Physics of Binary Star Evolution

PRINCETON SERIES IN ASTROPHYSICS

EDITED BY DAVID N. SPERGEL

For a full list of titles in the series, go to https://press.princeton.edu/series/princeton-series-in-astrophysics

Galactic Dynamics, Second Edition, *by James Binney and Scott Tremaine*

Dynamical Evolution of Globular Clusters, *by Lyman S. Spitzer, Jr.*

Supernovae and Nucleosynthesis: An Investigation of the History of Matter, from the Big Bang to the Present, *by David Arnett*

Unsolved Problems in Astrophysics, *edited by John N. Bahcall and Jeremiah P. Ostriker*

Galactic Astronomy, *by James Binney and Michael Merrifield*

Active Galactic Nuclei: From the Central Black Hole to the Galactic Environment, *by Julian H. Krolik*

Plasma Physics for Astrophysics, *by Russell M. Kulsrud*

Conversations on Electric and Magnetic Fields in the Cosmos, *by Eugene N. Parker*

High-Energy Astrophysics, *by Fulvio Melia*

Stellar Spectral Classification, *by Christopher J. Corbally and Richard O. Gray*

Extrasolar Planetary Atmospheres and the Physics of Modeling, *by Sara Seager*

Physics of the Interstellar and Intergalactic Medium, *by Bruce T. Draine*

The First Galaxies in the Universe, *by Abraham Loeb and Steven R. Furlanetto*

Theory of Stellar Atmospheres: An Introduction to Astrophysical Non-equilibrium Quantitative Spectroscopic Analysis, *by Ivan Hubeny and Dimitri Mihalos*

Exoplanetary Atmospheres: Theoretical Concepts and Foundations, *by Kevin Heng*

Magnetic Reconnection: A Modern Synthesis of Theory, Experiment, and Observations, *by Masaaki Yamada*

Physics of Binary Star Evolution: From Stars to X-ray Binaries and Gravitational Wave Sources, *by Thomas M. Tauris and Edward P. J. van den Heuvel*

Physics of Binary Star Evolution

From Stars to X-ray Binaries and Gravitational Wave Sources

Thomas M. Tauris
Edward P. J. van den Heuvel

PRINCETON UNIVERSITY PRESS
PRINCETON AND OXFORD

Published by Princeton University Press
41 William Street, Princeton, New Jersey 08540
99 Banbury Road, Oxford, OX2 6JX

press.princeton.edu

Library of Congress Control Number: 2022932836

ISBN 9780691179070
ISBN (pbk.) 9780691179087
ISBN (e-book) 9780691239262

British Library Cataloging-in-Publication Data is available

Editorial: Ingrid Gnerlich, Whitney Rauenhorst
Cover Design: Wanda España
Production: Jacqueline Poirier
Publicity: Matthew Taylor, Charlotte Coyne

Cover Credit: An artist's impression of gravitational waves generated by binary neutron stars. R. Hurt/Caltech/MIT/LIGO Laboratory

This book has been composed in Times

Printed on acid-free paper. ∞

Printed in the United States of America

10 9 8 7 6 5 4 3 2 1

Contents

Preface

The majority of all stars are members of a binary system. The evolution of such binary stars and their subsequent production of pairs of compact objects in tight orbits, such as double neutron stars and double black holes, play a central role in modern astrophysics. Binary evolution leads to the formation of different types of violent cosmic events such as novae, supernova explosions, gamma-ray bursts, mass transfer and accretion processes in X-ray binaries, and the formation of exotic radio millisecond pulsars. In some cases, the binary systems terminate as spectacular collisions between neutron stars and/or black holes. These collisions lead to powerful emission of gravitational waves, as detected by LIGO[1] since 2015. The coming decade is expected to reveal a large number of discoveries of binary compact systems, as well as their progenitors and merger remnants, from major instruments such as the radio Square-Kilometre Array; the gravitational wave observatories LIGO–Virgo–KAGRA–IndIGO and LISA; the astrometric space observatory Gaia; the James Webb Space Telescope; and the X-ray space observatories eXTP, STROBE-X, and Athena. In this light, it is important to have a modern textbook on the physics of binary stars evolution, from ordinary stars to X-ray binaries and gravitational wave sources. The scope of this book is that the reader (student or educated expert) will learn the physics of binary interactions, from stellar birth to compact objects, and relate this knowledge to the latest observations. The reader will learn about stellar structure and evolution and detailed binary interactions covering a broad range of phenomena, including mass transfer and orbital evolution, formation and accretion onto compact objects (white dwarfs, neutron stars, and black holes), and their observational properties. Exercises are provided throughout the book.

It has been a privilege for us to write this book, and we hope you enjoy it.

Thomas Tauris & Ed van den Heuvel — June 2022
(Aalborg / Aarhus / Bonn) (Amsterdam)

[1] See the list of Acronyms that is in the the back of the book.

Preface

Chapter One

Introduction

The Role of Binary Star Evolution in Astrophysics

Many key astrophysical objects and phenomena are related to the evolution of binary stars. This holds, for example, for the formation of the brightest X-ray sources in the sky and the formation of double neutron stars (NSs) and black holes (BHs), the mergers of which produce the strongest bursts of energy anywhere in the observable Universe, measurable on Earth with gravitational wave (GW) detectors.[1] Binary interactions also play a role for the origin of most supernovae (SNe), all nova explosions and short gamma-ray bursts, and the formation of millisecond radio pulsars and a large variety of stars with peculiar chemical abundances, such as barium stars and carbon-enriched metal poor stars.

It has been long realized, from the fact that most stars are members of binary systems (Abt & Levy, 1976; Bonnell et al., 2003), that binary evolution must play a key role in stellar evolution (e.g., van den Heuvel, 1994a). This early and important awareness has even been strengthened further by the findings of, for example, Chini et al. (2011) and Sana et al. (2012) that practically all massive stars are found in binaries with orbits such that at some stage in their evolution, the far majority of these stars will interact with each other. This implies that binary interactions dominate the evolution of massive stars.

The first ideas about the evolution of binary systems originated in the 1950s. They were largely inspired by the surprising characteristics of Algol-type eclipsing binary systems (see also Section 2.4). Here, *Algol-type* means physically similar to the Algol system, which consists of an unevolved B8V[2] main-sequence star with a mass of $3.2\,M_\odot$ together with an evolved but less massive subgiant companion star of spectral type K3 IV of mass $0.7\,M_\odot$. This situation, with the more evolved star having the smaller mass of the two, is just opposite to what one would expect on the grounds of stellar evolution, as stars of larger mass are expected to have shorter lives than stars of smaller mass do. In a binary—where both stars were born at the same time—one would therefore expect the star of larger mass to be in a more advanced stage of evolution at any time than that of its companion of smaller mass.

---

[1] The first ever detected BH merger event GW150914 (Abbott et al., 2016d) released an energy of $3\,M_\odot\,c^2$ within a fraction of a second, thereby outshining all stars in the Universe for a brief moment.

[2] B8V refers to a star of spectroscopic type B8 and luminosity class V.

This is what is called the *Algol paradox*. Crawford (1955) was the first to realize that this paradoxical situation can be explained if one assumes that large-scale mass transfer can take place during the evolution of a binary system: Crawford hypothesized that the subgiant components in Algol-type binaries were originally the more massive components of these systems. As the more massive star evolved faster than its less massive companion did, it was the first one to exhaust the hydrogen fuel in its core and evolve into a giant star with a much expanded envelope. The presence of the close companion, however, prevented such an evolution: when the outer layers of the expanding (sub)giant came under the gravitational influence of the companion, they were captured by this smaller star (the *accretor*), causing the accretor to increase in mass at the expense of the (sub)giant (the *donor*). The (sub)giant transferred so much of its mass that it was finally able to restabilize its internal structure. At that moment, it became the less massive of the two stars, and the originally less massive star became the most massive one of the pair.

The first attempt to carry out a real calculation of this type of evolution with mass transfer was by Morton (1960). He demonstrated the correctness of Crawford's conjecture that mass transfer, once it begins, continues until the (sub)giant has become the less massive star of the system. In his calculations, however, Morton still assumed that the orbital period of the system does not change during the mass transfer. This is not correct because, if one assumes that the total mass of the system is conserved, one also expects the total angular momentum of the system to be conserved. The total angular momentum is, in good approximation, equal to the orbital angular momentum (as the rotational angular momentum of the two stars is usually much smaller than the orbital one). Conservation of the orbital angular momentum implies that during the mass transfer the orbital period and separation change in a well-determined way, which will be explained in Chapter 4. The evolution of close binaries in this more realistic approach was first calculated, independently of one another, by Paczyński (1966), Kippenhahn & Weigert (1967), and Plavec (1967). Their work was the foundation of all subsequent work on the evolution of close binary systems. Furthermore, the discovery of the first celestial X-ray source in 1962 and of the X-ray binaries in 1971–1972—which earned Riccardo Giacconi the 2002 Physics Nobel Prize—has given a great stimulus to the research in this field. The X-ray binaries consist of a normal star together with a compact object: a NS or a BH, which are the end states of evolution of massive stars. The X-rays from these binaries, which are observed with space-borne detectors, are generated by the accretion of matter onto the compact star, which is captured from the outer layers of the companion star. During the accretion this gas, falling inward in the extremely strong gravitational field of the compact star, is heated by the release of gravitational potential energy to temperatures above 10^6 K, causing it to emit X-rays.

Without the occurrence of extensive mass transfer from the original primary star (the progenitor of the compact object) to the secondary star, most of these systems could not have survived the SN explosion of the primary star in which the NS or BH was formed (van den Heuvel & Heise, 1972; Tutukov & Yungelson, 1973b), because the probability of the post-SN orbit to remain bound depends on the relative mass loss from the system during the SN (see e.g. Section 4.3.10 and Chapter 13).

Since the early 1970s, the realization of the importance of accretion of matter onto a compact star (NS, BH, or white dwarf [WD]) as an energy source in many types of binaries, ranging from X-ray binaries to cataclysmic variables (CVs) and symbiotic stars, has been a further important source of inspiration for new research on the structure and evolution of close binary systems. The discoveries since 1974 of binary radio pulsars, of which at least 20 are double NSs (DNSs; for a recent review, see Tauris et al., 2017), have revealed many interesting properties, including the many relativistic effects that are measurable in them with unprecedented precision (e.g., Taylor & Weisberg, 1989; Taylor, 1992; Kaspi & Kramer, 2016).

These discoveries have created a new and fundamental branch of relativistic binary star astrophysics, which among other things has produced the most accurate measurements of masses of any stellar objects so far (Chapter 14). The measured rate of orbital decay of the first-discovered double NS, PSR B1913+16, is in exact agreement with the decay rate predicted from the emission of GWs according to the general theory of relativity. The highly precise detection of this and other relativistic effects earned the discoverers of this binary pulsar, Russell Hulse and Joseph Taylor, the 1993 Physics Nobel Prize. Later-discovered DNSs, particularly the double pulsar system J0737−3039, have further refined the up to five tests of relativity allowed by these systems (Section 14.7.1) to almost incredible precision (Wex, 2014; Kramer et al., 2021).

The amazing discoveries of the merger event of a double BH, starting with GW-150914 in September 2015 (Abbott et al., 2016d), and of a DNS, in August 2017 (Abbott et al., 2017c), have revealed the ultimate final destiny of a massive binary star system and demonstrated the production of strong bursts of GWs, that are observable on Earth. This earned the LIGO pioneers Rainer Weiss, Kip Thorne, and Barry Barish the 2017 Physics Nobel Prize. But how do ordinary stars born in a binary system end up as two NSs or BHs that finish as a final single BH remnant? To answer this question, we need to follow the binary system through a long chain of exotic binary interactions (Fig. 1.1), involving mass transfer between the stellar components, SNe, and relativistic effects. That the orbits of evolved massive stars in binaries will shrink to the very small sizes observed for the double compact objects was predicted before these objects were discovered (van den Heuvel & De Loore, 1973).

Radio pulsars are some of the most intriguing astrophysical objects, among which a sizable fraction of binaries is found with quite special characteristics that give important information on binary evolution. To understand the continuously growing diversity of observed radio pulsars (see Fig. 1.2), it is necessary to link their properties to the stellar and binary evolution of their progenitors (Bhattacharya & van den Heuvel, 1991).

The detection of close binary pulsars and of the merger events of double BHs and NSs have further increased the interest in the evolution of binary systems. These discoveries demonstrated that, even though binaries may have undergone several stages of mass transfer during their evolution, where up to 90% of their original mass and >95% of their original orbital angular momentum is lost from the system, and despite having experienced two SN explosions, the two stars might still survive as a (very close) binary system with two compact objects.

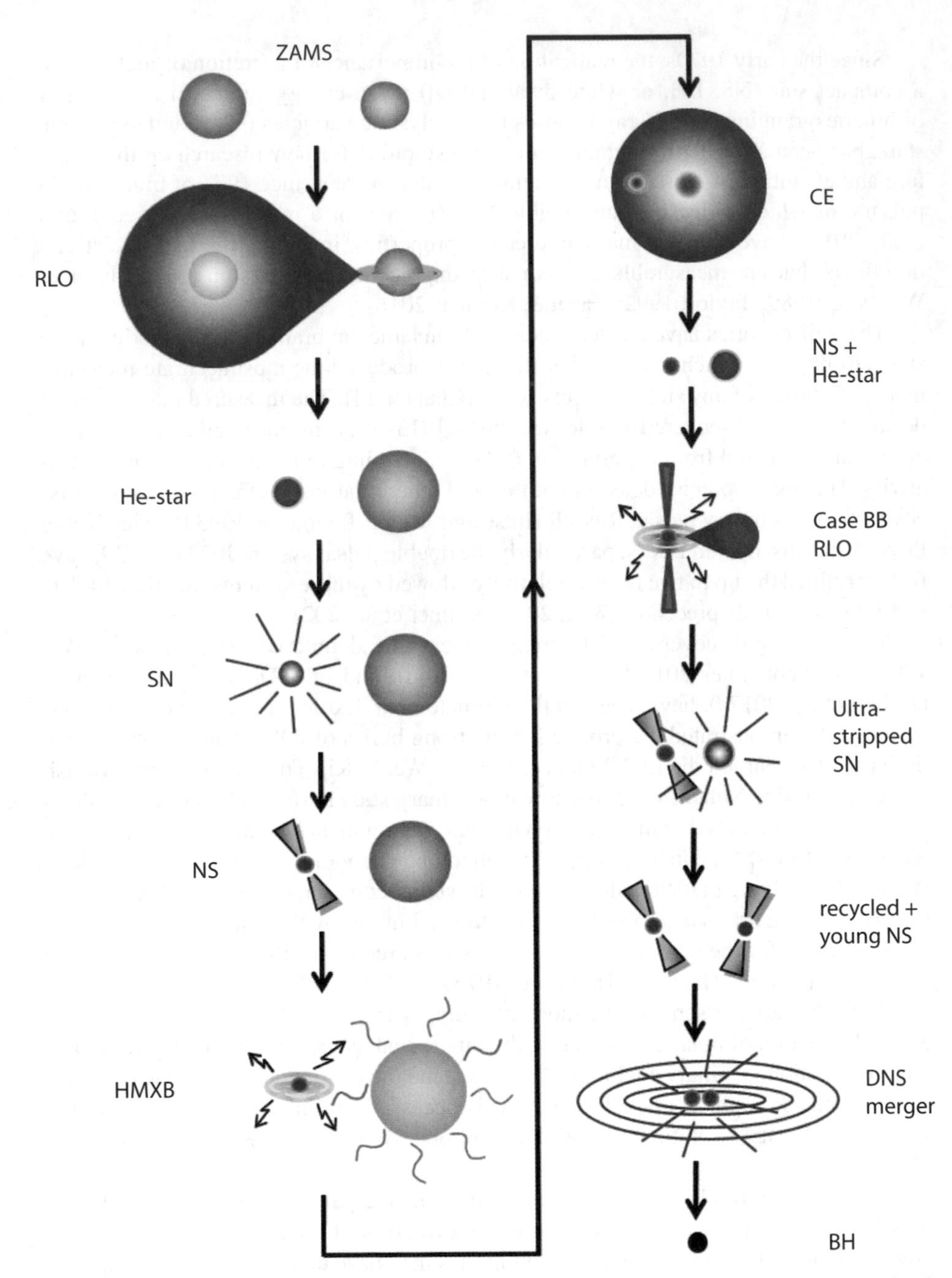

Figure 1.1. Formation model of a close double neutron star (DNS) system as final product of the evolution of a massive close binary. The DNS system may eventually merge and leave a solitary BH remnant. The same model, scaled up to higher initial stellar masses, is one of the main scenarios to explain the formation of double BHs (see Chapters 10, 12, and 15). After Tauris et al. (2017).

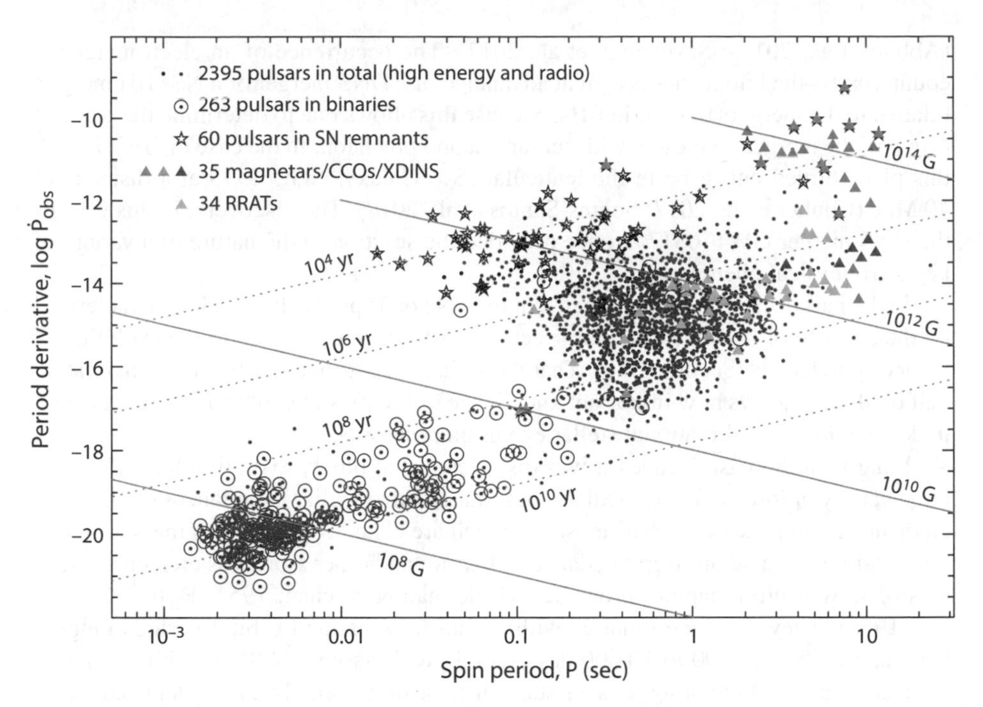

Figure 1.2. Observed population of ~2,400 radio and high-energy pulsars with measured values of both spin period (P) and period derivative ($\dot{P}$). The variety of pulsars has flourished immensely since their discovery in 1967 and continues to be a major science driver in modern astrophysics. Binary pulsars (blue circles) dominate among the fast-spinning millisecond pulsars. CCOs are central compact objects of SN remnants, XDINs are X-ray dim isolated NSs, and RRATs are rotating radio transients (see Chapter 14). Data taken from the *ATNF Pulsar Catalogue* in June 2021 (Manchester et al., 2005, https://www.atnf.csiro.au/research/pulsar/psrcat).

The existence of the close double NSs and close double BHs has also demonstrated that a precise knowledge of the physics of binary evolution is of vital importance for understanding fundamental astrophysics as diverse as the generation of the strongest bursts of GW radiation, the production of gamma-ray bursts, and the synthesis of heavy r-process (or rapid neutron-capture) elements in the Universe. The latter was predicted to be produced by the merging of double NSs or NS+BH binaries (Lattimer & Schramm, 1976) and expected to be observable as a so-called kilonova optical-infrared eruption (Metzger et al., 2010; Berger, 2014).

These predictions have been beautifully confirmed by the spectroscopic study of the optical-infrared transient that accompanied the DNS merger event GW170817 (Abbott et al., 2017c, 2017d, 2017e). The mergers of double compact object binaries with at least one NS also had been predicted to produce gamma-ray bursts (Paczynski, 1986; Eichler et al., 1989), which was confirmed by the Fermi and INTEGRAL missions by the detection of a short gamma-ray burst, following within 2 sec after GW170817

(Abbott et al., 2017a; Savchenko et al., 2017). The occurrence of an electromagnetic counterpart—the kilonova—is a great advantage of a DNS merger or a NS+BH merger, relative to the mergers of double BHs, because this allows one to determine the place of origin of the merger on the sky with sub-arcsecond precision. In the case of GW170817, this place turned out to be in the lenticular (S0) Galaxy. NGC 4993 at a distance of 40 Mpc (Coulter et al., 2017; Soares-Santos et al., 2017). The discovery, in this way, of the DNS merger GW170817 at once provided the solution to the nature of a variety of key astrophysical phenomena.

In the past decades it was realized that the SNe of Types Ia, Ib, and Ic (characterized by the absence of hydrogen in their spectra) most likely are related to the evolution of binary systems (see Section 2.10). That these three types together form approximately half of all SNe shows how important knowledge of binary star evolution is for a general understanding of observational stellar evolution processes.

Finally, in the past decades it was discovered that binary evolution has affected a large variety of low- and intermediate-mass stars with peculiar abundances of chemical elements. Examples are the barium stars, which are G and K giants with masses up to a few solar masses and an overabundance of barium and other s-process elements (products of slow neutron-capture processes) (Bidelman & Keenan, 1951; Boffin & Jorissen, 1988). They are wide binaries with eccentric orbits and orbital periods ranging from approximately 100 to 10,000 days (Boffin & Jorissen, 1988). The binary nature of these stars, of blue stragglers in star clusters, of extremely metal-depleted post–asymptotic giant branch stars, of carbon-enhanced metal-poor (CEMP) stars, and of $>80\%$ of nuclei of planetary nebulae was discovered in past decades. These discoveries show that the evolution of $\sim 50\%$ of all low- and intermediate-mass stars is expected to be affected by binary evolution (Abate et al., 2015). Because lower-mass stars are far more abundant in galaxies than massive stars are, the products of low- and intermediate-mass binaries are expected to vastly outnumber the products of the evolution of massive binaries. However, in this book we will concentrate mainly on the physics and evolution of the more massive binaries that produce NSs and BHs and on the origin of Type Ia SNe, which are caused by exploding WDs and are of crucial importance for cosmology.

In the last decades it was realized that stars are found often, even in triple systems or higher-order multiple systems. Observational estimates suggest that approximately 20%–30% of all binary stars are members of triple systems (Tokovinin et al., 2006; Rappaport et al., 2013). These systems may remain bound with a long-term stability in a hierarchical structure (a close inner binary with a third star in relatively distant orbit). Since the 1970s, a number of stability criteria for a triple system have been proposed (see Mikkola, 2008, for an overview) that enable predictions for the long-term stability of triple systems. Iben & Tutukov (1999) estimated that in $\sim 70\%$ of the triple systems, the inner binary is close enough that the most massive star will evolve to fill its Roche lobe. Furthermore, in $\sim 15\%$ of the triple star systems, the outer third (tertiary) star may even also fill its Roche lobe at some point, possibly leading to disintegration or production of rare configurations with three degenerate objects in the same system. In 2014, Ransom et al. (2014) reported the remarkable discovery of a triple-system pulsar (PSR J0337+1715), which is exactly the first example of such an exotic system—a NS orbited by two WDs. (For a formation and evolution scenario for this complex system, which must have survived a SN explosion and at least three stages of mass transfer between the stellar components, see our model in Tauris & van den Heuvel [2014].) Besides specific triple-star interactions, such as Kozai-Lidov resonances (Kozai, 1962; Lidov, 1962) and the above-mentioned dynamical stability considerations, most interactions between stars in triples are similar to the interactions between binary stars. For this reason, we focus only on binary stars in this book.[3]

The 2020s and 2030s are expected to reveal a large number of discoveries of new double compact object systems, as well as their progenitors and merger remnants. This field of astrophysics is strongly driven by investments in new big-science instruments. The Square-Kilometer Array (SKA) is expected to increase the number of known radio pulsars by a factor of 5 to 10, thus resulting in a total of >100 known DNS systems (Keane et al., 2015). The Five-hundred-meter Aperture Spherical Telescope is also expected to contribute a significant number of new radio pulsars (Smits et al., 2009; Nan et al., 2011), including new discoveries of pulsar binaries. High-mass X-ray binaries (HMXBs), the anticipated progenitors of double compact object systems containing NSs and BHs, are continuously being discovered with ongoing X-ray missions (INTEGRAL, Swift, XMM-Newton, and Chandra [see e.g., Chaty, 2013]). New and upcoming space-borne X-ray telescopes such as eXTP, STROBE-X, and Athena are expected to produce further discoveries of these systems. Hence, we are currently in an epoch when a large wealth of new information on exotic binaries is becoming available. In light of this, it is important to explore and understand the formation and evolution of such binary systems in more detail. Earlier textbooks on binary evolution are those of Shore et al. (1994), Hilditch (2001), and Eggleton (2006), to which we refer for further reading. For an earlier book on physics and evolution of relativistic objects in binaries, we refer to Colpi et al. (2009).

This book is organized as follows. In Chapter 2, we give a brief history of the discovery of the different types of binary systems and, where appropriate, summarize their importance in modern astrophysics. In Chapter 3, we consider how the orbital parameters of spectroscopic and eclipsing binaries are measured and, how from these measurements information is obtained about the masses and radii of the stars. We give an overview of the thus-derived masses of stars of different spectral types.

In Chapter 4, we consider basic aspects of the celestial mechanics of binary systems, the meaning and limitations of the Roche-lobe concept, as well as the changes in orbital period and binary separation that are induced by various processes of mass loss and mass transfer in binary systems. We first consider the somewhat idealized "conservative" evolution, in which the total mass and orbital angular momentum of the binary are expected to be conserved during the evolution, followed by the more realistic and exotic types of close binary evolution, so-called non-conservative evolution, in which large losses of mass and orbital angular momentum from the systems are taken into account. A treatment of common envelopes and the orbital evolution during the

[3]Formation of stellar binaries via triple- or quadruple-star dynamical interactions in dense cluster, however, are discussed in Chapter 12. For a recent review on the evolution of destabilized triple systems, see Toonen et al. (2022).

dynamically unstable in-spiral phase are described as well. Finally, we briefly discuss the Eddington accretion limit as well as accretion disks.

In Chapter 5, we describe the observed properties and the general classifications of the various types of interacting binary systems that do not contain NSs or BHs, concentrating mostly on systems in which at least one component is an evolved star, that is, a (sub)giant, or a WD.

In Chapters 6 and 7, we describe the observed properties of X-ray binaries: HMXBs as well as low-mass X-ray binaries (LMXBs), including mass determination of the accreting NSs and BHs. Regarding HMXBs, we discuss Be-star X-ray binaries, supergiant X-ray binaries, and, for example, stellar wind accretion and the Corbet diagram. We also discuss the recently discovered class of pulsating ultra-luminous X-ray sources. For LMXBs, we discuss the various types, including the systems with BHs and the symbiotic X-ray binaries with accreting NSs.

In Chapter 8, we give an overview of the evolution of single stars (with a special focus on the final evolution of massive stars). In Chapter 9, we apply this knowledge to the evolution of binaries in general. Here, we also discuss in detail the various cases of mass transfer (including mass loss) and orbital stability analysis, and we end with a comparison between the outcomes of single vs. binary star evolution.

This knowledge is crucial for understanding the formation and evolution of X-ray binaries, which are the subjects of Chapters 10 (HMXBs) and 11 (LMXBs). We discuss the final stages of HMXBs (including Wolf-Rayet [WR] star binaries), with or without a common envelope, leading to the formation of double NS/BH binaries. For LMXBs and CVs, we also discuss the mechanisms driving the mass transfer in LMXBs and CVs. These concern the internal evolution of the companion star, as well as the loss of orbital angular momentum due to emission of GWs and/or a magnetically coupled stellar wind, or a combination of these. We also discuss the final mass-transfer stage from WDs in very tight binaries, in the so-called AM Canum Venaticorum (double WD) systems and the ultra-compact X-ray binary sources (typically WD+NS systems).

Binaries with compact objects can also be formed by the dynamical evolution of dense star clusters. This is the subject of Chapter 12.

Chapter 13 concerns SNe in binaries. We first discuss the evolution leading to the thermonuclear Type Ia SNe, triggered by the accretion of matter by a WD, and we also consider accretion-induced collapse (AIC) of massive WDs. We subsequently discuss the origin of Type Ib/Ic SNe and the evidence derived from theoretical computations of the late stages of close binary evolution, including ultra-stripped SNe, combined with observations of SN light curves and known Galactic post-SN DNS systems. We discuss the evidence for momentum kicks imparted onto compact objects during various SN events (iron core collapse and electron-capture SNe) and examine the kinematical effects of these kicks on the resulting compact-object binaries.

In Chapter 14, using the results of the previous chapters, the final stages of the formation of binary and millisecond radio pulsars is studied; in other words, the transition from X-ray binaries to radio pulsars. We review the recycling of pulsars in detail, including the accretion torques at work. We review the rich diversity of resulting millisecond pulsar binaries and their component masses (theoretical expectations and measurements via relativistic effects). Finally, we revisit the formation of double NS (DNS) systems.

Chapter 15 is devoted to GW astrophysics. We cover the basic physics of GW emission and detection. We discuss formation channels of both high-frequency (LIGO–Virgo–KAGRA–IndIGO) and low-frequency (LISA and TianQin) GW sources. We review the signals expected from extragalactic merging double BHs and NSs, including the electromagnetic counterparts of these mergers, and the Galactic WD and NS binaries as continuous GW emission sources. The different models for the formation of the double BHs are discussed in depth in light of the latest observations of "ordinary" and exotic events from the LIGO–Virgo-KAGRA network.

Chapter 16 discusses the subject of binary population synthesis with an emphasis on methodology and statistics. Two examples are highlighted that illustrate a synthetic open star cluster population of binaries and the differences between estimates of empirical vs. theoretical double NS (DNS) merger rates.

Chapter Two

Historical Notes on Binary Star Discoveries

2.1 VISUAL BINARIES AND THE UNIVERSAL VALIDITY OF THE LAWS OF PHYSICS

On January 7, 1617, Benedictine monk and astronomer Benedetto Castelli, a former pupil of Galileo Galilei, discovered that the second magnitude star Mizar (ζ Ursae Majoris) of the Big Dipper, seen through a telescope, consists of two stars very close to each other, differing by approximately two magnitudes (a factor of $\sim$6) in brightness. Castelli, who had succeeded Galilei as professor of mathematics in Pisa, immediately wrote to Galilei about his discovery, and shortly later Galilei himself recorded that he also had observed this double star.[1] This was the first discovered so-called visual binary (double star). *Visual* means here that one can really see through a telescope that there are two stars. Viewing such a pair, one may wonder whether the two stars are really physically connected, moving around each other in gravitationally bound orbits, or whether they are two stars at different radial distances along the line of sight that have no physical connection, but happen to be seen close to each other on the sky. In the first case, one speaks of a *physical binary* and in the second case of an *optical binary*. The components of Mizar, called Mizar A and Mizar B, are separated on the sky by 14.4$''$ (the symbol $''$ means 1 sec of arc = 1/60 of an arc minute = 1/3,600 of a degree).

In the 17th and 18th centuries, the use of the telescope led to the discovery of many hundreds of visual binaries, often very close pairs. In 1767, the geologist-reverend John Michell (1724–1793) was the first to mathematically show that the observed frequency of close pairs is far larger than one would expect if the pairs were chance coincidences of stars randomly spread over the sky. On this basis, he concluded that most visual double stars must be physical pairs. At that time, no one had yet observed the orbital movement of a visual pair. (Michell was also the first to put forward, in 1790, the concept of BHs, an idea independently conceived around the same time by French mathematician Pierre-Simon Laplace [1749–1827].)

The first one to observe the orbital movement of a visual binary was Sir William Herschell (1738–1824). In 1803, in a paper published in the *Philosophical Transactions*

[1] In most literature, the discovery of the binary character of Mizar is ascribed to Jesuit-astronomer G. B. Riccioli in 1650. However, recent studies have revealed the much earlier discovery by Castelli, see "A New View of Mizar," https://www.leosondra.cz/en/

of the Royal Society, he mentions that in 1802 he had noticed that Castor B, the fainter component of the visual binary Castor (α Geminorum), had moved with respect to the brighter component Castor A, relative to its position recorded by British astronomers James Bradley and James Pound in 1719. The latter ones had noticed that the line connecting the two components of Castor passed through the stars σ and κ Geminorum. Herschell, who had started observing visual binaries in 1779, saw that since 1719 the orientation of this connecting line had rotated over a considerable angle on the sky.

This observation makes clear that the periods in which the stars in visual binaries orbit one another must often be counted in centuries; in the case of Castor A and Castor B, which are separated by $2''$, the period is 445 yr. The shortest periods recorded for visual binaries are on the of order of several decades. Therefore, observations by several generations of astronomers are generally required to determine the orbits of visual binaries. At present many thousands of visual binaries are being followed regularly, and astrometry satellites such as Gaia will multiply this number by orders of magnitude.

Herschell's very important discovery showed that Newton's law of gravity is valid not only on Earth and in the Solar System, but also throughout the Universe. This was the first physical law for which this universal validity was discovered. Later, in 1860, German physicists Gustav Kirchhoff and Robert Bunsen discovered that every chemical element has its own spectral lines and that the same lines also are present in the spectrum of the Sun, which had been recorded early in the 19th century by Bavarian genius instrument builder Joseph von Fraunhofer (1787–1826). Subsequently, English amateur astronomer William Huggins (1824–1910) showed that these lines also are present in the spectra of stars. This all showed that the Universe consists of the same elements that are found on Earth and that the physical laws discovered on Earth, which determine the behavior of light and matter, are apparently valid throughout the Universe. Although to many physicists this may seem obvious, it is only thanks to the astronomical observations by Herschell, von Fraunhofer, Huggins, and others that we know that this really is the case.

Visual double stars have provided the first means to measure the masses of stars, by applying Newton's concept of gravity via Kepler's laws of orbital motion (Exercise 2.1). Figure 2.1 shows as an example the relative (apparent) and absolute orbits, respectively, of the components Sirius A and Sirius B of the brightest star of the sky. These two stars move around their common center of mass in only 49.94 yr.

2.2 ASTROMETRIC BINARIES

Wide binaries, in which one of the two stars is too faint to be observable, can still be recognized by carefully following the proper motion of the visible component. The center of mass of a binary will move among the neighboring stars on the sky in a straight line, while the observable star orbits this center, so its proper motion on the sky will not be straight but will show a periodic "wiggle" (see bottom-right frame of Fig. 2.1). Sirius is the first star for which this was discovered; in 1834, the German astronomer Friedrich Wilhelm Bessel (1784–1846) noticed that its proper motion does not follow a straight line. This led him to conclude in 1844 that Sirius must have an unseen companion.

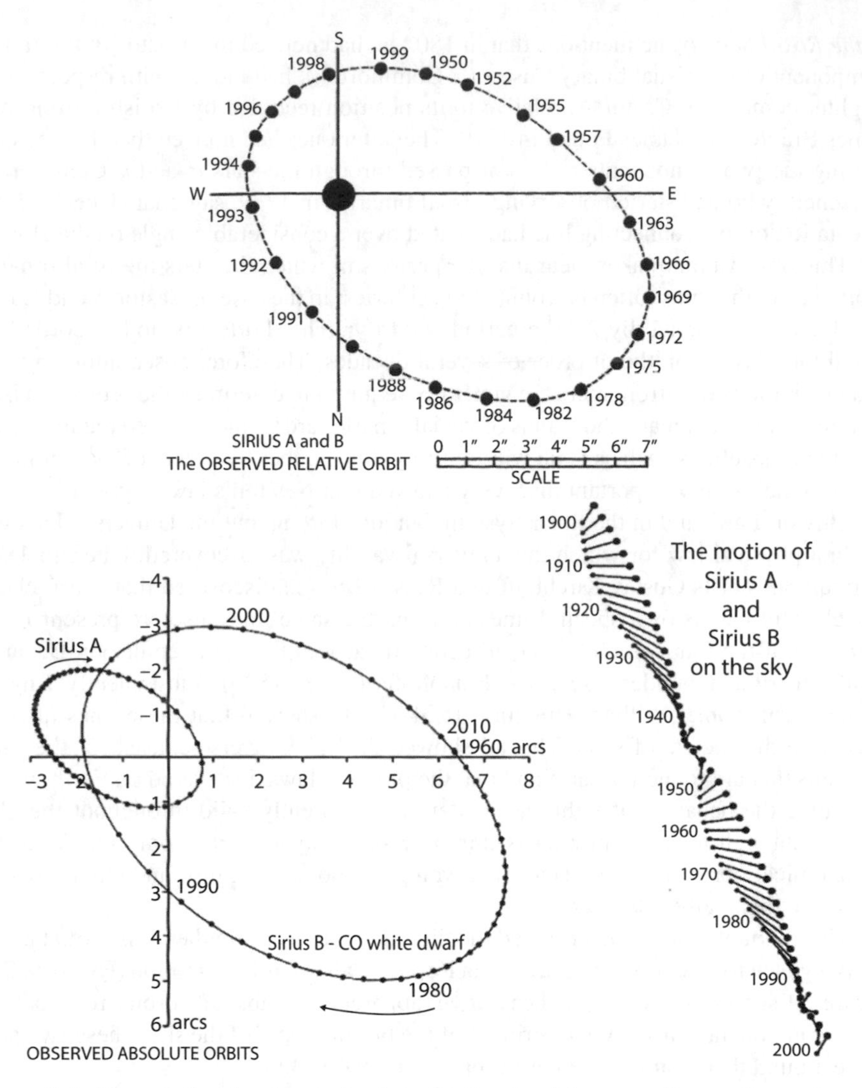

Figure 2.1. The relative (top) and absolute (bottom) orbits, as well as paths on the sky, of the components A and B of Sirius. Adapted from (top) the Lunar and Planetary Laboratory (http://vega.lpl.arizona.edu/sirius/A4.html) and (bottom) a figure by Scot Aaron after Lippincott (1961).

Indeed, in 1862, American telescope maker Alvan Graham Clark (1832–1897), who used Sirius as a light source to test the 47-cm refracting telescope he had made for Dearborn Observatory of Northwestern University, discovered the companion Sirius B of Sirius, which is ~10,000 times fainter than Sirius A. Clark had discovered the first WD star, which is a compact object. In this book, we shall discuss many WD binary systems, with a plethora of companion stars.

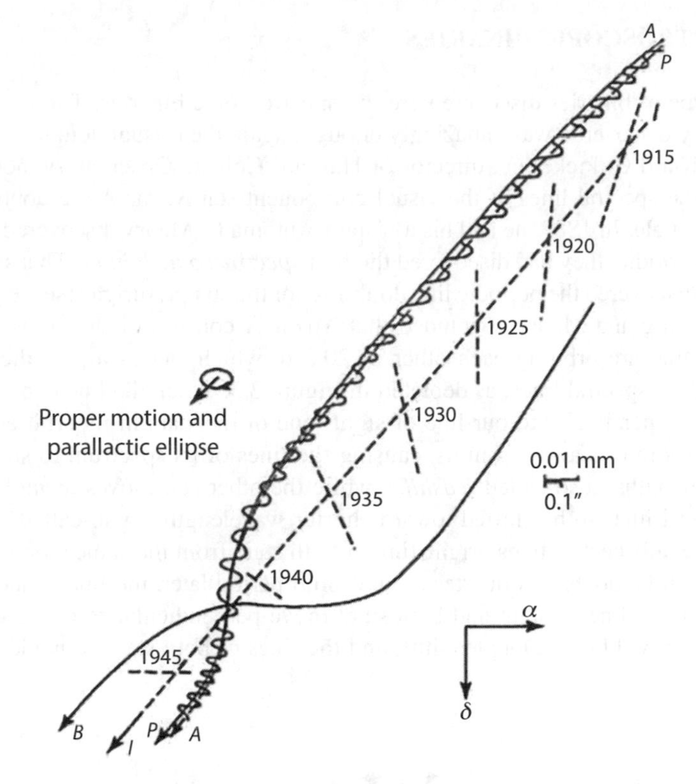

Figure 2.2. Geocentric astrometric path on the sky of 99 Herculis, showing its binary motion (stars A and B) and yearly parallactic motion on the sky. For the latter effect, see the oscillatory movement along track P of the primary component, star A. The orbital period of 99 Her is $P_{\rm orb} = 56.3$ yr. After Binnendijk (1960).

Since 1862, the motions of Sirius A and Sirius B on the sky have been carefully followed, leading to the orbits depicted in Figure 2.1. The bottom-right frame of this figure also shows the paths of both stars on the sky. In fact, the motion on the sky of the visible star of an astrometric binary is still somewhat more complex than a line with a wiggle with the binary period, because the position also shows a small yearly periodicity due to the parallax of the star (caused by the annual motion of the Earth around the Sun), as depicted in Figure 2.2. This figure shows the geocentric path on the sky of the photocenter of 99 Her, relative to its center of mass (see dashed line, *I*). The data were derived by Peter van de Kamp (1901–1995) from photographs taken by him at Sproul Observatory (Swarthmore College, Pennsylvania). Over the years, in this way, van de Kamp discovered many astrometric binaries, often with invisible components with sub-stellar masses ($<0.08\ M_\odot$), meaning that they are *brown dwarfs*, or very large planets, with masses of a few tens of times the mass of Jupiter (see, e.g., van de Kamp, 1981).

2.3 SPECTROSCOPIC BINARIES

The third type of binaries discovered are the spectroscopic binaries. The first spectroscopic binary discovered was, amazingly enough, again the (visual double) star Mizar. In 1887, Edward C. Pickering, director of Harvard College Observatory, noticed that sometimes the spectral lines of the visual component star Mizar A are double and at other times single. In 1889, he and his assistant Antonia C. Maury discovered this doubling to be periodic: they had discovered the first *spectroscopic binary*. That same year, Maury also discovered the periodic line doubling for the star β Aurigae (see, e.g., Sobel, 2016). Pickering and Maury concluded that Mizar A consists of two almost equally bright stars that are orbiting each other in 20.5 d, which causes a periodic Doppler shifting of their spectral lines, as depicted in Figure 2.3. When the line connecting the two stars is perpendicular to our line of sight, one of the stars in its orbit around the center of mass moves away from us, causing the lines of its spectrum to shift toward longer wavelengths (a so-called *redshift*), while the other star moves toward us, causing its spectral lines to be shifted toward shorter wavelengths (a so-called *blueshift*). At this phase, all spectral lines originating in both stars from the same atomic electron transitions will be double. A quarter of an orbital period later, the line connecting the stars is along our line of sight and both stars move perpendicular to our line of sight such that there will be no Doppler shift, and the lines of both stars coincide: one now

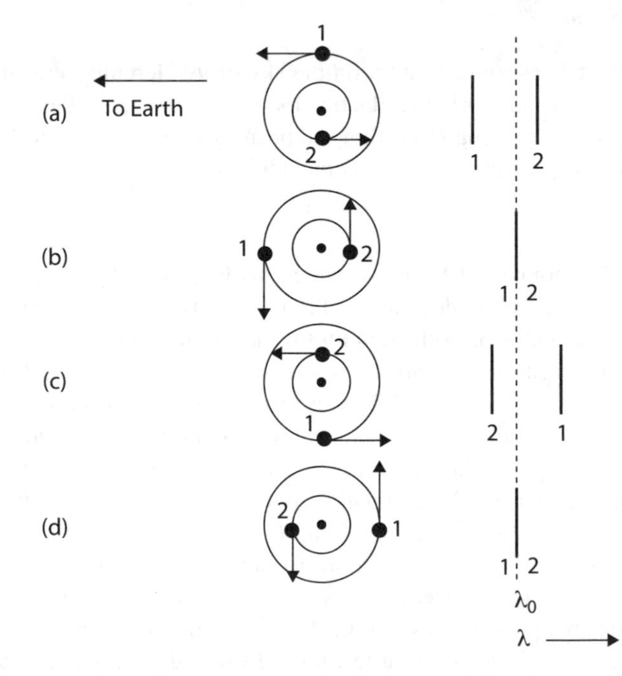

Figure 2.3. The Doppler effect causes a periodic line doubling in a spectroscopic binary with two almost equally bright components.

observes all spectral lines to be single. A quarter period later they will be double again, and so on.

The Doppler shift, $\Delta\lambda$ of the wavelength, λ of a spectral line is related to the radial velocity $v_{\rm rad}$ (the component of the velocity along our line of sight to the star) by the Doppler equation, which is valid for velocities $v \ll c$:

$$\frac{\Delta\lambda}{\lambda} = \frac{v_{\rm rad}}{c}. \tag{2.1}$$

Spectroscopic binaries that show this periodic doubling of the lines are called *double-lined* spectroscopic binaries (SB2). It is also possible that the companion of a star is too faint for its spectrum to be seen against the light of the star. If that is the case, one observes the spectral lines of the star to periodically shift toward the red and the blue. One then speaks of a *single-lined* spectroscopic binary (SB1). In 1908, Edwin B. Frost at Yerkes Observatory discovered that Mizar B is a spectroscopic binary, with a period of 175.6 d. Thus, Mizar is itself a quadruple stellar system. Moreover, Mizar's visible naked eye companion Alcor, which is two magnitudes fainter and $12'$ (12 arc minutes) away from it, is less than a light year distant from Mizar and has the same proper motion, which means that the two stars have the same origin. It is suspected that Alcor is loosely gravitationally bound to Mizar, with an orbital period of ~800,000 yr. Alcor itself has a close visual M-dwarf companion, with an orbital size of a few tens of astronomical units (AU). Therefore, the visible pair Mizar-Alcor is in fact a system of six stars!

The case is the same with Castor. The bright first magnitude pair Castor AB has a faint third companion, Castor C, that is separated from Castor AB by $73''$ and has an apparent visual magnitude of 10. All three visual components of Castor are spectroscopic binaries. Castor $A_{1,2}$ has an orbital period of 9.2128 d, Castor $B_{1,2}$ of 2.93 d, and Castor $C_{1,2}$ of only 0.813 d. Inspection of the *Bright Star Catalogue* (Hoffleit & Jaschek, 1991), which lists the properties of the ~9,000 naked-eye stars (that is, stars brighter than apparent visual magnitude of 6.5), shows that among these stars such multiple systems are quite common.

2.4 ECLIPSING BINARIES

With very close spectroscopic binaries, such as Castor B and Castor C, one may wonder what the probability is that one of the stars during its orbital motion passes in front of the other one, causing an eclipse. The inclination, i of the orbital plane is defined as the angle between the orbital plane and the plane of the sky (see Fig. 3.3). Therefore, the angle ϕ between the line of sight and the orbital plane is given by $\phi = 90° - i$. If the radii of the two stars are R_1 and R_2, and a is the orbital radius (assuming the orbit to be circular), a partial eclipse will occur if

$$\sin\phi < \sin\phi_0 = \frac{(R_1 + R_2)}{a}, \tag{2.2}$$

and a total eclipse will occur if

$$\sin\phi < \sin\phi_1 = \frac{(R_1 - R_2)}{a}, \tag{2.3}$$

as one can easily verify. If one assumes that the angles ϕ between the line of sight and the orbital planes of binaries are randomly distributed, then the probability of having an angle between 0 and ϕ_0 is the fractional surface area of the half sphere between the pole and the circle at an angle ϕ_0 from the pole. This area is $2\pi\,(1 - \cos\phi_0)$, while the area of the half sphere is 2π. So, the probability of having $\phi < \phi_0$ (and thus an eclipse) is

$$f_{\text{eclipse}} = (1 - \cos\phi_0). \tag{2.4}$$

Assuming ϕ_0 to be small, then

$$\sin\phi_0 \simeq \phi_0 \simeq \frac{(R_1 + R_2)}{a}. \tag{2.5}$$

Then according to the usual series expansion

$$\cos\phi_0 \simeq 1 - \frac{(R_1 + R_2)^2}{2a^2}. \tag{2.6}$$

So, in good approximation

$$f_{\text{eclipse}} = \frac{(R_1 + R_2)^2}{2a^2}. \tag{2.7}$$

One observes from Eq. (2.7) that, because $R_1 + R_2 < a$, the probability of an eclipse will be largest for (1) dwarf stars that are very close to each other, and (2) wide binaries, if at least one of the stars is a giant.

These are indeed the two most common kinds of eclipsing binaries known. For a general comprehensive review of the basics of geometry and physics of eclipses, we refer to Winn (2010).

In 1669, Italian astronomer Montanari noticed that the star Algol (β Persei, the second brightest star of the constellation Perseus) varies in brightness by approximately a factor of 3. The name Algol is derived from the Arabic "Al Ghul," the devil. This implies that most probably Arab astronomers already centuries earlier had noticed the "devilish" variability of this star. In 1783, English amateur astronomer John Goodricke (1764–1786), who was both deaf and mute, discovered that Algol's variability is regular with a period of 2.875 d. The presently best-determined period is 2 d; 20 hr; 49 min. Goodricke already proposed that this periodicity is due to eclipses. This was confirmed more than a century later, in 1889, by Pickering, who discovered that Algol is a single-lined spectroscopic binary with the same period as that of its light variations. So, Algol was the first-discovered eclipsing binary. (See also the description of the "Algol paradox" in Chapter 1.)

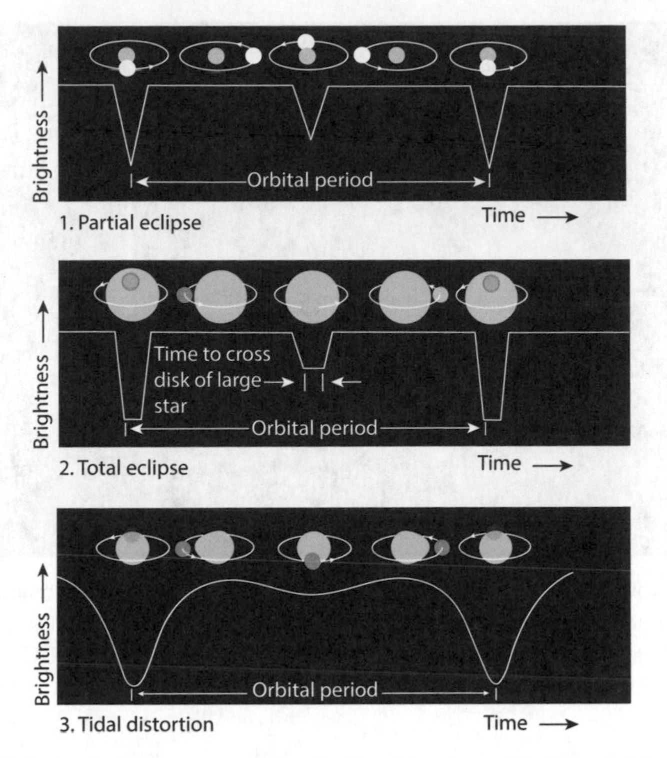

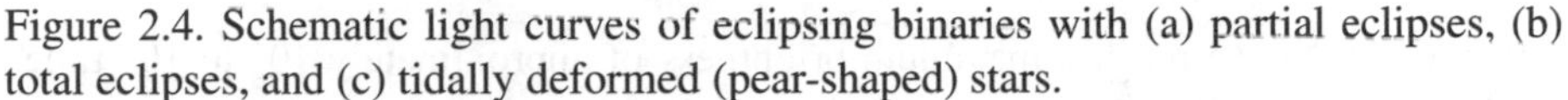

Figure 2.4. Schematic light curves of eclipsing binaries with (a) partial eclipses, (b) total eclipses, and (c) tidally deformed (pear-shaped) stars.

Figure 2.4 shows the schematic light curves of different types of eclipsing binaries. In the Algol system, the eclipses are only partial. Goodricke, in his short life,[2] also discovered the periodic nature of the light variations of the eclipsing binary β Lyrae and of the star δ Cephei. The latter is a pulsating star that has given its name to the important class of Cepheid variables, or Cepheids. Due to the tight relation between the pulsation period and intrinsic brightness, known as Leavitt's law (Leavitt & Pickering, 1912), Cepheids are "standard candles" and among the primary distance indicators in the Universe (Weinberg et al., 2013).

2.5 THE DISCOVERY OF THE BINARY NATURE OF NOVAE AND OTHER CATACLYSMIC VARIABLES

Novae ("new stars") are stars that suddenly appear in the sky and shine bright for several days to weeks and then disappear in the course of a few months. (They should not be confused with SNe which are many orders of magnitude more energetic, see

[2]He died at the age of 21 yr from pneumonia contracted during intense astronomical observations on cold winter nights.

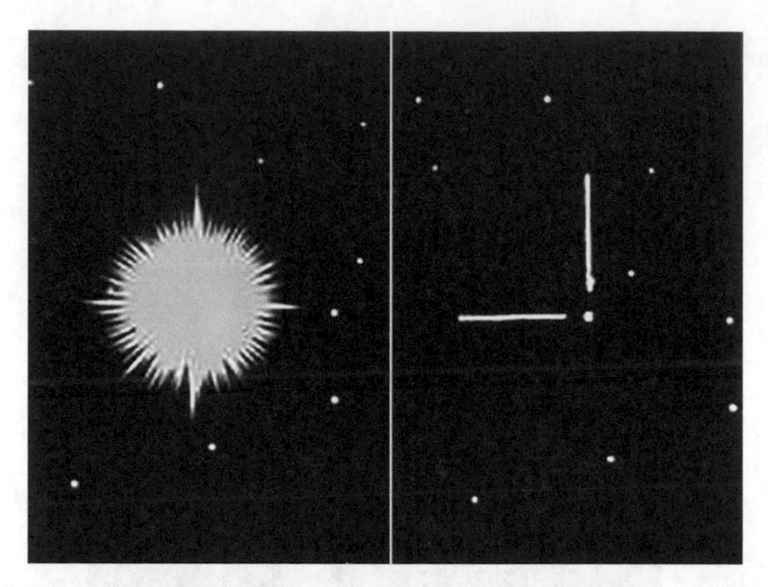

Figure 2.5. Nova Herculis 1934, during its outburst in 1934, when it reached apparent visual magnitude 1.4 and two months later (right) at apparent visual magnitude 12. The little variable star left after the explosion is called DQ Herculis. © UC Regents/Lick Observatory.

Chapter 13). Novae reach a maximum brightness of approximately 10^3 to 10^5 times the luminosity of the Sun ($L_\odot$). It has been determined that they are not new stars, because after they have faded, a faint star remains, about as luminous as the Sun, which was already present before the nova outburst.

In 1954, American astronomer Merle F. Walker, using photoelectric photometry with the 100-inch Mount Wilson telescope, discovered that the faint blue variable star DQ Herculis, located at the position of the Nova Herculis 1934 (Fig. 2.5), is an eclipsing binary with a period of 4 hr; 39 min (Walker, 1954). This *slow nova* was, in 1934, for several weeks one of the brightest stars in the sky, with an apparent visual magnitude of 1.4. In addition, Walker discovered that the brightness of DQ Her oscillates by 0.2 magnitude with a period of 71 sec. Such a short oscillation period indicates that the star must have a very high density, as the pulsation period of a star is inversely proportional to the square root of its mean mass density through the relation (see Chapter 8)

$$P_{\text{pulse}} \simeq 50\,\text{min}\left(\frac{\overline{\rho}_\odot}{\overline{\rho}}\right)^{1/2}, \tag{2.8}$$

where $\overline{\rho}_\odot$ and $\overline{\rho}$ denote the mean mass densities of the Sun and the star, respectively. As the mean density of the Sun is $1.4\,\text{g}\,\text{cm}^{-3}$, the 71 sec oscillation period of DQ Her, if it is indeed a pulsation, would indicate a mean density of this star of $3{,}500\,\text{g}\,\text{cm}^{-3}$. This is very much larger than the densities of normal stars that are powered by nuclear fusion. Such stars (even pure helium stars) never have densities larger than a few hundred g cm^{-3}.

The only stars much denser than normal stars known in 1954 were the WDs which have densities of order 10^6 to 10^7 g cm^{-3}. Nevertheless, it became clear after Walker's discovery that DQ Her must be a WD, *rotating* with a period of 71 sec, for the following reasons. WDs have, according to Eq. (2.8), pulsation periods of order 1 to 3 sec, which is much faster than the 71 sec observed in DQ Her. Therefore, DQ Her cannot be a pulsating WD. However, Eq. (2.8) also is (within a factor of order π, see Exercise 2.2) the equation for the shortest possible spin period allowed for a star. If a WD would spin with a period shorter than a few seconds, it will be ripped apart by centrifugal forces, as these become larger than the gravity that holds the star together. Although DQ Her cannot be a pulsating WD, there is no problem with it being a spinning WD, as 71 sec is much slower than the shortest allowed spin period of a WD. DQ Her is a rotating WD with a strong magnetic field, and it has two bright hot spots above its magnetic poles, which do not coincide with its rotation axis. We see the hot spots passing by with its 71 sec spin period, producing the oscillation of its light.

From the duration of the eclipses, one finds that the companion of the WD in DQ Her is a normal solar-like star, which is pear-shaped and filling its Roche lobe (critical equipotential surface, see Chapter 4), and is transferring matter to the WD. This matter spirals inward through a so-called accretion disk before reaching the surface of the WD. The hot accretion disk and the WD are the main light sources in the system, which produce the blue light and the emission-line spectrum of DQ Her.

Figure 2.6 schematically depicts a model similar to that of the DQ Her system, as derived from the observations by Walker in 1954. Subsequently, it was found that all novae are such binary systems in which one of the stars is a WD. Their orbital periods are in most cases between 1.3 and 12 hr. (More details about these systems are given in Section 5.4.) In a few cases, novae are also found in wider binaries with a subgiant or giant companion star. The WD appears to be the seat of the nova explosion, in which the binary system temporarily becomes 10^3 to 10^5 times brighter than normal. WDs are "dead" stars: that is, they are the burned-out remnant cores of stars that started their lives with a mass less than about 8 times that of the Sun (i.e., $M < 8\,M_\odot$). Their diameter is $\sim$100 times smaller than that of the Sun (i.e., they are Earth-size objects), but their mass is of the same order as that of the Sun ($\sim$330,000 times the mass of the Earth). They consist mostly of carbon and oxygen, products of the fusion of helium in their progenitor stars.

In the nova binaries, the normal star transfers matter (consisting mostly of hydrogen), from its outer layers to the WD (Fig. 2.6), and the nova outburst is due to the ignition of nuclear fusion in the hydrogen layer that is accumulating on the surface of the WD. After a critical amount of hydrogen has been accumulated, due to compression and temperature increase, unstable hydrogen fusion ignites under degenerate conditions. This leads to a thermonuclear explosion, in which much of the accreted matter is ejected, creating temporarily an expanded photosphere visible as the bright optical nova.

Later it was found that also dwarf novae occur in similar systems. Their outbursts are much less energetic (the stars become approximately 10 to 300 times brighter than

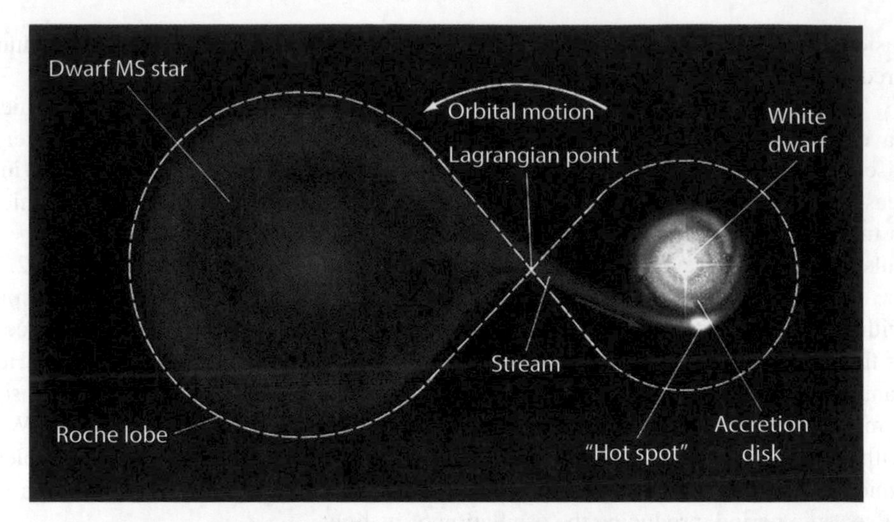

Figure 2.6. In a nova or dwarf nova binary, a normal star with a mass less than or similar to that of the Sun fills its critical equipotential surface (Roche lobe, see Chapter 4) and transfers matter to a WD star. The matter flows to the WD through an accretion disk. Where the gas stream hits the disk, a hot spot forms. Adapted after www.as.arizona.edu.

normal) and repeat on timescales between days and decades. These outbursts are caused by thermal-viscous instabilities in the accretion disk of matter surrounding the WD (Fig. 2.6). Novae and dwarf novae as a group are called CVs.

2.6 THE DISCOVERY OF THE BINARY NATURE OF THE BRIGHTEST X-RAY SOURCES IN THE SKY

In 1962, Riccardo Giacconi and colleagues, with an X-ray detector on-board a parabolic rocket flight, discovered Scorpius X-1 (Sco X-1), the brightest point X-ray source in the sky (Giacconi et al., 1962). This source, at a distance of hundreds of light years, is so bright in X-rays that when it rises above the horizon, the degree of ionization of the Earth's ionosphere increases markedly. In the 1960s, several dozen more point X-ray sources were discovered with rocket and balloon experiments, but their nature remained unclear. Figure 2.7 depicts the sky distribution of the brightest cosmic X-ray sources, clearly showing that these belong to our Milky Way Galaxy. Sandage et al. (1966) identified Sco X-1 with a faint blue star, with an emission-line spectrum, somewhat resembling that of a CV system. Shklovskii (1967) suggested that this source is a binary system like a CV, but with a NS instead of a WD as the accreting object. Thanks to the 1970 launch of the Uhuru satellite, which allowed long and uninterrupted observations of the strong celestial X-ray sources, the nature of these sources became clear: they are binaries consisting of a normal star and a NS or a BH, as will be explained in the following two sections. (For a most interesting personal account of the history of these discoveries, see Giacconi, 2005.)

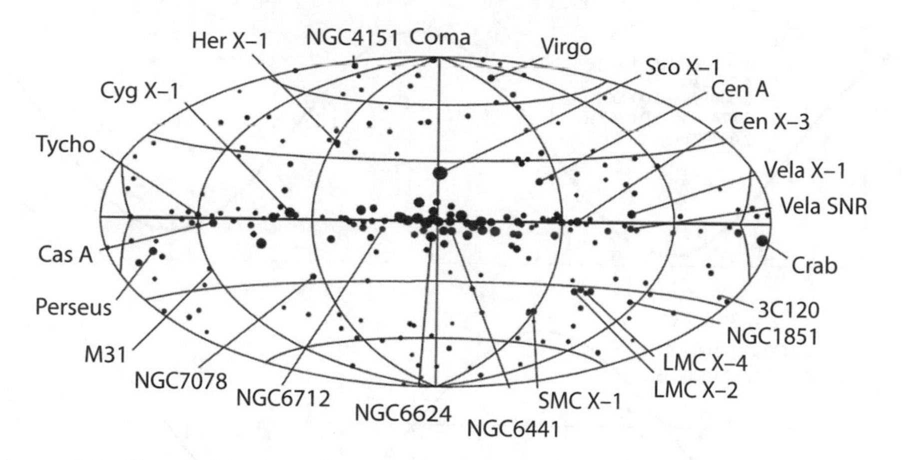

Figure 2.7. Sky distribution of the brightest cosmic X-ray sources, as recorded by UK+NASA's Ariel 5 X-ray satellite (Warwick et al., 1981; McHardy et al., 1981) (Aitoff equal area Galactic coordinate plot). Sources are represented by circles, with size indicating the strength of the source as recorded by the detector. The horizontal axis is along the Milky Way, and the center of the picture is the Milky Way center. The brightest source of the sky, above the center, is Sco X-1, a LMXB with an orbital period of 19.2 hr. The great majority of the bright circles along the Galactic plane are X-ray binaries; some of the most prominent sources are indicated by name: LMC, Large Magellanic Cloud; SMC, Small Magellanic Cloud; NGC 6624, globular cluster. Coma and Virgo are clusters of galaxies, like the Perseus Cluster. After Pounds (2020).

2.7 CENTAURUS X-3: DISCOVERY OF THE FIRST NEUTRON STAR X-RAY BINARY

Early in 1971, using the first X-ray satellite, Uhuru, developed by Riccardo Giacconi and his team, it was discovered that the source Centaurus X-3 (Cen X-3) shows regular X-ray pulsations with a period of 4.84 sec, clearly indicating that this must be a NS (for NS properties, see more in Chapters 6 and 14). However, it was found that this period is not steady; on some days the pulse period is slightly shorter, and on others it is slightly longer. Also, in the spring of 1971, it was found that on some days the X-ray source had completely disappeared, to be seen again the next day.

In November 1971, Ethan Schreier of the Uhuru team discovered that all this variability in Cen X-3 occurs with a regular periodicity of 2.087 d, and that this pulsating X-ray source is member of an eclipsing binary system with this orbital period (Schreier et al., 1972). The X-ray source is moving in a circular orbit around a normal massive star, which eclipses it every orbit for ~0.49 d, and the pulse period variations are due to the Doppler effect of the pulsar moving in its orbit around this star, as depicted in Figures. 2.8 and 2.9. The Doppler variations of the pulse period make this system a spectroscopic binary, and the projected radial velocity amplitude of the X-ray pulsar in its circular orbit, as measured from these variations, is $v \sin i = 415.1 \pm 0.4\,\mathrm{km\,s^{-1}}$,

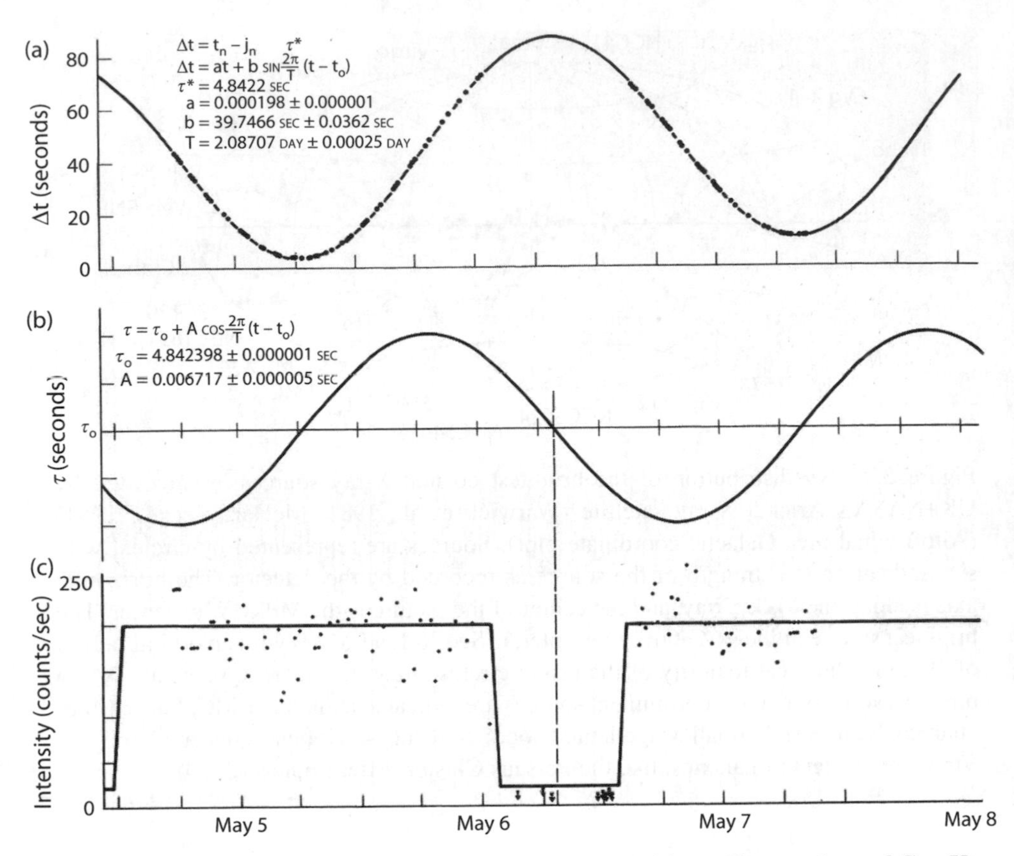

Figure 2.8. (a) Observed Doppler time delays of the 4.84 sec X-ray pulses of Cen X-3. (b) Pulse period as a function of time derived from curve (a). (c) Source intensity phased with the 2.087 d orbital period. Each dot represents the mean source intensity in a 0.1 d time interval. After Schreier et al. (1972).

where v is the orbital velocity. In addition, it is an eclipsing binary, such that a measure of the size of the normal massive companion star can be derived (Exercise 6.2). (For updated measurements and characteristics of Cen X-3, see, e.g., Shirke et al. [2021] and references therein.) How the X-ray pulses are generated by accretion of matter onto a magnetized NS is explained in Chapter 6.

2.8 CYGNUS X-1: DISCOVERY OF THE FIRST BLACK HOLE X-RAY BINARY

Around the same time as the discovery of the binary nature of Cen X-3, Bolton (1971) and Webster & Murdin (1972) noticed that in the X-ray error box (the roughly determined position on the sky) of the strong X-ray source Cyg X-1, there is a ninth magnitude blue supergiant star, which they found to be a single-lined spectroscopic binary

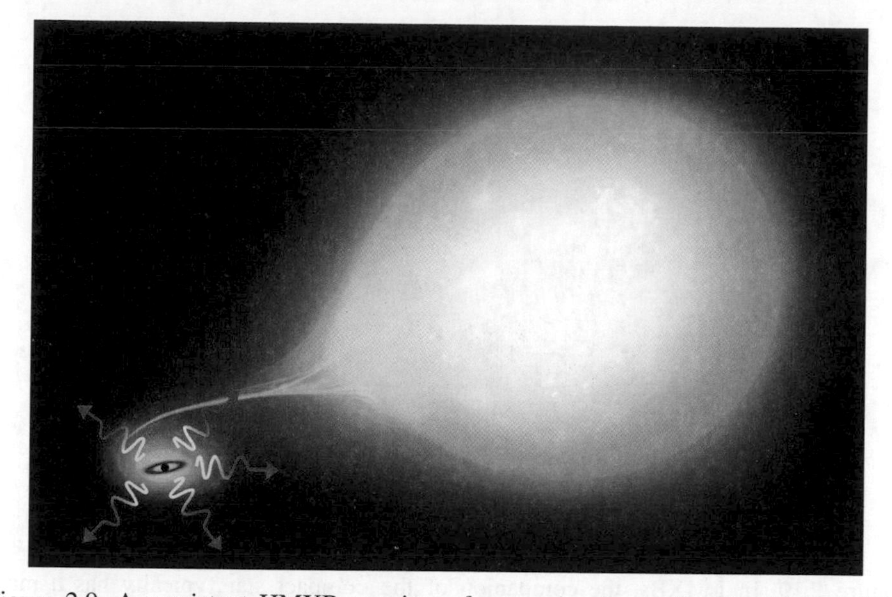

Figure 2.9. A persistent HMXB consists of a massive (typically $M > 15\,M_{\odot}$) blue supergiant O- or B-type star and a NS or BH accretor. Matter from the atmosphere of the supergiant flows toward the compact star through an accretion disk. Due to the friction of the in-spiraling matter, the inner part of this disk is heated to a temperature of order 10 million K, causing it to emit X-rays. In the case of a NS, also matter falling on the NS surface produces much X-rays. Credit: modified after original work by Mark Garlick, © Mark Garlick.

with a period of 5.6 d. The radial velocity amplitude of this star is $\sim 72\,\mathrm{km\,s^{-1}}$, and combined with the fact that the X-ray source is not eclipsed, and assuming a normal mass of $\sim 20\,M_{\odot}$ for the blue supergiant (spectral type O9.7Iab), they found the invisible companion to have mass of $>5\,M_{\odot}$ (see Section 6.1). As this is above the maximum allowed mass of $\sim 3\,M_{\odot}$ for NSs (Nauenberg & Chapline, 1973; Kalogera & Baym, 1996), they suggested that Cyg X-1 is a BH. Slightly later, in 1971, a very accurate radio position of Cyg X-1 was determined, which precisely coincided with the position of the blue supergiant star (Braes & Miley, 1971). Subsequently, it was found that changes in the radio brightness coincided in time with changes in the X-ray brightness of the X-ray source, which confirmed that the X-ray source indeed coincides with the blue supergiant star (Tananbaum et al., 1972), making clear that Cyg X-1 must indeed be a BH. The most recent determination of its mass is $M_{\mathrm{BH}} = 21.1 \pm 2.2\,M_{\odot}$ (Miller-Jones et al., 2021).

Thus, as of 1972, we have known that massive binaries with NS as well BH components do exist in nature. It was subsequently found that apart from these HMXBs, in which the companion of the compact star is an early-type star more massive than approximately $10-15\,M_{\odot}$, there is a second main category of luminous Galactic X-ray binaries, in which the companion star typically has a mass less than approximately

Figure 2.10. In LMXBs, the companion of the compact star typically has a mass $\lesssim 1.5\,M_\odot$. Like in CVs, the companion star feeds material to the compact star by Roche-lobe overflow (see Chapter 4). The compact star in these systems can either be a NS or a BH. In systems with a NS, the very long-lasting disk accretion can lead to accelerating the spin rate of the NS to more than 100 times per second, such that a millisecond X-ray pulsar results, as was briefly explained in the text and will be discussed in much detail in Chapter 14. Credit: Dana Berry/NASA GSFC.

$1.5\,M_\odot$. These are the LMXBs. Sco X-1 belongs to this second category and has an orbital period of 19.2 hr and a slightly evolved low-mass companion star. Figures 2.9 and 2.10 give an artist's impression of these two main categories of X-ray binaries. As will be explained in Chapter 6, the enormous X-ray energy output of these systems, typically approximately $10^3 - 10^5\,L_\odot$, is generated by the simple process of accretion ("falling inward") of mass onto the NS or BH. In the case of a NS, this simple accretion of an amount of mass m onto the star, releases of order $GMm/R = 0.1\,mc^2$ of gravitational potential energy, which is converted into kinetic energy (heat) and is approximately 20 times more efficient than the release of energy by nuclear fusion of the same amount of hydrogen. In the case of a fast spinning BH, as much as $0.42\,mc^2$ can be released before the matter disappears into the hole (see Section 7.6). Accretion of matter onto a NS or BH is the most efficient bulk electromagnetic energy generation process known in the Universe.

A further very important discovery of the Uhuru satellite team was that about a dozen bright LMXBs are present in globular clusters (GC; Gursky, 1973), the oldest ensembles of stars in our Galaxy, with ages more than 11 billion yr (11 Gyr). The total mass of all more than 150 Galactic globular clusters together is less than one thousandth of the mass of the Galaxy. As there are not more than about 50 bright LMXBs with similar X-ray luminosities in our Galaxy, this discovery showed that per unit mass the incidence of luminous LMXBs is on an order of 100 to 1,000 times higher in globular clusters than in the Galaxy at large (Gursky, 1973; Clark, 1975; Katz, 1975). This very high incidence of LMXBs can be understood by gravitational capture processes in the very dense cores of the clusters where, at star densities of $>10^5\,\mathrm{pc}^{-3}$, old NSs can capture a normal star as a companion. This capture process can take place in various ways: for example, by tidal “collisions” between NSs and normal cluster dwarf stars (see Fabian et al., 1975, for early work on this process), by direct collisions between NSs and giant stars (Sutantyo, 1975b), and by three-body processes, through which NSs can from time to time swap companions (Hut, 1983; Hut & Bahcall, 1983). We will return to this subject in Chapter 12.

2.9 THE DISCOVERY OF THE EXISTENCE OF DOUBLE NSs AND DOUBLE BHs

The detection by LIGO/Virgo of bursts of GWs generated by the mergers of double BHs and NSs was one of the greatest breakthroughs in physics of the past half-century (Abbott et al., 2016d, Abbott et al., 2017c, Abbott et al., 2021d). Until these discoveries, the existence of very close double BHs was not known—although the possibility of the existence of wider such systems had been predicted from theoretical arguments on the outcome of evolution of massive binary stars (Tutukov & Yungelson, 1973a; Bisnovatyi-Kogan & Komberg, 1974). On the other hand, very close double NSs (DNSs) had been discovered already in 1974 (Hulse & Taylor, 1975), and how such systems originate was long understood (see Flannery & van den Heuvel 1975; see also reviews by van den Heuvel, 1976, 1977; and Bhattacharya & van den Heuvel 1991). Presently, $\sim$20 DNS systems are known in our Galaxy (Tauris et al., 2017), half of them with orbital periods of only a fraction of a day. The reason why their orbital periods are so short has been long understood too (van den Heuvel & De Loore, 1973; van den Heuvel, 1976). Early studies of the relative detection rates of merging double BH/NS systems either predicted a negligible (Portegies Zwart & Yungelson, 1998) or comparable (Tutukov & Yungelson, 1993b) contribution of double BHs compared to that of DNSs, although some researchers had already anticipated that mergers of double BHs were to completely dominate the detection rate (Lipunov et al., 1997; Belczynski et al., 2002; Voss & Tauris, 2003), and thus a double BH merger was most likely to be detected first. Most of the models for the formation of double NSs and double BHs are basically very similar. Both are the later evolutionary products of HMXBs (as depicted in Fig. 1.1), involving a deep in-spiral phase after the massive companion star started to strongly expand due to its internal evolution. The only difference is that the double NSs originate from HMXBs with somewhat lower masses and double BHs from systems with rather high initial masses (see, e.g., Bogomazov, 2014; Belczynski et al., 2016; van den Heuvel et al., 2017). For the double BHs with component masses exceeding $25\,M_{\odot}$, alternative specific models can be conceived that have no analogues for the DNSs, as they work only for extremely close and massive binaries in metal-poor environments (de Mink & Mandel, 2016; Marchant et al., 2016). Finally, double compact objects can be produced via dynamical exchange interactions in dense

Type II SNe, they are core-collapse SNe of (initially) massive stars. The difference between Type Ib and Ic SNe is that Type Ib SNe have helium in their spectra and Type Ic do not. (A subclass of Type Ic SNe, after some time, do begin to show helium lines also. These are indicated as Ib,c.)

It appears that Type Ib SNe are due to the collapse of the naked helium core of a massive star that has been stripped of its hydrogen-rich envelope; in Type Ic SNe, also the helium envelope of the collapsing core has been stripped off before the core-collapse occurred. One conceivable way in which the hydrogen (and later the helium) envelope of a massive star can be stripped off is most likely by mass transfer in a binary system. Therefore, to make Type Ib or Ic SNe, very probably binary systems are required.

Very massive stars (mass more thatn $\sim 25\ M_{\odot}$) may also lose their hydrogen-rich envelopes during the evolution because of their strong stellar winds and possibly also because of instabilities in the red supergiant phase. They will then leave only the helium core as a so-called Wolf-Rayet (WR) star. With this *Conti scenario* for the formation of WR-stars, also single helium stars can be produced, which then may produce a single SN of Type Ib or Ic. Smartt (2009) and Smartt et al. (2017) find no evidence for Type Ib/c coming from very massive stars; these possibly could disappear without a SN, thus directly forming a BH. However, because very massive stars are quite rare, the absence, so far, of Type Ib/c SNe from massive star regions may also be an effect of small-number statistics.

Type Ia SNe appear to form a separate, distinct, and homogeneous class, with very characteristic light curves and spectra, with no helium, but strong lines of silicon, sodium, calcium, nickel, and other metals. They are very luminous SNe and occur in all kinds of galaxies. Thanks to their high luminosity and homogeneous behavior, they are excellent "standard candles" for cosmology, and it is thanks to these SNe that the acceleration of the expansion of the Universe has been discovered (Perlmutter et al., 1999; Riess et al., 1998). They are the only kind of SNe that occur in elliptical galaxies, in which star formation has ceased billions of years ago, and which therefore presently contain only low-mass stars like our Sun. For this reason, Type Ia SNe cannot be related to the deaths of massive stars. The general idea nowadays is that Type Ia SNe are thermonuclear explosions of carbon-oxygen WDs (CO WDs). Hoyle & Fowler (1960) were the first to realize that WDs consisting of carbon and oxygen still contain a large amount of nuclear fuel that, if ignited, will blow up the entire WD in a giant catastrophic thermonuclear explosion. The reason why this fusion will proceed explosively is that WDs consist of electron-degenerate matter. In such matter, the pressure is independent of the temperature. Normal stars consist in good approximation of ideal gas. In an ideal gas, an increase in temperature leads to an increase in pressure, which will make the star expand, and this expansion against gravity costs energy and leads to cooling. So, if in a star consisting of normal matter, the nuclear fusion produces too much energy, this will lead to extra heating, resulting in a pressure increase, which leads to cooling, slowing down the nuclear fusion again. So, the nuclear fusion in a normal star is stable thanks to this negative feedback reaction. In a WD, however, this negative feedback reaction is absent, and thus when nuclear fusion ignites, it will create extra heat, leading to a temperature rise, leading to enhanced fusion, leading to a further temperature rise, etc. Therefore, the entire fusion process, when started, will get completely out of hand, leading to the destruction of the entire star in a gigantic thermonuclear explosion. The final fusion product ^{56}Ni will decay to ^{56}Co on a timescale of 6.1 d, and ^{56}Co subsequently decays in 77.7 d to ^{56}Fe. The latter decay powers the Type Ia SN light curve for up to a few months.

One way to make a WD to explode, is if its mass increases to the so-called Chandrasekhar limit of $\sim 1.4\,M_{\odot}$, which is the maximum mass allowed for a stable WD. When the mass of a CO WD exceeds this limit, the star collapses, which will set off the ignition of the fusion of carbon nuclei to neon, sodium, and magnesium and of oxygen to silicon, sulpher, and other elements. As described, this fusion will completely run out of hand, causing the entire star to be converted into heavier elements with ^{56}Ni as the final fusion product.

The homogeneous behavior of the Type Ia SNe, in combination with their spectra, has originally led to the idea that they are to be identified with the thermonuclear explosions of WDs that have reached the Chandrasekhar mass limit. However, later on it was realized that a WD may also undergo a thermonuclear explosion before its mass reaches this limit. This can happen when a degenerate accreted layer of H/He on top of the original WD ignites fusion. The resulting explosion of this layer may send a shock wave into the WD that may trigger it to explode entirely or partly. These *edge-lit detonations* may also be visible as a SN Ia. These are the so-called sub-Chandrasekhar explosions. It has been found, as we will see in Chapter 13, that these explosions also may lead to the observed homogeneous behavior of SNe Ia. In both these SN Ia models, some mechanism is required to make the WD grow in mass. In elliptical galaxies, this mechanism should be able to start to operate only billions of years after the WD was formed. Because the amount of mass that a WD can accumulate from interstellar gas clouds over billions of years is negligible, the only conceivable way to make it grow in mass is by accretion of matter from a companion star in a close binary system. This companion star can either be a normal solar-like star, like in the CVs (see Fig. 2.6), or another WD. The first case is called the *single-degenerate* (SD) model and the second case the *double-degenerate* (DD) model. In the DD model, first proposed by Webbink (1984) and Iben & Tutukov (1984), the double WD system should have been born with an orbital period of less than half a day, such that by emission of GWs the orbit can shrink on a timescale less than the Hubble time, causing the two WDs to merge and explode. In the SD model, the low-mass normal companion star should be able to transfer mass at such a rate that nova explosions (which would prevent the WD to grow) will be avoided. This will require quite special conditions, as will be explained in Chapter 13. The two possible models for the Type Ia SNe are sketched in Figure 2.11. In Chapters 10, 11, and 13 we will consider the detailed binary evolution leading to SNe of Types Ia, Ib, and Ic.

2.12 BINARY NATURE OF BLUE STRAGGLERS, BARIUM STARS, AND PECULIAR POST-AGB STARS

As mentioned in Chapter 1, barium stars are G and K giants with masses up to a few solar masses, with an overabundance of barium and other s-process elements (products of slow neutron capture processes). They are wide binaries with eccentric orbits and

Figure 2.11. The two possible models for producing a Type Ia SN explosion: the SD model (right) and the DD model (left). In the SD model, the mass-donor companion of the CO WD is a normal star; in the DD model, it is another WD in a very narrow orbit. In the DD model, the WD merges with its companion due to GW radiation, which cause the stars to spiral toward each other and merge, leading to either a degenerate star with mass above the Chandrasekhar limit or a sub-Chandrasekhar WD that undergoes a double-detonation explosion (Sim et al., 2010; Shen et al., 2018; Shen et al., 2021). Credit: NASA/CXC/M. Weiss.

orbital periods ranging from approximately 100 to 10,000 d (Boffin & Jorissen, 1988). The binary nature of these stars, as well as of that of the blue stragglers in star clusters, extremely metal-depleted post–asymptotic giant branch (post-AGB) stars, carbon-enhanced metal-poor (CEMP) stars and of >80% of nuclei of planetary nebulae, was discovered in the last few decades. These discoveries show that the evolution of ~50% of all low- and intermediate-mass stars is expected to be affected by binary evolution (Abate et al., 2015). For further descriptions of these objects, we refer to Chapter 5.

For high-mass stars, the fraction of stars that exchange mass with a companion is even higher, >70% (Chini et al., 2011, 2012; Sana et al., 2012). Thus, we can safely conclude that to understand general, as well as more exotic, aspects of observed properties of stars, binary evolution and its aftermath must be understood in detail. This will be our task in this book.

EXERCISES

Exercise 2.1. *In this exercise, we consider the masses and luminosities of Sirius A and B.*

System data from observations:

Orbital period: 49.94 yr.

Parallax: $0.377''$.

Apparent magnitudes (bolometric) for A and B: -1.55 *and* $+5.69$.

Apparent semi-major axis of relative orbit: $7.62''$.

Ratio of orbits a_A/a_B: *0.466.*

Inclination: $\sim 0°$.

Absolute bolometric magnitude of the Sun: $+4.74$.

In addition, we need the following relationship between apparent bolometric magnitude, m_{bol}, *and absolute bolometric magnitude,* M_{bol}:

$$M_{bol} - m_{bol} = 5 + 5 \log p, \tag{2.9}$$

where p is the parallax in arcsec (″), and in terms of stellar luminosities:

$$M_{bol} - M_{bol,\odot} = -2.5 \log \left(\frac{L}{L_\odot} \right). \tag{2.10}$$

a. *What is the distance to Sirius in parsecs?*
b. *Calculate the absolute bolometric magnitudes for Sirius A and B.*
c. *Determine the luminosities of Sirius A and B expressed in* $L_\odot$.
d. *Determine the orbital semi-major axis in AU.*
e. *What is the mass ratio* M_A/M_B?
f. *And what is the total mass* $M = M_A + M_B$?
g. *Calculate the individual masses* M_A *and* M_B.
h. *What are the mass-to-light ratios* M/L *of Sirius A and B?*

Exercise 2.2. *Consider a star with radius, R, and mass, M. A stability criterion, leading to the limiting value of the maximum rotation of the star, must be that the magnitude of the centrifugal force must be smaller than the magnitude of the gravitational force acting on, for example, a proton at its surface in order not to dissolve the star, that is:* $F_{centrifugal} < F_{gravitational}$.

a. *Show that this simple criterion leads to the following constraint on the mean mass density:*

$$\overline{\rho} > \frac{3\pi}{GP^2}, \tag{2.11}$$

where P is the spin period of the star.

b. *Estimate the minimum spin period for the Sun ($M = 1.0\, M_\odot$ and $R = 696{,}000$ km).*

c. *Estimate the minimum spin period for a WD (assume $M = 0.7\, M_\odot$ and $R = 5{,}000$ km).*

d. *Estimate the minimum spin period for a NS (assume $M = 1.4\, M_\odot$ and $R = 12$ km).*

Chapter Three

Orbits and Masses of Spectroscopic Binaries

3.1 SOME BASICS ABOUT BINARY ORBITS

The classical view of how to construct a binary star system is based on the well-known work of Johannes Kepler and Isaac Newton, taking its basis from mass, orbital radius, and orbital energy. Kepler's three laws for planetary motion in the Solar System were modified by Newton after realizing the concept of a barycenter.[1] With Newton's modification, we can apply these laws to any gravitationally bound two-body system, thus also a binary star system (see illustrations in Fig. 3.1):

1. The orbit of each star is an ellipse with the center of mass (c.m., or the *barycenter*) as one common focus for both ellipses. These orbits around the barycenter are called the *absolute* orbits of the two stars.
2. A line connecting the two stars in a reference frame fixed on one star at a focal point (i.e., the *relative* orbit) sweeps over equal areas in equal intervals of time (the same law holds in each absolute orbit).
3. The square of the orbital period is proportional to the cube of the semi-major axis of the relative orbit.

It is from this third law that Newton derived his $1/r^2$ law of gravity. Without the work of Kepler (which he derived from Tycho Brahe's measurements of the orbit of Mars) and Huygens' equation for the centripetal acceleration, v^2/r, Newton would not have been able to derive this law.

The absolute and relative orbits have the same orbital period ($P = P_1 = P_2$) and elliptic shape (i.e., the same eccentricity, $e = e_1 = e_2$, defined in Eq. (3.1)). The point of the ellipse of closest approach of the stars is called the *periastron* and the point of largest distance the *apastron*. The line connecting periastron and apastron is called the *major axis* of the ellipse. The distance between the center of the ellipse and either

[1]Kepler thought that the Sun was the center of the Solar System, and that the Sun was in the focus of the ellipses of the planetary orbits. He did not know that the planets actually move around the barycenter of the Solar System. For example, for an isolated Sun-Earth orbit, the barycenter would be located 449 km from the center of the Sun, whereas for an isolated Sun-Jupiter orbit, the barycenter would be located just outside the surface of the Sun. The idea of barycenter came from Newton, thanks to his discovery of the law of gravity. He was the first to realize that two gravitating masses move around a common barycenter.

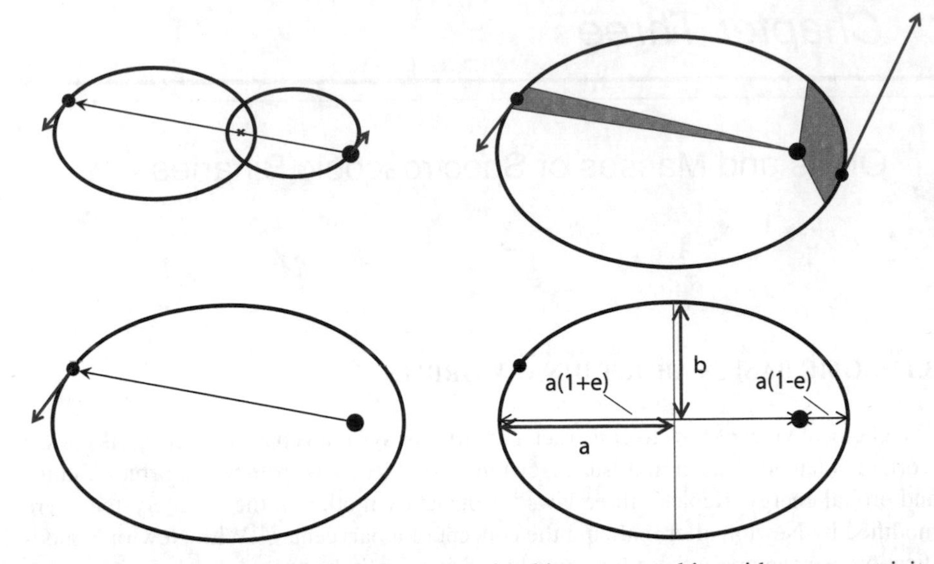

Figure 3.1. Absolute (barycentric) and relative binary star orbits with an eccentricity of $e \simeq 0.75$. The absolute orbits of the two stars are described around the barycenter (top left). The relative orbit (bottom left and both right panels) is the orbit that one star describes relative to the other one. The orbital velocity (red vector) is largest at periastron and smallest at apastron (separation between stars being $a(1-e)$ and $a(1+e)$, respectively), as dictated by Kepler's second law (top right). The semi-major, a, and semi-minor, b, axes are indicated (bottom right) and $e = \sqrt{1 - b^2/a^2}$.

periastron or apastron is the *semi-major axis*, a, of the ellipse (see Fig. 3.1). The semi-minor axis is denoted by b. Obviously, if the orbits are circular ($e = 0$), the semi-major axis (and semi-minor axis) is the radius of the circle of the relative orbit.

The periastron separation between the two stars is $a(1-e)$, and the apastron separation between the two stars is $a(1+e)$. The distance from the center of the ellipse to one of the focal points is ae. The sum of the semi-major (semi-minor) axes of the orbits of the two stars in the barycentric reference frame is $a = a_1 + a_2$ ($b = b_1 + b_2$). It should be noticed that the two stars and the c.m. always lie on a straight line during the orbital motion. Finally, the orbital eccentricity is given by

$$e = \sqrt{1 - \frac{b^2}{a^2}}. \tag{3.1}$$

It follows from Kepler's second law that the ratio between two areas swept out by one star is equal to the ratio between the two time intervals

$$\frac{A_1}{A_2} = \frac{\Delta t_1}{\Delta t_2}. \tag{3.2}$$

This relation describes orbital velocity as greatest at periastron and lowest at apastron, as depicted in Figure 3.1 for two equal time intervals ($\Delta t_1 = \Delta t_2$).

Notice that according to Newton's $1/r^2$ law of gravity, it also follows that the orbit of a binary will be an ellipse. The ratio of the semi-major axes in the barycenter reference frame, a_1 and a_2, is given by

$$\frac{a_1}{a_2} = \frac{M_2}{M_1}, \tag{3.3}$$

where M_1 and M_2 are the masses of the two stars. (In the same way that a heavier weight must be placed closer to the fulcrum of a balance beam, the heavier star must be closer to the barycenter—but this relation also follows from conservation of momentum: $M_1\vec{v}_1 + M_2\vec{v}_2 = 0 \Leftrightarrow M_1 v_1 = M_2 v_2$.)

The orbital energy ($E_{\rm orb} = E_{\rm pot} + E_{\rm kin}$) is given by

$$E_{\rm orb} = -\frac{GM_1M_2}{2a} = -\frac{GM_1M_2}{r} + \frac{1}{2}\mu v^2, \tag{3.4}$$

where $\mu \equiv M_1M_2/(M_1+M_2)$ is the reduced mass, and from this expression it is trivial to find the relative velocity of the two stars at any given position:

$$v = \sqrt{GM\left(\frac{2}{r} - \frac{1}{a}\right)}, \tag{3.5}$$

where $M = M_1 + M_2$ is the total mass of the system. For a circular orbit ($a = r$), Eq. (3.5) reduces to the well-known expression $v = \sqrt{GM/r}$. It is evident, that for an eccentric system, v will vary strongly during the orbital motion of the stars. This is illustrated in Figure 3.2, which shows the dependence of v on the orbital phase for three different eccentricities.

For simplicity, consider now a binary system with a circular orbit ($a = r$). We can combine the laws from Newton's mechanics for the force of gravity:

$$F_{\rm grav} = G\,\frac{M_1\,M_2}{a^2}, \tag{3.6}$$

and the centripetal force from the orbital motion:

$$F_{\rm centripetal} = \mu a\,\Omega^2, \tag{3.7}$$

where G is the gravitational constant, r is the distance between the two stars, and $\Omega = 2\pi/P$ is here the angular velocity, to derive Kepler's third law (Exercise 3.1):

$$\Omega^2 = \frac{GM}{a^3}. \tag{3.8}$$

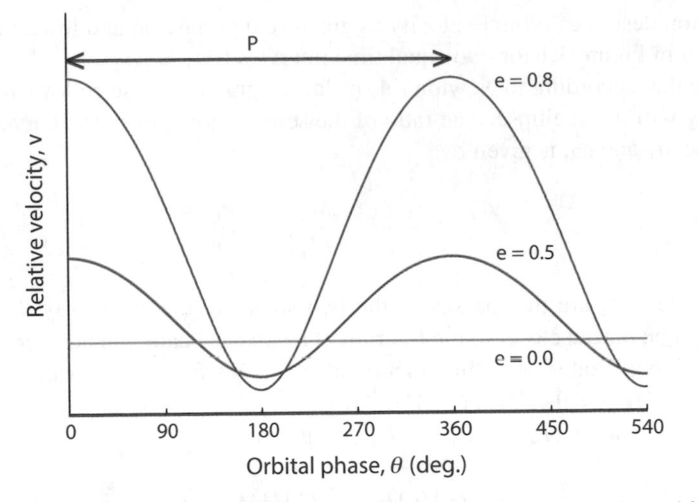

Figure 3.2. Relative velocity between two stars in systems with similar orbital period, but different eccentricities, as a function of orbital phase (true anomaly). See also Figure 3.5 for the absolute velocity of one of the components for different angles between our line of sight and the direction of the semi-major axis.

3.2 ORBIT DETERMINATION

Because interacting binaries have relatively short orbital periods, they will in general be observed as spectroscopic and/or eclipsing binaries, not as visual binaries. For this reason, we restrict ourselves here to the determination of the orbital elements of spectroscopic binaries. We now consider the motion of one of the components. There are in total six orbital elements that define the orbit of a binary: a, e, i, ω, T, and ϖ, the first five of which will be defined here. For spectroscopic binaries, the sixth one, ϖ (the orientation of the line of nodes on the plane of the sky), is of no importance and will not be considered here.

Figure 3.3 illustrates the geometry of the situation. The plane of the sky is here the XY-plane, and our line of sight to the star is the Z-axis, which is perpendicular to the XY-plane. We take the central focal point of the orbital ellipse to be located on the plane of the sky. The orbital plane of the binary cuts the plane of the sky along the line of nodes, AB, and is inclined with respect to the plane of the sky by the inclination angle, i. The ascending node, A, is the place where the star, crossing the plane of the sky, moves away from us, and in the descending node, B, it moves toward us. (Often in the literature the ascending node is indicated with the symbol Ω, and the descending node with the same symbol upside down; to avoid confusion with the use of the same symbol for angular velocity, we will in this book not use Ω and its upside-down counterpart for indicating the nodes.) One observes in the figure, the periastron, P, of the orbit. The angle ω, measured in the orbital plane from the ascending node, A, to P in the direction of the orbital motion, is called the *length of the periastron*.

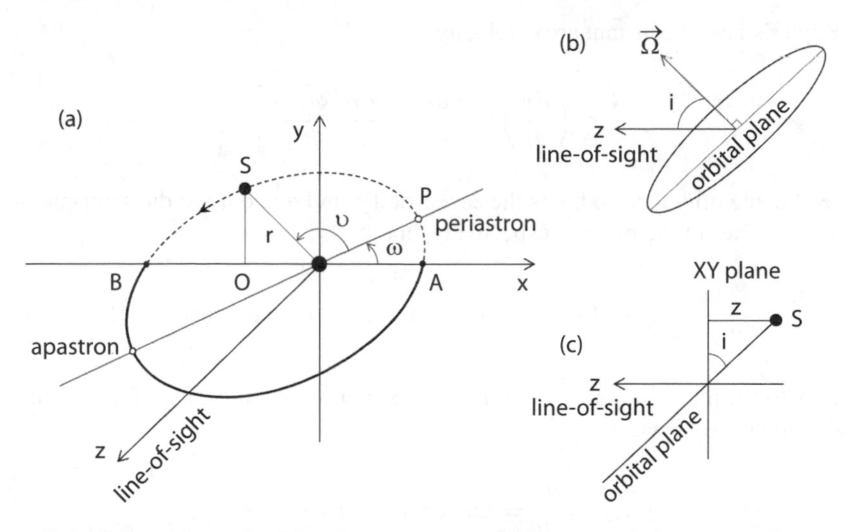

Figure 3.3. Orientation of orbit of a spectroscopic binary relative to an observer on Earth, with its various parameters indicated. The panels (a), (b) and (c) display the same orbit seen from different directions. The XY-plane is the plane of the sky; Z-direction is toward Earth. S is the position of the star in its orbit.

Further orbital elements are the semi-major axis, a; the eccentricity, e; and the time, T (in Julian Dates), at which the star passed through periastron. The true anomaly, v, of the star is the angle in the orbital plane between the periastron, P, and the actual position, S, of the star in its orbit, which is measured from the periastron in the direction of the orbital motion.

The distance of the star behind and in front of the plane of the sky is z or $-z$, respectively, and the radial velocity of the star is dz/dt. Consider now the line OS, drawn in the orbital plane perpendicular to the line of nodes, AB. It will be clear that

$$z = OS \cdot \sin i = r \sin i \cdot \sin(v + \omega), \tag{3.9}$$

where r is the radius vector from the focal point to the position S of the star.

Hence

$$\frac{dz}{dt} = \left(r \cos(v + \omega) \cdot \frac{dv}{dt} + \frac{dr}{dt} \cdot \sin(v + \omega) \right) \sin i. \tag{3.10}$$

Using the relation between the radius vector and the true anomaly:

$$r = \frac{a\,(1 - e^2)}{1 + e \cos v} \tag{3.11}$$

and Kepler's law of constant areal velocity:

$$\frac{1}{2} r^2 \left(\frac{dv}{dt} \right) = \frac{\pi ab}{P} = \frac{\pi\, a^2 \sqrt{1-e^2}}{P}, \tag{3.12}$$

where P is the orbital period, e is the eccentricity, and a and b are the semi-major and semi-minor axes of the binary, respectively, one finds

$$\frac{dz}{dt} = \frac{2\pi a \sin i}{P\sqrt{1-e^2}} \cdot [(1 + e \cos v) \cdot \cos(v + \omega) + e \sin v \cdot \sin(v + \omega)]. \tag{3.13}$$

It is easy to see that after applying some trigonometric identities, Eq. (3.13) results in a radial velocity (Exercise 3.2):

$$v_{\rm rad} = \frac{dz}{dt} = K\,[e \cos \omega + \cos(v + \omega)], \tag{3.14}$$

where

$$K = \frac{2\pi a \sin i}{P\sqrt{1-e^2}}, \tag{3.15}$$

which is known as the *semi-amplitude* of the radial velocity orbit (i.e., projected along the line of sight to the observer). This quantity can be measured for each of the two stars individually, that is, $K_1(a_1)$ and $K_2(a_2)$, where a_i is the semi-major axis of the orbit of star i in the c.m. reference frame.

The radial velocity of the star with respect to the c.m. of the Solar System (which is close to the edge of the Sun) is

$$\frac{d\zeta}{dt} = \gamma + K\,[e \cos \omega + \cos(v + \omega)], \tag{3.16}$$

where γ is the radial velocity of the c.m. of the binary with respect to the c.m. of the Solar System. (In observing from Earth, one must still correct the measured radial velocity of the star for the orbital motion of Earth around the c.m. of the Solar System with a velocity of $\sim$30 km s^{-1}. Application of this *solar correction* is a standard procedure in every observatory when measuring stellar radial velocities.)

The task of the astronomer is to determine, from a series of measurements of the radial velocities $d\zeta/dt$ with respect to the c.m. of the Solar System, the parameters that describe the orbit of the binary: P, e, ω, K, γ, and T. The orbital period, P, in days, follows quite easily from the recurrence times of the same radial velocities. And γ, as average of all measurements of the radial velocities, is also quite easily obtained in good approximation. The shape of the radial velocity curve already gives an indication of the values of the eccentricity, e and of ω (e.g., see Figs. 3.5 and 3.6 for shapes of radial velocity curves as a function of e and ω), which one can use as a starting value to begin the orbital analysis. The value of ω also yields a starting value of T. With these starting values one can find all orbital parameters by iteration. The average angular velocity in the orbit is

$$\mu_p = \frac{2\pi}{P} \text{ rad s}^{-1}. \tag{3.17}$$

The *mean anomaly*, M, is given by

$$M = \mu_p (t - T), \tag{3.18}$$

where the observation time, t, and the time of periastron passage, T, are expressed in Julian days. Equation (3.18) yields values of M for all observations. The mean anomaly, M, and the true anomaly, v, are related through the eccentric anomaly, E:

$$M = E - e \sin E, \tag{3.19}$$

where

$$\tan\left(\frac{E}{2}\right) = \sqrt{\frac{1-e}{1+e}} \tan\left(\frac{v}{2}\right). \tag{3.20}$$

There are well-known numerical methods for solving Eqs. (3.19) and (3.20) for obtaining v, when M is known and using an estimated value of e. The numerical program is aimed at obtaining, by an iterative procedure, a least-squares definitive fit to the observed radial velocity data, which yields the best fitting values of e, ω, and T. See, for example, Katoh et al. (2013).

3.2.1 Orbital Passage Time near Periastron

It is often of interest to calculate the relative amount of time that a binary system spends near periastron passage, for example, for the purpose of planning X-ray observations related to mass-transfer between stellar components near periastron passage in a transient X-ray binary system. The more eccentric the orbit, the shorter is the epoch near periastron passage. For a binary system with a reference frame fixed on one of the stellar components, we derived (Tauris et al., 1999) an analytical expression for the fraction of the orbital period the companion star spends between orbital phases θ_1 and θ_2 (see Fig. 3.4):

$$\left(\frac{\delta P_{\text{orb}}}{P_{\text{orb}}}\right) = 1 - \frac{1}{2\pi}\left[\frac{\sqrt{1-e^2}\, e \sin\theta}{1 - e\cos\theta} + \tan^{-1}\sqrt{\frac{1+e}{1-e}} \tan\left(\frac{\theta}{2}\right)\right]_{\theta_1}^{\theta_2}. \tag{3.21}$$

As an example, consider a hypothetical X-ray binary[2] that has an orbital period of 16.6 d and an eccentricity of $e = 0.94 \pm 0.04$. Using Eq. (3.21), one can find that it only

[2] A large eccentricity of $e = 0.94 \pm 0.04$ was derived by Tauris et al. (1999) for Cir X-1 from SN simulations based on a large radial velocity of $\sim$430 km s^{-1}, erroneously reported by Johnston et al. (1999). The current estimate of the eccentricity is much smaller ($e \sim 0.4$) and the companion mass in Cir X-1 is estimated to be 4–10 $M_\odot$ (Johnston et al., 2016).

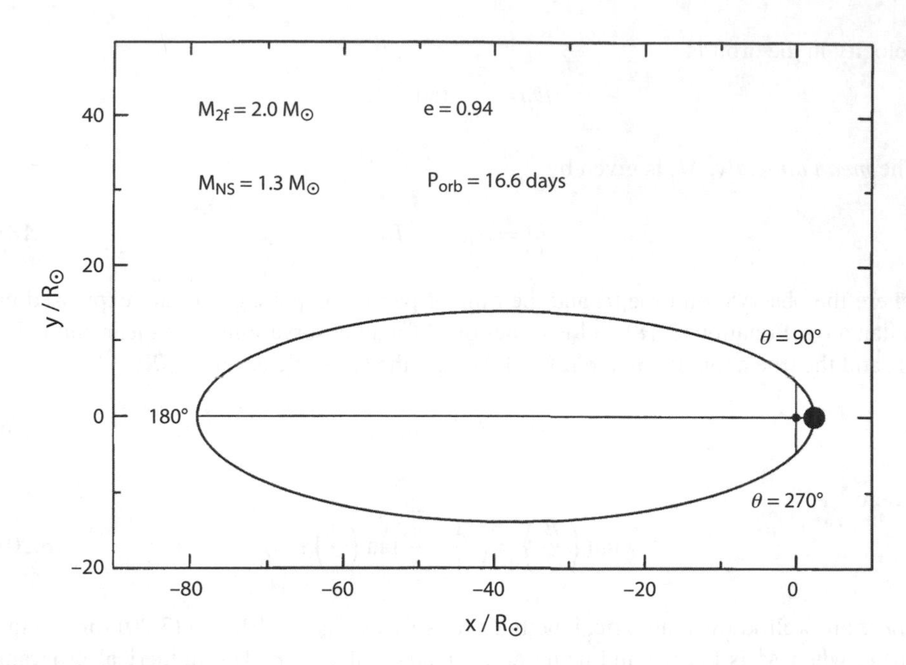

Figure 3.4. Orbital geometry of a hypothetical X-ray binary in a reference frame fixed on the NS accretor. The companion star is assumed to have a mass of $2.0\,M_\odot$ and $e = 0.94$. The periastron passage $\theta_1 = 270°$ ($-90°$) to $\theta_2 = 90°$ only lasts ~ 3.5 hr, which is a small fraction of the full orbital period of 16.6 d. After Tauris et al. (1999).

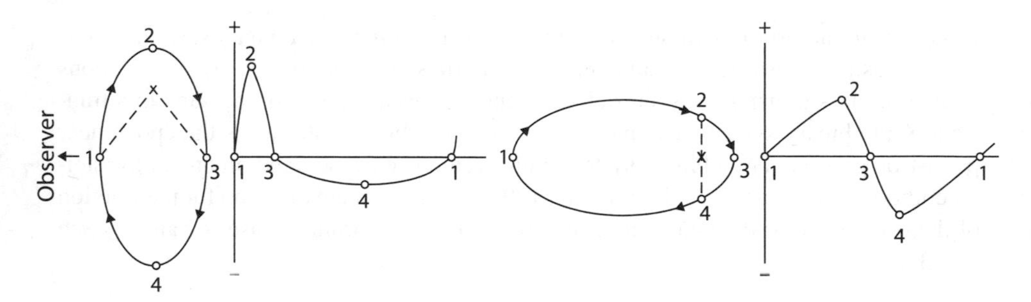

Figure 3.5. For systems with the same elliptic orbit, the shape of the radial velocity curve observed from Earth depends strongly on the value of the length of periastron ω. It is assumed that the line of sight to Earth is in the orbital plane. (Left) The image shows the radial velocity curve for a binary with $\omega = 0°$ (we look perpendicular to the major axis of the system). Due to Kepler's law of equal areas, between points 1, 2, and 3 the star moves quickly and reaches a high positive peak velocity; from points 3 to 4 to 1, it moves slowly with a low velocity amplitude. (Right) The same system has $\omega = 90°$ (we look along the major axis) and shows a saw tooth shape radial velocity curve.

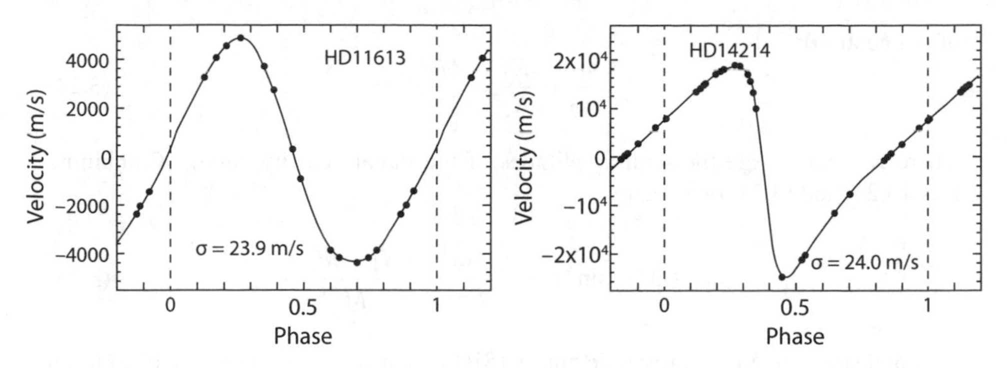

Figure 3.6. Two examples of well-determined radial velocity orbits. The dots are the observations and the curves are the best-fit radial velocity orbits. The orbital parameters of HD 11613 are $P = 836.8826$ d, $e = 0.1285$, $\omega = 63.61°$, $K_1 = 4.635\,\mathrm{km\,s^{-1}}$; those of HD 14214 are $P = 93.2828$ d, $e = 0.52414$, $\omega = 104.03°$, $K_1 = 19.264\,\mathrm{km\,s^{-1}}$. From these data, the mass functions can be determined (Exercise 3.4). After Katoh et al. (2013). (©) AAS. Reproduced with permission.

takes ∼3.5 hr for the companion star to move from $\theta_1 = 270°$ ($-90°$) to $\theta_2 = 90°$ (see Fig. 3.4). Hence, the companion star spends >99% of its time between these angles: $90° < \theta < 270°$. Observations must be planned very carefully to catch the periastron passage where, for example, the effects of companion heating owing to X-ray irradiation are most severe. Note, the expression in Eq. (3.21) only depends on the eccentricity and the angular width of the considered orbital phase.

3.3 DETERMINATION OF STELLAR MASSES

The projected semi-major axis $a \sin i$ of the orbit follows from Eq. (3.15) once P, e, and K are known. When P is in days and K in km s^{-1}, one has

$$a \sin i = 1.375 \times 10^4\,\mathrm{km} \quad K\,P\,\sqrt{1 - e^2}. \tag{3.22}$$

From Kepler's third law one has

$$M_1 + M_2 = \frac{(a_1 + a_2)^3}{P^2}. \tag{3.23}$$

Here a_1 and a_2 are expressed in astronomical units, P is in years, and M_1 and M_2 are in solar masses. When the spectra of both stars are visible, one has (from conservation of momentum):[3]

$$\frac{a_1}{a_2} = \frac{K_1}{K_2} = \frac{M_2}{M_1}, \tag{3.24}$$

where K_1 and K_2 are the semi-amplitudes of the radial velocity curves. Combining Eqs. (3.23) and (3.24) one obtains

$$(M_1 + M_2) \sin^3 i = \frac{a_1^3 \sin^3 i}{P^2} \frac{(M_1 + M_2)^3}{M_2^3}. \tag{3.25}$$

For a single-lined spectroscopic binary (SB1), $a_1 \sin i$, according to Eq. (3.22), can be found from the radial velocity curve of star 1, such that with Eq. (3.25) one obtains the so-called *mass function*, $f(M_1, M_2, i)$, given by (see Exercise 3.3)

$$f(M_1, M_2, i) \equiv \frac{(M_2 \sin i)^3}{(M_1 + M_2)^2} = \frac{v_1^3 P}{2\pi G} \tag{3.26}$$

$$= 1.036 \times 10^{-7} M_\odot \quad K_1^3 P (1 - e^2)^{3/2}, \tag{3.27}$$

where v_1 is the projected orbital velocity amplitude of star 1 (parallel to the line of sight), and here in the second expression, again, the orbital period, P, is in days, and the semi-amplitude of the radial velocity, K_1, is in km s^{-1}.

When two spectra are visible (one for each star), one finds (see Exercise 3.3)

$$M_2 \sin^3 i = 1.036 \times 10^{-7} M_\odot \quad K_1 (K_1 + K_2)^2 P(1 - e^2)^{3/2} \tag{3.28}$$

and

$$M_1 \sin^3 i = 1.036 \times 10^{-7} M_\odot \quad K_2 (K_1 + K_2)^2 P(1 - e^2)^{3/2}. \tag{3.29}$$

The inclination angle, i, can be solved only if the system is also eclipsing. For this reason, double-lined spectroscopic binaries that also are eclipsing are our most valuable sources for precise determination of stellar masses. These binaries can be found, for example, in the catalogs of Popper (1980), Andersen (1991), Moe & Di Stefano (2015), and Zasche et al. (2020), and in the list of accurate masses and radii of normal stars by Torres et al. (2010). Orbital elements of spectroscopic binaries can be found in the "Ninth Catalogue of Spectroscopic Binary Orbits" (Pourbaix et al., 2004) and in Katoh et al. (2013).

3.4 MASSES OF UNEVOLVED MAIN-SEQUENCE STARS

Table 3.1 lists the best-determined values of masses, effective temperatures, and bolometric luminosities (in units of the bolometric solar luminosity, $L_\odot$) of main-sequence stars (Morgan-Keenan luminosity class V) as a function of spectral type, as we derived

[3]Note, also the following relations: $a = a_1 + a_2$, $M = M_1 + M_2$, and $K = K_1 + K_2$.

Table 3.1. Average Masses, Effective Temperatures and Luminosities of Main-Sequence Stars

Spectral type	Mass ($M_\odot$)	$T_{\rm eff}$ (K)	$\log L/L_\odot$	Spectral type	Mass ($M_\odot$)	$T_{\rm eff}$ (K)	$\log L/L_\odot$
O7V(1)	27.27 ± 0.55	38,000	5.09 ± 0.02	A2−4V(6)	2.05 ± 0.20	8,650	1.36 ± 0.20
O8V(2)	22.10 ± 0.5	34,000	4.97 ± 0.05	A5−8V(10)	1.96 ± 0.40	8,000	1.25 ± 0.50
O9V(2)	18 ± 1.0	33,200	4.72 ± 0.07	F0V(3)	1.64 ± 0.20	7,090	0.86 ± 0.30
O9.5V(2)	16.5 ± 0.3	30,500	4.63 ± 0.01	F1−2V(4)	1.49 ± 0.14	6,840	0.78 ± 0.16
B0.5V(4)	13.5 ± 1.7	29,200	4.35 ± 0.20	F3−4V(5)	1.38 ± 0.20	6,650	0.60 ± 0.15
B1V(3)	12.2 ± 2.0	28,100	4.18 ± 0.01	F5V(12)	1.34 ± 0.15	6,460	0.53 ± 0.20
B2V(3)	9.3 ± 0.9	24,300	3.73 ± 0.15	F6V(2)	1.32 ± 0.15	6,200	0.47 ± 0.20
B2.5V(3)	6.03 ± 0.05	18,800	3.18 ± 0.03	F7−9V(14)	1.28 ± 0.20	6,180	0.55 ± 0.25
B3V(2)	5.7 ± 0.50	18,300	3.23 ± 0.06	G0V(2)	1.16 ± 0.02	5,960	0.25 ± 0.15
B3.5V(2)	5.12 ± 0.23	17,800	2.85 ± 0.10	G1−3V(5)	1.08 ± 0.03	5,680	0.28 ± 0.12
B4V(3)	4.96 ± 0.36	16,700	2.81 ± 0.15	G5V(6)	1.02 ± 0.10	5,680	0.12 ± 0.09
B5V(5)	4.84 ± 0.43	15,800	2.75 ± 0.15	G8−9V(5)	0.93 ± 0.02	5,300	-0.19 ± 0.07
B6−7V(3)	4.02 ± 0.10	14,900	2.44 ± 0.06	K0V(2)	0.92 ± 0.04	5,235	-0.17 ± 0.13
B8V(3)	3.25 ± 0.40	11,800	$2{,}04 \pm 0.25$	K1−2V(2)	0.89 ± 0.04	5,110	-0.36 ± 0.05
B9V(4)	2.97 ± 0.40	10,700	2.00 ± 0.40	K3V(3)	0.83 ± 0.07	4,825	-0.47 ± 0.10
B9.5V(2)	2.56 ± 0.24	10,100	1.60 ± 0.12	M1V(4)	0.60 ± 0.01	3,860	-1.12 ± 0.04
A0V(3)	2.24 ± 0.30	10,100	1.46 ± 0.18	M3.5V(2)	0.42 ± 0.02	3,140	-1.83 ± 0.05
A1V(5)	2.15 ± 0.20	9,380	1.47 ± 0.20	M4.5V(2)	0.22 ± 0.01	3,125	-2.28 ± 0.03

Derived from the best-studied double-lined eclipsing spectroscopic binaries, from the catalog of Torres et al. (2010). Numbers between parentheses after the spectral types indicate the number of stars in the catalog from which we derived the data.

from double-lined eclipsing spectroscopic binaries in the catalog of Torres et al. (2010). The uncertainty ranges of the masses and luminosities indicated at each spectral type were derived from the ranges of masses and luminosities measured at each spectral type. The luminosities were derived from the measured stellar radii and effective temperatures. Here it should be realized that a spectral type is not a quantitatively measured parameter, rather it is determined by visual inspection of spectra by experts in stellar spectroscopy. This always implies some subjectivity. Also, the same spectral type may not always correspond to exactly the same evolutionary stage of the star. For example, a star classified as A0V may be close to the zero-age main sequence (ZAMS), or it may be already halfway through its core hydrogen-burning phase (see the evolutionary tracks of single stars in Chapter 8). In the first case, its measured mass and luminosity will be smaller than in the latter case. In the latter case, it probably started out as a B9.5V star on the ZAMS. And if it would be near the end of core hydrogen burning, its mass and luminosity would still be higher, as in this case it started out like a B9V star on the ZAMS. This is the reason why the masses and luminosities of binary components, which have been assigned the same spectral type, may differ considerably. For certain spectral types, there are only very few measured masses and luminosities of binary components available. In such a case, the observed range of masses and luminosities for that spectral type, as indicated in Table 3.1, may be artificially small. This is, for example, the case for spectral types O7V to O9.5V in the table. For others, such as F0V and F5V, many binaries of those spectral types are available, with a considerable spread in the measured masses and luminosities, such that the indicated ranges of these parameters in the table may be relatively large. Therefore, the ranges indicated in the table are not probable errors, instead they are just due to the observed ranges in mass and luminosity of the available binaries at these spectral types, for which these quantities could be measured by Torres et al. (2010).

If one plots the observed luminosities of Table 3.1 against the stellar masses, in a log-log diagram, one observes that there is a clear relation between the mass and luminosity of the main-sequence stars, which is called the *mass-luminosity relation*. For masses larger than 1.3 $M_\odot$, this relation in the log-log diagram is a straight line with slope 3.5, which means that for masses greater than 1.3 $M_\odot$, the following holds:

$$\frac{L}{L_\odot} = \left(\frac{M}{M_\odot}\right)^{3.5}. \tag{3.30}$$

3.5 THE MOST MASSIVE STARS

Masses of O2–O6 main-sequence stars are not accurately known from eclipsing spectroscopic binaries. But these stars, with masses upward of 30 $M_\odot$, are known to exist in considerable numbers in young star clusters in our Galaxy and the Magellanic Clouds, two small satellite galaxies of our Galaxy on the southern sky. In the Galaxy, these stars are found, for example, in the star clusters NGC 3603, Pismis 24, Westerlund 1, and the Quintuplet Cluster near the Galactic center. They range in bolometric luminosity from $6 \times 10^5\ L_\odot$ for the O3.5 star Pismis 24-1SE, to $> 4 \times 10^6\ L_\odot$ for the Pistol Star in the

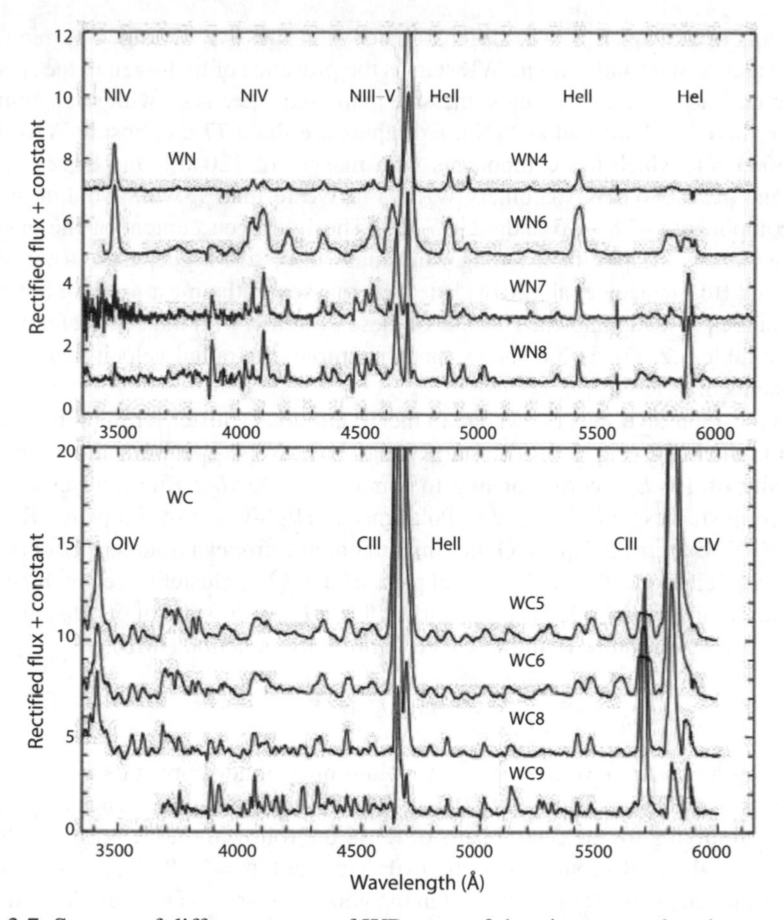

Figure 3.7. Spectra of different types of WR-stars of the nitrogen and carbon varieties. A series of WN-type spectra (top) and of WC-type spectra (bottom). Adapted after Crowther (2007).

Quintuplet Cluster (Figer et al., 1998). (Here $L_\odot$ indicates the bolometric luminosity of the Sun.) According to the theoretical mass-luminosity relation for hydrogen-burning stars (see Chapter 8), such stars must have masses ranging from 60 to >200 $M_\odot$. However, even more massive stars are found in the Large Magellanic Cloud (LMC).

The most massive main-sequence stars can have bolometric luminosities up to $\sim 10^7\ L_\odot$. Due to their enormous luminosities, in combination with their effective temperatures of ∼50,000 K, the radiation pressure in the outer layers of these stars is so large that these layers are forced to flow out in the form of an extremely strong stellar wind. These stellar winds produce strong emission lines in the spectra of these stars, of hydrogen, ionized helium, carbon, and nitrogen. This makes that their spectra resemble those of WR-stars (Fig. 3.7), which are very luminous helium stars, that are the remnant helium cores of massive stars that have lost their entire hydrogen-rich envelopes (for details see Chapter 5). The difference between massive hydrogen-burning upper main-sequence stars and genuine WR stars is the presence of hydrogen in the spectra of the former. The spectra of the most massive main-sequence stars with strong emission lines are therefore indicated as WNh. Examples are the 3.77 d eclipsing WNh-binary NGC 3603-A1, which has components with masses of $120\,M_\odot$ and $80\,M_\odot$, respectively, and the 3.686 d WNha binary WR20a in Westerlund 1, with two almost equal-mass components of 83 ± 5 and $82\pm5\,M_\odot$. The hydrogen content of the envelopes of these stars is, despite their strong wind mass loss, still higher than 40%. Martins (2015) and Bogomazov et al. (2018) listed the masses of the most massive binary systems that are presently known. The data of these systems, with their key references, are listed in Table 3.2. Figure 3.8 shows, as an example, the radial velocity curves of the components of WR20a.

The most massive stars known are in the 30 Doradus Cluster R136 in the Tarantula Nebula in the LMC (Fig. 3.9). The star R136a1 has a WNh spectrum and a bolometric luminosity of $10^7\,L_\odot$, corresponding to a mass of $300\,M_\odot$. Khorrami et al. (2017) found, from studies with the Spectro-Polarimetric High-Contrast Exoplanet Research, or SPHERE, Extreme Adaptive Optics instrument at European Southern Observatory's Very Large Telescope, that in the central parts of the R136 cluster there are three to five stars with a mass of $\sim 300\,M_\odot$, ~ 20 stars of $125\,M_\odot$, ~ 70 stars of $50\,M_\odot$, and ~ 200 stars of $20\,M_\odot$. The slope of the so-called *initial mass function* (IMF),

$$\Psi(M) = C\,M^{-\alpha} \tag{3.31}$$

(where C is a constant), which represents the number of newborn stars as a function of mass, is here in R136 as low as -1.90 (i.e., $\alpha = 1.90$), for an assumed age of the cluster of 1 Myr, implying for the central parts of R136 the following IMF: $\Psi(M) \propto M^{-1.90}$. Schneider et al. (2018) found the same IMF exponent, $\alpha = 1.90^{+0.37}_{-0.26}$, from studies of massive stars ($15-200\,M_\odot$) further out in the young cluster 30 Doradus. A similar IMF slope of $\alpha = 1.90$ was found for the massive Arches Cluster near the Galactic center by Figer (2005) and Kim et al. (2006). If IMFs with such low exponents are representative for the massive-star formation history (however, see Schootemeijer et al., 2021) within the observable LIGO–Virgo-KAGRA volume of GWs of the local Universe, then this will have a significant impact on the detected merger rate of double NSs and BHs (Kruckow et al., 2018).

3.6 FALSIFICATION OF RADIAL VELOCITY CURVES

We largely follow here what Underhill (1966) remarked about possible deformations of the radial velocity curves. The implicit assumption in determining the radial velocity curve of a spectroscopic binary is that the Doppler displacements of the spectral lines give a true measure of the radial velocity of the c.m. of the star. This will be true if the star is a uniformly illuminated circular disk and if there are no systematic motions of the gas in the stellar atmosphere with respect to the c.m. of the star.

Table 3.2. Masses of the Most Massive Binaries Known in the Galaxy and of R145 in the LMC.

Name	P_{orb} (d)	Spectral types	Masses ($M_\odot$)	Key reference
WR 20a	3.686	WNha/O3If + WNha/O3If	$(83 \pm 5, 82 \pm 5)$	Bonanos et al. (2004), Rauw et al. (2004)
WR 21a	31.672	O3/WNha + O3V	$(103.6 \pm 10.2, 58.3 \pm 3.7)$	Tramper et al. (2016)
WR 22	80.336	WR + O	$(55.3 \pm 7.3, 20.6 \pm 1.7)$	Rauw et al. (1996)
NGC 3603-A1	3.772	WNh6 + WNh6	$(116 \pm 31, 89 \pm 16)$	Schnurr et al. (2008)
HDE 311884	6.34	WNh6 + O5V	$(\sim 51, \sim 60)$	van der Hucht (2001)
R145	158.76	WRh + O	$(\geq 116 \pm 33, \geq 48 \pm 20)$	Schnurr et al. (2009)

Stars with spectral type WNh have very strong emission lines, resembling the spectra of WR-stars. But unlike the classical WR-stars, they do have hydrogen in their spectra, indicating that they are very massive main-sequence stars (like very early O-stars) characterized by very strong stellar-wind mass loss. Spectrum Of indicates O-stars with strong emission lines. See the text for further explanations.

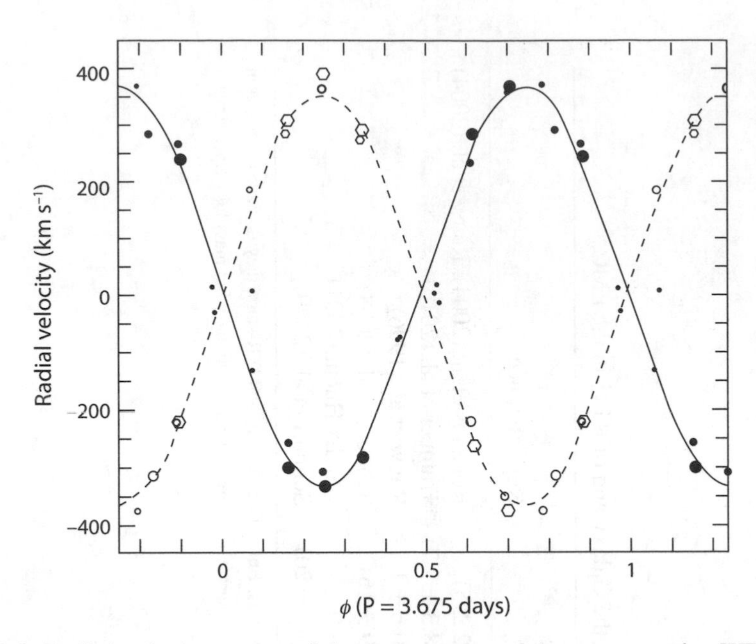

Figure 3.8. Radial velocity curves of the components of the very massive WR binary WR20a, with an orbital period of 3.675 d. After Rauw et al. (2004). © ESO. Reproduced with permission.

The following five types of disturbances of the radial velocity curves have been recognized: (1) the rotation effect, (2) the presence of gas streams in the system, (3) the reflection effect, (4) the heating effect, and (5) the deformation effect. That the radial velocity curve does not give a correct representation of the real line of sight motions of the star is most easily recognized from the fact that the value of the orbital eccentricity derived from the radial velocity curve differs from that derived from the eclipsing binary light curve. For example, one may see from the eclipses that the orbit must be circular, as the primary and secondary eclipses of the system differ exactly 0.50 in orbital phase, whereas still the radial velocity curve yields a clear orbital eccentricity different from zero.

An example of this is shown in Figure 3.10, which depicts the observed radial velocity curve of the B8 component of the circular-orbit Algol-type eclipsing binary U Cephei (after Struve & Huang, 1958). Here one sees that the radial velocity curve is not sinusoidal, as expected for a circular orbit, and also clearly shows the rotation effect. To begin with the latter, the rotational disturbance appears at the time of the eclipse. When the star is rapidly rotating around an axis perpendicular to the orbital plane, in the same direction as the orbital motion, then, on entering the eclipse, when half of its disk is eclipsed, the still-visible half of the stellar disk will have a mean rotational velocity away from the observer, which will cause an extra spike in the measured radial velocity just before the eclipse. Similarly, when coming out of the eclipse, the same half of the star is now covered by the companion, such that the now visible half will have a mean rotational velocity toward us, causing a negative spike in the measured radial velocity,

Figure 3.9. Tarantula Nebula in the LMC is the nearest giant star-forming region. In the upper-right is the very young massive star cluster R136. The bright star in the middle-left is the very massive binary R145. Source: NASA, Hubble Space Telescope.

just after the eclipse. The rotation effect is also called the Rossiter-McLaughlin effect, after its discoverers (Rossiter, 1924; McLaughlin, 1924).

The non-sinusoidal nature of the radial velocity curve of U Cephei suggests an orbital eccentricity between 0.2 and 0.5, while Hardie (1950) found from the eclipses that the eccentricity is certainly below 0.1. Hardie found that the hydrogen absorption lines in the spectrum are asymmetric, the core being displaced with respect to the wings. The rather sharp cores were found by him to be due to a gas stream in the system, while the underlying stellar hydrogen lines produce the wings, which yield radial velocities giving a very low orbital eccentricity, similar to that derived from the eclipses. The reflection effect and the heating effect in a close binary are both due to the fact that the light of the companion illuminates the half of the star directed toward it. The pure reflection makes this half of the star already brighter than its other half, but the irradiation also heats the atmosphere of this half of the star, making it still brighter.

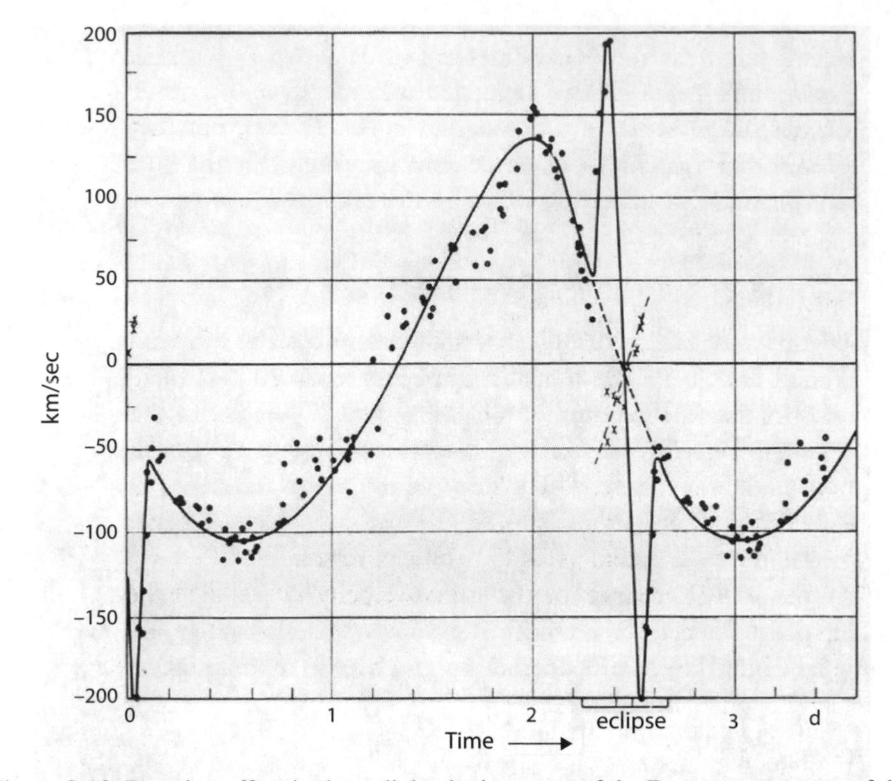

Figure 3.10. Rotation effect in the radial velocity curve of the B-type component of the Algol-type eclipsing binary U Cephei. This effect, also called the *Rossiter-McLaughlin effect* (after Rossiter, 1924; McLaughlin, 1924), is due to the fact that once half of the star is eclipsed by its companion, the rotational velocity of the non-eclipsed half of the star adds an extra velocity to the observed orbital radial velocity of the star. This produces just before the eclipse a spike in the radial velocity away from us, and just after the eclipse a spike in the radial velocity toward us. Also, the shape of the radial velocity curve suggests that the orbit is eccentric. However, from the eclipse light curve, we know that the orbit is circular. The apparent eccentricity of the radial velocity curve is due to gas streams in the system (see text). After Struve & Huang (1958).

These combined effects cause the center of light of the star to not coincide with its c.m., leading to a deformation of the measured radial velocity curve. In addition, due to the rotation of the illuminated star, this brightening of half of the star will produce a rotational disturbance of the radial velocity curve.

An important case where all these deformation effects have a crucial influence on the determination of stellar masses, is that of the so-called black widow pulsars (see Chapter 14), such as PSR 1957+20 (van Kerkwijk et al., 2011). The inferred high mass of this NS is indeed dependent on the light-curve modelling. See also the discussion and the interesting change of conclusion in Romani et al. (2015), on the black-widow pulsar PSR J1311−3430. This NS mass went down from 2.7 to 1.8 $M_{\odot}$, compared to earlier results in Romani et al. (2012). Also in HMXBs, the radial velocity curves are deformed by the above-mentioned effects, which is clearly illustrated by the shapes of the optical light curves of the deformed massive donor stars (see, e.g., Fig. 6.15). These effects must therefore be carefully taken into account in the determination of the masses of the NSs or BHs in these systems (see Chapter 6).

3.7 THE INCIDENCE OF INTERACTING BINARIES AND THEIR ORBITAL DISTRIBUTIONS AND MASSES

3.7.1 Introduction

In the late 1970s, Abt & Levy (1976, 1978) estimated that about half of the main-sequence stars of spectral type B2 to K have stellar companions with orbital periods ≤ 10 yr. At some stage of their evolution, the components of such systems will interact with one another. For the stars more massive than early B-type stars, the incidence of interacting binaries is even higher: it was found that these stars, if at all, rarely live in isolation (Blaauw 1961; van Albada 1968; Mason et al., 2009; Chini et al., 2011; Chini et al., 2012; Sana et al., 2012; Kobulnicky et al., 2014; Sana et al., 2014; Maíz Apellániz et al., 2016). Sana et al. (2012) found that that $>70\%$ of massive O-type stars are members of a binary system so close that the two stars will exchange mass, or even merge, before any of the two stars explode as a SN. (These observed binary fractions should be corrected for incompleteness, as systems with small mass ratios are hard to detect.) This leads to the conclusion that binary interaction among stars is the rule and single star evolution is the exception (Boffin et al., 2017).

Although the precise reasons for why star formation has a high preference for the formation of binaries is not known, a likely reason—such as for the formation of planetary systems—is that this is the simplest way for a contracting gas cloud to get rid of its excess angular momentum. After the formation of binaries in relatively wide orbits, the dynamical stellar interactions in young star clusters will harden the binaries. Indeed, a recent study (Ramírez-Tannus et al., 2021) showed that the velocity dispersion of very young clusters of massive stars rapidly increases with age, which most likely is due to the hardening of the binary orbits by dynamical interactions.

Since 2000, there has been remarkable progress in what we know from observations about the incidence and the statistical distributions of orbital sizes and mass ratios of binaries, in large part thanks to new techniques, such as interferometry. Excellent reviews of our present knowledge in this field were given by Duchêne & Kraus (2013), Moe & Di Stefano (2017), Moe (2019). Here, we will follow the review of Moe (2019).

3.7.2 The Fraction of Stellar Binaries as a Function of Mass

Moe (2019) defines the binary fraction, $F_{\rm bin}(M_1)$, as the fraction of main-sequence primaries of mass M_1 with at least one unevolved companion with $q = M_{\rm comp}/M_1 > 0.1$. He defines the multiplicity frequency $F_{\rm multi}(M_1)$, which is the average number of unevolved companions with $q > 0.1$ per primary. The binary fraction and the

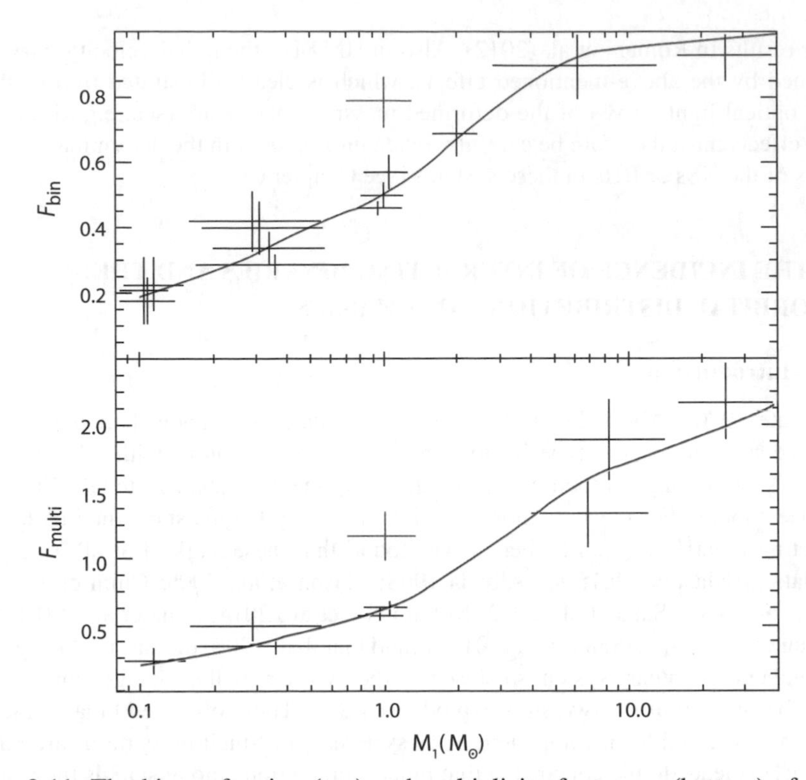

Figure 3.11. The binary fraction (top) and multiplicity frequency (bottom) of main-sequence stars as a function of the primary mass. The curves were derived from integrating the period distributions of inner binaries (top) and of all companions (bottom), based on the analytic fit presented by Moe & Di Stefano (2017). The binary fraction and multiplicity frequency of solar type pre-main-sequence stars (red crosses) are higher than for their main-sequence counterparts in the field, but less than for OB-stars. References for the data points are given in the caption of Figure 3.12 and in the original paper by Moe (2019).

multiplicity frequency are depicted in Figure 3.11. From the references mentioned in the caption of Figure 3.12, it appears that F_{bin} increases with increasing mass M_1, from 0.35−0.40 for early M-dwarfs to 0.46 ± 0.02 for solar-type stars, 0.69 ± 0.07 for A-type primaries, >0.80 for B-type primaries, and to >0.85 for O-type primaries.

After correction for selection effects, the multiplicity frequency F_{multi} is 1.35 ± 0.25 for B-type primaries (Rizzuto et al., 2013), 1.9 ± 0.3 for early B-type primaries (Abt et al., 1990), and 2.2 ± 0.3 for O-type primaries (Sana et al., 2012). The multiplicity factor F_{multi} of $\simeq 2$ for O-type and early B-type stars implies that the majority of massive stars is found in triples and higher-order multiples. On the other hand, only $\sim$13% of the solar-type main-sequence primaries are in triples or higher-order multiples (Raghavan et al., 2010; Tokovinin, 2014). The lower frame of Figure 3.11 depicts the multiplicity frequency as a function of main-sequence primary star mass.

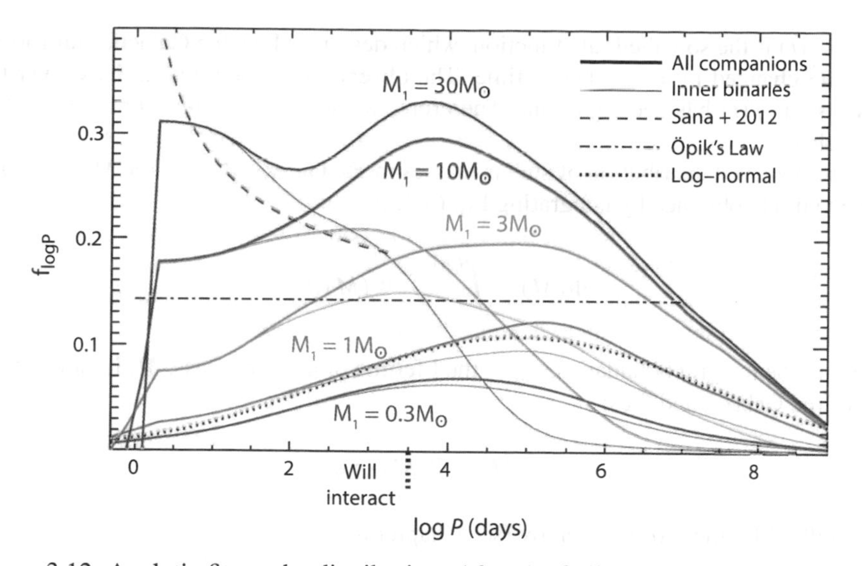

Figure 3.12. Analytic fits to the distributions ($f_{\log P}$) of all companions (thick curves) and inner binaries (thin curves) per decade of orbital period, for different primary star masses, M_1 (Moe & Di Stefano, 2017). Integrating the thick and thin curves provide the multiplicity frequency, $F_{\rm multi}(M_1)$ and binary star fraction $F_{\rm bin}(M_1)$, respectively. Field solar-type main-sequence binaries follow a log-normal period distribution that peaks at $\log P \simeq 4.9$ (dotted line: Duquennoy & Mayor, 1991; Raghavan et al., 2010). Companions to B-type main-sequence primaries approximately obey Öpik's law (Öpik, 1924), that is, a uniform distribution in log P (dash-dotted line: Abt et al., 1990; Kobulnicky & Fryer, 2007; Kouwenhoven et al., 2007). Inner companions to O-type main-sequence primaries are skewed significantly toward very short periods (dashed line: Sana et al., 2012). Close binaries with $\log P \leq 3.5$ (left of the thick dotted vertical line) will eventually interact via RLO. Figure and caption after Moe (2019).

3.7.3 Distribution of Primary Star Masses in Binaries

It is mostly assumed that the birth distribution of primary[4] star masses in binaries follows the distribution function of initial masses of isolated stars. This distribution is expected to be given by the general IMF (Eq. [3.31]), where the classic IMF, derived by Salpeter (1955) from observations, has $\alpha = 2.35$. (For stars with $M < M_\odot$, the exponent is expected to be smaller; see, e.g., Kroupa [2001]).

The *star formation rate* as a function of stellar mass, M is given by

$$\frac{dN(M)}{dt} = \Psi(M) \cdot f(t), \tag{3.32}$$

[4]The initially most massive of the two stellar components.

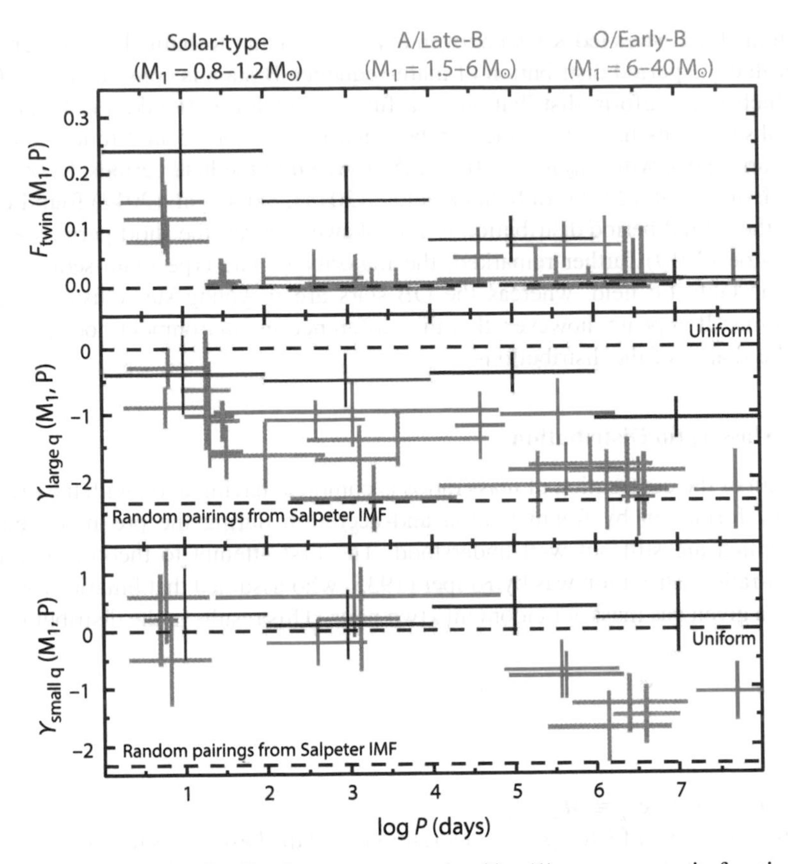

Figure 3.13. Mass-ratio distribution as parametrized by (1) an excess twin fraction $\mathcal{F}_{\rm twin}$ (top), (2) a power-law slope $\gamma_{{\rm large}\,q}$ across large mass ratios $q = 0.3 - 1$ (middle), and (3) a power-law slope $\gamma_{{\rm small}\,q}$ across small mass ratios $q = 0.1 - 0.3$ (bottom), all as a function of orbital period and colored according to primary masses. Close binaries follow a uniform mass-ratio distribution ($\gamma_{{\rm large}\,q} = \gamma_{{\rm small}\,q} = 0.0$) with a small excess twin fraction $\mathcal{F}_{\rm twin} = 0.1 - 0.2$, while wider binaries become increasingly weighted toward smaller mass ratios, especially those with massive primaries. Wide companions to early-type main-sequence stars are significantly skewed toward extreme mass ratios, but their mass-ratio distribution is still mildly discrepant from random pairings drawn from a Salpeter IMF ($\gamma_{{\rm large}\,q} = \gamma_{{\rm small}\,q} = -2.35$). Figure and caption after Moe (2019).

mass-ratio frequencies. Moe & Di Stefano (2017), however, found this to be too simple, and therefore they adopted a three-parameter model (a modified broken power-law probability distribution) as follows: a power-law slope $\gamma_{{\rm small}\,q}$ across small mass ratios, $q = 0.1 - 0.3$, a power-law slope $\gamma_{{\rm large}\,q}$ across large mass ratios $q = 0.3 - 1.0$, and an excess $\mathcal{F}_{\rm twin}$ of twins, thats is: systems with mass ratios near unity. These parameters appear all to be a function of primary mass M_1 and orbital period P. The results are represented in Figure 3.13, which is quite complex and shows the following.

For solar-type binaries, there is a modest excess twin fraction, $\mathcal{F}_{\rm twin} \simeq 0.20$ at short orbital periods, $P < 100$ d (Tokovinin, 2000; Halbwachs et al., 2003). Solar-type (Raghavan et al., 2010) and A-type (De Rosa et al., 2014) binaries with intermediate orbital separations, $a \simeq 1 - 100$ AU, show a smaller but significant excess twin fraction of $\simeq 0.05 - 0.10$. At wider separations, $a > 200$ AU, $\mathcal{F}_{\rm twin} \simeq 0$ (Lépine & Bongiorno, 2007; Raghavan et al., 2010). For more massive primaries, the twin fraction is negligible beyond $P > 20$ d. Only at very short periods, $P < 20$ d, do early-type main-sequence binaries exhibit a small twin fraction $\mathcal{F}_{\rm twin} \simeq 0.10$ (Pinsonneault & Stanek, 2006; Moe & Di Stefano, 2013, 2017).

Close companions ($\log P_{\rm d} \lesssim 2 - 3$, where P is in days) to both solar-type and early-type primaries, roughly follow a uniform mass-ratio distribution, $\gamma_{{\rm large}\,q} = \gamma_{{\rm small}\,q} = 0.0$ (Abt et al., 1990; Raghavan et al., 2010; Sana et al., 2012). For solar-type stars with intermediate separations, $a \simeq 10$ AU, the mass-ratio distribution broadly peaks at $q = 0.3$, as found by Duquennoy & Mayor (1991), giving $\gamma_{{\rm large}\,q} \simeq -0.5$ and $\gamma_{{\rm small}\,q} = 0.5$. At $a > 200$ AU, the mass ratio of solar-type binaries becomes weighted toward smaller mass ratios, $q = 0.3$ ($\gamma_{{\rm large}\,q} = -1.0$) and flattens for $q < 0.3$.

For early-type binaries, the power-law parameters also decrease with increasing separation, but much stronger, as can be seen in Figure 3.13. For A-type and late B-type primaries, at intermediate separations, $a \simeq 1 - 100$ AU, one has $\gamma_{{\rm large}\,q} = -1.0$, $\gamma_{{\rm small}\,q} = 0.0$ (Shatsky & Tokovinin, 2002; Tokovinin, 2000; De Rosa et al., 2014; Gullikson et al., 2016; Murphy et al., 2018), which then decrease to -2.0 and -1.0, respectively, for $a > 500$ AU (De Rosa et al., 2014).

More massive binaries are skewed even more toward smaller mass ratios at intermediate and wide separations. For early B- and O-type primaries, the power-law slopes are $\gamma_{{\rm large}\,q} = -1.5$ and $\gamma_{{\rm small}\,q} = 0.0$, in the range $a = 1 - 10$ AU (Abt et al., 1990; Rizzuto et al., 2013; Sana et al., 2014; Evans et al., 2015; Moe & Di Stefano, 2015), and $\gamma_{{\rm large}\,q} = -2.0$ and $\gamma_{{\rm small}\,q} = -1.5$ at wide separations, $a > 100$ AU (Abt et al., 1990; Peter et al., 2012; Sana et al., 2014).

EXERCISES

Exercise 3.1.

a. *Show that Kepler's third law can be expressed conveniently as:*

$$P \simeq \frac{a^{3/2}}{8.626\,\sqrt{M}} \tag{3.38}$$

where P is in units of days, a is in $R_\odot$ and M is in $M_\odot$.

b. *Derive Eq. (3.23).*

Exercise 3.2. *Derive Eqs. (3.13) and (3.14).*

Exercise 3.3. *Derive Eq. (3.28).*

Exercise 3.4.

a. *Determine the mass functions of the two binaries depicted in Figure 3.6.*
b. *Determine from these mass functions a lower limit to the masses of the companions (M_2) of these two stars, given that the estimated masses of the primary stars are $M_1 = 0.66\,M_\odot$ (HD 11613) and $M_1 = 1.05\,M_\odot$ (HD 14214), based on stellar mass-temperature relations for main-sequence stars.*

Exercise 3.5. *The young radio pulsar PSR J1755−2550 is orbiting an invisible companion star in a binary with an orbital period of $P_{\rm orb} = 9.7$ d and an eccentricity of $e = 0.09$. The inferred mass function is $f = 0.02117\,M_\odot$. Ng et al. (2018) argued that the young pulsar is either orbiting a massive WD with a mass of $M_{\rm WD} = 1.0 - 1.35\,M_\odot$ or an old NS with a mass of $M_{\rm NS} = 1.30 - 1.60\,M_\odot$. For now, we shall assume that the mass of the young pulsar itself is $M_{\rm PSR} = 1.30\,M_\odot$.*

a. *Assuming that the true nature of the companion is indeed one of the two above-mentioned options (i.e., a WD or a NS), calculate the probability that the binary is a double NS system.*
   (Hint: there is a one-to-one correspondence between the mass of the unseen companion star, M_2, and the orbital inclination angle of the system, i. One can use the fact that the probability $P(M_2 > X)$ is equal to the probability $P(i < i_X)$, which is given by $P = 1 - \cos i_X$.)
b. *Show that the result in question a. changes by <0.5% when considering a young pulsar mass of $M_{\rm PSR} = 1.15 - 1.45\,M_\odot$.*

Exercise 3.6. *Reproduce Figure 3.2 using Eqs. (3.5) and (3.11).*

Chapter Four

Mass Transfer and Mass Loss in Binary Systems

4.1 ROCHE EQUIPOTENTIALS

Consider two stars with masses M_1 and M_2 that are moving in circular orbits around their common center of mass c.m. In the approximation introduced by the French mathematician and astronomer Eduard Roche (1820–1883)—the so-called Roche approximation—the two stars are treated as point masses, and they are assumed to be rotating synchronously with their orbital motion, that is, in so-called co-rotation. Under these conditions, the total potential—gravitational plus centrifugal—at position vector, $\vec{r}$, in a coordinate system that is co-rotating with the binary and has its center in the c.m., can be written as

$$\Phi = -\frac{GM_1}{|\vec{r}-\vec{r}_1|} - \frac{GM_2}{|\vec{r}-\vec{r}_2|} - \frac{1}{2}\left(\vec{\Omega}\times\vec{r}\right)^2. \tag{4.1}$$

Here, $\vec{r}_1$ and $\vec{r}_2$ denote the position vectors of the two star centers and $\vec{\Omega}$ is the angular velocity vector of the binary system. If we now indicate by $\vec{v}$ the velocity of a fluid element in one of the stars relative to the rotating frame, the equation of motion for this element will be

$$\frac{\partial \vec{v}}{\partial t} + (\vec{v}\cdot\nabla)\vec{v} = -\nabla\Phi - \frac{1}{\rho}\nabla P - 2\vec{\Omega}\times\vec{v}, \tag{4.2}$$

where Φ is given by Eq. (4.1), and ρ and P are the local density and the pressure, respectively. The last term on the right-hand side of Eq. (4.2) is the Coriolis acceleration. The angular velocity, $\vec{\Omega}$, is related to the separation, a, of the binary components (here assuming a circular orbit with orbital period, P) by Kepler's third law (see Section 3.2):

$$\Omega = \frac{2\pi}{P} = \sqrt{\frac{G\,(M_1+M_2)}{a^3}}. \tag{4.3}$$

In unevolved close binaries with orbital periods shorter than 10 d, the stars are generally observed to be in co-rotation with the orbital motion. This is due to the strong tidal forces that occur when the star is not rotating synchronously: in that case, the difference in rotational and orbital frequency makes the star pulsate with a frequency equal to this frequency difference. The internal viscosity in the star will act to strongly

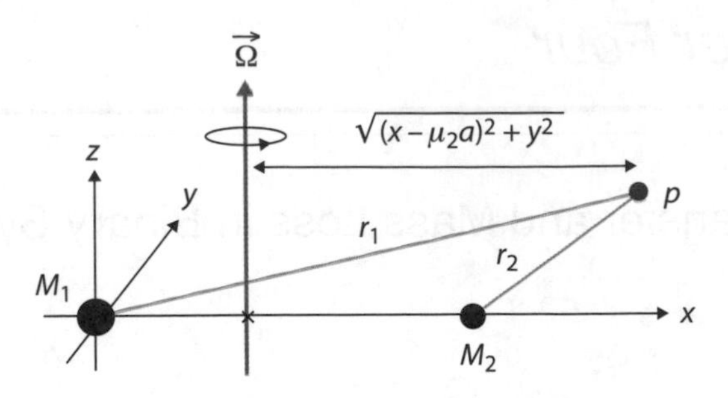

Figure 4.1. Schematic illustration of probing the Roche potential (Eq. 4.5) in a Cartesian coordinate system with origin in star 1.

dampen these pulsations, until the star rotates synchronously with the orbital motion. (In Section 4.7, the physics of tidal synchronization is considered in more detail.)

Co-rotation implies that in the co-rotating frame the stars are standing still, and thus the gases in their interiors are not moving, so in Eq. (4.2) $\vec{v} = 0$. This implies

$$\nabla P = -\rho \nabla \Phi. \tag{4.4}$$

Hence, surfaces of equal pressure (isobars) coincide with equipotential surfaces, which means that the pressure is a function only of the potential: $P = P(\Phi)$. Also, because $\rho = |dP(\Phi)/d\Phi|$, which depends only on Φ, surfaces of equal density coincide with equipotential surfaces, where Φ is a constant. This implies that, in the Roche approximation, the shape of the binary components will be given by the shape created by the Roche equipotential surfaces.

To determine the shapes of these surfaces, it is convenient to work in a Cartesian coordinate system with the origin at the center of the primary component, the x-axis in the direction of the companion star and the z-axis along the direction of the vector $\vec{\Omega}$, thus perpendicular to the orbital plane (Fig. 4.1). In this coordinate system the equipotentials take the form

$$-\frac{G M_1}{\sqrt{x^2 + y^2 + z^2}} - \frac{G M_2}{\sqrt{(x-a)^2 + y^2 + z^2}} - \frac{1}{2}\Omega^2((x - \mu_2 a)^2 + y^2) = \text{const}, \tag{4.5}$$

where $\mu_2 = M_2/(M_1 + M_2)$. From Eq. (4.5) it is clear that the shape of the equipotentials is determined by the mass ratio, q (defined here as M_1/M_2; see Fig. 4.1), while the scale is determined by the binary separation, a.

An example of the equipotentials in the orbital plane ($z = 0$) is given in Figure 4.2, while a three-dimensional (3D) plot of the potential is presented in Figure 4.3. The qualitative properties of the equipotential surfaces are as follows. Close to the center of each star, where the gravity of that star dominates, the equipotential surfaces are almost spherical. When one moves further away from the stellar center, two effects become important: (1) the tidal effect of the companion causes an elongation in the direction of the companion (along the x-axis) and (2) a flattening due to the centrifugal forces. As a result, the surfaces are distorted such that their largest dimension is along the line connecting the centers.

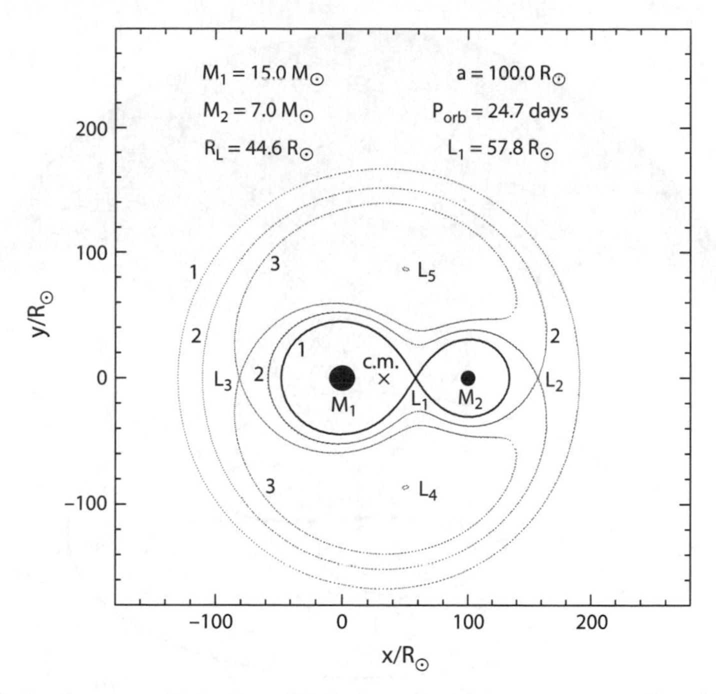

Figure 4.2. Roche potential in the orbital plane for a binary system with a 15 $M_{\odot}$ and a 7 $M_{\odot}$ star. The Lagrangian points L_1–L_5 are indicated, as well as equipotentials in order of increasing Φ.

A critical equipotential surface, from the point of view of binary star evolution, is the "eight-shaped" surface through the first Lagrangian point, L_1. The two pear-shaped volumes enclosed by this critical surface are called the Roche lobes of the two stars. A binary in which neither of the stars is filling these Roche lobes is called *detached*. If one of the two stars fills its Roche lobe but the other one not, one calls it *semi-detached*, and when both stars fill their Roche lobes, one speaks of a *contact system* (see Fig. 4.4; see also Section 5.2).

A binary system usually starts out being detached, that is, the stars are filling an equipotential surface well inside their Roche lobes. As a result of nuclear evolution, the most massive star evolves first, expands, and may fill its Roche lobe to initiate mass transfer. It is also possible that losses of angular momentum, for example, via magnetic braking and/or gravitational radiation, cause the orbital separation (and thus the Roche lobe) to shrink, which may force a star to fill its Roche lobe.

When a star fills its Roche lobe, matter can freely flow toward the companion star, as on the equipotential surface through L_1 matter needs no energy to flow from one side to the other side. The transfer of material is driven by the pressure difference across L_1 (see Section 4.5 for more details on the physics of mass transfer via L_1).

As a measure of the size of the Roche lobe of a star, one generally uses the so-called Roche radius, R_L, which is the radius of a sphere with the same volume as that of the stellar Roche lobe. The ratio $r_L = R_L/a$, where a is the orbital separation, depends only on the mass ratio, q (defined as donor star mass divided by accretor star mass).

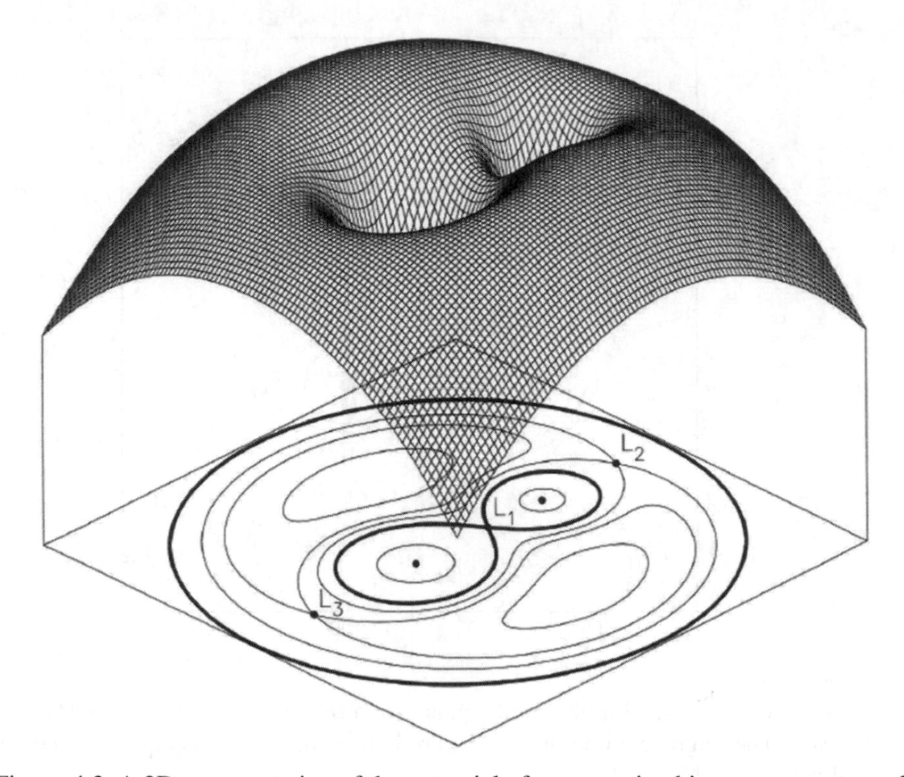

Figure 4.3. A 3D representation of the potential of a co-rotating binary system, according to the Roche model. In this representation, the value of the potential is plotted in the vertical direction. Each horizontal plane corresponds to a different potential; therefore the cuts of the different planes through the potential correspond to different equipotentials. The down-sloping of the potential beyond the L_2 and L_3 points is due to the centrifugal acceleration of the rotating coordinate system and will not occur in real binary systems; see also Figure 4.5. Credit: Martin Heemskerk, University of Amsterdam.

A convenient approximation formula for $r_L(q)$, for all values of $q = M_{\rm donor}/M_{\rm accretor}$ and accurate to within 1%, is

$$r_L = \frac{0.49\, q^{2/3}}{0.6\, q^{2/3} + \ln(1+q^{1/3})} \tag{4.6}$$

from Eggleton (1983).

For discussions on general numerical solutions, see Leahy & Leahy (2015). Other convenient approximation formulae are from Paczyński (1971b):

$$r_L = \begin{cases} 0.38 + 0.20 \log q & \text{for} \quad 0.523 \le q < 20 \\ 0.462 \left(\dfrac{q}{1+q}\right)^{1/3} & \text{for} \quad 0 < q < 0.523 \end{cases} . \tag{4.7}$$

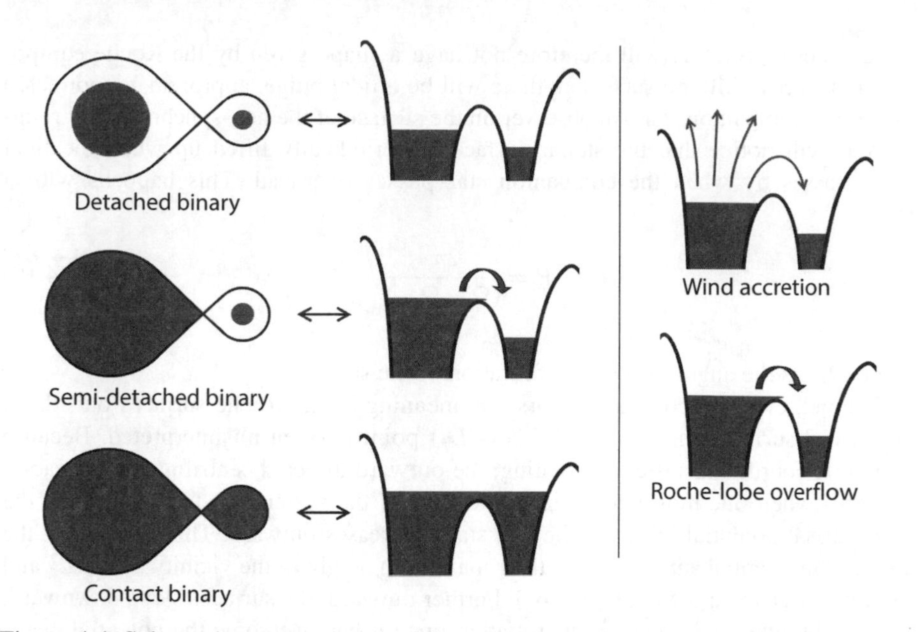

Figure 4.4. Schematic representation of the situation of *detached* (top), *semi-detached* (middle), and *contact binaries* (bottom) in terms of the position of the stellar surfaces with respect to the heights of the potential wells. The right column illustrates two cases of accretion.

The distance from the center of the donor star to L_1 is given by, to an error within 1%:

$$R_{L1} = \frac{a}{(1.0015 + q^{-0.4056)})} \qquad \text{for} \quad 1.0 \leq q \leq 25.0 \tag{4.8}$$

(Silber, 1992). Equation (4.7) together with Eq. (4.3) implies that for a Roche-lobe filling donor star, the average mass density, to an error within 1%, is given by

$$\langle \rho_2 \rangle = \frac{110.6\,\text{g}\,\text{cm}^{-3}}{(P/\text{hr})^2} = \frac{78.4\,\rho_\odot}{(P/\text{hr})^2} \qquad \text{for} \quad 0 < q < 0.8, \tag{4.9}$$

where P is the orbital period (here stated in hours), and $\langle \rho_2 \rangle$ is the mean mass density of the Roche-lobe filling star (here star 2): $\langle \rho_2 \rangle = M_2/(4\pi R_2^3/3)$. Thus, we see that a solar-like donor star would fill its Roche lobe for $P = 8.85$ hr, which is independent of the mass of the companion star, as one can easily verify. In Section 11.3, we discuss these relations for less massive dwarf donors.

4.2 LIMITATIONS IN THE CONCEPT OF ROCHE EQUIPOTENTIALS

It is important to notice that the Roche approximation has several limitations. First of all, it holds only for stars that are co-rotating with the orbital motion. This is because the Roche equipotential surfaces are defined in a co-rotating coordinate system. Stars that are not co-rotating will therefore not have a shape given by the Roche equipotential surfaces. Still, on such stars there will be a tidal bulge, approximately directed toward the companion star. An observer on the surface of the non-synchronously rotating star will notice that the stellar surface is periodically lifted up when the tidal bulge passes by when the companion star passes overhead. This happens with a period

$$P = \frac{2\pi}{(\Omega - \Omega_{\rm rot})}, \tag{4.10}$$

where $\Omega_{\rm rot}$ is the angular velocity of rotation of the star.

Second, even for co-rotating stars, the meaning of the Roche surfaces outside of the critical surfaces through the L_2 (and L_3) point is often misinterpreted. Because the whole coordinate system is rotating, the outward directed centrifugal acceleration increases when one moves away from the c.m. of the system. At the same time, the gravitational potential felt from the two stars decreases outward. This makes that the Roche equipotential surfaces reach their maximum height in the vicinity of the L_2 and L_3 points, as can be seen in Figure 4.3. Further outward, the surfaces slope downward. This might give the impression that matter, once it has overcome the potential peaks near L_2 and L_3, will move outward by itself, sliding down along the equipotential surfaces, and leave the system. This, however, is not correct. Matter will only do so if it is kept in co-rotation with the system out to large distances. That can only be achieved if the matter would be sliding along stiff rods fixed in the two co-rotating stars. However, no such rods exist in real stars (possibly, magnetic fields might play such a role in some cases).

Let us therefore consider what will really happen when the two stars overflow their Roche lobes and start filling one of the common equipotential surfaces outside the critical surface through L_1. Here, co-rotation is still possible as matter, which surrounds both stars, still belongs to the two stars. If, however, the stars keep swelling, there will come a point when they will fill the equipotential surface through the L_2 point. Matter will now flow out through L_2 and is no longer connected to the two stars: it may form a ring around the system, as indeed often is observed in real binary systems.

If the Roche model could be extended infinitely beyond L_2, matter would now simply be centrifuged out of the system, which is impossible, as the matter leaving L_2 is no longer forced to co-rotate. The particles flowing out through L_2 into a ring will only feel the gravitational attraction of the two stars and therefore are still bound to the system. Indeed, seen from a large distance, the two stars can be considered as one point mass of mass $M_1 + M_2$.

Thus seen from a distance, the system is deep inside a gravitational potential well, and a particle released in the vicinity of the system will fall toward the two stars and will certainly not be centrifuged out of the system. The preceding paragraphs imply that, in fact, in Figure 4.3, the parts of the equipotential surfaces beyond the surface through L_2 should be ignored. The real equipotential surfaces beyond L_2 keep sloping upward when one moves outward, and not downward as in Figure 4.2. Figure 4.5 schematically illustrates this more realistic situation.

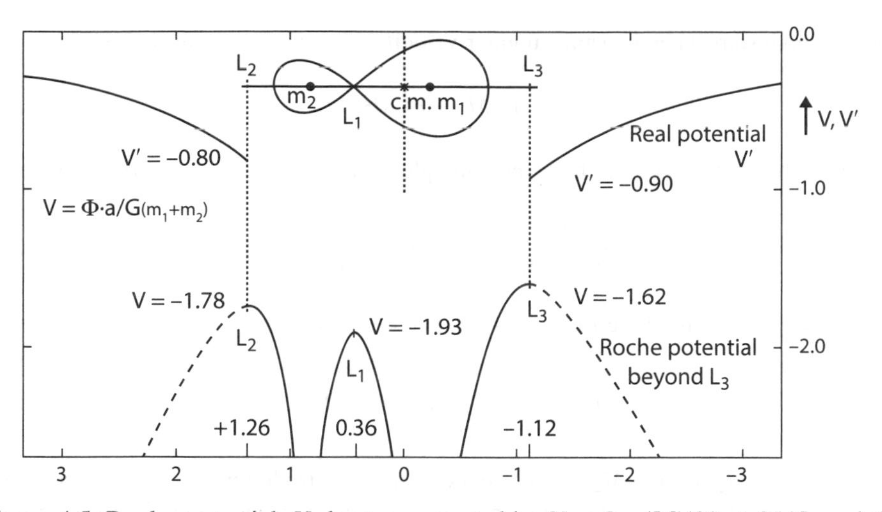

Figure 4.5. Roche potential, V, here represented by $V = \Phi\, a/[G(M_1 + M_2)]$, and the real potential, V', shown (schcmatically) around a binary system. Real potentials are fully drawn curves. The system depicted here has $M_2/M_1 = 1/3$. The Roche potential is defined in a rotating coordinate system; the rotational contribution $-\Omega^2\,(x - a/4)^2/2$ becomes more and more negative at increasing distances, r, from the center of gravity. However, beyond the points L_2 and L_3, the Roche potentials no longer have meaning (see text). The real potential, V', beyond L_2 and L_3 is the sum of the gravitational potentials of the two stars, which are revolving around their common center of gravity. After van den Heuvel (1994a).

These considerations are important if one wishes to study what happens to the orbit of the binary when matter with angular momentum leaves the system, for example, through the L_2 point. This will be the topic of the next section.

4.3 ORBITAL CHANGES DUE TO MASS TRANSFER AND MASS LOSS IN BINARY SYSTEMS

In this section, we consider the effects on the orbits of binaries produced by the exchange of matter between the stellar components and by the loss of mass and angular momentum from the systems. Following an introduction to the problem and the orbital balance equation, we consider some general *modes of mass loss* identified in the important paper by Huang (1963); see also Kruszewski (1964), van den Heuvel (1994a), and Soberman et al. (1997).

4.3.1 General Equations

In a general binary system with two stars of masses M_1 and M_2, orbiting in circular orbits of radii of r_1 and r_2, the total orbital angular momentum is given by ($\vec{J} = \vec{r} \times \vec{p}$):

$$J = J_1 + J_2 = r_1\, M_1\, v_1 + r_2\, M_2\, v_2. \tag{4.11}$$

Next, we introduce the orbital angular momentum balance equation, and afterward we demonstrate a few simple and relevant examples of such mass loss modes.

4.3.2 Mass Loss from the Second Lagrangian Point, L_2

In most cases of mass transfer in X-ray binaries, analysis has been done for mass transfer/loss via the inner Lagrangian point, L_1. In some cases of extreme mass transfer via L_1, such as for systems with donor stars substantially more massive than the accretors, RLO via L_1 might not be enough to provide an efficient mass-transfer rate and the donor will extend far beyond its Roche lobe to reach the equipotential surface passing through L_2 (Fig. 4.6). However, contrary to when mass passes through L_1, the material that crosses L_2 takes away a large amount of orbital angular momentum from the binary. In cases where the outer layers of the donor reach L_2 (or the donor obtains a volume equivalent to that of the equipotential lobe passing through L_2), it is expected that the binary orbit would shrink rapidly. Because that is an effective sink of orbital angular momentum, this situation is generally considered an onset of dynamical instability.

Using numerical methods, one can calculate. for example, the distance between the c.m. of the donor star and L_2 (D_{L2}), as well as the volume-equivalent radius (R_{L2}) of a donor star whose volume equals that of the L_2 equipotential surface. These values were calculated by Misra et al. (2020) in units of the Roche-lobe radius of the donor star, R_L, as a function of the mass ratio, q (Fig. 4.7). Slightly rewriting their fitting formulae to follow our definition of $q \equiv M_{\rm donor}/M_{\rm acc}$ yields

$$\frac{D_{L2}}{R_L} = \begin{cases} 3.334\, q^{-0.514}\, e^{-0.052/q} + 1.308 & \text{for } q \geq 1 \\ -0.040\, q^{-0.866}\, e^{-0.040/q} + 1.883 & \text{for } q \leq 1 \end{cases} \tag{4.27}$$

$$\frac{R_{L2}}{R_L} = \begin{cases} 0.784\, q^{-1.05}\, e^{-0.188/q} + 1.004 & \text{for } q \geq 1 \\ 0.290\, q^{-0.829}\, e^{-0.016/q} + 1.362 & \text{for } q < 1 \end{cases} . \tag{4.28}$$

The relative errors between numerical calculations and the corresponding fits obtained by Misra et al. (2020) are $<1\%$. Note, D_{L2} is discontinuous for $q=1$ and thus there are two solutions for D_{L2} when $q=1$ (i.e., due to symmetry in this case, the two Lagrangian points L_2 and L_3 overlap). A criterion for stable RLO is thus that $R_{\rm donor} < \min\{R_{L2},\ D_{L2}\}$ for the entire phase of mass transfer.

4.3.3 The Orbital Angular Momentum Balance Equation

Consider an interacting binary system with a donor star (with mass M_2) and an accretor (with mass M_1), that is, a total mass: $M = M_1 + M_2$. (In most of this book, we will define star 2 to be the donor and star 1 to be the accretor.) Any exchange and loss of mass in such a binary system will lead to perturbations of the orbital dynamics via

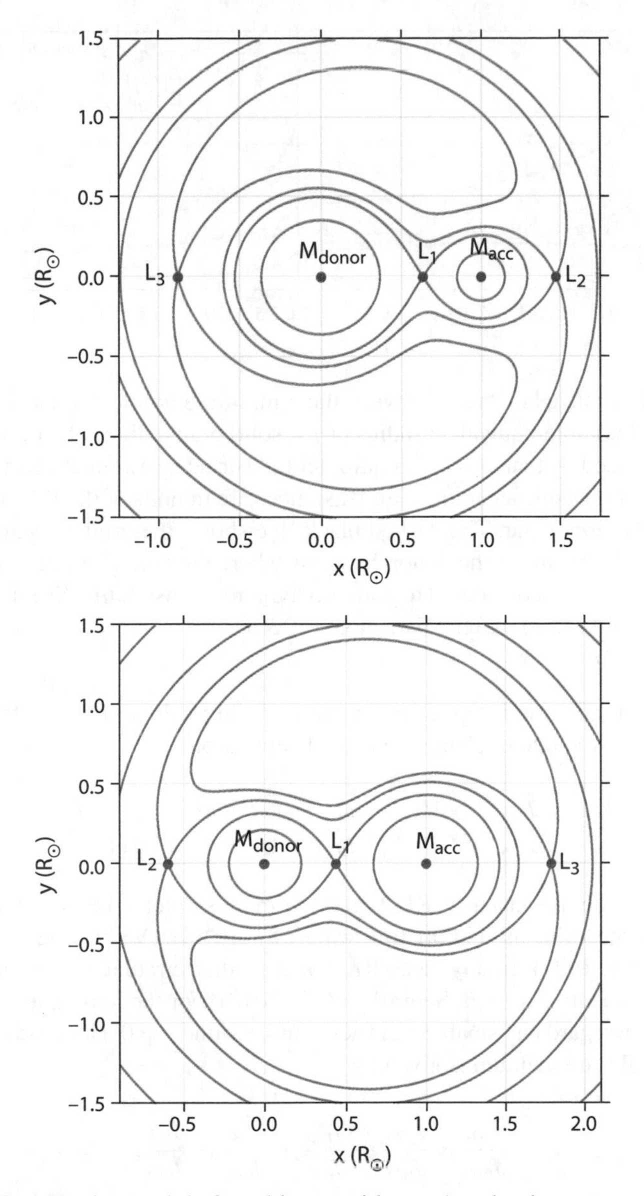

Figure 4.6. (Top) Equipotentials for a binary with $q > 1$ and point-mass components of $M_{\rm donor} = 5.0\, M_\odot$ and $M_{\rm acc} = 1.3\, M_\odot$. (Bottom) Equipotentials for a binary with $q < 1$ and point-mass components of $M_{\rm donor} = 0.7\, M_\odot$ and $M_{\rm acc} = 1.3\, M_\odot$. In both cases the orbital separation is $a = 1.0\, R_\odot$. The projected equipotential surfaces passing through the Lagrangian points L_1, L_2, and L_3 are shown. The locations of L_2 and L_3 are swapped when a binary evolves from $q > 1$ to $q < 1$; see also Figure 4.7. After Misra et al. (2020).

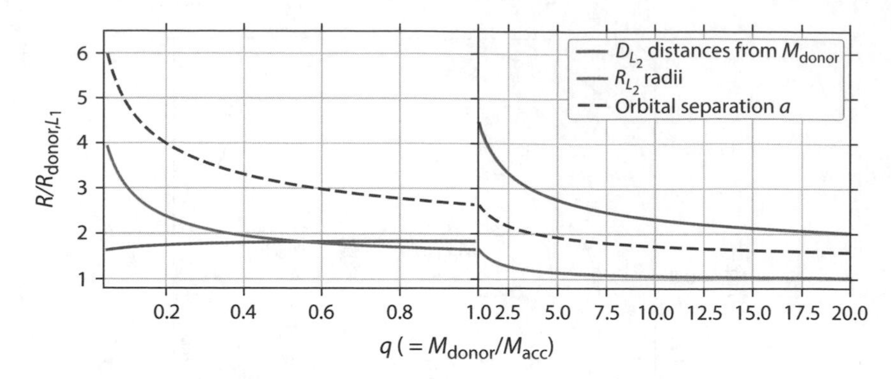

Figure 4.7. Normalized distance between the c.m. of the donor star and L_2 (D_{L2}, solid blue line), and volume-equivalent radius (R_{L2}, solid orange line) of a donor star whose volume is enclosed by that of the L_2 equipotential surface, as a function of mass ratio of donor to accretor. Radii, separations and distances are in units of the Roche-lobe radius, $\mathrm{R}_{\mathrm{donor},L_1}$ of the donor star. The dashed black line shows the orbital separation in units of the Roche-lobe radius of the donor. Systems where the donor's radius exceeds any of these limiting radii are considered to undergo dynamical instability. See fitting formulae in Eqs. (4.27) and (4.28). After Misra et al. (2020).

changes in the total orbital angular momentum. A simple logarithmic differentiation of Eq. (4.16) yields the rate of change in orbital separation:

$$\frac{\dot{a}}{a} = 2\,\frac{\dot{J}_{\mathrm{orb}}}{J_{\mathrm{orb}}} - 2\,\frac{\dot{M}_1}{M_1} - 2\,\frac{\dot{M}_2}{M_2} + \frac{\dot{M}_1 + \dot{M}_2}{M_1 + M_2} + 2\,\frac{e\,\dot{e}}{(1-e^2)}. \tag{4.29}$$

Tidal effects acting on a near-RLO (giant or main-sequence) donor star are believed to circularize the orbit on a short timescale of $\sim 10^4$ yr (Verbunt & Phinney, 1995; Claret & Cunha, 1997). In any case, RLO will further circularize any pre-RLO orbit with residual eccentricity (e.g., Sepinsky et al., 2010). For the remainder of this section we therefore disregard any small eccentricity and assume $e = 0$. Hence, the total change in orbital angular momentum is given by

$$\frac{\dot{J}_{\mathrm{orb}}}{J_{\mathrm{orb}}} = \frac{\dot{J}_{\mathrm{gwr}}}{J_{\mathrm{orb}}} + \frac{\dot{J}_{\mathrm{mb}}}{J_{\mathrm{orb}}} + \frac{\dot{J}_{\mathrm{ls}}}{J_{\mathrm{orb}}} + \frac{\dot{J}_{\mathrm{ml}}}{J_{\mathrm{orb}}}. \tag{4.30}$$

The first term on the right-hand side of this equation gives the change in orbital angular momentum due to gravitational wave (GW) radiation (Landau & Lifshitz, 1971). The second term in Eq. (4.30) arises due to so-called magnetic braking (e.g., Verbunt & Zwaan, 1981; Rappaport et al., 1983; van der Sluys et al., 2005; Istrate et al., 2014). This mechanism is only relevant for magnetically active stars (i.e., low-mass stars with convective envelopes), such as in LMXBs and CVs. Both of the terms $\dot{J}_{\mathrm{gwr}}$ and $\dot{J}_{\mathrm{mb}}$ are discussed in Chapter 11. The third term ($\dot{J}_{\mathrm{ls}}/J_{\mathrm{orb}}$) on the right-hand side of Eq. (4.30) describes possible tidal spin-orbit exchanges of angular momentum between the orbit and the rotation of the donor star due to its expansion or contraction.

Gravitational quadrupole variations can also contribute to this term and are discussed in Section 4.3.11. Finally, the last term on the right-hand side of Eq. (4.30) represents the change in orbital angular momentum caused by mass loss from the binary system.

Equations (4.29) and (4.30) together constitute what is called "the angular momentum balance equation." The mass-loss term in Eq. (4.30) is usually the dominant term in the orbital angular momentum balance equation and its effect depends on the chosen mode of mass loss.

In the following sections, we consider angular momentum loss only as a result of mass loss and mass exchange between the stellar components, that is, $\dot{J}_{\rm orb} = \dot{J}_{\rm ml}$. This is usually a good approximation in binaries that exchange mass on a short (thermal) timescale or in binaries that are not too tight (i.e., if their orbital periods are larger than a few days and the effects of GW radiation, magnetic braking, and spin-orbit interactions are negligible). We will derive a simple prescription to calculate the orbital changes as a result of different modes of mass exchange and mass loss. These modes include direct fast wind mass loss, RLO, isotropic re-emission, and a circumbinary disk. Common-envelope evolution will be discussed in Section 4.8. In all cases, we will parameterize the changes only as a function of the initial and final mass ratios of the binary as well as a few other simple parameters describing the physics of the given mode of mass loss.

4.3.4 Direct Fast Wind (Jeans' Mode)

Assume the donor star (here star 2) is losing mass in the form of a direct fast wind and neglect any accretion on the companion star. Such mass loss is called *Jeans' mode* in Huang's nomenclature. In this case, it is a reasonably good assumption that the mass lost from the system is carrying the specific angular momentum of the donor star. Hence, we have ($dM_2 < 0$)

$$dJ_{\rm orb} = \frac{J_2}{M_2}\, dM_2 = \frac{M_1 J_{\rm orb}}{M_2 M}\, dM_2, \tag{4.31}$$

where we have used the simple relation

$$J_1 M_1 = J_2 M_2 = \mu J_{\rm orb}, \tag{4.32}$$

where J_1 and J_2 are the orbital angular momenta of star 1 and star 2, respectively.

Inserting Eq. (4.31) into Eq. (4.29) and applying $\dot{M}_1 = 0$, we find

$$\frac{\dot{a}}{a} = 2\frac{M_1 \dot{M}_2}{M_2\, M} - 2\frac{\dot{M}_2}{M_2} + \frac{\dot{M}_2}{M} = -\frac{\dot{M}_2}{M} = -\frac{\dot{M}}{M}. \tag{4.33}$$

A simple integration (Exercise 4.2) of this expression yields

$$\frac{a}{a_0} = \frac{M_0}{M} = \frac{q_0 + 1}{q + 1}, \tag{4.34}$$

where the index 0 refers to initial values (before the mass loss) and $q = M_2/M_1$ is the stellar mass ratio. Using Kepler's third law, we can also write the change in orbital period $\dot{P}/P = -2\dot{M}/M$ or

$$\frac{P}{P_0} = \left(\frac{M_0}{M}\right)^2 = \left(\frac{q_0+1}{q+1}\right)^2. \tag{4.35}$$

The previous two equations for stellar wind mass loss are also valid if both stars would be losing mass by a stellar wind.

Verbunt (1990) considered a different assumption: he assumed that the wind material is leaving the orbit with the specific orbital angular momentum of the binary:

$$d\,J_{\rm orb} = \frac{J_{\rm orb}}{M_1+M_2}\,dM_2. \tag{4.36}$$

Under this special assumption one obtains

$$\frac{a}{a_0} = \left(\frac{M}{M_0}\right)^3 \left(\frac{M_{20}}{M_2}\right)^2 = \left(\frac{q+1}{q_0+1}\right)^3 \left(\frac{q_0}{q}\right)^2. \tag{4.37}$$

4.3.5 Wind Accretion

In some binaries the donor star is losing mass in the form of an intense wind, and the compact object is observed to accrete a certain fraction of this material. For example, this scenario is seen in binaries containing a WR-star and a compact object, for example, Cyg X-3 (see Section 6.5.3). Assume the mass fraction $(1-\alpha)$ of the wind lost from the donor is accreted onto the companion star. We then find

$$\frac{a}{a_0} = \left(\frac{q}{q_0}\right)^{2\,(\alpha-1)} \left(\frac{q_0+1}{q+1}\right) \left(\frac{(1-\alpha)q+1}{(1-\alpha)q_0+1}\right)^{3+2(1-\alpha)}. \tag{4.38}$$

It is trivial to see that this equation is equal to Eq. (4.34) for $\alpha \rightarrow 1$.

4.3.6 Conservative Roche-lobe Overflow

Conservative RLO is mass transfer from a donor star to an accretor where both the total mass and the orbital angular momentum remain constant. This evolution is an example of Huang's *slow mode*. Considering the orbital angular momentum (Eq. [4.16]) of a circular binary,

$$J_{\rm orb} = \sqrt{\frac{G\,M_1^2 M_2^2\,a}{M_1+M_2}}, \tag{4.39}$$

and comparing initial and final conditions while applying $J_{\rm orb,0} = J_{\rm orb}$ and $M_{10} + M_{20} = M_1 + M_2$, we find (see Exercise 4.5)

$$\frac{a}{a_0} = \left(\frac{M_{10}\,M_{20}}{M_1\,M_2}\right)^2 = \left(\frac{q_0}{q}\right)^2 \left(\frac{q+1}{q_0+1}\right)^4 \tag{4.40}$$

or in terms of the orbital period changes:

$$\frac{P}{P_0} = \left(\frac{M_{10}\,M_{20}}{M_1\,M_2}\right)^3 = \left(\frac{q_0}{q}\right)^3 \left(\frac{q+1}{q_0+1}\right)^6. \tag{4.41}$$

From these equations, it is clear that the orbit widens as long as the donor star is less massive than the accreting star ($q < 1$), and the orbit shrinks when the donor is more massive than the accretor ($q > 1$). This is easily seen by differentiating Eq. (4.40) to obtain

$$-\frac{\partial \ln a}{\partial \ln q} = 2 - \frac{4q}{q+1}. \tag{4.42}$$

The sign of this quantity is a measure of whether the orbit expands (when positive) or the orbit shrinks (when negative), given that $\partial \ln q < 0$ for any value of q and the solution for $(-\partial \ln a/\partial \ln q) = 0$ is $q = 1$ (see plotted examples in the top panel of Fig. 4.8).

4.3.7 Isotropic Re-emission Model

The idea behind the isotropic re-emission model, first introduced by van den Heuvel & De Loore (1973) and explored in detail in Bhattacharya & van den Heuvel (1991), is that matter flows over from a donor star (M_2) to an accreting compact object (M_1) in a conservative manner and thereafter a certain fraction, β, of this matter is ejected from the vicinity of the compact object with the specific angular momentum of the compact object (for example in a jet as illustrated in Fig. 4.9 and observed in SS433, see Chapter 6). Hence, in this model we can write the change in orbital angular momentum as

$$d\,J_{\rm orb} = \frac{J_1}{M_1}\,\beta\,dM_2 = \frac{\mu}{M_1^2}\,J_{\rm orb}\,\beta\,dM_2 \tag{4.43}$$

or

$$\frac{\dot{J}_{\rm orb}}{J_{\rm orb}} = \frac{\mu}{M_1^2}\,\beta\,\dot{M}_2 = \frac{\beta\,q^2}{1+q}\,\frac{\dot{M}_2}{M_2}. \tag{4.44}$$

Keep in mind that $(1-\beta)$ is the fraction of baryonic matter[1] that is accreted onto the compact object. We can now integrate Eq. (4.29) for $e = 0$ by inserting Eq. (4.44) as the first term on the left-hand side, and we obtain (Tauris, 1996)

$$\frac{a}{a_0} = \left(\frac{q_0(1-\beta)+1}{q(1-\beta)+1}\right)^{\frac{3\beta-5}{1-\beta}} \left(\frac{q_0+1}{q+1}\right) \left(\frac{q_0}{q}\right)^2. \tag{4.45}$$

---

[1] When the material falls down the deep gravitational potential well ($V_{\rm grav} = -GM/R$) of the accreting compact object, gravitational binding energy is released. See also Section 4.9.

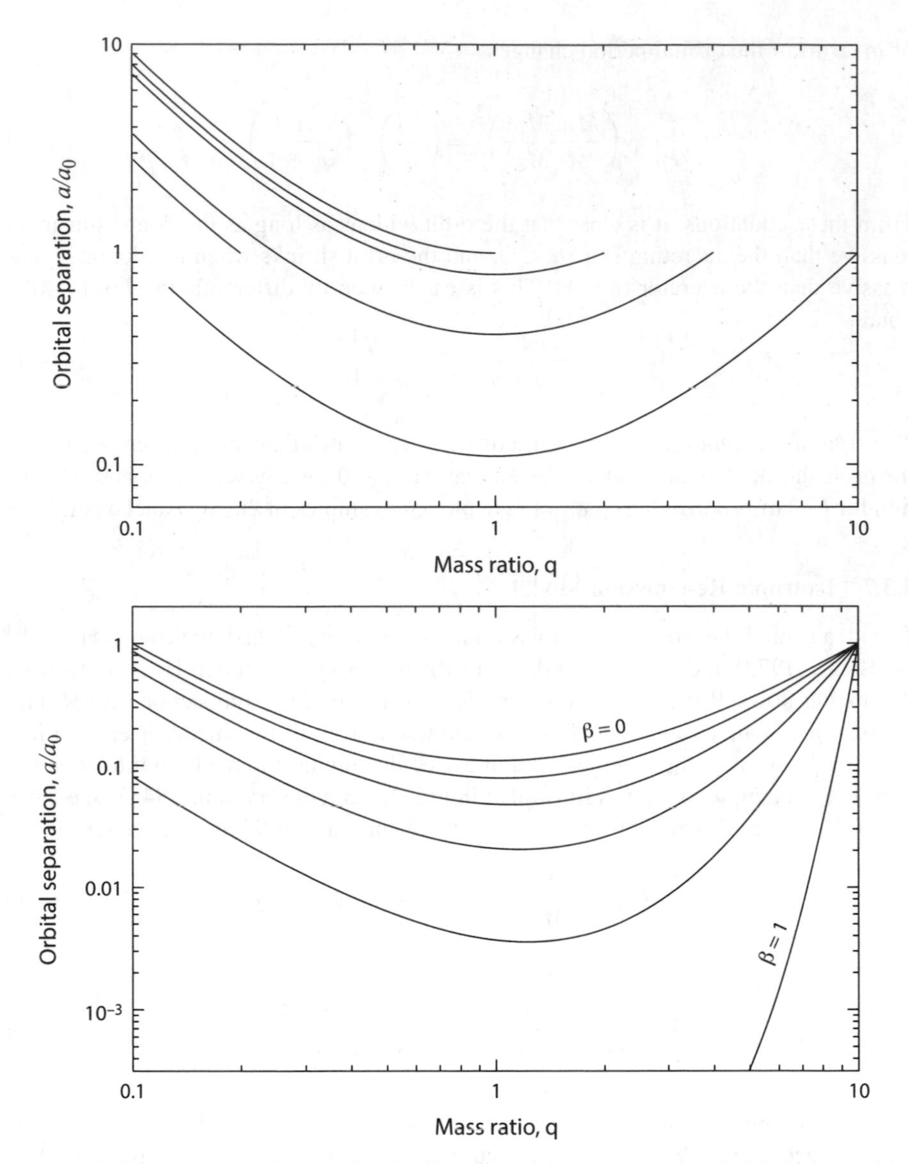

Figure 4.8. Orbital separation tracks for binaries with mass transfer. Orbital separation, a (in units of the initial orbital separation, a_0), is plotted as a function of the decreasing mass ratio, q. The evolution is from right to left. (Top) Conservative RLO ($\beta = 0$) and initial mass ratios of $q_0 = 0.2$, 0.6, 1.0, 2.0, 4.0, and 10 (see Eq. [4.40]). (Bottom) Mass transfer for $q_0 = 10$ (e.g., a HMXB) with isotropic re-emission with various degrees of mass loss from the accretor using $\beta = 0.0$, 0.2, 0.4, 0.6, 0.8, and 1.0 (see Eq. [4.45]). Many of these tracks lead to common envelopes (and possibly mergers) due to extreme orbital shrinkage.

Figure 4.9. Artist's impression of an X-ray binary. Isotropic re-emission via jets and disk winds are expected in many IMXBs and HMXBs with high mass-transfer rates. Credit: NASA/CMC/M.Weiss.

When applying Eq. (4.45) to, for example, intermediate mass or high-mass X-ray binary systems (HMXB, Fig. 4.9), where the accretor is a NS or a BH, the accretion rates onto these compact stars is limited by the Eddington limit (see Section 4.9) to a rate $\dot{M}_{\rm Edd} \simeq 1.8 \times 10^{-8}\, M_\odot\ {\rm yr}^{-1}$ for the case of a NS. The donor star, on the other hand, transfers matter on the thermal timescale of its envelope (for definition of the thermal timescale: see Chapter 8), which is between 10^4 and 10^6 yr, leading to mass-transfer rates two to four orders of magnitude greater than the Eddington rate. So, the bulk of the transferred matter will be blown away by the radiation pressure of the accreting compact star. In this case, $\beta > 0.99$. It is therefore interesting to consider Eq. (4.45) in the limit where $\beta \to 1$:

$$\lim_{\beta \to 1} \left(\frac{a}{a_0} \right) = \left(\frac{q_0 + 1}{q + 1} \right) \left(\frac{q_0}{q} \right)^2 e^{2(q - q_0)}. \tag{4.46}$$

According to Kepler's third law, $P^2 \propto a^3/M$; hence we can rewrite the preceding equations to yield the change in orbital period (see also King, Schenker et al., 2001):

$$\frac{P}{P_0} = \left(\frac{q_0(1-\beta)+1}{q(1-\beta)+1} \right)^{\frac{5\beta - 8}{1-\beta}} \left(\frac{q_0 + 1}{q + 1} \right)^2 \left(\frac{q_0}{q} \right)^3 \tag{4.47}$$

$$\lim_{\beta \to 1} \left(\frac{P}{P_0} \right) = \left(\frac{q_0 + 1}{q + 1} \right)^2 \left(\frac{q_0}{q} \right)^3 e^{3(q - q_0)}. \tag{4.48}$$

As we shall see later, it can be quite convenient to define the mass-ratio parameter, $k = q_0/q$ (Tauris et al., 2011). In this case, we can rewrite the latter equation:

$$\lim_{\beta \to 1} \left(\frac{P}{P_0} \right) = \left(\frac{kq+1}{q+1} \right)^2 k^3 \, e^{3q(1-k)}. \tag{4.49}$$

4.3.8 Mass Loss from a Circumbinary Disk

Mass loss from a binary system via a circumbinary disk is referred to as the *intermediate mode* by Huang (1963), van den Heuvel & De Loore (1973), van den Heuvel (1994a), and Soberman et al. (1997). Here it is assumed that the ejected particles have enough energy to overcome the attraction of the individual components and escape through L_2 (or L_3) to form a co-planar ring revolving around the entire system, that is, a circumbinary disk. Such circumbinary disks may be produced from either matter outflow during RLO or from the late stage of a common envelope evolution.

Through tidal interactions between the binary system and the circumbinary disk, orbital angular momentum is extracted from the binary orbit. Circumbinary disks are therefore often suggested to be at work in binary systems in which one cannot otherwise account for the observed orbital period decay (e.g., Chen & Li, 2015; Chen & Podsiadlowski, 2017; Jiang et al., 2017).

In first approximation, the particles in the circumbinary disk (of mass $M_{\rm disk}$) can be considered as moving under the gravitational attraction of a mass $M_1 + M_2$ ($M_{\rm disk} \ll M_1 + M_2$). The disk will probably be circular due to the frequent particle collisions. To be stable, the disk must be large with respect to the binary system. Otherwise, its orbit will fragment due to time-dependent tidal forces, which are comparable to the central force. If the circumbinary disk has a radius of $a_{\rm disk} = \gamma^2 \, a$ (where a is the semi-major axis of the binary system), a typical value assumed in the literature for a stable circumbinary disk is $\gamma^2 = 2.25$ (Soberman et al., 1997).

Assume that that a fraction, δ, of the mass lost from star 2 is transferred to the circumbinary disk and ejected from there with the specific orbital angular momentum of the disk. In this case, we have for the change in orbital angular momentum of the binary:

$$d\,J_{\rm orb} = \frac{J_{\rm disk}}{M_{\rm disk}} \, \delta \, dM_2 = \gamma \, \frac{J_{\rm orb}}{\mu} \, \delta \, dM_2, \tag{4.50}$$

where we have used the relation

$$\frac{J_{\rm disk}}{M_{\rm disk}} = \gamma \, \frac{J_{\rm orb}}{\mu}. \tag{4.51}$$

Hence we find

$$\frac{\dot{J}_{\rm orb}}{J_{\rm orb}} = \frac{\delta \, \gamma}{\mu} \, \dot{M}_2. \tag{4.52}$$

Inserting Eq. (4.52) into Eq. (4.29) and applying $\dot{M}_1 = 0$ we find

$$\frac{\dot{a}}{a} = \left(2\frac{\delta\gamma}{\mu} - \frac{2}{M_2} + \frac{1}{M}\right)\dot{M}_2. \tag{4.53}$$

Integration of this expression yields

$$\frac{a}{a_0} = \left(\frac{q}{q_0}\right)^{2(\delta\gamma-1)}\left(\frac{q+1}{q_0+1}\right)e^{2\delta\gamma(q-q_0)}. \tag{4.54}$$

This example is not physical because we have not accounted for the remaining fraction of the material leaving the donor star. Thus, the previous expression is only valid for $\delta = 1$ ($\alpha = \beta = 0$), under which conditions

$$\frac{a}{a_0} = \left(\frac{q}{q_0}\right)^{2(\gamma-1)}\left(\frac{q+1}{q_0+1}\right)e^{2\gamma(q-q_0)}. \tag{4.55}$$

The circumbinary disk (or toroid) and the binary system also interact via a gravitational torque that can be even more efficient in extracting angular momentum from the binary. In this case, it can be shown that $\dot{J}_{\rm orb} \propto \tau_{\rm vis}^{-1/3}$ (Spruit & Taam, 2001), where $\dot{J}_{\rm orb}$ is the torque and $\tau_{\rm vis}$ is the viscous timescale at the inner edge of the disk. For details of resonant and non-resonant interactions between a binary system and its circumbinary disk, see, for example, Artymowicz & Lubow (1994). Eccentricity pumping is a mechanism related to circumbinary disks (e.g., Lubow & Artymowicz, 1996) that might explain the eccentric orbits in post-AGB star binaries (Dermine et al., 2013), in systems such as the barium stars (see Section 5.4.2) and in some peculiar eccentric binary millisecond pulsar systems (Antoniadis, 2014).

4.3.9 The General Case

The last term ($\dot{J}_{\rm ml}$) on the right-hand side of Eq. (4.30) represents the change in orbital angular momentum caused by mass transfer/loss from the binary system. This quantity is usually the dominant term in the orbital angular momentum balance equation. It is often a good approximation to assume $\dot{J}_{\rm gwr}, \dot{J}_{\rm mb}, \dot{J}_{\rm ls} \ll \dot{J}_{\rm ml}$, especially during a relatively short-lasting RLO with a high mass-transfer rate. Thus far, we have considered different individual modes of mass transfer/loss from a binary system. Combining these effects in a single prescription yields its total effect given by

$$\frac{\dot{J}_{\rm ml}}{J_{\rm orb}} = \frac{\alpha + \beta q^2 + \delta\gamma(1+q)^2}{1+q}\,\frac{\dot{M}_2}{M_2}, \tag{4.56}$$

where α, β, and δ are the fractions of mass lost from the donor in the form of a direct fast wind, the mass ejected from the vicinity of the accretor, and from a circumbinary co-planar disk (with radius, $a_r = \gamma^2 a$), respectively (see Section 4.3.8). The accretion efficiency of the accreting star (star 1) is thus given by $\epsilon = 1 - \alpha - \beta - \delta$, or equivalently

$$\partial M_1 = -(1 - \alpha - \beta - \delta)\,\partial M_2, \tag{4.57}$$

where $\partial M_2 < 0$ (as usual, M_2 refers to the donor star mass). These factors will be functions of time as the binary system evolves during the mass-transfer phase. Assuming, however, that α, β, and δ are constant in time, Tauris & van den Heuvel (2006) demonstrated that the general solution for calculating the change in orbital separation during the RLO phase can be obtained by analytical integration of the orbital angular momentum balance equation (Eq. 4.29). The result is

$$\frac{a}{a_0} = \Gamma_{ls} \left(\frac{q}{q_0}\right)^{2\,(\alpha+\gamma\delta-1)} \left(\frac{q+1}{q_0+1}\right)^{\frac{-\alpha-\beta+\delta}{1-\epsilon}} \left(\frac{\epsilon q+1}{\epsilon q_0+1}\right)^{3+2\frac{\alpha\epsilon^2+\beta+\gamma\delta(1-\epsilon)^2}{\epsilon(1-\epsilon)}}, \tag{4.58}$$

where the subscript 0 denotes initial values and Γ_{ls} is a factor of order unity to account for any tidal spin-orbit couplings ($\dot{J}_{\rm ls}$). We remind the reader that $q \equiv M_{\rm donor}/M_{\rm accretor}$.

4.3.10 Mass Loss from Spherically Symmetric SNe

In Section 4.3.4, we considered the orbital widening due to stellar wind mass loss. Another variant of this Jeans' mode is a sudden spherically symmetric mass ejection from one component caused by a SN explosion. As the ejection velocity of the SN shell is of the order of 10^4 km s^{-1}, approximately two orders of magnitude larger than the typical orbital velocities of the two stars, it is a good approximation to assume instantaneous mass ejection. Such an event can drastically change the orbital parameters of the system, and in case the system remains bound after the SN, a recoil is imparted that can give rise to a significant systemic velocity. Thus the post-SN system may propagate far distances from the local standard of rest in which it was born. Alternatively, if the binary system is disrupted, the two stars will become runaway objects with velocities of the order of their original orbital velocities (see Chapter 13 for a detailed description).

This mode of mass ejection was studied extensively by Blaauw (1961) and Boersma (1961) to examine whether this mechanism may explain the origin of the runaway OB-stars—see discussions in van den Heuvel (1985) and Stone (1991). This mode is of great importance also in the formation of X-ray binaries, for which we know one stellar component underwent a SN explosion. In the following section, we consider the kinematic impact on the post-SN orbit for the idealized case of a symmetric SN (i.e., without a kick). A more general and in-depth treatment is given in Chapter 13.

We make the following simplifying assumptions:

- There is an originally circular orbit with radius, a_0.
- There is an instantaneous spherically symmetric SN explosion (i.e., no kick).
- There are point-mass stars, so the effects of the SN shell impact can be neglected.
- The motion of the stars can be neglected until the ejecta decouples gravitationally from the binary (when the SN shell passes by the companion star).

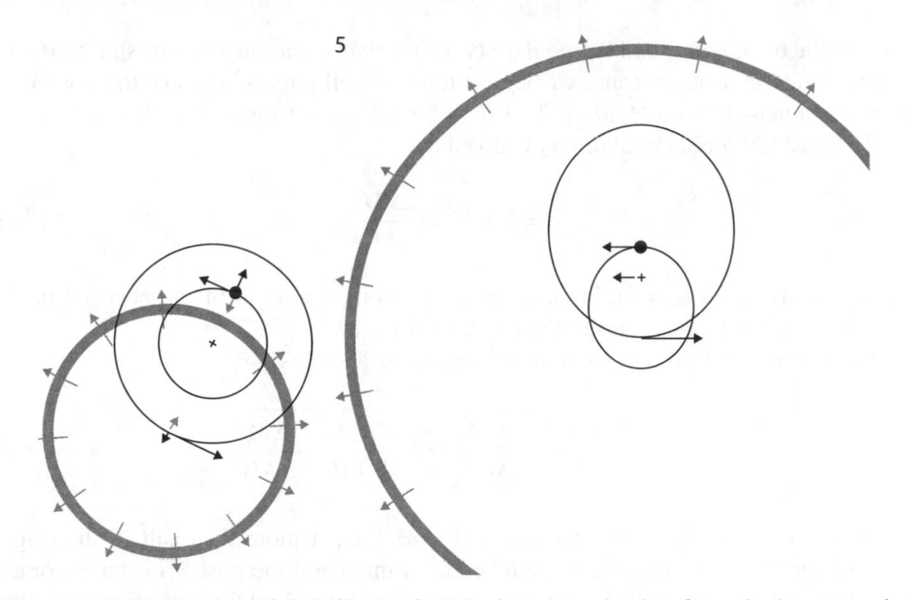

Figure 4.10. Illustration of the effects of explosive mass ejection from the least massive binary component on the binary orbits (i.e., the *Blaauw effect*). The companion star only "notices" the mass loss once the SN shell has passed by (right): it then suddenly feels less gravitational attraction from the exploding star than before, causing it to move into a wider and eccentric orbit. As the SN shell carries momentum caused by the orbital motion of the exploding star, the remaining post-SN binary will get a recoil velocity in the opposite direction. Unlike this illustration, for computational purposes we assume no orbital motion in the short time interval between the SN explosion (left) and the SN shell decoupling (right). After Gursky & van den Heuvel (1975).

Then we follow Flannery & van den Heuvel (1975) to calculate the properties of the post-SN system. As a result of the explosion, the total mass of the system is suddenly reduced, which causes an instantaneous reduction in the attraction between the stellar components. Furthermore, the suddenly unbalanced part of the centrifugal accelerations of the orbital motion of the stars will produce a wider and eccentric post-SN orbit (see Fig. 4.10).

Considering the difference in the total energy of the system before and after the SN explosion, one can derive (Exercise 4.10) the following:

$$\frac{a}{a_0} = \frac{M}{M - \Delta M}, \tag{4.59}$$

where $\Delta M = M_0 - M = M_{10} - M_1$ is the instantaneous mass loss from star 1 with pre-SN mass, M_{10}, and post-SN (gravitational) mass,[2] M_1. Here, M and M_0 denote the total

[2] In addition to the baryonic mass loss of the ejected material, the newly formed NS or BH is compact and releases gravitational binding energy from the core collapse.

mass of the binary after and before the SN, respectively, and the mass of star 2, M_2, is assumed to be constant because we neglect the SN shell impact. We note that for small amounts of mass loss ($\Delta M/M_0 \ll 1$), Eq. (4.59) takes the form of Eq. (4.34).

The post-SN orbital eccentricity is given by

$$e = 1 - \frac{a_0}{a} = \frac{\Delta M}{M} \tag{4.60}$$

(recall, we assumed $e_0 = 0$), because the periastron separation of the post-SN orbit, $a(1-e)$, is equal to the radius of the pre-SN orbit, a_0.

From Kepler's third law, we have $P^2 \propto a^3/M$, and thus

$$\frac{P}{P_0} = M\frac{\sqrt{M^2 - \Delta M^2}}{(M-\Delta M)^2} = M\sqrt{\frac{M+\Delta M}{(M-\Delta M)^3}}\,. \tag{4.61}$$

Equations (4.59)–(4.61) show that for $\Delta M > M$ (i.e., if more than half of the original total mass, M_0, is lost), the system becomes unbound: the post-SN orbit becomes hyperbolic ($e > 1$ and $a < 0$). This can be derived from the virial theorem (Exercise 4.9).

In case the post-SN binary circularizes efficiently due to tides (e.g., for a post-SN system with a giant companion star close to filling its Roche lobe), then for conservation of orbital angular momentum (Sutantyo, 1974), the semi-major axis will decrease by a factor of $(1-e^2)$, such that for a re-circularized post-SN orbit, we have $a/a_0 = M_0/M$, which is similar to Eq. (4.34). For the re-circularized post-SN orbit, it is easy to show that $P/P_0 = (M_0/M)^2$ (Exercise 4.11).

From conservation of momentum, the recoil (systemic runaway) velocity of the c.m. of the post-SN system is simply given by

$$v_{\rm sys} = \frac{\Delta M\, v_1}{M}, \tag{4.62}$$

where v_1 is the orbital velocity of the exploding star in the c.m. reference frame and given by

$$v_1 = \frac{M_2}{M}\, v_{\rm rel}, \tag{4.63}$$

and $v_{\rm rel} = \sqrt{GM_0/a_0}$ is the pre-SN relative velocity between the two stars.

4.3.11 Gravitational Quadrupole Moment Variations and Additional Spin-orbit Couplings

The third term ($\dot{J}_{\rm ls}/J_{\rm orb}$) on the right-hand side of Eq. (4.30) describes possible exchange of angular momentum between the orbit and the donor star. In the following, we briefly discuss two cases: (1) tidal interactions caused by radial expansion or contraction of a star in a close binary, and (2) the Applegate mechanism (with an extension).

The tidal torque in LMXBs can be determined by considering the effect of turbulent viscosity in the convective envelope of the donor star on the equilibrium tide (Terquem et al., 1998). By calculating the pre-RLO spin-orbit couplings in LMXBs, it can be shown (Tauris, 2001) that the sole effect of nuclear expansion of a (sub)giant donor in a tight binary will lead to an orbital period *decrease* of ~10%, prior to the onset of the RLO, as a result of tidal interactions. This effect is most efficient for binaries with $2 < P_{\rm orb} < 5$ d. In wide binaries, the tidal torque is weak (Section 4.7), and in more narrow orbits, the donor star does not expand very much. However, if the effect of magnetic braking (Section 11.3.2) is included in the calculations prior to RLO, this contribution will completely dominate the loss of orbital angular momentum, assuming its corresponding torque is relatively strong.

Spin-orbit couplings in X-ray binaries can help to stabilize the mass-transfer processes in intermediate-mass X-ray binaries (IMXBs) with $2-5\,M_\odot$ radiative donor stars (Tauris & Savonije, 2001). In such systems, the effect of pumping spin angular momentum into the orbit is clearly seen in the calculations as a result of a contracting mass-losing star in a tidally locked system. This causes the orbit to widen (or actually shrink less) and thereby survive the, otherwise dynamically unstable, mass transfer. The tidal effects in eccentric high-mass binary systems are discussed in Witte & Savonije (1999b) and Witte (2001), for example. See Section 4.7 for more details on tides.

The Applegate Mechanism

In the gravitational quadrupole coupling model (GQC model; Applegate, 1992; Applegate & Shaham, 1994), magnetic activity is driven by energy flows in convective layers of the companion star that is being irradiated by its close NS companion. This, in combination with wind mass loss, results in a torque on its spin that holds it slightly out of synchronous rotation, causing tidal dissipation of energy and heating of the star. The resultant time-dependent gravitational quadrupole moment (e.g., variations of the oblateness) causes modulation of the orbital period on a short (in principle), dynamical timescale (which is another name for the pulsation timescale, given by Eq. [8.11]; see also Chapter 8). The orbital separation of the two stars will shrink if there is an increase in the quadrupole moment of the companion star, and the orbit will widen if the quadrupole moment decreases. In Applegate's model, total orbital angular momentum is assumed to remain constant and the NS is treated as a point mass.

The variable quadrupole moment, ΔQ, is caused by cyclic spin-up and spin-down of the outer layers of the companion star, which leads to orbital period changes equal to (Applegate & Shaham, 1994)

$$\frac{\Delta P}{P} = -9\left(\frac{R_2}{a}\right)^2 \frac{\Delta Q}{M_2\,R_2}. \tag{4.64}$$

A redistribution of spin angular momentum to a thin shell of radius, $R_{\rm shell}$, and mass, $M_{\rm shell}$, rotating with angular velocity, Ω, in the gravitational field of a point mass, M_2, will change the angular velocity of the shell by an amount $\Delta\Omega$ and cause the quadrupole moment of the star to change by

$$\Delta Q = \frac{2}{9}\,\frac{M_{\mathrm{shell}}\,R_2^5}{G\,M_2}\,\Omega\,\Delta\Omega. \tag{4.65}$$

The torque needed to redistribute the angular momentum is speculated to be exerted by a mean subsurface magnetic field of several kG. Such a B-field could, for example, be driven by a dynamo caused by convection due to irradiation of the companion star by the pulsar wind.

The Applegate model is one of a number that have been applied to explain the cyclic modulation in the orbital period of magnetically active close binaries (see Section 4.4). For example, the GQC mechanism has been applied successfully by Applegate & Shaham (1994) to explain the orbital period variations in the eclipsing radio MSP binary, PSR B1957+20 (the original so-called black widow system [Arzoumanian et al., 1994]). In addition, it was proposed as the main mechanism for producing the orbital period variations of another black widow MSP, PSR J2051−0827 (e.g., Doroshenko et al., 2001; Lazaridis et al., 2011; Shaifullah et al., 2016; see illustration in Fig. 4.11). Although variations in the gravitational quadrupole moment are seen in such binary MSP systems, their effect on the *long-term* orbital evolution is limited in X-ray binaries. The reason is that in X-ray binaries with moderate or high mass-transfer rates, the dominant contribution to $\dot{J}_{\mathrm{orb}}$ is related to mass loss, which is the fourth term on the right-hand side of Eq. (4.30).

In Applegate's model, a rather large fraction of the stellar luminosity is required for the GQC operation. Empirical evidence for this would be to detect changes in the stellar luminosity in phase with the orbital period modulation. In the alternative model of Lanza & Rodonò (e.g., Lanza et al., 1998; Lanza & Rodonò, 1999), however, a change in the azimuthal magnetic field intensity of the companion star produces a change in the quadrupole moment by perturbing the oblateness (and thus the effective centrifugal acceleration).

Through spin-orbit couplings, the quadrupole of a rapidly rotating companion leads to apsidal motion and precession of the binary orbit. This in turn causes a variation of the longitude of periastron and of the projected semi-major axis of the pulsar, x, according to (Smarr & Blandford, 1976; Wex, 1998):

$$\frac{dx}{dt} = x\,\Omega\,\tilde{Q}\,\cot i\,\sin\Phi\,\frac{1}{2}\sin(2\theta), \tag{4.66}$$

where i is the orbital inclination of the binary system, θ is the angle between the spin and the orbital angular momentum, and Φ is the longitude of the ascending node with respect to the invariable plane (i.e., the plane perpendicular to the total angular momentum vector), see Figure 4.11. The dimensionless quadrupole moment, $\tilde{Q}$, is related to the quadrupole moment, Q, by

$$\tilde{Q} = \frac{9}{2}\,\frac{Q}{M_2\,a^2}. \tag{4.67}$$

For arbitrary values of θ, one obtains for combined GQC effects and spin-orbit couplings averaged over the orbit (Lazaridis et al., 2011):

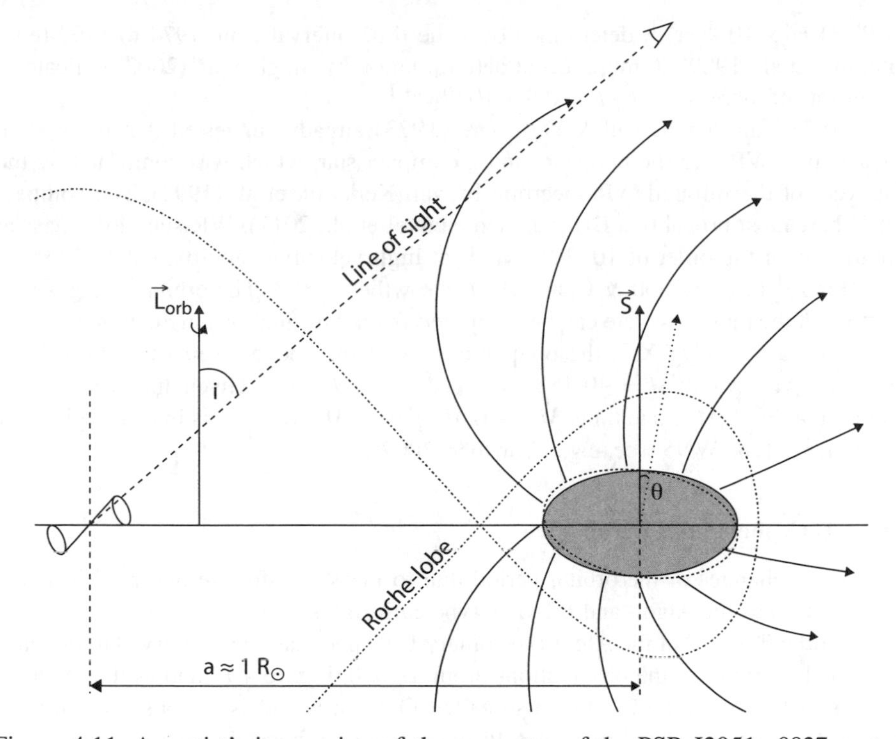

Figure 4.11. An artist's impression of the geometry of the PSR J2051−0827 system (not drawn to scale). The pulsar and its companion star are separated by $\sim 1\ R_\odot$ and the orbital inclination angle is approximately 40°. The pulsar wind interacts and pushes the wind of the companion out, resulting in a bow shock–like denser region of charged particles, which causes the system to be eclipsed at given orbital phases. Changes in the oblateness of the tidally locked companion star, possibly in combination with alterations in the direction of the spin axis, θ, cause changes in the quadrupole moment that affect the timing of the pulsar. After Lazaridis et al. (2011).

$$\frac{\Delta P}{P} = -2\,\Delta\tilde{Q}\left(1 - \frac{3}{2}\sin^2\theta\right). \tag{4.68}$$

Observational evidence in PSR J2051−0827 for variations in θ with time is discussed in the next section.

4.4 OBSERVATIONAL EXAMPLES

4.4.1 Cyg X-3

An example of a star that continuously loses mass in the form of a high-velocity spherically-symmetric stellar wind is the donor star in the X-ray binary Cyg X-3, which has an orbital period of 4.8 hr, which is increasing rapidly at a rate

$\dot{P}/P = 1.66 \times 10^{-6}\,\mathrm{yr}^{-1}$, determined over the time interval from 1974 to 1992 (e.g., Kitamoto et al., 1989). A more recent determination by Singh et al. (2002) indicates a slower rate of increase of $\dot{P}/P = 0.9 \times 10^{-6}\,\mathrm{yr}^{-1}$.

In 1973, van den Heuvel & De Loore (1973) already suggested that the system consists of a WR-star (helium star) and a compact star, which was confirmed by the discovery of the infrared WR spectrum by van Kerkwijk et al. (1992). The compact star is here most probably a BH (van den Heuvel et al., 2017). WR-stars lose mass at a high rate, of the order of $10^{-5}\,M_\odot\,\mathrm{yr}^{-1}$, at high velocities: approximately 2,000 to 3,000 $\mathrm{km\,s}^{-1}$ (e.g., Abbott & Conti, 1987; Crowther, 2007). The orbital changes as a function of the mass-loss rate can be computed from the equations in Section 4.3.4.

In the case of Cyg X-3, these equations, with the above-given value of $\dot{P}/P = 0.9 \times 10^{-6}\,\mathrm{yr}^{-1}$ yield: $\dot{M} = -0.45 \times 10^{-6}\,(M/M_\odot)\,M_\odot\,\mathrm{yr}^{-1}$. Given that $\dot{M} = \dot{M}_2$ and assuming $M = 10\,M_\odot$, one finds $\dot{M} = -0.45 \times 10^{-5}\,M_\odot\,\mathrm{yr}^{-1}$, as is indeed typical for a WR-star of type WN5 (see, e.g., Crowther, 2007).

4.4.2 U Cephei and β Lyrae

Continuous changes of the orbital period due to mass transfer are observed in many binaries, such as the Algol- and β Lyrae–type eclipsing systems.

The orbital period of the Algol-type binary U Cephei has been observed to increase since the beginning of the observations around 1880. Figure 4.12 shows these period changes of U Cep in the form of a so-called O–C curve (O is the observed time of eclipse, C is the calculated time of eclipse, assuming a constant binary period). The system consists of a B8Ve primary star (the accretor) and a G8III secondary star (the donor), with masses of $M_1 = 3.8\,M_\odot$ and $M_2 = 1.9\,M_\odot$, respectively. On the adopted zero-point date HJD 2434195,5750 (which was around the year 1950) the orbital period was 2.4929935 d, while the period in 1909 was 2.4929005 d. (The present period is well above 2.4931 d.) The figure shows that around 1950 the eclipses came already late by $\sim$0.9 d (21.6 hr) relative to the 1880 ephemeris. The rate of period increase is 1.4×10^{-8} d per cycle, corresponding to $\dot{P} = 5.6 \times 10^{-9}$ (Selam & Demircan, 1999). For a linear increase of the period with time, the deviation of the eclipse time from the one predicted for a fixed orbital period increases quadratically with time (i.e., $\propto E^2$, where E is the epoch number), hence the O–C curve will be a parabola, as indicated by the dotted curve in Figure 4.12. Similarly, Kuiper (1941) determined for the β Lyrae system: $\dot{P} = 6 \times 10^{-7}$ (see also Kreiner, 1981; Harmanec, 2002).

For conservative RLO, taking the logarithmic differentiation of Eq. (4.41) and isolating for the mass-transfer rate of the donor star yields

$$\dot{M}_2 = \frac{1}{3}\left(\frac{M_2\,M_1}{M_2 - M_1}\right)\frac{\dot{P}}{P}. \tag{4.69}$$

For U Cep, with the above-given masses and $P = 2.5$ d, the observed value of $\dot{P}$ yields $\dot{M}_2 = -1.036 \times 10^{-6}\,M_\odot\,\mathrm{yr}^{-1}$. For β Lyrae ($P = 12.9$ d), with $M_2 = 2\,M_\odot$, $M_1 = 12\,M_\odot$, the measured value of $\dot{P} = 6 \times 10^{-7}$ yields $\dot{M}_2 = -1.359 \times 10^{-5}$

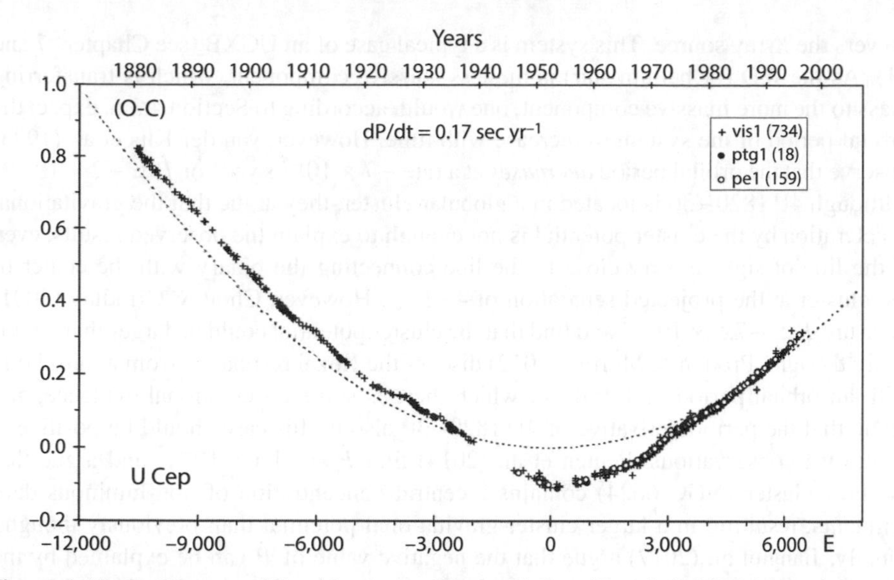

Figure 4.12. O–C curve of the Algol-type binary U Cep ($P = 2.4929935$ d on the zero-date HJD 2434195,5750) until 1995. The scale on the vertical axis is in days. The dotted curve is the best-fit parabola through the visual (vis), photographic (ptg), and photoelectric (pe) data. While in 1909 the orbital period was 2.4929005 d, the present orbital period is well above 2.4931 d. (Deviations from a linear increase of the period with time, as represented by the dotted parabola, indicate that sometimes there are decelerations and accelerations in the mass-transfer rate.) After Selam & Demircan (1999).

$M_\odot\,\mathrm{yr}^{-1}$. We will examine the causes of these very large rates of mass transfer in Chapter 9.

4.4.3 Peculiar Systems—Spin-Orbit Couplings in Action

There are peculiar systems that do not seem to fit to mass loss via fast winds or mass transfer via RLO, even though mass transfer or rapid mass loss is taking place in these systems. Examples include the ultra-compact X-ray binary (UCXB) 4U 1820−30 in the globular cluster NGC 6624, and the two black widow binary MSPs, PSR 1957+20 and PSR J2015−0827, in the Galactic disk.

4U 1820−30

4U 1820−30 is a brilliant example of a complex binary system that has intrigued astrophysicists for decades. The system has an X-ray luminosity of $2 - 10 \times 10^{37}$ erg s^{-1} and an orbital period of only 11 min (Stella et al., 1987). Hence, this binary clearly must consist of a NS (the X-ray source) and an $\sim 0.2\,M_\odot$ low-mass WD (or helium star), as mass donor, that is overflowing its Roche lobe and provides the mass accretion that powers the X-ray source. This system is a typical case of an UCXB (see Chapters 7 and 11). As the WD (or helium star) is the less massive component, which is transferring mass to the more massive component, one would, according to Section 4.3.6, expect the orbital period of the system to increase with time. However, van der Klis et al. (1993) observe that the orbital period *decreases* at a rate -7×10^{-5} s yr^{-1} or $\dot{P} \simeq -2 \times 10^{-12}$. Although 4U 1820$-$30 is located in a globular cluster, they argue that the gravitational acceleration by the cluster potential is not enough to explain the observed results, even if the line of sight is very close to the line connecting the binary with the center of the cluster at the projected separation of $4'' \pm 1''$. However, Chou & Grindlay (2001) measure $\dot{P} \simeq -7.5 \times 10^{-13}$ and find that the cluster potential could be larger than previously thought. Prodan & Murray (2012) discuss the Kozai resonance from a third body with an orbital period of 170 d, for which there is some observational evidence, and argue that the period derivative of 4U 1820$-$30 also in this case should be positive—contrary to observations. Peuten et al. (2014) find $\dot{P} = -1.1 \times 10^{-12}$ and argue that the host cluster (NGC 6624) contains a central concentration of non-luminous dark remnants, resulting in a larger cluster gravitational potential than previously thought. Finally, Jiang et al. (2017) argue that the negative value of $\dot{P}$ can be explained by the known superburst events that are caused by runaway thermal nuclear burning on the NS surface, which may eject material from the binary and feed a circumbinary disk once every ~10,000 yr. As is evident from this one example, the nature of binary (or multiple) stars can be truly puzzling.

PSR 1957+20

The PSR 1957+20 system consists of a 1.6 ms radio pulsar that is eclipsed every 9.2 hr by a red dwarf companion with a mass of $\sim 0.025\, M_\odot$ (Fruchter et al., 1988; Fruchter et al., 1990). This system has been optically identified, and the red dwarf shows a very large heating effect (van Paradijs et al., 1988; van Kerkwijk et al., 2011). The shape of the radio eclipse shows that the companion is being evaporated by the high-energy radiation from the pulsar that is impinging on its surface. The companion has a comet-like tail of gas directed away from the pulsar that is produced by the evaporation of its envelope. At large distances (of the order of a light-year), the system is surrounded by an Hα nebula that is produced by the matter evaporated from the companion (Kulkarni & Hester, 1988).

In view of the rapid evaporative mass loss from the companion, one would, according to the fast wind mode (see Section 4.3.4), expect the orbital period to increase with time. However, it was first observed to *decrease* at a rate $\dot{P} = -3.6(\pm 0.6) \times 10^{-11}$ (Ryba & Taylor, 1991). This observation gave rise to much theoretical work (e.g., Tavani & Brookshaw, 1992), such as simulations of mass outflow via a circumbinary disk (Section 4.3.8), to explain efficient loss of orbital angular momentum. A few years later, the orbital period was reported to *increase* with a rate of $\dot{P} = 1.47(\pm 0.08) \times 10^{-11}$ (Arzoumanian et al., 1994). Hence, the previous measurement of a large negative orbital period derivative reflected only the short-term behavior of the system during the early observations; the orbital period derivative is apparently undergoing quasi-cyclic orbital period variations that are similar to those found in other close binaries such as Algol and RS CVn. It is therefore thought that the $\sim 0.025\, M_\odot$ companion to

PSR B1957+20 is non-degenerate, convective, and magnetically active—see discussion on the Applegate mechanism in Section 4.3.11.

As a twist in the tail, PSR 1957+20 has been suggested to host a NS with a mass of $2.4 \pm 0.12\, M_\odot$ (van Kerkwijk et al., 2011). The estimates of the velocity amplitudes and the mass ratio between the stellar components (crucial for the resulting derived NS mass), however, are complicated. Corrections need to be made because the velocity amplitude is that of the center of light, which, because of the irradiation, is shifted toward the pulsar relative to the c.m; see also Section 3.6. A NS mass of $2.4\, M_\odot$ would have a profound impact on the equation of state of NS matter.

PSR J2051−0827

The PSR J2051−0827 system (Stappers et al., 1996) consists of a radio MSP with a spin period of 4.5 ms and an $\sim 0.05\, M_\odot$ semi-degenerate dwarf companion with an orbital period of 2.4 hr. Understanding the orbital evolution of this binary has proven to be rather challenging. In Figure 4.13, the observed changes in orbital period, P, differ significantly from those of the projected semi-major axis of the pulsar, x (Lazaridis et al., 2011; Shaifullah et al., 2016), which is given by $x = a_{\rm psr} \sin i / c$, where $a_{\rm psr}$ is the semi-major of the pulsar orbit around the c.m. and i is the orbital inclination angle.

In Section 4.3.11, we investigated a mostly stable companion star with an outer shell responsible for the changing quadrupole moment via the Applegate mechanism. An alternative (or additional) explanation, however, is needed for the non-correlated variations in P and x, as seen in Figure 4.13. Lazaridis et al. (2011) propose that while the observed variations in P could be provided by limited quadrupole changes, an overall small tilt of the angle between the stellar spin axis and the orbital angular momentum vector, θ, may cause a precession of both the star and the orbit. This would allow the entire quadrupole moment of the star, $\tilde{Q}$, to cause variations in x. Lazaridis et al. conclude that, with small changes in the orientation and magnitude of the quadrupole moment, they can explain the observed parameter changes. However, they do not know of any physical mechanism that would cause the required tilt of the star or its rapid changes. Hence, this system remains puzzling in nature.

The exact nature of the companion star in PSR J2051−0827, and those of all black widow systems (including PSR B1957+20, which we have discussed), are also not well known. Because the pulsars in these systems suffer from eclipses of their radio signals, it is clear that the companions are non- or semi-degenerate stars that suffer from irradiation-driven mass loss (Kluzniak et al., 1988; Ruderman et al., 1989; Chen et al., 2013). Their nature is most likely that of a brown dwarf-like star or a semi-degenerate helium remnant—the leftover from the irradiation process of a former He WD companion.

The size (radius) of the irradiated companion star is difficult to determine accurately. Using Hubble Space Telescope observations, Stappers et al. (2001) determine two different values for the filling factor (the ratio of the volume-equivalent radius of the companion star to its Roche lobe) of 0.43 and 0.95, corresponding to stellar radii of $\sim 0.064\, R_\odot$ and $0.14\, R_\odot$, respectively. The "dark" side of the irradiated companion was detected, and the R magnitude at minimum is ~ 26, while the difference between the side heated by the impinging pulsar radiation and the unirradiated side is ~ 3.3 mag. Finally, they estimated the orbital inclination angle to be approximately $i \simeq 40°$.

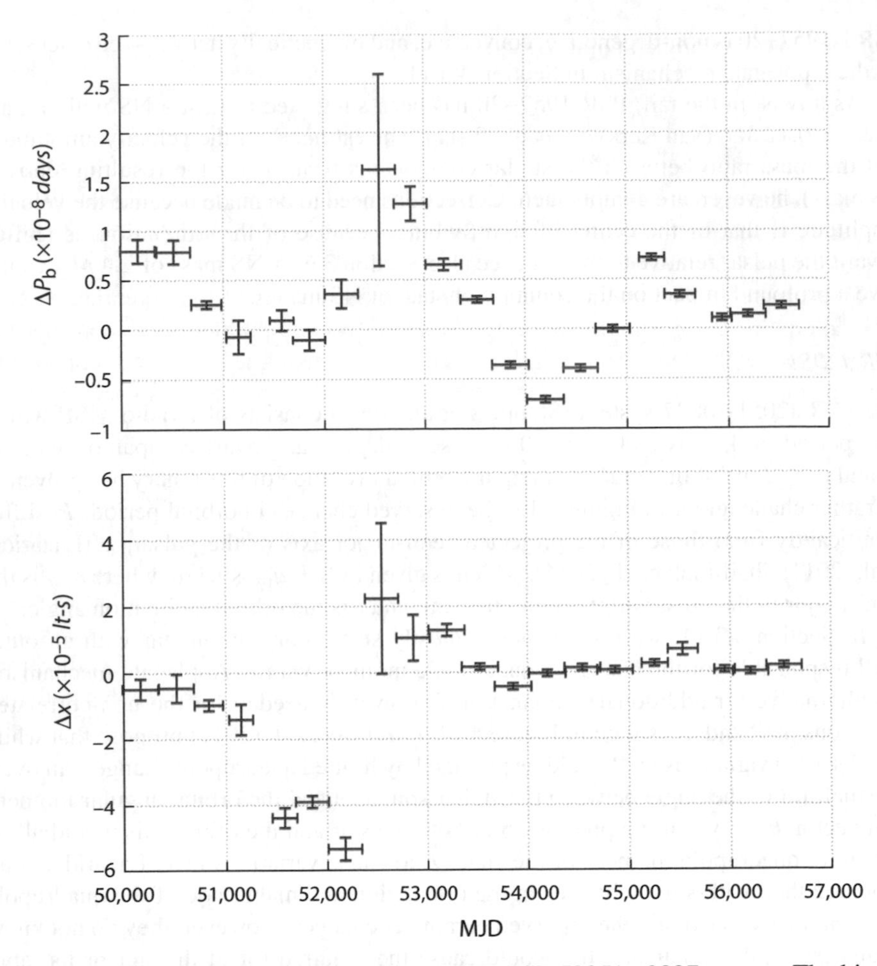

Figure 4.13. The puzzling orbital changes of the PSR J2051−0827 system. The binary consists of a radio millisecond pulsar with an $\sim$0.05 $M_\odot$ (possibly semi-degenerate) dwarf companion. The plots show changes in orbital period, ΔP (top), and projected semi-major axis of the pulsar, Δx (bottom), as measured by fitting P and x to $\sim$20 yr of observational data for eras of length 365 d (with an overlap of 30 d). The data for ΔP and Δx are somewhat correlated for the first 10 yr, whereas no correlation is seen in the last 10 yr of data. After Shaifullah et al. (2016).

4.5 BASIC PHYSICS OF MASS TRANSFER VIA L_1

The process of mass transfer was studied in early works by Kippenhahn & Weigert (1967), Jedrzjec (1969), Savonije (1979), and Meyer & Meyer-Hofmeister (1983). The rate of mass transfer from the Roche-lobe filling donor star, $\dot{M}$, can be estimated by integrating the mass-flux density ($\rho\, c_s$) over the cross section, Q, of the stream nozzle near the inner Lagrangian point, L_1:

$$\dot{M} = \int \rho\, c_S \frac{dQ}{d\phi}\, d\phi \simeq \rho\, c_s\, Q, \tag{4.70}$$

where ρ is the mass density, c_s is the local speed of sound (thermal velocity), ϕ is the gravitational potential, $dQ/d\phi$ is the differential cross section, and Q is the effective cross section of the flow at L_1.

To estimate Q, one may consider a series of expansions of the potential, ϕ, near its saddle point L_1. A simplification to the hydrodynamic problem is to consider the Roche potential in the plane through L_1 perpendicular to a line ($y=z=0$) connecting the two stars (e.g., Livio, 1994). Hence, $L_1 = L_1(x_1, 0, 0)$ and the distance, x_1, from the donor star to L_1 is given by Eq. (4.8). Consequently, $\nabla\phi = 0$ at L_1, which allows us to write, using a series expansion at L_1,

$$\Delta\phi(x_1, y, 0) \equiv \phi(x_1, y, 0) - \phi(x_1, 0, 0) \approx \frac{1}{2} \frac{\partial^2 \phi}{\partial y^2}\bigg|_{L_1} y^2 \simeq \frac{1}{2}\, \Omega^2\, y^2, \tag{4.71}$$

where the last part, $\partial^2\phi/\partial y^2 \simeq \Omega^2$, follows from Eq. (4.5).

The width of the nozzle is roughly $\sqrt{Q}$ and can be estimated from the fact that a gas element in the stream should have a kinetic energy sufficient to overcome the potential difference across it, from point $(x_1, \sqrt{Q}, 0)$ to $(x_1, 0, 0)$, and thus $C_S^2 \simeq 2\,\Delta\phi(x_1, \sqrt{Q}, 0)$ or

$$C_S^2 \simeq \Omega^2\, Q, \tag{4.72}$$

which then yields (via [Eq. 4.70])

$$\dot{M} \simeq \frac{\rho\, C_S^3}{\Omega^2}\bigg|_{L_1}. \tag{4.73}$$

We now consider two cases for which we estimate the mass-transfer rate: (1) a star with a radius inside the Roche lobe (beginning atmospheric RLO), and (2) a star with a radius overfilling its Roche lobe.

4.5.1 Atmospheric Roche-lobe Overflow

According to the canonical criterion for mass transfer (Kippenhahn & Weigert, 1967), one simply has to ask whether the donor star radius, R is larger or smaller than its Roche-lobe radius, R_L. Hence, one makes the implicit assumption that the edge of the star is infinitely sharp and therefore that the mass transfer starts/ends abruptly rather than following a gradual transition. This criterion can be improved by taking the finite scale height of the stellar atmosphere properly into account.

What one generally defines as the radius of a star is the radius of its photosphere, $R_{\rm ph}$, which is the radius where the star becomes transparent (at the optical depth,

$\tau = 2/3$). Above its photosphere, the star will still have an atmosphere with an exponentially decreasing density, with scale height, H, such that at the radius, R_L, of the Roche lobe we have approximately

$$\rho(R_L) = \rho_{\rm ph}\, e^{-(R_L - R_{\rm ph})/H}, \tag{4.74}$$

where $\rho_{\rm ph}$ is the density of the photosphere. There will then be outflow of matter with density, $\rho(R_L)$, from the atmosphere toward the companion. This is what one calls (beginning) atmospheric RLO (Savonije, 1978, 1979; Ritter, 1988). To obtain the mass-transfer rate along L_1 for this atmospheric RLO, one should insert this density into Eq. (4.73).

The case of (beginning) atmospheric RLO is important in HMXBs before the radius of the donor star begins to exceed that of its Roche lobe. Atmospheric RLO then may produce accretion rates of the order of $10^{-10} - 10^{-8}\, M_\odot\,{\rm yr}^{-1}$ onto a NS or BH companion, which is sufficient to make them a strong X-ray source for periods of the order of $10^4 - 10^5$ yr (Savonije, 1978, 1979, 1983).

Once the photospheric radius of the star exceeds that of its Roche lobe, real (full-blown) RLO sets in, which in HMXBs typically leads to mass-transfer rates $>10^{-4}$ $M_\odot\,{\rm yr}^{-1}$, which will completely obscure the X-ray source. For a NS accretor and a typical $10 - 25\, M_\odot$ donor star in a HMXB, the large mass ratio makes the RLO dynamically unstable as the orbit shrinks efficiently (Fig. 4.8), leading to the onset of a common envelope (CE) (Section 4.8). See discussions on stability of RLO in Sections 4.5.2–4.5.5 and further details in Chapter 9.

Breakdown of a Mass-Transfer Scheme for Giant Donors

As discussed in the previous section, the mass loss from the donor star can be modelled as a stationary isothermal, subsonic flow of gas that reaches sound velocity near the nozzle at the first Lagrangian point, L_1. The accompanying mass-transfer rate is given by Eq. (4.73), or in the version of Ritter (1988):

$$|\dot{M}| = \rho\, c_s\, Q \simeq \frac{1}{\sqrt{e}}\, \rho_{\rm ph}\, c_s\, Q\, e^{-\Delta R/H_p}, \tag{4.75}$$

where $c_s = \sqrt{kT/(\mu m_{\rm H})}$ is the isothermal sound speed, Q is (again) the effective cross section of the flow at L_1 (see Meyer & Meyer-Hofmeister, 1983), and $\Delta R \equiv R_L - R$. The last parameter,

$$H_p = \frac{kT\,R^2}{\mu m_{\rm H}\, GM}, \tag{4.76}$$

is the pressure scale height of the stellar atmosphere (μ is the mean molecular weight and $m_{\rm H}$ is the mass of the hydrogen atom). This scheme was developed to study the turn-on (turn-off) of mass transfer in nearly semi-detached systems, that is, so-called *optically thin* mass transfer for which $R < R_L$ (see also D'Antona et al., 1989;

Kolb & Ritter, 1990).[3] However, the mass-transfer algorithm in Eq. (4.75) was derived for low-mass main-sequence donor stars in CV binaries for which $H_p \ll R$, and therefore mass transfer was only assumed to occur for $\Delta R \ll R$.

For giant stars this picture has to change (Pastetter & Ritter, 1989). These stars with low surface gravities often have $H_p/R \simeq 0.04$. As a result, Tauris et al. (2013) find that using the above-mentioned prescription for giant stars can in some cases cause mass-transfer rates $>10^{-7}\,M_\odot\,\text{yr}^{-1}$ even for $\Delta R = 0.3\,R$ (i.e., while the donor star is still *underfilling* its Roche lobe by $\sim$23% in radius). Hence, for giant star donors, the assumptions behind the original Ritter scheme break down.

4.5.2 Stability Criteria for Mass Transfer

The stability and nature of the mass transfer is very important in binary stellar evolution. It depends on the response of the mass-losing donor star and of the Roche lobe—see Soberman et al. (1997) for a review. If the mass transfer proceeds on a short timescale (thermal or dynamical) the system is unlikely to be observed during this short phase; whereas if the mass transfer proceeds on a nuclear timescale it is still able to sustain a high enough accretion rate onto the NS or BH for the system to be observed as an X-ray source for a long time.

When the donor star fills its Roche lobe and is perturbed by removal of mass, it falls out of hydrostatic and thermal equilibrium. In the process of re-establishing equilibrium the star will either grow or shrink—first on a dynamical (sound-crossing) timescale, and then on a slower thermal (Kelvin-Helmholtz) timescale. But also the Roche lobe changes in response to the mass transfer/loss. As long as the donor star's Roche lobe continues to enclose the star, the mass transfer is stable. Otherwise it is unstable and proceeds on a dynamical timescale. Hence the question of stability is determined by a comparison of the exponents in power-law fits of radius to mass, $R \sim M^\zeta$, for the donor star and the Roche lobe, respectively:

$$\zeta_{\text{donor}} \equiv \frac{\partial \ln R_2}{\partial \ln M_2} \quad \wedge \quad \zeta_L \equiv \frac{\partial \ln R_L}{\partial \ln M_2}, \tag{4.77}$$

where R_2 and M_2 refer to the mass-losing donor star. Given $R_2 = R_L$ (the condition at the onset of RLO) the initial stability criteria becomes

$$\zeta_L \leq \zeta_{\text{donor}}, \tag{4.78}$$

where ζ_{donor} is the adiabatic or thermal (or somewhere in between) response of the donor star to mass loss. Note that the stability might change during the mass-transfer phase so that initially stable systems become unstable, or vice versa, later in the evolution (Kalogera & Webbink, 1996). The radius of the donor is a function of time and

[3] It should be noted that ΔR and the mass ratio, q, are defined differently in Ritter (1988) and Kolb & Ritter (1990). The first paper has a typo in the last term in Eq. (A8), which should be $f_2^{-3}(q)$.

mass and thus

$$\dot{R}_2 = \left.\frac{\partial R_2}{\partial t}\right|_{M_2} + R_2\,\zeta_{\rm donor}\frac{\dot{M}_2}{M_2} \tag{4.79}$$

$$\dot{R}_L = \left.\frac{\partial R_L}{\partial t}\right|_{M_2} + R_L\,\zeta_L\frac{\dot{M}_2}{M_2}. \tag{4.80}$$

The second terms on the right-hand sides follow from Eq. (4.77); the first term of Eq. (4.79) is due to expansion of the donor star as a result of nuclear burning (e.g., shell hydrogen burning on the red giant branch [RGB]), and the first term in Eq. (4.80) represents changes in R_L that are not caused by mass transfer—such as orbital decay due to GW radiation and tidal spin-orbit couplings. Tidal couplings act to synchronize the orbit whenever the rotation of the donor is perturbed (e.g., as a result of magnetic braking or an increase in the moment of inertia while the donor expands). The mass-loss rate of the donor can be found as a self-consistent solution to Eqs. (4.79) and (4.80), assuming $\dot{R}_2 = \dot{R}_L$ for stable mass transfer.

For a simplified treatment based on polytropic equation-of-state donor stars, see Section 11.3.2. For discussions on the stability of mass transfer via the second Lagrangian point, L_2, see Section 4.3.2.

4.5.3 Response of the Roche Lobe to Mass Transfer/Loss

To study the dynamical evolution of an X-ray binary let us consider the cases where tidal interactions and GW radiation can be neglected. We shall also assume that the amount of mass lost directly from the donor star in the form of a fast wind, or via a circumbinary toroid, is negligible compared to the flow of material transfered via the Roche lobe. Hence we have $\dot{J}_{\rm gwr} = \dot{J}_{\rm mb} = \dot{J}_{\rm ls} = 0$ and $\dot{J}_{\rm ml}/J_{\rm orb} = \beta q^2/(1+q) \times (\dot{M}_2/M_2)$. This corresponds to the mode of *isotropic re-emission* where matter flows over from the donor star onto the compact accretor (NS or BH) in a conservative way, before a fraction, β, of this material is ejected isotropically from the system with the specific angular momentum of the accretor: $\dot{M}_1 = -(1-\beta)\dot{M}_2$, $dJ_{\rm orb} = (J_1/M_1) \times \beta dM_2$, and $J_1 = (M_2/M)\,J_{\rm orb}$. In the general-case formalism outlined in Section 4.3.9, this corresponds to $\alpha = \delta = 0$ and $\Gamma_{ls} = 1$. This is actually a good approximation for many real systems (with orbital periods larger than a few days). Following Tauris & Savonije (1999), one can then combine Eqs. (4.6), (4.57), and (4.58) and obtain an analytical expression for ζ_L:

$$\begin{aligned}\zeta_L = \frac{\partial \ln R_L}{\partial \ln M_2} &= \left(\frac{\partial \ln a}{\partial \ln q} + \frac{\partial \ln (R_L/a)}{\partial \ln q}\right)\frac{\partial \ln q}{\partial \ln M_2},\\ &= [1+(1-\beta)q]\psi\; + \;(5-3\beta)q,\end{aligned} \tag{4.81}$$

where

$$\psi = \left[-\frac{4}{3} - \frac{q}{1+q} - \frac{2/5 + 1/3\,q^{-1/3}(1+q^{1/3})^{-1}}{0.6 + q^{-2/3}\ln(1+q^{1/3})}\right]. \tag{4.82}$$

In the limiting case where $q \to 0$ (when the accretor is much heavier than the donor star; for example, in a soft X-ray transient system hosting a BH),

$$\lim_{q \to 0} \zeta_L = -5/3. \tag{4.83}$$

The behavior of $\zeta_L(q, \beta)$ for different X-ray binaries is plotted in Figure 4.14. This figure is quite useful to get an idea of the stability of the mass transfer when comparing with ζ of the donor star. We see that in general the Roche lobe, R_L, increases ($\zeta_L < 0$) when material is transferred from a relatively light donor to a heavy accretor ($q < 1$). In this situation the mass transfer will be stable. Correspondingly R_L decreases ($\zeta_L > 0$) when material is transferred from a heavy donor to a lighter accretor ($q > 1$). In this case the mass transfer has the potential to be dynamically unstable. This behavior can be understood from the bottom panel of Figure 4.14 where we plot

$$-\partial \ln(a)/\partial \ln(q) = 2 + \frac{q}{q+1} + q\frac{3\beta - 5}{q(1-\beta)+1} \tag{4.84}$$

as a function of q. The sign of this quantity is important because it tells whether the orbit expands or contracts in response to mass transfer (note $\partial q < 0$). It is noticed that the orbit always expands when $q < 1$ and it always decreases when $q > 1.28$ [solving $\partial \ln(a)/\partial \ln(q) = 0$ for $\beta = 1$ yields $q = (1 + \sqrt{17})/4 \approx 1.28$]. If $\beta > 0$ the orbit can still expand for $1 < q < 1.28$. There is a point at $q = 3/2$ where $\partial \ln(a)/\partial \ln(q) = 2/5$, independent of β. It should also be mentioned for the curious reader that if $\beta > 0$ then, in some cases, it is actually possible for a binary to decrease its separation, a, while increasing $P_{\rm orb}$ at the same time!

4.5.4 Response of the Mass-losing Star—The Effect on the Binary

The radius of the mass-losing donor star will respond to mass loss. Therefore, to obtain a full stability analysis of the mass-transfer process it is important to know whether the donor star expands (or contracts) in response to mass loss. This is determined by the stellar structure (i.e., temperature gradient and entropy) of the envelope at the onset of the RLO. A summary of early works on the stability criteria in binary systems with donor stars treated as simple polytropic equation-of-state models is given in Section 11.3.2.

Donor Stars with a Radiative Envelope

Donor stars with radiative envelopes (Case A RLO, i.e., mass transfer during hydrogen core burning—see Chapter 9 for definitions of cases of RLO) or only slightly convective envelopes (early Case B RLO) will usually shrink, or keep a roughly constant radius, in response to mass loss. Therefore, these stars will give rise to a dynamically stable mass-transfer phase if the mass ratio, q, is not too large. Tauris et al. (2000) demonstrated in detail that even IMXB systems with such unevolved or slightly evolved donor stars of mass $2 < M_2/M_\odot < 5$ are able to survive extreme mass transfer

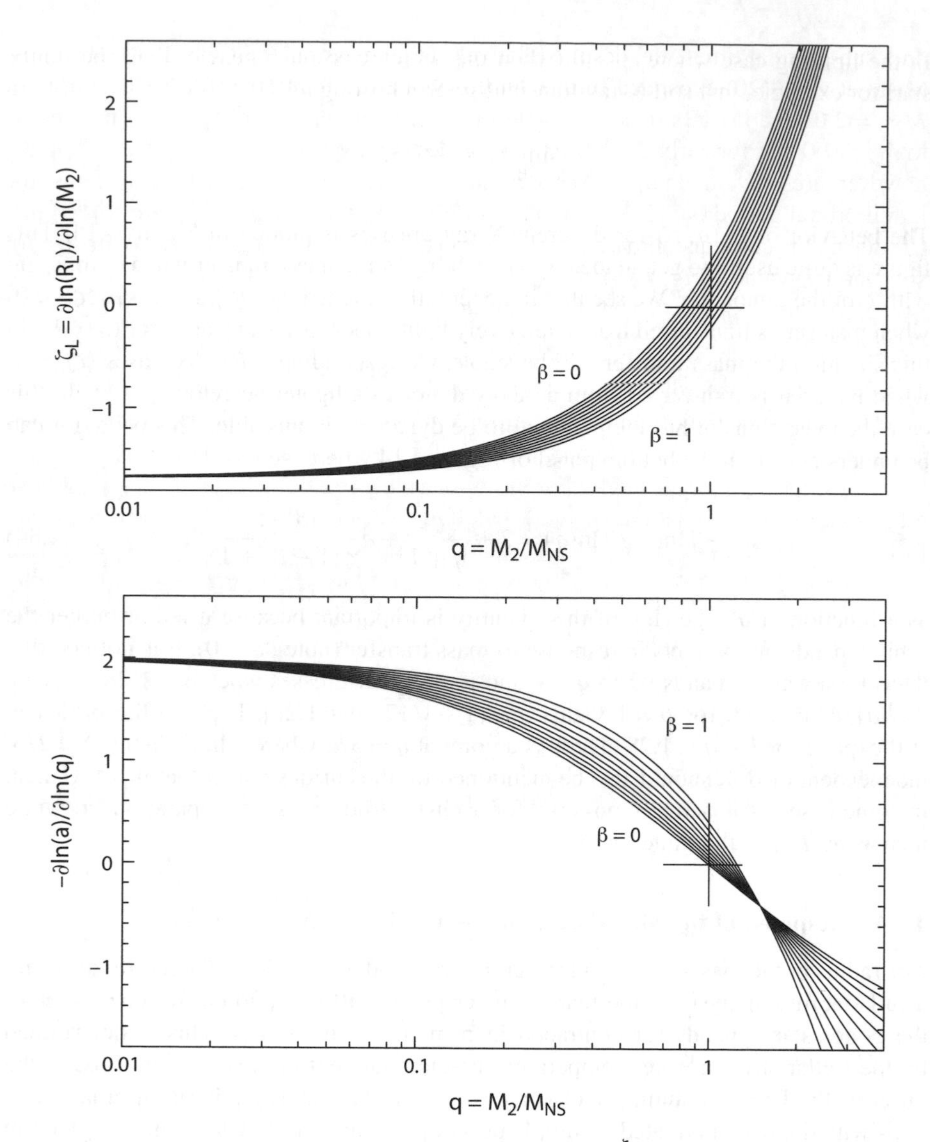

Figure 4.14. (Top) The Roche-radius exponent ($R_L \propto M_2^{\zeta_L}$) for LMXBs as a function of q and β. The different curves correspond to different constant values of β in steps of 0.1. Tidal effects are not taken into account and the mass loss is according to the isotropic re-emission model (i.e., $\alpha = 0$, $\delta = 0$). A cross is shown to highlight the cases of $q = 1$ and $\zeta_L = 0$. (Bottom) We have plotted $-\partial \ln a / \partial \ln q$ as a function of q. When this quantity is positive the orbit widens. This is the case when $q \leq 1$. For more massive donor stars ($q > 1$) the orbit shrinks in response to mass transfer. Because M_2 (and hence q) is decreasing with time, the evolution during the mass-transfer phase follows these curves from right to left, although β need not be constant. After Tauris & Savonije (1999).

on a sub-thermal timescale, despite relatively large mass ratios of $q \simeq 2-3$. Recently, Misra et al. (2020) expanded further on this work and found that for NS accretors of $M_{\rm NS} = 2.0\,M_\odot$, IMXBs undergo stable mass transfer for donor stars with masses up to $M_2 \lesssim 7.0\,M_\odot$ for early Case B RLO (Fig. 11.24). IMXBs undergoing Case A RLO, however, are only stable up to $M_2 \lesssim 5.5\,M_\odot$. The reason is that in these systems, the initial orbital period is $\lesssim 2$ d at the onset of RLO, and thus their orbits are forced in a CE during the subsequent mass transfer because binary orbits always shrink in size for such large q-values (Fig. 4.14).

Donor Stars with a Convective Envelope

The thermal response of a donor star with a deep convective envelope is much more radical. It expands rapidly in response to mass loss due to the super-adiabatic temperature gradient in its giant envelope. This is clearly an unstable situation if the Roche lobe does not grow accordingly. For systems with $q \gtrsim 2.0$, the orbital shrinking is so efficient that, in combination with an expanding convective donor, it always leads to the formation of a CE and (dynamically unstable) spiral-in evolution. Calculations by Tauris & Savonije (1999) and Podsiadlowski et al. (2002) showed that whereas all LMXBs with donor stars $M_2 \lesssim 1.8\,M_\odot$ (and initial $1.3\,M_\odot$ NS accretors) will undergo stable mass transfer, systems with $M_2 > 2.0\,M_\odot$ will be unstable for giant star donors (late Case B or Case C RLO). Hence, this is most likely the destiny for all HMXBs, as well as all IMXBs with fairly wide orbits ($P_{\rm orb} \gtrsim 30$ d).

Further discussions on the interplay between stellar evolution and binary interactions are given in Chapter 9.

4.5.5 Donor Stars Overfilling Their Roche Lobes

If a donor star fills its Roche lobe and undergoes full-blown RLO, the mass-transfer rate can be rather high—depending on the mass ratio between the two stars (as discussed in Section 4.5.3) and the stellar structure in the envelope of the donor star (Chapter 8). In particular, for donor stars more massive than the accretor and with a radiative envelope, very large rates of mass transfer will ensue, on the order of the thermal timescale. For example, for $2-5\,M_\odot$ donor stars in IMXBs, the mass-transfer rates can reach $10^{-5}-10^{-4}\,M_\odot\,{\rm yr}^{-1}$, and yet these systems still remain dynamically stable despite $q \simeq 1.5-4$ (Tauris et al., 2000; Podsiadlowski et al., 2002; Shao & Li, 2012; Pavlovskii & Ivanova, 2015; Misra et al., 2020).

The situation is different when the donor is less massive than the accretor as is the case in CVs and LMXBs.[4] In this case, the donor can slightly keep overfilling its Roche lobe, leading to modest rates of mass transfer that are sufficient to fill an accretion disk around a WD, NS, or BH and turn the latter two types of objects into strong X-ray sources (Fig. 4.15).

[4] Although, highly super-Eddington mass-transfer rates ($10^{-4}\,M_\odot\,{\rm yr}^{-1}$) are still possible, on a thermal timescale, for evolved low-mass stars on the RGB (Tauris & Savonije, 1999; Podsiadlowski et al., 2002).

Figure 4.15. Artist's impression of the BH binary system V404 Cygni with a beautiful accretion disk and jets. The temperature profile across the disk can be used to determine the spin rate of the BH. Credit: Gabriel Pérez Díaz, SMM, Instituto de Astrofisica de Canarias.

We now redefine $\Delta R \equiv R - R_L$ as the degree by which the star *overfills* its Roche lobe, for the radius of which we can take Eq. (4.6). In the expression for the mass-transfer rate given by Eq. (4.73), we can combine an expression for the sound speed, $c_s = \sqrt{\gamma P/\rho}$ (γ is the adiabatic index) with a general polytropic equation of state (i.e., where the pressure is given by $P = K\,\rho^\gamma$) and express

$$\rho \propto c_s^{\frac{2}{\gamma-1}}, \tag{4.85}$$

which then yields

$$\dot{M} \propto \frac{1}{\Omega^2}\, c_s^{\frac{3\gamma-1}{\gamma-1}}. \tag{4.86}$$

We aim at expressing $\dot{M}$ as a function of $\Delta R/R$. There are a couple of ways to do this in an approximate manner. One method is to consider the gradient of the potential, ϕ, through L_1 at a point on the surface of the donor star far away from L_1. Here, $\phi \simeq -GM/R$, and thus we can write $\Delta\phi/\Delta R = GM/R^2$, or

$$\Delta\phi = \frac{GM}{R}\,\frac{\Delta R}{R}, \tag{4.87}$$

which can simply be combined with $c_s^2 = 2\,\Delta\phi$ to yield

$$c_s \propto \sqrt{\frac{\Delta R}{R}}. \tag{4.88}$$

Another method (Exercise 4.12) is simply to assume $\Delta R \sim H_p$, which also leads to this scaling relation. Now, we can simply insert this expression into Eq. (4.86) to obtain

$$\dot{M} \propto \left(\frac{\Delta R}{R}\right)^{\frac{3\gamma-1}{2(\gamma-1)}}. \tag{4.89}$$

Applying, for example, $\gamma = 5/3$, which is relevant for stars with convective envelopes (red giants and low-mass main-sequence stars), yields

$$\dot{M} \propto \left(\frac{\Delta R}{R}\right)^3. \tag{4.90}$$

A more careful analysis by Paczyński & Sienkiewicz (1972), results in

$$\dot{M} \simeq -A\, M \Omega \left(\frac{\Delta R}{R}\right)^3, \tag{4.91}$$

where A is a dimensionless coefficient that depends on the density profile of the outer layers of the mass-losing star and slightly on the mass ratio of the system. For typical systems, A is of the order of a few. Two-dimensional numerical hydrodynamic calculations of mass flow along L_1 by Edwards & Pringle (1987) have verified Eq. (4.91).

Following an idea in Onno Pols' lecture notes we can rewrite Eq. (4.91):

$$\frac{\Delta R}{R} \propto \left(\frac{\dot{M}}{M}\, P_{\rm orb}\right)^{1/3} = \left(\frac{P_{\rm orb}}{\tau_{\rm mt}}\right)^{1/3}, \tag{4.92}$$

where $\tau_{\rm mt}$ is the mass-transfer timescale. It is seen that the mass-transfer rate is very sensitive to any increase in ΔR (also for other values of the adiabatic index, γ), and it is evident that for mass transfer proceeding on a nuclear or a thermal timescale, $\Delta R/R \lesssim 0.01$. Hence, the donor stars usually only evolve to overfill their Roche lobes by a very small amount. However, as we shall see later, this is not always the case.

As another example, one finds (Livio, 1994) that to obtain a mass-flow rate of $10^{16}\,{\rm g\,s^{-1}}$ ($\sim 1.6\times10^{-10}\,M_\odot\,{\rm yr^{-1}}$) in a CV with a 3 hr orbital period, only requires $\Delta R/R \simeq 3\times10^{-5}$. For a Roche-lobe filling companion of $0.5\,M_\odot$, with a radius of $0.5\,R_\odot$, this means an overfill $\Delta R \sim 10$ km. This overfill can be caused by either evolutionary expansion of the star on a nuclear timescale or shrinking of the orbit by orbital angular momentum loss, for example, by emission of gravitational radiation or by magnetic braking, as described in Chapter 11.

The effect of irradiation feedback is a cause of uncertainty in the calculations of the mass-transfer rate via RLO. In X-ray and pulsar binaries, the flux of hard photons and/or relativistic particles from the NS accretor impinging on the surface layers of the donor star will cause changes in the mass-transfer rate (e.g., Büning & Ritter, 2004). However, the impact and the modelling of this effect, often leading to cyclic accretion, is still unclear. Investigations by Benvenuto et al. (2012) on the evolution of

UCXBs suggests that the inclusion of irradiation feedback is not very significant for the long-term evolution of a compact binary and its final properties. Furthermore, for the wide-orbit LMXBs with giant donors, Ritter (2008) argues that irradiation-driven mass-transfer cycles cannot occur because these systems are transient because of accretion disk instabilities.

4.6 ACCRETION DISKS

In a semi-detached system with RLO, gas flowing through L_1 has too much angular momentum to fall directly onto the surface of the accreting star. Hence, the gas forms an accretion disk around the mass-gaining star, through which the gas particles slowly spiral in due to viscosity before being accreted (see Fig. 4.15 for an illustration). We now discuss this process in a bit more detail. The standard steady state accretion disk model was developed by Shakura & Sunyaev (1973) and subsequently refinements were added by a variety of authors. Time-dependent disks were studied by Lynden-Bell & Pringle (1974) and gas streams by Lubow & Shu (1975). For more detailed discussions, see, for example, the textbook by Frank et al. (2002).

4.6.1 Particle Trajectories and Disk Formation

When the donor star fills its Roche lobe, the gas can escape from the atmosphere of the donor at L_1 and move into the Roche lobe of the accretor because of the pressure difference across L_1. The flow resembles the escape of gas through a nozzle into a vacuum. The flow velocity is approximately the thermal velocity of the atoms, roughly equal to the sound speed in the atmosphere of the donor star. The fact that the gas funneling through L_1 is supersonic simplifies the calculation of the stream trajectory, which can be treated as a set of test particles in more or less ballistic orbits. If a particle starts with almost zero velocity ($\dot{r} = 0$) on the Roche lobe after passing through L_1, it does not have sufficient energy to cross the lobe at any other point. The trajectory lies entirely within the Roche lobe of the accretor, and whenever the particle approaches the lobe it does so with low velocity. Thus the stream trajectory can be found from integrating the equations of motion for a particle in the rotating binary frame. In conserving energy along the trajectory, a particle obeys the following equation:

$$\frac{1}{2}\dot{r}^2 + \phi(r) = \text{constant}. \tag{4.93}$$

The stream of gas particles from the donor will follow an orbit reaching a minimum distance, d_{min}, from the c.m. of the accretor, which scales with the orbital separation, a:

$$d_{\text{min}} = 0.0488\, q^{-0.464}\, a \qquad \text{for} \quad 0.05 < q < 1. \tag{4.94}$$

This expression is obtained from trajectory computations and approximated to 1% accuracy.

As the thermal speed with which the particles are launched from L_1 is very small, the total energy (potential plus kinetic) of the particles will be that corresponding to the equipotential through L_1. This means that, after sweeping around the accretor, the trajectory of the particle stream can just reach the Roche lobe of the accretor, and from there will fall inward again, as depicted in Figure 4.16. In falling back from the Roche lobe, the particle stream in its trajectory will now collide (possibly after several orbits) with the stream coming from L_1 (see Flannery, 1975; Lubow & Shu, 1975). This happens for two reasons: (1) the presence of the donor star causes the stream of gas particles to precess slowly in their motion back and forth through the Roche lobe of the accretor, and (2) there is deflection that is caused by the Coriolis effect. As a continuous stream is trying to follow this orbit it will therefore collide with itself, resulting in dissipation of energy via shocks (Fig. 4.16). Thus, the stream entering from L_1 will only hit the accreting star directly if it has a stellar radius $R > d_{\min}$ (which is unlikely for a compact object accretor); otherwise it continues in its orbit and collides with itself, forming an accretion disk around the companion. Given that a circular orbit has the least energy for a given orbital angular momentum, the dissipation within the stream will tend to produce a ring of gas.

In the disk, there will be various dissipative processes (e.g., collisions of gas particles, shocks, viscous dissipation etc.). Eventually, some of the energy of the gas particles is radiated away, causing the gas particles to sink deeper into the gravitational potential of the accretor. This requires that the gas particles lose orbital angular momentum. In the absence of external torques, loss of angular momentum proceeds via internal torques. However, the timescale for redistributing the orbital angular momentum is usually much longer than the timescale of radiative cooling and the dynamical (orbital) timescale. Therefore, the gas only slowly spirals inward, while remaining approximately in circular orbits. The net outward transfer of angular momentum implies that the disk is broadened both outward and inward compared to its initial location at $R_{\rm circ}$.

At the outer edge of the disk some other process must finally remove the angular momentum. It is likely that some of lost angular momentum is fed back into the binary system through tides exerted on the outer disk by the donor star. For compact object accretors, the inner edge of the disk is truncated: either by the strong B-field of a NS (Section 14.6.2) or by the location of the innermost stable circular orbit, $R_{\rm isco}$, around a BH (see also Section 4.9).

4.6.2 Location of the Accretion Disk

Here, we derive the initial location (radius) of the ring of gas building up the accretion disk around a central compact object accretor. Flowing out from the L_1 point, the gas accelerates toward the accreting star, which in the case of a CV is a WD and in X-ray binaries is a NS or a BH. Because these compact objects have dimensions much smaller than the size of the binary system, which is at least of the order of a solar radius, the gas will acquire velocities when falling toward the compact object of mass, M, which will be of magnitude

$$v \simeq \sqrt{\frac{GM}{r}} > \sqrt{\frac{GM}{a}} \simeq 440\sqrt{\left(\frac{M}{M_\odot}\right)\left(\frac{R_\odot}{a}\right)}\ \mathrm{km\,s^{-1}}. \tag{4.95}$$

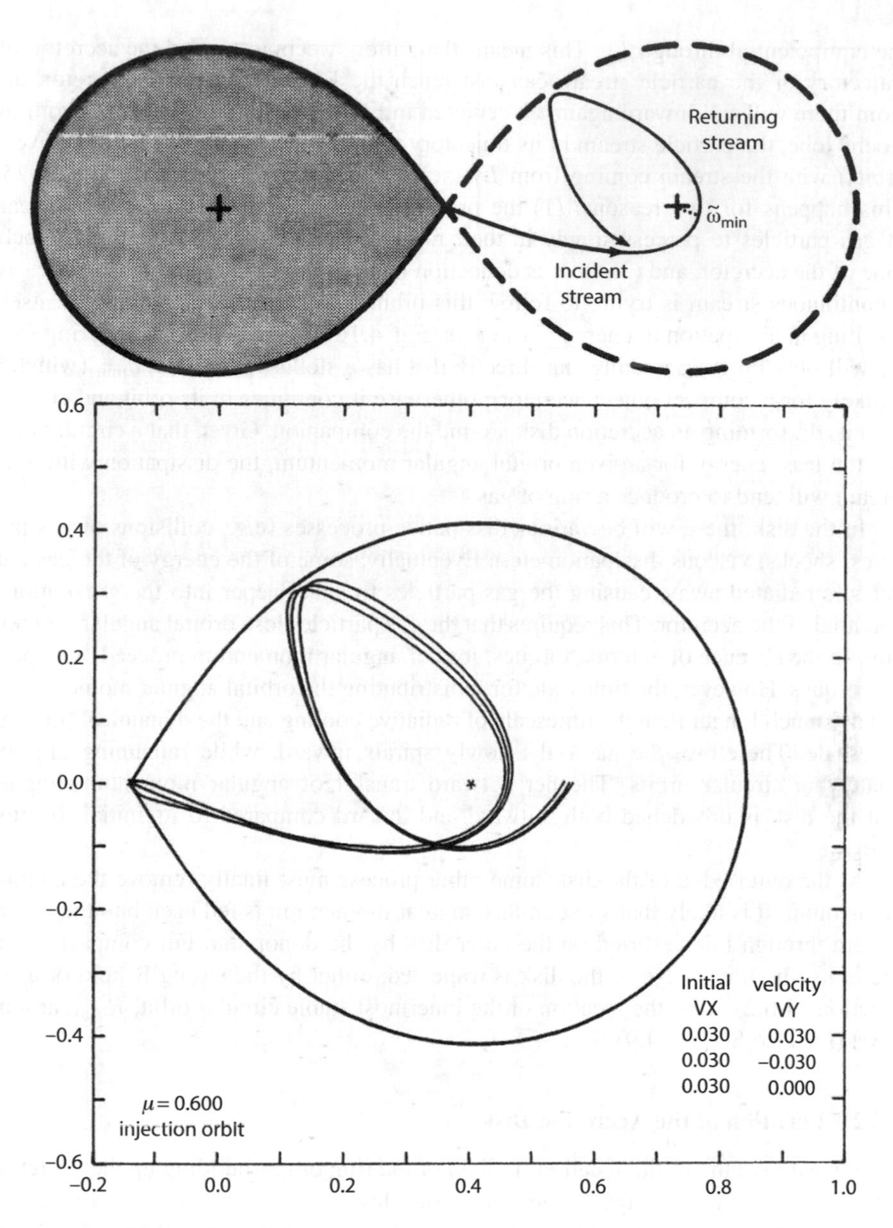

Figure 4.16. Illustration of the accretion stream flow of particles. (Top) The returning stream collides with the incident stream, creating shocks and dissipation of energy (from Lubow & Shu, 1975). The minimum distance between the accretion stream and the accretor (see Eq. [4.94]) is here denoted by $\tilde{\omega}_{\rm min}$. (Bottom) Particle trajectories of the accretion stream in the orbital plane, after passing L_1 with a low velocity (in units of Ωa), for a system with $q = 3/2$ (using our definition of q). After Flannery (1975).

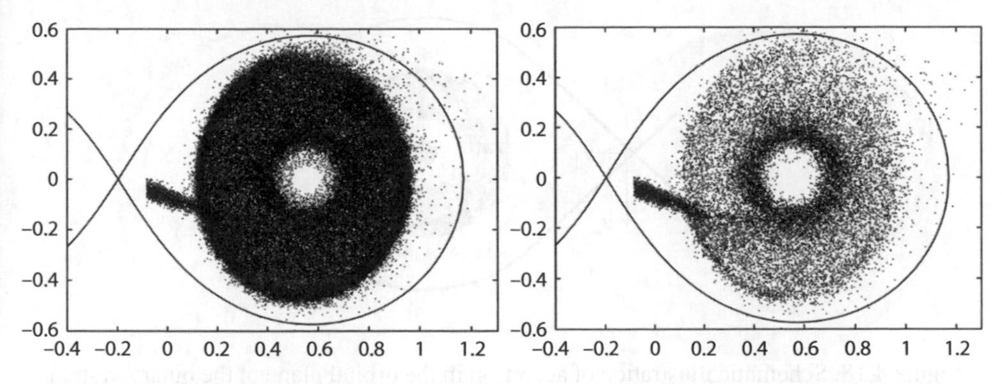

Figure 4.17. These 3-D simulations show the accretion flow in DQ Her, an intermediate polar with a mass ratio of $q=2/3$. (Left) All particles; (right) stream particles. After Kunze et al. (2001).

The speed of sound of the gas at a typical temperature of $\sim 10^4$ K of the outer layers of the donor star in a CV or an X-ray binary, however, will only be

$$c_s = \sqrt{\frac{\gamma k T}{\mu m_{\rm H}}} \sim 10\sqrt{\frac{T}{10^4\,{\rm K}}}\ {\rm km\,s^{-1}}. \tag{4.96}$$

Therefore, the Mach number of the flow after the matter leaves the L_1 nozzle grows to approximately 10–100 and becomes highly supersonic. There are two effects of this: (1) after leaving L_1, the gas forms a relatively narrow stream (as the expansion timescale of the stream is much longer than the flow timescale), and (2) along the stream, pressure forces can be neglected, and the path of the stream can be calculated as though the stream consists of test particles that follow ballistic trajectories, as mentioned in Section 4.6.1.

In deriving the location of the accretion disk, we follow Frank et al. (2002). For the accretor, the gas stream appears to move almost orthogonally to the line connecting the two stars via L_1. To be more specific, let $v_\parallel$ and $v_\perp$ be the components of the stream velocity in a non-rotating reference frame, parallel and perpendicular to the line connecting the two stars. Thus we have

$$v_\parallel \sim c_s \qquad \text{and} \qquad v_\perp \sim x_{\rm acc}\,\Omega, \tag{4.97}$$

where c_s is the sound speed in the envelope of the donor star (extending to L_1), and $x_{\rm acc}$ is the distance from L_1 to the center of the accretor (i.e., $x_{\rm acc} = a - R_{L1}$, where R_{L1} is given by Eq. [4.8]). Generally $v_\perp \gg v_\parallel$ and the gas particles crossing L_1 will therefore have a large specific angular momentum, which prevents direct accretion onto the accretor.

The location of the center of the initial ring (disk) can easily be found because the angular momentum of the gas particles is conserved during their flow toward the accretor while dissipating energy. Hence, the size of this ring, $R_{\rm circ}$, is found by equating

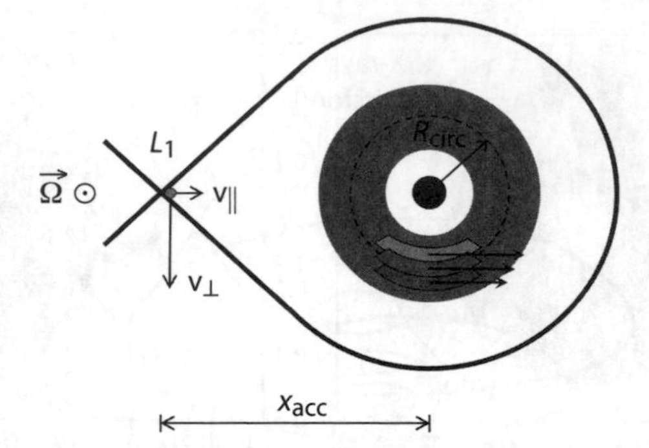

Figure 4.18. Schematic illustration of accretion in the orbital plane of the binary system. A gas element passing through L_1 has a nearly orthogonal velocity component ($v_{\perp} \gg v_{\|}$) as seen from the accretor. Conservation of specific angular momentum means that $x_{acc}\, v_{\perp} \simeq R_{circ}\, v_K(R_{circ})$. When the acretion disk forms, it spreads inward and outward from R_{circ}. Shear (differential rotation) in the disk causes angular momentum to be transported outward while the gas particles lose energy and slowly spiral inward in almost circular orbits.

the specific orbital angular momentum at the ring to that at L_1 (which is $x_{acc}\, v_{\perp}$, see Fig. 4.18), that is,

$$R_{circ}\, v_K(R_{circ}) \simeq \Omega\, x_{acc}^2, \tag{4.98}$$

where the Keplerian velocity for a circular orbit at a distance, r, from the accretor is

$$v_K(r) = \sqrt{\frac{GM_{acc}}{r}}, \tag{4.99}$$

and where M_{acc} is the mass of the accretor. Equation (4.98) can simply be rewritten (Exercise 4.13) as

$$\frac{R_{circ}}{a} = (1+q)\left(\frac{x_{acc}}{a}\right)^4. \tag{4.100}$$

Numerical evaluations of this expression yield (e.g., Plavec & Kratochvil, 1964)

$$\frac{R_{circ}}{a} = \begin{cases} (1+q)\,(0.5 - 0.227\log q)^4 & \text{for} \quad q < 1, \text{ or} \\ 0.0859\, q^{-0.426} & \text{for} \quad 0.05 < q < 1 \end{cases}, \tag{4.101}$$

where the latter expression is accurate to within 1% and which compared to Eq. (4.94) yields

$$R_{circ} \sim 1.75\, d_{min}. \tag{4.102}$$

For typical CVs and LMXBs, one finds from Eq. (4.101) that R_{circ} is smaller than the Roche-lobe radius of the compact object by a factor of 2 to 3.

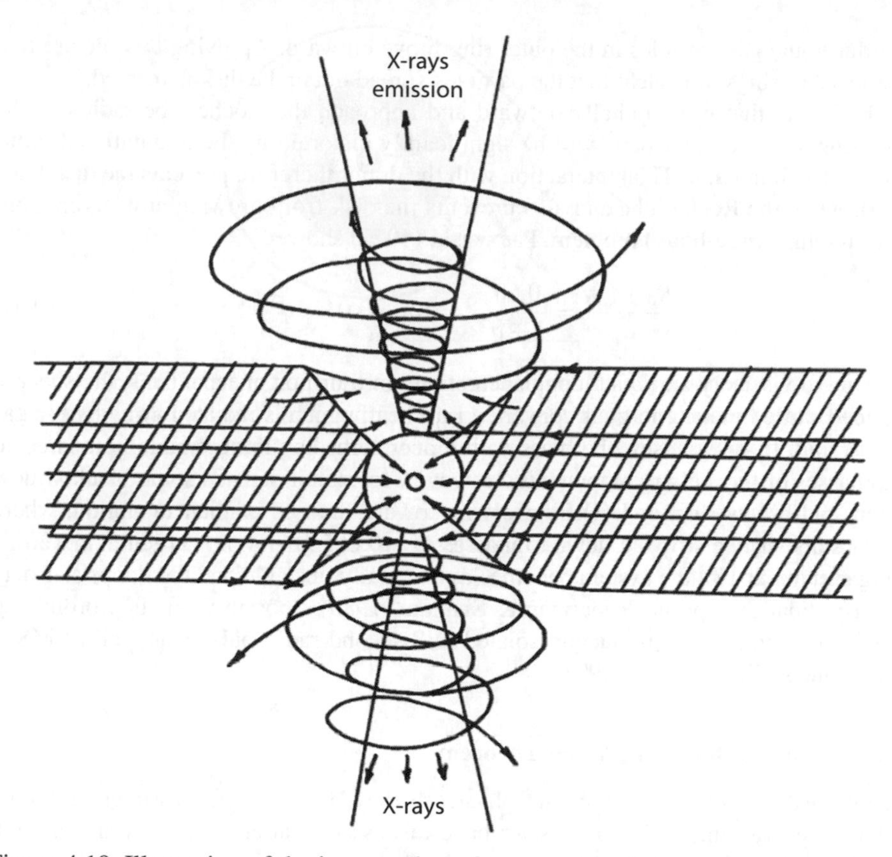

Figure 4.19. Illustration of the inner region of an accretion disk and an emission beam of X-rays. After Shakura & Sunyaev (1973).

4.6.3 Differential Rotation and Maximum Disk Size

Within the ring formed, differential rotation (shear) across a radial distance, Δr, is given by

$$\Delta v = \frac{dv}{dr}\,\Delta r = \frac{d}{dr}\sqrt{\frac{GM}{r}}\,\Delta r = -\frac{v}{2}\frac{\Delta r}{r}. \tag{4.103}$$

Friction acts to oppose the shear and causes the ring to spread inward and outward. (The precise cause of the viscosity is still being disputed. Most probably it is due to the combination of small-scale turbulence and magnetic fields in the disk.) This spreading of the disk is understood by considering a set of concentric rings centered on the accreting star (Fig. 4.18). A given ring rotates faster than its neighboring ring located further out. As a result, friction between the two rings will try to slow down the inner ring and speed up the outer ring. Hence, angular momentum is transferred from the inner ring to the outer ring. As the gas particles in the inner ring are forced to remain in Keplerian orbits despite their loss of angular momentum, they must move further *inward.*

Similarly, the gas particles in the outer ring move outward. Applying this picture to a whole set of rings, it is clear that the particles spread out and a disk is formed.

For disks that grow radially outward and approach the Roche-lobe radius of the accreting star, the outer parts will be significantly distorted by the gravitational influence of the donor star. Tidal interaction with the donor therefore prevents the disk from overflowing the Roche lobe and also prevents the disk from growing any larger. Considering this (three-body) problem, Paczynski (1977) showed

$$\frac{R_{\rm disk,max}}{a} = \frac{0.60}{1+q} \qquad \text{for} \quad 0.03 < q < 1. \tag{4.104}$$

Due to viscosity and other angular momentum transport mechanisms, the shear is reduced while kinetic energy is converted into heat, which is radiated away by the gas of the ring. In most cases, the mass of the accretion disk is negligible compared to the masses of the stellar components. However, for donor stars in some black widow binary pulsar systems, or UCXBs evolving toward the end of their evolution, where donor stars will be stripped down to masses of $\sim 0.005\,M_\odot$ (van Haaften et al., 2012; Sengar et al., 2017), the system becomes dynamically unstable and the donor star may undergo tidal disruption (Ruderman & Shaham, 1985)—possibly this is a promising formation scenario for producing isolated millisecond radio pulsars as well as MSPs with planets (Phillips et al., 1993).

4.6.4 Heating Due to the Virial Theorem

Accretion disks are important in astrophysics as they efficiently transform gravitational potential energy into radiation. As we have discussed, matter spirals inward through the disk due to viscosity. The reason why the disk gets hot in this process is basically due to the virial theorem. Matter moving in a bound Keplerian orbit obeys the virial theorem:

$$E_{\rm pot} + 2\,E_{\rm kin} = 0, \tag{4.105}$$

where $E_{\rm pot}$ and $E_{\rm kin}$ denote the potential and kinetic energy, respectively, of a given amount of disk matter, dM.

When dM moves inward to a smaller Keplerian orbit, its potential energy, $-GMdM/r$, decreases (becomes more negative), where r is the orbital radius and M is the mass of the accretor. According to the virial theorem, the change of the kinetic energy of dM due to its new orbit is given by $dE_{\rm kin} = -dE_{\rm pot}/2$, which implies that half of the lost potential energy is converted into an increase of the orbital kinetic energy and the other half is converted into heat by the friction in the disk and radiated away. Thus, the gradual inward motion originating from friction produced by viscosity causes the disk to be heated more and more. According to the virial theorem, the total energy, $E_{\rm tot}$ of dM is

$$E_{\rm tot} = E_{\rm pot} + E_{\rm kin} = \frac{1}{2} E_{\rm pot} < 0. \tag{4.106}$$

This means that when dM moves inward, its *total amount of energy decreases* because of energy losses by radiation, caused by conversion of potential energy into heat. The closer the disk matter comes to the central compact object (i.e., the smaller the radius r), the larger the loss of potential energy per unit length decrease in r, and the hotter the disk will become and the stronger it will radiate. In the case of NSs and BHs, the inner parts of the disk, just outside the compact object, can reach temperatures of the order of 10^7 to 10^8 K, such that they radiate mainly in the X-ray band of the spectrum. This temperature profile across the disk is quite useful and can be used to determine the spin rate of the BH (McClintock et al., 2014); see Section 7.6.

We note here that the reason of the heating of the disk is basically the same as the reason of the heating of a space probe when it enters the Earth's atmosphere. The friction with the air causes its orbital motion to be braked, such that it moves into a lower Keplerian orbit. In this lower orbit, however, its orbital speed is higher: the braking causes its speed to *increase*! This causes still higher friction, which causes the space probe to get hot. And the more it is braked, the hotter it gets. This is the reason why space ships that bring astronauts back to Earth need a ceramic heat shield, because otherwise the outside wall of the ship would heat up to temperatures far above 1,000°C.

4.6.5 Accretion Disk Instabilities

As a result of the possibility of thermal-viscous instabilities in accretion disks (van Paradijs, 1996; Dubus et al., 2001; Lasota et al., 2008), not all X-ray binaries are detectable as long-term stable (persistent) sources. Some X-ray binaries are transient (unstable) sources that are often only detectable when they experience an outburst that is caused by such accretion disk instabilities. Whether the disk suffers from thermal-viscous instabilities depends on the mass-transfer rate and the size of the disk.

At low mass-transfer rates, thermal-viscous instabilities result from a relatively large and sudden local increase in the opacity and the viscosity of the disk material (Osaki, 1974). The high viscosity causes a much higher accretion rate onto the central compact object, which is called an *outburst*. Basically, all the accumulated material in the accretion disk is suddenly dumped toward the compact object at a very high rate, making the source very luminous in X-rays. Outbursts are alternated in cycles (see schematic illustration in Fig. 4.20) by low-viscosity stages during which the disk builds up again. Stable behavior only persists if the entire disk has a homogeneous degree of ionization. X-ray binaries can therefore have a stable disk if the mass-transfer rate is sufficiently high to keep the entire disk ionized through X-ray irradiation. Thus a critical mass transfer exists that depends on the orbital period (disk size) of the system, separating transient and persistent systems. For example, for an irradiated disk with a pure composition of helium (particularly applicable to UCXBs), Lasota et al. (2008) found

$$\dot{M}_{\rm crit} = 2.4\times10^{-12}\left(\frac{0.60}{(1+q)^{2/3}}\right)^{2.51}\left(\frac{P_{\rm orb}}{\rm min}\right)^{1.67} M_\odot\,{\rm yr}^{-1} \tag{4.107}$$

$$\simeq 6.7\times10^{-13}\,(1+q)^{-5/3}\left(\frac{P_{\rm orb}}{\rm min}\right)^{1.67} M_\odot\,{\rm yr}^{-1}, \tag{4.108}$$

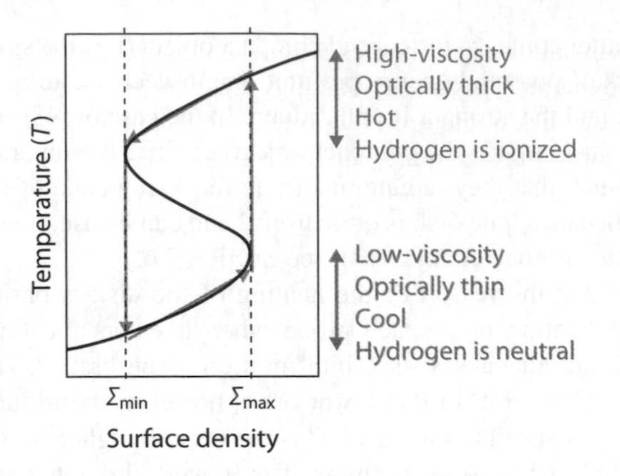

Figure 4.20. Illustration of thermal-viscous instability cycles in hydrogen-rich accretion disks. The plot shows the repeating pattern (so-called *S-curves*) in the surface density–temperature diagram. After M. Kimura, University of Kyoto.

assuming an irradiated disk temperature of

$$T_{\rm irr} = \left(\frac{C}{10^{-3}} \frac{\dot{M} c^2}{4\pi R^2 \sigma} \right)^{1/4} \tag{4.109}$$

(where C is an irradiation constant), although the value of $T_{\rm irr}$ depends on disk geometry, albedo, radiative efficiency, irradiation spectrum, and so on. From Eq. (4.108) we can see that an UCXB with a relatively large orbital period is expected to have $\dot{M} < \dot{M}_{\rm crit}$, such that these systems should be transients.

For a non-irradiated disk, the value of $\dot{M}_{\rm crit}$ (in Eq. [4.107]) is roughly four times larger. For a disk containing even a small amount of hydrogen, the value of $\dot{M}_{\rm crit}$ is significantly smaller. The value of $\dot{M}_{\rm crit}$ is reduced by a factor of the order of 10 if $X = 0.1$ ($Y = 0.9$), and this value is further reduced by another factor of ~ 2 if $X = 0.7$ ($Y = 0.3$). The accretion disk mass is given by

$$M_{\rm disk} = \int_{R_{\rm in}}^{R_{\rm out}} 2\pi R\, \Sigma\, dR, \tag{4.110}$$

where Σ is the surface density, and $R_{\rm in}$ and $R_{\rm out}$ are the inner and outer disk radii, respectively. The timescale on which this disk rebuilds after a collapse is roughly $M_{\rm disk}/\dot{M}$.

To test this thermal-viscous disk instability model, Coriat et al. (2012) selected a sample of 52 persistent and transient NS and BH X-ray binaries and found that their data are in very good agreement with theoretical expectations that the observed persistent (transient) systems do lie in the appropriate stable (unstable) region of parameter space predicted by the theoretical models.

Another disk instability for binaries with $q \ll 1$, unrelated to the thermal-viscous instability, is the dynamical instability that may arise due to limited feedback of angular momentum from the disk to the orbit (see, e.g., van Haaften et al., 2012) and that may cause the donor star to become tidally disrupted.

Final Notes on Accretion Disks

An omitted aspect of accretion disks in binaries is the relativistic Lense-Thirring effect (or *frame dragging*) that occurs in systems which have a tilted accretion disk around a spinning compact object (Bardeen & Petterson, 1975; Armitage & Natarajan, 1999; Middleton, 2016). Such tilted accretion disks experience a torque due to the Lense-Thirring effect, which leads to precession of the inner disk and a warped disk structure. It has been suggested that this effect might be responsible for some low-frequency quasi-periodic oscillations (QPOs) observed in the X-ray light curves of NS and BH X-ray binary systems (Belloni & Stella, 2014).

Furthermore, when considering RLO in very tight binaries (e.g. double WD systems, see Section 5.4) there is the possibility that the accreting WDs are large compared to the orbital separations. This makes it likely that the mass-transfer stream can hit the accretor directly and causes a loss of angular momentum from the orbit which can destabilize the mass transfer (Marsh et al., 2004; Kremer et al., 2017).

Finally, we end this section on accretion disks by mentioning that they also play a large role in star formation (Hartmann, 1998; Krumholz et al., 2009) and in the physics of active galactic nuclei (Urry & Padovani, 1995).

4.6.6 Spherical Accretion on Black Holes

Apart from disk accretion, in which a large part of the loss of potential energy of the inspiralling accreted matter is efficiently converted into radiation energy before the matter disappears into the BH, one may also consider the case of purely spherical accretion onto a BH. This will happen when the accreted matter has no (or very little) angular momentum, as sometimes is the case with wind accretion in a binary (see Chapter 7).

In two important papers, Park (1990a, 1990b) makes a detailed study of spherical accretion onto BHs. Although these papers are aimed at supermassive BHs, the results also apply to stellar BHs (see, e.g., Nobili et al., 1991). These studies showed that spherical accretion is expected to be radiatively very inefficient, which means that even at very high accretion rates the amounts of X- and γ-rays produced may be very modest, much smaller than in the case of disk accretion. Park (1990b) obtains relativistic radiation hydrodynamics solutions for spherical accretion and finds that the main radiative processes in this case are relativistic bremsstrahlung and Comptonization. We give here only some of the main results of Park's work; readers should refer to his original papers for details.

Park introduced the following parameters/quantities:

$$\dot{M}_{\rm Edd} \equiv \frac{L_{\rm Edd}}{c^2} \simeq 0.22\,\mu_e\,M_8 \quad M_\odot\,{\rm yr}^{-1}, \tag{4.111}$$

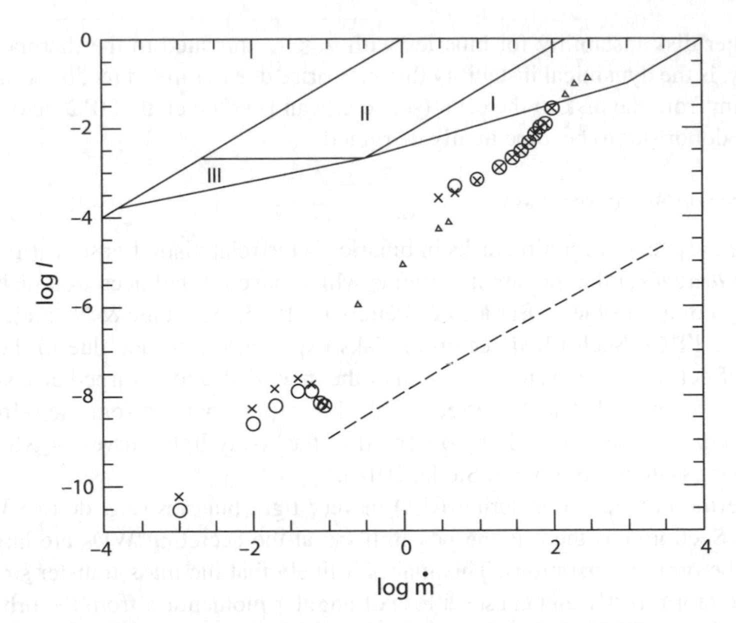

Figure 4.21. The electromagnetic luminosity l (in units of the Eddington luminosity) for spherical accretion onto a BH, as a function of the accretion rate $\dot{m}$ (in units of the Eddington accretion rate). Circles are the results from Park (1990b). Models from Park (1990a) are crosses (high temperature) and the dashed line (low temperature). Small triangles are models by Wandel et al. (1984). There are two regions of self-consistent solutions as described in the text. The regions where the flow becomes time-dependent by pre-heating (Ostriker et al., 1976) are the regions I, II, and III (top) bounded by dashed lines. After Park (1990b).

where $L_{\rm Edd}$ is the Eddington luminosity (see Section 4.9) given by

$$L_{\rm Edd} = \frac{4\pi\, GMc\, \mu_e m_p}{\sigma_T} \simeq 1.3 \times 10^{46}\, \mu_e\, M_8 \quad {\rm erg\, s^{-1}} \tag{4.112}$$

where M_8 is the BH mass in units of $10^8\ M_\odot$, μ_e is the mean molecular weight per electron, m_p is the proton mass, and σ_T is the electron's Thomson scattering cross section. Furthermore, he defined $l \equiv L/L_{\rm Edd}$, $\dot{m} \equiv \dot{M}/\dot{M}_{\rm Edd}$, and $\epsilon \equiv l/\dot{m} = L/\dot{M}c^2$ (radiation efficiency).

Park (1990b) found two self-consistent regions of stable spherical accretion for $\dot{m} \leq 0.1$ and for $5 \leq \dot{m} \leq 30$, as shown in Figure 4.21. The figure shows the radiative inefficiency of the accretion. For example, at $\log l = -1.5$ it is seen that $\log \dot{m} = 2$, which means that at $L = 10^{-1.5}\, L_{\rm Edd}$, the mass-accretion rate is $10^2\, \dot{M}_{\rm Edd}$, and hence the luminosity is only $L = 10^{-3.5}\, \dot{M}c^2$ ($\epsilon = 10^{-3.5}$). And at $L = 10^{-8}\, L_{\rm Edd}$, the energy production per unit mass of accreted material is only $\epsilon = 10^{-6.5}$.

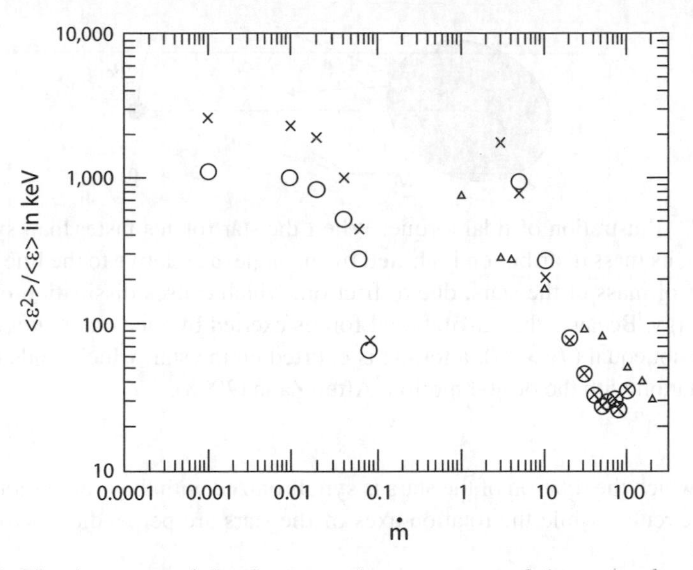

Figure 4.22. The (energy-weighted) mean photon energy of the emitted radiation for the models of Park (1990b). The meanings of the symbols are the same as those in Figure 4.21. After Park (1990b).

The mean photon energies for these two branches of stable solutions are shown in Figure 4.22. One observes that the photons are emitted typically in the energy range of approximately 20–3000 keV, which is in the hard X-ray and γ-ray parts of the spectrum.

4.7 TIDAL EVOLUTION IN BINARY SYSTEMS

For excellent reviews of the problem of tidal evolution in binary systems and star-planet systems, we refer to Zahn (2008) and Ogilvie (2014). For an overview of the role of tides in the evolution of populations of binaries, we refer to Hurley et al. (2002).

Here, we give only a brief summary of the essentials of the problem, largely based on the review by Zahn (2008). The simplest problem is that of two detached stars with no loss of mass or angular momentum by winds or gravitational waves. The total mass and angular momentum of such a system are conserved. The gravitational interaction between the stars will raise tidal bulges on them, roughly in the direction of the companion star, and if the rotation of the stars is not synchronized with the orbital motion, the motion of the tidal bulges over the stellar surface will cause friction and thus dissipation of the rotational kinetic energy of the star. If the orbit is eccentric, the heights of the tidal bulges will vary during the orbital motion, causing additional dissipation of kinetic energy. Tidal interaction thus leads to the exchange of kinetic energy and angular momentum between the rotation of the stars and their orbital motion. Due to the dissipation of kinetic energy, the system will evolve toward a stage of minimum kinetic

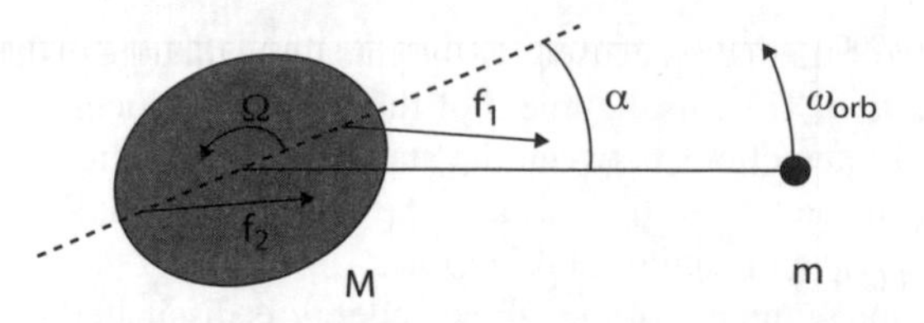

Figure 4.23. Illustration of tidal torque. When the star rotates faster than synchronous ($\Omega > \omega_{\rm orb}$), its mass distribution is shifted by an angle, α, relative to the line connecting the centers of mass of the stars, due to friction, which causes dissipation of rotational kinetic energy. Because the gravitational forces exerted by the companion on the tidal bulges are not equal ($f_1 > f_2$), a torque is exerted on the star, which tends to synchronize its rotation with the orbital motion. After Zahn (2008).

energy in which the rotation of the stars is synchronized with the orbital motion and the orbits are circular, while the rotation axes of the stars are perpendicular to the orbital plane.

The timescale on which this final state is reached depends on the strength of the tidal forces, which depends strongly on the orbital separation of the stars. Indeed, observations of solar-type spectroscopic binaries with ages $t \leq 1$ Gyr and orbital periods $P_{\rm orb} \leq 6$–7 d have circular orbits (Mazeh, 2008); while for older systems, the lower-limiting period for circular orbits increases with age (Mathieu et al., 2004; Mazeh, 2008). Besides the orbital separation, the precise physical processes that cause the dissipation (the tidal friction) are also an important factor that determine the timescales of orbital synchronization and circularization. The subject of tidal evolution goes back to the work of Darwin (1879) on tidal evolution of the Earth-Moon system, but its application to stars started only since the mid-1960s with the works of Zahn (1966a, 1966b, 1977), Alexander (1973), and Counselman (1973).

4.7.1 The Equilibrium Tide

Here, we follow Zahn (2008) and Hut (1981). Each star raises tides on the surface of the other one. We simplify the configuration to a system consisting of a primary star and a point-mass secondary, as depicted in Figure 4.23. This configuration is very similar to that of a system in which one of the stars is a WD, NS, or BH and therefore applies to X-ray binaries and CVs. We start from the assumption that the (primary) star is in hydrostatic equilibrium and without dissipation adjusts instantaneously to the tidal forces exerted by the companion; thus its shape follows equipotential surfaces. For this case, the shapes of the tidal bulges can be calculated, and the tidal bulges will be aligned with the line connecting the centers of the stars. If the rotation of the star is non-synchronous, dissipation mechanisms will cause the tidal bulges to deviate from this instantaneous equipotential state. Assume that the tidal bulges will in first approximation keep their original equilibrium shape, but that due to the non-synchronous rotation they will have become misaligned with respect to the line connecting the centers of the stars, as depicted in Figure 4.23. This is what is called the *equilibrium tide*, and the calculation of the forces caused by the misaligned bulges is the so-called *weak friction approximation*. The misalignment of the bulges produces a torque component in the gravitational interactions between the stars, which leads to exchange of angular momentum between rotation of the star and the orbital motion, as well as dissipation of kinetic energy of rotation and orbital motion.

Even without misalignment of the bulges, energy is dissipated if the amplitudes of the tides lag with respect to the equipotential surfaces, although in this case no angular momentum is exchanged. In all cases, the orbital parameters will change and either asymptotically approach an equilibrium state or will lead to an accelerated spiral-in. The latter, known as *Darwin instability* (Darwin, 1879), is at work if in the final coplanar, synchronized circular orbit the ratio of orbital angular momentum and rotational angular momentum of the star, $J_{\rm orb}/J_{\rm rot} = \omega_{\rm orb}/\Omega < 3$ (Counselman, 1973; Hut, 1980). In this case, no final equilibrium state is possible and spiral-in and coalescence are unavoidable.

We now estimate the timescales for synchronization and circularization in the weak friction approximation and consider a primary star of mass, M_1, and the point-mass secondary of mass, m_2, at a distance, d, in a circular orbit (see Fig. 4.23). The relative elevation, $\delta R_1/R_1$, of the tidal bulges is approximately given by the ratio of the differential acceleration exerted on the bulges and the surface gravity, g:

$$\frac{\delta R_1}{R_1} \simeq \frac{Gm_2R_1/d^3}{GM_1/R_1^2} = \frac{m_2}{M_1}\left(\frac{R_1}{d}\right)^3. \tag{4.113}$$

If the primary has constant mass density, the mass of the bulges δM_1 is of the order of $\rho_1 R_1^2 \delta R_1 \approx M_1 \delta R_1/R_1$. (In reality, the mass is lower, as the density decreases outward in the star). The tidal bulges produce a quadrupole gravitational field that causes apsidal motion (e.g., Schwarzschild, 1958; Mazeh, 2008); see Zasche et al. (2020) and Figure 4.24 for recent measurements of apsidal motion (precession) of 162 eccentric eclipsing binaries in the LMC. The tidal bulges, which exert forces f_1 and f_2 on the companion star (see Fig. 4.23), are assumed to be misaligned by an angle α with respect to the line connecting the stellar centers.

For the calculation of the effects of the equilibrium tide, we continue following Hut (1981) and Zahn (2008). Neglecting factors on the order of unity, the magnitude of the torque can be estimated as

$$\begin{aligned} N \simeq (f_2 - f_1)\,R_1 \sin\alpha &= -\delta M_1\left(\frac{Gm_2R_1}{d^3}\right) R_1 \sin\alpha \\ &= -\frac{Gm_2^2}{R_1}\left(\frac{R_1}{d}\right)^6 \sin\alpha. \end{aligned} \tag{4.114}$$

The angle α is a function of the difference between the rotational angular velocity, Ω, and the orbital angular velocity, $\omega_{\rm orb}$, and decreases when these angular velocities approach one another. In the simple *weak friction* case, $\alpha \propto (\Omega - \omega_{\rm orb})$. It further depends on the physical processes that cause the tidal friction (i.e., that dissipate the

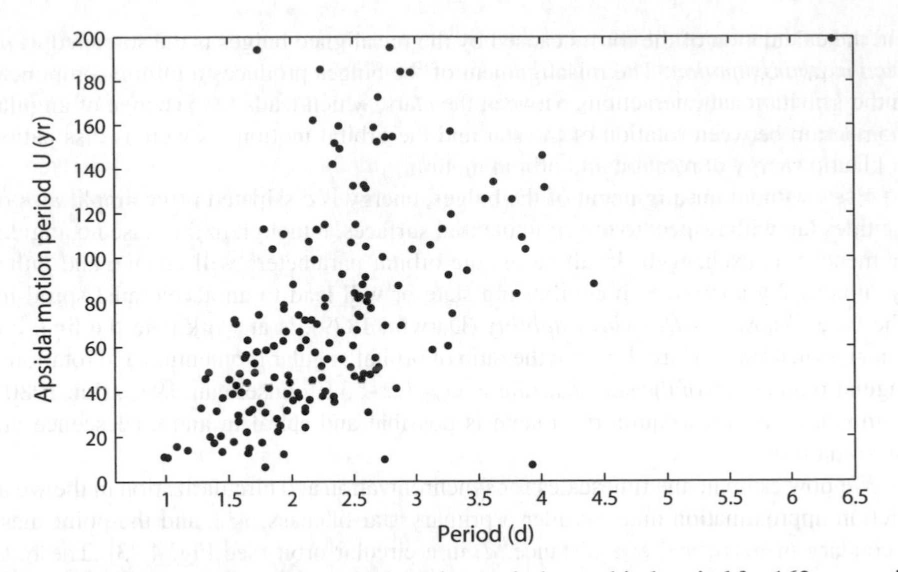

Figure 4.24. Diagram showing apsidal motion period vs. orbital period for 162 eccentric eclipsing binaries in the LMC. An apsidal period is the time interval required for an orbit to precess through 360°. After Zasche et al. (2020).

kinetic energy), which are characterized by a dissipation time, $\tau_{\rm diss}$, which leads to

$$\alpha = \frac{(\Omega - \omega_{\rm orb})}{\tau_{\rm diss}} \left(\frac{R_1^3}{GM_1} \right), \tag{4.115}$$

where α has been made dimensionless by dividing by the square of the free-fall time, which is the natural dynamical time of the star (Section 8.1.3). The tidal torque now becomes

$$N = -\frac{(\Omega - \omega_{\rm orb})}{\tau_{\rm diss}} q^2 M_1 R_1^2 \left(\frac{R_1}{d} \right)^6, \tag{4.116}$$

where the mass ratio is $q = m_2/M_1$. The weak friction approximation was originally derived for solid elastic bodies like planets and satellites, but in first approximation can also be applied to fluid bodies like stars, assuming that the dissipation is viscosity-like and does not depend on the tidal frequency $(\Omega - \omega_{\rm orb})$. (Note that in radiative cores or envelopes this condition may not be fulfilled, as oscillations may be excited by the tides, leading to enhanced dissipation.) The correct expression for the tidal torque for stars is then given by Eq. (4.116), and from this one can derive the equation for the synchronization time:

$$\frac{1}{t_{\rm sync}} = -\frac{1}{I\Omega} \frac{d(I\Omega)}{dt} = -\frac{N}{I\Omega} = \frac{1}{\tau_{\rm diss}} \frac{(\Omega - \omega_{\rm orb})}{\Omega} q^2 \frac{M_1 R_1^2}{I} \left(\frac{R_1}{a} \right)^6, \tag{4.117}$$

where I is the moment of inertia of the primary star, with spin angular momentum, $I\Omega$. The torque, N, has been averaged here over the elliptic orbit with semi-major axis a. The dissipation time, assuming viscous dissipation, is of the order of $R_1^2/\langle\nu\rangle$, where $\langle\nu\rangle$ is the average kinematic viscosity.

The previous expression is valid only for a circular orbit. Hut (1981) derived an expression for the torque averaged over an elliptic orbit and found that the torque does not vanish for $\Omega = \omega_{\rm orb}$, but for

$$\frac{\Omega}{\omega_{\rm orb}} = \frac{1 + \frac{15}{2}e^2 + \frac{45}{8}e^4 + \frac{5}{16}e^6}{(1-e^2)^{3/2}(1+3e^2+\frac{3}{8}e^4)}. \tag{4.118}$$

Here *pseudo-synchronized* equilibrium is achieved; for this ratio of angular velocities there is no longer a net spin-up or spin-down tidal torque summarized over the orbit. When rotation is pseudo-synchronized, the still present torque (which varies over the elliptic orbit) continues to circularize the orbit by reducing the orbital eccentricity, e. For the circularization time, one can thus derive the following expression (see, e.g., Hut, 1980, 1981):

$$\frac{1}{t_{\rm circ}} = -\frac{d\ln e}{dt} = \frac{1}{\tau_{\rm diss}}\left(9 - \frac{11}{2}\frac{\Omega}{\omega_{\rm orb}}\right) q(1+q)\left(\frac{R_1}{a}\right)^8. \tag{4.119}$$

One sees from Eqs. (4.117) and (4.119) that synchronization goes much faster than circularization of the orbit. This is because the orbital angular momentum ($\sim M_1 a^2 \omega_{\rm orb}$) is, in general, much larger than the rotational angular momentum of the star(s) ($M_1 R_1^2 \Omega$). Darwin (1879) was the first to derive that the eccentricity *increases* when $\Omega/\omega_{\rm orb} \geq 18/11$. (That result also follows immediately by solving for the right-hand side of Eq. [4.119] being equal to 0.) Such a situation may arise in binaries when a star overflows its Roche lobe and transfers mass to its companion via an accretion disk, thereby spinning up the rotation rate of the accretor. This mass-receiving companion may then spin much faster than the orbital angular velocity, and if the radius of the companion is temporarily inflated, for example, due to the energy release by accretion, tidal forces may be strong and pump up the eccentricity. Possibly, this may be an explanation for the surprisingly high orbital eccentricities of the barium stars, which are thought to be post–mass-transfer stars whose companion now has become a WD (see Chapter 5).

As noted by Zahn (2008), the assumption that the angular velocity is constant throughout the star needs not be correct, as the torque varies with depth. Taking this, as well as the resulting transport of angular momentum throughout the star, into account may complicate matters considerably, but we will not discuss these complications in this book.

4.7.2 Dissipation in Convective Zones: Turbulent Convection

In stellar interiors, the normal viscosity due to microscopic processes is very small and has no importance for tidal processes. However, as noted by Zahn (1977), turbulence can be a major cause of viscosity in stars. The kinetic energy of the large-scale flows, induced by the tides, cascades down to smaller and smaller eddies until it is converted to heat by viscous friction. The force acting on the tidal flow can be expressed as a turbulent viscosity, $\nu_t = v\,l$, where v is the average vertical velocity of the turbulent eddies and l is the mixing length (i.e., the average vertical distance travelled by the eddy). The tidal torque can then be written as the integral of the turbulent viscosity over the stellar interior, and the dissipation time is then proportional to the convective turnover time

$$\frac{1}{\tau_{\rm diss}} = \frac{6\,\lambda_2}{\tau_{\rm conv}}, \tag{4.120}$$

where

$$\tau_{\rm conv} = \left(\frac{M_1 R_1^2}{L_1}\right)^{1/3} \tag{4.121}$$

and where the dimensionless constant, λ_2, is given by

$$\frac{\lambda_2}{\tau_{\rm conv}} = \frac{336}{5}\pi\,\frac{R_1}{M_1}\int x^8 \rho_1 \nu_t\, dx. \tag{4.122}$$

Here L_1 is the luminosity of the star, and for deriving these equations it is assumed that this luminosity is entirely transported by convection; furthermore, ρ_1 is the mass density and $x = r/R_1$ is the dimensionless radial coordinate. The resulting dissipation time is very short—in the Sun, $\tau_{\rm conv} = 0.435$ yr—and as a result the turbulent convection is the main dissipation mechanism of the tides (Zahn, 1966b). Assuming that the entire stellar luminosity is transported by convection and the star is fully convective, one derives $\lambda_2 = 0.019\,l^{4/3}$, where l is the classical mixing length parameter (Zahn, 1989). In stars with convective cores (such as all massive stars), tidal dissipation by turbulent convection is much smaller, as the dissipation timescales as $\tau_{\rm diss} \propto (r_{\rm conv}/R_1)^{-7}$, where $r_{\rm conv}$ is the radius of the convective core.

4.7.3 Dissipation in Radiative Regions: The Dynamical Tide

In massive stars, with radiative envelopes, the tidal dissipation is mainly caused by the dynamical excitation by the tides of different types of oscillatory modes—acoustic modes, internal gravity modes, and inertial modes—where the restoring force is the compressibility of the gas, the buoyancy of the stably stratified gas, or the Coriolis force in the rotating star, respectively. Most of the work so far has been concerned with the gravity modes, for which the buoyancy provides the restoring force. The dissipation of these modes is by *radiative damping*, which is the conversion of their energy into radiation. We only give the resulting equations for the tidal synchronization and circularization, as derived by Zahn (1975); see the original paper for the derivation.

For the synchronization time, for a uniform rotation of the star, one finds

$$\frac{1}{\tau_{\rm sync}} = -\frac{d}{dt}\left(\frac{2(\Omega-\omega_{\rm orb})}{\omega_{\rm orb}}\right)^{-5/3}$$
$$= 5\left(\frac{GM_1}{R_1^3}\right)^{1/2} q^2(1+q)^{5/6}\left(\frac{M_1R_1^2}{I}\right)E_2\left(\frac{R_1}{a}\right)^{17/2}, \quad (4.123)$$

and for the circularization time,

$$\frac{1}{\tau_{\rm circ}} = -\frac{d\ln e}{dt} = \frac{21}{2}\left(\frac{GM_1}{R_1^3}\right)^{1/2} q(1+q)^{11/6}E_2\left(\frac{R_1}{a}\right)^{21/2}. \quad (4.124)$$

Here E_2 is a parameter that measures the coupling between the tidal potential and the gravity mode. It depends strongly on the size of the convective core and thus on the mass of the star. Zahn (1975) gives its expression and Claret & Cunha (1997) tabulate it for various stellar models. For a 10 $M_\odot$ ZAMS star, its value is $\sim 10^{-6}$.

Importantly, Witte & Savonije (1999a; 1999b; 2002) and Savonije & Witte (2002) have pointed out that in the spectrum of eigen modes of the star, often frequencies are present coinciding with the tidal frequency, such that resonances may be exited. This causes much larger radiative dissipation, and thus much faster tidal synchronization and circularization. See the review of this subject by Savonije (2008).

Recently, Justesen & Albrecht (2021) compiled and investigated a catalogue of eclipsing binaries from the TESS, and the Kepler mission. They found a clear dependency of stellar temperature and orbital separation in the eccentricities of close binaries. While the eccentricity distribution in their sample of convective binaries is in excellent agreement with the predictions from the equilibrium tide, they also found that some binaries with radiative envelopes and temperatures between 6,250 K and 10,000 K may be tidally circularized significantly more efficiently than is usually assumed from dynamical tide models.

Finally, in Chapter 15, we will see that tides on WR-stars in tight, massive binaries (e.g., Kushnir et al., 2016) play an important role for the expected spins of the BH components in double BH mergers as detected by the LIGO network of GW detectors.

4.8 COMMON ENVELOPES

A very important stage, and possibly the most uncertain aspect, of the evolution of close binaries is the formation of a common envelope (CE) and its subsequent ejection. This phase is often triggered by an unstable, runaway mass transfer (see Chapters 4 and 9), causing the companion star to be engulfed in the envelope of the donor star. The capture of the companion is then accompanied by the creation of a drag force, arising from its motion through the CE, which leads to dissipation of orbital angular momentum (the spiral-in process) and deposition of orbital energy in the envelope. Hence, the global outcome of a CE phase is reduction of the binary separation, often by more than a factor of 100, and possibly ejection of the envelope. Alternatively, in case the envelope is not ejected, the spiral-in continues and causes the companion star to coalesce (merge) with the core of the donor star.

There is strong evidence of such past orbital shrinkage (i.e., similar to the expected outcome of a CE phase) in a number of observed tight binary systems containing a pair of compact objects, with orbital periods of a few hours or less. Examples include binaries with pulsars and/or WDs: PSR 1913+16, PSR J0348+0432 and J0651+2844; the AM Canum Venaticorum (AM CVn) systems (Nelemans, Yungelson et al., 2001); and obviously also the progenitor systems of the recently discovered GW mergers.[5] In all these systems it is clear that the precursor of the last-formed degenerate star must have achieved a radius much larger than the current orbital separation. CEs are also anticipated to be involved in the formation of the progenitor systems of SNe Ia as well as long γ–ray bursts (the latter require the collapse of a stellar core with rapid spin and a lost envelope). Finally, planets might be captured and eject the CEs of low-mass giant stars, which thereby produce isolated low-mass He WDs (Soker, 1998; Nelemans & Tauris, 1998).

There is also strong evidence that the nuclei of planetary nebulae and other types of close WD binaries are the results of CE evolution (see Section 5.4 and references therein).

There are many uncertainties involved in calculations of the onset of a CE and the outcome of the spiral-in phase. The evolution is often tidally unstable and the angular momentum transfer, dissipation of orbital energy, and structural changes of the donor star take place on a very short timescale ($\sim 10^3$ yr). As a result, the final post-CE separation between the two stars is difficult to predict as a result of our poor understanding of all the complex physical processes involved in the envelope ejection. All results from population synthesis including CE evolution should therefore be taken with a grain of salt.

Although the earliest considerations of CE evolution originates from the 1970s (Paczyński, 1976; van den Heuvel, 1976; Webbink, 1976), even modern 3D-hydrodynamical simulations (see examples in Figs. 4.26 and 4.27) have difficulties ejecting the envelope and securing deep spiral-in (Taam & Sandquist, 2000; Passy et al., 2012; Ricker & Taam, 2012; Nandez et al., 2014; Ohlmann et al., 2016; Moreno et al., 2021). Such simulations may use different techniques, such as Eulerian adaptive mesh refinement (Sandquist et al., 1998; Ricker & Taam, 2008; Passy et al., 2012; MacLeod & Ramirez-Ruiz, 2015b; Staff et al., 2016; MacLeod, Antoni et al., 2017; Iaconi et al., 2017, 2018; Chamandy et al., 2018; De et al., 2020), moving meshes (Ohlmann et al., 2016, 2017; Prust & Chang, 2019; Moreno et al., 2021), and smoothed particle hydrodynamics (Rasio & Livio, 1996; Lombardi et al., 2006; Passy et al., 2012; Nandez et al., 2015; Nandez & Ivanova, 2016; Reichardt et al., 2019).

[5] We discuss the origin of the GW mergers in Chapter 15.

Such computations are troublesome because of the huge ranges involved in both length scales and timescales. The range in timescale is of the order of 10^{10} (from the dynamical timescale of a few seconds of an in-spiralling NS to the thermal timescale of a few thousand years in the envelope) and the range in space coordinates is of the order of 10^8 (from the $\sim$10 km size of an in-spiralling NS to the $\sim$1,000 $R_\odot$ size of a giant star envelope). However, recent progress overcoming these issues and apparently succeeding with CE ejection has been demonstrated by Law-Smith et al. (2020) who investigated 3D-hydrodynamical simulations of a CE between a red supergiant and a NS. Also promising are the first general relativistic simulations of the accretion flow onto a (non-rotating) BH moving supersonically in a medium with regular but different density gradients have been carried out (Cruz-Osorio & Rezzolla, 2020), thereby also mimicking conditions in the envelope of a red supergiant in a HMXB. Nevertheless, our limited basic understanding of the physics involved remains the main cause of the problem. An interesting finding by Clayton et al. (2017) was that the CE evolution where a red giant envelope is involved may lead to pulsational instability of the giant envelope, leading to episodic mass ejection events. For general reviews on CE evolution, see, for example, Iben & Livio (1993), Taam & Sandquist (2000), Podsiadlowski (2001), Taam & Ricker (2010), Ivanova et al. (2020), and Ivanova, Justhan, Chen et al. (2013).

4.8.1 Onset and Main Phases of CE Evolution

The onset of a CE phase is believed to be connected with one of the following situations:

- dynamical unstable (runaway) RLO mass transfer
- occurrence of a Darwin instability
- expansion of the accreting star

As we have shown in in Section 4.5.2 and Figure 4.8, and to be discussed in further detail in Chapter 9, mass transfer from a massive donor to a less massive accretor often results in effective orbital shrinking, such that the donor star will continuously overfill its Roche lobe more and more as the mass transfer proceeds. At some point the companion star is physically captured by the envelope of the donor star as the orbit keeps shrinking (MacLeod et al., 2018). Another possible cause for the onset of a CE is the Darwin instability (Darwin, 1879). If the sum of the spin angular momenta of the two stars, in synchronous rotation with the orbital motion, exceeds 1/3 of the orbital angular momentum ($|\vec{J}_1| + |\vec{J}_2| > 1/3\,|\vec{L}|$), then the orbit will shrink in a catastrophic manner as tidal forces try to keep the system synchronous.

Finally, a CE may arise if the accretor cannot handle the amount of material it receives, such that it reacts by swelling up until it fills its own Roche lobe and a contact binary is formed. As a rule of thumb, this will be the case for mass transfer that is fast compared to the thermal timescale of the envelope of the accretor (i.e., if $\tau_{\rm MT} < \tau_{\rm th}$).

The CE event can be divided into a number of distinct phases, where each phase operates on its own timescale (Podsiadlowski, 2001):

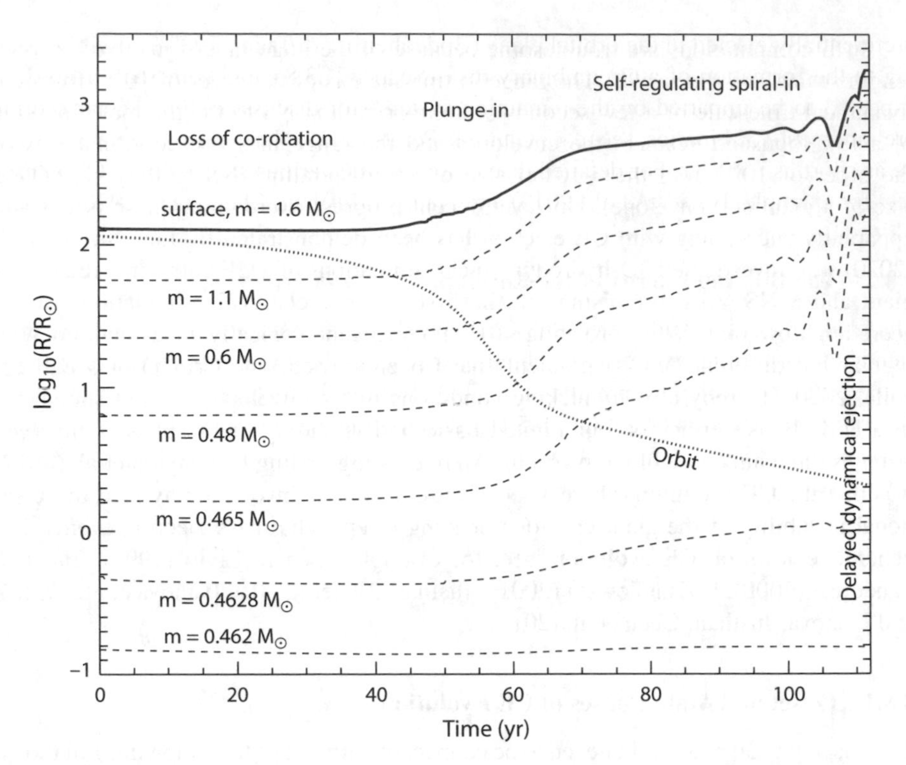

Figure 4.25. The main phases of a CE event prior to the envelope ejection or the merger. This example is for a 1.6 $M_\odot$ red giant and a 0.30 $M_\odot$ WD. The dashed lines represent locations at fixed mass coordinates, the solid blue line represents the location of the stellar surface, and the red dotted line shows the location of the in-spiralling WD. After Ivanova, Justhan, Chen et al. (2013).

I: loss of co-rotation
II: plunge-in
III: slow (self-regulating) spiral-in
IV: envelope ejection

These phases are illustrated in Figure 4.25. The binary first loses co-rotation (which may take a few hundred years) before it enters the rapid spiral-in phase, during which frictional and gravitational drag forces cause orbital energy to be extracted and deposited in the envelope. In this stage, which proceeds on a dynamical timescale of the donor star (i.e., a few orbital periods), the major part of the orbital decay takes place. The plunge-in is followed by a slower phase where the (almost) stripped core adapts to the rapid loss of its envelope via expansion or contraction of the remaining shell on a thermal timescale of the remaining layer. If the remaining core expands, the mass transfer and spiral-in will reactivate, and in this way the CE may rebuild in several cycles. Eventually the envelope is ejected or the system will coalesce when either the companion star or the exposed core of the donor overfills its Roche lobe. The CE material is preferentially ejected in the orbital plane (Bodenheimer & Taam, 1984), possibly leading to the formation of a circumbinary disk (Section 4.3.8), and some eccentricity is expected to be imparted on the remaining system (for discussions on eccentric orbit pre- and post-CE binaries, see, e.g., Glanz & Perets, 2021). The entire CE process may be as short as $\sim 10^3$ yr. For detailed discussions of the various stages of the CE phase, see, for example, Ivanova et al. (2013) and Ivanova, Justhan, Chen et al. (2020).

4.8.2 The Binding Energy of the Envelope

The ability to successfully eject the CE depends on its binding energy and the efficiency of converting the released orbital energy from the spiral-in into kinetic energy in the envelope, which eventually causes the envelope to be lifted and ejected. Using the energy prescription (Webbink, 1984) for estimating the outcome of a CE and spiral-in evolution, one can define the efficiency, $\alpha_{\rm CE}$, of converting released orbital energy, $\Delta E_{\rm orb}$, into ejection of the envelope by

$$E_{\rm bind} \equiv \alpha_{\rm CE}\, \Delta E_{\rm orb}, \tag{4.125}$$

where the total binding energy of the envelope, $E_{\rm bind}$, is calculated from the sum of the gravitational binding energy and the internal thermodynamic energy (Han et al., 1994, 1995),

$$E_{\rm bind} = \int_{M_{\rm core}}^{M_{\rm donor}} \left(-\frac{GM(r)}{r} + \alpha_{\rm th}\, U \right) dm, \tag{4.126}$$

and where $M(r)$ is the mass within radial coordinate, r, of the donor star with total mass, $M_{\rm donor}$; core mass, $M_{\rm core}$; and envelope mass, $M_{\rm env} \equiv M_{\rm donor} - M_{\rm core}$. The mass coordinate of $M_{\rm core}$ is sometimes referred to as the *bifurcation point*, which separates the ejected envelope from the remaining exposed core—either a helium core (from Case B RLO)[6] or a CO core (for Case BB and Case C RLO). In the preceding equation, the U term involves the basic specific thermal energy for a simple perfect gas ($3\mathcal{R}T/2\mu$), the specific energy of radiation ($aT^4/3\rho$), as well as terms due to ionization of atoms and dissociation of molecules and the Fermi energy of a degenerate electron gas. The value of $\alpha_{\rm th}$ depends on the details of the ejection process, which is very uncertain. Values of $\alpha_{\rm th}$ equal to 0 and 1 correspond to maximum and minimum envelope binding energies, respectively.

The total change in orbital energy is simply given by (here star 1 being the accretor as usual)

$$\Delta E_{\rm orb} = -\frac{G M_{\rm core} M_1}{2\,a} + \frac{G M_{\rm donor} M_1}{2\,a_0}, \tag{4.127}$$

where a_0 is the initial separation at the onset of the RLO and a is the final orbital separation after the CE phase. Given that $a \ll a_0$, we can approximate $\Delta E_{\rm orb} \simeq -G M_{\rm core} M_1/2a$.

A structure parameter, λ, was introduced by de Kool (1990) as a numerical factor that depends on the stellar density distribution, such that the binding energy of the

[6] See Chapter 9 for more on the different cases of RLO.

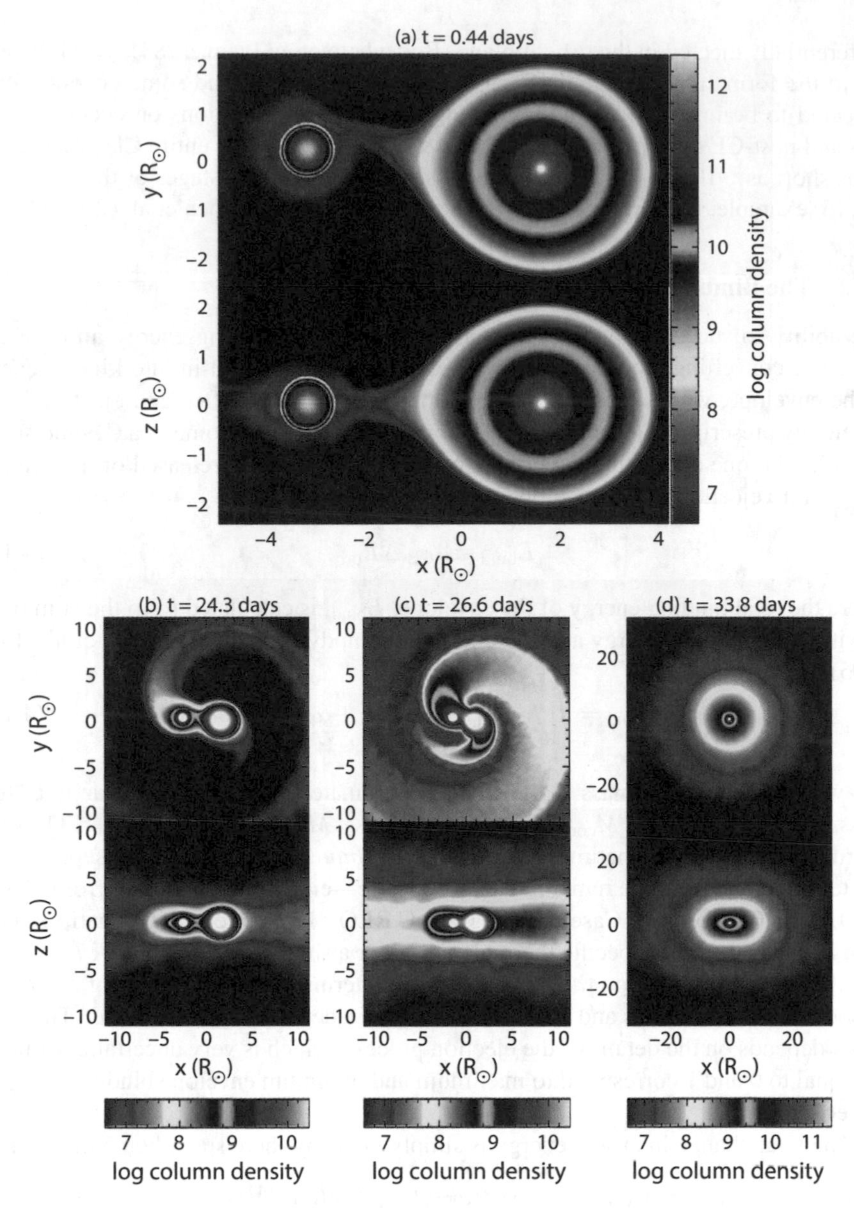

Figure 4.26. Simulation of evolution of a CE with a $1.2\,M_\odot$ giant and a $0.6\,M_\odot$ main-sequence star, seen from above (top) and along (bottom) the orbital plane. It is seen that the majority of the mass ejection is in the orbital plane, while here the end result is a merger event. The smoothed-particle hydrodynamics simulation with 2.2×10^5 particles was performed by J. Lombardi and R. Scruggs. After Ivanova, Justhan, Chen et al. (2013).

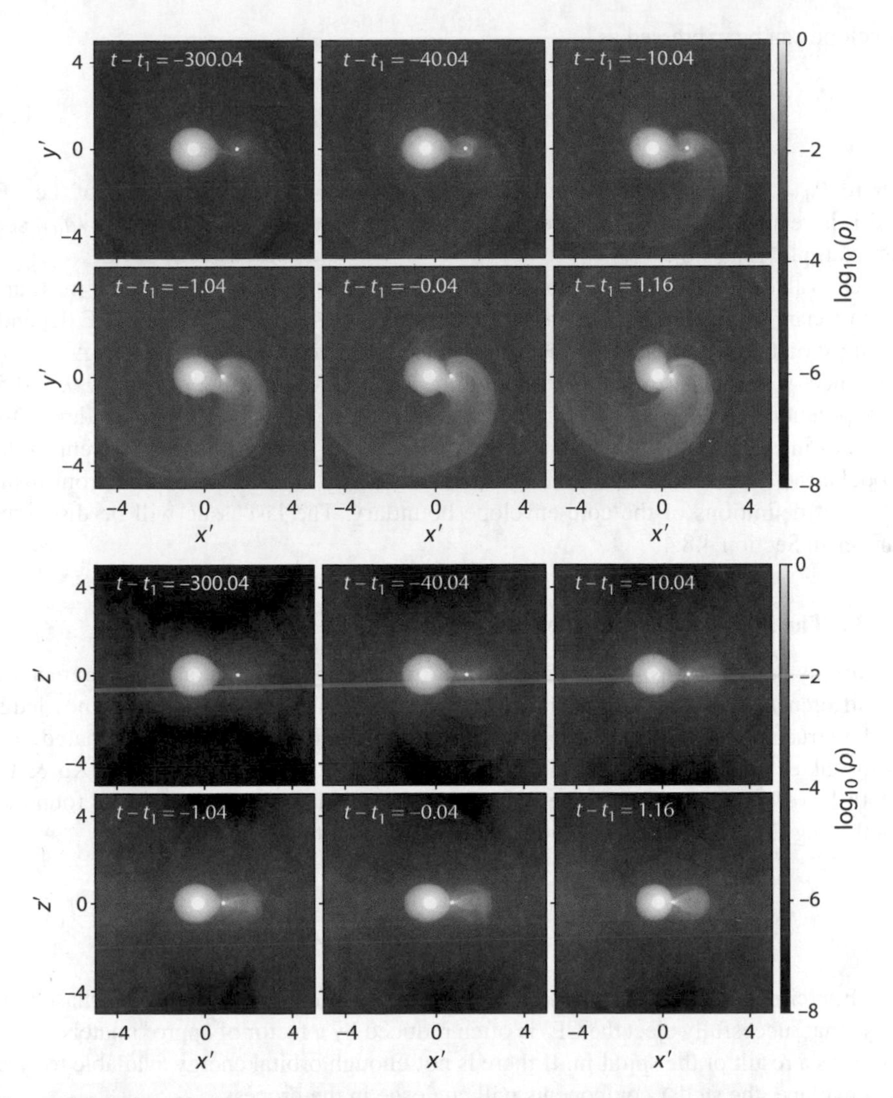

Figure 4.27. Simulation of the early in-spiral leading to the plunge-in phase. Shown here are slices of gas density through the orbital plane (top) and perpendicular to the plane (bottom). Time t_1 marks the onset of the merger where $a < R$. Time is in units of the dynamical timescale of the primary star. Common features in the flow include exchange of mass from the donor envelope toward the accretor and then into tails that extend from the Lagrangian points, L_2 and L_3. After MacLeod et al. (2018).

envelope can be expressed as

$$E_{\rm bind} \equiv -\frac{G M_{\rm donor} M_{\rm env}}{\lambda\, R_{\rm donor}}, \tag{4.128}$$

where $R_{\rm donor}$ is simply the Roche-lobe radius of the donor star at the onset of the CE, and it is related to a_0 via the dimensionless Roche-lobe radius, $r_L = (R_{\rm donor}/a_0)$, see, for example, Eq. (4.6).

The value of λ varies by several orders of magnitude, depending on stellar mass and evolutionary status (Fig. 4.28). Therefore, the ability to successfully eject a CE depends strongly on the evolutionary status of the donor star at the onset of the CE. As a consequence, we emphasize that one should not use a constant value of λ (e.g., $\lambda = 0.5$) in population synthesis calculations because the outcome can be very misleading. Differences in calculated λ-values may arise, for example, from the use of different stellar models, the degree of available recombination energy, and, in particular, from using different definitions of the core-envelope boundary. The last issue will be discussed further in Section 4.8.4.

4.8.3 The Post-CE Orbital Separation

By simply equating Eqs. (4.126) and (4.128), one is able to calculate the parameter λ for different evolutionary stages of a given star (Dewi & Tauris, 2000). From knowledge of the structure of the donor star at the onset of the CE, $M_{\rm core}$ can be estimated, and using tables[7] of λ available in the literature (Dewi & Tauris, 2000, 2001; Xu & Li, 2010; Loveridge et al., 2011), the final post-CE orbital separation, a can be found by combining Eqs. (4.125), (4.127) and (4.128). The result is

$$\frac{a}{a_0} = \left(\frac{M_{\rm core} M_1}{M_{\rm donor}}\right) \left(M_1 + \frac{2\, M_{\rm env}}{\alpha_{\rm CE}\, \lambda\, r_L}\right)^{-1} \tag{4.129}$$

(see Exercise 4.14). The orbital separation of the binaries surviving the CE phase (i.e., those that succesfully eject the CE) is often reduced by a factor of approximately 100 – 1,000 as a result of the spiral-in. If there is not enough orbital energy available to eject the envelope, the stellar components will coalesce in the process.

4.8.4 CE Ejection: Dependence on λ and Accretor Mass

For low- and intermediate-mass stars, $\lambda < 1$ on the RGB and $\lambda \gg 1$ on the AGB (especially for stars with $M < 6\,M_\odot$). Hence, the envelopes of such donor stars on the AGB are easily ejected because $|E_{\rm bind}|$ is small, resulting in only a relatively modest decrease

[7] Kruckow et al. (2016) argued, however, that for reliable results, λ must be evaluated using the same stellar evolution code as the one applied to determine $M_{\rm core}$, because λ is very sensitive to the mass coordinate of $M_{\rm core}$.

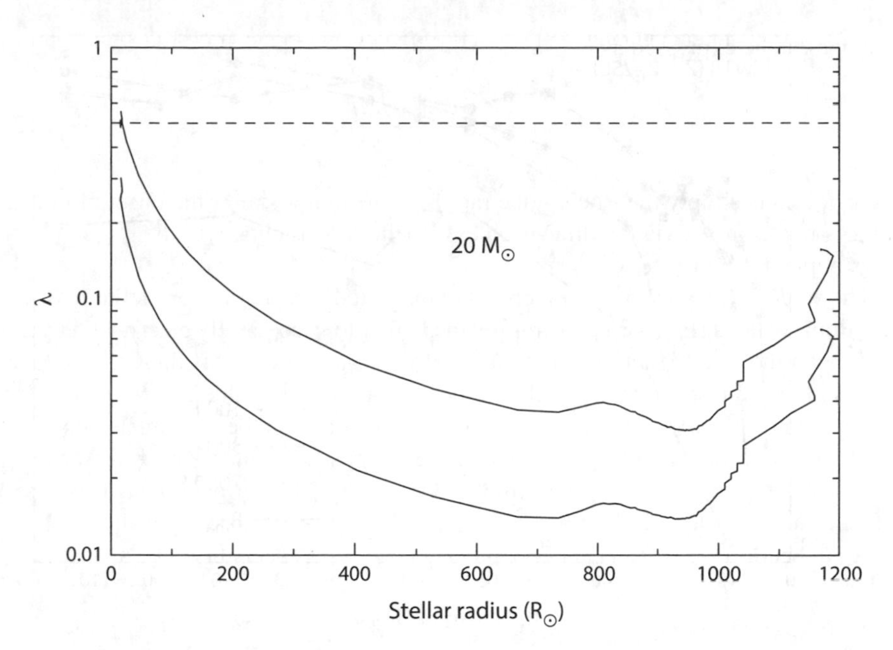

Figure 4.28. The λ-parameter for a 20 $M_{\odot}$ star as a function of its stellar radius. The upper curve includes internal thermodynamic energy ($\alpha_{\rm th} = 1$), whereas the lower curve is based on the sole gravitational binding energy ($\alpha_{\rm th} = 0$); see Eq. (4.126). There is roughly a factor of 2 in difference between the λ-curves in accordance with the virial theorem. It is a common misconception to use a constant value of $\lambda = 0.5$ (marked by the dashed line) for population synthesis. After Voss & Tauris (2003).

in orbital separation from a weak spiral-in. For more massive stars ($M > 10\,M_{\odot}$), $\lambda < 0.1 - 0.01$ (see Fig. 4.28), and for such stars $|E_{\rm bind}|$ is large, mainly due to the large envelope mass, although it usually becomes significantly smaller in the giant star stage (depending on metallicity).

Final Fate of CEs in Post-HMXB Systems

The large value of $|E_{\rm bind}|$ for massive star envelopes has the important consequence that many HMXBs with NSs will not survive the CE phase to later produce double NS (DNS) systems, because the envelope is too tightly bound to be ejected (Taam, 1996; Taam & Sandquist, 2000; Kruckow et al., 2016; Tauris et al., 2017). Therefore, only HMXBs with NSs and $P_{\rm orb} \gtrsim 1$ yr are able to eject the CE. Instead, most of such HMXB systems will coalesce and produce Thorne-Żytkow objects (Thorne & Żytkow, 1975, 1977), see Section 10.7.4. These objects, which have a NS embedded in the center of a red supergiant, are difficult to distinguish from normal cool red supergiants, and it is presently not clear whether they actually exist. Based on theoretical arguments, their birth rate in the Galaxy is expected to be quite high ($2 \times 10^{-4}\,{\rm yr}^{-1}$) (Podsiadlowski et al., 1995). It is also possible that the merger of a NS or a BH with

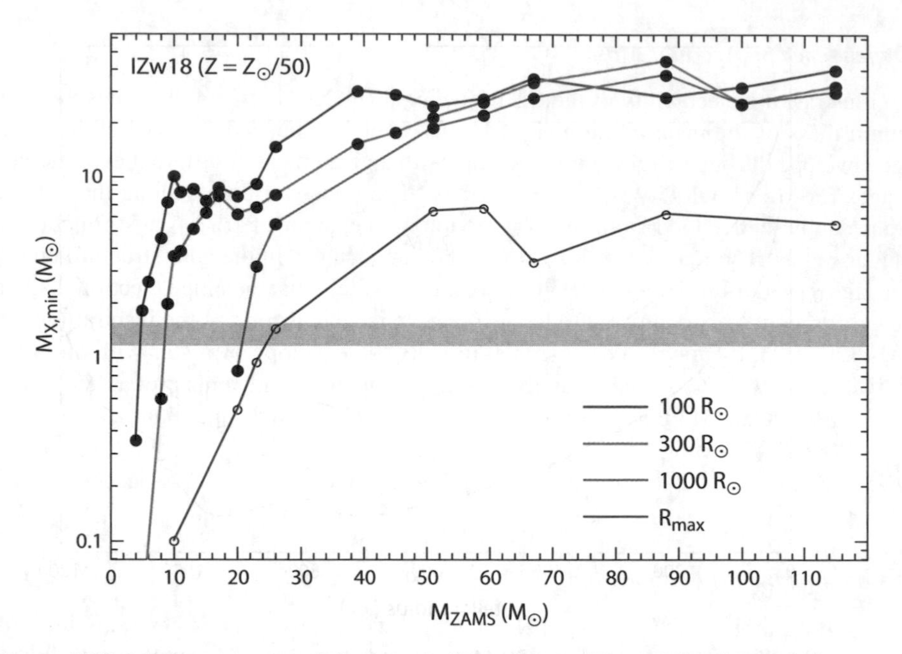

Figure 4.29. Minimum mass of the in-spiralling object, $M_{X,min}$, that is needed to expel the envelope during a CE evolution with $\alpha_{CE} = 1$, for a given donor star radius as indicated by the colored lines, as a function of ZAMS mass of the donor star, M_{ZAMS}, in a low-metallicity environment ($Z = Z_{\odot}/50$). R_{max} is the maximum radial extent during the stellar evolution. The gray band between 1.17 and 1.56 $M_{\odot}$ indicates masses of NSs observed in DNS systems (Martinez et al., 2015). Applying more realistic efficiencies of $\alpha_{CE} < 1$ would require higher values of $M_{X,min}$ and shift all plotted curves upward. The plotted lines also depend on metallicity. After Kruckow et al. (2016).

the core of a massive star may trigger a merger-driven explosion (Schrøder et al., 2020, and references therein). As a direct consequence of many HMXBs with NSs coalescing during the CE phase, the estimated Galactic GW merger rate of DNS binaries is relatively small (Voss & Tauris, 2003). BHs in HMXBs can more easily eject the CE because they are more massive and therefore release more orbital energy as they spiral-in.

We note that only evolved (i.e., giant) donor stars have relatively small values of $|E_{bind}|$, which is often necessary to ensure a successful CE ejection. This condition requires wide pre-CE orbits such that a donor star only fills its Roche lobe and initiates a CE when it is in an evolved stage. This means that the first mass-transfer phase, from the primary star (the progenitor of the NS or BH) to the (less evolved) secondary star, must have been dynamically stable in order for the system to remain wide (or, in case of an unstable RLO, the orbital separation could only have been slightly reduced). Later in Chapter 10, we discuss HMXBs with BHs that may avoid the formation of a CE, because their RLO can be stable.

Dependence on Accretor Mass

To illustrate the ejectability of the CE following a HMXB, Figure 4.29 shows the minimum mass of the inspiralling object, $M_{X,\min}$, which is needed to successfully expel the envelope during a CE evolution of stars with a given mass at different evolutionary stages for low-metallicity stars ($Z = Z_\odot/50$). The scatter of points along the colored lines can be understood from the non-monotonic behavior of $|E_{\rm bind}|$ as a function of stellar radius. The reason for this is that there are changes in the core structure during the stellar evolution (Kruckow et al., 2016). It is evident that envelope ejection is facilitated for giant stars compared to less evolved stars, and as long as the inspiralling BH masses are large enough. For example, to eject the envelope of a star with an initial ZAMS mass of 50 $M_\odot$ requires an inspiralling companion with a mass of at least 6 $M_\odot$ (thus a BH for a HMXB system) in case the donor star is a red supergiant with a radius of ~3,000 $R_\odot$. However, if the donor star is less evolved with $R < 1{,}000\,R_\odot$ then it is seen in Figure 4.29 that $M_{X,\min} \gtrsim 20\,M_\odot$ (i.e., a fairly massive BH is needed to eject the envelope).

Location of the Bifurcation Point

One of the major problems in our understanding of CE ejection is the difficulty in localizing the physical point of envelope ejection, that is, the bifurcation point that separates the ejected envelope from the remaining core. Where does the spiral-in stop? Three main categories proposed for determining the bifurcation point are nuclear energy generation, chemical composition, and thermodynamic quantities (Tauris & Dewi, 2001; Podsiadlowski et al., 2003; Ivanova, 2011; Kruckow et al., 2016; Vigna-Gómez et al., 2022). For example, Kruckow et al. (2016) demonstrated that for an 88 $M_\odot$ star (with a metallicity, $Z = Z_\odot/50$), a spread in the location of the core boundary in mass coordinates of ~4 $M_\odot$ (<8% of the remaining core mass) leads to a corresponding spread in radius coordinates (and thus a spread in final post-CE orbital separation) of an astonishing factor of 500. This extreme example confirms the difficulty in making precise predictions for the GW merger rates (Abadie et al., 2010) and illustrates again that all quoted rates from simulations (let alone other uncertain effects, e.g., SN kicks) should be taken with a huge grain of salt. Recently, Marchant et al. (2021) suggested a method to self-consistently determine the core-envelope boundary in a CE. However, further investigations are needed to judge the applicancy and validity of this proposed method.

The calculated envelope binding energies and λ-values are also affected by the choice of the core convective overshooting parameter applied in stellar models, $\delta_{\rm OV}$. For example, for a 20 $M_\odot$ star, Kruckow et al. (2016) found that when it has evolved to a radius of $R = 1{,}200\,R_\odot$ then $|E_{\rm bind}|$ can be almost a factor of 10 smaller (and λ a factor of 10 larger) using $\delta_{\rm OV} = 0.0$ compared to $\delta_{\rm OV} = 0.335$. The corresponding core masses are approximately 5.9 $M_\odot$ and 7.2 $M_\odot$, respectively.

4.8.5 CE Ejection: Dependence on $\alpha_{\rm CE}$

It is important to stress that in the ($\alpha_{\rm CE}$, λ)-formalism discussed here, the CE ejection efficiency parameter must be $0 < \alpha_{\rm CE} < 1$. It is a misconception when $\alpha_{\rm CE}$ is sometimes

issue, as well as the question of implicitly assuming angular momentum conservation in the (α_{CE}, λ)-formalism, are needed and possibly require new development of 3D hydrodynamical simulations.

4.8.6 Hypercritical Accretion onto a NS in a CE?

It has been argued that a NS undergoing in-spiral in a CE might experience hypercritical accretion and thereby collapse into a BH (e.g., Chevalier, 1993). For hypercritical accretion to occur, the photon radiation must be trapped in the inward flow of accreted material. This requires high accretion rates ($\sim 10^{-3}\ M_\odot\ \mathrm{yr}^{-1}$) such that the temperature at the base of an accreting NS is sufficiently high to drive neutrino emission. Because neutrinos can escape the NS without interacting with the infalling material, the classic Eddington accretion limit derived for photon pressure (Section 4.9) does not apply.

Assuming the work of Chevalier (1993) to be correct, to explain the existence of double NS systems an alternative formation scenario without CE evolution was invented: the *double core scenario* (Brown, 1995; Bethe & Brown, 1998; Dewi et al., 2006). In this scenario, two stars with an initial mass ratio close to unity evolve in parallel and reach the giant stage roughly at the same time. Therefore, when the CE forms it will embed both stars in their giant stages (or as a giant star and a helium core), thereby avoiding the formation of a CE with a NS.

Nevertheless, it was already argued a few decades ago that hypercritical accretion might be inhibited by rotation (Chevalier, 1996) and strong outflows from the accretion disk (Armitage & Livio, 2000). Indeed, in more recent studies of hydrodynamical simulations, MacLeod & Ramirez-Ruiz (2015a, 2015b) found that a compact object such as a NS embedded in a CE only accretes a very modest amount of material during its in-spiral as a result of a density gradient across its accretion radius, which strongly limits accretion by imposing a net angular momentum to the flow around the NS. This conclusion supports earlier work by Ricker & Taam (2012) who also found that the true accretion rate of the accreting star is much less than that predicted by the Bondi-Hoyle prescription.

Finally, from an empirical point of view the idea of hypercritical accretion seems difficult to conciliate with observations of a number of very tight-orbit binary pulsars with massive WDs or NSs. Such systems show clear evidence of having evolved through a stage with a NS embedded in a CE (van den Heuvel & Taam, 1984), and yet these NSs often have masses only slightly greater than their expected birth masses (Tauris et al., 2017). This would not be the case if they had accreted efficiently at a very large rate in a CE.

4.8.7 Direct Observational Evidence for a CE: Red Novae

Given the short duration of a CE phase (of the order of 10^3 yr) (Meyer & Meyer-Hofmeister, 1979; Podsiadlowski, 2001), it is an extremely rare event to detect. Nevertheless, a number of candidates for direct observations of CE events have been proposed. The prime candidates are the luminous red nova (LRN) transients. In particular, V1309 Sco (a V838 Mon-class event) seems to be the most promising case so far

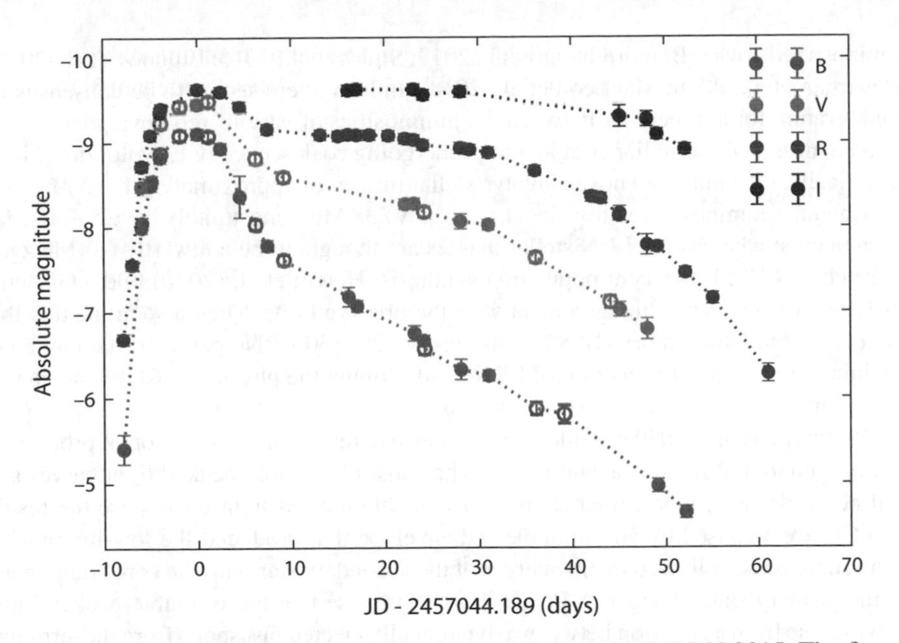

Figure 4.30. Absolute light curve of the optical outburst of M31LRN 2015. The figure is constructed assuming $E(B-V)=0.15$ mag, and error bars show measurement and distance errors while ignoring reddening error. The similarity of this light curve to other luminous red novae transients, such as V838 Mon, identifies M31LRN 2015 as a stellar merger. After MacLeod, Macias et al. (2017); see further details of the data and discussions thereof in this reference.

for an active CE event (or merger) being caught in action (Tylenda et al., 2011; Ivanova, Justhan, Chen et al. 2013). It was also the first star to provide conclusive evidence that contact binary systems end their evolution in a stellar merger. Another example of an observed LRN is M31 2015, shown in Figure 4.30, whose final progenitor binary system has been investigated and simulated in detail (e.g., MacLeod, Macias et al., 2017).

The energy involved in producing these red novae events is order-of-magnitude comparable to the likely orbital energy release from a CE phase or the binding energy of the envelope ($\sim 10^{47}$ erg) (Bond et al., 2003; Kulkarni et al., 2007). Observational examples of this class include M85 OT 2006-1 (Kulkarni et al., 2007; Ofek et al., 2008), NGC300 OT 2008 (Bond et al., 2009), PTF 10fqs (Kasliwal et al., 2011), M31 RV (Bond, 2011), and M31LRN 2015 (MacLeod, Macias et al., 2017). See also Stritzinger et al. (2020) for further discussions on observational features, including the candidate LRN AT 2014ej. The rate of similar events has been estimated to be as high as 20% of the core-collapse SN rate (Thompson et al., 2009).

It has been suggested that LRNe (i.e., gap transients with absolute magnitudes brighter than $M_V \sim -13$ and often possessing double-peaked light curves) might simply be more scaled-up versions of normal red novae (typically fainter than $M_V \sim -10$), thus resulting from merger events involving more massive stars compared to the less luminous red novae (Blagorodnova et al., 2017; Smith et al., 2016; Lipunov et al., 2017; Mauerhan et al., 2018; MacLeod et al., 2018). Indeed, there seems to be consensus in the literature for a direct link between the luminosities of various red nova phenomena and the masses of the stellar components undergoing coalescence (CE evolution). More specifically, the fainter red novae involve stellar masses of approximately $1 - 5\,M_{\odot}$; the intermediate-luminosity red transients, such as V838 Mon, most likely have $5 - 10\,M_{\odot}$ components; whereas the LRN stellar masses are thought to be a few $10\,M_{\odot}$ (Metzger & Pejcha, 2017). In a recent population synthesis, Howitt et al. (2020) find a Galactic LRN rate of $\sim 0.2\,\mathrm{yr}^{-1}$, in agreement with the observed rate. They also argue that the Vera C. Rubin Observatory (LSST) will observe 20–750 LRNe per year, thereby constraining the luminosity function of LRNe and probing the physics of CE events in the near future.

Pastorello et al. (2019) provide a comprehensive review and discussion of properties and progenitors of red novae and LRNe. The causes of double-peaked light curves are still being debated. One explanation is that double-peaked light curves are the result of a CE event caused by an initial ejected envelope that produced the low-luminosity light-curve peak, followed by the merger of the secondary star onto the core component of the primary star. Metzger & Pejcha (2017) propose that these double-peaked light curves arise from a collision between a dynamically ejected fast shell (from the terminal stage of the CE merger) and pre-existing slow equatorial circumbinary material shed from the plunge-in of the secondary star during the early phase of in-spiral. That is, the first optical peak arises through cooling envelope emission, whereas subsequent radiative shocks in the equatorial plane powers the second light-curve peak. In this scenario, different phenomenologies between various red gap transients can possibly be explained by different viewing angles of observers with respect to the orbital plane of the progenitor binary (see Fig. 2 of Metzger & Pejcha, 2017).

Glebbeek et al. (2013) studied the evolution of stellar mergers formed by a collision involving massive stars. They concluded that mass loss from the merger event is generally small ($<10\%$ of the total mass for equal-mass star mergers at the end of the main sequence) and that little hydrogen is mixed into the core of the merger product. Nevertheless, such an amount of mass loss may be sufficient to explain the double-peaked light curves in LRN events.

We note that a link has been suggested between LRNe and core-collapse SNe. Pastorello et al. (2019) argued that the LRN UGC5460−2010 OT1 may be an outburst precursor of the Type IIn-P SN2011ht, in a scenario where an evolved massive star likely exploded shortly ($\sim 300\,\mathrm{d}$) after a CE event. However, for an in-spiral system that survives as a binary with two stellar components after the CE ejection, we find this hypothesis to be very unlikely: it would require fine-tuning in stellar nuclear evolution such that the core of the (more massive) primary star undergoes core collapse coincidentally at the (almost) exact time the star is caught in a short-lasting CE phase.

Finally, it has been argued that ALMA detections of so-called "water fountains" ($<4\,M_{\odot}$ stars that have ejected a significant fraction of their mass over less than a few hundred years) may be directly related to CE ejection (Khouri et al., 2021).

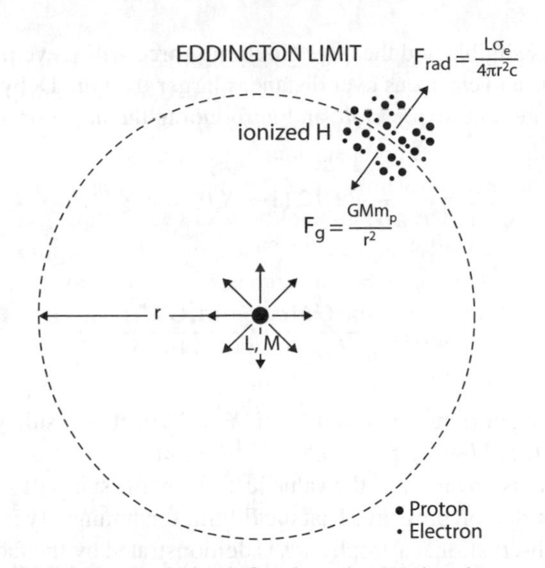

Figure 4.31. The Eddington limit is defined as the luminosity at which the radiation pressure on ionized gas (in this case pure hydrogen) around the star is just balanced by the gravitational attraction of the star (explanation in the text). After van den Heuvel (1994a).

4.9 EDDINGTON ACCRETION LIMIT

A critical luminosity, the Eddington luminosity (L_{Edd}), sets an upper limit to the accretion luminosity, L_{acc}, of a spherically accreting compact object, because for $L_{\mathrm{acc}} > L_{\mathrm{Edd}}$ further accretion of matter will be inhibited by the radiation pressure. The classical Eddington luminosity, first calculated by Eddington around 1920, is simply derived by equating the outward radiation pressure force on ionized plasma near the stellar surface with the gravitational force exerted by the star on this plasma (see Fig. 4.31):

$$|\vec{F}_{\mathrm{rad}}| = |\vec{F}_{\mathrm{grav}}| \quad \Longleftrightarrow \quad \frac{L_{\mathrm{Edd}}\,\sigma_T}{4\pi r^2\,c} = \frac{GMm_p\mu_e}{r^2}, \tag{4.130}$$

where σ_T is the Thomson scattering cross section of an electron (the dominant opacity source in an ionized plasma at low density), and $\mu_e \simeq 2/(1+X)$ is the mean molecular weight per electron, and X is the hydrogen mass fraction. Here σ_T is given by

$$\sigma_T = \frac{8\pi}{3}\left(\frac{e^2}{m_e c^2}\right)^2 \quad \mathrm{cm}^2. \tag{4.131}$$

Equation (4.130) for L_{Edd} follows from the fact that an ion (which is several thousand times heavier than an electron) is attracted by the force of gravity while the electron associated with the ion feels the outward radiation force. The gravitational force on the electron is negligible and the huge Coulomb force will prevent any charge separation between ions and electrons over distances larger than the Debye length.

Rewriting the previous expression, and introducing the mean opacity,

$$\kappa = \frac{\sigma_T}{m_p\,\mu_e} \simeq 0.2\,(1+X) \quad \mathrm{cm^2\,g^{-1}}, \tag{4.132}$$

one obtains

$$L_{\mathrm{Edd}} = \frac{4\pi\,GMc}{\kappa} \simeq \frac{4\pi\,GMc}{0.2\,(1+X)}. \tag{4.133}$$

For a typical hydrogen-rich donor star with $X = 0.70$, this result yields a valuc of $L_{\mathrm{Edd}} = 1.5 \times 10^{38}\,(M/M_\odot)\,\mathrm{erg\,s^{-1}} = 3.8 \times 10^4\,(M/M_\odot)\,L_\odot$.

If the luminosity is greater than the value just given, the star will start blowing away its outer layers by radiation pressure. That the Eddington luminosity indeed plays a very important role in observational astrophysics is demonstrated by the fact that none of the persistent accreting X-ray sources in our Galaxy considerably exceeds the Eddington limit. The apparent exceptions are the newly discovered pulsating ultra-luminous X-ray sources, which we discuss further in Chapter 7.

For a compact object accreting material from its companion star, the resulting luminosity is the sum of the contribution from released gravitational binding energy (as the accreted material falls deep into the gravitational potential of the compact object) and that of nuclear burning of the accreted material at the surface of the compact object, in other words,

$$L = \epsilon\,\dot{M} \qquad \text{and} \qquad \epsilon = \epsilon_{\mathrm{nuc}} + \epsilon_{\mathrm{grav}}, \tag{4.134}$$

where ϵ is the energy production per unit mass of accreted material. Hence, the Eddington mass-accretion rate can be defined as

$$\dot{M}_{\mathrm{Edd}} = \frac{4\pi\,GMc}{\kappa\,\epsilon} \simeq \frac{4\pi\,GMc}{0.2\,(1+X)\,\epsilon}. \tag{4.135}$$

For accretion of hydrogen and helium, the value of ϵ_{nuc} is approximately $6.4 \times 10^{18}\,\mathrm{erg\,g^{-1}}$ and $7.5 \times 10^{17}\,\mathrm{erg\,g^{-1}}$, respectively (the former value being based on $4\,^1\mathrm{H} \rightarrow {}^4\mathrm{He}$, and the latter value depending on the final $^{12}\mathrm{C}/^{16}\mathrm{O}$ ratio). Note, for the complete transition of hydrogen to iron, or helium to iron, the energy yield is $\sim 8.4 \times 10^{18}\,\mathrm{erg\,g^{-1}}$ (8.8 MeV/nucleon) or $2.0 \times 10^{18}\,\mathrm{erg\,g^{-1}}$, respectively.

Whereas for a WD accretor, the dominant contribution to the accretion luminosity is given by ϵ_{nuc}, the nuclear burning is negligible for NS and BH accretors (i.e., $\epsilon \simeq \epsilon_{\mathrm{grav}} \gg \epsilon_{\mathrm{nuc}}$, where $\epsilon_{\mathrm{grav}} = GM/R$). Hence, for NS and BH accretors, we can express $L = GM\dot{M}/R \equiv \eta\dot{M}c^2$, where η is of the order of 0.1 to 0.4 (see the following paragraphs). Based on this information, we can derive limits on the maximum accretion rates for all compact objects.

4.9.1 White Dwarf Accretor

For a WD accretor, $\eta \sim 10^{-4}$ and the Eddington accretion limit is roughly

$$\dot{M}_{\rm Edd} = \begin{cases} 3.6 \times 10^{-7}\, M_\odot\, {\rm yr}^{-1}\, (M/M_\odot) & \text{for hydrogen-rich matter } (X=0.7) \\ 4.4 \times 10^{-6}\, M_\odot\, {\rm yr}^{-1}\, (M/M_\odot) & \text{for accretion of helium} \end{cases} \tag{4.136}$$

4.9.2 Neutron Star Accretor

For a NS accretor, $\eta \simeq 0.15$–0.20 and the Eddington accretion limit is roughly

$$\dot{M}_{\rm Edd} = \frac{4\pi\, Rc}{0.2\,(1+X)} = \begin{cases} 2.1 \times 10^{-8}\, M_\odot\, {\rm yr}^{-1} & \text{for hydrogen-rich matter } (X=0.7) \\ 3.6 \times 10^{-8}\, M_\odot\, {\rm yr}^{-1} & \text{for accretion of helium} \end{cases} \tag{4.137}$$

for an assumed NS radius of 12 km. There is some uncertainty of the order of 20% depending on the assumed equation of state and the mass of the NS. The corresponding Eddington luminosity (often emitted almost entirely in X-rays) is then $L_{\rm Edd} = (1-3) \times 10^{38}\ {\rm erg\ s}^{-1}$, as mentioned earlier. See also Chapter 6 for further observational details.

4.9.3 Black Hole Accretor

For a BH accretor, $\eta \simeq 0.04$–0.42, depending on its spin. We derive these values in Section 7.6 where we also discuss how to measure the spin of a BH in an X-ray binary. Here we briefly summarize the result. The critical radius for the radiation of the released gravitational binding energy is the location of the innermost stable circular orbit, $R_{\rm ISCO}$, roughly corresponding to the inner edge of the accretion disk. Beyond (inside) this point, the accreted material is simply advected across the event horizon of the BH without producing further radiation. Defining the dimensionless spin parameter, $a_* \equiv Jc/(GM^2)$, which takes the values $a_* = \{-1, 0, 1\}$ for a retrograde orbit around a maximally spinning BH, a non-spinning BH, and a prograde orbit around a maximally spinning BH. The corresponding values of the released gravitational binding energy in units of the rest mass energy are $\eta = \{1-\sqrt{25/27},\ 1-\sqrt{8/9},\ 1-\sqrt{1/3}\} \simeq \{0.038, 0.057, 0.42\}$, respectively (Section 7.6). The locations of $R_{\rm ISCO}$ (in units of GM/c^2) are $R_{\rm ISCO} = \{9, 6, 1\}$, respectively, whereas the Schwarzschild radii (or event horizons) are $R_{\rm H} = \{1, 2, 1\}$ for $a_* = \{-1, 0, 1\}$ (Section 7.6). The radius of interest for an Eddington-limited accreting BH is therefore that of the ISCO, and we can finally express

$$\dot{M}_{\rm Edd} = \frac{4\pi\, R_{\rm ISCO}\, c}{0.20(1+X)} = \alpha \cdot 4.4 \times 10^{-9} \left(\frac{M/M_\odot}{1+X}\right) M_\odot\, {\rm yr}^{-1}, \tag{4.138}$$

where $\alpha = \{9, 6, 1\}$ for $a_* = \{-1, 0, 1\}$, see Table 7.2.

EXERCISES

Exercise 4.1.

a. *Verify Eq. (4.9).*

b. *The binary radio pulsar J1719−1438 has an orbital period of 2.18 hr. Its companion star with a mass of $M_2 \simeq 0.001 - 0.004\, M_\odot$ is not filling its Roche lobe (i.e., $R_2 < R_{\rm L1}$), which means that $\langle \rho_2 \rangle$ is larger than was given by Eq. (4.9). Notice that this density limit is independent of the orbital inclination angle and the pulsar mass. Verify the finding of Bailes et al. (2011) who suggested that PSR J1719−1438 may host a "diamond" planet.*

It seems clear that this system is the remnant of an UCXB (see Section 11.4), and although the chemical composition of the companion might be helium rather than carbon (thus a diamond core), it is certainly a dense object and most probably also crystallized.

Exercise 4.2. *Follow the recipe in Section 4.3.4 and derive Eq. (4.34).*

Exercise 4.3. *Consider a binary system with two stars of masses $M_1 = 2\,M_2$ and M_2, respectively, and assume $\Delta M = \frac{1}{2} M_2$ is lost in a fast stellar wind from star 1.*

a. *Calculate the resulting widening of the orbit, a/a_0.*

b. *Show that the change in orbital angular momentum is $J/J_0 = \mu/\mu_0$ and demonstrate that J decreases even though the orbit is widening.*

c. *Calculate J/J_0 assuming ΔM is lost from star 2.*

d. *If the mass is lost from star 2, show that $\dfrac{J}{J_0} = \dfrac{\frac{1}{q_0}+1}{\frac{1}{q}+1}$, where $q < q_0$.*

e. *Calculate J/J_0 assuming ΔM is lost from star 1.*

f. *Explain why there is a difference in the answers to parts c and e.*

Exercise 4.4. *In very wide orbit LMXBs ($P_{\rm orb} > 100\,{\rm d}$), the orbital separation will always* widen *prior to RLO because the stellar wind mass loss becomes very important for such giant stars. To quantize this effect, one can apply the Reimers (1975) wind mass-loss rate:*

$$\dot{M}_{\rm wind} = -4 \times 10^{13}\, \eta_{RW} \left(\frac{L\,R}{M} \right) \quad M_\odot\,{\rm yr}^{-1}, \tag{4.139}$$

where the luminosity, radius, and mass of the mass-losing star are in solar units and $\eta_{RW} \simeq 0.5$ is a mass-loss parameter. Give a rough estimate of the typical widening of the orbit of such a system prior to RLO by considering appropriate values of L, R, and M and for the time duration of a low-mass giant star on the RGB. Compare Eqs. (4.34) and (4.37) for the widening of the orbit.

Exercise 4.5. *For conservative mass transfer* $\dot{M}_1 = -\dot{M}_2$ *and* $\dot{J}_{\rm orb} = 0$. *Show that*

$$\frac{\dot{a}}{a} = -2\left(1 - \frac{M_2}{M_1}\right)\frac{\dot{M}_2}{M_2} \tag{4.140}$$

and derive Eqs. (4.40) and (4.41).

Exercise 4.6. *Occasionally, the definition of mass ratio is inverted (i.e.,* $q' \equiv 1/q$ *and* $q'_0 \equiv 1/q_0$*). Show that with this mass-ratio inversion, Eq. (4.41) remains invariant, whereas Eq. (4.48) takes the form*

$$\lim_{\beta\to 1}\left(\frac{P}{P_0}\right) = \left(\frac{q_0+1}{q+1}\right)^2\left(\frac{q}{q_0}\right)^5 e^{3(q_0-q)/qq_0}. \tag{4.141}$$

Exercise 4.7. *Show that the general solution (Eq. 4.58) to the orbital angular momentum equation transforms into*

a. *Equation (4.34)* *for* $\alpha \to 1$.
b. *Equation (4.40)* *for* $\alpha = \beta = \delta \to 0$.
c. *Equation (4.45)* *for* $\alpha = \delta \to 0$ *and* $\beta > 0$.
d. *Equation (4.55)* *for* $\delta \to 1$.

Exercise 4.8. *Show that for isotropic re-emission with a mass-loss rate of* $\beta\dot{M}_2$, *and assuming* $\alpha = \delta = 0$, *the rate of loss of orbital angular momentum is given by*

$$\dot{J}_{\rm ml} = a_1^2\Omega\,\beta\dot{M}_2, \tag{4.142}$$

where a_1 *is the distance from the accreting compact object (star 1) to the c.m.*

Exercise 4.9. *Show by means of the virial theorem, that if at least half of the total mass of a binary system is suddenly lost in a SN, the system becomes unbound.*

Exercise 4.10. *Show that for a symmetric SN, the changes in semi-major axis and eccentricity are given by*

$$\frac{a}{a_0} = \frac{M}{M - \Delta M} \qquad\qquad e = \frac{\Delta M}{M}, \tag{4.143}$$

where a *and* a_0 *denote the post- and pre-SN semi-major axes, respectively,* M *is the post-SN total mass of the binary, and* ΔM *is the amount of mass lost in the SN. Assume the explosion to be instantaneous and that the pre-SN orbit is circular.*

Hint: consider the ratio of the orbital (total) energy before and after the SN, where $E_{\rm orb} = E_{\rm pot} + E_{\rm kin}$, *and where* $E_{\rm orb} = -GM_1M_2/2a$, $E_{\rm pot} = -GM_1M_2/r$, *and*

$E_{\rm kin} = 1/2\, \mu\, v_{\rm rel}^2$. *A circular pre-SN orbit means that* $a_0 = r$. *Notice, after the SN, the periastron separation of the new orbit,* $q = a(1-e)$ *is equal to the orbital separation between the two stars at the moment of the explosion* ($r = a_0$).

Exercise 4.11.

a. *Show that the semi-major axis of a post-SN orbit (assuming a symmetric SN with no kick) that re-circularizes due to tidal forces is related to the pre-SN semi-major axis according to an expression similar to that of Eq. (4.34).*

b. *Show also that*

$$\frac{P_{\rm circ}}{P_0} = \left(\frac{M_0}{M}\right)^2, \tag{4.144}$$

where $P_{\rm circ}$ *and* P_0 *denote the post-SN/circularization and pre-SN orbital period, respectively, and* M *and* M_0 *are the post- and pre-SN total mass of the binary, respectively.*

Exercise 4.12. *Assume* $\Delta R = H_p/R$ *and derive Eq. (4.88).*

Exercise 4.13. *Derive Eq. (4.100).*

Exercise 4.14. *Derive Eq. (4.129). Try to insert some realistic numbers for the various quantities (search the literature) and estimate* (a/a_0).

Exercise 4.15. *Kepler's nightmare: Discuss under which conditions it is possible to observe a binary system increasing its orbital period while the orbital size, a, shrinks.*

Exercise 4.16. *Reproduce Figure 4.8.*

Exercise 4.17. *Derive Eqs. (4.45) and (4.47).*

Exercise 4.18.

a. *Show that by eliminating* $d\epsilon/\epsilon$ *from Eqs. (4.21)–(4.23), one can write*

$$\frac{dP}{P} = -2\,\frac{d(M_1+M_2)}{M_1+M_2} + 3\,\frac{dj}{j} + 3\,\frac{e\,de}{(1-e^2)}, \tag{4.145}$$

where j *is the orbital angular momentum per unit reduced mass, that is,*

$$j = J\,\frac{(M_1+M_2)}{M_1 M_2} \tag{4.146}$$

and J *is the orbital angular momentum of the system.*

b. *Derive from Eq. (4.146) an expression for dj/j and show that this implies that for circular orbits, Eq. (4.145) can be rewritten as*

$$\frac{dP}{P} = 3\frac{dJ}{J} + \frac{d(M_1+M_2)}{M_1+M_2} - 3\frac{dM_1}{M_1} - 3\frac{dM_2}{M_2}. \tag{4.147}$$

Exercise 4.19. *Consider the case for a circular binary where ejected particles form a circumbinary ring around the system, in which they move in Keplerian orbits around the c.m. of the system, in the same direction as the orbital motion. Assume that the radius of the ring, $a_{\rm ring}$, is larger than the orbital radius, a, of the binary. Here, we call the orbital angular momentum per unit mass of the ring $h_{\rm ring}$ and that of the binary h, so*

$$h = \frac{M_1 M_2}{(M_1+M_2)^2}\sqrt{G(M_1+M_2)\,a}. \tag{4.148}$$

a. *Show that*

$$h_{\rm ring} = \sqrt{G(M_1+M_2)\,a_{\rm ring}}. \tag{4.149}$$

b. *Show that $h_{\rm ring} > h$.*

c. *The decrease of the orbital angular momentum of the binary due to the escape of matter to the ring is equal to $dJ = h_{\rm ring}\,d(M_1+M_2)$, where*

$$J = \frac{M_1 M_2}{M_1+M_2}\sqrt{G(M_1+M_2)\,a}. \tag{4.150}$$

Show now that Eq. (4.147) transforms into

$$\frac{dP}{P} = (1+3\Gamma)\frac{d(M_1+M_2)}{M_1+M_2} - 3\frac{dM_1}{M_1} - 3\frac{dM_2}{M_2}, \tag{4.151}$$

where

$$\Gamma \equiv \frac{(M_1+M_2)^2}{M_1 M_2}\sqrt{\frac{a_{\rm ring}}{a}} = \frac{h_{\rm ring}}{h}. \tag{4.152}$$

Consider now a system with $M_1 = M_2$, $dM_1 = dM_2$, and $a_{\rm ring} = 1.2\,a$.

d. *Show that for this system the following holds:*

$$\frac{dP}{P} \simeq 8.2\frac{d(M_1+M_2)}{M_1+M_2}. \tag{4.153}$$

We notice that because $\Gamma \simeq 4.4$, formation of a circumbinary ring acts as an efficient sink for orbital angular momentum, that is, its drains J efficiently.

Exercise 4.20. *We consider stellar wind matter that is accelerated from one of the binary stars due to radiation pressure. While moving in the binary system, the wind* *particles also experience gravitational interactions with the two stars. We consider now the case where, when the particle has reached a distance, a, from the c.m. of the system, a fraction, η, of the total energy required for its complete expulsion from the binary is provided by (through gravitational interactions) the orbital energy, E, of the system (the remaining required energy coming from the radiation pressure). We assume that the orbit remains circular. The decrease of the orbital energy due to the expulsion of a mass element, dM_1, is given by*

$$dE = \eta \frac{G(M_1 + M_2)}{a} dM_1, \tag{4.154}$$

where

$$E = -\frac{GM_1 M_2}{2a}, \tag{4.155}$$

such that

$$\frac{dE}{E} = -2\eta \frac{(M_1 + M_2)}{M_1 M_2} dM_1. \tag{4.156}$$

a. *Show that if M_2 is constant, the change of the orbital radius, a, of the system is given by*

$$\frac{a}{a_0} = \left(\frac{M_1}{M_{10}}\right)^{(1+2\eta)} e^{2\eta \frac{(M_1 - M_{10})}{M_2}}, \tag{4.157}$$

where index 0 denotes the initial values of a and M_1.

b. *Assume $\eta = 0.2$ and $M_2 = 0.5\, M_{10}$. Calculate how much the orbital radius will shrink when M_1 decreases from M_{10} to $0.5\, M_{10}$.*

Chapter Five

Observed Binaries with Non-degenerate or White Dwarf Components

5.1 INTRODUCTION

Interacting binaries are almost always spectroscopic or eclipsing binary systems, and most of our fundamental knowledge on masses and radii of stars have been derived from these systems. Here, we describe the properties of the various important types of interacting binaries found in nature that contain non-degenerate or WD components. Observed binaries that contain interacting NSs or BHs (i.e. X-ray binaries) are reviewed in Chapters 6 and 7.

Interacting binaries can be classified in various ways, such as by their *observational characteristics* or the *evolutionary status* of their components. As we are primarily interested here in the evolutionary aspects, we will concentrate on an overview of systems in terms of the second type of classification, which will be given in the following sections. Because the nomenclature of systems is often derived from their observational characteristics, we briefly describe here the much-used observational classification in terms of the degree to which one or both components fill their critical Roche surfaces (i.e., Roche lobes), as depicted in Figure 5.1. In this classification, four categories of systems are distinguished as follows:

1. Detached systems: neither component fills its Roche lobe.
2. Semi-detached systems: one of the two stars fills its Roche lobe. In short-period systems, this is almost always the less-massive component, like in classical Algol systems. In wide systems, like ζ or ϵ Aurigae, it often is the more massive star that is (almost) filling its Roche lobe.
3. Contact systems: both stars are filling their Roche lobes and have a common outer envelope.
4. Evolved systems of unclear type: it is unclear whether one of the components fills its Roche lobe.

5.1.1 Important Observed Categories of Systems

In more detail, one can characterize these four categories of systems in the following way:

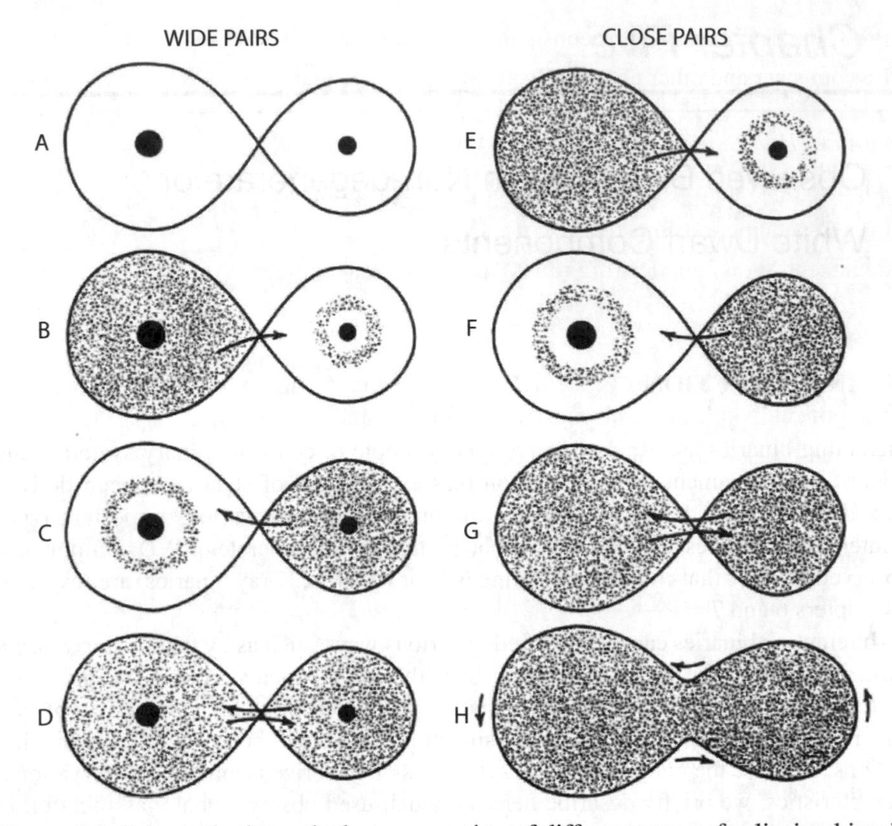

Figure 5.1. Schematic theoretical representation of different types of eclipsing binaries (for systems with an adopted mass ratio of 3:2). In supergiant systems B and C, the low-density outer atmosphere (chromosphere) of one star fills one of the lobes, and the other star has a gaseous ring. System D is typical for some supergiant systems: the low-density atmospheres of both stars fill both lobes. In system E the more massive star fills its Roche lobe, and the other star has a gaseous ring. Such systems are rare. Systems like F, on the other hand, are common, for example, Algol and U Cep. System G represents massive contact systems, and system H represents stellar pairs like W Ursa Majoris. Arrows show the direction of the flow of gases between the stars. After Payne-Gaposchkin & Haramundanis (1970).

Detached Systems

- All visual binaries.
- Non-evolved spectroscopic and eclipsing binaries (both stars are on the main sequence); these are the systems from which the masses and other data in Tables 3.1 and 5.1 were derived.
- Double WDs (DWDs).
- Double NSs (DNSs) and other binary radio pulsars.

- Pre-CVs: detached systems consisting of a red dwarf and a WD.
- Barium stars and other chemically peculiar G- or K-giants, with WD companions.

Semi-detached Systems

- Algol-type systems (orbital periods of the order of days).
- CVs (orbital periods mostly $\leq$0.5 d).
- Persistent X-ray binaries (high-mass and low-mass donor stars).

Contact Systems

- W Ursa Majoris binaries (systems of low total mass, with two F, G, or K dwarfs as components; orbital periods mostly between 0.2 and 0.5 d).
- β Lyrae-type binaries: high-mass contact systems consisting of two stars of relatively early spectral type: O, B, or A; orbital periods days to weeks.

Evolved Systems of Unclear Type

- ζ Aurigae systems (orbital periods of the order of years to decades).
- Symbiotic stars (orbital periods of the order of years).

Figure 5.1 gives a schematic observational classification of eclipsing binaries after Payne-Gaposchkin & Haramundanis (1970). Figure 5.2 shows characteristic light curves of a semi-detached eclipsing system (Algol-type) and of two types of contact systems (β Lyrae-type and W Ursa Majoris-type). For a detailed description of each of these types of systems and of their characteristic light curves we refer to the general literature (e.g., Tsesevich, 1973).

Here, we define binaries as interacting if they are so close that during some stage of their evolution at least one of the components will overflow (or has overflowed) its Roche lobe and will transfer (or has transferred) mass to its companion. In Chapter 9 we will give details on the maximum orbital periods, as a function of stellar mass, up to which stars can still be considered as interacting. Red Supergiants (RSGs) or stars on the AGB can reach radii as large as 5 AU (or $\sim$1,000 $R_\odot$), so these maximum orbital periods can be of the order of several decades. Hence, all binaries with orbital periods shorter than a few decades can be considered as (potentially) interacting, even when at present both components are still main-sequence stars that are deep inside their Roche lobes.

Because evolution is the topic of this book, we now consider the various types of interacting binaries classified according to their evolutionary status. We divide the systems into the following four categories:

1. Unevolved systems
2. Evolved systems with non-degenerate components
3. Evolved systems with one or two WD components
4. Evolved systems with one or two NS or BH components.

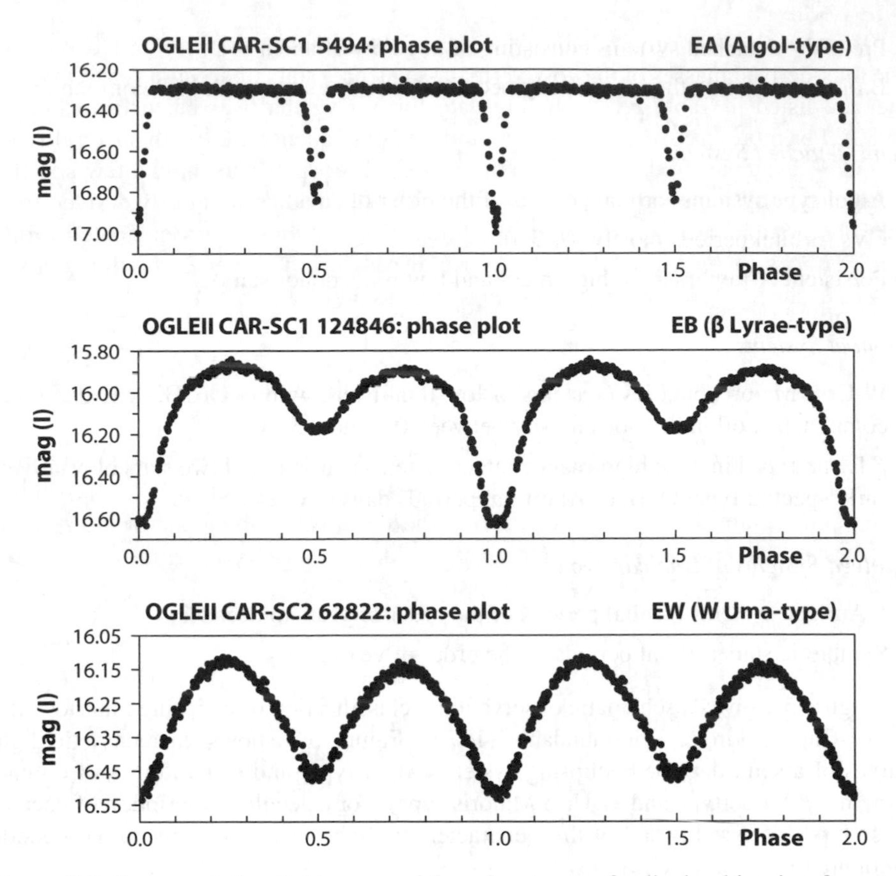

Figure 5.2. Representative lightcurves of three key types of eclipsing binaries, from top to bottom: Algol type (EA), β Lyrae type (EB) and W UMa type (EW); from OGLE-II observations (Szymanski, 2005) and taken from Hümmerich et al. (2013) who plotted these curves. The vertical scale is stellar magnitudes in color I. After Hümmerich et al. (2013). See also Figure 2.4.

The first three categories of systems are discussed in this chapter, in Sections 5.2 to 5.4, and the last category in Chapters 6 and 7. To get an impression of the observational appearance of the different types of systems, in this chapter we list characteristic examples of each type of system.

5.2 UNEVOLVED SYSTEMS

These are systems in which both components are still unevolved main-sequence stars. Among these are systems that are both eclipsing and spectroscopic binaries. Such systems can be completely solved for the masses, radii, effective temperatures, and luminosities of the components, as described in Chapter 3. It is from these systems that the mass-luminosity relation of main-sequence stars (Eq. [3.30]) has been derived.

Torres et al. (2010) give reliable data of unevolved stars, derived from such systems. The thus-derived masses of unevolved (main-sequence) stars of spectral type O7V and later are listed in Table 5.1, including data for Y Cyg that was taken from Popper (1980). For more massive stars, which on the main sequence start with spectral types O3–O6V, examples of data are given in Table 3.2. For such stars, only a few spectroscopic and eclipsing binary data are available, and the masses are mostly derived from their observed absolute luminosities, by fitting these luminosities to calculated evolutionary tracks of very massive stars. As we discussed in Chapter 3, the thus-derived masses of the most massive stars can be as large as 300 $M_{\odot}$.

5.3 EVOLVED SYSTEMS WITH NON-DEGENERATE COMPONENTS

These systems are either close systems (orbital periods less than a few weeks), in which one of the components has evolved into a sub-giant (luminosity class IV or III) and/or shows considerable signs of mass loss, or wide systems, in which one of the components has evolved into a yellow or red giant or supergiant. We consider these two types separately.

5.3.1 Close-orbit Evolved Systems

Tables 5.2–5.4 show some characteristic examples of three varieties of systems in this group. These are the hot evolved close systems, the classical Algol systems, and the WR binaries. We briefly comment on each of these three categories.

Hot Evolved Close Binaries

In Table 5.2 there are two systems of special interest. The first one concerns the system BD+40°4220 (V729 Cyg), which consists of two almost identical early Of-type stars of almost equal luminosity (the notation f indicates the presence of emission lines in the spectrum, which is characteristic for a strong stellar wind; the mass-loss rate of each star is $5 \times 10^{-6}\, M_{\odot}\,\mathrm{yr}^{-1}$). Despite their similar luminosities, the stars differ by more than a factor of 3 in mass! The 15.6 $M_{\odot}$ O7f star has the luminosity of an (at least) 45 $M_{\odot}$ star. As noticed by Bohannan & Conti (1976), this system must be on its way to becoming a WR binary: the 15.6 $M_{\odot}$ O7f star is expected to be the helium core of a star originally approximately three times more massive, still surrounded by a thin hydrogen rich envelope of at most 1–3 $M_{\odot}$. When this hydrogen envelope is completely removed by the stellar wind, the helium core will be left as a WR-star. The second system of interest is V Puppis, a naked-eye star of apparent visual magnitude of $\sim 4^m$. Its binary period is only 1.4545 d, but its eclipse period shows a cyclic variation with a period of 5.47 yr (Qian et al., 2008), indicating the presence of a third component with a mass $\geq 10.4\, M_{\odot}$. As no trace of light of this massive third star has been observed, this star has been suggested to be a BH (Qian et al., 2008). This would mean that V Pup could be the first naked-eye star with a BH companion.

Table 5.1. Some Characteristic Examples of Detached Unevolved Double-lined Eclipsing and Spectroscopic Binaries with Well-determined Physical Parameters

Name	Spectral type	P_{orb} (d)	M_1 ($M_\odot$)	M_2 ($M_\odot$)	$\log L_1$ ($L_\odot$)	$\log L_2$ ($L_\odot$)
V3903 Sgr	O7V+O7V	1.74	27.27 (±0.55)	19.01 (±0.44)	5.088 (±0.087)	4.658 (±0.088)
EM Car	O8V+O8V	3.41	22.83 (±0.32)	21.38 (±0.33)	5.021 (±0.104)	4.922 (±0.104)
V478 Cyg	O9.5V+O9.5V	2.88	16.62 (±0.33)	16.27 (±0.33)	4.631 (±0.058)	4.635 (±0.057)
Y Cyg	O9.8V+O9.8V	3.00	16.70 (±0.50)	16.70 (±0.50)	4.45 (±0.11)	4.45 (±0.11)
V578 Mon	B1V+B2V	2.41	14.50 (±0.12)	10.26 (±0.08)	4.285 (±0.045)	3.888 (±0.045)
V539 Ara	B3V+B4V	3.17	6.24 (±0.07)	5.31 (±0.06)	3.293 (±0.051)	2.955 (±0.055)
U Oph	B5V+B6V	1.68	5.72 (±0.09)	4.74 (±0.07)	2.90 (±0.03)	2.71 (±0.03)
V451 Oph	B9V+A0V	2.20	2.77 (±0.06)	2.35 (±0.05)	1.93 (±0.13)	1.53 (±0.09)
RS Cha	A8V+A8V	1.67	1.854 (±0.016)	1.817 (±0.018)	1.24 (±0.05)	1.24 (±0.05)
FS Mon	F2V+F4V	1.91	1.63 (±0.01)	1.46 (±0.01)	0.89 (±0.03)	0.64 (±0.03)
V636 Cen	G1V+K2V	4.28	1.019 (±0.005)	0.855 (±0.003)	0.053 (±0.025)	−0.431 (±0.035)

Sources: Except for Y Cyg (Popper, 1980), the data are from the catalog of Torres et al. (2010).

Table 5.2. Examples of Hot Evolved, in Most Cases Semi-detached, Systems

Name	Spectral type	P_{orb} (d)	M_1 ($M_\odot$)	M_2 ($M_\odot$)	R_1 ($R_\odot$)	R_2 ($R_\odot$)	$\log L_1$ ($L_\odot$)	$\log L_2$ ($L_\odot$)
BD+40°4220[1]	O6f+O7f	6.6	51.8	15.6	27	15.5	6.1	5.9
LY Aur[2]	O9II+O9III	4.0	25.5	14.5	16.1	12.6	5.33	5.13
AO Cas[2]	O8V(f)+O9.2II	3.52	15.59	9.65	4.61	9.43	5.06	4.82
V Pup[3]	B1V+B3	1.45	14.86	7.76	6.18	4.90	4.34	4.04
MP Cen[4]	B3V+B6-7	2.99	11.4	4.4	7.7	6.6	3.8	3.0
β Lyrae[2]	B7V+BII(?)	12.925	13.16	2.97	6.0	15.2	4.42	3.87
V356 Sgr[5,6]	B2-3+A2II	8.93	12.1	4.7	6.0	14.0	3.45	3.00
Z Vul[2]	B4V+A5III	2.46	5.4	2.3	4.5	4.6	3.45	2.20

Sources: [1]Bohannan & Conti (1976), Rauw et al. (1999). [2]SIMBAD. [3]Andersen et al. (1983). [4]Terrell et al. (2005). [5]Dominis et al. (2005). [6]Popper (1980).

V Pup, MP Cen, and V356 Sgr have β Lyrae-like light curves.

Table 5.3. Examples of Classical Semi-detached Algol-type Eclipsing Binaries

Name	Spectral type	P_{orb} (d)	M_1 ($M_\odot$)	M_2 ($M_\odot$)	R_1 ($R_\odot$)	R_2 ($R_\odot$)	$\log L_1$ ($L_\odot$)	$\log L_2$ ($L_\odot$)
RY Per[1]	B4V+F7II–III	6.86	6.24	1.69	4.06	8.10	3.0	1.9
RS Vul[2,3]	B5V+G1III	4.48	6.59	1.76	4.71	5.84	2.9	1.5
U Cep[4]	B8V+G8III	2.49	3.6	1.9	2.4	4.4	1.92	1.03
β Per (Algol)[4]	B8V+K3IV	2.87	3.7	0.81	2.9	3.5	2.34	0.66
RT Per[4]	F2V+G5–8IV	0.85	1.7	0.4	1.41	1.27	0.64	−0.03
ST Per[4]	A3V+K1–2IV	2.85	3.3	0.5	2.32	2.95	1.51	0.78
TX UMa[4]	B8V+G0III–IV	3.06	4.8	1.2	2.83	4.24	2.30	1.17
XZ And[4]	A0–1V+G5IV	1.36	3.2	1.3	2.4	2.0	1.57	0.68

Sources: [1]Barai et al. (2004). [2]Holmgren (1989). [3]Richards et al. (2010). [4]Selam & Demircan (1999) and references therein.

Note: For Algol a more recent estimate has revealed component masses of ~3.2 and 0.7 $M_\odot$, respectively (Baron et al., 2012).

In the 1970s, V Pup was surprisingly detected as a weak Uhuru X-ray source, which is otherwise rather unusual for this type of binary (see references in Qian et al., 2008). The presence of a BH, accreting from the stellar wind of the inner binary, might be a way to solve this problem. However, Maccarone et al. (2009) suggested that the X-ray emission may also be due to the colliding winds of the components of V Pup itself. In addition, there still is a possibility that the massive companion is not a BH, but just itself a close binary consisting of two main-sequence stars of ~5 $M_\odot$ each. Due to the

Table 5.4. Examples of Galactic WR Binaries with Well-determined Orbital Parameters

Name	Spectral type	$P_{\rm orb}$ (d)	$M_{\rm WR}$ ($M_\odot$)	$M_{\rm O}$ ($M_\odot$)	Ref.
WR 155 = CQ Cep	WN6o+O9II-I	1.64	24	30	
WR 151 = CX Cep	WN4o+O5V	2.13	26.4	36.0	1
WR 139 = V444 Cyg	WN5o+O6III-V	4.21	12.4	28.4	2
WR 47 = HD 311884	WN6o+O5V	6.24	51	60	
WR 42 = HD 97152	W7C+O7V	7.89	14	23	
WR 21 = HD 90657	WN5o+O4-6	8.26	19	37	
WR 79 = HD 152270	WC7+O5-8	8.89	11	29	
WR 62a = SMS NPL 13	WN4-5o+O5.5-6V	9.15	22	41	3
WR 127 = HD 186943	WN5o+O9.5V	9.56	13	24	4
WR 9 = HD 63099	WC5+O7	14.31	9	32	
WR 30 = HD 94305	WC6+O6-8	18.81	16	34	
WR 141 = HD 193928	WN5o+O5V-III	21.69	45	30	
WR 113 = HD 168206	WC8+O8-9IV	29.70	13	27	
WR 11 = γ^2 Vel	WC8+O7.5III	78.53	9.5	30	

Source: [1]Hutton et al. (2009). [2]Hirv et al. (2006). [3]Collado et al. (2015). [4]de La Chevrotière et al. (2011). Where no reference is given, the data are from the van der Hucht (2001) catalog, to which we refer for the original references. Systems are listed in order of increasing orbital periods. The indication WNo means that hydrogen lines are absent in the WN spectrum.

mass-luminosity relation, the luminosity of such a binary is ~6 times lower than that of a 10 $M_\odot$ main-sequence star. This would make the dark companion ~20 times fainter than V Pup itself, and therefore its light would not be seen in the spectra of V Pup. For this reason, the presence of a massive unseen companion does not always need to point to the presence of a BH. The same holds for the alleged presence of a BH of rather unusual mass in the 83 d red-giant binary 2MASS J02515658+4359220 (Thompson et al., 2019). We pointed out previously (van den Heuvel & Tauris, 2020) that instead of a BH, the unseen companion of the red giant may be a close binary consisting of two K0-4V main-sequence stars. It should be kept in mind that triple systems consisting of a close binary and a distant third star are very common. Approximately 20% of all wide binaries are such triples. Also wide binaries composed of two close twins are not very rare.

Classical Algol Systems

These are the systems for which the Algol paradox was noticed by Crawford (1955), see Chapter 2, which means that the more evolved sub-giant component, which is filling its Roche lobe, is *less* massive than its unevolved companion. The solution to this paradox is that in these systems large-scale mass transfer has taken place, like in several of the

Figure 5.3. WR 31a and its wind-induced nebula is an example of many WR-stars that are surrounded by a nebular shell. Credit: ESA/NASA Hubble Space Telescope/Judy Schmidt.

hot evolved systems of Table 5.2. Table 5.3 lists some characteristic examples of classical semi-detached Algol-type systems. (An excellent early overview of these systems and their evolutionary origins was given by De Greve [1986].)

Wolf-Rayet Binaries

WR-stars are characterized by very prominent emission-line spectra with either strong nitrogen lines (WN-stars) or strong carbon lines (WC-stars) (see Fig. 3.7). In the classical WR-stars, lines of hydrogen are absent, but helium emission lines are prominent. The emission spectrum arises in the very dense stellar wind of the WR-star, with a mass-loss rate of the order of $10^{-5}\ M_{\odot}\ \mathrm{yr}^{-1}$ (Sander & Vink, 2020). The strong winds of WR-stars often produce wind-induced nebulae surrounding the star (Fig. 5.3). WR binaries consist of a WR-star together with an O-type star; the latter being unevolved in most cases and more massive than the WR-star. The WR-star is in a more advanced stage of evolution than its O-type companion. This is evidenced by the fact that the classical WR-stars—which are different from the so-called WNh-stars (see Table 3.2 and the following discussion)—despite the absence of hydrogen in their spectra, nevertheless have bolometric luminosities similar to those of their more massive O-type companions. The general consensus is that WN- and WC-stars are the remnant helium cores of the originally more massive components of their binaries (see, e.g., Crowther,

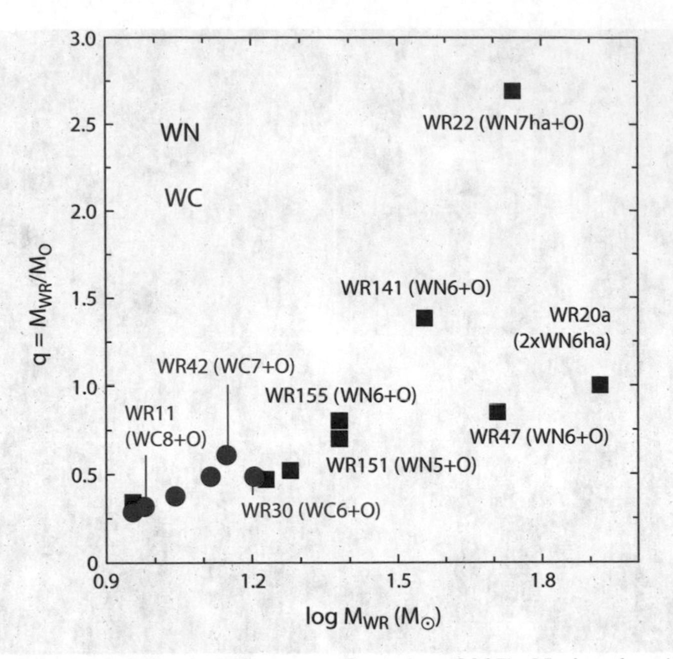

Figure 5.4. Masses of Galactic WR-stars (Crowther, 2007). Notice that (1) with one exception (WR 141) the WN-stars (without hydrogen in their spectra) always are less massive than their O-type companions; (2) WNh-stars (with hydrogen in their spectra) tend to be more massive than their O-type companions; and (3) WC-stars tend to be less massive than WN-stars and have lower binary mass ratios.

2007), as was first pointed out by Paczyński (1967c). In view of their larger masses, their progenitors evolved faster than their (at that time less massive) companions, lost their hydrogen-rich envelopes (either to their companions via RLO or due to strong stellar wind mass loss), and now are already in the core-helium-burning phase, while their O-type companions are still in the core-hydrogen-burning phase. The helium burning in their cores produces carbon and oxygen. It is thought that the WR-stars begin their evolution as WN-stars (the nitrogen having been produced by the CNO cycle during core-hydrogen burning). When the WN-star has lost a large part of its envelope due to wind mass loss, the carbon produced by the helium burning becomes visible at the surface, and the star turns into a WC-star. Still later, even the oxygen produced by the helium burning may come to the surface. The star then becomes a WO-star, a very rare type of WR-star. The evolutionary sequence therefore is thought to be WN $\longrightarrow$ WC $\longrightarrow$ WO (Crowther, 2007; Langer, 2012).

This evolutionary picture fits with the observation that in general WC+O binaries have lower mass ratios than the WN+O systems, as can be seen in Figure 5.4 and Table 5.4. The special group of WNh-stars that do show hydrogen in their spectra are physically very different from the genuine WN-stars. They are extremely massive and luminous hydrogen-rich stars, still in the phase of core-hydrogen burning. They are an extension of the earliest O-type main-sequence stars to still higher masses, as indicated in Table 3.2. Due to their very high wind mass-loss rates, they have very dense winds, which lead to the formation of an emission-line spectrum resembling that of the hydrogen-free WR-stars, the difference being also hydrogen emission lines are visible. To distinguish the genuine WN-stars from the WNh-stars, one nowadays indicates the genuine hydrogen-free WN-stars also as WNo. In Table 5.4, this nomenclature has been followed and several characteristic examples of WR binaries are listed. The Galactic Wolf Rayet Catalogue of WR-stars is updated regularly.[1]

Because the winds are driven by radiation pressure, which depends on the opacity of the stellar material, which is largely determined by the heavy element abundances, the wind mass-loss rate depends on the metallicity, Z (abundance of elements heavier than helium), of the stellar material: the higher Z, the higher the mass-loss rate. The mass-loss rate of O-stars and WR-stars is roughly proportional to Z^m, where $m = 0.8\ (\pm 0.2)$ (Mokiem et al., 2007; Crowther, 2007); see also Shenar et al. (2020) and references therein. In the case of the WR-stars, this leads to different ratios of the numbers of WN- and WC-stars in regions of different metallicity. This ratio is $\sim$1 in our Galaxy (the Milky Way, MW), but 5 in the LMC, where $Z = 0.5\ Z_{\rm MW}$, and $\sim$10 in the SMC, where $Z = 0.2\ Z_{\rm MW}$. (For an up-to-date compilation of observationally determined WR wind mass-loss rates, see Yoon [2017].)

Apart from formation by mass-stripping in binaries, it is thought that, because O-stars themselves have strong radiation-driven stellar winds, (single) WR-stars can also be formed by self-stripping of the hydrogen envelope by wind mass loss in single O-stars, the so-called Conti scenario—see, for example, Meynet et al. (2011) and Shenar et al. (2020) and references therein; see also Section 8.1.7 where this scenario is discussed in some detail. Because the rate of self-stripping by radiation-driven stellar winds of massive stars is expected to be metal-dependent, and the WR-stars are descendants of supergiants, one might expect the ratios of the numbers of WR-stars and red supergiants in galaxies to be metal-dependent. This is, however, not observed (see Fig. 8.16).

In our Galaxy and the LMC, the number ratio of WR- to O-stars is 0.15, but in the metal-deficient SMC it is only 0.01. The deficiency of WR-stars in the SMC is often ascribed as being caused by the metal-dependence of the mass-loss rate of O-star winds. It is thought that at low metallicity, most O-stars do not lose enough mass to turn into WR-stars, which would explain the low WR incidence in the SMC. In addition, at lower metallicity, a higher luminosity is needed to produce a strong radiation-driven WR wind. Therefore, the lower mass limit for single helium stars to show up as WR-stars is higher at lower metallicity, as was recently established by Shenar et al. (2020). They found that the lower mass limits for helium stars for becoming WR-stars in the Galaxy, the LMC, and the SMC are $7.5\ M_\odot$, $11\ M_\odot$, and $17\ M_\odot$, respectively. This factor may have contributed considerably to the low incidence of WR-stars in the SMC.

Still, large WR populations are nevertheless inferred also in very metal-deficient galaxies such as IZw18 and SBS 0335−052E, where the presence of strong nebular

[1] https://pacrowther.staff.shef.ac.uk/WRcat/.

Table 5.5. The Class of ζ Aurigae Stars and Other Giants of Eclipses

Name	Spectral type	V (mag)	P_{orb} (d)	ecc	Eclipse (d)	(M_g, M_2) $(M_\odot)$	(R_g, R_2) $(R_\odot)$
ζ Aur[1,2]	K4Ib+B5V	3.8	972	0.38	37	(5.8, 4.8)	(154, ?)
31 Cyg[1,2]	K4Ib+B3-4	3.8	3,784	0.21	61	(11.7, 7.1)	(197, ?)
32 Cyg[1,2]	K5Ib+B6V	4.0	1,148	0.30	Grazing	(9.7, 4.8)	(175, ?)
VV Cep[1,3]	M2Iab+B0-2	4.9	7,430	0.35	233	(18.2, 18.6)	(1,050, ?)
22 Vul[1,2]	G7Ib+B8.5V	5.2	249	0.0	8	(–, –)	(77, ?)
HR 6902[1,2]	G9IIb+B8.5V	5.7	385	0.31	3.8	(–, 2.95)	(33, ?)
HR 2554[1,2]	G7II+A1V	4.4	195.3	0.0	Partial	(3.1, 2.0)	(31.3, ?)
τ Per[1,2]	G8III+A4V	4.0	1,561	0.73	Partial	(–, –)	(15.5, ?)
γ Per[1.2]	G8II-III+A1V	2.9	5,328	0.79	7.3	(–, –)	(24.1, 3.2)
HD 223971[1,2]	G7III+F2III	6.6	50.1	0.0	≤ 1	(–, –)	(6.8, 3.2)
ϵ Aur[1,4]	F0Iab+disk	2.98	9,896	0.23	720 Partial*	(–, –)	(160, 1,400)

Sources: [1]Griffin & Ake (2015). [2]Griffin et al. (2015). [3]Bennett & Bauer (2015). [4]Stencel (2015), see also work by Ake & Griffin (2015).

*This is ≥ 400 d of total eclipse.

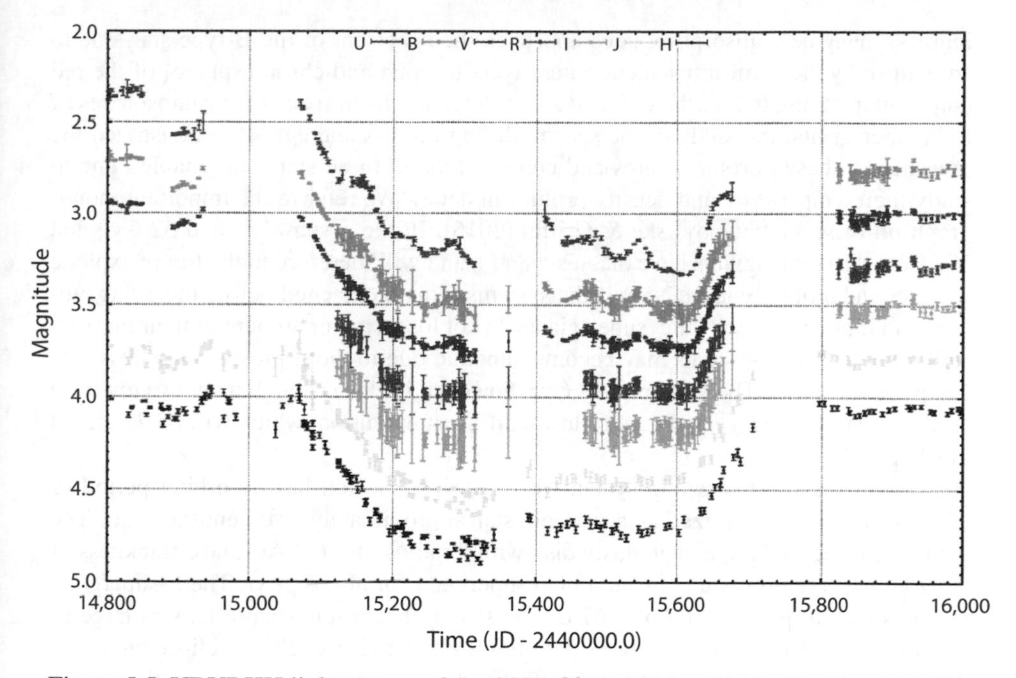

Figure 5.5. UBVRIJH light curves of the 2009–2011 eclipse of ϵ Aurigae (arranged in downward sequence). Reproduced with permission from Kloppenborg et al. (2012).

singly ionized helium (He II) ($\lambda = 4,686$ Å) is observed, which indirectly indicates the presence of hot weak-lined WN-stars, which emit much ionizing ultraviolet (UV) radiation (Crowther, 2007, and references therein). These WR-rich blue compact dwarf galaxies are even called WR galaxies (Vacca & Conti, 1992). These large WR populations might possibly be due to binary evolution, instead of resulting from wind mass loss, as binary evolution can easily remove the envelope of the most massive star of the binary. Indeed, the observed binary frequency among WR-stars in metal-poor and metal-rich galaxies is similar: ~40% (Crowther, 2007).

5.3.2 Wide-orbit Evolved Systems

These are the giant and supergiant Algol systems, the so-called ζ Aurigae systems: eclipsing binaries with periods of years to decades, in which one of the components is a red giant or supergiant star. Table 5.5 lists characteristic examples. With the exception of HD 223971 they are all naked-eye stars. The system of γ Persei is, after β Per (Algol), the second-brightest eclipsing binary of the Northern sky. Also ϵ and ζ Aurigae, 31 and 32 Cyg, and τ Per are bright naked-eye stars. These systems belong to the most marvellous types of eclipsing binaries, which are of interest to professional and amateur astronomers alike. In the ζ Aurigae systems, to which also 31 and 32 Cyg belong, the companion is a hot unevolved B-type main-sequence star. The long duration of the eclipses, of the order of months, allow one to carefully monitor the phases of ingress and egress, both spectroscopically and photometrically. Before and during ingress, many new absorption lines appear in the spectrum of the B-type star, due to an eclipse by the semi-transparent outer layers (corona and chromosphere) of the red (super)giant. Long before the UV- and X-ray detection from space of chromospheres of red (super)giants, the study of the spectra during ingress and egress demonstrated the presence of these chromospheres and coronae around these stars and enabled one to study their temperature and density profiles in detail. We refer to the important monograph on these systems by Ake & Griffin (2015). In the systems 31 and 32 Cyg and VV Cep, the (super)giants have masses $>8\ M_\odot$ and will therefore in the future explode as SNe, and probably leave NSs. If the systems remain detached before the SN explosion and if prior to the SN the (super)giants do not lose a major amount of their mass by stellar winds, these systems may become unbound due to the explosive mass loss (see Chapters 4 and 13). The system of 32 Cyg, however, is close enough to evolve toward a common envelope stage and may follow a different evolution, which will be described in Chapter 10.

One of the most puzzling systems is ϵ Aurigae, which has an orbital period of 27.1 yr and was recognized as a variable star approximately two centuries ago. The eclipsing object is here a huge dusty disk with a radius of ~6.4 AU and a thickness of 0.5 AU, which eclipses the visible F0Iab supergiant for almost 2 yr. The F-supergiant pulsates with a quasi-period of ~67 d. The system has a semi-major axis as large as that of the orbit of Neptune. Figure 5.5 shows its last (2009–2011) eclipse curves in different colors; see Kloppenborg et al. (2012). For a detailed discussion of the system and its possible evolutionary status, see Stencel (2015). At the center-time of the eclipse the light of the system temporarily increases. During the 1982–1984 and 2009–2011 eclipses, its brightness increased during mid-eclipse by ~0.25 magnitudes (~25%), indicating that the disk has a central hole, through which we see some light of the F-supergiant. The fact that no light of a central star of the disk is seen has led to the suggestion (Lissauer & Backman, 1984; Eggleton & Pringle, 1985) that the central object is a close binary consisting of two main-sequence B- or A-stars that, due to the mass-luminosity relation, will together emit only a few percent of the light of the F-supergiant. For example, with its mass function of $2.51\ M_\odot$ (Stencel, 2015), and assuming the F-star to have a mass of $20\ M_\odot$, the central object of the disk will have a mass of $14.4\ M_\odot$, and if it consists of two B-type main-sequence stars of $7.2\ M_\odot$, its luminosity will be only 6% of that of the F-supergiant. If the F-star has a mass of $30\ M_\odot$, then the central object has a mass of $18\ M_\odot$; if the central object consists of two stars of $9\ M_\odot$, then its total luminosity will be only 2.8% of that of the supergiant. Such masses of the F-star would mean that the system is <5–8 Myr old. On such a timescale, a protoplanetary disk around a young star may still survive. There are, however, also other possible models of this system, and they are described by Stencel (2015).

5.4 SYSTEMS WITH ONE OR TWO WHITE DWARFS

The nearest known WD is Sirius B, at 8.6 ly. It is the smaller component of the Sirius binary star, and has approximately the same mass as the Sun (see Fig. 2.1). The astronomical world was astonished when it was first recognized in 1917, based on its

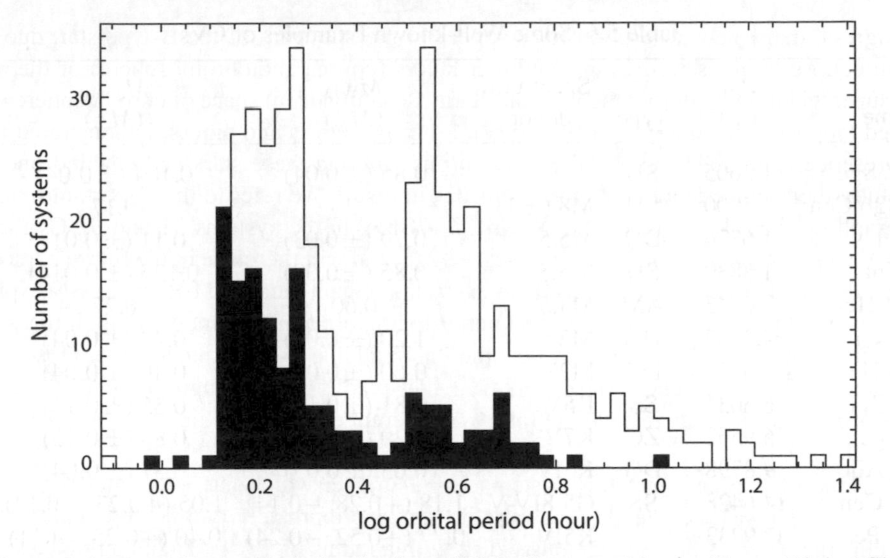

Figure 5.6. The orbital period distribution of CVs discovered in the Sloane Digital Sky Survey (SDSS black histogram, after Southworth et al., 2012), and of known non-SDSS CVs (white histogram, from the catalog of Ritter & Kolb, 2003). See also Gänsicke et al. (2009) and Figure 11.13.

faintness and relatively high effective temperature as determined in 1915 by Walter Adams on Mount Wilson, that its radius is of the order of that of the Earth, such that the mean mass density of Sirius B is $\geq 300\,\mathrm{Kg\,cm^{-3}}$. Furthermore, because a WD no longer undergoes fusion reactions it remained a mystery (at the time) how this star could avoid collapsing due to gravity. The solution of electron degeneracy pressure was only found after the introduction of quantum mechanics. In 1922, Willem Luyten coined the name WD, given the high temperature and compactness of these objects. In the following section, we describe WDs in binaries. Here one may also distinguish between close- and wide-orbit systems.

5.4.1 Close-orbit WD Systems

Cataclysmic Variables (Novae, Dwarf Novae)

In most cases, these systems consist of an unevolved low-mass main-sequence star (red dwarf) that fills its Roche lobe and a WD accretor. The main-sequence star is usually less massive than the WD and is transferring mass to the latter, thereby giving rise to explosive eruptions from the WD surface, that is, novae and dwarf novae (for reviews, see, e.g., Starrfield et al., 2016; Chomiuk et al., 2021). The discovery of novae and further observational characteristics are discussed in Section 2.5.

The WD in a CV binary is surrounded by an accretion disk that is the main light source in the system. The orbital periods are generally short, ranging mostly from approximately 80 min to 12 hr. The orbital period distribution shows a characteristic

Table 5.6. Some Well-known Examples of CVs

Name	P_{orb} (hr)	Type	Spectrum donor	M_{WD} ($M_\odot$)	M_{donor} ($M_\odot$)	Ref.
WZ Sge	1.3605	SU		0.85 ($\pm$0.04)	0.104 ($\pm$0.04)	1
V 436 Cen	1.5000	SU	M8 ($\pm$1)		0.17	2
EX Hya	1.6376	DQ	M5.5	0.79 ($\pm$0.03)	0.11 ($\pm$0.01)	2
Z Cha	1.7830	SU	M5.5	0.85 ($\pm$0.09)	0.125 ($\pm$0.014)	2
AM Her	3.0943	AM	M4.5	0.60	0.27	3
U Gem	4.2457	UG	M2	1.20 ($\pm$0.05)	0.42 ($\pm$0.04)	4
DQ Her	4.6469	DQ	M2V	0.60 ($\pm$0.07)	0.40 ($\pm$0.04)	5
SS Cyg	6.6031	SS	K5V	0.81 ($\pm$0.19)	0.55 ($\pm$0.15)	6
Z Cam	6.9562	ZC	K7V	1.20 ($\pm$0.20)	0.85 ($\pm$0.15)	
AE Aqr	9.8798	DQ	K5IV	0.63 ($\pm$0.05)	0.37 ($\pm$0.04)	7
BV Cen	14.6428	SS	G5-8IV-V	1.18 (+0.28, −0.14)	1.05 (+0.23, −0.14)	8
GK Per	47.9233		K5IV	0.77 (+0.52, −0.24)	0.40 (+0.25, −011)	5

Sources: [1]Steeghs et al. (2007). [2]Hamilton et al. (2011). [3]Hessman et al. (2000). [4]Echevarría et al. (2007). [5]Harrison et al. (2013). [6]Schreiber & Lasota (2007). [7]Echevarría et al. (2008). [8]Watson et al. (2007).

Note: DQ Her and GK Per were novae that erupted in 1934 and 1901, respectively. The other systems are dwarf novae with various characteristic types of outburst light curves: AM, AM Her-type; DQ, DQ Her-type; SS, SS Cyg-type; SU, SU UMa-type; UG, U Gem-type; ZC, Z Cam-type. AM and DQ systems contain magnetic WDs and are indicated as *polars* and *intermediate polars*, respectively.

"period gap" between 2 and 3 hr (see Fig. 5.6), where very few systems are found. Furthermore, there is a sharp lower cutoff of the orbital periods at ~80 min. Very few CVs are found below this cutoff. Table 5.6 lists the orbital parameters of a few well-known CV systems.[2] Unless more recent references are given, the data in the table are taken from the catalog given in footnote 2. For a comprehensive review on CV systems, see Knigge et al. (2011). In Chapter 11, we return to CVs and their period gap.

Pre-CVs and Double Nuclei of PNe

Pre-CVs are systems resembling the CV binaries but differing from those by the fact that the non-degenerate star is not (yet) filling its Roche lobe. When this star during its evolution expands or the orbital separation decreases, for example, by gravitational radiation (see Chapter 4), these systems may turn into CV binaries. This will in most cases take several billion years. Closely related to these systems are the double nuclei of planetary nebulae (PNe, see Fig. 5.7), which clearly are an earlier evolutionary phase of the pre-CVs. In the double nuclei of PNe, the compact stars are genuine PN nuclei: extremely hot degenerate stars of spectral type (sub-)dwarf-O, which after cooling will become WDs. Up to ~80% of the PN nuclei may be binaries (Jones, 2019). When

[2]A complete online catalog of CVs, updated until February 1, 2007, is available at https://archive.stsci.edu/prepds/cvcat/.

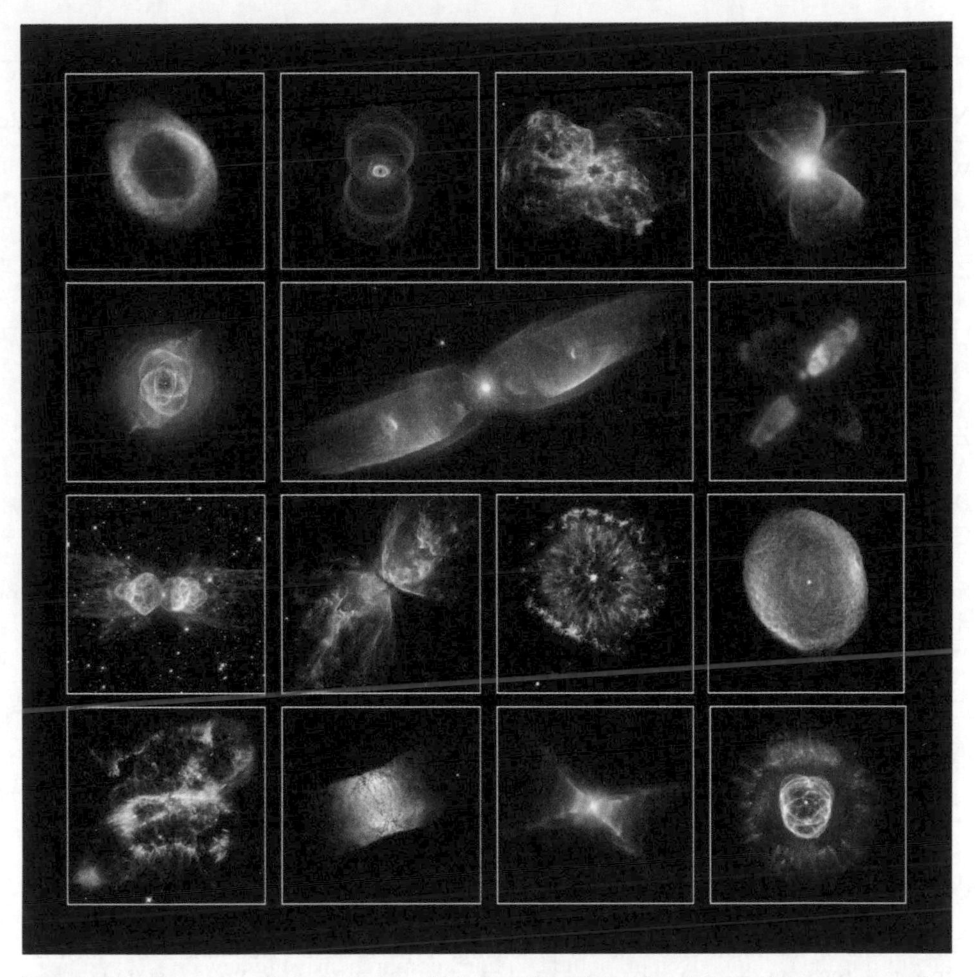

Figure 5.7. Peculiar shapes of most PNe are thought to be due to the presence of a companion star or planets around the hot central star of the PN. Credit: Hubble Space Telescope (NASA/ESA/STScI).

the systems are close, the one-sided heating of the main-sequence star by the hot PN nucleus produces a detectable periodic light variation of the main sequence star, which yields the orbital period (depicted in Fig. 5.8). Table 5.7 lists some examples of both these categories, taken from Ritter & Kolb (2003) and more recent references that are indicated in the table. Three of the pre-CVs belong to the Hyades open star cluster, indicating that several red giants in binaries in this cluster have entered CE evolution and have produced these very close systems (as explained in Chapters 4 and 11. The light of the A-giant in the system V651 Mon varies in a rather irregular way (Kato et al., 2001). The bloated appearance of several of the companions of the hot pre-WDs in PNe cannot be explained by heating alone. In addition to heating, the bloated appearance is thought to be due to recent mass accretion from the CE that was ejected when the two stars spiralled toward each other.

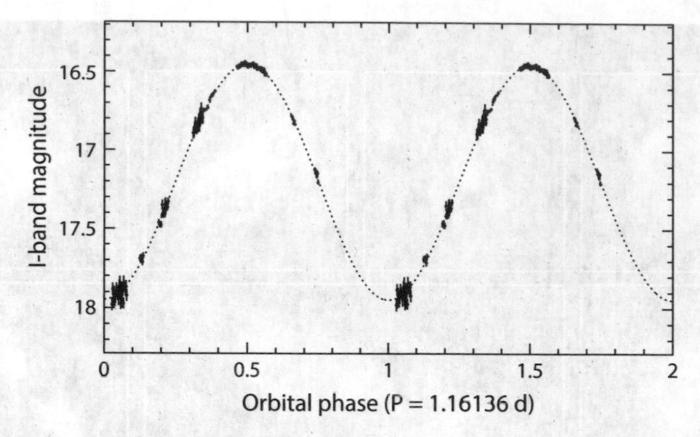

Figure 5.8. Infrared light curve of the central binary of the planetary nebula IPHASXJ94359.5+170901. The light variation is due to one-sided heating of the normal companion star by the hot central pre-WD that produced the nebula. After Corradi et al. (2011).

A special case is the double (pre-)WD nucleus of PN Henize 2−428 (discovered by Santander-García et al. [2015]) with a total mass greater than the Chandrasekhar limit; see Table 5.8. This system has an orbital period of 4.2 hr, causing it to merge by GW losses within 700 Myr.

Double WDs

DWDs are important for a variety of reasons. Close systems are strong sources of GWs that ultimately may be detected, for example, by the space-borne GW mission LISA (Nelemans et al., 2001; Amaro-Seoane et al., 2017; Amaro-Seoane & Andrews et al., 2022; see also Chapter 15). Furthermore, the mergers of close-orbit DWDs are thought to be one of the main channels for producing Type Ia SNe (Webbink 1984; Iben & Tutukov 1984; Maoz et al., 2014; see also Chapter 13). To merge within a Hubble time, systems should start out with orbital periods shorter than ~12 hr. Table 5.8 lists a few examples of close-orbit DWD binaries.

The discovery of DWDs goes back to the late 1980s (L870−2 with $P_{\rm orb} = 1.56$ d [Saffer et al., 1988]) and the first sources were initially dominated by low-mass ($\lesssim 0.4\, M_\odot$) He WDs that cannot be produced from single star evolution within a Hubble time. These sources were thus the targets for radial-velocity searches for binarity[3] (e.g., Bragaglia et al., 1990; Marsh et al., 1995). By the year 2000, ~180 WDs had been checked for radial-velocity variations yielding a sample of 18 DWDs with $P_{\rm orb} < 6.3$ d (Marsh, 2000). Later, more massive DWDs were found using the Sloan Digital Sky Survey (SDSS) and the ESO SPY survey (Napiwotzki et al., 2020, and references therein) which benefits from Gaia data.

[3] An interesting puzzle was therefore the discovery of isolated low-mass He WDs (Maxted & Marsh, 1998); see Nelemans & Tauris (1998) for a possible explanation.

Table 5.7. Examples of Pre-cataclysmic variables (top) and Double Nuclei of planetary nebulae

Name of pre-CV	P_{orb} (d)	Spectral type	M_{WD} ($M_\odot$)	$M_{companion}$ ($M_\odot$)	Remarks
SDSS J1205−0242[1]	0.049	WD+BD	0.39 ± 0.02	0.049 ± 0.006	
SDSS J1231+0041[1]	0.050	WD+?	0.56 ± 0.07	≤ 0.095	
NN Ser[2]	0.1301	WD+dMe	0.535 ± 0.012	0.111 ± 0.004	
PG 0308−	0.2891	DA4+M2-3V	—	—	
V471 Tau[2,3]	0.5212	DA2+K0V	0.80	0.79	Hyades
HZ 9[4,5]	0.5643	DA2+dM4.5e	0.56	0.20	Hyades
G UMa[4,5]	0.6675	DA4+dM2	≥ 0.40	0.36	Hyades
HW Vir[6]	0.1167	sdO+dM	0.50 ± 0.04	0.14 ± 0.01	
AA Dor[6]	0.2614	sdO+dM?	0.50 ± 0.03	0.086 ± 0.005	
Name of double nuclei	**P_{orb} (d)**	**Spectral type**	**$M_{subdwarf}$ ($M_\odot$)**	**$M_{companion}$ ($M_\odot$)**	**PN**
KV Vel[6]	0.3571	sdO+dM?	0.63 ± 0.03	0.23 ± 0.01	DS1
UU Sge[7]	0.4651	sdO+K5	—	—	Abell 63
V477 Lyr[7]	0.4717	SdO9+K	—	—	Abell 46
V664 Cas[8,9]	0.5817	sdO+F5-K5	0.6	0.4−1.1	HFG 1
No name[10]	1.1614	sdO+?	0.5−1.0	1.0?	IPHASXJ1943*
No name[8]	2.29	sdO+?	0.48−0.61	0.13−0.26	EC11575−1845
IN Com[9]	5.9(?)	sdO+G5III-IV	—	—	LoTr5
V651 Mon[11]	16.089	sdO+A0III	—	—	NGC 2346

Sources: [1]Parsons et al. (2017). [2]Parsons et al. (2010). [3]Kundra & Hric (2011). [4]Bleach et al. (2002b). [5]Bleach et al. (2002a). [6]Hilditch et al. (1996). [7]Kiss et al. (2000). [8]Exter et al. (2005). [9]Montez et al. (2010). [10]Corradi et al. (2011). [11]Kato et al. (2001). If no reference is given, the data are from Ritter & Kolb (2003).

*IPHASXJ194359.5+170901.

BD is an acronym for brown dwarf.

Table 5.8. Examples of DWDs with Short Orbital Periods

Name	$P_{\rm orb}$ (min)	M_1 ($M_\odot$)	M_2 ($M_\odot$)	Ref.
ZTF J1539+5027	6.9	0.210 ± 0.01	0.610 ± 0.02	1
J0651+2844	12.8	0.247 ± 0.01	0.49 ± 0.02	2
SDSS J093506.92+441107.0	19.9	0.312 ± 0.019	0.75 ± 0.24	3
SDSS J163030.58+423305.7	39.8	0.298 ± 0.019	0.76 ± 0.24	3
SDSS J092345.59+302805.0	64.7	0.275 ± 0.015	0.76 ± 0.23	3
CD −30° 11223	71.5	0.54 ± 0.02	0.79 ± 0.01	3
CSS 41177	167.1	0.38 ± 0.02	0.32 ± 0.01	4
2331+290	239.7	~ 0.39	>0.32	5
Henize 2−428	253.2	~ 0.88	~ 0.88	6
L870−2	2240.3	~ 0.35	~ 0.39	7

Sources: [1]Burdge et al. (2019). [2]Hermes et al. (2012). [3]Kupfer et al. (2018). [4]Bours et al. (2014). [5]Marsh et al. (1995). [6]Santander-García et al. (2015). [7]Saffer et al. (1988).

These examples include the double (pre-)WD nucleus of PN Henize 2−428, the hot subdwarf system CD −30° 11223, and the first discovered DWD: L870−2 with $P_{\rm orb} = 1.56$ d.

Since 2010, many more DWDs were discovered with increasingly sophisticated instrumentation and, in particular, resulting from the successful Extremely Low-Mass WD survey (Brown et al., 2010, 2020) that uses color-color diagrams to target a part of the parameter space that is occupied by (subdwarf) B-stars, as well as very low mass ($\lesssim 0.2\,M_\odot$) proto-WDs that are still approaching the cooling track and are thus relatively large and bright (Istrate et al., 2014, 2016).

Since 2017, observational campaigns using the Zwicky Transient Facility (ZTF) have discovered several cases of DWDs with orbital periods <1 hr. Three of the new eclipsing DWDs discovered by ZTF (e.g., Burdge et al., 2020) are ZTF J1539+5027 ($P_{\rm orb} = 6.91$ min; see Fig. 5.9), ZTF J2243+5242 ($P_{\rm orb} = 8.80$ min), and ZTF J0538 +1953 ($P_{\rm orb} = 14.4$ min).

Apart from these close-orbit detached DWDs, there are two other kinds of *interacting* WD binaries with extremely short orbital periods: (1) the AM Canum Venaticorum (AM CVn) stars, which are semi-detached binaries consisting of a Roche-lobe filling donor (at least partly degenerate) and a more massive WD—see further description in next section; and (2) the UCXBs, which consist of a Roche-lobe filling WD or a partly degenerate star and a NS accretor. The latter systems, which have typical orbital periods between 11 and 50 min (Heinke et al., 2013), will be described in Chapter 11.

AM CVn Systems

AM CVn stars consist of a WD accreting from a hydrogen-deficient star (or WD) companion (Warner, 1995; Nelemans et al., 2001; Solheim, 2010; Levitan et al., 2015). Their formation history involves at least one phase of CE evolution (cf. Fig. 15.36).

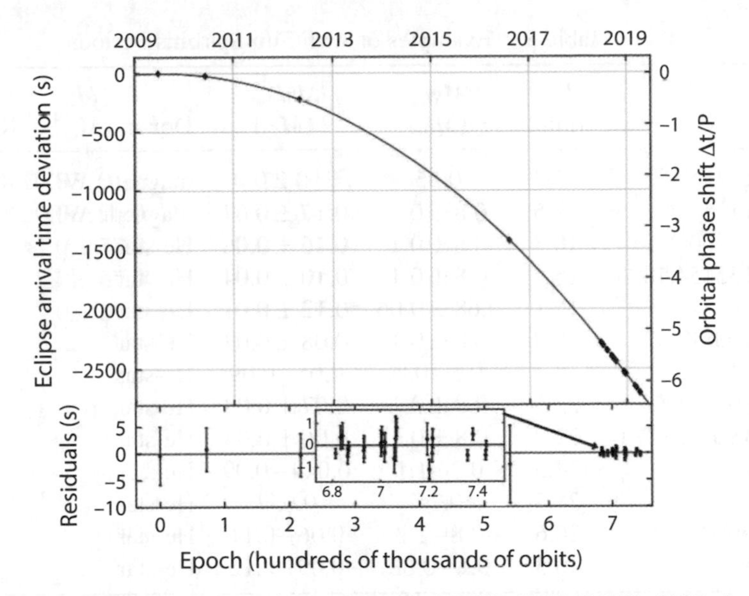

Figure 5.9. Orbital decay of the 6.91 min period DWD ZTF J1539+5027, due to GW radiation, as a function of time. Plotted is a second-order polynomial fit to the deviation of the measured eclipse times as a function of time, compared to a system with constant orbital period. The orbital decay rate inferred is consistent with that expected from general relativity. The relative velocity of the two WD components is $1,180\,\mathrm{km\,s^{-1}}$. After Burdge et al. (2019).

The active mass transfer in current AM CVns is initiated due to orbital damping caused by GW radiation at orbital periods of typically $5-20$ min (depending on the nature and the temperature of the companion star). The mass-transfer rate is determined by a competition between orbital angular momentum loss through emission of GWs and orbital widening due to RLO from the less massive donor star to the more massive WD accretor. Some examples of AM CVn systems are listed in Table 5.9.

There are currently $\geq$70 AM CVn binaries known in the Milky Way (Ramsay et al., 2018; van Roestel et al., 2021), and yet their formation pathways are still a puzzle (Green et al., 2018). Their compact orbits and the lack of hydrogen in their spectra, led to three different proposed formation channels (Solheim, 2010, and references therein): (1) the donor is a low-mass He WD; (2) the donor is a semi-degenerate hydrogen-stripped, helium-burning star (e.g., main-sequence helium star, or hot subdwarf); or (3) the donor is a helium-rich core of a low-mass main-sequence star that has not undergone helium-burning.

The best way to fully constrain binary parameters and the nature of the donor stars is to study eclipsing AM CVns (Burdge et al., 2020). Recent results from such systems revealed that the donor stars are likely larger and more massive than previously was assumed (Copperwheat et al., 2011; Ramsay et al., 2018), implying a semi-degenerate donor is looking more likely for such systems, unless the donor star is a low-mass He WD that can remain bloated on a Gyr timescale (Istrate et al., 2014).

Table 5.9. Examples of AM CVn Systems

Name	P_{orb} (min)	M_{WD} ($M_\odot$)	M_{donor} ($M_\odot$)	Donor	Remarks
HM Cnc[1]	5.4	0.55	0.27	magnetic WD	X-rays
V407 Vul[2]	9.5	0.8 ± 0.1	0.17 ± 0.07	magnetic WD	X-rays
ES Cet	10.3	0.8 ± 0.1	0.16 ± 0.06	He-star	
SDSS J135154.46	15.7	0.8 ± 0.1	0.10 ± 0.04	He-star	
AM CVn	17.1	0.68 ± 0.06	0.12 ± 0.01	He-star	
SDSS J190817.07	18.0	0.8 ± 0.1	0.08 ± 0.03	He-star	
HP Lib	18.4	$0.5-0.8$	$0.05-0.09$	He-star	
PTF J191905.19	22.5	0.8 ± 0.1	0.07 ± 0.03	He-star	
CXOGBS J175107.6	22.9	0.8 ± 0.1	0.06 ± 0.03	He-star	
CR Boo	24.5	$0.7-1.1$	$0.044-0.09$	He-star	
KL Dra	25.5	0.76	0.057	He-star	
V803 Cen	26.6	$0.8-1.2$	$0.06-0.11$	He-star	
GP Com	46.5	$0.5-0.68$	$0.09-0.12$	He-star	

Sources: [1]Roelofs et al. (2010). [2]Cropper et al. (1998). Most data in the table are from Kupfer et al. (2018) and references therein; see also Solheim (2010).
The helium-star companions may be degenerate (i.e., He WDs) or partly degenerate.

The exact number density of AM CVns in the Milky Way is still uncertain—mainly because there are open questions regarding their previous CE phase and the stability of their current RLO to a WD accretor. Future transient sky surveys, such as using the Vera C. Rubin Observatory, would likely have improved success in detecting short-period AM CVn systems with the implementation of appropriate very short cadence intervals.

If the system survives the onset of the semi-detached phase, a stable accreting AM CVn binary is formed in which the orbital separation widens shortly after onset of RLO, and the system evolves to longer orbital periods (see Figs. 15.25 and 15.35) on a Gyr timescale. For this reason, AM CVn sources are prime candidates for being bright sources in GWs, and thus there are high expectations for LISA (Chapter 15) to detect many of these sources. When they reach an orbital period of $\sim$60 min (after a few Gyr), the donor star has been stripped down to $\sim$5 jupiter masses (Tauris, 2018). These systems have been hypothesised to be possible progenitors of faint thermonuclear explosions (flashes, or Type .Ia SNe [Nelemans et al., 2001; Bildsten et al., 2007]).

Magnetic WDs

The origin of magnetic fields in WDs remains an unsolved problem in stellar astrophysics. One reason for this is that the very different fractions of strong-magnetic-field WDs (exceeding 1 MG) in close binaries cannot easily be reproduced by current knowledge of their intimate evolutionary links. While strong-magnetic-field WDs are very rare ($<2\%$) among detached binary WDs that are younger than $\sim$1 Gyr (Parsons

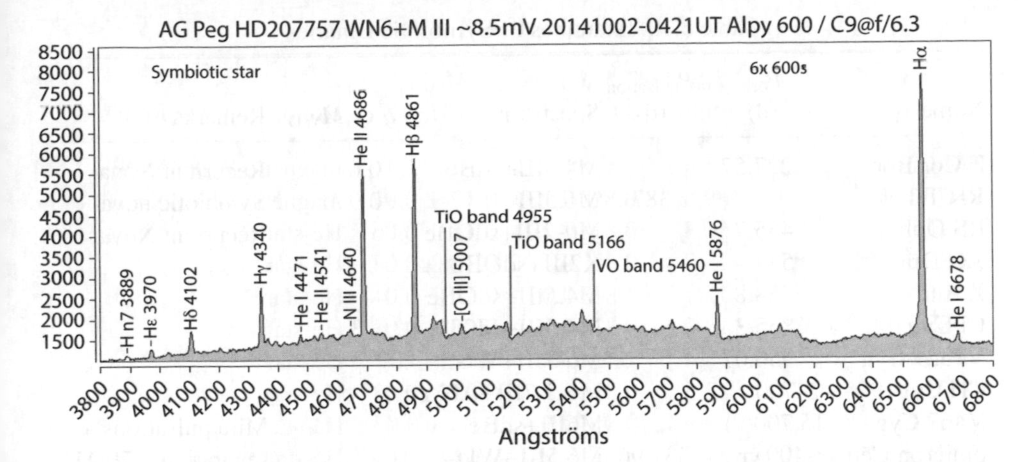

Figure 5.10. The spectrum of the symbiotic star AG Peg is characteristic for the emission-line spectra observed in symbiotic stars. The spectrum resembles that of a WR-star of type WN6, but it is due to the accreting WD in the system. Note, some old novae and some nuclei of PNe have spectra that mimic the emission spectra of massive WR-stars. After a figure from Jim Ferreira, Livermore, CA.

et al., 2021), they represent almost 1/3 of the WDs in semi-detached CVs in which the WD accretes from a low-mass star companion (Ferrario et al., 2015; Pala et al., 2020). Schreiber et al. (2021) recently presented new binary star models and demonstrated that a crystallization- and rotation-driven dynamo similar to those working in planets and low-mass stars can generate strong magnetic fields in the WDs in CVs, which may explain their large fraction among the observed population. Furthermore, they argued that when the magnetic field generated in the WDs connects with that of the secondary stars, synchronization torques and reduced angular momentum loss cause the binary to detach for a relatively short period of time, thereby explaining the few known strong-magnetic-field WDs in detached binaries, such as AR Sco, as we will discuss now.

A New Class of Radio-pulsating WD Binary Star

In 2016, Marsh et al. (2016) discovered a peculiar WD. The star, AR Scorpii (AR Sco), was classified in the early 1970s as a δ-Scuti star, a common variety of a periodic variable A-type star. The observations of Marsh et al. (2016) revealed, however, a 3.56 hr orbital period close binary, pulsing in brightness on a periodicity of 1.97 min. The latter period increases with time and is interpreted as the spin period of a magnetic WD, spinning down due to magnetodipole radiation (Chapter 14). Hence, it appears that AR Sco is a *WD pulsar!* There are no signs of accretion in the system, and although the pulsations are driven by the WD's spin, they probably originate from the cool star companion via some interaction with the magnetosphere of the WD. It will be interesting to see whether further such WD pulsars will be discovered in the near future.

Table 5.10. Examples of Symbiotic Binaries

Name	P_{orb} (d)	$P_{pulsation}$ (d)	Spectrum	q or M_{WD}	Remarks
T Cor Bor	227.57		M4.5III+sdBe	0.6	Recurrent Nova
RR Tel		387	M6.5III	0.9 $M_{\odot}$	Symbiotic nova
RS Oph	455.7		M0-2III+sdOBe		Recurrent Nova
AG Dra	544		K2III+sdOBe		
Z And	758.8		M4.5III+sdOBe		
CI Cyg	855.3		M4.5III+sdOBe	3	
V1329 Cyg	950.07		M6-7III+WN5		
CH Cyg	5,700		M6.5-7III+A0Veq		
V467 Cyg	15,700(?)		M6III+sdBe		Mira pulsations
omicron Ceti (Mira)	~400 yr	331.96	M6.5III+WD		separation ~70 AU

Sources: After Belczyński et al. (2000).

5.4.2 Wide-orbit WD Systems

Here we briefly discuss symbiotic binaries and barium stars and other chemically peculiar binaries.

Symbiotic Binaries

Symbiotic stars are recognized by their spectra that are a combination of the spectrum of a cool giant and a hot blue star with a strong emission-line spectrum (see Fig. 5.10). They are binaries that consist of a red-giant donor star, probably with mass $\leq 3\, M_{\odot}$, and, in most cases, a hot blue WD-like accreting companion star. Examples of symbiotic binaries are listed in Table 5.10. In some cases, the giant is an AGB star; also, in some cases, the blue companion is a main-sequence star accretor. The hot temperature of the companion is due to the accretion of matter from the wind of the giant or by RLO from this star or by a mixture of these two processes. The winds from the giants are slow, with velocities of a few tens of km s^{-1}, and the wind mass-loss rates are of the order of $10^{-6}\, M_{\odot}\, \mathrm{yr}^{-1}$, leading to accretion rates on the companion of $\sim 10^{-7}\, M_{\odot}\, \mathrm{yr}^{-1}$. For RLO, accretion rates (or mass-transfer rates) anywhere between 10^{-7} and $10^{-3}\, M_{\odot}\, \mathrm{yr}^{-1}$ can be reached. The red giants typically have luminosities in the range 100 to 1,000 $L_{\odot}$. Reviews of the properties of symbiotic stars can be found, for example, in Kenyon (1986), Belczyński et al. (2000), and Sokoloski et al. (2016).

The fact that the WDs in symbiotic binaries show a hot subdwarf O-(sdO) star or subdwarf B (sdB) star spectrum does not necessarily imply that they are young objects that were formed only recently, like the nuclei of PNe. Rather, their hot appearance is due to the accretion of the matter from the envelopes of their giant companions, which causes the WDs to become very hot and luminous. These stars, as well as the chromospheres of the red giants, can be hot enough to become X-ray sources, as can be

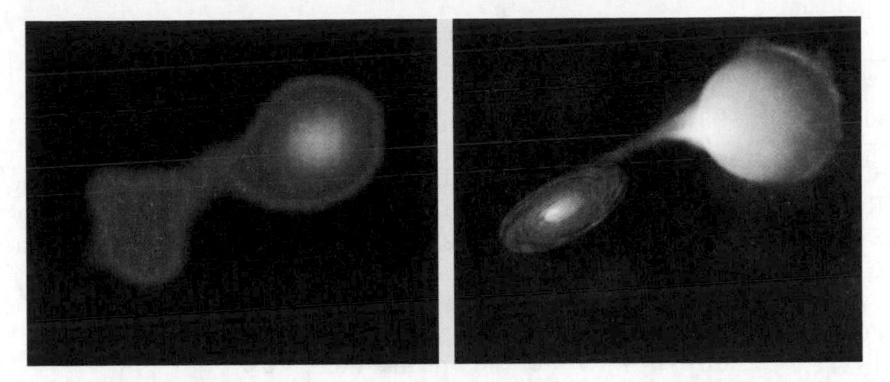

Figure 5.11. (Left) Chandra X-ray image of the symbiotic star Mira Ceti. (Right) Artist's impression of the Mira Ceti system: the accreting WD-star, at a distance of 70 AU, is surrounded by a hot X-ray emitting accretion disk. Credit: NASA/CXC/SAO/M. Karovska et al./M. Weiss.

seen in the case of Mira (o Ceti; see Fig. 5.11). Note, in some sense the ζ Aurigae stars and VV Cep systems might also be considered an extension of the group of symbiotic stars to the massive stars (see Table 5.5). However, in the usual terminology, they are considered a separate class, because they never contain a WD component.

Chemically Polluted Binaries: Barium Stars, CH-Stars, and Carbon-enhanced Metal-poor (CEMP) Stars

As mentioned in Chapter 2, barium stars are G- and K-giants, whose spectra indicate an overabundance of s-process elements (elements produced by slow neutron captures) by the presence of the $\lambda = 455.4$ nm line of singly ionized barium (Ba II). Barium stars also show enhanced bands of the molecules CH, CN, and C_2, indicating an enhanced carbon abundance. It has been found that all barium stars are in wide binaries, with orbital periods between 100 and 10,000 d, and often with considerable eccentricity (e.g., Jorissen et al., 1998). The enhanced s-process elements cannot have been produced in the interiors of the present G-, and K-giants in these systems, as they are in a too early stage of evolution for this. The only way to understand their enhanced abundances is that when these stars were still on the main sequence, their companion stars were thermally pulsating giant stars (AGB stars), which produced the s-process elements (for ways in which this element production in AGB stars takes place, see, e.g., Busso et al. [1999]). By the thermal pulses, enhanced by a stellar wind, they transferred the s-enriched matter from their envelopes to the main-sequence companion stars, enriching them with these elements. The AGB companions subsequently terminated their evolution as WDs. When, many tens of millions of years later, the s-process-enriched main-sequence stars left the main sequence and evolved into G- and K-giants, they became the presently observed barium stars. A companion of a barium star has never been seen, which is consistent with them being WDs, and this model, originally proposed by Boffin & Jorissen (1988), has now been generally accepted. For reviews, see Izzard

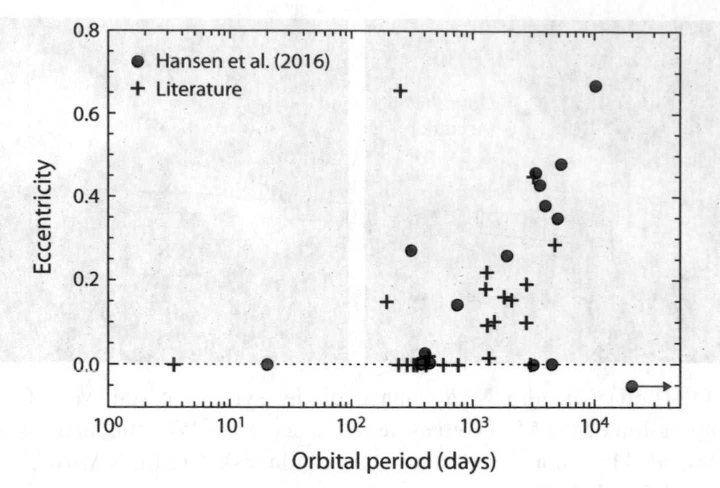

Figure 5.12. The orbital eccentricity vs. orbital period relation for barium stars and CEMP stars. After Abate et al. (2018) and Hansen et al. (2016).

et al. (2010) and more recently Abate et al. (2018). Two other classes of chemically-enriched giant stars—the CH stars and the carbon-enriched metal-poor (CEMP) stars—have been found to be 100% in binaries. The CH stars have typical metallicities of $0.1\ Z_{\odot}$, and the CEMP stars, which are in the halo of the Galaxy, have $Z = 0.01\ Z_{\odot}$. While the barium stars make up $\sim$1% of the giants in the Galactic disk, the CH stars make up 2% of the giants with $Z = 0.1\ Z_{\odot}$, and the CEMP stars make up 6 to 20% of the halo giants (Pols, private communication, 2018). The relatively high eccentricities of all these types of binaries (Fig. 5.12) are quite remarkable (see, e.g., Karakas et al., 2000; Izzard et al., 2010; Abate et al., 2018).

In wide low-mass binaries with a red giant or AGB star, the effects of wind mass transfer and torques on the orbit due to this transfer may become large (e.g., Jahanara et al., 2005; Saladino et al., 2018, 2019; Saladino & Pols, 2019) and may lead to drastic orbital changes, possibly leading to these eccentricities.

5.4.3 Luminous Super-soft Binary X-ray Sources

This is a special class of WD binaries that can have close as well as wide orbits. For reviews we refer to Kahabka & van den Heuvel (1997), Kahabka & van den Heuvel (2006), and Kato (2010). The supersoft sources (SSS) were recognized with the ROSAT X-ray Observatory, by Trümper et al. (1991), as an important new class of intrinsically bright X-ray sources (see also Greiner et al., 1991). In fact, four of them had already been found in the Magellanic Clouds by the Einstein Observatory circa 1980, but they had not been recognized as a separate new class (Long et al., 1981; Seward & Mitchell, 1981). A careful analysis of the ROSAT data on the first LMC sources showed that while the X-ray luminosities can be as high as the Eddington limit (they range from approximately 10^{36} to 10^{38} erg s^{-1}), their X-ray spectra are extremely soft, peaking in the range of 20 to 100 eV, corresponding to black-body temperatures of 10^5 to 10^6 K.

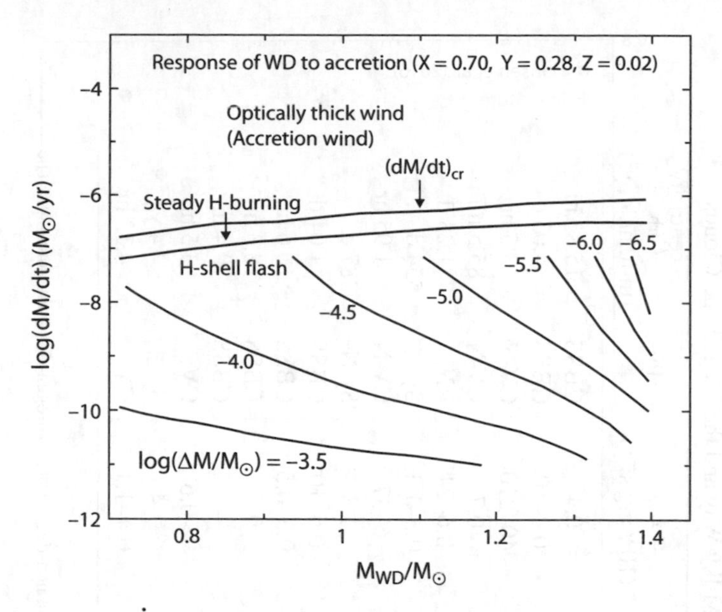

Figure 5.13. $M_{\rm WD} - \dot{M}_{\rm acc}$ plane, with regimes of optically thick winds (if the WD is single, a red-giant envelope forms here), steady surface nuclear burning and nova-like flashes. The ΔM values for the nova-like flashes indicate envelope masses, for a given accretion rate, at which burning is ignited. After Hachisu & Kato (2001); adapted from Nomoto (1982).

This is approximately two orders of magnitude lower than for a classical X-ray binary that contains an accreting NS or BH (see Chapter 6). Approximately 40 SSS have been discovered by ROSAT, 16 in the Andromeda Galaxy (M31), approximately a dozen in the Magellanic Clouds, 10 in our own Galaxy, and one in NGC 55. Since then, several dozen SSS have been discovered with the X-ray observatories BeppoSAX, Chandra, and XMM (X-ray Multi-Mirror Mission)-Newton, mostly in external galaxies. A catalog of SSS is given by Greiner (2000a). In view of the very large interstellar extinction of soft X-rays due to neutral hydrogen, the sources in other galaxies can only be observed when they are near the outer edge of the interstellar hydrogen layer, at the outside of a Galaxy. Taking this into account, the total number of sources in M31 and our Galaxy is estimated to be approximately two orders of magnitude larger than the observed numbers, that is, of the order of 10^3 (Rappaport et al., 1994). The SSS therefore form a major population of highly luminous X-ray sources in spiral galaxies like our own. The same holds for irregular galaxies like the Magellanic Clouds. From the observed luminosities and black-body temperatures of SSS, one infers effective stellar radii comparable to those of WDs. Their observed characteristics are consistent with those of WDs that are steadily or cyclically burning hydrogen-rich matter accreted onto the surface at a rate of the order of $10^{-7}\,M_\odot\,{\rm yr}^{-1}$ (van den Heuvel et al., 1992). When WDs accrete hydrogen at this rate, steady nuclear burning of the accreted matter ensues (see, e.g., Nomoto, 1982; Nomoto et al., 1984, and references therein). Figure 5.13 depicts the different surface hydrogen-burning regimes of WDs

Table 5.11. Examples of Optically Identified Binary SSS in the Milky Way and the Magellanic Clouds

Galaxy	Name	Alias	V_{mag}	T_{bb} (eV)	L_X (10^{37} erg s^{-1})	Type	P_{orb} (min, hr, d, yr)
MW	RX J0019.8+2156	QR And	12.4	25–37	0.4	CBSS	15.85 hr
	RX J0925.7−4758	MR Vel	17.1	75–94	0.2–10	CBSS	4.03 d
	V382 Vel	N Vel 1999	2.8–8.0	34–48	40–210	CV-N	3.5 hr
	GQ Mus	N Mus 1983	18	38–43	0.7	CV-N	85.5 min
	U Sco	BD−17 4554	8–19	74–76	0.6–24	RN	1.23 d
	AG Dra	BD+67 922	8.3–9.8	10–15	0.3–1.1	Sy	554 d
	V1974 Cyg	N Cyg 1992	4.4–17	34–51	6–37	CV-N	1.95 hr
	RR Tel	N Tel 1948	6.7–11	12	1.3	Sy N	387 d
LMC	CAL 83	LHG 83	16.2–17.1	28–50	0.6–10	CBSS	1.04 d
	CAL 87	LHG 87	19–21	55–76	0.3–0.5	CBSS	10.6 hr
	RX J0513.9−6951	HV 5682	16.5–17.5	52	9.5	CBSS	18.3 hr
	RX J0527.8−6954	HV 2554	—	18	—	CBSS	9.4 hr
	RX J0537.7−7034	—	19.7	18–30	>0.6	CV	3.5 hr
SMC	RX J0048.4−7332	SMC 3	15.0	30–35	≥ 4	Sy N	~4.4 yr
	1E 0035.4−7230	SMC 13	20.1–21.5	27–48	0.4–1.1	CV	4.13 hr

Source: After Kahabka & van den Heuvel (2006).

The examples consist of a WD with steady surface hydrogen-burning, together with a non-degenerate hydrogen-rich companion star. The different types are defined in the text.

CV-N, cataclysmic variable nova; Sy N, symbiotic system nova.

as a function of $M_{\rm WD}$ and $\dot{M}_{\rm acc}$ (see also Kato, 2010). In the optically thick wind regime hydrogen is also burning steadily. Below the steady hydrogen-burning band, the accreted hydrogen burns in flashes. This is what happens in novae. The ΔM values indicate envelope masses (for a given accretion rate) at which burning is ignited. The accretion rates required for steady burning can be supplied by mass transfer on a thermal timescale from a close companion that is more massive than the WD accretor, typically $1.0-3.0\,M_\odot$ (van den Heuvel et al., 1992). These are the so-called close-binary super-soft sources (CBSS). Alternatively, the donor star may also be low-mass red (sub-)giant, these are the symbiotic systems (Sy), among which there are symbiotic recurrent novae (RNe) such as RS Oph and T CorBor. Steady burning can also occur in a postnova phase (these are the CV-type SSS), but only for relatively short timescales, up to decades (see references in Kahabka & van den Heuvel, 2006). All these systems are binaries, but also some single SSS have been observed, presumably highly evolved stars on their way to the WD stage, such as PG 1159 stars and the nuclei of some PNe (see Reinsch et al., 2002). It has been suggested, based on similarities in optical variability, that a number of unusual CVs, such as V Sge, T Pyx and WX Cen, also belong to the SSS class (e.g., Greiner, 2000b). Table 5.11 lists characteristic examples of binary SSS.

EXERCISES

Exercise 5.1.

a. *Use data in Table 5.3 to calculate the Roche-lobe radius of each of the two stars in β Per (Algol) and TX UMa. State which star fills its Roche lobe.*

b. *Both systems are eclipsing binaries, which means that their orbital inclination is not too far from $i \sim 90°$. Consider now β Per (Algol). As in most semi-detached systems, the orbit is circularized ($e = 0$) due to strong tidal forces. Estimate the mass function of the binary from the given stellar masses and compare it to that obtained from the derived orbital velocity of star 1 (assuming again $i \sim 90°$; see Section 3.3).*

Chapter Six

Observed Binaries with Accreting Neutron Stars and Black Holes: X-ray Binaries

In this chapter, we continue the review of interacting binaries and now consider the observed properties of binaries containing accreting NSs or BHs. These systems are the X-ray binaries. Other systems containing NSs and BHs are the binary radio pulsars, BH+NS binaries, and double BHs. Chapter 14 is devoted to the binary radio pulsars. The known properties of double BHs (and also BH+NS and NS+NS systems) derived from their merger events are reviewed in Chapter 15 on GWs.

6.1 DISCOVERY OF THE NS AND BH CHARACTER OF BRIGHT GALACTIC X-RAY SOURCES

The approximately 200 brightest Galactic point X-ray sources with fluxes $> 10^{-10}$ erg $cm^{-2}\,s^{-1}$ in the energy range of 1–10 keV above the Earth's atmosphere show a clear concentration toward the Galactic center and toward the Galactic plane (Figs. 2.7 and 6.1), indicating that they mostly belong to our Galaxy. A few strong sources are also found in the two nearby satellite systems of our Galaxy, the LMC and SMC. Fainter point sources have also been discovered in other nearby galaxies, such as M31 (the Andromeda Galaxy), showing distributions in these galaxies similar to the distribution of the strong point sources in our Galaxy, indicating that they are of a similar nature to the sources in our Galaxy (e.g., Fabbiano, 2006). For the strong sources around the Galactic center and in the LMC and SMC a fair estimate of the distance can be given as $\sim$8.5 kpc and $\sim$60 kpc, respectively. Together with the X-ray fluxes measured on Earth, this yields typical X-ray luminosities of $10^{34} - 10^{38}$ erg s^{-1}, that is, between 2.5 and 2.5×10^4 times the total energy output of the Sun, concentrated in the X-ray part of the electromagnetic spectrum. As mentioned in Chapter 2, also about a dozen bright sources are found in globular clusters (GCs) in our Galaxy (Fig. 6.1), for which the distances are known. These yield source luminosities of typically $10^{35} - 10^{38}$ erg s^{-1}. The same holds for the about two dozen sources in the GCs of M31 (Fabbiano, 2006).

Zeldovich & Guseynov (1966) were the first to suggest that the strong Galactic point X-ray sources that had been discovered since 1962 (Giacconi et al., 1962), with instruments carried above the atmosphere by rockets or to great heights by stratosphere balloons, are accreting NSs in binary systems. Slightly earlier, it had been pointed out by Salpeter (1964) and Zel'dovich (1964) that when gas is accreted by a BH, it is heated so much that a large part of its loss of gravitational potential energy can be

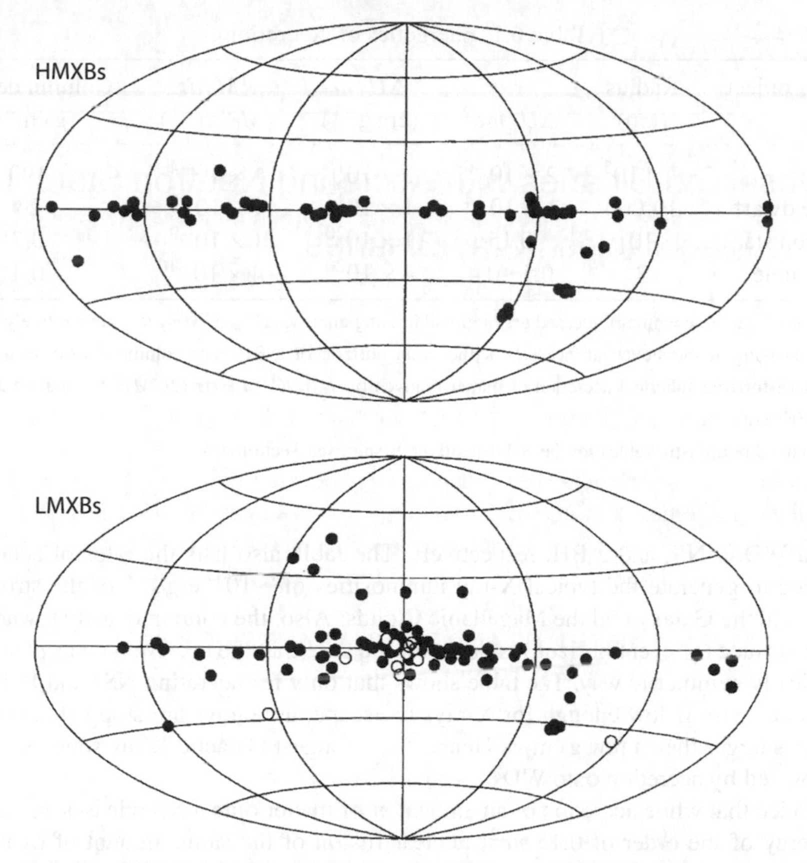

Figure 6.1. Sky distribution in Galactic coordinates of HMXBs (upper panel) and LMXBs (lower panel) derived by van Paradijs (1998) on the basis of his catalog of X-ray binaries (van Paradijs, 1995). The 27 LMXBs within 2° from the Galactic center, known at the time this figure was made, were not included to avoid congestion. The horizontal axis is along the Galactic plane, and the center of the figure is the Galactic center. Among the LMXBs, the globular cluster X-ray sources are indicated by open circles. Notice the strong concentration of the HMXBs to the Galactic plane, characteristic for a very young stellar population. On the other hand, the LMXBs are distributed much wider around the Galactic plane and strongly concentrated toward the Galactic center, as is characteristic for a much older stellar population. See also Figure 2.7 and van Paradijs (1998).

emitted in the form of electromagnetic radiation before the gas disappears into the hole. But confirmation of these ideas had to wait until the launch of the first X-ray satellite, Uhuru, in 1971 by the team of Riccardo Giacconi.

Typically an amount of matter, m, accreted on a NS or a BH can in this way generate an amount of radiation in the range of 0.1–0.4 mc^2 (see Section 4.9). Table 6.1 lists the amounts of gravitational potential energy released per unit mass by accretion onto the

Table 6.1. Energetics of Accretion

Stellar object $1\,M_\odot$	Radius (km)	$\Delta U/mc^2$	$\Delta U/m$ (erg g^{-1})	dM/dt ($M_\odot$ yr^{-1})	Column density (g cm^{-2})
Sun	7×10^5	2×10^{-6}	2×10^{15}	1×10^{-4}	190
White dwarf	10,000	1×10^{-4}	1×10^{17}	1×10^{-6}	23
Neutron star	10	0.15	1×10^{20}	1×10^{-9}	0.7
Black hole*	3	0.1~0.4	4×10^{20}	4×10^{-10}	0.4

The approximate (Newtonian) released gravitational binding energy, $\Delta U \equiv GMm/R$, refers to freely falling matter, arriving at the stellar surface (R) or the event horizon of a BH. The column density to the stellar surface for free spherical accretion of matter is given by $N = R^{1/2}L_X/[\pi(2GM)^{3/2}]$, and we assume $L_X = 10^{37}$ erg s^{-1}.

*The general relativistic values for the BH depend on its spin, see Section 4.9.

Sun, a WD, a NS, and a BH, respectively. The table also lists the rates of accretion required to generate the typical X-ray luminosities of $\sim10^{37}$ erg s^{-1} of the strongest sources in the Galaxy and the Magellanic Clouds. Also, the column density toward the stellar surface (or event horizon) is listed for the case that the accretion takes place in a spherically symmetric way. The table shows that only for accreting NSs and BHs, the column density is low enough for X-rays to escape, as X-rays are stopped at column densities larger than a few g cm^{-2}. Hence, the strongest Galactic X-ray sources cannot be powered by accretion onto WDs.

Notice that while accretion of an amount m of matter onto a NS releases an amount of energy of the order of $0.15\,mc^2$; nuclear fusion of the same amount of hydrogen releases only $0.007\,mc^2$. So, the simple process of accretion of matter onto a NS is energetically $\sim$20 times more efficient than hydrogen fusion! In the case of a BH, between 0.04 and as much as $0.42\,mc^2$ (depending on its spin, see Table 7.2) can be released before the matter disappears into the hole. Apart from pure matter-antimatter annihilation, accretion of matter onto a NS or a BH is the most efficient energy generation process known in the Universe. One also notices that accretion of matter onto a WD releases $\sim$1,000 times less energy than onto a NS, and therefore $\sim$60 times less energy than hydrogen fusion. Hydrogen fusion on the surface of a WD is therefore energetically much more efficient than energy release by accretion; it powers the X-ray emission of the so-called super-soft X-ray sources (van den Heuvel et al., 1992); see Chapter 5.

The accreting NS model was nicely confirmed in 1971 with the discovery (Schreier et al., 1972) that the source Cen X-3 is regularly pulsing with a period of 4.84 sec (thus it is a NS) and is a member of an eclipsing binary system. The regular X-ray eclipses have a duration of 0.488 d and repeat every 2.087 d. As mentioned in Section 2.7, the pulse period shows a sinusoidal Doppler modulation with the same 2.087 d period and is in phase with the X-ray eclipses, indicating that the X-ray pulsar is moving in a circular orbit with a projected velocity of 415.1 km s^{-1} (Fig. 2.8). This velocity, together with the orbital period, allows one to determine the mass function, $f(M)$, of

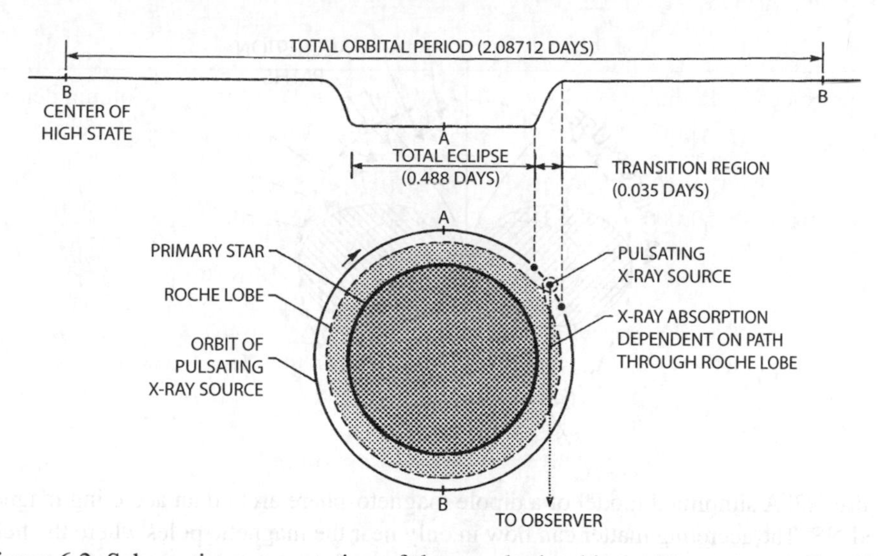

Figure 6.2. Schematic representation of the occultating binary X-ray system Cen X-3, with detected X-ray flux (top) as a function of orbital phase. After Tananbaum (1973).

this binary (see Chapter 3):

$$f(M) = \frac{(M_{\rm opt}\ \sin i)^3}{(M_{\rm X} + M_{\rm opt})^2}$$
$$= 1.036 \times 10^{-7}\, M_\odot \quad K_{\rm X}^3\, P_{\rm orb}\, (1 - e^2)^{3/2} = 15.6\, M_\odot, \tag{6.1}$$

where $M_{\rm opt}$ and $M_{\rm X}$ denote the masses of the optical and the X-ray star, respectively; i is the inclination of the orbital plane with respect to the plane of the sky; $P_{\rm orb}$ is the orbital period (in days); e is eccentricity; and $K_{\rm X}$ is the observed radial velocity amplitude (in km s^{-1}) of the X-ray pulsar. Because $M_{\rm X} \geq 0$ and $\sin i \leq 1$, the mass function yields immediately a lower limit to the mass of the optical star: $M_{\rm opt} \geq 15.6\, M_\odot$. Thus, the X-ray pulsar is moving in a very close orbit around a massive companion star, which, from the duration of the eclipses, has a radius of >70% of the orbital radius of the X-ray source (see Fig. 6.2). The pulsed character of the X-ray emission is due to the rotation of the NS: this star has a magnetic field, which channels the accreting matter toward the magnetic poles, where two hot X-ray–emitting gas columns are formed (Fig. 6.3). Due to the rotation of the star, we see these columns under different angles, which causes the received X-rays to be modulated with the NS spin period.

At present >100 of these X-ray pulsars in HMXBs are known. Particularly the small Galaxy SMC is abnormally rich in such HMXBs, practically all of the B-emission (Be)/X-ray type (for a description of this type, see Chapter 7). There are ~100 HMXBs known in the SMC, ~70 of which are NS-Be/X-ray binaries (Coe & Kirk, 2015). The large number of HMXBs in the SMC is remarkable because it is comparable to the number of HMXBs in the Milky Way, even though the mass ratio between these two

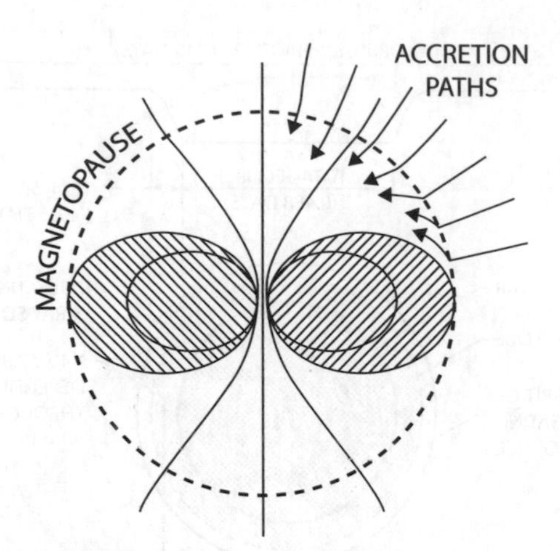

Figure 6.3. A simplified model of a dipole magnetosphere around an accreting magnetized NS. The accreting matter can flow in only near the magnetic poles where the field lines are "open", leading to the formation of two hot X-ray emitting accretion columns, one above each pole. The NS rotates around an axis that is inclined with respect to the magnetic axis; this gives rise to the pulsed appearance of the X-ray emission. After Davidson & Ostriker (1973).

galaxies is ~1:100. The reason for the exceptional abundance of HMXBs in the SMC is attributed to a burst of star formation ~60 Myr ago (Harris & Zaritsky, 2004). By contrast, the few known HMXBs in the LMC is in line with its stellar mass content when compared with that of the Milky Way. We discuss in much more detail, the different subclasses of observed HMXBs in Chapter 7.

Figure 6.4 depicts the distribution of the known HMXBs in the Galactic plane, together with the positions of associations of young massive OB-stars (the blue circles). Apart from these HMXBs, there is a second class of X-ray binaries, in which the donor star has a mass $\lesssim 1.5\,M_\odot$. The latter are called low-mass X-ray binaries (LMXBs).

Figure 6.5 sketches the two main classes of X-ray binaries: HMXBs and LMXBs. Table 6.2 lists some key examples of regularly pulsating HMXBs.

6.1.1 Cygnus X-1: Discovery of the First BH X-ray Binary

Around the same time as the discovery of the binary nature of Cen X-3, Bolton (1971) and Webster & Murdin (1972) independently noticed that in the error box (the region on the sky in which with 3-σ certainty the source is located) of the strong X-ray source Cyg X-1, there is a 9th magnitude blue supergiant star, which coincides within 1 arcsec with a new radio source that had appeared in this error box, discovered by Braes & Miley (1971) and Hjellming & Wade (1971). They found this supergiant to be a single-lined spectroscopic binary with a period of 5.6 d. The radial velocity amplitude of this star is $\sim$72 km s^{-1}, and this combined with the fact that the X-ray source is not eclipsing (implying an inclination angle $i \leq 70°$), and assuming a normal mass of approximately

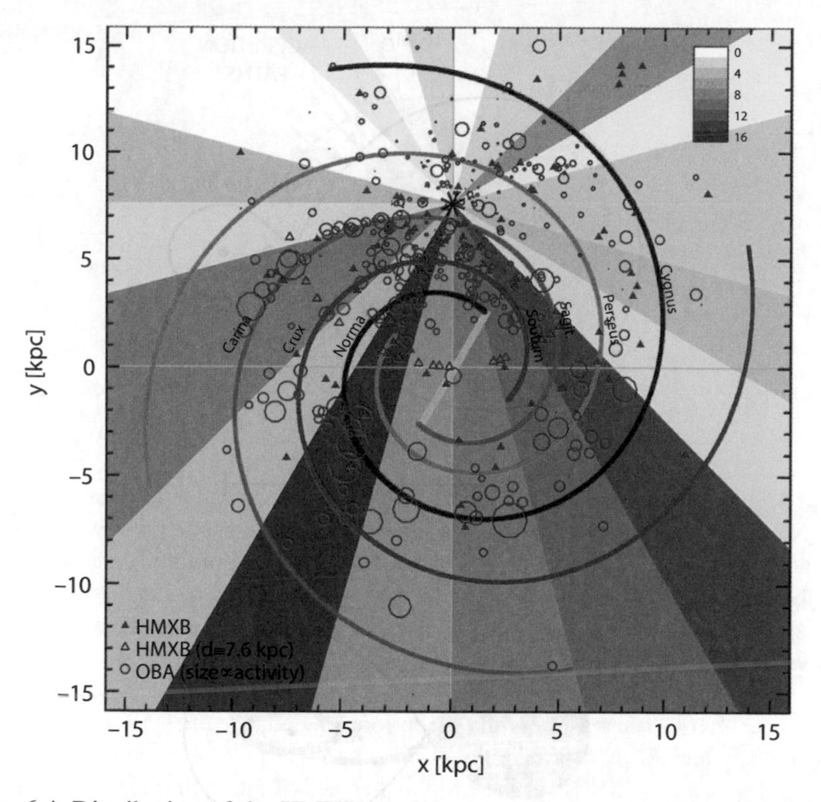

Figure 6.4. Distribution of the HMXBs and young stellar groups (blue circles) in the Galaxy. For systems without a distance estimate, a distance of 7.6 kpc is assumed. The location of the Sun is marked by an asterisk. See also Kretschmar et al. (2019). Figure from Arash Bodaghee (private communication 2021).

$M_{\rm opt} \simeq 20\,M_\odot$ for the blue supergiant (spectral type O9.7Iab), one finds the invisible companion to have mass of $M_{\rm X} > 5.55\,M_\odot$, as one can see from the mass function:[1]

$$f(M) = \frac{(M_{\rm X}\,\sin i)^3}{(M_{\rm opt} + M_{\rm X})^2} = 0.2163\,M_\odot. \tag{6.2}$$

As $5.55\,M_\odot$ is greater than the upper mass limit of $\sim 3.0\,M_\odot$ allowed for a NS—see Nauenberg & Chapline (1973) and Kalogera & Baym (1996)—Webster & Murdin (1972) and Bolton (1971) suggested that Cyg X-1 is a BH. Shortly later, in 1972, it was discovered that the X-ray source underwent a large intensity and spectral change exactly at the time that the new radio source in its (large) error box had appeared,

[1]Note the difference between the expression of the mass function here and in Eq. (6.1). Here, the mass function is based on observations of the supergiant; whereas in the previous version, the mass function is based on observations of the pulsating X-ray source.

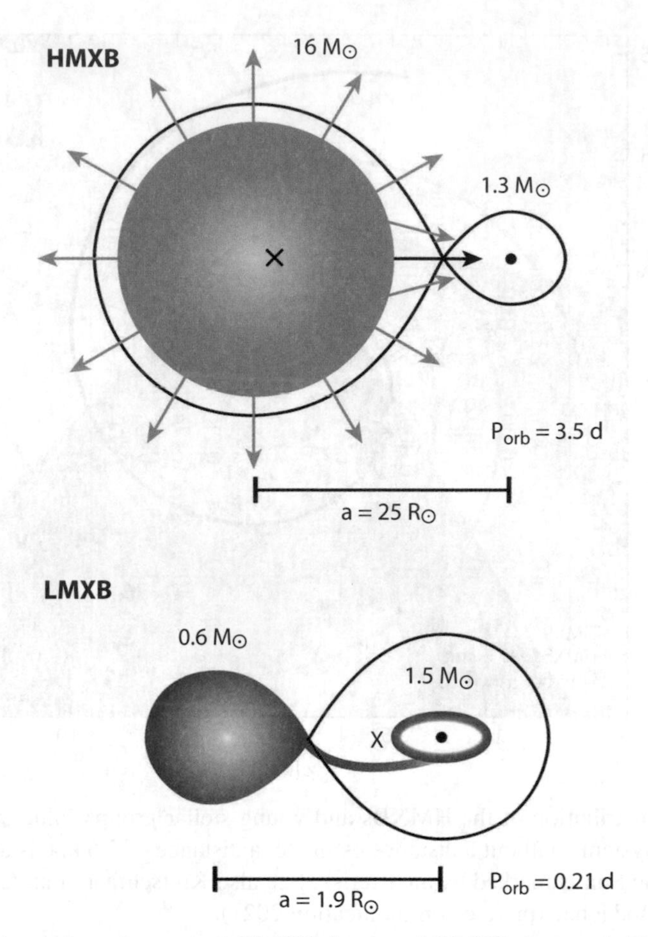

Figure 6.5. Examples of a typical HMXB (top) and LMXB (bottom). The NS in the HMXB has a strong magnetic field ($>10^{11}$ G) and is fed by a strong high-velocity stellar wind and/or by beginning atmospheric RLO. The NS in a LMXB in most cases has a weak magnetic field ($<10^{9}$ G) and is surrounded by an accretion disk that is fed by RLO. There is also observational evidence for HMXBs and LMXBs harboring BHs.

proving that the X-ray source and the radio source are the same object (Tananbaum et al., 1972). This means that the X-ray source is exactly—within 1 arcsec—at the same position as the blue supergiant star. This made clear that Cyg X-1 must be a BH. So, since 1972, we know that X-ray binaries can harbor NSs as well as BHs. We will return to the BH X-ray binaries as a separate category in Section 6.5.

The most recent mass determination of Cyg X-1 is $M_{\rm BH} = 21.1 \pm 2.2\, M_\odot$ (Miller-Jones et al., 2021). Before that, the commonly accepted BH mass estimate was $14.8 \pm 1.0\, M_\odot$ (Orosz et al., 2011), which is quite a surprise given the quoted error bars that correspond to a change of the result by many standard deviations. This is a reminder that astrophysical measurements can be very complex and sometimes the

Table 6.2. Examples of Pulsating HMXBs

Name	$P_{\rm spin}$ (s)	$P_{\rm orb}$ (d)	$a_{\rm X} \sin i$ (light-sec)	$f(M)$ $(M_\odot)$	$L_{\rm X}$ $(10^{35}\ {\rm erg\ s^{-1}})$	Ecc.	Type
A0538-66	0.069	16.65	—	—	8×10^3	0.7	B2IIe transient
SMC X-1	0.717	3.892	53.46	10.84	5×10^3	≤0.007	Sg B0Ib
4U 0115+63	3.61	24.309	140.13	5.007	3×10^2	0.3402	B0.2Ve
V0332+53	4.38	34.25	48	0.101	4	0.37	B0-1Ve transient
Cen X-3	4.84	2.087	39.664	15.386	5×10^2	0.0008	Sg O6.5II-III
LMC X-4	13.5	1.408	26	9.4	7×10^3	≤0.02	Sg O7III-V
2S1417-62	17.6	—	—	—	—	0.45	B1Ve transient
A0535+26	104	111	500	20	2×10^2	0.3	B0Ve-transient
GX 304-1	272	133	500	—	30	≥0.5	B1Ve transient
4U 0900-40*	283	8.964	113.0	19.29	20	0.092	Sg B0.5Ib
4U 1907+09	438	8.376	83	8.8	4×10^2	0.28	Sg O8-9Ia
4U 1538-52	529	3.728	52.8	11.4	40	0.174	Sg B0Iab
GX 301-2	696	41.508	371.2	31.9	30	0.47	Sg B1-1.5Ia
X Per	835	250	454	1.61	1	0.11	O9.5Ve steady

Source: Data from the catalog of Liu et al., 2006.

Ecc, eccentricity.

*Vela X-1.

uncertainties are much larger than the formal error bars indicate (see also discussions in Section 6.6.2).

6.2 TWO TYPES OF PERSISTENT STRONG X-RAY SOURCES: HMXBs AND LMXBs

As we already mentioned, it was also found in the early 1970s that, apart from the HMXBs, in which the companion of the compact star is an early-type star more massive than $\sim 10\,M_{\odot}$, there is a second main category of luminous Galactic X-ray binaries: the LMXBs, in which the companion star has a mass $\lesssim 1.5\,M_{\odot}$. The strongest X-ray source of the sky, Sco X-1, is a LMXB. It has an orbital period of 19 hr and a slightly evolved low-mass companion star (Gottlieb et al., 1975). Like most LMXBs, Sco X-1 does not show regular X-ray pulsations, although there is much evidence—as we will explain later—that, like, in Cen X-3, the accretor is here a NS. A complete list of the 166 LMXBs in our Galaxy and the LMC is given in the review of Sazonov et al. (2020), which is a rich source of information about LMXBs and their properties. Table 6.3 lists the system parameters of several dozens of well-known LMXBs. Among them there is quite a number in which the compact accreting star is a BH. We will come back to the properties of the ~ 20 Galactic BH systems later in this chapter and in the following chapter.

As of 2022, about 74 LMXBs have measured orbital periods in the range between ~ 11 min–1160 d (most are shown in Figure 6.8). For LMXBs with a NS accretor, about 26 spin periods have been found between ~ 1.6 ms–7.7 s. Most LMXBs are transients and X-ray bursters (see following paragraphs). There are well over 100 HMXBs known in our Galaxy, about half of which have their orbital periods measured with values in the range ~ 0.2–262 d (Fig. 6.8). In roughly 66 HMXBs, a NS accretor spin is measured between ~ 33 ms–4 hr.

Except for the system of Hercules X-1, in which the donor star is $\sim 2\,M_{\odot}$, there are no persistent X-ray binary systems observed with donor stars in the mass range between $\sim 1.5\,M_{\odot}$ and $\sim 10\,M_{\odot}$ (see discussion on intermediate-mass X-ray binaries below). HMXBs and LMXBs are, therefore, two clearly separated categories of systems. They indeed differ in several important phenomenological characteristics, listed in Table 6.4, and graphically illustrated in Figures 6.1 and 6.5–6.8.

First, in the HMXBs, the companion of the X-ray source is a luminous early-type star, of spectral type O or B (like in the Cen X-3 and Cyg X-1 systems), with a mass typically between 10 and 40 $M_{\odot}$. In the LMXBs, it is a faint star of mass $\lesssim 1.5\,M_{\odot}$. In most LMXBs, the stellar spectrum is not visible as the light of the systems is dominated by that of the accretion disk around the compact star.

Second, in space, the HMXBs show a strong concentration toward the Galactic plane (but not toward the Galactic center), as is typical for a very young stellar population (Fig. 6.1). This fits with the fact that the companion stars in these systems have $M \geq 10\,M_{\odot}$. Such stars do not live longer than 20 Myr, so the HMXBs and their NS or BH accretors must have ages less than this value. On the other hand, the LMXBs

Table 6.3. Orbital Periods of a Number of LMXBs

Source	P_{orb} (hr)	Nature of modulation
X1820−303	0.19	X-ray (GC)
X1850−087	0.34	UV (GC)
X1626−673	0.7 (pulsar)	(optical + Doppler effect pulsar)
X1832−330	0.73	UV (GC)
X1916−053	0.83	X-ray, optical
SAX J1808.4−366	2.0 (pulsar)	Doppler radial velocity
MAXI J1659−152	2.414 (BH)	X-ray dips
X1323−619	2.9	X-ray dips
X1636−536	3.8	Eclipsing
X0748−676	3.8	Eclipsing (trans.)
X1254−609	3.9	X-ray dips
X1728−169	4.2	Optical
X1755−338	4.4	X-ray dips
X1735−444	4.6	Optical
J0422+32	5.1 (BH)	Optical radial velocity
X2129+470	5.2	Optical
X1822−371	5.6	Eclipse
J2123−058	6.0	Eclipse
Nova Vel 93	6.9 (BH)	Optical radial velocity
4U 1658−298	7.2	X-ray dip
A0620−00	7.8 (BH)	Optical radial velocity
G2000+26	8.3 (BH)	Optical radial velocity
A1742−289	8.4	Eclipse
4U 1957+115	9.3	Optical
Nova Mus 91	10.4 (BH)	Optical radial velocity
Nova Oph 77	12.5 (BH)	Optical radial velocity
Cen X-4	15.1 (trans.)	Optical radial velocity
4U 2127+119	17.1	Eclipse (GC)
Aql X-1	19	Optical
Sco X-1	19.2	Optical
X1624−490	21	X-ray dips
Nova Sco 94	62.6 (BH)	Optical radial velocity
V404 Cyg	155.4 (BH)	Optical radial velocity
2S0921−630	?216	Eclipse
Cyg X-2	235	Optical radial velocity
J1744−28	283 (pulsar)	Doppler radial velocity
GX1+4	7300 or 27800	Doppler effect pulsar

Sources: Data from Liu et al. (2007) and Charles & Coe (2010).

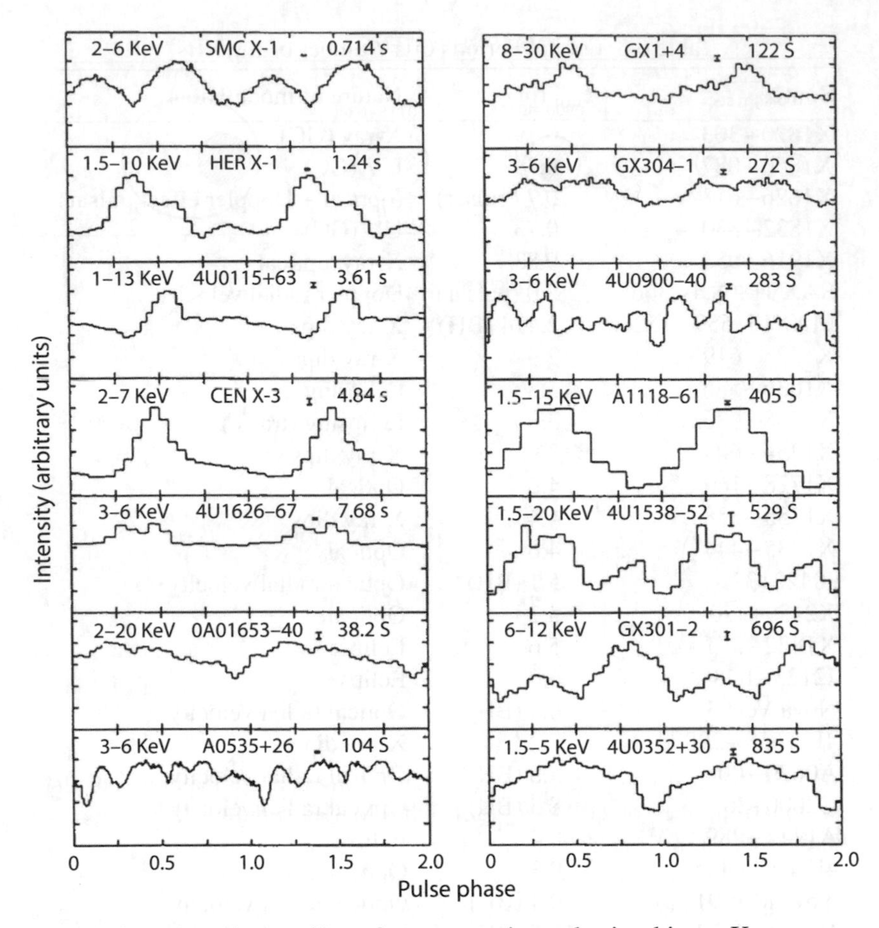

Figure 6.6. Sample pulse profiles of representative pulsating binary X-ray sources in several energy bands, observed by a variety of satellites. Notice that in some cases, such as 4U 0900−40 (Vela X-1) and A0535+46, the magnetic field of the NS is more complex than a dipole. After Rappaport & van den Heuvel (1982).

show a strong concentration toward the Galactic center, and a much broader distribution around the Galactic plane, as is typical for a stellar population with an age of the order of 5–10 Gyr. Indeed, a sizeable number (∼10%) of the ∼130 brightest Galactic LMXBs is found in globular clusters (GCs) which are the oldest objects in our Galaxy, with ages 11–13 Gyr.

Third, each group has its own characteristic type of time variability. In most HMXBs, the X-ray source is a regular X-ray pulsar with a spin period between ∼0.1 and 10^3 s, indicating that the NS has a strong magnetic field (dipole surface strength in the range of $10^{11.5}$–$10^{13.5}$ G), as is supported by the detection of cyclotron features in the X-ray spectra of many of these sources [Staubert et al., 2019]). Some characteristic pulse profiles are depicted in Figure 6.6. On the other hand, very few LMXBs show regular X-ray pulsations with periods in the same range (some exceptions are Her X-1, with a

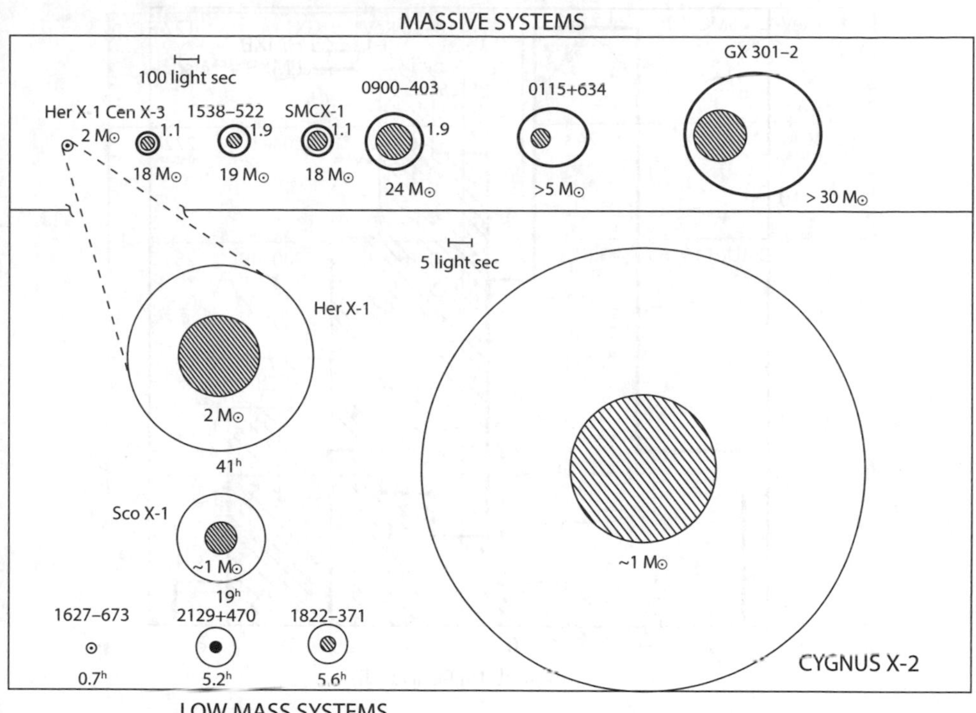

Figure 6.7. Sketch, to scale, of the sizes of the orbits and companions (shaded) of a representative number of HMXBs (top) and LMXBs (bottom). The geometry is based on an estimate of the companion mass, derived either from Doppler tracking or from its optical characteristics. As indicated in the figure, Her X-1 can be considered an intermediate-mass X-ray binary. In the wide-orbit LMXBs, the companion is an evolved star (low-mass subgiant). After van den Heuvel (1994a).

period of 1.24 s; 4U 1627−67, with a period of 7.68 s; and GX 1+4, with a period of 122 s), but many of them show thermonuclear X-ray flashes—so-called Type I X-ray bursts—of the matter accreted on the NS surface. As thermonuclear flashes are suppressed when the magnetic field strengths are $>10^{11}$ G (e.g., Strohmayer & Bildsten, 2006, and references therein), the surface magnetic fields of the NSs in these old systems are expected to be weaker than this value. That the burst systems contain NSs has been demonstrated convincingly by (1) the study of the time evolution of the spectra of the X-ray bursts (e.g., see Chapter 7 and the reviews by Lewin et al., 1995; Strohmayer & Bildsten, 2006) and, (2) more directly, by the detection of regular millisecond X-ray pulsations in several tens of LMXBs (Wijnands & van der Klis, 1998; van der Klis, 2006). The physics of X-ray bursts and of millisecond variability of the X-ray emission of NS–LMXBs is described in more detail in Chapter 7.

Fourth, the X-ray spectra of the HMXBs are, on average, harder ($kT \geq 10$ keV) than those of the LMXBs ($kT \leq 10$ keV). This is thought to be due to the presence of strong magnetic fields in the NSs of HMXBs and weak fields in the NSs of LMXBs.

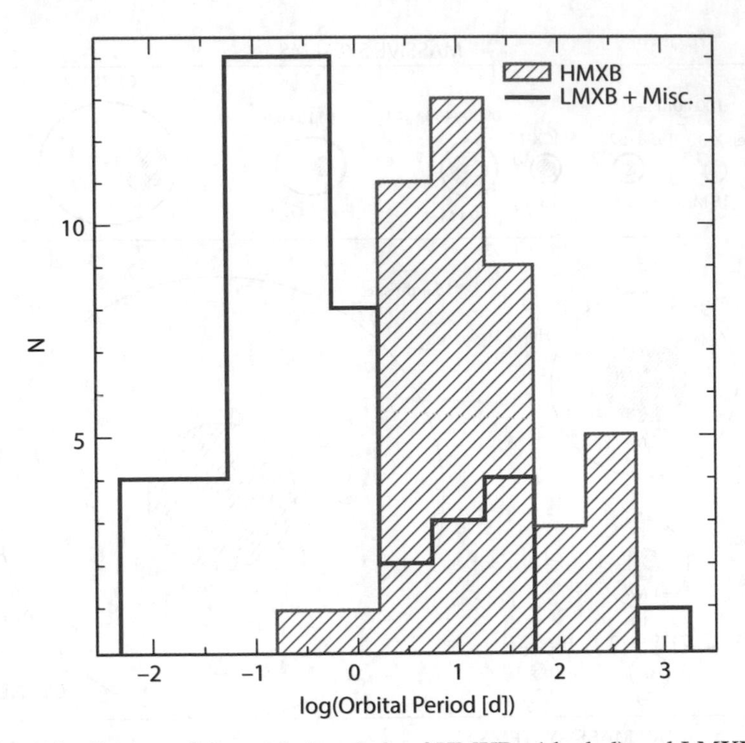

Figure 6.8. Distribution of the orbital periods of HMXBs (shaded) and LMXBs (red), from the catalogs of Liu et al. (2000, 2001), supplemented by new sources discovered by INTEGRAL until 2007. After Bodaghee et al. (2007).

Apart from these characteristic differences between LMXBs and HMXBs, also the distributions of their orbital periods is quite different (see Figs. 6.7 and 6.8). One observes that for LMXBs there is a very wide range of orbital periods, from approximately 11 min to several hundred days. For HMXBs the range is much smaller, from approximately 1.5 to a few hundred days (the WR-star X-ray binaries are included here). The much higher value of the lower limit to the orbital periods of HMXBs is because hydrogen-rich massive companion stars have much larger radii than low-mass stars do, such that the restriction that they should be smaller than their Roche lobes forbids very short orbital periods. On the other hand, hydrogen-rich stars in LMXBs cannot allow orbital periods much shorter than ~80 min (the *period minimum of CVs*), which means that systems with orbital periods below this value, cannot have hydrogen-rich donors: they are helium stars or WDs, as will be further explained in Chapters 7 and 11.

6.3 HMXBs AND LMXBs vs. IMXBs

We have reviewed the properties of the two main classes of observed X-ray binaries, the HMXBs and the LMXBs. Nature, however, also produces IMXBs, the progenitors of binary radio pulsars with CO WD companions, for example, which are described in

Table 6.4. The Two Main Classes of Strong Galactic Binary X-ray Sources

HMXBs	LMXBs
• Optical counterparts massive and luminous early-type stars of spectrum OB, $L_{\rm opt}/L_{\rm X} \geq 1$. • Concentrated in space toward Galactic plane: young stellar population with age $\leq 2 \times 10^7$ yr. • Type of variability: mostly regular pulses in range $P = 0.1 - 10^3$ s; no thermonuclear X-ray bursts. • Relatively hard X-ray spectra: $kT \geq 10\,{\rm keV}$.	• Faint blue optical counterparts, $L_{\rm opt}/L_{\rm X} \leq 1$. • Concentrated in space toward Galactic center, fairly widespread around Galactic plane; old stellar population with age $(5-13) \times 10^9$ yr. • Type of variability: thermonuclear X-ray bursts, kHz QPOs and regular millisecond X-ray pulsations (only a few sources with $P = 1 - 10^3$ s). • Relatively soft X-ray spectra: $kT \leq 10\,{\rm keV}$.

QPOs: quasi-periodic oscillations.

Chapter 14 (van den Heuvel, 1994b; Tauris et al., 2000; Podsiadlowski & Rappaport, 2000; Tauris et al., 2011, 2012; Lazarus et al., 2014). These systems typically must have had donor star masses between 2 and 6 $M_\odot$. The reason that IMXBs are rarely detected is due to a pure selection effect. Whereas HMXBs and LMXBs are observable as persistent X-ray sources due to a strong donor stellar wind (producing a bright X-ray source) or as a result of long-term RLO, respectively, the donor stars in IMXBs only have relatively weak stellar wind mass-loss rates, and once they fill their Roche lobes, the resultant X-ray phase is dynamically unstable or short-lived (typically 1–2 Myr [Tauris et al., 2000]) compared to of the order of 100 Myr to several Gyr for LMXBs (Tauris & Savonije, 1999; Podsiadlowski et al., 2002), causing them to be much rarer than LMXBs. In this section, we elaborate further on this explanation.

6.3.1 Why the Two Major Classes Exist: HMXBs and LMXBs

To produce a steady long-lived binary X-ray source with a typical X-ray luminosity in the observed range of $\sim 10^{34}–10^{38}\,{\rm erg\,s^{-1}}$, the companion star must offer a steady mass-transfer rate between $\sim 10^{-12}$ and $10^{-8}\,M_\odot\,{\rm yr}^{-1}$ to the NS or BH, for a reasonably long time interval.

If the companion provides continuously a mass-transfer rate of more than the Eddington limit ($\dot{M}_{\rm Edd}$; see Section 4.9), the source is likely to be choked, as the compact star cannot accept all matter transferred to it. The excess inflowing matter will pile up around the source and absorb the X-rays and degrade them to radiation at longer wavelengths, in other words, one will no longer observe an X-ray source. (An example of a source that shows such choking when too much mass is surrounding it is the pulsating X-ray binary Cen X-3, which shows extended "off" periods, which are preceded by ever increasing X-ray absorption, indicating an increasing mass-accretion rate; see, e.g., Giacconi [1975].) To understand why we observe the two major classes of X-ray binaries, HMXBs and LMXBs, with companion masses of $\gtrsim 15\,M_\odot$ and $\lesssim 1.5\,M_\odot$, respectively, and hardly any donor mass in between, let us follow the argument by van den Heuvel (1975) and assume that in Nature compact stars are born with companion stars of any kind of mass. Which companions will then be able to provide a steady accretion rate onto the compact star in the suitable range? As we have already alluded to, systems with donor stars in the range of 1.5–6 $M_\odot$ are believed to exist, and they are suggested to be the progenitors of the radio pulsars with massive WD companions. To understand why they rarely (if ever) show up as X-ray sources, we ask the following questions: In which companion mass[2] ranges do we expect the mass-accretion rates provided by the companion to be much higher than the Eddington rate? And in which companion mass ranges is a steady mass-accretion rate not higher than the Eddington rate expected to occur?

To answer these questions, one has to realize that companions can lose mass at a large rate in two ways, namely via:

1. Roche-lobe overflow. The donor star in this case fills its Roche lobe and transfers mass through the L_1 point to its companion (Chapter 4).
2. A strong, high-velocity stellar wind: In this case, the star does not need to fill its Roche lobe (see also Section 7.2).

We now consider these two cases in more detail (and in addition, the variable mass loss from Be stars).

6.3.2 Roche-lobe Overflow

In this case, the mass transfer takes place at low velocities and, in principle (but not in practice, see Section 14.6), all the transferred matter can be accreted by the compact star. As shown in Chapter 4, if the mass-losing star is more massive than its companion, the orbit of the system and the Roche lobe of the donor star shrink. This will either lead to an unstable type of mass transfer (the star becomes larger than its Roche lobe and loses mass at an accelerated pace, leading to a common envelope (CE), see Section 4.8) or the system remains dynamically stable with rapid mass transfer ensuing on a timescale equal to (roughly) the thermal timescale of the mass-losing star (see Chapter 8):

$$\tau_{\rm th} = \frac{GM^2}{RL} = 16\,{\rm Myr}\ \frac{(M/M_\odot)^2}{(R/R_\odot)\,(L/L_\odot)}. \tag{6.3}$$

Using the mass-luminosity relation and the mass-radius relation for main-sequence stars, $L \simeq M^{3.5}$ and $R \simeq M^{0.7}$, one obtains

$$\tau_{\rm th} \simeq 16\,{\rm Myr}\ (M/M_\odot)^{-2.2}. \tag{6.4}$$

[2] As discussed in Chapter 9, it also depends on the evolutionary state of the donor star.

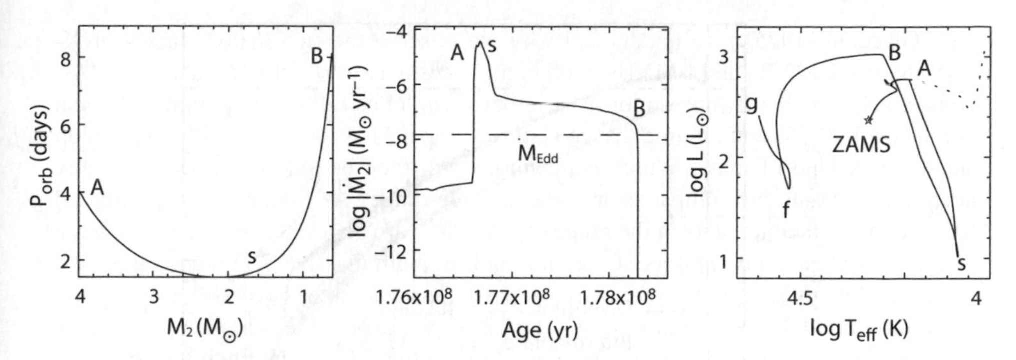

Figure 6.9. Calculation of an IMXB system with an initial donor star mass of $M_2 = 4.0\,M_\odot$ and orbital period of $P_{\rm orb} = 4.0$ d. (Left) Evolution of $P_{\rm orb}$ as a function of M_2 (evolution proceeds to the right). (Middle) Mass-loss rate of the donor as a function of its age since the ZAMS. (Right) Evolution of the mass-losing donor (solid line) in an HR diagram. The dotted line represents the evolutionary track of a single $4.0\,M_\odot$ star. After Tauris et al. (2000).

As most of the stellar envelope mass ($\simeq 0.7 - 0.8\,M$) will be transferred on this timescale, one therefore has

$$\dot{M}_{\rm RLO} \simeq 4 \times 10^{-8}\,M_\odot\,{\rm yr}^{-1}\,(M/M_\odot)^{3.2}. \tag{6.5}$$

Hence, we see that for intermediate-mass donor stars ($2 - 6\,M_\odot$), $\dot{M}_{\rm RLO}$ will be of the order of a few $10^{-7} - 10^{-5}\,M_\odot\,{\rm yr}^{-1}$, and the X-ray source will likely be obscured by a too large mass-transfer rate. Furthermore, as can be inferred from Eq. (6.4), the duration of the X-ray phase is at most a few Myr for such IMXBs (in agreement with detailed computations in, e.g., Tauris et al. [2000]; see Fig. 6.9). Thus, there are two reasons for selection effects *against* observing IMXBs.

Equation (6.5) also shows that, for a NS accretor, thermal-timescale mass-transfer rates below the Eddington limit can only be obtained for donor stars with $M \lesssim M_\odot$. For donor star masses $M \lesssim M_\odot$, however, Eq. (6.5) is not valid, because the mass-losing star is no longer more massive than the NS, as was the condition for deriving this expression for thermal-timescale mass transfer. For donor stars less massive than the NS accretor (which have measured masses between 1.2 and $2.1\,M_\odot$, see Fig. 14.32), mass transfer will cause the orbit to widen, so the slow evolutionary expansion of the star, on a nuclear timescale, will drive the mass transfer. (For the sake of argument, we will consider here only this type of mass transfer, although other mechanisms for driving the mass transfer may come into play in tight orbits, such as gravitational radiation losses and magnetic braking, driving the two stars toward each other; see Chapters 9 and 11). The nuclear timescale (Chapter 8) is roughly

$$\tau_{\rm nuc} = 10\,{\rm Gyr}\,\frac{(M/M_\odot)}{(L/L_\odot)}. \tag{6.6}$$

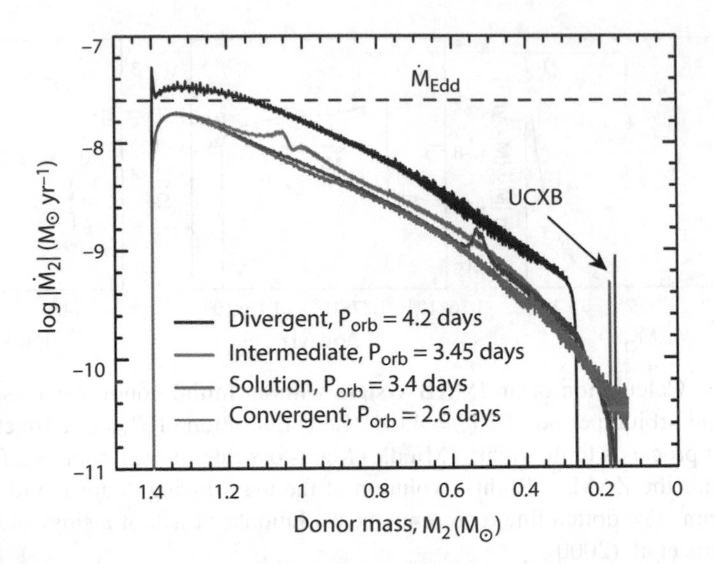

Figure 6.10. Mass-transfer rate vs. decreasing donor star mass for LMXBs near the so-called bifurcation period. The systems accumulate with time at low masses where the mass-transfer rate is smallest. The converging system does not detach at all. A few detached systems eventually evolve into an UCXB (Section 7.8.4) when the WD fills its Roche lobe. After Istrate et al. (2014).

Applying again the mass-luminosity relation, $L \simeq M^{3.5}$, one finds that the mass-transfer rate becomes

$$\dot{M}_{\mathrm{nuc}} \simeq 0.8 \times 10^{-10}\ M_{\odot}\ \mathrm{yr}^{-1}\ (M/M_{\odot})^{3.5}. \tag{6.7}$$

Indeed, one finds that in systems with donor masses $M \leq M_{\mathrm{NS}}$, mass-transfer rates lower than the Eddington rate are obtained, typically in the range of a few $10^{-10}\ M_{\odot}\mathrm{yr}^{-1}$ or less, lasting for billions of years (see example in Fig. 6.10). The larger the initial P_{orb}, the higher the mass-transfer rate and the shorter the RLO episode will be (e.g., Tauris & Savonije, 1999; Podsiadlowski et al., 2002). For LMXBs in wide orbits with giant donor stars, it becomes less than a few Myr. In these systems, the mass-transfer rate exceeds the Eddington limit as a consequence of the deep convective envelopes of the giants (see Chapter 9).

So, to conclude, RLO can only produce steady long-lived binary X-ray sources with $\dot{M}_{\mathrm{RLO}} < \dot{M}_{\mathrm{Edd}}$ if the donor star is a main-sequence star (or subgiant) and has a mass less than or close to the mass of the NS accretor. This is the simple explanation for the existence of the observed class of LMXBs.

6.3.3 Accretion from a Stellar Wind

The stellar winds of early-type stars are driven by the radiation pressure in the outer layers of the stars. The wind mass-loss rate of such luminous stars is subject to large uncertainties. The observational and theoretical estimated values depend on the

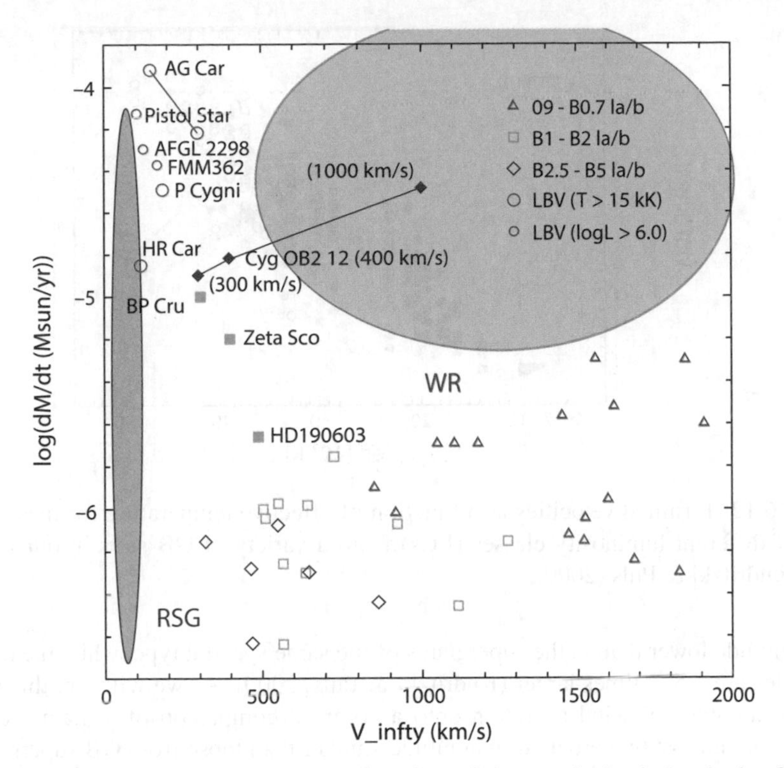

Figure 6.11. Wind mass-loss rate as a function of the terminal wind velocity for various types of stars in our Galaxy (i.e., for solar metallicity). Circles are luminous blue variables (LBVs), while triangles, squares, and diamonds are OB supergiants of various masses. The position of red supergiants is shown by the left red ellipse and that of WR-stars by the top-right blue region. After Martins (2015) and Clark et al. (2012).

modelling of the terminal wind velocity, clumping effects, metallicity, and absorption properties in the local environment (Kudritzki & Puls, 2000; Vink et al., 2001; Repolust et al., 2004; Crowther et al., 2006; Mokiem et al., 2007; Markova & Puls, 2008; Langer, 2012; Šurlan et al., 2013; Chaty, 2013; Martins, 2015; Falanga et al., 2015).

Strong stellar winds, with mass-loss rates exceeding $10^{-6}\,M_{\odot}\,\mathrm{yr}^{-1}$, are only found in stars more massive than $\sim 15\,M_{\odot}$, as depicted schematically in Figure 6.11. The plot shows that Galactic O9 to B0.7 blue supergiants (luminosity classes Ia/b) have high mass-loss rates of the order of $(0.5-5)\times 10^{-6}\,M_{\odot}\,\mathrm{yr}^{-1}$ and high terminal wind velocities between 1,000 and 2,000 km s^{-1}, while B1–B2 supergiants have mass-loss rates of $(0.2-2)\times 10^{-6}\,M_{\odot}\,\mathrm{yr}^{-1}$ and terminal wind velocities between 400 and 1,200 km s^{-1}; still later type supergiants (B2.5–B5) have mass-loss rates of $(0.2-1)\times 10^{-6}\,M_{\odot}\,\mathrm{yr}^{-1}$ and terminal wind velocities between 300 and 900 km s^{-1}.

Figure 6.12 shows that the terminal wind velocity is a strong function of the effective temperature of the star and depends also on whether the star is on the main sequence (luminosity class V) or is a giant (luminosity class III) or supergiant (luminosity class I or II). In the OB-type main-sequence stars, the wind mass-loss rates are at least an order

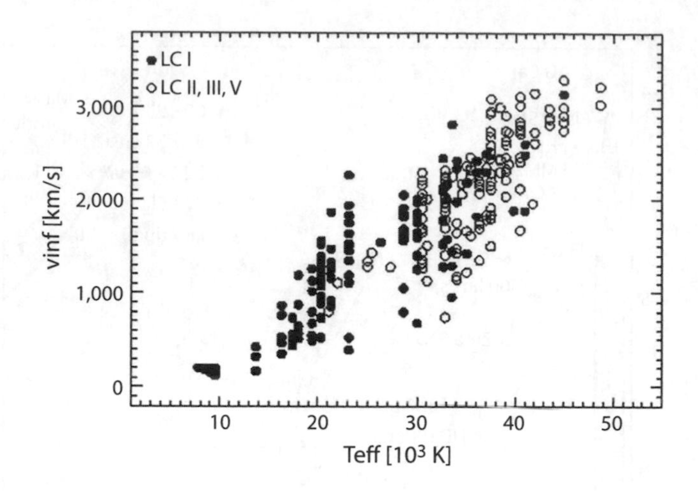

Figure 6.12. Terminal velocities as a function of effective temperature for massive hot stars of different luminosity classes (LCs) from a variety of OB-stars in our Galaxy. After Kudritzki & Puls (2000).

of magnitude lower than in the supergiants of the same spectral type, while the outflow velocities are 1.5–2 times larger (Kudritzki & Puls, 2000). As we will see, this means that the amounts of wind accretion onto a compact companion of a main-sequence OB-star are almost two orders of magnitude smaller than those from OB supergiants.

As the wind outflow velocities observed in these massive and luminous stars are very high, the bulk of the wind matter will flow past a NS companion without being affected by its presence, as is depicted in Figure 6.13. Only a tiny fraction of the wind matter will be captured. To estimate the accretion rate in HMXBs due to stellar winds, and the total amount of material accreted by a NS, we now follow Tauris et al. (2017). We shall begin by assuming a wind mass-loss rate of a typical OB-star companion in a HMXB of $|\dot{M}_{2,\mathrm{wind}}| \simeq 10^{-6} - 10^{-5}\, M_\odot\,\mathrm{yr}^{-1}$ (assuming that the effect of X-ray irradiation on the wind mass-loss rate of the companion star is negligible). The wind velocity can be approximated by the escape velocity at the stellar surface: $v_{\mathrm{wind}} \simeq v_{\mathrm{esc}} = \sqrt{2GM_2/R_2}$, where M_2 and R_2 are the companion star mass and radius, and G is the constant of gravity. Moreover, we assume *Bondi-Hoyle–type accretion* to be valid, given the highly supersonic flow of the stellar wind from the companion star. The gravitational-capture radius of the accreting NS (also called the Bondi-Hoyle radius, after the authors who first studied this type of accretion of gas captured by the gravity of an object [Bondi & Hoyle, 1944] (see also Section 7.2 on more details of the accretion process), is then given by

$$r_{\mathrm{acc}} = \frac{2\,GM_{\mathrm{NS}}}{(v_{\mathrm{rel}}^2 + c_s^2)}, \tag{6.8}$$

where $c_s = \sqrt{\gamma P/\rho} \approx 11\sqrt{T/(10^4\,K)}\,\mathrm{km\,s^{-1}}$ is the local sound speed of the ambient medium (γ is the adiabatic index, P is the pressure, ρ is the mass density, and T is

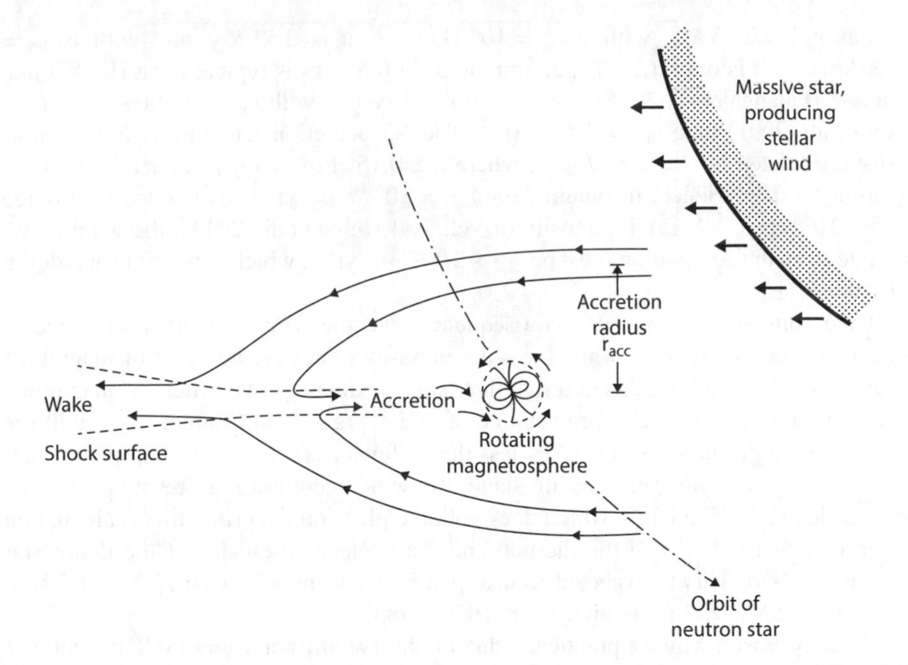

Figure 6.13. Streamlines of stellar-wind material, in the frame of an accreting NS. Relative dimensions are not to scale. After Davidson & Ostriker (1973).

the temperature). The Bondi-Hoyle treatment of accretion is, in fact, a refinement of the Hoyle-Lyttleton representation (Hoyle & Lyttleton, 1939) of the wind accretion process, in which the wind is assumed to consist of free-moving particles instead of a gas (see Exercise 6.5 for the Hoyle-Lyttleton treatment, which yields a result similar to that of the Bondi-Hoyle treatment). The relative velocity between the wind and the NS ($\vec{v}_{\rm rel} = \vec{v}_{\rm wind} + \vec{v}_{\rm orb}$) is usually dominated by the wind velocity compared to the orbital velocity, that is, $v_{\rm wind} \gg v_{\rm orb}$ (except for very tight and/or highly eccentric binaries). Similarly, we have $v_{\rm wind} \gg c_s$. The resulting accretion rate onto the NS, $\dot{M}_{\rm NS} \approx \pi r_{\rm acc}^2 \rho\, v_{\rm wind}$ can be combined with the continuity equation, $\rho = |\dot{M}_{2,\rm wind}|/(4\pi a^2\, v_{\rm wind})$, to yield

$$\dot{M}_{\rm NS} \approx \frac{(GM_{\rm NS})^2\,|\dot{M}_{2,\rm wind}|}{a^2 v_{\rm wind}^4} \equiv f_{\rm acc} \cdot |\dot{M}_{2,\rm wind}|, \tag{6.9}$$

where a is the orbital separation. An estimate from this equation indicates that, in wind-driven scenarios, $f_{\rm acc} = 10^{-3} - 10^{-4}$; in other words, $\dot{M}_{\rm NS}$ is typically $10^3 - 10^4$ times lower than $|\dot{M}_{2,\rm wind}|$, and for helium-star winds the ratio is even lower by an order of magnitude. Hence, we adopt a typical NS accretion rate of $\dot{M}_{\rm NS} \approx 10^{-9}\, M_\odot\,{\rm yr}^{-1}$, corresponding to an X-ray luminosity of the order $L_{\rm X} \approx 10^{37}\,{\rm erg\,s^{-1}}$.

Equations (6.8) and (6.9) show that the captured fraction of the wind mass-loss rate depends very strongly on the wind velocity: it is inversely proportional to the fourth power of this velocity. For example, with $v_{\rm wind} = 500\,{\rm km\,s^{-1}}$ and a $1.4\,M_\odot$ NS accretor, one has $r_{acc} = 2.13\,R_{\odot}$; with $v_{wind} = 1000\,km\,s^{-1}$, it is $0.53\,R_{\odot}$; and with $v_{wind} =$ $2{,}000\,km\,s^{-1}$, it is only $0.13\,R_{\odot}$. Assuming $a = 60\,R_{\odot}$, as is typical for a HMXB like Vela X-1 (Falanga et al., 2015), one finds for this system with $v_{wind} = 600\,km\,s^{-1}$ (van Loon et al., 2001) that $f_{acc} = 1.5 \times 10^{-4}$. The NS accretion rate and its X-ray luminosity are related by $L_X = \epsilon\,\dot{M}_{NS}c^2$, where $\epsilon \simeq 0.15$ (Eq. [4.134]), depending on the equation of state. Hence, to obtain $\dot{M}_{NS} \geq 6 \times 10^{-10}\,M_{\odot}\,yr^{-1}$, as needed to provide the $5 \times 10^{36}\,erg\,s^{-1}$ X-ray luminosity of Vela X-1 (Fürst et al., 2014), the wind mass-loss rate from the companion must be $\geq 4 \times 10^{-6}\,M_{\odot}\,yr^{-1}$, which is possible according to Figure 6.11.

If the companion is an OB main-sequence star, the wind velocities are twice as large as for the supergiants, while the wind mass-loss rates are an order of magnitude lower, which leads to an accretion rate that is approximately two orders of magnitude lower, and thus a very weak X-ray source of $\sim 10^{33}\,erg\,s^{-1}$. In Chapter 9, we will see that in close high-mass systems (P_{orb} less than a few days), where the companion star is still in the core-hydrogen burning stage, the wind accretion may be supplemented by so-called beginning RLO, which does not take place on a thermal timescale, but on a timescale that is between the thermal and the nuclear timescales of the donor star (Savonije, 1978). This is expected to take place in systems like Cen X-3, LMC X-4, and possibly SMC X-1 (Savonije, 1978, 1979, 1983).

The only reason why we practically do not observe any persistent IMXBs (Her X-1 and LMC X-3 are the few exceptions) is that these intermediate-mass companions have no suitable mode of mass transfer available to produce a steady long-lived strong X-ray source with an X-ray luminosity below the Eddington limit. Their winds are too weak, but on the other hand, RLO produces far too much accretion (greater than the Eddington limit). We thus see that the existence of the two main classes of strong persistent binary X-ray sources in Nature is due to a selection effect (van den Heuvel, 1975). The sample of compact objects in binaries observed on the X-ray sky is therefore a highly biased sample of the real population of binaries with compact objects in the Galaxy. From these discussions we reach the following conclusion (van den Heuvel, 1975):

If one assumes that NSs (and BHs) can be formed as companions of stars of any mass, one expects to find in Nature only two classes of persistent (long-lived) binary X-ray sources:

1. *Systems with a low-mass ($\lesssim 1.5\,M_{\odot}$) companion star, in which the accretion onto the NS is due to RLO.*
2. *Systems with a high-mass ($\gtrsim 15\,M_{\odot}$) companion star, in which the NS is accreting from the strong stellar wind of a blue (super)giant companion.*

6.4 DETERMINATIONS OF NS MASSES IN X-RAY BINARIES

Because the eclipsing and pulsating HMXBs are double-lined spectroscopic binaries (SB2), for which the orbital inclination angle can be determined, they enable one to determine the masses of NSs. Together with the binary radio pulsars, in which the precise measurement of several relativistic effects allow one to determine NS masses

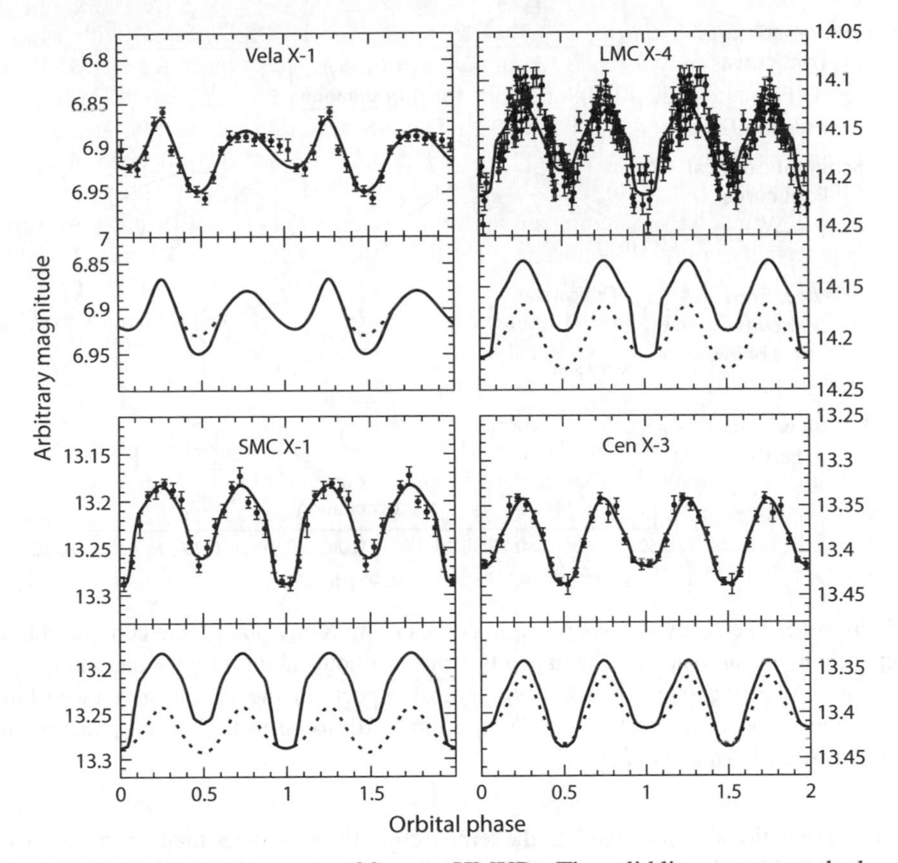

Figure 6.14. Optical light curves of four sg-HMXBs. The solid lines represent the best-fit model light curves for each system, all of which include the presence of an accretion disk around the NS, according to Rawls et al. (2011). The dotted lines depict the models with the light from the accretion disk subtracted, for comparison. The light curves of Vela X-1, SMC X-1, and Cen X-3 are all V-band data (Pojmanski, 2002; Priedhorsky & Holt, 1987; van Paradijs et al., 1983), while the curve for LMC X-4 is B-band data (Ilovaisky et al., 1984). All data are phased relative to the time of the X-ray eclipse. After Rawls et al. (2011).

(see Chapter 14), they are our source of empirically determined masses of NSs. The first-ever NS mass that was determined in this way is that of the supergiant HMXB (sg-HMXB) pulsar Vela X-1 (4U 0900−40), which was determined by van Paradijs et al. (1976).

In the HMXBs, the information about the inclination of the orbital plane is obtained from the regular photometric variability of the optical star. Because this star is nearly filling its Roche lobe, it is pear-shaped and will show a double-wave optical light curve (two maxima and two minima per orbital revolution) as schematically explained in Figure 6.15. As the stellar surface facing the NS is more extended (or pear-shaped) due

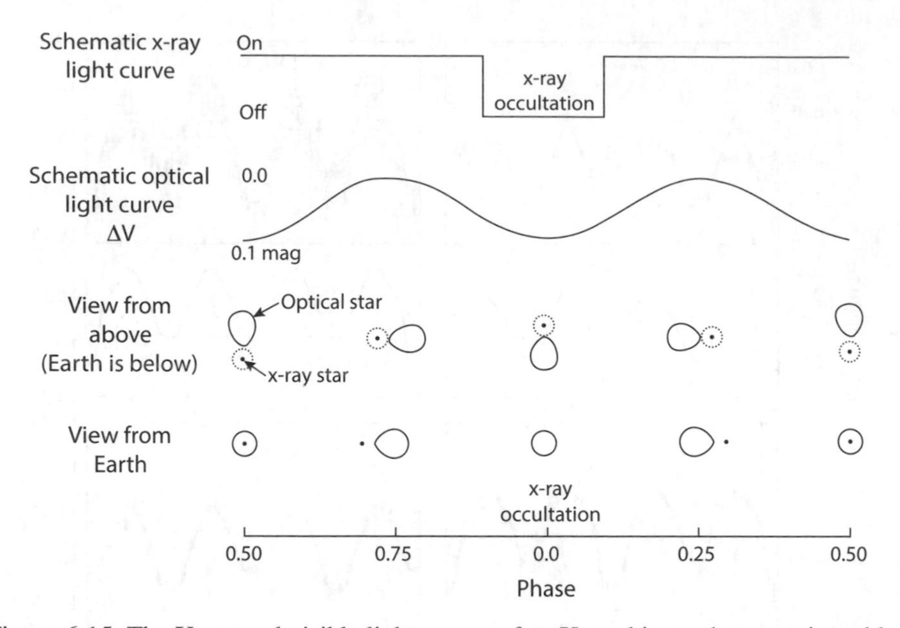

Figure 6.15. The X-ray and visible light curves of an X-ray binary that contains a blue supergiant star are shown schematically (top), without taking gravity darkening into account. Also illustrated (bottom) are several aspects of the system as viewed both from above and from the Earth, assuming the Earth to be in the orbital plane of the system. After Forman (1974).

to the gravitational pull of the NS, the temperature there is lower than on the opposite side; hence this side of the star is darker ("gravity darkening") than the opposite side. As a result, the light minimum at phase 0.50 is deeper than that at phase 0. In Figure 6.15 this effect is not taken into account, but the dotted curves in Figure 6.14 show the full theoretical ellipsoidal light curve, with its two unequal minima. Still, as this figure shows, these theoretical ellipsoidal light curves do not accurately fit the observations for these systems. To explain the observed light curves, one has to take into account that (1) on the side of the X-ray source, the surface of the optical star is heated by the X-rays, and (2) that the accretion disk around the NS also is a light source. The fully drawn curves in Figure 6.14 include these two effects. The amplitude of the light variations depends on the orbital inclination. Therefore, the precise model fitting of the observed light curve yields the value of the orbital inclination. Once this inclination is known, one can then simulate as well as possible the observed radial velocity curve of the donor star, by fitting it to models of this gravitationally deformed and one-sided heated star.

This then yields the real radial velocity curve of the mass center of this star, which together with the Doppler radial velocity curve of the X-ray pulsar companion yields the masses of the two stars (see Fig. 6.16). This type of mass determination of both components of HMXBs by light velocity and radial velocity curve fitting has been brought to highest perfection in the past decades by Jerome Orosz and his collaborators. Rawls

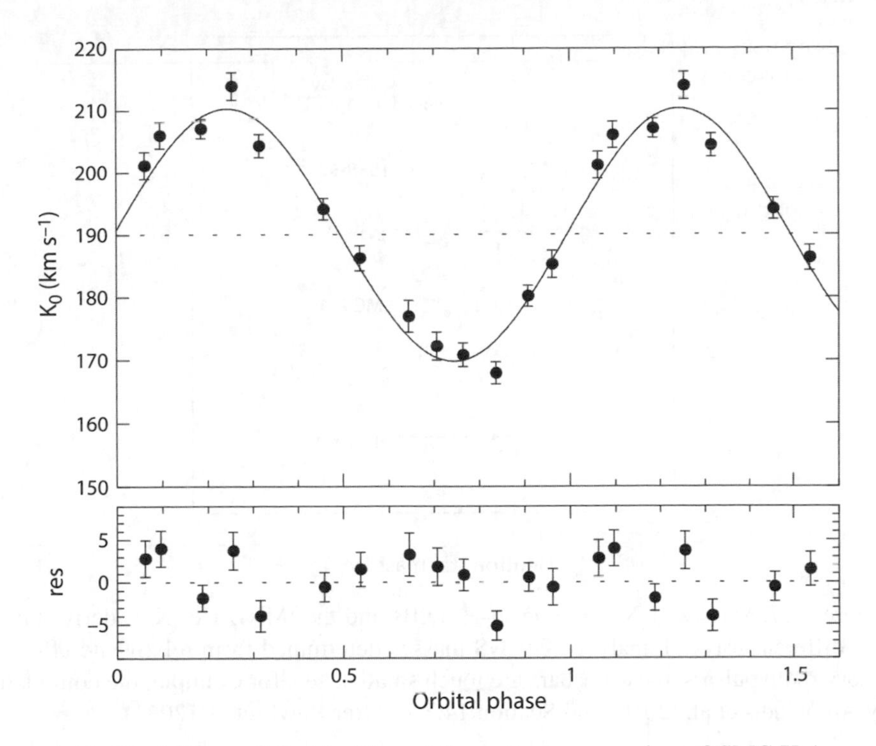

Figure 6.16. (Top) The radial velocities of the optical companion of SMC X-1, measured in one hydrogen line (Balmer $10 \rightarrow 2$), by van der Meer et al. (2007), together with the best fitting radial velocity curve for its circular orbit. (Bottom) The residuals of each measurement with respect to this curve are shown. One-sigma error bars are indicated. The radial velocity amplitude is $20.2 \pm 1.2\,\mathrm{km\,s^{-1}}$, which together with the mass function of the X-ray pulsar and the system inclination of 67 ± 5 degrees leads to a NS mass of $1.06 \pm 0.1\,M_{\odot}$. After Kaper et al. (2006).

et al. (2011) used two different models for obtaining the best fit to the observations, which gave slightly different results, as depicted in Figure 6.17. The figure shows that, except for the case of 4U 1538−52, whose orbital eccentricity is uncertain, for the HMXBs the masses determined in the two ways are very similar. The high mass of the NS Vela X-1 ($1.79 \pm 0.16\,M_{\odot}$) stands out, as was already known because of the work of Barziv et al. (2001).

6.5 BH X-RAY BINARIES

In a number of the non-pulsating HMXBs and LMXBs, the radial velocity amplitude of the optical star is so large that the mass of the compact star is in all likelihood $>3.0\,M_{\odot}$, which is expected to be the absolute upper limit to the mass of a NS (for example, see the references given in Section 2.8). One therefore expects the compact stars in these

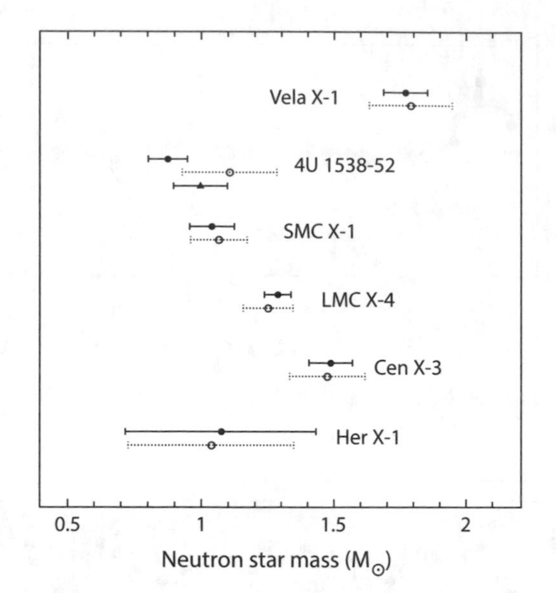

Figure 6.17. Masses of NSs in five sg-HMXBs and the IMXB Her X-1, derived using two different ways of analysis. For NS masses determined from relativistic effects in binary radio pulsars, the error bars are much smaller, see, for example, the compilation by Antoniadis et al. (2016) and Section 14.7.1. After Rawls et al. (2011).

systems to be BHs. The X-ray sources in the BH-LMXB systems show characteristics that are quite different from those of the regularly pulsing sources and "normal" LMXBs that we know to contain NSs, such as the X-ray bursters and accreting millisecond X-ray pulsars. The clearest of these characteristics is an X-ray spectrum that either has a very hard tail, to energies beyond 100 keV, or is super-soft (softer than usually observed in LMXBs). Some sources, such as the HMXB Cyg X-1, are found to switch between these two states, others are always found in one of these two states (see the reviews by Fender & Belloni, 2004; McClintock & Remillard, 2006). On the basis of their very hard X-ray spectral tails, Barnard et al. (2014) identified ~50 Chandra sources in M31 as BH candidates, of which about a dozen are located in globular clusters.

A very important group of the BH X-ray binaries is that of the so-called *soft X-ray transients* or *X-ray novae*, sources that suddenly appear in the sky as a strong soft X-ray source, and after the outburst disappear after a few weeks to months (see Fig. 6.18). Among these transients, the so-called FRED sources have fast rise and exponential decay, and the FRFT sources have fast rise and flat top, as illustrated in Figure 6.18. The class of X-ray novae was first recognized and put on the map as an important new class of X-ray transients by Tanaka and colleagues, working with the Japanese X-ray satellite Ginga (see reviews by Tanaka & Lewin, 1995; McClintock & Remillard, 2006; Motta et al., 2021). During the outburst, a quite bright optical object is also seen. After the disappearance of the X-ray source and the bright optical object (a bright accretion disk), one often finds at the position of the transient a low-mass star, that

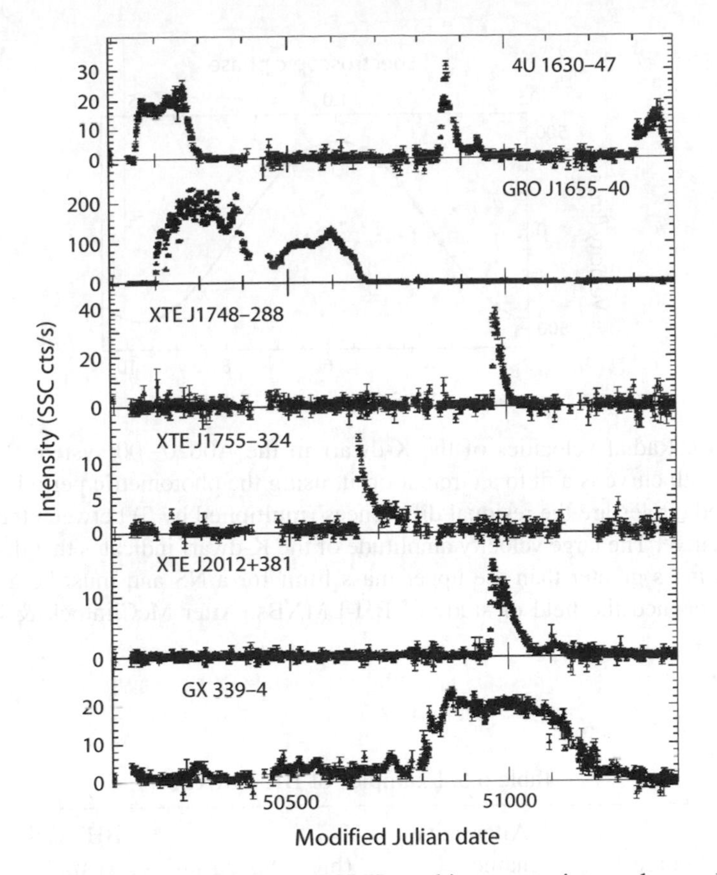

Figure 6.18. X-ray light curves of six BH X-ray binary transients, observed with the All-Sky Monitor (ASM) of NASA's Rossi X-ray Timing Explorer satellite (RXTE). After Bradt et al. (2000).

is a single-lined spectroscopic binary with a very large radial velocity amplitude. The first-discovered case of this type was the source A0620−00 (Nova Monocerotis 1975), where the optical star is a K-dwarf, which was found by McClintock & Remillard (1986) to be a spectroscopic binary with an orbital period of 8 hr and a radial velocity amplitude of almost 450 km s^{-1} (Fig. 6.19). Even if the mass of the K-dwarf is zero this already leads to a mass of the compact star of $>3.0\,M_\odot$. By 2006, already 15 of such X-ray novae BH systems, with low-mass donor stars, had been discovered, making them the largest class of BH X-ray binaries. At the time of writing, >30 of these systems are known.

Table 6.5 lists 15 examples of these systems: 8 with very short orbital periods (<12.5 hr) and M- and K-dwarf companions, and 7 systems with wider orbits and low-mass subgiant (luminosity class IV) or giant (luminosity class III) donor stars. One of the latter systems, V1033 Sco (Nova Scorpii 1994 = GRO J1655-40) deserves special attention. It was discovered by Israelian et al. (1999) that the spectrum of its F-subgiant

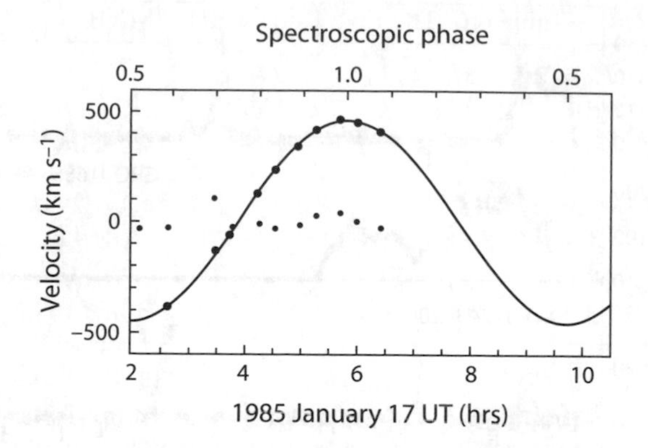

Figure 6.19. Radial velocities of the K-dwarf in the A0620−00 system (large solid dots). Smooth curve is a fit to a circular orbit, using the photometric period of the star. Small filled circles are the residual differences (multiplied by 5) between the data and the fitted curve. The large velocity amplitude of the K-dwarf indicates that the compact star has a mass greater than the upper mass limit for a NS and must be a BH. This discovery opened the field of study of BH-LMXBs. After McClintock & Remillard (1986).

Table 6.5. Examples of BH-LMXBs

Source	Alternative name	$P_{\rm orb}$ (hr)	Spectrum	BH mass ($M_\odot$)
0422+32	V518 Per	5.1	M2V	3.2−13.2
0620−003	V616 Mon	7.8	K4V	3.3−12.9
1009−45	MM Vel	6.8	K7/M0V	6.3−8.0
1118+480	KV UMa	4.1	K5/M0V	6.5−7.2
1124−684	GU Mus	10.4	K3/K5V	6.5−8.2
1543−475	IL Lupi	26.8	A2V	7.4−11.4
1550−564	V381 Nor	37.0	G8−K8IV	8.4−10.8
1655−40	V1033 Sco	62.9	F3−F5IV	6.0−6.6
1659−487	V821 Ara	42.1	—	—
1705−250	V2107 Oph	12.5	K3/7V	5.6−8.3
1819.3−2525	V4641 Sgr	67.6	B9III	6.8−7.4
1859+226	V406 Vul	9.2	—	7.6−12.0
1915+105	V1487 Aql	804.0	K/MIII	10.0−18.0
2000+251	QZ Vul	8.3	K3/K7V	7.1−7.8
2023+338	V404 Cyg	155.3	K0III	10.1−13.4

After McClintock & Remillard (2006).

Table 6.6. The Five Known BH-HMXBs

Source	P_{orb} (d)	M_{donor} ($M_\odot$)	M_{BH} ($M_\odot$)	Reference
Cyg X-1	5.6	41 ($\pm$ 7)	21.2 ($\pm$ 2.2)	Miller-Jones et al. (2021)
LMC X-1	3.9	31.8 ($\pm$ 3.5)	10.9 ($\pm$ 1.4)	Orosz et al. (2009)
LMC X-3	1.7	3.6 ($\pm$ 0.6)	7.0 ($\pm$ 0.6)	Orosz et al. (2014)
MCW 656	$\sim$60	$\sim$13	4.7 ($\pm$ 0.9)	Casares et al. (2014)
M33 X-7	3.45	70 ($\pm$ 7)	15.7 ($\pm$ 1.5)	Orosz et al. (2007)

Updated after van den Heuvel (2019).

companion star shows extraordinarily strong lines of the alpha elements oxygen, magnesium, silicon, and sulfur, indicating overabundances by a factor of 6 to 10 of these elements in the atmosphere of this star. This is seen as evidence that the formation of the BH in this system has been accompanied by a SN explosion. Further evidence for an explosion in this system is its high systemic velocity ($114 \pm 19\,\text{km s}^{-1}$) relative to its local Galactic rest frame (Brandt et al., 1995). In scenarios for BH formation, it is possible that the BH forms after an intermediate stage of short duration in which initially a meta-stable NS is formed. A few seconds later, due to fallback, the NS accretes much matter and collapses to a BH (see e.g., Sukhbold et al. [2018]).

While several tens of BH systems are known among the LMXBs, among the HMXBs only five BH systems are known: Cyg X-1, LMC X-1, LMC X-3, M33 X-7, and MCW 656. M33 X-7 consists of a 70 $M_\odot$ O-star and a 15.65 $M_\odot$ BH in a 3.45 d orbit (see Orosz et al., 2007; Valsecchi et al., 2010). MCW 656 is the only known BH-Be/X-ray binary. Table 6.6 lists parameters of these five systems.

Excellent overviews of our knowledge of the BH X-ray binaries and their specific observed characteristics—such as the presence of relativistic jets and so-called *soft X-ray transient* outbursts or "X-ray novae"—are given by Tanaka & Lewin (1995); Remillard & McClintock (2006); McClintock & Remillard (2006); Fender (2006); Li (2015); Mirabel (2017) and Motta (2021). Since its start in 2009, the Japanese Monitor of All-sky X-ray Image (MAXI) mission on the International Space Station discovered many new BH candidates, for example MAXI J1659−152, MAXI J1543−564, MAXI J1836−194, MAXI J1305−704, MAXI J1910−057/Swift J1910.2−0546, and MAXI J1828−249. The light curves of several X-ray novae are shown in Figure 6.18. For details of the behavior of the different sources, such as the presence or absence of a plateau phase in the light curve, and the possible causes of these behavioral differences, we refer to Maccarone et al. (2014).

Figure 6.20 depicts the relative sizes of 22 BH X-ray binaries, and Figure 6.21 depicts graphically examples of the masses of stellar BHs and NSs, including the ones measured from the GW signals of merging BHs and NSs measured by LIGO–Virgo until September 2020 (see Chapter 15 for an update). This figure shows that (1) the masses of NSs tend to cluster around $1.4 - 1.5\,M_\odot$, although there are also NSs with masses of $\sim 2.0\,M_\odot$ (see also Fig. 14.32), and (2) the masses of the stellar BHs in X-ray binaries and three of the merging double BH binaries cluster between

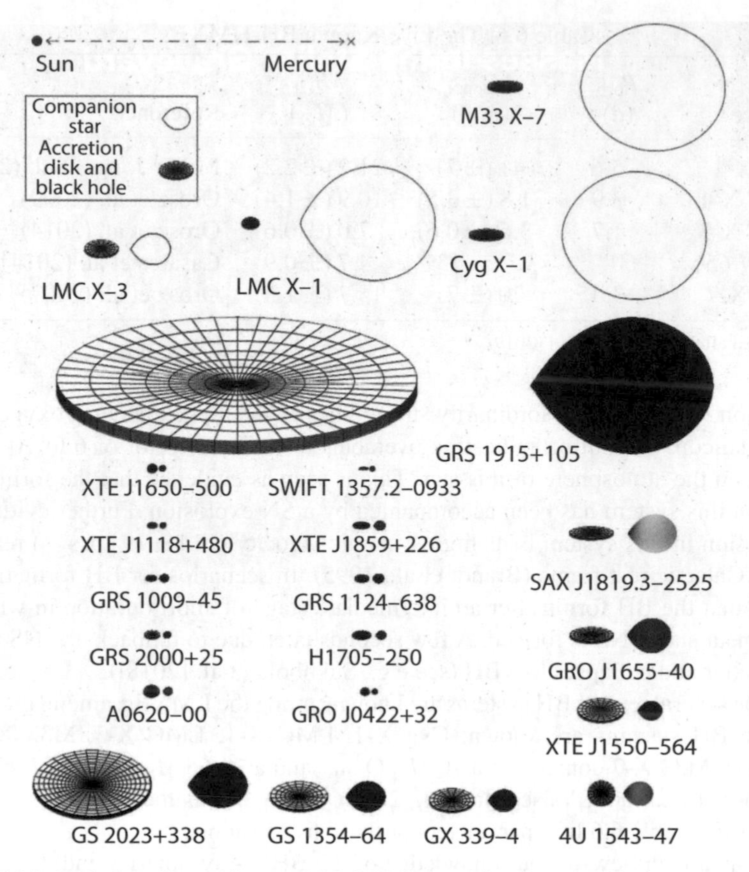

Figure 6.20. Relative dimensions of 22 BH X-ray binaries and their accretion disks. Alternative names of sources: GRS 1124−683 = Nova GUMus; GS 2023+338 = V404 Cyg; GRO J1655−40 = V1033 Sco; GRS 1915+105 = V1487 Aql; GRO J0422+32 = V518 Per. Credit: Jerome A. Orosz (private communication, 2021).

5 and $\sim 20\,M_{\odot}$, while the other merging double BHs had components with masses between 20 and $\sim 100\,M_{\odot}$ (GW190426 and GW190521 being most massive). Jonker et al. (2021) recently argued for observational selection effects, implying that the current sample of confirmed BH-LMXBs is biased against the most massive BHs. We will consider the origins of merging BH masses in Chapter 15.

6.5.1 Microquasars: X-ray Binaries with Strong Radio Emission

Microquasars are X-ray binaries that, like quasars (supermassive BHs in the nuclei of galaxies), exhibit two-sided relativistic jet outflows producing double radio structures (see, e.g., the review by Fender, 2006). The most striking example is the transient BH X-ray binary system of GRS 1915+105 (see Fig. 6.22), which during transient outbursts shows superluminal expansion of two-sided radio lobes, exactly as is observed

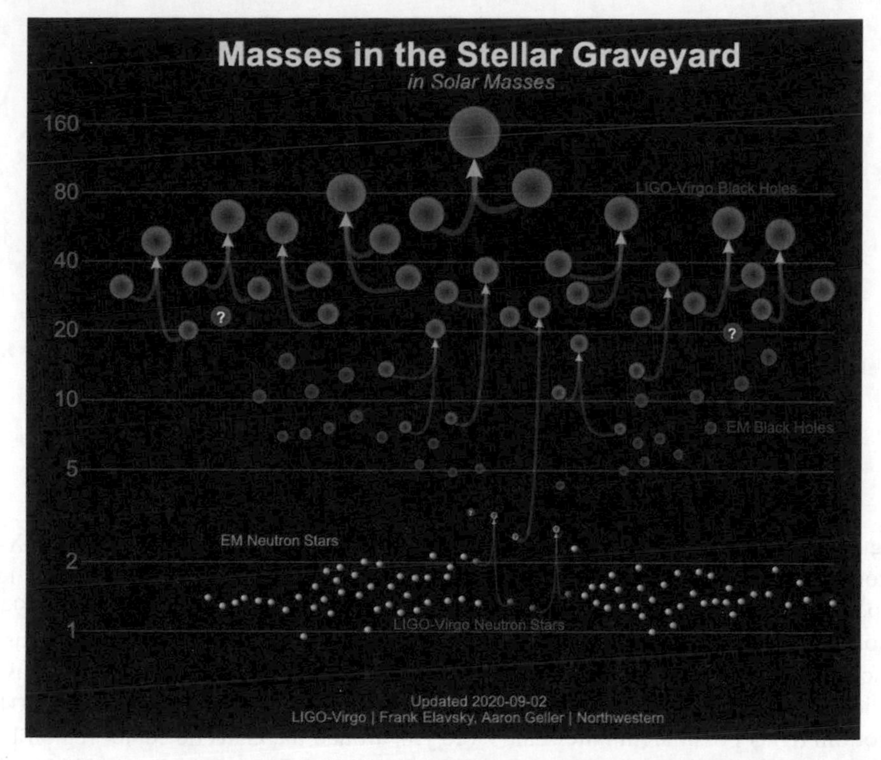

Figure 6.21. Measured masses of NSs and stellar-mass BHs. The electromagnetic (EM) BHs (magenta) are the 24 ones known from X-ray binaries, and the EM NS masses (yellow) are from radio pulsars and X-ray binaries. The 13 LIGO–Virgo BH pairs (blue) were determined from the GW signals of 12 double BH mergers and one BH+NS merger (GW190814). The LIGO–Virgo NS masses (orange) resulted from the GW signals of the double NS (DNS) mergers GW170817 and GW190425. Error bars (particularly large in X-ray binaries) are not shown and all precisely determined NS masses are between 1.2 and 2.1 $M_{\odot}$. Credit: Frank Elavsky and Aaron Geller (Northwestern University, September 2020). See Figure 15.13 for more GW sources.

in quasars (e.g., Fender & Belloni, 2004). On the other hand, the steady BH–HMXB Cyg X-1 shows a small steady relativistic radio jet (top-left panel of Fig. 6.22), although the large and faint double radio source in its surroundings discovered by Gallo et al. (2005) (see Fig. 6.23) shows that in the past it has undergone relativistic mass-ejection episodes. Two microquasars that deserve special attention are SS433 and Cyg X-3 (and other WR X-ray binaries), which we will now consider in more detail.

6.5.2 SS433

This is a peculiar 14th magnitude emission-line star at a distance of ~5 kpc, which coincides with the X-ray source A 1909+04. It is centrally located in the huge radio shell Westerhout 50 (Fig. 6.24), which resembles a SN remnant and is also visible in

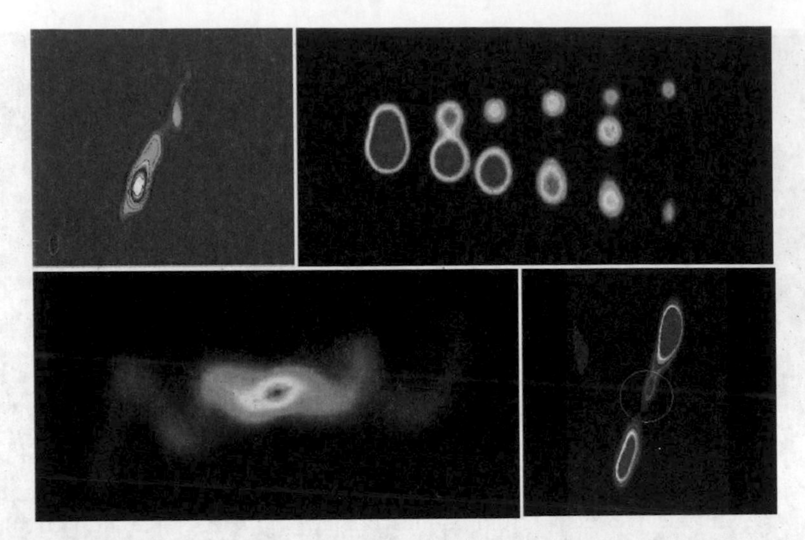

Figure 6.22. (Top left) Milliarcsecond-scale radio jet from the BH-HMXB Cyg X-1 (from Stirling et al., 2001). (Top right) Transient arcseacond-scale radio jets from the superluminal Galactic jet source GRS 1915+105 (from Mirabel & Rodríguez, 1994). (Bottom left) Arcsecond-scale radio jets from SS433, the second Galactic microquasar discovered. The binary orbit is almost edge-on; the precessing disk of SS433 causes its jets to trace a "corkscrew" in the sky every 162 d (from Blundell & Bowler, 2004). (Bottom right) Fossil arcminute-scale radio jets around the Galactic center region BH in 1E140.7−2042 from Mirabel et al. (1992). Combined figure after Gallo (2010).

X-rays. Two oppositely directed X-ray beams extend from SS433 to the shell, which is elongated in the direction of the beams. In 1978, Margon announced the discovery of a number of broad emission lines in the spectrum of SS433 that change drastically in wavelength from night to night (Margon, 1983, 1984). He found that these changing lines are blue-shifted and red-shifted components of Balmer and He-I (neutral helium) lines. In any given spectrum, all red-shifted components share the same redshift ($\Delta\lambda/\lambda$), and all blue-shifted components share the same blueshift. Figure 6.25 shows a sample spectrum. For extensive reviews of SS433, see Fabrika (2004) and Cherepashchuk et al. (2020).

The amplitudes of the Doppler shifts of the red-shifted lines range up to 50,000 $\mathrm{km\,s^{-1}}$, and those for the blue-shifted lines, up to 30,000 $\mathrm{km\,s^{-1}}$. Most surprisingly, these Doppler shifts were found to vary in a cyclical way, with a period of 162.375 d (see Fig. 6.26 and the review by Fabrika [2004]).

During the cycle, the mean Doppler shift of the red- and blue-shifted components of each line remains constant at $z = 0.035$ (see Fig. 6.26). This was explained by Fabian & Rees (1979), who pointed out that narrowly collimated beams ejected in opposite directions from a central object, with constant velocity of $0.26c$, could explain the data. The variations in the observed Doppler velocities would in this model be due to a change in the inclination angle of the beams with respect to the line of sight. The beam

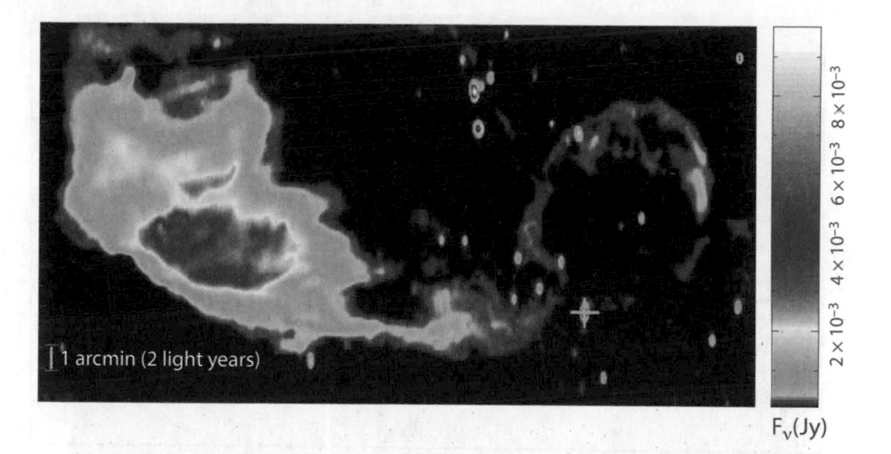

Figure 6.23. Jet-powered radio nebula around Cyg X-1, with size in excess of 10 ly, discovered with the Westerbork Radio Telescope at 1.4 MHz. Notice in particular the "bubble" blown by the jet on the right-hand side of the figure. After Gallo et al. (2005).

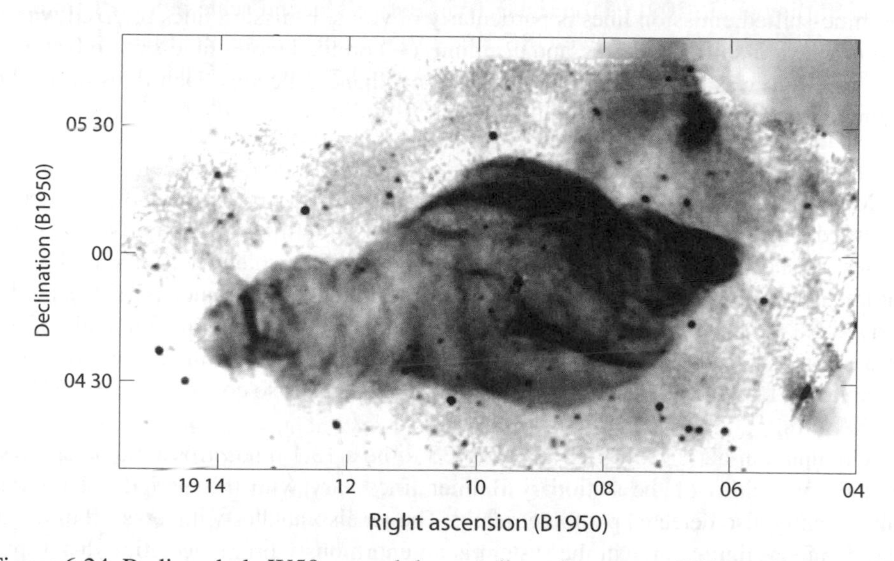

Figure 6.24. Radio nebula W50 around the peculiar microquasar SS433. The two-sided "ears" of the nebula are along the direction of the axis of the precessing relativisitic jets of SS433. Total length of nebula along main axis is ~200 pc. After Dubner et al. (1998).

velocity $v = 0.26c$ was derived from the mean redshift of the blue- and red-shifted lines, which in this model is simply due to the transverse Doppler effect, and amounts to $z = 0.5\,(v/c)^2$, where v is the velocity of the beams (see Milgrom, 1979).

Milgrom (1979) suggested the 162 d period and explained it in terms of a model in which the beams precess around an axis that is inclined to the line of sight (see Fig. 6.27). This model has since been confirmed by other observations (see reviews

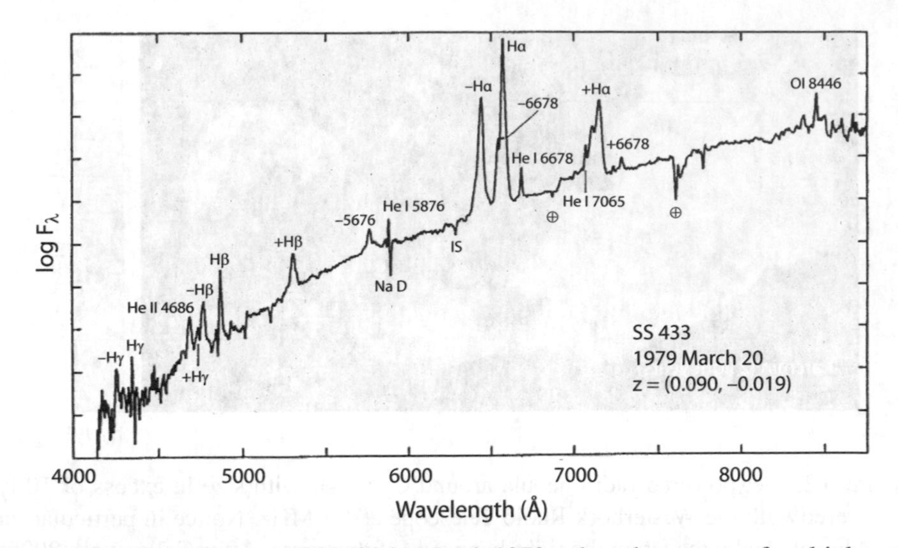

Figure 6.25. The spectrum of SS433 in March 1979, when the pattern of multiple red- and blue-shifted emission lines is particularly obvious. Emission lines prefixed with a plus (+) are red-shifted features, and the minus (−) prefix denotes blue-shifted features. The spectrum was obtained with the Lick 3 m Shane reflector. Each division in the ordinate is 0.83 mag. After Margon (1983).

by Margon, 1984; Fabrika, 2004). The beams and their precession have also been observed with radio telescopes, using very long baseline interferometry (VLBI) techniques, as shown in the lower-left picture of Figure 6.22, which shows the corkscrew nature of the SS433 outflow. Table 6.7 lists the kinematic parameters of the model derived from the stellar observations (Fabrika, 2004); i is the inclination angle of the precession axis to the line of sight, and θ is the opening angle of the precession cone. An amazing fact is that the outflow velocity of the beams is so constant: $v = 0.2647c = 78{,}000\,\mathrm{km\,s^{-1}}$.

Crampton et al. (1980) discovered SS433 to be a 13.1 d spectroscopic binary system: the velocities of the stationary Balmer lines vary with this period, which was subsequently also detected photometrically. The jet also nutates with a period of 6.3 d. The dominant light source in the system is an enormously bright accretion disk (optical luminosity of $>10^4\,L_\odot$), which produces the stationary emission lines, and the beams are expected to be ejected perpendicular to the disk. Thus the precession of the beams is due to the precession of the disk, which was first detected photometrically by Cherepashchuk (1981); see the reviews by Fabrika (2004) and Cherepashchuk et al. (2020). Finally, the orbital eccentricity of SS443 has been measured to be $e = 0.05 \pm 0.01$, which lends support to the slaved-disk model (Cherepashchuk et al., 2021).

Hillwig & Gies (2008) detected absorption lines that they ascribe to a A3−7I supergiant companion star, which in the optical does not emit more than 20% of the light of the system. From radial velocity variations they derived a mass of $12.3 \pm 3.3\,M_\odot$ for the A-supergiant and $4.3 \pm 0.8\,M_\odot$ for the compact companion. Because the latter value is

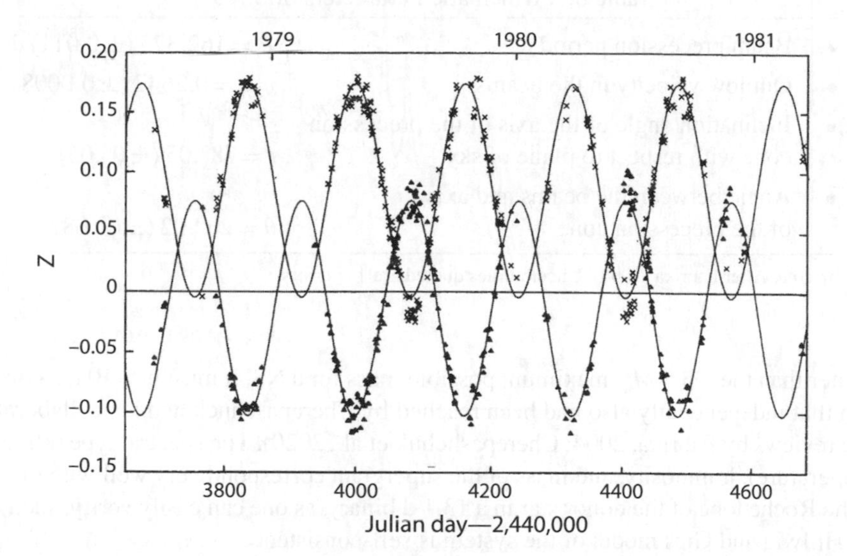

Figure 6.26. The values of the redshifts and blueshifts of SS433 over an almost 3 yr period. Displayed are close to 500 separate Doppler-shift values obtained from 300 nights. Most of the data have been obtained by Margon and University of California colleagues, and a few points are from the literature. The solid line is the best fit to the kinematic model, with free parameter values given in Table 6.7. After Margon (1984).

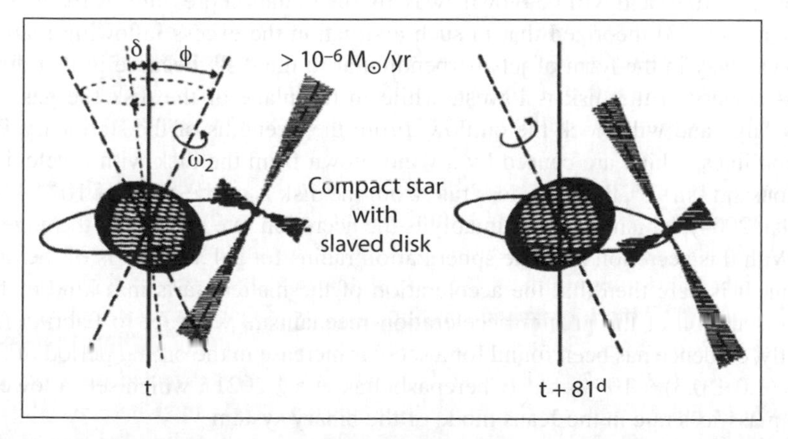

Figure 6.27. Outline of the "slaved-disk" model for the 162 d precession of the beams of SS433. After van den Heuvel et al. (1980).

Table 6.7. Kinematic Parameters of SS433

• Beam precession period	$P = 162.375\ (\pm 0.011)$ d
• Outflow velocity in the beams	$v/c = 0.2647\ (\pm 0.0008)$
• Inclination angle of the axis of the precession cone with respect to plane of sky	$i = 78^\circ.05\ (\pm 0^\circ.05)$
• Angle between the beams and axis of the precession cone	$\theta = 20^\circ.02\ (\pm 0^\circ.08)$

Source: After Fabrika (2004). Uncertainties quoted are 1 σ errors.

greater than the $\sim$3.0 $M_\odot$ maximum possible mass for a NS, it must be a BH, a conclusion that independently also had been reached by Cherepaschuck and his collaborators (see reviews by Fabrika, 2004; Cherepashchuk et al., 2020). The spectral type (effective temperature), luminosity, and mass of the supergiant correspond very well with the size of the Roche lobe of the donor star in a 13.1 d binary, as one can easily verify, such that the Hillwig and Gies model of the system is self-consistent.

The mass-loss rate in the beams is $\sim 10^{-6}\ M_\odot\ \mathrm{yr}^{-1}$, corresponding to the kinetic luminosity $L_{\mathrm{kin}} = 10^{39}\ \mathrm{erg\ s}^{-1}$, which is close to the Eddington luminosity of the 4 $M_\odot$ compact star. This suggests that the jets are radiatively driven by super-Eddington accretion. If the accretion rate of matter entering the disk is far greater than the Eddington accretion rate, then at a distance from the compact star equal to the *spherization radius*, $R_{\mathrm{sph}} = (\dot{M}/\dot{M}_{\mathrm{Edd}})\ R_{\mathrm{Sch}}$, where R_{Sch} is the Schwarzschild radius (Shakura & Sunyaev, 1973), the accretion luminosity reaches the Eddington limit, and matter cannot flow in further and will be blown away by the radiation pressure. Already Shakura & Sunyaev (1973) theorized that in such a situation the excess inflowing matter will be blown away in the form of jets perpendicular to the disk, because in that direction the gas density in the disk is lowest, while in the plane of the disk the gas density is very large and will block the outflow. From the strengths of the stationary Balmer emission lines, which are caused by a wind blown from the disk with a velocity of a few thousand $\mathrm{km\ s}^{-1}$, the mass-loss rate from the disk is of the order of $10^{-4}\ M_\odot\ \mathrm{yr}^{-1}$ (Fabrika, 2004), which then presumably is the accretion rate arriving at the edge of the disk. With this accretion rate, the spherization radius for a 4 $M_\odot$ BH is of the order of 10^{10} cm. It is here then that the acceleration of the matter starts in a kind of funnel. For a discussion of the precise acceleration mechanism, we refer to Fabrika (2004). Recently, evidence has been found for a secular increase in the orbital period of SS433, $\dot{P}_{\mathrm{orb}} = (1.0 \pm 0.3) \times 10^{-7}\ \mathrm{s\ s}^{-1}$ (Cherepashchuk et al., 2021), which sets a lower limit on the mass-loss rate in the Jeans mode of the binary system.

SS433 is unique among the microquasars because (1) the outflowing matter is neutral hydrogen and helium gas, and (2) the outflow velocity in the jets is so extremely constant and equal to 0.2647 c. In the other microquasars, like GRS1915+105, Cyg X-1, and Cyg X-3, the jets are highly relativistic with velocities very close to the speed of light, and presumably the jets consist of relativistic electrons accelerated in magnetic fields and radiating synchrotron radiation. This is all very different from the case of

SS433. The main reason for this difference seems to be that the accretion rate of SS433 is many orders of magnitude higher than that of these other microquasars, which only flare up when temporarily their accretion rate exceeds the Eddington limit.

Milgrom et al. (1982) noticed that the value $\beta = v/c = 0.2647$ is very close to 0.25, with a special relativisitic correction. The value $\beta = 0.25$ is a special quantity in the hydrogen atom (and in all hydrogen-like ions): it is the relative energy difference between the Lyman limit and Lyman-alpha line. Milgrom (1979) suggested that the value $\beta = 0.25$ is due to radiative acceleration by Lyman-alpha absorption with line-locking: the radiative acceleration being due to a continuous radiation flux with a spectrum with a lower cutoff wavelength at the Lyman limit. In this case, the matter cannot be accelerated further than when the Doppler-shifted Lyman continuum edge is seen by the hydrogen atom as the wavelength of Lyman-alpha. At that moment the velocity reached by the accelerated gas is $v = 0.25c$ plus a special relativistic correction. Conditions for line-locking are (1) the underlying continuum flux is strong enough for the radiation pressure to overcome gravity, which clearly is the case in SS433; (2) the dominant momentum transfer is due to Lyman-line absorption by some hydrogenic ion; and (3) the continuum flux falls off sharply at wavelengths shorter than the Lyman edge. We thus see that any hydrogenic ion can do the acceleration up to $\beta = 0.25$, and so can atoms more massive than hydrogen when stripped from their outer electrons by X-ray absorption. This line-locking mechanism not only explains the constant velocity of the jets of SS433, but also the fact that we see neutral hydrogen in the jets. We fully agree with Fabrika (2004) that line-locking is the most plausible explanation of the highly peculiar properties of the SS433 jets.

As mentioned already, Cherepashchuk et al. (2020) have argued that the mass of the relativistic component in SS433 is $M_X > 7\,M_\odot$ (thus a BH). Combined with an estimated optical component (donor) mass of $M_2 > 12\,M_\odot$ (Hillwig et al., 2004), the relatively small binary mass ratio ($q < 2$) in SS433 allows us to understand why there is no common envelope in this binary at the current evolutionary stage and why the system remains semi-detached, transferring mass by stable RLO (van den Heuvel et al., 2017). See further discussions in Section 10.12.

6.5.3 Cyg X-3 and Other Wolf-Rayet X-ray Binaries

Cyg X-3 is the first-discovered microquasar. It has an orbital period of only 4.8 hr and a quite unique sinus-like light curve in X-rays as well as γ-rays (Fig. 6.28). It can give huge radio outbursts at irregular times. The first-discovered events of this type were the 1972 radio outbursts of 2–11 September and 18–25 September, which made it for several weeks the brightest radio source in the sky. These outbursts triggered observing campaigns at many wavelengths, the results of which were published in an issue of *Nature Physical Sciences*, entirely devoted to this source.[3] One of the discoveries was its huge infrared emission with a 4.8 hr modulation (Becklin et al., 1972). The spectral evolution of the radio emission during the outbursts resembled exactly that of a quasar

[3] *Nature Physical Sciences*, volume 239, October 1972.

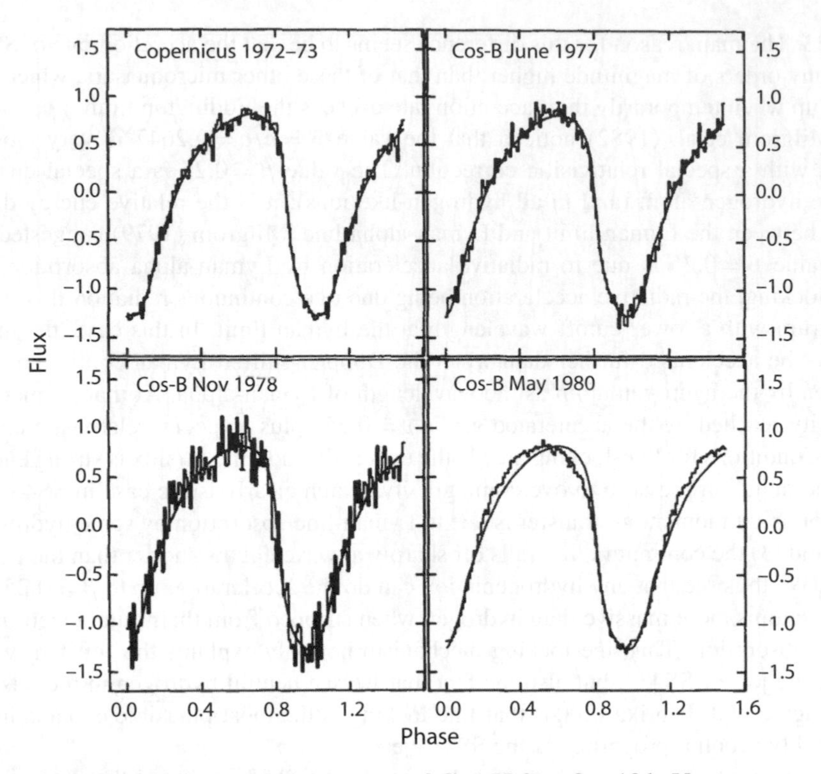

Figure 6.28. The mean X-ray light curve of Cyg X-3 at 2 – 12 keV energy, sampled over 9 yr by the Copernicus and Celestial Observation Satellite B (COS-B) spacecrafts. After Bonnet-Bidaud & van der Klis (1981).

outburst at cosmological distance: it is a synchrotron emission of an expanding cloud of relativistic electrons with magnetic fields, the so-called *van der Laan model* for quasar outbursts (Hjellming & Balick, 1972a, 1972b; Hjellming, 1973). The source is right in the middle of the Galactic plane, and during the radio ourbursts the H I 21 cm absorption lines from three spiral arms between us and the source are visible, yielding a distance of ~10 kpc. The visual extinction in its direction is at least 20 magnitudes, so no optical counterpart is known. Radio VLBI observations show that, like SS433, it has (in this case, small) radio jets, presumably due to relativistically outflowing matter. Based on theory of binary evolution, van den Heuvel & De Loore (1973) argued that the system is the later evolutionary stage of a HMXB and consists of a helium star and a compact object. This was confirmed in 1991 by the work of van Kerkwijk et al. (1992), who discovered that the infrared spectrum of the system shows only helium and nitrogen lines, no hydrogen, and is very similar to that of a nitrogen-type WR-star of type WN7. Its absolute infrared luminosity is indeed consistent with it being a Population I WR-star (see Figs. 3.7 and 6.29), in other words, it is a helium star with a mass of the order of 10 $M_{\odot}$.

The large wind mass-loss rate observed in WR-stars gives a consistent explanation for the gradual increase of their orbital period on a timescale of the order of 10^6 yr (see Sections 4.3.4 and 5.3). So far, Cyg X-3 is the only WR X-ray binary known in

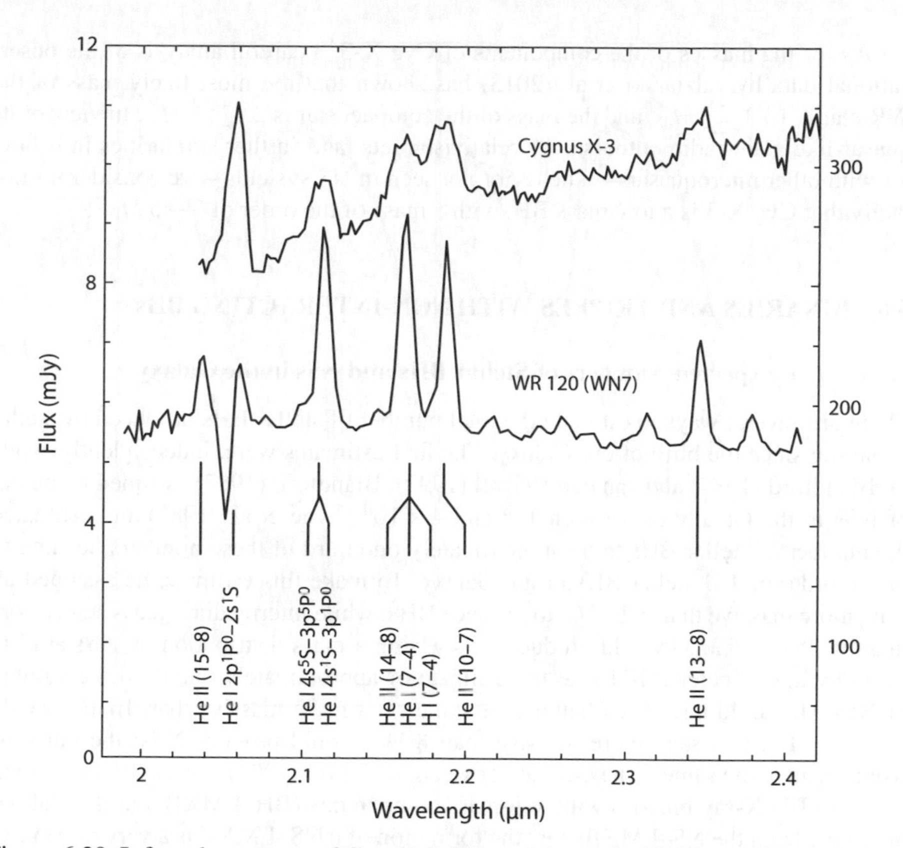

Figure 6.29. Infrared spectrum of Cyg X-3 (left scale) and the WN7-star WR 120 (right scale, arbitrary units), showing that the companion of Cyg X-3 is a WR-star of type close to WN7. After van Kerkwijk et al. (1992).

our Galaxy, but in recent years, several more WR X-ray binaries have been discovered in other galaxies. The X-ray light curve of one of them, the Circinus Galaxy source CG X-1 with an orbital period of 7.2 hr, is very similar to that of Cyg X-3 (Esposito et al., 2015). It seems very likely that the compact stars in all these systems are BHs, as argued on theoretical grounds by van den Heuvel et al. (2017). Table 6.8 lists the properties of the six known WR X-ray binaries. It is clear that their orbital periods tend to be very short: of the order of 1 d or less. If indeed the compact stars in these systems are BHs, these binaries are excellent progenitors of close double BH systems that will merge within a Hubble time (Esposito et al., 2015; van den Heuvel et al., 2017). It should be stressed, however, that the measured accretor masses in these systems are highly uncertain, as the radial velocity curves derived from the emission lines of the WR-stars, on which these mass estimates were based, are almost 90° out of phase with the radial velocity curves expected from the eclipse curves of the systems. Therefore, these observed radial velocity curves cannot represent the actual radial velocities of the WR-stars, but more likely are those of bow-shock structures in the WR winds (Maccarone et al., 2014; Laycock et al., 2015).

As for the masses of the components of Cyg X-3, a careful analysis of the observational data by Zdziarski et al. (2013) has shown that the most likely mass of the WR-star is $10.3^{+3.9}_{-2.8}\,M_\odot$, and the mass of the compact star is $2.4^{+2.4}_{-1.1}\,M_\odot$. In view of its quasar-like giant radio outbursts, its relativistic jets, and further similarities in behavior with other microquasars—a behavior not seen in NS systems—we consider it most likely that Cyg X-3 is a low-mass BH, with a mass of the order of $3 - 5\,M_\odot$.

6.6 BINARIES AND TRIPLES WITH NON-INTERACTING BHs

6.6.1 The Expected Numbers of Stellar BHs and NSs in the Galaxy

There are several ways to estimate the total number of stellar BHs produced by stellar evolution since the birth of our Galaxy. The first estimates were, independently, made by Blandford (1987) and van den Heuvel (1992). Blandford (1987) assumed a number of NSs in the Galaxy of between 10^8 and 4×10^8 (Lyne et al., 1985) and estimated the number of stellar BHs to be approximately one third of these numbers, leading to of the order of 10^8 stellar BHs in the Galaxy. To make this estimate, he assumed all stars more massive than $\sim 10\,M_\odot$ to produce BHs, while intermediate-mass stars (more massive than $\sim 5\,M_\odot$) would produce NSs. Using a mass distribution of stars at birth given by the Salpeter IMF led to the estimate of approximately one third the number of NSs. (It would have been better to assume stars more massive than $16\,M_\odot$ would become BHs, and stars more massive than $8\,M_\odot$ would produce NSs; the outcome would remain the same). The estimate of van den Heuvel (1992) was based on the incidence of BH X-ray binaries with low-mass companions (BH–LMXBs) in the Galaxy. We know from the NS–LMXBs that the formation of a NS–LMXB is a very rare event: less than one in 10^5 NSs in the Galaxy ends up in a NS–LMXB (see Chapter 11). Assuming the same to be true for stellar BHs in relation to BH–LMXB (because, similar to the case of a NS–LMXB, it is very difficult to retain a low-mass companion next to an exploding massive star), one may use the estimated Galactic number of presently active BH–LMXBs and their estimated lifetime to make an estimate of the total number of stellar BHs present in the Galaxy. This estimate led to at least 10^8 stellar BHs in the Galaxy (van den Heuvel, 1992).

The estimate by Blandford (1987) may have been on the low side, for the following reasons. The present metallicity of stars in the Galactic disk is approximately $Z = 0.015$, which with a disk mass of $\sim 10^{11}\,M_\odot$ corresponds to $\sim 1.5 \times 10^9\,M_\odot$ of elements heavier than helium synthesized in the lifetime of the Galaxy. Assuming that on average a NS-producing SN explosion ejects $1.5\,M_\odot$ of elements heavier than helium, one finds the Galactic number of NSs to be 10^9. Assuming again the number of stellar BHs to be one third that of NSs, one finds the number of BHs to be $\sim 3 \times 10^8$.

More recently, using detailed binary population synthesis calculations, Olejak et al. (2020) made more refined estimates of the Galactic numbers of stellar-mass BHs and of the numbers of BH+BH and BH+NS systems (as well as of the expected Galactic rate of mergers of the latter systems). They found a number of 1.2×10^8 single BHs with an average mass of $14\,M_\odot$ and 9×10^6 BHs in binary systems with an average mass

Table 6.8. The WR Star X-ray Binaries

Galaxy	Source	P_{orb} (hr)	M_{WR} ($M_\odot$)	M_X ($M_\odot$)
Milky Way	Cyg X-3	4.8	8 – 12	≥3(?)
NGC 4490	CXOU J123030.3+413853	6.4	—	—
Circinus	CG X-1	7.2	—	—
NGC 253	CXOU J004732.0−251722.1	14.5	—	—
NGC 300	X-1	32.8	26	17 ± 4
IC 10	X-1	34.9	35	33(?)
M101	ULX-1	196.8	19	20(?)

The data are taken from Zdziarski et al. (2013), Esposito et al. (2015), Binder et al. (2021), and references therein.

The mass of the WR star (M_{WR}) in Cyg X-3 is estimated from its infrared luminosity (see text). The mass estimates of the compact stars (M_X) in WR X-ray binaries are very uncertain—as examples, see Laycock et al. (2015) and Binder et al. (2021) for the cases of IC 10 and NGC 300 X-1, respectively. On the basis of binary evolution, thesc compact stars are expected to be BHs.

of 19 $M_\odot$ in the Galaxy. In view of the large uncertainties in the outcomes of binary population synthesis models (see Chapter 16), the last-mentioned numbers have a large uncertainty.

6.6.2 Non-interacting BHs in Binaries and Triple Systems

In view of the large expected numbers of stellar-mass BHs in the Galaxy, and the high incidence of binaries and multiple systems, one may expect that the number of binaries or multiple systems in which one of the stars is a BH will be large. A BH in a binary will only be easily recognized if it is accreting matter from a companion star such that it becomes an X-ray source. Olejak et al. (2020), Wiktorowicz et al. (2020), and Yungelson et al. (2020) pointed out that most of the BHs in binaries will not be interacting (i.e., not accreting matter from their companion) and therefore cannot be detected in X-rays.

The presence of these so-called non-interacting BHs can only be detected on the basis of large-amplitude radial velocity variations of the companion star, or periodic astrometric motions of the companion star, detectable with ESA's Gaia astrometry satellite (Wiktorowicz et al., 2020). If such studies show the presence of an invisible companion with a mass ≥3.0 $M_\odot$, one will have a candidate for a non-interacting BH. Wiktorowicz et al. (2020) estimate that ≤14 such non-interacting BH binaries may be detectable by means of radial velocity studies, using the Large Sky Area Multi-Object Fibre Spectroscopic Telescope (or LAMOST), and approximately 40 – 340 by using astrometry with Gaia. Still, the measurement of solely the mass of an unseen star may not be a sufficient criterion, as can be shown by the following example (related to the interpretation of the unseen companion of the red-giant star 2MASS J05215658

+4359220; see van den Heuvel & Tauris [2020]). Assume that a normal main-sequence companion is detectable if it emits >2% of the light of the brightest star in the binary. Suppose that the mass of a main-sequence star that emits this 2% is M_2. Then, if no light of a companion is detected, one will conclude that any main-sequence companion of the main star has a mass $\leq M_2$. If now the unseen companion has a mass of, say, $1.4 \times M_2$, and this mass is $\geq 3.0\,M_\odot$, one will conclude that the unseen star cannot be a normal light-emitting star and must be a BH. However, in this reasoning it was assumed that the unseen companion is a *single* main-sequence star. In binaries with orbital periods $\geq 30-40$ d, a companion may itself also be a close binary: the full system may be a hierarchical triple! A "star" of mass $1.4 \times M_2$ will in that case consist of, for example, two normal main-sequence stars, each of mass $0.7 \times M_2$, which—because of the mass–luminosity relation ($L \propto M^{3.5}$)—will each have a luminosity of $0.7^{3.5} = 0.29$ times the luminosity of a star of mass M_2. So, a binary of two such stars has the luminosity of 0.58 times the luminosity of a star of mass M_2 and will therefore be undetectable against the light of the main star, even though the mass of this unseen companion is $1.4 \times M_2$. A mass of the unseen star $\geq 3.0\,M_\odot$ is therefore not yet proof that we are dealing here with a BH. (This triple star possibility for dark companions was first put forward by Trimble & Thorne [1969].)

Detection of large radial velocity variations caused by an unseen companion was the first method proposed for detection of the presence of a BH (Guseinov & Zel'dovich, 1966; Trimble & Thorne, 1969). This method was applied recently by Khokhlov et al. (2018), Liu et al. (2019), Thompson et al. (2019), and Rivinius et al. (2020), who claimed to have detected massive unseen companions that are BHs. Already prior to this, Qian et al. (2008) had discovered that the bright naked-eye eclipsing binary V Pup (spectral types B1V+B3, $P_{\rm orb} = 1.45$ d; see Table 5.2) has an unseen companion of $\geq 10.3\,M_\odot$ in a 5.47 yr orbit. As no light of this companion could be detected, Qian et al. suggested it to be a BH.

It has turned out that four of these five claimed BHs probably do not exist. Only the case of the companion of the Be-star AS 386 studied by Khokhlov et al. (2018) has so far survived closer scrutiny. The most spectacular case was that by Liu et al. (2019) who claimed that the young Be-star LB-1 has a 70 $M_\odot$ BH companion in a circular orbit with $P_{\rm orb} = 79$ d. Soon after its publication, it was shown by several authors (e.g., Simón-Díaz et al., 2020; Irrgang et al., 2020; Shenar et al., 2020 and El-Badry & Quataert, 2021) that the analysis by Liu et al. (2019) of the complex emission line spectra of the Be-star was flawed. Careful reanalysis of the spectra showed that the companion is a UV-bright stripped star with a mass of only $\sim 1.1\,M_\odot$ (Irrgang et al., 2020; Shenar et al., 2020), but see Lennon et al. (2021).[4]

Likewise, the alleged triple system HR 6819 (again a Be-star), in which the inner binary with an orbital period of 40.33 d was suggested to contain a BH of mass $\geq 4\,M_\odot$ (Rivinius et al., 2020), was immediately shown by Bodensteiner et al. (2020) and El-Badry & Quataert (2021) not to be a triple system but just a binary in which the Be-star has a stripped low-mass companion of only $\sim 0.4\,M_\odot$.

---

[4]This case shows that even papers that passed the very strict refereeing system of the journal *Nature* (probably involving three or four referees) can still be flawed.

The third disappearing BH case is that of the red-giant 2MASS J05215658+435922, which was proposed by Thompson et al. (2019) to have a $\geq 3.3\,M_\odot$ BH in a 83 d orbit around an $\sim 3\,M_\odot$ red giant. van den Heuvel & Tauris (2020) showed that an alternative interpretation is possible here in terms of a system consisting of a $1\,M_\odot$ red giant with an unseen close binary system of mass $1.8\,M_\odot$, which consists of two $0.9\,M_\odot$ K-dwarfs that together emit only 0.5% of the light of the red giant, and is therefore undetectable.

A similar interpretation can also be applied to the $10\,M_\odot$ unseen companion in a 5.5 yr orbit of the eclipsing binary V Pup. If this companion consists of two $5\,M_\odot$ main-sequence stars, the light of this pair is undetectable against the very bright binary V Pup, as the reader can easily verify.

In three of these five above-mentioned claimed BH cases, the primary star is a Be-star, a type of star for which the interpretation of the spectral lines (emission lines with often variable absorption components) is notoriously difficult. Here observers who are not sufficiently cautious or experienced in the interpretation of the spectra of these stars—which are often found in binaries—may easily go astray.

In the case of the claim of the $70\,M_\odot$ BH companion of the young massive Be-star with solar metallicity, the impossibility of this case should have been immediately clear to people with knowledge of stellar evolution and of the observations of the winds of massive stars. This is because during the past four decades it had become clear from careful observations of the stellar wind mass-loss rates of massive stars in the Galaxy and the Magellanic Clouds that at solar-like metallicity, the winds of massive stars are so strong that by the time the core collapses, the stars have lost so much mass that no BHs with masses above about $20\,M_\odot$ can form (see discussions in Chapters 8 and 9). Should one throw away four decades of careful observational work of hundreds of astronomers worldwide because of one new observational paper published on hard-to-interpret spectra of one Be-star? The answer is clearly "of course not."

The study of these four cases shows that one should be extremely cautious with believing the claims that an unseen companion is a BH, particularly so long as one has not exhaustively studied alternative interpretations of the observations. For further discussion on this important topic of false positive detections of BH candidates, see Clavel et al. (2021).

EXERCISES

Exercise 6.1. *In our Galaxy there are* $\sim$*100 bright X-ray sources with fluxes,* $F_X > 10^{-10}\,\mathrm{erg\,cm^{-2}\,s^{-1}}$ *in the energy range of* $1-10\,\mathrm{keV}$*. These fluxes correspond to source luminosities of* $L_X \approx 10^{34}-10^{38}\,\mathrm{erg\,s^{-1}}$*. These luminosities are roughly equivalent to the rate at which gravitational potential energy is released from matter falling down the deep potential well of a compact object (see Table 6.1).*

a. *Calculate the accretion rate,* $\dot{M}$ *(in* $M_\odot\,\mathrm{yr}^{-1}$*), needed to feed a* $1.4\,M_\odot$ *NS with a detected X-ray flux,* $F_X = 2.8\times10^{-9}\,\mathrm{erg\,cm^{-2}\,s^{-1}}$*, located at a distance of 7.0 kpc. Assume a NS radius of 12 km.*
b. *Does this accretion rate exceed the Eddington limit for a NS? (See Section 4.9.)*

Exercise 6.2. *Using the observed radial velocity amplitude and orbital period, calculate the lower limit of the mass of the massive companion of Cen X-3 (see Section 2.7), by assuming that the X-ray pulsar has a mass of 1 or 2* $M_\odot$.

Notice, the mass function is given by Eq. (6.1), for example, and derived in Chapter 3.

Exercise 6.3. *The system of Cen X-3 is considered again. Observed data are*

- *Orbital period,* $P_{\rm orb} = 2.087$ d. *Eclipse duration,* $\Delta t = 0.488$ d.
- *The orbit is circular (i.e.,* $e \simeq 0$*).*
- *Mass function,* $f(M) = \dfrac{(M_{\rm opt} \sin i)^3}{(M_{\rm X} + M_{\rm opt})^2} = 15.6\, M_\odot$ *(see Eq. 6.1).*

Here $M_{\rm X}$ *is the mass of the NS (X-ray source);* $M_{\rm opt}$ *is the mass of the optical companion star and* i *is the inclination of orbital plane (defined as the angle between the orbital plane and the plane of the sky; e.g.,* $i = 90°$ *if our line of sight is parallel to the orbital plane). In the following, we assume the optical star to be spherical with radius* $R_{\rm opt}$*, and the NS to be a point source in a relative circular orbit with radius* a.

a. *Show that, if the values of* $R_{\rm opt}$ *and* a *are known, one finds for an orbital inclination angle,* $i \leq i_{\rm persist} = \cos^{-1}(R_{\rm opt}/a)$*, the emitted emission from the NS is persistent with no eclipses observed.*

b. *Show that for an orbit seen edge on (*$i = 90°$*), the ratio of the radius of the optical star,* $R_{\rm opt}$*, and the radius of the relative orbit,* a*, is given by*

$$\frac{R_{\rm opt}}{a} = \sin\left(\frac{\Delta t}{P_{\rm orb}}\pi\right), \tag{6.10}$$

where Δt *is the duration of the eclipse.*

One can show (e.g., van den Heuvel & Heise, 1972; Winn, 2010) that, if Δt *and* $P_{\rm orb}$ *are known, the ratio of the radius of the optical star,* $R_{\rm opt}$*, and the radius of the relative circular orbit,* a*, is for all values of* $i > 0$ *given by*

$$\frac{R_{\rm opt}}{a} = \sqrt{1 - \cos^2\left(\frac{\Delta t}{P_{\rm orb}}\pi\right)\sin^2 i}. \tag{6.11}$$

c. *Check the validity of the preceding equation in the limits where* $i = i_{\rm persist}$ *(i.e.,* $\Delta t \simeq 0$*) and for* $i = 90°$ *and compare with questions a and b.*

d. *Using Kepler's third law and the mass function, show that for Cen X-3 one can write*

$$R_{\rm opt} \simeq 4.21 \left(\frac{P_{\rm orb}^{2/3}\sqrt{M_{\rm opt}\sin i}}{15.6^{1/6}}\right)\sqrt{1 - \cos^2\left(\frac{\Delta t}{P_{\rm orb}}\pi\right)\sin^2 i}, \tag{6.12}$$

where $P_{\rm orb}$ *is measured in days,* $M_{\rm opt}$ *in* $M_\odot$*, and* $R_{\rm opt}$ *in* $R_\odot$.

e. *If the mass of the NS in Cen X-3 is* $M_X = 1.4\,M_\odot$*, find (using the mass function) the values of* M_{opt} *for* $i = 90°$, $75°$, *and* $60°$.

f. *Argue that for a given measured eclipse fraction of* $(\Delta t / P_{orb})$*, the smallest possible value of* R_{opt} *is obtained for* $i = 90°$*. Then calculate the values of* R_{opt}*, in units of* $R_\odot$*, for* $i = 90°$, $75°$, *and* $60°$.

g. *Derive Eq. (6.11).*
Hint: draw the projection on the sky of the spherical star and the orbit of the NS for the case that $i < 90°$*, such that the projection of the orbit on the sky is an ellipse. The length cord on the sky of the curved eclipsed part of the orbit is a cord of the circle with radius,* R_{opt}*, that represents the star. But it is also equal to the cord of the circle that spans in the real circular orbit the angle* $(\pi\,\Delta t / P_{orb})$*. Therefore, calculate the value of this cord both on the projection of the star and in the orbit of the NS and set these two values to be equal to each other. This gives Eq. (6.11).*

Notice that, because of tidal and centrifugal forces, the Cen X-3 stellar companion is not spherical in shape, but instead rather flattened. Hence, Eq. (6.11) is not exactly valid and this explains why, shortly after the discovery of Cen X-3, van den Heuvel & Heise (1972) argued for a NS mass in this system only of the order of $0.5\,M_\odot$*. The most recent measurements of Cen X-3 reveal that the X-ray pulsar has a mass of* $M_X = 1.49 \pm 0.08\,M_\odot$*, and the optical counterpart has a mass of* $M_{opt} = 20.5 \pm 0.7\,M_\odot$ *and a radius of* $R_{opt} = 11.4 \pm 0.7\,R_\odot$*. See, for example, Shirke et al. (2021) for an updated spin measurement and references therein.*

Exercise 6.4.

a. *Calculate for both systems in Figure 6.5, the distance from the center of the donor star to the c.m.*

b. *Calculate the Roche-lobe radius* (R_L) *of the donor star in both systems.*

c. *Calculate the distance from the center of the donor star to the first Lagrangian point* (L_1) *in both systems.*

d. *Write a computer script to plot the Roche lobes and compare with Figure 6.5.*

Exercise 6.5. *An alternative way to calculate the accretion radius of a compact star in a stellar wind is the following, after Hoyle and Lyttleton (Hoyle & Lyttleton, 1939), as depicted in Figure 6.30. Consider the wind to consist of free particles moving in the X-direction. To be captured by a point mass with mass,* M*, the orbit of the passing particle must be deflected by the gravity of* M*, so much that the particle, after it collides with a particle from the opposite side at the same distance, retains a velocity in the X-direction that is not larger than the escape velocity from* M *at the collision point. In this case the colliding particles will dissipate their transverse velocity components into heat, which is radiated away, and then fall to the compact star and be captured. Particles deflected less will retain an X-velocity larger than the escape velocity and flow away, although with a somewhat reduced speed.*

Show that for the trajectory to be deflected such that it is captured, a particle with a velocity v_{rel} *with respect to the compact star has to pass this star within a distance*

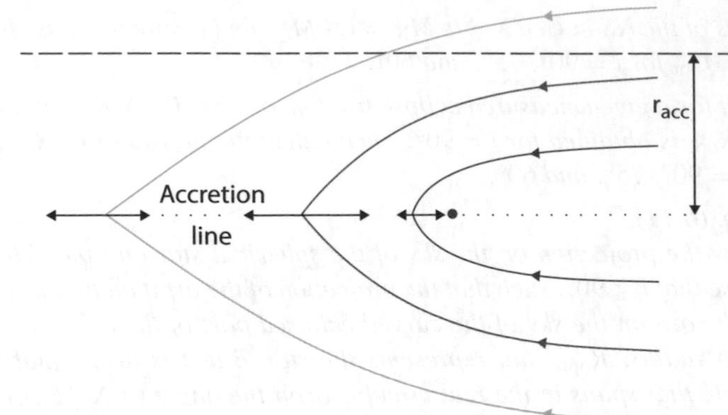

Figure 6.30. Hoyle-Lyttleton treatment of wind accretion. The wind consists of a parallel stream of free particles, moving in the left direction (X-direction). The star is the small red circle. The accretion radius is defined as the largest distance from the star where, after deflection of the particle orbit due to the gravity of the star and collision with a particle from the other side, it retains a velocity in the X-direction that equals the escape velocity from the star. All the material passing within the accretion radius will be gravitationally captured by the star. After Boffin (2015).

$r_{\rm acc}$ *given by*

$$v_{\rm rel} = v_{\rm esc}(r_{\rm acc}) = \sqrt{2\,GM_{\rm X}/r_{\rm acc}}. \tag{6.13}$$

Hence,

$$r_{\rm acc} = 2\,GM_{\rm X}/v_{\rm rel}^2, \tag{6.14}$$

where $M_{\rm X}$ is the mass of the compact star. Thus particles will be captured that pass through a circle of radius $r_{\rm acc}$ around the star. Notice that this expression for the accretion radius is basically the same as the accretion radius for Bondi-Hoyle accretion.

Chapter Seven

Observed Properties of X-ray Binaries in More Detail

7.1 HIGH-MASS X-RAY BINARIES IN MORE DETAIL

7.1.1 Supergiant HMXBs vs. Be-HMXBs

It appears that the HMXBs fall into two main categories that differ in a number of important characteristics:

1. Persistent so-called standard HMXBs, in which the massive OB-type companion of the X-ray source is a blue supergiant star that is close to filling its Roche lobe, as evidenced by the double-wave optical light curve (see Fig. 7.1).
2. Be/X-ray binaries.

Table 6.2 lists a number of characteristic examples of both categories and Figure 7.2 depicts four characteristic differences between the two groups:

1. In standard supergiant-HMXBs (sg-HMXBs), the companions are evolved blue supergiant stars with radii of 10–30 $R_\odot$, optical luminosity $L_{\rm opt} \geq 10^5\ L_\odot$, and an initial mass (derived from their luminosities by using evolutionary tracks $>20\ M_\odot$). On the other hand, the optical components of Be/X-ray binaries are generally unevolved stars of spectral type O9Ve to B2Ve (in some cases they are of luminosity class IV: subgiants). Such stars have relatively small radii, $\lesssim$5–10 $R_\odot$; absolute luminosities $\lesssim 3 \times 10^4\ L_\odot$; and masses between 8 and 20 $M_\odot$.
2. Standard HMXBs tend to have short orbital periods, mostly between 1.4 and 10 d, whereas the Be/X-ray binaries have orbital periods in the range of 15 d to several years; most of them have periods between 3 and 12 months.
3. Characteristics 1 and 2 explain why the standard systems in many cases exhibit X-ray eclipses and always show double-wave ellipsoidal light variations, indicating that their optical components are close to filling their Roche lobes. The mass transfer in these systems is thought to be due to the strong, enhanced stellar winds of the supergiants and/or beginning atmospheric RLO, as is explained in Section 4.5.1. On the other hand, characteristics 1 and 2, in combination with the absence of X-ray eclipses and of periodic ellipsoidal light variations in the Be/X-ray binaries, indicate that the Be-stars in these systems reside deep inside their Roche lobes. This rules out RLO or enhanced winds as sources of the mass accretion.

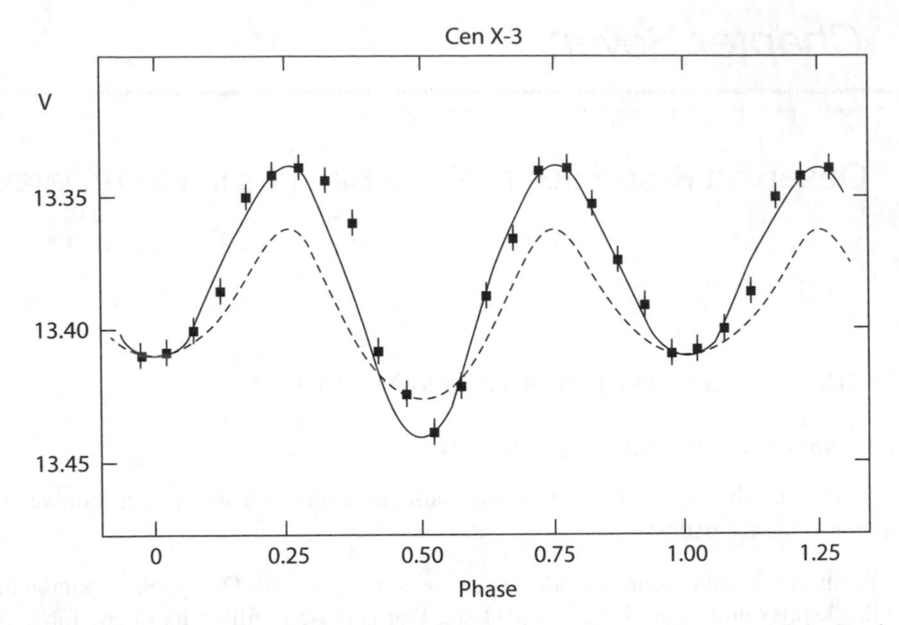

Figure 7.1. Observed light curve of Cen X-3 in comparison with two models. Dashed line includes only ellipsoidal variations, and solid line includes both ellipsoidal variations and an accretion disk. After Tjemkes et al. (1986).

4. The *standard* systems are persistent X-ray sources (i.e., they are always *on*). On the other hand, many of the Be sources are so-called *transients*, which means that they may be completely unobservable (*off*) for long periods of time, ranging from months to many years, and then suddenly turn on as a strong X-ray source for a duration of weeks to months. Because RLO is ruled out here as a source of mass accretion, the sudden mass transfer must be due here to the intrinsic mass-losing properties of the Be component, as was first argued by Maraschi et al. (1976).

Be-stars show two types of intrinsic mass loss: (1) by a weak stellar wind and (2) by mass ejection from the equatorial regions of the star. All Be-stars are very rapid rotators and, as a result, highly flattened (ellisoidal) stars. The irregular mass-ejection events from their equatorial regions are thought to be rotationally driven, possibly by excitation of non-radial oscillations (e.g., Slettebak, 1988; Baade et al., 2018). The weak winds of the Be-stars can only generate X-ray luminosities of the order of 10^{31} to 10^{34} erg s^{-1} at most (cf. Section 6.3.3), which are of the same order as observed for the nearby weak Be/X-ray binaries X Per and γ Cas. On the other hand, the transient outbursts with $L_X = 10^{37}$ to 10^{39} erg s^{-1} cannot be explained in this way. They must be due to the equatorial outbursts of mass ejection that are characteristic for Be-stars (Maraschi et al., 1976). These outbursts are already long known to occur at completely erratic moments (Underhill, 1966). Figure 7.3 shows as a characteristic example of the intensity variations of the strong pulsating transient source V0332+53, which had on stages only in 2015, 2005, 1983/84, and 1973/74 and was off for the decades in

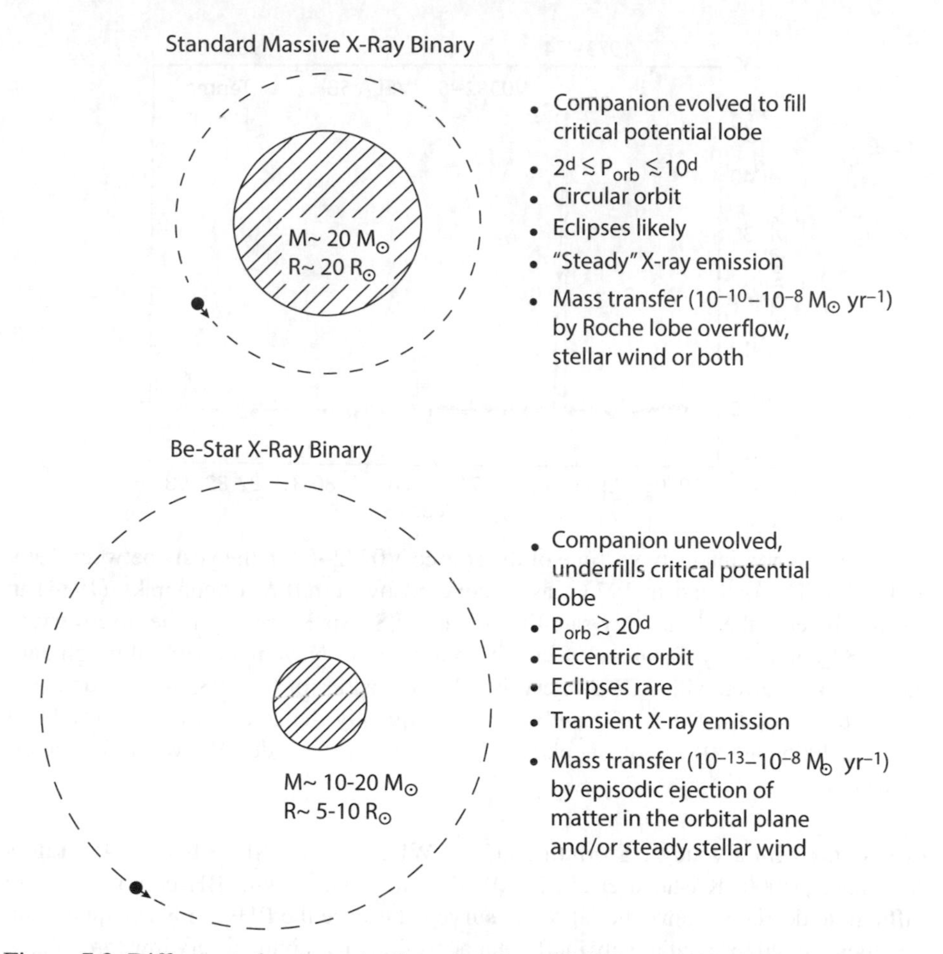

Figure 7.2. Differences between the "standard" sg-HMXBs and the Be/X-ray binaries. After Rappaport & van den Heuvel (1982).

between. Between November 14, 1983, and January 24, 1984, it had three on stages of ~1 week, recurring approximately every 34 d, separated by off intervals. The orbit is moderately eccentric, and its periodic turn-ons in 1983/84 occurred near periastron passage. Since then the system went through outbursts again in 2005 and in 2015 (June–October). From the last outburst the orbital period and eccentricity were derived by Doppler tracking of the pulse period to be $P = 33.85$ d and $e = 0.37$, respectively (Doroshenko et al., 2016). Figure 7.4 gives a schematic picture of how the periodic X-ray flares of a Be/X-ray binary during a phase of equatorial mass ejection from the Be-star are thought to arise.

As mentioned in Chapter 6, the SMC is particularly rich in Be/X-ray binaries, with ~100 detected systems of this type. This is presumably due to a burst of star formation ~60 Myr ago. A handful of X-ray sources in the SMC are identified as WD-Be/X-ray binaries (Coe et al., 2020; Kennea et al., 2021). These binaries are progenitors

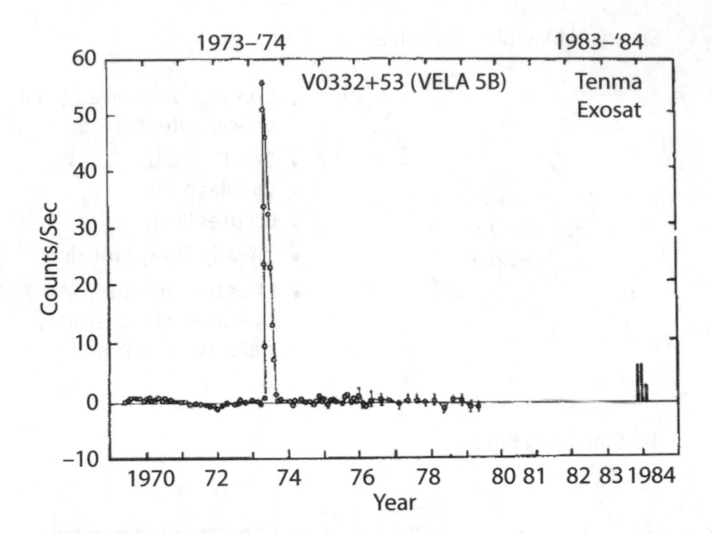

Figure 7.3. X-ray intensity vs. time of the source V0332+53 in the years between 1969 and 1985. The outburst in 1973 was discovered by Terrell & Priedhorsky (1984) in the database of the Vela γ-ray satellites of the U.S. Air Force, after the discovery of the 1983 outburst by Tanaka (1983). The outbursts in November 1983 through January 1984 occurred with $\sim$34 d intervals. The intensities of the 1983/84 outbursts are drawn to the scale of the 1973 outburst. Subsequently the source went into outburst again in 2005 and 2015 (Doroshenko et al., 2016). After van den Heuvel & Rappaport (1987).

of systems with a young NS orbiting an old WD, such as PSR J1141−6545 (Tauris & Sennels, 2000; Krishnan et al., 2020). Be-star binaries with BH companions are difficult to detect by conventional X-ray surveys because the BHs are X-ray quiescent, i.e., they are fed by a radiatively inefficient accretion flow giving a very low luminosity, such as $<1.6 \times 10^{-7}$ times the Eddington luminosity for the MW BH-Be/X-ray binary MWC 656 (Casares et al., 2014). This is the only system of this type known.

A couple of Be/X-ray binaries have been found in SN remnants in our satellite galaxies: the SMC and the LMC (e.g., Gvaramadze et al., 2021; Maitra et al., 2021). These systems must therefore be very young.

7.1.2 Highly Obscured HMXBs

Two new subclasses of sg-HMXBs were discovered in the past decades: the highly obscured HMXBs and the supergiant fast X-ray transients (SFXTs). Until 2003, only $\sim$10 sg-HMXBs were known: seven pulsating ones and two BH ones (Cyg X-1 and LMC X-1). Only seven of them are in our Galaxy. Since its launch in October 2002, the European γ-ray satellite INTEGRAL, with its IBIS (Imager on Board the INTEGRAL Satellite) instrument, which is sensitive in the hard X-ray (i.e., soft γ-ray) spectral range between 17 and 100 keV, discovered almost two dozen new sg-HMXBs in our Galaxy. These had been missed by earlier X-ray satellites that always operated at softer X-ray

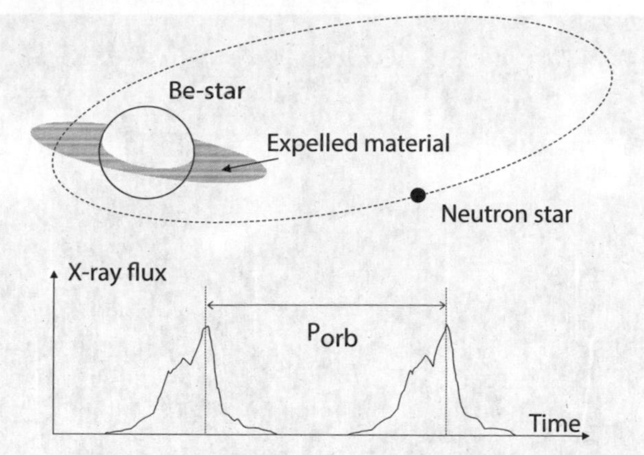

Figure 7.4. Schematic model of a Be/X-ray binary system such as A0535+26 and V0332+53. The NS moves in an eccentric orbit around the Be-star, which is much smaller than its Roche lobe at any position in the orbit. The rapidly rotating Be-star is temporarily surrounded by matter expelled in its equatorial plane (misaligned with the orbital plane), which forms a circumstellar disk of gas. Near its periastron passage, the NS enters this circumstellar matter and the resultant accretion produces an X-ray outburst, lasting several days to weeks. After Tauris & van den Heuvel (2006).

energies, below about 10 keV. These softer X-rays are stopped by a neutral-hydrogen column density toward the X-ray source in excess of $\simeq 10^{23}$ atoms cm^{-2}, while the harder X-rays, >17 keV are still able to reach us through these column densities. Excellent reviews of the HMXBs studied by INTEGRAL were given by Walter et al. (2015) and Kretschmar et al. (2019). The newly discovered INTEGRAL sources appear to be highly obscured, as first discovered by Walter et al. (2003), and only observable in hard X-rays. The obscuring matter is in the binary system and its surroundings. This is clear because in the well-studied systems of this type one observes the amount of obscuration to vary with the binary orbital phase, as is illustrated in Figures 7.5 and 7.6 for the system of IGR J17252−3616.

An interesting feature of this variation with orbital phase is that numerical hydrodynamic simulations of the wind accretion process show that the amplitude of the obscuration variability depends on the mass of the NS. This enabled Manousakis et al. (2012) to determine that the NS in IGR J17252−3616 has a mass of between 1.9 and 2.0 $M_\odot$, as is shown in Figure 7.6. The new HMXBs tend to harbor X-ray pulsars, mostly with pulse periods of hundreds of seconds, and their companions are late O- and early B-supergiants.

The Extremely Obscured Source IGR J16318−4848: A Special Case

This is the first highly obscured sg-HMXB discovered by INTEGRAL (Walter et al., 2003). At the same time, it is also the most extreme obscured system; it is very different from all the other obscured sg-HMXBs discovered by INTEGRAL. It therefore should be considered as a separate class by itself.

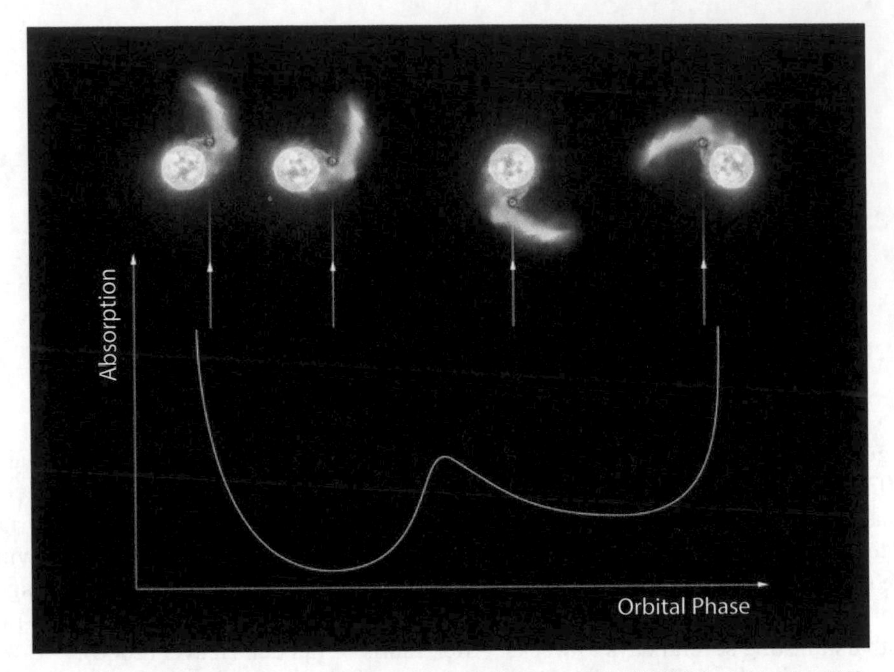

Figure 7.5. Illustration of the eclipsing X-ray binary IGR J17252−3616 in different orbital phases. The system consists of a NS accreting from the stellar wind of its companion, a blue supergiant star. The orbital period is 9.74 d and the pulse period is 404 s. It is a highly obscured system of the type of which INTEGRAL discovered many. The top pictures show the high-density wake in the wind of the donor star at four orbital phases, as derived from numerical-hydrodynamic simulations of the wind-accretion process. Credit: ESA: https://sci.esa.int.

Contrary to the other obscured sg-HMXBs, which have O- and B-supergiant donor stars, its optical counterpart is a B[e]-supergiant (pronounced "B-bracket-e"; not to be confused with the Be-stars). The notation B[e], which was coined in 1976 by Peter Conti (see Zickgraf, 1998), concerns B-stars characterized by a large infrared excess and with emission lines of neutral and low-ionization atoms such as Fe-II and [Fe-II] (the latter ones are forbidden lines). The optical spectrum is further dominated by emission lines of H-I and He-I, often with P-Cygni profiles, indicating strong stellar wind mass loss, though at a relatively low velocity of $\sim$300–400 km s^{-1}. The effective temperatures, derived from their optical spectral energy distributions, are in the range of B0 to B8 stars. Photospheric features in these stars can be masked by their dense winds (see, e.g., de Freitas Pacheco, 1998). The B[e]-stars form a mixed group of objects, ranging from pre-main-sequence stars to luminous supergiants. The latter ones form a rare separate class, first recognized in the LMC and SMC, with luminosities often between 10^5 and $10^6\,L_\odot$ or even greater (Zickgraf, 1998). The spectral types of the most luminous ones ($L \approx 10^6\,L_\odot$) are B(0–2)Ia to Ia$^+$ (Zickgraf, 1998). It appears likely that their large wind mass loss is driven by their enormous luminosities, but the

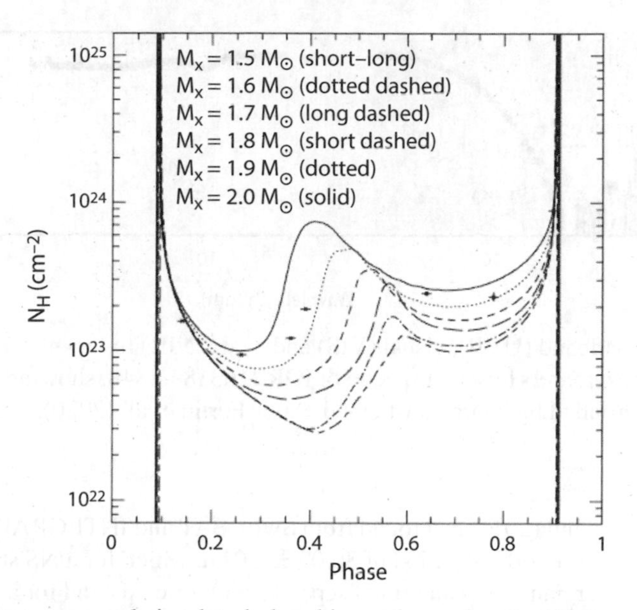

Figure 7.6. Time-averaged simulated absorbing column density (N_{H}) as a function of orbital phase in the system IGR J17252−3616, for various assumed NS masses. The stellar wind is characterized by $\dot{M} = 10^{-6}\, M_{\odot}\, \mathrm{yr}^{-1}$ and $v_{\mathrm{term}} = 500\,\mathrm{km\,s}^{-1}$. The orbital radius is 1.75 times the radius of the supergiant. After Manousakis et al. (2012).

origin of the large amount of circumstellar dust, often in the form of disk- and ring-like structures, moving in Keplerian orbits (Kraus, 2016), which produces the IR-excess, is not well understood. Because yellow hypergiants (YHGs), which have similar luminosities, also may show the presence of circumstellar ring-like structures, it has been speculated that the B[e]-supergiants are at a later evolutionary stage of these stars (Aret et al., 2017), but this is by no means certain; it could also be the other way around. For a further discussion of the B[e]-stars and their possible evolutionary connections with other luminous stars, see Chapter 8.

Detailed study of the B[e]-supergiant optical counterpart of the X-ray source IGR J16318−4848 (Chaty, 2011; Chaty & Rahoui, 2012; Fortin et al., 2020) showed that its spectral energy distribution has an enormous infrared excess (see Fig. 7.7). Its infrared spectrum shows lines and features of, for example, neon, polycyclic aromatic hydrocarbons (PAHs), and silicon, proving that the extra absorbing component surrounding the system is made up of dust and cold gas. By fitting the optical to mid-infrared spectra with a sophisticated aspheric disk model developed for Herbig Ae-Be–stars, and adapted to sg-B[e]–stars, Fortin et al. (2020) showed that the supergiant star is surrounded by a hot rim of dust at 6,740 K, at a distance of ~1.5 AU, along with a hot dusty viscous outer disk component at an inner temperature of ~1,374 K, extending to ~6 AU.

The star was found to have an effective temperature of ~20,000 K and a radius of approximately 34–36 $R_{\odot}$, if its mass is 25 $M_{\odot}$, and approximately 75–80 $R_{\odot}$ if its mass

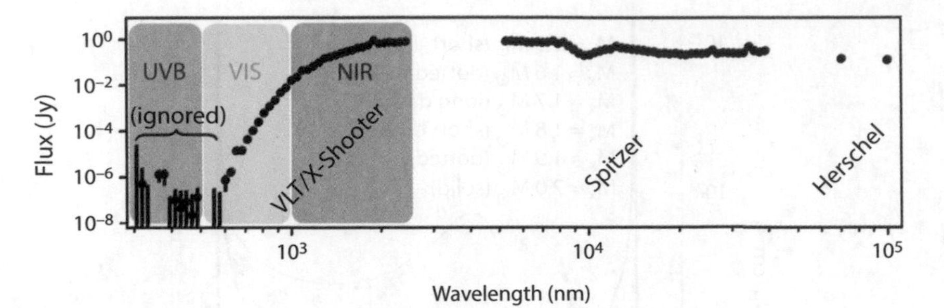

Figure 7.7. Broadband (UVB), visual (VIS) and near- (NIR) to far-infrared, ESO-VLT/X-shooter, Spitzer, and Herschell spectra of IGR J16318−4848, showing that the entire system is enshrouded by a large dust cloud. After Fortin et al. (2020).

is 50 $M_\odot$. Iyer & Paul (2017) had found from Swift-BAT and INTEGRAL observations that the system has an orbital period of 80.09 ± 0.01 d, which for a NS star companion and a 25 $M_\odot$ supergiant gives an orbital separation of 1.06 AU and for a 50 $M_\odot$ supergiant, 1.34 AU. As the hot inner rim of the disk is located somewhat beyond the latter value, the compact star orbits inside the inner cavity of the disk (Fortin et al., 2020). The outside rim of the cool disk is located at ∼6 AU. As the stellar radius is much smaller than the orbital separation, the X-ray source must be powered by wind accretion. Fortin et al. (2020) found that the supergiant is helium-rich: its He/H ratio is 3.7 times that of the Sun. This, together with the fact that the star is deep inside its Roche lobe, implies that its strong stellar wind must already have removed a considerable part of its initial hydrogen-rich envelope. IGR J16318-4848 is not the only sg-B[e] HMXB. The other one is CI Cam, which also appears to have quite a long orbital period of 19.4 d (Bartlett et al., 2013). The difference between the two systems seems to be that, while the former one is seen practically edge-on (high orbital inclination), CI Cam is seen more pole-on (Fortin et al., 2020). The fact that in both systems the dusty disk coincides with the orbital plane of the compact companion star strongly suggests that, as in the less obscured system of IGR J17252−3616 (see Figs. 7.5 and 7.6), the dense slow wind of the B[e]-star (wind velocity on the order of 400 km s^{-1} or less) produces a dense spiral-shaped wake behind the compact star. It appears that this out-spiralling dense wake then has produced the disk in the orbital plane, beyond the orbit of the compact star. The cooling of the disk matter has then caused the dust condensation.

The fact that the disk is in the orbital plane (which does not need to be the equatorial plane of the supergiant, because the compact star may have received a random kick at its birth) is therefore, in our opinion, a strong indication that the presence of the compact star has been instrumental in creating this disk surrounding its orbit. It appears that the dense and absorbing circumstellar material (disk and envelope) enshrouds the entire binary system, as sketched in Figure 7.8. From its X-ray emission, we know that there is a NS or BH in this system. Its supergiant companion must have been ejecting this huge envelope of gas and dust already for quite some time, and the compact star is apparently accreting from this reservoir of matter.

Figure 7.8. Sketch of the possible configuration of the system IGR J16318−4848. The model is a massive NS binary with a 25 $M_\odot$ donor in a nearly circular orbit of 80 d period. This nearly edge-on system is enshrouded by a large dust disk in the orbital plane, through which only hard X-rays are able to escape. After Chaty (2013).

Judging from the properties and the huge luminosities of the B[e]-supergiants in the Magellanic Clouds, it may well be that the large mass loss of the star is radiatively driven by its huge luminosity itself (e.g., that it is close to the Eddington limit) and is not related to the binary character. An alternative is that the compact companion has played a major role in ejecting the envelope matter, by some form of binary interaction. If indeed B[e]-stars originate from YHGs, as Aret et al. (2017) have suggested, the donor star will, in its YHG phase, have engulfed the compact star, leading to some kind of CE configuration. We would now be observing the system after the CE interaction.

7.1.3 Supergiant Fast X-ray Transients

Apart from the highly obscured sources, INTEGRAL discovered another new class of sg-HMXBs: the SFXTs, which generally have X-ray luminosities $<10^{34}$ erg s^{-1}, but at irregular times may show sudden outbursts to X-ray luminosities of 10^{36}–10^{37} erg s^{-1}, lasting on the order of a few hours. The total energy release in such a flare is 10^{38}–10^{40} erg, corresponding to a total accreted amount of 10^{18}–10^{20} g. Regular pulsation periods have been measured for approximately half of the systems, and these fall in the same range as for the other sg-HMXBs (for reviews, see Sidoli, 2012; Postnov et al., 2012).

In a recent investigation of stellar and wind properties of HMXBs, Hainich et al. (2020) claim that there is no systematic difference between the wind parameters of the

Table 7.1. Spectral Types of the Identified Donor Stars of SFXTs and Persistent sg-HMXBs Discovered by the INTEGRAL Satellite

SFXT donors	Persistent sg-HMXB donors
O8.5Ib-II(f)p	BN0.7Ib
O8Ia(fpe)	ON9.7Iab
O8.5I	B1Ia
O9.5I/B0.5I	Bsg-e
O8.5Ib	B0.5sg-e
O8.5Iab(f)	O8I
O8Iab(f)	B0.5Ib
O9Ib	O8.5I
B0.5Iab	B1Ib
B0I	OB
O9.5I	Early BI
B0.5-1Iab	B0.5Ia
Early BIa	—

Source: Taken from Table A.1 of Kretschmar et al. (2019).

Indications f or e point at the presence of different types of emission lines in the spectra; indication p means that the spectrum has additional peculiarities. These indications are assigned by spectroscopic observers from visual inspection of the spectra. It should be remembered that symbol I indicates supergiants, where the luminosity increases from Ib to Ia. How observers classify such supergiants, and for the precise evolutionary state of these supergiants, will be discussed in Chapter 10.

donor stars in SFXTs and those of persistent HMXBs. Furthermore, they argue that the SFXTs in their sample are characterized by high orbital eccentricities and that the wind velocities at the position of the NSs, and thus the accretion rates, are strongly dependent on the orbital phase. This would suggest that the orbital eccentricity might be the decisive factor for the distinction between SFXTs and persistent HMXBs.

However, this certainly is not the full story because, although at first glance the blue supergiant donors of the SFXTs and persistent sg-HMXBs appear very similar, there is a systematic difference between the *evolutionary states* of the supergiant donor stars of SFXTs and persistent wind-fed sg-HMXBs. This difference can be seen from the lists of spectral types of the supergiant donors in the two types of systems listed in Table 7.1. This table lists the spectral types of the identified donor stars of SFXTs and persistent wind-fed sg-HMXBs discovered by ESA's INTEGRAL satellite. Because these systems were all discovered by the same satellite, in a similar way, this can be considered to be a representative sample.

The sample in the table consists of 13 SFXT systems and 12 persistent systems. The table shows that the spectral types of the supergiant companions of the 13 SFXTs are systematically earlier than those of the companions in the persistent sg-HMXBs. Of the 13 companions of the SFXTs, nine are of spectral type OI, and four of type BI; whereas only three of the 12 companions of persistent sg-HMXBs are of type OI, eight are of type BI, and one is of type OB. The difference is even more striking if one compares the numbers of companions of the earliest types (O8–8.5I) and the latest types (B0.5–1I): among the SFXTs, there are six companions of types O8–8.5I, and two of type B0.5–1I; whereas among the persistent sg-HMXBs, there are two companions of type O8–8.5I and six companions of type B0.5–1I. These differences imply that the supergiant companions of the SFXTs systematically are less evolved, hotter and have smaller radii than those of the persistent sg-HMXBs. Therefore their wind velocities will be larger, and the stars will be deeper inside their Roche lobes than in the persistent sg-HMXBs. That the companions of SFXTs can be very deep inside their Roche lobes is illustrated by the fact that several of the SFXT systems have very long orbital periods: IGR J18483, IGR J16465−4507, and IGR J10100−5655 have orbital periods of 18.55 d, 30.32 d, and 165 d, respectively, so here the supergiant donors are indeed very deep inside their Roche lobes. But still these NSs are able to accumulate matter captured from the winds of their donors, which then is accreted suddenly in bursts lasting several hours. For possible models why the matter is not accreted here continuously, but released to the NS in bursts, we refer to the review of Kretschmar et al. (2019). In most models, the accumulation is thought to take place outside the boundary of the magnetosphere of the NS.

Because the blue supergiant donor stars of the SFXTs are on average clearly in an earlier evolutionary stage than those of the persistent wind-fed sg-HMXBs are, and these donor stars tend to be relatively deep inside their Roche lobes, it seems likely to us that the SFXTs can be seen as an earlier evolutionary stage of the sg-HMXBs. Additional support for this idea is that the spin periods of the SFXTs are systematically shorter than those of the persistent wind-fed sg-HMXBs, as can be seen in the Corbet diagram of Figure 7.18. When the radii of their donors grow and begin to approach those of their Roche lobes, one would expect SFXT binaries to turn into persistent wind-fed sg-HMXBs (while at the same time, the spin rates of the NSs are further decreased due to enhanced torque interactions at this stage).

7.1.4 Massive Gamma-ray Binaries

The first discovered γ-ray–emitting high-mass binary is LSI +61°303. It is a peculiar Be-star with a 25 d eccentric orbit, which shows radio and γ-ray outbursts recurring with the orbital period. It appears that the outbursts take place near periastron. Since the discovery of this system, four more γ-ray–emitting high-mass binaries have been discovered. The TeV γ-ray emitters: LS 5039, an O-star with an eccentric orbit with $P_{\rm orb} = 3.9$ d; HESS J0632+057, HESS J1832−093 (which coincides with the SN remnant [SNR] G22.7−02, see [Mori et al. (2017)]); and finally the very wide eccentric-orbit Be-star/radio pulsar system PSR B1259−63, which becomes a γ-ray source near periastron passage. Figure 7.9 depicts the relative sizes of the orbits and massive stars in four of these systems investigated by Dubus (2013). The γ-ray emission in these systems is thought to be due to the presence of a very active young pulsar. This is known for the system of PSR B1259−63, and very likely for the system of HESS J1832, as this system is located in a SNR. The ultra-relativistic electrons and positrons of the pulsar wind of the young pulsar will, by the inverse Compton effect, transform optical and UV photons of the companion star into γ-ray photons. Also, the annihilation of

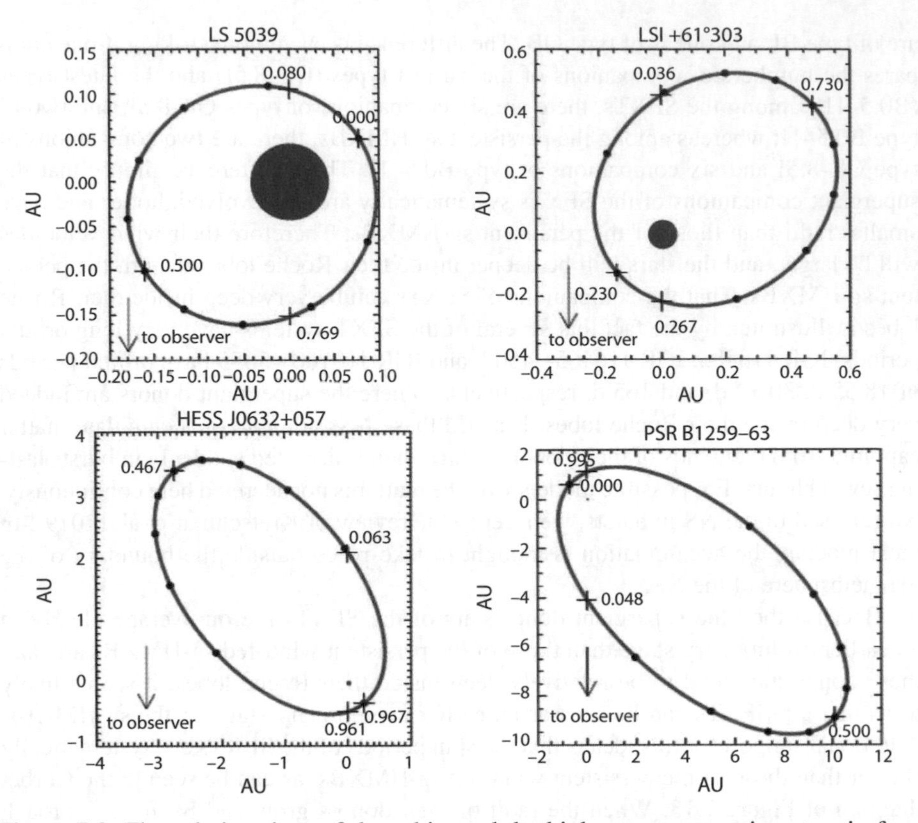

Figure 7.9. The relative sizes of the orbits and the high-mass companion stars in four γ-ray binaries. Three of these systems have wide and eccentric orbits, and we expect that they later evolve into Be/X-ray binaries. The system LS 5039, on the other hand, has a smaller orbit ($P_{\rm orb} = 3.9$ d) and we expect that it becomes a sg-HMXB. After Dubus (2013).

the positrons and the interaction of the ultra-relativistic pulsar wind particles with the stellar wind of the companion leads to the production of γ-rays. All these effects are thought to be at play here. The fact that a young fast-spinning newborn NS is present in these systems means that these systems are the progenitors of what later in life will become the HMXBs. Three of the systems in Figure 7.9 have wide and eccentric orbits and are expected to become Be/X-ray binaries. The system LS 5039 is expected to become a sg-HMXB. For further details on these γ-ray systems, we refer to the reviews by Dubus (2013) and Mori et al. (2017).

7.1.5 Fast Radio Bursts: A HMXB Connection?

The mysterious fast radio bursts (FRBs) are among the unanswered new puzzles in astrophysics in the first two decades of the 21st century. FRBs were discovered in 2007 by Lorimer et al. (2007), but the announcement of the first FRB (the *Lorimer burst*) received fairly little attention in the community.[1] This has changed dramatically since then, especially after the discovery of the first repeating source, FRB 121102 (Spitler et al., 2016). FRBs are millisecond-timescale bright (0.1 – 100 Jy ms fluence) radio transients that originate at cosmological distances (except for FRB 200428). The origin of FRBs is unknown, though a plethora of models invoking sources from magnetars to binary NS mergers have been proposed. The link to NSs has recently been strengthened further from detections of multi-component millisecond-timescale pulse-profile FRBs. For reviews on FRBs, see Cordes & Chatterjee (2019) and Petroff et al. (2019, 2021).

Magnetars (young NSs with extremely high B-fields, see Sections 10.3.3 and 14.1, Turolla et al., 2015; Kaspi & Beloborodov, 2017) show magnetically driven enhancements of their thermal and non-thermal emission, referred to as outbursts (Beloborodov & Li, 2016; Coti Zelati et al., 2018). In addition, short bursts and flares in X-rays and γ-rays, believed to involve the magnetosphere, are also observable magnetar activities, which are possibly triggered by their interior dynamics and sudden crustal deformations (star quakes). It is mainly these flare activities that are hypothesized to link magnetars to FRBs (e.g., Popov & Postnov, 2007, 2013; Dehman et al., 2020).

The magnetar model for the origin of FRBs received a boost from the detection of a pair of bursts from FRB 200428 (Bochenek et al., 2020; Kirsten et al., 2020). This source was detected from the same area of sky as the Galactic magnetar SGR 1935+2154, suggesting an origin *within* the MW. Indeed, the dispersion measure of the radio signal is too low to have originated anywhere outside our own Galaxy. Despite being thousands of times less intrinsically bright than previously observed (extragalactic) FRBs, its comparative proximity rendered it the most powerful FRB yet detected, reaching a peak flux of approximately $10^3 - 10^5$ Jy (or a fluence of $>1.5 \times 10^6$ Jy ms) and thus comparable to the brightness of the radio sources Cassiopeia A and Cygnus A at the same frequencies. This established magnetars as at least one ultimate source of FRBs. However, another strong candidate FRB source relates to binary NSs, most likely HMXB systems.

One example of an FRB with a potential binary NS origin is FRB 20180916B (Andersen et al., 2019). This source is repeating, showing an activity period of 16.35 d, and is located significantly offset with respect to nearby star-forming regions in its host Galaxy (Tendulkar et al., 2021). At a distance of 149 Mpc, it is by far the closest-known extragalactic FRB with a robust host Galaxy association. Based on their derived constraint on the upper limit of Hα luminosity in the vicinity of the source, Tendulkar et al. (2021) concluded that possible stellar companions to FRB 20180916B should be of a cooler, less massive spectral type than O6 V.

They also argued that because FRB 20180916B is 250 pc away (in projected distance) from the brightest young star forming region, it would need 800 kyr to 7 Myr to traverse the observed distance from its presumed birth site, assuming high velocities of NSs inherited from their formation in an asymmetric SN (see Chapter 13). This timescale is inconsistent with the true active ages of magnetars ($\lesssim$10 kyr) (Turolla

[1] The discovery paper merely received a handful of citations within its first year.

et al., 2015; Kaspi & Beloborodov, 2017). Rather, the inferred age and observed separation are compatible with the ages of HMXBs and γ-ray binaries, and their separations from the nearest OB associations (Coleiro & Chaty, 2013). On this basis, Tendulkar et al. (2021) suggested that the system could be a HMXB where the FRBs are possibly generated through an interaction between the companion wind and NS magnetosphere (the "cosmic comb" model; see, e.g., Ioka & Zhang, 2020). Pleunis et al. (2021) discuss several aspects (especially dispersion measure variations) in the context of interacting binary models for FRB 20180916B, which we briefly summarize in the following.

A number of interacting binary models have been proposed for the origin of FRBs (e.g., Ioka & Zhang, 2020; Lyutikov et al., 2020; Du et al., 2021) in which the FRBs are being produced by a highly magnetized NS whose magnetosphere is "combed" by the ionized wind of a massive companion star (i.e., a HMXB origin). This interaction could possibly lead to magnetic reconnection events, which have been proposed as a source of FRBs (Lyutikov & Popov, 2020). Pleunis et al. (2021) argue that special geometric conditions can explain why we do not see FRBs from Galactic HMXBs: the FRBs may only be visible within a funnel where the strong pulsar wind shields against the companion's wind, which is otherwise opaque to induced Compton or Raman scattering for FRB emission (see illustration in Fig. 7.10). The observed windows of observable burst activity (lasting for approximately 4–5 d in the case of FRB 20180916B) then correspond to when this funnel, and the induced magnetic tail of the NS, are pointed toward Earth. This special viewing geometry may thus explain why we do not see sources of bright radio bursts among the $\sim$200 Galactic HMXBs. Additional suggested evidence for possible magnetars in HMXBs comes from studies of NS spin evolution.

While a model with a highly magnetized NS in an interacting HMXB system can plausibly explain many of the observed FRB phenomena to date (Pleunis et al., 2021), several authors have argued that the 16.35 d activity period of FRB 20180916B may instead correspond to the rotational period (Beniamini et al., 2020) or precession period (e.g., Zanazzi & Lai, 2020) of an isolated magnetar.

Most recently, it was discovered by Nimmo et al. (2021) that the repeating FRB 20200120E (Bhardwaj et al., 2021) can produce isolated shots of emission as short as $\sim$60 ns in duration, with extreme brightness temperatures, comparable to "nano-shots" from the Crab pulsar. Comparing both the range of timescales and luminosities, Nimmo et al. (2021) find that FRB 20200120E bridges the gap between known Galactic young pulsars and magnetars and the much more distant extragalactic FRBs. However, shortly thereafter it was shown by Kirsten et al. (2022) that FRB 20200120E is associated with a globular cluster in the M81 galactic system (40 times closer than any other known extragalactic FRB). This is a very important discovery, because such globular clusters host old stellar populations (Chapter 12) and therefore this association challenges FRB models that invoke magnetars because these are young NSs formed recently in a core-collapse SN of a massive star (Chapter 13). To explain this puzzle, Kirsten et al. (2022) propose that FRB 20200120E is indeed a young, highly magnetized NS that formed via either accretion-induced collapse (AIC) of an old WD (Section 11.1.3) or via merger of compact stars in a binary system (Section 11.1.3; see Fig. 11.7). To us, these two possibilities seem the most likely ones. Alternative scenarios suggested by Kirsten et al.

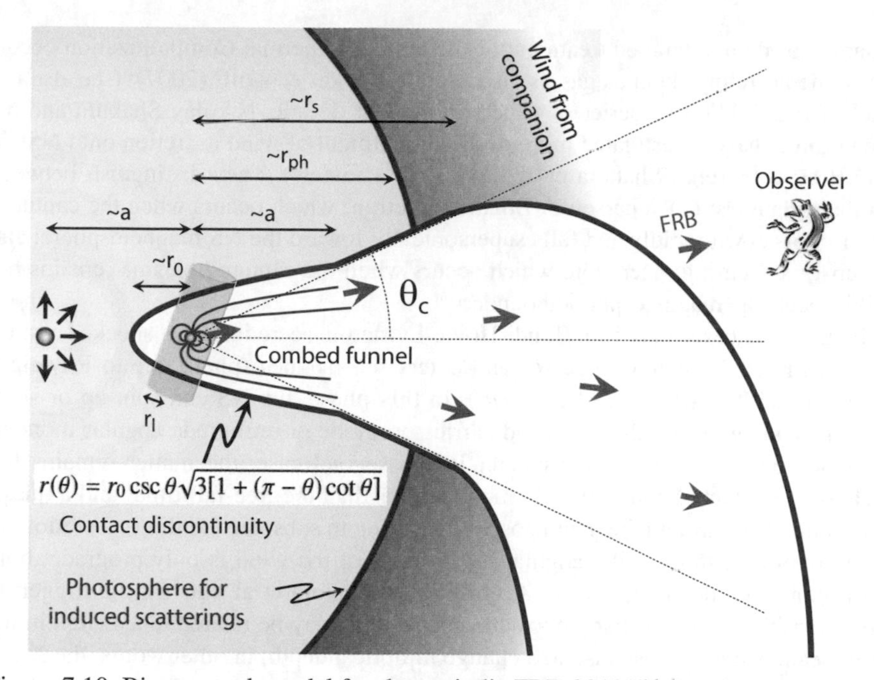

Figure 7.10. Binary comb model for the periodic FRB 20180916B. The FRBs are produced by a highly magnetized pulsar, whose magnetic field is "combed" by the strong wind from a companion star, for example, as in a HMXB system. The pulsar wind retains a clear funnel in the companion's wind that is otherwise opaque to induced Compton or Raman scatterings for repeating FRB emission. The 16.35 d activity period of FRB 20180916B corresponds in this model to the orbital period of the HMXB, and the 4–5 d active window corresponds to the time when the funnel points toward Earth. The wind from the FRB pulsar creates a clean funnel with a half-opening angle θ_c, which is combed by the wind from the companion. For the several characteristic length scales of this model, see the original publication. After Ioka & Zhang (2020).

(2022) are giant pulses from MSPs and compact binary X-ray systems, some either interacting through winds or magnetic fields or via an accretion disk.

We conclude that the FRB mystery remains unsolved and certainly not all FRB phenomena are anchored in HMXBs. Giving the current strong science focus in this area, the origin of FRBs will hopefully soon be settled.

7.2 STELLAR WIND ACCRETION IN MORE DETAIL

In Section 6.3.3, we gave a simplified treatment of wind accretion, as pioneered by Davidson & Ostriker (1973) and depicted in Figure 6.13. In the following, we go into somewhat more detail when considering wind accretion in a HMXB system. An example of a theoretical model for the spectral formation process in accretion-powered X-ray pulsars, based on a detailed treatment of the bulk and thermal Comptonization occurring in the accreting shocked gas, is discussed in Becker & Wolff (2007); (see also top panel of Fig. 7.11). In a series of papers in the last decade, Nikolay Shakura and his collaborators have investigated more detailed treatment of wind accretion onto NSs in HMXB systems (e.g., Shakura et al., 2015, for a review). They distinguish between two main regimes: (1) supersonic (Bondi) accretion, which occurs when the captured matter cools down rapidly and falls supersonically toward the NS magnetosphere; and (2) subsonic (settling) accretion, which occurs when the captured plasma remains hot until it meets the magnetospheric boundary.

In the first case, classical Bondi-Hoyle-Lyttleton accretion, the shocked matter cools down via Compton processes and enters the magnetosphere due to Rayleigh-Taylor instabilities (Arons & Lea, 1976). In this phase, the NS can spin-up or spin-down depending on whether the wind carries prograde or retrograde angular momentum when entering the magnetosphere. In the second case, the matter remains hot (because the plasma-cooling time is much longer than the free-fall time) and a quasi-static shell forms around the magnetosphere leading to subsonic (settling) accretion. In this case, even if the specific angular momentum of the wind is only prograde, both spin-up and spin-down can occur. According to Shakura et al. (2015), the triggering of the transition from supersonic to subsonic regime may be related to a switch in the X-ray beam pattern in response to a change in optical depth, in other words, the X-ray beam pattern changes with decreasing X-ray luminosity (near 4×10^{36} erg s^{-1}) from a fan beam to a pencil beam (Fig. 7.11, bottom panel). This hypothesis is supported by pulse profile observations of Vela X-1 in different energy bands (Doroshenko et al., 2011).

7.2.1 Bondi-Hoyle-Lyttleton Accretion

Given the supersonic flow of the stellar wind from the companion star, we first assume *Bondi-Hoyle-Lyttleton-like accretion* to be valid. The wind velocity from the companion star, such as a blue supergiant, can be approximated by the escape velocity at the stellar surface: $v_{\rm wind} \simeq v_{\rm esc} = \sqrt{2GM_2/R_2}$, where M_2 and R_2 are the mass and radius of the companion star, and G is the constant of gravity. The gravitational-capture radius of the accreting NS (also called the *Bondi-Hoyle-Lyttleton radius*, after the authors who first studied this type of accretion of gas captured by the gravity of an object) is then given by

$$r_{\rm grav} = \frac{2\,GM_{\rm NS}}{(v_{\rm rel}^2 + c_s^2)}, \tag{7.1}$$

where $c_s = \sqrt{\gamma P/\rho} \approx 11\,\sqrt{T/(10^4\,K)}$ km s^{-1} is the local sound speed of the ambient medium (γ is the adiabatic index, P is the pressure, ρ is the mass density, and T is the temperature). The relative velocity between the wind and the NS ($\vec{v}_{\rm rel} = \vec{v}_{\rm wind} + \vec{v}_{\rm orb}$) is usually dominated by the wind velocity compared to the orbital velocity, that is, $v_{\rm wind} \gg v_{\rm orb}$ (except for very tight and/or highly eccentric binaries; see, e.g., Hainich et al. [2020]).

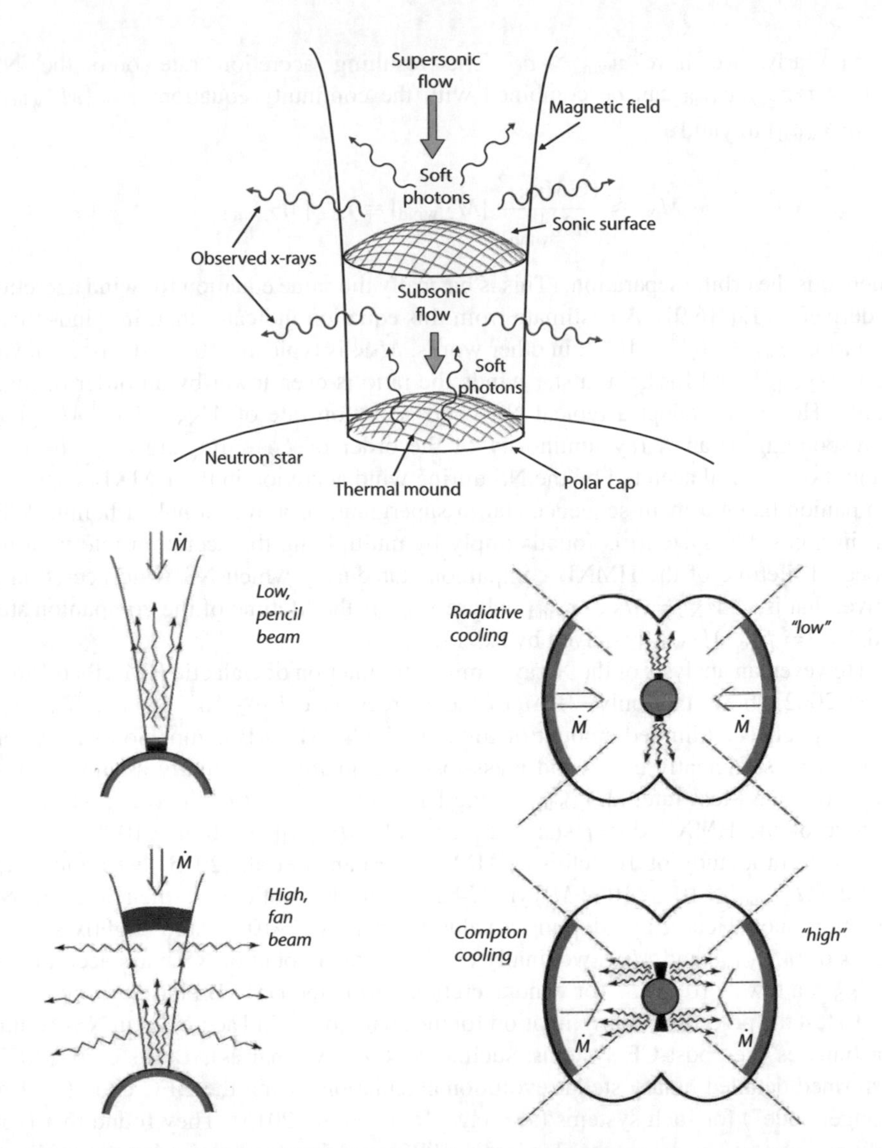

Figure 7.11. (Top) Schematic illustration of gas accreting onto the magnetic polar cap of a NS. Seed photons are created throughout the column via bremsstrahlung and cyclotron emission, and additional black-body seed photons are emitted from the surface of the thermal mound near the base of the column. After Becker & Wolff (2007). (Bottom) Schematic illustration of the transition from Compton ("high state") to radiative ("low state") plasma cooling regimes, due to the switch from fan (bottom) to pencil (top) X-ray beam pattern with decreasing mass-accretion rate. After Shakura et al. (2015).

Similarly, we have $v_{\rm wind} \gg c_s$. The resulting accretion rate onto the NS, $\dot{M}_{\rm NS} \approx \pi r_{\rm grav}^2 \rho\, v_{\rm wind}$ can be combined with the continuity equation: $\rho = |\dot{M}_{2,\rm wind}|/(4\pi a^2\, v_{\rm wind})$ to yield

$$\dot{M}_{\rm NS} \approx \frac{(GM_{\rm NS})^2}{a^2 v_{\rm wind}^4}\, |\dot{M}_{2,\rm wind}| \equiv f_{\rm acc}\, |\dot{M}_{2,\rm wind}|, \tag{7.2}$$

where a is the orbital separation. (This is basically the same equation for wind accretion as derived in Eq. [6.9]). An estimate from this equation indicates that, in wind-driven scenarios, $f_{\rm acc} = 10^{-3} - 10^{-4}$, in other words, $\dot{M}_{\rm NS}$ is typically $10^3 - 10^4$ times lower than $|\dot{M}_{2,\rm wind}|$, and for helium star winds the ratio is even lower by an order of magnitude. Hence, we adopt a typical NS wind accretion rate of $\dot{M}_{\rm NS} \approx 10^{-9}\, M_\odot\, {\rm yr}^{-1}$, corresponding to an X-ray luminosity of the order of $L_X \approx 10^{37}\, {\rm erg\, s}^{-1}$. The total amount of material accreted by the NS during wind accretion in the HMXB stage (the companion being a main-sequence star, a supergiant, or a even a naked helium/WR-star in a post-CE system) is found simply by multiplying the accretion rate with the expected lifetime of the HMXB companion star during which NS wind accretion is active: that is, $\Delta M_{\rm NS} = \dot{M}_{\rm NS} \cdot \tau_{\rm wind}$, where $\tau_{\rm wind}$ is the lifetime of the companion star, and $\dot{M}_{\rm NS} = f_{\rm acc}\, |\dot{M}_{2,\rm wind}|$ is given by Eq. (7.2).

However, an analysis of the X-ray luminosity function of Galactic HMXBs (Grimm et al., 2002) shows that only $\sim$10% of the sources have $L_X \geq 10^{37}\, {\rm erg\, s}^{-1}$. This fact probably reflects a limited amount of time in which a HMXB companion star is able to deliver a sufficiently high wind mass-loss rate and/or our initially assumed (average) wind mass-loss rate, $|\dot{M}_{2,\rm wind}|$, is too large. During most of the (main-sequence) lifetime of the HMXB donor star, the value of $|\dot{M}_{2,\rm wind}|$ is often $< 10^{-6}\, M_\odot\, {\rm yr}^{-1}$. Indeed, from a study of 10 eclipsing HMXBs, Falanga et al. (2015) list typical values of $|\dot{M}_{2,\rm wind}| \approx 10^{-7}$–$10^{-6}\, M_\odot\, {\rm yr}^{-1}$, based on observations of their average NS accretion rates. Hence, by adopting a value for $\tau_{\rm wind}$ of $\sim 10^7$ yr and slightly smaller values of $|\dot{M}_{2,\rm wind}|$ and $\dot{M}_{\rm NS}$, we finally obtain a total amount of NS mass accretion of $\Delta M_{\rm NS} \approx$ a few $\times\, 10^{-3}\, M_\odot$ for wind accretion during the HMXB phase.

To test the preceding approximation for the analogous wind accretion in NS–helium star binaries (i.e., post-CE systems, such as WR X-ray binaries), Tauris et al. (2017) performed detailed binary stellar evolution calculations using the BEC code (i.e., the "Langer code") for such systems (see, e.g., Tauris et al., 2015). They found that typically $f_{\rm acc} \simeq$ a few $\times\, 10^{-4}$ (see also Israel, 1996), and they obtained values of $\Delta M_{\rm NS} < 4 \times 10^{-4}\, M_\odot$ when integrating throughout the wind accretion phase of the NS–helium star binaries. Hence, NSs accrete very little wind material in this phase of their evolution. Combined with the small amount of wind accretion in the previous (pre-CE) HMXB phase (see above), this has the important consequence that observed NS masses in both HMXB and in double NS (DNS) systems are close to their birth masses (Tauris et al., 2017; see also Chapter 14).

The structured nature of the stellar wind of a supergiant star (i.e., a wind where cool dense clumps are embedded in a photoionized gas) makes it difficult to derive an estimate of its mass-loss rate that is more accurate than a factor of a few (e.g.,

Martinez-Nunez et al., 2017). Another caveat is that the presence of a NS causes hydrodynamical interactions of the wind flow in close binaries (Manousakis et al., 2012). However, it is also possible that a strong reduction of the long-term mass-accretion rate is produced by the effect of a strong magnetic field and/or a fast spin of the NS, which could lead to magnetic and/or centrifugal barriers, respectively (Bozzo et al., 2008, 2016), see Section 7.3. Such complications are not taken into account in our simplified Bondi-Hoyle-Lyttleton-like treatment of the accretion process. For a semi-analytical treatment to wind accretion in NS sg-HMXBs, we refer to Bozzo et al. (2021).

Before we discuss the four basic stages of accretion (Section 7.3), we end the description of NS wind accretion by summarizing current ideas of "settling accretion."

7.2.2 "Settling Accretion" and Efficient Spin-down of NSs to Very Slow Rotation

In a series of important papers, Shakura and collaborators (e.g., Shakura et al., 2012, 2014, 2015, 2018) have analyzed how matter that is captured from the wind of a companion, and that is infalling spherically and symmetrically toward the NS, enters the magnetosphere of this rotating star. They found that there are two main accretion regimes, depending on the X-ray luminosity, L_X, of the source.

For X-ray luminosities $L_X \geq 4 \times 10^{36}$ erg s^{-1}, the infalling matter, heated by the liberation of gravitational energy, cools down rapidly due to Compton cooling by X-ray irradiation. The matter in this case falls supersonically toward the NS, until it is stopped at the Alfvén radius, R_A, and from there enters the magnetosphere of the NS via Kelvin-Helmholtz instabilities at the Alfvén surface. The Alfvén radius is the distance from the NS center at which the magnetic pressure of the dipole magnetic field of the NS equals the ram pressure of the infalling matter (see Eq. [7.11]). This case of supersonic (Bondi) accretion was discussed in Section 7.2.1.

For $L_X \leq 4 \times 10^{36}$ erg s^{-1}, the infalling matter is unable to cool and the matter falls in subsonically and forms a hot quasi-spherical stationary shell around the NS magnetosphere. The actual accretion rate onto the NS is now determined by the ability of the plasma to enter the magnetosphere, which will proceed by the Rayleigh-Taylor instability (a dense "liquid" on top of a low-density "liquid": the magnetic field). This regime is called *settling* accretion, and the presence of the hot quasi-spherical shell around the magnetosphere causes large friction with the magnetosphere, leading to rapid spin-down of the NS to a very long spin period, as is observed in most HMXBs. This frictional spin-down occurs therefore when the accretion rate (and thus the X-ray luminosity) is low, because it is in this regime that the hot quasi-spherical shell forms outside the Alfvén boundary, the radius of which is again given by Eq. (7.11), although sometimes the radius is expressed with a somewhat different scaling value of the constant because the radial velocity of matter in the shell is smaller than the free-fall velocity (see, e.g., Shakura & Postnov, 2017, for details).

This rotational braking mechanism introduced by Shakura and collaborators is quite similar to that proposed by Kundt (1976). In the Kundt model the torque is produced by the magnetic field lines of the rotating star being swept back by the matter accreting onto the magnetosphere boundary. The presence of a hot shell outside the Alfvén boundary, which may further accelerate the spin-down rate, was not foreseen in the Kundt model. Shakura and collaborators (e.g., Shakura & Postnov, 2017) furthermore propose that in the low-accreting regime plasma can keep piling up in the hot shell around the magnetosphere, but the plasma is not able to enter the magnetosphere. Only when the mass of the shell exceeds a certain threshold, the Rayleigh-Taylor instability may set in, causing all matter of the shell to be suddenly accreted, leading to a very bright X-ray flare with a duration of a few hours. They suggest that this may be the cause of the outbursts of the SFXTs discovered by INTEGRAL.

7.3 SPIN EVOLUTION OF NEUTRON STARS

7.3.1 Four Basic Stages of Accretion

An accreting NS will experience a number of different stages while interacting and accumulating material from a nearby companion star (see, e.g., the excellent review in (Ghosh, 2007). Consider a case where a young NS (accretor) is in a binary system with a companion star (donor) of mass M_*, having a wind mass-loss rate of $\dot{M}_*$. The orbital separation is assumed to be a. In the following, we describe the basic long-term evolution of the spin of such a NS, depending on various critical radii, using a somewhat simplistic approach in four stages. Then in Section 7.3.2, we briefly summarize the aspects of a more advanced treatment and also discuss the effects of the Roche-lobe decoupling phase on the NS spin.

Initially, the young pulsar is assumed to have relatively fast spin (i.e., Ω is large) and a strong B-field typical of young pulsars (say, $B \sim 10^{13}$ G)—see the location of young pulsars in the $P\dot{P}$-diagram (Fig. 1.2). We then follow the evolution and the accretion stages of the NS while its values of Ω and B decreases over time; see also pioneering work by Lamb et al. (1973).

1) Isolated Pulsar ($r_{\rm stop} > r_{\rm grav}$)

Right after the NS is born in a SN, it is very energetic and the wind plasma from the companion star is stopped by the pressure of magnetodipole radiation from the pulsar at a location, $r_{\rm stop}$, outside the radius of gravitational capture of the NS, $r_{\rm grav}$. Hence, the NS does not even "feel" the presence of its companion star (aside from its gravitational pull keeping the two stars orbiting each other), that is, it evolves as an *isolated* pulsar. To estimate the value of $r_{\rm stop}$, we simply equate the outward pressure of magnetodipole radiation to that of the incoming ram pressure ($P_{\rm dipole} \simeq P_{\rm ram}$). That is,

$$P_{\rm dipole} \simeq P_{\rm ram} \tag{7.3}$$

$$\frac{\dot{E}_{\rm dipole}}{4\pi r_{\rm stop}^2 c} \simeq \frac{1}{2}\rho_{\rm w} v_{\rm w}^2, \tag{7.4}$$

where $\rho_{\rm w}$ and $v_{\rm w}$ represent the mass density and the speed of the wind at the location of $r_{\rm stop}$. Applying Eq. (14.13) for $\dot{E}_{\rm dipole}$ where $|\ddot{\mu}| \simeq BR^3\Omega^2$ (R being the NS radius), and the continuity equation,

$$\rho_{\rm w} = \frac{\dot{M}_*}{4\pi\, a^2 v_{\rm w}}, \tag{7.5}$$

one obtains (Exercise 7.2):

$$r_{\rm stop} = \sqrt{\frac{4B^2 R^6 \Omega^4 a^2}{3c^4 v_{\rm w} \dot{M}_*}}. \tag{7.6}$$

Whereas the gravitational capture radius of the NS (given by $r_{\rm grav} \sim 2GM/v_{\rm w}^2$, see also Eq. [7.1] for the general case) is independent of the spin of the NS, we see that $r_{\rm stop} \propto B\Omega^2$. Therefore, after a while, the initially energetic and rapidly spinning pulsar (i.e., large Ω) loses rotational energy due to emission of magnetodipole waves and the value of Ω decreases, whereby the pulsar enters the regime where $r_{\rm stop} < r_{\rm grav}$.

2) Gunn-Ostriker Mechanism ($r_{\rm stop} < r_{\rm grav}$ and $r_{\rm A} > r_{\rm lc}$)

At this stage $r_{\rm stop} < r_{\rm grav}$, however, the magnetospheric boundary (i.e., roughly the Alfvén radius, $r_{\rm A}$) is located outside the light cylinder, $r_{\rm lc}$, as the young NS is still spinning relatively fast. Because matter cannot couple to the magnetosphere (that would require co-rotation with a speed $v > c$), it is accelerated to relativistic energies by magnetodipole waves (i.e., the *Gunn-Ostriker mechanism* or the ejector phase). The location (distance from the NS center) of the light cylinder is simply given by

$$r_{\rm lc} = \frac{c}{\Omega} \simeq 48\,{\rm km} \cdot P_{\rm ms}, \tag{7.7}$$

where $P_{\rm ms}$ is the NS spin period in milliseconds.

As mentioned earlier, the location of the Alfvén radius, $r_{\rm A}$, is found by equating the magnetic energy density to the incoming ram pressure of the wind:

$$\frac{B(r_{\rm A})^2}{8\pi} \simeq \frac{1}{2}\rho_{\rm w} v_{\rm ff}^2, \tag{7.8}$$

where for a perfect dipole

$$B(r_{\rm A}) = B\left(\frac{R}{r_{\rm A}}\right)^3 \tag{7.9}$$

and the wind speed is roughly the free-fall velocity inside the gravitational capture radius, that is, $v_{\rm ff} = \sqrt{2GM/r_{\rm A}}$, and the value of $\rho_{\rm w}$ is again obtained from the continuity equation (Eq. [7.5]) but with $v_{\rm w} = v_{\rm ff}$. Assuming that the NS wind accretion rate can be estimated from simple solid angle accretion geometry,

$$\dot{M}_{\rm NS} = \frac{\pi r_{\rm A}^2}{4\pi\, a^2}\,\dot{M}_*, \tag{7.10}$$

we finally obtain (Exercise 7.3)

$$r_{\mathrm{A}} = \left(\frac{1}{32} \frac{B^4 R^{12}}{G M \dot{M}_{\mathrm{NS}}^2} \right)^{1/7} \simeq 32\,\mathrm{km} \cdot \left(\frac{B_8^4 R_{12}^{12}}{M_{1.4} \dot{M}_{-10}^2} \right)^{1/7}, \tag{7.11}$$

where B_8 is the NS B-field strength in units of 10^8 G, R_{12} is the NS radius in units of 12 km, $M_{1.4}$ is the NS mass in units of 1.4 $M_{\odot}$, and $\dot{M}_{-10}$ is the NS accretion rate in units of $10^{-10}\,M_{\odot}\,\mathrm{yr}^{-1}$. Note, uncertainties in the acccretion geometry and infall velocity of the material lead to uncertainties in Eq. (7.11) of at least a factor of 2. For a slightly stronger NS surface magnetic flux density of $B = 10^9$ G, keeping the other values constant, the Alfvén radius is located at 120 km. We notice that $r_{\mathrm{A}} \propto B^{4/7}\,\dot{M}_{\mathrm{NS}}^{-2/7}$ and that the size of the magnetosphere is controlled by the battle between the strength of the magnetic field and the incoming ram pressure.

As the pulsar continues to lose rotational energy, the value of Ω decreases further and causes the light cylinder to move outward and beyond the magnetospheric boundary, thereby allowing material to enter the magnetosphere.

3) Propeller Phase ($r_{\mathrm{co}} < r_{\mathrm{A}} < r_{\mathrm{lc}}$)

At this stage, infalling matter can couple to the magnetosphere. However, the magnetospheric boundary is still outside the co-rotation radius of the spinning NS, that is, the radius where matter travels in stable circular Keplerian orbits:

$$r_{\mathrm{co}} = \left(\frac{GM}{\Omega^2} \right)^{1/3} \simeq 17\,\mathrm{km} \cdot P_{\mathrm{ms}}^{2/3}\,M_{1.4}^{1/3} \tag{7.12}$$

and therefore the centrifugal force acting on the incoming matter exceeds the gravitational force, whereby material piles up near the magnetospheric boundary and creates a strong braking torque, N, acting on it:

$$N = \dot{J}_{\mathrm{spin}} \approx \frac{\partial}{\partial t} \left(m\, r_{\mathrm{A}}^2 \Omega_{\mathrm{K}} \right) = \dot{M}_{\mathrm{NS}} \sqrt{G M r_{\mathrm{A}}}, \tag{7.13}$$

where m is the mass of an incoming test particle, and its angular momentum with respect to the NS can be expressed as $\vec{J} = |\vec{r} \times \vec{p}|$.

Given that the X-ray luminosity, L_{X}, $\propto \dot{M}_{\mathrm{NS}}$, it is trivial to show that $N \propto L_{\mathrm{X}}^{6/7}$. Thus, the brighter the X-ray source, the stronger the braking torque (or spin-up torque, see below). The braking torque acting on the NS causes the NS spin rate (Ω) to decrease and thus the co-rotation radius to increase until $r_{\mathrm{co}} > r_{\mathrm{A}}$.

4) Accretion Phase ($r_{\mathrm{A}} < r_{\mathrm{co}}$)

In this regime, efficient accretion and spin-up of the NS is finally possible. The accretion torque can be written as

$$N = \dot{M}_{\mathrm{NS}} \sqrt{G M r_{\mathrm{A}}}\,\xi, \tag{7.14}$$

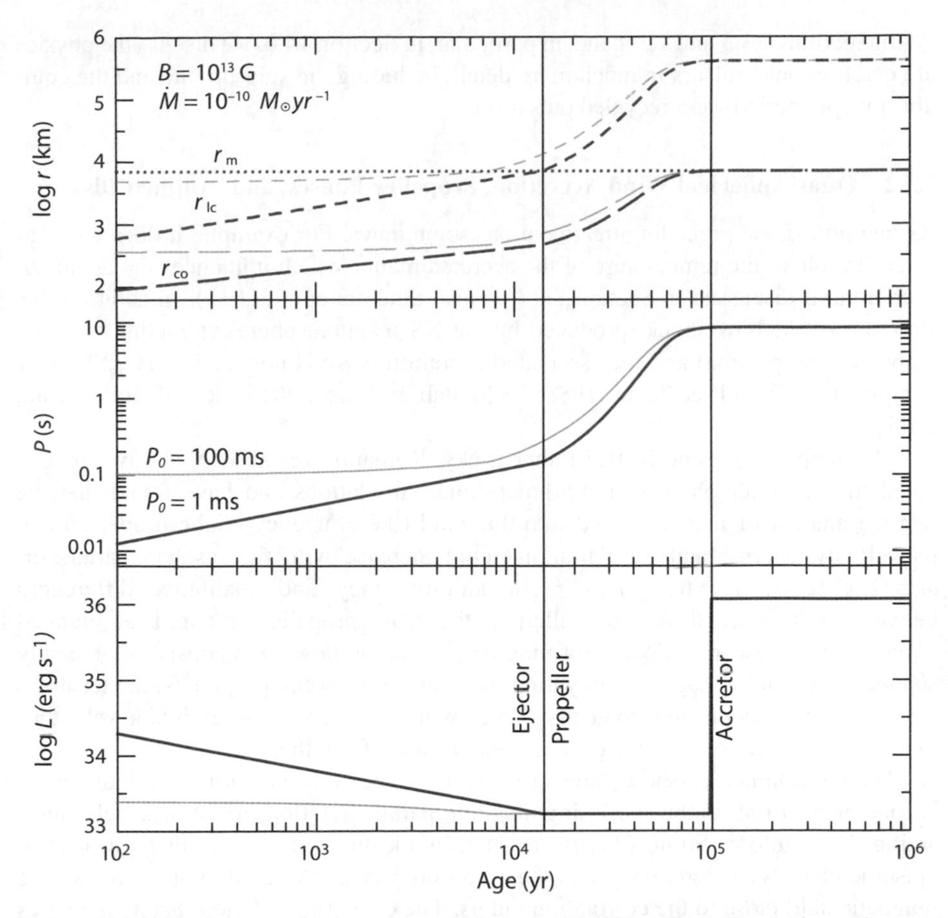

Figure 7.12. Evolution of a newborn NS in a HMXB. (Top) Evolution of magnetosphere radius ($r_{\rm mag} = r_{\rm m}$, dotted), light cylinder radius ($r_{\rm lc}$, short-dashed), and co-rotation radius ($r_{\rm co}$, long-dashed) for constant surface magnetic field $B = 10^{13}$ G, constant mass-accretion rate $\dot{M}_{\rm NS} = 10^{-10}\ M_\odot\ {\rm yr}^{-1}$, and initial spin periods $P_0 = 1$ ms (dark green) and 100 ms (light green). Vertical lines separate ejector ($r_{\rm mag} > r_{\rm lc}$), propeller ($r_{\rm co} < r_{\rm mag} < r_{\rm lc}$), and accretor/spin-equilibrium phases ($r_{\rm mag} \le r_{\rm co}$). (Center) Spin period evolution (see original paper for details). (Bottom) Accretion luminosity from liberated gravitational binding energy ($L = GM_{\rm NS}\dot{M}_{\rm NS}/r$). After Ho, Wijngaarden, et al. (2020).

where ξ is an efficiency parameter determined by the geometry and flow of the accreted material (typically $\xi \simeq 1$ for RLO; $\xi \simeq 0.01 \sim 0.1$ for wind accretion without disk formation).

The magnetospheric radius, $r_{\rm mag}$, is often assumed to be roughly equal to the Alfvén radius, $r_{\rm A}$. It may vary by a factor of the order of unity such that $r_{\rm mag} = \phi \cdot r_{\rm A}$, where ϕ is $0.5 - 1.4$ (Ghosh & Lamb, 1992; Wang, 1997; D'Angelo & Spruit, 2010). Numerical examples of the spin evolution of young NSs in HMXBs are given, for example, in Ho, Wijngaarden, et al. (2020); see also Figure 7.12.

In Sections 7.3.3 and 7.3.4 and, in particular, in Section 14.6, we discuss the physics of accretion onto pulsars in much more detail, including the spin-up line and the equilibrium spin period for a recycled pulsar.

7.3.2 Quasi-spherical Wind Accretion, Propeller Phases, and Trapped Disks

As mentioned, the preceding treatment has some flaws: For example, it does not consider the role of the temperature of the accreted matter which influences the details of accretion, as discussed in Section 7.2 (see also Shakura et al., 2015). In addition, the structure of the bow shock (produced by the NS magnetosphere interacting with the ambient wind plasma) and the associated accretion wake is non-stationary and rather complicated (Fryxell & Taam, 1988; El Mellah & Casse, 2015; de Val-Borro et al., 2017).

The propeller regime is also rather tricky. Romanova et al. (2018) have investigated the propeller phase in multidimensional simulations and have found that the relative amount of matter ejected into the wind (the efficiency of the propeller) and its velocity, energy, and angular momentum increase with the fastness parameter: $\omega \equiv \Omega_{\rm NS}/\Omega_{\rm K}(r_{\rm mag}) = (r_{\rm mag}/r_{\rm co})^{3/2}$. In addition, they find qualitative differences between the strong and weak propellers: in the strong propellers, matter is accelerated higher than the escape velocity, forming large-scale outflows that consist of conically shaped winds and a magnetic (Poynting flux) jet. In the weak propellers, matter flows into a more widely opened, conically shaped wind with sub- or super-escape velocities that may partly fall back to the disk at some distance from the NS.

The unexplained episodic flares of some accreting NSs show quasi-periodic oscillations or recurrent outbursts. Using a simple parametrization of the disk-field interaction, D'Angelo & Spruit (2010) presented work on an instability that can lead to episodic outbursts (accretion) when the accretion disk is truncated by the star's strong magnetic field close to the co-rotation radius. The cycle time of these bursts increases with a decreasing accretion rate. Their solutions show that the usually assumed propeller state, in which mass is ejected from the system, need not occur even at very low accretion rates. In additional work, D'Angelo & Spruit (2012) showed that disks accreting onto the magnetosphere of a NS can end up in a trapped state, in which the inner edge of the disk stays near the co-rotation radius, even at low and varying accretion rates. The accretion in these trapped states can be steady or cyclic. D'Angelo & Spruit (2012) demonstrated how trapped states evolve from both non-accreting and fully accreting initial conditions, and they calculated the effects of cyclic accretion on the spin evolution of the NS.

7.3.3 Accretion Disk and Additional Torques

When matter is transferred by (beginning) RLO, the accreted gas from a binary companion possess large specific angular momentum. In this case, contrary to wind accretion, the flow of gas onto the NS is not (quasi-)spherical, but leads to the formation of an accretion disk (Section 4.6) where excess angular momentum is transported outward by (turbulent-enhanced) viscous stresses, as discussed by, for example,

Shapiro & Teukolsky (1983) and Frank et al. (2002). Depending on the mass-transfer rate, the opacity of the accreted material, and the temperature of the disk, the geometric shape and flow of the material may take a variety of forms (thick disk, thin disk, slim disk, torus-like, advection-dominated accretion flow). Popular models of the inner disk (Ghosh & Lamb, 1992) include optically thin/thick disks that can be dominated by either gas or radiation pressure.

The exact expression for the spin-up line of recycled pulsars in the $P\dot{P}$-diagram (Section 14.6.3) also depends on the assumed model for the inner disk, mainly as a result of the magnetosphere boundary, which depends on the characteristics of the inner disk. Close to the NS surface, the magnetic field is strong enough that the magnetic stresses truncate the Keplerian disk and the plasma is channeled along field lines to accrete onto the surface of the NS. At the inner edge of the disk, the magnetic field interacts directly with the disk material over some finite region. The physics of this transition zone from Keplerian disk to magnetospheric flow is important and determines the angular momentum exchange from the differential rotation between the disk and the NS. Apparently the resultant accretion torque, acting on the NS, which is calculated using detailed models of the magnetosphere–disk interaction, does not deviate much from simple expressions assuming idealized, spherical accretion and Newtonian dynamics. For example, Ghosh & Lamb (1992) considered it a fortuitous coincidence that the equilibrium spin period calculated under simple assumptions of spherical flow resembles the more detailed models of an optically thick, gas pressure–dominated inner accretion disk.

The Magnetosphere–Disk Interaction

The exchange of angular momentum at the magnetospheric boundary eventually leads to a gain of NS spin angular momentum. In addition to material torque (the dominant term), the accretion torque acting on the spinning NS has a contribution from both magnetic stress and viscous stress. The total torque, $N = \dot{J}_\star \equiv (\partial/\partial t)(I\Omega)$, where $J_\star$ is the NS spin angular momentum, Ω is its angular velocity, and $I \simeq 1$–2×10^{45} g cm^2 is its moment of inertia.

In numerical calculations of the propeller phase, for example, one should include the effect of additional spin-down torques acting on the NS due to both magnetic field drag on the accretion disk (Rappaport et al., 2004) as well as magnetic dipole radiation (see Eq. [14.13]), although these effects are usually not dominant. The magnetic stress in the disk is related to the critical fastness parameter, ω_c (Ghosh & Lamb, 1979b; Ghosh, 2007). Here, we follow Tauris (2012) and write the total spin torque as

$$N_{\rm total} = n(\omega)\left(\dot{M}\sqrt{GMr_{\rm mag}}\,\xi + \frac{\mu^2}{9r_{\rm mag}^3}\right) - \frac{\dot{E}_{\rm dipole}}{\Omega}, \tag{7.15}$$

where

$$n(\omega) = \tanh\left(\frac{1-\omega}{\delta_\omega}\right) \tag{7.16}$$

is a dimensionless function, depending on the fastness parameter, $\omega = \Omega_\star / \Omega_K(r_{mag}) = (r_{mag}/r_{co})^{3/2}$, which is introduced to model a gradual torque change in a transition zone near the magnetospheric boundary. The width of this zone has been shown to be small (Spruit & Taam, 1993), corresponding to $\delta_\omega \ll 1$ and a step function-like behavior $n(\omega) = \pm 1$. In the numerical modelling of Tauris (2012) for the Roche-lobe decoupling phase (Section 14.6.8), we used $\delta_\omega = 0.002$, $\xi = 1$, $r_{disk} = r_{mag}$. For further discussions on the final stages of the X-ray phase, the radio ejection phase and transitional MSPs, see Section 14.3.1.

7.3.4 The Equilibrium Spin Period

In the following, we briefly summarize the concept of an equilibrium spin period, P_{eq}. For a detailed review of the spin-up and recycling process of old NSs, see Section 14.6. The equilibrium spin period refers to the situation where the angular velocity of the accreting NS equals the Keplerian angular velocity of co-rotating matter located at the magnetospheric boundary. For accreting MSPs (see, e.g., original work by van den Heuvel, 1977; Bhattacharya & van den Heuvel, 1991), P_{eq} can be expressed as

$$P_{eq} \simeq 1.40\,\text{ms} \quad B_8^{6/7} \left(\frac{\dot{M}}{0.1\,\dot{M}_{Edd}} \right)^{-3/7} \left(\frac{M}{1.4\,M_\odot} \right)^{-5/7} R_{13}^{18/7}, \tag{7.17}$$

where B_8 is the dipole magnetic flux density at the NS surface in units of 10^8 G, R_{13} is the NS radius in units of 13 km, and $\dot{M}_{Edd}$ is the Eddington accretion limit for a NS (a few $10^{-8}\ M_\odot\ \text{yr}^{-1}$, see Section 4.9). Hence, 1.40 ms is the shortest spin period that a magnetized and accreting NS can reach for $B_8 = 1$ and $\dot{M} = 0.1\,\dot{M}_{Edd}$. Were the NS to rotate faster than with P_{eq}, the matter would be centrifugally swung out of the magnetosphere and accretion would be prevented (however, see discussions in Sections 7.2.2 and 7.3.2). On the other hand, for $P \geq P_{eq}$, matter can enter the magnetosphere and accretion is in principle possible. This is illustrated in Figure 7.13.

The absolutely shortest rotation period possible for a NS has been under debate for many years (see also Section 14.6.9 on sub-ms MSPs that are shorter than a millisecond). In principle, for $B_8 = 1$ and $\dot{M} = 1.0\,\dot{M}_{Edd}$, one would naively expect the possibility of an ~0.5 ms MSP. However, such high accretion rates are rare on a long timescale (if they are at all possible for a timescale of greater than a few Myr), and often in systems with high mass-transfer rates the NS does not reach P_{eq} during the X-ray phase (see Section 14.6.6 on the spin-relaxation timescale).

For HMXBs, the NS B-fields are still fairly large (e.g., Taani et al., 2018) because these young NSs have not accreted much material (Section 7.2.1). For example, for $B = 10^{12}$ G ($B_8 = 10^4$), $\dot{M} = \dot{M}_{Edd}$ and assuming $R_{13} = 1$, the equilibrium spin period in Eq. (7.17) is estimated to be 1.4 s. Thus, the strongly-magnetized NS in SMC X-1 is spinning close to its minimum possible spin period, as it is accreting at a rate close to the Eddington rate ($L_X = 2 \times 10^{38}\ \text{erg s}^{-1}$) and has $P = 0.7$ s. (Notice, a factor of 2 in the value of P_{eq} is easily explained by slightly changing the assumed NS radius or NS mass.)

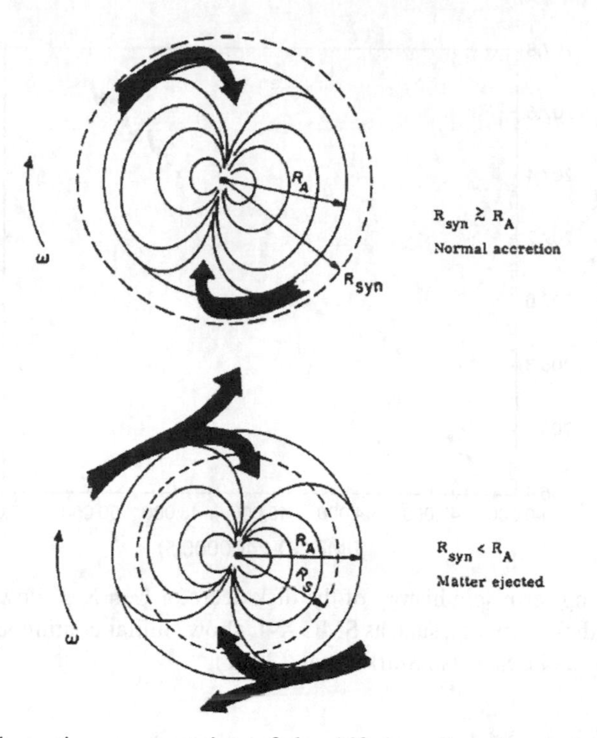

Figure 7.13. Schematic representation of the Alfvén radius, R_A, and the co-rotation radius (here denoted R_{syn}), of a rotating, accreting, and magnetized NS. R_A depends on the accretion rate, $\dot{M}$, mass M, and dipole magnetic field strength B (at the stellar surface) of the NS; R_{syn} depends on the rotation period P and mass of the NS. When $R_A = R_{syn}$ the NS spins at its equilibrium spin period P_{eq}. If it rotates equal to or slower than this rate, accretion is possible. If it rotates faster, the centrifugal forces on matter entering the magnetosphere are larger than the gravitational attraction by the NS, which will swing this matter out, and accretion is prevented. After Schreier (1977).

7.3.5 Spin Evolution as a Diagnostic of Mass-transfer Type

Table 6.2 shows a selection of the spin periods of NSs in HMXBs (see also Fig. 7.30). A striking fact observed in both the table and the figure is that, contrary to the radio pulsars, most X-ray pulsars tend to have very long spin periods, of the order of hundreds of seconds. Spin periods of the order of seconds are found only in systems where, from optical and UV observations, there is clear evidence of the presence of an accretion disk around the compact star. These are systems in which a large part of the mass transfer is due to beginning RLO: the systems of Her X-1, Cen X-3, SMC X-1, OAO 1657−415, and LMC X-4. Such systems are expected to be powered by a combination of stellar wind and beginning atmospheric RLO, as first pointed out by Savonije (1978, 1983). The spin periods of the accreting X-ray pulsars are observed to vary on relatively short timescales; see, for example, the important review by Nagase (1989). Figures 7.14–7.16 show the spin histories of a number of well-known binary X-ray pulsars over time spans

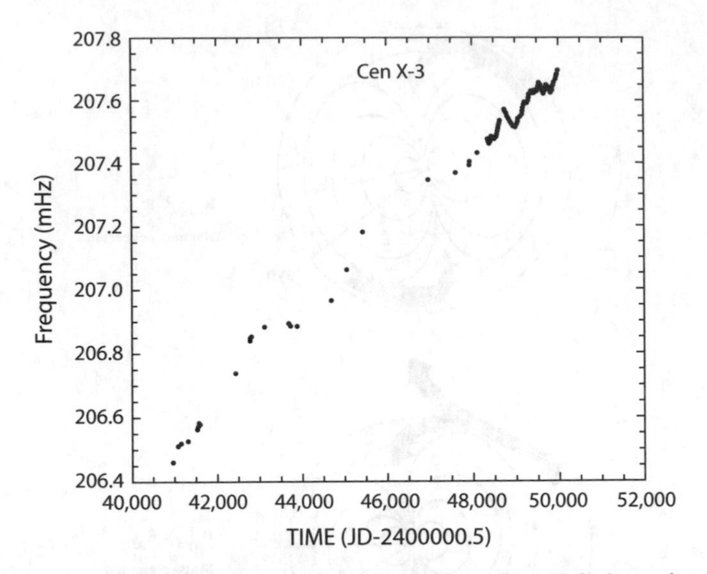

Figure 7.14. Long-term spin history of the disk accretor Cen X-3 shows its continuous spin-up. Other disk accretors, such as SMC X-1, show similar continuous spin-up. After Bildsten et al. (1997). See also Shirke et al. (2021).

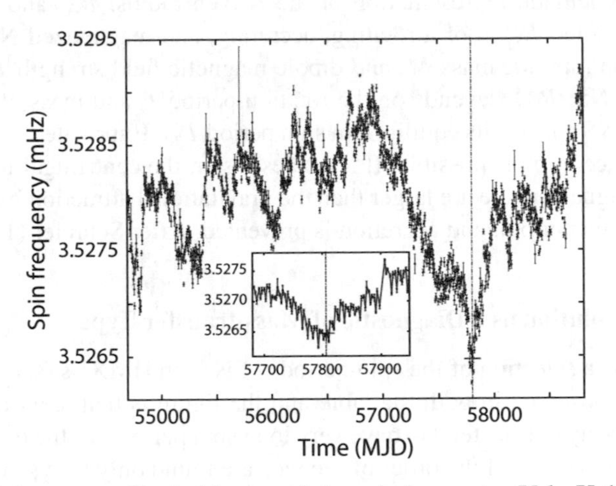

Figure 7.15. Eight years of spin history of the wind accretor Vela X-1 (4U 0900-40; pulse period 283 s) measured by Fermi/GBM, showing its continued irregular epochs of spin-up and spin-down. Inset shows enlarged 200 d epoch of spin-down and spin-up around the time marked by the red cross. These irregular spin-up and spin-down epochs, with the spin period hovering around the same average value, are characteristic for all wind accretors with blue supergiant companions. After Liao et al. (2020).

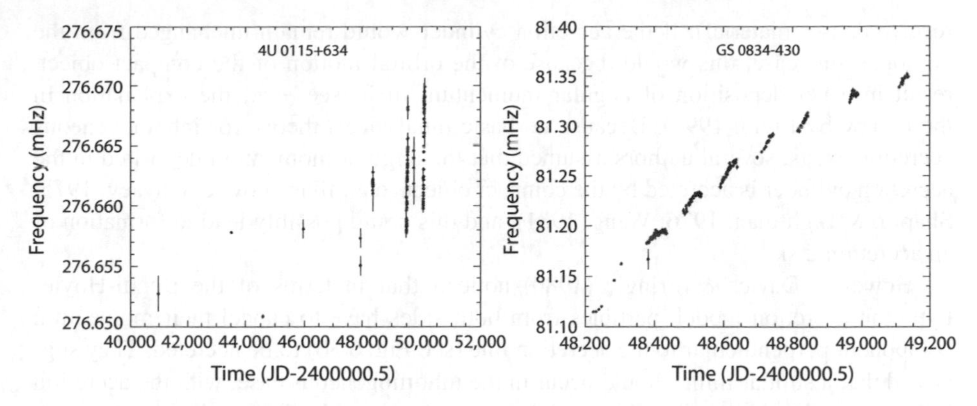

Figure 7.16. Long-term spin history of two Be/X-ray binaries. During transient outbursts, these pulsars spin-up; during quiescence periods between outbursts, they spin-down. After Bildsten et al. (1997).

ranging from 3 yr to several decades. Figure 7.14 and the review by Nagase (1989) show that the sources in which an accretion disk is observed, such as Cen X-3 (and also Her X-1 and OAO 1657−415), experience a secular decrease of the spin period, although from time to time this *spin-up* is interrupted for brief periods of time, and sometimes briefly reversed. The timescales for the secular spin-up are relatively short, of the order of 10^4 yr (Cen X-3, SMC X-1, OAO 1657−415) to $\sim 10^5$ yr (Her X-1). (For Cen X-3 the most recent determination of the secular spin-up timescale is 7709 ± 58 yr [Shirke et al., 2021].) This secular spin-up is due to the angular momentum fed to the NS by the accretion disk matter. On the other hand, Figure 7.15 shows that sources that are clearly accreting from a wind, such as Vela X-1 (4U 0900-40), show erratic variations of the spin period, sometimes spinning up and then again spinning down. On average, their spin periods fluctuate around a mean value of hundreds of seconds. We know that these sources are accreting from a wind because often the orbits are eccentric, such as $e \simeq 0.09$ for Vela X-1 and $e \simeq 0.5$ for GX 301-2, so the companions cannot be overflowing their Roche lobes. As shown in Figure 7.16, the Be/X-ray binaries tend to have very long pulse periods and show alternating spin-up and spin-down. However, during their transient outbursts they tend to show a continuous spin-up, as is clearly seen in GS 0834−430 and EXO 2030+375, while between outbursts, when the Be-companion has quieted down and does not eject matter from its equatorial regions, the NS spins down again, as seen, for example, in 4U 1145−819 (e.g., Nagase, 1989).

7.3.6 Disk Formation and Instabilities in Stellar Wind Accretion

Accretion from a stellar wind is generally not homogeneous. Even in the absence of clumping, just the density (or velocity) gradient in the wind results in a difference in density (velocity) of the material that enters the accretion cylinder's cross section (points of the cross section that are closer to the donor star are expected to receive denser material). If the accretion cylinder would remain unchanged from the homogeneous case, this would, because of the orbital motion of the compact object, result in a net deposition of angular momentum in it (see, e.g., the explanation in the review by Livio, 1994). Because no basic (analytical) theory for inhomogeneous accretion exists, several authors assumed that the angular momentum deposited in the accretion cylinder is accreted by the compact object (e.g., Illarionov & Sunyaev, 1975; Shapiro & Lightman, 1976; Wang, 1981), and this could possibly lead to formation of an accretion disk.

However, Davies & Pringle (1980) noticed that in terms of the Bondi-Hoyle-Lyttleton accretion model, particles from both sides have to cancel their momentum component perpendicular to the accretion line (see Fig. 6.30) to be accreted. They suggested that a similar thing should occur in the inhomogeneous case, with the accretion line being somewhat displaced toward the lower density side. They pointed out that, as a result, the conditions imposed on the material to be accreted conflict with the idea that angular momentum can be accreted at all. They showed with a simplified 2D model that the angular momentum accreted by an orbiting compact object is indeed zero. Soker & Livio (1984) showed that in the 3D case, the angular momentum accreted can be slightly different from zero.

These analytical calculations were followed by numerical hydrodynamic calculations of wind accretion in binaries that showed that the accreted angular momentum is indeed much smaller than expected from the above-mentioned analytical calculations of angular momentum deposition in the accretion cylinder. High-resolution 2D hydrodynamical wind accretion simulations by Matsuda et al. (1987), Taam & Fryxell (1988), and Fryxell & Taam (1988) showed that the wind accretion did not reach a steady state but exhibits a "flip-flop" behavior, in which the accretion shock cone oscillates from side to side, accompanied by short periods of disk formation. Subsequently, more precise 2D and 3D simulations for various circumstances and accreting-star sizes by Matsuda et al. (1992) showed that this instability arises particularly when the accretor is small in size compared to the accretion radius, as is the case with a NS or BH. Figure 7.17 shows an example of this flip-flop behavior of the accretion cone.

Accretion disks were discussed in Section 4.6. For a disk to form, the specific angular momentum of the accreted material must be sufficient to enable it to enter a Keplerian circular orbit around the magnetosphere (for magnetized NSs). This can be expressed by the condition $j > (GM_{\rm NS}r_{\rm A})^{1/2}$, where the radius of the magnetosphere, $r_{\rm A}$, is given by Eq. (7.11). For the wind velocities and mass-loss rates of blue supergiants in HMXBs, this condition leads, with the flip-flopping cone, to alternating formation of small prograde and retrograde disks. This may explain the spin behavior of the slow-spinning wind accretors in HMXBs, such as that of Vela X-1 depicted in Figure 7.15.

Apart from pure wind accretion and pure RLO, there also may occur an intermediate mode of accretion, called *wind RLO* (see e.g., El Mellah et al., 2019). This mode of accretion is thought particularly to be important in systems with relatively low wind velocities.

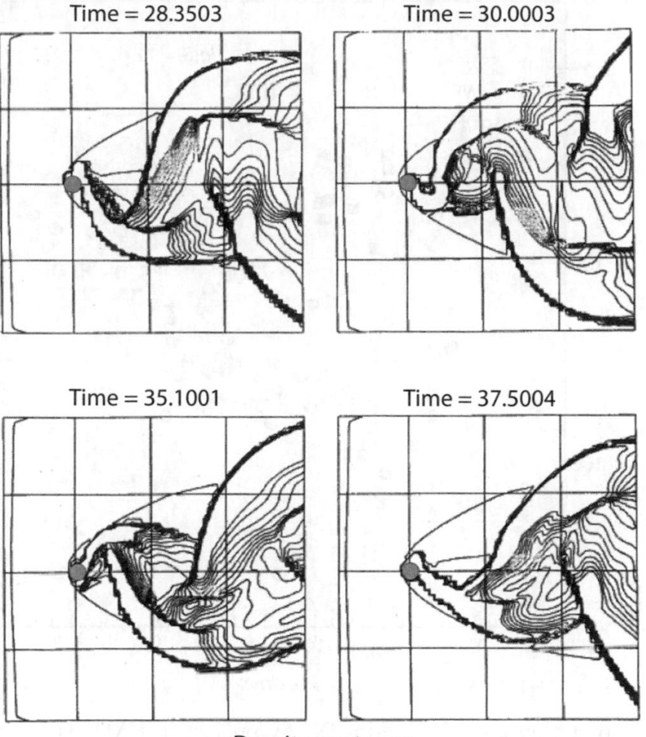

Figure 7.17. Density contours in a nearly isothermal numerical hydrodynamical calculation of wind accretion by a compact object (blue circle), showing flip-flop behavior. After Matsuda et al. (1992).

7.4 THE CORBET DIAGRAM FOR PULSATING HMXBs

In the so-called Corbet diagram, introduced by Corbet (1984), the spin period of accreting X-ray pulsars is plotted as a function of the orbital period, as shown in Figure 7.18, after Kretschmar et al. (2019). (The accreting MSPs are not included in this diagram.) One observes here that there are three distinct regions in the diagram: wind-accreting sg-HMXBs occupy the upper-left part of the diagram, the disk-accreting ones are located in the lower-left corner, and the Be/X-ray binaries exhibit a more or less linear relation between spin period and orbital period, in a band stretching from upper right to lower left in the diagram. Furthermore, the SFXTs tend to be located mostly in the same region as the Be/X-ray binaries. The diagram shows that in the sg-HMXBs and the Be/X-ray binaries with $P_{\rm orb} \geq 100$ d, the spin periods of the NSs are very long, in the range of 10^2–10^4 s. These NSs are extremely slow rotators. As they are born with spin periods <1 s (see Section 14.1 and also Fig. 1.2), their rotation has been braked to a very slow value within the lifetime of the companion star, which generally is $\leq 10^7$ yr.

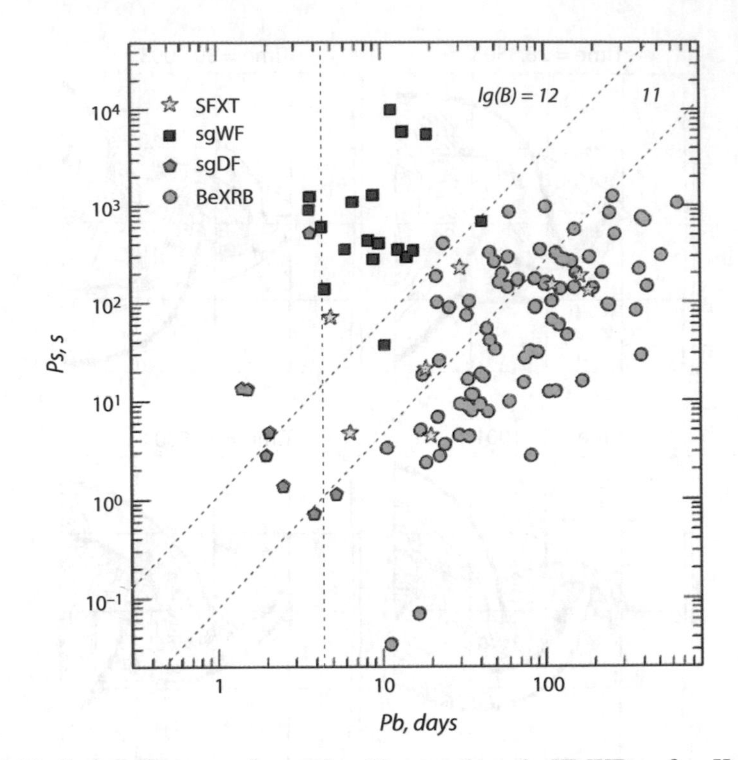

Figure 7.18. Corbet diagram of accreting X-ray pulsars in HMXBs, after Kretschmar et al. (2019). Magenta squares are persistent wind-fed sg-HMXBs (denoted as sgWF). The green pentagons are disk-accreting persistent HMXBs (denoted as sgDF; among these are also some ULXs in external galaxies). The yellow stars are the SFXTs. The blue circles are the Be/X-ray binaries. Below the blue dashed lines, quasi-spherical accretion from the stellar wind for two different magnetic field strengths, B, is inhibited by the centrifugal barrier (see text), assuming a wind speed of 800 km s^{-1}. Credit: Sergei A. Grebenev.

This strong braking cannot be explained by magnetodipole radiation (Section 14.1.3) and is thought to be due to the interaction between the NS's magnetic field and the accretion of wind matter from the companion star at a rather low rate. As shown in Figure 7.14, the spin periods of the disk-accreting NSs show a secular continuous decrease, due to the disk angular momentum accreted by the NS. The disks in these systems are produced by (beginning) RLO; see, for example, Savonije (1978, 1983). This continuous spin-up leads to the short pulse periods observed in the lower-left corner of the Corbet diagram. On the other hand, the spin period of a supergiant or a Be-source irregularly somewhat increases and decreases, but keeps hovering around the same mean value, showing that on average no angular momentum is fed to the NS by the stellar wind.

For an interpretation of the Corbet diagram in terms of stellar (binary) evolution, see Chapter 10.

7.5 ORBITAL CHANGES DUE TO TORQUES BY STELLAR WIND ACCRETION, MASS LOSS, AND TIDES

For the effects of stellar wind torques on the orbital evolution of binaries, we refer to Schrøder et al. (2021) and references therein. In particular, in wide low-mass binaries with a red-giant star such effects may become large (e.g., Jahanara et al., 2005; Saladino et al., 2018; Saladino et al., 2019; Saladino & Pols, 2019) and may lead to drastic orbital shrinking and even coalescence. On the other hand, Schrøder et al. (2021) find that for HMXBs such as Vela X-1 and Cyg X-1, the effects of wind torques in combination with wind mass loss may make the orbits expand on a timescale of a few tens of Myr. However, the orbital angular momentum of these systems also changes due to tidal torques that pump angular momentum from the orbit into the rotation of the donor star. This effect causes the orbits of these systems to shrink on a timescale of the order of the tidal circularization timescale that, using the equation for tidal circularization given by Hurley et al. (2002), yields a timescale of 9×10^3 yr for Cyg X-1 and of 2×10^5 yr for Vela X-1 (Schrøder et al., 2021). For HMXBs therefore the orbital shrinking by tidal effects is expected to dominate completely over the orbital expansion due to wind torques and mass loss. Indeed, for Vela X-1, the observed orbital period derivative is negative (Falanga et al., 2015), confirming that tidal effects are dominant here.

7.6 MEASURING BH SPINS IN X-RAY BINARIES

In Section 4.9, we introduced the spin of BHs and defined the dimensionless spin parameter, $a_* \equiv Jc/(GM^2)$, which takes the values $a_* = \{-1, 0, 1\}$ for a retrograde orbit around a maximally spinning BH, a non-spinning BH, and a prograde orbit around a maximally spinning BH, respectively. But how is the BH spin actually measured in X-ray binaries? For detailed reviews on BH spins and their observations, we refer to McClintock et al. (2014), Middleton (2016), and Reynolds (2021), for example. Here, we briefly summarize the results.

Three different techniques have been proposed, of which two are widely used. One method is based on modelling the profile of the relativistically broadened iron-emission line, which is a feature in the disk reflection spectrum (Fabian et al., 1989; Laor, 1991). The other main method appeals to modelling the accretion disk spectral continuum, also referred to as the *continuum-fitting method* (Zhang et al., 1997; McClintock et al., 2014), which will briefly be described in this chapter. Finally, entirely based on X-ray timing of GRO J1655−40, Motta et al. (2014) demonstrated that three frequencies of the quasi-periodic oscillations (QPOs) and of the broad band noise components and their variations match accurately the strong field general relativistic frequencies of particle motion in the close vicinity of the innermost stable circular orbit, as predicted by the relativistic precession model. Thus, these frequency measurements allowed Motta et al. (2014) to solve exactly the system of equations to obtain a simultaneous measurement of the mass and spin of the BH with very small error bars. They obtained a BH mass ($M_{BH} = 5.31 \pm 0.07\, M_\odot$) that is fully consistent with the value obtained from optical/near-infrared dynamical studies. The spin that they obtained

($a_* = 0.290 \pm 0.003$), however, is inconsistent with the estimates coming from either of the two above-mentioned methods: the Fe Kα line method or the continuum-fitting method. Unfortunately, this new method is limited to sources with three QPOs, which currently prevents further tests in other BH binaries.

Before describing the continuum-fitting model and the measured BH spins in X-ray binaries, we first follow Shapiro & Teukolsky (1983) and Reynolds (2021) and review a few basic equations describing the physics of the spin of BHs.

7.6.1 Basic Physics of a Spinning BH

Within the framework of general relativity (GR), the exact description of an isolated, spinning, and uncharged BH is given by the Kerr metric (Kerr, 1963; Misner et al., 1973). The geometry of the space-time around a BH is simply described by its mass, M, and spin angular momentum, J. Section 4.9 introduced the dimensionless spin parameter, $a_* \equiv Jc/(GM^2)$, and this parameter enables the description of several other BH physical parameters of interest. Using the Boyer-Lindquist coordinates (t, r, θ, ϕ), which are a straightforward generalization of spherical polar coordinates to space-time structure (Boyer & Lindquist, 1967), the line element of the Kerr metric is given by

$$ds^2 = -\left(1 - \frac{2Mr}{\Sigma}\right) dt^2 - \frac{4a_* M^2 r \sin^2\theta}{\Sigma} dt\, d\phi + \frac{\Sigma}{\Delta} dr^2 \tag{7.18}$$

$$+ \Sigma\, d\theta^2 + \left(r^2 + a_*^2 M^2 + \frac{2a_*^2 M^3 r \sin^2\theta}{\Sigma}\right) \sin^2\theta\, d\phi^2,$$

where $\Delta \equiv r^2 - 2Mr + a_*^2 M^2$ and $\Sigma = r^2 + a_*^2 M^2 \cos^2\theta$ and using the default setting of $G = c = 1$.

Setting $a_* = 0$ yields the Schwarzschild metric for a non-spinning BH. The $dt\, d\phi$ term shows the rotational frame-dragging properties of a spinning BH. The event horizon, $R_{\rm H}$, is found when the dr^2 term in the line element becomes singular, that is, when $\Delta = 0$, or

$$R_{\rm H\pm} = (1 \pm \sqrt{1 - a_*^2})\, M, \tag{7.19}$$

where only $R_{\rm H+}$ has an astrophysical meaning. We find here immediately the result that $R_{\rm H+} = \{M,\ 2M,\ M\}$ for $a_* = \{-1,\ 0,\ 1\}$.

The effective potential felt by particles orbiting a BH is given by

$$V \equiv E^2\,(r^3 + a_*^2 M^2 r + 2a_*^2 M^3) - 4a_* M^2 E\, l - (r - 2M)\, l^2 - m^2 r\, \Delta, \tag{7.20}$$

where E, l, and m are the binding energy, orbital angular momentum and particle mass (Shapiro & Teukolsky, 1983). For circular orbits, the specific energy and the specific orbital angular momentum are

$$\frac{E}{m} = \frac{r^2 - 2Mr \pm a_* \sqrt{M^3 r}}{r\sqrt{r^2 - 3Mr \pm 2a_* \sqrt{M^3 r}}} \tag{7.21}$$

$$\frac{l}{m} = \frac{\sqrt{Mr}\,(r^2 \mp 2a_* \sqrt{M^3 r} + a_*^2 M^2)}{r\sqrt{r^2 - 3Mr \pm 2a_* M}}. \tag{7.22}$$

In GR, circular orbits can exist from $r = \infty$ to $r = R_{\rm ph}$ (the photon radius, see below), and these can be both bound or unbound. But not all bound orbits are stable. Circular orbits are only stable outside of a critical radius, $R_{\rm ISCO}$ known as the innermost stable circular orbit (ISCO). This radius is roughly equal to the inner edge of the accretion disk (depending on the disk geometry). Beyond (meaning inside) this point, the accreted material is advected across the event horizon of the BH without producing further radiation. Using the stability criterion $\partial^2 V/\partial r^2$ yields (after some algebra [Bardeen et al., 1972])

$$R_{\rm ISCO} = \left(3 + Z_2 \mp [(3 - Z_1)(3 + Z_1 + 2Z_2)]^{1/2}\right) M, \tag{7.23}$$

where the orbits considered are restricted to the $\theta = \pi/2$ plane; the $\mp$ sign is for particles in prograde/retrograde orbits, respectively; and Z_1 and Z_2 are defined by

$$Z_1 \equiv 1 + \left(1 - a_*^2\right)^{1/3} \left[(1 + a_*)^{1/3} + (1 - a_*)^{1/3}\right] \tag{7.24}$$

$$Z_2 \equiv \left(3a_*^2 + Z_1^2\right)^{1/2}. \tag{7.25}$$

The binding energy at the ISCO is important for determining the accretion luminosity when matter is transferred onto a BH. It can be shown that at the location $r = R_{\rm ISCO}$, a_* and E and m are related according to

$$a_* = \frac{4\sqrt{2}\sqrt{1 - (E/m)^2} - 2E/m}{3\sqrt{3}\,(1 - (E/m)^2)}, \tag{7.26}$$

which means that $E/m = \{\sqrt{25/27}, \sqrt{8/9}, \sqrt{1/3}\}$ for $a_* = \{-1, 0, 1\}$. We can now find the maximum gravitational binding energy release[2] (in units of the rest mass) for a particle falling into the BH, simply as $\eta = 1 - E/m$, and thus we finally obtain $\eta = \{0.0378, 0.057, 0.423\}$ for $a_* = \{-1, 0, 1\}$; see Table 7.2. This means that between 4% and 42% of the rest mass energy of infalling matter can be released by accretion onto a BH. In comparison, accretion onto a NS gives rise to an energy release of

[2]Defining the binding energy to be 0 at the event horizon.

Table 7.2. Radii of Interest for a Spinning and Accreting BH (see also Fig. 7.19)

a_*	-1	0	1
$R_{\rm ISCO}$	9	6	1
$R_{\rm H+}$	1	2	1
η	0.038	0.057	0.423

The location of the ISCO, $R_{\rm ISCO}$, and the event horizon (Schwarzschild) radius, $R_{\rm H+}$, in units of $GM/c^2 \simeq 1.5\,{\rm km}$ $(M/M_\odot)$ given as a function of the BH spin parameter, a_* (see text). The bottom row gives the fraction of released gravitational binding energy in units of rest mass energy.

approximately 10%–15%, depending on its exact EoS (equation of state, almost independent of its spin rate).

The event horizon marks the sphere of no return for any infalling matter or light. No signal emitted from $r < R_{\rm H+}$ is able to propagate outward. For a non-spinning BH ($a_* = 0$), this event horizon, the Schwarzschild radius, is given by $R_{\rm Sch} = 2GM/c^2 \simeq 3\,{\rm km}\,(M_{\rm BH}/M_\odot)$. Formally, at the very center of the BH, there is a location where the space-time curvature diverges, that is, a space-time singularity.

Spin introduces an interesting phenomenon in space-time surrounding a BH: inertial frames of reference are dragged along into rotation around the BH. As a result of this, if a particle orbits the BH in a plane that is not aligned with the BH spin (i.e., if the orbit does not lie in the $\theta = \pi/2$ plane), the orbital plane will undergo Lense-Thirring precession with a frequency (Reynolds, 2021) of

$$\Omega_{\rm LT} = 2a_* \frac{(GM)^2}{c^5 r^3} \,. \tag{7.27}$$

Note, that NSs and WDs are also able to produce significant frame-dragging effects. Recently, evidence for frame-dragging was demonstrated in a NS+WD system by the binary radio pulsar PSR J1141$-$6545 (Krishnan et al., 2020).

The combined effects of Lense-Thirring precession and internal disk viscosity acts to align the mid-plane of the inner region of the accretion disk with the equatorial plane of the BH (the *Bardeen-Petterson effect*, Bardeen & Petterson 1975) and thereby, over time, acts to align the BH spin axis with that of the orbital angular momentum of the binary system.

Very close to a rapidly spinning BH, frame-dragging effects may become extreme. Inside the *ergosphere*, space-time is dragged along causing matter and even photons to co-rotate in the same sense as the BH, as seen by a distant observer. The outer boundary of the ergosphere, called the *static limit*, is defined as where a test particle moving at $v = c$ would compensate frame-dragging and appear stationary for a distant observer. In principle, rotational energy of the BH can be extracted from the ergosphere by the

Penrose process (Penrose, 1969). Penrose demonstrated through a series of thought experiments that physical processes (e.g., particle-particle scattering) interior to the static limit can tap into and extract the spin energy of the BH. (Hence, the name of this region, the ergosphere, *ergon* being a Greek word for "work.")

Considerations based on BH thermodynamics reveal that the (extractable) energy associated with BH spin is given by (Misner et al., 1973)

$$E_{\rm spin} = \left[1 - \frac{1}{2}\left(\left[1+\sqrt{1-a_*^2}\right]^2 + a_*^2\right)^{1/2}\right] Mc^2. \tag{7.28}$$

For a maximally spinning BH ($|a_*| = 1$), this yields $E_{\rm spin} = (1 - 1/\sqrt{2})Mc^2$, which is an amazing 29% of the rest mass energy of the BH.

Following Reynolds (2021), we now come to an important realization, namely that BH spin can be an astrophysically important energy source. The processes envisaged in Penrose's original work require quite violent particle interactions within the ergosphere. As a result, high-energy particle acceleration and γ-ray emission is possible. More attention, however, has been given to the magnetic version, the *Blandford-Znajek mechanism*, in which relativistic jets extract energy and angular momentum from the BH via an accretion disk with a strong poloidal magnetic field (Blandford & Znajek, 1977)—the poloidal B-field is twisted by frame-dragging, thereby producing an outgoing Poynting flux along twin jets. This mechanism is believed to be the driving mechanism of long γ-ray bursts (Piran, 2004)—the violent deaths (hypernovae or collapsars) of extremely massive stars (Woosley, 1993; MacFadyen & Woosley, 1999; see also Chapter 13).

The rate of extracting rotational energy of the BH, that is, the power, scales as

$$\dot{E}_{\rm spin} \propto \Phi_{\rm tot}^2 \Omega_{\rm H}^2, \tag{7.29}$$

where $\Phi_{\rm tot} \propto BM^2$ is the total poloidal magnetic flux in the jet, and

$$\Omega_{\rm H} = \frac{c^3}{2GM}\left(\frac{a_*}{1+\sqrt{1-a_*^2}}\right) \tag{7.30}$$

is the angular velocity of the BH at the event horizon (Tchekhovskoy et al., 2010). For Cyg X-1, for example, we thus estimate its spin period to be $P_{\rm H} = 1.3$ ms. For observational evidence that supports Eq. (7.29), see, for example, McClintock et al. (2014).

We end this section by defining yet another couple of characteristic BH radii. As mentioned, in BH space-times, circular orbits are only stable outside the ISCO. However, on energetic grounds and dependent on the geometry of the accretion disk (Frank et al., 2002), marginally bound orbits exist down to a radius:

$$R_{\rm mb} = \left(2 \mp a_* + 2\sqrt{1 \mp a_*}\right) M, \tag{7.31}$$

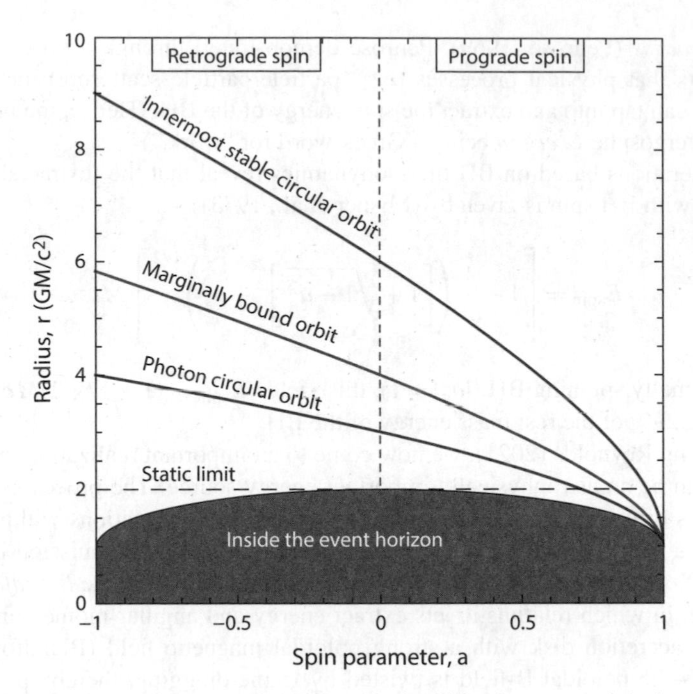

Figure 7.19. Locations of five special BH orbits/radii in the equatorial plane of a spinning BH as a function of the dimensionless spin parameter, a_*. See text for explanation. After Reynolds (2021).

where, again, the $\mp$ sign is for particles in prograde/retrograde orbits and is restricted to orbits within the $\theta = \pi/2$ plane (Bardeen et al., 1972).

Inside of the marginally bound orbit is a special location where photons can orbit the BH, the photon circular orbit. This radius is also spin-dependent and given by (Bardeen et al., 1972)

$$R_{\rm ph} = 2\left(1 + \cos\left[\frac{2}{3}\cos^{-1}(\mp a_*)\right]\right) M. \tag{7.32}$$

This radius can be found as a singularity in Eqs. (7.21) and (7.22) (solving for the denominators being equal to 0). In other words, a photon sphere is a spherical region of space where gravity is strong enough that photons are forced to travel in orbits. Figure 7.19 illustrates the relative location of the five special BH orbits/radii introduced in this section.

7.6.2 BH Spins from the Continuum-fitting Model

The (thermal) continuum-fitting method (Zhang et al., 1997; McClintock et al., 2014) is based on the fact that the measurable temperature at the inner edge of the accretion disk, and thus the location of the ISCO, is a function of the BH spin (Bardeen et al.,

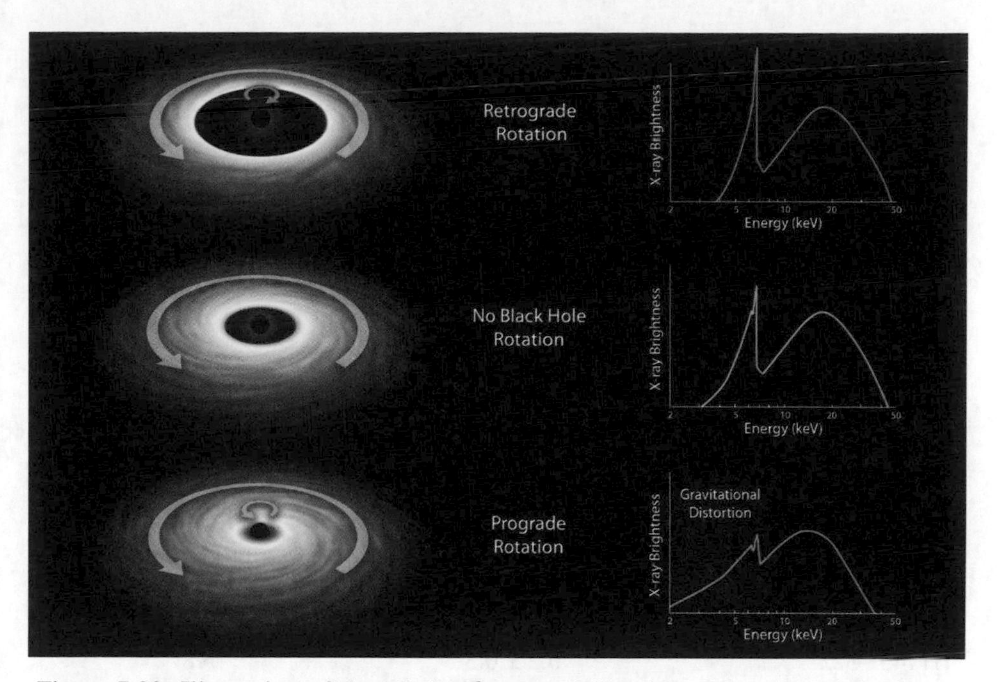

Figure 7.20. Illustration of the relation between BH spin and location of the innermost stable circular orbit (ISCO) which is determined from the accretion disk spectrum in BH X-ray binaries. Credit: NASA/Jet Propulsion Laboratory (JPL)-Caltech.

1972). The reason for this is that the location of the inner disk determines the amount of released gravitational binding energy extracted from each gram of accreting matter, and thus the heating of the disk—with prograde accretion onto a rapidly rotating BH yielding the highest temperatures. More specifically, the disk spectrum (peak emission and temperature) depends on $R_{\rm ISCO}$ (see Fig. 7.20 for an illustration) and can be determined from continuum-fitting techniques within the context of a specific accretion disk model, which depends on the mass-accretion rate of the BH.

In principle, the method is simple: it is analogous to measuring the radius of a star whose flux, temperature, and distance are known. By this analogy, it is clear that it is essential to know the luminosity of the accretion disk. Determining this luminosity, however, involves uncertainties from estimating the source distance, as well as the disk inclination. Additionally, one must know the BH mass to scale $R_{\rm ISCO}$ and thereby determine a_*.

Table 7.3 shows examples of the measured BH spins in X-ray binaries using the continuum-fitting method. It is seen that, whereas the BH spins in the persistent (wind-fed) X-ray sources are rather high ($a_* > 0.8$), the BH spins in the transient (RLO) X-ray binaries cover a wide range of BH spins ($a_* \simeq 0.1$–1). It has been suggested, based on evolutionary considerations, that the high BH spins in the persistent sources must be natal (see further discussions in Section 15.8). As shown in Figure 7.21, BH spins

Table 7.3. Measured Values of BH Spins (a_*) Using the Continuum-fitting Method

Source	BH spin (a_*)	BH mass ($M_{BH}/M_{\odot}$)
Persistent X-ray binaries		
Cyg X-1	>0.9985*	21.2 ± 2.2
LMC X-1	$0.92^{+0.05}_{-0.07}$	10.9 ± 1.4
IC 10 X-1	$0.85^{+0.04}_{-0.07}$	15 (assumed)
M33 X-7	0.84 ± 0.05	15.65 ± 1.45
Transient X-ray binaries		
GRS 1915+105	>0.95	10.1 ± 0.6
MAXI J1803−298	0.991 ± 0.001	?
4U 1543−47	0.80 ± 0.10	9.4 ± 1.0
GRO J1655−40	0.70 ± 0.10	6.3 ± 0.5
Nova Mus 1991	$0.63^{+0.16}_{-0.19}$	11.0 ± 1.8
XTE J1550−564	$0.34^{+0.20}_{-0.28}$	9.1 ± 0.6
LMC X-3	<0.3	7.6 ± 1.6
H1743−322	0.2 ± 0.3	~8
MAXI J1820+070	$0.13^{+0.07}_{-0.10}$	8.5 ± 0.7
A0620−00	0.12 ± 0.19	6.6 ± 0.25

Source: After McClintock et al. (2014), Zhao et al. (2021), Reynolds (2021), Feng et al. (2021), Miller-Jones et al. (2021), and references therein.

*Assuming aligned spin ($\delta = 15^{\circ}$ yields $a_* = 0.9696$; Miller-Jones et al., 2021).

obtained using the other main method of modelling, based on the relativistically broadened iron-emission line, sometimes yield quite different results.

A word of caution: in the continuum-fitting method, models of BH accretion disks and their expected emitted radiation is needed to predict the expected continuum spectrum. The physics of these models involves a number of assumptions, not all of them well-verified. Therefore, the uncertainties of the predicted spectra may be considerable, but it is not known how large they are.

7.7 ULTRA-LUMINOUS X-RAY BINARIES

A puzzling discovery made by the Einstein Observatory is the existence of off-nucleus X-ray point sources with luminosities $>10^{39}$ erg s^{-1} (Long & van Speybroeck, 1983), sometimes even reaching 10^{41} erg s^{-1}, assuming isotropic radiation (Gao et al., 2003). These extragalactic ultra-luminous X-ray (ULX) sources are difficult to explain in terms of accreting compact objects in HMXBs, as the Eddington limits for NSs and stellar mass BHs are less than a few times 10^{38} erg s^{-1} and a few times 10^{39} erg s^{-1},

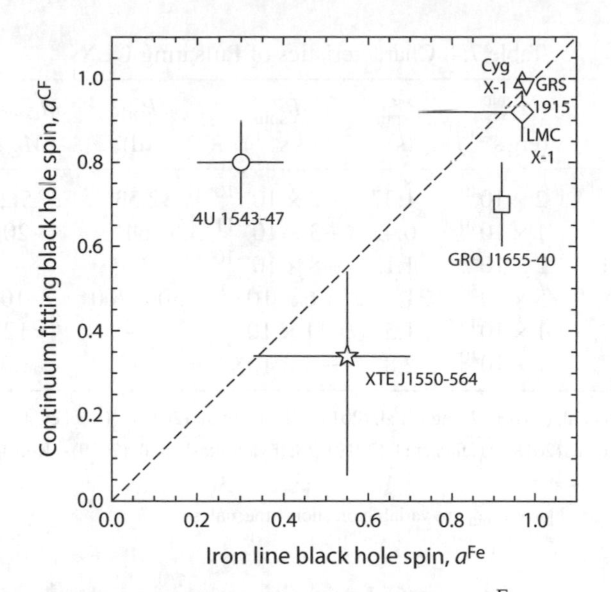

Figure 7.21. BH spin measurements from the iron-line (a^{Fe}) vs. continuum-fitting (a^{CF}) techniques. Plotted error bars are indicating the 68% confidence level. Spin measurements for, in particular, 4U 1543−47, but also GRO J1655−40, and marginally XTE J1550−564, are in disagreement, as these points do not lie on the dashed diagonal that indicates agreement between the two measurement techniques. After Salvesen & Miller (2021).

respectively.[3] One possibility to explain these high luminosities is to consider the existence of a population of intermediate-mass BHs, with masses between 10^2 and $10^5\ M_\odot$, possibly arising from the collapse of primordial stars (Madau & Rees, 2001) or formed in dense globular clusters (e.g., Miller & Hamilton, 2002a).

If the accreting objects in ULXs have a stellar origin, there are various potential explanations for their high luminosities. For instance, beaming of the radiation emitted would imply that the actual full-sky luminosity of these sources is much lower, so that ULXs could consist of BHs with masses $<10\ M_\odot$ accreting close to the Eddington rate. This beaming could be a purely geometrical effect (King, Davies, et al., 2001) or the result of relativistic beaming (Körding et al., 2002). For example, in ULX systems, a low to moderate amount of beaming can be present due to the presence of outflows (King et al., 2017). The intrinsic luminosity ($L_{X,\rm int}$) of a system can be expressed as a function of a beaming parameter, b:

$$L_{X,\rm int} = \frac{L_X}{b} \simeq \frac{GM\dot{M}}{R}, \qquad b \geq 1, \tag{7.33}$$

[3]BHs of masses $M_{\rm BH} > 30\ M_\odot$ (reaching luminosities $L_X > 4 \times 10^{39}$ erg s^{-1} without exceeding their Eddington luminosities) may form via chemically homogeneous evolution in very massive close binaries at low metallicity (Marchant et al., 2017).

Table 7.4. Characteristics of Pulsating ULXs

Source	$L_X^{\max}$ (erg s^{-1})	$P_{\rm spin}^*$ (s)	$\dot{P}_{\rm spin}^*$ (s s^{-1})	$P_{\rm orb}$ (d)	$M_{\rm donor}$ ($M_\odot$)	References
M82 X-2	2×10^{40}	1.37	-2×10^{-10}	2.53	>5.2	1
NGC 7793 P13	1×10^{40}	0.42	-3×10^{-11}	64	~20	2,3,4
NGC 5907 X-1	2×10^{41}	1.13	-8×10^{-10}	2.5	—	5
NGC 300 ULX-1	5×10^{39}	31	-6×10^{-7}	300–700	8–10	6,7
NGC 1313 X-2	1×10^{40}	1.5	-1×10^{-8}	—	<12	8
M51 ULX-7	1×10^{40}	2.8	-3×10^{-9}	—	>8.3	9

Sources: [1]Bachetti et al. (2014). [2]Motch et al. (2014). [3]Israel et al. (2017). [4]Israel et al. (2017). [5]Fürst et al. (2016). [6]Carpano et al. (2018). [7]Heida et al. (2019). [8]Sathyaprakash et al. (2019). [9]Rodríguez Castillo et al. (2020).

*Note, $P_{\rm spin}$ and particularly $\dot{P}_{\rm spin}$ are variable on short timescales.

where L_X is the measured isotropic luminosity derived from the observed X-ray flux, F_X. If beaming is indeed present, the derived mass-accretion rate and B-field should yield an inner disk radius smaller than the spherization radius, where the disk thickness becomes comparable to its radius (Shakura & Sunyaev, 1973; Poutanen et al., 2007), and the disk luminosity becomes equal to the Eddington luminosity.

So is the actual accretion rate onto the compact object super-Eddington or not? According to Shakura & Sunyaev (1973), it seems that any super-Eddington mass-transfer rate from a companion star would blow up the accretion disk to be geometrically thick and this helps keeping the accretion rate at each radius below the Eddington limit. On the other hand, another potential mechanism is the photon-bubble instability (Begelman, 2002; Ruszkowski & Begelman, 2003), which acts on radiation-dominated accretion disks and produces clumping, that could cause photons to be radiated away through low-density regions, allowing for overall accretion rates exceeding the Eddington limit by an order of magnitude.

Recent discoveries, however, reveal that several ULXs show X-ray pulsations (Table 7.4), which is only expected if the compact accretor is a NS (Kaaret et al., 2017). At the same time, these X-ray pulsating ULXs show long-term X-ray luminosities that are clearly in the ULX regime ($L_X > 10^{39}$ erg s^{-1}). Examples include the first discovered source M82 X-2 (Bachetti et al., 2014), which is in a 2.5 d orbit with a companion more massive than 5.2 $M_\odot$, and the 0.42 s X-ray pulsar NGC 7793 P13 (Israel, Papitto et al., 2017), which has an X-ray luminosity of ~500 times the Eddington limit if one assumes the emission to be isotropic. All the pulsating ULXs are highly variable sources (Kaaret et al., 2017). They exhibit a luminous phase with peak luminosity of $L_X \simeq 10^{40}$–10^{41} erg s^{-1} and flux variation of a factor of ~10 during which pulsations are sometimes detected, along with a faint phase with fluxes below $L_X \simeq 10^{37}$–10^{38} erg s^{-1}. This high level of variability, is not observed in the usual non-pulsating ULX population.

That the Eddington limit (Chapter 4) plays a very important role in Nature is demonstrated by the fact that none of the persistent pulsating accreting X-ray sources known in our Galaxy, in the Magellanic Clouds, and other nearby galaxies considerably exceeds the Eddington limit for a 1.4 $M_\odot$ NS for longer than a few weeks. Pulsating transient sources, such as A0538−66 in the LMC, may sometimes exceed the Eddington limit by an order of magnitude for short periods of time, possibly due to accretion disk instabilities. As a consequence, the existence of extragalactic pulsating ULXs exceeding the Eddington limit for spherical accretion was at first ascribed to highly non-spherical accretion where the X-rays are emitted highly beamed, sideways from the accretion column above the magnetic poles of the NS. In such a case the beam flux, if it were emitted by an entire spherical surface, would give the impression of a highly super-Eddington source (similar to the beaming idea proposed for BH ULXs). Super-Eddington accretion by a factor of a few was already suggested by Abramowicz et al. (1980) via a thick accretion disk with a central funnel.

Whereas beaming effects might help to explain the very large X-ray luminosities, some NSs in ULXs have been shown to experience an extreme spin-up on a short timescale—for example, NGC 7793 P13, which has spun up from ~1.4 to 1.1 s in just one decade (Israel, Papitto et al., 2017). Providing such a strong torque responsible for the spin-up only seems possible if the NS actually *accreted* at a rate much higher than the classical Eddington limit (but see King & Lasota, 2019). Furthermore, measurements from ionization nebulae around some ULXs appear to confirm the isotropic estimate of their luminosities (Pakull & Mirioni, 2003). For other pulsating ULXs, such as NGC 1313 X-2, it is argued that most of the accreted material has been expelled over the lifetime of the ULX, favoring physical models including strong winds and/or jets for NS ULXs (Sathyaprakash et al., 2019).

One way to understand high intrinsic luminosities and accretion rates onto NSs is to invoke magnetic fields with a strength $\gg 10^{13}$ G (i.e., magnetar B-field strengths), which could reduce the radiative opacity (electron scattering cross section) of the accreted matter and thus raise the Eddington limit significantly (Basko & Sunyaev, 1976; Paczynski, 1992; Dall'Osso et al., 2015; Mushtukov et al., 2015). In an alternative scenario (Israel, Belfiore et al., 2017; Chashkina et al., 2017), the presence of a strong multi-polar B-field close to the surface of the NS, possibly coupled with a modest degree of beaming, appears as a reasonable way out of the problem of requiring a very large dipole B-field component. In this scenario, a more standard dipolar B-field component of approximately $10^{12} - 10^{13}$ G dominates at large distances over the multi-polar component, the effect of which is limited to the region close to the accretion column base near the surface of the NS. The presence of a corona supported by strong magnetic fields could also help to reduce the radiation pressure and allow super-Eddington accretion (Socrates & Davis, 2006). A remaining puzzle, however, is how the strong B-fields in these models can remain strong without decaying (Bhattacharya, 2002), given the expected vast amount of material accreted by such a NS—unless the duration of the accretion phase is very short and not a long-term effect (a solution that might be important for understanding the nature of these sources). Gao & Li (2021) find that pulsating ULXs exhibit two branches of dipole magnetic field solutions, broadly

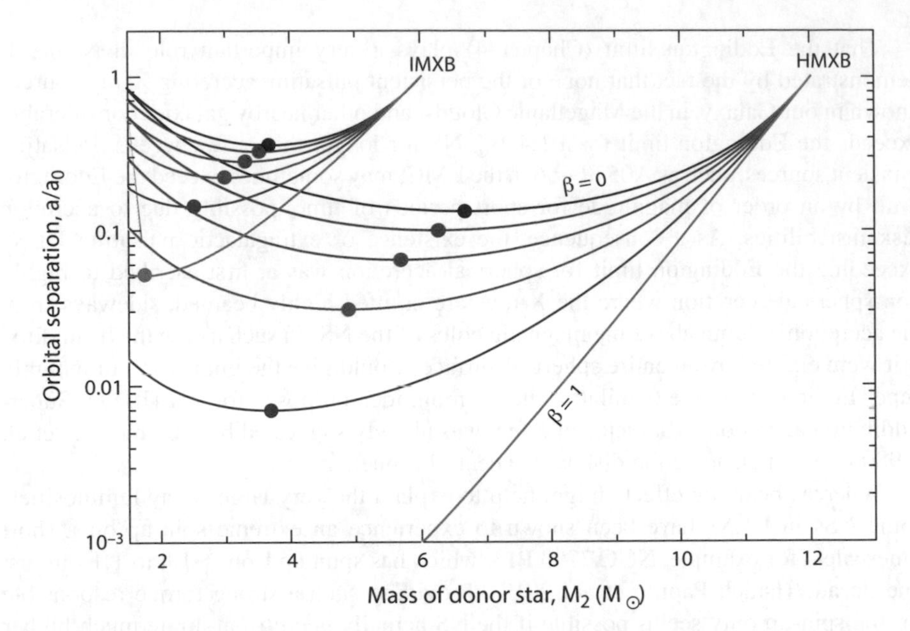

Figure 7.22. Orbital evolution of IMXBs and HMXBs based on the isotropic re-emission model (Chapter 4). The plot shows the decrease in orbital separation (in units of the pre-RLO orbital separation, a_0) as a function of decreasing donor star mass, for an IMXB system ($M_2 = 6\,M_\odot$) and a HMXB system ($M_2 = 12\,M_\odot$). The evolution is along tracks from right to left. In both cases, an initial accreting NS mass of $1.35\,M_\odot$ is assumed. Minimum values of (a/a_0) are shown with circles for different values of β in steps of 0.2. While IMXBs would be more stable on a longer timescale compared to HMXBs, several NS ULXs are found to host a massive companion star. This fact challenges our understanding of their origin. After Tauris et al. (2017).

distributed and centered on roughly $B \sim 10^{10}$ G and $B \sim 10^{13}$ G. In this picture, the low B-field solutions correspond to the state where the NSs are far away from spin equilibrium, and the high B-field solutions are close to spin equilibrium (see Section 7.3). For an additional study on the B-fields of pulsating ULXs, see Chen X., et al. (2021).

7.7.1 Origin of Pulsating ULXs

The discovery of pulsating ULXs poses an interesting challenge to modern accretion physics and binary star evolution. Tauris et al. (2017) and Misra et al. (2020) have suggested that ULXs hosting NSs may be explained by intermediate-mass donor stars ($\sim 2-6\,M_\odot$); they have based their suggestions on the IMXB models of, for example, Tauris et al. (2000), Podsiadlowski et al. (2002), and Shao & Li (2012). The rationale being that solving the orbital angular momentum balance equation (Section 4.3.3) reveals that the orbital evolution of HMXB donors would drive the systems into contact, whereby the systems should be dynamically unstable on a short timescale (see Fig. 7.22). The (thermal) timescale on which such a HMXB becomes dynamically unstable after initiating RLO (and presumably forming a CE) is of the order of 10^2–10^4 yr (Savonije, 1978), depending on the orbital period at RLO. In comparison, the IMXB systems are typically able to provide mass-transfer rates, $|\dot{M}_2| > 10^{-5}\,M_\odot\,\mathrm{yr}^{-1}$ for a few $10^5 - 10^6$ yr (e.g., Tauris et al., 2000, 2011), followed by RLO at a lower rate for approximately $10^6 - 10^7$ yr, depending on the orbital period. Therefore, based on the preceding arguments, IMXBs are good candidates to explain the nature of ULXs hosting an accreting NS.

However, the donor star in the pulsating ULX source NGC 5907 X-1 is constrained to be $\gtrsim 10\,M_\odot$ (Israel, Belfiore et al., 2017) and its orbital period is found to be $P_{\mathrm{orb}} = 5.3^{+2.0}_{-0.9}$ d. Motch et al. (2014) determined the orbital period of the ULX NGC 7793 P13 to $\sim$64 d, which was afterward discovered to be a pulsating ULX with a mass donor being a B9Ia supergiant of $\sim 20\,M_\odot$ (Israel, Papitto et al., 2017). And recently, NGC 300 ULX-1 was found to have a red supergiant donor star ($8 - 10\,M_\odot$) and an orbital period of approximately $300 - 700$ d (Heida et al., 2019). Except for being ultraluminous, the parameters of these three pulsating ULX systems are reminiscent of those of the sg-HMXBs discussed earlier. Also M51 ULX-7 has an OB-giant donor star of mass $> 8.3\,M_\odot$ (Rodríguez Castillo et al., 2020). An HMXB origin of pulsating ULXs has been suggested by Eksi et al. (2015), Kluzniak & Lasota (2015), Shao & Li (2015), and Mushtukov et al. (2015), including Be-star HMXBs (Karino & Miller, 2016; Christodoulou et al., 2017). Further evidence for a HMXB origin comes from the Chandra discovery of a population of ULXs with lifetimes $\lesssim$10 Myr in the Cartwheel Galaxy (King, 2004). However, these ULX sources were not detected to be pulsating and might therefore be hosting accreting BHs; if so, their orbital evolution is much more stable. It has been argued that the current population of known pulsating ULXs is merely the tip of the iceberg and that a relatively large fraction of ULXs host a NS rather than a BH accretor (King et al., 2017).

If the majority of pulsating ULXs are indeed HMXBs (rather than IMXBs), the challenge is to explain long-term accretion at a rate of $\dot{M}_{\mathrm{NS}} \gg \dot{M}_{\mathrm{Edd}}$ and *how* such a system could avoid a delayed dynamical instability. Quast et al. (2019) investigated how the proximity of supergiant donor stars to the Eddington limit, and their advanced evolutionary stage, may influence the evolution of ULXs with supergiant donor stars. They constructed models of massive donor stars with different internal hydrogen/helium gradients and exposed them to mass loss to probe the response of the stellar radius. They concluded that, under certain conditions, stable nuclear timescale mass transfer in sg-HMXBs is possible, even with a NS accretor and mass ratios > 20. In their binary evolution models, the donor stars rapidly decrease their thermal equilibrium radius and can therefore cope with the inevitably strong orbital contraction imposed by the high mass ratio.

Whatever the formation channel is, one should keep in mind that usually only one ULX source is seen in a given Galaxy. That is, the formation channel may be a rare event, allowing for somewhat more exotic binary possibilities. Alternatively, these pulsating ULXs could simply be explained by standard HMXBs where the donor stars are at the early onset of thermal timescale RLO. Even though such systems become dynamically unstable approximately $10^2 - 10^4$ yr after initiating RLO (Savonije, 1978), there could still be one or two such systems observable during that time epoch in a typical MW-like Galaxy that contains approximately 100 to 200 HMXB sources.

7.8 LOW-MASS X-RAY BINARIES IN MORE DETAIL

7.8.1 Scope of This Section

Here, we describe in more detail the characteristic types of variability of the NS-LMXBs, which were already briefly mentioned in Section 6.2, and discuss their possible causes. We also describe here three special groups of LMXBs that stand out from the bulk of the "standard" LMXBs in the Galactic disk with core-hydrogen–burning donor stars. These are (1) the globular cluster X-ray sources; (2) the ultra-compact LMXBs, which have hydrogen-poor donor stars; and (3) the symbiotic LMXBs, in which the donor star is a low-mass red-giant star, except in one case, where it is a quite massive red supergiant.

7.8.2 Categories of Variability of NS-LMXBs

Two categories can be distinguished characterized by (1) variations in the X-ray flux and/or the optical brightness related to the orbital motion of the stars, and (2) variations directly related to the intrinsic luminosity of the X-ray source itself. We start by discussing the first category, followed by a discussion of the second category.

Table 6.3 lists the orbital periods of several dozens of LMXBs, which have been determined in the course of many decades. In most cases, the determination of this period is quite difficult, because the light of the system is dominated by that of the bright accretion disk, compared to which the donor star is almost always invisible. Only if our line of sight almost coincides with the orbital plane do regular eclipses occur. In most cases, however, no orbital periodicity of the X-rays is seen. In some cases one sees repeating faint X-ray dips due to shallow eclipses of the X-ray source by gas steams in the system. In some cases an optical periodicity is seen due to the one-sided heating of the donor star. This is, for example, the case in Sco X-1. If the source is transient, and goes into an "off" state, one observes ellipsoidal light variations of the Roche-lobe filling donor star, which yields the orbital period. This is particularly the case in many of the BH-LMXBs, because most of these are transients. Contrary to the case of the CVs (Fig. 5.6), the distribution of the orbital periods of the LMXBs does not show a period gap, nor does it have a lower-limit cutoff at an orbital period at 80 min. Systems with periods below this limit we call ultra-compact X-ray binaries (*UCXBs*), which will be described later in Section 7.8.4.

As to the second category, characterized by variations directly related to the intrinsic luminosity of the X-ray source itself, LMXBs show two common broad classes of variability. The first of these is *thermonuclear X-ray bursts*. These are bright sudden outbursts of X-rays from NSs in LMXBs, typically reaching the Eddington limit and lasting in total several tens of seconds, with a decay timescale of the order of 5 s (see Fig. 7.23). An excellent review of these bursts, also called Type I X-ray bursts, and of their causes, is given by Strohmayer & Bildsten (2006), as well as by Ghosh (2007).

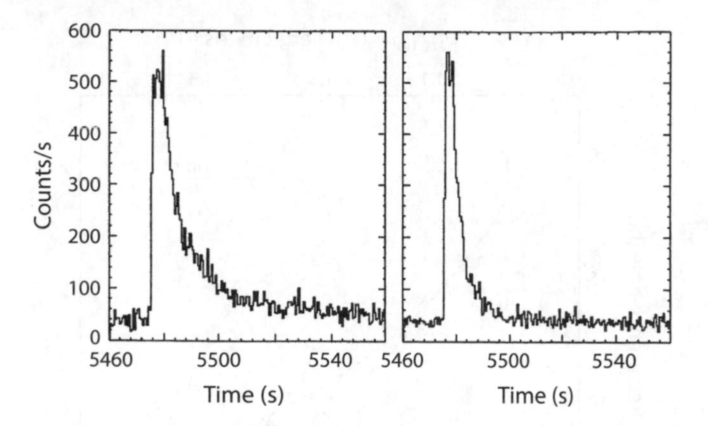

Figure 7.23. Type I X-ray burst from 4U 1702−429 observed with EXOSAT in the 1.2 − 5.3 keV band (left) and the 5.3 − 19 keV band (right). Time is in seconds since April 8, 1986, UT 03:31:31. The softening of the X-ray burst spectrum during the decay is visible as a relatively long tail of the low-energy burst tail: the black-body temperature of the X-ray spectrum decreases with time during the burst. From T. Oosterbroek (unpublished).

The other broad class among the intrinsic luminosity variation sources is due to *periodic or quasi-periodic variations* of the X-ray flux on millisecond timescales. These are observed in NSs as well as BHs. For excellent overviews of this type of variability, see van der Klis (2006) and Ghosh (2007). We discuss here briefly the main aspects of these two classes of intrinsic variability sources and restrict ourselves to the systems with NSs.

Thermonuclear X-ray Bursts

Also called Type I X-ray bursts,[4] these are seen in ∼50% of all NS-LMXBs. They are recurrent phenomena, and the typical recurrence time can be hours to weeks or months. It varies from source to source and is on average shorter for the intrinsically brighter sources. This can be seen very well during a transient accretion outburst of a source. One then observes the recurrence time of Type I outbursts to increase when the accretion rate decreases (Fig. 7.24).

Type I X-ray bursts were discovered in 1975 by Grindlay and Heise, working with the Astronomical Netherlands Satellite (ANS), when observing the LMXB 4U 1820−30 in the globular cluster NGC 6624 (Grindlay et al., 1976). The burst spectrum appears to be in good approximation of black-body radiation, with a temperature that decreases with time during the burst (Swank et al., 1977). Figure 7.23 shows as an example a

[4] Apart from the Type I bursts, which are very common, there are also Type II bursts, which are very rare. They are seen in only a few sources and have a very different spectral behavior with time than the Type I bursts do. They are not of thermonuclear origin and most probably are related to magnetospheric instabilities, see, for example, Lewin et al. (1993). We will not discuss these bursts here.

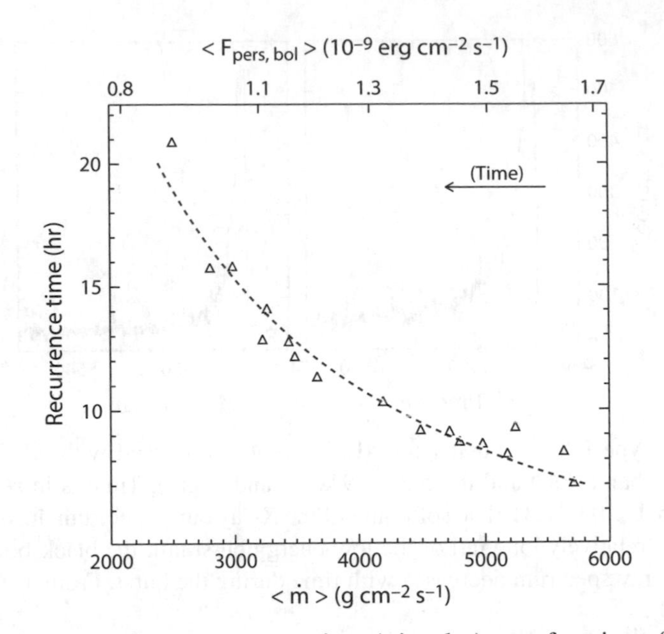

Figure 7.24. Observed burst recurrence time (triangles) as a function of local mass-accretion rate (the corresponding flux is shown in the upper X-axis) for the accreting X-ray MSP IGR J17511−3057 during the decay of a transient outburst of this source. The dashed line is a best-fit power-law model. The recurrence time increases with time roughly as $\langle F_{\rm pers,bol}\rangle^{-1.1}$, where $\langle F_{\rm pers,bol}\rangle$ is the averaged persistent flux between the bursts. After Falanga et al. (2011).

Type I X-ray burst from 4U 1702−429 as observed by ESA's European X-ray Observatory Satellite (EXOSAT), in two energy bands. It clearly shows the faster decay in the higher energy band. It was found by van Paradijs (1978) that the observed X-ray luminosity of a Type I burst decays, in good approximation, during the burst as

$$L_{\rm X} = A\,\sigma\,T_{\rm eff}^4, \tag{7.34}$$

where A is a constant, $T_{\rm eff}$ is the effective temperature, and σ is the Stefan-Boltzmann constant. This means that we are observing a cooling surface of constant area. For sources with a known distance, one can determine the value of A. van Paradijs (1978) investigated LMXBs around the Galactic center, which have a distance of ~8 kpc and found A to be of the order of $10\,\mathrm{km}^2$. One thus appears to be dealing here with a cooling spot on the NS surface with a radius of a few kilometers. It seems reasonable to assume that this spot is related to a magnetic pole, to which accreted matter is channeled by the magnetic field.

Already before the discovery of Type I X-ray burst was published, Hansen & van Horn (1975) had shown that burning of the hydrogen and helium accumulated on the NS surface occurs in radially thin shells that are susceptible to an instability of thin shell helium burning that had been discovered by Schwarzschild & Härm (1965). After the discovery of Type I bursts, it was quickly realized by theorists that it is this type of instability that causes these bursts (Woosley & Taam, 1976; Maraschi & Cavaliere, 1977; Joss, 1977, 1978; Lamb & Lamb, 1978). For a brief historical overview see Lewin et al. (1993). The generally accepted picture for the Type I bursts is that the accreted hydrogen on the surface of the NS is converted to helium on the hot ($T > 10^7$ K) surface of the accreting NS by steady burning through the CNO cycle. The bursts are then flashes of helium burning, which convert the helium to carbon and oxygen. In the late 1990s, a rare class of very long bursts was discovered (later called "intermediate bursts"), lasting from several minutes to a quarter of an hour. These were discovered with the Wide Field X-ray cameras of the Italian-Dutch BeppoSAX satellite (Jager et al., 1997) in combination with the ASM of NASA's Rossi X-ray Timing Explorer (RXTE) (Levine et al., 1996). Still more recently, so called "superbursts" were discovered, which sometimes last up to a day.

These intermediate bursts and superbursts (see, e.g., Ubertini et al., 1999) are thought to be due to mixed hydrogen+helium burning at very high temperatures ($\sim 10^8$ K), which gives rise to the rapid proton fusion process, during which a large variety of proton-rich nuclei are synthesized, up to heavy elements such as Sn, Sb, and Te (see Strohmayer & Bildsten, 2006, and references therein). The helium burning produces ^{12}C that is used in the carbon cycle for high-temperature hydrogen fusion. It is thought that in carbon-rich matter, the duration of the hydrogen fusion can be much extended, which may be the cause of the superbursts. For a review see Sazonov et al. (2020).

Variability of LMXBs on Millisecond Timescales and kHz QPOs

The X-ray flux of all X-ray binaries is powered by gravitational potential energy release of the accreted matter, which is emitted near the inner region of the accretion flow and, if present, at the NS surface. For a compact object with a size of about $\sim$10 km, $\sim$90% of the energy is released in the inner 100 km of the accretion flow, which for weak magnetic field NSs and BHs is the inner region of an accretion disk. If the magnetic dipole field is strong ($B \sim 10^{12}$G), the Alfvén surface, where the magnetic pressure of the magnetic field begins to dominate the ram pressure of the inflowing matter (Section 7.3.1), is located at a distance of the order of 10^3 km from the NS. Therefore, for $r < 10^3$ km, the matter flows inward along the magnetic field lines toward the magnetic poles where hot X-ray emitting gas columns form, see, for example, Figure 7.11. The rotation of the NS then causes the observed X-ray flux to be modulated by the spin period of the NS: one observes an X-ray pulsar. For weak dipole magnetic fields of $B \sim 10^8$ G (similar to those observed in radio MSPs), the Alfvén surface for accretion rates in a wide range of 10^{-11}–$10^{-8}\, M_\odot\, \mathrm{yr}^{-1}$ is located just above, or at, the NS surface (see Eq. 7.11). This means that the Keplerian velocity of disk matter orbiting at the inner edge of the accretion disk (near the Alfvén surface) is of the order of $v_\mathrm{K} = \sqrt{GM/R} \lesssim 0.5\, c$, where M and R are the mass ($1.2 - 2.0\, M_\odot$) and radius (approximately $10 - 13$ km) of the NS, respectively. The dynamical timescale of the matter moving through the emitting region is short: $t_\mathrm{dyn} \simeq \sqrt{r^3/(GM)}$, which is $\simeq$0.1 ms at $r = 10$ km and 2 ms at $r = 100$ km, for a typical NS.

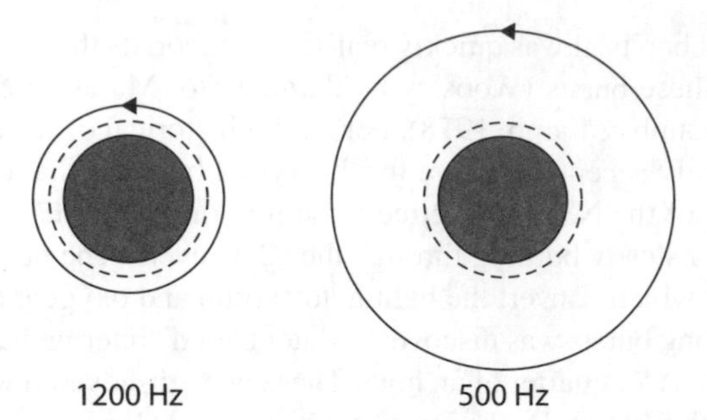

Figure 7.25. Orbital frequencies of particles orbiting a 10 km radius NS of 1.4 $M_\odot$ at two distances. The dashed circle indicates the ISCO (see Section 7.6). Closer to the NS, matter is driven inward due to a general relativistic instability. After van der Klis (2000).

Because the Keplerian period at the inner edge of the accretion disk, just above the NS surface (Fig. 7.25), is of the order of 1 ms, the NS may by the accretion have been spun-up to a rotation period of the order of 1 ms, as was predicted by Radhakrishnan & Srinivasan (1982) and Alpar et al. (1982). This prediction was confirmed in 1998, when Wijnands & van der Klis (1998), using the RXTE satellite, discovered regular X-ray pulsations with a frequency of 401 Hz in the LMXB SAX J1808−3658. This discovery of the first accreting X-ray millisecond pulsar (AXMSP) was followed by many others, discovered in particular by Chakrabarty, Strohmayer, and colleagues at the Massachusetts Institute of Technology, and van der Klis and colleagues in Amsterdam. Table 7.5 lists some well-known examples of these AXMSPs.

The LMXBs that harbor weak magnetic field NSs are, because the inner parts of the accretion disk are located just above the NS surface, the systems in which one expects, apart from regular millisecond X-ray pulsations, to find all other kinds of variability on millisecond timescales, produced by the orbital motion of blobs of disk matter with periods of this order. Indeed, already before the 1998 discovery of the first AXMSP, van der Klis et al. (1996) discovered millisecond timescale QPOs of the entire X-ray luminosity of LMXBs, with kHz frequencies, the so-called *kHz QPOs*. The first LMXB in which these kHz QPOs were discovered was Sco X-1, the brightest X-ray source in the sky. Fourier analysis of the X-ray flux of this source shows two strong kHz QPO peaks (Fig. 7.26). These double-peaked kHz QPOs appear to be common in LMXBs.

It is observed that for a given source, the frequencies ν_u and ν_l of the upper and lower kHz QPO peaks, as well as the frequency separation $\Delta\nu$ between the peaks, depend on the X-ray luminosity of the source. The frequencies of the peaks increase when the X-ray luminosity (i.e., accretion rate) of a source increases. Above a certain X-ray luminosity (the saturation luminosity), this increase levels off, as can be seen by the example of the source 4U 1820−30 in Figure 7.27. With increasing X-ray luminosity, the frequency separation of the peaks for some sources remains approximately constant; while for others, such as Sco X-1, the frequency separation decreases. For understanding this complex behavior of the kHz QPOs of LMXBs, a variety of

Table 7.5. Accreting X-ray Millisecond Pulsars (AXMSPs) in LMXBs

Source name	$\nu_{\rm spin}$ (Hz)	$P_{\rm orb}$ (hr)	f_X ($M_\odot$)	$M_{\rm comp}^{\rm min}$ ($M_\odot$)	Companion
IGR J00291+5934	599	2.46	2.8×10^{-5}	0.039	BD
Aql X-1	550	18.95	N/A	~0.6	MS
Swift J1749.4−2807	518	8.82	5.5×10^{-2}	0.59	MS
SAX J1748.9−2021	442	8.77	4.8×10^{-4}	0.1	MS/subgiant
XTE J1751−305	435	0.71	1.3×10^{-6}	0.014	He WD
IGR J17498−2921	401	3.84	2.0×10^{-3}	0.17	MS
SAX J1808.4−3658	401	2.01	3.8×10^{-5}	0.043	BD
HETE J1900.1−2455	377	1.39	2.0×10^{-6}	0.016	BD
IGR J174944−3030*	376	1.25	1.4×10^{-6}	0.014	He/CO WD
XTE J1814−338	314	4.27	2.0×10^{-3}	0.17	MS
IGR J18245−2452	254	11.03	2.3×10^{-3}	0.17	MS
IGR J17511−3057	245	3.47	1.1×10^{-3}	0.13	MS
NGC 6440 X-2	206	0.95	1.6×10^{-7}	0.0067	He WD
XTE J1807−294	190	0.67	1.5×10^{-7}	0.0066	CO WD
XTE J0929−314	185	0.73	2.9×10^{-7}	0.0083	CO WD
Swift J1756.9−2508	182	0.91	1.6×10^{-7}	0.007	He WD

Source: Updated after Patruno & Watts (2021).

Listed quantities are their spin frequencies, orbital periods, mass functions, minimum companion donor star masses (assuming $i = 90°$, $M_{\rm NS} = 1.4\,M_\odot$), and types.

*Companion star type acronyms. MS, main-sequence star; N/A, not available; BD: brown dwarf; He WD: helium white dwarf; CO WD: carbon-oxygen white dwarf.

*Data from Ng, Ray et al. (2021).

theoretical models have been proposed, all based on the fact that the matter that causes these phenomena is moving so close to the NS that its motion is dominated by strong-field general relativistic effects (see, e.g., Ghosh, 2007).

An initially popular model for the kHz QPOs was one in which the general relativistic Lense-Thirring precession is involved, due to frame-dragging, produced by the rapid spin of the NS. However, none of the models proposed so far are able to explain all the peculiarities that are observed in the behavior of the kHz QPOs. For example, there seems to be a relation of these QPOs with the spin frequency of the NS, but this relation is not the same for all sources. For example, the frequency difference between the two QPO peaks in SAX J1808−3658 is ~195 Hz, which is approximately half of its spin frequency of 401 Hz. The same is the case for 4U 1636−53, which has a spin frequency of 581 Hz, and a separation of 250–320 Hz between the two peak QPO frequencies. On the other hand, 4U 1702−429 has a spin frequency of 330 Hz, which is practically the same as the frequency separation of 315 − 344 Hz between its two QPO peaks. So, unfortunately, there is not one clear and simple relation between the spin

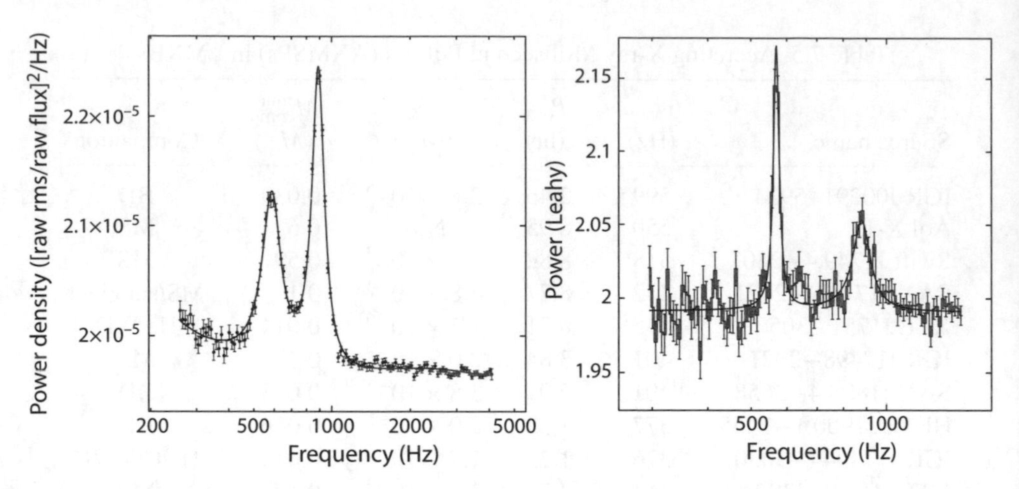

Figure 7.26. Twin kHz QPO peaks of Sco X-1 (left) (van der Klis et al., 1997) and 4U 1608−52 (right) (Méndez et al., 1998).

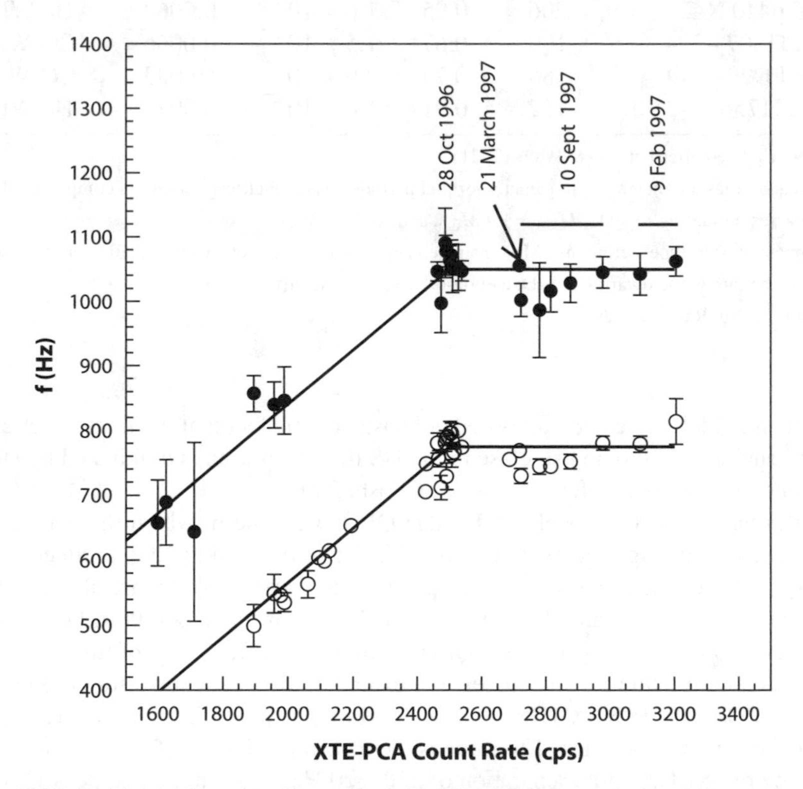

Figure 7.27. Evidence for a leveling off of the kHz QPO frequencies with count rate in 4U 1820−30. (cps: counts per second). After Zhang et al. (1998).

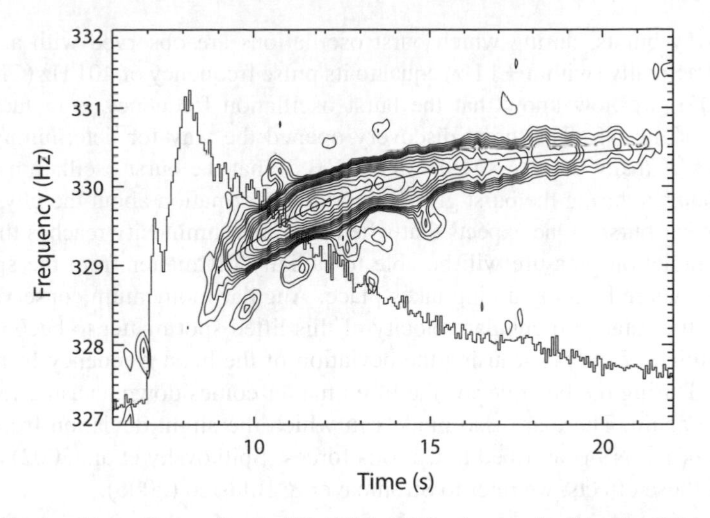

Figure 7.28. An X-ray burst from 4U 1702−429 observed with the PCA on board NASA's RXTE satellite. Shown are the contours of constant power spectral density as a function of frequency and time. The solid curve shows the best-fitting exponential model. The burst profile is also shown. After Strohmayer & Markwardt (1999).

frequency and the frequency separation of the QPO peaks, as had been hoped for. van Doesburgh & van der Klis (2017) found for systems in which the NS spin is known that there is no clear relation between the kHz QPO frequencies themselves and the spin frequency of the NS, which seems to rule out Lense-Thirring precession as the direct cause of these QPOs.

Other models involve, for example, the presence of a two-armed spiral structure in the accretion disk, which is prone to radial oscillations that can produce the two frequency peaks (Kato & Machida, 2020). Changes in the accretion rate move the inner edge of the disk (Alfvén radius), changing the shape of the spiral structures and the correlated frequencies.

A discussion of all the models proposed to explain the QPO observations goes beyond the scope of this book. For the two main types of kHz QPO sources, the Z sources and the atoll sources, and the kHz QPO interpretations according to the beat-frequency model, or the relativistic precession model, we refer to the review by van der Klis (2000). For further reviews of the extensive literature on the subject, see also van der Klis (2006) and Wang (2016).

Burst Oscillations and NS Spin

Strohmayer and colleagues discovered that often during thermonuclear X-ray bursts from LMXBs, (almost) regular X-ray pulsations of the entire X-ray flux are observed also, which during the burst slightly increase in frequency by a few Hz. Figure 7.28 shows as an example such burst oscillations observed in the LMXB 4U 1702−429. Thanks to the fact that the first-discovered AXMSP, SAX J1808−3856, also shows

Type I X-ray bursts, during which burst oscillations are observed with a frequency which is practically (within ± 1 Hz) equal to its pulse frequency of 401 Hz (Chakrabarty et al., 2003), we now know that the burst oscillation frequency is in fact the spin frequency of the NS. This great discovery opened the way for determining the spin frequencies of many more AXMSPs in LMXBs. That the burst oscillation frequency slightly changes during the burst gives important information about the physics of the thermonuclear bursts. One expects that when the burst luminosity reaches the Eddington limit, radiation pressure will be able to radially lift matter from the spot on the NS surface where helium burning takes place. Angular momentum conservation may then cause the rotational angular velocity of this lifted spot matter to become slightly lower than that of the NS, causing the deviation of the burst frequency from the spin frequency. During the burst decay, the lifted matter comes down, causing its spin rate to increase again. There are also models in which the slight deviation from the spin frequency of the NS is ascribed to Coriolis forces (Spitkovsky et al., 2002). For a discussion of these effects, we refer to Strohmayer & Bildsten (2006).

7.8.3 The Globular Cluster Sources

A group of LMXBs that deserves special attention is that of the GC X-ray sources, which are indicated in Figure 6.1. The incidence of LMXBs in GCs is (per unit mass) $\sim 10^3$ times larger than in the Galaxy as a whole (Gursky, 1973). With a total mass of $\sim 2 \times 10^7\ M_\odot$, the $\sim$120 GCs in our Galaxy (which has a stellar mass of $\sim 10^{11}\ M_\odot$) contain $\sim$10% of all strong LMXB sources of the Galaxy. The same is true for M31 and other galaxies (e.g., Fabbiano, 2006; Verbunt & Lewin, 2006). This abnormally high incidence of LMXBs in GCs is due to a special formation mechanism that can operate in GCs, but not in the Galactic disk: binary formation by close encounters either by tidal captures or by exchange collisions. The latter are binary-single star interactions that cause one binary component to be exchanged by a passing NS (see the review by Verbunt & Lewin, 2006). The dynamical formation mechanisms of LMXBs and other compact star binaries in GCs will be discussed in Chapter 12.

Apart from LMXBs with high X-ray luminosities, which have been known since the early 1970s, the sensitive X-ray observatories ROSAT, Chandra, and XMM-Newton discovered a considerable population of low-luminosity X-ray sources in GCs, which are of diverse types: NS binaries with low accretion rates, radio MSPs, and also magnetically active WDs in CV systems (for details, see Verbunt & Lewin, 2006). The reason why GCs are particularly favorable places for binary formation by the above-mentioned stellar encounter processes is their combination of very high stellar densities (central densities up to $10^4\ \mathrm{pc}^{-3}$) with relatively low stellar velocities (a few tens of $\mathrm{km\,s}^{-1}$). The discovery of many LMXBs in GCs was the first evidence for the presence of a population of old NSs in the Galaxy. Some of these may date from the early days of our Galaxy, $\sim$13 Gyr ago, but there is also evidence, from the presence of apparently young radio pulsars in some GCs (see Sections 11.1.3, 12.2, and 14.5), that at present NSs may still be forming in these clusters, presumably due to accretion-induced collapses of accreting massive WDs in binaries (see Chapter 11). In GCs, one also finds in particular many so-called UCXBs.

Table 7.6. Examples of ultra-compact X-ray binaries (UCXBs)

Source name	$P_{\rm orb}$ (min)	Location	P/T
4U 1738−34	10.8	Bulge	P
4U 1820−30	11.4	GC	P
4U 1850−08	13 or 20.6	Field	P
2S 0918-549	17.4	Field	P
4U 1543−624	18.2	Field	P
M15 X-2	22.6	GC	P
4U 2129+11	23.0	Field	P
47 Tuc X-9	28.2	GC	T
IGR J17062−6143	38.0	Field	T
1RXS J180408−342058	40.0	Bulge	T
XTE J1807−294	40.1	Bulge	T
4U 1626−67	42.0	Field	P
XTE J0929−314	43.6	Field	T
NGC 6652B	44.0	GC	T
MAXI J0911−655	44.3	GC	T
4U 1916−053	50.0	Field	P
4U 0614+091	50.0	Field	P
Swift J1756.9−2508	54.7	Bulge	T
XB 1832−330 (NGC 6652)	55.0	GC	P
NGC 6440 X-2	57.3	GC	T

Sources: After van Haaften et al. (2012), Cartwright et al. (2013), Tudor et al. (2018), Di Salvo & Sanna (2020), and references in these articles.

P. persistent source; T. transient source.

7.8.4 Ultra-compact X-ray Binaries

UCXBs are systems with orbital periods $\lesssim$1 hr. There are almost 30 of such LMXBs known, and many of them are located in GCs. The systems with the shortest orbital periods are 4U 1738−34 and 4U 1820−30 (which is in GC NGC 6624), with $P_{\rm orb} =$ 10.8 and 11.4 min, respectively. These systems are smaller in size than the diameter of the Sun. Table 7.6 lists 20 of the UCXBs that are presently known.

It is known from the properties of CVs that the mass donors in UCXBs are rarely expected to be normal hydrogen-rich stars with nuclear fusion, because systems with such hydrogen-rich companions are not expected to have orbital periods much shorter than the period minimum of CVs, which is ~60−80 min (see Fig. 11.13). The companions in UCXBs are therefore mostly expected to be very helium-rich and/or (partly) degenerate, that is, brown dwarfs or WDs (Rappaport et al., 1982; Di Salvo & Sanna, 2020). The shortest orbital periods of the systems in Table 7.6, between 10 and ~20 min, are those of 4U 1738−34 (10.8 min), 4U 1820−30 (11 min), 4U 1850−087 (13 or

Table 7.7. Examples of Pulsating Symbiotic X-ray Binaries

Source name	P_{spin} (s)	P_{orb} (d)	L_X (erg s^{-1})	Distance (kpc)
Sct X-1	112	111	2×10^{34}	$\geq$4
GX1+4	$\simeq$140	$\simeq$300*	$10^{35} - 10^{36}$	4.3
4U 1700+2	404	—	—	—
3XMM J1819−1706	408	—	2.8×10^{34}	—
1RXS J1804311−273932	494	—	—	—
2XMM J1740−2905	626	—	—	—
IGR J16358−4726	5,880	—	$3 \times 10^{32} - 2 \times 10^{36}$	5.5 or 12.5
IGR J17329−2731	6,680	—	—	2.7
4U 1954+31	$\simeq$19,100	$\geq$400	1.6×10^{33}	3.3

Sources: After Yungelson et al. (2019) and Staubert et al. (2019).

With the exception of 4U 1954+31, where the companion is a red supergiant with a mass of $\sim 9\,M_{\odot}$ (Hinkle et al., 2020), the companion stars are red giants with masses similar to that of the Sun. In total, about a dozen such systems are known. Here we list their spin (pulse) periods, orbital periods, X-ray luminosities, and distances, in so far as known.

*Various orbital periods have been reported: 1,161 d, or 295 ± 70 d, or 304 d.

21 min), 2S 0918−549 (17 min), and 4U 1543−624 (18 min), but there are still more UCXBs known in this period range. Among the UCXBs in GCs are 4U 1820−30, M15 X-2, NGC 6440 X-2, 47 Tuc X-9, MAXI J0911−655, XB 1832−330, and NGC 6652B. The latter two systems are both in NGC 6652. For further observational aspects of UCXBs, we refer to Nelemans et al. (2010) and Heinke et al. (2013), for example. Theoretical work and detailed discussions on the formation and evolution of UCXBs, linking LMXBs to UCXBs and their descendants, are given in Section 11.4. In Chapter 15, we further review UCXBs as important GW sources.

7.8.5 The Symbiotic X-ray Binaries: Red Giants with Slowly Rotating NSs

On the other end of the range of orbital periods of LMXBs, are the accreting X-ray pulsars with low-mass red-giant (M-class) donor stars. In recent years, these were recognized as an important new class of X-ray binaries, the so-called *symbiotic X-ray binaries* (Corbet et al., 2008; Yungelson et al., 2019), of which now $\gtrsim$13 sources are known. Nine of these sources are known to be regularly pulsing; these are listed in Table 7.7. One of these systems, 4U $1954+31$, is not a LMXB: it was recently found, thanks to a Gaia distance determination, to harbor a red supergiant with a mass of $\sim 9 M_{\odot}$ (Hinkle et al., 2020).

Of the LMXB symbiotic systems (see Table 7.7), GX $1+4$ has been known for the longest time; it was already discovered in the 1960s with balloon experiments Figure 7.29 shows its spin history of more than four decades. While for the first 15 yr

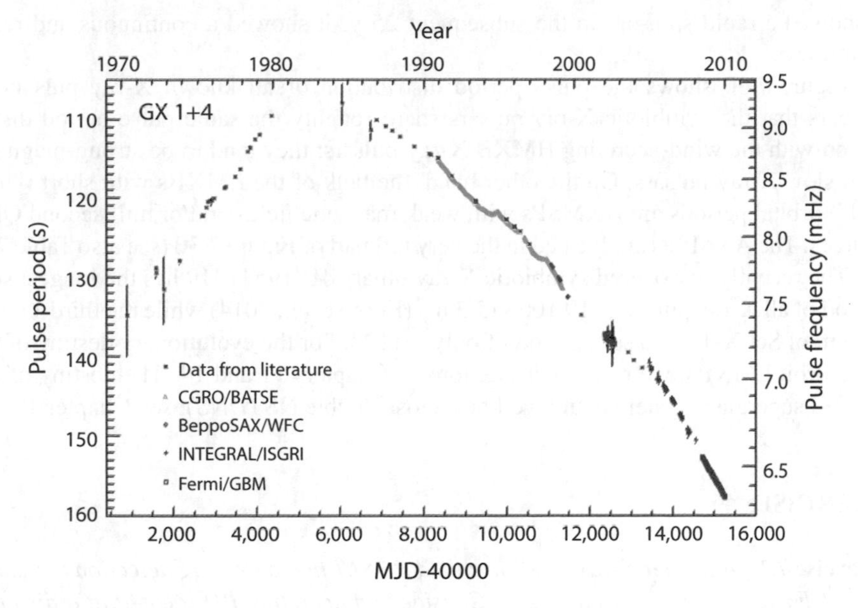

Figure 7.29. More than 40 yr of spin history of the symbiotic X-ray binary pulsar GX 1 + 4. After González-Galán et al. (2012).

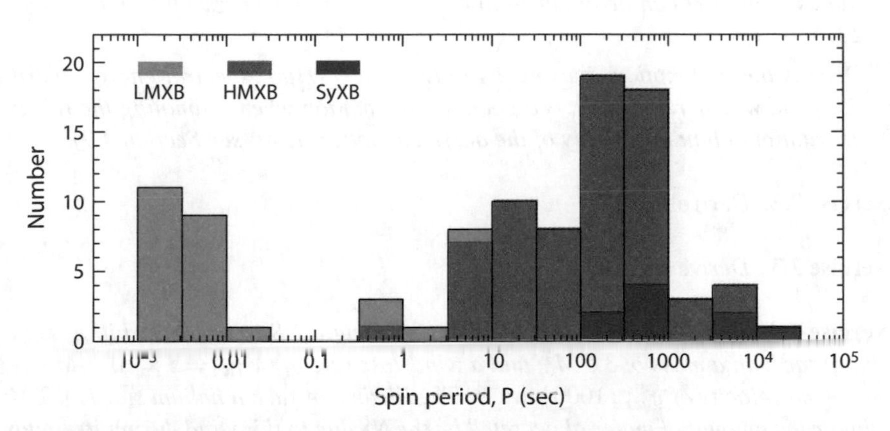

Figure 7.30. Pulse (spin) period distribution of 96 accreting X-ray pulsars known until 2021. Data for sources with periods of $P \geq 0.01$ s are from Staubert et al. (2019) and Yungelson et al. (2019). Data for accreting AXMSPs ($P \leq 0.01$ s) are from Di Salvo & Sanna (2020). Blue: HMXBs, green: LMXBs, and red: symbiotic X-ray binaries (SyXBs).

it showed a rapid spin-up, in the subsequent 25 yr it showed a continuous and rapid spin-down.

Figure 7.30 shows the pulse period distribution of all known X-ray pulsars. It appears that the symbiotic X-ray pulsars share roughly the same pulse-period distribution with the wind-accreting HMXB X-ray pulsars: they tend to be strong-magnetic field slow X-ray pulsars. On the other hand, the bulk of the LMXBs with short ("normal") orbital periods are AXMSPs with weak magnetic fields and/or millisecond QPO sources. The AXMSPs are located in the very left part of Figure 7.30 (see also Table 7.5).

The recently discovered symbiotic X-ray binary 4U 1954+319 has the longest spin period of all X-ray pulsars,[5] 19,100 s (5.3 hr) (Enoto et al., 2014) while the third-known system of Sct X-1 has a spin period of only $\sim$112 s. For the evolutionary destiny of the wide-orbit LMXBs, we refer to discussions in Chapters 11 and 14. The destiny of the one red supergiant system could well be a close double NS (DNS); see Chapter 10.

EXERCISES

Exercise 7.1. *A former X-ray satellite called ROSAT had a limiting detection threshold flux of* $F_X \simeq 5 \times 10^{-13}\,\mathrm{erg\,cm^{-2}\,s^{-1}}$. *Consider an accreting BH located at a distance of 1 kpc.*

a. *Show that the accretion luminosity of an object accreting matter with a rate of* $\dot{M}$ *and radius R is given by* $L \simeq GM\dot{M}/R$.
b. *Calculate the accretion rate needed for ROSAT to detect this BH. (Notice: the radius of the BH can be assumed to be given by its Schwarzschild radius:* $R_{\mathrm{Sch}} = 2GM/c^2$*).*
c. *Discuss the assumption of using* $F_X = L/(4\pi d^2)$ *(Hint: spectral energy distribution) and whether* $R = R_{\mathrm{Sch}}$ *is a good approximation when estimating the released gravitational binding energy of the accreted matter (Hint: see Section 4.9).*

Exercise 7.2. *Derive Eq. 7.6.*

Exercise 7.3. *Derive Eq. 7.11.*

Exercise 7.4. *Consider a naked helium star orbiting a NS in a tight orbit. Assume a helium star with a mass of* $3.5\,M_\odot$ *and a wind-loss rate of* $|\dot{M}_{\mathrm{He}}| = 5 \times 10^{-7}\,M_\odot\,\mathrm{yr}^{-1}$ *and a wind velocity of* $v_w \simeq 1000\,\mathrm{km\,s^{-1}}$. *The lifetime of such a helium star is* <2 *Myr. Estimate the amount of material accreted by the NS due to this wind during its lifetime, if the orbital period is 2.0 d and the NS mass is* $1.35\,M_\odot$.

Exercise 7.5. *Reproduce Figure 7.22.*

[5] This is the slowest spin measured for any of the >3,000 known NSs.

Chapter Eight

Evolution of Single Stars

8.1 OVERVIEW OF THE EVOLUTION OF SINGLE STARS

For an overview of the evolution of single stars, we refer to the textbooks on the subject, such as those by Clayton (1968), Cox & Giuli (1968), Weiss et al. (2004), Kippenhahn & Weigert (1990), Hansen et al. (2004), Iben (2013a, 2013b), Kippenhahn et al. (2013), and Eldridge & Tout (2019). Excellent course lecture notes on the subject can also be found online.[1]

8.1.1 Why Stars Shine and Evolve: The Virial Theorem

A non-rotating globe of ideal monatomic gas without energy sources and in hydrostatic equilibrium obeys the simplest form of the virial theorem:

$$2E_{\rm th} + E_{\rm pot} = 0, \tag{8.1}$$

where $E_{\rm th}$ is the thermal energy content of the globe, and $E_{\rm pot}$ is its gravitational potential energy (binding energy). This equation can be derived from the equation of hydrostatic equilibrium, as can be found in the above-mentioned textbooks and lecture notes.

$E_{\rm th}$ is given by

$$E_{\rm th} = \frac{3}{2} N k \overline{T} = \frac{3}{2} M (\mathcal{R}/\mu) \overline{T}, \tag{8.2}$$

where N is the number of particles in the star; k is the Boltzmann constant; $\overline{T}$ is the mean temperature of the gas; M is the mass of the globe; $\mathcal{R} = k/m_{\rm H}$ the ideal gas constant, and μ is the mean particle mass in units of the mass $m_{\rm H}$ of the hydrogen atom.

$E_{\rm pot}$ is given by

$$E_{\rm pot} = -\alpha\, GM^2/R, \tag{8.3}$$

where R is the stellar radius; G is the constant of gravity; and α is a constant of proportionality of order unity, which depends on the density distribution in the star (see,

[1] Such as those by O. Pols: https://www.astro.ru.nl/~onnop/education/stev_utrecht_notes/.

e.g., Cox & Giuli, 1968; Weiss et al., 2004). Substitution of Eqs. (8.2) and (8.3) into Eq. (8.1) yields

$$\overline{T} = \alpha \left(\frac{GM}{R}\right)\left(\frac{\mu}{3\mathcal{R}}\right), \tag{8.4}$$

which shows that for a given stellar mass, the mean internal temperature depends only on the stellar radius and increases when the star contracts. The preceding equations imply that a globe of ideal gas in hydrostatic equilibrium and without energy sources must contract indefinitely, or until it ceases to obey the ideal gas law, for the following reasons. Its temperature is >0 K, which implies that it will radiate and thus lose energy. This energy loss happens at the expense of its total energy content that, using the virial theorem, is given by

$$E_{\rm tot} = E_{\rm th} + E_{\rm pot} = \frac{1}{2} E_{\rm pot} = -\alpha \frac{GM^2}{2R}. \tag{8.5}$$

Because $E_{\rm tot} < 0$, a decrease in $E_{\rm tot}$ implies, according to this equation that R decreases when the star loses energy. The heat losses from its surface therefore force the star to contract. But, according to Eq. (8.4) this contraction implies that the mean internal temperature of the star $\overline{T}$ will increase. Thus, while the star tries to cool itself by radiating away energy from its surface, it gets hotter instead of cooler! The star has a *negative heat capacity*. Clearly, this cannot be a stable situation: the gas globe, starting out with a very large radius as part of an interstellar cloud, is forced by radiation losses into a vicious circle of contraction and heating. The more it radiates to cool itself, the more it contracts and the hotter it gets, and the more it is forced to go on radiating. It is this *vicious virial circle* (depicted in Fig. 8.1) that in a nutshell explains how and why an interstellar globe of gas, once it has started contracting, must finally end its life as a compact object—a BH, a NS, or a WD—or a brown dwarf in case the central temperature never increases sufficiently high ($\sim 10^7$ K) to ignite hydrogen fusion and make a star.

In the meantime, the globe may spend millions or billions of years in intermediate stages, which are called the *main sequence* and the *giant branch*, etc. These stages are, however, only temporary "stations" in the contraction history of the globe. It is important to realize that stars do not shine because they burn nuclear fuel. They shine because they are hot, and they obtained their high internal temperature due to their history of gravitational contraction, not due to nuclear burning. Eqs. (8.1) and (8.5) show that for an ideal monatomic gas globe, half of the gravitational potential energy lost during the contraction of the globe is converted into thermal energy, and the other half is radiated away.

Thus far we have considered only the case of an ideal monatomic gas, i.e. with an adiabatic index of $\gamma = C_P/C_V = 5/3$, where C_P and C_V denote the specific heats at constant pressure and constant volume, respectively. This is an excellent approximation for globes of non-degenerate ionized hydrogen and helium, which all normal stars are. It is useful, however, also to have the preceding equations for other values of γ. The generalized forms of Eqs. (8.1), (8.2), and (8.5) are, respectively, as one can easily

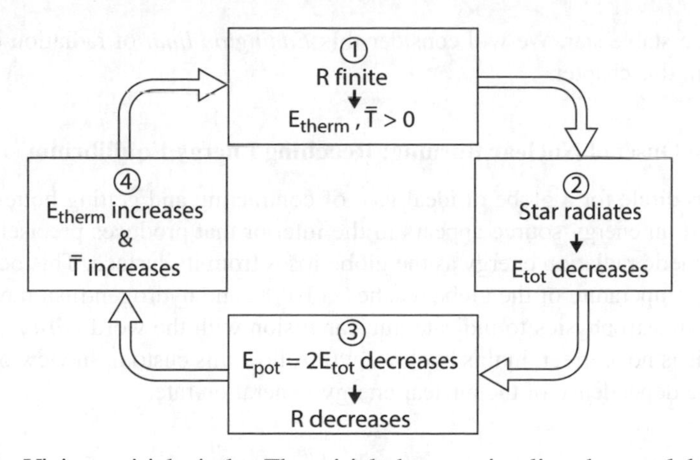

Figure 8.1. Vicious virial circle. The virial theorem implies that a globe of ideal monatomic gas ($\gamma = 5/3$) in hydrostatic equilibrium, once it has a finite radius, is forced to contract indefinitely. According to the virial theorem, because of its finite radius the star has a positive mean temperature $\overline{T}$ (1), and therefore it radiates, causing its total energy content to decrease (2), causing its potential energy and radius to decrease (3), causing the star's mean temperature $\overline{T}$ to increase (4), causing the star to radiate even stronger (1), causing it to contract further, etc.

derive (see, e.g., Weiss et al., 2004),

$$3(\gamma - 1)E_{\rm th} + E_{\rm pot} = 0, \tag{8.6}$$

$$E_{\rm th} = \frac{1}{\gamma - 1} Nk\overline{T} = \frac{1}{\gamma - 1} M(\mathcal{R}/\mu)\overline{T}, \tag{8.7}$$

and

$$E_{\rm tot} = E_{\rm th} + E_{\rm pot} = \frac{3\gamma - 4}{3(\gamma - 1)} E_{\rm pot}. \tag{8.8}$$

To have a gravitationally bound star, the total energy must be negative, which, because $E_{\rm pot} < 0$, implies that γ must be $>4/3$. For $\gamma \leq 4/3$, the star cannot be in hydrostatic equilibrium, in other words, it collapses or it explodes. This is important, for example, when the star becomes fully convective, such as early in the pre-main-sequence phase, and undergoes the Hayashi collapse. Or when, as a WD, it becomes relativistically degenerate when reaching the Chandrasekhar mass limit ($\sim$1.4 $M_\odot$) and collapses to a NS (or possibly explode in a SN Ia). In both these cases, $\gamma = 4/3$ and hydrostatic equilibrium is impossible. Also, when stars have a very high mass, they have a very high luminosity; as a result, their internal pressure is dominated by the pressure of the *photon gas* in their interior, which has $\gamma = 4/3$. Such stars are also close to getting unstable: this sets an upper limit of a few hundred solar masses to the mass of of a stable star. We will consider this *Eddington limit* of radiation-dominated stars later in this chapter.

8.1.2 The Onset of Nuclear Burning: Reaching Energy Equilibrium

The vicious circle for a globe of ideal gas, of contracting and getting hotter, can only be broken if an energy source appears in the interior that produces precisely as much electromagnetic radiation energy as the globe loses from its surface. This occurs when the central temperature of the globe reaches $\sim 10^7$ K, and hydrogen fusion begins. It is customary in astrophysics to indicate nuclear fusion with the word *burning*; although literally this is not correct, in this book we will follow this custom. In view of the steep temperature dependence of the nuclear energy generation rate,

$$L_{\rm nuc} \sim T^n, \tag{8.9}$$

where n has a value ranging from 4 (for the p-p reactions) to 20 (for the CNO cycle), a slight further contraction and temperature rise causes an enormous increase in the $L_{\rm nuc}$, such that the star will be able to reach an equilibrium situation at which

$$L_{{\rm nuc},\gamma} = L_{\rm out}, \tag{8.10}$$

where $L_{\rm out}$ is the heat loss at the outer surface of the star, and $L_{{\rm nuc},\gamma}$ is the photon part of the nuclear luminosity (Fig. 8.2). A part of the nuclear luminosity, $L_{{\rm nuc},\nu}$, is lost in the form of neutrinos, which freely escape into space. When Eq. (8.10) is fulfilled, the contraction stops. The nuclear burning equilibrium that now is reached is stable, because the virial theorem works as a safety valve. If too much heat were to be produced by the nuclear burning, the internal temperature would rise, causing an increase in gas pressure, which would lead to expansion and cooling of the star.

8.1.3 The Three Basic Stellar Timescales

There are three timescales for single stars that are important for binary evolution and for the mass and angular momentum loss processes associated with this evolution. We therefore give them here. In order of increasing length they are (1) the dynamical or pulsation timescale, (2) the thermal or Kelvin-Helmholtz timescale, or (3) the nuclear timescale.

i) The Dynamical Timescale or Pulsation Timescale

This is the timescale on which the star restores a perturbation of its hydrostatic equilibrium (see Chapter 2). It is the time required for a sound wave (pressure disturbance) with velocity c_s to cross the stellar radius R:

$$\tau_{\rm dyn} = \frac{R}{c_s} \simeq 50\,{\rm min}\left(\frac{\overline{\rho}_\odot}{\overline{\rho}}\right)^{1/2}, \tag{8.11}$$

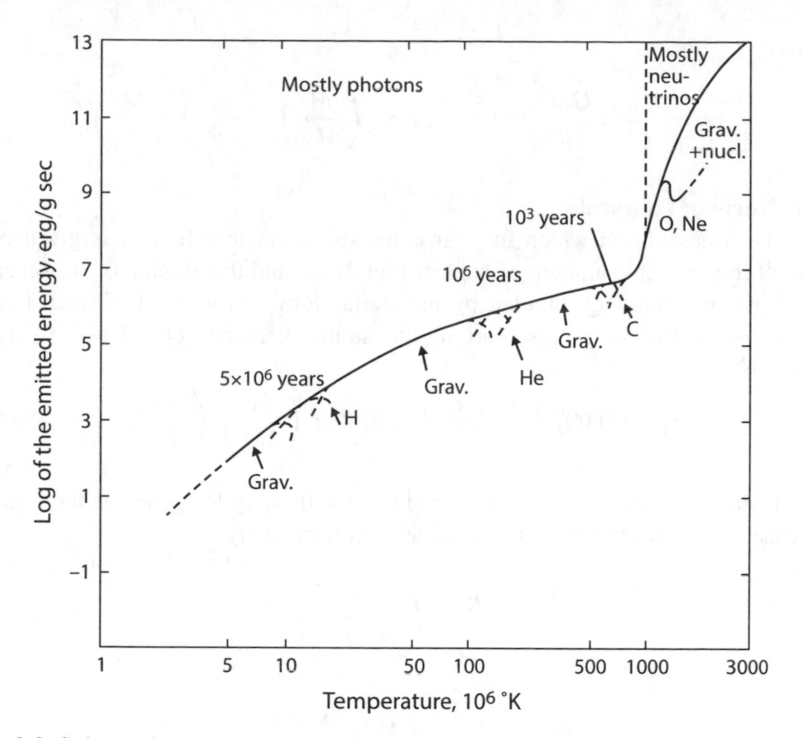

Figure 8.2. Schematic representation of the evolution of the stellar luminosity as a function of the central temperature. The total luminosity on the *photon* branch is equal to the outer luminosity (heat radiation) of the star. It is provided by alternating stages of gravitational contraction and nuclear burning until carbon burning. During carbon burning and beyond, most of the energy generated in the interior escapes in the form of neutrinos, causing the duration of these stages to become very short, and the neutrino luminosity to become very large. The timescales are for a star of 25 $M_\odot$ (see, e.g., Weaver et al., 1978; Weaver & Woosley, 1980). (For the duration of the different burning stages of a 15 $M_\odot$ star, see Table 8.1). Adapted after Reeves (1968).

where $\overline{\rho}_\odot = 1.4\,\mathrm{g\,cm^{-3}}$ and represents the mean mass density of the Sun, and $\overline{\rho}$ indicates the mean mass density of the star. (For an ideal gas, $c_s = \sqrt{\gamma\, P/\rho} = \sqrt{\gamma\, kT/\mu m_u}$, and by using the average temperature of the star given by Eq. [8.4], Eq. [8.11] for the dynamical timescale follows.)[2]

ii) The Thermal or Kelvin-Helmholtz Timescale

This is the timescale on which the star reacts when its radiative energy loss and its energy generation are not in equilibrium. This is, for example, the case during the pre-main-sequence contraction, as well as after the exhaustion of a nuclear fuel. It is basically the timescale on which it emits its thermal energy content E_{th} at its present

[2]See Exercise 8.1 for an alternative derivation based on the free-fall timescale.

luminosity L:

$$\tau_{\rm th} = \frac{E_{\rm th}}{L} = \frac{GM^2}{2RL} \simeq 1.6 \times 10^7\,{\rm yr} \left(\frac{M}{M_\odot}\right)^2 \left(\frac{R_\odot}{R}\right) \left(\frac{L_\odot}{L}\right). \tag{8.12}$$

iii) The Nuclear Timescale

This is the timescale on which the star exhausts its nuclear fuel. It is given by the product of the available amount of nuclear fuel $M_{\rm core}$ and the amount of fusion energy released per unit mass Q, divided by the stellar luminosity. For hydrogen burning, which accounts for the largest part of the stellar lifetime, $Q = 0.007\,c^2$, and the timescale is

$$\tau_{\rm nuc} = 0.007 \left(\frac{M_{\rm core} c^2}{L}\right) \simeq 10^{10}\,{\rm yr} \left(\frac{M}{M_\odot}\right) \left(\frac{L_\odot}{L}\right). \tag{8.13}$$

For main-sequence (hydrogen-burning) stars with $M \geq M_\odot$, one has the following approximate relations between mass, radius, and luminosity:

$$\frac{R}{R_\odot} = \left(\frac{M}{M_\odot}\right)^{0.7} \tag{8.14}$$

and

$$\frac{L}{L_\odot} = \left(\frac{M}{M_\odot}\right)^{3.5}. \tag{8.15}$$

With these relations, the three timescales can all be expressed as only a function of stellar mass, leading to the following simple relations:

$$\tau_{\rm dyn} = 50\,{\rm min} \left(\frac{M}{M_\odot}\right)^{0.55}, \tag{8.16}$$

$$\tau_{\rm th} = 1.6 \times 10^7\,{\rm yr} \left(\frac{M}{M_\odot}\right)^{-2.2}, \tag{8.17}$$

and

$$\tau_{\rm nuc} = 10^{10}\,{\rm yr} \left(\frac{M}{M_\odot}\right)^{-2.5}. \tag{8.18}$$

8.1.4 Evolution beyond Hydrogen Burning

The onset of nuclear burning puts a temporary halt on the gravitational contraction of the gas. However, this stage is only effective until the time when hydrogen is exhausted in the core. The further evolution of the star after the exhaustion of the hydrogen fuel in its burning core can be described again by applying the virial theorem, as follows. When the hydrogen in the core is exhausted, $L_{\rm nuc}$ drops to 0. However, the star keeps

Table 8.1. Properties of Nuclear Burning Stages in a 15 $M_\odot$ Star

Burning stage	T_c (10^9 K)	ρ_c (g cm^{-3})	Fuel	Main products	Timescale
Hydrogen	0.035	5.8	H	He	1.1×10^7 yr
Helium	0.18	1.4×10^3	He	C, O	2.0×10^6 yr
Carbon	0.83	2.4×10^5	C	O, Ne	2.0×10^3 yr
Neon	1.6	7.2×10^6	Ne	O, Mg	0.7 yr
Oxygen	1.9	6.7×10^6	O, Mg	Si, S	2.6 yr
Silicon	3.3	4.3×10^7	Si, S	Fe, Ni	18 d

Source: After Woosley et al. (2002).

T_c and ρ_c are the central temperature and mass density, respectively. The internal evolution of this star is depicted in Figure 8.18.

shining because its outer layers are hot, so L_{out} does not decrease right away. The central thermal energy content of the star declines, causing the internal pressure support to decrease, such that the star starts contracting again. This causes, according to the virial theorem, its central temperature to increase: the star resumes its vicious circle of contraction and rising of its central temperature. The contraction of the central parts of the star goes on until a central temperature is reached at which the next nuclear fuel is ignited. This is helium, which burns at temperatures of $1 - 2 \times 10^8$ K. Information about the burning stages through which the stellar interior subsequently passes is given in Table 8.1 for a star of 15 $M_\odot$. It should be noted that for the radius R in Eq. (8.4) one should read the radius of the dense burning core. The outer mantle of the star, which in most cases comprises ~70%–80% of the stellar mass, does not take part in the nuclear burning. In the stages beyond hydrogen burning, they have a tendency to expand, while the core further contracts. The reasons for this envelope expansion are complex and have to do with the steepening of the (negative) temperature gradient between center and limb of the star and the decrease of the mean molecular mass of the stellar material from center to limb. The presence of the outer layers hardly affects the value of E_{pot} in Eq. (8.3), as the largest contribution to the binding energy is made by the dense core, due to the small radius of this core. See, for example, Kippenhahn & Weigert (1990) and Iben (2013a, 2013b).

In Table 8.1, it is seen that for massive stars the duration of the stages beyond carbon burning is very short, because most of the nuclear energy generated in the interior is liberated now in the form of neutrinos. These represent a *leak of energy from the interior* as they freely escape without any thermal interaction with the stellar gas. This causes an acceleration of the nuclear burning, as only the photon fraction of the liberated energy is of interest for maintaining the hydrostatic equilibrium and for preventing the star from contracting. Beyond carbon burning, the fraction of the energy lost in the form of neutrinos becomes so large that the duration of these stages becomes almost negligible.

For the final evolution of stars beyond hydrogen burning, three mass regimes can be distinguished: $M \geq 12 M_\odot$, $M \leq 8 M_\odot$, and $M = 8 - 12 M_\odot$.

i) Massive Stars, $M \geq 12 M_{\odot}$

These stars pass without complications through the sequence of burning stages listed in Table 8.1. At the end of their lives, they produce a collapsing iron core, leaving a NS or a BH. This core collapse may be accompanied by a SN explosion. These stars are discussed in much more detail in Section 8.2 and in Chapter 13.

ii) Lower-mass Stars, $M \leq 8 M_{\odot}$

Here the occurrence of degeneracy in the core causes complications with the ignition of nuclear fuels. Because for an (electron-)degenerate gas, the pressure depends only on the density but not on the temperature ($P = K_1 \rho^{5/3}$ in the non-relativistic case and $P = K_2 \rho^{4/3}$ in the relativistic degenerate case, where K_1 and K_2 are constants and ρ is the mass density [see Cox & Giuli, 1968; Weiss et al., 2004]), degenerate cores have no safety valve. When a nuclear fuel ignites in a degenerate gas, the temperature will rise, due to the liberation of nuclear energy, but because the pressure is independent of the temperature, the pressure will not "feel" this temperature rise and no expansion and stabilizing cooling (due to the virial theorem) will occur. As the rate of the nuclear energy generation depends on a high power of the temperature, the temperature rise produced by nuclear burning will cause the luminosity to rise enormously (up to $10^{11} L_{\odot}$ for a few seconds).

As a result, the ignition of nuclear burning in a degenerate gas causes a runaway nuclear energy generation and a runaway rise in the internal temperature, producing a so-called *flash*. Only when the temperature becomes so large that the boundary between a degenerate and an ideal gas is crossed (i.e., when degeneracy is lifted; see the double-dashed band in Fig. 8.3), the gas will be able to expand and cool. In stars with $M < 2.3 M_{\odot}$, the helium core becomes degenerate during hydrogen shell burning, and when M_{He} reaches $\sim 0.47 M_{\odot}$, helium ignites with a flash; the temperature rises to $\sim 10^9$ K, at which the degeneracy is removed. The helium flash is not violent enough to cause disruption of the entire star (cf. Cox & Giuli, 1968; Weiss et al., 2004). Stars with masses in the range of $2.3 M_{\odot}$ to $\sim 8 M_{\odot}$ ignite carbon with a flash.

Figure 8.3 depicts the evolutionary tracks of the central temperature and density of stars of different masses. Only greater than $\sim 8 M_{\odot}$ is carbon ignited sufficiently close to the boundary of the non-degenerate region that the star can avoid the carbon flash. For stars in the mass range of $3 - 8 M_{\odot}$, the carbon flash may, in principle, generate sufficient energy ($\sim 10^{51}$ erg) to disrupt the entire star in a so-called *carbon-deflagration SN*. The carbon is converted into ^{56}Ni which decays in 120 d to ^{56}Fe. It is, however, very likely that in single stars in this mass range the hydrogen-rich envelope has already been lost during helium-shell burning when the star is on the AGB and also has a hydrogen-burning shell. This double shell burning causes the star to experience thermal pulses. These together with pulsational instability of its red-supergiant (RSG) envelope may cause the star to eject its hydrogen-rich envelope, such that only the degenerate carbon core is left as a CO WD. In this way, in single stars with masses in the range of $2.3 M_{\odot}$ to $\sim 8 M_{\odot}$, the carbon-ignition stage is never (or very rarely) reached. The idea is that the carbon ignition in degenerate CO cores can probably only be reached in binary systems, when the CO core that lost its envelope has become a CO WD. When this star receives matter from its companion star, this may lead to carbon ignition and complete

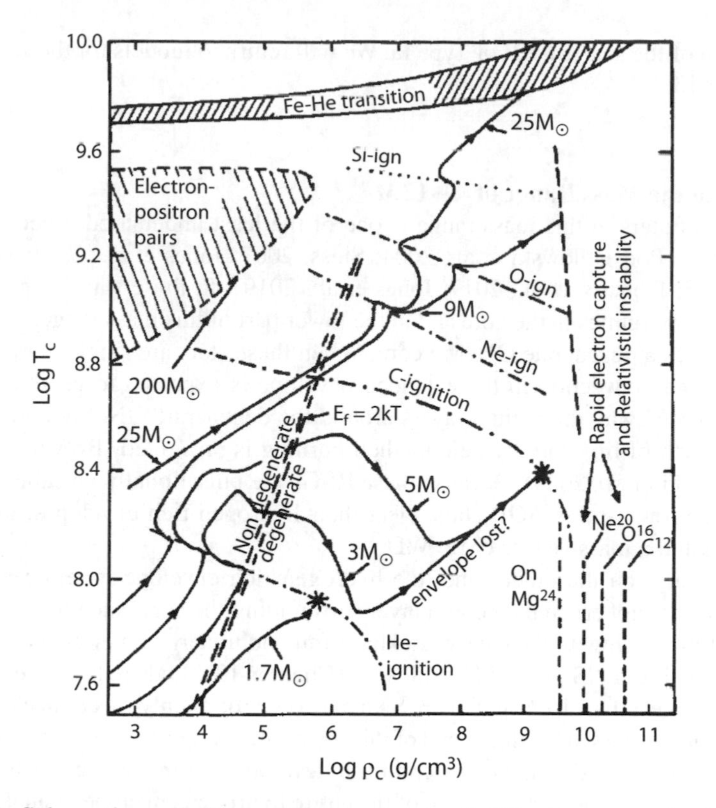

Figure 8.3. Schematic evolutionary tracks of stars of different masses in the central temperature vs. central mass density diagram. Dash-dotted lines indicate approximate loci of ignition of nuclear fuels. Shaded areas indicate regions of instability due to pair production (upper left) and photo-disintegration of iron (top). Dashed lines at far right indicate places of instability due to electron captures and relativistic instability. Solid lines are (slightly simplified) evolutionary tracks of the centers of stars of 1.7, 3, 5, and 9 $M_\odot$ after Maeder & Meynet (1989); 25 $M_\odot$ after Weaver et al. (1978) and Weaver & Woosley (1980); and 200 $M_\odot$ after Mazurek & Wheeler (1980). PISNe triggered by electron-positron production give way to collapse due to iron–photo-disintegration when the initial ZAMS masses drop below $\sim$120 $M_\odot$. (The mass ranges indicated here are for stars without stellar wind mass loss.) Photo-disintegration of iron gives way to electron capture on lighter elements for masses below $\sim$12 $M_\odot$; stars $< 8\,M_\odot$ either completely disrupt on carbon ignition, or—most likely—lose their envelopes before this happens. Adapted after Trimble (1982).

destruction of the star in a SN of Type Ia. We will return to models for these explosions in Chapter 13.

iii) Stars in the Mass Range of $8-12\,M_\odot$

The death of stars in this mass range is one of the least understood aspects of stellar physics (e.g., Podsiadlowski et al., 2004; Siess, 2007; Jones et al., 2013; Woosley & Heger, 2015; Doherty et al., 2017; Jones et al., 2019). In these stars, non-degenerate carbon burning ignites in the core and in the lower part of this mass range, up to $\sim 10-11\,M_\odot$, leaves a degenerate ONeMg core. When these stars are single, they are on the AGB, and the convection in the extended envelope is expected to penetrate into the core and erode away the helium layers around the degenerate ONeMg core, such that further growth of this core by helium-shell burning is prevented (Brown et al., 2001; Podsiadlowski et al., 2004). Because these RSGs are pulsationally unstable, they will, just like the lower-mass AGB stars, eject their hydrogen-rich envelopes, in this case leaving behind a massive (ONeMg) WD.[3]

In a binary, on the other hand, the hydrogen rich envelope is removed by mass transfer and cannot erode the helium layer surrounding the core, such that this ONeMg core can further grow to the Chandrasekhar limit and undergo an electron-capture collapse, producing a NS and a SN explosion (Chapter 13). Podsiadlowski et al. (2004) therefore suggest that electron-capture SNe may only (or mainly) occur in close binaries.

Some of the stars in the upper end of this mass interval ignite neon and oxygen burning off-center, and a degenerate silicon flash develops that in some cases is so violent that it could lead to the early ejection of the entire hydrogen envelope of an isolated star (Woosley & Heger, 2015). The precise mass ranges for the different types of final fate of the stars in this mass regime appears to depend quite strongly on metallicity. Even at solar metallicity, some studies find that an iron core collapse may be the outcome of stars with initial masses down to $\sim 9.5 M_\odot$.

Recent studies by Chanlaridis et al. (2022) of stripped-envelope stars, with initial ZAMS masses between $\sim 7-11\,M_\odot$, show that their final outcome may, however, be a SN Ia because of explosive oxygen burning in degenerate cores with central densities $\lesssim 10^{9.6}\,\mathrm{g\,cm^{-3}}$. Chanlaridis et al. conclude that the amount of residual carbon retained after core carbon burning plays a critical role for the final destiny: Chandrasekhar-mass cores with a central carbon mass fraction, $X_C \gtrsim 0.004$ result in ONe SNe Ia, while those with lower carbon mass fractions become ECSNe.

After this brief discussion of the overall evolution after core hydrogen burning and the final fate of stars, we now consider the precisely computed evolution of single stars in the different mass ranges in more detail.

[3] ONeMg WDs have masses $\gtrsim 1.05\,M_\odot$ and are also referred to as ONe WDs because ^{23}Na, which is produced in carbon burning, may also be the third most abundant isotope, rather than ^{24}Mg, (Siess, 2006). Based on updated electron-capture rates on ^{20}Ne (Kirsebom et al., 2019), so-called thermonuclear electron capture SNe, that is, incomplete explosions of degenerate ONeMg cores by oxygen deflagration, have been suggested to leave behind bound ONeFe WD remnants (Jones et al., 2019; Zha et al., 2019).

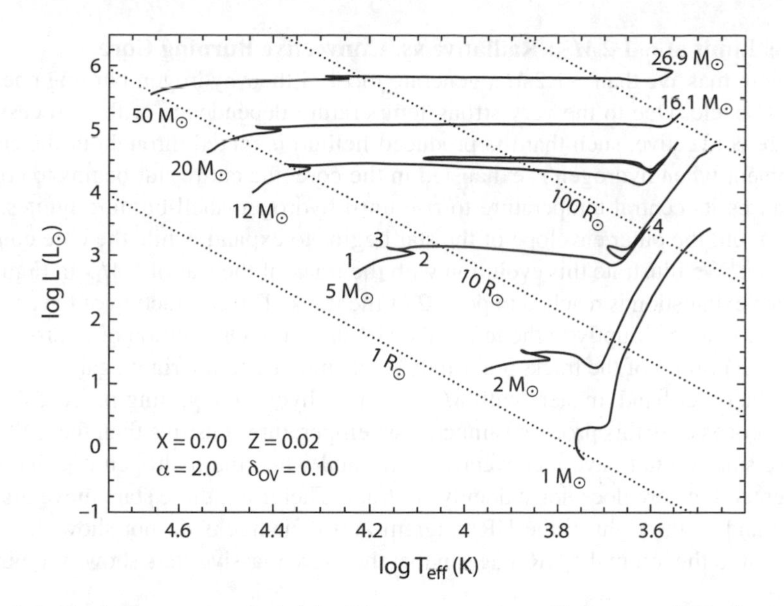

Figure 8.4. Evolutionary tracks in the Hertzsprung-Russell diagram (HR diagram) of non-rotating single stars of 1.0 to 50 $M_\odot$ together with lines of constant radii (in solar radii, dotted lines). Tracks were calculated by us using Eggleton's evolutionary code (see Tauris & van den Heuvel, 2006). For the two most massive stars, effects of mass loss by stellar wind are included, and their final masses are stated at the end of their tracks.

8.1.5 Further Details of the Evolution of Single Stars

The evolution of a star depends, apart from its mass, on its chemical abundances and rotation. In the following, we consider the evolution of stars with a chemical composition of 70% hydrogen, 28% helium, and 2% heavier elements (i.e., $Z = 0.02$), which is roughly similar to the composition of the Sun and stars in the disk of the MW (Population I stars). We first discuss the evolution without rotation and include some of these effects later in this chapter.

Figure 8.4 depicts how for stars of different masses the observable stellar parameters—luminosity (L), radius (R), and effective surface temperature (T_{eff})—change during the evolution of the stellar interior. Because $L = 4\pi R^2 \sigma T_{\mathrm{eff}}^4$, only two of these parameters are independent of one another. The figure depicts the evolutionary tracks in the Hertzsprung-Russell diagram (HR diagram) of single non-rotating stars with masses ranging from 1.0 $M_\odot$ to 50 $M_\odot$. For stars more massive than $\sim$15 $M_\odot$, mass loss by stellar winds becomes important, as can be seen from the tracks of the two most massive stars in the figure. We will consider these mass loss effects in more detail later in Sections 8.1.6 and 8.1.7, and concentrate here on the main aspects of the internal evolution of stars of different masses.

It appears that for stars in the phases of core hydrogen burning and hydrogen shell burning there are three critical mass limits (for solar metallicity):

1) Mass Limit at $\sim 1.2\,M_\odot$: Radiative vs. Convective Burning Core

Stars more massive than $\sim$1.2 $M_\odot$ generate most of their hydrogen-burning energy by the CNO cycle. Due to the very strong temperature dependence of this process, their cores are convective, such that the produced helium is mixed throughout the core. At the moment when hydrogen is exhausted in the core, the entire star begins to contract. This causes its central temperature to rise until hydrogen shell-burning ignites. From this point on, the outer envelope of the star begins to expand, while the core continues to contract. We illustrate this evolution with the track of the star of 5 $M_\odot$ in Figure 8.4: hydrogen exhaustion is reached in point 2 of the track. The contraction of the entire star causes it to move abruptly to the left in the HR diagram. On ignition of hydrogen-shell burning, in point 3 of the track, it abruptly starts moving to the right again.

On the other hand, in stars with $M \leq 1.2\,M_\odot$, hydrogen burning proceeds through the p-p process. As this process is much less temperature sensitive than the CNO cycle is, these stars do not have a convective core, and when the hydrogen is exhausted in their center, the core does not suddenly contract. Therefore, these stars move gradually upward and to the right in the HR diagram and their tracks do not show the sudden movement to the left in the HR diagram that the more massive stars show (see Fig. 8.4).

2) Mass Limit at $\sim 1.5\,M_\odot$: Convective vs. Radiative Envelope

At masses less than $\sim$1.5 $M_\odot$ main-sequence stars have convective outer envelopes, so higher than this mass limit the envelope becomes radiative. On the main sequence, this mass limit corresponds to spectral type of approximately F5V. Stars with convective envelopes have magnetic activity, like the Sun does. Such stars therefore have magnetically coupled stellar winds, which due to the magnetic field co-rotate with the star out to a certain distance, and therefore carry off the star's angular momentum. This *magnetic braking* is the reason why main-sequence stars of spectral type later than approximately F5V are observed to be slow rotators (Schatzman, 1962; Mestel, 1968; Skumanich, 1972; Meibom et al., 2015; see also Chapter 4), whereas main-sequence stars with earlier spectral types are rapid rotators, as is illustrated in Figure 8.9. It is interesting to note that main-sequence stars with masses between 1.2 and $\sim$1.5 $M_\odot$ have a convective core as well as a convective outer envelope, with a radiative layer in between.

3) Mass Limit at $\sim 2.3\,M_\odot$: Degnerate vs. Non-degenerate Helium Core

In stars less massive than $\sim$2.3 $M_\odot$, after hydrogen-shell ignition, the contracting helium core becomes degenerate, and the hydrogen-burning shell around the core generates the entire stellar luminosity. While the mass of its degenerate helium core gradually grows due to the hydrogen-shell burning, the star slowly climbs upward along the RGB, until it reaches helium ignition with a flash, when its helium core mass has grown to $\sim$0.47 $M_\odot$. Along the RGB, the stellar luminosity is only a function of the mass M_c of the degenerate helium core and is given by $L = 2.3 \times 10^5\,L_\odot\,(M_c/M_\odot)^6$. The star here climbs up along the Hayashi line that is appropriate for its mass. For stars with masses between 1.0 and 2.3 $M_\odot$ the Hayashi lines are very close to each other, such that there is basically one core mass vs. radius relation for these subgiants and giants (see also Chapter 11). For all stars less massive than $\sim$2.3 $M_\odot$ the helium ignition happens at the

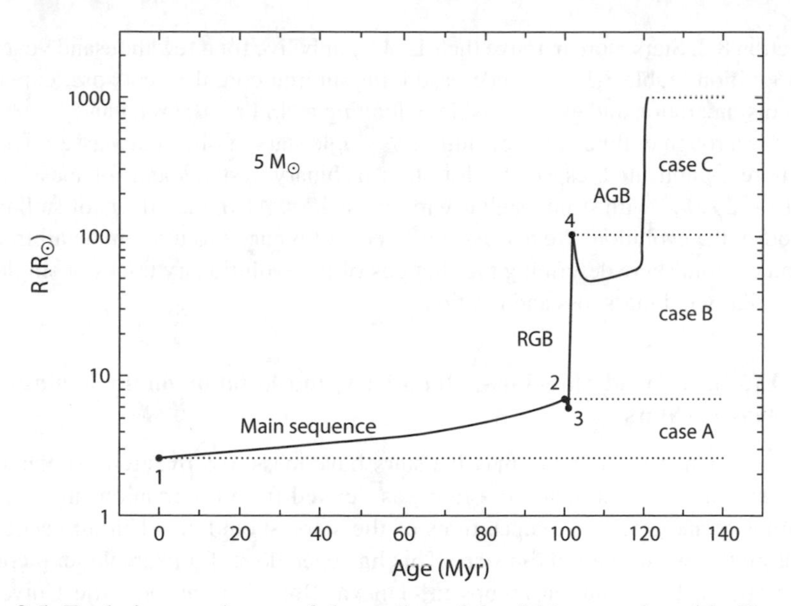

Figure 8.5. Evolutionary change of the radius of the 5 $M_\odot$ star plotted in Figure 8.4. The ranges of radii for mass transfer to a companion in a binary system according to the RLO Cases A, B, and C are indicated—see Chapter 9 for an explanation. (See also Tauris & van den Heuvel, 2006).

same core mass of 0.47 $M_\odot$. During helium burning, these stars, after an excursion to the left, again climb up along the giant branch, this time the AGB.

In the stars more massive than ~2.3 $M_\odot$, during hydrogen-shell burning the core does not become degenerate and continues to contract, while the outer layers of the star continue to expand until helium burning ignites in the core at a temperature of ~10^8 K, at point 4 of the tracks in Figures 8.4 and 8.5. At this time, the star has become a red giant, with a very dense core and a very large outer radius. During core-helium burning, hydrogen is burning in a shell around the core; during this phase, the star describes a loop in the HR diagram. The stars with $M > 2.3\ M_\odot$ move from hydrogen-shell ignition to core-helium ignition on a thermal timescale and describe the helium-burning loop on a nuclear timescale. During helium-shell burning, the outer radius of the star again expands, and the star becomes a RSG.

As described in Section 8.1.4, stars with masses up to ~8 $M_\odot$ then have a degenerate CO core and are located on the AGB, where they experience double shell burning (in hydrogen and helium shells), and probably lose their hydrogen-rich envelope due to pulsational instability, leaving a CO WD. Stars more massive than 8 $M_\odot$ ignite carbon burning and also become RSGs. In the mass range of approximately 8 – 10(12) $M_\odot$, the CO core is non-degenerate. The further evolution of such stars, with an ONeMg core produced by carbon burning and located on the AGB (RSGs), may in the end produce an ONeMg WD or possibly lead to collapse by electron capture producing a NS (see Fig. 8.3). However, their final evolution is complex as described in Section 8.1.4

and Section 8.2. Stars more massive than 12 $M_\odot$ only live for a few thousand years after carbon ignition (Table 8.1) and once producing an iron core they collapse, experience an iron disintegration and explode as SNe, leaving a NS or a BH remnant.

This is a rough outline of the evolution of single stars of different masses. The reality is more complicated, especially for stars in binary systems and for massive stars ($M \geq 15-20\, M_\odot$) with strong stellar wind mass loss. Also, the effects of stellar rotation modify the evolution. We discuss the effects of binary evolution in detail in Chapter 9 and continue here describing the changes of the evolutionary tracks of single stars due to stellar wind mass loss and rotation.

8.1.6 Effects of Wind Mass Loss, Metallicity, and Rotation on the Evolution of Massive Stars

As shown in Chapters 5 and 6, massive stars have mass loss by strong stellar winds. For this reason, the wind mass-loss rates, as derived from observations as a function of luminosity and effective temperatures of the stars, should be taken into account in calculating the evolution of these stars. This has been done, for example, in increasing detail by the stellar evolution groups in Geneva, Bonn, and at the Vrije Universiteit Brussels. We give here as examples the results of some of the extensive investigations by the Geneva group, led by Maeder and Meynet, as these nicely show the evolutionary tracks over a large range of stellar masses (0.8–120 $M_\odot$) and for different metallicities, including wind mass loss.

Figures 8.6 and 8.7 show the tracks calculated without stellar rotation (Schaller et al., 1992), for stars with $Z=0.02$ and 0.001. This investigation shows very clearly the large effects of wind mass loss on the evolutionary tracks of stars more massive than $\sim 20\, M_\odot$. It was later found that the wind mass-loss rates used by Schaller et al. (1992) are somewhat too high. However, the refined calculations with slightly different wind mass-loss rates for non-rotating stars with a slightly different solar metallicity ($Z=0.014$), by Ekström et al. (2012), and for low metallicity ($Z=0.002$), by Georgy et al. (2013), respectively, show qualitatively quite similar results. Therefore, the results of Schaller et al. (1992) for non-rotating stars are certainly not outdated. Contrary to the figures of the non-rotating tracks in the HR diagrams of Ekström et al. (2012) and Georgy et al. (2013), the tracks by Schaller et al. (1992) nicely depict the different regions of hydrogen and helium burning in the HR diagram, therefore, we prefer them here to discuss the evolution of stars of different masses. The stellar tracks calculated with rotation by Ekström et al. (2012) and Georgy et al. (2013) for $Z=0.014$ and 0.002, respectively, are somewhat different, however, as the example of Figure 8.9 for $Z=0.014$ shows.

The interesting effects of rotation on the evolution of massive stars have been investigated intensively since 2000 by the Geneva group (e.g., Maeder & Meynet, 2000a, 2000b; Meynet & Maeder, 2000, 2002, 2003, 2005; Hirschi et al., 2005), and for binaries in particular by the group of Langer in Bonn (and before that in Utrecht) (e.g., Heger et al., 2000; Heger & Langer, 2000; Yoon & Langer, 2005; Hunter et al., 2008; Brott et al., 2011; de Mink et al., 2013; Marchant et al., 2016; Sen et al., 2022). Very fast rotation can have a significant effect on the radii of stars via chemical mixing

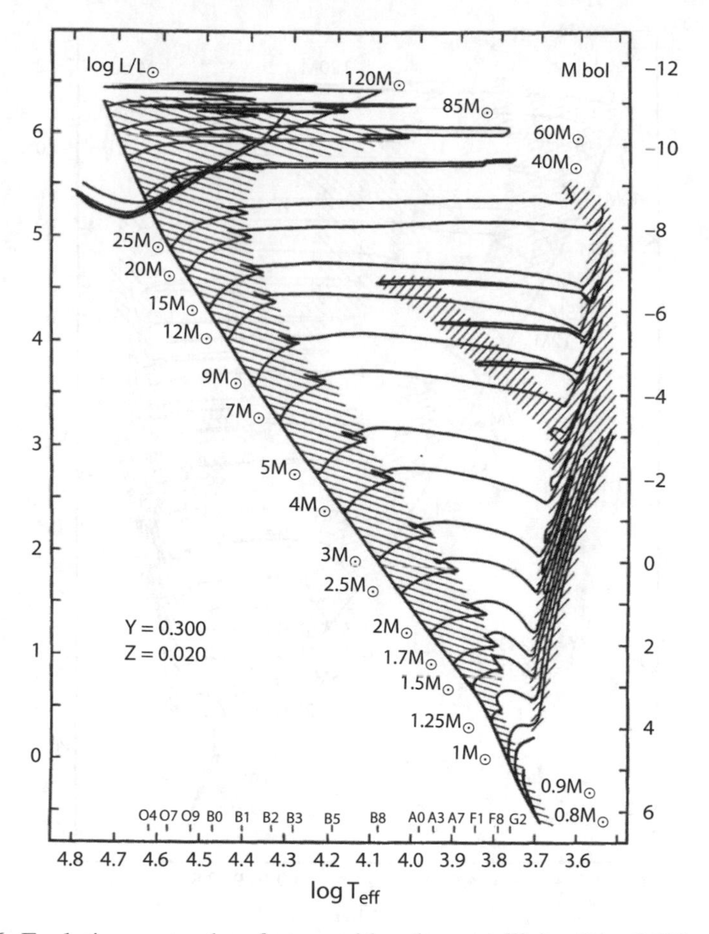

Figure 8.6. Evolutionary tracks of stars with solar metallicity ($Z = 0.02$), calculated without rotation. The shaded areas are the regions of long-lasting hydrogen or helium burning in the HR diagram. The tracks are modified if somewhat different wind mass-loss rates are assumed or if effects of significant rotation are included, for example, see Figure 8.9. After Schaller et al. (1992).

processes. The helium produced by hydrogen burning in the core can, due to rotational mixing, be mixed throughout the star and, in extreme cases, prevent such stars from becoming giants. Instead, these stars evolve blueward in the HR diagram and turn into pure helium stars (WR-stars), as illustrated in Figure 8.11.

The importance of this rotational effect is that it can turn a star of relatively low mass, such as 15 $M_\odot$, into a WR-star (a helium star more massive than 8 $M_\odot$), while for a non-rotating star with wind mass loss, only the helium core can turn into a WR-star, which would require a progenitor with a mass of at least $\sim$25 $M_\odot$. Furthermore, the rotational effects may, as Figures 8.9 and 8.11 show, prevent massive stars from moving toward the right in the HR diagram and becoming RSGs.

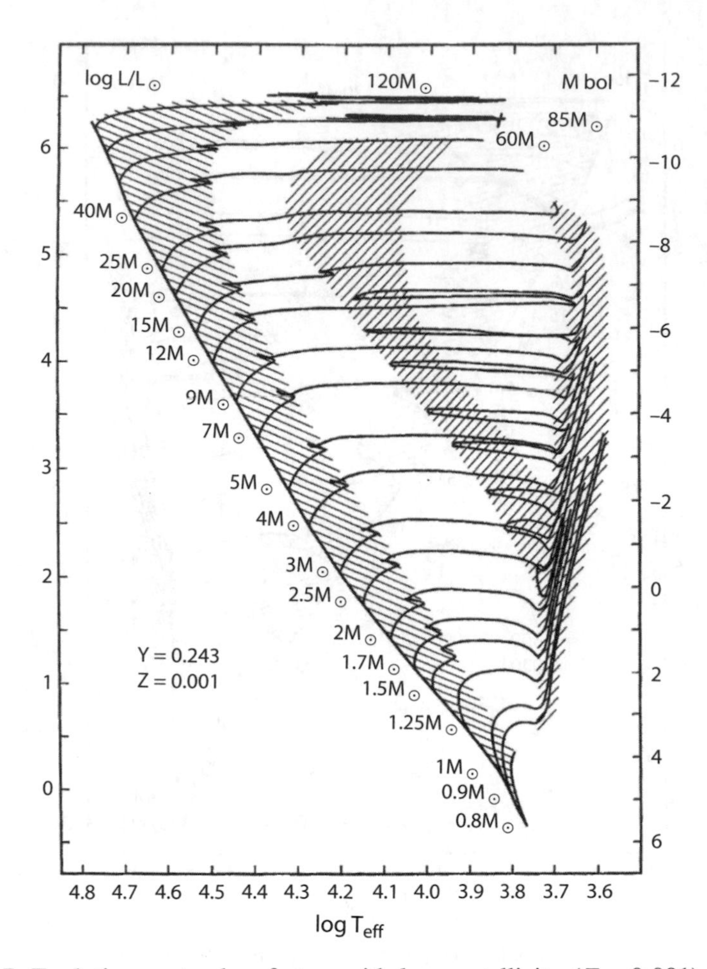

Figure 8.7. Evolutionary tracks of stars with low metallicity ($Z = 0.001$), calculated without rotation. The shaded areas are the main regions of core-hydrogen, hydrogen-shell (for low mass stars), core-helium, and helium-shell burning. After Schaller et al. (1992).

As to the effects of the metallicity Z on the wind mass-loss rates, from observations of the mass-loss rates of hot massive stars in other galaxies, with different Z-values (e.g., the LMC and SMC, which have $Z = 0.4\ Z_\odot$ and $Z = 0.2\ Z_\odot$, respectively [Hunter et al., 2007]), Mokiem et al. (2007) found the empirical relation

$$\dot{M}_{\text{wind}} \propto Z^{0.83 \pm 0.16} \tag{8.19}$$

for which theoretical explanations also exist (see, e.g., Conti et al., 2008). Garcia et al. (2014) found for the low-metallicity Galaxy IC 1613 ($Z = 0.1\ Z_\odot$) mass-loss rates and terminal wind velocites similar to those in the SMC. Earlier empirical relations had a value of the exponent α of Z in the preceding relation of order $\alpha = 0.5$ (Kudritzki & Puls, 2000).

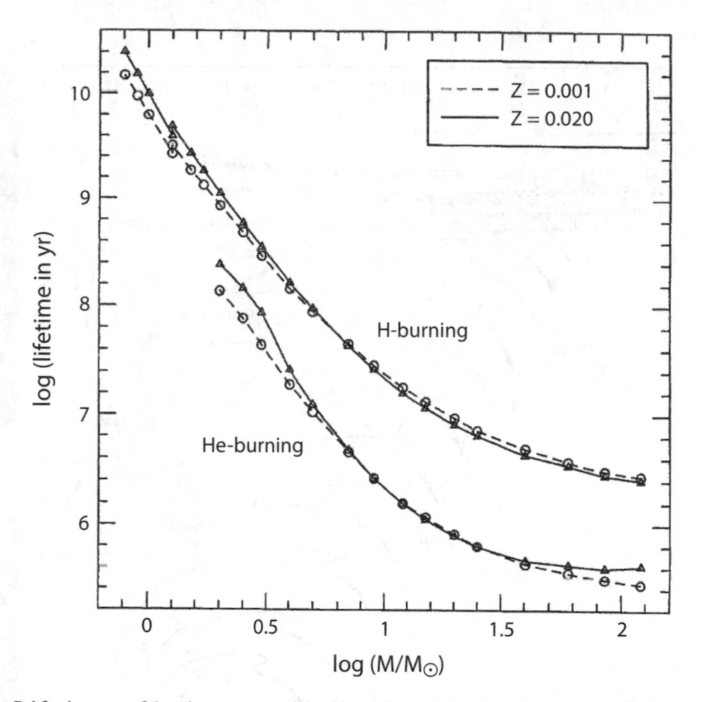

Figure 8.8. Lifetimes of hydrogen and helium burning for the stars of the evolutionary computations in Figures 8.6 and 8.7, for high and low metallicity, respectively (after Schaller et al., 1992). These relations remain valid in good approximation also for evolution with rotation and with somewhat different stellar wind mass-loss rates (see, e.g., Ekström et al., 2012; Georgy et al., 2013).

Because the tracks of Schaller et al. (1992) nicely show the regions of hydrogen and helium burning in the HR diagram for massive stars with mass loss, for different metallicities, they are well suited for discussing the qualitative effects of wind mass-loss and metallicity on the evolution of the most massive stars. As to the quantitative outcome regarding the final pre-SN masses of the stars as a function of their original mass one should use the results of the computations by Ekström et al. (2012), which are depicted in Figure 8.10. This figure shows that indeed the calculations with improved wind mass-loss rates, with and without rotation, produce considerably larger final masses of the massive stars than those of Schaller et al. (1992). Finally, the effects of metallicity on the lifetimes of hydrogen and helium burning of massive stars are rather minor as can be seen in Figure 8.8.

The computations of Ekström et al. (2012) show that the non-rotating (rotating) star of 120 $M_\odot$ has a mass of 63.6 $M_\odot$ (34.6 $M_\odot$) at the end of its hydrogen-burning phase and 31 $M_\odot$ (19 $M_\odot$) at the end of its helium-burning phase. At the beginning of helium burning, these stars already have lost their entire hydrogen-rich envelopes and have become WR-stars, which they remain throughout helium burning. Their final mass values could still be somewhat larger than depicted in Figure 8.10, as the mass-loss rates of WR-stars have since been found from observations to be somewhat lower than the

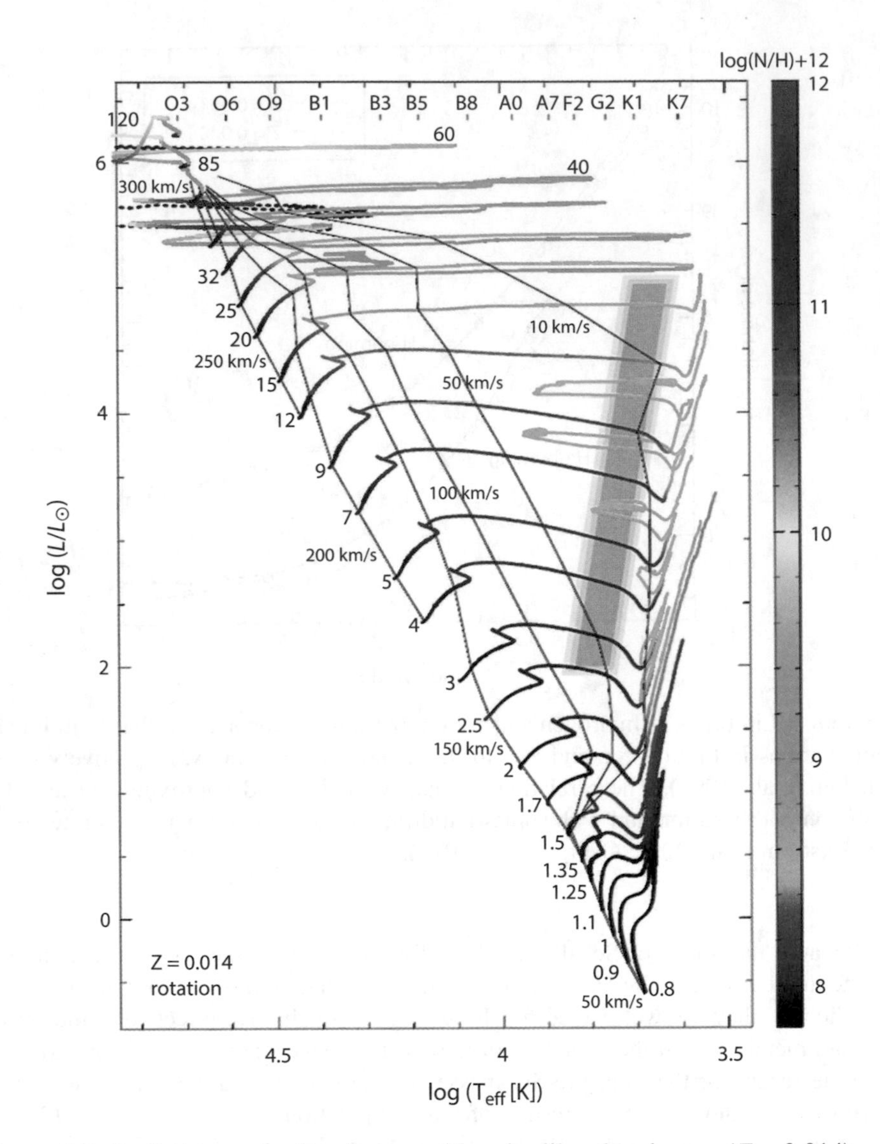

Figure 8.9. Evolutionary tracks of stars with solar-like abundances ($Z = 0.014$) and rotation. The assumed zero-age rotational velocities are indicated at the left at the beginning of the tracks. These velocities are the average observed rotational velocities of main-sequences stars in the different mass ranges. Because main-sequence stars with masses $< 1.5\, M_{\odot}$ are observed to rotate very slowly, they were assumed to start non-rotating. The thin lines indicate the computed rotational velocities at later evolution stages. The gray band indicates the Cepheid pulsational instability strip. The colors indicate the (N/H) ratios in the envelopes of the stars. When hydrogen is depleted in the envelope due to stellar wind mass loss, and nitrogen produced by the CNO cycle is enriched in the envelope, (N/H) becomes large. After Ekström et al. (2012).

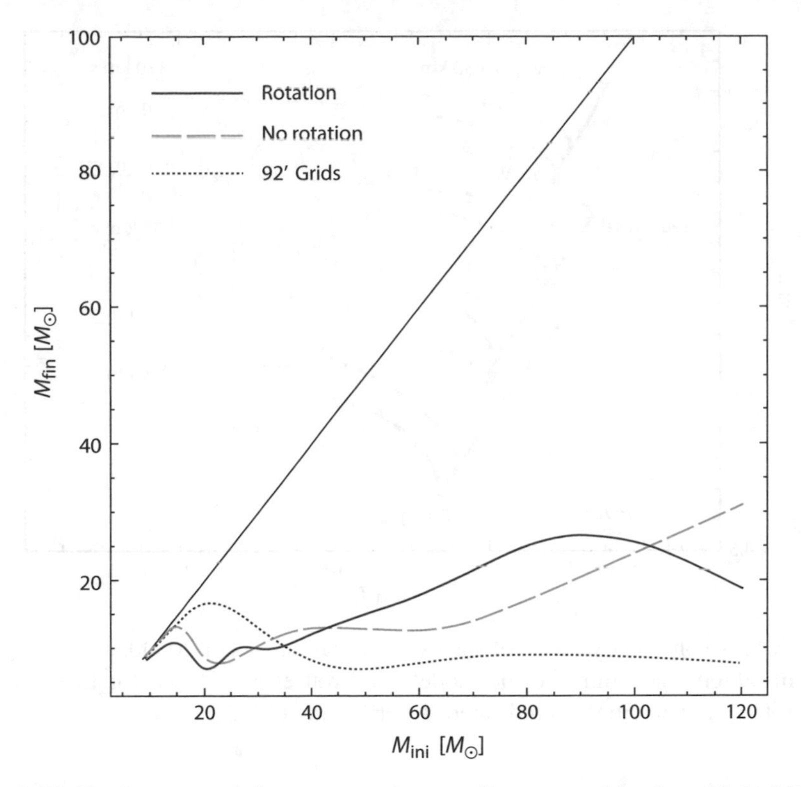

Figure 8.10. Final masses, at the moment of core collapse, as a function of initial ZAMS stellar mass for $Z = 0.014$. Green dashed curve indicates the mass without rotation; blue curve shows the mass with rotation. The red dotted curve (labelled 92' Grids) represents the results of Schaller et al. (1992) for $Z = 0.02$, which used higher wind mass-loss rates, particularly for stars more massive than 30 $M_\odot$. After Ekström et al. (2012). (It can be concluded that the final mass is very dependent on the assumed stellar wind mass-loss rate and thus on metallicity. Indeed, for rotating models with metallicity similar to that of the SMC ($Z = 0.2\,Z_\odot$), the final masses can be approximately 1.5 to 2 times larger than for $Z = Z_\odot$; see Meynet & Maeder [2005].)

ones used here, perhaps by as much as a factor of 2 (see, e.g., Shenar et al., 2020, and the discussion in Chapter 10).

Figures 8.6 and 8.7 show that for stars with wind mass loss, the following three conclusions can be drawn: (1) To the right of the main sequence, there are two locations of helium-burning stars in the HR diagram: as RSGs ($\log T_{\rm eff} \leq 3.7$) and as yellow and white (spectral types A–G) supergiants ($\log T_{\rm eff} = 3.7{-}4.1$), respectively. (2) At high metallicity, the yellow and white supergiants occur for masses up to $\sim$12 $M_\odot$ and RSGs ($\log T_{\rm eff} = 3.5{-}3.7$, spectral types K and M) up to $\sim$25 $M_\odot$. (Note, while these figures place the maximum mass limit for RSGs near 25 $M_\odot$, other recent works, see, e.g., Fig 8.12, place the observational limit near 40 $M_\odot$, although in this figure, except for a few cases, the RSGs have masses $<$25–30 $M_\odot$.) (3) At low metallicity, the

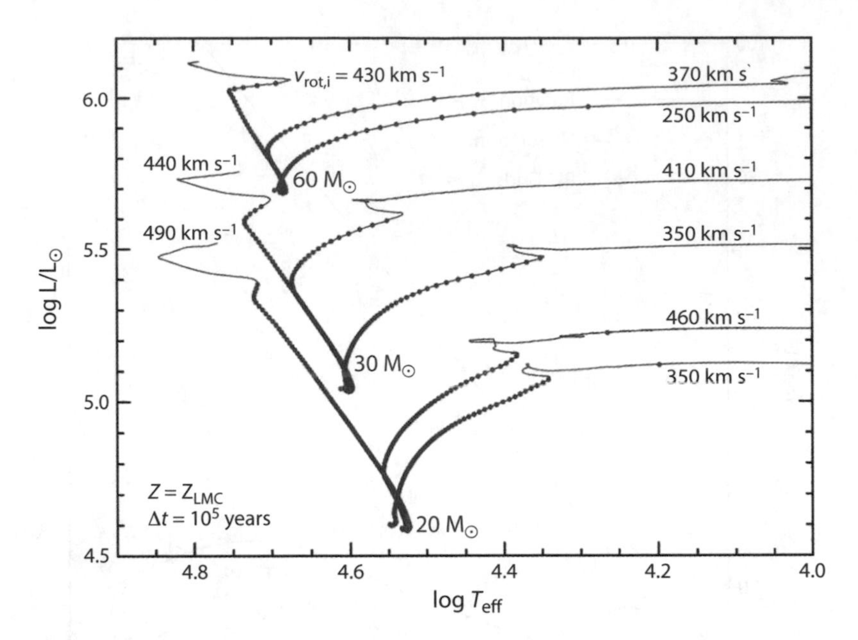

Figure 8.11. Evolutionary tracks of stars with initial masses of 20, 30, and 60 $M_\odot$ with LMC metallicity, according to the models of Brott et al. (2011), for three different initial rotational velocities, as indicated. After Langer (2012).

wind mass-loss effects are smaller, which implies that the helium-burning supergiants (spectral types B3–G; colors: blue, white, and yellow) occur up to $\sim$60 $M_\odot$, but RSGs (spectral types K and M; $\log T_{\rm eff} = 3.5-3.7$) still occur only up to only $\lesssim$25 $M_\odot$. The fact that, according to these calculations with wind mass loss, stars more massive than $\sim$25 $M_\odot$ are not expected to become red giants or supergiants, has important consequences for the evolution of such stars in binaries as we will see in Chapters 9 and 10.

Figure 8.12 depicts the positions in the HR diagram of the supergiants in the M31 (Andromeda) Galaxy as observed by Humphreys et al. (2017). (For M33, the figure determined by these authors is quite similar; see also Gordon & Humphreys [2019].) The figure shows that indeed, as predicted by the evolutionary tracks with mass loss and rotation as computed by the Geneva group (and, e.g., by Eldridge & Tout, 2004, and the group of Langer in Bonn), stars with masses larger than $\gtrsim$40 $M_\odot$ do not become RSGs (notice that the number of RSGs with luminosities corresponding to stars with masses of $>$25 $M_\odot$ is very small). In fact, in our Galaxy and the Magellanic Clouds, no RSGs with masses $\gtrsim$25 $M_\odot$ are seen (Fitzpatrick & Garmany, 1990; Levesque et al., 2006; Georgy et al., 2013; Massey, 2002), even though the metallicity of the SMC is only 0.2 $Z_\odot$, such that one naively would expect lower wind mass-loss rates. The Schaller tracks for low metallicity indeed predict no RSGs $\gtrsim$25 $M_\odot$, in accordance with the observational upper limits to the luminosities of RSGs of $\log(L/L_\odot) = 5.2-5.3$ (Levesque et al., 2005). Similar results were obtained by Massey et al. (2017) for RSGs in other galaxies in the Local Group. The dashed line in Figure 8.12 is the so-called

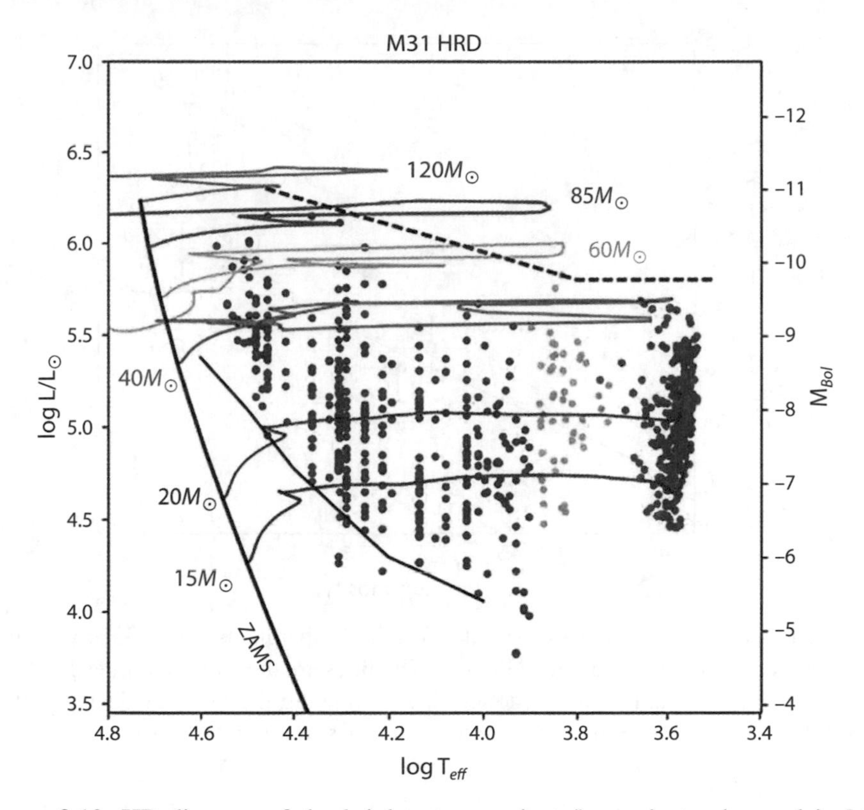

Figure 8.12. HR diagram of the brightest supergiants/hypergiants observed in M31 together with evolutionary tracks for solar metallicity (with stellar wind mass loss, but without rotation) from Ekström et al. (2012). Observed OBA-stars are blue, FG-supergiants are yellow, and RSGs (K and M spectra) are red. Notice that the observed red supergiants have masses $\leq 40\ M_\odot$, while there are only a few greater than approximately $25 - 30\ M_\odot$. The same is true in M33 and the LMC. The upper envelope of observed stars indicated by the dashed line is called the Humphreys-Davidson limit. After Humphreys et al. (2017).

Humphreys-Davidson limit (Humphreys & Davidson, 1994): the observational upper limit to the luminosity of stars as a function of effective temperature.

Many of the red and yellow-white supergiants in M31 and the other Local Group galaxies are variable and have high mass-loss rates (Gordon et al., 2016; Neuhäuser et al., 2019). They are systematically more reddenend than the blue supergiants in their neighborhood, indicating the presence of circumstellar dust that is indicative of high mass loss from these stars (Massey et al., 2005). The mass-loss rates from RSGs can be $>10^{-4}\ M_\odot\ \mathrm{yr}^{-1}$ (Neuhäuser et al., 2019), as depicted in Figure 8.13. With such mass-loss rates, stars with masses $>40\ M_\odot$ will lose their hydrogen-rich envelopes in a very short time. About the evolution and fate of RSGs, as a function of metallicity, see also Georgy et al. (2013) and Ekström et al. (2013).

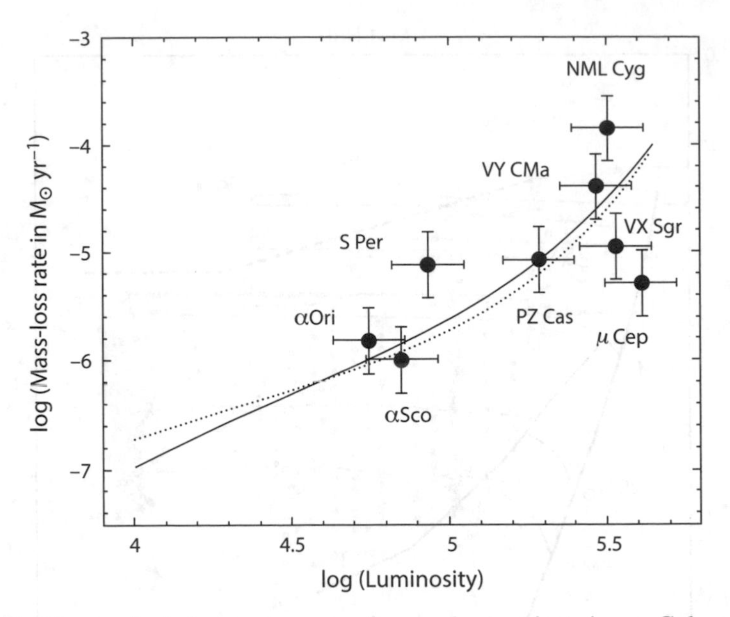

Figure 8.13. Observed wind mass-loss rates from red supergiants in our Galaxy, derived from observed amounts of circumstellar gas. The lines are the mass-loss rates for $T_{\rm eff} =$ 4,000 K (solid line) and 3,500 K (dotted line) following the work of de Jager et al. (1988). Luminosity is in units of $L_\odot$. After Mauron & Josselin (2011).

The Effects Solely of Metallicity on the Stellar Radius

It is important to notice that without including the effects of rotation and stellar wind mass loss, metallicity influences the values of the radii of the stars. Table 8.2 shows the pre-SN radii for stars with masses between 12 and 40 $M_\odot$, for a range of metallicities, computed without wind mass loss. Table 8.2 shows that for $Z = 0$ and very low Z, the radii of massive stars up to the time of the SN explosion remain small. For $0 \leq Z \leq 10^{-2}\ Z_\odot$, for most masses, the pre-SN stars are blue supergiants. Only for higher metallicities do they become RSGs. In principle, these results can be important for the formation of BHs in binaries with mass transfer, because whether mass transfer will occur depends critically on the radius evolution of stars.

As we already noticed, for $Z \geq 0.1\ Z_\odot$, when stellar wind mass loss is included, the radii of stars more massive than $\sim$40 $M_\odot$ never become as large as those of RSGs (see, e.g., Figs. 8.7, 8.6, and 8.12). Instead, these massive stars in the MW and other Local Group galaxies are observed with temperatures $T_{\rm eff} > 20{,}000$ K (Humphreys & Davidson, 1994; Castro et al., 2014). These stars are the so-called *luminous blue variables* (LBVs), and it has been suggested that these are to be identified with massive stars that while on their way to the RSG stage have been shedding their envelopes and are now becoming WR-stars (the "Conti scenario" for producing LBVs and single WR-stars)—see, for example, the tracks in Figures 8.9 and 8.12 that show that a 60 $M_\odot$ star with MW metallicity does not expand $\gtrsim$160 $R_\odot$ (and see also Fig. 8.15). For other similar track calculations, see, for example, Kruckow et al. (2016).

Table 8.2. Pre-SN Radii of Massive Stars as a Function of Metallicity

Mass ($M_\odot$)	$Z=0$	$10^{-4}\,Z_\odot$	$0.01\,Z_\odot$	$0.1\,Z_\odot$	$0.033\,Z_\odot$
12	0.140	2.64	2.79	2.83	2.55
13	0.0898	0.694	2.98	3.01	2.80
15	0.0674	0.181	0.709	3.44	3.35
18	0.0696	0.144	0.309	1,79	4.17
20	0.0795	0.215	0.866	3.42	4.88
22	0.0890	0.186	0.617	4.76	5.19
25	0.0916	0.239	0.809	2.14	5.91
30	0.113	0.316	1.72	6.89	6.87
35	0.174	0.563	5.35	8.50	7.78
40	0.196	0.625	7.96	9.52	(28)

Source: Woosley (2015, Lecture notes No. 7).
Stellar radii are given in units of 10^{13} cm $\simeq 140\,R_\odot$ and computed without stellar wind mass loss and rotation.

For a recent detailed paper about the effects of metallicity, rotation, and the variation of several model physical parameters on the evolution of the stellar radii, we refer to Klencki et al. (2020). This paper is specifically aimed at calculating the evolution of massive binary systems, with primary star masses in the range of 10–80 $M_\odot$.

We now briefly discuss the possible nature of these LBV stars. It has been suggested that they are a crucial evolutionary stage for either single massive stars on their way to becoming WR-stars (i.e., the Conti scenario), or they might be mass gainers from a previous interaction with a companion star. Both hypotheses will be discussed in Section 8.1.7.

8.1.7 The Question of the LBVs and Their Origin

Figure 8.9 shows that stars at solar metallicity with masses between approximately 25 $M_\odot$ and 120 $M_\odot$ throughout their evolution with mass loss (and rotation) have a luminosity in the range of approximately $2\times10^5 - 2\times10^6\,L_\odot$. This is roughly the range of luminosities of the LBVs, as can be seen in Figure 8.14. This figure shows that a few LBVs, such as η Carinae, can reach even higher luminosities of the order of $\log L/L_\odot > 6.5$, implying that their masses are $>120\,M_\odot$. At quiescence, the LBVs are located in the diagonal strip at high $T_{\rm eff}$ (blue), while during their LBV outbursts, which happen at irregular times, they move toward cooler spectral types, approximately $T_{\rm eff}\sim 7{,}500$ K (white-yellowish color), and eject quite a lot of mass. In some spectacular cases, such as η Carinae, the amount of mass ejected during an outburst may be as large as several solar masses. The precise causes of these outbursts of mass loss are not known. But it is known from theoretical studies that stars in this top part of the HR diagram may be pulsationally unstable (Saio, 2011). Studies of stellar dynamical instabilities in this part of the HR diagram, by de Jager et al. (2001), had already suggested that there may be two regions of instability: a *blue* region and a *yellow-red* region.

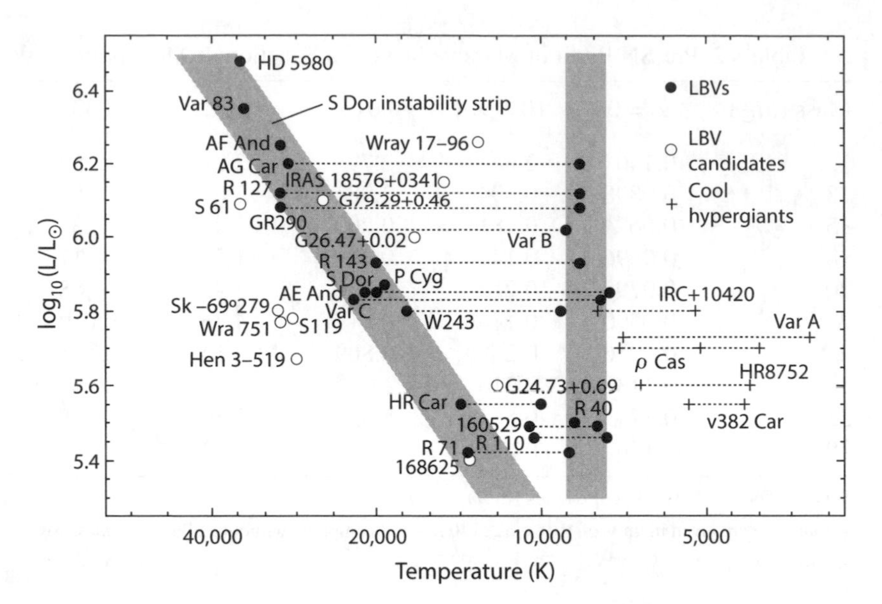

Figure 8.14. The upper HR diagram of LBVs and some LBV candidates. The most massive LBV candidates such as η Carinae and the Pistol Star are off the top of the diagram. At quiescence, the LBVs reside in the diagonal strip. During the LBV outbursts, they make an excursion to the right where they reside in the vertical strip at an effective temperature of $\sim$7,500 K (white-yellowish color). After Smith et al. (2004).

Until recently, the *standard model* for LBV instability, first suggested by Peter Conti (see, e.g., Abbott & Conti, 1987), was that these mass ejection episodes are due to such internal (pulsational) instabilities in very luminous single stars. According to this standard model, single stars more massive than approximately $25 - 40\, M_{\odot}$ will, due to the radiation pressure produced by their enormous luminosities, have stellar winds so strong that when they reach the RSG stage, assisted by pulsational instability, they blow away their envelopes. Such evolution is illustrated by the above-mentioned Geneva evolutionary tracks calculated by, for example, Ekström et al. (2012). According to this model, LBVs just are the evolutionary stage in which single massive stars are shedding their hydrogen-rich envelopes, to become a WR-star. For a description of this standard model, see, for example, Meynet et al. (2011).

Figure 8.15 shows that indeed the positions in the HR diagram of LBVs at quiescence fall in the region crossed by the tracks of very massive stars. Still, these massive single star evolutionary tracks with mass loss do not predict why the LBVs at quiescence should be located in the diagonal strip in the HR diagram where they are found, and they do not indicate why their excursion to the right in the HR diagram terminates in the vertical strip indicated in Figure 8.14. Their location in the vertical strip is suggestive for a kind of internal (pulsational?) instability that occurs when stars enter this strip, as suggested by de Jager et al. (2001).

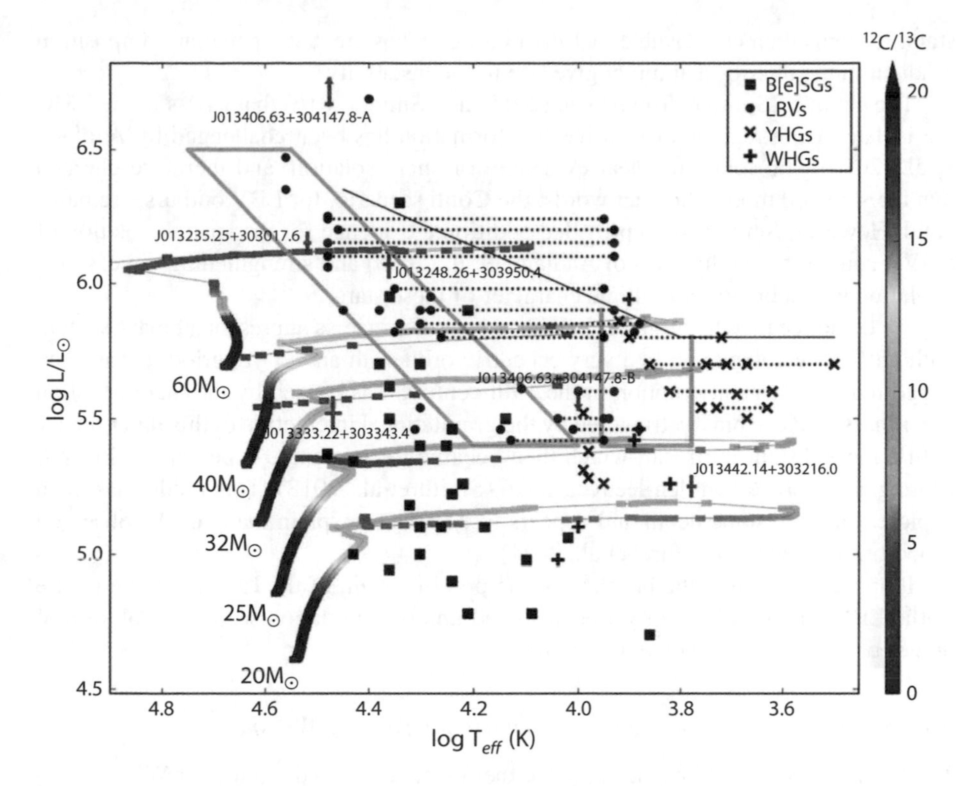

Figure 8.15. The positions in the HR diagram of LBVs, yellow hyper-giants YHGs, white hyper-giants (WHGs), and B[e]-supergiants, from studies in Local Group galaxies. Evolutionary tracks of stars of 20–60 $M_\odot$ with solar metallicity and rotation (40% critical, initially) computed by Ekström et al. (2012) are indicated. The thick black line is the Humphreys-Davidson limit. The gray lines mark the boundaries of the alledged S Dor instability strip, and of the cooler Yellow Void, where only few YHGs are found. The red symbols are six objects in M33. The dotted lines mark the transition phases for several LBVs and YHGs. After Kourniotis et al. (2018).

However, Smith & Tombleson (2015) and Smith (2016) put forward arguments suggesting that LBVs have originated from binaries, and that the outbursts are due to their binary history (that may have affected their internal structure). They base this suggestion on the fact that they found LBVs in space to be quite isolated from regions where groups of massive stars are found. This would then suggest that the LBVs have moved away from such regions with substantial runaway velocities.

If indeed all LBVs would have originated in binaries, this instability might possibly arise in binary components that have undergone considerable mass gain due to mass transfer. It is interesting to note that according to the simulations by Renzo et al. (2019), ~86% of all massive post-mass-transfer binaries are disrupted in the SN explosion of the initially most massive star, such that these mass gainers will be single runaway stars. It seems then conceivable that the mass gain has created a peculiar composition gradient in the interior that might give rise to the instability.

The claim by Smith & Tombleson (2015) and Smith (2016) that LBVs in the LMC are isolated from regions of massive star formation has been challenged by Aadland et al. (2018) who found no clear evidence for such isolation, and therefore claimed that the standard model, in other words, the Conti scenario, for LBV outbursts remains valid. However, Smith (2019) provided additional evidence for the spatial isolation of LBVs, countering the findings of Aadland et al. (2018) and strengthening the case for a relation with a binary or multiple character of these stars.

We do notice that the brightest LBV known, η Carinae, is at present a binary system with an O-type companion in a very eccentric orbit with an $\sim$5 yr period. It has been suggested that its giant eruption in the 19th century was caused by the merging of an inner massive close binary, triggered by the gravitational interaction of this inner binary with a more distant third star, which then became the present O-type companion star (Portegies Zwart & van den Heuvel, 2016; Smith et al., 2018). This model has been explored in quite some detail and appears to be able to explain many of the observed properties of the system (Hirai et al., 2021).

It appears to us that the problem of the precise origin of the LBVs and the nature of the LBV outbursts has not yet been solved and that a relation with the evolution of binary or multiple stars cannot be excluded.

Are LBVs Really Needed in the Conti Scenario for Forming WR-stars?

The basic thesis of the Conti scenario for the formation of hydrogen-poor WR-stars by single star evolution is that very massive single post-main-sequence stars rapidly lose their hydrogen-rich envelopes and turn into WR-stars. The loss of their envelopes is, in this scenario, due to the enormous luminosity of a star more massive than approximately $25-40\,M_{\odot}$, which causes stellar-wind mass loss at rates so large that on the way to the RGB, or on this branch itself, the evolutionary tracks of the stars turn around and the stars become WR-stars. The evolutionary calculations with realistic wind mass-loss rates (and also rotation) by the Geneva and Bonn groups and others (see, e.g., Figs. 8.6, 8.7, and 8.9), show that this is indeed expected to happen to single massive stars. Notice that a LBV stage is not necessarily assumed to play a role in these calculations. Therefore, the Conti scenario for forming hydrogen-poor WR-stars by single massive star evolution remains valid, also without invoking a LBV stage.

It just happens to be the case that the tracks of massive luminous stars with strong wind mass loss cross the region of the HR diagram where the LBVs are located, too. But this does not necessarily mean that the LBVs are causally related to these evolutionary tracks. It could also be that the LBVs are an evolutionary stage of massive binaries, which happen just by accident to be in the same part of the HR diagram crossed by the tracks of single massive stars with strong winds, on their way to the WR stage. We thus conclude that the LBV stage is not a necessary ingredient of the Conti scenario for forming single WR-stars. Possibly, it has nothing to do with it.

We notice here that apart from the LBVs, there is another group of highly luminous stars with very large mass-loss rates in the upper part of the HR diagram of the LMC,

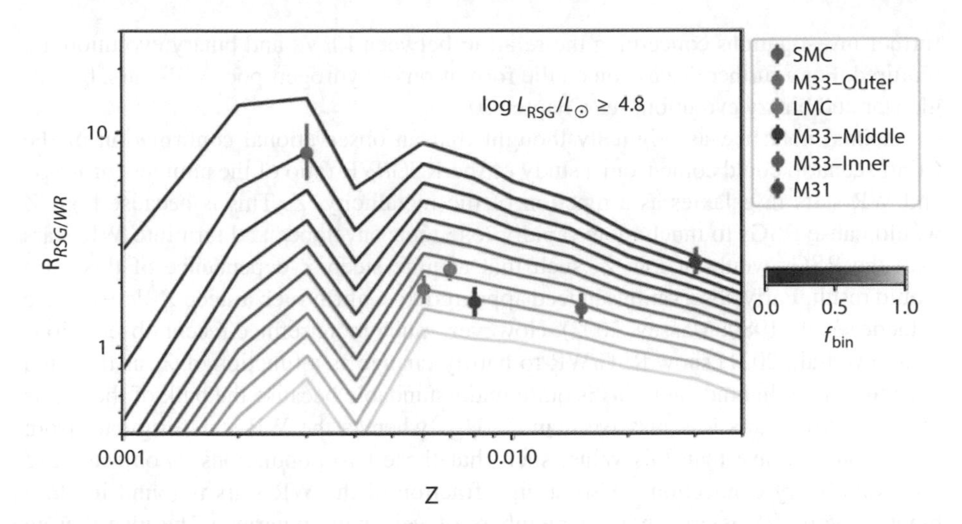

Figure 8.16. Observed RSG/WR ratios (points) in different galaxies as a function of Z. The figure shows no clear Z-dependence of this ratio. The curves are predicted by the Binary Population and Spectral Synthesis (BPASS) (Stanway et al., 2020) binary population synthesis evolutionary tracks. The color range of the BPASS models corresponds to interpolating between models with no binaries ($f_{bin} = 0$) and models with all stars in binaries ($f_{bin} = 1$). After P. Massey (private communication, 2021).

SMC, and other Local Group galaxies. These are the B[e]-supergiants, *B-bracket-e* stars which were briefly described in Chapter 7, in connection with the extremely obscured sg-HMXBs IGR J16318−4848 and CI Cam. They have luminosities typically in the range between $10^{4.5}$ and more than $10^6\ L_\odot$ (Zickgraf, 1998) and are enshrouded by a thick infrared-emitting disk of gas and dust. We saw in Chapter 7 that in these sg-HMXBs the B[e] donor stars are deep inside their Roche lobes, such that their large wind mass-loss rates cannot have been triggered by the binary nature, but must be intrinsic to the stars. The large helium abundance in the envelope of the donor of IGR J16318−4848 indicates that this star has already lost a substantial part of its hydrogen-rich envelope and might be on its way to becoming a WR-star. The positions in the HR diagram of B[e]-supergiants and YHGs in Local Group galaxies are indicated in Figure 8.15, and illustrate that, if the LBV stage would indeed not be a phase in the evolution of single massive stars, as Smith & Tombleson (2015) and Smith (2016, 2019) propose, these stars might possibly be an alternative intermediate phase of massive stars on their way to becoming (single) hydrogen-poor WR-stars.

On the other hand, an evolutionary sequence in which the supergiant B[e]-stars as well as the LBVs play a role in the evolution of single massive stars was proposed by Stothers & Chin (1996). These authors indicated the possibility of hydrogen-poor and hydrogen-rich B[e]-stars to occur before and after LBV and RSG phases of evolving single massive stars. See also Morris et al. (1997) and Marston & McCollum (2008).

Thus far, we conclude that at present the precise evolutionary relation between LBVs and the formation of hydrogen-poor WR-stars is still controversial and that further investigations concerning the relation between LBVs and binary evolution are required. For a further discussion on the formation of hydrogen-poor WR-stars, by single star and binary evolution, see Chapter 10.

A final note: it was originally thought that an observational confirmation of the Conti scenario could come from a study of the RSG/WR ratio of the numbers of RSGs and WR-stars in galaxies as a function of the metallicity, Z. This is because high Z would cause RSGs to much more rapidly lose their envelopes and turn into WR-stars than the RSGs would at low Z, such that a quite steep Z-dependence of this ratio would result. Early observations indeed appeared to confirm such a steep Z-dependence (Maeder et al., 1980; Massey, 2002). However, much more refined recent observations (Massey et al., 2021) show RSG/WR to hardly change as a function of Z, as depicted in Figure 8.16. In hindsight, this is quite understandable because the bulk of the RSGs originate from stars less massive than 25 $M_{\odot}$, whereas the WR-stars originate from stars more massive than this value, such that these two populations of objects have no evolutionary connection. (Also, a high fraction of the WR-stars is found in close binaries while RSGs are single or members of very wide binaries.) The idea that an observed clear Z-dependence of RSG/WR could provide a confirmation of the Conti scenario therefore proved elusive.

8.1.8 Final Evolution of Stars in the Mass Range of $1 - 8\,M_{\odot}$: Competition between Mass Loss and Carbon Ignition; Evidence from WDs in Open Clusters

When stars in this mass range are approaching carbon ignition, they are on the AGB appearing as bright red (super)giants. Such stars tend to be pulsationally unstable (they show pulsations of Mira type, with periods of hundreds of days) and lose mass by strong stellar winds with rates up to $10^{-4}\,M_{\odot}\,\mathrm{yr}^{-1}$. WDs are found in open stellar clusters, such as the Pleiades, that still contain main-sequence stars with masses as large as $8 \pm 1\,M_{\odot}$ (Weidemann, 1990; Kalirai et al., 2008). This indicates that single stars as massive as at least $8\,M_{\odot}$ may still terminate life as a WD; in other words, these single stars are apparently able to shed their envelopes in the AGB phase before carbon ignites. This confines the possible mass range for carbon deflagration "SNe" of single stars to at most $7 - 8\,M_{\odot}$. It is possible, and even likely, that single stars never reach this phase, and that SNe due to carbon ignition in single stars do not occur in nature (however, see also Antoniadis et al., 2020).

Observed WDs can be grouped in different spectral classes, which appear to be related with the masses of their progenitors. This classification is as follows: WDs with hydrogen-dominated atmospheres are assigned the primary spectral classification DA. They make up ~80% of all observed WDs. The WDs with spectral classification DB have neutral helium in their spectra and are the second most frequent class in number (~15%–20%). WDs with ionized helium in their atmosphere as called DO. Carbon-dominated atmosphere WDs are hot, with temperatures >15,000 K, and are called DQ (but they only constitute ~0.1% of all WDs). Some WDs are assigned DAB, which means a hydrogen-rich and neutral helium-rich atmosphere is present. Pulsating variable WDs are assigned with an extra "V" in their classification (e.g., DAV, DBV), while magnetic WDs without detectable polarization are assigned with an extra "P" or "H", respectively.

From the observed and theoretically predicted relations between the initial stellar mass and the final WD mass for low- and intermediate-mass stars, it appears that stars up to $\sim 2\,M_\odot$ (which are the bulk of the WD progenitors) leave DA WDs with masses of approximately 0.55–0.60 $M_\odot$. These WDs consist mainly of carbon and oxygen and still have a very thin hydrogen skin. Stars more massive than $\sim 3.0\,M_\odot$ leave WDs with masses in the range $\sim$0.70–1.0 $M_\odot$ (Kalirai et al., 2008), consisting of carbon and oxygen with a thin layer of helium (plus sometimes traces of hydrogen) on top: these are the DB or DAB WDs. When they are still young and hot they may be of type DO. In the initial mass range of $2-3\,M_\odot$, the WD remnants are expected to have masses between $\sim$0.6 and 0.7 $M_\odot$ and are probably of type DAB.

8.2 FINAL EVOLUTION AND CORE COLLAPSE OF STARS MORE MASSIVE THAN 8 $M_\odot$

8.2.1 Electron-capture Collapse vs. Iron-core Collapse

In the mass range between 8 and $\sim 10-12\,M_\odot$ (it is not precisely known if the upper limit is 10 or possibly even 12 $M_\odot$), carbon ignition is nonviolent, but the ONeMg core that is formed (the oxygen originating from helium burning, the neon and magnesium from carbon burning) is degenerate. Figure 8.17 shows the last part of the internal evolution of such a core of 2.4 $M_\odot$ (which is a helium star) computed by Nomoto (1984). (For recent calculations of the internal evolution of helium stars with masses between 1.6 and 120 $M_\odot$, showing evolution of lower-mass helium stars roughly similar to that in Fig. 8.17, see Woosley [2019].) When the mass of the ONeMg core approaches the Chandrasekhar limit, the density becomes so high that the threshold for electron captures by the magnesium, neon, and oxygen nuclei in this core is crossed (Langanke et al., 2021; Kirsebom et al., 2019), as is indicated in Figure 8.3. This causes the onset of core collapse, as the removal of the (most energetic) electrons causes the pressure of the electron-degenerate gas to drop, as was first pointed out by Miyaji et al. (1979) and further elaborated in important papers by Nomoto (1984, 1987). At the time of the collapse, neon and oxygen ignition may already have been passed, as the track of the 9 $M_\odot$ star in Figure 8.3 shows, and a small iron core may have formed that collapses to a NS in an electron-capture SN (see also Nomoto et al., 1991). Because this collapse is due to reaching the Chandrasekhar limit (which here is $\sim 1.4\,M_\odot$), the NSs produced by this mechanism are expected to have a baryonic mass equal to the Chandrasekhar mass. As the loss of gravitational binding energy in the formation of a NS is $\sim 0.10\,Mc^2$ (depending on the exact equation of state; see Lattimer & Yahil [1989]), the gravitational mass[4] of a NS produced by electron-capture collapse is expected to be $\sim 1.25\,M_\odot$.

[4]Using $M_{\rm NS}^{\rm grav} = M_{\rm NS}^{\rm baryonic} - \Delta M_{\rm NS}^{\rm bind}$, where $\Delta M_{\rm NS}^{\rm bind} \simeq 0.084\,M_\odot\,(M_{\rm NS}^{\rm grav}/M_\odot)^2$.

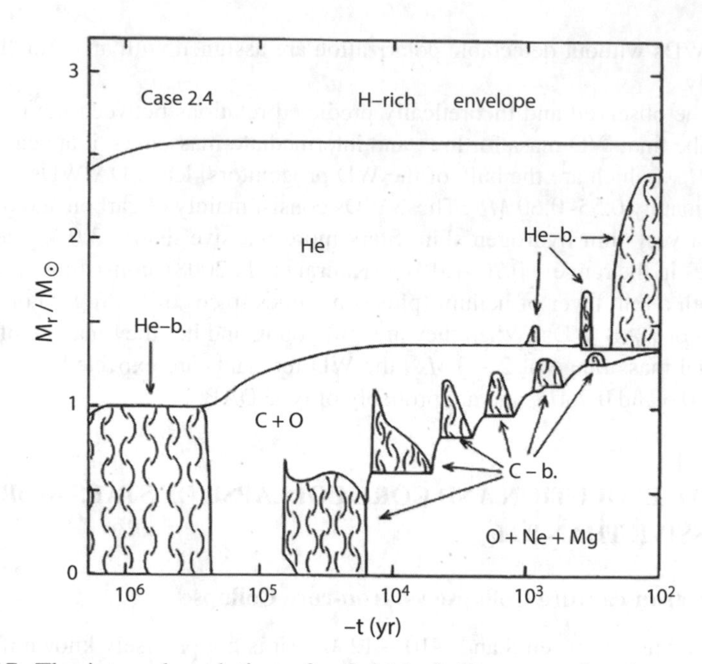

Figure 8.17. The internal evolution of a 2.4 $M_\odot$ helium core forming a degenerate ONeMg core, which evolves to electron-capture collapse. Shown is remaining time in years until core collapse. After Nomoto (1984).

For stars more massive than about ~10–12 $M_\odot$, an iron core of mass >1.2 $M_\odot$ is formed that collapses as a result of the iron-helium transition (photo-disintegration of iron), as indicated in Figure 8.3 and schematically described in Figure 8.20; see also Hillebrandt (1987). Detailed investigations in the 2010s by Jones et al. (2013) and Woosley & Heger (2015) provide evidence of an initial threshold mass of ~10 $M_\odot$ for producing an iron core collapse at solar metallicity. For other metallicities, the threshold mass may vary (Ibeling & Heger, 2013). The physics of iron-core collapse and of the SNe that may be produced by this process will be briefly summarized in the next two sub-sections (see also Chapter 13).

For a discussion of the present state of our knowledge about electron-capture collapses in the mass range 8 to ~12 $M_\odot$, where the final evolution is still not well understood, see Section 8.1.4 and, for example, Poelarends et al. (2008), Jones et al. (2013), Woosley & Heger (2015), Schwab et al. (2015), Schwab et al. (2017), Doherty et al. (2017), Kirsebom et al. (2019), Zha et al. (2019), Chanlaridis et al. (2022), and the references therein. Because electron-capture collapse is expected to be particularly important for the formation of NSs in binary systems, including double NSs (DNSs), we will return to this subject in more detail in Chapter 13. As we will see in that chapter (and as was mentioned in Section 8.1.4), there may be a strong preference for electron-capture collapse to occur in interacting binaries, rather than in single stars (Podsiadlowski et al., 2004).

Hydrodynamic simulations of electron-capture core collapses (as well as of collapses of highly stripped iron cores) show that these collapses are expected to proceed very symmetrically, such that the NSs formed in these collapses are not expected to receive a kick velocity of more than a few tens of km s^{-1} at birth (Janka, 2017, and references therein). They are therefore expected to be quite easily retained in binaries. On the other hand, Janka (2017) shows that collapses of more massive iron cores are expected to impart a large range of kick velocities onto the NSs formed in them. For a detailed discussion about NS kicks and their effects on binaries, see Chapter 13.

8.2.2 Internal Evolution of Stars More Massive than $\gtrsim 12\,M_{\odot}$

The internal evolution of a star more massive than $\sim 12\,M_{\odot}$ is well illustrated by the evolution of the 15 $M_{\odot}$ star computed by Woosley et al. (2002). The durations of the subsequent burning stages of this star are listed in Table 8.1.

Figure 8.18 depicts the internal evolution of the star during these burning stages, until core collapse, and Figure 8.19 depicts the evolution of its central temperature and density through these burning stages. The short duration of the stages after the start of carbon burning is due to the large neutrino energy losses. Figure 8.18 shows that starting from central carbon ignition, there follow three stages of convective carbon shell burning, while partly simultaneously neon is ignited in the core, followed by convective core oxygen burning and two stages of convective oxygen shell burning, followed by convective core silicon burning and shell silicon burning. It is important to note here that the computed number of shell burning stages for each element is model dependent. All stars more massive than $\sim 10-12\,M_{\odot}$ show similarly complex core evolution before final core collapse.

The reasons for the collapse of the iron core can be described as follows. As one observes in Figures 8.3 and 8.19, the cores of stars with ZAMS masses $\geq 12\,M_{\odot}$, during and beyond carbon burning are not degenerate (they are located above the degeneracy region in the temperature vs. density diagram). For this reason, their collapse is not directly related to the Chandrasekhar mass, as is the case for electron-capture collapse, but is related to the precise physical processes taking place in a collapsing iron core, which may result in NS masses different from, and in some cases much larger than, the baryonic Chandrasekhar mass (and in some cases, even less than the Chandrasekhar mass [Timmes et al., 1996]).

Indeed, in some HMXBs, there are observed NS (birth) masses much larger than the Chandrasekhar mass, for example, $1.77 \pm 0.08\,M_{\odot}$ for Vela X-1 (Rawls et al., 2011) and $1.9-2.0\,M_{\odot}$ for IGR J17252-3616 (Manousakis et al., 2012), respectively. As the amount of accretion experienced by NSs in HMXBs is negligible (see Section 7.2), these masses are the birth masses of these NSs.

The processes leading to the collapse of the iron core to a NS can be briefly summarized as follows. ^{56}Fe, produced by ^{28}Si fusion, is the most stable nucleus known in the periodic system, with the highest nuclear binding energy (mass deficiency) per baryon (proton or neutron).

As a result, fusion of iron nuclei to heavier elements requires energy (endothermic) instead of producing energy (exothermic). Iron therefore is the end product of all nuclear fusion reactions in the star. The iron core can no longer produce energy by fusion to keep up the pressure and is forced to keep contracting continuously, converting gravitational potential energy into heat, as depicted in Figure 8.2.

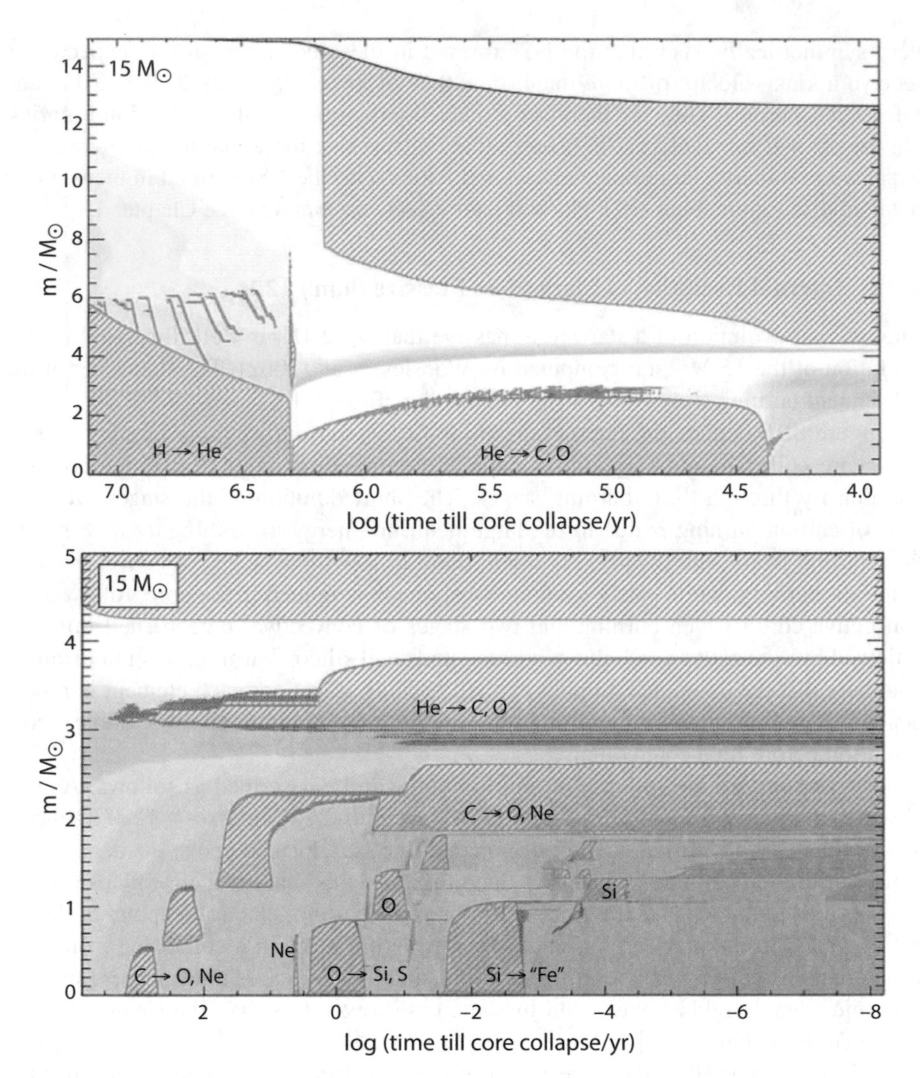

Figure 8.18. Kippenhahn diagram of the evolution of a 15 $M_\odot$ star (mass coordinate vs. logarithmic time in years until core collapse), showing convective regions (cross hatching) and nuclear-burning intensity (blue shading) during central hydrogen and helium burning (top panel) and during the late stages in the inner 5 $M_\odot$ of the star (bottom panel). A complex series of convective core-burning and shell-burning domains occur due to carbon burning (approximately $\log t \simeq 3$), neon burning ($\log t \simeq 0.6$), oxygen burning ($\log t \simeq 0$), and silicon burning ($\log t \simeq -2$). After Woosley et al. (2002).

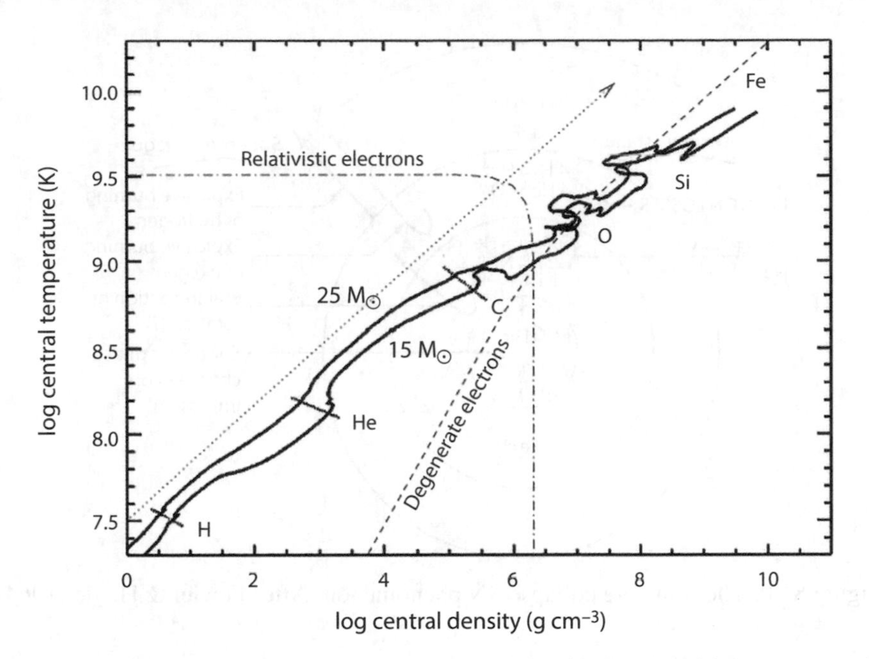

Figure 8.19. Evolution of central temperature and density in 15 and 25 $M_\odot$ stars with $Z = 0.02$ through all nuclear burning stages up to iron-core collapse, from the calculations by Woosley et al. (2002). The dashed line indicates where electrons become degenerate and the dash-dotted line where they become relativistic. The dotted line with arrow indicates the trend $T \sim \rho^{1/3}$ as expected from homologous contraction. Non-monotonic behavior is seen whenever nuclear fuels are ignited and a convective core is formed. Adapted by Pols (lecture notes, 2011) from Woosley et al. (2002).

The contraction starts at the end of silicon burning when the core temperature is $\sim 8 \times 10^9$ K, and the density is $\sim 4 \times 10^9$ g cm^{-3}. The contraction during its later stages is accelerated due to the photo-inonization of nuclei. At a temperature of $\sim 10^{11}$ K, the thermal gamma-rays break down the iron and other nuclei into alpha particles (helium nuclei) and neutrons:

$$\gamma + {}^{56}\mathrm{Fe} = 13\,\alpha + 4\,n - 124\,\mathrm{MeV}. \tag{8.20}$$

As this reaction requires energy at the expense of the thermal energy, it forces the core to contract even faster.

The next step is the breaking down of the alpha particles into protons and neutrons, which happens at a temperature of $\sim 2 \times 10^{11}$ K:

$$\gamma + {}^{4}\mathrm{He} = 2p + 2n - 28\,\mathrm{MeV}. \tag{8.21}$$

Because there is a larger number of particles at the right-hand side of this equation (higher entropy), an equilibrium state results that consists of protons and neutrons.

One observes that in this way the release of gravitational potential energy by the contraction of the core has broken down all the heavier nuclei that had been built up by fusion during the entire preceding lifetime of the star. By the time the core becomes

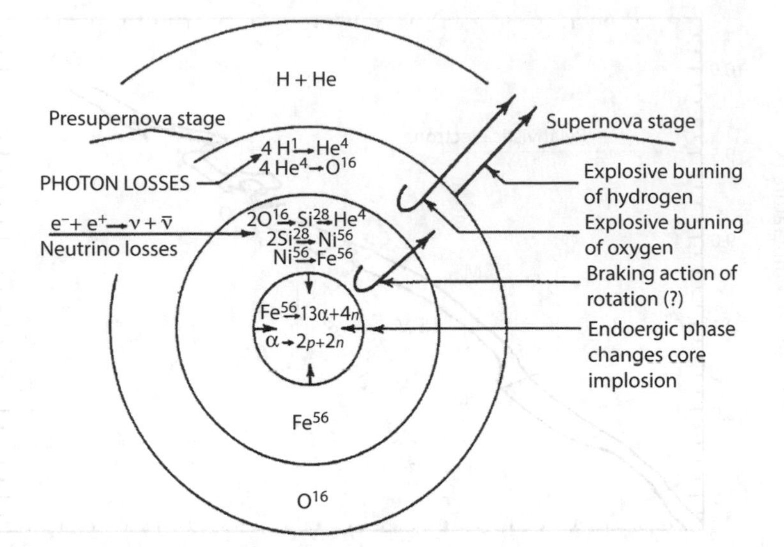

Figure 8.20. The iron-core collapse SN phenomenon. After Fowler & Hoyle (1964).

a NS, the total gravitational energy released is of the order of $0.1\,M_{\rm NS}\,c^2$, which is $\sim 3 \times 10^{53}$ erg. This exceeds by >1 order of magnitude the nuclear energy released during the entire lifetime of the star.

At the time of the last photo-ionization reaction, the pressure that supports the core against gravity is provided by the gas of electrons and protons (present in equal numbers) plus neutrons. At some point during the contraction, the electrons are captured by the protons, producing neutrons by the reaction

$$p + e^- = n + \nu_e. \tag{8.22}$$

This vanishing of the pressure causes the contraction to accelerate to free fall and the formation of the NS. During the free fall the matter reaches a velocity comparable to the escape velocity from a NS surface, which is close to $0.5\,c$. When the infalling matter of the inner part of the core reaches nuclear density, the repulsion due to the strong nuclear force halts the collapse, but due to the large infall velocity the matter gets compressed, causing it to bounce back, sending a shock wave into the still infalling matter of the outer core. It has turned out that the shock energy of $\sim 10^{51}$ erg is just sufficient to photo-ionize the infalling outer core matter and convert it into neutrons, but it is not enough energy to produce an explosion.

Most of the $\sim 3 \times 10^{53}$ erg in gravitational energy released in the collapse is converted into heat that is converted into neutrinos due to plasma processes and other neutrino-generating reactions. The total number of neutrinos produced largely outnumbers the $\sim 10^{57}$ neutrinos released by the capture of the electrons by the protons in the above-mentioned reaction.

As the cross section of neutrinos for interaction with matter is extremely small, most of the neutrinos escape into space, carrying off the bulk of the released energy.

Still, because the matter of the outer core is very dense, some neutrinos will interact with this matter on their way out and may deposit their energy here. To eject the matter of the infalling outer core and the envelope around it, requires some 10^{51} erg, which is the observed SN explosion energy (kinetic energy of the ejecta).

Much work of the 2010s on 2D and 3D hydrodynamic collapse models, particularly by the groups in Munich (Sukhbold et al., 2016; Summa et al., 2016; Summa et al., 2018) and Princeton (Burrows et al. 2019; Burrows et al., 2020), has demonstrated that the neutrinos can indeed cause accumulation of energy below the accretion shock until there is enough energy to drive an explosion. This takes ~500 ms. This is the *delayed neutrino-driven explosion*. If the mechanism fails, a BH will be formed. For a global comparison of core-collapse SN (CCSN) simulations, see, for example, O'Connor et al. (2018). As an alternative, it has also been proposed that the SN ejection can be driven by a fast-spinning magnetized newly formed NS. This mechanism, suggested by Ostriker & Gunn (1971), works only if the newborn NS is spinning with a period of a couple of ms and has a dipole field of $>10^{12}$ G.

8.2.3 Which Core Collapses Will Lead to SNe and NSs, and Which Ones to BHs?

As pointed out by Sukhbold et al. (2016), Sukhbold et al. (2018), Ebinger et al. (2019), and Curtis et al. (2019), observational estimates of explosion energies and pre-explosion masses are available for a number of CCSNe. A key example is SN 1987a, for which the ejected masses of ^{56}Ni to ^{58}Ni and of ^{44}Ti are available. Also, some information about failed SNe, which presumably leave BHs, is available: for example, Adams et al. (2017) found that a 25 $M_{\odot}$ ZAMS mass RSG star has disappeared without exploding. In Table 8.3 we display a rough mapping between initial ZAMS masses of single massive stars and their final outcome as NSs or BHs. However, it has become clear during the 2010s, and as we now discuss in more detail, that the picture of a one-to-one mapping is too simplified and reality is more complex.

As we have mentioned, understanding the complex problem of the explosion mechanism of CCSNe still has not fully been solved. Nevertheless, the 2D and 3D CCSN simulations that became available in the period about 2010 to 2020 (e.g., Janka et al., 2016; Müller et al., 2019; Burrows et al., 2020, and references therein) have provided important new insights to the various factors that may lead to a successful explosion. While it was originally thought that there is a one-to-one relation between ZAMS stellar mass and outcome of the core collapse, works from the 2010s through the early 2020s (Ugliano et al., 2012; Pejcha & Thompson, 2015; Müller et al., 2016; Sukhbold et al., 2016; Sukhbold et al., 2018; Ebinger et al., 2019; Curtis et al., 2019; Ebinger et al., 2020) have shown that there is no such linear relationship. These authors used a large set of stellar models using various one-dimensional neutrino-transport models with artificially enhanced neutrino heating for the explosion, calibrated by connecting with what we know about SN1987A and, in some cases, other observational constraints. For example, Müller et al. (2016) cautioned that SN1987A might not provide a good calibration point because this SN might be the result of a binary merger (Podsiadlowski et al., 1990). The modelling is based on the approach by Ugliano et al.

(2012) for simulating neutrino-driven SNe in spherical symmetry; see also Ertl et al. (2016).

It appears that due to the complex structure of the various burning shells in the pre-collapse stellar core (as depicted in Fig. 8.18), the ZAMS mass is not necessarily a good predictor of whether a successful SN explosion will take place. Several authors have investigated the connection between the final pre-collapse structure of the inner parts of the massive star and the outcome of the neutrino-driven explosions in effective explosion models (O'Connor & Ott, 2011; Ugliano et al., 2012). O'Connor & Ott (2011) showed that the so-called *compactness* ξ_M of the pre-collapse stellar core of mass M is an important parameter for predicting which cores will lead to a successful explosion and which cores will not.

The compactness of the pre-explosion core is defined as

$$\xi_M = \frac{M/M_\odot}{R(M)/1{,}000\,\mathrm{km}}, \tag{8.23}$$

where M is a given pre-explosion core mass and $R(M)$ is the radius of this mass. Ebinger et al. (2019), Curtis et al. (2019), and Ebinger et al. (2020) use $M = 2\,M_\odot$, while Sukhbold et al. (2016), Sukhbold et al. (2018) use $M = 2.5\,M_\odot$, which both appear to be adequate choices. The compactness characterizes the innermost structure of the pre-explosion models, which is strongly influenced by the complex pre-explosion burning regions. (Note that, although O-shell burning can now be modelled routinely in 3D just before the very moment of core collapse [e.g., Müller (2020) and references therein], the progenitor star models are only calculated in 1D up until a few minutes before the SN.) Ebinger et al. (2019) find that for approximately $\xi_{2.0} \lesssim 0.4$, models tend to explode successfully and leave a NS (the real mass dependency of $\xi_{2.0}$ for exploding stars is somewhat more complex, see the original paper), whereas for larger compactness values the cores tend to collapse to BHs.

We now describe results from some of the presently best simulations that used the compactness criterion for studying the fate of massive stars. As remarked by Woosley (2019), for example, the compactness parameter is a too simple representation of core structure, but still it is known to correlate with other measures of explodability (e.g., Sukhbold et al., 2018). The inclusion of effects due to rotation and magnetic fields on core structure evolution is at a very early stage, and to study large grids of models would require extremely large amounts of computer time, which, at the time of writing, is not possible. Despite these limitations, impressive results have been achieved, which we now briefly summarize.

The results from the extensive core-collapse simulations by Müller et al. (2016), Sukhbold et al. (2016), Sukhbold et al. (2018), Ebinger et al. (2019), and Ebinger et al. (2020) show that, as a function of ZAMS mass, there are islands of "explodability." For example, for solar metallicity ($Z = Z_\odot$), these computations show that in the ZAMS mass range[5] of $22 - 25\,M_\odot$, the core compactness is very high, and the outcome of

[5]Müller et al. (2016) find slightly lower ZAMS mass range of $20 - 23\,M_\odot$ for the first region of BH formation, because of updated neutrino rates for the advanced burning stages.

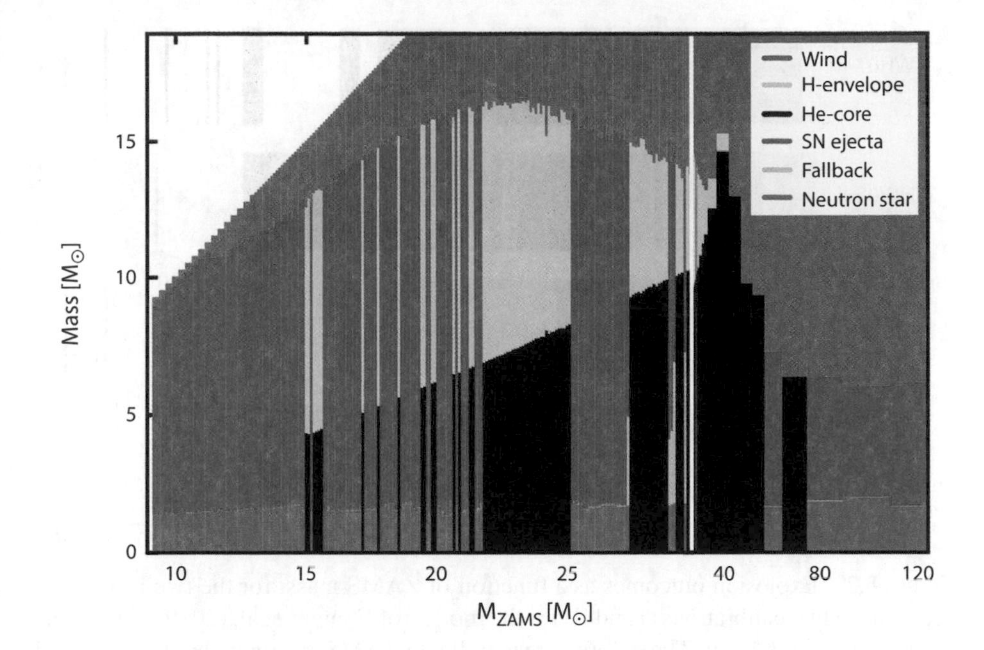

Figure 8.21. Results from CCSNe for single stars with ZAMS masses of $9-120\,M_{\odot}$ for solar metallicity. Black indicates BH remnants and the height of the vertical bars indicates BH mass, which is assumed to be equal to that of the helium core. Green represents NS remnants. Light-blue represents BHs formed by fallback. Orange represents SN ejecta, and yellow represents the hydrogen rich envelope. Gray represents mass lost by wind: stars more massive than $20\,M_{\odot}$ lose a large part of their mass by wind. New here is that there is no one-to-one relation between ZAMS mass and outcome of the core collapse: even a $15\,M_{\odot}$ star can produce a BH, while an $80\,M_{\odot}$ star can still leave a NS. (It should be noted that for single stars with masses larger than $\sim 30\,M_{\odot}$, the wind mass-loss rates are probably considerable overestimates, and thus the outcomes are uncertain.) Further explanation in the text. After Sukhbold et al. (2016).

core collapse is in most cases the formation of a BH, whereas for masses $<22\,M_{\odot}$, most collapses are successful and lead to NSs. This is illustrated in Figures 8.21 and 8.22. The large majority of successful explosions leave a NS, but some, by fallback, leave a BH. The great majority of the BHs are in these simulations formed by direct collapse, not by fallback, as Figure 8.21 illustrates.

One sees from Figure 8.21 that a few stars with masses as low as $15\,M_{\odot}$ still terminate as a BH. The figures result from the calculation of models for thousands of ZAMS mass values and show that there seems to be some randomness in the precise outcome of the core collapses. This is due to interactions between burning layers just in the pre-collapse stage (Sukhbold et al., 2018), which may be different when the ZAMS mass changes by only a small amount.

Ebinger et al. (2019) as well as Sukhbold et al. (2016) based their calculations for $Z = Z_{\odot}$ on two sets of input models, from Woosley et al. (2002) (which Ebinger

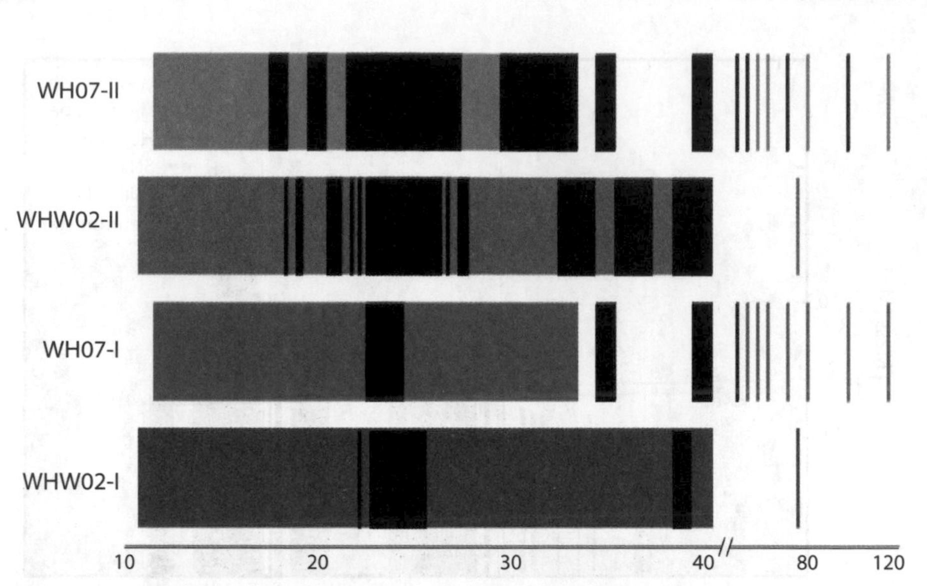

Figure 8.22. Explosion outcomes as a function of ZAMS mass, for the two progenitor sets and the two calibrations (I and II) for the models of Ebinger et al. (2019); all models are for solar metallicity. The colored areas indicate ZAMS intervals that leave behind NSs, and the black areas indicate BH remnants. Notice that with the standard calibration (I) only stars with ZAMS masses between ~22 and 25 $M_{\odot}$ leave BHs while most other stars with $M \leq 40\, M_{\odot}$ leave NSs. On the other hand, with the second calibration (II), BHs are already formed from stars greater than ~ 17 $M_{\odot}$ and in a broad region between ~21 and about ~28 $M_{\odot}$. The differences between these outcomes and those of Müller et al. (2016) and those of Sukhbold et al. (2016) and Sukhbold et al. (2018) may primarily be due to somewhat different assumed stellar wind mass-loss rates (see text), but other input factors may also play some role. After Ebinger et al. (2019).

and colleagues call the WHW02 series of models) and from Woosley & Heger (2007) (which Ebinger and colleagues call the WH07 models, which are illustrated by Fig. 8.22). Furthermore, in their simulations, Ebinger and colleagues made use of the availability of quite extensive observational information on energies and pre-explosion masses of CCSNe to calibrate their theoretical models. They used two possible types of calibrations; of these two, the *standard* calibration, which uses core compactness $\xi_{2.0}$, gives a good fit to all observed CCSN observations. These models are indicated as WHW02–I and WH07–I. The *second* calibration uses $\xi_{1.75}$ and these models are indicated as WHW02–II and WH07–II. Figure 8.24 illustrates the successful fit of the model outcomes of the standard calibration with the CCSN observations. The results of Ebinger et al. (2019) and Sukhbold et al. (2016) imply that single stars of solar metallicity with masses $\gtrsim 30\, M_{\odot}$ mostly will leave NSs as remnants and relatively rarely will leave BHs. It is very likely, however, that this result is due to the assumed large stellar wind mass-loss rates of massive stars. This holds in particular for the winds of massive helium stars (WR-stars) used in the input models taken from Woosley et al. (2002) and

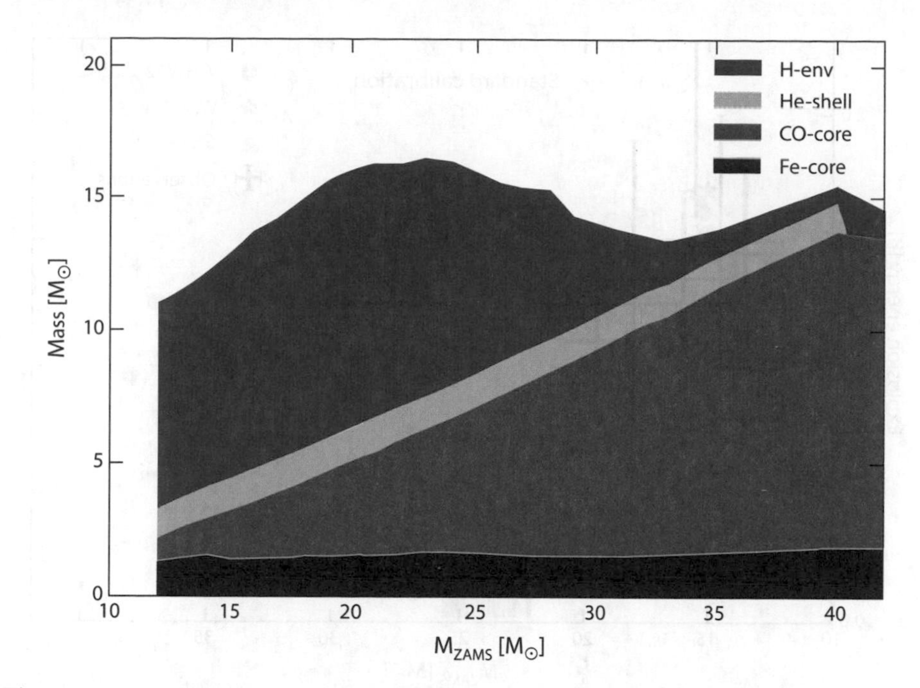

Figure 8.23. Iron-core mass (black), CO-core mass (blue), helium-shell mass (yellow), and hydrogen envelope (red) are shown as functions of ZAMS mass at the onset of core collapse for the progenitor models computed by Woosley & Heger (2007), which were used for the WH07 core collapse models of Ebinger et al. (2019). After Ebinger et al. (2019).

Woosley & Heger (2007). These models used wind mass-loss rates given by estimates from Wellstein & Langer (1999), which for massive WR-stars (helium stars more massive than $\sim 8\,M_\odot$) in later years have been found to be overestimated by a factor that can be as large as an order of magnitude—mainly due to clumping effects (e.g., Nugis & Lamers, 2000; Yoon, 2017; Sander & Vink, 2020). The effects of the large wind mass-loss rates assumed by Ebinger et al. (2019) and Sukhbold et al. (2016) for ZAMS masses of $\geq 30\,M_\odot$ and solar metallicity can be clearly seen in Figures 8.21 and 8.23. More recently, Sukhbold et al. (2018) and Woosley (2019) used revised empirical wind mass-loss rates. Woosley (2019) used the WR mass-loss rates compiled by Yoon (2017) to calculate the final evolution and fate of massive stars. He finds that for helium stars originating from stars with ZAMS masses $\gtrsim 30\,M_\odot$ the fractions terminating as NSs and BHs are quite different from those derived by Ebinger et al. (2019) and Sukhbold et al. (2016) and that notably more stars finish as BHs. On the other hand for stars with ZAMS masses $\lesssim 30\,M_\odot$, wind mass-loss rates are only modest and no substantial revisions of wind loss rates are needed such that the results from Ebinger et al. (2019) as well as from Sukhbold et al. (2016) remain valid for these relatively lower mass stars.

The following gives an illustration of the effects of using different wind mass-loss recipes for massive stars. In the models of Ebinger et al. (2019), in total, taking into

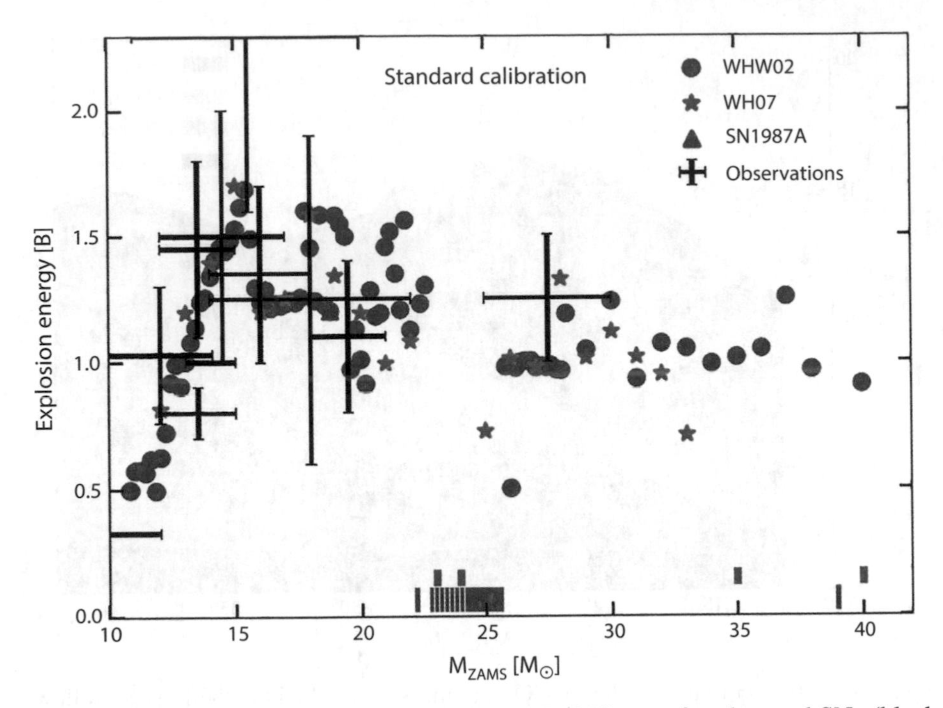

Figure 8.24. Explosion energies as a function of ZAMS mass for observed SNe (black crosses with error bars) for pre-explosion models from Woosley et al. (2002) (blue circles) and Woosley & Heger (2007) (green stars) and computed by Ebinger et al. (2019) for their standard calibration. The vertical dashes at the bottom of the figure indicate masses for which BHs were formed. The explosion energy unit is Bethe, $B \equiv 10^{51}$ erg. After Ebinger et al. (2019).

account the shape of the IMF in the simulations for $Z = Z_\odot$ and the standard calibration, the core collapses of single stars in the ZAMS mass range of 8–150 $M_\odot$ leave BHs between 5% (for the WHW02–I models) and 8% (for the WH07–I models) of the cases; for the WHW02–II and the WH07–II models, these percentages are ~16% and 21%, respectively. On the other hand, for the simulations by Sukhbold et al. (2018) this fraction is 35% for all stars with ZAMS mass of >9 $M_\odot$ and 92% for all stars more massive than 18 $M_\odot$.

In the standard models of Ebinger et al. (2019) for $Z = Z_\odot$, the BH masses produced from stars in the ZAMS mass range of 20 – 30 $M_\odot$ peak at ~14 $M_\odot$ for ZAMS mass of ~25 $M_\odot$. However, when the model is weighted with the IMF, the average BH mass is consistent with the average observed value in X-ray binaries of 7.8(± 1.2) $M_\odot$ obtained by Özel et al. (2010), if it is assumed that the entire helium core collapses to a BH. The average mass of the BHs produced in the simulations of Sukhbold et al. (2018), weighted with the IMF distribution of ZAMS masses and assuming that the entire helium core collapses into a BH, is 6.6 $M_\odot$.

Although there are differences in the predictions of BH production between these two sets of models, the predicted average BH masses do not differ much. This is because the outcomes for ZAMS masses of $<30\,M_\odot$ are quite similar, while due to the IMF, these masses are weighted the most when determining the average BH mass. The remaining differences must mostly be due to the fact that, as a result of different wind-mass loss recipes, in the models of Ebinger et al. (2019) the stars $>30\,M_\odot$ have hardly any hydrogen envelope left, whereas in the models of Sukhbold et al. (2018) there still is a significant hydrogen envelope left. For further discussions of *islands of explodability* in the mass spectrum (leaving NS remnants rather than BHs), see, for example, Patton & Sukhbold (2020).

It should still be kept in mind that in spite of the great advances in core-collapse modelling, the above-described models are not yet definitive solutions to the core-collapse problem, because they are not fully 3D and do not include rotation and magnetic fields, which we know from NS observations to be quite crucial parameters. (However, see recent work by Bugli et al. [2021], who performed 3D simulations of core-collapse SNe with realistic magnetic structures, such as quadrupolar fields and a tilted dipolar field.) For this reason, the judgment about which of these simulations are the most realistic ones should come from a comparison of their predictions with observations.

With respect to the observed massive merging BHs observed by LIGO, it is important to realize that the BH masses predicted by evolution of stars of very low metallicity ($Z < 10^{-3}\,Z_\odot$) can reach considerably larger masses, approaching $M_{\rm BH} \simeq M_{\rm ZAMS}$, because stellar wind mass loss is very weak for such small metallicities. See further discussion in Chapter 15.

We close this section with an insightful diagram giving an overview of the final products of the evolution of massive single stars as a function of ZAMS mass and metallicity given by Heger et al. (2003), as depicted in Figure 8.25. Although the detailed results of Ebinger et al. (2019) and Sukhbold et al. (2018) are somewhat different, this figure still gives a simple approximate overview of how ZAMS mass and metallicity determine the outcome of the evolution. It should be kept in mind, however, that the wind mass-loss rates of stars more massive than $\sim 30\,M_\odot$ that were used for constructing the figure were overestimated by at least a factor of two.

8.2.4 About the Stochastic Relation between ZAMS Stellar Mass and Resulting NS Mass

The results of the above-described core-collapse simulations by Müller et al. (2016), Sukhbold et al. (2018), and Ebinger et al. (2019), for the mass range up to $\sim 27\,M_\odot$, are globally not greatly different, despite slightly different definitions of the compactness parameter. In both types of simulations, there is a large scatter in the compactness parameter of the pre-SN stellar cores in the region between ZAMS masses of 14 and $19\,M_\odot$. Also between approximately 22 and $25\,M_\odot$, where the compactness reaches a peak, there is still quite some scatter, as is clear from Figure 8.26.

In the 1990s, Timmes et al. (1996) noticed that at $\sim 19\,M_\odot$, coming from the lower masses, there is the transition in resulting iron core mass from *convective* central carbon burning to *radiative* central carbon burning at the higher masses. At the higher masses, the energy produced by carbon burning does not exceed the neutrino losses in the star's

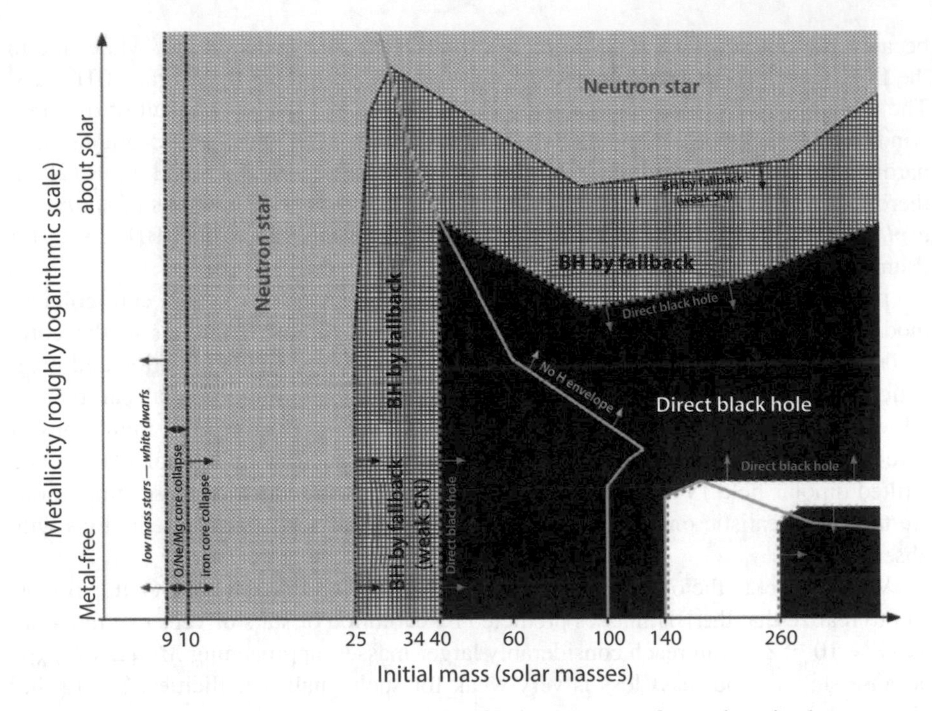

Figure 8.25. Approximate overview of the final remnants of massive single stars as a function of ZAMS mass and metallicity. Although for some values the depicted outcomes differ slightly from those computed more recently, for example, by Sukhbold et al. (2016), Sukhbold et al. (2018), and Ebinger et al. (2019), this diagram gives a simple and insightful impression of how the type of final product roughly depends on initial stellar mass and metallicity. In the very left part of the diagram, WD remnants are produced and for slightly larger ZAMS masses NSs are produced by electron-capture collapse. In the lower-right white part of the diagram, the entire star explodes in a PISN (Section 15.5.1) and no remnant is left. After Heger et al. (2003).

center. As a result, the core shrinks and moves very quickly to oxygen burning, and this causes the burning in shells to behave differently from the burning that is happening below this transition mass. As remarked by Sukhbold et al. (2018), this seems to cause the drastic transition in compactness near 20 $M_{\odot}$, as seen in Figure 8.26.

It is important to stress also that the SN explosion itself comes with an element of stochasticity. For example, the actual magnitude of a NS kick depends on the explosion asymmetry, which develops stochastically (Wongwathanarat et al., 2013). The statistical probability distribution of this stochastic asymmetry should therefore be determined using numerous explosion simulations.

Because of the stochastic outcome of the collapse process, as a function of ZAMS mass, one can only predict statistically, on the basis of computing the evolution of a large number of stars of different masses until the final core collapse, what is the probability of stars in a certain mass range to finish as a NS or as a BH. Sukhbold

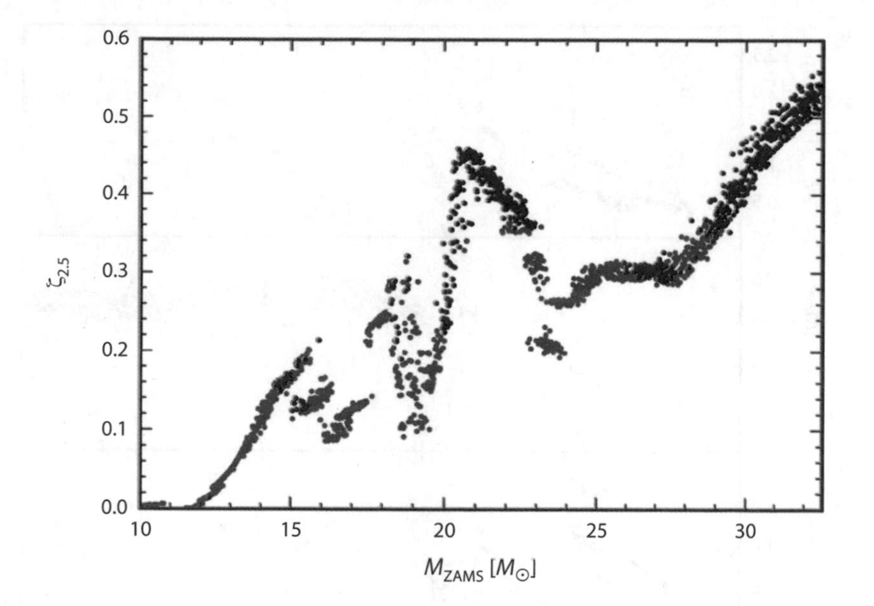

Figure 8.26. Compactness parameter of solar metallicity models with ZAMS masses between 12 and 32.5 $M_\odot$, color-coded according to the success or failure of the explosion, from Müller et al. (2016). Red symbols denote successful explosions according to the semi-analytical criterion of these authors. If the criterion formulated by Ertl et al. (2016) would be used, some more BHs are produced. Black dots denote BHs formed by direct collapse. Blue dots denote models where shock revival is initiated but the explosion finally fails and a BH forms by fall back. (See also Fig. 15 of Sukhbold et al. [2018].) After Müller et al. (2016).

et al. (2018) did this by computing the evolution of 4,000 stars of solar metallicity with masses between 12 and 60 $M_\odot$. The fraction of BHs that they predicted were already mentioned. They furthermore found that the average gravitational mass of the NSs produced is 1.38 $M_\odot$, although there is a large spread in the obtained NS masses, plotted as a function of the initial ZAMS mass. Such a plot, derived from the computations by Müller et al. (2016) is depicted in Figure 8.27, where one observes that in these models (and those of Sukhbold et al. [2018]) the largest NS gravitational masses, between 1.8 and 2.0 $M_\odot$, surprisingly, are produced in the relatively low-mass ZAMS range of approximately $14-15$ $M_\odot$, where, in these simulations, some BHs are also produced.

These large NS masses appear to be related with the presence of a significant second oxygen-burning shell. Furthermore, in these models, the stars in the ZAMS mass range of $22-27$ $M_\odot$ that do not collapse to BHs produce fairly massive NSs, with gravitational masses of $\sim$1.65 $M_\odot$. In binaries, the initial ZAMS masses for producing these massive NSs will be larger because, due to envelope stripping, the helium cores of their primary stars cannot further grow in mass, contrary to the case of single stars. So, a primary star in a binary of $\sim$20 $M_\odot$ may leave a helium star with the same mass as the final core of an $\sim$$14-15$ $M_\odot$ single star.

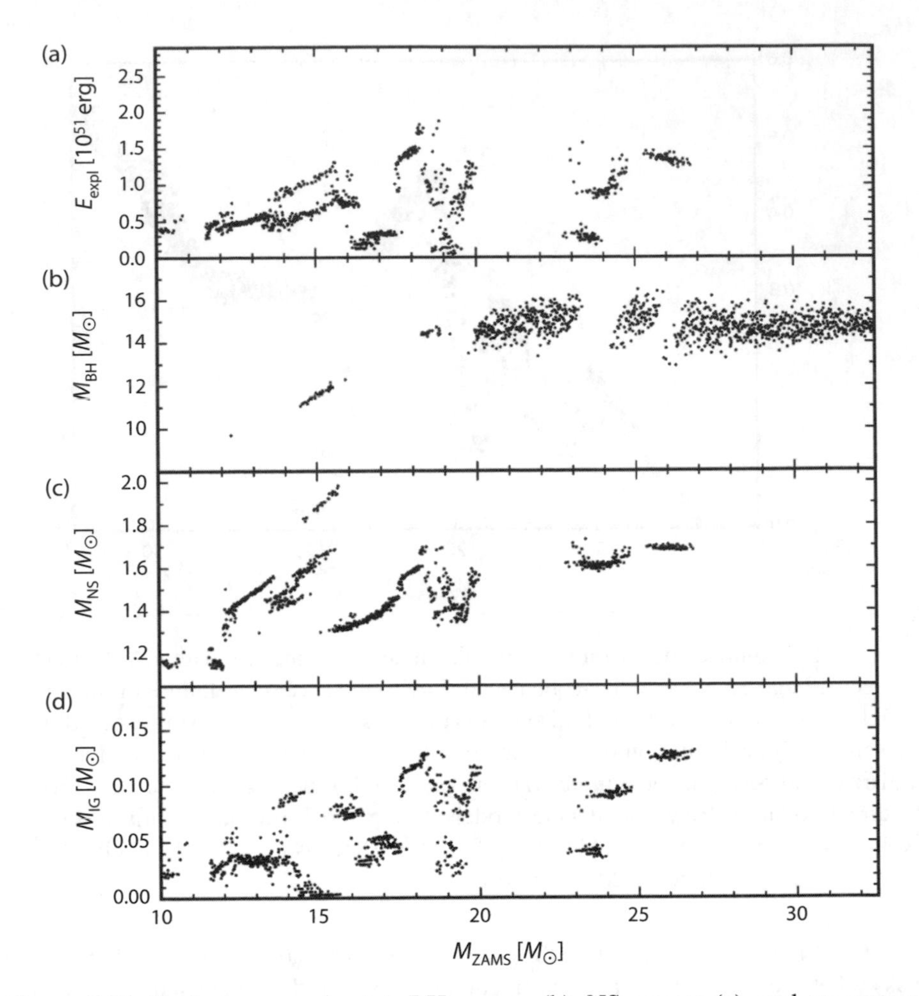

Figure 8.27. Explosion energies (a), BH masses (b), NS masses (c), and amounts of iron-group elements produced (d), resulting from the models computed by Müller et al. (2016) for stars in the mass range of 10–32.5 $M_{\odot}$ with normal stellar wind mass loss. The masses are the gravitational masses of the NSs and BHs. The most massive NSs on this plot come from models with $M_{ZAMS} \simeq 14-15\, M_{\odot}$, which are the most massive models with a significant second oxygen-burning shell. The existence of multiple branches of core models results in a large range of NS masses being accessible by stars of nearly the same ZAMS mass (see also Fig. 17 of Sukhbold et al. [2018]). The gap near 11 $M_{\odot}$ is due to the lack of models around this mass, not due to the formation of BHs. After Müller et al. (2016).

For further discussions on SN progenitor models and hydrodynamics of CCSNe, we refer to the recent review by Müller (2020). In Chapter 13 we discuss SNe in more general terms and particularly focus on SNe in binaries.

8.2.5 Failed SNe

Thus far, we considered formation of NSs and BHs via successful SNe from the gravitational collapse of massive stars. Numerical studies agree that a progenitor with a more compact inner core is likely to fail to revive after the bounce shock (O'Connor & Ott, 2011). These failed SNe will be observed as massive stars suddenly disappearing in the sky on BH formation, but the complete picture of their observational signatures is not yet known. There may still be some observational features to detect. The outcome may well depend on the rotation of the collapsing star (Pan et al., 2021). Failed SNe from stars with rapid or moderate rotation are expected to leave accretion disks around the BHs, from which transients such as γ-ray bursts, fast luminous transients, or intermediate-luminosity red transients are generated (e.g., Woosley, 1993; MacFadyen & Woosley, 1999; Kashiyama & Quataert, 2015; Quataert et al., 2019; Tsuna et al., 2020), depending on the physical conditions of, and near, the progenitor star (e.g., blue supergiant, stripped star, or WR-star). Collapsing progenitor stars with slow rotation may generate a sound pulse from the sudden release of gravitational binding energy ($\sim$10% of the core's rest mass) due to neutrinos escaping during the proto-NS phase prior to BH formation. This sound pulse can produce a shock and unbind the outer envelope of the star on shock breakout (e.g., Nadezhin, 1980; Lovegrove & Woosley, 2013; Coughlin et al., 2018).

8.3 EVOLUTION OF HELIUM STARS

Helium stars are basically naked cores of former hydrogen-rich stars that have lost their envelope due to mass loss in a strong stellar wind or via RLO to a companion star. A pioneering study on such helium stars is Paczyński (1971a), followed, for example, by the studies of Biermann & Kippenhahn (1971), Arnett, (1972a, 1972b, 1974), De Greve & De Loore (1976), Arnett (1978), Iben & Tutukov (1985), Habets, (1986a, 1986b), Dewi et al. (2002), Dewi & Pols (2003), Ivanova et al. (2003), Tauris et al. (2013), and Tauris et al. (2015). Most of these works focused on low- and intermediate-mass helium stars in binaries. If a helium star is more massive than $\sim 8\,M_{\odot}$ it is often observable as a WR-star (Crowther, 2007), as discussed in Chapters 5 and 6.

For a recent study of the general evolution and fate of single helium stars with masses of $1.6 - 120\,M_{\odot}$, see Woosley (2019). Since the evolution of helium stars less massive than $\sim 5\,M_{\odot}$ is quite complex, with effects of degeneracy and radius expansion etc., and since also the evolution of such lower-mass helium cores inside stars with hydrogen-rich envelopes may lead to complications, we separately discuss in the following sub-sections the evolution of lower-mass helium stars and that of the more massive helium stars.

8.3.1 Evolution of Helium Stars $\lesssim 5\,M_\odot$

For low-mass stars, the evolution of the helium core in post-main-sequence stars is practically independent of the presence of an extended hydrogen-rich envelope. However, for more massive stars ($> 2.3\,M_\odot$) the evolution of an *embedded* (or *clothed*) core of an isolated star differs from that of a *naked* helium star—see Section 13.4.3 and the list of arguments given in, for example, Brown et al. (2001), Podsiadlowski et al. (2004), Tauris et al. (2011), and references therein.

A simple relation between radii of helium stars on the helium-star ZAMS and their masses (for a chemical composition of $Y = 0.97$ and $Z = 0.03$) was given in one of our earlier publications (Tauris & van den Heuvel, 2006), based on a fit to calculations by O. Pols (2002, private communication):

$$R_{\rm He} = 0.212\,(M_{\rm He}/M_\odot)^{0.654}\ R_\odot. \tag{8.24}$$

It is important to realize that helium cores in binaries have residual envelopes of hydrogen when they detach from RLO. This has important effects on the subsequent radial evolution of both massive pre-SN stars (e.g., Laplace et al., 2020) as well as low-mass helium stars with only tiny ($< 0.01\,M_\odot$) envelopes of hydrogen (e.g., Han et al., 2002).

As discussed earlier in this chapter, prior to about year 2000, wind mass-loss rates of the more massive helium stars (WR-stars) were considerably overestimated. Since then more realistic values of the mass-loss rates of these stars were estimated, for example, by Nugis & Lamers (2000); Yoon (2017); Sander & Vink (2020).

A reasonable analytic fit to the wind mass-loss rates of helium stars is, for example, given by Dewi et al. (2002):

$$|\dot{M}_{\rm He,\,wind}| = \begin{cases} 4.0\times10^{-37}\,(L/L_\odot)^{6.8} & M_\odot\,{\rm yr}^{-1} \quad \text{for } \log(L/L_\odot) < 4.5 \\ 2.8\times10^{-13}\,(L/L_\odot)^{1.5} & M_\odot\,{\rm yr}^{-1} \quad \text{for } \log(L/L_\odot) \ge 4.5 \end{cases}. \tag{8.25}$$

A precise knowledge of the wind mass-loss rates of the higher-mass helium stars is important for determining the threshold mass for core collapse into a BH (see our detailed discussions in the previous Section 8.2.3, as well as in Chapters 9 and 13). Woosley (2019) uses in his studies of helium star evolution the wind mass-loss rates given by Yoon (2017), which are expected to be quite reliable.

Figures 8.28 and 8.29 show the evolutionary tracks of helium stars with masses between 0.5 and 28 $M_\odot$ in the HR diagram, and the internal evolution of a 3.2 $M_\odot$ helium star, respectively. One observes in Figure 8.28 that, while having radii of not much more than 0.5 $R_\odot$ on the He-ZAMS, during their later evolution the radii of helium stars less massive than $\sim$3.5 $M_\odot$ expand to quite large dimensions, which may cause such a star in a binary to overfill its Roche lobe and lose considerable mass by mass transfer, as we shall discuss in the next chapters in more detail. The 3.2 $M_\odot$ helium star in Figure 8.29 corresponds to the naked core of a hydrogen-rich star of $\sim$12 $M_\odot$. The figure shows that the heavy-element (metal) core of the star has a mass of $\sim$2.2 $M_\odot$, which means that it certainly will undergo core collapse and leave a NS

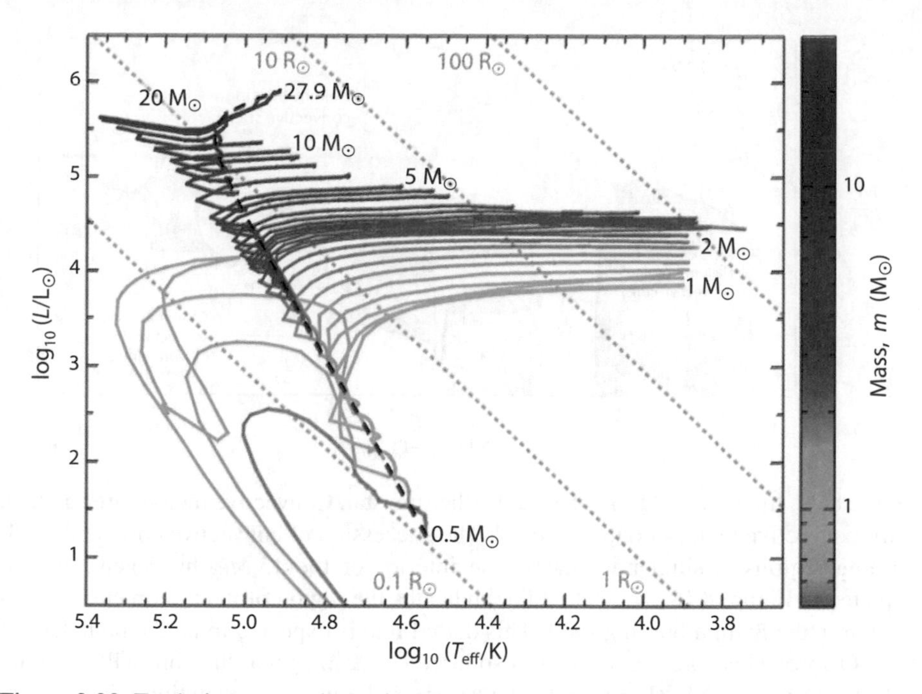

Figure 8.28. Evolutionary tracks in the HR diagram of helium stars with MW metallicity as calculated by Kruckow et al. (2018). The mass along the tracks is color-coded. The gray dotted lines indicate the stellar radii. The black dashed line is the He-ZAMS. One observes that the envelopes of the lower-mass helium stars expand to very large radii (see also Fig. 8.30). In a binary, this leads to so-called Case BB mass transfer, as discussed in Chapters 9 and 10. After Kruckow et al. (2018).

(see also Woosley, 2019). The evolution of binaries with very low-mass helium stars ($0.5 - 1.4\ M_\odot$) is discussed in Section 11.3.9.

Habets (1986b) found that from carbon ignition onward, the evolution of the central parts of helium stars more massive than $\sim 2\ M_\odot$ follows the same pattern alternating convective central and shell burning as seen in hydrogen-rich stars, as depicted in Figure 8.18. As the mass of the metal cores of helium stars more massive than $\sim 2.2\ M_\odot$ is larger that the Chandrasekhar mass ($\sim 1.4\ M_\odot$), Habets (1986a) expected that the final remnants of helium stars more massive than $\sim 2.2\ M_\odot$ will be NSs or BHs. This can be compared with the results of Tauris et al. (2015), discussed below, and Woosley (2019). If helium stars do not lose their envelopes, Woosley (2019) found that at mass $1.9\ M_\odot$ electron capture collapse sets in, leading to a SN and formation of a NS. He also found that if the helium stars lose their envelopes (e.g., by binary mass transfer), then at masses $\lesssim 2.4\ M_\odot$, core collapse by electron capture can occur. He further found helium stars more massive than $\sim 2.4\ M_\odot$ to develop collapsing iron cores, with or without envelope loss. (In fact, the precise evolution of the cores of the helium stars in the mass range of 1.9 to $\sim 3.2\ M_\odot$ is very complex, due to the occurrence of degeneracy, flashes and electron capture, and will be discussed in more detail in Chapter 13.)

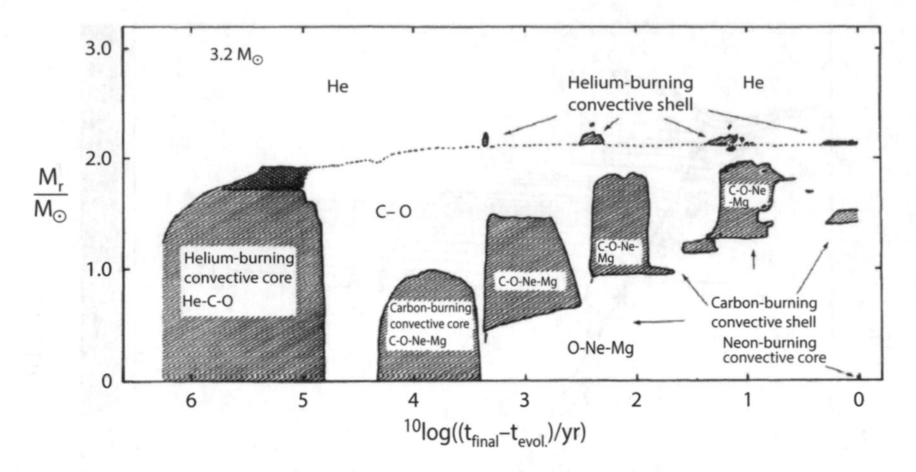

Figure 8.29. Internal evolution of a 3.2 $M_{\odot}$ helium star. Convective regions are hatched, semi-convective regions doubly hatched. The succession of convective core- and shell-burning regions is similar to that in the interior of the 15 $M_{\odot}$ hydrogen-rich star depicted in Figure 8.18. The dotted line indicates the region of maximum energy generation in the helium-burning shell. This dotted line is expected to be the boundary of the CO core. This core mass becomes so large ($\geq 2.2\,M_{\odot}$) that the core will certainly collapse to form a NS. (For recent calculations of helium star evolution with roughly similar results, see Woosley [2019].) After Habets (1986b).

Figure 8.30 shows the evolution of naked helium stars with initial masses between $2.6-4.5\,M_{\odot}$. We calculated these models using the Binary Evolution Code (BEC), developed by Norbert Langer and his group, assuming a chemical composition of $Y=0.98$ and $Z=0.02$, a mixing-length parameter of $\alpha=l/H_{\rm p}=2.0$, and a stellar-wind prescription adopted from Wellstein & Langer (1999) (for these lower-mass helium stars that wind prescription is expected to be adequate). For further description of calculations similar to those presented here, see Tauris et al. (2015). It is seen in the lower panel that helium stars with masses $M_{\rm He}>3.2\,M_{\odot}$ do not expand as much as their lower mass equivalents do. The substantial radial expansion of helium stars depicted here with $M_{\rm He}=2.6-3.2\,M_{\odot}$ (some even expand to giant dimensions exceeding 100 $R_{\odot}$) is very important for our later discussions regarding Case BB RLO (Chapter 10) and the formation of DNS systems via electron-capture SNe and iron-core collapse SNe (Chapter 13).

For comparison, Figure 8.31 shows the early computations of Habets (1986a) on the evolution of the radii of helium stars with masses from 2.0 to 4.0 $M_{\odot}$, up to oxygen ignition (at point E). The helium stars depicted in this figure, and in the same mass range in Figures 8.28 and 8.30, have only moderate wind mass-loss rates and the very high effective temperatures during most of their life mean that they spend most of their life as far-UV emitters. Such UV-emitting moderate-mass helium stars are expected to be formed in considerable numbers by binary evolution, as we will discuss in the remainder of this book. Götberg et al. (2018) and Smith et al. (2018) identified a

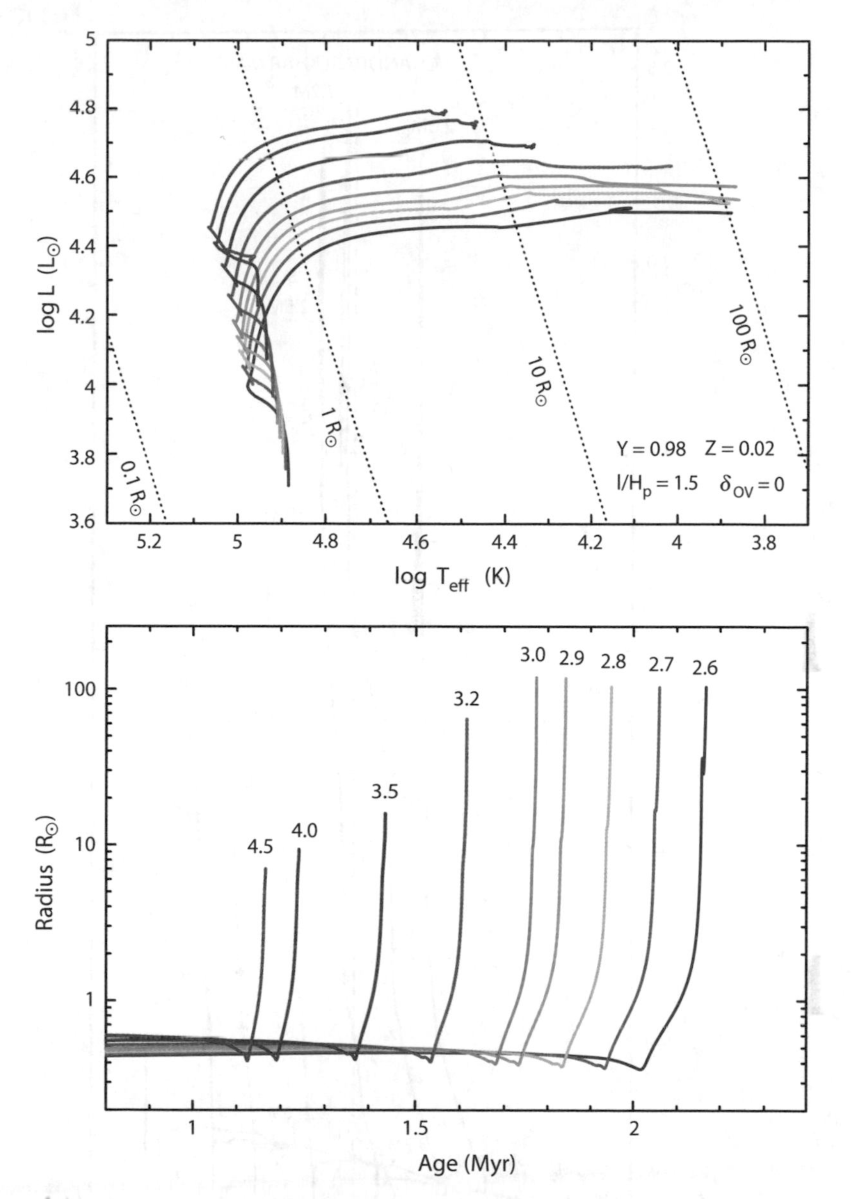

Figure 8.30. HR diagram (upper) and radius vs. age (lower) of the evolution of an intermediate-mass range of single, naked helium stars calculated with the BEC code of Norbert Langer, see text. Their initial masses are $M_{\mathrm{He}} = 2.6 - 4.5\ M_{\odot}$. (Color code for each track is given in the lower panel.)

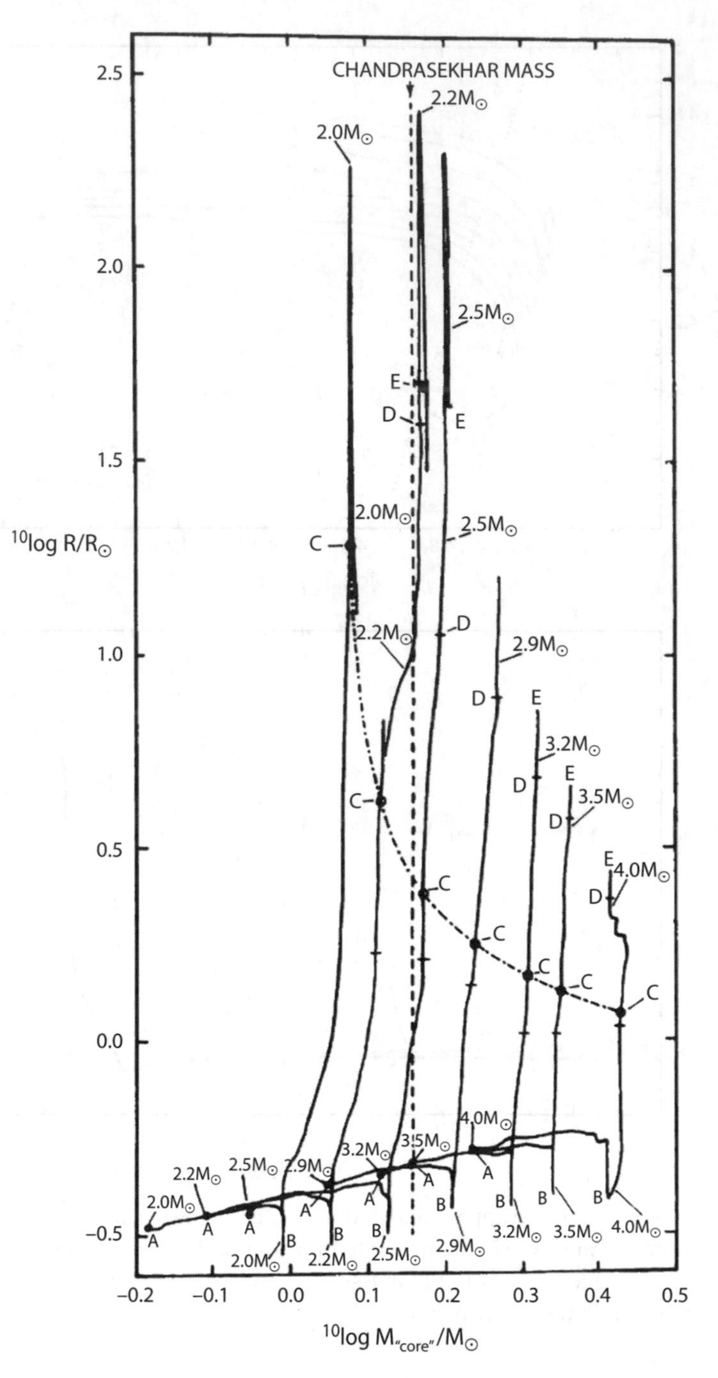

Figure 8.31. (see caption at bottom of the next page)

number of known objects with such helium stars, an important new discovery, which will also be described in Chapter 9. Helium stars more massive than $\sim 3.5\,M_\odot$ never evolve to large radii, as can be seen in Figure 8.30, which means that in binaries they rarely overflow their Roche lobes.

Our calculations of single helium stars in Figure 8.30, using the BEC code, result in iron-core collapse SNe for $M_{\rm He} \geq 2.8\,M_\odot$, electron-capture SNe for $2.7 \geq M_{\rm He} \geq 2.6\,M_\odot$, and ONeMg WDs just below this limit (i.e., $M_{\rm He} \leq 2.5\,M_\odot$)—see further details in Tauris et al. (2015). Compared to the pioneering models of Habets (1986a), a further remarkable difference is that our newer models peak in radial expansion for masses near $3.0\,M_\odot$, whereas the models of Habets (1986a) peak in radial expansion for helium stars with initial masses near $2.2\,M_\odot$. This means that a much larger fraction of our helium stars in post-HMXB/CE binaries would evolve to become ultra-stripped prior to the second SN, which again will affect the properties of the newly formed DNS systems and their prospects to merge within a Hubble time.

8.3.2 Evolution of Helium Stars with Masses from $5\,M_\odot$ to $\sim 45\,M_\odot$ (and Above)

Figure 8.32 shows that there is a clear relation between both helium-core mass and CO-core mass and initial ZAMS stellar mass for the stars studied by Sukhbold et al. (2018) that is quite independent of the adopted stellar wind mass-loss rate. The helium-core mass as a function of the initial ZAMS mass can be approximated here by

$$M_{\rm He} = 6.46\,M_\odot \left(\frac{M_{\rm ZAMS}}{20\,M_\odot}\right)^{1.27}. \tag{8.26}$$

(Other analytical approximation formulas for the helium-core mass, after Woosley [2019] are given in Chapter 9.) It has turned out that the final evolution of helium stars more massive than $\sim 40\,M_\odot$ proceeds quite differently from the more moderate mass ones below this value. The very massive ones evolve to SNe involving the pair-instability process. We will therefore consider the evolution of helium stars in these two mass ranges separately.

Helium stars more massive than $\sim 8\,M_\odot$ are known from observations as WR-stars, thanks to their very prominent emission lines of helium, nitrogen, and carbon that are

---

Figure 8.31. Core mass vs. radius relation of helium stars in the mass range of 2.0–4.0 $M_\odot$ undergoing carbon and neon ignition (points C and D on the tracks) and in several cases to oxygen ignition (points E on the tracks). During convective core-helium burning, between points A and B on the tracks, the core mass is defined as the mass of the convective helium-burning core plus adjacent semi-convective regions. After convective core-helium burning, the core mass is defined as the mass of the core (consisting of carbon, oxygen, and possibly neon and magnesium, and some helium at the outer core boundary) that is bounded by the helium-burning shell with maximum energy generation. The circles mark the onset of either central or off-center convective carbon burning (at points C). The dash-dotted line connects these circles. After Habets (1986b).

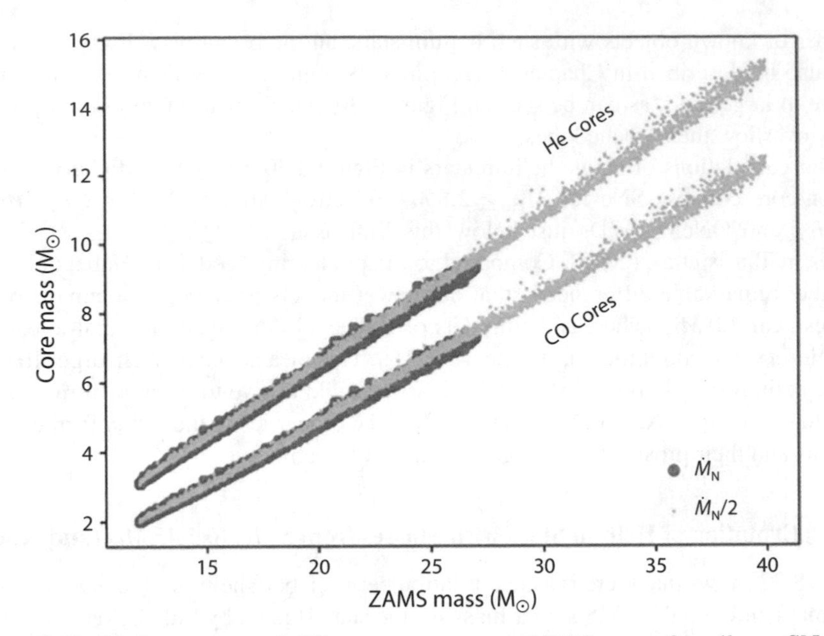

Figure 8.32. Helium-core and CO-core masses vs. ZAMS mass for all pre-SN stars from sets with normal stellar wind mass loss and half of normal stellar wind mass loss, for 4,000 models. Despite significant variations in mass loss, the final helium and CO cores are well determined by the star's initial mass and a standard choice of stellar physics. After Sukhbold et al. (2018).

produced by their very strong stellar winds (see Chapter 5). As Table 5.4 shows, the most massive known hydrogen-poor WR-stars have masses up to $\sim 45 - 50\, M_\odot$. As to final evolution and expected remnants of helium stars in different mass ranges, we refer to the detailed discussions at the end of Chapter 9 and in Chapter 13.

8.3.3 Evolution of Massive Helium Stars, Pair Instability, and BH Formation

As Figure 8.3 shows, hydrogen-rich stars with masses greater than $\sim 100\, M_\odot$ begin to enter the pair-instability region of the log T_c vs. log ρ_c diagram. In this region stars become unstable before oxygen ignition can set in. The resulting pair-instability SNe (PISNe) are caused by violent ignition of carbon and oxygen, following the initial collapse triggered by (e^-, e^+)–pair creation from high-energy photons, which thereby remove pressure support in the core. This instability leads to one or more thermonuclear explosions. Up to a certain initial stellar mass, these explosions are expected to eject considerable amounts of mass before the BH forms, in a so-called pulsational pair-instability SN (PPISN) or, above a certain mass limit, completely disrupt the star, in a regular PISN, leaving no compact remnant behind. The PPISNe/PISNe are important because they are expected to constrain the mass spectrum and leave a mass gap among the massive stellar BHs (roughly between 45 and 125 $M_\odot$), which can be tested with LIGO observations of BH mergers. See further discussions on PISNe and detected BH merger masses in Chapters 13 and 15.

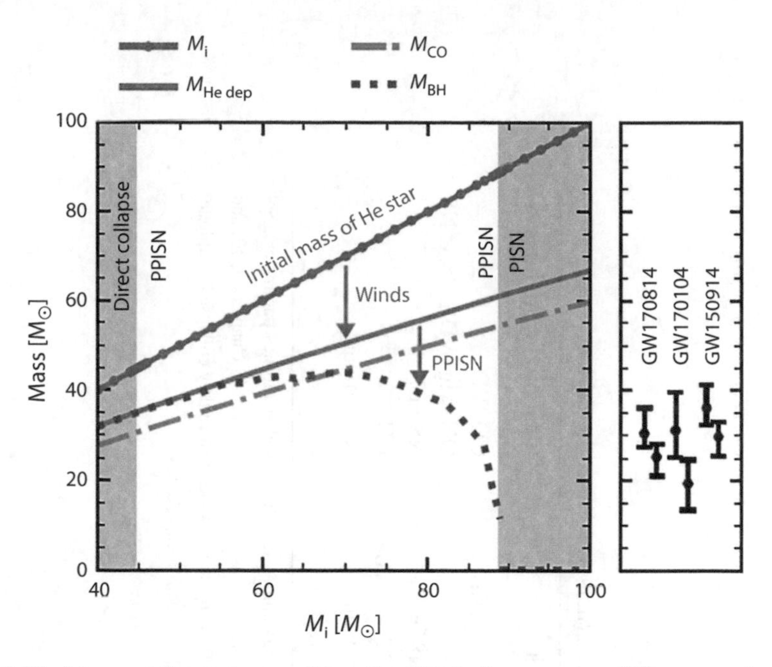

Figure 8.33. Masses of metal-poor ($Z = Z_\odot/10$) helium stars at different evolutionary stages for pulsational pair-instability supernova (PPISN) models as a function of the initial helium star mass, M_i. $M_{\rm He\ dep}$ and $M_{\rm CO}$ are the total and CO-core masses at helium depletion, while $M_{\rm BH}$ is the final mass of the BH formed. For reference, the component masses of three relatively massive binary BHs detected in the LIGO-Virgo network observing runs O1 and O2 are shown. Individual dots in the blue line indicate individual simulations that were performed. After Marchant et al. (2019). Note, since this publication, significantly more massive BH component masses have been discovered in the O3 run; see Chapter 15.

Woosley (2017) pioneered the study of PPISNs and found that for pure helium stars of zero metallicity, the mass range where PPISNe occur is approximately 40 to 90 $M_\odot$, which would correspond roughly to an initial hydrogen-rich star mass range[6] of approximately 80 to 200 $M_\odot$. These studies were followed-up by Marchant et al. (2019), who considered the PPISNe in helium stars with metallicity $Z = 0.1\ Z_\odot$. Their results confirmed the helium-star mass range for PPISNe found by Woosley (2017), as depicted in Figure 8.33. The pair-instability pulsations cause considerable pre-SN mass loss, particularly at the high-mass end of the PPISN mass range, leading to the final BH masses indicated by the dotted curve in the figure. One observes here, and from the tables in Marchant et al. (2019), that PPISNe can produce BH masses up to $\sim$44 $M_\odot$. For helium stars more massive than 90 $M_\odot$, the pair-instability explosion leaves no compact remnant, up to helium star masses of $\sim$240 $M_\odot$, after which again BHs are expected to be produced, with masses upward of $\sim$120 $M_\odot$. Thus, there is an

[6]The exact mass boundaries are strongly dependent on metallicity and assumed stellar physics and may therefore vary throughout this book.

Table 8.3. Final Evolutionary Products of Helium Stars as a Function of Mass

Helium-core mass	ZAMS mass	Final product	Final product in binary
$1.4 - 1.7\,M_\odot$	$3 - 8\,M_\odot$	CO WD (or C-flash)	CO WD
$1.7 - 2.5\,M_\odot$	$8 - 10(12)\,M_\odot$	ONeMg WD	ONeMg WD/ECSN: NS
$>2.5\,M_\odot$	$>10\,(12)\,M_\odot$	Fe-CCSN: NS or BH	NS or BH
$45 - 100\,M_\odot$	$60 - 200\,M_\odot$	PPISN: BH $12 - 45\,M_\odot$	BH
$100 - 240\,M_\odot$	$200 - 400\,M_\odot$	PISN: no remnant	No remnant
$>240\,M_\odot$	$>400\,M_\odot$	PISN: BH $> 120\,M_\odot$	BH

This is to be taken only as a rough guide, as the outcome depends on, for example, metallicity, stellar physics parameters, SN physics, and orbital period of binaries—see discussions and references throughout this chapter, as well as in Chapters 9, 13, and 15. It should be noted that the massive helium stars lose about half of their mass by stellar winds before they collapse or explode. This is the reason why a $\sim$240 $M_\odot$ helium star leaves a $\sim$120 $M_\odot$ BH. ECSN: electron-capture SN.

expected mass gap in the spectrum of BH masses produced from isolated and binary star evolution. Mergers of lower-mass BHs in dense clusters (Chapter 12), however, may be able to fill this mass gap. For discussions on the exact location of the BH mass gap expected from isolated and binary star evolution, we refer to Marchant et al. (2019), Renzo et al. (2020), and Woosley & Heger (2021), for example.

8.3.4 Mapping Initial Stellar Masses to NS/BH Remnants

Table 8.3 summarizes the above-described evolution of helium stars and their expected final remnants. This can be taken as a rough guideline for the mapping between ZAMS stars, helium stars, and final compact objects left behind. As we have explained, reality is more complex—in particular, the mapping to ZAMS masses (which depends on metallicity and details of stellar structure evolution), the question of isolated vs. close binary star evolution, details of the final shell-burning stages up to core collapse, and the SN itself. More detailed discussions of final binary star evolution, BHs, and (P)PISNe are given in Chapters 9, 13, and 15.

EXERCISES

Exercise 8.1.

a. *Derive Eq. (8.11).*

b. *An often used alternative derivation of the dynamical timescale follows by considering the free-fall timescale, that is, the time it takes for a particle to freely fall in the gravitational field of the star over a distance equal to the stellar radius. Show that*

$$\tau_{\rm dyn} \equiv \left(\frac{2R}{g}\right)^{1/2} \simeq \left(\frac{2R^3}{GM}\right)^{1/2} \simeq 40\,\text{min}\left(\frac{\bar{\rho}_\odot}{\bar{\rho}}\right)^{1/2}. \tag{8.27}$$

Exercise 8.2. *Based on Figure 8.5, discuss whether one will be most likely to observe RLO of Case A, B, or C (defined in Section 9.2.1)? Is the answer different for more massive stars?*

Exercise 8.3. *From a close inspection of Figures 8.6 and 8.7, determine whether low-metallicity stars tend to have smaller or larger radii than high-metallicity stars? (Hint: compare, e.g., 12 $M_\odot$ or 20 $M_\odot$ tracks.) Discuss the reason for the result.*

Exercise 8.4. *Estimate the difference in wind mass-loss rate between a $Z = 0.02$ star and a $Z = 0.001$ star.*

Chapter Nine

Stellar Evolution in Binaries

9.1 HISTORICAL INTRODUCTION: IMPORTANCE OF MASS TRANSFER

9.1.1 Algol Binaries

The ideas of how binary systems evolve with mass exchange have been largely inspired by the surprising characteristics of Algol-type eclipsing binary systems, which were described in Chapter 5. Such systems consist of an unevolved main-sequence star, in the case of Algol, a B8V star of $3.2\,M_{\odot}$, together with a less massive subgiant (i.e., more evolved) companion star, in the case of Algol, a $0.7\,M_{\odot}$ star of spectral type K3IV. This situation, with the more evolved star having the smaller mass of the two, is opposite what one would expect on evolutionary grounds, as stars of larger mass are expected to live shorter and thus at any time to be in a more advanced stage of evolution than stars of smaller mass are.

This is what is called the *Algol paradox*. Crawford (1955) was the first to realize that this paradoxical situation can be explained if one assumes that large-scale mass transfer can take place during the evolution of a binary system: he hypothesized that the subgiant components in Algol systems were originally the more massive components of these systems. Because the more massive star evolved faster than the less massive companion did, it was the first to have evolved away from the main sequence toward the giant branch. The presence of the close companion, however, prevented such an evolution: when the outer layers of the expanding (sub)giant star came under the gravitational influence of the companion, they were captured by that star, causing the latter to increase its mass at the expense of the (sub)giant. The (sub)giant transferred so much of its mass that it was finally able to restabilize its internal structure. At that moment it had become the less massive of the two stars.

The first attempt to carry out a real calculation of this type of evolution with mass transfer was by Morton (1960). He demonstrated the correctness of Crawford's conjecture that mass transfer, once it begins, continues until the more evolved star has become the less massive of the two. This was a very important finding, which is the basis of all later work on close binary evolution. In his calculations, however, he still assumed that the orbital period of the system does not change during the mass transfer. He just kept it fixed. This is not correct because, if one assumes the total mass is to be conserved (as he did), then the total angular momentum of the system should be conserved too. Because by far the largest part of this angular momentum is the orbital angular momentum, a good approximation is to keep the orbital angular momentum fixed. This implies that the orbital period will change during the mass transfer in a well-determined way, as was described in Chapter 4. The first to calculate, independently of one another, the evolution of close binaries in this more realistic approach were Paczyński (1966, 1967a,b), Kippenhahn & Weigert (1967), and Plavec (1967). Their works were the foundation of all subsequent works on the evolution of close binary systems.

9.1.2 X-ray Binaries and GW Sources

The discoveries of X-ray binaries in 1971/1972 and of binary radio pulsars in 1974 provided a great stimulus to further research in this field. Without the occurrence of extensive mass transfer from the original primary star to the secondary star, the progenitor systems of many of the X-ray binaries could not have survived the SN of the primary star in which the (first) NS was formed (van den Heuvel & Heise, 1972; Tutukov & Yungelson, 1973b), see Chapter 10. In the case of DNSs, the same holds for the formation event of the second NS (Flannery & van den Heuvel, 1975). For earlier reviews of the subject, particularly of the formation of relativistic binaries with compact star components, we refer to van den Heuvel (1974, 1976), Bhattacharya & van den Heuvel (1991), van den Heuvel (1994a), Tauris & van den Heuvel (2006), Colpi et al. (2009), and Postnov & Yungelson (2014).

Another class of astrophysical sources that are the outcome of close binary evolution is GW sources. After the first direct GW detection by LIGO in 2015 (Abbott et al., 2016d), the field of binary compact star research accelerated further to become a dominant field in astrophysics in the 2020s. With the scheduled launch of the space-borne GW detectors in about a decade (e.g., LISA, [Laser Interferometer Space Antenna] see Amaro-Seoane et al. [2017]; Amaro-Seoane, Andrews et al. [2022]), this field will grow even further in the years to come. GW astrophysics is discussed in Chapter 15.

9.2 EVOLUTION OF THE STELLAR RADIUS AND CASES OF MASS TRANSFER

In the following, we refer to the evolutionary tracks of single stars, such as those as depicted in Figure 8.4. The stellar mass intervals which we quote below are based on ZAMS stars of solar-like chemical composition ($X = 0.70$, $Z = 0.02$). As an example of the evolution of a so-called *intermediate-mass* star, which is characteristic for all stars in the mass range of approximately 2.3–15 $M_{\odot}$, the evolution of the stellar radius of a 5 $M_{\odot}$ star is plotted in Figure 8.5. Evolution of such stars was described in detail in Chapter 8 and here we briefly summarize the important evolutionary phases. Starting from the bottom of Figure 8.5, the numbers along the track subsequently indicate the radius at (1) the ZAMS, (2) core-hydrogen exhaustion, (3) hydrogen-shell ignition, and (4) core-helium ignition, respectively. All intermediate-mass stars cross the Hertzsprung gap (between corresponding points 3 and the base of the red giant branch (RGB) below point 4; see Fig. 8.4) rapidly, on a thermal timescale, until core-helium ignition at the tip of the RGB, after which they describe the helium-burning loop in the HR diagram on a nuclear timescale. After central helium exhaustion, the radius increases again and the star climbs up along the AGB. Stars up to $\sim 8\, M_\odot$ develop degenerate CO cores at this point, and stars in the mass range of $\sim 8 - 10(12)\, M_\odot$ ignite carbon burning and develop degenerate ONeMg cores on the AGB. Beyond $\sim 10(12)\, M_\odot$, stars go through all burning stages and develop iron cores (Jones et al., 2013; Woosley & Heger, 2015).

The evolution of stars with masses $<2.3\, M_\odot$ takes a somewhat different course (see Section 8.1.5) because the helium core becomes degenerate during hydrogen-shell burning, and this burning shell generates the entire luminosity of the star. While the degenerate helium core mass grows, the star climbs up along the RGB until it reaches helium ignition with a flash when the core mass reaches $0.47\, M_\odot$. The evolution we have just described depends only slightly on the initial chemical composition and on the effects of convective overshooting.

Massive stars $>15\, M_\odot$, similar to intermediate-mass stars, expand rapidly—on a thermal timescale—when crossing the Hertzsprung gap, and up to $25\, M_\odot$ burn helium as red supergiants (RSGs), with a slightly expanding radius. Stars with masses $\gtrsim 25\, M_\odot$ develop very strong stellar wind mass loss, as seen in Figures 8.6, 8.7 and 8.9. This means that in such stars most or all the hydrogen-rich envelope may be removed by the time helium ignites. The radius evolution of these massive stars therefore becomes more complex than that of the intermediate-mass stars. In what follows we will, for the moment, ignore these complications.

9.2.1 Nomenclature of Cases of Mass Transfer

Based on the evolution of the stellar radius with time as depicted in Figure 8.5, Kippenhahn & Weigert (1967) defined three cases of mass transfer (known as RLO) in close binaries:

- Case A: the binary is very tight, such that the radius of the evolving star already during core-hydrogen burning becomes larger than the radius of its Roche lobe.
- Case B: the system is wider, such that the star fills its Roche lobe after leaving the main sequence, but before reaching helium ignition.
- Case C: the system is so wide that the star fills its Roche lobe only after core-helium exhaustion, when it has developed a carbon-oxygen core.

These cases are illustrated in Figure 9.1

Further subdivision of evolutionary cases of RLO are introduced in the literature with a nomenclature that describes the later stages of binary star interactions with multiple stages of RLO. For example, RLO Case BB refers to mass transfer from a naked helium star that lost its hydrogen-rich envelope in an earlier episode of mass transfer. Strictly speaking, this stage can be further subdivided into Case BA, Case BB, and Case BC RLO, for mass transfer initiated during core-helium burning (Case BA), helium-shell burning (Case BB), and beyond ignition of carbon burning (Case BC).

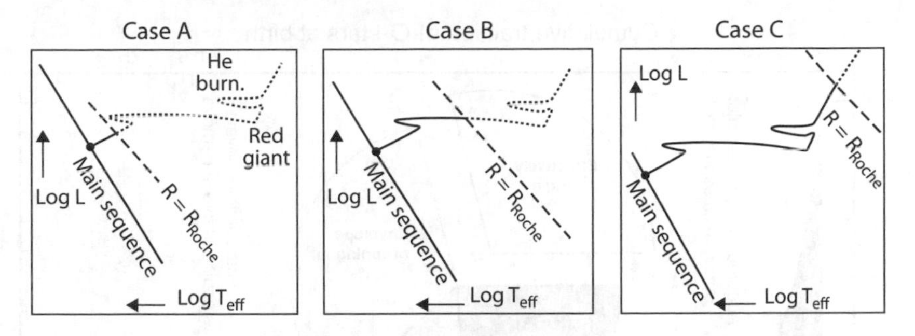

Figure 9.1. The three basic cases of mass transfer in close binaries, defined as Case A, Case B, and Case C RLO, illustrated at the evolutionary track of an intermediate-mass star in the HR diagram. After Kippenhahn & Weigert (1967).

Other examples of nomenclature for composite stages of mass transfer by RLO include, for example, Case AB RLO, which is a second stage of mass transfer during hydrogen-shell burning (B), following an initial stage of RLO while the donor star was burning hydrogen in its core (A) (see Fig. 14.17); or Case BBB RLO, which is mass transfer during carbon-shell burning (B) of a stripped helium star (BB) (see, for example, Fig. 10.17).

From the distribution of orbital periods of binaries (see Chapter 3), it appears that for low- and intermediate-mass binaries, Case A RLO is relatively rare: of the order of 10% of all mass-exchange binaries (for massive binaries, greater than $\sim 20\,M_\odot$ the fraction is significantly higher). On the other hand, Case B RLO is very common: more than half of all observed unevolved spectroscopic binaries will evolve according to this case. Because of strong selection effects against the detection of long-period spectroscopic binaries, however, the precise incidence of Case C RLO is still not well known, but it is expected to be high as well. Figure 9.2 provides a schematic representation of binary interactions among unevolved O-type stars (stars with masses $\gtrsim 20\,M_\odot$). According to the studies of Chini et al. (2011), Chini et al. (2012), and Sana et al. (2012), >70% of all O-stars are members of interacting binaries. The binary fraction of stars is increasing with their masses as was shown in Figure 3.11; see also further discussions in Chapter 3.

For a system with a primary star mass of $5\,M_\odot$ and a mass ratio (donor star mass divided by accretor star mass) of $q = 2.0$, Case A occurs only for orbital periods < 1.5 d, whereas Case B RLO occurs for orbital periods between ~ 1.5 and 87 d, and Case C RLO for all greater orbital periods.

Figure 9.3 illustrates how the orbital period ranges for the Cases A, B, and C RLO depend on the initial primary mass, for an assumed mass ratio of $q = 2.0$. The various symbols indicate the orbital periods at which the primary star fills its Roche lobe, for stars with masses ranging from 1.0 to $16\,M_\odot$. The figure shows, for example, that for a $12\,M_\odot$ star, Case B RLO occurs for all systems with orbital periods between 2.5 and about ~ 600 d.

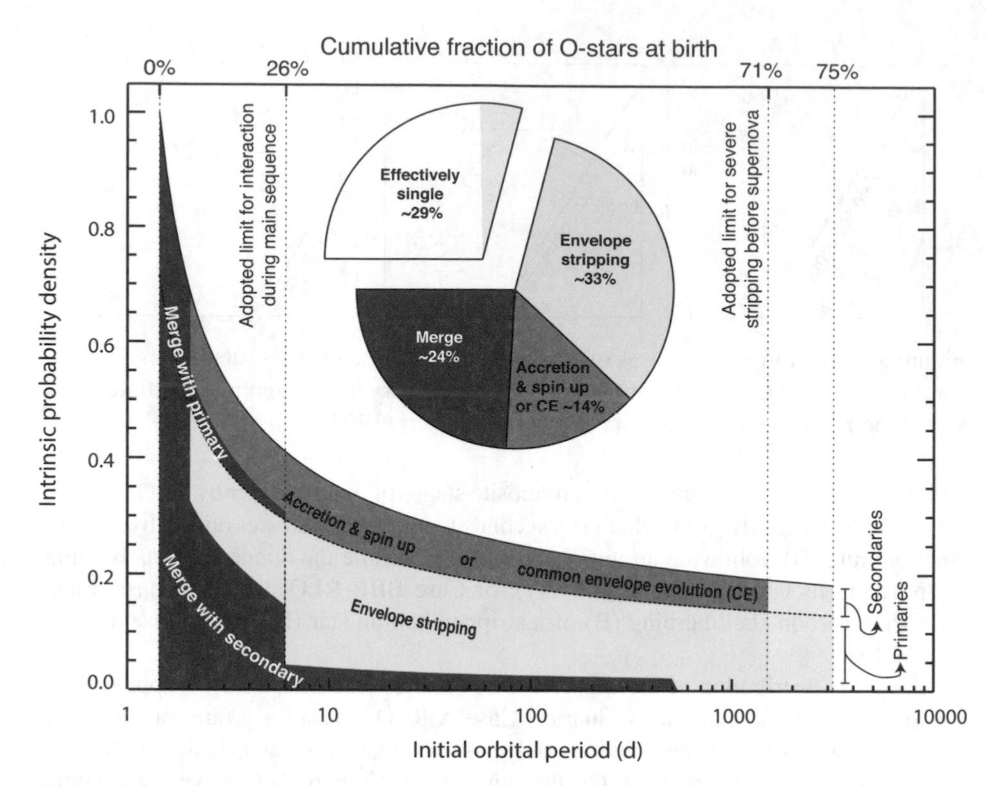

Figure 9.2. Schematic representation of the relative importance of different binary interaction processes among unevolved O-type stars. All percentages are expressed in terms of the fraction of all stars born as O-type stars, including the single O-stars and the O-stars in binaries, either as the initially more massive component (the primary) or as the less massive one (the secondary). The solid curve gives the adopted intrinsic distribution of orbital periods, normalized to unity at the minimum period considered, as a function of initial orbital period in days. The dotted curve separates the contributions from O-type primary and secondary stars. The colored areas indicate the fractions of systems that are expected to merge (red) or experience stripping (yellow) or accretion/CE evolution (orange). The pie chart compares the fraction of stars born as O-stars that are effectively single (i.e., single [white] or in wide binaries with little or no interaction effects [light green]; 29% combined) with those that experience significant binary interaction (71% combined). The binaries that undergo accretion or CE (14%) are the progenitors (if they survive the SN) of X-ray binaries that host a NS or BH. After Sana et al. (2012).

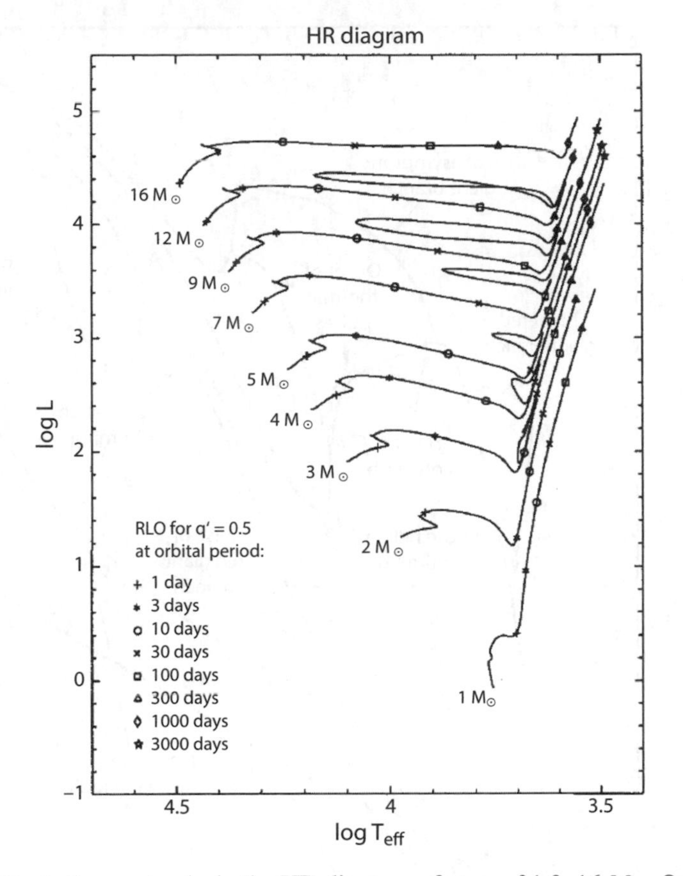

Figure 9.3. Evolutionary tracks in the HR diagram of stars of 1.0–16 $M_{\odot}$. On the tracks are indicated the radii at which the stars fill their Roche lobes in binaries with the indicated orbital periods, for $q = M_1/M_2 = 2.0$. (Beware of the inverse definition of mass ratio in the plot legend, $q' \equiv 1/q = 0.5$.) After Pols (1993).

9.2.2 The Webbink Diagram

The *Webbink diagram* allows one to immediately observe for what ranges of primary star radii the different Cases A, B, and C RLO occur. The Webbink diagram is depicted in Figures 9.4 and 9.5, after Ge et al. (2015) (see also Ge, Webbink, Chen, Han, 2020; Ge, Webbink, Han, 2020), Figure 9.4 shows the stellar radii as a function of stellar mass at various evolutionary stages. Curves of different evolutionary stages—such as ZAMS, tip of RGB, tip of AGB, etc.—are indicated in the diagram. From this diagram one can derive the orbital period ranges for close binary evolution according to the RLO Cases A, B, and C. Figure 9.5 from Ge et al. (2015), depicts these period ranges for a mass ratio of $q = 1$. The dashed line in both figures (base of the giant branch) is the boundary above which stars have deep convective envelopes, such that the mass transfer from the primary star to the secondary star is dynamically unstable (see Ge et al., 2015, Ge, Webbink, Chen, Han, 2020; Ge, Webbink, Han, 2020, and Sections 9.3.5 and 9.3.7).

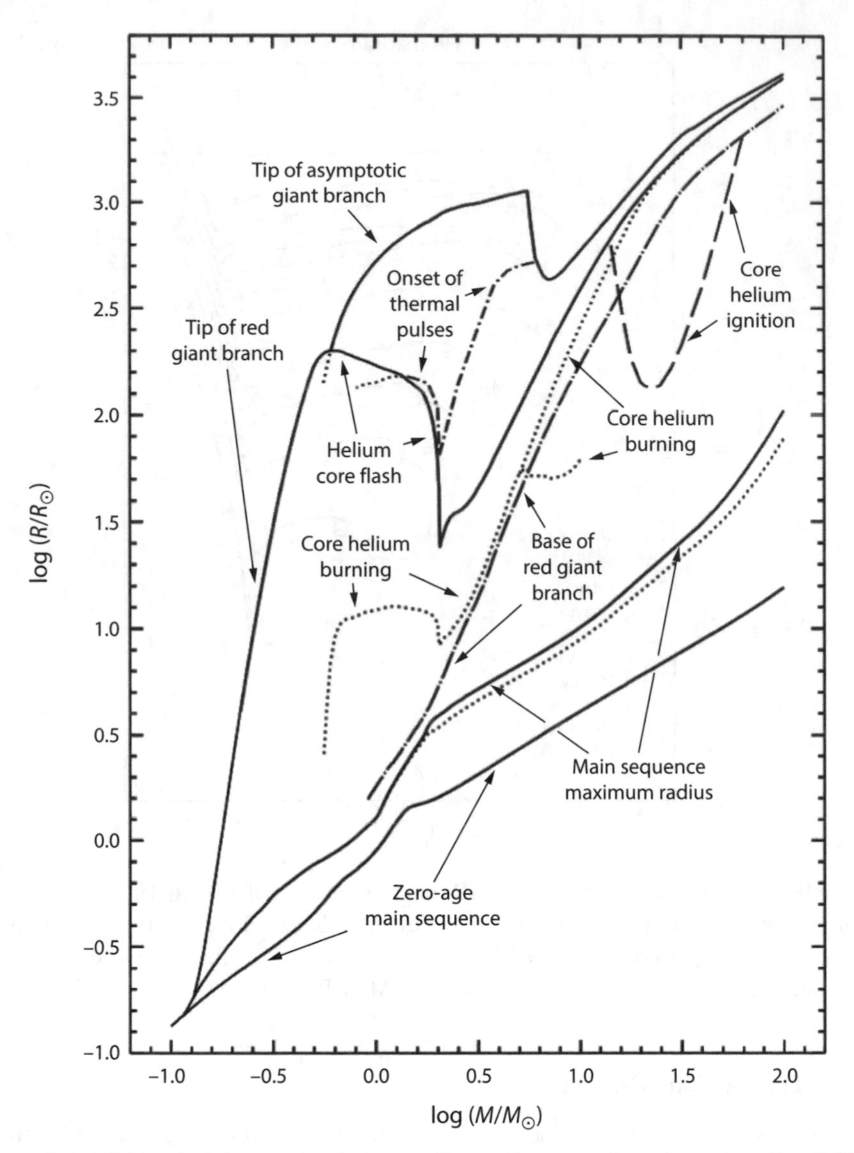

Figure 9.4. Webbink diagram depicting stellar radius as a function of stellar ZAMS mass at various evolutionary stages for stars with Population I (solar) metallicity. With help of this diagram, one can determine for which orbital periods, as a function of stellar mass, the binary evolutionary RLO Cases A, B, and C occur. See examples for a mass ratio of $q = 1$ in Figure 9.5. Further explanation in the text. After Ge et al. (2015).

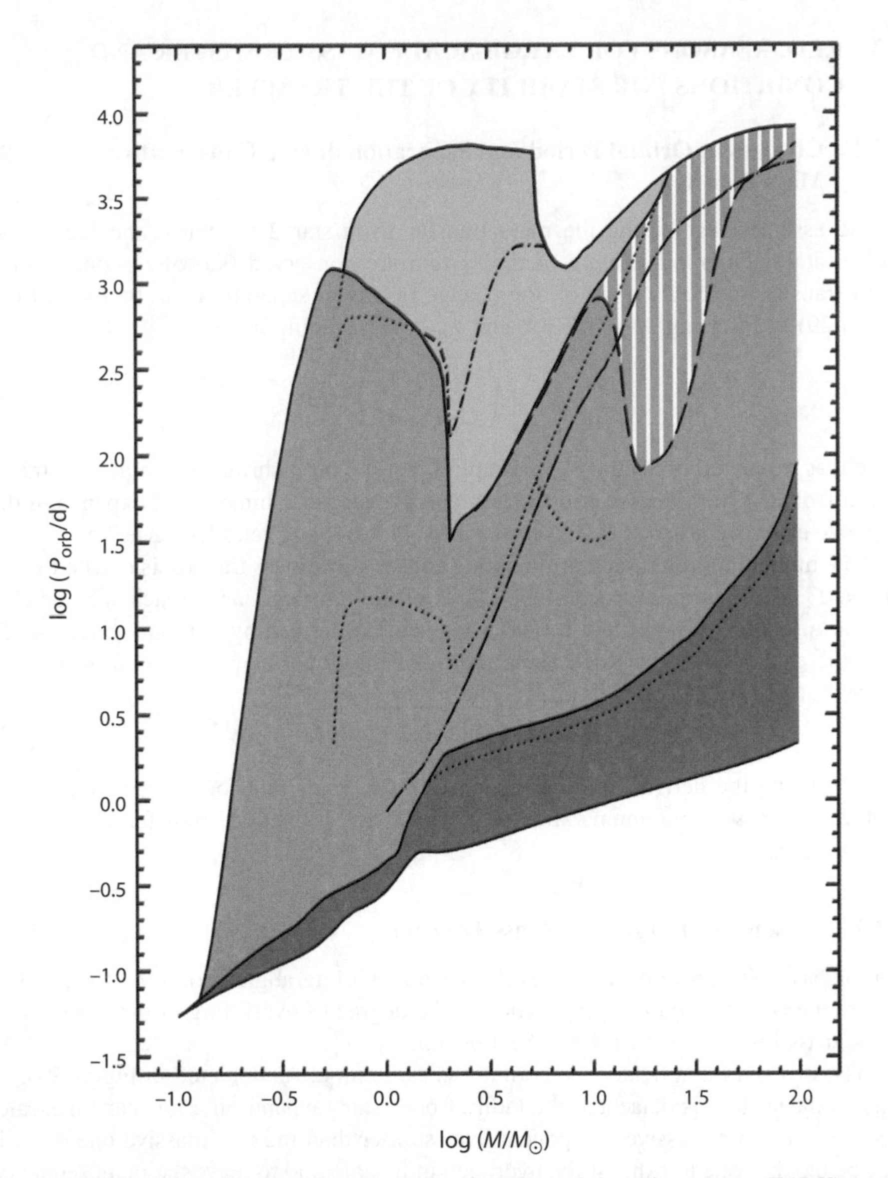

Figure 9.5. Webbink diagram depicting as a function of the primary star ZAMS mass, for which orbital periods in binaries with mass ratio $q = 1$, the system will evolve according to the RLO Cases A, B, and C. Darkest color is Case A, medium color is Case B, and the lightest color is Case C. In the hatched region, for primary masses larger than $\sim 15\, M_{\odot}$, the evolution proceeds similar to Case B at lower primary masses, even though formally the evolution does not fit the original Kippenhahn definition of Case B. The meaning of the different curves is the same as was defined in Figure 9.4. The diagram is for stars with Population I (solar) metallicity. Modified in 2021 by H. Ge after Ge et al. (2015).

9.3 RLO: REASONS FOR LARGE-SCALE MASS TRANSFER AND CONDITIONS FOR STABILITY OF THE TRANSFER

9.3.1 Changes in Orbital Period and Separation during Conservative Mass Transfer

If one assumes that during the mass transfer from star 2 to star 1, the total mass and orbital angular momentum of the system are conserved (so-called *conservative* mass transfer, see Section 4.3.6), the change in orbital separation can be found from Eq. (4.29) by inserting $\dot{M}_1 + \dot{M}_2 = 0$ and $\dot{J}_{\rm orb} = 0$, resulting in

$$\frac{da}{a} = -2\left(\frac{1}{M_2} - \frac{1}{M_1}\right) dM_2, \tag{9.1}$$

which, as explained in Section 4.3.6, implies that the orbit shrinks when mass is transferred from the more massive to the less massive star in a binary, and expands in the opposite case, because $dM_2 < 0$; see also Eqs. (4.40)–(4.42) and Exercise 4.5.

The minimum orbital separation during conservative mass transfer is reached when $M_1 = M_2$. The minimum Roche-lobe radius of the primary star is reached somewhat later at a smaller value of $q < 1$. This value can be derived by combining Eq. (4.40) with the equation for the Roche-lobe radius of the donor star, such as from the top expression of Eq. (4.7):

$$R_{\rm L} = a\ (0.38 + 0.20 \log q)\,. \tag{9.2}$$

By setting the derivative of $R_{\rm L}$ equal to 0, one finds that for conservative evolution, $R_{\rm L}$ reaches its minimum value for $q = 0.78$, and if the mass transfer continues, it increases again.

9.3.2 Reasons for Large-scale Mass Transfer

For the basic physics of the mass transfer via the first Lagrangian point, L_1 (i.e., RLO), and the rate of mass transfer, in relation to the degree of overfilling of the Roche lobe of the mass-losing star, we refer to Section 4.5.

The two stars in a binary are born at the same time. As depicted in Figure 9.6, as long as the system is detached, the radii of both stars expand on a nuclear timescale. Because the more massive component lives shorter than the less massive one does, it will be the first one to exhaust the hydrogen in its core and to leave the main sequence and expand to overfill its Roche lobe, causing it to start transferring mass to its companion. As in the conservative case, the mass transfer from the more massive to the less massive star makes the orbit shrink, and the onset of mass transfer then may lead to an unstable situation: while the star wishes to expand, its Roche lobe shrinks, which will lead to an enhanced tendency to transfer mass. Equilibrium can be restored after the mass-losing star has become the less massive one of the two, such that when it transfers further matter, its Roche lobe expands (see Section 4.5.3 and Fig. 4.14 for details). In this way, the expansion of the Roche lobe now can keep pace with the further evolutionary expansion of the star, on a nuclear timescale.

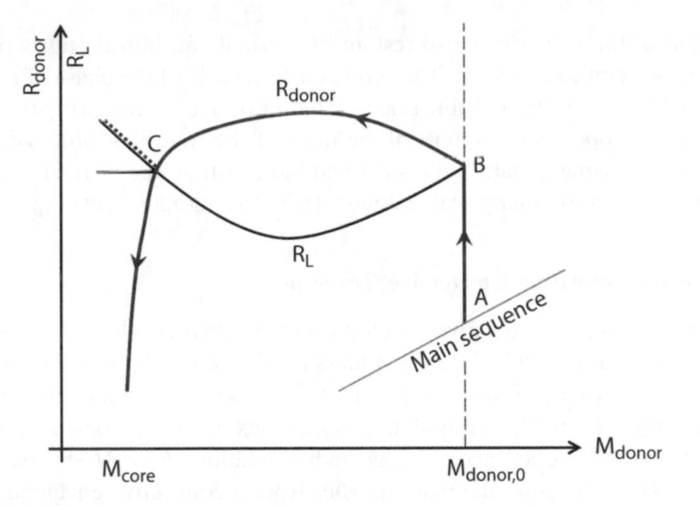

Figure 9.6. Schematic variation of donor star radius, $R_{\rm donor}$ (solid red line), and its Roche-lobe radius, R_L (thin black line), before/during/after mass transfer from a primary with a radiative envelope to the secondary, for a binary evolving according to Case B RLO (mass transfer starting after the star has left the main sequence, but before core-helium ignition). $R_{\rm donor}$ expands on a nuclear timescale while evolving from the main sequence (point A) until RLO is initiated at point B when $R_{\rm donor} > R_{\rm L}$. During RLO (point B to point C), the donor star adapts to its thermal-equilibrium radius, which significantly exceeds $R_{\rm L}$, giving rise to a high mass-transfer rate on a thermal timescale. Only after the primary has become less massive than the secondary (Fig. 4.14) will R_L begin to increase. Consequently, equilibrium is not possible before the primary has lost so much mass that it has become the less massive component. At point C, RLO ceases and leaves an almost naked helium core of mass, $M_{\rm core}$, which may contract to the helium main sequence. If it still has a thin hydrogen envelope, it may continue somewhat further (dotted line) to expand filling its Roche lobe, on a nuclear timescale, and consuming the residual hydrogen by hydrogen-shell burning. See also Fig. 14.17.

The timescale of the unstable mass-transfer stage depends on the physical state of the envelope of the mass-losing star: radiative or convective. We now first consider these two cases and then discuss the precise criteria for stability of the mass transfer.

9.3.3 Donor Star with a Radiative Envelope

When the star has a radiative envelope, and mass is taken away, its first reaction is to restore its hydrostatic equilibrium: this causes it to shrink, which occurs on a dynamical (pulsational) timescale. However, the mass loss causes the thermal equilibrium of the envelope to be disturbed, and it turns out that to restore this equilibrium the star has to expand. This is depicted in Figure 9.6, in which the thermal equilibrium radius, $R_{\rm eq}$, and the Roche-lobe radius, R_L, of the more massive (primary) component of a binary are plotted as a function of the mass M_1 of the mass-losing primary star. The expansion of the radius of the primary star to restore its thermal equilibrium takes place on the thermal timescale of its envelope. Therefore, the timescale of the phase of mass transfer is of the order of the thermal timescale of the envelope of the primary star, which, because the envelope contains most of the mass of the star, is of order of the thermal timescale of the primary star. Figures 8.5 and 9.6 illustrate the different timescales of the evolution of a close binary with a donor star with a radiative envelope.

9.3.4 Donor Star with a Convective Envelope

Contrary to the case of a radiative envelope, a convective envelope does not shrink to restore its hydrostatic equilibrium when mass is taken away from it. Just the opposite, it expands, on a dynamical timescale. This is because stars with deep adiabatic convective envelopes ($\gamma = 5/3$) are polytropes of index 1.5 (like non-relativistic degenerate stars), which have an inverted mass-radius relation: $R \propto M^{-1/3}$ (see, e.g., Cox & Giuli, 1968). As a result, the mass transfer from a convective envelope tends to be violently unstable: the envelope expands on a dynamical timescale, leading to mass transfer on a very short timescale. Because the companion star needs its own thermal timescale to adapt its radius to the large amount of mass dumped onto it, it will often be unable to accommodate this huge mass transfer, and the result will be the formation of a CE around the binary system, as explained in Section 4.8, leading to highly non-conservative evolution with significant in-spiral and possibly envelope ejection.

9.3.5 Stability Criteria for Mass Transfer: I

The preceding qualitatively described reasons for stable and unstable mass transfer can be refined and put into a more quantitative form. In Section 4.5.2, we discussed the stability of mass transfer in view of the mass-radius relation of stars and their Roche lobes. The first one to derive such quantitative criteria was Hjellming (1989). Here we summarize the arguments and largely follow the description of the stability criteria as given by Pols (2011, chapters 6–8).

As explained in Chapter 4, one can derive stability criteria by comparing how the stellar radius reacts to the mass loss and how the Roche-lobe radius reacts to the mass loss. Figure 9.7 depicts the behavior of the Roche-lobe radius and the stellar radius as a function of the donor star mass for an assumed conservative mass transfer of a binary with total mass of 2.0 $M_\odot$ and two evolutionary tracks: for an initial donor mass $M_\mathrm{d} = 1.0\,M_\odot$ (starting at point A) and $M_\mathrm{d} = 1.4\,M_\odot$ (starting at point D). Like in Figure 9.6, the stars evolve vertically upward in the diagram on a nuclear timescale until the onset of RLO, which starts in the points B and E, respectively. On mass loss, the donor radius may behave as given by the dotted line through point B or the dashed line through point E. In the first case, after losing a small amount of mass $\delta M_\mathrm{d} < 0$, the radius R_d of the donor star has become slightly smaller than the Roche-lobe radius, R_L. So, strictly speaking, the donor would detach from its Roche lobe and has to re-expand before RLO can continue. Therefore, this is the condition for stable mass transfer. In reality, however, because RLO is not dictated by a strict on/off contact, but instead by a gradual increase in mass transfer when $R_\mathrm{d} \simeq R_L$ (see discussion on finite scale

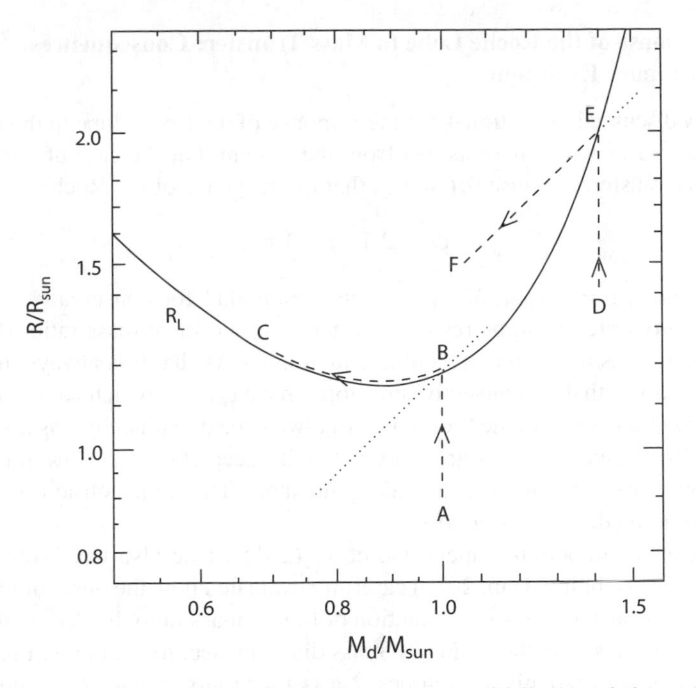

Figure 9.7. Behavior of the Roche-lobe radius (R_L, solid line) and the stellar radius (dashed lines, for two cases) as a function of the donor mass. The Roche-lobe radius curve is for conservative mass transfer in a binary with a total mass of $2.0\,M_\odot$. The dotted lines show a fiducial stellar mass-radius relation that, for example purposes, is taken to be $R_\mathrm{d} \propto M_\mathrm{d}$ (i.e., $\zeta_* = 1$). Further explanation in the text (figure from Pols, 2011, chapters 6–8).

height of the stellar atmosphere in Section 4.5), the RLO proceeds continuously but at a slow pace following the radius evolution of the donor star. On the other hand, in the second case, where the mass transfer starts at point E, after losing $\delta M_\mathrm{d} < 0$, R_d has become larger than R_L, and the mass-transfer rate, which depends very strongly on $\delta R = R_\mathrm{d} - R_L$ (Section 4.5), increases, leading to more mass loss and a larger δR. This is, at first sight, a runaway situation leading to unstable mass transfer (however, see Sections 9.3.6 and 9.3.7). It thus appears that to check for stability of the mass transfer, one has to compare the slope of the stellar mass-radius relation with that of the Roche-lobe mass-radius relation. These slopes are expressed by the mass-radius exponents (see also Section 4.5.2):

$$\zeta_* \equiv \frac{\partial \ln R_\mathrm{d}}{\partial \ln M_\mathrm{d}} \qquad \wedge \qquad \zeta_L \equiv \frac{\partial \ln R_L}{\partial \ln M_\mathrm{d}}, \tag{9.3}$$

where $\zeta_* \geq \zeta_L$ implies stability and $\zeta_* < \zeta_L$ implies instability of the RLO. As mentioned in Sections 9.3.2–9.3.4, however, the situation is complicated by the fact that stars react to perturbations, such as mass loss, on very different timescales.

9.3.6 Response of the Roche Lobe to Mass Transfer: Consequences for Binary Evolution

We already discussed in Section 4.5.3 the response of the Roche lobe to the mass transfer for a few cases where mass is lost from the system. For the case of purely conservative mass transfer, Exercise 9.1 shows that the response of the Roche lobe becomes

$$\zeta_L = 2.13\,q - 1.67, \tag{9.4}$$

where, as usual, $q = M_{\rm donor}/M_{\rm accretor}$. This means that for conservative mass transfer the stability criteria can be rewritten in terms of a critical mass ratio. Because the initial mass ratio is $q > 1$, the preceding equation shows that it is always the case that $\zeta_L > 0.46$. Stars with deep convective envelopes have $\zeta_{\rm ad} \leq 0$, which means that the first stage of mass transfer from these donors will always be dynamically unstable. Becausc in Case C RLO the donor stars are red giants with deep convective envelopes, Case C RLO is always expected to be dynamically unstable. The same is true for late Case B RLO with intermediate-mass donors.

In a series of important papers, Ge et al. (2015), (see also Ge, Webbink, Chen, Han, 2020; Ge, Webbink, Han, 2020) carefully evaluated how the onset of dynamically unstable mass transfer varies, as a function of binary mass ratio divided by the location of the mass-losing star in the radius vs. mass diagram; see, for example, Figure 9.4.

For donors with radiative envelopes, $\zeta_{\rm ad}$ is much larger than zero, and RLO will only become dynamically unstable if the donor is much more massive than the accretor. For $q < 4$, mass transfer is dynamically stable and will take place on a thermal timescale (see, for example, Tauris et al., 2000; Hurley et al., 2000; Pavlovskii et al., 2017). Only if the donor fills its Roche lobe very close to the ZAMS ($\zeta_{\rm eq} = 0.6 - 1.0$, as defined in Section 9.3.7) in a very close binary with nearly equal masses, can one expect stable nuclear timescale mass transfer. For this reason, thermal timescale mass transfer is expected to occur in most Case A binaries, as well as in Case B binaries of intermediate and high mass with periods such that mass transfer occurs not too late during the crossing of the Hertzsprung gap.

In the case of non-conservative mass transfer, such as RLO with mass loss from the system described in Section 4.3, ζ_L will depend on the precise mode of loss of mass and angular momentum, and again its dependence on q can be derived (see Soberman et al., 1997, for a detailed description). We now describe the stability criteria in more detail.

9.3.7 Stability Criteria for Mass Transfer: II

If a star suddenly loses mass, it reacts by readjusting its structure and thereby its radius to recover equilibrium. However, on mass loss, both the *hydrostatic* and the *thermal* equilibrium of the star are disturbed, and they recover on different timescales. We therefore need to redefine the mass-radius exponent, ζ_*, for each of the two different cases.

Hydrostatic readjustment happens on the star's dynamical timescale, which is much shorter than its thermal timescale (see Chapter 8). Therefore the initial dynamical response will be almost adiabatic. For the question of the *dynamical stability* of the mass transfer, one therefore must consider the *adiabatic* response of the stellar radius to the mass loss. This can be expressed as

$$\zeta_{\rm ad} \equiv \left(\frac{\partial \ln R_{\rm d}}{\partial \ln M_{\rm d}} \right)_{\rm ad} . \tag{9.5}$$

The criterion for dynamical stability of the mass transfer then becomes $\zeta_{\rm ad} \geq \zeta_L$. If this criterion is fulfilled the star will, on mass loss, shrink within its Roche lobe on a dynamical timescale and is able to recover its hydrostatic equilibrium. In that case, the slower thermal readjustment of the donor becomes relevant. On the thermal (Kelvin-Helmholtz) timescale it will attempt to recover its thermal equilibrium radius appropriate to its new mass $M_{\rm d} + \delta M_{\rm d}$ (where $\delta M_{\rm d} < 0$), and the change of its thermal equilibrium radius can be expressed as $(\delta R/R_{\rm d})_{\rm eq} = \zeta_{\rm eq}(\delta M_{\rm d}/M_{\rm d})$, or

$$\zeta_{\rm eq} \equiv \left(\frac{\partial \ln R_{\rm d}}{\partial \ln M_{\rm d}} \right)_{\rm eq} . \tag{9.6}$$

If, in addition to $\zeta_{\rm ad} \geq \zeta_L$, also $\zeta_{\rm eq} \geq \zeta_L$, then the new equilibrium radius will be smaller than the Roche radius and we will have the condition for *secularly stable* mass transfer. In the intermediate case $\zeta_{\rm ad} \geq \zeta_L > \zeta_{\rm eq}$, thermal readjustment of the donor keeps pushing it to overfill its Roche lobe. Mass transfer then proceeds on the donor's thermal timescale, as illustrated in Figures 9.8 and 9.6.

Based on these stability considerations one can, with regard to stability of the mass transfer, distinguish three types of mass transfer, corresponding to different timescales. (These classifications are independent of the division into the Cases A, B, and C that we discussed in Section 9.2.1 and that are illustrated in Fig. 9.1.)

1. Stable mass transfer: if $\zeta_L \leq \min\,(\zeta_{\rm ad},\ \zeta_{\rm eq})$.
   This corresponds to mass transfer starting at point B in Figure 9.7. The donor remains in thermal equilibrium and the mass transfer is driven by the nuclear evolution of the donor. Initially, both R_L and $R_{\rm d}$ may decrease, but in the course of time both quantities will expand while the mass transfer is still driven by nuclear evolution (e.g., at point C).

   The transfer can also be driven by orbital shrinking due to loss of orbital angular momentum, $J_{\rm orb}$ (from magnetic braking or GW radiation), in which case R_L shrinks. In this situation both R_L and $R_{\rm d}$ may shrink. The mass ratio, q, also plays a key role for the evolution of R_L (i.e., shrinking or widening of the orbit in response to mass transfer/loss), and thus a competition may arise between the shrinkage of the orbit due to loss of $J_{\rm orb}$ and the widening of the orbit due to RLO from a less massive donor star to a more massive accretor star. The result in this case is often that the orbit shrinks at a reduced rate (unless the mass ratio $q \ll 1$, in which case the system always widens if the mass-transfer rate is high enough). Evolution with RLO driven by loss of $J_{\rm orb}$ applies, for example, to LMXBs, CVs, UCXBs, and AM CVns (see Sections 11.3 and 11.4).

2. Thermal timescale mass transfer: if $\zeta_{ad} \geq \zeta_L > \zeta_{eq}$.
Mass transfer is dynamically stable, but driven by the thermal re-expansion of the donor. This corresponds to the mass transfer starting at point E of Figure 9.7. Initially, the mass-transfer rate increases, but then stabilizes at a value determined by the donor's thermal timescale:

$$|\dot{M}_{\rm d}^{\max}| \simeq M_{\rm d}/\tau_{\rm th}. \tag{9.7}$$

This case of mass transfer is sometimes called *thermally unstable* mass transfer, but the mass transfer here is not really unstable: it is *self regulating*. If $|\dot{M}_{\rm d}|$ would fall below the value of the preceding equation, the donor will be allowed to expand to regain thermal equilibrium, leading to an increased rate of mass transfer. On the other hand, if $|\dot{M}_{\rm d}|$ were much larger than the preceding thermal timescale value, the donor would react by adiabatically shrinking inside its Roche lobe. Thus, the radius excess $\delta R/R_{\rm d}$ adjusts itself to maintain a thermal-timescale mass-transfer rate. As shown in Section 4.5, this implies $\delta R/R_{\rm d} < 0.01$. Hence, the donor radius closely follows the Roche-lobe radius, but $R_{\rm d} < R_{\rm eq}$.
3. Dynamically unstable mass transfer: if $\zeta_L > \zeta_{ad}$.
The adiabatic response of the donor is unable to keep it within its Roche lobe, leading to ever-increasing mass-transfer rates. As discussed in Section 9.3.4, this occurs if the donor has a deep convective envelope. Detailed calculations show that the mass transfer accelerates to a timescale between the thermal and dynamical timescales of the donor. This tends to lead to the formation of a CE in which the companion and the dense core of the donor spiral toward each other, and the envelope may be ejected (see Section 4.8). Only in cases where the donor star is not significantly more massive than the accretor, and significant mass loss (isotropic re-emission) takes place, can a CE be avoided and the system may survive. This is the case in many wide-orbit LMXB and IMXB systems, even though the donor stars in these systems have a deep convective envelope (Tauris & Savonije, 1999; Tauris et al., 2000; Misra et al., 2020). In HMXBs or IMXBs where $q \gg 1$, the destiny is always dynamically unstable mass transfer and formation of a CE, regardless of the nature of the donor star's envelope.

9.4 RESULTS OF CALCULATIONS OF BINARY EVOLUTION WITH CONSERVATIVE MASS TRANSFER

Before we discuss results of non-conservative binary evolution, which is expected to be common in nature, it is instructive to first consider some results of conservative binary evolution, as this provides us with a guide and insightful approximations of the expected outcomes of close binary evolution with mass transfer. We restrict ourselves here to discussing outcomes of Case A and Case B evolution, because Case C is expected to always lead to non-conservative evolution, which will be discussed in Section 9.5. Part of the examples we give here are from the pioneering works of the group of Kippenhahn and Weigert (then at the University of Göttingen), Paczyński in

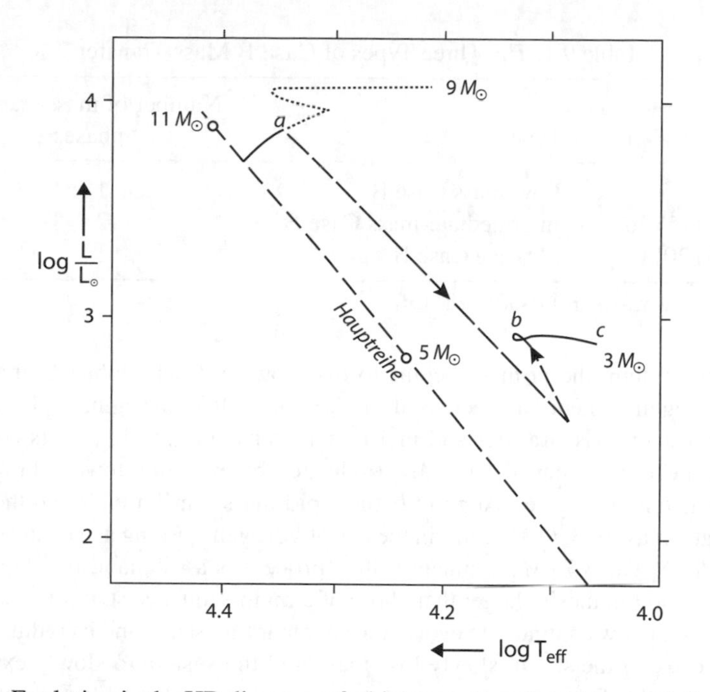

Figure 9.8. Evolution in the HR diagram of a binary undergoing Case A RLO with component masses of $9+5\,M_\odot$ and chemical abundances of $X=0.602$ and $Z=0.044$. ("Hauptreihe" means: main sequence.) The primary star begins on the ZAMS and evolves to the point *a* where it begins to overflow its Roche lobe (if it had been single, it would have followed the dotted evolutionary track). This leads to rapid transfer of matter and the donor star moves down in the HR diagram along the dashed line until at point *b* the rapid mass transfer (on a thermal timescale of the primary) ends, and a phase of slow mass transfer, on a nuclear timescale, follows. The secondary star has grown to almost $11\,M_\odot$ by the end of the mass transfer, which is assumed to be conservative, and has moved upward along the main sequence as indicated. A further description of the evolution of the system is given in the text. After Kippenhahn & Weigert (1967).

Warsaw and Plavec in Praha, who, independent of each other, in the 1960s, pioneered the computation of the evolution of close binaries.

9.4.1 Case A RLO

One of the first close binary evolution calculations ever carried out is shown in the HR diagram in Figure 9.8. This calculation by Kippenhahn & Weigert (1967) depicts the evolution of the primary star in a close binary system with initial component masses of $9+5\,M_\odot$ and an initial orbital period of 1.20 d, for initial chemical composition $X=0.60$ and $Z=0.044$. (The dotted line shows the evolutionary track that the $9\,M_\odot$ star would have followed if it had been single.) At point *a*, 1.25×10^7 yr after the

Table 9.1. The Three Types of Case B Mass Transfer

M_{donor} ($M_\odot$)	Type	Number of mass-transfer phases
$\lesssim$2.6	Low-mass Case B	1
2.6 – (15) 20	Intermediate-mass Case B	2
$\gtrsim$(15) 20	Massive Case B	1

The limiting masses are for solar metallicity.

birth of the system, the primary begins to overflow its Roche lobe. At that time its central hydrogen content has decreased to $X_c = 0.35$. It now begins a phase of rapid thermal-timescale mass transfer and in 10^4 yr it transfers 5.27 $M_\odot$ to its companion, which now becomes a star of 10.27 $M_\odot$ (indicated by evolution toward 11 $M_\odot$ on the main sequence in the figure). At point *b*, the rapid mass transfer ends and the star now stabilizes at a mass of 3.73 $M_\odot$, still in the core-hydrogen-burning stage, and still filling its Roche lobe. The 3.73 $M_\odot$ remnant of this process is a subgiant, in other words, it has a radius and luminosity larger than those of a main-sequence star of the same mass. From here on, it now continues to evolve on a nuclear timescale, and its radius gradually increases, causing the star to slowly lose mass and the system to slowly expand on a nuclear timescale. The calculations were terminated at $t = 3 \times 10^7$ yr (point *c*), when the primary mass had decreased to 3.02 $M_\odot$, and the primary was still in the core-hydrogen-burning phase, and the accretor had a mass of 10.98 $M_\odot$.

Between points *b* and *c* (and also beyond *c*), the system nearly resembles the massive semi-detached Algol-type close binaries listed in Tables 5.2 and 5.3, such as RY Per and RS Vul, as well as MP Cen and Z Vul. In all these systems, the Roche-lobe filling component is the less massive of the two and is a subgiant that is over-luminous for its mass. As Figure 9.8 shows, the same is true for the post-mass-transfer system: the subgiant, with its mass decreasing from 3.73 to 3.03 $M_\odot$ has a luminosity similar to that of an ~5 $M_\odot$ main-sequence star.

9.4.2 Case B RLO

For systems evolving according to Case B mass transfer, there appear to be three possible kinds of outcome depending on the initial mass (M_{donor}) of the primary star (Table 9.1): the so-called low-mass Case B (for solar metallicity: $M_{donor} \leq 2.6\,M_\odot$), the intermediate-mass Case B, and the massive Case B.

The lower dichotomy near $M_{donor} = 2.6\,M_\odot$ reflects that in low-mass stars ($\leq$2.3 $M_\odot$) the post-main-sequence evolution proceeds differently from that of stars of higher mass, as described in Section 8.1.5. (In binaries, as a result of mass transfer, the limiting mass of the donor star up to which a degenerate helium core develops is somewhat increased to ~2.6 $M_\odot$.) The helium cores in these post-main-sequence stars become degenerate, and their entire luminosity is generated by hydrogen-shell burning. During their slow ascent along the giant branch, the mass of their degenerate core grows, and when they reach the tip of the giant branch, and their helium core mass reaches 0.47 $M_\odot$, they ignite helium burning with a flash. On the other hand, in stars more massive than 2.3 $M_\odot$ ($\sim$2.6 $M_\odot$ for donor stars in a binary), the helium core contracts and heats when the stars cross the Hertzsprung gap, and they immediately ignite core-helium burning and become giants powered by core-helium burning supplemented by hydrogen-shell burning, as described in Chapter 8.

The upper dichotomy for $M_{\rm donor} \simeq 15-20\,M_\odot$ (for solar metallicity) distinguishes between the intermediate-mass Case B stars that undergo a second phase of mass transfer, so-called Case BB RLO (strictly speaking, Cases BA, BB, or BC; see Section 10.13), before these donor stars terminate their evolution, and the massive Case B stars where this second mass-transfer phase is absent.

Low-mass Case B RLO

Figures 9.9 and 9.10 as well as Table 9.2 show another historic example of early binary stellar evolution with conservative RLO, namely the first-ever calculated evolution of a low-mass Case B system by Kippenhahn et al. (1967). The system initially consisted of stars of masses $2+1\,M_\odot$, with an orbital period of 1.14 d. We now summarize its evolution in detail.

After 5.75×10^8 yr, the primary star has left the main sequence and begins to overflow its Roche lobe (point D). Now a phase of fast (thermal timescale) mass transfer ensues, lasting 6×10^6 yr, after which (point F) the mass of the primary has been reduced to 0.96 $M_\odot$, while that of the secondary has increased to 2.04 $M_\odot$. After this, a long phase ($\sim10^8$ yr) of nuclear timescale mass transfer sets in, during which the degenerate helium-core mass gradually grows and the Roche-lobe-filling subgiant gradually expands and transfers mass, roughly at a rate of $10^{-8}\,M_\odot\,{\rm yr}^{-1}$. Energy is generated in the subgiant by hydrogen-shell burning. At 6.35×10^8 yr after the birth of the system (point G), the primary mass has been reduced to 0.28 $M_\odot$ and it detaches slightly from its Roche lobe. The star now consists of a 0.25 $M_\odot$ degenerate helium core and a 0.03 $M_\odot$ hydrogen-rich envelope. After a temporary contraction phase, the primary star expands again and refills its Roche lobe at point I ($t=6.68\times10^8$ yr). Hydrogen-shell burning continues to provide the luminosity and drive the mass transfer. After this final phase of RLO, lasting 13.7 Myr, during which the hydrogen-rich envelope is removed, at point K the donor star detaches from RLO and it begins to contract to finally form a 0.26 $M_\odot$ He WD (with a tiny hydrogen-rich envelope of $<10^{-3}\,M_\odot$) that settles on the cooling track (evolving along the track between points N and O). This calculation is the *first one ever* in which stellar evolution calculations straightforwardly produced a WD.

We see from this example that low-mass Case B RLO produces long-lived lower-mass classical Algol-type binaries such as Algol itself and U Cephei (see Tables 9.2 and 5.3). Contrary to the higher-mass Algol-type systems that result from Case A evolution, such as Z Vul, in which the subgiant still is in the core-hydrogen burning phase, here the subgiant has a degenerate helium core and its luminosity is generated by hydrogen-shell burning. Many of the known classical Algols are of this type.

In the above-described Case A and low-mass Case B evolutionary models, which produced classical Algol-type binaries, conservative evolution was assumed. In comparing the parameters of observed Algol-type binaries with those predicted by theoretical models, it has been found that while several systems fitted well with conservative evolution, for several other systems the observed parameters could only be

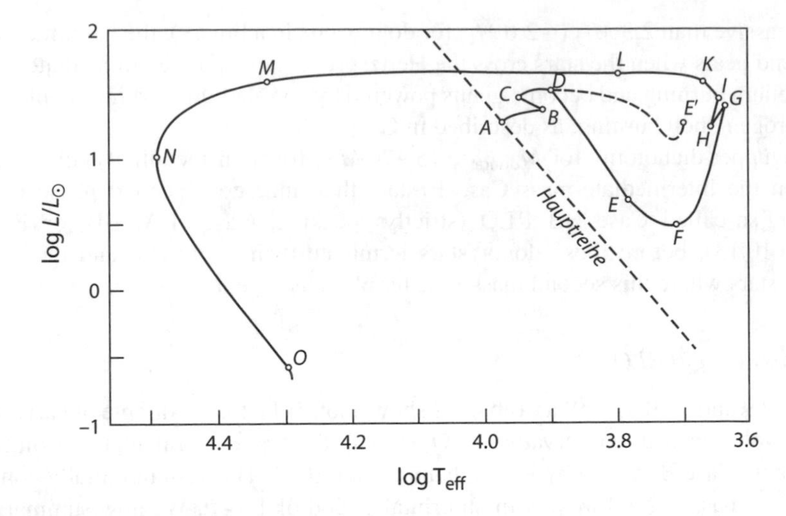

Figure 9.9. Evolution in the HR diagram of a low-mass Case B RLO donor star of $2\,M_{\odot}$ in orbit with a $1\,M_{\odot}$ accretor and an initial orbital period $P = 1.14$ d and chemical composition of $X = 0.602$ and $Z = 0.044$. ("Hauptreihe" means: main sequence). The mass-transfer evolution is assumed to follow conservative RLO. At point D, the mass transfer begins. At point F, the mass ratio of the system is practically reversed ($q = 0.47$). The mass transfer is now driven by the expansion of the envelope due to hydrogen-shell burning. At point G, the mass transfer temporarily stops, and between G and H, the system is slightly detached with the primary still being a subgiant. After point H, the mass transfer resumes, still driven by hydrogen-shell burning around the degenerate helium core. At point K, the primary star detaches from the Roche lobe, and its hydrogen-envelope begins to shrink. At point O, the star becomes a WD. The orbital period at points K to O is 23.74 d. The system parameters at various evolutionary stages are given in Table 9.2 and summarized in the text. Figure 9.10 depicts how various system parameters evolved in the course of time. The deep drop in luminosity between points D and F is due to the fact that here the nuclear energy generated in the interior is temporarily used to drive the mass transfer and is not radiated away. After Kippenhahn et al. (1967).

explained if during the evolution a considerable amount of mass and orbital angular momentum had been lost from the systems. So, the evolution of the latter systems did not proceed conservatively. These precise analyses were carried out particularly by the group at the Free University, Brussels. See, for example, van Rensbergen & De Greve (2016, 2020) and references therein. We will return to this problem in Section 9.5 on non-conservative evolution.

Intermediate-mass and Massive Case B RLO

For stars with initial masses $M_1 > 2.6\,M_{\odot}$ in binaries, the helium core in the post-main-sequence star is sufficiently massive ($>0.4\,M_{\odot}$) to avoid degeneracy and to contract immediately to helium ignition. It appears that practically independent of the mass

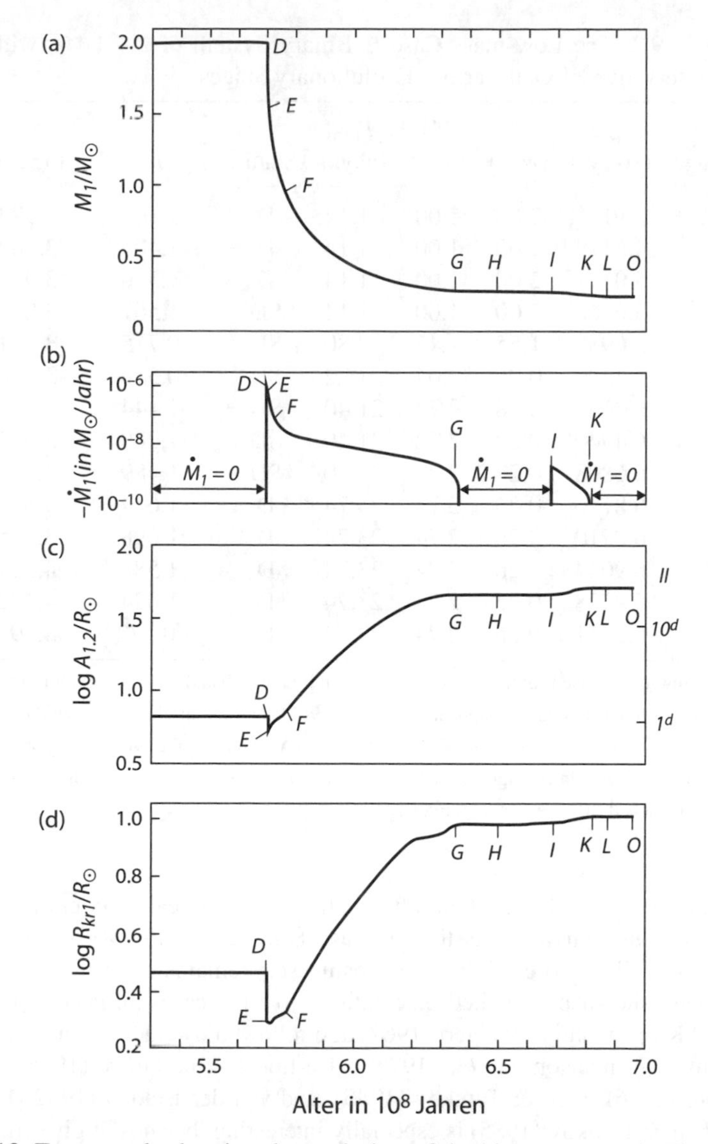

Figure 9.10. Four panels showing the evolution of the binary system in Figure 9.9: (a) donor star mass, (b) mass-transfer rate, (c) orbital separation and (d) Roche-lobe radius of donor star. "Alter in Jahren" means: Age in years. After Kippenhahn et al. (1967).

of the envelope, the thermal equilibrium radius of a star with a helium-burning core and a hydrogen-rich envelope is *always* very large, as can be seen from Figure 9.11. This figure shows the radii of stars with a helium-burning core mass of $0.8\,M_\odot$ and a total mass ranging from 0.8 to $5\,M_\odot$. The figure shows that for a total mass of $>0.9\,M_\odot$ (i.e., a hydrogen-rich envelope mass of $>0.1\,M_\odot$), the stars all have giant-like radii. From this it follows that in a mass-transfer binary such a star can only be

Table 9.2. The Low-mass Case B Binary System of $2+1\,M_{\odot}$ with Conservative RLO at Various Evolutionary Stages

Stage	Age (10^8 yr)	M_1 ($M_{\odot}$)	M_2 ($M_{\odot}$)	$P_{\rm orb}$ (days)	Status	$\log(L/L_{\odot})$	$\log T_{\rm eff}$
A	0	2.00	1.00	1.14	D	1.307	3.972
B	4.8179	2.00	1.00	1.14	D	1.405	3.909
C	5.0591	2.00	1.00	1.14	D	1.500	3.947
D	5.6968	2.00	1.00	1.14	SD	1.567	3.899
E	5.6989	1.55	1.45	0.80	SD	0.715	3.781
F	5.7509	0.96	2.04	1.22	SD	0.518	3.710
G	6.3538	0.28	2.72	21.40	SD	1.444	3.634
H	6.4940	0.28	2.72	21.40	SD	1.280	3.653
I	6.6808	0.28	2.72	21.40	SD	1.489	3.642
K	6.8176	0.26	2.74	23.74	D	1.634	3.667
L	6.8710	0.26	2.74	23.74	D	1.698	3.795
M	6.9044	0.26	2.74	23.74	D	1.588	4.329
N	6.9328	0.26	2.74	23.74	D	1.024	4.492
O	6.9651	0.26	2.74	23.74	D	−0.572	4.297

This low-mass Case B binary system is shown in Figures 9.9 and 9.10. The evolutionary stages were calculated by Kippenhahn et al. (1967). The initial chemical abundances were $X=0.602$, $Y=0.354$, and $Z=0.044$. Letters A–O refer to the evolutionary stages (points) indicated in the figures. Status refers to whether the system is detached (D) or semi-detached (SD; i.e., undergoing RLO).

made to stay inside its Roche lobe after it has lost practically its entire hydrogen-rich envelope. One thus expects that in Case B systems with $M_1 > 2.6\,M_{\odot}$ practically only the helium core of the mass-losing star remains after the mass transfer. This is indeed shown by detailed calculations for this case, initiated by Paczyński (1967c) and Kippenhahn & Weigert (1967) and followed by calculations by, for example, Tutukov & Yungelson, (1973a, 1973b), De Loore & de Grève (1976), De Greve & De Loore (1976), Iben & Tutukov (1985), and van der Linden (1982, 1987). The paper by Iben & Tutukov (1985) is especially interesting, because it gives the detailed evolution of initial primary stars between 3 and $12\,M_{\odot}$, up to the end of its life as a WD.

Thanks to the increase of computing power and improved input physics, many new models of such stars have been calculated in more recent years. Grids of such evolutionary models of stripped stars can be found in, for example, Eldridge et al. (2008), Eldridge et al. (2013), Yoon et al. (2010), Yoon et al. (2017), and references therein.

Figures 9.12 and 9.13 depict examples of such grids calculated by Götberg et al. (2018). It should be noted that these models are not fully conservative, because stellar wind mass loss was taken into account. Figure 9.12 shows the evolution of the primary stars of four Case B systems with an initial solar composition of $X=0.70$, $Z=0.014$;

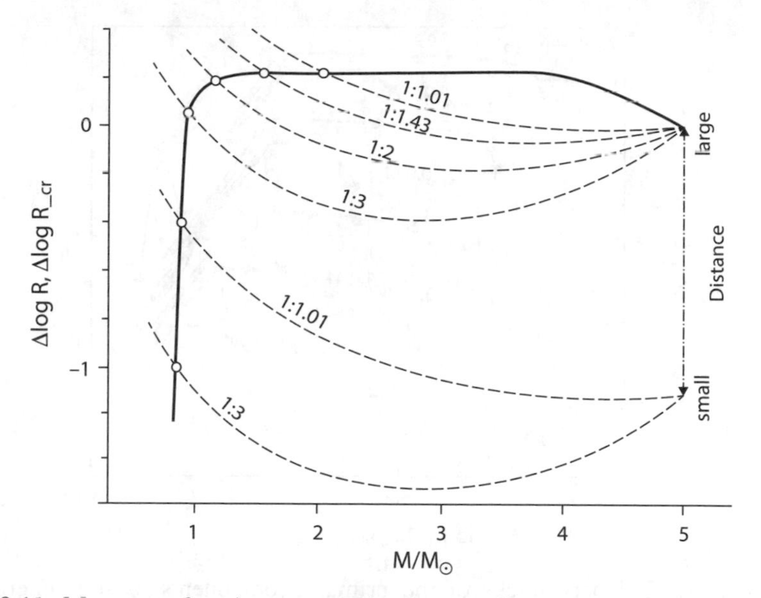

Figure 9.11. Mass transfer via conservative intermediate-mass Case B RLO, for an initial donor star mass of 5 $M_{\odot}$ (the originally more massive component; solar abundance), for different mass ratios of secondary and primary stars (from 1:1.01 to 1:3) for different initial separations of the components. Solid curve: $\Delta \log R$ plotted as a function of the donor star mass, M, where R is the possible thermally adjusted remnant radius of the star. (The radius difference is the difference between the thermally adjusted radius and the radius at the onset of the mass transfer.) Dashed curves: similar curves for changes in its Roche-lobe radius ($\Delta \log R_{cr}$). For different initial separations of the components, the $R_{\rm cr}$ curves are shifted parallel up and down. For a given case, the mass transfer stops at the intersection of the corresponding dashed curve and the solid curve (open circles). After Giannone et al. (1968).

mass ratio of $q = 1.25$; and initial primary star masses of 2.4, 4.0, 9.0, and 18.2 $M_{\odot}$, respectively. The initial orbital periods were 6.4, 8.8, 15.2, and 31.7 d, respectively.

During the rapid mass transfer, the luminosity of the primary goes down significantly, as can be seen at the tracks in the HR diagram. This is because most (80%–90%) of the nuclear energy generated in the interior is used to drive the expansion of the envelope and the mass transfer. More extreme cases of this effect are sometimes seen in Case BB RLO, where the central helium burning can temporarily shut off as a result of the decreasing core temperature (see Section 10.13.2). As long as the primary is more massive than the secondary, the mass-transfer causes the orbit to shrink, which boosts the mass-transfer rate $\dot{M}$. The maximum value of $\dot{M}$ occurs at minimum luminosity and is due to the formation of a small convective zone in the envelope of the primary, which causes enhanced mass transfer even after the mass ratio has been reversed. Then follows a phase when the thermal equilibrium is restored and the mass-transfer rate drops to a low value. The timescale of the fast mass-transfer phase is of the order of the thermal timescale, as can be seen, for example, in van der Linden (1987).

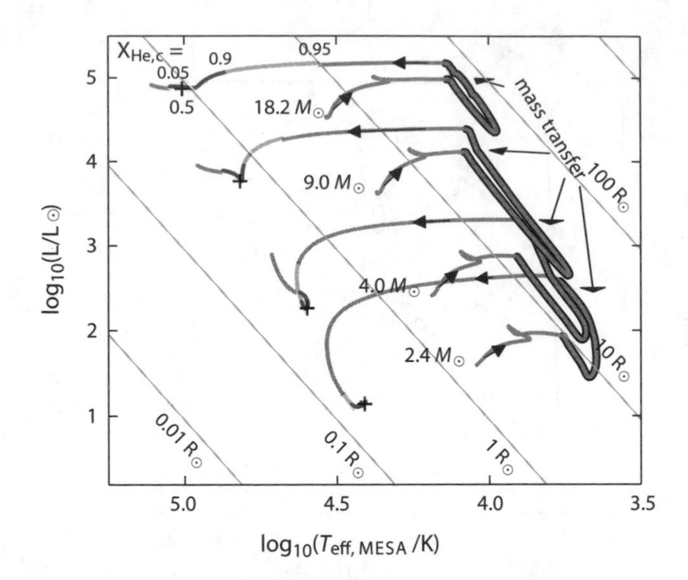

Figure 9.12. Evolutionary tracks of the primary components of four intermediate Case B systems in the HR diagram for solar metallicity and conservative mass transfer. Indicated at the beginning of each track are the initial stellar masses. The initial mass ratios of primary and secondary stars are 1.25. The masses of the final stripped (helium) stars are 0.44, 0.80, 2.5, and 6.7 $M_\odot$, respectively. The thermal-timescale mass-transfer phase is marked by the thick gray lines. Helium-core burning is marked with yellow and dark orange, corresponding to central helium mass fractions in the ranges $0.95 > X_{\rm He,c} > 0.90$ and $0.90 > X_{\rm He,c} > 0.05$, respectively. The models have solar metallicity ($Z = 0.014$). After Götberg et al. (2018).

As Figure 9.12 shows, the remnants of the donor stars of solar metallicity are in all cases the helium cores of the stars (with only a very thin remaining hydrogen envelope), which in the HR diagram are situated close to the pure helium star main sequence. The remnant masses range from 0.44 to 6.7 $M_\odot$. The 0.44 $M_\odot$ star still reaches helium burning.

Götberg et al. (2018) computed the evolution of a large grid of similar Case B binaries for different metallicities: $Z = 0.014$, 0.006, 0.002, and 0.0002. The masses and properties of the resulting stripped stars, as a function of initial donor star mass, are depicted in Figure 9.13. One observes in this figure that the masses of the stripped stars do not depend much on metallicity, although the mass and, as a consequence, the radius of the thin hydrogen-rich envelope (panel b) do. Because in these models mass loss by stellar winds was included, they are not fully conservative. As the wind mass-loss rates are quite modest here the differences with conservative evolution are only minor. For discussions on their applied stellar wind mass-loss rates, see the original paper (Götberg et al., 2018).

The presence of a considerable hydrogen-rich envelope in stripped stars of low metallicity (as depicted in Fig. 9.13b) is important for the further evolution of binaries

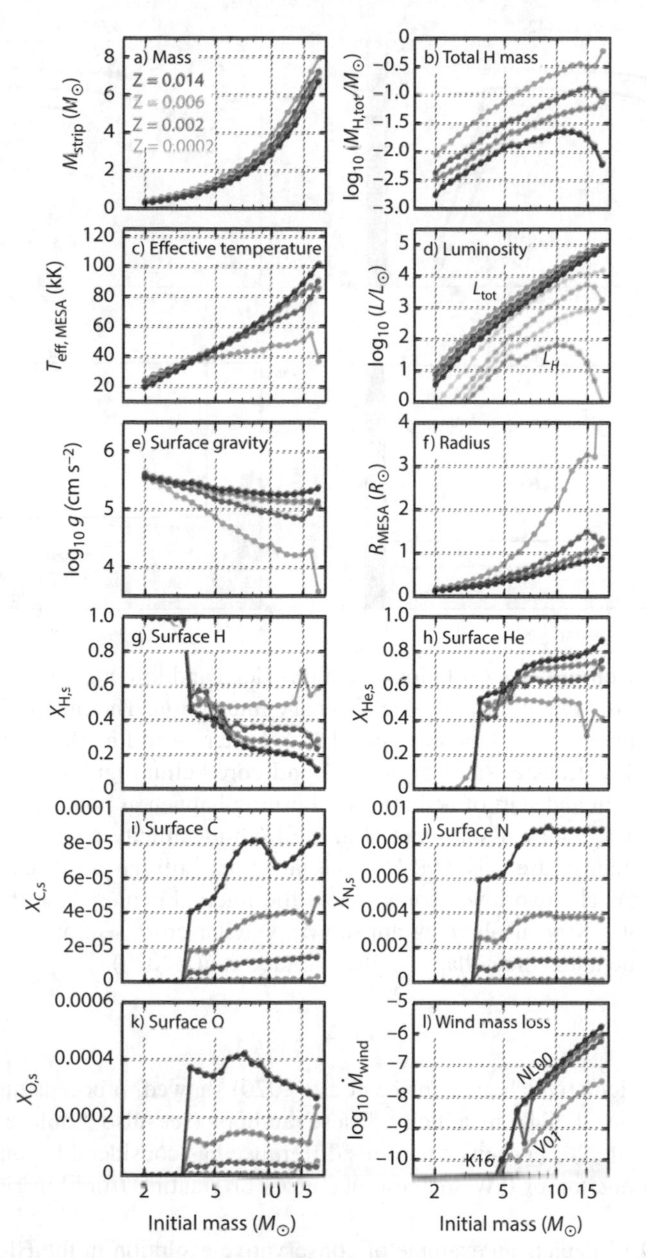

Figure 9.13. Properties of stripped (helium) stars in binaries as a function of initial mass for different metallicities, computed using MESA. Panel (b) shows the mass of the very thin hydrogen plus helium envelope left after the RLO. Panel (d) shows the total luminosity L_{tot} as well as the luminosity L_H of the hydrogen-burning shell, in lighter colors. The labels in panel (l) denote the applied wind mass-loss rates (see original paper). After Götberg et al. (2018).

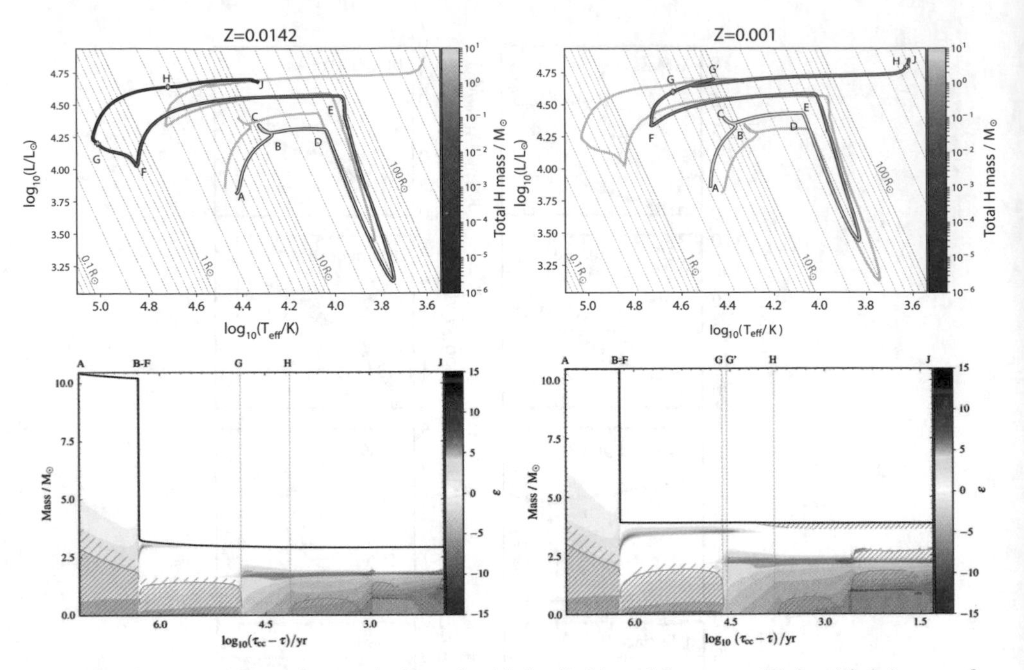

Figure 9.14. Conservative evolution of a high- (left) and low-metallicity (right) star of initially 10.5 $M_\odot$ that is stripped due to binary interaction. The initial binary period is 25 d and the initial companion mass is 8.4 $M_\odot$. Letters A to J mark different evolutionary points. Mass transfer starts at point D and core helium ignition at point F. Core-helium depletion and start of helium shell-burning happen at point G (carbon ignition at point H). The total mass at point H in the high-metallicity case is 2.98 $M_\odot$ and in the low-metallicity case it is 3.90 $M_\odot$ (the latter one still has a substantial hydrogen-rich envelope). The two lower panels show the internal evolution of the primary star in the form of a Kippenhahn diagram showing stellar cross section as function of the remaining time until core collapse. After Laplace et al. (2020).

of low metallicity, which, as Laplace et al. (2020) showed, proceeds quite differently from evolution at high metallicity. These authors (see also Laplace et al., 2021; Vartanyan et al., 2021) showed that this difference has considerable consequences for SNe and production of GW source progenitors originating from binaries of different metallicities.

Figure 9.14 depicts an example of conservative evolution in the HR diagram of a 10.5 $M_\odot$ donor star in Case B binaries with metallicities $Z = 0.0142$ and 0.001. The systems have initial orbital period $P_{\rm orb} = 25$ d and an 8.4 $M_\odot$ companion star. The figure shows that in the low-metallicity case, a considerably more massive He and CO core is produced than in the high-metallicity case. These results are expected to have important implications for the outcome of binary population synthesis evolution calculations for the production of SNe and DNSs (Laplace et al., 2020).

Results of Massive Case B RLO

In the traditional textbook view of Case B RLO, the donor star loses nearly its entire hydrogen-rich envelope in a short-lived phase of thermal-timescale mass exchange ($\lesssim 10^4$ yr). This is because the donor is thought to usually be a rapidly-expanding hydrogen-shell burning giant. This view is, however, not true for massive binaries in low-metallicity environments. It was recently shown by Klencki et al. (2021) that in such environments two types of post-main-sequence mass transfer may result: (1) mass exchange on the long nuclear timescale that continues until the end of the core-helium burning phase, and (2) rapid mass transfer leading to detached binaries with mass-losers that are only partially stripped of their envelopes.

Keeping this behavior of the evolution in the low-metallicity case in mind, we now discuss the evolution in the high (solar) metallicity case. Here, in massive Case B systems, with initial donor masses of $\geq 15-20\,M_\odot$ in metal-rich environments (like the MW), the donor star leaves a helium star with a mass of $\gtrsim 3.2-3.5 M_\odot$, which in later life only expands modestly and therefore often does not experience further RLO. If the exposed helium stars are more massive than $\sim 8\,M_\odot$, they are observed as WR-stars which have very strong stellar winds (see discussions in Chapters 5 and 10). The final evolution of intermediate-mass and massive Case B binaries leading to the formation of HMXBs is discussed in Chapter 10.

9.4.3 Response of the Secondary to Mass Transfer

The mass-transfer process in binary systems has a significant impact on the further evolution of the donor star just as accretion onto the companion star has important consequences for its structure and evolution. Initially, while undergoing thermal readjustment and (often) expansion in response to bulk accretion (i.e., *rejuvenation*), the accretor may spin-up to a critical rotation limit—a threshold value beyond which mass shedding, due to increased centrifugal forces resulting from a gain in angular momentum, may set in (Packet, 1981; Petrovic et al., 2005; de Mink et al., 2013). Thus, unlike the general case for a compact object accretor (X-ray binary), a non-degenerate mass gainer cannot be treated accurately as a point mass, and its evolution should be resolved with an advanced binary stellar evolution code, for example, Modules for Experiments in Stellar Astrophysics (MESA[1] Paxton et al., 2013, 2015, 2019.)

9.4.4 Further Evolution of Helium Stars in Intermediate-mass Case B Binaries: Case BB RLO

As was mentioned in Section 9.2.1, after the first phase of Case A, B, or C RLO, a second or even third phase of mass transfer may occur. For Case B systems, this will occur for systems with initial primary masses between $\sim 2.6\,M_\odot$ and $15-20\,M_\odot$ (we call these *intermediate-mass Case B systems*), for the following reasons. In Section 8.3.1, we saw for helium stars with masses $\leq 3.5\,M_\odot$, during helium-shell burning the outer

[1] http://mesa.sourceforge.net/

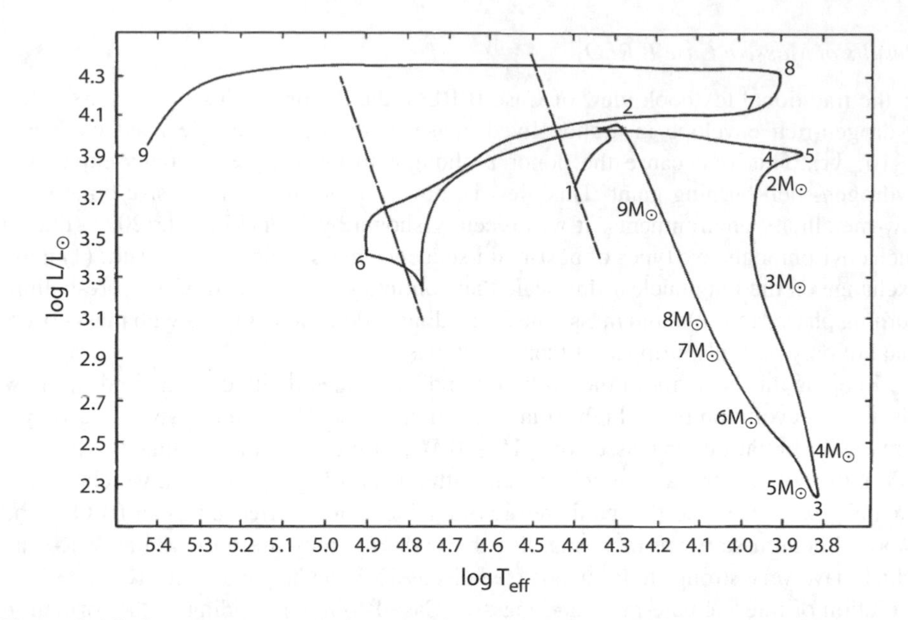

Figure 9.15. Conservative evolution of a binary of $10.0 + 8.0\, M_\odot$, $P_{\rm orb} = 3.15$ d, undergoing Case B and Case BB RLO. The figure shows the evolution of the donor (primary) star in the HR diagram. The two dashed diagonal lines are the hydrogen and the helium ZAMSs. The system first goes through a Case B mass-transfer episode (up to point 5), after which the primary star leaves a $1.66\, M_\odot$ helium star in an orbit with $P_{\rm orb} = 81.07$ d. During helium-shell burning, the helium star evolves into a giant (point 7) and, via Case BB RLO, transfers $0.54\, M_\odot$ to the secondary star and finishes as a $1.12\, M_\odot$ CO WD, in an orbit with $P_{\rm orb} = 236.44$ d. This was the first ever calculation of Case BB mass transfer and of the formation of a CO WD. After De Greve & De Loore (1976). See also De Loore & de Grève (1976).

radius of the star expands to giant dimensions, as is also visible in the evolutionary track of the $2.98\, M_\odot$ mass-transfer remnant in the the top-left diagram of Figure 9.14. This means that during the later evolution of the intermediate-mass Case B binary with a relatively short orbital period, a second phase of mass transfer may occur. Delgado & Thomas (1981) were the first to call attention to the occurrence of such a second mass-transfer phase and gave it the name Case BB RLO (see also Section 10.13). For helium stars more massive than $\sim 3.5\, M_\odot$ and high metallicity, this radius expansion is strongly reduced or absent, implying that for such Case B systems with primaries more massive than $\sim 15\, M_\odot$, subsequent Case BB mass transfer will not occur.

Even before Case BB RLO was formally defined by Delgado & Thomas (1981), De Greve & De Loore (1976) had calculated the evolution of such a system until the end. They computed the evolution of a system of $10 + 8\, M_\odot$ and initial orbital period of 3.15 d, which is depicted in Figure 9.15. After the first mass transfer, the secondary (accretor), now a star of $16.34\, M_\odot$, is rejuvenated and restarts its evolution practically on the ZAMS. The orbital period is then 81.07 d, and the helium core left by the primary is $1.66\,M_\odot$ on the helium main sequence. This star terminates core-helium $\sim 2.2 \times 10^6$ yr after the mass transfer. It has a contracting degenerate CO core of mass $0.94\,M_\odot$ and moves to the giant branch to overflow its Roche lobe again. Case BB mass transfer at a rate of $2 \times 10^{-5}\,M_\odot\,\mathrm{yr}^{-1}$ takes place for 3×10^4 yr, leaving a WD remnant of $1.12\,M_\odot$ in a 236 d orbit around a $16.88\,M_\odot$ star. This will be a very rapidly rotating B0.5V main-sequence star, presumably a Be-star, as will be argued in Section 9.6.1 and Chapter 10. This was the first ever calculation in which a CO WD was directly produced by stellar evolution.

Many similar calculations of the evolution of intermediate-mass binaries, through Case B and Case BB RLO, up to the final remnant of the primary star as a WD, were presented, for example, by Iben & Tutukov (1985), which we particularly recommend to the reader.

9.5 EXAMPLES OF NON-CONSERVATIVE MASS TRANSFER

9.5.1 Modelling the Post-Algol System V106

In Section 9.4.2 when we discussed the formation of Algol-type semi-detached binaries, we noted that many of the observed systems show signs of losses of mass and orbital angular momentum during their evolution (van Rensbergen & De Greve, 2016, 2020). An example of non-conservative low-mass Case B RLO leading to an Algol binary is the star V106 in the old open cluster NGC 6791. It consists of a $1.67\,M_\odot$ blue straggler star (BSS) orbiting a $0.182\,M_\odot$ extremely low mass (ELM) helium WD in a tight binary with an orbital period of $P_{\rm orb} = 1.45$ d Brogaard et al., 2018. In this system, the still bloated (proto-)ELM WD was originally the more massive of the two stars in the binary. V106 is a non-eclipsing version of the EL CVn systems that contain an A- or F-type dwarf star and an ELM He WD (Chen et al., 2017). These systems are also characterized by obeying the WD mass-orbital period relation (Tauris & Savonije, 1999).

A MESA modelling of the evolution of the V106 system is shown in Figure 9.16. The age of V106 is calculated to be ~ 8.42 Gyr. The initial binary consisted of two ZAMS stars with component masses of $1.15\,M_\odot$ and $0.80\,M_\odot$ and an orbital period of $P_{\rm orb} = 3.42$ d. These stars then evolved via magnetic braking and early Case B RLO (initiated when $P_{\rm orb} = 0.82$ d) to produce the current ELM He WD and the BSS, respectively. The fact that V106 is a member of a star cluster, whose properties are well determined from observations, can be exploited to obtain even more information about the system and constrain uncertain physics in the applied binary evolution models.

An interesting finding by Brogaard et al. (2018) is that in this system the RLO only terminated completely ~ 40 Myr ago, and that the system will come into contact, producing a merger, in ~ 80 Myr when the BSS will fill its Roche lobe. A merger event is unavoidable from this future RLO because it will be dynamically unstable due to the large mass ratio between the BSS donor star and the ELM WD accretor. These constraints, as well as the input parameters behind the numerical model, are based on the present observables of the system: M, R, and $T_{\rm eff}$ of the two stars, as well as their age, t, and $P_{\rm orb}$.

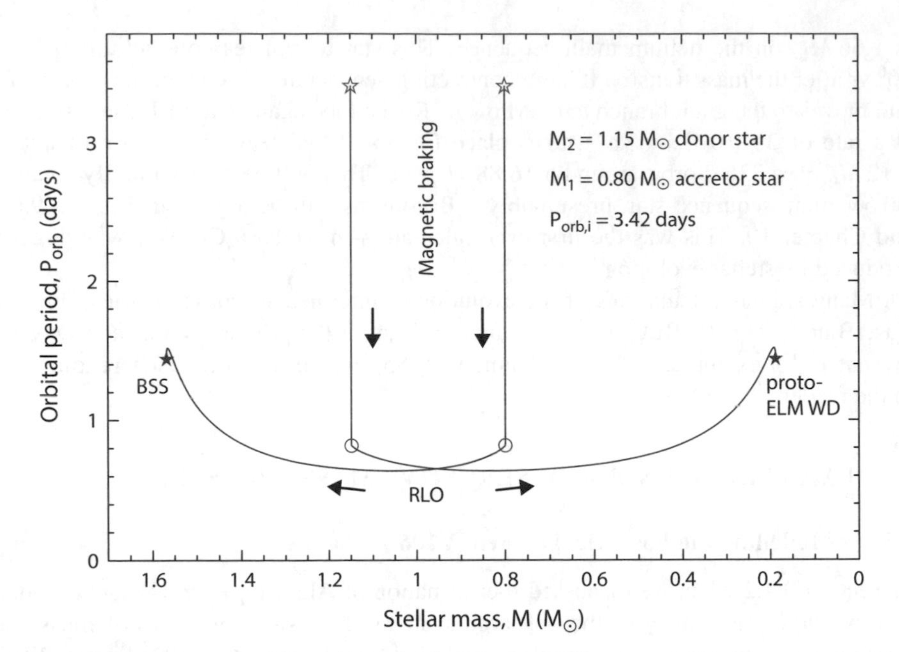

Figure 9.16. Formation model of the post-Algol V106 system via mass transfer in a binary. The initial configuration is two ZAMS stars (open star symbols) of masses 1.15 and 0.80 $M_\odot$, in a circular orbit with an orbital period of 3.42 d. The 1.15 $M_\odot$ donor star evolves to become the present secondary star ($\sim$0.18 $M_\odot$ proto-ELM WD), while the 0.80 $M_\odot$ star accretes material and produces the present primary star ($\sim$1.6 $M_\odot$, BSS); see solid star symbols. As a result of magnetic braking, the orbital period decreased to $\sim$0.82 d prior to RLO (open circles). After Brogaard et al. (2018).

The modelling of the evolution of this system is non-conservative due to both magnetic braking (which converts orbital angular momentum into stellar spin angular momentum) and mass loss from the system. Further details of the model are provided in Figure 9.17, which shows the mass-transfer rate (top panel) and the evolution of the donor star (i.e., the present secondary star that is now an ELM He WD) in the ($T_{\rm eff}$, $\log g$) diagram in the bottom panel. In this model, it was assumed ad hoc that 20% of the transferred material was lost from the system, that is, $\beta = 0.20$ which means that this fraction of the transferred material is ejected from the accretor, and thus is carrying the specific orbital angular momentum of the accretor (see Section 4.3.7). However, this β value is quite uncertain and could easily be somewhere in the interval $\beta = 0.05 \sim 0.25$. The β value depends, for example, on the mass-transfer rate and the saturation of accretor star spin rate (when it reaches critical rotation, it will start to shed some the incoming mass transferred from the donor star). Finally, we notice that a system analogous to V106, but with a massive WD companion, is the post-CV system LAMOST J0140355+392651 (El-Badry et al., 2021).

We see from these examples (V106 and the previous conservative model of Kippenhahn et al. [1967], presented in Figures 9.9 and 9.10) that both conservative and

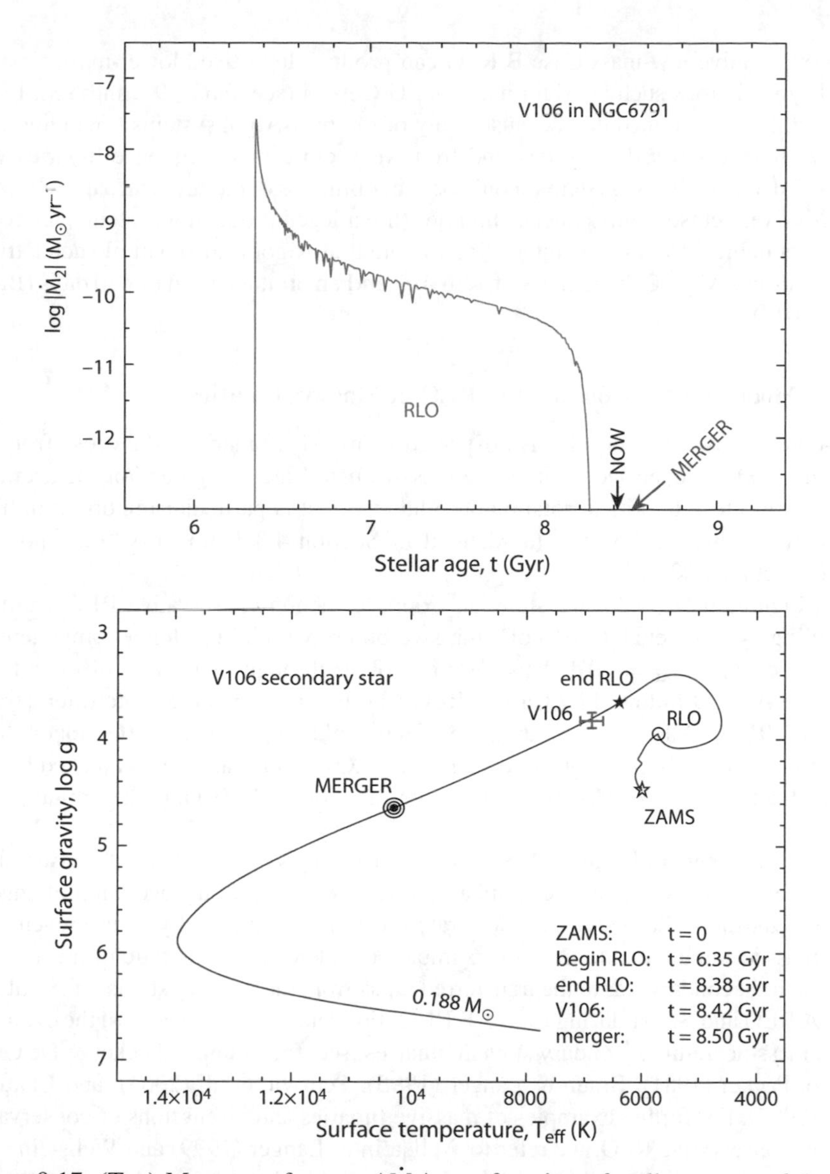

Figure 9.17. (Top) Mass-transfer rate, $|\dot{M}_2|$ as a function of stellar age, t, of the former donor star in V106 (i.e., the present secondary star; see Figure 9.16). The RLO ceased completely at $t \simeq 8.38$ Gyr, ~40 Myr ago. The present age of the modelled system is $t = 8.42$ Gyr (marked by "now"), in agreement with current age estimates of NGC 6791. It is anticipated that the system will merge in ~80 Myr when the primary star (BSS) fills its Roche lobe. Some numerical noise is seen in the calculated values of $|\dot{M}_2|$. (Bottom) Past and future evolution of the secondary star in V106 in the ($T_{\rm eff}$, $\log g$) diagram, according to the model. The current location of the observed $0.182 \pm 0.006\, M_\odot$ secondary star in V106 is plotted with a green cross (error bars shown) on top of the track for a $0.188\, M_\odot$ proto-ELM WD. V106 is expected to initiate a merger event in ~80 Myr. After Brogaard et al. (2018).

non-conservative low-mass Case B RLO can produce long-lived lower-mass classical Algol-type binaries such as Algol itself and U Cephei (see Tables 9.2 and 5.3). Lagos et al. (2020) have argued that because many of the post-Algol systems (including ones that we have discussed) may descend from very close main-sequence binaries with $P_{\rm orb} < 3$ d, most of these systems should be inner binaries of hierarchical triples because >95% of very close main-sequence binaries (the alleged progenitor systems) are found to be hierarchical triples. It is interesting to note that Algol is also part of such a triple. The third star, Algol C, has a mass of 1.76 $M_\odot$ and an orbital period of 680 days (Baron et al., 2012).

9.5.2 Modelling Non-conservative RLO in Massive Binaries

Non-conservative evolution is also expected in massive binaries. Mass loss from the system ($\beta \neq 0$) is expected if either the mass-transfer rate is high and/or the accreting star evolves close to critical rotation. The latter effect is particular important in high-mass binaries where β-values (as defined in Section 4.3.7) up to 0.90 are possible (Petrovic et al., 2005).

In Figures 9.18–9.20, we show an example of non-conservative RLO from the modelling by Yoon et al. (2010) of a massive binary with initial stellar components of $18 + 17\,M_\odot$ and $P_{\rm orb} = 4.0$ d. This system evolves through two stages of mass transfer: Case A RLO followed by Case B RLO (the latter stage is therefore often termed Case AB RLO). A Kippenhahn diagram of the evolution of the 18.0 $M_\odot$ donor star is shown in Figure 9.18. After the two stages of RLO, the primary star is a naked helium star with a mass of 3.79 $M_\odot$ (with a CO core mass of 2.14 $M_\odot$) and the orbital period is 29.7 d.

The lower panel of Figure 9.19 shows that the mass transfer is not conservative. When the secondary star reaches critical rotation as a result of the accretion of angular momentum, the stellar wind mass-loss rate increases so drastically as to prevent efficient mass accumulation. In the case example considered here, the ratio of the accreted mass in the secondary star to the transferred mass from the primary star is ~0.83 during Case A RLO and ~0.41 during Case AB RLO. For detailed discussions on the evolution of the mass-accreting secondary stars in binaries, see, for example, Packet & De Greve (1979), Packet (1981), Braun & Langer (1995), Petrovic et al. (2005), and Cantiello et al. (2007). For further examples of massive binaries and discussions of conservative vs. non-conservative RLO, we refer to Wellstein & Langer (1999) and Wellstein et al. (2001), for example. Detailed models of massive Algols and their descendants were studied by Sen et al. (2022).

To help determine the fate of the donor star in the binary system calculated by Yoon et al. (2010), Figure 9.20 shows the mean specific angular momentum as a function of evolutionary time of the innermost 3.0 $M_\odot$ and 1.4 $M_\odot$ of the donor star in Figures 9.18 and 9.19. It is seen that angular momentum in the core of the donor (primary) star is mostly removed during the mass-transfer phases (Case A and Case AB RLO). The decrease of the core angular momentum during Case A RLO results from the synchronization that occurs even when the orbit is rapidly widened due to mass

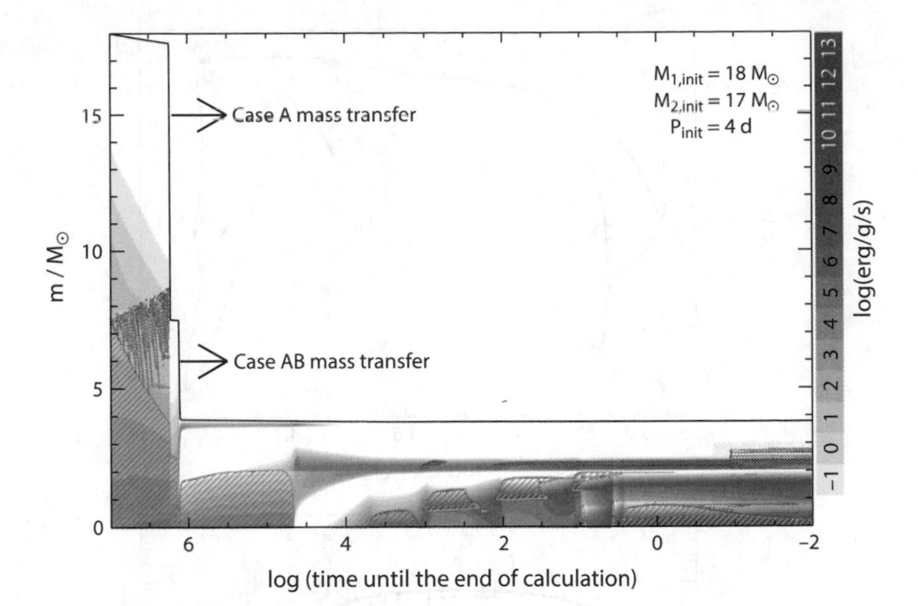

Figure 9.18. Kippenhahn diagram of the evolution of a 18.0 $M_\odot$ donor star in a binary with a 17.0 $M_\odot$ companion star with an initial orbital period of 4.0 d. The hatched lines and the red dots denote convective layers and semi-convective layers, respectively. The different shades give the nuclear energy generation rate, for which the scale is shown on the right-hand side. The surface of the star is marked by the topmost solid line. After Yoon et al. (2010).

exchange. During Case AB RLO, the further increase of the orbital separation significantly weakens the role of synchronization for the redistribution of angular momentum inside the donor star. However, both the Spruit-Tayler dynamo (a magneto-hydrodynamical model of angular momentum transport in radiative zones of differentially rotating stars; see, e.g., Spruit [2002]) and the mass loss lead to rapid braking of the thermally contracting helium core. Further significant core braking by the Spruit-Tayler dynamo occurs during the CO core contraction phase, and the mean specific angular momentum of the innermost 1.4 $M_\odot$ becomes about 2.5×10^{14} cm^2 s^{-1} at the neon burning phase, as seen in Figure 9.20.

Yoon et al. (2010) argue that in comparison to models of single stars, this amount of angular momentum retained in the core of the primary star at the pre-SN stage is almost similar. This value is smaller by one or two orders of magnitude than what is necessary to make long γ-ray bursts by magnetar or collapsar formation, or very energetic SNe (hypernovae) powered by rapid rotation and strong magnetic fields (e.g., Burrows et al., 2007), although it may suffice to produce MSPs (Heger et al., 2005). Yoon et al. (2010) suggest that the destiny of the modelled donor star is instead a Type Ib SN given the rather thick helium envelope with a very thin hydrogen layer, but it might perhaps also appear as a Type IIb SN.

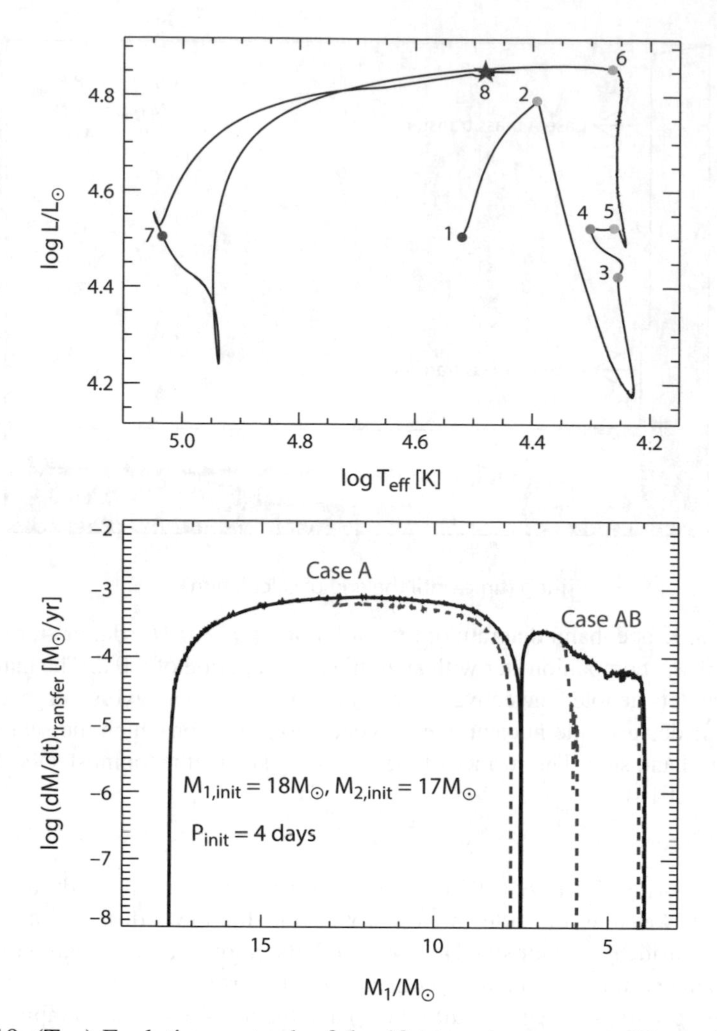

Figure 9.19. (Top) Evolutionary track of the 18 $M_\odot$ primary star in Figure 9.18. The solid circles on the track mark different evolutionary epochs as the following: (1) ZAMS, (2) onset of Case A RLO, (3) end of Case A RLO, (4) core hydrogen exhaustion, (5) onset of Case AB RLO, (6) end of Case AB RLO, (7) core helium exhaustion, and (8) neon burning (end of calculation). (Bottom) Mass-transfer rates from the primary star (solid line) and mass-accretion rates onto the secondary star (dashed line) during Case A and Case AB RLO as a function of the primary star mass. After Yoon et al. (2010).

9.5.3 Case C RLO

In Case C RLO, the primary star has a deep convective envelope when it begins to overflow its Roche lobe. When it loses mass, such an envelope expands on a dynamical timescale. Therefore, the onset of RLO will in most cases lead to very large rates of

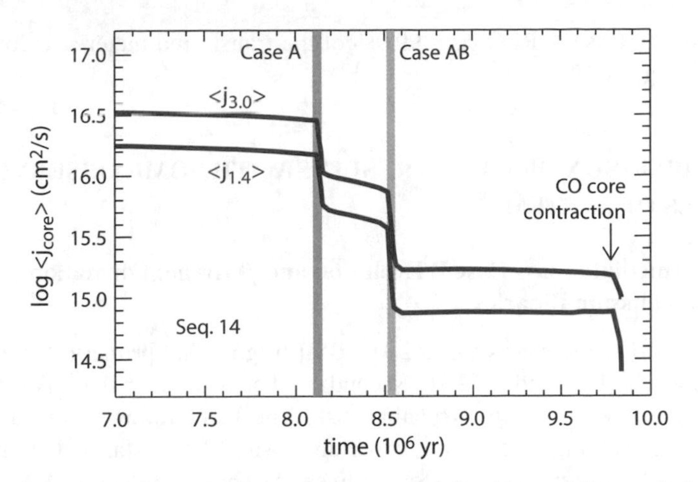

Figure 9.20. Mean specific angular momentum as a function of the evolutionary time of the innermost 3.0 $M_\odot$ and 1.4 $M_\odot$ of the primary star in Figures 9.18 and 9.19. The time spans for Case A and AB RLO are color-coded green and yellow, respectively. After Yoon et al. (2010).

mass transfer, which the companion star will be unable to accommodate. In most cases, the result will be the formation of a CE, (see Section 4.8) and large mass ejection from the system, while the companion star spirals inward very deeply. As the timescale for in-spiral will be very short, the companion will have no time to accrete much mass and will emerge from the CE stage basically unchanged. Because in Case C, the core of the donor (primary) star had time to grow through previous hydrogen- and helium-shell burning, the core of such a star will be (much) more massive than that of a Case B donor star with the same ZAMS mass. Its core mass will be similar to that of a single star (see Section 9.7). After Case C RLO and CE-evolution, a binary will result consisting of this relatively massive core plus the unchanged companion star, in most cases in a relatively narrow orbit. If the core is sufficiently massive, it will collapse to a NS or a BH and, if the system is not disrupted in the accompanying SN explosion, a system will result consisting of a NS or BH plus a low- or intermediate-mass companion star. Such systems may resemble a LMXB or an IMXB, with an accreting NS or BH. If the mass of the core of the donor star is too low to collapse to a NS or BH, the final system will consist of WD plus a low- or intermediate-mass companion star. Such systems resemble the CVs and the super-soft X-ray binaries described in Chapter 5. Much work on the formation of such WD binaries was done by Iben, Tutukov, and Yungelson. For the formation and evolution of LMXBs and IMXBs, we refer to Chapter 11.

9.5.4 Non-conservative RLO in X-ray Binaries

As a result of the Eddington accretion limit (Section 4.9), X-ray binaries with high mass-transfer rates often experience extremely non-conservative evolution because the compact objects simply cannot accrete at the same rate as mass is being transferred toward them. Such an example is given in Section 14.4.2, where $\beta \simeq 0.999$ during phase A1 of the Case A RLO (i.e., 99.9% of the transferred material is lost from the system; see Fig. 14.17).

9.6 COMPARISON OF CASE B RESULTS WITH SOME OBSERVED TYPES OF SYSTEMS

9.6.1 Intermediate-mass Case B Evolution and β Lyrae: Formation of B-emission Binaries

The system of β Lyrae consists of a 2 $M_\odot$ B8II bright giant primary star that is overflowing its Roche lobe and a 12 $M_\odot$ secondary that is dimmer than the primary and has the appearance of a rapidly rotating disk. The latter follows from the shape of the light curve (see Figure 5.2), the underluminosity of this star and from the enormous rotational broadening of its spectral lines. As shown in Section 4.4.2, the mass-transfer rate in the system, derived from the rate of change of its orbital period, is $1.36 \times 10^{-5}\ M_\odot\ \mathrm{yr}^{-1}$, which implies that the B8II giant is still transferring mass to its companion. Judging from its large luminosity, it consists mainly of the helium core of the initial primary surrounded by a low-mass hydrogen-rich envelope and is close to the end of its mass-transfer phase.

Most remarkable is that in the near-infrared spectrum of β Lyrae, the absorption lines of CI, which normally are strong in a B8II giant, are absent (see Fig. 9.21), which indicates that carbon is underabundant by a factor of $\geq$25. From the strengths of its helium lines, one finds a composition ratio $Y_{\mathrm{He}}/X_{\mathrm{H}} = 1.5$, which is five times larger than normal. These two facts together imply that the atmospheric material of the primary has been nuclearly processed by the CNO cycle: the material was part of the hydrogen-burning core of the original primary star. It is well-known that during the main-sequence evolution of intermediate-mass and massive stars, the size of the convective core, in which hydrogen-burning proceeds by the CNO cycle, gradually shrinks, leaving behind a region enhanced in helium and depleted in carbon. From this information, Tomkin & Lambert (1987) conclude that the B8II primary star consists of a 1.6 $M_\odot$ helium core surrounded by a 0.4 $M_\odot$ hydrogen rich envelope. When the latter envelope will have been transferred, ~30,000 yr from now, the system will consist of a 1.6 $M_\odot$ helium star and a 12.4 $M_\odot$ companion with an orbital period of ~20 d.

After the disk matter surrounding the companion has settled, this star will become a very rapidly rotating early B-type main-sequence star (spectral type of approximately B2V, see, e.g., Packet & De Greve, 1979). Because very rapidly rotating main-sequence early B-stars in general have the characteristics of Be-stars, shedding some mass from their equatorial regions from time to time (see Chapters 7 and 10, and Fig. 7.4), we expect this star to look like a B2V-emission star (van der Linden, 1987; Habets, 1985, 1987). Thus, ~30,000 yr from now, the system will consist of a B2Ve star with an invisible (because $T_{\mathrm{eff}} \geq 10^5$ K) helium star with an orbital period of $P_{\mathrm{orb}} \simeq 20$ d.

The original mass of the primary star was $\sim 9 - 10\ M_\odot$, which implies that the companion originally had a mass of $\sim 4 - 5\ M_\odot$. In view of the low mass of the helium star that will form in the β Lyrae system, the primary star here will later go into Case BB

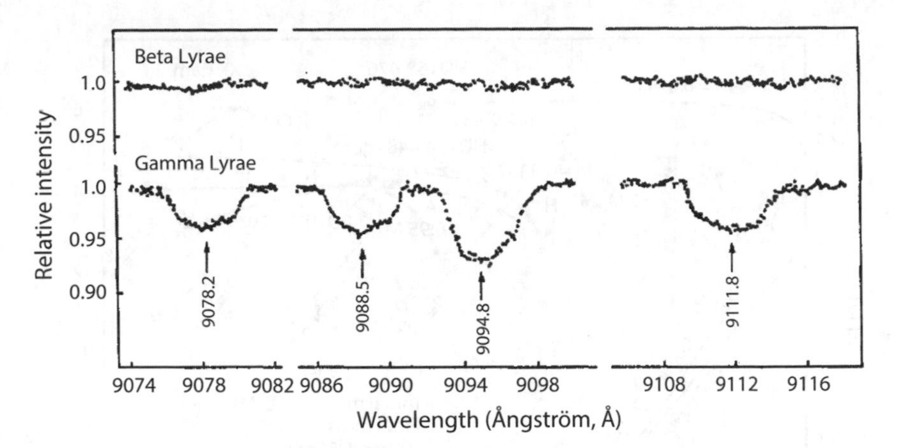

Figure 9.21. The 9,100 Å absorption lines of neutral carbon in β and γ Lyrae. γ is of normal chemical composition, but otherwise similar to β. The complete absence of the lines in the latter indicates an unusually low abundance of carbon in the B8II primary, just as would be expected if it is the evolved core of what was once a more massive star. Further explanation in the text. After Tomkin & Lambert (1987).

mass transfer and, like in the systems studied by De Greve & De Loore (1976) and Iben & Tutukov (1985), will then probably leave a CO WD or ONeMg WD.

9.6.2 R Cor Bor Stars and Helium Stars

Iben & Tutukov (1985) have pointed out that the the evolutionary tracks in the HR diagram of the (almost pure) helium-star remnants of the primary (donor) stars of intermediate-mass close binaries pass through the regions of the HR diagram where extremely hydrogen-deficient giant stars are found. These stars are rare; only a couple dozen of them known. They are the pulsating R Coronae Borealis (abbreviated R Cor Bor or R CrB) stars and the non-pulsating helium stars. For a detailed description of the properties of these stars and how these were discovered, we refer to Iben & Tutukov (1985). They have hydrogen deficiencies (with respect to iron) of a factor of 10^{-5} to 10^{-4}, and an overabundance of CNO by a factor of $10 - 30$. Figure 9.22 shows the positions of several of these stars in the HR diagram, together with the evolutionary tracks of a 6.95 $M_{\odot}$ primary star that evolved according to Case B and Case BB RLO. It is clear from this diagram that the locations in the HR diagram and other properties of the R Cor Bor and helium stars fit well with the tracks in the HR diagram of the RLO remnants of intermediate-mass Case B mass transfer. (For the alternative possibility that R Cor Bor stars are instead the remnants of the merger between a He WD and a CO WD, see recent work by Schwab [2019].)

9.6.3 Identification of Binaries with an Intermediate-mass Core-helium-burning Stripped Star Component

As mentioned in Chapter 5, helium stars (in the phase of core-helium burning) with masses upward of $\sim 8\, M_{\odot}$ are recognized as WR-stars. Because of their very high

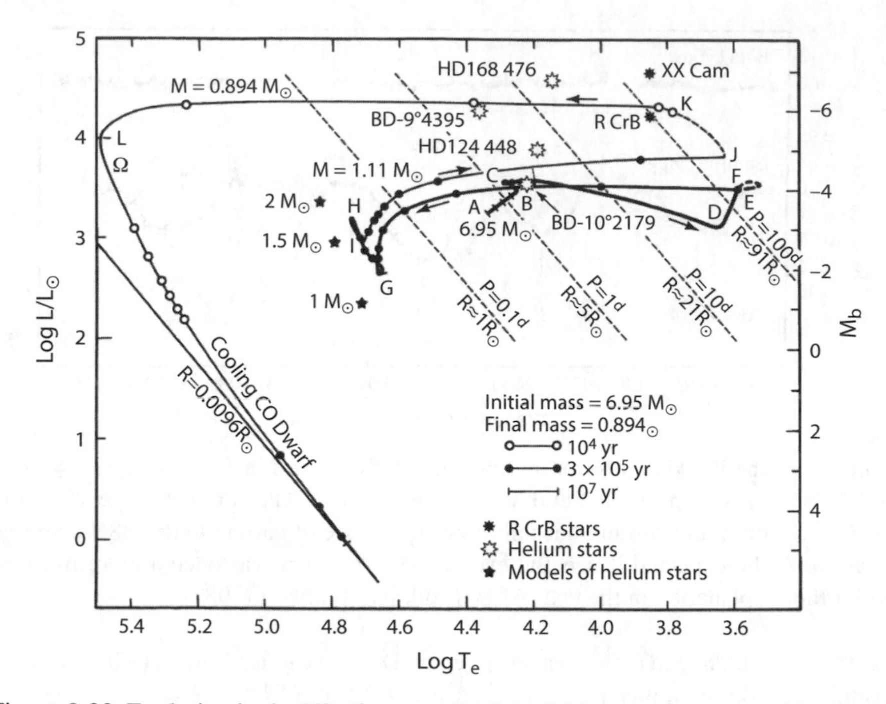

Figure 9.22. Evolution in the HR diagram of a Case B binary component with an initial mass of 6.95 $M_\odot$ and $X = 0.70$ and $Z = 0.02$. The positions of pure helium stars are given by the five-pointed stars (taken from Paczyński, 1971a). Lines of constant orbital period and Roche-lobe radius are for binaries with a mass ratio $q = 1.0$. The temperature of the CO shell reaches a maximum at the point Ω. Mass loss occurs along the dashed parts of the track. Solid eight-pointed stars give positions of R Cor Bor stars; open eight-pointed stars are helium stars. After Iben & Tutukov (1985).

luminosites, these stars have very strong radiation-driven stellar winds, which produce the characteristic WR emission-line spectra in the optical part of the spectrum, making it easy to recognize them in optical surveys.

On the other hand, core-helium-burning helium stars of lower mass, on or near the helium main sequence, appear to have much weaker winds and do not produce clear emission spectra; therefore, they are hard to recognize in the optical part of the spectrum. As Figures 9.12 through 9.15 and 9.22 show, for high metallicities their effective temperatures are of the order of $0.5 - 1 \times 10^5$ K, such that their energy output peaks in the UV part of the spectrum and is highly ionizing for neutral hydrogen. Figure 9.13 shows that the effective temperatures and thus also the ionizing power of a stripped star varies as a function of metallicity.

Because of the high incidence of close binaries among O- and B-type stars, and the fact that a large fraction of such systems will evolve according to Case B RLO, one expects the products of this evolution—stripped-star (helium-star) binaries with an intermediate-mass main-sequence companion—to be very common. This was recognized already in the late 1960s. Still, until quite recently, very few systems with an intermediate-mass stripped star were recognized. This is well understandable because for systems that evolved conservatively, recognition of the stripped star is difficult as the companion star generally is a luminous early B-type main-sequence star that in the optical part of the spectrum outshines the stripped star. The observationally confirmed sample of such systems, discovered in the past 40 yr, consists of a handful of very hot subdwarfs: ϕ Per, 59 Cyg, 60 Cyg, HR 2142, HD 55606, and FY CMa, which all have rapidly rotating Be-type companions (see references in Götberg et al., 2018, and Chojnowski et al., 2018). The systems tend to be characterized by a UV-color excess, which is expected because of the high effective temperatures of the helium stars. As Götberg et al. (2018) pointed out, systems that are most easily recognized in the optical spectrum are those with relatively large stripped-star masses and relatively low companion masses.

Systems with a large mass ratio of initial primary and secondary stars and a wide orbit will have evolved highly non-conservatively, with large mass loss from the system, in a CE phase. Here, the secondary will hardly have gained any mass. It will then be a relatively low-mass star with a low luminosity, such that the helium star can outshine it and become observable. For example, this is the case in system HD 45166 that consists of an $\sim$4.5 $M_{\odot}$ stripped star with a WR-like spectrum and a B5V main-sequence star of approximately the same mass (Groh et al., 2008). The stripped star is clearly visible in the composite spectrum (Steiner & Oliveira, 2005), with emission lines such as HeII 4686 Å. The latter are due to the wind of the helium star, which is at least an order of magnitude weaker than that of normal classical WR-stars, but still produces an emission spectrum. Systems with similarly relatively low-mass companions to the stripped stars were recently identified as bright UV binaries in the LMC (see Fig. 9.23). They are of two types: so-called WN3/O3 stars (the magenta diamonds) and UV OGLE eclipsing binaries. The latter are eclipsing binaries detected in the Optical Gravitational Lensing Experiment (OGLE) survey in the V band (Udalski et al., 2008) and their UV flux was measured in the UVM2 band of the Swift satellite.[2] Among the UV OGLE eclipsing binaries in Figure 9.23 there are three different categories: (1) helium stars (dark blue circles), which must have much less luminous companions that are presumably low- or intermediate-mass main-sequence stars outshone by the helium star; (2) systems with B-type spectra (the light-blue circles), where the helium star is less bright than its B-type main-sequence companion; and (3) systems with composite spectra (medium-blue circles), where the helium star and its main-sequence companion must be similar in luminosity. As the figure shows, the stripped stars (helium stars) are mostly less massive than 7 $M_{\odot}$ and therefore will not be observed as classical WR-stars. The WN3/O3 stars are a particularly interesting group that we will now discuss.

9.6.4 The WN3/O3 Stars: Stripped Stars Originating from Intermediate-mass Binaries?

By precisely computing model atmospheres of the stripped stars, including wind outflows, Smith et al. (2018) found that they are expected to span a continuous range

[2] See references in Götberg (2018).

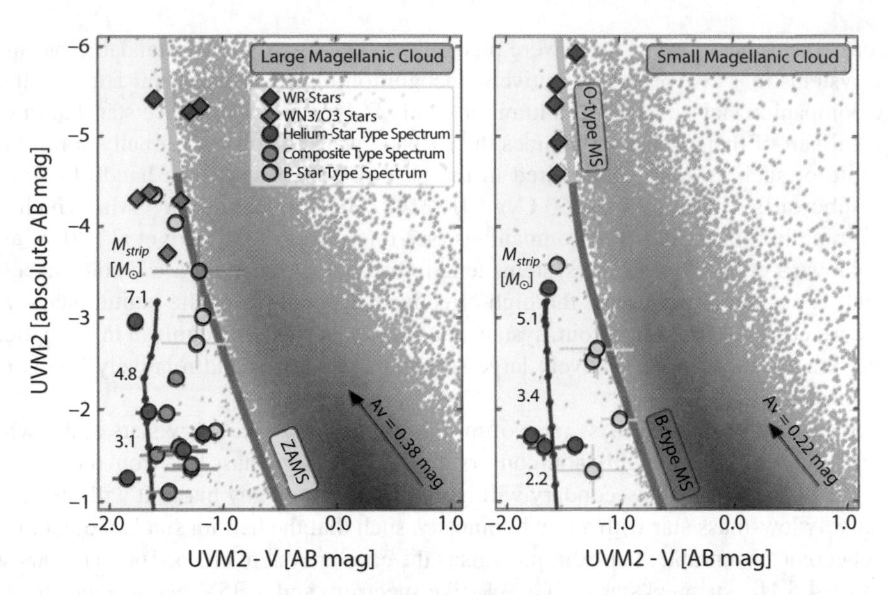

Figure 9.23. UV color-magnitude diagram of identified UV-bright LMC and SMC stars from UVM2 data from the Swift Ultraviolet Survey of the Magellanic Clouds (SUMaC) and V data from the OGLE survey (Udalski et al., 2008), corrected for foreground extinction. The candidate stripped-star systems spend most of their lifetimes in the region depicted to the left of the main sequence. The different kinds of stripped stars have been indicated. (See previous figures and the text for explanations of the types.) All these stars are binaries that were detected by OGLE. Models of single stripped stars are marked by the dark almost vertical curves with the most massive one in the LMC being 7.1 $M_{\odot}$. With a companion star, one expects the composite color of the system to be redder, and there should therefore also be systems between the location of the single stripped models and the ZAMS, as is shown in the figure. Credit: M. R. Drout, Y. Goetberg, and B. Ludwig in Drout et al. (in preparation, 2021).

of spectral types, from WR-like spectra characterized by emission lines formed in the winds of the more massive and metal-rich stripped stars, to subdwarf-like spectra dominated by absorption features resulting from the photosphere of stripped stars with transparent outflows. They further predicted the existence of a hybrid intermediate class of spectra showing a combination of absorption and emission lines, very similar to those observed for the recently discovered new class of WN3/O3 stars.

The WN3/O3 stars form a new class of peculiar objects, distinctly different from the classical massive and highly luminous WR+O binaries. This new class was recognized by Massey et al. (2014) and Massey et al. (2015, 2017) in surveys for WR-stars in the Magellanic Clouds (SMC, LMC). These authors pointed out that the stars are far too faint for their spectra to simply be spectra of WR+O3V pairs, and that such pairings are also unlikely because the lifetimes of O3V stars are too short to allow the companion to evolve into a WN-star. The spectra of the WN3/O3 stars, when compared to model atmosphere spectra, show very high effective temperatures ($T_{\rm eff} = 80{,}000-100{,}000$ K);

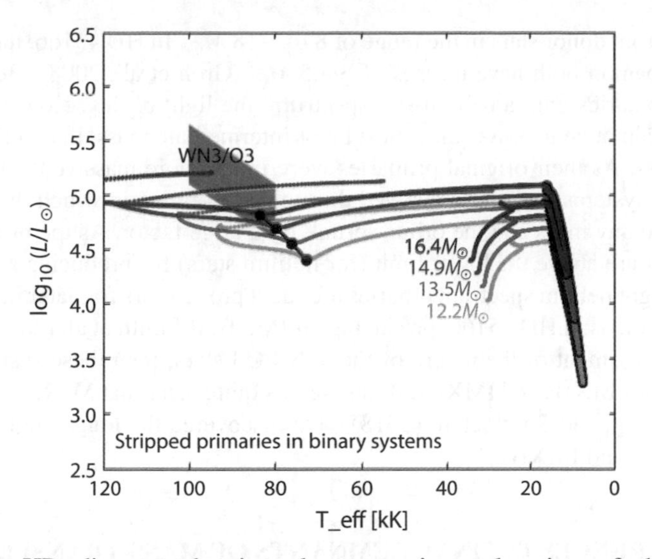

Figure 9.24. HR diagram showing the approximate location of the observed WN3/O3 stars shaded in green. The figure shows the tracks of primary stars in binary systems (solar-like metallicity), and the black dots show the location of the stripped stars halfway through core-helium burning. Labels indicate initial masses at the onset of hydrogen burning. After Smith et al. (2018).

hydrogen depletion; a factor of 10 nitrogen enhancement; luminosities of $\sim 10^5\,L_\odot$; and mass-loss rates of $\approx 10^{-6}\,M_\odot\,\mathrm{yr}^{-1}$, which are more than an order of magnitude lower than for classical WR-stars (Massey et al., 2014; Neugent et al., 2017). In visual light, the WN3/O3 stars are several magnitudes fainter than the classical WN-stars are (see Fig. 9.23).

Smith et al. (2018) show that as to their locations in the LMC, these stars are extremely isolated from the regions where the normal massive O- and WR-stars are found, which indicates that they are not evolutionary related to the populations of very massive stars in the LMC. They argue, in our view convincingly, that according to their locations in the HR diagram, these stars are most probably stripped stars in binaries originating from initial donor (primary) stars with masses between ~ 12 and $\sim 18\,M_\odot$ (see Fig. 9.24). The reason why the companion star is not seen is, most probably, that it has a relatively low mass, say $2-5\,M_\odot$, such that its luminosity is too low to be detected. The WN3/O3 systems will then be the result of highly non-conservative (CE) evolution, initiated by late Case B or Case C RLO, in which the low-mass secondary star hardly accreted any matter.

9.6.5 The Stripped UV-bright LMC Stars in Binaries and HD 45166: Progenitors of LMXBs and IMXBs?

The UV-bright OGLE binaries in the LMC, indicated in Figure 9.23 are, according to their luminosity, stripped stars (helium stars) with masses in the range of 3–7 $M_\odot$, originating from donor stars in the range of 8 to $\sim$18 $M_\odot$. In HD 45166, the helium star and its companion both have masses of $\sim$4.5 $M_\odot$ (Groh et al., 2008). Because in the UV OGLE binaries with a helium-star spectrum, the light of the secondary star is not seen, these cannot be massive: they must be of intermediate mass, like in HD 45166, or of lower mass. As their original primaries were much more massive than their secondaries, these systems must have evolved through a CE phase in which the secondaries hardly gained any mass and the orbits shrunk by a large factor. As the masses of these stripped stars are above the lower limit (for helium stars) for producing a NS or a BH, these UV-bright helium spectrum binaries are ideal progenitors for, later in life, producing a LMXB or, as in HD 45166, producing an IMXB. If Smith et al. (2018) are correct with their assessment of the nature of the WN3/O3 stars, then these systems are also progenitors of LMXBs and IMXBs. It thus seems quite clear that M. R. Drout, Y. Goetberg, B. Ludwig, and Smith et al. (2018) have discovered the long-sought progenitors of the LMXBs and IMXBs.

9.7 DIFFERENCES IN FINAL REMNANTS OF MASS-TRANSFER BINARIES AND SINGLE STARS

In Table 8.3, we display a rough mapping between initial ZAMS masses, helium star masses, and final compact objects left behind. This mapping, however, is strongly dependent on whether a star evolves as an isolated star or in a close orbit, where it is subject to mass loss early in its evolution. The differences in stellar evolution between single and binary stars have been highlighted in a number of papers, for example, Podsiadlowski et al. (1992), Brown et al. (2001), Podsiadlowski et al. (2004), and most recently by Woosley (2019).

9.7.1 Key Differences between Final Remnants of Case B and Single Star/Case C Evolution

It turns out that for the same initial primary star mass, the final outcomes of these two types of binary evolution can be very different. This is due to the fact that in Case B, the mass of the remnant of the primary star is that of the helium core at helium ignition (about the same core mass as when the star left the main sequence). On the other hand, in Case C it is the mass of the core at the time of carbon ignition, which is much larger because during core-helium burning the core still grows, like in a single star, by hydrogen-shell burning. For the same ZAMS primary star mass, this Case C core mass is basically equal to that of a single star of this mass at carbon ignition. As the time between carbon ignition and core collapse is very short, the core mass does not grow after carbon ignition. Therefore, the core mass at this stage is a very good approximation of the core mass at the time of core collapse. Figure 9.25 shows these different core masses as a function of initial primary star mass for solar metallicity. One immediately sees in this figure that the helium core masses left after Case B RLO are much smaller than the final core masses of single stars. The figure allows one to determine

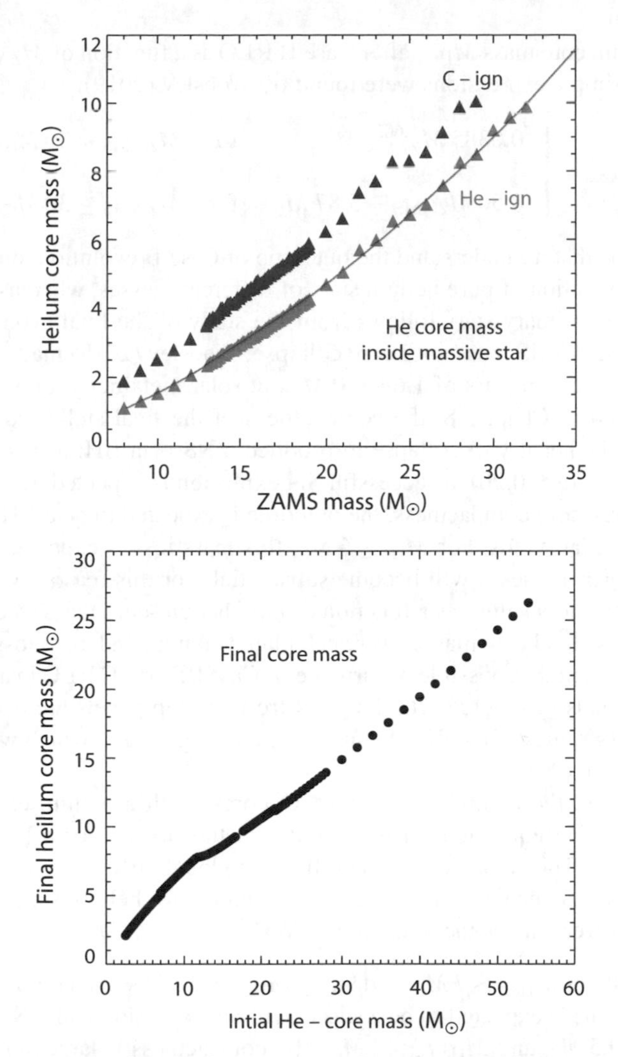

Figure 9.25. (Top) Mass of the helium core as a function of ZAMS mass, at the times of helium ignition and carbon ignition, for single stars at solar metallicity. These progenitor stars do not lose all their hydrogen envelope prior to core collapse. The upper (red) and lower (green) rows of triangles are core masses after carbon and helium ignition, respectively. The upper row thus also reflects the core mass at onset of core collapse. (Bottom) Final pre-SN helium-core mass as a function of initial naked helium-core mass, assuming a standard wind mass-loss prescription. The inflection $\sim$11 $M_{\odot}$ reflects the uncovering of the CO core by wind mass loss. After Woosley (2019).

the initial helium core mass $M_{\mathrm{He,i}}$ after Case B RLO as a function of M_{ZAMS}, for which the following simple expressions were found by Woosley (2019):

$$M_{\mathrm{He,i}} = \begin{cases} 0.0385\, M_{\mathrm{ZAMS}}^{1.603}\, M_\odot & \text{for} \quad M_{\mathrm{ZAMS}} < 30\, M_\odot \\ 0.50\, M_{\mathrm{ZAMS}} - 5.87\, M_\odot & \text{for} \quad M_{\mathrm{ZAMS}} \geq 30\, M_\odot \end{cases} . \tag{9.8}$$

The result is that, to understand the outcome of Case B evolution, one has to study the complete evolution of pure helium stars of different masses, whereas the final evolution of Case C primary stars follows from the study of the final evolution of single stars of the same ZAMS mass until core collapse. Woosley (2019) made a full study of the evolution of helium stars of $1.6-120\, M_\odot$, at solar metallicity, until core collapse. As we discussed in Chapter 8, the compactness of the final stellar core is expected to determine whether it will collapse to produce a NS or a BH: at low values of the core compactness ($\xi \lesssim 0.20$), a successful SN explosion is expected, resulting in a NS, whereas for large core compactness, the outcome is expected to be a BH. Helium stars lose mass by stellar winds. For $M_{\mathrm{He}} \leq 6\, M_\odot$ this mass loss is expected to be negligible, but for larger masses it will become substantial. For this reason, Woosley (2019) gives the compactness values as a function of *final* helium-core mass (see top panels of Fig. 9.26). (These final core masses were calculated taking wind mass loss into account, but they did not include a possible occurrence of Case BB or BC RLO in a close binary.) In the lower panels of Figure 9.26, the plots from the top panels are overplotted with the compactness values of the final helium cores of single stars, which were calculated by Sukhbold et al. (2018).

As mentioned, these single star final CO cores (with a helium envelope) are in good approximation equivalent in mass and structure to the cores left behind from Case C evolution. This figure shows the following global differences between the final outcomes of Case B and Case C (single star) evolution, for helium stars with a mass of $\gtrsim 1.8\, M_\odot$ (for lower masses the remnant is a WD):

- Case B RLO: For $M_{\mathrm{He,f}} \lesssim 7\, M_\odot$ and $9.5 \lesssim M_{\mathrm{He,f}} \lesssim 12\, M_\odot$, the compactness is small and the outcome is expected to be a successful SN explosion and a NS remnant. For $M_{\mathrm{He,f}} \simeq 7-9.5\, M_\odot$ and $M_{\mathrm{He,f}} \gtrsim 12\, M_\odot$, the compactness is large, and the outcome is expected to be a BH remnant.
- Case C RLO (and single stars): For $M_{\mathrm{He,f}} \lesssim 6.5\, M_\odot$, between islands of low compactness where successful SNe and NSs are expected, there are islands of high compactness around $4.3\, M_\odot$, $5.1\, M_\odot$, $5.6\, M_\odot$, and $M_{\mathrm{He,f}} \simeq 6.5-7.5\, M_\odot$, where BHs are expected to be produced, as they are for $M_{\mathrm{He,f}} \gtrsim 10\, M_\odot$; while for $M_{\mathrm{He,f}} \simeq 7-10\, M_\odot$, NSs may still be produced.

We see from Figure 9.26 and the preceding discussion that in Case B RLO, only for quite high final helium-core masses, $M_{\mathrm{He,f}} = 7-10\, M_\odot$, BHs can be produced; while for $M_{\mathrm{He,f}} \lesssim 7\, M_\odot$, the result will always be a successful SN and production of a NS. According to Figure 9.25, a final helium-core mass of $M_{\mathrm{He,f}} = 7\, M_\odot$ means an initial helium-core mass of $M_{\mathrm{He,i}} \simeq 11\, M_\odot$, which means a $\sim 35\, M_\odot$ ZAMS primary star. Thus, we see that all helium stars (WR-stars) originating through Case B evolution from

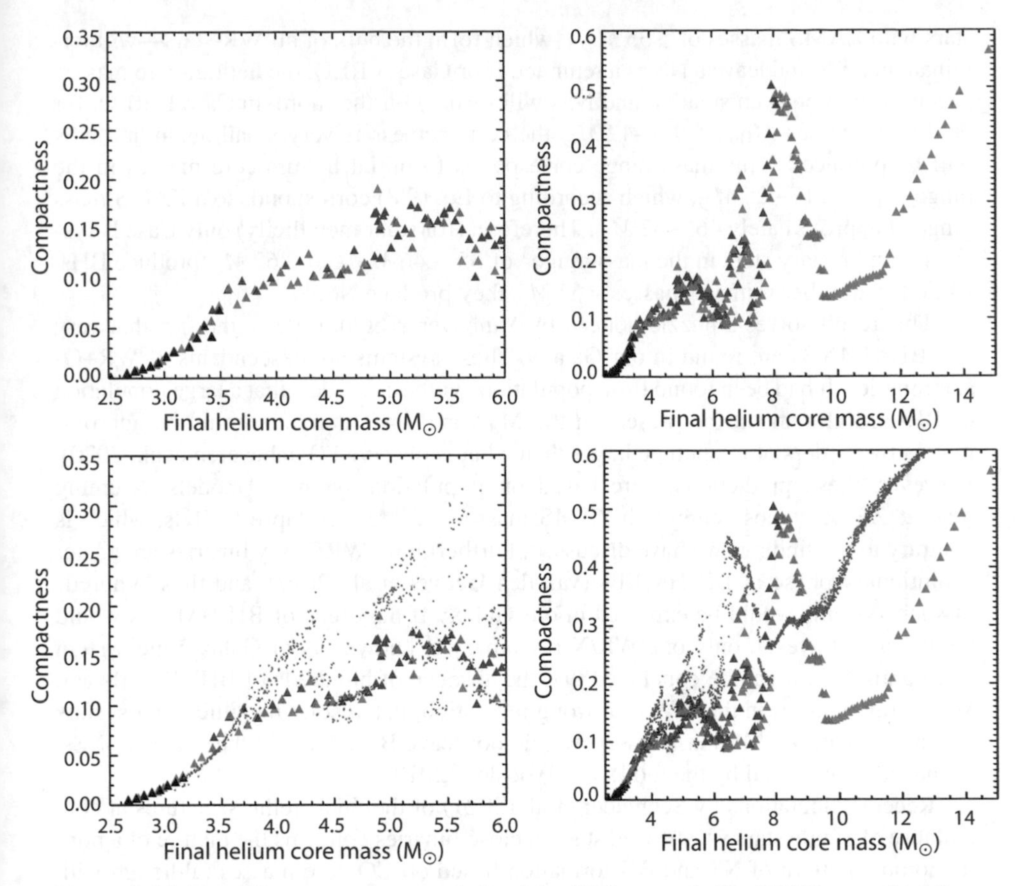

Figure 9.26. Pre-SN compactness as defined by O'Connor & Ott (2011), as a function of final helium core mass. (Top) Results for naked helium stars (Case B remnants). The different color points indicate variation in stellar physics and mass-loss rate used in the models. Green points are the newest models of Woosley (2019), with standard nuclear physics. Blue points used a reduced surface boundary pressure. Black points used a large nuclear network. Red points used 1.5 times the standard wind mass-loss rate. The left-hand panel is a zoom-in of the right-hand panel. (Bottom) The same results as in the top panels, overplotted with the $\dot{M}/2$ case for final helium cores of hydrogen-rich single star models of Sukhbold et al. (2018): gold points (see their Fig. 8). Note the large number of high compactness parameters $<6\,M_{\odot}$. The peak of compactness for single stars is at a final helium core mass of $6.5\,M_{\odot}$. For the new set approximating Case B evolution, it is at $8\,M_{\odot}$. The minimum above the first peak now has a much lower compactness than for the single stars and is shifted to high pre-SN masses, which will still leave NSs. After Woosley (2019).

stars with ZAMS masses of $\lesssim 35\,M_\odot$—which form the bulk of all WR-stars—will terminate in a SN and leave a NS as a remnant. (For Case A RLO, the helium-core masses produced will be even smaller, and NSs will result.) Furthermore, in Case B RLO, for final core masses $M_{\rm He,f} \simeq 9.5-12\,M_\odot$ the compactness is very small again and NSs will be produced. This mass range corresponds to initial helium-core masses in the range $M_{\rm He,i} \simeq 17-25\,M_\odot$, which according to Eq. (9.8) corresponds to a ZAMS mass range of approximately $46 - 62\,M_\odot$. Therefore, (for solar metallicity) only Case B systems with primary stars in the mass range[3] of $35 - 46\,M_\odot$ and $>62\,M_\odot$ produce BHs, while for all other primary masses $<62\,M_\odot$ they produce NSs.

This result solves a puzzle noticed by Vanbeveren et al. (2020): the fact that very few BH-HMXBs are found in our Galaxy. These systems are descendants of WR+O-star binaries. It had been found from population synthesis models that a large population of BH-HMXBs should be present in the MW, and their number should be approximately two orders of magnitude larger than what is observed (Vanbeveren et al., 2020). However, these predictions were based on population synthesis models assuming that, at $Z = Z_\odot$, most stars with ZAMS masses $>30\,M_\odot$ collapse to BHs, which is contrary to the findings we have discussed. Furthermore, WR/X-ray binaries are a later evolutionary phase of BH-HMXBs (van den Heuvel et al., 2017), and thus hundreds of such systems would be expected in our Galaxy if hundreds of BH-HMXBs would be present. However, only one WR/X-ray binary is known in our Galaxy: the system of Cygnus X-3. It can be concluded that the absence of hundreds of BH-HMXBs and WR/X-ray binaries in the MW is a strong indication that at solar metallicity most stars in binaries with ZAMS masses $>30\,M_\odot$ do not leave BHs, but end their lives as NSs, as indeed is predicted by the models of Woosley (2019).

Recent calculations by Schneider et al. (2021) of the final stellar structures at core collapse of single stars vs. stripped stars in close binaries confirms the picture of a non-monotonic pattern of NS and BH formation based on CO core masses (although with somewhat different mass intervals compared to those of Woosley [2019]). They find that stripped binary stars can have systematically different pre-SN structures compared to genuine single stars and thus also different SN outcomes. In terms of initial masses, they find that single stars of $>35\,M_\odot$ all form BHs, while this transition is only at $\sim 70\,M_\odot$ in stripped stars.

We should also mention here that the magnetar CXOU J164710.2−455216 has been found in the very young massive star cluster Westerlund 1, which contains 24 WR-stars, many blue and red and 6 yellow supergiants. The progenitor of this NS must have been more massive than the progenitors of the WR-stars (these progenitors were stars more massive than $\sim 30\,M_\odot$). Hence, here a star that was more massive than $30\,M_\odot$ left a NS as remnant. (It may well be that it is the remnant of a WR-star in a binary, although to date, no magnetars have been found in a binary system; see Section 10.3.3.)

Contrary to Case B evolution, as will be further discussed in Chapter 10, Case C evolution can produce BHs already from initial primary stars with ZAMS masses as

[3]The exact mass boundaries are strongly dependent on metallicity and assumed stellar physics and may therefore vary throughout this book.

low as 15 $M_\odot$ (final helium-core masses of $\sim$5 $M_\odot$). As we will see in Chapter 10, this is very important for understanding the formation of the BH-LMXBs (soft X-ray transients). We thus see that the new theoretical findings of final stellar structure and SNe simulations are of great importance for understanding the formation of X-ray binaries in general, as well as BH/NS GW mergers.

9.7.2 The Formation and Evolution of X-ray Binaries: The Next Two Chapters

We see from the previous sub-section that the evolution of Case B and Case C RLO (and the much rarer Case A as well) is closely connected with the formation and evolution of X-ray binaries. To properly cover the formation and evolution of these systems, we consider the formation and evolution of the HMXBs and LMXBs separately in Chapters 10 and 11.

9.8 SLOWLY ROTATING MAGNETIC MAIN-SEQUENCE STARS: THE PRODUCTS OF MERGERS?

A special type of products of early binary evolution are the products of mergers of two normal stars. As will be discussed in Chapter 16, in star clusters the products of mergers of normal stars are expected to be quite common and may constitute $\sim$10% of the stars. The merger products of main-sequence stars deserve special attention because in recent years it has been suggested that the magnetic main-sequence stars, which make up $\sim$10% of all main-sequence stars with radiative envelopes in the spectral ranges A, B, and O, are products of mergers of two (pre–)main-sequence stars (Ferrario et al., 2009; Schneider et al., 2016; Schneider et al., 2019). In these references it was proposed that these stars acquired their magnetic fields through dynamo processes that occur during and shortly after the merger of two (proto-)stars. Because a merger product absorbed most of the orbital angular momentum of the progenitor binary, it will be an extremely rapid rotator—rotating near break-up speed at its formation. Such extremely rapid rotators will be highly flattened and strongly differentially rotating. This differential rotation will cause small poloidal magnetic fields that always are present in stellar matter, to be stretched out and amplified into strong toroidal magnetic fields, which by buoyancy will be turned into strong poloidal fields, which will be further amplified to still stronger toroidal fields, creating even stronger poloidal fields, etc. This would then lead to the observed strong frozen-in magnetic fields present in the outer layers of the magnetic Ap and Bp stars, which are concentrated in the spectral region A2 to approximately B5 on the main sequence (stellar masses between 1.6 and $\sim$5 $M_\odot$). These A- and B-type stars with highly peculiar overabundances of certain iron-like and heavier elements have been known to be magnetic stars since the 1950s. They have frozen-in steady surface magnetic fields with strengths typically between 500 G and 35,000 G. (For a more detailed explanation of how these strong magnetic fields in merger products may be generated, see Wickramasinghe et al. [2014] and Schneider et al. [2019].) It is well-known that the observed overabundances of heavy elements in these stars are only an atmospheric phenomenon and not part of their interiors. This is clear from the fact that the luminosities of these stars in visual binaries and clusters are quite normal for their spectral types and effective temperatures, indicating that they obey the normal mass-luminosity relation. In the 2000s and 2010s, also strong magnetic fields have been found in main-sequence stars more massive than 5 $M_\odot$: the O- and early B-stars (Fossati et al., 2016). Apart from their strong surface magnetic fields, these magnetic stars have as many as four other characteristics that make them stand out from the other main-sequence stars in the same spectral range:

1. Slow rotation. While normal A- and B-type main-sequence stars typically have projected equatorial rotation velocities, $v \sin i$ of 100–400 km s^{-1}, the magnetic stars (although presumably born rapid rotators) tend to have $v \sin i$ typically $<50\,\mathrm{km\,s^{-1}}$, although a few rotate faster, and many others much slower.
2. Absence of spectroscopic binaries. While the spectroscopic binary percentage among normal A- and B-type main-sequence stars is $\sim$50% or more (see Chapter 3), there is among the hundreds of magnetic Ap and Bp stars only one spectroscopic binary known (Abt's star, HD 98088), which has a magnetically locked companion that co-rotates with the orbital motion (Abt et al., 1968).
3. For the magnetic stars in the spectral range B5 to approximately A2, large atmospheric overabundances of rare earth elements (in particular dysprosium, gadolinium, and europium) and/or elements such as strontium, silicon, chromium, and other iron-like elements are found. They fall mostly into two classes: the hotter ones as Si-stars and the cooler (early A-)types as Eu-Cr-Sr stars. The rare earth elements can be overabundant by factors as large as 10^3 or even higher. The overabundant elements tend to be concentrated in spots, presumably around magnetic poles. For example, in the extremely slowly rotating star HR 465 there is one spot in which chromium and manganese and other iron-peak elements are concentrated, and another spot in which the rare earth elements, such as neodymium, samarium, gadolinium, and dysprosium are concentrated (Rice, 1988).
4. BSSs. As was noticed by van den Heuvel (1968b), in open star clusters, such as Coma, the α Per cluster and the UMa Moving Cluster, the chemically peculiar Ap/Bp stars tend to be blue stragglers. He ascribed this fact to binary mass exchange, suggested that these stars have a WD companion that is very hard to detect (thus explaining the apparent absence of spectroscopic binaries), and said the peculiar abundances were due to a nuclear evolutionary history of the companion. However, it soon became clear that this idea is untenable because some visual binaries were discovered in which one component is an Ap star while the visual companion is still a pre–main-sequence Herbig Ae star (e.g., the naked-eye Ap star HR 6000 with its Herbig Ae companion HR 5999 [Tjin A Djie et al., 1982]), showing that some Ap stars are extremely young and have just arrived on the main sequence.

The merger model of Ferrario et al. (2009) nicely explains both the absence of spectroscopic binaries (Schneider et al., 2016) and the blue straggler character in star clusters, assuming that the merging process occurs not necessarily only in the pre–main-sequence phase, but may also occur after the stars have arrived on the main sequence. Furthermore, the slow rotation of the magnetic stars is also understandable (Fossati et al., 2016). This is because all stars have stellar winds, and in a magnetic star the wind matter will be forced by the magnetic field to co-rotate with the star out to quite a large distance (the Alfvén radius), such that it will very efficiently carry off the rotational angular momentum of the stars (see also Keszthelyi et al., 2020). However, a few magnetic stars have exceedingly slow rotation, such as the magnetic Ap stars HR 465 and HD 166473, which have rotation periods of 21.7 and 10.5 yr, respectively (Mathys et al., 2020). Also some magnetic O-stars rotate extremely slowly (Sundqvist et al., 2013). It seems very difficult to attain such slow rotation within the star's lifetime by a magnetically coupled stellar wind (unless the wind mass-loss rate in the pre-main-sequence stage is very large). Here other mechanisms for rotational slowdown have been proposed, such as magnetic coupling to a pre-stellar cloud (Spruit, 2018).

9.8.1 The Class of Slowly Rotating B and A Main-sequence Stars

van den Heuvel (1966, 1968a) discovered that all along the main sequence, in the field as well as in OB associations, at spectral types between O8 and A2 where the stars have radiative envelopes, there is a distinct class of slowly rotating stars, which comprises approximately 20%–25% of the main-sequence stars. This class clearly stands out from the normal rapidly rotating stars in the same spectral region: the distribution of the rotational velocities is double-peaked, one peak between 0 and $\sim$50 km s^{-1}, the other between 90 and $\sim$250 km s^{-1}. Between spectral types of approximately A2 and B5, more than half the stars in this slowly rotating class are of the magnetic Ap/Bp type, which suggests that in the remaining spectral classes between B5 and O8 a large part of the slow rotators is related to the magnetic Ap/Bp stars but were not recognized as such. Indeed, in the 2000s and 2010s, it was discovered that among the O- and early B-stars there are stars with strong magnetic fields and these are slow rotators. Ferrario et al. (2009) and Schneider et al. (2016) have suggested that all these stars, like the magnetic Ap/Bp stars, are merger products. The fact that not all the slowly rotating A2 to B5 stars are peculiar may be an evolutionary effect: the magnetic fields may decay on timescales of the order of 10^8 yr, and the peculiar heavy elements in their atmospheres may then be able to diffuse down into the stellar interior, causing the surface abundance anomalies to disappear. In that case, a slowly rotating normal A2–B5 star will result. (The slowly rotating A-type blue straggler HD 73666 in the Praesepe cluster [Fossati et al., 2010] may be such an object.) The merger suggestion provides a nice explanation of the origin of the class of slowly rotating main-sequence stars. It makes all the peculiar characteristics of the slowly rotating class fall in place, including the presence of strong magnetic fields, its blue straggler character in open clusters, and the absence of spectroscopic binaries. A remaining problem is how to explain the peculiar atmospheric abundances of certain groups of elements in the magnetic stars. We propose a brief speculative explanation for this phenomenon.

9.8.2 On the Possible Origins of the Abundance Anomalies in the Magnetic Ap/Bp Stars

It is well known that the atoms of the rare earth elements, which are highly overabundant in the atmospheres of the magnetic Ap/Bp stars, have peculiar magnetic moments. The same holds for several of the iron-group elements. These peculiar magnetic properties are the reason why for the construction of very strong magnets, industry uses solids containing rare earths and iron-group elements. For example, for the strong magnets needed in many technical applications, the rare earth elements dysprosium and gadolinium, which are particularly prominent in many Ap/Bp stars, are frequently used in combination with iron or iron-related elements. The peculiar magnetic properties of these two groups of elements imply that in proto-planetary disks that surround a young strongly magnetized star, the rocks that condense out of these disks and contain rare earth and iron-group elements will preferentially—with respect to other types of rocks—experience strong magnetic forces that will couple them to the magnetic field of the star. This may cause these magnetic rocks to slide selectively (with respect to other kinds of rocks) along the magnetic field lines to the stellar surface and to pollute the stellar atmosphere with these elements. It seems to us that this may be a plausible way to explain why just these elements with peculiar magnetic properties are so overabundant in the atmospheres of the magnetic A- and B-stars and concentrated in large spots around the magnetic poles. The strong frozen-in magnetic fields in the stellar atmosphere will prevent these atoms from rapidly diffusing inward, as long as the strong magnetic fields are present. Only when these gradually decay might the surface abundance anomalies of these stars gradually disappear. Because rocks contain large amounts of silicon, such a model for producing the abundance anomalies of the magnetic Ap/Bp stars will also explain the overabundance of silicon observed in the atmospheres of most of these stars.

EXERCISES

Exercise 9.1. *Derive Eq. (9.4).*

Exercise 9.2. *Use a binary stellar evolution code (e.g., MESA) to compute the evolution of an Algol-type binary and compare to the results in Section 9.5.1.*

Exercise 9.3. *Consider the fusion of hydrogen.*

a. Calculate the net amount of energy liberated in the fusion of 1 Kg of hydrogen to helium.

b. Calculate the percentage of energy liberated during hydrogen fusion in units of the rest mass energy of the processed hydrogen.

Consider accretion onto a NS (with a mass of $1.4\,M_\odot$ *and a radius of 12 km).*

c. *Estimate the amount of (gravitational potential) energy liberated if 1 liter of Danish beer were poured onto a NS from far away.*

d. *Calculate the percentage of energy liberated during the accretion process in units of the rest mass energy of the beer.*

e. *Answer questions c and d for the Sun, a WD ($M_{\rm WD} = 0.7\, M_\odot$, $R_{\rm WD} = 10{,}000$ km), and a BH ($M_{\rm BH} = 10\, M_\odot$, $R_{\rm BH} = 30$ km).*

Chapter Ten

Formation and Evolution of High-mass X-ray Binaries

10.1 INTRODUCTION: HMXBs ARE NORMAL PRODUCTS OF MASSIVE BINARY STAR EVOLUTION

Later, in Section 10.6, we will see that the bright persistent standard sg-HMXBs have, with a lifetime of the order of 10^5 yr and a number of the order of 10^2 in the Galaxy, a formation rate of the order of 10^{-3} yr^{-1}. The Be/X-ray binaries are transients, which are *on* only a fraction of the time, and thus they must outnumber the sg-HMXBs by at least one order of magnitude. This can be understood, for example, by considering the SMC Galaxy, which has a mass of $\sim$0.03 times the mass of our Galaxy, but yet contains >100 Be/X-ray binaries (see, e.g., Haberl & Sturm, 2016). Translated to our Galaxy, this implies some 3,000 Be/X-ray binaries in the MW. Assuming them to live $\sim 5 \times 10^6$ yr, their formation rate must be 0.6×10^{-3} yr^{-1}. This leads to a total HMXB formation rate of $\sim 1.6 \times 10^{-3}$ yr^{-1}, which is about one order of magnitude lower than the NS formation rate. Hence, of the order of 1 of every 10 NSs ends up in a HMXB, which means that the HMXBs are not exotic objects but quite normal products of massive star evolution.

10.2 FORMATION OF SUPERGIANT HIGH-MASS X-RAY BINARIES

The formation of sg-HMXBs follows the final evolution of massive Case B systems and WR binaries. For primary stars more massive than approximately 15–20 $M_\odot$, the evolution through Case B mass transfer proceeds in a way similar to that of the Case B systems with primary masses between 2.3 and 15 $M_\odot$ (Section 10.3), the only difference being that the helium stars that are produced have masses $\gtrsim 3.5 - 4\, M_\odot$ and thus expand to smaller radii (see Fig. 8.31), such that they will not go through Case BB mass transfer before they collapse to a NS or a BH. Thereafter, in general, these systems evolve to become sg-HMXBs.

Pioneering calculations of massive Case B evolution were made by Paczyński (1967a) and Kippenhahn & Weigert (1967). Paczyński (1967a) was the first to point out that the systems resulting from this evolution closely resemble WR binaries. Later such systems are expected to produce HMXBs, as was first pointed out by van den Heuvel & Heise (1972) and Tutukov & Yungelson (1973a, 1973b); see also van den Heuvel (1973). As an example, Figure 10.1 shows the evolution in the HR diagram of

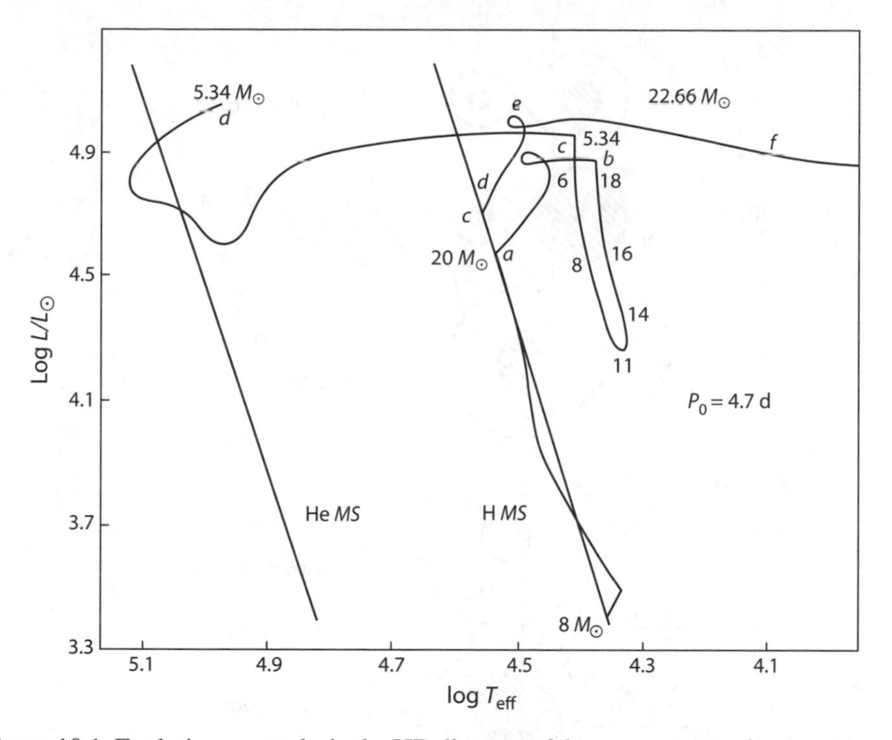

Figure 10.1. Evolutionary tracks in the HR diagram of the components of a close binary with initial masses of 8 and 20 $M_\odot$ and an initial orbital period of 4.70 d, computed by De Loore & De Greve (1975), with the assumptions of conservation of mass and orbital angular momentum. This system later in life becomes a sg-HMXB. Letters correspond to the evolutionary phases indicated in Figure 10.2. The first phase of mass transfer begins in b and terminates in c. In d the primary, a 5.34 $M_\odot$ helium star, explodes in a SN and is assumed to leave a 2 $M_\odot$ NS. In f, the secondary reaches its Roche lobe and the second phase of mass transfer begins. Approximately 10^5 yr prior to this, the secondary is a blue supergiant with a strong stellar wind, which turns the NS into a strong X-ray source. Some values of the masses of primary and secondary are listed along the tracks. After De Loore & De Greve (1975).

a Population I (solar composition) binary with initial masses of 20 and 8 $M_\odot$ and an initial orbital period of 4.70 d to become a sg-HMXB, as computed by De Loore & De Greve (1975). The evolution is here presumed to proceed conservatively.

Figure 10.2 graphically depicts the evolution of this system. We now describe this evolution in the HR diagram: at $t = 6.17 \times 10^6$ yr (stage b) the primary begins to overflow its Roche lobe, and after 3×10^4 yr a close binary is left consisting of a 5.34 $M_\odot$ helium star and a 22.66 $M_\odot$ companion in a 10.86 d orbit (stage c). Such a system, consisting of two components differing about a factor four in mass, but with about the same bolometric luminosity, has a resemblance to the characteristics of the WR-binaries, although the mass of the helium star here is still somewhat too low for such

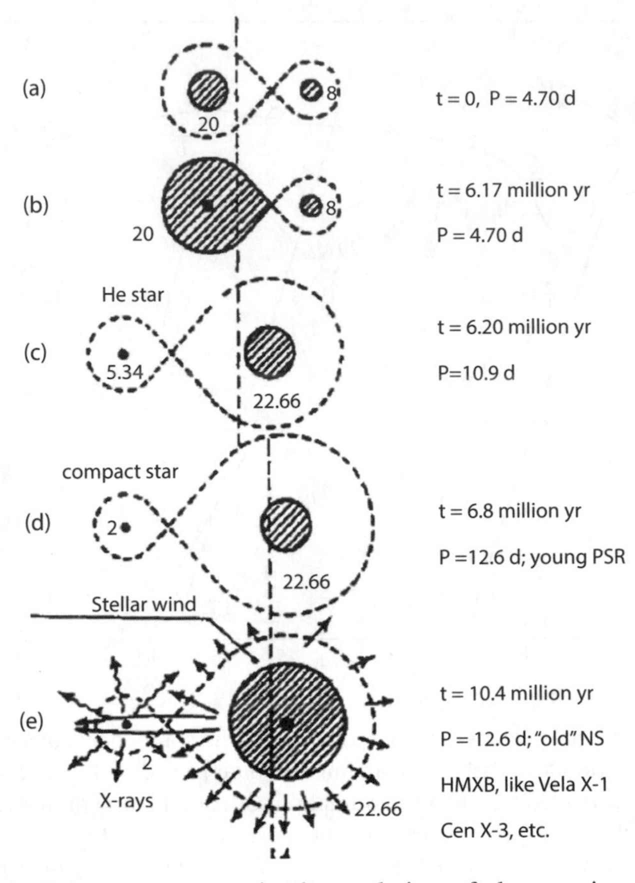

Figure 10.2. Subsequent stages in the evolution of the massive close binary of Figure 10.1, producing a sg-HMXB. Ages and orbital periods of the system are indicated. It is assumed that the SN explosion of the helium star leaves a 2 $M_\odot$ NS, which becomes an X-ray source when the companion star has left the main sequence and has become a blue supergiant with a strong stellar wind. The SN mass ejection is assumed to have been spherically symmetric (no kick). This symmetric ejection of 3.34 $M_\odot$ will impart a runaway velocity of 39.8 km s^{-1} to the system. This is indicated by the displacement after the SN of the vertical axis through the c.m. of the system. Illustration from van den Heuvel (1976) after evolutionary calculations by De Loore & De Greve (1975).

a star, as WR-stars have masses $\gtrsim 8\, M_\odot$ (see Chapters 5 and 8). (Just like massive helium-burning helium stars, WR-stars have in general little or no hydrogen in their atmospheres and are overluminous for their masses by a factor of the order of 30 as compared with the normal mass-luminosity relation for massive hydrogen-rich stars.) The evolutionary calculations predict for massive helium stars a small radius of the order of $\lesssim 2\, R_\odot$, as shown in Figure 8.31. However, in the case of WR-stars, due to their heavy stellar wind-mass loss, at rates of the order of $10^{-5}\,M_\odot$ yr, their stellar winds are opaque to optical radiation out to distances of the order of 5–10 $R_\odot$. The surface of optical depth $\tau = 2/3$ in the wind is located at approximately such radial distances. Therefore, WR-stars have a pseudo-photosphere (not a real surface) located at this distance, which causes the effective temperatures of WR-stars to be only approximately $(4-5)\times 10^4$ K lower than the temperatures at the real radius of the helium star.

At point d on the track in the HR diagram, the helium star explodes as a SN. Here it is assumed it leaves behind a NS of 2 $M_\odot$. Because observations show that NSs receive a kick velocity at birth, here also the effects on the orbit of a 100 km s^{-1} kick velocity in the orbital plane were calculated for two velocities at which the 3.34 $M_\odot$ SN shell was ejected: 10^4 and 2×10^4 km s^{-1}. Figure 10.3 shows how the kick affects the post-SN orbital eccentricity of the system. In the case of no kick (spherically symmetric mass ejection), the sudden ejection of 3.34 $M_\odot$ in the SN explosion imparts a runaway velocity of 39.8 km s^{-1} to the c.m. of the system: that is, the system becomes a runaway star. In case a kick is imparted onto the NS, the resulting systemic velocity will be somewhat different but often of similar order unless the kick is very large. One thus expects the sg-HMXBs to be runaway stars (van den Heuvel & Heise, 1972). This is indicated in Figure 10.2 by the displacement after the SN of the position of the vertical axis through the c.m. of the system. Kinematic effects of SNe in binaries are discussed in detail in Chapter 13.

In Figure 10.2 the mass ejection was assumed to be spherically symmetric. In stage e of this figure, the companion star has left the main sequence and has become a blue supergiant star with a radius close to that of its Roche lobe. The wind mass-loss rate is now larger than on the main sequence, and the wind velocity is lower, such that the accretion rate onto the NS is much larger than on the main sequence (the rate is inversely proportional to the fourth power of the wind velocity, cf. Section 6.3.3), and the system becomes a sg-HMXB. The duration of this phase will be relatively short and is estimated to be of the order of 10^5 yr (see later in this chapter; the duration depends on whether the supergiant is still in a late phase of core-hydrogen burning or already on its way to core-helium burning, crossing the Hertzsprung gap). As its radius tends to expand due to its internal evolution, the supergiant will at a certain point overflow its Roche lobe, and a very large rate of mass transfer will ensue, which quenches the X-ray emission, and will cause the system to enter CE evolution and spiral-in toward a much narrower orbit (see Section 4.8).

10.3 FORMATION OF B-EMISSION (Be)/X-RAY BINARIES

The formation of Be/X-ray binaries follows the final evolution of Case B systems with primary masses between 8 and 15 $M_\odot$. The helium cores left by the primaries in these close systems are expected to be in the mass range of 2.2–3.5 $M_\odot$, causing them to also leave NSs or possibly BHs. These naked helium stars may later in life experience a sufficiently large radius expansion, causing them to undergo Case BB mass transfer, leaving only small stripped metal cores at the time of the collapse.

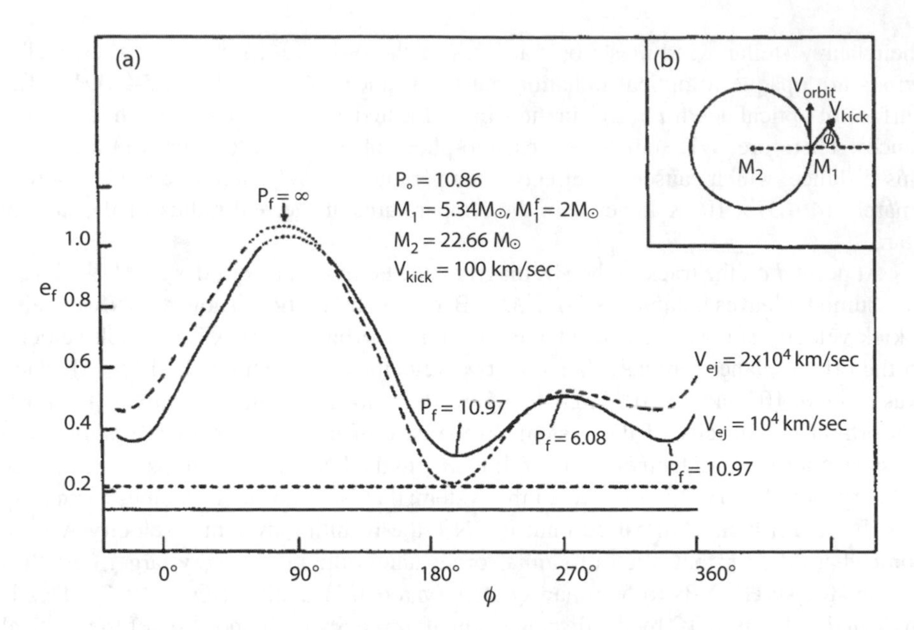

Figure 10.3. The effects of asymmetric mass ejection on the post-SN orbital eccentricity of the system of Figure 10.2, for two values of the ejection velocity of the SN shell. It is assumed that the NS received a kick in the orbital plane with a velocity of 100 km s^{-1}, directed under an angle ϕ as defined in the inset. The effects of impact of the SN shell were taken into account (hence the dependence on the SN ejecta velocity). The two straight horizontal lines indicate the eccentricities resulting from symmetric mass ejection. The system is disrupted for ϕ close to 90°. Orbital periods are in days. After van den Heuvel (1976).

This second mass-transfer phase (Case BB RLO) from the lighter helium star to its early B-type companion (which will mostly be in the mass range of 10–20 $M_\odot$), will further widen the orbit, such that a relatively wide early-type binary will result prior to core collapse of the primary star. The mass transfer to the B-type star accretor will lead to formation of an accretion disk around this star, like in the β Lyrae system, and the accretion of this high-angular-momentum matter from the disk will efficiently speed up the rotation of the B-star, such that it will become a very rapid rotator (Packet, 1981) of the Be type (see Chapter 9). After the SN collapse of the core of the naked helium star, the system, if it remains bound, will resemble a B-emission (Be)/X-ray binary (as pointed out, for example, by De Loore & Sutantyo, 1984).

Figure 10.4 depicts the evolution to the formation of such a system, which originated from a $13 + 6.5\,M_\odot$ binary with a period of 2.58 d, as calculated by Habets (1985). The evolution of the system during all mass-transfer phases is assumed to take place conservatively and proceeds as follows, and is illustrated in the figure.

The Case B (first) mass transfer begins at $t = 1.6 \times 10^7$ yr and results in a binary with component masses of $2.5 + 17.0\,M_\odot$ in a 20.29 d orbit. Two million years later, the Case BB (second) mass transfer begins, which takes 2×10^3 yr, leaving a $2.2\,M_\odot$ remnant plus a $17.3\,M_\odot$ main-sequence star in a 25.1 d orbit. Figure 10.5 depicts the

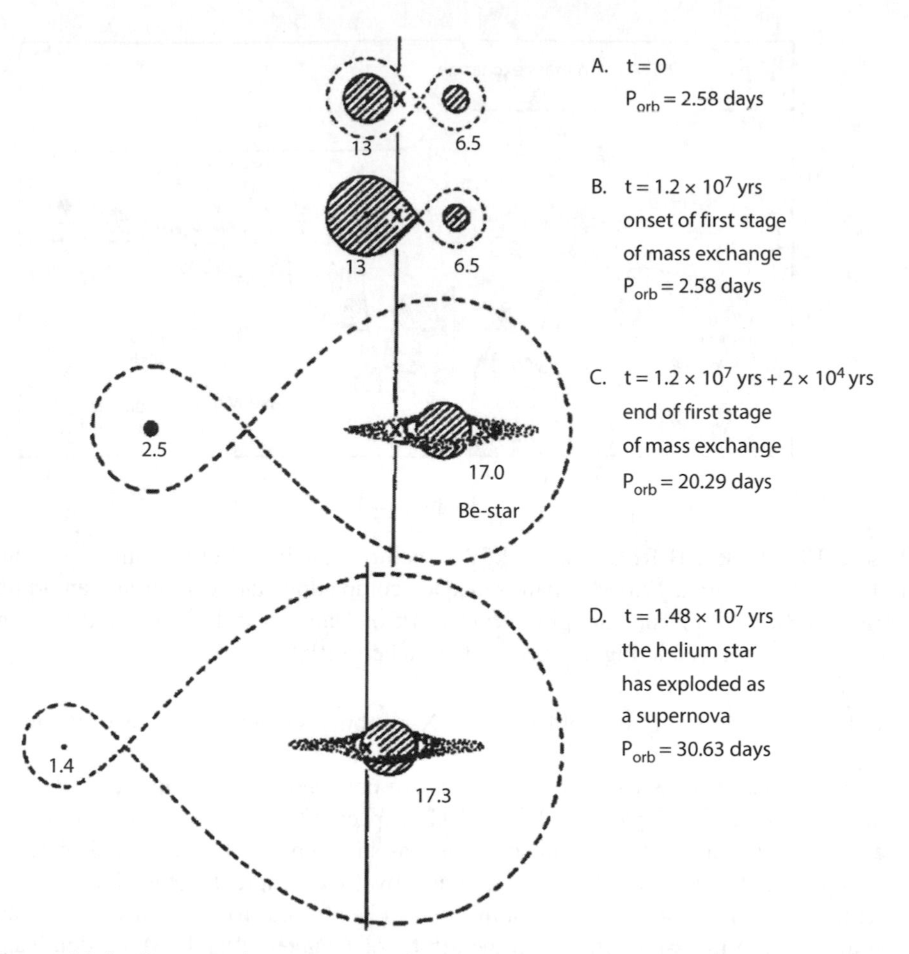

Figure 10.4. Conservative mass-transfer scenario for the formation of a Be/X-ray binary, which here originated from a close pair of B-type main-sequence stars (masses $13 + 6.5\ M_{\odot}$). The numbers indicate mass in solar masses. After the mass transfer, the $17.0\ M_{\odot}$ early B-star secondary is expected to have a circumstellar disk or shell related to its rapid rotation, induced by previous disk accretion of matter with high angular momentum. Prior to the collapse of the core of the primary star, a phase of Case BB mass transfer occurred, which reduced the mass of this helium star to $2.2\ M_{\odot}$ at the time of the collapse, such that a SN mass ejection of only $0.8\ M_{\odot}$ ensued. Modified after Habets (1986a).

internal evolution of the $2.5\ M_{\odot}$ helium star companion and shows its mass reduction to $2.2\ M_{\odot}$ during the Case BB mass transfer prior to its core collapse. The core collapse occurs 4×10^3 yr after the end of the Case BB mass transfer and is here assumed to leave a $1.4\ M_{\odot}$ NS. Assuming the SN mass ejection to be spherically symmetric, the orbital period increases to 30.63 d, the orbital eccentricity becomes 0.043, and the system obtains a runaway velocity of $\sim 10\ \mathrm{km\ s^{-1}}$ (which in Fig. 10.4 is indicated by the displacement of the vertical line through the c.m. of the system).

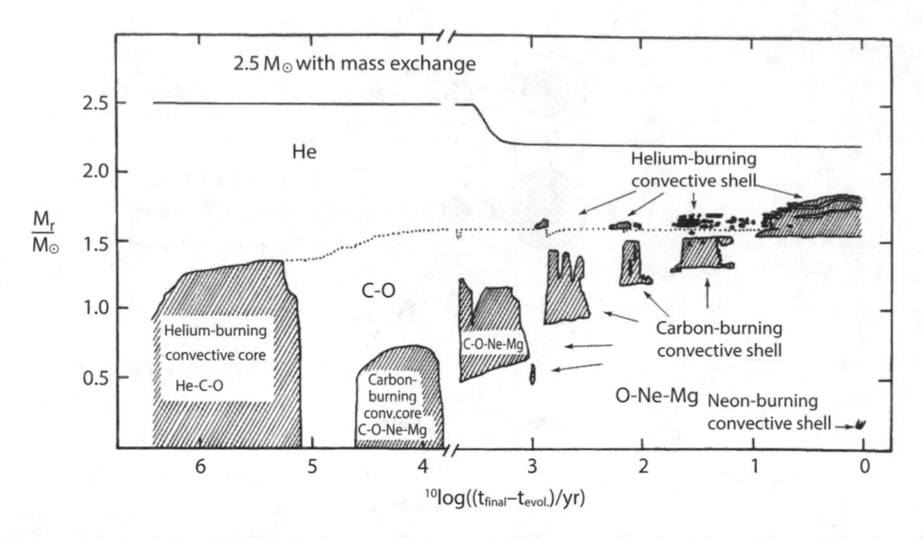

Figure 10.5. Case BB RLO from a 2.5 $M_\odot$ helium star in the close binary depicted in Figure 10.4, with a 17.0 $M_\odot$ main-sequence companion star accretor and an orbital period of $P_{\rm orb} = 20.29$ d. This pioneering work of Habets can be compared to more recent work displayed in Figure 10.21. After Habets (1986a).

The final system closely resembles the Be/X-ray binaries described, for example, in Chapter 7.

The impact of the SN shell onto the massive companion is, like in the case of the more massive system depicted in Figure 10.2, expected to have a very small effect on the companion star itself and on the orbit, as was shown by the hydrodynamic calculations by Fryxell & Arnett (1981), and later also by, for example, Hirai et al. (2018) and references therein. A small amount of matter will be ablated from the massive star, but overall the pure mass-loss effects on the orbit (see Chapters 4 and 13) are dominant. (The same is not necessarily true if the companion is a low-mass star [see e.g., Taam & Fryxell, 1984; Liu et al., 2015], although here, too, in most cases the impact effects are smaller than one might have expected.) In view of this, it is justified in the formation of HMXBs to neglect the shell impact effects.

10.3.1 Orbital Eccentricities of Be/X-ray Binaries and Kicks

The system depicted in Figure 10.4 obtained an eccentricity of only 0.043 due to an assumed symmetric ejection of 0.8 $M_\odot$ when the 1.4 $M_\odot$ NS formed. Even if the exploded star in this system would have had a mass of 3 $M_\odot$, the eccentricity imparted in a spherically symmetric SN explosion would have been <0.1. Observationally, there is indeed a group of Be/X-ray binaries with almost circular orbits, as discovered by Pfahl et al. (2002). These systems must have resulted from nearly spherically symmetric SN mass ejection.

However, as the table of HMXBs (Table 6.2) in Chapter 6 shows, many Be/X-ray binaries have orbital eccentricities in the range of 0.3–0.7, while the binary radio pulsar with an early B-type companion star, PSR 1259−63, even has an orbital eccentricity of $e = 0.97$. Clearly, to obtain eccentricities in the range of 0.3–0.97, the NSs in these high-eccentricity systems must have received a considerable kick velocity at birth, depending on the orbital periods, in the range of 50–200 km s^{-1} (see van den Heuvel, 1994a, and see also Section 13.8).).

To obtain eccentricities between 0.3 and 0.97 from a SN with symmetric mass ejection, the mass of the exploding helium star, producing a 1.4 $M_\odot$ NS and for a B-star companion of 15 $M_\odot$ in a system somewhat similar to that of Figure 10.4, must have been between 6.3 $M_\odot$ and 17.3 $M_\odot$ (Eq. 4.60). In systems with such massive helium stars, the initial primary stars would have been very massive (between 22 and ~40 $M_\odot$). Such massive stars are quite rare, while the Be/X-ray binaries just constitute the most abundant class of HMXBs, implying that, due to the shape of the IMF, they must originate from relatively moderate-mass primary stars, typically in the range of 8–15 $M_\odot$. So, based on this argument alone, such massive helium star progenitors of their NSs are ruled out. (In addition, a fraction of massive stars between 22 and ~40 $M_\odot$ are likely to leave behind BHs as remnants; see Section 8.2.3.) The conclusion must be that the high eccentricities of more than half of all observed Be/X-ray binaries can only be explained by intrinsic SN kicks imparted to their NSs at birth and therefore provide direct proof of such kicks (van den Heuvel, 1994a).

However, that there is also a sizeable group of Be/X-ray binaries with almost circular orbits, the prime example being the X Persi system (Pfahl et al., 2002), means that there is a class of helium stars for which the collapse of the stripped core imparts a negligible kick velocity to the NS. As we will see in Chapter 14, among the double NSs (DNSs) there is a group with very low-eccentricity orbits, which indicates that the second-born NS in these systems received hardly any kick at birth. We now consider briefly possible ways in which the two groups of Be/X-ray binaries may have formed.

10.3.2 Non-conservative Evolution, Dichotomous Kicks and Formation of the Two Groups of Be/X-ray Binaries

Podsiadlowski et al. (2004) have proposed the model depicted in Figure 10.6 for the formation of the two groups of Be/X-ray binaries, in which they took into account that binary evolution in many cases will not proceed conservatively. The two systems considered here have the same starting masses of the stellar components: 14 $M_\odot$ + 10 $M_\odot$, but different initial binary periods, leading to very different evolution. In the left system, with an initial orbital period of 500 d, the primary star grows to become a red giant in which the core is able to grow to a 3.9 $M_\odot$ helium star. The onset of the mass transfer from the red giant's deep convective envelope is assumed to be highly unstable and leads to the formation of a common envelope (CE, see Chapter 4), in which the 10 $M_\odot$ companion rapidly spirals in very deeply, leading to ejection of the CE, a final orbital period of only 2 d, and no mass gain of the secondary. In the right-hand system, with initial orbital period of 100 d, the helium core is 2.4 $M_\odot$ when stable, but non-conservative, Case B mass transfer occurs, in which 6.0 $M_\odot$ of the primary's envelope is captured by the secondary, 5.6 $M_\odot$ is lost from the system, and the orbital period is apparently assumed to remain at 100 d. The latter could, in principle, be achieved by

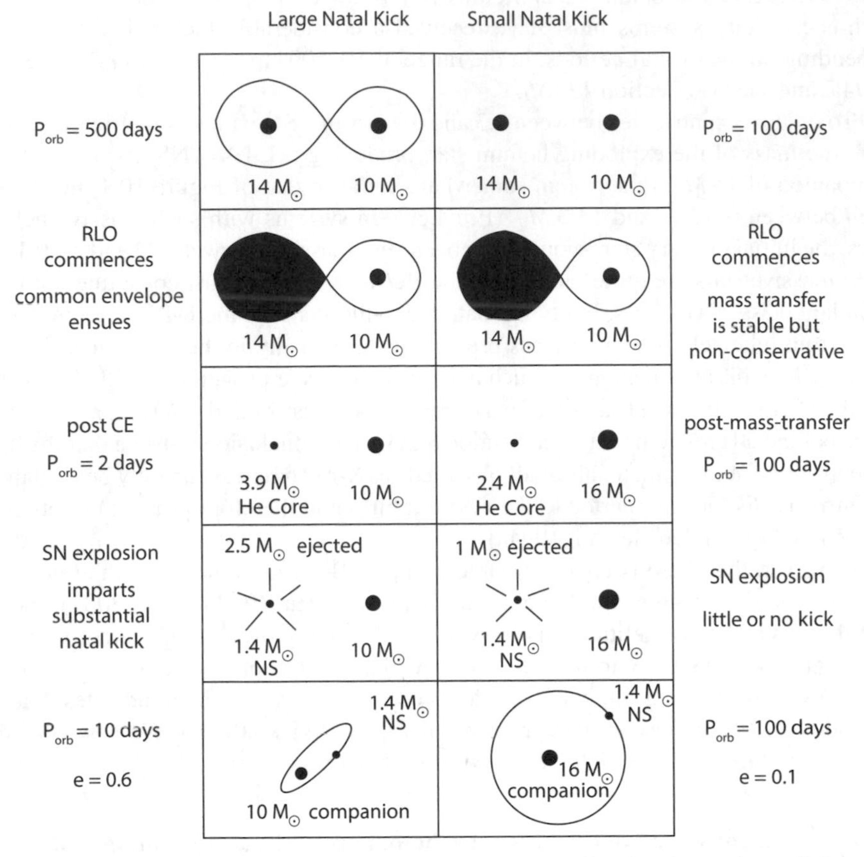

Figure 10.6. Non-conservative mass-transfer scenarios for the formation of the two classes of Be/X-ray binaries, according to Podsiadlowski et al. (2004), taking into account that lower-mass helium stars ($M_{\rm He} \sim 2.4\,M_\odot$) may undergo electron-capture core collapse, which imparts hardly any kick velocity to the NS, whereas a more massive helium star ($3.9\,M_\odot$) undergoes iron-core collapse, which imparts a large kick velocity. In the originally very wide system in the left frames, mass transfer starts in the red-giant phase, leading to CE evolution in which the orbit shrinks by a large factor and the $10\,M_\odot$ companion does not gain any mass. The large kick imparted to the NS leads then to a very eccentric Be/X-ray binary. The system in the right-hand frames evolves non-conservatively but via stable RLO. The companion does gain mass, and the electron-capture core collapse of the ONeMg core of the helium star imparts hardly any kick, leading to an almost circular orbit of the Be/X-ray binary. Figure adapted from Podsiadlowski et al. (2004). (Note, the assumption of a constant value of $P_{\rm orb} = 100\,$d stated at all stages in the right-side column is not correct, and we believe it must be a mistake or typo; see also Exercise 10.3.)

making very special assumptions about the orbital angular momentum carried off by the 5.6 $M_\odot$ of material lost from the system (see Exercise 10.3). However, assuming that the orbital period does not change at all during the entire evolution (including the SN event) is not correct.

Because lower-mass helium stars ($M_{\rm He} \sim 2.5\, M_\odot$) are expected to undergo electron-capture core collapse (see Chapter 13), in which hardly any kick is imparted onto the NS, the right-hand system terminates as a relatively wide Be/X-ray binary with an almost circular orbit. On the other hand, the 3.9 $M_\odot$ helium star in the left-hand system undergoes iron-core collapse, which often imparts a large kick velocity, leading to a very eccentric Be/X-ray binary system with an orbital period of about 10 d.

Pfahl et al. (2002) and Podsiadlowski et al. (2004) carried out extensive binary population evolution studies implementing these scenarios for the formation of the Be/X-ray binaries and showed that in this way a good agreement with the observed characteristics of the two groups of systems can be achieved. Further support for a kick dichotomy came from a study of observational data of Be/X-ray binaries, in which Knigge et al. (2011) reached the conclusion that the Be/X-ray binaries are composed of two distinct subpopulations with different characteristic spin periods, orbital periods, and orbital eccentricities. They argued that these two subpopulations are most probably associated with the two distinct types of NS-forming SNe (i.e., iron-core collapse and electron-capture SNe). This explanation seems very plausible to us. However, because the mass range for electron-capture SNe might be rather small (see Section 13.4), it is perhaps possible as an alternative that the high-kick and low-kick NS groups both are the result of iron-core collapses with different amounts of envelope stripping (see discussions in Section 13.7).

Recent SN modelling suggests a broad range of kick velocities (from $<10\,{\rm km\,s^{-1}}$ to $>1000\,{\rm km\,s^{-1}}$) imparted on newborn NSs from iron-core collapse SNe alone (see further discussions in Chapter 13). In general, there seems for these iron cores to be a correlation between the mass of the exploding (stripped) star and the resulting kick, which might provide an alternative explanation for why helium star masses play an important role for explaining the observed properties of the Be/X-ray binaries. It remains to be shown, however, that mass-related kicks of iron cores may indeed lead to two distinct groups of kicks.

10.3.3 Absence of Magnetars in HMXBs

Without doubt the most dramatic and radiatively unpredictable type of NSs are the so-called *magnetars*—a population of very young and the most strongly magnetized of all NSs, with surface dipole B-fields in the regime of $10^{13}-10^{15}$ G (Fig. 1.2). These objects, of which ~30 are currently known in our Galaxy (Olausen & Kaspi, 2014),[1] have as their hallmark the emission of bright, occasional super-Eddington, bursts of X-rays and soft γ-rays. A sizeable fraction of the magnetars is located in SN remnants, which implies that they are among the youngest of all known NSs. The energy source

[1] See the McGill Online Magnetar Catalog at https://www.physics.mcgill.ca/~pulsar/magnetar/main.html

of their outbursts, as well as for the luminous X-ray pulsations observed from many magnetars, is believed to be the decay of an intense internal magnetic field (Viganò et al., 2013), which causes stresses and fractures of the stellar crust (Thompson & Duncan, 1995, 1996). The strengths of these fields are inferred from a variety of different arguments (see Thompson & Duncan, 1996; Rea et al., 2010) but arguably most commonly from the spin period and period derivative measured from the X-ray pulsations via the conventional dipole spin-down formula (Eq. 14.4), which often yields dipolar B-fields $> 10^{14}$ G, as first discovered by Kouveliotou et al. (1998). This breakthrough discovery definitely established the existence of their enormous magnetic fields. For reviews on magnetars, see, for example, Turolla et al. (2015); Kaspi & Beloborodov (2017).

Outstanding questions regarding magnetars include the origin of such high-magnetic fields (Damour & Taylor, 1992), what fraction of the NS population is born as magnetars (Keane & Kramer, 2008), and whether there is an evolutionary relationship between magnetars and other NSs, such as high-magnetic-field radio pulsars. In such a picture (e.g., the grand unification of NSs [GUNS], Kaspi [2010]), the differences between the distinct types are mainly caused by discrepancies in age and initial magnitude of the B-field. The latter may depend on the degree of SN fallback (Zhong et al., 2021). While initial magnetar spin periods of the order of milliseconds may be required at birth for producing long γ-ray bursts, superluminous SNe, and FRBs (e.g., Section 7.1.5 and Dall'Osso & Stella [2022]), modelling of the spin period evolution of magnetars to reproduce the currently observed population does not require birth spin periods <100 ms (Jawor & Tauris, 2022).

Interestingly, none of the known magnetars is a member of a binary system.[2] It has been suggested (Schneider et al., 2019) that their progenitors are stars with strong and large-scale surface magnetic fields. Such stars are thought to be the product of merged massive stars (see Chapters 4 and 9). This merger hypothesis thus would provide a simple explanation for the lack of magnetars in binary systems (unless they formed in a triple system, but that can probably only account for a few rare cases). It is believed that approximately 10%–20% of all massive stars are the product of a previous merger process (see Chapters 4 and 9), and given the Galactic rate of core-collapse SNe of about 2 per century (Diehl et al., 2006), the resulting birth rate of magnetars is close to that inferred from NS population studies of $0.3^{+1.2}_{-0.3}$ per century (Keane & Kramer, 2008); however see also Jawor & Tauris (2022) for higher magnetar birth rates based on synthetic populations.

10.4 WR BINARIES, HMXBs, AND RUNAWAY STARS

The evolution of the system of Figure 10.2 into a sg-HMXB is straightforward. However, in the evolution of massive binaries there are complications due to stellar wind-mass loss that we have so far ignored. This will not qualitatively alter the evolutionary

[2] Although Klus et al. (2014) and Shi et al. (2015) have suggested that magnetars in general may populate a group of HMXBs observed in the SMC, the B-field estimates remain controversial in those cases (Ikhsanov & Mereghetti, 2015).

picture, but it will change the numerical values of the masses of the components (and also the orbital period) at various evolutionary stages. But the main line of the evolution will remain the same. The more massive the stars are, the stronger the wind mass-loss effects will be. The most massive observed evolved binaries are the WR binaries, such as those listed in Table 10.1. As described earlier in Chapters 5 and 8, WR-stars are helium stars with, at the metallicity of our Galaxy, masses generally upward from $\sim 8\,M_{\odot}$. They are considered to have lost their hydrogen-rich envelopes either by strong stellar wind-mass loss (single WR-stars: Conti scenario) or by binary mass transfer; see discussions in Section 8.1.7. They have large stellar wind mass-loss rates of the order of $10^{-5}\,M_{\odot}\,\mathrm{yr}^{-1}$. As they live for approximately $(3-5)\times 10^{5}$ yr, they will lose a considerable amount of mass during their lifetimes (Shenar et al., 2020).

We made some toy calculations for the expected further evolution of the WR-binaries of Table 10.1 into binaries consisting of an O-star and a compact star and have listed the results in the table. As mentioned in Section 9.7.1, Woosley (2019) finds that helium stars in the initial mass range between 14 and 22 $M_{\odot}$ reach final masses (after wind mass loss) between 8 and 11 $M_{\odot}$ and are expected to explode, presumably leaving NSs. Because most observed WR-stars have masses in the range of approximately $8-22\,M_{\odot}$, we thus expect a sizeable fraction of them to undergo a successful SN explosion and leave a NS. In our calculations for Table 10.1, we considered two cases: (i) the core collapses to form a 1.4 $M_{\odot}$ NS and the rest of the envelope mass is ejected in a spherically symmetric SN explosion; we assume here the mass of the WR-star at the time of the explosion to be 9 $M_{\odot}$ for WR-stars with initial masses $\leq 20\,M_{\odot}$, and half of the initial WR-star mass for larger masses; and (ii) the star at the time of core collapse has the same mass as in case (i), but the star collapses to form a BH. In case (ii), we assumed the entire remaining mass of the WR-star to collapse to the BH, such that no mass is ejected in a SN, and no kicks are imparted to the BH (i.e., no loss of gravitational binding energy). We have assumed that the original orbits are circular, and that the orbital period remained unchanged until the core collapses. With these assumptions, we calculated the effect of the SN mass ejection on the orbital period and eccentricity and the systemic runaway velocity v_{sys} imparted to the c.m. of the system by the SN mass ejection. In the case that a BH forms, because of these assumptions for this case, the orbital period, eccentricity, and system velocity do not change. We have listed in Table 10.1 for case (i): the final orbital period $P_{\mathrm{orb}}^{\mathrm{f}}$, eccentricity e, and the system's runaway velocity v_{sys}. For case (ii), the values of $P_{\mathrm{orb}}^{\mathrm{f}}$ remained constant and $v_{\mathrm{sys}}=0$, and we just listed the masses of the BH and its O-type companion.

Table 10.1 shows that for the case of NS remnants, the orbital periods range from 3.5 d to 149 d, with the majority in the range of 3.5–15 d, as observed in most of the sg-HMXBs. One observes that due to the SN mass ejection, runaway velocities mostly between 45 and 140 $\mathrm{km\,s^{-1}}$ will be imparted on the systems. One further observes that in about half of the systems in Table 10.1, the O-type companion stars are more massive than 30 $M_{\odot}$. This is the case in only two of the known sg-HMXBs in our Galaxy: 4U 1700−37 and GX 301−2 (4U 1223−62), which have as companions the $>30\,M_{\odot}$ O6.5f star HD 153919 (Clark et al., 2002) and the $38-68\,M_{\odot}$ B1.5Iae supergiant Wray 977, respectively. In both systems, the X-ray source is a NS, because in 4U 1700−37 the compact object has a mass of $\sim 2\,M_{\odot}$ and in GX 301−2 it is an X-ray pulsar. That very massive original primary stars still left a NS might be seen as

Table 10.1. Examples of Evolutionary Products of WR Binaries with Well-determined Orbital Parameters

Name	Spectral type	P_{orb} (d)	M_{WR} ($M_\odot$)	M_O ($M_\odot$)	P^f_{orb} (d)	e	v_{sys} (km s^{-1})	$M_{BH} + M_O$ ($M_\odot$)
CQ Cep	WN6o+O9II-I	1.64	24	30	3.5	0.34	151.7	12+30
CX Cep	WN4o+O5V	2.127	26.4	36.0	4.3	0.32	140.4	13.2+36.0
V444 Cyg	WN5o+O6III-V	4.212	12.4	28.4	7.4	0.26	86.4	9+28.2
WR 47	WN6o+O5V	6.239	51	60	15.6	0.39	140.6	25.5+60
WR 42	W7C+O7V	7.886	14	23	15.8	0.31	76.2	9+23
WR 21	WN5o+O4-6	8.255	19	37	12.6	0.20	60.2	9+37
WR 79	WC7+O5-8	8.891	11	29	15.3	0.25	66.1	9+29
WR 62a	WN4-5o+O5.5-6V	9.145	22	41	14.9	0.23	68.0	11+41
WR 127	WN5o+O9.5V	9.555	17	36	14.7	0.20	110.2	9+36
HD 63099	WC5+O7	14.305	9	32	23.4	0.23	53.8	9+32
HD 94305	WC6+O6-8	18.82	16	34	29.8	0.21	47.7	9+34
HD 193928	WN5o+O5V-III	21.690	45	30	149.3	0.67	110.0	22.5+30
HD 168206	WC8+O8-9IV	29.704	13	27	53.4	0.27	45.7	9+27
γ^2 Vel(WR 11)	WC8+O7.5III	78.53	9.5	30	132.6	0.24	31.5	9+30

Source: The observed systems are taken from the catalog of van der Hucht (2001). Systems are listed in order of increasing orbital periods. The indication WNo means that hydrogen lines are absent in the WR spectrum. For the final evolution of the systems, we considered two cases, as described in the main text. The values of P^f_{orb}, e, and v_{sys} are for case (i).

a confirmation of the work of Woosley (2019). It is possible that a few of the most massive WR-stars in Table 10.1 will leave BHs. These will then be progenitors of BH sg-HMXBs, such as Cyg X-1 and LMC X-1. Note, that after tidal circularization of the orbits, the orbital periods of the post-SN systems will be 10%–25% shorter than the $P^{\mathrm{f}}_{\mathrm{orb}}$ values in Table 10.1.

10.4.1 Runaway OB-stars

In 1954, Blaauw & Morgan (1954) discovered that the naked-eye stars AE Aur (O9.5V) and μ Colombae (B0V), which are in the Northern and Southern sky, respectively, are moving away from the Orion region with almost oppositely directed velocities of 127 km s^{-1}, as depicted in Figure 10.7. The stars must have left the Orion Nebula region ~2.6 Myr ago. Somewhat later Blaauw discovered that also the naked-eye star 53 Ari had left Orion with a similar speed in a different direction a few Myr ago (e.g., see Blaauw, 1956). Blaauw and Morgan named these stars *runaway stars*, which has become their accepted name. Practically no runaway stars are found of spectral type later than B2V. Extensive observational studies of the runaway OB-stars were made, for example, by Gies & Bolton (1986), Gies (1987), and Stone (1991). For an excellent recent overview of the literature on observations of runaway stars and of theoretical work on models of their origins, see Renzo et al. (2019). One generally calls an OB-star a runaway star if it has an excess space velocity of >30 km s^{-1} with respect to its local rest frame in the Galaxy. Two possible models have been suggested for the origins of these stars: the *binary SN model* suggested by Blaauw (1961), and the *dynamical interaction model*, suggested by Poveda et al. (1967). In the first model, the most massive component of the binary (which has the shortest lifetime) explodes as a SN and disrupts the binary. This will happen in the case of a symmetric SN (i.e., assuming no kick) where the exploding star ejects more than half the total mass of the system. As a consequence of the disrupted binary, the unevolved companion leaves the binary with a velocity of the order of its orbital velocity (of typically ~100 km s^{-1}; Blaauw, 1961). In the second model, the runaway star received its high velocity due to dynamical interactions between single stars and/or between close binaries and single stars in dense young star clusters (see, e.g., Portegies Zwart et al., 2010, and references therein). Such dense small clusters consisting of massive stars of spectral type O and early B are indeed found in OB associations such as those located in Orion.

10.4.2 Binary Evolution and Runaway OB-stars

For an up-to-date review of the history and present state of the subject, we refer to Renzo et al. (2019). We see in Table 10.1 that if the remnants of the WR-stars are NSs, the systems after the SN explosion of the WR-star are expected to have runaway velocities in the range of 30–140 km s^{-1}. Such velocities are similar to the typical runaway velocities observed for O-type runaway stars. The spectroscopic binary character of a massive O-star with a NS companion will be difficult to detect, as the velocity amplitude of the O-star will often be below the detection limit of 15–20 km s^{-1} for these broad-lined stars.

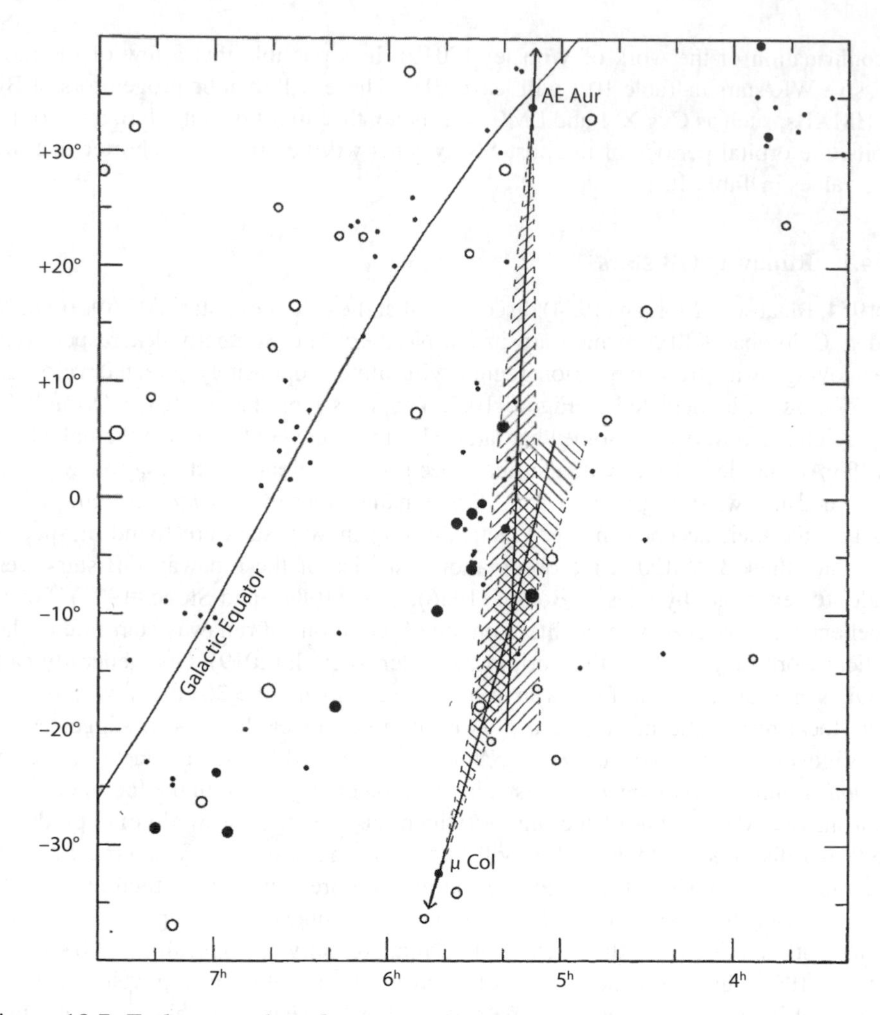

Figure 10.7. Early examples of runaway OB-stars. The naked-eye stars AE Aurigae (O9.5V) on the Northern sky and μ Colombae (B0V) on the Southern sky were found by Blaauw & Morgan (1954) to move in almost opposite directions away from the Orion region, with a velocity of 127 km s^{-1}. Arrows show the directions of their proper motions, corrected for the solar motion. Shaded areas show the uncertainties of their directions in 1954. Stars brighter than apparent visual magnitude of 3.5 are represented by large dots (OB-stars) and open circles. Smaller dots are fainter OB-stars. The two runaway stars left the Orion Nebula region ~2.6 Myr ago. Later, also the naked-eye star 53 Ari was found to move away from Orion with a similarly high velocity (Blaauw, 1956). The horizontal coordinates are right ascension, and the vertical ones are declination. After Blaauw & Morgan (1954).

Another factor is that we assumed in the calculations for Table 10.1 that the SN-mass ejection was spherically symmetric, so no kicks of the newborn NSs were included. However, as we discuss in Chapter 13, NSs formed by iron-core collapses are often imparted a large kick velocity at birth, typically in the range of $100 - 500\,\mathrm{km\,s^{-1}}$. These kicks are expected to disrupt a fair fraction of the systems. However, the orbital velocities of the WR-stars in the first eight systems in the table are between 250 and $450\,\mathrm{km\,s^{-1}}$. To disrupt a system with a forward kick (in the direction of the orbital motion), and under the special assumptions assumed with no mass loss during the formation of the BH, a kick is needed with a magnitude that is $(\sqrt{2} - 1)$ times the orbital velocity. The kicks must then be between 100 and $180\,\mathrm{km\,s^{-1}}$. Kicks in other directions need to be larger to cause disruption, and kicks in backward direction relative to the orbital motion (50% of all kicks, if kick directions are random) have the smallest chance of disrupting the systems. Therefore, as a first-order rough estimate, we expect $>50\%$ of the WR-systems in Table 10.1 to survive the SN-mass ejection combined with (moderate) kicks.

Recently, from careful binary population synthesis evolution calculations, summarizing over the observed distributions of primary star masses, initial orbital periods, and mass ratios and assuming a distribution of birth kick velocities of NSs similar to that observed for radio pulsars, Renzo et al. (2019) find that $86^{+11}_{-22}\%$ of all binaries are disrupted in the first SN. That most of the WR binaries of our table are not expected to be disrupted may be a selection effect: these systems are, because of their considerably higher masses than average for post-mass-transfer binaries, not a representative sample of the total population of these binaries.

Still, even if a large kick is imparted to the NS and the system is disrupted, the O-stars (or B-stars in similar systems) will often be ejected from their binary with a velocity of the order of their orbital velocity in the pre-SN binary (although deviations can be quite large depending on kick direction; Tauris & Takens, 1998). Although these ejection velocities of the single runaway OB-stars can in some cases be significantly larger (even producing hyper-velocity stars; Tauris, 2015), they are often comparable to those post-SN systemic velocities of surviving binaries listed in Table 10.1. Therefore, both the binary and single O-stars resulting after the SN in a WR binary similar to those in Table 10.1, will be genuine runaway stars, given that the typical velocity dispersion of ZAMS O-stars in OB associations is only $\sim 10\,\mathrm{km\,s^{-1}}$.

In his original paper on the binary-SN origin of runaway OB-stars Blaauw (1961) did not take mass transfer in binaries into account, because this was not known at the time. In his original model it was therefore always the most massive star that exploded, and the systems were disrupted.

After he knew about mass transfer, Blaauw (1993) showed that the seven best-studied runaway OB-stars (for all of which the parent association [loose star cluster] from which they were ejected is known) indeed show traces of prior close-binary mass transfer. These traces are (1) a higher than normal helium abundance; (2) very rapid rotation in 6 of the 7 cases, which is expected after mass transfer in a binary, such as in the Be/X-ray binaries (the rapid rotation of runway OB-stars was first noticed by Conti & Ebbets, 1977); and (3) in the HR diagram of their parent associations, these runaway stars are *blue stragglers*, just as would be expected from close-binary mass transfer, as

Figure 10.8. SN remnant Semeis 147 contains the B0.3V runaway star HD 37424 with an excess transverse velocity of $74 \pm 7.5\,\mathrm{km\,s^{-1}}$ and the pulsar PSR J0538+2817 (the + symbol) with an excess transverse velocity of $383 \pm 1\,\mathrm{km\,s^{-1}}$, both directed away from the center of the SN remnant (the X symbol), as discovered by Dinçel et al. (2015). As explained in the text, these two objects originated 30,000 yr ago from a post-mass-transfer close binary that was disrupted in the SN explosion. Credit: B. Dincel.

this mass transfer will have increased the mass of the OB-star, as well as rejuvenated it. Most of these seven runaway stars are O-stars, and it thus seems logical to associate them with WR+O progenitor binaries such as the ones in Table 10.1.

Example of Evidence for a Binary SN Origin of a Single Runaway Star

Dinçel et al. (2015) discovered that in the SN remnant Semeis 147 there is the B0.3V runaway star HD 37424 with an excess transverse velocity with respect to its local rest frame of $74 \pm 7.5\,\mathrm{km\,s^{-1}}$, plus a high-velocity pulsar, PSR J0538+2817, with an excess transverse velocity of $383 \pm 1\,\mathrm{km\,s^{-1}}$. Both these velocities are directed away from the center of the SN remnant (see Fig. 10.8). The pulsar's velocity and distance from the center lead to a kinematic age of 30,000 yr, when both stars originated in the center of the SN remnant. The high velocity of the B0.3V-star indicates that the pre-SN system had a relatively short orbital period, less than $\lesssim 15$ days, because only the disruption of a close system can have produced its large runaway velocity (unless unlikely and extreme fine-tuning in the kick direction from the newborn NS has accelerated the runaway star in its escape trajectory Tauris & Takens, 1998; Tauris, 2015). The short orbital period means that prior to the SN there must have been extensive mass transfer in the system, and that at the time of the explosion, the exploding star was the less massive component of the system. That the system still was disrupted means that the disruption can only have been due to the high kick velocity imparted onto the NS at its birth. This is direct proof of high kick velocities imparted onto NSs (see also Chapter 13). At the same time, it is *direct proof* of the Blaauw mechanism for producing runaway OB-stars.

10.4.3 The Runaway Character of HMXBs

van den Heuvel & Heise (1972) and Tutukov & Yungelson (1973b) predicted HMXBs to be runaway stars. It was subsequently indeed found that the HMXBs as a group share several of the characteristics of the runaway OB-stars (van den Heuvel, 1985; van Oijen, 1989, namely): (i) they are practically never found in OB associations, and (ii) they have a significantly wider Galactic z-distribution than the normal OB-stars do.

Since then the runaway velocities of several well-known individual sg-HMXBs have been measured, of which the key examples are the relatively nearby systems Vela X-1 (4U 0900−40) and 4U 1700−37, which both have very bright optical companions of apparent visual magnitudes of 7.0 and 6.5, respectively. Also, the 3.73 d orbital period sg-HMXB 4U 1538−52 has the 9th magnitude optical companion QV Norma, which has an excess radial velocity of 84 km s^{-1} with respect to its Galactic rest frame; this system is located 280 pc above the Galactic plane, which is far too high for a massive young star, illustrating its runaway character.

The excess transverse velocity of 45 km s^{-1} of Vela X-1 (4U 0900-40; orbital period 8.97 d) was measured thanks to the discovery by Kaper et al. (1997) of the bow shock in front of this source as it ploughs through the interstellar medium, which is depicted in Figure 10.9.

The excess space velocity of 66 km s^{-1} of the sg-HMXB 4U 1700−37 (orbital period 3.4 d) with respect to the Sco OB1 association, from which it originated, resulted from proper motion measurements with the Hipparcos astrometry satellite by Ankay et al. (2001), which were significantly refined recently by van der Meij et al. (2021) on the basis of data obtained with the *Gaia* astrometry satellite. The latter results are depicted in Figure 10.10.

These measured excess space velocities of sg-HMXBs are in the normal range for runaway OB-stars and fit well with the velocities expected for descendants of WR binaries given in Table 10.1.

For the Be/X-ray binaries, the expected runaway velocities are much smaller. This is due to a combination of their wider orbits, lower masses, and relatively small amounts of mass ejected in the SN. Furthermore, generally, the wider orbits mean that the pre-SN orbits were wide also, which therefore only allows a system to remain bound if the imparted NS kick is small. Thus, the resulting systemic runaway velocity will be small too. Indeed the measured excess velocities here mostly do not exceed 10 km s^{-1} (van den Heuvel et al., 2000). Also the work of Renzo et al. (2019) indicates that the runaway velocities of intermediate-mass post-SN systems are small (they become so-called *walkaway stars*).

10.5 STABILITY OF MASS TRANSFER IN HMXBs

From numerical modelling of IMXBs including spin-orbit interactions, stable mass transfer is found for main-sequence and Hertzsprung-gap donor stars less than ~3 times more massive than the accreting compact object (Tauris et al., 2000; Misra et al., 2020).

Figure 10.9. Bow shock in the interstellar medium produced by the ionizing radiation together with the motion of the 7th visual magnitude sg-HMXB Vela X-1 (4U 0900−40) with an excess transverse velocity of 45 km s^{-1} with respect to its local rest frame in the Galaxy. The cross indicates the location of Vela X-1. The bow shock shows that this HMXB is a runaway star. After Kaper et al. (1997).

These donor stars, which have radiative or not too deep convective envelopes, undergo an initial rapid phase of mass transfer on a (sub)thermal timescale (see Section 11.6). It has been found that the same is true for HMXBs (Pavlovskii et al., 2017; van den Heuvel et al., 2017). Once the donor stars in IMXBs and HMXBs develop deep convective envelopes, however, the mass transfer usually become dynamically unstable (Savonije, 1978, 1979).

Quast et al. (2019) explored the stability of sg-HMXBs with very high donor masses (up to 20 times more massive than the NS accretors) and found that an ultra-luminous X-ray (ULX) phase could be long lasting ($\sim$0.4 × 10^6 years) for such an extreme mass ratio, if the super-giant star has a H/He gradient in the layers beneath its surface. They demonstrated that such systems can evolve on a nuclear timescale with a BH or NS accretor. In their binary evolution models, the donor stars rapidly decrease their thermal equilibrium radius and can therefore cope with the inevitably strong orbital contraction imposed by such a high mass ratio. These binaries could be post-CE systems, where the super-giant donor star has lost most of its hydrogen-rich envelope and the remaining one is enriched in helium. Recent 1D hydrodynamical simulations of a CE phase between a super-giant donor and a NS predict the formation of such binary configurations (Fragos et al., 2019). The final fate of the models by Quast et al. (2019), however,

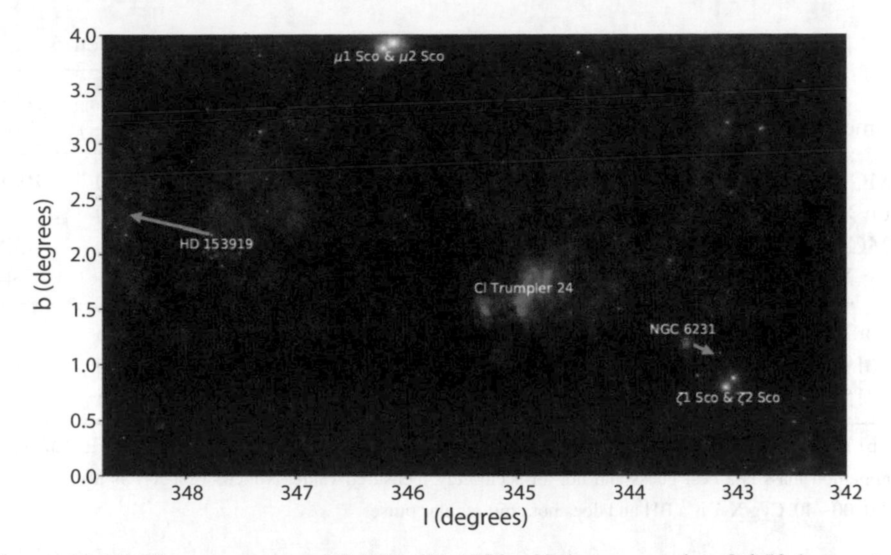

Figure 10.10. The runaway sg-HMXB 4U 1700−37 has a massive O6.5f donor star, HD 153919, which is almost visible to the naked eye (apparent visual magnitude of 6.5). Gaia proper motion measurements by van der Meij et al. (2021) showed that this is a runaway star, which originated in the young massive-star cluster NGC 6231 in the center of the OB association Sco OB1, confirming earlier findings by Ankay et al. (2001) on the basis of Hipparcos proper motions. The arrows show the distances travelled by the HMXB and the cluster in 0.5 Myr. Open cluster Cl Trumpler 24 and four naked eye Scorpius stars are indicated in the figure (source: Aladin). 4U 1700−37 moves in space with a velocity of 66 km s^{-1} relative to the cluster, and left the cluster 2.2($\pm$ 0.1) Myr ago. The coordinates are Galactic longitude (l) and latitude (b). After van der Meij et al. (2021).

is that following this ULX phase the mass transfer most likely becomes dynamically unstable and the system enters a CE phase.

10.6 THE X-RAY LIFETIME AND FORMATION RATE OF THE BLUE SUPERGIANT HMXBs

10.6.1 Mechanisms Driving the Accretion: Two Classes of Persistent sg-HMXBs

In the Corbet diagram (Fig. 7.18), in which the spin period of the NS is plotted against the orbital period of HMXBs, one observes that there are two distinct classes of persistent sg-HMXBs.

The first is a small class of systems located in the lower left part of the diagram, which are characterized by the following: i) a short pulse period, of the order of 1 s; ii) circular orbit; iii) a steadily decreasing pulse period, indicative of accretion from a disk; iv) a Roche-lobe–filling donor star; v) the presence of a tertiary period that

Table 10.2. Two Types of sg-HMXBs: Disk Accretors and Wind Accretors

Name	Spectral type	$P_{\rm orb}$ (d)	$P_{\rm pulse}$ (s)	e	$R_{\rm L,peri}$ $(R_\odot)$	$M_{\rm comp}$ $(M_\odot)$	P_3 (d)
SMC X-1	B0Ib	3.892	0.717	$\leq$0.007	17.6	20	40–60
Cen X-3	O6.5II−IIIV	2.087	4.84	0.0008	11.6	20	—
LMC X-4	O7 II−V	1.408	13.5	$\leq$0.02	10.7	30	30
Cyg X-1	O9.7Iab	5.60	—	0.002	22.9	41	294
Vela X-1	B0.5Ib	8.964	283	0.092	31.1	25	—
4U 1907+09	O8−9Ia	8.376	438	0.28	24.4	30	—
4U 1538−52	B0Iab	3.728	529	0.174	14.9	25	—

(Top) Disk accretors; (bottom) wind accretors. Except in the case of Cyg X-1 ($M_{\rm comp} = 41 \pm 7\, M_\odot$), the companion mass is a best guess and not an accurately measured value. Notice, Vela X-1 is also known as 4U 0900−40. Cyg X-1 is a BH and does not emit regular pulses.

is much longer than the orbital period. The upper part of Table 10.2 lists the known examples of this class. Also, the pulsating ULXs in external galaxies belong to this class (see Chapter 7). Furthermore, we have tentatively also included the BH X-ray binary Cyg X-1 in this class, as it also has a circular orbit and a tertiary period that is indicative of the presence of a precessing accretion disk. The (big) difference between this system and the NS systems is the large mass (21 $M_\odot$) of the compact star in this system (Miller-Jones et al., 2021). This large mass of the accretor enables stable RLO in this system, as will be explained in Section 10.12. For NS accretors in HMXBs, an epoch of stable RLO is much more difficult to achieve, but still potentially possible, as will be argued in later in this sub-section.

The second class is the main class of persistent sg-HMXBs: the wind-accreting systems. These are characterized by i) very long pulse periods, ranging from 10^2 to 10^4 s; ii) erratic pulse period variations (alternating spin-up and spin-down episodes, the pulse period hovering around an average value with no systematic trend), indicative of accretion from a wind (see Chapter 7); iii) orbits with still substantial eccentricity; and iv) donor stars that do not fill the Roche lobe, not even at periastron. The lower part of Table 10.2 lists a few key examples of this class.

In the first class of systems, the presence of an accretion disk and steady decrease of the pulse period indicates that the mass transfer is driven by RLO. However, as we have seen in Chapter 9, fully developed RLO proceeds on the thermal timescale of the donor's envelope, which is $<10^4$ yr, leading to a mass-transfer rate of the order of $10^{-3}\, M_\odot\, {\rm yr}^{-1}$, which would completely drown the X-ray source and lead to CE formation. However, as described in Chapter 4, it was shown by Savonije (1978, 1979, 1983) that when the donor star almost fills its Roche lobe, there can be a phase of *beginning atmospheric RLO*, in which the mass-transfer rate is still below the Eddington rate of the accretor. For donor stars that are still in the core-hydrogen burning stage—like the donors of Cen X-3, LMC X-4, and most probably also SMC X-1 and Cyg X-1—this phase may last of the order of 10^4 yr (which is the observed spin-up timescale of the X-ray pulsar Cen X-3; Shirke et al., 2021).

Furthermore, Quast et al. (2019) show that for systems with NS accretors and core-hydrogen burning donors there can even be a much longer phase of stable RLO at a rate of the order of $10^{-5}\,M_\odot\,\mathrm{yr}^{-1}$, lasting $\sim 2.5 \times 10^5$ yr, before fully developed RLO begins. (This stable RLO phase occurs if the envelopes of the donor stars have an enhanced helium abundance, as is indeed observed in these systems.) Quast et al. (2019) show that the mass-transfer rate of the order of $10^{-5}\,M_\odot\,\mathrm{yr}^{-1}$ fits well with the observed rates of decrease of the orbital periods of the RLO HMXBs (Falanga et al., 2015) (see Fig. 10.11), as well as with the inferred accretion rates in the pulsating ULXs in external galaxies. It thus appears that the systems with NSs in the upper part of Table 10.2 may possibly live as long as $\sim 2.5 \times 10^5$ yr under certain circumstances. In the system of Cyg X-1, stable fully developed RLO is in principle possible, as the mass ratio of donor and accretor is well below 3. However, as this would lead to a very large mass-transfer rate (on a thermal timescale of the donor), which is not observed here, this cannot yet be the case in the present Cyg X-1 system. Most probably it is just in the beginning atmospheric RLO phase, which may last for thousands of years.

Before we can consider the lifetime of the class of wind-accreting sg-HMXBs, we must establish in which evolutionary stage the donor stars in the wind-driven systems are. These supergiant donors are always of very early type, ranging from O6I to B1.5I. We now first must examine what this classification *early-type blue supergiant* really means in terms of stellar evolution.

10.6.2 The Evolutionary State of Blue Supergiants of Type Earlier than B1.5I

The classification *blue supergiant* is an *observational spectroscopic classification*. When observers notice that the spectral lines of a luminous early-type star are quite sharp, they classify this star as a giant or a supergiant. Sharp spectral lines indicate that there is a low-density atmosphere (little pressure broadening of the lines), which is evidence for low atmospheric gravity that is much lower than for a ZAMS star, and therefore a large radius.

Figures 8.6, 8.7, and 8.9 show that stars of approximately 20 and 25 $M_\odot$ begin on the ZAMS as spectral types O7V and O6V main-sequence stars, respectively; and near the end of core-hydrogen burning they have reached spectral types of approximately B1 and B0.5, respectively.

As the evolutionary tracks in Figure 8.12 show, stars of $\sim 20\,M_\odot$ at their latest stages of core-hydrogen burning are already classified by observers as *blue supergiants*, while stars of 40 $M_\odot$ that are beyond some 65% of their core-hydrogen burning stage are already classified as such supergiants (see also discussion in Chapter 8). We thus see that what is called a *blue supergiant* by a spectroscopic observer can, at masses $\geq 20\,M_\odot$, still be in a late phase of core-hydrogen burning. It does not need to be a post-main-sequence star, as the supergiants of spectral type beyond B1.5I probably are.

During core-hydrogen burning, the radius of the star increases only slowly with time. Much slower than after the end of core-hydrogen burning, when the radius very rapidly expands by a large factor in only a few thousand years. This is shown in

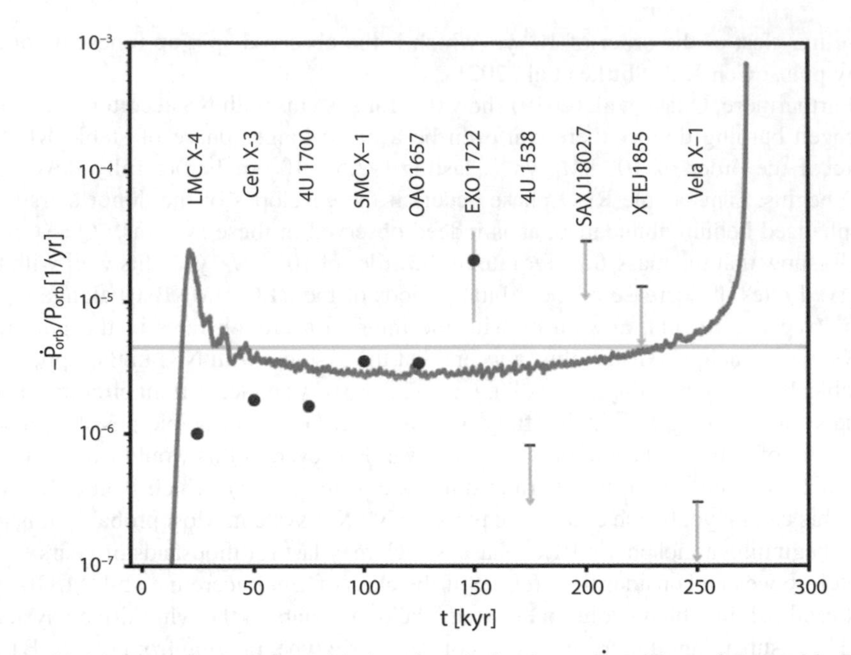

Figure 10.11. Evolution of the orbital period derivative $-\dot{P}_{\rm orb}/P_{\rm orb}$ of the stable RLO model calculations by Quast et al. (2019), together with an analytical estimate for the same model (the horizontal line) by these authors. The ZAMS mass of the donor was $60\,M_\odot$, which due to stellar wind-mass loss had decreased to $33.6\,M_\odot$ at the onset of the mass transfer, when the orbital period was 9.1 d. The NS has a mass of $2.0\,M_\odot$. After a brief switch-on state, the mass-transfer rate stabilized at $\sim 3 \times 10^{-6}\,M_\odot\,{\rm yr}^{-1}$ for 250,000 yr, while the orbital period decreased to 1.4 d. Overplotted are the measured orbital period decay rates of 10 sg-HMXBs as inferred by Falanga et al. (2015) who give the measured values with 1σ error bars (blue dots with error bars; for most of these dots the sizes of the bars are smaller than the size of the dot) and four upper limits (blue arrows). It should be noted that the time axis only has a meaning for the orange theoretical curve of the period evolution of the binary, not for the period derivatives of the observed sources. It should also be noted that the systems with upper limits are wind accretors. Most of the other systems are disk accretors with RLO. After Quast et al. (2019).

Figure 10.12, which depicts the evolution of the stellar radius from the ZAMS beyond core-hydrogen exhaustion of stars of $20\,M_\odot$, $25\,M_\odot$, and $40\,M_\odot$, as calculated by Schaller et al. (1992) for solar metallicity, taking wind mass loss into account. (In later calculations of massive star evolution, such as those by Brott et al. [2011] and Ekström et al. [2012], also with rotation, the radius expansion for these masses is very similar.) One observes that for the $20\,M_\odot$ star the radius increases from $5.75\,R_\odot$ on the ZAMS to $19.59\,R_\odot$ at core-hydrogen exhaustion. For the $25\,M_\odot$ star, it increases from $6.49\,R_\odot$ on the ZAMS to $21.53\,R_\odot$ at core-hydrogen exhaustion, and for the $40\,M_\odot$ star, it increases from $8.49\,R_\odot$ on the ZAMS to $41.31\,R_\odot$ at core-hydrogen exhaustion,

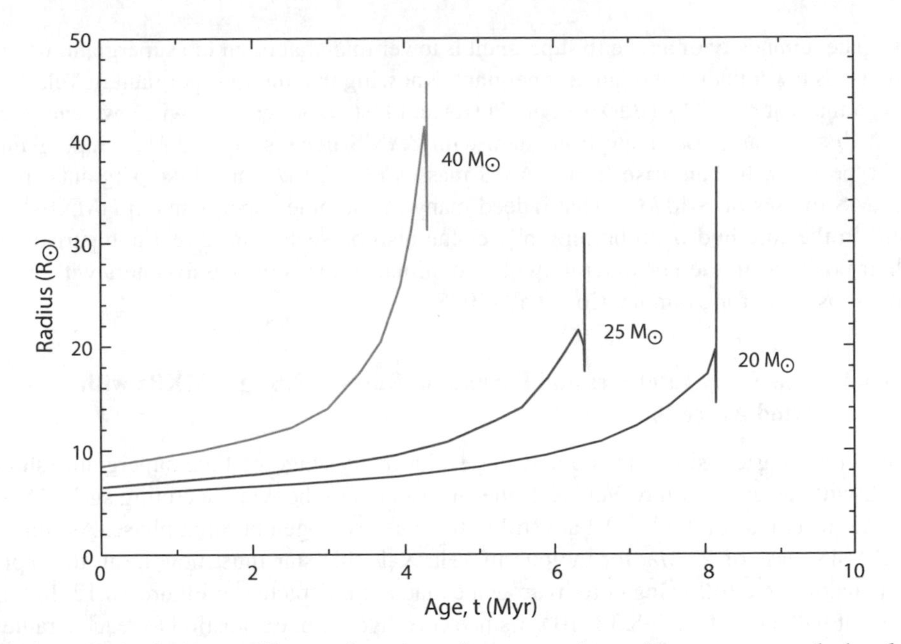

Figure 10.12. Stellar radii as a function of age, on the main sequence and shortly thereafter, for three models of Schaller et al. (1992) of ZAMS masses: 20, 25, and 40 $M_{\odot}$.

where the mass of the latter star has been reduced to 36.23 $M_{\odot}$ due to stellar winds. (For the two lighter stars, the mass loss on the main sequence is close to negligible.) For the 20 $M_{\odot}$ star, the radius doubles during the last 25% of hydrogen-burning; for the 25 $M_{\odot}$ star, this is the case in the last 20% of hydrogen-burning; and for the 40 $M_{\odot}$ star, during the last 12% of hydrogen-burning. One thus observes that the more massive the star, the shorter the time in which its radius doubles. (Thus, the more massive the star, the closer to core-hydrogen exhaustion the fast radial expansion occurs.)

Now the question arises: At what radius do observers begin to call a hot massive star a *blue supergiant*? Here the sg-HMXBs provide a good guide, because in these systems we know the size of the Roche lobe, and we know that their supergiants are not far from filling the Roche lobe, as we see regular ellipsoidal photometric variations (see Chapter 6). Assuming their radius to be close to that of the Roche lobe at periastron, we can see for the systems in Table 10.2 that the O9.7Iab supergiant of 41 ± 7 $M_{\odot}$ in the Cyg X-1 system (Miller-Jones et al., 2021) has a radius of $\sim$23 $R_{\odot}$, which is slightly larger than the $\sim$17 $R_{\odot}$ radius of the B0Ib supergiant in SMC X-1, while the B0Iab supergiant in 4U 1538−52 has a radius not larger than 15 $R_{\odot}$. We see that these companion stars with radii of only 15 − 23 $R_{\odot}$ are called supergiants while their given radii imply that they certainly are still in the core-hydrogen burning stage. For the donor star of Cyg X-1 this argument was put forward by Ziolkowski (2012). On the other hand, the radius of the B0.5Ib supergiant in Vela X-1 (4U 0900−40) is $\sim$30 $R_{\odot}$ and the O8-9Ia supergiant in 4U 1907+09 is $\lesssim$24 $R_{\odot}$.

The luminosity of a type Ib supergiant is lower than that of an Iab supergiant, which in turn is lower than that of an Ia supergiant. Knowing that the Ib supergiant in Vela X-1 has a mass of $\sim 23\,M_\odot$ (Barziv et al., 2001) and lost some mass by wind, we can infer that Ib supergaints originate from stars with ZAMS masses of $\sim 25\,M_\odot$. Supergiants of type Iab will then arise from ZAMS masses of $\sim 30\,M_\odot$ and Ia supergiants from ZAMS masses of $\sim 40\,M_\odot$. That indeed many of the blue supergiants in HMXBs are still in the core-hydrogen burning phase, can also be seen from direct comparison of their positions in the HR diagram with evolutionary tracks of massive stars with wind mass loss, see, for example, Cox et al. (2005).

10.6.3 The X-ray Lifetimes and Formation Rates of NS sg-HMXBs with Wind Accretion

The preceding considerations about the evolutionary stage of blue supergiants show that, with the exception of Vela X-1, the supergiants in the wind accreting sg-HMXBs in the lower part of Table 10.2 are still in the core-hydrogen burning phase. Assuming a ZAMS mass of $25\,M_\odot$ for the one in Vela X-1, this star must now be in the rapid expansion stage following core-hydrogen exhaustion, depicted in Figure 10.12. In this case, it will have taken $\sim 8.3 \times 10^4$ yr since core-hydrogen exhaustion to reach a radius of $30\,R_\odot$ (the radius at exhaustion was $21.5\,R_\odot$, after which it shrunk to $17.5\,R_\odot$ in 7.8×10^4 yr, to rapidly expand in 5,000 yr to $30\,R_\odot$). Thus, the X-ray lifetime of Vela X-1 cannot have been longer than $\sim 9 \times 10^4$ yr, and the same will be the case for the wider sg-HMXB systems, with orbital periods longer than approximately 6–7 d.

Because in all persistent sg-HMXBs, the donor star shows clear ellipsoidal light curves with an amplitude of ≥ 0.05 magnitudes in V, it appears that to become a persistent wind-accreting sg-HMXB, the radius of the donor star must be quite close to that of its Roche lobe at periastron. The reason for this requirement is likely as follows: If the radius is close to that of the Roche lobe, the surface gravity in the direction of the first Lagrangian point, L_1 (Chapter 4), is reduced such that matter in that direction can escape more easily than for a single star in the same evolutionary stage. This will cause the stellar wind mass-loss rate in that direction to be enhanced, producing a relatively high accretion rate on the compact star. At the same time, X-ray ionization of the wind atoms is expected to slow down the radiative acceleration of the wind matter, as is shown by the fact that the winds of the donors in wind-accreting sg-HMXBs are systematically slower than those of single supergiant stars of the same luminosity and effective temperature (see, e.g., Fig. 5.16 in Charles & Coe, 2010). Both these effects enhance the accretion rate and X-ray luminosity. It thus seems likely from these observational facts that to become a persistent wind-accreting sg-HMXB, the radius of the donor star must be close to that of its Roche lobe at periastron. We will assume that this means that the stellar radius must be $\geq 0.9\,R_L$ at periastron. The lifetime as a strong X-ray source is then the time needed to increase its radius from 0.9 to 1.0 times the Roche-lobe radius at periastron. This assumption allows us to calculate the lifetimes as a wind-accreting sg-HMXB for the systems in the lower part of Table 10.2 (with the exception of Vela X-1, for the reasons mentioned) by comparison with the radius-evolution tracks of stars of $20 - 40\,M_\odot$ in Figure 10.12.

Assuming for 4U 1907+09 a ZAMS mass of 30 $M_\odot$, we find for this system an X-ray lifetime similar to that of Vela X-1, so $\sim 10^5$ yr. For 4U 1538−52, assuming a ZAMS mass for the supergiant of 25 $M_\odot$, the lifetime is $\sim 3 \times 10^5$ yr. On the other hand, because Cyg X-1 has relativistic radio jets, it must have an accretion disk and therefore cannot be a pure wind accretor.[3] As argued earlier in this section, it must be powered by beginning atmospheric RLO. Because its 41 $M_\odot$ donor is still in the core-hydrogen burning phase (see Fig. 10.12), its evolutionary expansion proceeds very slowly and may last for at least several thousands of years, up to perhaps 10^4 yrs.

We thus see that wind-accreting sg-HMXBs with orbital periods less than ~6 d live for $\sim 2.5 \times 10^5$ yr and those with longer orbital periods, for $\sim 10^5$ yr, whereas the disk accreting systems have relatively short lifetimes, of the order of $<10^4$ years.

In the Corbet diagram (Fig. 7.18), we see that there are five persistent sg-HMXBs plus two SFXTs with orbital periods <6 d, plus another six SFXTs that, despite their longer orbital periods, have O-type supergiant companions, which therefore must be core-hydrogen burning stars. In total there are 13 sg-HMXBs with a lifetime of the order of 2.5×10^5 yr. On the other hand, there are also 13 persistent sg-HMXBs in that diagram with orbital periods >6 d, which will live for $\sim 10^5$ yr.

The total number of wind-accreting sg-HMXBs plus SFXTs known in the Galaxy is ~30 (Kretschmar et al., 2019). However, as Figure 6.4 shows, the part of the Galaxy opposite from us with respect to the Galactic center is very poorly sampled for these systems. Assuming the sample within 7 kpc from the Sun, which contains most of the known wind-accreting sg-HMXBs, to be complete and assuming a radius of the massive star-forming region in the Galactic disk to have a radius of 12 kpc, the total number of wind-driven sg-HMXBs with NSs in the Galaxy will be ~90. Based on this discussion, we expect roughly half of them to have a lifetime of 2.5×10^5 yr, and the rest to have a lifetime of 10^5 yr. This leads to a birthrate of 1.8×10^{-4} yr^{-1} plus 4.5×10^{-4} yr^{-1}, yielding a total birthrate of wind-accreting sg-HMXBs of 6.3×10^{-4} yr^{-1} in the Galaxy.

In comparison, the RLO systems are extremely rare: Cen X-3 and possibly OAO 1657−415 and Cyg X-1 are the only ones known in our Galaxy. So their contribution to the birthrate of sg-HMXBs is negligible.

With a Salpeter IMF, the birth and death rate of stars more massive than 20 $M_\odot$ is ~4 times lower than that of stars more massive than 8 $M_\odot$. So, 1/5 of all NSs are expected to result from stars with masses $\geq 20\ M_\odot$ (assuming naively that the same relative fraction of these stars leave NSs as remnants). Assuming a Galactic core-collapse SN (CCSN) rate of 10^{-2} yr^{-1}, the core-collapse rate of stars that are $\geq 20\ M_\odot$ will be 2×10^{-3} yr^{-1}. Because at least half of these massive stars are primaries of interacting binaries, the interacting binary core-collapse rate in interacting binaries with primaries more massive than 20 $M_\odot$ is expected to be $\sim 10^{-3}$ yr^{-1}.

This rate is 1.6 times larger than the wind-accreting sg-HMXB formation rate in the Galaxy that we previously calculated. This would mean, taking these results at

[3]Some wind-fed HMXBs have been suggested to produce accretion disks (e.g., Karino et al., 2019). See also the arguments given in Section 7.3.6 against such a possibility.

face value, that ∼60% of the interacting binaries with initial primary masses $>20\,M_\odot$ survived the SN explosion in the system and ∼40% of the systems have been disrupted.

The theoretically expected disruption rate of interacting binaries in the first SN event was calculated, for example, by Renzo et al. (2019), in which the effects of kick velocities imparted to the NSs in their birth events are taken into account. Renzo et al. (2019) predicted a disruption rate of 88^{+11}_{-22} % for all systems with initial primary masses that are $\geq 15\,M_\odot$, so a disruption rate of >66%. However, we must realize that our above-mentioned estimates of formation rates of sg-HMXBs still involve quite some uncertainties and may easily be off by a factor of 2. So, it is too early to conclude whether the order of 2/3 of the potential progenitor systems were disrupted, as Renzo et al. (2019) predicted. For an illustration of which orbital periods and kick velocities will result in a survival probability of ∼25% in the first SN explosion, subsequently leading to HMXBs, see Figure 13.24.

10.6.4 Formation Rate and Expected Total Number of Be/X-ray Binaries

Based on the results of Renzo et al. (2019), one would expect only approximately one quarter of the binaries with primary stars more massive than $8\,M_\odot$ to survive the first SN explosion. Assuming half of all stars in this mass range to be in binaries, one expects the formation rate of NS binaries with primary stars in the mass range of $8-20\,M_\odot$ to be roughly $1/8$ of the SN rate in this mass range, which is itself $4/5$ of the total SN rate. So the formation rate of NSs from primary stars of $8-20\,M_\odot$ in binaries is expected to be of the order of 10% the Galactic SN rate. Assuming the latter to be $10^{-2}\,\mathrm{yr}^{-1}$, one expects the formation rate of NS binaries from binaries with primaries in the mass range of $8-20\,M_\odot$ to be $10^{-3}\,\mathrm{yr}^{-1}$. Assuming these systems to become Be/X-ray binaries, in which the donor star lives for $\sim 10^7$ yr, one expects of the order of 10^4 potential Be/X-ray binary sources to be present in the Galaxy. If one were to put the upper limit to the primary mass for forming a Be/X-ray binary at a lower value, such as $15\,M_\odot$, the number would come out about a factor 2 lower, so $\sim 5\times 10^3$.

From the observational side, estimating the real Galactic number of Be/X-ray binaries is difficult, as these systems are transients, which go into outbursts at erratic and unpredictable times. A possible way to make a rough observational estimate of their Galactic number is to compare with the total number of such systems observed in a small external Galaxy: the SMC. This Galaxy has of the order of 100 recognized Be/X-ray binaries (Haberl & Sturm, 2016). Taking into account that not all Be/X-ray binaries in the SMC will have been recognized, we assume its real number to be twice that, so ∼200. Because the mass of the SMC is ∼3% of the mass of the MW, one would—assuming the incidence of Be/X-ray binaries in both galaxies to be the same—expect the Galactic number of Be/X-ray binaries to be $\sim 6\times 10^3$. Taking into account the roughness of these estimates, this number appears to agree quite well with the preceding theoretical estimate of $\sim 0.5-1\times 10^4$.

10.7 HIGHLY NON-CONSERVATIVE EVOLUTION AND FORMATION OF VERY CLOSE RELATIVISTIC BINARIES

We saw in Chapter 9 for Algol- and β Lyrae-type binaries, and in the present chapter for WR binaries and HMXBs, that the evolutionary history can be understood reasonably well in terms of a history with an assumed (more or less) *conservative* mass transfer; that is, conservation of both mass and orbital angular momentum. We know that these conservative assumptions are only approximate ones, because during the evolution of massive stars, these stars may lose a significant fraction of their mass by stellar winds. This mass loss can particularly substantial in the case of WR-stars. Also some mass is lost in the SN explosion of the most evolved component.

On the other hand, there are also many types of evolved close binaries, discussed in Chapters 5 and 6, for which there is very clear evidence that they have lost a very large fraction of their original mass and orbital angular momentum during their evolution. For such systems, the conservative approach certainly does not lead to an understanding of how these systems formed. Examples of such systems are:

i) *Close DNSs* such as PSR B1913+16 ($P_{\rm orb} = 7.75$ hr, $e = 0.61$). To form a pulsar in a binary, one has to start out with a star more massive than $8 - 10\, M_{\odot}$, and to form two NSs, the original binary must have had a total mass of at least $\sim 15\, M_{\odot}$ (while the final total mass of the DNS system is $\lesssim 3\, M_{\odot}$). Also, these binaries must have started out with much larger initial orbital periods, because otherwise they could not have encompassed progenitor stars producing a massive enough helium core to make a NS. So, they must have lost $>80\%$ of their original mass and $>95\%$ of their original orbital angular momentum in the course of their evolution.

ii) *LMXBs*. LMXBs with NSs now have total masses $<2.5\, M_{\odot}$, while the progenitor of the NS must have had a mass $\gtrsim 8 - 9\, M_{\odot}$. Clearly, the low-mass ($\lesssim 1.0\, M_{\odot}$) non-degenerate companion star hardly gained any mass. The very short orbital periods (the shortest one is 11 min, and most of them have orbital periods between 2 and 12 hr) means that also here the bulk of the original orbital angular momentum has been lost.

iii) *CVs, double WDs, and double cores of planetary nebulae*. In most cases, the WDs in CVs have masses in the range of $0.5 - 1.2\, M_{\odot}$. To form such a CO WD (and in some cases ONeMg WD), the progenitor must have been able to evolve to the AGB and must have had a mass of $\geq 2 - 3\, M_{\odot}$. This implies initial orbital radii of the order of one AU or more, whereas the present orbital radii are only a few $R_{\odot}$

These systems must therefore have lost an enormous amount of orbital angular momentum, and also quite some mass, as was first pointed out by Paczyński (1976) and J. P. Ostriker (private communication to Paczyński, mentioned in Paczyński [1976]). Clear examples that also quite some mass has been lost are the pre-CV systems in thc Hyades listed in Table 5.7. Their orbital periods are only ~ 0.5 d; their WDs have masses between 0.4 and 0.8 $M_{\odot}$; and their total system masses range from 0.8 to 1.6 $M_{\odot}$. As the main-sequence turn-off point of the Hyades cluster is

$\sim 2\,M_\odot$, the WDs have originated from stars more massive than $2\,M_\odot$ that have evolved to red giants or AGB stars and that lost more than $1.2\,M_\odot$ of material thereafter. In these systems, as well as in the double nuclei of planetary nebulae and the LMXBs, the masses of the non-degenerate companion stars, mostly M- and K-dwarf main-sequence stars, are so low (mostly between 0.1 and $1.0\,M_\odot$) that these stars can hardly have captured any mass from their initially much more massive companion stars.

Similar arguments of large losses of mass and orbital angular momentum apply also to the DWDs listed in Table 5.8 and the AM CVn systems listed in Table 5.9.

10.7.1 Reasons for Highly Non-conservative Evolution

Such evolution occurs primarily in high mass-ratio binaries, that is, when two non-degenerate components of a binary differ very much in mass, or when one of the components is a compact object (WD, NS, or BH) that is unable to accept all the mass that is being transferred to it by a normal non-degenerate companion star. This is, for example, the case in HMXBs that have a NS accretor or in binaries with a massive star donor and an accreting WD companion. In these cases, the mass transfer is dynamically unstable and formation of a CE is unavoidable for the reasons described in detail in Section 4.8.

On the other hand, in the case of a HMXB with a BH accretor, RLO does not have to lead to the formation of a CE if the mass ratio $q \equiv M_{\rm donor}/M_{\rm accretor} \lesssim 3$ and the envelope of the donor star is in radiative equilibrium (see van den Heuvel et al., 2017, and Section 9.3, and references therein). In this case, the evolution of the system is still highly non-conservative, as the Eddington limit prevents the BH from being able to accept all the mass transferred to it by RLO. However, the system remains dynamically stable, and it may eject most of the mass transferred to it in the form of relativistic jets. This is presumably what is happening in the system of SS 433 (van den Heuvel et al., 2017; Cherepashchuk et al., 2020). We will return to this case later in Section 10.12.

10.7.2 The Outcome of CE Evolution of HMXBs

As we saw in Section 4.8, assuming circular orbits, the change of the orbital separation a with respect to thc initial orbital radius a_0 during the CE spiral-in can be expressed as

$$\frac{a}{a_0} = \left(\frac{M_{\rm core} M_1}{M_{\rm donor}}\right) \left(M_1 + \frac{2\,M_{\rm env}}{\alpha_{\rm CE}\,\lambda\, r_L}\right)^{-1}, \qquad (10.1)$$

where $M_{\rm donor}$, $M_{\rm core}$, and $M_{\rm env}$ are the masses of the donor, its core (consisting of helium or heavier elements), and its hydrogen-rich envelope, respectively; M_1 is the mass of the engulfed companion star (in HMXBs, the accreting compact object); λ is a parameter related to the density profile of the donor's envelope; and $\alpha_{\rm CE}$ is the efficiency parameter of the CE process (see Section 4.8 for details).

The preceding equation enables one to quickly estimate how deep a system with certain values of λ and α_{CE} will spiral-in and whether it will be able to survive this in-spiral as a binary or whether it will coalesce in the process.

Let us assume, as an example, $\lambda = \alpha_{CE} = r_L = 0.5$, $M_{donor} = 20\,M_\odot$, $M_{core} = 6\,M_\odot$, and $M_1 = 1.5\,M_\odot$, as in a typical NS-HMXB. Eq. (10.1) then yields $a/a_0 = 1/500$: the orbit shrinks by a factor 500! If α_{CE} and λ are both a factor of 2 larger (or smaller), and all other parameters are the same, the orbit shrinks by a factor of $\sim$125 (or $\sim$2,000). Based on these assumed values, one sees that the values of λ and α_{CE} play an important role in determining the degree of orbital shrinking of the system.

As mentioned in Section 4.8.4, for low- and intermediate-mass stars, $\lambda \leq 1$ on the RGB, but $\lambda \gg 1$ on the AGB (especially if $M \leq 6\,M_\odot$ [Dewi & Tauris, 2000]). For more massive stars, $10 \lesssim M \lesssim 20\,M_\odot$, the gravitational binding energy of the envelope is larger and typically $0.01 < \lambda < 0.1$ (Dewi & Tauris, 2001). For these stars, as they finally evolve to super-AGB stars, or RSGs, their λ values increase by a factor of $\geq$5. This has been confirmed by detailed numerical 1D and 2D hydrodynamic simulations by, for example, Taam (1996); see the review by Taam & Sandquist (2000). This means that for HMXBs with NSs, only systems with long orbital periods ($\gtrsim$1 yr) can survive the CE in-spiral and eject the envelope (Taam et al., 1978; Terman et al., 1995).

Figure 10.15 shows the value of λ for a variety of massive stars at two very different metallicities. It can be seen that the assumption sometimes used in the literature of a constant value of λ for all stars is erroneous and will yield unreliable results when applied in population synthesis studies.

10.7.3 Final Evolution of Wide NS-HMXBs

The arguments presented by Taam et al. (1978), Terman et al. (1995), Taam (1996), Taam & Sandquist (2000), and Kruckow et al. (2016) show that only NS-HMXBs with orbital periods longer than about one year will survive the CE and in-spiral phase that ensues when the massive donor star begins to overflow its Roche lobe and transfer mass at a thermal-timescale rate of the order of 10^{-4} to $10^{-3}\,M_\odot\,\mathrm{yr}^{-1}$ to the NS. Such systems will then, after the hydrogen-rich envelope of the massive star has been lost, consist of the (naked) helium core of the donor star, plus the NS, in an orbit with a radius of typically a few $R_\odot$ and an orbital period of a few hours.

Systems of this type will resemble the WR X-ray binary Cyg X-3, with an orbital period of 4.8 hr (van den Heuvel & De Loore, 1973) (although this system probably contains a BH instead of a NS and its WR-star must have a mass $\gtrsim 8\,M_\odot$). The only HMXBs known with orbital periods long enough to survive the CE are the Be/X-ray binaries (Fig. 7.18). These have relatively moderate donor masses, mostly in the range of 8–15 $M_\odot$. The helium stars in such a post-CE system will then typically have a mass in the range of 2.6–3.5 $M_\odot$, and in most cases will, after a final Case BB mass-transfer episode, explode as a SN, leaving a second NS (for the more detailed description of this final evolution, see Section 10.13). If the system remains bound, it will be a DNS system with an eccentric orbit; if it is disrupted, two runaway NSs will have formed: one old and recycled NS and one newborn NS, as depicted in Figure 10.13.

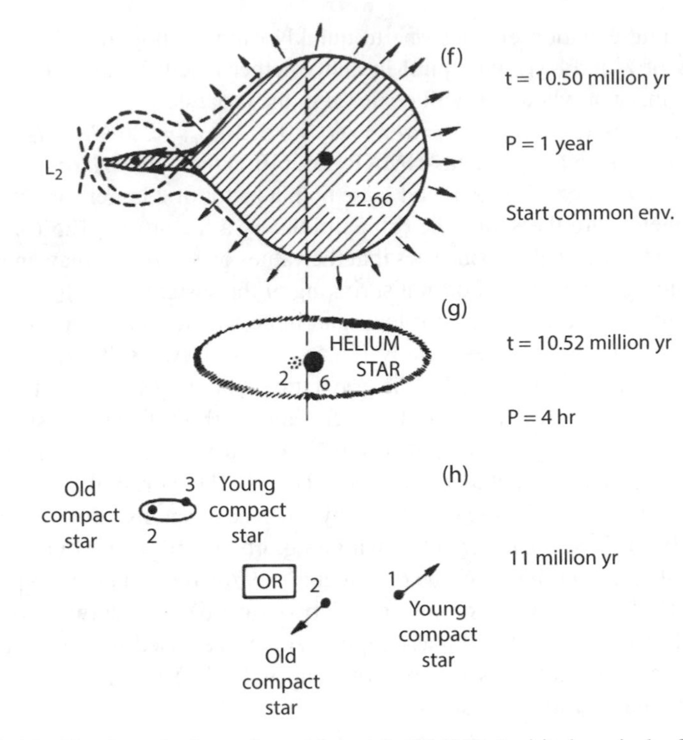

Figure 10.13. Final evolution of a wide-orbit HMXB (orbital period of 1 yr) with a 2 $M_\odot$ NS, resembling an Oe-Be/X-ray binary. The RLO following the HMXB phase leads to CE evolution and a deep in-spiral of the NS, producing a helium star–NS binary with an orbital period of typically <1 d. Explosion of the helium star then either produces a DNS system in a close eccentric orbit, consisting of one young and one old (recycled) NS or two runaway NSs. If the system survives the explosion and the orbital period of the DNS is shorter than ~1 d, it may merge within a Hubble time and produce a very massive NS or, more likely, a BH. (The letters (f), (g), and (h) in the figure are a continuation of the letters of the evolutionary phases in Figure 10.2. The same model, starting from a wide BH-HMXB, with a higher-mass donor star, may produce a close double BH (e.g., Bogomazov, 2014; Belczynski et al., 2016), as described later in this chapter, as well as in Chapter 15; see Fig. 15.10). After van den Heuvel (1976).

Very interestingly, recently Hinkle et al. (2020) discovered that the symbiotic X-ray binary 4U 1954 + 31 is composed of a RSG with a mass of ~9 $M_\odot$ and a very slowly rotating X-ray pulsar (see Table 7.7). This system resembles a later evolution stage of a wide-orbit Be/X-ray binary, and appears to be an excellent candidate for later potentially producing a close DNS after CE evolution, in-spiral, and a second SN explosion.

This CE and spiral-in model for the formation of close DNSs was put forward by van den Heuvel (1976) as an improvement on earlier RLO and spiral-in models by van den Heuvel & De Loore (1973), Flannery & van den Heuvel (1975), and De Loore et al. (1975). Although the model provides a correct qualitative insight in the final evolution of wide-orbit HMXBs, the real quantitative picture of the formation of different kinds of DNSs is much more complex and depends on the precise initial binary parameters and the assumed physical parameters of binary mass transfer and CE evolution. This quantitative picture, as is required if one wishes to make quantitative predictions about the formation, properties, and merger rate of DNSs, is described in detail in Section 10.13 and in Chapters 13 and 14.

10.7.4 Final Evolution of NS-HMXBs with Short Orbital Periods: Formation of TZOs?

Practically none of the sg-HMXBs with NS accretors will survive the in-spiral (Kruckow et al., 2016). In fact, coalescence during the CE in-spiral is the expected fate for the majority of known HMXBs shown in the Corbet diagram (Fig. 7.18). In these systems the NS will spiral into the core of the donor star, possibly leading to the the formation of a Thorne-Żytkow Object (TŻO) (see Thorne & Żytkow, 1977), which is a red supergiant with a NS core, as described in Section 4.8.4. For an overview of the models of such stars, we refer to Podsiadlowski et al. (1995) and references therein. Due to nuclear burning in shells surrounding the NS, the formation of proton-rich isotopes of a variety of elements, such as lithium, rubidium, and molybdenum, is expected (Biehle, 1991; Cannon, 1993). Because the envelope of the object is fully convective, these elements are expected to show up in the spectra of these RSGs, which would allow these RSGs to be identified as TŻOs. Alternatively, it has been suggested that the merger of the NS and the core may trigger an explosion (Schrøder et al., 2020, and references therein). More research is needed to confirm whether this may really occur.

Because star formation in the Galaxy is expected to be in a steady state, the formation rate of TŻOs in the MW is expected to be the same as the formation rate of the sg-HMXBs plus Be/X-ray binaries with orbital periods shorter than $\sim$1 yr. This rate is uncertain but, as we saw in Sections 10.6.3 and 10.6.4, is of the order of $10^{-3}\ \mathrm{yr}^{-1}$. In other words, the rate is probably somewhere between $\sim 2 \times 10^{-4}\ \mathrm{yr}^{-1}$ and $10^{-2}\ \mathrm{yr}^{-1}$, where the lower limit is estimated by Podsiadlowski et al. (1995) and the (hard conservative) upper limit is set by the rate of CC SNe. The lifetimes of TŻOs are also uncertain. Probably, such RSGs have very strong mass loss by stellar winds. Still, if they are not subject to pulsational instabilities, they might easily live for 10^5 yr. In such a case, their number in the Galaxy would be $\sim$100.

TŻOs will appear as very cool RSGs. Because these are difficult to distinguish from normal RSGs, it is presently not clear whether they actually exist and so far, no TŻOs have been identified with certainty. Depending on the uncertain lifetime of this phase (which is limited, e.g., by wind mass loss), a few to 10% of all RSGs with a luminosity comparable to or above the Eddington limit for a NS ($L_{\mathrm{Edd}} \sim 10^5\ L_{\odot}$) may harbor NS cores. Despite numerous searches, to date only one candidate TŻO has been identified (HV 2112 in the SMC; Levesque et al., 2014) and its interpretation as a TŻO has remained controversial (Maccarone & de Mink, 2016). This suggests that the lifetime of the TŻO phase (if it exists) is much shorter than previously was estimated. The properties of exploding massive TŻOs, caused by a collapse anticipated to follow exhaustion of nuclear burning near the NS surface, were investigated by Moriya & Blinnikov (2021).

10.7.5 Summary: Comparison of the Predictions of the HMXB Evolution Models with Observations

The above-described models for the formation and later evolution of HMXBs made five predictions:

1. HMXBs will be runaway stars (van den Heuvel & Heise, 1972; Tutukov & Yungelson, 1973a).
2. During their later evolution, HMXBs will spiral-in significantly, either by orbital angular momentum loss by "isotropic re-emission" (van den Heuvel & De Loore, 1973) or by CE evolution (van den Heuvel, 1976), producing very close binaries consisting of a helium star and a NS.
3. There will be OB-binaries with a young energetic pulsar companion (van den Heuvel, 1974).
4. Because NSs may receive a large kick at birth, there may be systems that are disrupted in the first SN explosion (Flannery & van den Heuvel, 1975).
5. A sizeable fraction of all HMXBs will spiral-in completely, to form a TŻO (Thorne & Żytkow, 1977).

As we have seen in this chapter and in Chapter 7, the first four of these predictions have been confirmed by the observations. We saw in this chapter that HMXBs are runaway stars and that disrupted systems exist. We saw in particular that the sg-HMXBs have substantial runaway velocities, while the velocities of the Be/X-ray binaries are much smaller, just as predicted. Furthermore, the discovery of the close DNS systems with eccentric orbits has confirmed the spiral-in predictions (van den Heuvel & De Loore, 1973; van den Heuvel, 1974, 1976). Also, the discovery of the binary pulsars PSR B2159−63 (Johnston et al., 1992) and PSR J0045−7319 (Kaspi et al., 1994), as well as the gamma-ray binaries consisting of an O- or early B-type star and a gamma-ray emitting compact star (see Dubus [2013] and Chapter 7), have confirmed the existence OB-stars with a young pulsar companion.

The only prediction that, so far, has escaped confirmation is the existence of Thorne-Żytkow stars. What the fate is of the many HMXBs that later in life spiral-in completely remains a mystery.

10.8 FORMATION MODELS OF HMXBs DIFFERENT FROM CONSERVATIVE CASE B EVOLUTION

Although the conservative Case B evolution model for the formation of HMXBs described in Sections 10.2 and 10.3 globally explains the formation of the supergiant and Be-star HMXBs quite well, it is idealized and does not fit the formation of all known HMXBs—particularly not of systems that may have formed through another formation channel, such as Case A or Case C mass transfer, and not of systems for which there is evidence that in their history much mass and angular momentum has been lost. Figure 10.6 considers a few such models for the formation of Be/X-ray binaries with considerable mass and orbital angular momentum losses. Further examples of evolution with such losses are the WR+O binaries with very short orbital periods such as CQ and CX Cephei (orbital periods of only 1.64 and 2.127 d, respectively) and the HMXB Cen X-3 with its orbital period of only 2.08 d. We will now consider some alternative or non-conservative models for forming HMXBs and WR binaries in some more detail.

10.8.1 The Case of GX 301-2/Wray 977

This wide-orbit and eccentric HMXB system consists of the B1.5Iae blue supergiant star Wray 977, with a mass of $40 - 50\ M_{\odot}$, and the X-ray pulsar GX 301-2, with a spin period of 696 s. It has an orbital period of 41.508 d and an orbital eccentricity $e = 0.47$. The spectral type of its companion means that it belongs to brightest blue supergiants known and has emission lines in its spectrum, indicating strong stellar wind-mass loss. The mass function of the pulsar is $f = 31.9\ M_{\odot}$, and because it is not eclipsing, one finds that the mass of the supergiant is certainly $>40\ M_{\odot}$ and probably close to $50\ M_{\odot}$.

Ergma & van den Heuvel (1998) assumed that the progenitor system of this wide binary had evolved according to Case B RLO and found that in that case the progenitor of the pulsar (NS) must have started out with a high mass of the order of $50\ M_{\odot}$. They thought that this result then would enable them to set a lower limit of $50\ M_{\odot}$ for stars that form BHs in binaries. However, shortly later, Wellstein & Langer (1999) showed that the system might have evolved according to Case A RLO, in which case it could have started out as a close O-type binary with components of $26 + 25 M_{\odot}$ and an orbital period of 3.5 d. Figure 10.14 shows the evolutionary tracks of both components of this Case A system in the HR diagram. The dashed curve is the evolutionary track of the $26\ M_{\odot}$ primary star and the fully drawn curve is that of the $25\ M_{\odot}$ secondary. After the Case A mass transfer, the primary has a mass of $\sim 15\ M_{\odot}$ and the mass of its hydrogen-burning core has shrunk to $6.41\ M_{\odot}$. The primary continues to transfer matter, and when hydrogen is exhausted in the core, it initiates Case B RLO, at which time its helium core has a mass of $3.17\ M_{\odot}$. When this star explodes, it leaves a NS. Due to the mass transfer, the mass of the secondary grows to $40.5\ M_{\odot}$ at the time of the explosion of the primary. The orbital period has increased to 46.22 d just before the SN explosion in which the NS was formed and to 47.3 d after the (assumed symmetric) explosion.

This case of the system GX 301-2/Wray 977 shows that it is not possible to uniquely reconstruct the past evolution of a HMXB from its present system parameters. Wellstein & Langer (1999) provided a series of evolutionary models of Case B RLO as well as Case A+AB RLO (like the one given in the figure) that all can produce systems resembling this particular HMXB. The importance of this paper is that it shows that, contrary to what Ergma & van den Heuvel (1998) had thought, one cannot use this system to uniquely constrain the upper limit up to which stars in binaries still produce NSs (i.e., equivalent to the lower limit for producing a BH).

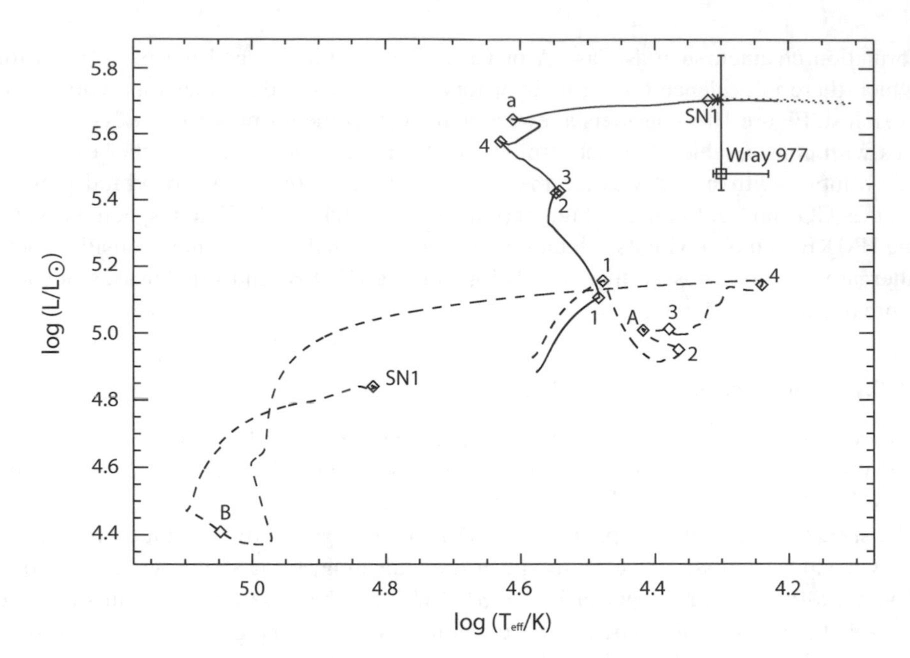

Figure 10.14. Evolution of a close binary with initial component masses of $26+25\,M_\odot$ and initial orbital period of 3.5 d. The system evolves according to Case A of binary evolution and terminates as a wide-orbit HMXB consisting of a NS and a $40\,M_\odot$ companion in a 47.3 d orbit, closely resembling the wide-orbit massive HMXB GX 301-2/ Wray 977. The dashed curve is the evolutionary track of the primary star, in which the numbers and letters have the following meanings: 1 indicates the beginning of Case A RLO; 2, the end of Case A RLO; 3, the beginning of Case AB; 4, end of Case AB; A/a, core-hydrogen exhaustion of the primary/secondary; B, end of the primary's core-helium burning. The position of Wray 977 according to Kaper and Najarro (private communication, 1999) is indicated by the square with error bars. After Wellstein & Langer (1999).

10.8.2 Non-conservative Formation of Short-period WR+O Binaries and HMXBs

We saw in Figure 10.6 that Podsiadlowski et al. (2004) found that for the formation of several types of Be/X-ray binaries highly non-conservative evolutionary models are needed. Similar arguments hold for the formation of close WR+O-star binaries and very close sg-HMXBs, such as Cen X-3 ($P_{\rm orb}=2.087$ d). We consider here as examples the WR binaries CQ Cephei and CX Cephei; the orbital periods and component masses of which are listed in Table 10.1. Their orbital periods of 1.64 and 2.127 d, respectively, are so short and the masses of their WR-stars (helium stars) are so large, $24\,M_\odot$ and $26.4\,M_\odot$, respectively, that these systems could never have evolved through simple conservative Case B or Case A mass transfer.

To reach their very short orbital periods, large amounts of orbital angular momentum and mass must have been lost in these systems. The WR-stars in these systems are the helium cores of progenitors with a mass of about 60 $M_{\odot}$. The companion O-stars in these systems have masses of 30 $M_{\odot}$ and 36 $M_{\odot}$, respectively. Exercise 10.1 shows that formation by conservative evolution is out of the question for CQ Cephei, and the same will hold for CX Cephei.

Because conservative close binary evolution is excluded for CQ and CX Cephei, one will have to turn to highly non-conservative evolution (e.g., CE evolution). This formation path is closely related to the general formation of short-period massive X-ray binaries such as Cen X-3, which consists of a blue supergiant with a mass of $\sim$18 $M_{\odot}$, and a NS, which may have a typical mass of $\sim$1.4 $M_{\odot}$ and an orbital period of 2.087 d. Also here the present orbital period excludes a conservative mass-transfer history during the formation of the system. Just like the two short-period WR+O-star binaries, a consistent non-conservative model is needed for the origin of this system (see also Exercises 10.1 and 10.2).

10.9 THE LOWER MASS LIMIT OF BINARY STARS FOR TERMINATING AS A BH

10.9.1 Single Stars vs. Binaries

We saw from the case of GX 301-2/Wray 977 that there is no simple way to derive the lower ZAMS mass limit for producing BHs from studying observed HMXBs. In Section 8.2.3, we saw that according to the work of Sukhbold et al. (2018) most single stars of solar metallicity leave BHs if their initial mass is larger than $\gtrsim$22 $M_{\odot}$, while even around 15 $M_{\odot}$ some single stars leave BHs; in the mass range of 22–25 $M_{\odot}$ these authors find all stars leave BHs. However, although Ebinger et al. (2019) find roughly similar results for single stars less massive than approximately 25 $M_{\odot}$, they find, for their calibration no. I that for masses $\gtrsim$27 $M_{\odot}$ most stars leave NSs, while Sukhbold et al. (2016), and Sukhbold et al. (2018) find most of these stars to terminate as BHs. (For their calibration II, Ebinger et al. [2019] find that for a ZAMS mass $\gtrsim$27 $M_{\odot}$, there are still several mass intervals where BHs are produced—see Fig. 8.22—while still the bulk of the stars leave NSs). This difference in outcomes for $M \gtrsim 27$ $M_{\odot}$ is presumably due to the larger stellar wind mass-loss rates adopted in the simulations of Ebinger et al. (2019), as described in Section 8.2.3.

For stars in binaries, the lower mass limit for leaving a BH is much higher than in single stars, as has been shown in a number of seminal papers by Langer, Wellstein, and collaborators (Wellstein & Langer, 1999, 2001; Brown et al., 2001; Fryer et al., 2002) and was recently further confirmed by the work of Woosley (2019), as discussed in Section 9.7. As we saw in Section 9.7, and also already from the case of GX 301-2/Wray 977, in binaries the situation is very different from that of single stars. Let us, for example, consider a primary star of a close binary with a mass of 30 $M_{\odot}$, which evolves according to Case B RLO of binary evolution. When this star overflows its Roche lobe, it transfers its hydrogen-rich envelope to its companion and leaves a helium core with a mass of $\sim 9.0\,M_\odot$, which is a WR-star. Such stars, now devoid of their hydrogen-rich envelope, have very strong stellar winds and in the course of time their masses decrease much by the wind mass loss, and they will terminate with a mass of $\sim 7\,M_\odot$ (Woosley, 2019) and will leave a NS instead of a BH (see, e.g., Fryer et al., 2002; Woosley, 2019). If this helium core would have kept its hydrogen-rich envelope, its mass would not have gone down but would have increased further by hydrogen-shell burning to $\sim 11\,M_\odot$, and the core of the $30\,M_\odot$ star would have collapsed to a BH. Thus, for binary components the lower mass limit for terminating as a BH is considerably higher than for single stars, as we saw in Section 9.7. This result was found already by Fryer et al. (2002), who investigated the fate of $60\,M_\odot$ primary stars in close binaries and found that, dependent on the adopted wind mass-loss rates of WR-stars, if they evolve according to Case A or Case B RLO, they may still leave NSs. For relatively small assumed WR wind mass-loss rates, they found that stars of an initial mass of $60\,M_\odot$ may leave BHs, however, with masses $\lesssim 10\,M_\odot$. They pointed out that they were therefore unable to explain the relatively large observed mass (at that time thought to be only $14.8\,M_\odot$) of the BH in the 5.6 d orbit HMXB Cyg X-1, by evolving an initial primary star of $60\,M_\odot$ with Case A or Case B evolution. They therefore suggested that this system is the result of Case C evolution. Although Case C evolution is nowadays thought to be very important for the formation of the BH X-ray binaries with a low-mass companion (the BH-LMXBs), as we will describe next, for Cyg X-1 there are also alternative formation scenarios possible, such as non-conservative Case A or Case B (e.g., Qin et al., 2019; Neijssel et al., 2021).

10.9.2 Case C Origin of BH-LMXBs vs. the Formation of Cyg X-1

As mentioned in Section 9.7, in wide-orbit binary systems that evolve according to Case C RLO, the primary star still evolves like a single star through the entire core-helium burning phase and overflows its Roche lobe only when it is on its way to core-carbon ignition. It thus evolves like a single star until carbon ignition and, at the onset of the mass transfer, has a massive core like that of a single star, which will be left behind after losing its hydrogen rich envelope by RLO. Because the lifetime of this massive core remnant after carbon ignition is very limited, it will not lose much mass by a wind before it reaches core collapse. Therefore, in Case C evolution the final outcome of the core collapse of the primary star is basically the same as that for a single star with the same initial mass. Single stars with solar metallicity in the mass range of 22–25 $M_\odot$ leave BHs that, according to Ebinger et al. (2019), have maximum masses of $\sim 14\,M_\odot$. Raithel et al. (2018), in work based on Sukhbold et al. (2016), give a peak of BH masses of $\sim 16\,M_\odot$ in the same ZAMS mass range for solar metallicity. (For any other ZAMS mass in the range of 15–100 $M_\odot$, they find for solar metallicity the final BH masses to be lower.) The Case C model for the origin of BH X-ray binaries was proposed by Brown et al. (1999), Brown et al. (2001a), Brown et al. (2001b); Kalogera (2001), and Fryer & Kalogera (2001).

The reason for proposing Case C models for the origin of BH X-ray binaries with low-mass donor stars came already quite some time ago from two observational constraints from observed BH systems regarding the BH mass limit in binaries. Portegies Zwart et al. (1997) found from population synthesis studies that the number of low-mass BH X-ray binaries in the Galaxy requires that the lower limit to the initial progenitor mass of the BHs in these binaries is $\leq 25\, M_{\odot}$. And Ergma & van den Heuvel (1998) argued that the observed orbital periods of less than $\sim$10 hr found in most low-mass BH systems are incompatible with a lower limit to the mass of the initial BH progenitor star of $>25\, M_{\odot}$. These observational constraints fit well with Case C evolution of the progenitor binaries, because we now know that many single stars with masses $\leq 25\, M_{\odot}$ leave BHs, and the same will be true for primary stars in Case C binaries. For a review of formation of BH-LMXBs, we refer to Li (2015). For a recent important population synthesis investigation aimed at explaining the formation of BH X-ray binaries see Shao & Li (2020). These authors also find that Case C evolution likely plays a key role in the formation of the BH-LMXBs and is also important for other BH X-ray binaries. Our conclusion is that for the formation of the BH-LMXBs, Case C evolution is the most likely formation mechanism.

For the BH-HMXBs, however, the situation is much less clear. As we have already discussed, it was recently found by Miller-Jones et al. (2021) that the mass of the BH in Cyg X-1 is $21\, M_{\odot}$ and its companion has a mass of $\sim 41\, M_{\odot}$, and these authors correctly concluded that the $21\, M_{\odot}$ mass of Cyg X-1 implies that the wind mass-loss rates used in the calculations of the evolution of stars of solar metallicity were too high.

Indeed, the fact that the computations of Ebinger et al. (2019), Sukhbold et al. (2016), and Raithel et al. (2018) never reach final BH masses $\geq 14\, M_{\odot}$ and $\geq 16\, M_{\odot}$, respectively, for ZAMS masses in the range of $15-100\, M_{\odot}$ (at solar metallicity) may be due to having used a too large stellar wind mass-loss rate for hydrogen-rich and helium-rich (WR) stars. For stars with ZAMS masses up to $\sim 25\, M_{\odot}$, where the wind mass-loss rates are only modest, this will not make much of a difference, but for the higher ZAMS masses (25–100 $M_{\odot}$), a considerable reduction in the wind mass-loss rate during the later evolution will make a large difference in the final BH mass. The model calculations of core compactness by Ebinger et al. (2019), Sukhbold et al. (2016), and Raithel et al. (2018) all used as starting models the models by Woosley et al. (2002) and Woosley & Heger (2007), which used for helium stars (WR-stars) the wind mass-loss rate in Wellstein & Langer (1999). As mentioned in Chapter 8, these estimates lead to unrealistically large mass loss from massive helium stars. For example, a star with a ZAMS mass of $60\, M_{\odot}$ with solar metallicity would leave a $26.8\, M_{\odot}$ helium star, which with the exaggerated mass-loss rate would finish as a pre-SN star of just $3\, M_{\odot}$

However, as shown by Nugis & Lamers (2000), if one includes the effects of clumping in the wind, the empirical mass-loss rates of massive WR-stars turn out to be almost an order of magnitude lower than those given by the estimates of Wellstein & Langer (1999). Woosley (2019) has included the reduced empirical WR mass-loss rates compiled by Yoon (2017) in his calculations of the final fate of massive helium stars. According to his calculations, the $26.8\, M_{\odot}$ helium star of Wellstein & Langer (1999) terminates as a $\sim 13\, M_{\odot}$ helium star with a $\sim 10\, M_{\odot}$ CO core, which collapses to form a BH.

Formation of Cyg X-1

Using the work of Woosley (2019), we now examine how the 21 $M_\odot$ BH in the Cyg X-1 system may have formed. We notice here that a WR binary system that comes close to being a progenitor of the Cyg X-1 system with its revised component masses is the binary WR 47, which consists of a 51 $M_\odot$ WN6 star and a 60 $M_\odot$ O5V star, in an orbit with $P_{\rm orb} = 6.24$ d (see Table 5.4). If we assume that the 51 $M_\odot$ WR-star in this system is just at the beginning of its helium-star evolution, then according to the calculations by Woosley (2019), it will at the end of its evolution have a mass of $\sim$25 $M_\odot$ with a CO core of $\sim$20.5 $M_\odot$. Thus, the WR-star in the WR 47 system may leave a BH of approximately 20–25 $M_\odot$ (some mass will be lost in released gravitational binding energy). By that time, due to wind mass loss, the mass of its O-type companion will have been reduced to $\sim$50 $M_\odot$. So, the resulting system may be slightly more massive than the Cyg X-1 system, but it will further be quite similar. So, it should be no surprise that a system like Cyg X-1 exists with a BH of mass 21 $M_\odot$. Whether a system like WR 47 has evolved according to Case A or Case B (or even, possibly, Case C) RLO is not so clear. In any case, systems of this type did not form by purely conservative evolution and may have lost quite some mass and orbital angular momentum already, as is shown in Exercises 10.1 and 10.2. An evolution resembling Case A or Case B RLO, but with considerable loss of mass and orbital angular momentum during the mass exchange, seems to us most likely for the systems of WR 47 and Cyg X-1. See also Neijssel et al. (2021).

10.10 FINAL EVOLUTION OF BH-HMXBs: TWO FORMATION CHANNELS FOR DOUBLE BHs

It appears that the final evolution of BH-HMXBs is more complex than that of the NS-HMXBs. This is due to the larger masses of the BHs ($\gtrsim$5 $M_\odot$) relative to NSs. As a result, the mass ratio between the donor star and the accretor in the BH-HMXBs is in general less extreme than in the NS-HMXBs. This opens the possibility that systems, when they enter RLO, do not necessarily have to enter CE evolution, like in the case of NS-HMXBs (see Fig. 10.13), but that certain systems can also enter stable RLO from the donor to the BH. There are, therefore, two different ways in which the further evolution of BH-HMXBs may proceed. Both of them finally may lead to the formation of close double BH or BH+NS binaries. These two ways are described in Sections 10.11 and 10.12.

The first way is from a *wide-orbit* BH-HMXB, via CE evolution and spiral-in, just as for the wide-orbit NS–HMXBs, for example, as depicted in Figure 10.13. This model is discussed in Section 10.11 where we also discuss the limitations of the CE evolution model for BH-HMXBs.

The second path for the final evolution of BH-HMXBs is from a relatively *close-orbit* BH-HMXB, via stable RLO. This model will be discussed further in Section 10.12.

Other formation channels of double BHs, and further details of the models presented here to produce the various double BH mergers detected by the LIGO-Virgo-KAGRA network of GW detectors, are described in Chapters 12 and 15.

10.11 FINAL EVOLUTION OF WIDE-ORBIT BH-HMXBs VIA CE EVOLUTION

This model is principally similar to the van den Heuvel (1976) model for producing close DNSs, depicted in Figure 10.13, and scaled up to higher initial component masses. This is the model put forward by, for example, Bogomazov (2014) and Belczynski et al. (2016) and investigated in many other studies over the past couple of decades. In this model, one must start from a very wide binary that, due to CE evolution, shrinks by a large factor, producing a system resembling the WR/X-ray binaries with orbital periods around one day. After the collapse of the WR-star, one is left with a double BH (or a BH+NS system) in a narrow orbit. Figure 15.10 graphically depicts this CE model for double BH formation.

Belczynski et al. (2016) found that to produce massive double BHs with components more massive than $30\,M_{\odot}$, such as the first double BHs observed by LIGO, the progenitor binaries must have had a metallicity $\lesssim 0.1\,Z_{\odot}$. For higher metallicities, the stellar-wind mass-loss rates are so high that that such massive remnants can never be left.

It is important to realize that the BH+main-sequence binary that finally produces here the close double BH is very wide with an initial separation of the order of 10 – 20 AU (orbital period of 3 – 7 yr), and prior to the CE formation when the secondary star evolves, the separation is even larger, as the orbit widens due to stellar wind-mass loss. The reason for the required initial wide orbit was given by Voss & Tauris (2003) and applies to formation all BH+BH, BH+NS, and NS+NS systems: for the systems to avoid merging in the CE and spiral-in phase when the secondary star evolves, the systems must be in a fairly wide orbit prior to the HMXB and subsequent CE phase. (This also means that these systems can only survive very small kicks in the first SN explosion to avoid disruption.) The wider the pre-CE orbit, the smaller the binding energy of the CE to be ejected (see Figs. 10.15 and 10.16), which facilitates the CE ejection and binary survival.

However, it is not yet certain that the outcome of the CE evolution of the first-formed BH in the envelope of its RSG companion will indeed result in deep in-spiral and formation of a short-period binary, as put forward in these models, for reasons that we will now briefly discuss.

10.11.1 Limitations of the CE Model for Forming Double BHs

A simple way to illustrate a common difficulty of the CE scenario to successfully eject the envelope and produce tight systems with massive stars is to consider the release of orbital energy during the in-spiral and compare to the binding energy of the donor star's envelope (Section 4.8). Detailed investigations by Kruckow et al. (2016), Klencki et al. (2021), and Marchant et al. (2021) clarify which massive binaries may be expected to survive a CE evolution without undergoing coalescence. However, although solutions exist, the range of solutions for a given binary is often quite narrow. Furthermore, the structure of massive giant stars is a hot topic that is currently debated aside from the physics of the hydrodynamic ejection process itself.

The reasons why the deep in-spiral of a BH in the envelope of a RSG companion remains an unsolved issue, and may possibly not be successful, are (at least) threefold:

i) Realistic 3D hydrodynamic and radiation-hydrodynamic CE simulations of such a system by Ricker et al. (2019) so far do not show the BH to spiral-in very deeply. The system still remains wide after simulating 50–80 yr of CE evolution, and only a relatively small amount of envelope mass (of the order of 10%) has been lost. The simulation will have to be continued to see whether a deeper spiral-in can be achieved (Ricker et al., 2019).

ii) It is not certain whether the postulated very massive RSGs of low metallicity, required for this scenario, will really form in Nature. This is because, as described in Chapter 8, RSGs with masses that are $\gtrsim 40\,M_\odot$ have never been observed in our Galaxy nor in galaxies with metallicities as low as that of the SMC ($0.2\,Z_\odot$). Strong stellar winds, rapid rotation, or dust formation, leading to rapid hydrogen envelope ejection, may prevent such supergiants from forming in Nature, even though theoretical models of such stars exist, as calculated by, for example, Klencki et al. (2021).

iii) If low-metallicity RSGs with masses that are $>40\,M_\odot$ do exist, as in the theoretical models of Klencki et al. (2021), there is the problem that the gravitational binding energy of the envelope of such a star, in most cases, is larger than the released drop in orbital energy of the binary during in-spiral, such that the latter amount is insufficient to drive off the CE, and the system will merge. Kruckow et al. (2016), and later also Klencki et al. (2021) and Marchant et al. (2021), show that there are only some relatively narrow windows of initial system parameters that may lead to a successful in-spiral and CE ejection.

We illustrate the last point with the following simple example. Figures 10.15 and 10.16 display the total envelope binding energies, $E_{\rm bind}$ (see Eq. 4.126), of massive stars as a function of their radii and metallicity. These binding energies can be compared to the orbital energies in the post-spiral-in tight binaries consisting of the core of a RSG plus a BH, which is the progenitor of a close double BH system. This comparison must show whether this spiral-in could have been successful. Consider a hypothetical post-spiral-in system consisting of a $30\,M_\odot$ stellar core (of an originally $60\,M_\odot$ RSG) plus a BH of $30\,M_\odot$, with an orbital period $P = 1.56$ d. The orbital energy of this system is $E_{\rm orb} = -GM_1M_2/2a = -7.7 \times 10^{49}$ erg, given that $M_1 = M_2 = 30\,M_\odot$, and $a = 22.1\,R_\odot$. Therefore, we can conclude that at most $\Delta E_{\rm orb} \simeq 7.7 \times 10^{49}$ erg has been released via in-spiral (the limiting value is in-spiral from infinity) and that this value therefore sets the upper limit for the value of $E_{\rm bind}$ to have a successful CE ejection (assuming 100% efficiency in converting released orbital energy via in-spiral into kinetic energy that may eject the CE, and neglecting the very small initial orbital energy of the wide binary and any widening of the post-CE system from winds). Figure 10.16 shows that it is difficult to find a progenitor solution for a $60\,M_\odot$ star that fulfills the criterion that $|E_{\rm bind}| < |\Delta E_{\rm orb}|$. According to Klencki et al. (2021), there may be a small window in evolution for late Case B or Case C RLO for a narrow metallicity abundance near $Z = 0.1\,Z_\odot$, where the binding energy of the envelope is just sufficiently

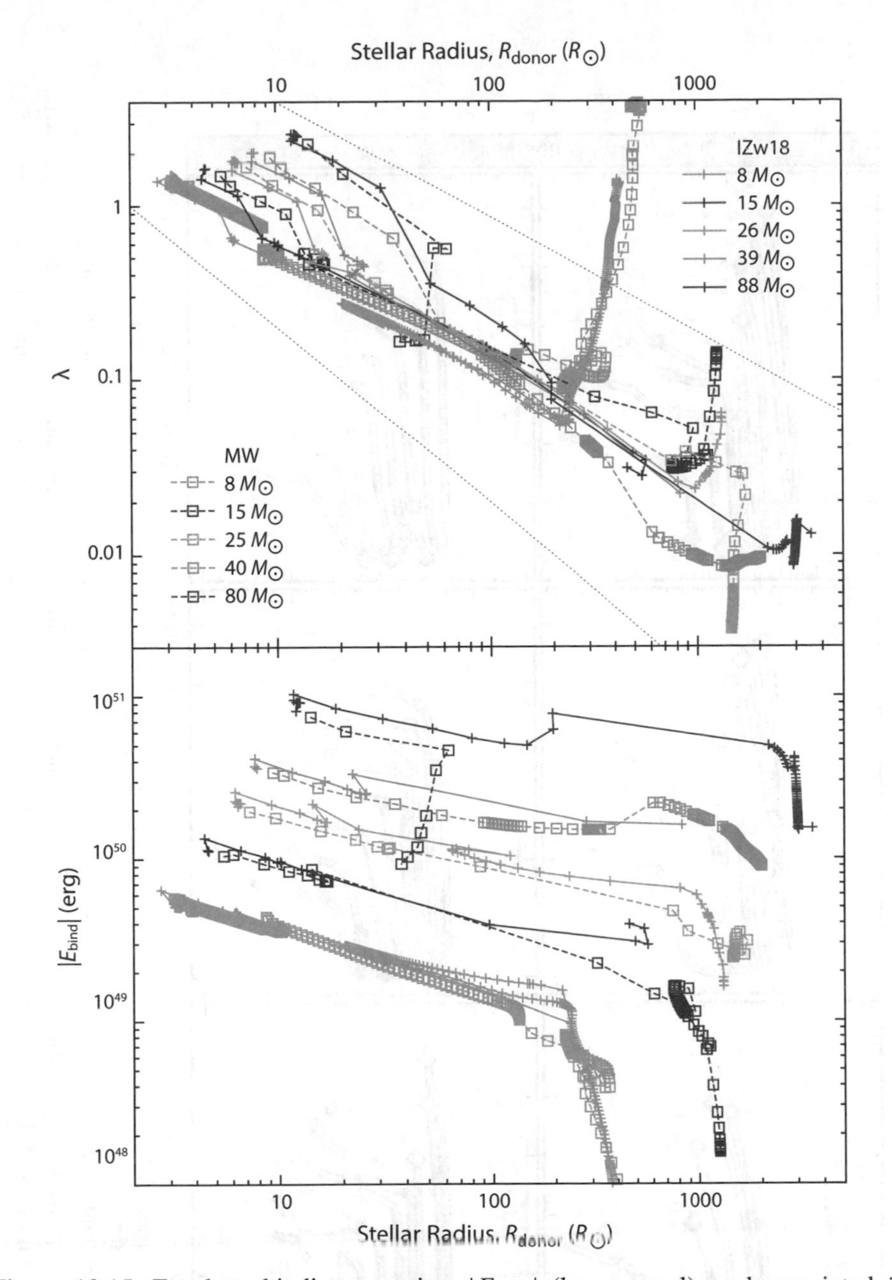

Figure 10.15. Envelope binding energies, $|E_{\rm bind}|$ (lower panel), and associated λ-values, (upper panel) as a function of total stellar radius for two sets of models with metallicities $Z = Z_\odot/50$ (full lines, crosses) and $Z = Z_{\rm MW}$ (MW, dashed lines, squares). Independent of mass and Z, and before reaching the giant stages ($R \lesssim 1{,}000\,R_\odot$), the λ-values almost follow a power law with an exponent between -2/3 and -1 (upper and lower gray lines, respectively). The exceptions are stars with initial masses $\gtrsim 60\,M_\odot$ at $Z = Z_{\rm MW}$ (dashed blue line), which either become LBV stars or have their envelopes stripped by enhanced wind mass loss. See original paper for further details. After Kruckow et al. (2016).

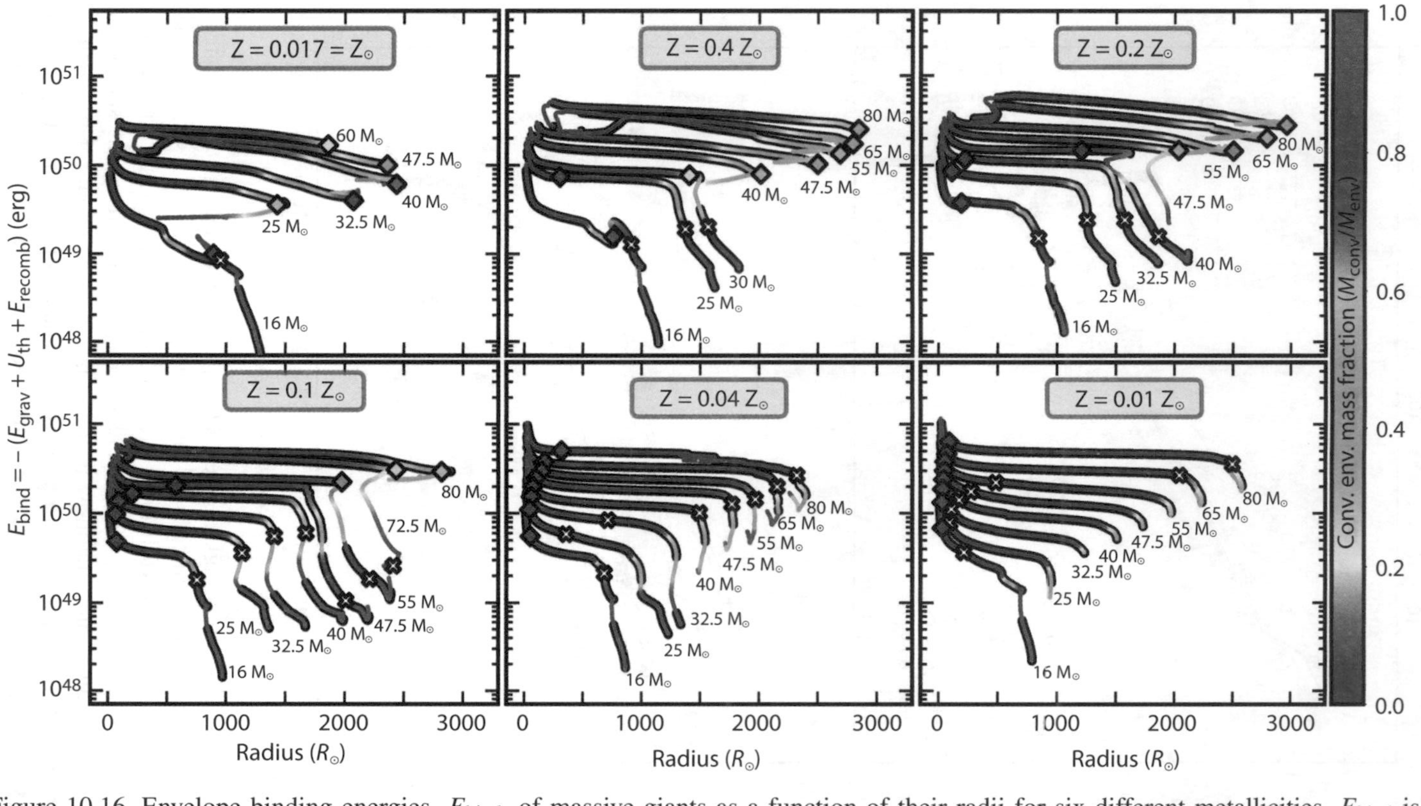

Figure 10.16. Envelope binding energies, $E_{\rm bind}$, of massive giants as a function of their radii for six different metallicities. $E_{\rm bind}$ is calculated as the sum of gravitational potential energy and internal energy (including the recombination terms). Colors indicate what fraction of the envelope mass is in the outer convective zone. Diamonds (same coloring) and white crosses mark the onset and end of core-helium burning, respectively. See original paper for further details. After Klencki et al. (2021).

small to be ejected ($R \simeq 2100-2400\, R_\odot$), and double BHs may be formed in this case. See also Figure 4.29 for similar energy considerations on the possibilities of ejecting CEs in X-ray binary systems in general. Finally, we refer to Exercise 10.2 for a similar difficult example of finding a solution to the CQ Cephei system (possibly a post-CE WR+O-star binary) discussed in Section 10.8.2.

The preceding example illustrates that much fine-tuning may be needed to find a solution, if one exists. Population synthesis (Chapter 16) can help to judge the statistical probability for any given formation channel, but it often lacks the detailed input physics to probe outliers in the observed population. Besides that, our knowledge on input physics for the models may still be rather incomplete.

10.12 FINAL EVOLUTION OF RELATIVELY CLOSE-ORBIT BH-HMXBs VIA STABLE RLO

10.12.1 Microquasars and SS433-like Systems: Stable RLO

Research in the 2010s has shown that the final evolution of BH-HMXB systems differs from that of the NS-HMXB systems in that it allows an important fraction of the systems to avoid CE evolution (e.g., Pavlovskii et al., 2017; van den Heuvel et al., 2017). These systems evolve with stable RLO, in which most of the overflowing matter is ejected from the accretion disk that surrounds the BH. This type of evolution holds for BH-HMXBs with orbital periods up to at least several months.

The first clear observational indication that such stable RLO evolution can take place is provided by the peculiar X-ray binary system of SS433. In view of the crucial importance of this system for the evolution of BH-HMXBs, we now describe its derived evolutionary state in some detail (see Section 6.5 for further discussions on microquasars and the nature of SS433).

As mentioned in Section 6.5.2, the SS433 system consists of a Roche-lobe–filling A4-7I supergiant donor star with an estimated mass of $12.3 \pm 3.3\, M_\odot$ and a luminosity of $\sim 3{,}800\, L_\odot$, plus a compact star with a mass of $4.3 \pm 0.8\, M_\odot$, in a 13.1 d orbital period binary (Hillwig & Gies, 2008). (Some authors favor higher masses for the components of SS433, though with the same mass ratio as given by Hillwig & Gies [2008]. See particularly Cherepashchuk et al. [2020] and references therein.) We will assume here the masses derived by Hillwig & Gies (2008). The $4.3\, M_\odot$ compact object in SS433 must be a BH, because causality does not allow NSs to have masses larger than $\sim 3\, M_\odot$ (Nauenberg & Chapline, 1973; Kalogera & Baym, 1996); also when the masses of the components of SS433 are larger than assumed here, the compact object will still be a BH (Cherepashchuk et al., 2020).

The BH is surrounded by an extended and luminous accretion disk, which is approximately an order of magnitude brighter than its A-supergiant companion. This disk ejects the famous precessing relativistic jets with a velocity of $0.265\, c$, in which neutral hydrogen is ejected at a rate of $\sim 10^{-6}\, M_\odot\, \mathrm{yr}^{-1}$. At the same time, in a strong disk wind with a velocity of $\sim 1{,}500\, \mathrm{km\, s}^{-1}$, mass is ejected at a rate of the order of $\sim 10^{-4}\, M_\odot\, \mathrm{yr}^{-1}$, as is seen in the form of the stationary Hα and broad absorption lines (Fabrika, 2004). The total mass loss from the disk is basically all the matter that the A-supergiant donor is transferring to the BH by RLO on its thermal timescale of $\sim 10^5$ yr (see also Begelman et al., 2006). The observed radiative accretion luminosity of the BH with its disk does not exceed the Eddington luminosity $L_{\rm Edd} = 6 \times 10^{38}$ erg s^{-1} of the BH (which corresponds to a real accretion rate onto the BH of the order of only a few times $10^{-8}\, M_\odot$ yr^{-1} [Section 4.9]; although when seen along the jets, the UV luminosity might be as large as 10^{40} erg s^{-1} [Fabrika, 2004], which would correspond to an accretion rate of the order of $10^{-7}\, M_\odot$ yr^{-1}).

This mass loss/transfer has been going on for thousands of years, as can be seen from the large radio nebula W50 that surrounds the system and has been produced by the precessing jets and the strong disk wind. Despite how long this transfer has been happening, the system is still dynamically stable and has not entered a CE stage. King & Begelman (1999) and King et al. (2000) argue that the reason why SS433 has not gone into a CE phase is because the A-supergiant star has a radiative envelope. If one takes away mass from a star with a radiative envelope, this envelope responds by shrinking on a dynamical timescale, followed by a re-expansion on the thermal timescale of the envelope. King et al. (2000) argue that as a result, this star can keep its radius close to that of its Roche lobe and will transfer matter to its BH companion on the thermal timescale of its envelope, without going into a CE phase. However, we know from Sections 4.5.4 and 9.3, that this mass transfer can still become unstable. It turns out that in the case of massive donors with a radiative envelope there is an extra condition for keeping the RLO stable, which was not mentioned in the preceding references: the thermal timescale mass transfer from a radiative donor envelope may become unstable, if the mass ratio of $q \equiv M_{\rm donor}/M_{\rm accretor}$ is larger than a value in the range of 3–4 (e.g., Tout et al., 1997; Hurley et al., 2000; Tauris et al., 2000; Pavlovskii & Ivanova, 2015; Pavlovskii et al., 2017).

For the sake of argument, we will assume here this limiting mass ratio to be $q_{\rm crit} \simeq 3.5$. For mass ratios $q > q_{\rm crit}$, the orbital decay of the system due to the mass transfer goes so fast that the shrinking of the donor star cannot keep pace with it, and the system will enter a CE phase (Section 4.8). SS433, however, has $q < q_{\rm crit}$ and therefore avoided going into a CE phase. Apart from systems having $q < 3.5$ and a donor star with a radiative envelope (including those with convective envelopes that are not too deep), the formation of a CE is unavoidable (except for very special circumstances, see Quast et al., 2019).

The A-supergiant donor star of SS433 with its mass of $12.3\, M_\odot$ evolved from a main-sequence early B-type star with a mass of approximately $14 - 15\, M_\odot$, and the orbital period of 13.1 d is typically that of a Be/X-ray binary. The only difference with other Be/X-ray binaries is that here the compact star is not a NS but a $4.3\, M_\odot$ BH. The simple reason why SS433 can stably survive this type of orbital decay for 10^4 to perhaps 10^5 yr is because of its unique combination (for a Be/X-ray binary progenitor) of a quite massive compact star and a relatively moderate-mass donor star, such that the mass ratio of donor and compact star is $q < 3.5$.

10.12.2 Orbital Evolution During SS433-like Mass Transfer

As shown by King et al. (2000) and Begelman et al. (2006), in the case of RLO from donor stars with a radiative envelope, a further condition for avoiding the formation of a CE is that the spherization radius, $R_{\rm sp}$, of the accreting compact object remains smaller than its Roche lobe, where $R_{\rm sp}$ is given by (Shakura & Sunyaev, 1973)

$$R_{\rm sp} = \frac{27}{4}\left(\frac{\dot{M}}{\dot{M}_{\rm Edd}}\right) R_{\odot}. \tag{10.2}$$

In this case, if the donor has a radiative envelope and a mass ratio $q < 3.5$, one expects the system to go into normal RLO evolution, similar to that of SS433. The *SS433 mode* of mass transfer is equivalent to what is also called *isotropic re-emission* (e.g., see Chapter 4 and van den Heuvel & De Loore [1973]; Bhattacharya & van den Heuvel [1991]; Soberman et al. [1997]; Tauris & Savonije [1999]; Tauris & van den Heuvel [2006]). With the SS433 mode of mass transfer, followed by mass loss from the accretion disk, which has the specific orbital angular momentum of the compact object, it is simple to calculate how the orbit of the system will change (see Section 4.3 and Eqs. (4.45)–(4.48) in Section 4.3.7 in particular). For SS433, the fraction of matter lost by the donor star that is accreted by the BH is of the order of $1-\beta = 10^{-4} - 10^{-3}$, meaning that $\beta \simeq 1$, in other words, practically all transferred matter is re-emitted from the vicinity of the BH.

In this limit, we found earlier that we can express the change in orbital period due to RLO as

$$\left(\frac{P}{P_0}\right) = \left(\frac{kq+1}{q+1}\right)^2 k^3\, e^{3q(1-k)}, \tag{10.3}$$

where $k = q_0/q$. In the case of SS433, assuming a 4.3 $M_{\odot}$ BH accretor currently orbited by an A-supergiant donor star of mass 12.3 $M_{\odot}$ and with a helium core of $\sim$3.5 $M_{\odot}$, means that at the end of the RLO phase $q = 0.81$, while at present $q_0 = 2.86$. Inserting these values into Eq. (10.3), and using a present orbital period of $P_0 = 13.1$ d, one finds that at the end of the RLO the orbital period of the system will be $P = 5.6$ d.

So SS433 will with these assumed component masses finish as a detached binary consisting of a 3.5 $M_{\odot}$ helium star and a 4.3 $M_{\odot}$ BH. The entire process will take place on the thermal timescale of the envelope of the A-supergiant, which is between 10^4 and 10^5 yr. The helium star in the resulting system may during helium-shell burning go through a second mass-transfer phase (Case BB RLO, Section 10.13) and finally explode as a SN, likely leaving a NS. If the system remains bound in response to the natal kick of the NS, a close eccentric binary will result, consisting of the present 4.3 $M_{\odot}$ BH plus a NS.

One may wonder what is so special about SS433 and why we do not see more SS433-like systems. van den Heuvel et al. (2017) propose that the answer is that the very unusual combination for a HMXB of a rather low donor mass (presently 12.3 $M_{\odot}$ and initially 14–15 $M_{\odot}$) plus a quite massive compact star (4.3 $M_{\odot}$). A 14–15 $M_{\odot}$ initial donor mass and an orbital period of $10-15$ d are typical for a Be/X-ray binary, which is the most common type of HMXB. There are $\sim$220 Be/X-ray binaries known in our Galaxy and the Magellanic Clouds (about half of which are in the SMC). In all but one of the known Be/X-ray binaries, the compact stars are NSs that have a typical mass of approximately $1.2-2.1\,M_{\odot}$. Only one Be/X-ray binary is known to harbor (probably) a BH companion with a mass of 3.6–6.9 $M_{\odot}$ (Casares et al., 2014). If the companion of a 14 $M_\odot$ Be-star is a 1.4 $M_\odot$ NS, then the mass ratio is $q=10$, and the formation of a CE is unavoidable. The two stars will merge, unless the orbital period is longer than ~1 yr (which is the case for only a small fraction of the Be/X-ray binaries). Only if the compact star has a mass of $\gtrsim 4\,M_\odot$ and the donor has a radiative envelope, the system may remain dynamically stable and survive the SS433-like mass-transfer process. (For a model for the formation and future evolution of the possible BH Be/X-ray binary, see Grudzinska et al., 2015). Therefore, of the 220 Be/X-ray binaries known, possibly only the one system with an (alleged) BH companion will in the future evolve like SS433 and survive the RLO. The Be/X-ray binaries with a NS accretor will evolve into a CE phase, and most of these systems have orbital periods that are too short to survive and eject the CE. Only a few, initially wide-orbit, systems will survive the in-spiral and produce a very close helium star–NS system. So, the birthrate of SS433-like systems is of the order of 0.5% of the birthrate of Be/X-ray binaries.

It should be noted that Cherepashchuk et al. (2021) have recently discovered the orbital eccentricity ($e=0.05\pm0.01$) and a secular increase in the orbital period of SS433 of $\dot{P}_{\rm orb}=(1.0\pm0.3)\times10^{-7}\,{\rm s\,s^{-1}}$. The latter value is very interesting as it sets a lower limit on the mass-loss rate from the system ($\sim7\times10^{-6}M_\odot\,{\rm yr}^{-1}$ in the Jeans mode) and also constrains the current mass ratio to be $q_0=M_X/M_V>0.8$. The orbital eccentricity means that the donor star has not yet been tidally synchronized, which lends support to the "slaved disk" model for the precession of its relativistic beams (Cherepashchuk et al., 2021).

10.12.3 Known WR+O Binaries, Stable RLO, and Production of Double BH Systems

Only five BH-HMXBs are known (see Table 6.6), of which Cyg X-1 is the best-known example. Additionally, in our Galaxy, there is the Be/X-ray binary MCW 656 (Casares et al., 2014). In the LMC, there are two sources: LMC X-1 and LMC X-3; in M33, there is the very massive system M33 X-7. In our Galaxy, one may further add the system of SS433, although it is a quite extraordinary object.

The only known BH-HMXB systems that probably are massive enough to produce in the future a double BH are Cyg X-1 and M33 X-7. Cyg X-1 harbors a 21 $M_\odot$ BH and a 41 $M_\odot$ companion in a 5.6 d orbit (Miller-Jones et al., 2021), and M33 X-7 harbors a 15.7 $M_\odot$ BH, moving around a 70 $M_\odot$ companion in a 3.45 d orbit (Orosz et al., 2007). A model of the evolutionary origin of M33 X-7 has been presented by Valsecchi et al. (2010), and for Cyg X-1 a formation model was described in Section 10.9.2. In both systems the donor star is massive enough to also leave a BH.

The rareness of BH-HMXBs with respect to BH-LMXBs, of which ~50 are presently known (Corral-Santana et al., 2016), is a selection effect: the $<2\,M_\odot$ donor stars in LMXBs are very long-lived ($>10^9$ yr), whereas those in HMXBs are very short-lived ($<5\times10^6$ yr). This makes the chance to observe a BH-LMXB, even though they are "on" only a fraction of the time, much greater than the chance to observe a BH-HMXB.

An important point to notice is that, if a HMXB donor star is massive enough to produce a BH, that is, it is more massive than ~15 $M_\odot$, and has a NS companion, stable SS433-like evolution is impossible. This is because at the start of the RLO, mass transfer with a very large mass ratio will become unstable and CE evolution will ensue, leading to the complete in-spiral of the NS and resulting in coalescence and formation of a TŻO, as discussed in Section 10.7.4. Therefore, the only HMXBs with relatively short orbital periods (such that the donor envelopes remain in, or close to, radiative equilibrium) that can survive as binaries and produce WR/X-ray binaries are the ones in which the accretor is a BH. This implies that the compact stars in the WR/X-ray binaries that result from this stable RLO evolution of HMXBs can only be BHs (van den Heuvel et al., 2017).

van den Heuvel et al. (2017) made exploratory (toy) calculations, in which they assumed the WR-stars in WR+O binaries to terminate life as BHs, and the same to happen, later on, with the WR descendants of their O-type companions. They showed that with these assumptions, a number of the well-known WR+O binaries (see Table 10.3) may well later in life become massive BH-HMXBs, which through stable RLO will later evolve into short-period WR/X-ray binaries and finally produce short-period double BHs (or, in some cases, BH+NS systems).[4]

The results of the calculations listed in Table 10.3 show that (at least) a fraction of the merging close double BHs is expected to descend from normal well-known WR+O binary systems like the ones listed in this table. These systems are therefore representative of a formation channel producing GW mergers detected by the LIGO-Virgo-KAGRA network. It is important to know the relative contributions of each of the proposed models for producing by binary evolution close-orbit double BHs that will merge within the age of the Universe. A study by Neijssel et al. (2019) of the evolution of stellar populations with different metallicities, and with a realistic fraction of binaries, indicates that the majority of the GW-detectable double BH systems that formed by binary evolution may have been produced by the stable RLO (SS433-like) model and only a minority by other binary evolution channels. This remains an open question according to other works.

In Chapter 15, we discuss the formation of GW sources in further detail along with other proposed formation channels for producing double BH systems, such as the dynamical formation channel discussed in Chapter 12.

10.13 REFINEMENT OF THE DNS FORMATION MODEL: CASE BB RLO IN POST-HMXB SYSTEMS

The outcome of the evolution of some of the HMXBs described in this chapter is that they will eventually evolve into double NS (DNS) systems, following a second SN; see illustrations in Figures 1.1 and 10.13 and see Section 14.9.1 for a résumé of DNS formation according to the standard model for isolated binaries. Here we highlight the effects of mass transfer onto the first-born NS and the details of the post-HMXB/post-CE evolution of donor stars with initial masses in the range of approximately 8–16 $M_{\odot}$.

[4]It should be kept in mind here that in reality not all WR-stars are expected to collapse to BHs, as this would overproduce BH-HMXBs, as mentioned in Section 9.7. There is quite strong evidence that many WR-stars may leave NSs, see also Woosley (2019).

Table 10.3. Anticipated Future Evolution of Seven Well-observed WR+O Binaries

		Observed WR+O		HMXB $_{RLO}$	WR-HMXB		Double BH		
Name	Spectrum	P_{orb} (day)	Masses ($M_\odot$)	Masses ($M_\odot$)	Masses ($M_\odot$)	P_{orb} (day)	Masses ($M_\odot$)	P_{orb} (day)	τ_{merge} (Gyr)
WR 127	WN5o + O9.5V	9.555	17 + 36	**9.6** + 33	**9.6** + 13.6	1.54∗	**9.6 + 7.0**	1.71	6.91
WR 21	WN5o + O4-6	8.255	19 + 37	**10.2** + 34	**10.2** + 14.1	1.64∗	**10.2 + 7.9**	1.77	6.50
WR 62a	WN4.5o + O5.5V	9.145	22 + 40.5	**10.8** + 37	**10.8** + 16.1	1.45∗	**10.8 + 8.8**	1.61	4.40
WR 42	WC7 + O7V	7.886	14 + 23	**10.4** + 22	**10.4** + 8.0	13.75	**10.4 + 4.5**	14.71	—
WR 47	WN6o + O5V	6.2393	51 + 60	**18.1** + 46	**18.1** + 25.8	7.96	**18.1 + 10.4**	8.59	—
WR 79	WC7 + O5-8	8.89	11 + 29	**9.0** + 27.4	**9.0** + 10.1	2.44	**9.0 + 5.4**	2.64	—
WR 11(CE)	WC8 + O7.5III	78.53	9.0 + 30	**7.8** + 28.5	**7.8** + 10.5	0.90	**7.8 + 5.6**	0.98	2.24

The evolution of these binaries was calculated by van den Heuvel et al. (2017) for the hypothetical case that all these WR-stars would collapse to a BH and the same would later be the case for their O-type companions. (Notice, from the review in Section 9.7, we know that in reality many of these WR and O-stars may also explode as SNe and leave behind NSs; however, certainly a few will leave BHs.) The orbital periods, spectral types, and masses of the components were taken from the catalog by van der Hucht (2001). The WR-stars and O-stars in these systems are sufficiently massive to leave BH remnants. The ways how the masses of these BHs (bold numbers) were calculated are explained in van den Heuvel et al. (2017). (Notice that the assumptions used in calculating these BH-HMXB masses are slightly different from those used in calculating the BH-HMXB masses in Table 10.1.) It is assumed for the top six systems that the resulting O+BH systems will spiral-in following the SS433 recipe. In the three first systems, the mass ratio of O-donor and BH is very close to 3.5, such that it is not fully certain that the mass transfer will be stable. To indicate this uncertainty, an asterisk (∗) was placed after the short orbital periods of the resulting WR-star HMXBs. Assuming the SS433-type mass transfer to be stable, the masses of the double BHs, with their orbital periods, indicated in the last columns, will result. The very last column lists the GW merger times of these systems (no times are given if the merger time is longer than the age of the Universe). In case where the SS433-type mass transfer would not be stable, the WR-star HMXBs indicated with ∗ will go into CE evolution; in which case none of these systems will survive. The bottom line gives the anticipated evolution of WR 11 (Gamma-2 Velorum), which will in the second phase of mass transfer go into CE evolution and produce a very short-period WR-star HMXB and double BH.

The latter stars produce post-CE helium stars with masses in the range of approximately 2.2–3.5 $M_{\odot}$. As we discussed in Chapter 9, during the later evolution the envelopes of such helium stars expand to giant dimensions, such that in a close binary they will go into a second case of mass transfer (Case BB RLO, as depicted in Fig. 10.5; see Chapter 9 for further details).

In the following, we consider further important examples of such post-CE/HMXBs that undergo Case BB RLO to a NS. The reason to pay so much attention to Case BB RLO is that, due to the shape of the IMF, the bulk of the DNSs are expected to originate from binaries with initial component masses in the range of 8–16 $M_{\odot}$. (See, e.g., the works by Dewi et al. [2002]; Dewi & Pols [2003]; Ivanova et al. [2003]; and Tauris et al. [2015] in which Case BB RLO to a NS was explored in increasing detail.)

For close binaries, this is the last phase of mass transfer and it is important for two reasons: i) it strips the outer layers of the helium star donor very efficiently, leaving an almost naked metal core prior to the SN explosion (Tauris et al., 2013, 2015), and ii) it causes the first-born NS in the binary to accrete material and become a recycled radio pulsar (Dewi et al., 2005; Tauris et al., 2015; Tauris et al., 2017). Both aspects will be discussed in more detail in Chapters 13 and 14, in relation to ultra-stripped SNe. For successful ultra-stripping models until CCSN, see Jiang et al. (2021).

10.13.1 A Detailed Example of Case BB RLO to a NS

As an example of a detailed calculation of the evolution of tight helium star–NS binaries, we show in Figure 10.17 that the HR diagram of a helium-star donor with initial mass $M_{\rm He} = 2.9\,M_{\odot}$, a NS mass of $M_{\rm NS} = 1.35\,M_{\odot}$, and orbital period $P_{\rm orb} = 0.10$ d. The points along the evolutionary track correspond to (A) helium star ZAMS ($t = 0$); (B) core helium exhaustion at $t = 1.75$ Myr, defining the bottom of the giant branch with shell helium burning; (C) onset of Case BB RLO at $t = 1.78$ Myr; (D_1) core-carbon burning during $t = 1.837 - 1.849$ Myr ($\Delta t \simeq 12{,}000$ yr), leading to radial contraction and Roche-lobe detachment (as a result of the mirror principle when the core expands); (D_2) consecutive ignitions of carbon-burning shells during $t = 1.850 - 1.854$ Myr and Roche-lobe detachment again; (E) maximum luminosity at $t = 1.854304$ Myr; (F) off-center ($m/M_{\odot} \simeq 0.5$) ignition of oxygen burning at $t = 1.854337$ Myr ($T_c = 9.1 \times 10^8$ K; $\rho_c = 7.1 \times 10^7\,{\rm g\,cm^{-3}}$), marked by a bullet. Shortly thereafter (point G) the binary orbit becomes dynamically unstable when $|\dot{M}_2| > 10^{-2}\,M_{\odot}\,{\rm yr^{-1}}$, and for technical reasons the model star was taken out of the binary to continue its final evolution as an isolated star ($t \geq 1.854337$ Myr). At this stage $P_{\rm orb} = 0.070$ d. After 3–4 yr of further stellar aging, the computer code broke down (<10 yr prior to core collapse). The mass-transfer rate of this evolution is plotted in the bottom panel of Figure 10.17. The chemical abundance profile and the Kippenhahn diagram of this ultra-stripped helium star donor are shown in Figures 10.18 and 13.14, respectively. Note that a similar episode of Case BB RLO from post-CE binaries can also produce a recycled NS in a binary with a massive WD companion (Tauris et al., 2012; Lazarus et al., 2014).

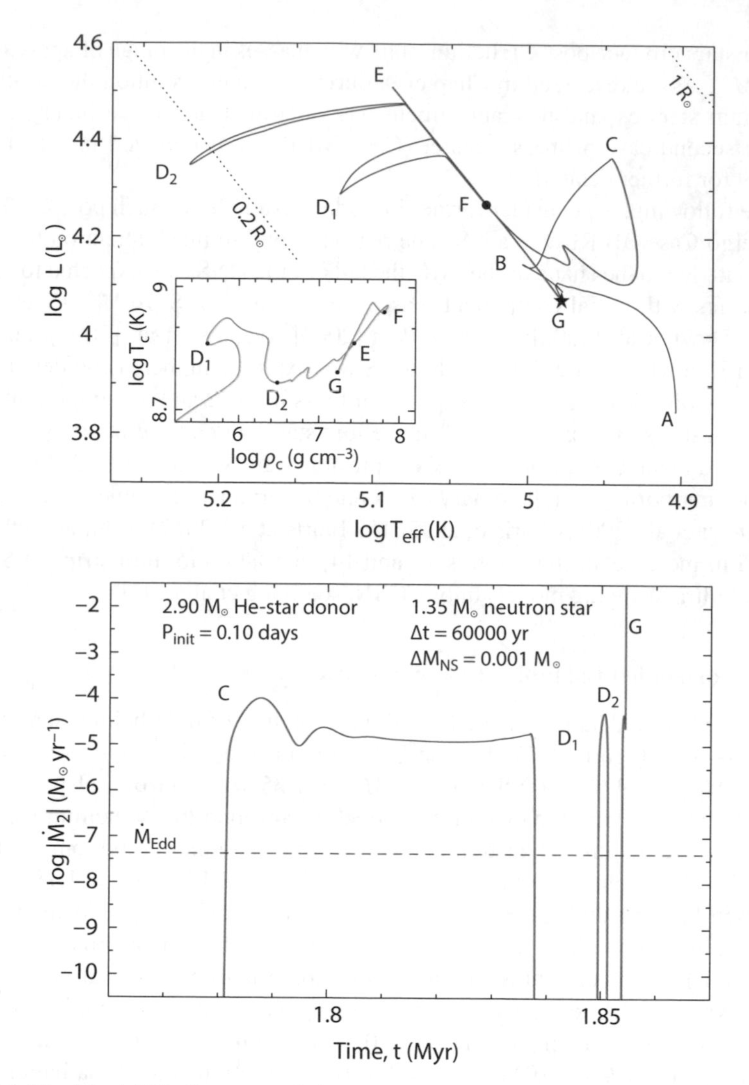

Figure 10.17. (Top) HR diagram of the evolution of a 2.9 $M_\odot$ helium star that loses mass to a NS companion star prior to the CCSN. The initial orbital period is 0.1 d. The red part of the track corresponds to a detached system and the blue part is marking RLO. See text for explanation of the evolutionary sequence A, B, C, D_1, D_2, E, F, and G. The inset shows the final evolution in the central temperature–central density plane. (Bottom) Mass-transfer rate as a function of stellar age for the same helium star. The major phase of Case BB mass transfer lasts for ~56,000 yr. The horizontal dashed line marks the Eddington accretion rate. Following stages of shell-carbon burning (Case BBB RLO, after a detached phase with central carbon burning, D_1), a rigorous helium-shell flash leads to the spike in point G and the computation ended. After Tauris et al. (2013).

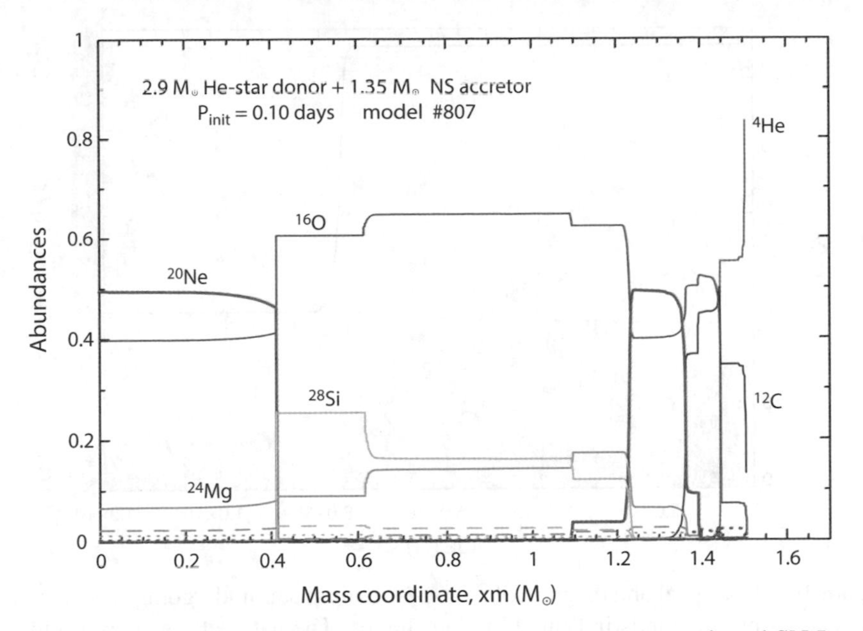

Figure 10.18. Chemical abundance structure of the 1.50 $M_\odot$ ultra-stripped SN Ic progenitor star resulting from Figure 10.17, ~10 yr prior to core collapse (during which oxygen and silicon burning produce an iron core while the core–envelope boundary remains frozen). At this early stage, the pre-collapsing star has a hybrid structure with an ONeMg inner core enclosed by a thick OMgSi outer core, which again is enclosed by shells of ONeMg and CO, and outermost a tiny envelope with 0.033 $M_\odot$ of helium. After Tauris et al. (2013).

10.13.2 Mass-transfer Rates during Case BB RLO to a NS

Figure 10.17 (bottom) shows a typical example of mass-transfer rate, $|\dot{M}_{\rm He}|$, as a function of time for a Case BB RLO leading to core collapse of an ultra-stripped star. Maximum values of $|\dot{M}_{\rm He}| \simeq 10^{-4}\ M_\odot\ {\rm yr}^{-1}$ (i.e., ~3–4 orders of magnitude above $\dot{M}_{\rm Edd}$) are similar for almost all calculations independent of the initial helium star mass and orbital period (Tauris et al., 2015). The total duration of the Case BB RLO varies: it is ~1,000 yr for Case BC RLO (i.e., RLO initiated during shell carbon burning), up to 100,000 yr for regular Case BB RLO; and as long as 50 Myr for Case BA RLO (i.e., RLO initiated during core helium burning), for example, for 2.5 $M_\odot$ helium stars producing ~0.9 $M_\odot$ CO WD remnants. In general, the duration of this mass-transfer phase *decreases* with increasing initial orbital period ($\Delta t_{\rm Case\ BA} > \Delta t_{\rm Case\ BB} > \Delta t_{\rm Case\ BC}$) and increasing helium star mass (e.g., Table 1 in Tauris et al., 2015).

Temporary Extinguishing of Central Helium Burning

The intense mass loss resulting from RLO can cause the central nuclear burning in the donor star to cease. An example of such a case is displayed in Figure 10.19, which shows a Kippenhahn diagram for the evolution of a 2.6 $M_\odot$ helium star, placed in a

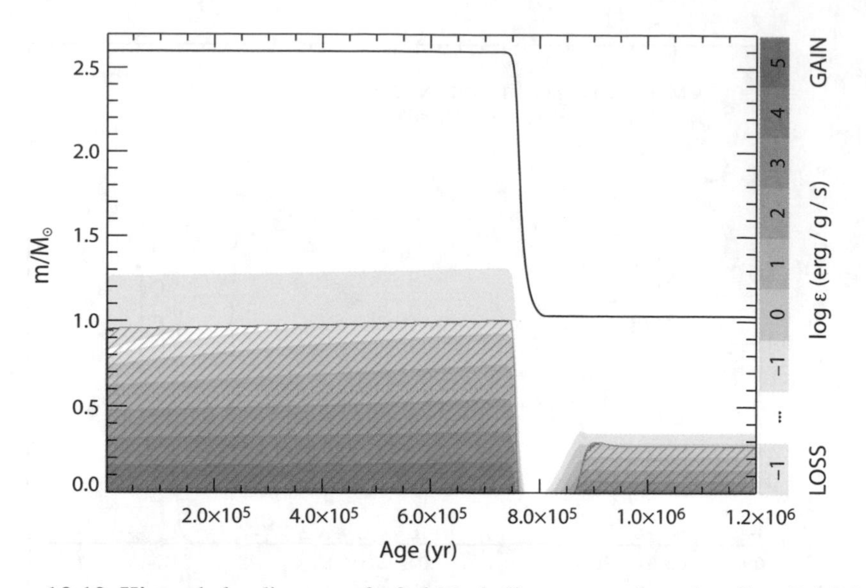

Figure 10.19. Kippenhahn diagram of a 2.6 $M_{\odot}$ helium star undergoing Case BA RLO. The energy production is indicated in blue shades. The hatched green color indicates convection. The central helium burning is temporarily completely shut off as a result of the intense mass loss from the outer layers of the donor star. Central helium burning is only reignited after the RLO has terminated and the central temperature has risen again. After Tauris et al. (2015).

tight orbit with a NS and $P_{\rm orb} = 0.06$ d. The helium star fills its Roche lobe during early core-helium burning (i.e., Case BA RLO). The resulting mass transfer causes the star to lose its outermost $\sim 1.6\, M_{\odot}$ of material in $<50{,}000$ yr. As a direct consequence of this mass loss, the star is driven out of thermal equilibrium. For the star to replace the lost envelope with material from further below and to remain in hydrostatic equilibrium, an endothermic expansion of its inner region (requiring work against gravity) causes the surface luminosity to decrease by >3 orders of magnitude. This effect also has an important consequence for the nuclear burning in the star. The core expansion causes the central temperature to decrease, and given the extreme dependence of the triple-α process on temperature, the result is that all fusion processes are completely quenched for $\sim 80{,}000$ yr (see further details in Tauris et al., 2015). Thereafter, when the central temperature has risen again, core-helium burning is reignited.

10.13.3 NS Recycling via Case BB RLO

Case BB RLO results in accretion onto the NS of 5×10^{-5} to $3 \times 10^{-3}\, M_{\odot}$ for Eddington-limited accretion, and up to $0.01\, M_{\odot}$ if allowing for accretion above the Eddington limit by a factor of 3 (Tauris et al., 2015). The gain in spin angular momentum of the accreting NS can be expressed as (Tauris et al., 2012)

$$\Delta J_{\star} = \int n\,(\omega, t)\, \dot{M}(t)\, \sqrt{G M(t)\, r_{\rm mag}(t)}\, \xi(t)\, dt, \qquad (10.4)$$

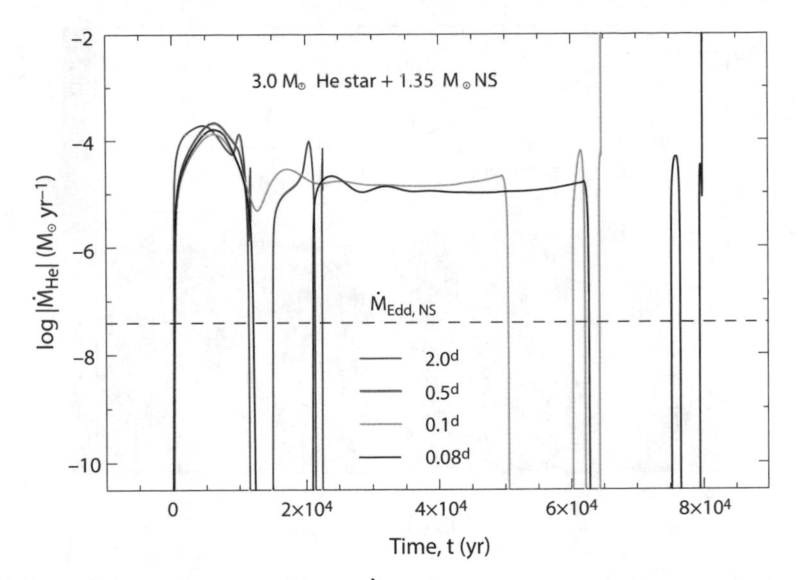

Figure 10.20. The mass-transfer rate, $|\dot{M}_{\rm He}|$, as a function of time for $3.0\,M_\odot$ helium star donors with different values of initial orbital period, $P_{\rm orb}$. The time $t=0$ is defined at RLO onset for all systems. For the two narrow systems, the calculations are terminated by a vigorous helium-shell flash following off-center oxygen ignition, similar to the case discussed in Tauris et al. (2013). It is clearly shown how the duration of the mass-transfer phase, Δt_x, increases with decreasing value of $P_{\rm orb}$, resulting in more rapidly spinning recycled pulsars in DNS systems in close orbits. In all cases shown the outcome is an ultra-stripped iron CCSN with final envelope mass $M_{\rm env,f}=0.04-0.23\,M_\odot$. After Tauris et al. (2015).

where n is the dimensionless accretion torque, ω is the fastness parameter, t is time, $\dot{M}$ is the NS mass accretion rate, G is the constant of gravity, $M(t)$ is the mass of the NS, $r_{\rm mag}$ is the radius of the magnetosphere, and ξ is a parameter depending on the geometric flow of the material ($\simeq$1 for accretion from a disk). This integral can roughly be approximated as a simple correlation between the equilibrium spin period in milliseconds, $P_{\rm ms}$, and the (minimum) amount of mass accreted, $\Delta M_{\rm eq}$ (see Chapter 14):

$$P_{\rm ms}=\frac{(M/M_\odot)^{1/4}}{(\Delta M_{\rm eq}/0.22\,M_\odot)^{3/4}}. \tag{10.5}$$

Hence, one can see that the above-mentioned range of mass accreted via Case BB RLO corresponds to spinning up the NS to a resulting spin period of $11-350$ ms (Tauris et al., 2015). For a comparison, the observed range of spin periods of recycled NSs in DNS systems is currently spanning $17-185$ ms (Stovall et al., 2018).

Some of the remaining questions to investigate in relation to this recycling process are the decay of the B-field of the accreting NS (for which we have evidence from observations, see, for example, Taam & van den Heuvel [1986]), its accretion efficiency, and how the excess material (transferred toward the NS at a highly super-Eddington

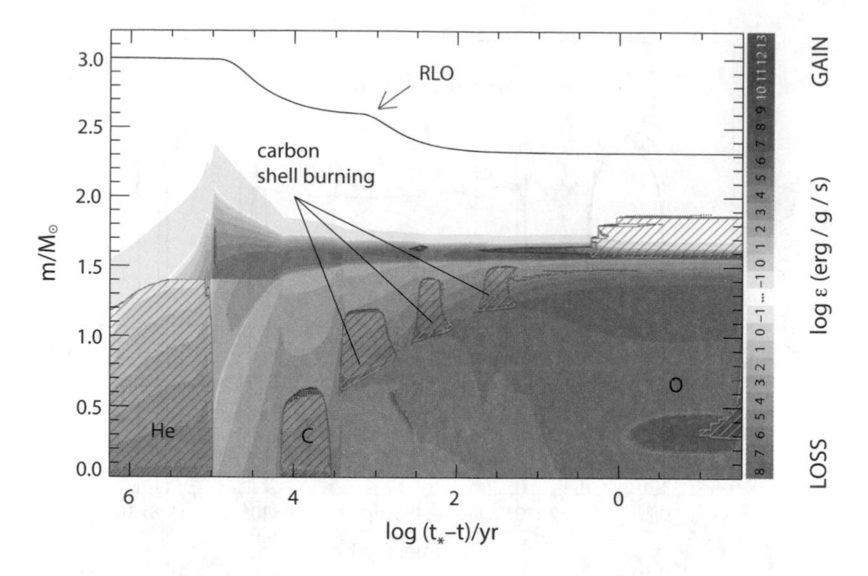

Figure 10.21. Kippenhahn diagram of a 3.0 $M_{\odot}$ helium star undergoing Case BB RLO to a NS in a wide binary with $P_{\mathrm{orb}} = 50$ d. The plot shows the evolving cross section of the star, in mass coordinate on the Y-axis, as a function of remaining calculated lifetime on the X-axis. The green hatched areas denote zones with convection and the intensity of the blue/purple color indicates the net energy-production rate, ε. The total mass of the star is shown by the solid black line. For this wide-orbit binary the onset of the RLO (see red arrow) will occur at a late evolutionary stage (shell-carbon burning) such that the star will undergo core collapse within ~1,000 yr, which is before the end of the RLO. Mass loss prior to RLO is due to a stellar wind. After Tauris et al. (2017).

rate from the donor star) is ejected from the vicinity of the NS. When discussing the recycling of, for example, the first-born NS in a DNS system, one must include potential phases of NS accretion prior to and after Case BB RLO. Tauris et al. (2017) argued that all taken together, the first-born NS in a DNS system is expected to have accreted less than ~0.02 $M_{\odot}$ (see also Chapter 14).

Given the relation between the amount of accreted material and the equilibrium spin rate of the recycled pulsar (see Eq. [10.5]), one can predict a correlation between the orbital period and the spin period of recycled pulsars in DNS systems (Chapter 14), because the amount of accreted mass is strongly dependent on the orbital period of the binary at the onset of Case BB RLO. Examples of detailed calculations of Case BB RLO showing how the duration of the mass-transfer phase decreases with increasing initial orbital period are shown in Figure 10.20. The wider the initial orbit of the NS–helium star system, the more evolved is the helium star when it fills its Roche lobe and the shorter is its remaining lifetime before it collapses and produces a SN explosion. Therefore, in wide binaries little mass is transferred to the NS prior to core collapse, and thus the recycling process is relatively ineffective. A sample case is illustrated in Figure 10.21 where the donor star only fills its Roche lobe during shell-carbon burning. The resulting spin period of the accreting NS will remain large, that is, it will be a *marginally recycled pulsar*. As an example, PSR J1930−1852 (Swiggum et al., 2015) has a large orbital period of 45 d and a slow spin period of 185 ms, as expected (see Chapter 14). We refer to Tauris et al. (2017) for detailed discussions on various aspects of the anticipated properties of NSs in DNS systems, including stellar evolution aspects and kinematical effects of the second SN explosion. A summary of these aspects is given in Section 14.9.

EXERCISES

Exercise 10.1. *Is a conservative Case B mass transfer history possible for CQ Cephei? CQ Cephei is a WR+O-star* ($24 + 30\,M_\odot$) *system (see Section 10.8.2). The* $60\,M_\odot$ *progenitor of the* $24\,M_\odot$ *helium star (WR-star) in this system must have had a very strong stellar wind. We therefore assume that at the onset of Case B mass transfer, its mass had been reduced to* $40\,M_\odot$ *due to wind mass loss. In the case of subsequent conservative RLO, it will then have transferred* $16\,M_\odot$ *to its O-star companion, which in that case must have had an initial mass of* $14\,M_\odot$*, to end with its present mass of* $30\,M_\odot$*.*

a. *Show that to end with its present orbital period of* 1.64 *d, its shortest orbital period during the conservative mass transfer would have been* 1.58 *d. (Hint: see Section 4.3.6.)*
b. *The mass-losing primary would at that time still have had a hydrogen-rich envelope of* $3.0\,M_\odot$*, and such a star has a radius of* $\sim 265\,R_\odot$*. Show that with an orbital period of* 1.58 *d, the entire system must have been deeply embedded inside the radius of the O-star, that is, it must have been in a CE phase.*
c. *There are further arguments against conservative RLO. Show that if it evolved conservatively, its initial orbital period would have been* 3.49 *d.*
d. *The* $40\,M_\odot$ *progenitor of the WR-star will have a radius of* $\sim 60\,R_\odot$ *when it leaves the main sequence. Show that with this orbital period, it must therefore have evolved according to early Case A, not Case B RLO.*

In early Case A RLO, however, this $40\,M_\odot$ *star could never produce a helium core as massive as in Case B RLO, as illustrated by Figure 10.14, and therefore it could never have reached the* $24\,M_\odot$ *WR-star mass of CQ Cephei. The conclusion is that CQ Cephei could not have formed via conservative Case B RLO.*

Exercise 10.2. *In Exercise 10.1 it was concluded that the WR+O-star binary CQ Cephei (Section 10.8.2) could not have formed via conservative Case B RLO. Here, we investigate whether this binary represents a post-CE system, that is, whether solutions exist for a CE scenario producing CQ Cephei. Assume that the progenitor of the* $24\,M_\odot$ *WR-star was a* $60\,M_\odot$ *ZAMS star, and assume furthermore a low-metallicity environment such that we ignore here stellar wind mass-loss prior to onset of a CE with its* $30\,M_\odot$ *O-star.*

a. *The present (post-CE) orbital period of CQ Cephei is $P_{\rm orb} = 1.64\,d$. Calculate its orbital separation.*

b. *Show that the orbital energy of the present system is -6.2×10^{49} erg.*
Therefore, we can conclude that at most $\Delta E_{\rm orb} \simeq 6.2 \times 10^{49}$ erg has been released via spiral-in and that this value therefore sets the upper limit for the value of $E_{\rm bind}$ to have a successful CE ejection (assuming 100% efficiency in converting released orbital energy via spiral-in into kinetic energy that may eject the envelope, and neglecting the very small initial orbital energy of the wide binary and any widening of the post-CE system from winds).

c. *Apply Figure 10.16 to find a progenitor solution (if any) for a 60 $M_\odot$ star that fulfills the criterion that $|E_{\rm bind}| < |\Delta E_{\rm orb}|$.*

Exercise 10.3. *The orbital period in the right-side column of Figure 10.6 is stated with a constant value of $P_{\rm orb} = 100\,d$. This must be a typo or a mistake.*

a. *Assuming that the system has $P_{\rm orb,0} = 100\,d$ at the time it initiates RLO, calculate the post-RLO value of $P_{\rm orb}$ using the isotropic re-emission model from Section 4.3.7, assuming $\alpha = \delta = 0$ and $\Gamma_{ls} = 1$ in Eq. (4.58).*

b. *Assume instead that, by some unspecified orbital angular momentum loss, the system has $P_{\rm orb} = 90\,d$ prior to the SN (and assume this value stayed roughly constant after the RLO). Investigate any possible solutions for the parameters $(\alpha, \beta, \delta, \gamma)$ of the RLO if the initial (pre-RLO) $P_{\rm orb,0}$ was somewhere between 100 and 200 d and $\Gamma_{ls} = 1.0$.*
(Hint: use numerical computations to explore any possible solutions by treating $(\alpha, \beta, \delta, \gamma)$ as free parameters, with the caveat that any solution may not be physical. Remember, $\alpha + \beta + \delta + \epsilon = 1$ (and use $\Gamma_{ls} = 1$) and that the mass ratios (q_0 and q as well as the accretion efficiency, ϵ, can be calculated from the given masses in the figure.)

c. *Calculate the change in $P_{\rm orb}$ as a consequence of the SN explosion, assuming a symmetric SN. (Hint: see Section 4.3.10.)*

Chapter Eleven

Formation and Evolution of Low-mass X-ray Binaries

In the first part of this chapter, we describe the possible formation mechanisms of LMXBs. We argue here that the formation of a LMXB is a very rare event and discuss the various proposed formation models. Subsequently, we discuss the mechanisms that drive the mass transfer in these systems, which turn out to be very long-lasting (timescales 10^8 to 10^9 yr) and have important implications for the precise further evolution of the systems. This leads us to the expected final products of this evolution, which in the case of NS-LMXBs are expected to be millisecond radio pulsars, in binaries as well as single, as will be further discussed in Chapter 14.

11.1 ORIGIN OF LMXBs WITH NEUTRON STARS

11.1.1 The Need for a Rare Formation Process

Since the mass of the donor star in a LMXB generally is smaller than that of the compact star, the mass transfer in these systems will be stable and will therefore take place on a long timescale: either the nuclear timescale of a star typically less massive of 1 $M_{\odot}$ ($\sim 10^{10}$ yr) or the timescale of orbital shrinking by angular momentum losses (see later in this section), which also is very long, of the order of 10^9 yr. With such a lifetime, the total number of steady bright ($> 10^{36}$ erg s^{-1}) LMXBs in our Galaxy of $\sim 10^2$ indicates that their formation rate in the Galaxy is of the order of 10^{-7} yr^{-1}. This is more than five orders of magnitude lower than the core-collapse supernova rate. So, not more than one in $\sim 10^5$ neutron stars (NSs) finished as a member of a bright steady LMXB. (A similar factor also holds for the stellar black holes (BHs) ending up in BH-LMXBs.) This is in great contrast to the bright standard HMXBs where of the order of *one out of every ten* NSs ends up in a HMXB (Chapter 10).

Thus, while the production of a HMXB is apparently an almost normal event for every high-mass close binary system, the production of a LMXB is clearly a most unusual event that occurs extremely rarely in Nature. It therefore requires very special conditions to take place. The following three processes have been suggested:

1. Common-envelope (CE) evolution of a fairly massive binary with a large initial mass ratio.
2. Accretion-induced collapse (AIC) of a massive WD reaching the critical Chandrasekhar mass limit ($\sim 1.4\, M_{\odot}$) in a close binary.

3. Origin from a HMXB with a much more distant low-mass main-sequence (G- or K-dwarf) companion (i.e., a triple star origin).

We briefly discuss each of these processes.

11.1.2 Formation via Common-envelope Evolution

This model, first proposed by Sutantyo (1975a) to explain the origin of the Her X-1 X-ray binary and subsequently worked out in further detail by Kalogera & Webbink (1996, 1998) and Kalogera (1997, 1998), has become one of the two key scenarios for producing NS-LMXBs (the other one being the AIC scenario). The Her X-1 system consists of an X-ray pulsar together with a $2\,M_{\odot}$ A-type donor star in a 1.70 d orbit. The system is located 3 kpc outside the Galactic plane. Nevertheless, it must be a fairly young object as a $2\,M_{\odot}$ star cannot live longer than $\sim 8 \times 10^{8}$ yr. Stars of that age belong to Population I and are always found within 200 pc from the Galactic plane. The system must therefore have originated near the Galactic plane, but to reach its present distance of 3 kpc above this plane, it must have been shot out of the plane with a velocity in excess of $120\,\mathrm{km\,s^{-1}}$. This large runaway velocity is undoubtedly due to the sudden mass ejection in the supernova (SN) in which the NS was created, possibly augmented by a birth kick to this NS in its formation event (Chapter 13). Verbunt et al. (1990) have shown that a system runaway velocity of this magnitude could never be reached if the NS was formed by the AIC of a white dwarf (WD), even if the NS received a kick of several hundreds of $\mathrm{km\,s^{-1}}$ at birth. They found that the kick required to give it the necessary velocity perpendicular to the Galactic plane either would have disrupted the system or would have shot the NS into its companion. Moreover, as we shall see later, AIC is not expected to impart a large kick to a NS.

For these reasons, for the formation of this system, the only model possible is the one suggested by Sutantyo (1975a, 1992), as depicted in Figure 11.1 (see also Fig. 11.2). In this CE model, the initial system consists of a fairly massive star ($\sim$12 to $15\,M_{\odot}$) together with a $2\,M_{\odot}$ secondary star in a fairly wide orbit ($P_{\mathrm{orb}} > 1$ yr). When the massive component becomes a giant, at an age of $\sim 10^{7}$ yr, the system enters a CE phase in which the $2\,M_{\odot}$ star spirals inward in the convective hydrogen-rich envelope of the massive giant and the envelope is ejected. The post-CE system consists of a helium star of between 3 and $4\,M_{\odot}$ and the $2\,M_{\odot}$ secondary star in an orbit with $P_{\mathrm{orb}} \simeq 0.5 - 1.0$ day. Some 1.5×10^{6} yr later, the helium star explodes as a SN (disregarding here a phase of Case BB RLO [Sections 9.2.1 and 10.13]). In a symmetric explosion, the system is not disrupted, but it receives a large runaway velocity due to the sudden mass ejection ($\Delta M = 1.6 - 2.6\,M_{\odot}$), and the orbit becomes highly eccentric. As the system is now 1.15×10^{7} yr old, and the $2\,M_{\odot}$ star remains on the main sequence for another 8×10^{8} yr, the orbit has ample time to circularize due to tidal forces, and the system has sufficient time to travel to 3 kpc outside the Galactic plane before the $2\,M_{\odot}$ begins to overflow its Roche lobe and the system becomes an X-ray source. As a matter of fact, the system had probably already completed several oscillations around the Galactic plane before it became an X-ray source.

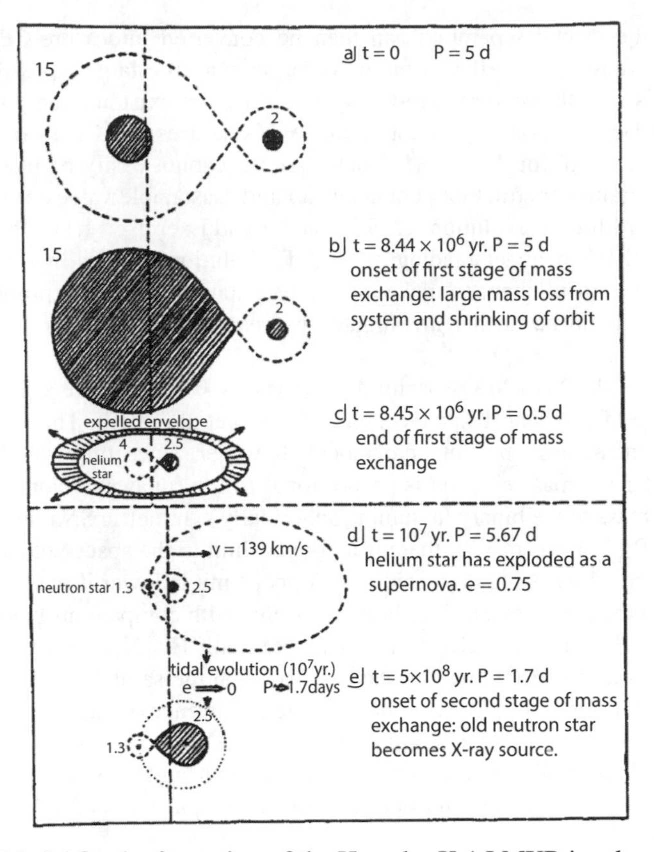

Figure 11.1. Model for the formation of the Hercules X-1 LMXB involves a CE phase in which the 2 $M_\odot$ secondary spiraled down into the envelope of its original massive primary companion. In this model, proposed by Sutantyo (1975a) as depicted by van den Heuvel (1976), the initial orbital period is 5d; it is nowadays thought that it should be longer than 1 yr.

Important subsequent work on LMXB formation, partly building forth on the Sutantyo model but in which birth kicks to the NS are included, was done by Kalogera and Webbink at the University of Illinois (Kalogera & Webbink, 1996, 1998; Kalogera, 1998). This shows that indeed LMXBs with lower mass donor stars than in the Her X-1 system can also be formed in this way. There may here, however, be difficulties in forming LMXBs with donor stars of initial masses below 1 $M_\odot$ for the following reasons (see van den Heuvel, 1994a): the minimum post spiral-in period of a system consisting of a helium star plus the low-mass companion (M_2) is given by the condition that the latter star should not overflow its Roche lobe. This yields for a given value of M_2 and of the helium core mass M_{He} a minimum orbital period. Using the equations for CE evolution given in Chapter 4, one can calculate for these mass combinations the ratio a/a_0 of the initial and final orbital separations (as $M_{\mathrm{He}} = M_{\mathrm{core}}$), and the envelope mass of the massive star is uniquely determined by $M_{\mathrm{env}} = M_1 - M_{\mathrm{He}}$. The thus obtained pre-CE value a_0 of the orbital separation can then be converted into a pre-CE Roche-lobe radius of the massive star. If this radius is larger than the largest possible red-giant radius for this star, the required pre-CE system does not exist and the post-CE binary with a donor star of mass M_2 cannot be made. If one uses this CE formalism, it does turn out that, indeed, for $M_2 \leq 2\, M_\odot$ such systems cannot easily be made for a realistic massive giant-star structure parameter (λ) and reasonable values of the efficiency parameter ($\alpha_{\rm CE}$) for CE evolution (cf. Section 4.8 and Exercise 11.1). However, it may very well be that our understanding of the CE evolution process in very wide binaries at present is insufficient to make a definitive statement on this problem. Clearly, this problem should be further investigated by realistic 3D radiation-hydrodynamics simulations.

The bulk of the NS-LMXBs (with donor masses $< 1\, M_\odot$) have z-distances to the Galactic plane of less than 1 kpc (see, e.g., Repetto et al., 2017). This cannot be used as an argument against the Sutantyo model, however, since the orbital eccentricity induced by the SN mass ejection is proportional to the runaway velocity imparted to the center of mass of the binary (assuming spherically symmetric SN mass ejection, see Section 4.3.10). For this reason, there is an upper limit to the space velocity that a system can obtain before being disrupted. This upper limit is lower if the mass M_2 of the companion to the NS is lower. Thus bound systems with companions below $1\, M_\odot$ will in general have lower space velocities than a system like Her X-1 with a $2\, M_\odot$ companion, such that these LMXBs will not be able to reach the same high z-distance above the Galactic plane as Her X-1. Also, the possible occurrence of large birth kicks to NSs in their formation events may not be able to carry LMXBs to z-distances beyond 1 kpc because large kicks will also have unbound the systems, and the systems that remained bound thanks to smaller kick did not get very large space velocities (Repetto et al., 2017).

A slightly updated standard formation path leading to production of LMXBs with donor masses $\leq 1\, M_\odot$, and their further evolution, is depicted in Figure 11.2. Most noteworthy is the inclusion of Case BB RLO prior to the SN in relatively tight systems. The further evolution of LMXBs leading to ultra-compact X-ray binaries (UCXBs) and radio millisecond pulsars (MSPs) is discussed in Section 11.4 and Chapter 14, respectively.

11.1.3 Accretion-induced Collapse (AIC)

Observational Evidence

The general question of the origin of NSs is closely related to many of their observable parameters: spin, B-field, age, space velocity, and the nature of their companion star. The observational evidence suggested in the literature for NSs formed via AIC can be categorized into different pieces described next (see Tauris et al., 2013, for a full review), including the formation of some odd NSs both in globular clusters (GCs, Chapter 12) and in the Galactic disk.

In Table 11.1, we list a number of apparently young and non-recycled NSs (characterized by slow spin and relatively high B-fields) that are found in GCs. The lifetime as

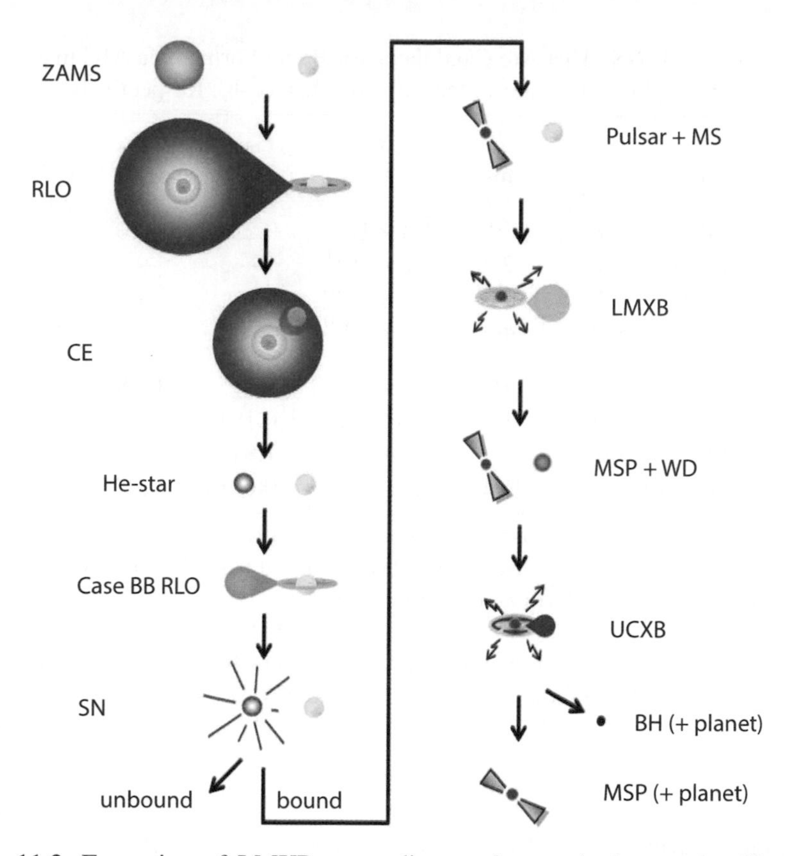

Figure 11.2. Formation of LMXBs according to the standard scenario. Close-orbit LMXBs may later evolve into an UCXB and eventually leave behind a planet companion, become an isolated MSP, or possibly collapse into a BH. More quantitative details of the evolution are given in Figure 14.10. Acronyms used in this figure—ZAMS: zero-age main sequence; RLO: Roche-lobe overflow (mass transfer); CE: common envelope; SN: supernova; LMXB: low-mass X-ray binary; MSP: millisecond pulsar; WD: white dwarf; UCXB: ultra-compact X-ray binary; BH: black hole.

an observable radio source is of the order of 100 Myr for a young (i.e., non-recycled) pulsar. Therefore, if these NSs had formed via iron-core collapse SNe, their existence in GCs would not only be unlikely for kinematic reasons (kicks imparted on such newborn NSs would be larger than the escape velocity of the GCs), it would simply be impossible given that the stellar progenitor lifetimes of SNe II and SNe Ib/c are less than a few times 10 Myr, much shorter than the age of the many Gyr old stellar populations in GCs. Similarly, the nuclear evolution timescales of stars undergoing electron capture SNe (EC SNe), which are expected to produce small kicks (Chapter 13), is of the order of 20–50 Myr, which is still short compared to the age of GCs, and for this reason also EC SNe cannot explain the existence of young NSs in GCs today. It

Table 11.1. NSs That Are Candidates for Being Formed via AIC in a Globular Cluster (a–d) or in the Galactic Disk (e–h), Respectively

Object	P ms	B^* G	$P_{\rm orb}$ days	$M^{**}_{\rm comp}$ $M_\odot$	Ref.
PSR B1718−19	1004	4.0×10^{11}	0.258	∼0.10	a
PSR J1745−20A	289	1.1×10^{11}	—	—	b
PSR J1820−30B	379	3.4×10^{10}	—	—	c
PSR J1823−3021C	406	9.5×10^{10}	—	—	d
GRO J1744−28	467	1.0×10^{13}	11.8	∼0.08	e
PSR J1744−3922	172	5.0×10^{9}	0.191	∼0.10	f
PSR B1831−00	521	2.0×10^{10}	1.81	∼0.08	g
4U 1626−67	7680	3.0×10^{12}	0.028	∼0.02	h

See text for explanations and discussion. After Tauris et al. (2013).

* B-field values calculated from Eq. (5) in Tauris et al. (2012), which includes a spin-down torque due to a plasma-filled magnetosphere.

** Median masses calculated for $i=60°$ and $M_{\rm NS}=1.35\,M_\odot$.

(a) Lyne et al. (1993); (b) Lyne et al. (1996); (c) Biggs et al. (1994); (d) Boyles et al. (2011); (e) van Paradijs et al. (1997); (f) Breton et al. (2007); (g) Sutantyo & Li (2000); (h) Yungelson et al. (2002).

is therefore clear that these NSs in GCs, if they are truly young[1], are formed via a different channel.

The evidence for AIC is found not only in GCs. In Table 11.1 we list a number of Galactic disk binary NS systems that are postulated candidates for having formed via AIC. A common feature of these NSs is a slow spin and a relatively high B-field and an ultra light ($\leq 0.10\,M_\odot$) companion star in a close orbit. The idea that the origin of some high B-field, slow spinning NSs (e.g., 4U 1626−67 and PSR B0820+08) is associated with AIC was originally suggested by Taam & van den Heuvel (1986). Although it was believed at that time that B-fields decay spontaneously on a timescale of only 50 Myr (and therefore these NSs could not have much larger ages), many of these sources remain good candidates for AIC today, even though it has been demonstrated that pulsar B-fields can remain high on much longer timescales (Kulkarni, 1986; Bhattacharya et al., 1992). One reason these NS systems remain good AIC candidates is that their companion stars have very small masses, indicating that a significant amount of material ($0.5-1.0\,M_\odot$) was transferred toward the compact object. The paradox is therefore that these NSs still have high B-fields and slow spins even though a significant mass transfer has occurred.

[1]For an alternative view, see Verbunt & Freire (2013) who argue that these NSs that appear to be young are not necessarily young.

There is solid observational evidence that the surface B-field strengths of NSs decrease with accretion (Taam & van den Heuvel, 1986; Shibazaki et al., 1989; van den Heuvel & Bitzaraki, 1994, 1995). The exact mechanism for this process is still unknown. It may be related to decay of crustal fields by ohmic dissipation and diffusion from heating via nuclear processing of accreted material (Romani, 1990; Geppert & Urpin, 1994; Konar & Bhattacharya, 1997), burial (screening) of the field (Zhang, 1998; Cumming et al., 2001; Payne & Melatos, 2007), or decay of core fields due to flux tube expulsion from the superfluid interior induced by rotational slowdown in the initial phases of mass accretion (Srinivasan et al., 1990); see also the review by Bhattacharya (2002). Even a small amount of material accreted may lead to significant B-field decay, contradicting the observational evidence that these binary NSs have accreted large amounts of material.

There are two possible solutions to the previously mentioned paradox of observing close binaries with ultra-light companions orbiting high B-field NSs: i) the relativistic pulsar wind of the young, fast-spinning, high B-field NS, during the first hundreds of thousands of years after the NS formed, has ablated a large amount of mass from the companion, that is, in the way as envisaged by van den Heuvel & van Paradijs (1988); or ii) these NSs were formed recently via AIC during the mass-transfer process in a binary. For normal binary evolution models (binary interactions in globular clusters are discussed in Chapter 12), the latter scenario requires fine-tuning of the AIC event to occur near the termination of the mass-transfer phase, which is important for preventing accretion of significant amounts of matter after the formation of the NS (resulting in low B-fields and fast spin). This issue was investigated by Tauris et al. (2013) who computed systems with main-sequence stars, helium stars, and giant stars as donors. They noted that a general problem with postulating that a given observed high-B-field NS was formed via AIC is that (as pointed out by Wijers [1997] and also demonstrated by Sutantyo & Li [2000]) it requires quite some fine-tuning to have the AIC occurring at the very final phases of the mass transfer to explain the high B-field of the NS. If the WD collapses earlier, the B-field of the newly formed NS (and its spin period, depending on its mass-accretion rate) should decrease significantly when the donor star refills its Roche lobe following the AIC, thereby evolving through a relatively long-lasting (up to 10^7-10^9 yr) post-AIC LMXB phase.

Figure 11.3 shows examples of mass transfer in pre-AIC binaries (i.e., resembling super-soft X-ray systems, see Section 5.4.3) with $2.5\,M_\odot$ main-sequence donor stars and a $1.2\,M_\odot$ ONeMg WD accretor (treated as a point mass). It is noted that by the time the WD reaches a critical Chandrasekhar mass and undergoes AIC, the donor star still has a remaining envelope mass of $\sim 0.8-1.0\,M_\odot$. This material will be transferred to the newborn NS once the donor star refills its Roche lobe after $\lesssim 10^5$ yr, following the instantaneous widening of the orbit due to the AIC process. Thus, in all the shown models, the newly formed NSs are able to accrete sufficient mass in the post-AIC LMXBs to become fully recycled MSPs and thereby become indistinguishable from MSPs formed via the standard LMXB channel. For main-sequence star donors, typically $0.03-0.3\,M_\odot$ will be accreted by the NS (Fig. 11.4), all traces of its origin will be erased, and the final NS cannot be distinguished from those recycled pulsars formed via the standard SN and CE channel. Even accretion of a few $0.01\,M_\odot$ is

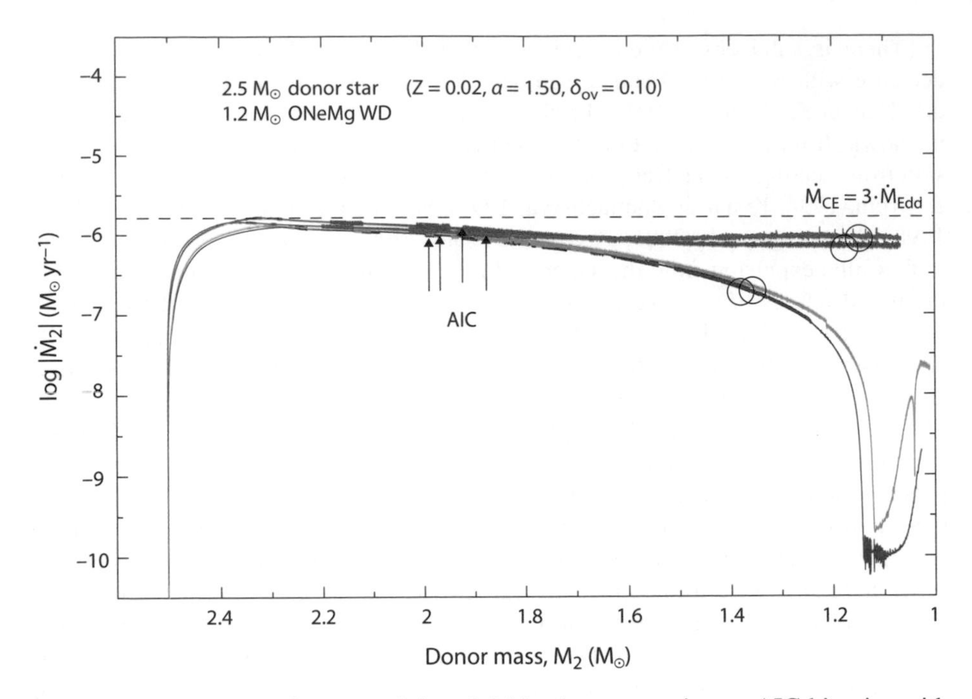

Figure 11.3. Mass-transfer rates of four 2.5 $M_\odot$ donor stars in pre-AIC binaries with initial orbital periods between 1.0 – 3.0 days leading to post-AIC LMXB systems. The initial mass of the WD is assumed to be 1.2 $M_\odot$. The arrows mark the collapse of the accreting WD (AIC) and the evolution from this point onward (ignored in this plot) is computed first by the widening of the orbit as a consequence of the AIC event, and thereafter by post-AIC accretion once the donor star refills its Roche lobe after $\lesssim 10^5$ yr. The open circles indicate hypothetical super-Chandrasekhar WD masses of 2.0 $M_\odot$. The dashed line is the adopted upper limit for stable mass transfer (here assumed to be $\dot{M}_{\rm WD,max} = 3\,\dot{M}_{\rm Edd}$, above which a CE is assumed to be formed). A slightly more massive donor star would lead to excessive mass-transfer rates and thus possibly not result in an AIC event. After Tauris et al. (2013).

enough to decrease the B-field significantly according to some models (e.g., Wijers, 1997; Zhang & Kojima, 2006, and references therein). In all model calculations of Tauris et al. (2013) the newborn NS undergoes post-AIC accretion, although in some rare cases, with giant-star or helium-star donors, less than $10^{-3}\,M_\odot$ is accreted (Fig. 11.4), enabling the NS to retain a high B-field.

Nevertheless, in our view, the presence of several young pulsars in GCs, in which star formation ceased over 10 Gyr ago, can only be explained by relatively recent AIC of massive WDs in binary systems. The situation is here, however, very different from normal binary evolution, since in GCs the binaries consisting of a massive WD and a normal main-sequence star usually formed by dynamical interactions (i.e., collisions; see Chapter 12). When such a close binary formed, the Roche lobe of the main-sequence star at the time of its formation may be smaller than the radius

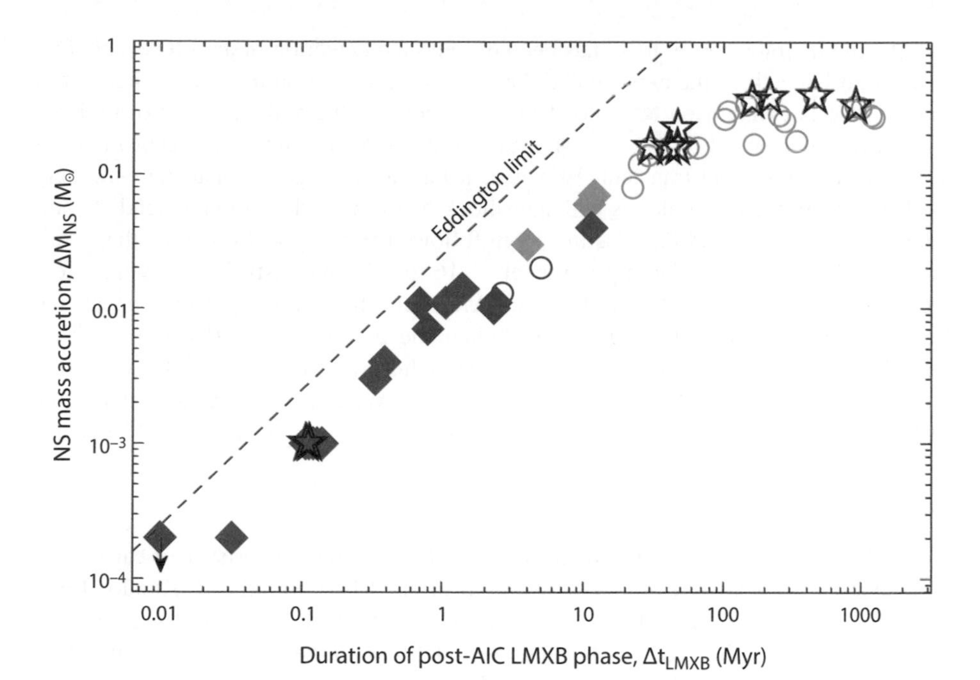

Figure 11.4. The amount of mass accreted by the NSs during the post-AIC LMXB phase, $\Delta M_{\rm NS}$, as a function of the duration of the post-AIC LMXB phase, $\Delta t_{\rm LMXB}$. The three different donor star progenitor classes (green, red, and blue) correspond to main-sequence, giant-star, and helium-star donors, respectively. Open circles are final MSPs with He WD companions, and solid diamonds indicate MSPs with (hybrid) CO WD companions. Symbols with a superimposed open black star correspond to AIC models where a kick was applied at NS birth. The calculations assumed an accretion efficiency of 30%. The deviation from a straight line is caused by the lower mass-transfer rates $|\dot{M}_2| \ll \dot{M}_{\rm Edd}$ in systems with long mass-transfer timescales. After Tauris et al. (2013).

of the star, such that a large mass-transfer rate may ensue, even while the companion has a low mass. This large mass-transfer rate then induces the AIC, leading to the formation of a young pulsar plus a low-mass companion, like in the GC pulsar binary PSR B1715−19. It also is possible, with an almost head-on collision, that the Roche lobe in the newly formed system would be very much smaller than the radius of the main-sequence star and the latter star completely destroyed, forming a massive disk around the WD—the mass transfer from this disk would trigger the AIC, leading to a single new GC pulsar resembling the three known single young GC pulsars. In all of these cases, the AIC is expected to be triggered by the mass transfer once the WD mass reaches the Chandrasekhar-mass limit of ~1.44 $M_\odot$ (or 1.48 $M_\odot$ for a rigidly rotating WD (Yoon & Langer, 2004); differential rotation can persist if the timescale of angular momentum transport is smaller than the accretion timescale in accreting WDs and leads to a critical mass that is significantly higher than the canonical Chandrasekhar mass (e.g., Yoon & Langer, 2004).

Despite its importance, there has been no reported detection of an AIC event. This is not surprising since the expected AIC rate is no more than at most a few percent of that of SNe Ia (Yungelson & Livio, 1998). In addition, the small amount of radioactive nickel synthesized ($\lesssim 10^{-3}\,M_{\odot}$; Dessart et al., 2006; Kitaura et al., 2006) implies a relatively dim optical transient. However, it has been suggested that the $\sim 10^{50}$ erg explosion from the AIC collides with and shock-heats the surface of an extended companion, creating an X-ray flash lasting ~ 1 hr followed by an optical signature that peaks at an absolute magnitude of approximately -16 to -18 and lasts for a few days to a week (Piro & Thompson, 2014). Moreover, AIC events have recently been suggested to be associated with the following observed phenomena: γ-ray bursts (Lyutikov & Toonen, 2017), X-ray transients (Yu et al., 2019), radio-bright transients (Moriya, 2019) and fast radio bursts (FRBs) (Margalit et al., 2019). We can therefore hope that some of the many high-cadence sky surveys will soon detect a strong candidate AIC event.

Theoretical Work

The AIC formation model was already suggested long ago by theorists (for early work and reviews, see Canal et al., 1990; Nomoto et al., 1991; Nomoto & Kondo, 1991; Nomoto & Yamaoka, 1992; Woosley et al., 1992; Timmes & Woosley, 1992). Specifically for LMXBs, the AIC NS formation model was suggested already in the mid-1970s by van den Heuvel (1976, 1977) and was further worked out by other groups. The theoretical work can be divided into two main branches: i) investigations of binary evolution, and ii) electron-capture and explosion physics. For binary evolution aspects, the great majority of work related to accreting WDs has been focusing on the much-related topic of CO WDs evolving toward SNe Ia, as described in Chapter 13. For work specifically aimed at AIC (either from the single or double-degenerate scenario), we refer to the works by, for example, Tauris et al. (2013); Wang (2018); Ruiter et al. (2019); Wang & Liu (2020), and references therein. For relatively recent papers related to the physics of the AIC, we encourage interested readers to see, for example, Shen & Bildsten (2007) for thermally stable nuclear burning on accreting WDs; Schwab et al. (2015) for thermal runaway during the evolution of ONeMg cores toward AIC; Brooks et al. (2017) for AIC with a helium-star donor; Jones et al. (2016, 2019) for the issue of ONeMg deflagration and total or partial thermonuclear disruption vs. NS formation via AIC or EC SNe; Kirsebom et al. (2019) and Suzuki et al. (2019) for electron-capture physics and EC SNe; Zha et al. (2019) for EC SNe vs. thermonuclear disruption; Leung et al. (2019) for AIC of dark matter admixed WDs; and Tauris & Janka (2019) for a debate on formation of low-mass NSs or Fe-WD remnants from partially thermonuclear EC SNe.

There is consensus in the field that an AIC event is not very different from an EC SN, and it is therefore expected that also NSs formed via AIC will receive a small kick (if any significant kick at all [Kitaura et al., 2006; Dessart et al., 2006]), and thus NSs formed via AIC will have a high chance of remaining inside a GC where the escape velocity is $< 50\,\mathrm{km\,s^{-1}}$. The theoretical understanding of the fate of dense stellar cores, however, is still rather uncertain and controversial. For example, for AICs and EC SNe, the question of thermonuclear disruption or not depends on the central density of the

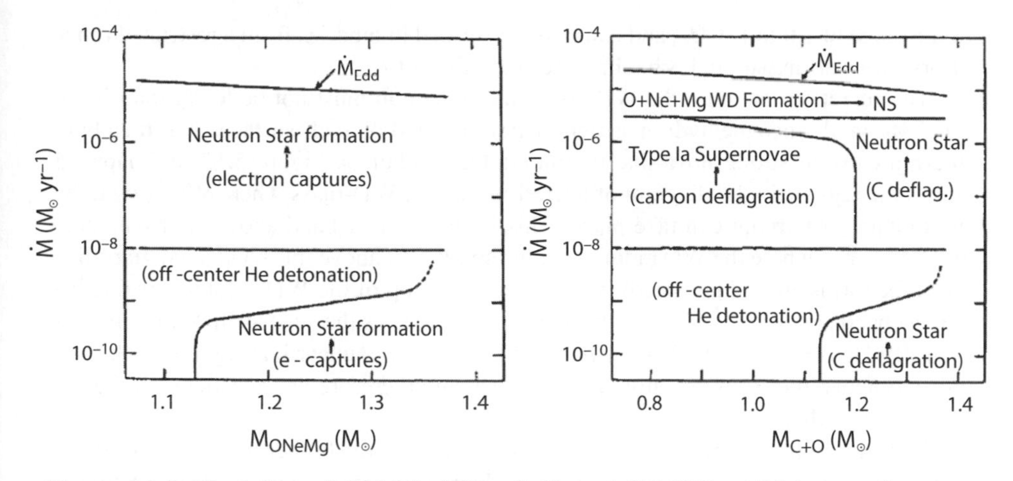

Figure 11.5. Final fate of ONeMg WDs (left) and CO WDs (right), as a function of accretion rate and initial WD mass, according to Nomoto & Kondo (1991). In ONeMg WDs, the collapse to a NS is triggered by electron captures on ^{24}Mg and ^{20}Ne. In CO WDs, in two regions of the diagram carbon deflagration is ignited at central densities as high as $\rho_c = 10^{10}$ g cm^{-3}, probably leading to NS formation by electron captures on oxygen. In CO WDs with masses below 1.2 $M_\odot$, and/or at low accretion rates, C-deflagration leads to the explosion of the CO WD as a SN Ia, which may or may not leave a small remnant behind. After Nomoto & Kondo (1991).

degenerate ONeMg core at oxygen ignition, which again depends on the core growth rate, convection and semi-convection, and the relevant microphysics such as electron capture rates and Coulomb effects.

Conditions for Growth of a WD via Mass Transfer

It appears that for AIC to really work and produce a NS, many boundary conditions have to be fulfilled. These include the mass and chemical composition of the WD, the mass-transfer rate from its companion star, and possibly the age of the WD, which affects the sinking of heavier elements (O,Ne,Mg) to its center and thus the electron-capture collapse. We now elaborate more on the basics of some of these conditions.

ONeMg WDs present the most favorable conditions for AIC: the required mass-transfer rate is at least 10^{-8} $M_\odot$ yr^{-1}, as depicted in Figure 11.5, after Nomoto & Kondo (1991). Also, massive CO WDs may, under certain circumstances, be triggered to undergo electron-capture collapse, but here very much fine-tuning is required since under most circumstances these stars will C-deflagrate and no, or only a small, remnant will be left. This is most likely what occurs in Type Ia SNe (see Fig. 11.5 and Chapter 13).

In view of the very special conditions required for AIC of CO WDs, ONeMg WDs present the most realistic possibility for NS formation via AIC in binaries, first elaborated in detail by Nomoto & Kondo (1991). Such WDs are formed with masses mostly in the range of 1.0 to 1.3 $M_\odot$ and can only be formed in binaries from a restricted range of primary star masses and orbital periods (see Fig. 11.6).

The accretion rate onto the WDs from a companion must not be lower than $0.5 \times 10^{-7}\ M_\odot\ \mathrm{yr}^{-1}$ since the hydrogen-rich matter accreted on their surfaces will then burn intermittently in strong nova-like thermonuclear outbursts. Figure 5.13 in Chapter 5 shows the regions in the diagram of accretion rate vs. WD-mass where WD growth by thermonuclear burning can take place. This is in a narrow band above approximately $10^{-7}\ M_\odot\ \mathrm{yr}^{-1}$ where the WD radius remains small, and above this band a red giant-like envelope forms around the steady-burning layer on top of the WD. Systems inside the narrow band above $10^{-7}\ M_\odot\ \mathrm{yr}^{-1}$ have been identified as the super-soft X-ray binaries (van den Heuvel et al., 1992; see Chapter 5). As stable nuclear burning takes place in those systems, the WDs steadily grow in mass and therefore are excellent candidates for reaching AIC.

Concerning the formation of a red-giant envelope at the higher mass-transfer rates: in a close binary there is no room for a red giant, so in systems with these higher mass-transfer rates, the envelope around the WD will exceed the Roche-lobe radius, and the envelope may eventually be expelled in the form of a high-velocity stellar wind (Hachisu & Kato, 2001), causing mass loss from the system possibly while the WD mass keeps growing at a small rate. For investigations of the evolution of these systems with a specific emphasis on the strong winds from the accreting WD, see Hachisu et al. (1996, 1999, 2012). Alternatively, at high mass-transfer rates, a CE forms that could potentially lead to a coalescence (Cassisi et al., 1998; Langer et al., 2000). Constraints on the wind mass loss from an accreting WD can be determined directly from radio observations of SN Ia remnants (Chomiuk et al., 2012).

The mass-transfer rates in excess of $10^{-7}\ M_\odot\ \mathrm{yr}^{-1}$ required for the WD mass to grow are never found in cataclysmic variables (CVs) with short orbital periods and low-mass main-sequence donors, which form the bulk of all CVs. In these systems, the observed relatively low mass-transfer rates in the range of $10^{-10} - 10^{-8}\ M_\odot\ \mathrm{yr}^{-1}$ lead to nova-like outbursts, in which probably all the accreted mass is lost, such that the WD mass will not experience a net growth. However, main-sequence star donors with masses $M_2 \simeq 1.3 - 2.7\ M_\odot$ (depending on metallicity and mass of the accreting WD) can easily deliver the required mass-transfer rate in excess of $10^{-7}\ M_\odot\ \mathrm{yr}^{-1}$ (see Fig. 11.3). The only CVs with low-mass donors ($M_2 \lesssim 1.3\ M_\odot$) in which WD growth may be possible are the few longer-period systems in which the donor is a subgiant or a giant star, and mass transfer is driven by internal nuclear evolution in these stars, such as in the Nova Per 1901 system ($P_{\mathrm{orb}} = 1.996803$ d) and the recurrent nova systems such as RR Tel and T Cor Bor, in which the donor star is a red giant. We will consider the evolution of such systems in a subsequent section. These systems will then eventually produce binaries consisting of a red (sub)giant and a NS, and may then later evolve into LMXBs with relatively wide orbits, such as Cyg X-2 and the symbiotic X-ray binaries.

From binary evolution population synthesis computations, one can, in principle, derive the rates of AIC for ONeMg WDs for different binary formation channels. For the channel with a main-sequence donor, as well as for the channel with a giant donor, the thus computed formation rate is $< 10^{-5}\ \mathrm{yr}^{-1}$ (S. Toonen, 2020, private communication; see also Yungelson & Livio, 1998). There are a number of other AIC channels,

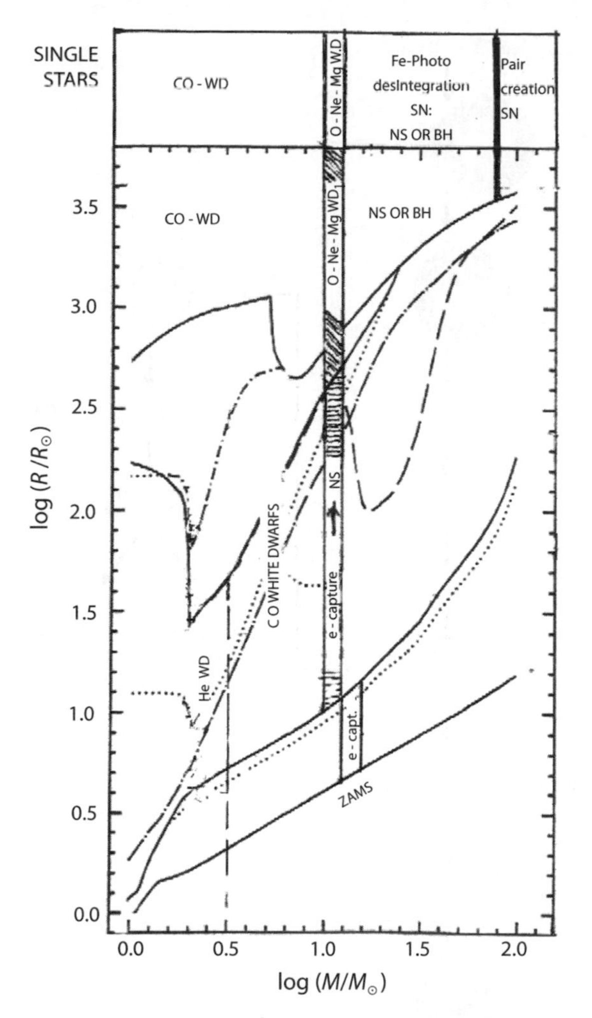

Figure 11.6. Schematic classification of expected final evolutionary states of primary stars of close binaries as a function of mass and radius of the Roche-lobe filling primary star at the onset of mass transfer, for solar metallicity. For a given binary mass ratio, the radius and mass of the Roche-lobe filling star can be converted into an orbital period (see Fig. 9.5). If mass transfer starts on the main sequence (Case A), higher primary star masses are required to produce the same remnants as in Cases B or C. At the top of the figure, the final remnants of single stars are indicated. In a subsequent stage of reverse mass transfer, a CO WD might be triggered to explode as a SN Ia (complete disruption). Some CO WDs with a mass close to the Chandrasekhar limit, as well as ONeMg WDs, may possibly be triggered to implode to form a NS (see text). Adapted after Ge et al. (2015).

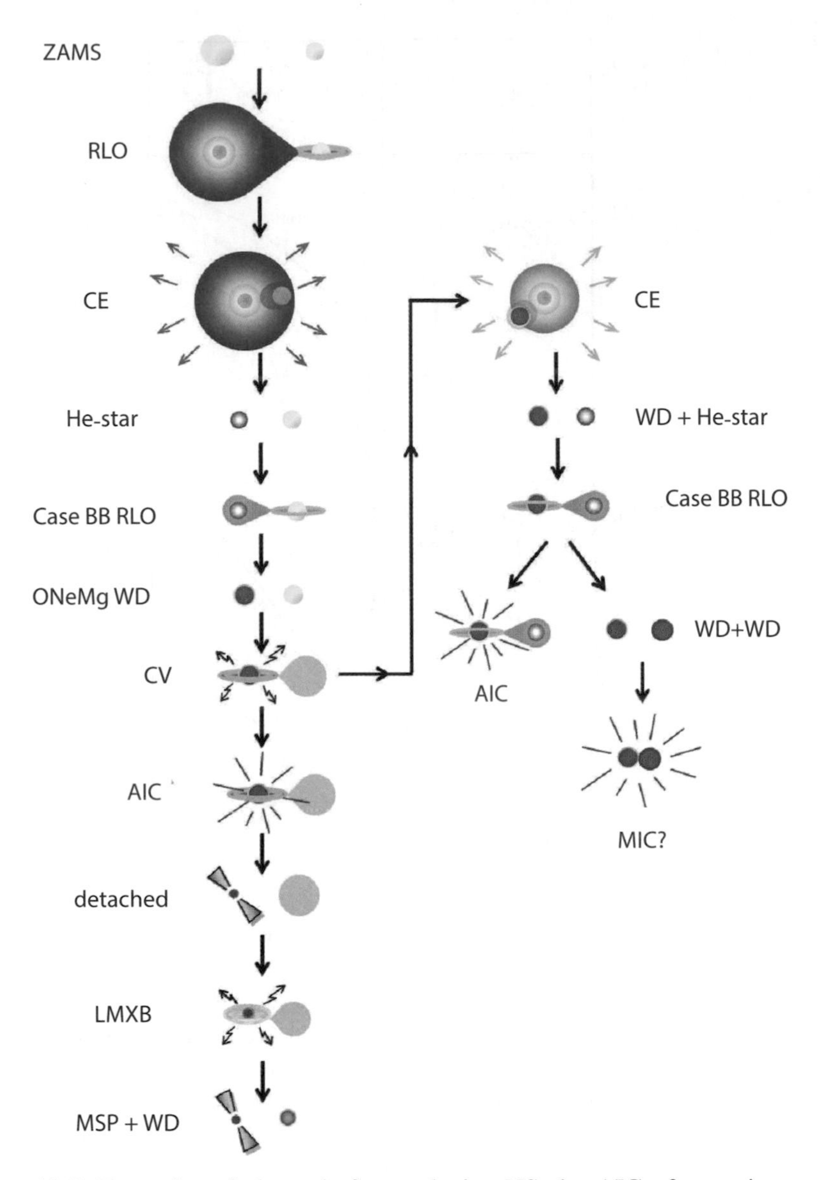

Figure 11.7. Examples of channels for producing NSs by AIC of accreting massive WDs with main-sequence, helium-star or WD companions. Post-AIC evolution is only shown for the case of a main-sequence companion star—see text. Acronyms used in this figure—ZAMS: zero-age main sequence; RLO: Roche-lobe overflow (mass transfer); CE: common envelope; WD: white dwarf; CV: cataclysmic variable; AIC: accretion-induced collapse; LMXB: low-mass X-ray binary; MSP: millisecond pulsar; MIC: merger-induced collapse.

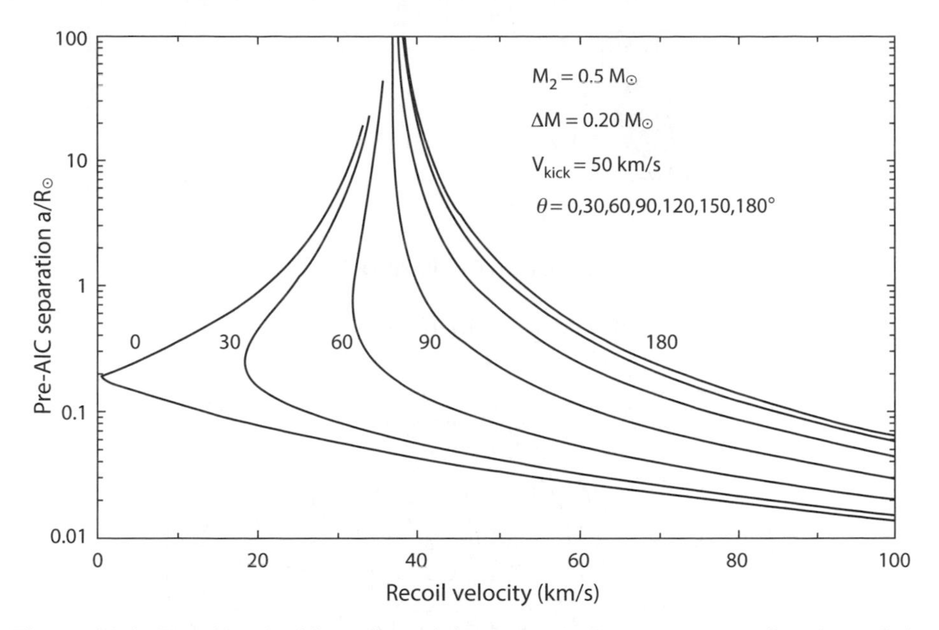

Figure 11.8. Recoil velocities of surviving post-AIC systems as a function of the pre-AIC orbital separation and direction of applied kick angle, $\theta = 0-180°$ (see Chapter 13 for details of geometry). The donor star is assumed to have a mass of $M_2 = 0.5\,M_\odot$ at the time of the AIC, and the assumed instantaneous mass loss from the collapsing $1.4\,M_\odot$ WD (from release of gravitational binding energy) is taken to be $\Delta M = 0.2\,M_\odot$. A kick of 50 km s^{-1} is applied. Main-sequence and giant donor stars require pre-AIC orbits $> R_\odot$ and $> 10\,R_\odot$, respectively. After Tauris & Bailes (1996).

that is, with a helium-star donor (Brooks et al., 2017) or a giant-star donor, but it is not clear if these will result in an observable LMXB. Figure 11.7 depicts examples of possible formation channels leading to AIC as well as a merger-induced collapse (MIC), which has also been suggested to produce a NS remnant, especially in globular clusters (Ivanova et al., 2008). Many more scenarios are discussed in the recent review by Wang & Liu (2020).

Early work on the dynamical effects of AIC on the surviving binary was made by Taam & Fryxell (1984) who carried out 2-D hydrodynamical calculations on the shell impact effect on a red dwarf-like star in a close binary. They found that if the ejection velocity is $< 1.5 \times 10^4$ km s^{-1} and the mass of the ejected shell is $< 0.1\,M_\odot$, companions with masses in the range $0.2 - 1.0\,M_\odot$ survive the explosion, and the systems are accelerated to velocities < 18 km s^{-1}. As this is smaller than the escape velocity from the core of a globular cluster, such systems are not expected to escape from the clusters. Moreover, there is, in fact, no reason to expect that in AIC formation of a NS any baryonic mass is ejected. Therefore, and since AIC is not expected to impart a birth velocity to a NS, it is indeed a most plausible mechanism to explain the existence of NSs in globular clusters. Even if AICs would produce NS kicks of 50 km s^{-1}, Tauris & Bailes (1996) demonstrated (Fig. 11.8) that also in this case the majority of post-AIC

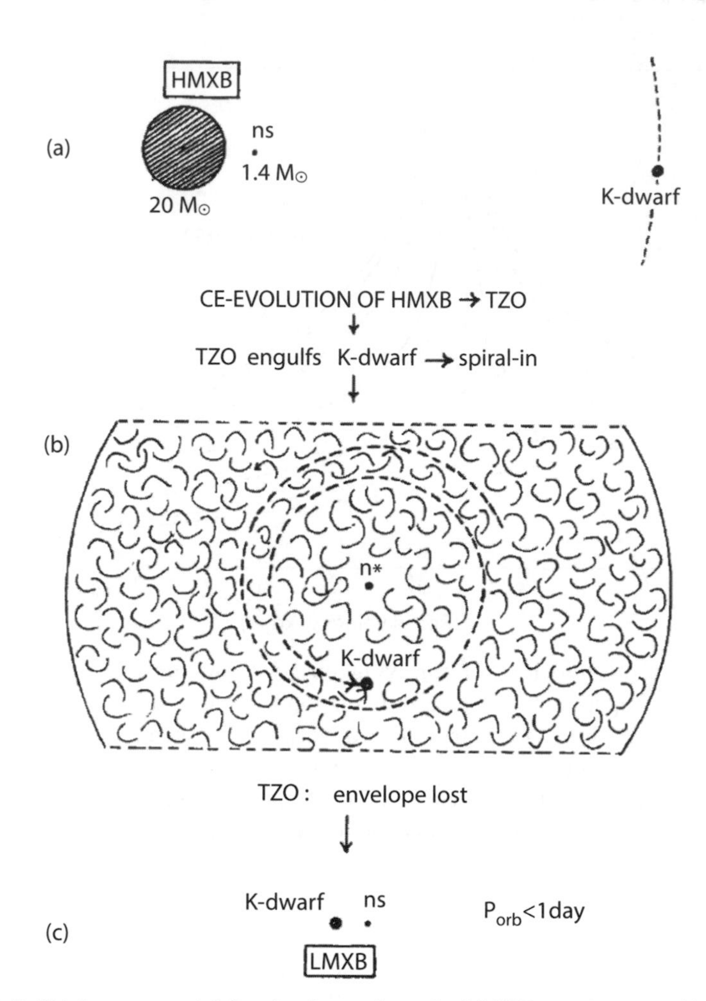

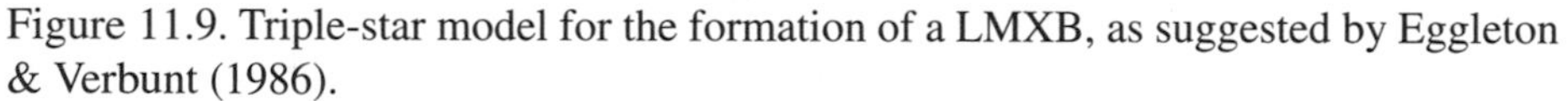
Figure 11.9. Triple-star model for the formation of a LMXB, as suggested by Eggleton & Verbunt (1986).

systems with main-sequence or giant-star companions would remain in the cluster. For recent literature on electron-capture collapse of ONeMg cores and on AIC of ONeMg WDs, we refer to Chapter 13.

11.1.4 Triple-star Model for Formation of LMXBs

This model was suggested by Eggleton & Verbunt (1986). It makes use of the fact that the later evolution of NS-HMXBs, with orbital periods shorter than approximately one year, leads to the formation of a Thorne-Żytkow object (TZO, Thorne & Żytkow, 1975, 1977): a red supergiant with a NS in its center, see Figure 11.9. So, if a relatively close-orbit HMXB has a low-mass ($\leq 1\ M_{\odot}$) tertiary companion at a distance of a few AU, this companion may be engulfed by the TZO's envelope and spiral down into this envelope as depicted in the figure. If the mass-ejection rate from the TZO envelope during the spiral-in is such that the envelope is lost just by the time that the orbital period of the low-mass star has decreased to less than a day, a close system will remain, consisting of a NS with a low-mass companion.

It is therefore clear that this model will require much fine-tuning, since for a star with a mass $\leq 1\ M_{\odot}$ the drop in orbital binding energy during the CE in-spiral is insufficient for removing the giant's envelope, unless it is close to its very maximum extent (see Fig. 4.29).

Only a tiny fraction of systems of the type depicted in Figure 11.9 are expected to be able to fulfill this condition, and, in most of the systems, the low-mass companion will spiral-in completely and only a single object will remain.

11.2 ORIGIN OF LMXBs WITH BLACK HOLES

All of the BH-LMXBs are transients, and, based on their observed numbers in the Solar neighborhood and their recurrence times, their total number in the Galaxy has been estimated to be between 100 and 1,000 (Tanaka & Lewin, 1995; McClintock & Remillard, 2006). Also LMC X-3 can considered to be a BH-LMXB, as here the B3V donor is less massive than the $> 5.9\ M_{\odot}$ BH. For the origin of the BH-LMXBs, of course, the WD AIC model is not possible, and neither is an AIC of a NS. The latter because the masses of all BHs in these systems are more likely $\geq 5\ M_{\odot}$. The low-mass companions (less than $\sim 2\ M_{\odot}$) do not have sufficient mass to have been able to increase the $\leq 2\ M_{\odot}$ mass of a NS to a $5\ M_{\odot}$ BH. Also, the triple-star model is not possible for formation of BH-LMXBs since TZOs cannot have a BH in their center. So, the only remaining model for forming BH-LMXBs is a model similar to the Sutantyo/Kalogera model for the formation of Her X-1 and NS-LMXBs via CE evolution.

11.2.1 Common-envelope Evolution of a Wide Massive Binary with an Extreme Initial Mass Ratio

To produce a BH-LMXB, the massive progenitor star must here be massive enough to leave a BH as remnant. From the evolutionary calculations of Ugliano et al. (2012); Pejcha & Thompson (2015); Sukhbold et al. (2016, 2018); Ebinger et al. (2020), it is apparent that already stars with ZAMS masses as low as $\sim 12 - 15\ M_{\odot}$ (for metallicity $Z = Z_{\odot}$) may occasionally leave BHs as remnants (see Chapter 8), while a gray zone exists for stars with ZAMS masses all the way up to $\sim 30\ M_{\odot}$, where the outcome can be either NSs or BHs, depending on the detailed core structure and explosion physics. So, the model for forming BH-LMXBs indeed is very similar to Sutantyo's formation model of Her X-1. Even the mass of the red giant in whose envelope the low-mass main-sequence companion spirals down during the CE phase is here quite similar, in the range of 12 to 30 $M_{\odot}$. The only difference is that here the core of the red giant collapses to a BH.

A problem, however, is that in a sizeable number of the BH-LMXBs the present donor star is an M- or K-dwarf with a mass below 1 $M_{\odot}$. Similar to the problem we encountered for the formation of the NS–LMXBs discussed previously, the big question is if the mass of such a dwarf star is too low to be able to drive off the CE of the red giant companion. Indeed, it may well be that only red giants that are already on the asymptotic giant branch, where the envelope has become very loosely gravitationally bound (see Fig. 4.29 and discussions in Section 10.11), are suitable for this kind of evolutionary model. Our understanding of the CE ejection process in very wide binaries is at present insufficient to make a more definitive statement on this problem, and more precise computations for this evolutionary phase are required.

11.3 MECHANISMS DRIVING MASS TRANSFER IN CLOSE-ORBIT LMXBs AND CVs

As the winds of low-mass stars are never very strong (except perhaps when they reach the asymptotic giant branch), in general, mass transfer by Roche-lobe overflow (RLO) is required to power these sources. The companion is here generally less massive than the NS (or BH), so the mass transfer will be stable since it leads to an increase of the orbital separation and Roche-lobe radius. The only exception is Her X-1, where the companion has a mass of $\sim 2\,M_\odot$. This system is in fact a boundary case between LMXBs and IMXBs (see also Chapter 6); here beginning atmospheric RLO can power the source for hundreds of thousands of years before it reaches the Eddington limit (see Savonije, 1978, 1983).

Assuming the donor star (star 2) to fill its Roche lobe, the orbital period $P_{\rm orb}$ is directly related to its average mass density, as was derived in Chapter 4. Eq. (4.9) can be rewritten as:

$$P_{\rm orb} \approx 9\,{\rm hr}\left(\frac{\rho_2}{\rho_\odot}\right)^{-1/2}, \tag{11.1}$$

where ρ_2 and $\rho_\odot$ are the mean mass density of the donor star and of the Sun, respectively. Equation (11.1) directly indicates what kind of mass donors fit in systems with known orbital periods. Three types of donor stars, given in table 11.2 and graphically

Table 11.2. Approximate mass-radius Relations and Orbital Periods for Roche-lobe Filling Low-mass ($\leq 1\,M_\odot$) Stars in Thermal Equilibrium

Main sequence	Helium main sequence	Degenerate star
$R_2 = M_2$	$R_2 = 0.2\,M_2$	$R_2 = 0.013\,(1+X)^{5/3}\,M_2^{-1/3}$
$P_{\rm orb} = 9\,{\rm hr}\cdot M_2$	$P_{\rm orb} = 0.9\,{\rm hr}\cdot M_2$	$P_{\rm orb} = 48\,{\rm sec}\cdot(1+X)^{5/2}\,M_2^{-1}$

During the mass transfer the thermal equilibrium may gradually be disturbed, such that deviations from these relations become possible). Donor star radii (R_2) and masses (M_2) are in units of $R_\odot$ and $M_\odot$, respectively. The fractional hydrogen abundance is indicated as X. After Verbunt (1990).

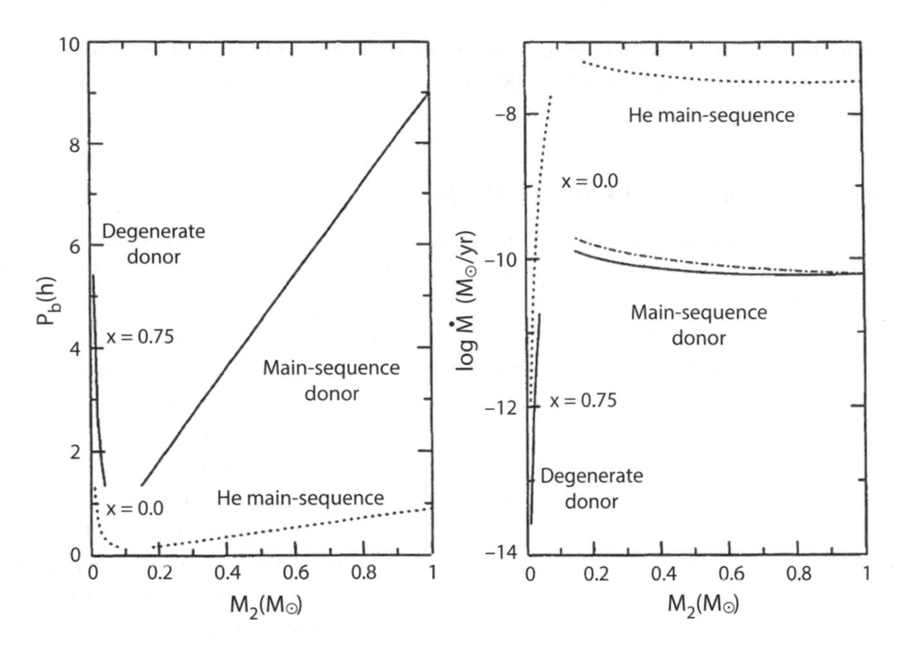

Figure 11.10. Orbital period (left panel) and mass-transfer rate (right panel) in LMXBs that evolve by means of angular momentum loss by GW radiation, as a function of decreasing mass of the donor star M_2 (evolution from right to left), with a $1.4\,M_\odot$ compact accretor. It is assumed that all transferred matter is accepted by the accretor (conservative mass transfer), except for the case of the dash-dotted curve, where it is assumed that half the transferred matter leaves the system with the specific orbital angular momentum of the main-sequence donor. Donors are assumed to be in thermal equilibrium. Helium stars ($X = 0$) have shorter orbital periods and higher mass-transfer rates than hydrogen-rich stars ($X = 0.75$). After Verbunt (1993).

represented in Figure 11.10, are possible for short-period (< 10 hr) systems. The table and the figure show that systems with orbital periods shorter than 80 min most probably contain helium stars or degenerate stars.

Binaries with initial orbital periods in the range of 80 min to 10 hr are expected to contain normal (hydrogen-rich) main-sequence stars, that is: unevolved solar-type stars. LMXBs with orbital periods longer than 10 hr cannot contain Roche-lobe filling unevolved (i.e., main-sequence) stars. If the donor is a star of $\sim 1\,M_\odot$ or less, as clearly is the case in systems such as Sco X-1 ($P_{\rm orb} = 0.78$ d), Cyg X-2 ($P_{\rm orb} = 9.8$ d), and X0921−63 ($P_{\rm orb} = 9.0$ d), it must be an evolved star, that is: a (sub-)giant. One usually assumes a NS mass of $\sim 1.4\,M_\odot$. The X-ray luminosities of the bright LMXBs are mostly in the range of $10^{36} - 10^{38}$ erg s^{-1} corresponding to an accretion rate of 10^{-10} to $10^{-8}\,M_\odot\,{\rm yr}^{-1}$. With a donor mass of $\sim 1\,M_\odot$, this allows for a lifetime of 100 Myr to 10 Gyr, if most of the donor star were to be consumed. This is some 3 to 5 orders of magnitude longer than the expected lifetime of a standard HMXB (Chapter 10). As the numbers of bright LMXBs and standard HMXBs in the Galaxy are similar to one another, this implies that the formation rate of bright LMXBs is some 3 to 5 orders of magnitude lower than that of the HMXBs, as we already mentioned at the beginning of this chapter.

11.3.1 Converging vs. Diverging LMXBs

Before considering how the observed mass-transfer rates can be obtained, we first summarize the two main evolutionary classes of LMXBs. It has been shown by Pylyser & Savonije (1988, 1989) that a critical orbital bifurcation period ($P_{\rm bif}$) exists at the onset of the RLO, separating the formation of *converging* LMXBs from *diverging* LMXBs. The converging systems evolve with decreasing $P_{\rm orb}$, roughly until the mass-losing component becomes degenerate and an ultra-compact binary is formed (see later in this section. and in Section 11.4), whereas in the diverging systems the orbit eventually expands until the mass-losing star has lost its hydrogen-rich envelope and a wide-orbit binary with a detached WD is formed (Fig. 14.13). The theoretically estimated value of $P_{\rm bif}$ is $\sim$1 day but depends strongly on the treatment of tidal interactions and the assumed strength of magnetic braking, which drains the system of orbital angular momentum (Verbunt & Zwaan, 1981; van der Sluys et al., 2005; Ma & Li, 2009; Istrate et al., 2014; Van et al., 2019). An illustration of the bifurcation effect is shown in Figure 11.11. The outcome of a LMXB evolution is therefore strongly dependent on the initial orbital period (Fig. 11.12). Istrate et al. (2014) argued that with current modelling, severe fine-tuning is needed to reproduce the observed population of resulting binary MSPs with orbital periods between 2 and 9 hr (i.e., systems driven by a combination of orbital angular momentum loss and internal evolution). The final fate of LMXBs is often radio MSPs: either wide-orbit MSPs with He WD companions or tight-orbit MSPs with low-mass ($< 0.20\,M_\odot$) He WDs (which may later evolve into UCXBs), ultra-light companions, planets, or solitary MSPs. The formation processes leading to MSPs are described in detail in Chapter 14.

11.3.2 Mass Transfer Driven by Losses of Orbital Angular Momentum —The Basics

In the following, we will largely stick to a basic analysis as outlined by Verbunt (1990). It is based on simple polytropic stellar mass-radius relations and disregards mass loss from the binary system and includes spin-orbit coupling through the magnetic braking assumptions (Verbunt & Zwaan, 1981). This is excellent for pedagogic reasons and demonstrates some key features of mass transfer in binaries. For examples of more realistic results based on modern stellar evolution codes, see, for example, Tauris et al. (2000); Podsiadlowski et al. (2002) and Misra et al. (2020).

For simplicity we thus tentatively assume that the total mass of the binary is conserved, that is, the mass lost by the donor (star 2) is accreted onto the compact companion, such that $\dot{M}_1 = -\dot{M}_2$. (For a straightforward generalization for the case of mass loss from the system, see, e.g., Rappaport et al. [1982]; the outcome is in most cases

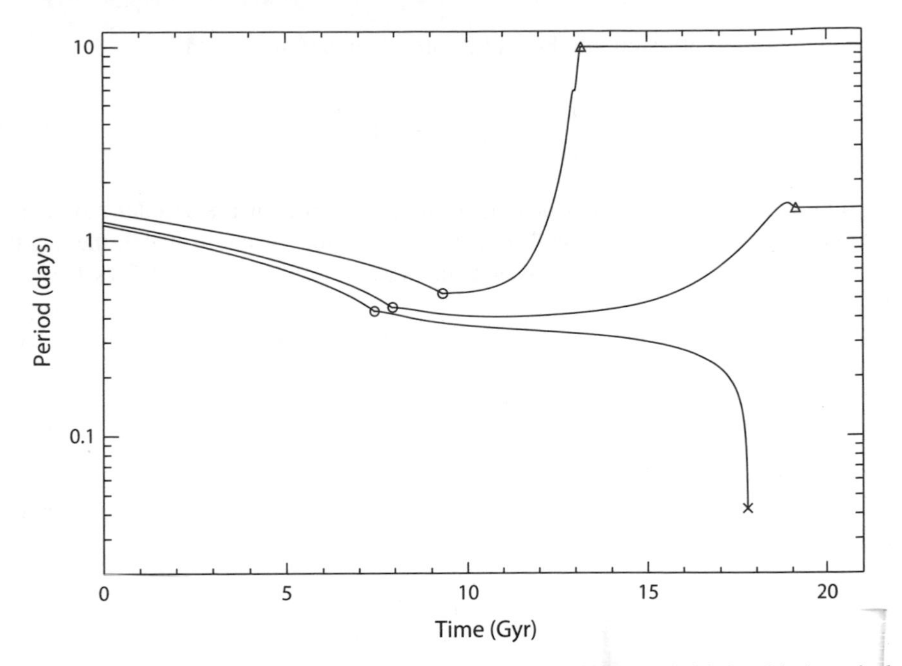

Figure 11.11. LMXB tracks calculated for slightly different initial orbital periods. The existence of a bifurcation period, separating converging from diverging systems, is clearly seen. The exact orbital evolution depends on the applied magnetic braking model. The mass of the donor star is $1.0\,M_{\odot}$. Using more massive donors ($1.1-1.5\,M_{\odot}$) will terminate the evolution within a Hubble time. After Ma & Li (2009).

not much different.) From Eq. (4.29) and assuming circular orbits one derives:

$$\frac{\dot{a}}{a} = 2\frac{\dot{J}_{\rm orb}}{J_{\rm orb}} - 2\left(1 - \frac{M_2}{M_1}\right)\frac{\dot{M}_2}{M_2}. \tag{11.2}$$

Since we designated the donor as star $2(\dot{M}_2 < 0)$ a condition for stable mass transfer is that $M_2 < M_1$, because in this case the orbital radius (and the Roche-lobe radius) increases when mass is being transferred. To examine whether the mass transfer is indeed stable, that is, if the condition is also sufficient, one has to compare the change in radius of the donor with that of its Roche lobe when mass is being transferred. The second expression of Eq. (4.7) for the Roche-lobe radius is:

$$\frac{R_L}{a} = 0.462\left(\frac{M_2}{M_1 + M_2}\right)^{1/3}. \tag{11.3}$$

Substituting this expression into Eq. (11.2) one obtains (Exercise 11.2):

$$\frac{\dot{R}_L}{R_L} = 2\,\frac{\dot{J}_{\rm orb}}{J_{\rm orb}} - 2\left(1 - \frac{M_2}{M_1}\right)\frac{\dot{M}_2}{M_2} + \frac{\dot{M}_2}{3\,M_2}, \tag{11.4}$$

which gives the change of the Roche-lobe radius in terms of the angular momentum loss rate from the system. We will now consider various ways in which the angular momentum can be lost from the system in the case of LMXBs and CVs (see also Chapter 4).

If one writes the mass-radius relation of the donor star as $R_2 = C\,M_2^n$, one will have:

$$\frac{\dot{R}_2}{R_2} = n\,\frac{\dot{M}_2}{M_2}. \tag{11.5}$$

To have stable mass transfer, R_2 should remain equal to R_L (and thus $\dot{R}_2 = \dot{R}_L$), which implies that Eqs. (11.4) and (11.5) together yield:

$$\frac{\dot{J}_{\rm orb}}{J_{\rm orb}} = \left(\frac{5}{6} + \frac{n}{2} - \frac{M_2}{M_1}\right)\frac{\dot{M}_2}{M_2}. \tag{11.6}$$

Since both GW radiation and magnetic braking cause the orbital angular momentum to decrease, in the case of mass transfer, the left side of this equation is negative and the same holds for $\dot{M}_2$. For stable mass transfer this implies that:

$$q = \frac{M_2}{M_1} < \frac{5}{6} + \frac{n}{2}. \tag{11.7}$$

For a given value of n, this sets a limit to the mass ratio above which the mass transfer is unstable. For example, for a non-relativistic degenerate star donor (as in an ultra-compact X-ray binary), one has $n = -1/3$, and $q = M_2/M_1 < 2/3$ is required for stable mass transfer (however, a more refined treatment yields smaller critical q-values [van Haaften et al., 2012; Bobrick et al., 2017]). On the other hand, for a main-sequence donor $n = 1$, and the mass transfer is found to be stable for $q = M_2/M_1 < 4/3$ (again here, a more realistic treatment shows that RLO is actually stable for much larger mass ratios, up to $q \lesssim 3$; see Section 11.6).

If the mechanism of orbital angular momentum loss is prescribed, the evolution of the rate of mass transfer can be calculated with Eq. (11.6). The two mechanisms that are generally accepted to drive the mass transfer in close ($P_{\rm orb} < 10\,$hr) systems are the orbital angular momentum loss by i) *gravitational wave (GW) radiation* and ii) *magnetic braking*, already briefly mentioned in Section 4.3.3.

In the case of angular momentum losses by GW radiation only, one has (Chapter 15):

$$\frac{\dot{J}_{\rm GW}}{J_{\rm orb}} = -\frac{32G^3}{5c^5}\,\frac{M_1 M_2 (M_1 + M_2)}{a^4}. \tag{11.8}$$

However, most often in tight systems with low-mass non-degenerate stars, the loss of orbital angular momentum is dominated by magnetic braking, which is a spin-orbit coupling we now discuss in more detail.

11.3.3 Magnetic Braking

Magnetic braking takes place when the Roche-lobe filling donor star is a dwarf (solar-like) main-sequence star with a convective envelope. Such stars are observed to have surface magnetic fields and hot coronae, with a temperature of several 10^6 K, that produce a stellar wind similar to the solar wind. Studies of solar-like stars of different ages (e.g., in the Pleiades and Hyades star clusters) show that their rotation is slowed down in the course of time. This is generally attributed to the fact that angular momentum is carried away by the stellar wind, which, out to a certain distance (5 to 10 stellar radii or more), is forced by the magnetic fields of the star to co-rotate with the star. This so-called *magnetically-coupled stellar wind* involves negligible mass loss (of the order of $10^{-13}\, M_\odot\, \mathrm{yr}^{-1}$) but provides an important drain on the spin angular momentum of the star (Mestel, 1968; Skumanich, 1972).

In a close binary, the spin angular momentum loss of a low-mass donor star ($\lesssim 1.5\, M_\odot$) by a magnetically coupled wind causes the rotation of the star to slow down. On the other hand, the tidal forces in a close system are strong and will continuously work to spin the star back up into co-rotation with the orbital revolution. This spin-up happens at the expense of the orbital angular momentum (i.e., there is a spin-orbit transfer of orbital angular momentum into spin angular momentum), and thus the magnetically coupled stellar wind indirectly produces a drain on the orbital angular momentum of the system, as was first realized for LMXBs and CVs by Verbunt & Zwaan (1981). As a result, the orbital separation decreases with time.

Using the observed magnetic braking laws for solar-type stars given by Skumanich (1972), one obtains numerically for the orbital angular momentum loss rate:

$$\dot{J}_{\mathrm{mb}} = -0.5 \times 10^{-28}\, f^{-2} I_2 R_2^2\, \omega^3, \tag{11.9}$$

where I_2 and R_2 are the moment of inertia and the radius of the donor star, ω, is the orbital angular velocity and f is a constant of order unity (Skumanich, 1972; Smith, 1979). I_2 is given by:

$$I_2 = k^2\, M_2 R_2^2, \tag{11.10}$$

where k is the *radius of gyration* of the star and M_2 its mass. A more generalized braking law for binaries with solar-like stars by Rappaport et al. (1983) yields:

$$\frac{\dot{J}_{\mathrm{mb}}}{J_{\mathrm{orb}}} = -3.8 \times 10^{-30}\, f\, \frac{R_\odot^4\, (R_2/R_\odot)^\gamma\, GM^2}{a^5\, M_{\mathrm{NS}}} \quad \mathrm{s}^{-1}, \tag{11.11}$$

where γ is the magnetic braking index ($\gamma \simeq 2-5$), which expresses the dependence of the braking on the radius of the donor star. This law assumes that the rotational slow-down of young G-stars involves the entire star and not just its convective envelope.

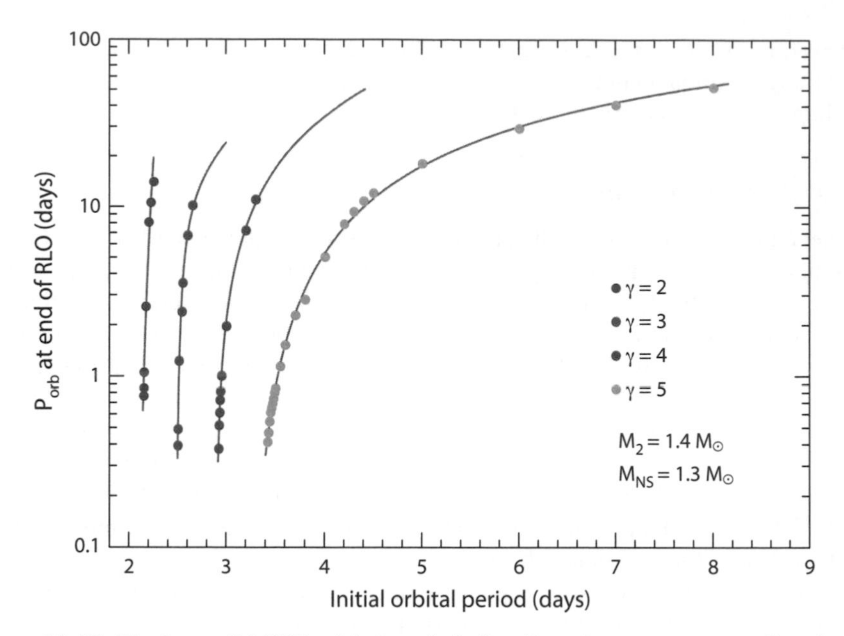

Figure 11.12. Final post-LMXB orbital periods for diverging systems as a function of initial orbital periods (roughly at the ZAMS of the donor star) for different values of the magnetic braking index ($\gamma \simeq 2-5$). It is seen how the orbital period fine-tuning problem becomes worse for smaller values of γ. After Istrate et al. (2014).

The importance of γ for the orbital evolution of LMXBs is shown in Figure 11.12. The net effect of applying the previously described prescription for magnetic braking is that the orbital period of relatively close ($P_{\rm orb} \simeq 2-5$ d) pre-LMXBs (or pre-Algol systems) typically decreases by a factor of three to five (depending on γ) prior to the RLO and onset of LMXB phase, that is, magnetic braking causes orbital decay and forces the donor star to fill its Roche lobe and initiate mass transfer already on the main sequence or early into the Hertzsprung gap—see Figure 9.16 for an analogue evolution in a pre-Algol system. For further and more recent investigations of magnetic braking laws, see, for example, Sills et al. (2000); van der Sluys et al. (2005); Knigge et al. (2011); and Istrate et al. (2014). Van et al. (2019) investigated modifications for the magnetic braking law and found that including a scaling of the magnetic field strength with the convective turnover time, and a scaling of magnetic braking with the wind mass-loss rate, can better reproduce persistent LMXBs and does a better job at reproducing transient LMXBs. On the other hand, it was demonstrated by Chen H.-L., et al. (2021) that this magnetic braking prescription has problems reproducing wide-orbit radio MSPs.

11.3.4 Results for Main-sequence Donors

Equating Eqs. (11.6) and (11.8) and eliminating a by using Eq. (11.3) together with the mass-radius relation $R_2 = C M_2{}^n$, one obtains the mass-transfer rate $\dot{M}_2$ as a function of M_1 and M_2, for the case of mass transfer *driven purely by GW radiation losses*. For a main-sequence donor, assuming $n = 1$ and $C = 1$ and applying $M_1 = 1.4\,M_\odot$ (NS accretor), one then obtains for M_2 in the range 0.1 to 1 $M_\odot$ a mass-transfer rate of the order of $10^{-10}\,M_\odot\,\mathrm{yr}^{-1}$, as depicted in Figure 11.10. This figure shows that a binary with an original donor mass of 1 $M_\odot$ will evolve from $P_{\mathrm{orb}} = 9\,\mathrm{hr}$ to $P_{\mathrm{orb}} < 3\,\mathrm{hr}$, when $M_2 < 0.3\,M_\odot$, and that the mass-transfer rate produced by GW radiation losses alone is always $\lesssim 10^{-10}\,M_\odot\,\mathrm{yr}^{-1}$. This will produce an X-ray luminosity of the order of $10^{36}\,\mathrm{erg\,s}^{-1}$.

A sizable number of observed LMXBs, however, have X-ray luminosities between 10^{37} and $10^{38}\,\mathrm{erg\,s}^{-1}$. To explain these values, additional angular momentum loss mechanisms should be considered (see Eq. 4.30), of which magnetic braking is the most plausible one, or these systems have more evolved donor stars. The first study of LMXBs to fully include magnetic braking was done by Rappaport et al. (1983), as mentioned in the last section. For a more complete study, see, for example, Tauris & Savonije (1999); Podsiadlowski et al. (2002); and Istrate et al. (2014). Figure 6.10 demonstrates such calculated values of $|\dot{M}_2|$ for both converging and diverging LMXBs. Here, at the onset of the mass transfer, the rate is very high, $>10^{-8}\,M_\odot\,\mathrm{yr}^{-1}$, but rapidly tapers off to a low rate ($\sim 10^{-10}\,M_\odot\,\mathrm{yr}^{-1}$) for a very long period of time (see also Fig. 11.17). The reason for the high values of $|\dot{M}_2|$ at the onset of RLO is that the donor star is initially driven out of thermal equilibrium. The central entropy is smaller in less massive main-sequence stars, and thus the mass-losing star has to radiate away extra energy to remain in thermal equilibrium. As a result, the donor star is initially inflated in size for roughly a thermal timescale (Eq. 8.12), thereby enhancing its mass-transfer rate.

11.3.5 The Period Gap for CV Binaries

For CVs there are very few systems with orbital periods between 2 and 3 hr (i.e., the *period gap*, see Fig. 11.13), whereas many systems are found with periods above and below this period gap. A plausible explanation of this gap was proposed by Rappaport et al. (1982); Spruit & Ritter (1983); and Verbunt (1984), involving that as a result of the mass-loss history the typically $\sim 0.3\,M_\odot$ donor star (at that epoch) deviates strongly from thermal equilibrium and has an inflated radius. It was argued that when M_2 has decreased to $\sim 0.3\,M_\odot$, the star becomes completely convective, upon which magnetic braking switches off[2] and the mass-transfer rate drops by an order of magnitude. This drop in the mass-transfer rate allows the donor to shrink back to thermal equilibrium, causing it to detach itself from the Roche lobe, upon which the mass transfer stops entirely. Further decay of the orbit due to GW radiation losses causes the orbit to shrink so far that the secondary star, with its reduced (thermal equilibrium) radius, once again overflows its Roche lobe. This happens when the orbital period is $\sim 2\,\mathrm{h}$. This magnetic-braking theory thus qualitatively explains why there is an orbital period gap for the population of CV binaries.

---

[2]This issue is still a matter of debate, see Section 11.3.6.

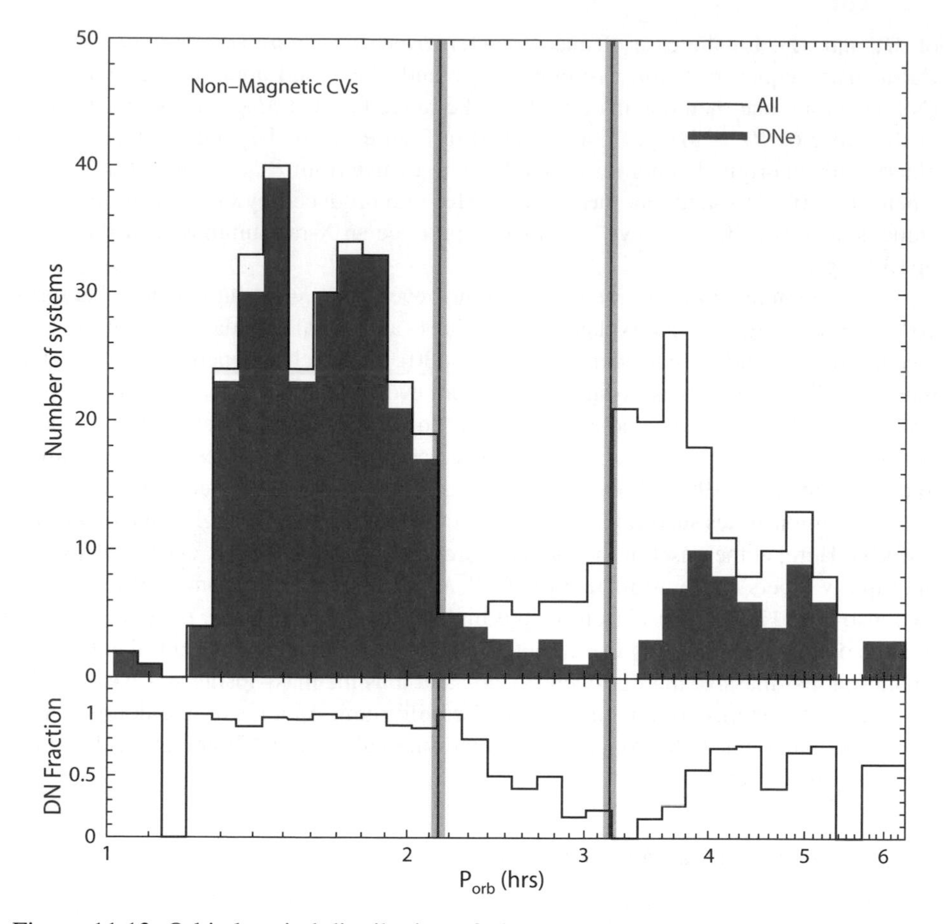

Figure 11.13. Orbital period distribution of observed (non-magnetic) CVs, and in red color the fraction of these systems which exhibit dwarf novae (DNe) eruptions. After Knigge et al. (2011). See also Figure 5.6.

11.3.6 The Period Minimum of CVs

The absence of CVs with periods below $P_{\rm orb} = 80$ min (see Fig. 11.13) has traditionally been explained by the reversal of the mass-radius relation around this period, at which point the hydrogen-rich donor, with its reduced mass, becomes degenerate and begins to follow a mass-radius relation of the form $R_2 \propto M_2^{-1/3}$ (see Fig. 11.10). For hydrogen-rich donors in close mass-transfer binaries, the minimum possible binary period was therefore argued to be around 80 min (Paczynski & Sienkiewicz, 1981). This minimum period can only be reached if the donor star is not driven far out of thermal equilibrium by the preceding mass loss. If this is the case, the minimum period becomes longer. It turns out that to reach the observed period minimum of $\sim$80 min, the orbital angular momentum loss by magnetic braking must be negligible in comparison to the angular momentum loss by GW radiation in secondaries $< 0.3\,M_\odot$ (Paczynski & Sienkiewicz, 1981; Rappaport et al., 1983). In this classical picture, after passing the orbital period minimum, the mass-loss rate in the out-spiraling systems (the *period bouncers*) drops to very small values, as can be seen in Figure 11.10, and the evolutionary timescale becomes longer than the Hubble time. These systems will therefore be difficult to observe as CVs.

The basic magnetic-braking scenario outlined in the previous paragraph has been recognized as the standard picture for 30 yr. However, it has been recognized for some time now that there may be issues with this picture. Knigge et al. (2011) and the references therein point to three problems challenging the standard picture: (1) the revised theoretically predicted minimum orbital period is only ~ 37 min (see the next paragraph), and thus substantially shorter than the observed one; (2) the standard model predicts that the Galactic CV population should be completely dominated by short-period systems and period bouncers, which is not compatible with observations even when selection effects are considered; (3) there is substantial evidence that fully convective low-mass stars (and perhaps even brown dwarfs) can sustain significant magnetic fields and thus should still operate a magnetic-braking mechanism. Thus we conclude that while magnetic braking must play an important role for explaining the orbital period distribution, including the period gap, additional ingredients are needed to fully explain the observed population of CVs, in particular, the observed orbital period minimum. For a detailed discussion, see the comprehensive review by Knigge et al. (2011).

The Real Minimum Orbital Period of Binaries with a Hydrogen-Rich Donor Star

As already alluded to in Section 4.1, in binary systems with RLO there is a relation between donor stellar structure and orbital period. Obviously, more dense donor stars can fit into tighter orbits. The minimum orbital period for non-degenerate hydrogen-rich donor stars has in recent years been found to be $P_{\rm orb}^{\rm min} \sim 37$ min (Nelson et al., 2018; Rappaport et al., 2021). Considering the transition with decreasing mass from dwarf stars to brown dwarfs and giant gas planets, the value of the mean mass density of the donor, $\langle\rho_2\rangle$ increases gradually from a few g cm^{-3} ($\langle\rho_\odot\rangle = 1.4$ g cm^{-3}), peaking above 200 g cm^{-3} for the most massive brown dwarfs ($M_2 \simeq 0.07\,M_\odot$), before declining back to a few g cm^{-3} for Jupiter-like objects. Thus, we expect the lowest-mass hydrogen-rich stars (at the transition limit to brown dwarfs) to be donors in the binary systems of the shortest orbital periods. We found earlier (Eq. 11.1) that for RLO: $P_{\rm orb} \propto \langle\rho_2\rangle^{-1/2}$. However, deriving the exact scaling factor in simple fitting formulae relating $P_{\rm orb}^{\rm min}$ and $\langle\rho_2\rangle$ can be somewhat tricky. As pointed out by Rappaport et al. (2021), brown dwarfs (and giant gas planets) are not substantially centrally concentrated to be described well by simple polytropic models of index $n = 3$ or $n = 3/2$. These objects are better described by $n = 1$ polytropes. For comparison, Chandrasekhar (1933) found that the ratios between central density and average density, $\rho_c/\langle\rho\rangle$, are: 54.2 ($n = 3$), 5.99 ($n = 3/2$), and 3.29 ($n = 1$); whereas for an incompressible fluid ($n = 0$) the ratio is obviously 1. Notice that the usual expressions for the Roche geometry and the critical equipotential surfaces are also derived assuming pure synchronous rotation—see Leahy & Leahy (2015) for discussions of rotational effects.

Consequential Angular Momentum Loss?

Despite the success of the basic model for CV formation and evolution, a number of discrepancies persist between observations and theoretical modelling related to WD masses, location of the orbital period minimum, and the space density of such systems—see, for example, Zorotovic & Schreiber (2020), Metzger et al. (2021), and references therein. It has been argued (Schreiber et al., 2016) that these tensions are alleviated if CVs experience an additional sink of orbital angular momentum loss beyond magnetic braking and GWs, referred to as *consequential angular momentum loss* (CAML). One proposed candidate is related to classical novae. Following such a thermonuclear outburst from the surface of the WD, the bloated WD envelope may engulf the companion star, causing a gas drag similar to the situation in a CE (Chapter 4). The outcome may lead to a merger of the low-mass star donor and the WD accretor (Metzger et al., 2021), thereby removing a number of sources from the CV population.

11.3.7 Model Calculations of CV and AM CVn Systems

A detailed calculation of tight binaries that evolve and produce, first, a CV system, then detach, before evolving to an AM CVn system is shown in Figure 11.14; see Tauris (2018) for further details. The progenitor of the first system (top panel) is a 1.40 $M_\odot$ main-sequence star orbiting a 0.7 $M_\odot$ CO WD (treated as a point mass) with an initial orbital period of 3.64 days. After orbital decay caused by magnetic braking and the subsequent CV phase, the system detaches again with a 0.162 $M_\odot$ He WD orbiting a 1.07 $M_\odot$ WD with an orbital period of 4.76 hr. Over the next 4.1 Gyr, this double WD system spirals in farther due to emission of low-frequency GWs with a constant chirp mass, $\mathcal{M} = 0.335\, M_\odot$ (Chapter 15), until the low-mass He WD fills its Roche lobe (at $P_{\rm orb} = 15.7$ min and $T_{\rm eff} = 9\,965$ K) and initiates mass transfer to the CO WD and the binary becomes observable as an AM CVn system. The final calculated model is composed of a $6.26 \times 10^{-3}\, M_\odot$ planet-like donor object, orbiting a 1.11 $M_\odot$ WD with an orbital period of 1.12 hr and a stellar age of 14.3 Gyr.

In the second model (bottom panel), the aim was to reproduce some of the known detached double WDs systems and AM CVn binaries that have lower WD masses (Kupfer et al., 2018) (i.e., total system masses below 0.9 $\rm M_\odot$). To achieve this, a progenitor system is chosen with a 1.40 $M_\odot$ MS star orbiting a 0.5 $M_\odot$ CO WD with an initial orbital period of 3.9482 days. Assuming rather inefficient accretion onto the WD ($\beta = 0.80$), the system detaches after the CV phase with a 0.160 $M_\odot$ He WD orbiting a 0.706 $M_\odot$ CO WD with an orbital period of 4.81 hr. Over the next 5.5 Gyr, the detached system spirals in further due to GW radiation with a constant chirp mass, $\mathcal{M} = 0.278\, M_\odot$, until the low-mass He WD fills its Roche lobe (at $P_{\rm orb} = 12.4$ min and $T_{\rm eff} = 8\,999$ K) and initiates mass transfer to the CO WD as an AM CVn system. The final calculated model is composed of a $6.81 \times 10^{-3}\, M_\odot$ planet-like donor object, orbiting a 0.736 $M_\odot$ WD with an orbital period of 1.06 hr and a stellar age of 17.4 Gyr.

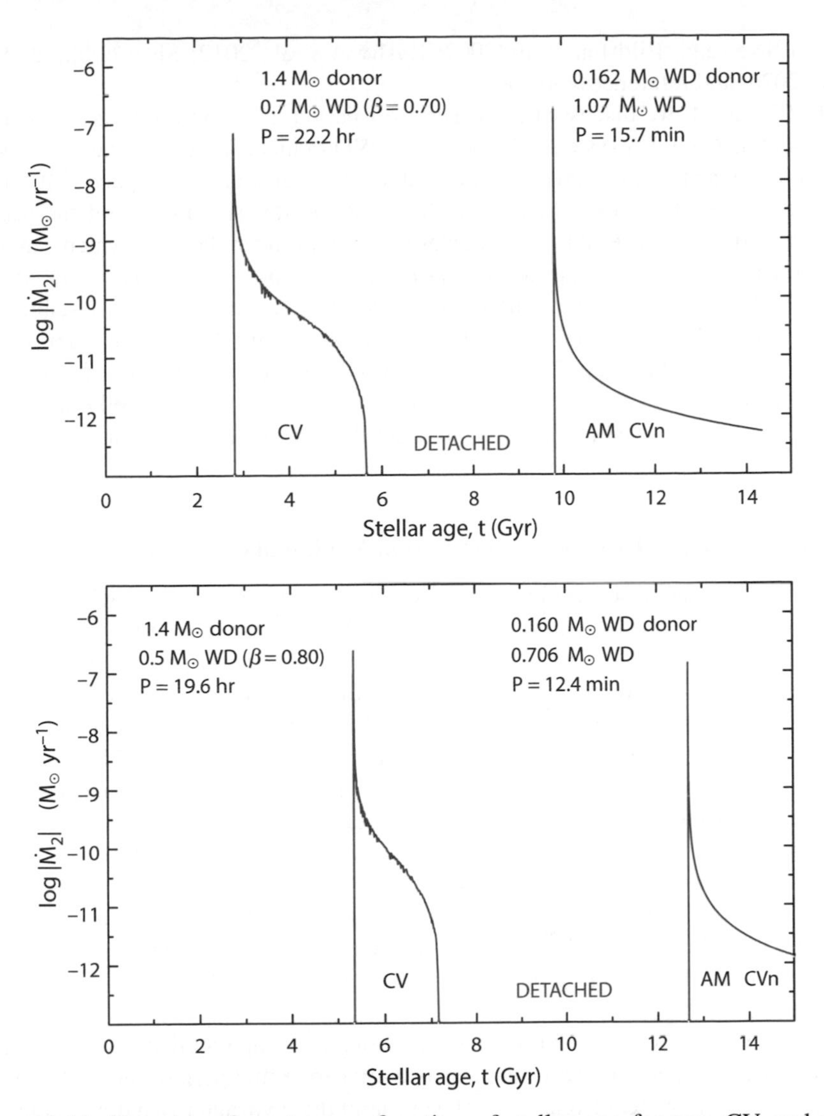

Figure 11.14. Mass-transfer rate as a function of stellar age for two CV and their descendant AM CVn systems. The donor and the accretor star masses, as well as the orbital period, are stated at the onset of each mass-transfer phase. After Tauris (2018).

The relation to GW sources for LISA for these systems is discussed in Tauris (2018); see also Chapter 15. Finally, it should be noted that these calculations did not include nuclear burning of the accreted helium material at the surface of the accreting CO WDs. Such systems may potentially undergo (double-detonation) helium-shell explosions, producing Ca-rich gap transients, which are an emerging population of faint and fast-evolving SNe identified by their conspicuous [Ca II] emission in nebular phase spectra SNe (e.g., Bildsten et al., 2007; Kasliwal et al., 2012; Shen et al., 2019; De et al., 2020; and references therein).

In Chapter 4, we discussed the general formation of accretion disks. It should be noted that in AM CVn systems, the accreting WD radius is large relative to the orbital separation. Marsh et al. (2004) investigated such systems and concluded that this has two effects: (1) it makes it likely that the mass-transfer stream can hit the accretor directly; and (2) it causes a loss of angular momentum from the orbit which can destabilize the mass transfer unless the angular momentum lost to the accretor can be transferred back to the orbit. They further concluded that the effect of the destabilization may significantly reduce the number of systems which survive mass transfer. Later, it was argued by Kaplan et al. (2012) that for AM CVn systems with (hot) extremely low-mass (ELM) He WD donors, the initial contraction of the ELM He WD due to mass loss allows for more stable mass transfer than that originally found for cold He WDs studied by Marsh et al. (2004).

11.3.8 The Period Distribution and Period Minimum of LMXBs

The previously described evolution of CV binaries, which start out with hydrogen-rich low-mass main-sequence donor stars of $\sim 1\,M_\odot$, appears not entirely applicable to the LMXBs. For orbital periods longer than ~ 3 hr, the orbital period distributions of CVs and LMXBs are quite similar, suggesting that in this period range both types of systems follow a similar evolutionary pattern. However, the virtual absence of LMXBs with periods between 80 min and 2 hr is striking, since CV binaries in this period range are numerous. For LMXBs, however, the situation at very short orbital periods is more complex than for CVs. Irradiation of the donor star by the large X-ray flux generated by the accretion onto the NS induces extra stellar wind mass loss from the donor, eroding this star at a faster pace. This makes the evolution of these systems more complex than that of the CVs, and the simple steady evolutionary model sketched in the previous section will no longer apply. (See also the discussion of *evaporation* of the companion star, as well as alternating periods of accretion and radio-pulsar action of the NS, in Chapter 14.)

We mentioned in Section 11.3.6 that it has long been thought that systems with normal hydrogen-rich donor stars cannot easily evolve to orbital periods much shorter than 80 min, but in more recent years it has been found that hydrogen-rich donors may produce smaller orbital period minima down to $\sim 40{-}65$ min (e.g., Podsiadlowski et al., 2002; Nelson et al., 2018; Rappaport et al., 2021). Binary systems with still much shorter periods, such as UCXBs and AM CVns, can therefore (in general) not contain hydrogen-rich donors (Rappaport et al., 1982). The donors in these very short period systems are expected to be (almost always, see Section 11.4) core-burning helium stars or hydrogen-poor degenerate or semi-degenerate stars. Thus, it is believed that both UCXBs and AM CVns generally have WD or subdwarf helium-star donors, and NS and WD accretors, respectively. UCXBs are discussed in more detail below. Figure 11.10 shows the mass-transfer rates and orbital periods of systems with core-burning helium stars (dotted lines). The observed AM CVns indeed appear to have helium-rich donors:

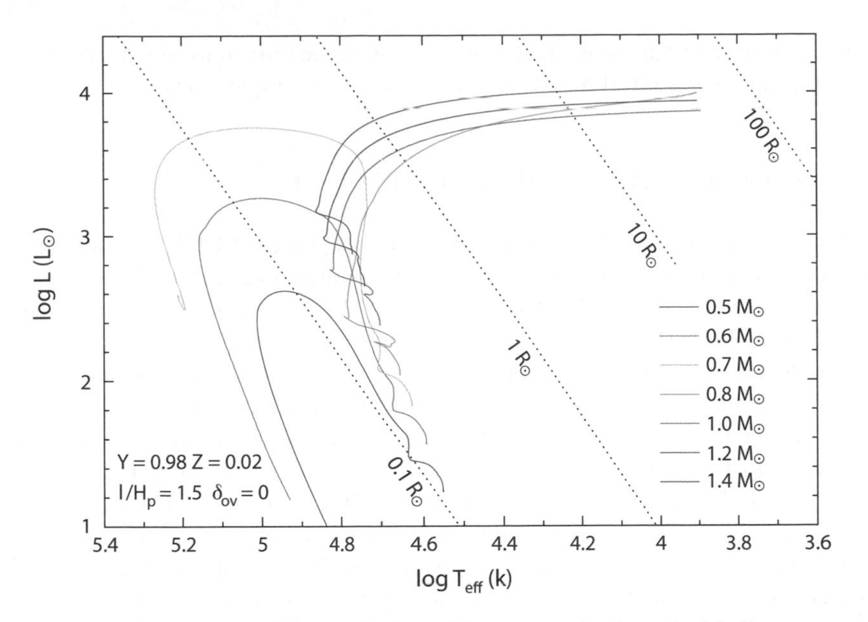

Figure 11.15. HR diagram of the evolution of low-mass single, naked helium stars calculated with the BEC code. The initial masses of the zero-age helium main-sequence stars are between $M_{\rm He} = 0.5 - 1.4\, M_{\odot}$. ($l/H_p$ is the mixing length in units of the pressure scale height; ov means: overshooting.)

either sdB (helium) stars or He WDs. Systems with such helium donors can reach a period minimum at around 6 min (e.g., Tauris, 2018).

11.3.9 Binaries with Low-mass Helium Stars

A star that evolves in a close binary system and loses its envelope via RLO may later be observed as a hot subluminous (or subdwarf) star, often referred to as sdB or sdO stars with typical masses of $\sim 0.5\, M_{\odot}$ (Han et al., 2002; Heber, 2016). These stars are thus stripped cores of former giant stars, and their further evolution, as well as that of their more massive cousins, is depicted in Figure 11.15, which shows their evolutionary tracks from the zero-age helium main sequence and onward. We calculated these tracks using the BEC code of Norbert Langer, assuming a chemical composition of $Y = 0.98$ and $Z = 0.02$ (with no envelope) and a mixing-length parameter of $\alpha = l/H_{\rm p} = 2.0$. It is remarkable for their further evolution that there is a clear bifurcation mass near $0.75\, M_{\odot}$, below which these naked helium stars will evolve bluewards in the HR diagram, and above which these stars will evolve redwards (i.e., re-expanding during helium shell burning to reappear as giant stars). This behavior is then directly related to the orbital periods of binaries for which one expects an additional phase of mass transfer—that is, the more massive of these helium stars being more likely to initiate a new phase of RLO in a binary system. We notice that the detailed evolution of the tracks near the bifurcation mass is dependent on the remaining amount of envelope mass (typically a few 0.001 $M_\odot$, from a more realistic computation).

11.4 FORMATION AND EVOLUTION OF UCXBs

In Chapter 7, we briefly introduced observational properties of UCXBs. These are generally defined as binaries with NS accretors and orbital periods less than ~60 min (Webbink, 1979; Nelson et al., 1986; Podsiadlowski et al., 2002; Nelemans et al., 2010; van Haaften et al., 2012; Heinke et al., 2013). As mentioned earlier, because of the compactness of UCXBs, the donor stars are usually constrained to be a WD, a semi-degenerate dwarf, or a helium star (Rappaport et al., 1982). Depending on the mass-transfer rate, the UCXBs are classified into two categories: persistent and transient sources. Until now, only 14 UCXBs have been confirmed (nine persistent, five transient), and an additional 14 candidates are known (Nelemans et al., 2010; Heinke et al., 2013). Therefore, we can infer that UCXBs are difficult to detect or represent a rare population. UCXBs are detected with different chemical compositions in the spectra of their accretion discs (e.g., H, He, C, O, and Ne; Nelemans et al., 2010). Explaining this diversity requires donor stars that have evolved to different levels of nuclear burning and interior degeneracy and, therefore, points to different formation scenarios of UCXBs. Since a large fraction of the UCXBs are found in globular clusters, some of these UCXB systems could also have formed via stellar exchange interactions (see Chapter 12). From orbital stability analysis, one can place constraints on the mass ratio between the two components in UCXBs. For a 1.4 $M_\odot$ NS accretor, van Haaften et al. (2012) found that only CO WDs with a mass of $\lesssim 0.4\,M_\odot$ lead to stable UCXB configurations (see also Chen et al., 2022). Hydrodynamical simulations (Fig. 11.16) suggest that this critical WD mass limit could be lower (Bobrick et al., 2017).

UCXBs can either form as a result of continuous RLO in tight-orbit LMXBs (e.g., Podsiadlowski et al., 2002; Istrate et al., 2014) or from detached binary millisecond pulsars (MSPs, Chapter 14) with He WD companions and $P_{\rm orb} \lesssim 9$ hr. The latter systems will come into contact within a Hubble time due to emission of GWs and thus become UCXBs. Istrate et al. (2014) have demonstrated the need for extreme fine-tuning of initial parameters (stellar mass and orbital period of the LMXB progenitor systems) to produce an UCXB from such a detached MSP system. This problem is closely related to our limited understanding of magnetic braking (see, e.g., investigations by Chen H.-L. et al. [2021]; Deng et al. [2021]). Early detailed evolutionary calculations to explain UCXBs, such as X2259+59, X1626−67, and X1916−05, have been carried out, for example, by Savonije et al. (1986). The first successful numerical calculations of UCXB formation that included finite-temperature effects of the WD donor and were able to probe the full mass-transfer phase, were performed by Sengar et al. (2017), Tauris (2018), using the MESA code. In the following we highlight some of their results.

To fully model realistic UCXBs, it is not sufficient to start with cold equation-of-state (EoS) polytropic WDs as donor stars. The reason is that finite-temperature (specific entropy) effects are important for the size and envelope structure of the WD,

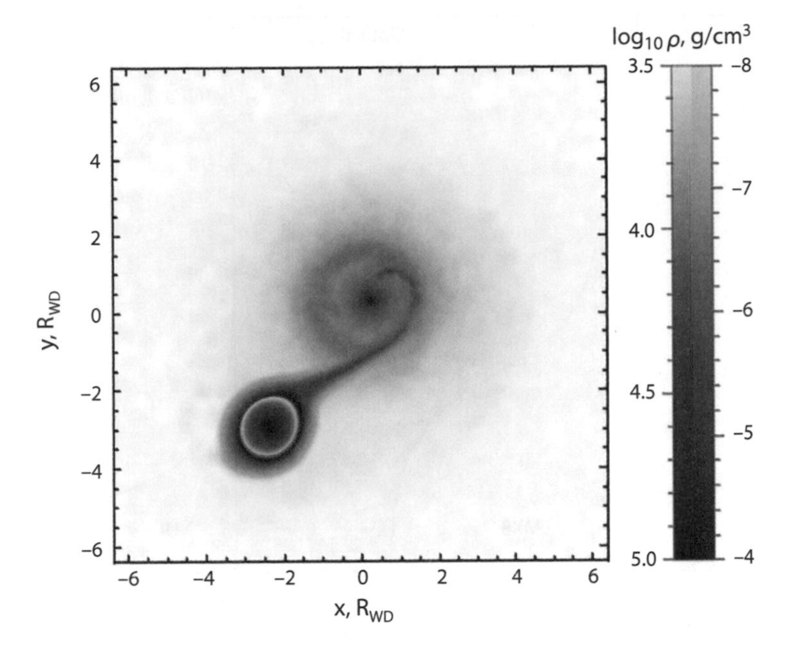

Figure 11.16. Hydrodynamical simulation of dynamical stability in an UCXB with a $0.15\,M_{\odot}$ WD donor star and a $1.4\,M_{\odot}$ accreting NS, in an orbit with an initial eccentricity of 0.04. The plot shows mass density in the orbital plane after 13 orbits of RLO. The coordinates x, y are in units of WD radius. Eccentric structures in the accretion disk reflect the complex character of the flow near the circularization radius. A strong density cusp appears near the NS. The envelope surrounding the binary is sparse, but its total mass is significant compared to that of the disk. After Bobrick et al. (2017).

which determines the mass-transfer rate and thus the observational properties of the UCXB systems. One must therefore begin the binary evolution modelling with a NS orbited by a main-sequence low-mass star (practically a ZAMS star, given that the lifetime of the NS progenitor, typically 10–20 Myr, is short compared to the ~ Gyr nuclear evolution timescale of the low-mass companion star). Figure 11.17 shows an example of the full evolution of such a NS+MS star binary: from a LMXB to a detached MSP, which becomes an UCXB until the termination of the calculation when the donor star has become a planet-like object after a Hubble time (14 Gyr). The red graph displays the mass-transfer rate, $|\dot{M}_2|$, as a function of stellar (or binary) age, t.

Following the work of Sengar et al. (2017), we now discuss the formation of UCXBs in more detail. Figure 11.18 shows six examples of binary evolutionary tracks leading to UCXBs. The donor stars evolve in this diagram from left to right along the tracks. The colors of the tracks correspond to different initial orbital periods of the progenitor systems (before magnetic braking evolves the systems to the LMXB stage). The upper four tracks in the top panel evolve to produce a detached He WD donor prior to

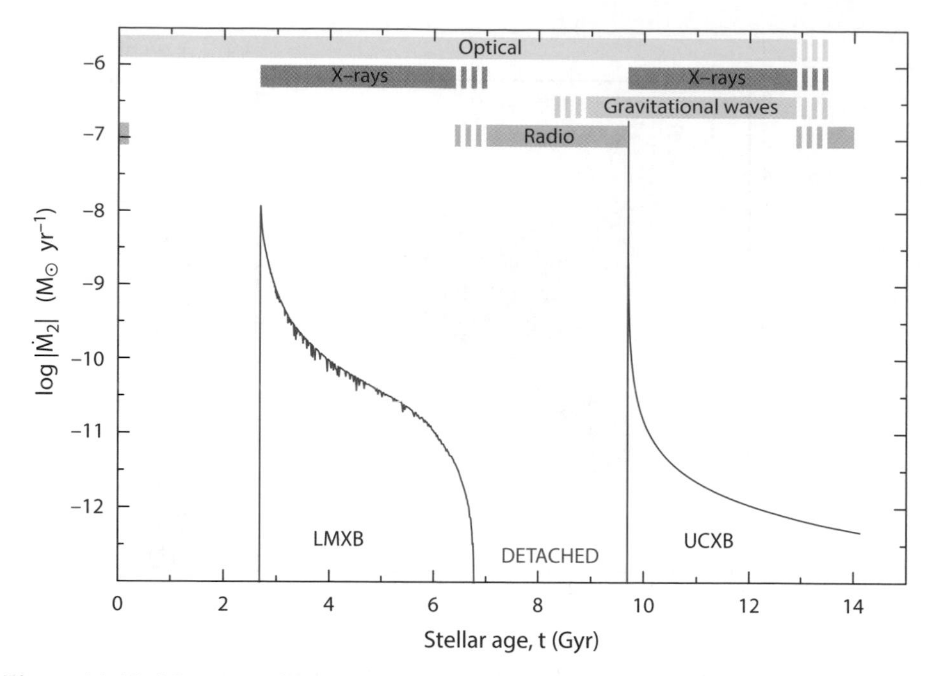

Figure 11.17. Mass-transfer rate of the donor star as a function of its stellar age. The initial MS star + NS binary has components of 1.40 $M_{\odot}$ and 1.30 $M_{\odot}$, respectively. The system evolves through two observable stages of mass transfer: a LMXB for 4 Gyr, followed by a detached phase lasting ~3 Gyr where the system is detectable as a radio MSP orbiting the He WD remnant of the donor star, until GW radiation brings the system into contact again, producing an UCXB. The color bars indicate detectability regimes. After Tauris (2018).

the UCXB phase, whereas the orange track marginally detaches before evolving to an UCXB, and the lower (red) track is for a system that produces a degenerate donor that still contains hydrogen and evolved continuously from mass transfer in a LMXB system to become an UCXB (i.e., without detachment in between). Observed UCXBs are overplotted and seen to accumulate to the right side of the plot where the evolutionary timescale is long due to the ever decreasing mass-transfer rate (see Fig. 11.17 for an example of $|\dot{M}_2|$ as a function of age). The lower panel shows the radii of the donor stars as a function of their masses, for the same tracks as in the upper panel (here the red track for the hydrogen-rich donor star is seen at the top-right corner). Initially, right after the onset of the UCXB stage, the donor stars decrease their radii until their residual hydrogen-rich envelope, of a few $10^{-3}\ M_{\odot}$, has been removed (e.g., Kaplan et al., 2012). Thereafter, they expand in response to mass loss and follow more or less the curves expected for a degenerate configuration. The red dashed and black dotted lines are theoretical mass-radius relations: $R_{\rm WD} = 0.013\ R_{\odot} \cdot (M_{\rm WD}/M_{\odot})^{-1/3}$ and that of Eggleton used in Rappaport et al. (1987).

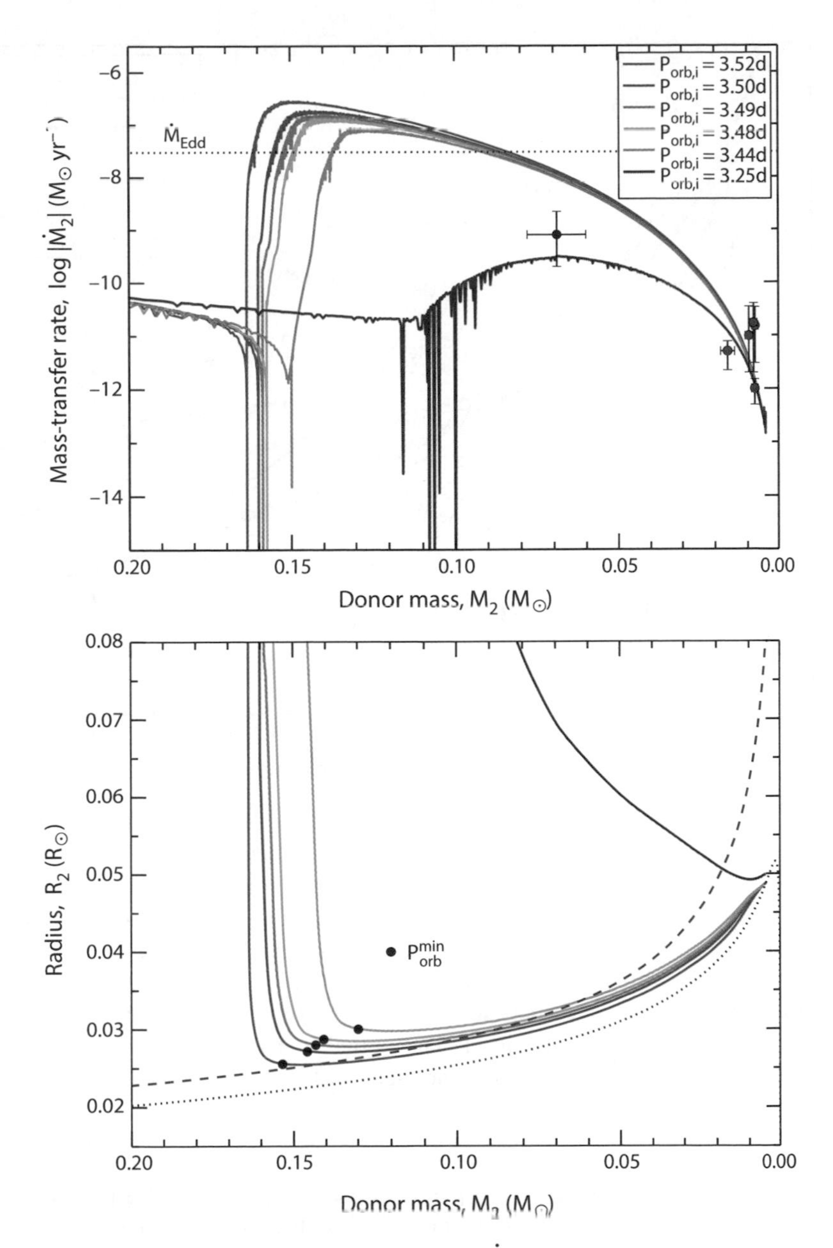

Figure 11.18. Upper panel: Mass-transfer rate, $|\dot{M}_2|$ as a function of donor star mass (M_2) for the UCXB evolutionary tracks shown in Figure 11.19. Evolution proceeds from left to right (decreasing donor mass). Crosses indicate observed UCXBs. Different colored tracks correspond to different initial orbital periods (see text). Lower panel: Donor star radius vs. mass for the same tracks. Black dots mark the orbital period minimum, P_{orb}^{min}. Further explanation is in the text. After Sengar et al. (2017).

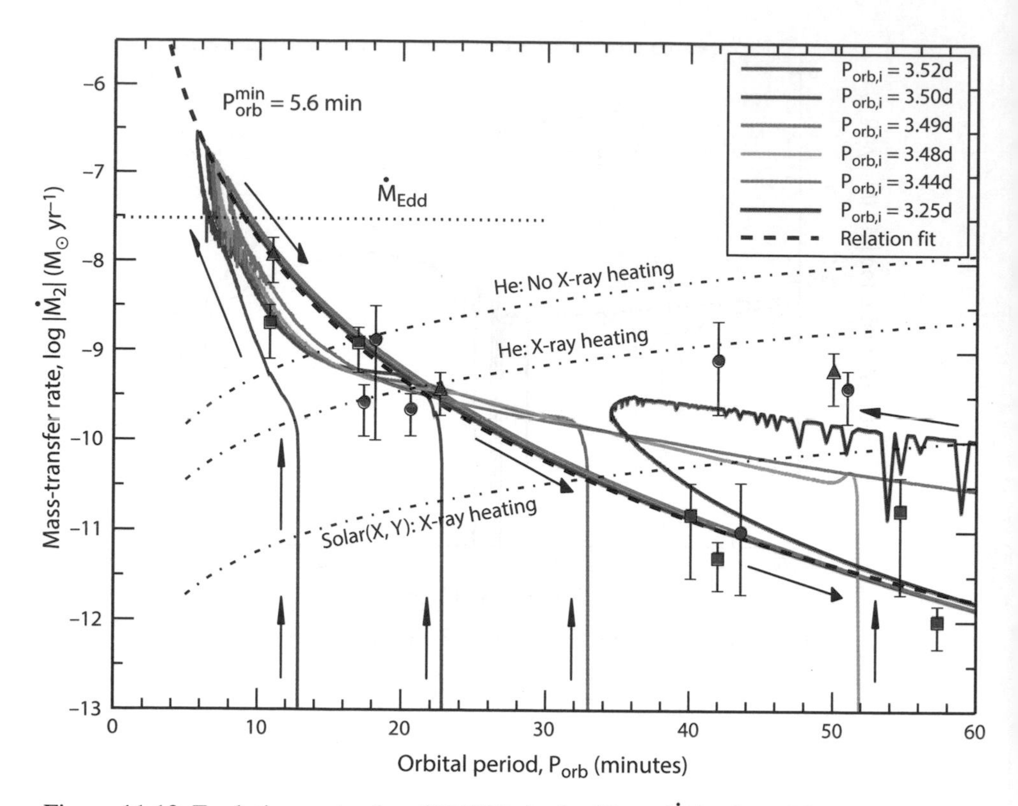

Figure 11.19. Evolutionary tracks of UCXBs in the ($P_{\rm orb}$, $\dot{M}_2$)–plane. The UCXB evolution (see black arrows for direction) begins on the vertical tracks, when the He WD initiates RLO, and decreases $P_{\rm orb}$ due to GWs while climbing up the *ascending* branch until the tip of the track at $P_{\rm orb}^{\rm min}$, before the system settles on the common *declining* branch (dashed line) while $P_{\rm orb}$ steadily increases on a Gyr timescale. Observed UCXBs are shown with red (persistent source) and blue (transient source) symbols. The dot-dashed lines indicate accretion disk instability thresholds, depending on X-ray heating and chemical composition. The dotted line is $\dot{M}_{\rm Edd}$ for a NS accreting helium. Additional details of the tracks are seen in Figure 11.18. After Sengar et al. (2017).

Figure 11.19 shows the ($P_{\rm orb}$, $\dot{M}_2$)–plane for the same UCXB tracks just described. Four important conclusions can be drawn from this figure (Sengar et al., 2017):

(1) It is evident that LMXBs with initial orbital periods, $P_{\rm orb,i}$, below a certain threshold value (depending on initial values of M_2, $M_{\rm NS}$, chemical composition, and treatment of magnetic braking) will never detach from RLO to produce a He WD. Their donor stars still possess a significant hydrogen content—even in their cores—and due to their very small nuclear-burning rates they still have a mixture of hydrogen and helium when they finally become degenerate near the orbital period minimum,

$P_{\rm orb}^{\rm min} \simeq 10-85$ min (Paczynski & Sienkiewicz, 1981; Rappaport et al., 1982; Nelson et al., 1986; Podsiadlowski et al., 2002; van der Sluys et al., 2005). These converging systems become hydrogen-rich UCXBs and are most likely the progenitor systems of the so-called black widow MSPs (Chapter 14). The minimum orbital period, $P_{\rm orb}^{\rm min}$, for the hydrogen-rich donor system (red track) is 35 min. This is contrary to the initially slightly wider-orbit LMXBs, which detach and form a He WD that cools off while decreasing in size such that much smaller values of $P_{\rm orb}^{\rm min} \simeq 5-7$ min are obtained in these systems.

(2) For UCXBs with He WD donors, the larger the value of $P_{\rm orb,i}$, the smaller is $P_{\rm orb}$ at the onset of the UCXB phase. The reason for this is that in wider binaries, He WDs have larger masses (Tauris & Savonije, 1999); and, more importantly, since in wider systems it takes a longer time for GWs to cause the He WDs to fill their Roche lobe, they will be less bloated (Istrate et al., 2014), that is, more compact (and colder) by the time they reach the onset of the UCXB phase.

(3) Sengar et al. (2017) identify a unique pattern in the tracks of these UCXBs (see black arrows for direction of evolution in this diagram). The UCXBs begin on the vertical tracks, when the He WD initiates RLO, and continue with decreasing $P_{\rm orb}$ due to GW losses, while climbing up the *ascending* branch until the tip of the track at $P_{\rm orb}^{\rm min} \simeq 5-7$ min. Following $P_{\rm orb}^{\rm min}$, which coincides with a maximum value of $|\dot{M}_2| \simeq 10\,\dot{M}_{\rm Edd}$ (see also Fig. 11.18), all systems settle on the common *declining* branch while $P_{\rm orb}$ steadily increases on a Gyr timescale, with the relation (Sengar et al., 2017):

$$\log|\dot{M}_2/M_\odot\,yr^{-1}| = -5.15 \cdot \log(P_{\rm orb}/{\rm min}) - 2.62. \qquad (11.12)$$

(4) The shape of the UCXBs tracks can be understood from the ongoing competition between GW radiation and orbital expansion caused by mass transfer/loss. The reason that the maximum value of $|\dot{M}_2|$ coincides with $P_{\rm orb}^{\rm min}$ is partly that the He WD donor stars are fully degenerate, which means that their mass-radius exponent is negative, whereby they expand in response to mass loss (Fig. 11.18). The onset of RLO leads not only to very high mass-transfer rates but also to an outward acceleration of the orbital size, as a result of the small mass ratio ($q \simeq 0.1$) between the two stars, such that at some point the rate of orbital expansion dominates over that of shrinking due to GW radiation. As the orbits widen further, the value of $|\dot{M}_2|$ decreases and the strength of the GW radiation levels off due to its steep dependence on orbital separation (Chapter 15), and the systems settle on the common declining branch while the orbit expands at a continuously slower pace. An analogy of the here described UCXB models can be made to RLO in double WD systems (Kaplan et al., 2012).

All the five observed UCXB transient systems likely populate the declining branch of the evolutionary tracks. Due to their wider orbits, the radius of their accretion disk is larger and its temperature lower, which causes thermal viscous instabilities and thus a transient behavior (Lasota et al., 2008). Heinke et al. (2013) suggested that this might be the reason why so relatively few UCXBs are seen in wide orbits with $P_{\rm orb} > 30$ min (keeping in mind that UCXBs should accumulate in wide orbits over time). The transient behavior allows radio MSPs to turn on, whereby the "radio ejection mechanism" (Burderi et al., 2002) can prevent further accretion. However, pulsar-wind irradiation of the donor may operate, possibly until $M_2 < 0.004\,M_\odot$ (if beaming is favorable), at which point the star is likely to undergo tidal disruption, potentially leaving behind pulsar planars (Radhakrishnan & Shukre, 1985; however, see also Priedhorsky & Verbunt, 1988; Martin et al., 2016). The evolutionary tracks of Sengar et al. (2017) are terminated just before that when $M_2 \simeq 0.005\,M_\odot$ (Jupiter masses) and $P_{\rm orb} = 70 - 80$ min (Fig. 11.18). Their computed M_2–R_2 tracks are in good agreement with (within 5% of) the adiabatic helium models of Deloye & Bildsten (2003) and the cold helium models of Zapolsky & Salpeter (1969). Interestingly, Sengar et al. (2017) find that their He WD donors never crystallize but remain Coulomb liquids with $\Gamma \equiv E_{\rm coulomb}/E_{\rm thermal} = Z^2e^2/(\langle r_{\rm ion}\rangle\, kT) \leq 35$. The final NS masses in their (post) UCXB systems are $\sim 1.7\,M_\odot$, reflecting the assumed NS birth mass ($1.3\,M_\odot$) and the accretion efficiency.

Analytical investigations by van Haaften et al. (2012a, 2012b) on the evolution of UCXBs reveal that these systems might evolve to final orbital periods of 100–110 min, thereby explaining the existence of the so-called diamond planet pulsar (Bailes et al., 2011).

A further consequence of the tight correlation between $P_{\rm orb}$ and $M_{\rm WD}$ in post-LMXB systems (see Sections 11.5 and 14.3) is that the initial WD donor mass in tight UCXBs, which evolve to become GW sources, are all found to be very close to $M_{\rm WD} \simeq 0.162 \pm 0.005\,M_\odot$ (Tauris, 2018; Chen et al., 2020, 2021). We shall return to UCXBs as GW sources in Chapter 15.

11.5 MECHANISMS DRIVING MASS TRANSFER IN WIDE-ORBIT LMXBs AND SYMBIOTIC BINARIES

As mentioned previously, there is a dichotomy in the evolution of close- and wide-orbit LMXBs that is partly related to the difference in the internal structure of the donor stars at different evolutionary stages. In Figure 11.20 is shown the difference in internal structure between the donor stars in converging and diverging systems. (Details of the binary system parameters are given in the figure caption.) While the radius of the unevolved star in the upper panel decreases in response to mass loss, because it has a positive mass-radius exponent (which enables dynamically stable evolution in decaying orbits), the giant-star donor in the lower panel expands its radius significantly while the star loses mass to its companion star due to a negative mass-radius exponent for convective envelope material. This expansion to giant dimensions remains intact despite the low mass of the star and its envelope during its later evolution.

In low-mass systems with $P_{\rm orb} \gtrsim 0.5 - 1$ day (depending on the strength of the magnetic braking), the Roche-lobe filling donor star has a too large radius to be unevolved:

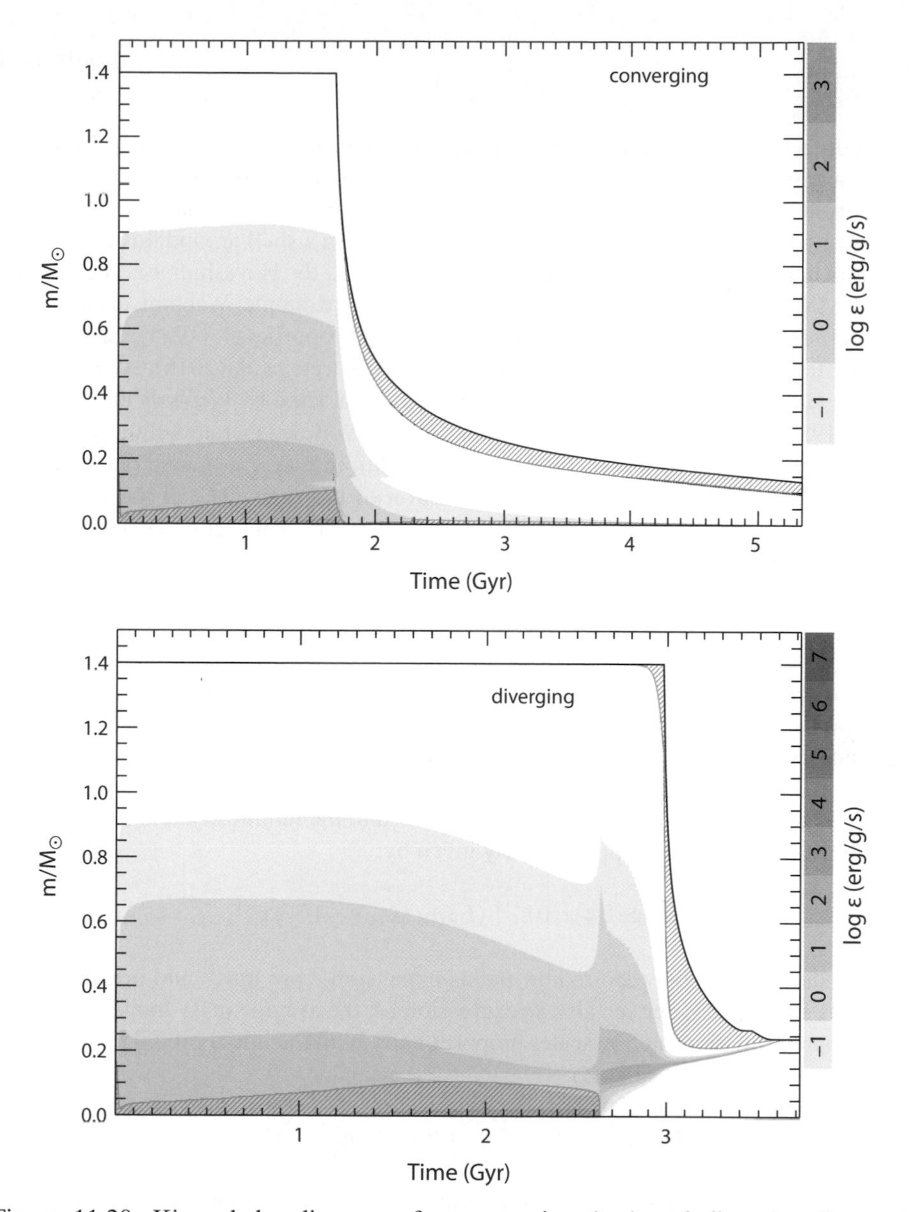

Figure 11.20. Kippenhahn diagram of a converging (top) and diverging (bottom) LMXB system, respectively, as calculated by Istrate et al. (2014). The following initial values were used: $M_2 = 1.4\,M_\odot$, $M_{\rm NS} = 1.3\,M_\odot$, $\gamma = 5$, and $P_{\rm orb,i} = 2.6$ and 4.2 days, respectively. In the converging system, the donor star experiences Case A RLO while the diverging system undergoes Case B RLO. The plots show cross-sections of the stars in mass-coordinates from the center to the surface of the star, along the y-axis, as a function of stellar age on the x-axis. The duration of the LMXB-phase is: "∞" (no detachment) and 600 Myr, respectively; and the final orbital period at termination of the calculation is: $\sim$2.6 hr and $\sim$8.0 days, respectively. The green hatched areas denote zones with convection (according to the Schwarzschild criterion), initially in the core and later in the envelope of the donor stars. The intensity of the blue color indicates the net energy-production rate ε; the hydrogen burning shell is clearly seen in the case of the diverging system at $m/M_\odot \simeq 0.2$. After Istrate et al. (2014).

the hydrogen in its core must be near or beyond exhaustion. In diverging LMXBs, the donor star initiates RLO after ignition of hydrogen shell burning (i.e., Case B RLO). In evolved low-mass stars ($< 2.3\,M_\odot$), the helium core becomes degenerate and the luminosity is entirely generated by hydrogen burning in a shell around this core, while the star climbs up along the red-giant branch (RGB, i.e., the Hayashi line). These wide-orbit LMXB systems have been studied in detail by, for example, Webbink et al. (1983); Taam (1983); Savonije (1987); Joss et al. (1987); Rappaport et al. (1995); Ergma et al. (1998); Tauris & Savonije (1999); Lin et al. (2011); Istrate et al. (2016). As Refsdal & Weigert (1971) showed, and was later demonstrated in detail by Webbink et al. (1983); Taam (1983) and others, in these low-mass giants with degenerate helium cores and energy generation by hydrogen-shell burning, the luminosity, $L_{\rm RGB}$, and radius, $R_{\rm RGB}$, are—for a given metallicity, Z—only a function of the mass, $M_{\rm c}$, of the degenerate core, and independent of the mass of the hydrogen-rich convective envelope of the star (see numerical examples in Table 14.2), and therefore one can formulate simple expressions as:

$$L_2 = f_1(M_{\rm c}, Z) \qquad \wedge \qquad R_2 = f_2(M_{\rm c}, Z), \tag{11.13}$$

where L_2 and R_2 are the luminosity and radius of the (sub)giant donor star.

It was demonstrated by Webbink et al. (1983) that for $0.16 \le M_{\rm c}/M_\odot \le 0.45$ the radii and luminosities of such stars can approximately be fitted by simple polynomial relations. Since for a star on the RGB, the luminosity is generated by hydrogen shell burning, the growth in core mass is directly related to the luminosity, and these authors found that for solar Z it is approximately given by:

$$\dot{M}_{\rm c} \simeq 1.4 \times 10^{-11}\,(L_{\rm RGB}/L_\odot) \quad M_\odot\,{\rm yr}^{-1}. \tag{11.14}$$

From their relations between radius, radius expansion, core mass, and orbital period, Webbink et al. (1983) derived also an expression for the average mass-transfer rate that, in first approximation, simply scales proportionally with the initial orbital period of the LMXB, roughly as (for solar Z):

$$\langle|\dot{M}_{\rm RGB}|\rangle \simeq 5.3 \times 10^{-10}\,(P_{\rm orb,i}/{\rm days}) \quad M_\odot\,{\rm yr}^{-1}. \tag{11.15}$$

Examples of applications of such fitting formulae are shown in Figure 11.21. Assuming that a LMXB typically transfers $\sim$0.75 $M_\odot$ of material, the mass-transfer timescale can be roughly estimated as:

$$\tau_{\rm LMXB} \simeq 1.4\,{\rm Gyr}\,({\rm days}/P_{\rm orb,i}). \tag{11.16}$$

For other donor masses and chemical compositions, the relations become slightly different. (For further discussions on these relations in wide-orbit LMXBs and their importance for binary radio pulsars, see Section 14.3.2.) Relations similar to those above can also be derived for low-mass stars with degenerate CO cores. These stars are

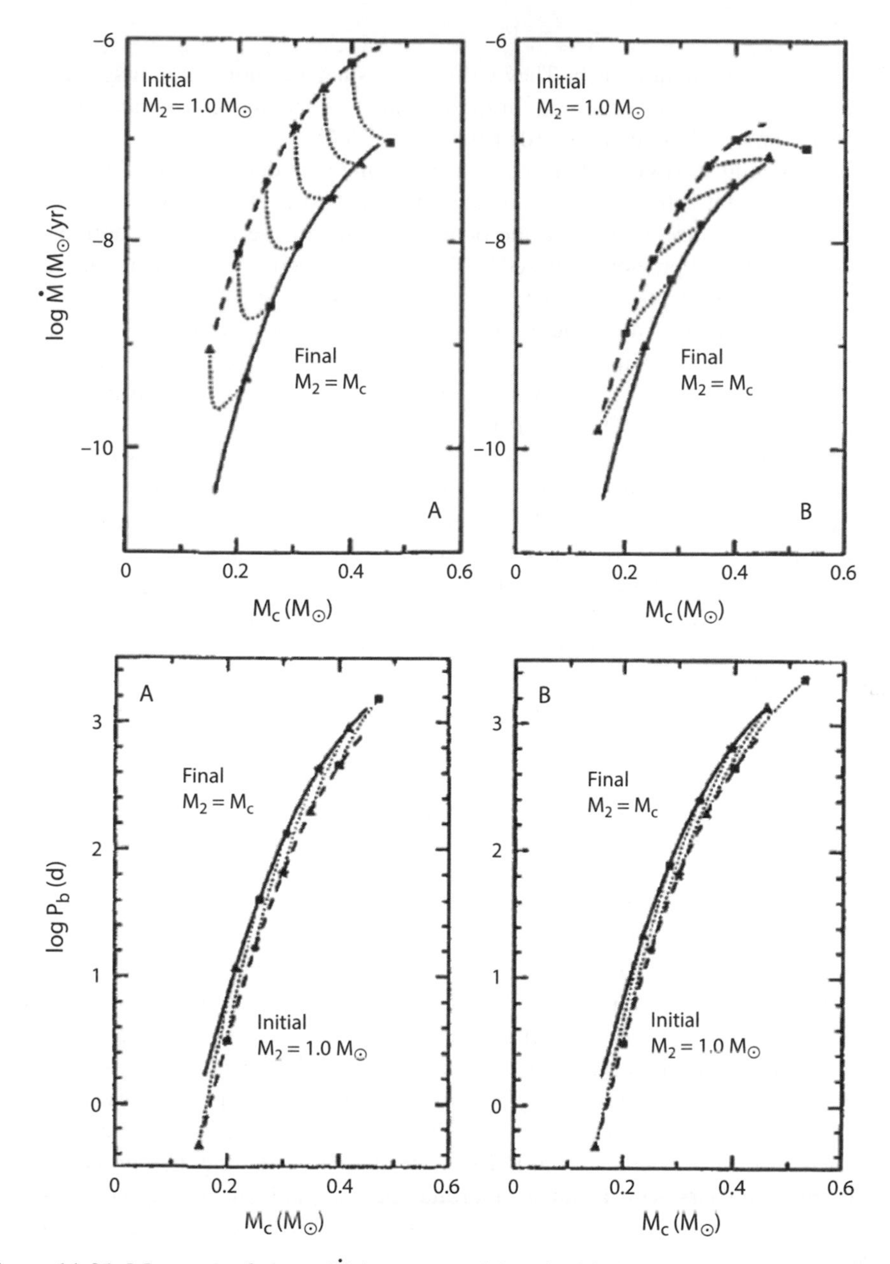

Figure 11.21. Mass-transfer rate $\dot{M}$ (upper panels) and orbital periods (lower panels) for wide-orbit LMXBs driven by the evolutionary expansion of the low-mass giant donor star, as a function of the core mass M_c of the giant. The curves are for a metallicity of $Z = 0.02$. The dashed line indicates the rates at the onset of the mass transfer, for a giant mass of 1 $M_\odot$. The full line indicates the end of the mass transfer, and dotted lines indicate the evolution during the mass transfer. Initial accreting compact object masses of 1.4 $M_\odot$ (left panels) and 7 $M_\odot$ (right panels) were adopted. The mass transfer is assumed to be conservative. (Left: after Verbunt [1993]. Right: after Bitzaraki & van den Heuvel [1993].)

on the asymptotic giant branch. They will occur as mass donors only in systems with orbital periods of a year or longer. In these systems, very high mass-transfer rates of order $10^{-7}\,M_{\odot}\,\mathrm{yr}^{-1}$ are expected (de Kool et al., 1986).

The widening LMXBs often experience a temporary detachment. This is caused by a transient contraction of the donor star when its hydrogen shell source encounters a slight chemical discontinuity at the outer boundary of the hydrogen burning shell (i.e., due to material previously mixed by the outer convection zone when it penetrated inward to its deepest point; see the diverging LMXB in Fig. 11.20 near $t = 3.4\,\mathrm{Gyr}$). Although this effect of a transient contraction of single low-mass stars evolving up the RGB has been known for many years (Thomas, 1967; Kippenhahn & Weigert, 1990), its application in LMXBs has surprisingly only been investigated in a few studies (Tauris & Savonije, 1999; D'Antona et al., 2006).

The LMXB evolution for giant donor stars discussed earlier applies equally well to symbiotic binaries where the accreting objects are WDs instead of NSs. The final general correlation (see Section 14.3.2) between orbital period and WD remnant mass of the low-mass giant donor star is independent of the mass and nature of the accretor (Exercise 11.3) and also independent of the accretion efficiency. However, for an *individual* binary system, the final orbital period and WD mass does depend on these quantities. An early example of computed LMXB evolution leading to formation of a binary MSP is shown Figure 11.22. Note that the initial NS mass is assumed to be $1.0\,M_{\odot}$, which is in conflict with the current understanding of stellar evolution and SN explosion physics.

Since the derivation of the previously mentioned polynomial fit relations for stellar evolution in binaries (e.g., Webbink et al., 1983) and pioneering computations by Pylyser & Savonije (1988, 1989) in the 1980s, much work has been done in this field. Nowadays, stellar evolution modelling (and thus binary evolution) is computed on advanced modern stellar evolution codes with the latest input physics.[3] An excellent example of such work by Lin et al. (2011) is demonstrated in Figure 11.23. Their aim was to explain the origin of the binary pulsar PSR J1614-2230, which harbors a massive NS and a $0.5\,M_{\odot}$ CO WD (thus originating from an IMXB system, see Section 14.4) in an orbit with a period of 8.7 days. The authors computed the evolution of 42,000 models of binaries that cover 60 initial donor masses over the range of 1 to $4\,M_{\odot}$ and, for each of these, 700 initial orbital periods over the range 10−250 hr. These results provide an excellent database for studying the evolution of a wide range of LMXBs and IMXBs as well as for finding the progenitors of binary radio pulsars with a range of orbital periods and masses of their NSs and WDs. The formation and properties of binary MSPs will be discussed in detail in Chapter 14.

[3]However, the CE stage still cannot be accounted for satisfactorily in such codes.

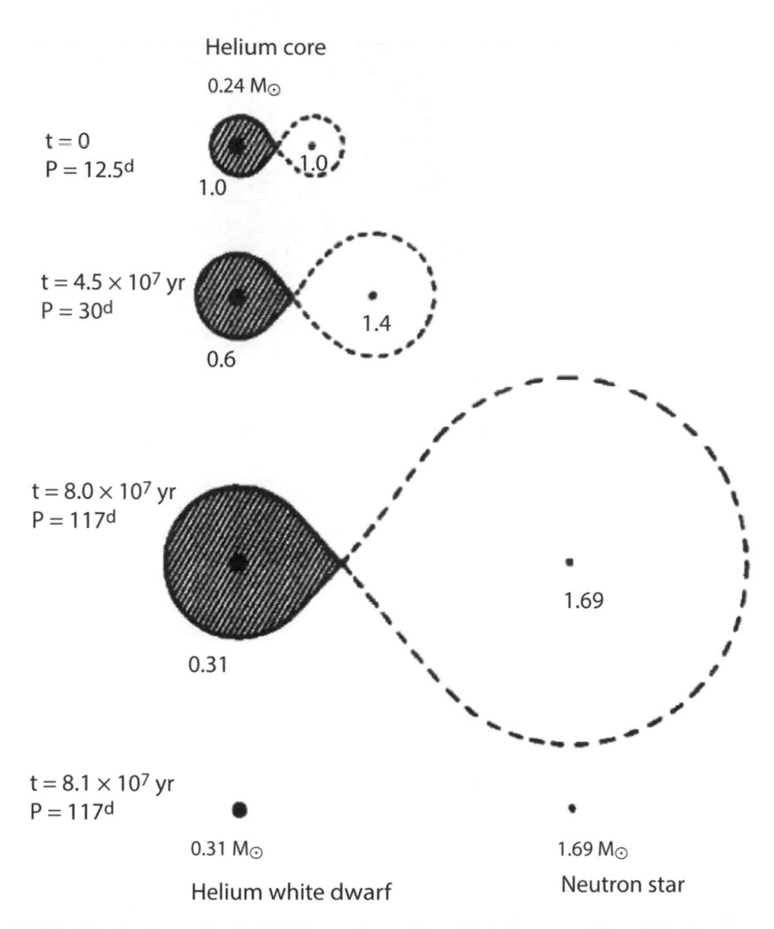

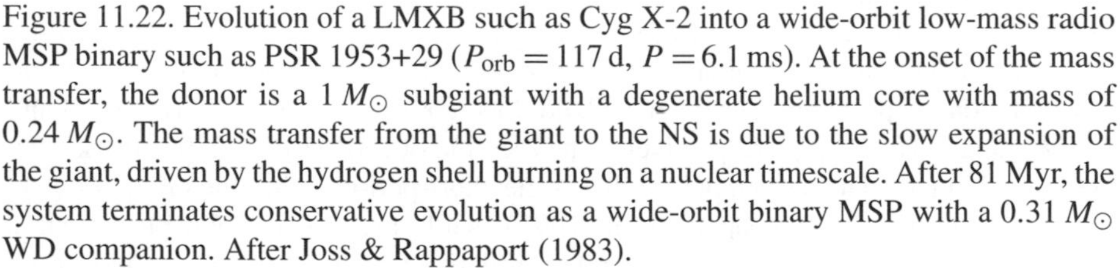
Figure 11.22. Evolution of a LMXB such as Cyg X-2 into a wide-orbit low-mass radio MSP binary such as PSR 1953+29 ($P_{\rm orb} = 117$ d, $P = 6.1$ ms). At the onset of the mass transfer, the donor is a 1 $M_\odot$ subgiant with a degenerate helium core with mass of 0.24 $M_\odot$. The mass transfer from the giant to the NS is due to the slow expansion of the giant, driven by the hydrogen shell burning on a nuclear timescale. After 81 Myr, the system terminates conservative evolution as a wide-orbit binary MSP with a 0.31 $M_\odot$ WD companion. After Joss & Rappaport (1983).

11.6 STABILITY OF MASS TRANSFER IN INTERMEDIATE-MASS AND HIGH-MASS X-RAY BINARIES

In Section 11.3.2, we summarized a classic investigation of orbital stability of RLO, based on simple polytropic approximations for the stellar structure, and found that for main-sequence donor stars, RLO is only dynamically stable for $q < 4/3$. However,

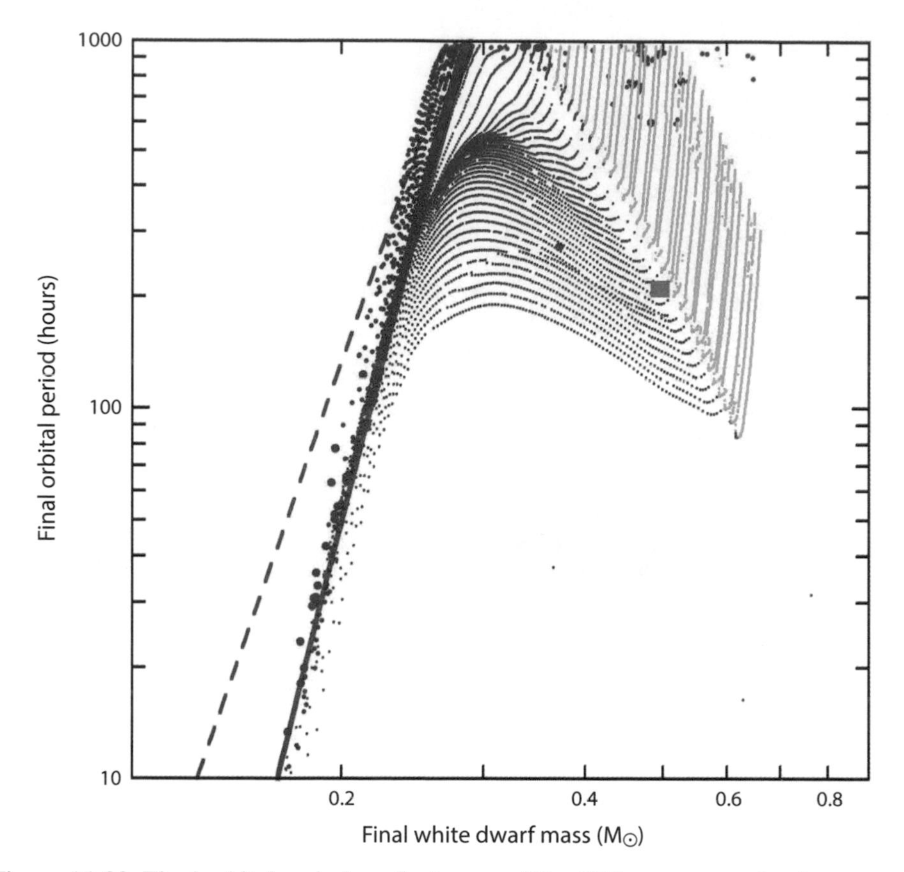

Figure 11.23. Final orbital period vs. final mass of the WD component for the approximately 14,000 systems, whose final orbital periods are in the range 10 to 1,000 hr (∼0.5−40 d), computed on the MESA code by Lin et al. (2011). The color coding of the dots is as follows: cyan and blue are for initial $M_2 > 2.2\,M_\odot$ and Cases B and AB evolution, respectively. The green and red dots represent initial M_2 between 1.4 and 2.2 $M_\odot$, and 1.0 to 1.4 $M_\odot$, respectively. The location of PSR J1614-2230 is marked by the orange square. The dashed purple line is the short-period end of the ($P_{\rm orb}$, $M_{\rm WD}$)–correlation derived by Rappaport et al. (1995), while the steeper solid purple curve (below the cluster of red dots) fits the relation derived by Tauris & Savonije (1999). After Lin et al. (2011).

using more realistic non-polytropic stellar models, and including spin-orbit interactions, stable mass transfer is found in IMXBs for main-sequence and Hertzsprung-gap donor stars more than 2.5 times more massive than the mass of the accreting NS (Tauris et al., 2000; Misra et al., 2020). These systems with relatively massive donors, which have radiative or thin convective envelopes, undergo an initial rapid phase of mass transfer on a (sub)thermal timescale. This rapid mass transfer remains stable for a substantial range of donor masses (up to 7 $M_\odot$) and orbital periods, as can be seen in Figure 11.24.

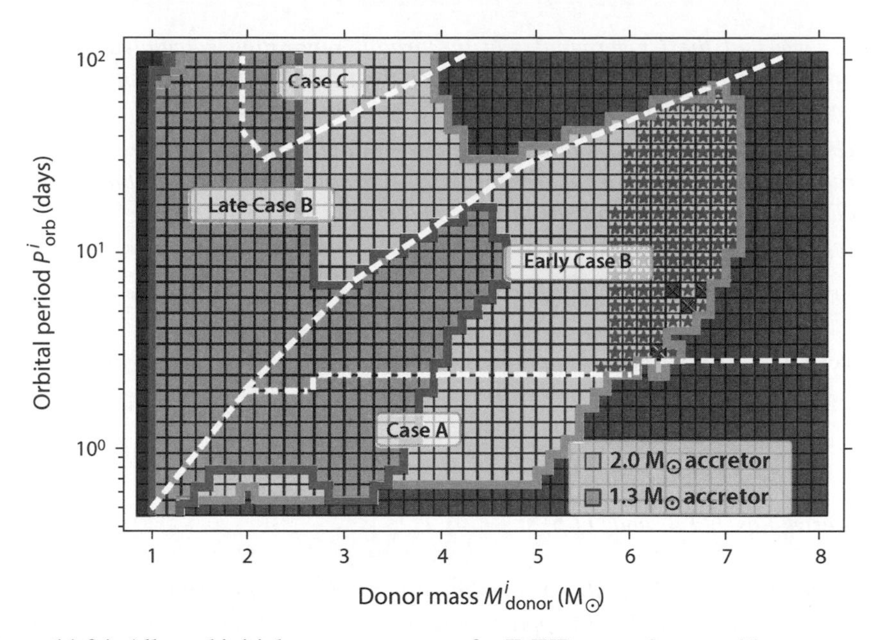

Figure 11.24. Allowed initial parameter space for IMXBs to undergo stable mass transfer for accreting NSs of mass $1.3\,M_\odot$ (red squares) and $2.0\,M_\odot$ (blue squares). The gray squares correspond to systems that encountered dynamical instability. The dashed white lines separate systems undergoing different cases of RLO. Green stars correspond to those systems that undergo a second mass-transfer phase from a stripped helium giant star, for example, Case BB RLO. After Misra et al. (2020).

It has been found that the same is true for HMXBs with donors with radiative envelopes for mass ratios (donor mass over accretor mass) of up to $q = 3$ (Pavlovskii et al., 2017). Once the donor stars in IMXBs/HMXBs develop deep convective envelopes, the mass transfer usually becomes dynamically unstable (see more detailed discussions in Chapters 7 and 10).

EXERCISES

Exercise 11.1. *Here we investigate the possibility of producing LMXBs via CE evolution. Follow the arguments in Section 11.1.2 (see also Section 4.8) and investigate the possibility of envelope ejection for a* $15\,M_\odot$ *giant donor star with a core mass of* $4.0\,M_\odot$*, assuming* $\lambda = 0.2$ *and* $\alpha_{CE} = 1$*, and a companion star mass of* $1.50\,M_\odot$.

a. *Start by calculating the minimum post-CE orbital separation to avoid RLO from the secondary star after in-spiral.*
   (Hint: For the radius of the secondary star, use Eq. 8.14).

b. *Assume that the CE terminates at the orbital separation calculated in (a). The orbital energy and that point is an upper limit to the released amount of orbital energy during CE in-spiral. Calculate this amount,* $\Delta E_{\rm orb}$.

c. *Using* $E_{\rm bind} = \alpha_{\rm CE} \Delta E_{\rm orb}$ *in combination with the* λ*-value, calculate the required radius of the* 15 $M_\odot$ *giant donor star* ($R_{\rm donor}$) *to obtain this envelope binding energy* ($E_{\rm bind}$). *Is this radius realistic for a* 15 $M_\odot$ *star?*

d. *What will be the outcome if* $\alpha_{\rm CE} = 0.3$?

e. *Repeat this whole exercise for a* 2.0 $M_\odot$ *companion star. Discuss the difference.*

Exercise 11.2. *Under the assumptions outlined in Section 11.3.2, show that:*

$$\frac{\dot{R}_{\rm L}}{R_{\rm L}} = \frac{\dot{a}}{a} + \frac{1}{3}\frac{\dot{M}_2}{M_2} \tag{11.17}$$

and derive Eqs. (11.4) and (11.6).

Exercise 11.3. *The* ($P_{\rm orb}$, $M_{\rm WD}$) *correlation for radio pulsars with He WD companions (see Eqs. 14.9 and 14.10) applies more broadly to all kinds of post-mass-transfer systems in which the donor star was a low-mass RGB star. Here we demonstrate that this correlation is* independent *of the mass of the accreting companion star (see also Xu et al. [2012] for numerical examples).*

a. *Consider a binary with a RGB donor star of* $M_2 = 1.4\,M_\odot$, $R_2 = 50\,R_\odot$ ($M_{\rm core} \simeq 0.34\,M_\odot$). *Assume an accreting object of mass:* $M_X = 1.4\,M_\odot$, $M_X = 14\,M_\odot$, *or* $M_X = 140\,M_\odot$, *and show that if the donor star fills its Roche lobe, the orbital periods of all three systems are identical to within a few percent.*

b. *Assume the following expression (see Eq. 11.3) to be approximately valid for* $q \leq 1$:

$$r_L = 0.462 \left(\frac{q}{1+q}\right)^{1/3},$$

where the mass ratio, $q \equiv M_2/M_X$. *Combine this expression with Kepler's third law and show that:*

$$P_{\rm orb} \propto R_2^{3/2}\, M_2^{-1/2}, \tag{11.18}$$

for example, the orbital period of the system is independent of the mass of the accretor. The dependence on M_2 *will disappear since* $R_2(M_{\rm core})$ *is independent of* M_2, *and eventually the donor star will be stripped down to* $M_2 \to M_{\rm core}$ *at the end of the RLO and the orbital period will be only a function of the core mass:* $P_{\rm orb}(M_{\rm core})$. *Notice that* $M_{\rm core}$ *will grow as the system widens during RLO and also that* individual *systems with different accretor masses,* M_X, *will end up at different locations on the final global* ($P_{\rm orb}$, $M_{\rm WD}$) *correlation (see, e.g., Fig. 14.13).*

Exercise 11.4. *Search for data or numerical relations on radius vs. mass of main-sequence stars below* $1\,M_\odot$*, brown dwarfs, and giant gas planets (down to* $\sim 1\,M_{\rm Jupiter}$*), and make a single plot of orbital period,* $P_{\rm orb}$*, as a function of donor mass,* M_2 *of a Roche-lobe filling donor star/brown dwarf/planet.*

a. *What is the minimum orbital period? (For which donor mass?)*

b. *Does it make a difference if the accretor is a* $1.2\,M_\odot$ *WD or a* $100\,M_\odot$ *BH? (Hint: you may take advantage of Eq. (4.9).*

Chapter Twelve

Dynamical Formation of Compact Star Binaries in Dense Star Clusters

12.1 INTRODUCTION

In the foregoing chapters, we described models for the formation of binaries containing NSs and BHs, including double NSs (DNSs) and double BHs, as products of the evolution of close binary systems. A fundamentally different formation model for these systems is by dynamical processes in dense star clusters. From the presence of NS low-mass X-ray binaries (NS-LMXBs), radio millisecond pulsars (MSPs), and three DNSs in globular star clusters, we know that indeed such processes occur in nature (see the arguments given in the next section). There are, therefore, good reasons to expect that also double BHs may be formed by such processes, for example, see Sigurdsson & Hernquist (1993), Portegies Zwart & McMillan (2000), Colpi & Devecchi (2009), Downing et al. (2010), and Ziosi et al. (2014).

During the first few tens of Myr of the life of a globular cluster (GC), BHs will be formed as the remnants of stars more massive than 15 to 20 $M_\odot$ (see Chapter 8). From the shape of the initial mass function, one expects that in a cluster with an initial mass $\gtrsim 10^5\,M_\odot$ hundreds of BHs will have formed, with masses typically in the range of 5 to 20 $M_\odot$. As most of the BHs will have formed by direct core collapse and little or no mass ejection, they are not expected to have received a large velocity kick at birth, and most of them are therefore retained by their clusters (see, however, also the last section of this chapter). From $\sim$100 Myr after the birth of the GC, when all non-degenerate stars more massive than 5 $M_\odot$ have moved off the main sequence and have terminated their evolution, the BHs will have become the most massive stars in the cluster. Because of N-body interactions, the most massive stars will sink to the cluster center, and the BHs will from this time on be concentrated in the cluster core where they will start interacting gravitationally and form binaries by three-body encounters.

In clusters with high central densities (*compact clusters*) these will be close (short-period) binaries. Due to exchange collisions with single BHs and other close BH binaries, these compact BH binaries and the single BHs will ultimately receive large recoil velocities and be ejected from their clusters (Kulkarni et al., 1993; Downing et al., 2010; Mapelli & Bressan, 2013; Breen & Heggie, 2013). As a result, compact GCs are expected to have lost practically all of their single and binary BHs (in theory, one single or binary BH may be left in such a cluster). On the other hand, in GCs that are born with very low central densities (*fluffy clusters*), the dynamical interactions produce only relatively wide-orbit binaries that, through exchange interactions with single

BHs and other wide-orbit BH binaries, will not receive large recoil velocities and are thus expected to be retained by their clusters. Indeed, it has been argued by Gieles et al. (2021) that such fluffy GCs, of which Palomar 5 is a key example, harbor a large population of BHs. In summary, one expects that in compact GCs many close-orbit double BHs have formed (which were ejected from their clusters) and that therefore the dynamical formation of double BHs in such clusters is a viable model for the formation of systems like GW150914 (Downing et al., 2010; Mapelli, 2016; Rodriguez et al., 2016; Parker et al., 2016). We will discuss these dynamical formation processes of double BHs later in this chapter. (Because the double BHs that formed in compact GCs are expected in most cases to have been ejected from their clusters, the later mergers of these systems are expected to occur mostly outside GCs; see Section 12.4.)

Before discussing in more detail the formation of DNSs and double BHs by dynamical processes, we first consider the formation processes of NS binaries with a noncompact companion in dense star clusters, because observations of X-ray binaries and binary radio pulsars in GCs have given us very much observational evidence pointing to the importance of dynamical formation of such binaries in dense star clusters (Fregeau et al., 2004).

In compact GCs, from which the BHs have been ejected, the NSs will become the most massive cluster members only after the normal stars more massive than 1.2 $M_\odot$ have left the main sequence and terminated their evolution as WDs, which is some 6 Gyr after the birth of the cluster. In a GC, which at present has an age of $\gtrsim$ 11 Gyr, the normal stars will have masses smaller than 0.95 $M_\odot$, and the NSs, with their masses mostly in the range 1.2 to 2.1 $M_\odot$, will be the most massive stars in the cluster. Due to dynamical encounters with other stars—leading to equipartition of kinetic energy—these NSs will have sunken to the center of the cluster. The same is true for the most massive main-sequence stars and WDs. On the other hand, due to encounters, the low-mass stars will gain speed and move to the outskirts of the cluster. In the central parts of the cluster, where the NSs and the most massive main-sequence stars and WDs reside, the star density is highest, making it an ideal place for close encounters and binary formation.

Thanks to the discovery of LMXBs in GCs in the 1970s, around the early 1990s, the basics of the formation of NS binaries by dynamical interactions had already become quite well understood. In the following sections, we first give a brief overview of the discovery and formation processes of NS binaries in GCs, largely following what was written on this subject in the review by Bhattacharya & van den Heuvel (1991) supplied with updated references, followed by discussions of formation processes of double BH systems, which later may become gravitational wave (GW) mergers. A variety of extensive reviews concerning different aspects of formation and evolution of X-ray binaries and binary MSPs in GCs has been available in the literature for more than three decades, for example, from the early works by Verbunt (1989, 1990, 1993); Verbunt & Hut (1987); and Phinney & Kulkarni (1994) to more recent and important work by Ivanova et al. (2008, 2010); Colpi & Devecchi (2009); and Benacquista & Downing (2013, and references therein). The reader is referred to these papers for more detailed information. For a popular science review of compact objects in GCs, see, for example, Maccarone & Knigge (2007).

12.2 OBSERVED COMPACT OBJECT BINARIES IN GLOBULAR CLUSTERS: X-RAY BINARIES AND RADIO PULSARS

GCs are compact, gravitationally bound stellar systems that move in extended orbits around the Galactic center and form a nearly spherical "halo" population (see, e.g., Ashman & Zepf, 1998; Binney & Tremaine, 2008). Such a spatial distribution, as well as the shape of their HR diagrams, indicates that the GCs are part of the oldest stellar population of the Galaxy, with ages in the range $11-13$ Gyr. The total number of GCs in the Galaxy is estimated to be $\lesssim 200$. Each of these clusters contains $\sim 10^5$ to 10^6 stars and practically no gas. Because there has been no recent star formation in these systems, all surviving stars in them are > 11 Gyr old, and therefore of low mass ($\leq 0.95\,M_\odot$). The more massive primordial stars will have evolved, leaving WD and NS remnants and, as mentioned earlier, also BHs. One of the early surprises in X-ray astronomy was the discovery of a relatively large population of bright X-ray sources in GCs (Gursky, 1973; private communication to van den Heuvel; Katz, 1975; Clark, 1975). All GCs together contain only $\leq 10^{-3}$ of the total mass of the Galactic disk, but $\sim 20\%$ of the bright LMXBs in the Galaxy.

This huge overabundance (per unit mass) of LMXBs in GCs was immediately attributed (Fabian et al., 1975; Sutantyo, 1975b) to an enhanced rate of binary production by stellar encounters that can occur in the clusters but are completely unimportant in the Galactic disk on account of low stellar density and large relative velocities of the stars here. In the next section, we will discuss the different ways in which encounters can form NS binaries in GCs. Like in the Galactic disk (see Chapters 7 and 11), one expects the LMXB evolution in GCs to lead to the formation of a significant number of rapidly spinning MSPs as the final product. Because of this expectation (Helfand, 1987), searches for radio pulsars in GCs have been undertaken, leading to the first detection of a MSP in the cluster M28 (Lyne et al., 1987), followed by many more such discoveries (e.g., Manchester et al., 1991; Camilo et al., 2000; Ransom et al., 2005; Freire et al., 2008; Cadelano et al., 2018; Ridolfi et al., 2022).

For example, in 1991, it was discovered by Manchester et al. (1991) that 10 MSPs reside in 47 Tucanae. This important discovery emphasized that GCs are breeding grounds for tight binaries with compact objects. Since then, many more MSPs were found in 47 Tucanae (Camilo et al., 2000) and in other GCs (Freire et al., 2008; Pan et al., 2021), in particular in Terzan 5, which hosts, at least, a total of 37 pulsars (Ransom et al., 2005; Cadelano et al., 2018). For recent MeerKAT discoveries of MSPs in GCs, see (Ridolfi et al. (2021, 2022).

At present, the count is some 250 pulsars in GCs,[1] the large majority of which are MSPs with weak magnetic fields (10^7-10^9 G). While this clearly can be interpreted as a dramatic confirmation of the standard expectations (i.e., the recycling model for forming MSPs, see Chapter 14), these discoveries have also raised some new problems regarding the detailed evolutionary routes leading to the formation of these pulsars. There may be some routes to binary MSPs that are characteristic only for globular star clusters and do not occur in the Galactic disk. We describe the possible origin

[1] https://www3.mpifr-bonn.mpg.de/staff/pfreire/GCpsr.html

of some specific type of sources in Section 14.3. Surprisingly, as already mentioned in Section 11.1.3, at least four young strong-magnetic-field radio pulsars have been found in GCs, one (PSR B1718−19 in NGC 6342) being a member of a close binary (Lyne et al., 1996). A simple calculation shows that with their short spin-down ages, the formation rate of these strong magnetic field pulsars must of the same order as that of the MSPs in GCs (Lyne et al., 1996). As they appear to have been formed recently, the only way to explain their existence in these very old stellar systems, in which massive stars have disappeared more than 10 Gyr ago, is formation by AIC of a massive WD in a binary system, as was described in Chapter 11.

One indeed expects that many WDs are present in GCs. Massive ONeMg WDs, with masses $\geq 1.2\,M_\odot$, will, like NSs, sink to the cluster center, where they can capture a non-degenerate companion by the dynamical processes described next for NSs. Mass transfer from the non-degenerate companion to the WD may then make the mass of the latter increase to near the Chandrasekhar mass, upon which electron captures will lead to collapse of the WD to a NS. Because e-capture collapse is not expected to induce a substantial birth kick to the NS (see Section 13.7), the system is expected to be retained in the cluster. Like any other core collapse, this collapse to a NS is expected to be a very violent process, releasing some 10^{53} erg—of which 99% is in the form of neutrinos—and producing an extremely hot and rapidly rotating young NS. This newly-formed NS is not expected to retain any memory of how it precisely formed. For this reason, this newly-formed NS is expected, like any other newly-formed NS, to have a strong magnetic field due to dynamo processes during the collapse and neutrino-driven convective overturn in the young differentially-rotating liquid proto-NS (e.g., Flowers & Ruderman, 1977; Dessart et al., 2006). (Some authors have postulated that electron-capture collapse would immediately produce a MSP with a weak magnetic field [Grindlay & Bailyn, 1988], but there is no physical model justifying why this should be the case.) These newly-formed pulsars will after some 50–100 Myr stop pulsing and disappear into the pulsar graveyard. Much later in life they may have sunken to the cluster core and captured a companion and be spun-up back to a short pulse period, the magnetic field decaying because of the accretion (see Chapters 11 and 14), thus becoming a MSP. A variety of extensive reviews concerning different aspects of formation and evolution of X-ray sources and radio pulsars in GCs is available in the literature, see the introduction of this chapter.

12.3 POSSIBLE FORMATION PROCESSES OF NS BINARIES IN GLOBULAR CLUSTERS

We give here a brief summary of the possible processes of dynamical formation of compact object binaries in dense star clusters. Formation of a bound system of two stars starting from an initial unbound configuration requires a sink of orbital energy, which can be achieved in four different ways:

1. In a collision (gravitational encounter) involving three stars, one of the objects may take up the excess energy, leaving the other two in a bound system. Because this process requires a simultaneous convergence of three independent trajectories, the

probability of this happening is mostly rather small for NSs. (For BHs it may under certain circumstances be a more viable way.)

2. Tidal capture: during a close passage of a compact star, a normal star can undergo a substantial tidal deformation at the cost of part of the relative kinetic energy of the orbit. Most of this energy will eventually be dissipated through oscillations and heating. If the amount of energy thus lost exceeds the total positive energy of the initial unbound orbit, a bound system will result. This mechanism is thought to be the most important binary formation route in GCs.
3. Direct physical collision. If the two stars do not coalesce in this process, still a binary may remain.
4. Exchange collisions: if a NS interacts with an already existing binary, an *exchange* may take place, in which the NS replaces one of the components of the original binary system.

We only consider here the three last-mentioned formation mechanisms.

12.3.1 Tidal Capture

The conditions necessary for tidal capture to occur, and the cross section for this process, were first computed by Fabian et al. (1975). Since then, several more refined calculations have been made that yielded very similar results. The main points are illustrated by the following example, after Verbunt & Hut (1987): Consider an interaction between a normal star of mass M (the target star) and a compact star (NS or WD) of mass m. The relative kinetic energy of the orbit at any time can be written as:

$$E_{\mathrm{kin}} = \frac{1}{2}\mu v^2 , \tag{12.1}$$

where $\mu \equiv Mm/(M+m)$ is the reduced mass, and v is the relative velocity of the two stars at that time. The initial total (positive) energy E_0 is given by Eq. (12.1) with $v = v_\infty$, the initial relative velocity at a large distance. At the closest distance of approach, d, the height, h, and mass, M_1, of the tidal bulge on the target star are:

$$h \simeq \left(\frac{m}{M}\right)\left(\frac{R^4}{d^3}\right) \tag{12.2}$$

$$M_1 \simeq k\left(\frac{h}{R}\right) M \simeq k\left(\frac{R}{d}\right)^3 m , \tag{12.3}$$

where R is the radius of the target star, and k is a constant (the gyration radius) that depends on the density distribution within the target star. For the deeply convective main-sequence stars in GCs, k is estimated to be ~0.14 (Motz, 1952). The tidal deformation of the compact star itself is negligible because of its small size and strong gravity. The potential energy of the tidal bulge is:

$$E_1 \simeq \frac{GMM_1}{R}\left(\frac{h}{R}\right) \simeq k\,\frac{Gm^2}{R}\left(\frac{R}{d}\right)^6 . \tag{12.4}$$

If a fraction, η, of this energy is dissipated, then the condition for tidal capture to occur can be written as $\eta E_1 > E_0$, that is,

$$d < \left[2\,\eta k\, G\left(\frac{m}{M}\right)\left(\frac{m+M}{R}\right)\right]^{1/6} R\, v_\infty^{-1/3} \tag{12.5}$$

$$d < 3.2\,R\left[\eta\left(\frac{k}{0.14}\right)\left(\frac{m}{M}\right)\left(\frac{m+M}{2\,M_\odot}\right)\left(\frac{R_\odot}{R}\right)\right]^{1/6}\left(\frac{v_\infty}{10\,\mathrm{km\,s^{-1}}}\right)^{-1/3} .$$

The normalizations in Eq. (12.5) are on parameters appropriate for stars in GCs. The small exponents make the expression very insensitive to all the parameters involved, and thus a safe rule of thumb is that the tidal capture occurs when the compact star approaches the target star within a distance of ~3 stellar radii. This result is valid, of course, only if the escape velocity from the surface of the target star is significantly larger than v_∞. This condition is satisfied in most cases, except for very bloated giant stars (such as AGB stars). In such cases, the upper limit to d for which capture can take place will be smaller than given by Eq. (12.5). However, the duration of such an evolutionary phase is so short that this will hardly make a difference in the net binary formation rate in a GC. Because the tidal effects are negligible until the two stars are very near closest approach, the relation between the distance of closest approach, d, and the impact parameter, b, can be obtained using conservative Newtonian dynamics:

$$b = d\sqrt{1 + \frac{2\,G(m+M)}{v_\infty^2 d}} . \tag{12.6}$$

For all situations of interest in GCs, the second term in the square root, which represents "gravitational focusing," is much larger than unity, and it is this effect that makes the cross section of this process considerable in spite of the very close approaches that are needed. The cross section can be written as:

$$\sigma = \pi b^2 \simeq \pi d\,\frac{2\,G(m+M)}{v_\infty^2} . \tag{12.7}$$

Of particular interest is the *linear* dependence of the cross section on d, also a consequence of the strong gravitational focusing. Because captures occur for $d \lesssim 3.2\,R$, this implies that at least one-third of the captures involve a direct collision between the two stars (Fabian et al., 1975). Direct collisions do not always need to produce a binary system (see Section 12.3.2). Using the cross section given by Eq. (12.7), one can write the rate of tidal capture of compact stars per unit volume as (see also Lee &

Ostriker, 1986):

$$\begin{aligned}\Gamma &= n_{\mathrm{c}} n\, v_\infty\, \sigma \\ &\simeq n_{\mathrm{c}} n\, \pi d\, \frac{2\,G(m+M)}{v_\infty} \\ &\simeq 4.0\times 10^{-10}\,\mathrm{yr}^{-1}\,\mathrm{pc}^{-3} \\ &\quad \times \left(\frac{n_{\mathrm{c}}}{10^{2}\,\mathrm{pc}^{-3}}\right)\left(\frac{n}{10^{4}\,\mathrm{pc}^{-3}}\right)\left(\frac{3.2\,R}{R_\odot}\right)\left(\frac{m+M}{2\,M_\odot}\right)\left(\frac{10\,\mathrm{km\,s}^{-1}}{v_\infty}\right),\end{aligned} \tag{12.8}$$

where n_{c} and n are the number density of compact stars and target stars, respectively. The net tidal capture rate in GCs can be obtained by integrating this equation over the cluster volume.

12.3.2 Direct Collisions

As mentioned previously, approximately one-third of the tidal capture events involve physical collisions between the two stars. The outcome of such a collision is not entirely certain; it depends to a large extent on the evolutionary state of the target star and several other poorly understood details. A trajectory with a very small distance of closest approach ($d \leq R/4$ for a polytropic star) would not have enough angular momentum to form a binary, and a merger of the two stars is expected (Lee & Ostriker, 1986). Numerical experiments of collisions between dust spheres show that a significant amount of angular momentum may be lost from the system by jet-like ejection of matter (Benz & Hills, 1987), which would indicate a merger outcome even for larger distances of closest approach. It has been suggested that a direct hit between a compact star and a main-sequence star results in the complete disruption of the main-sequence star, part of its matter forming a massive disk around the compact star (Krolik et al., 1984; Shara & Regev, 1986; Soker et al., 1987). Direct collisions between two main-sequence stars are likely to produce a single star with the combined mass of the two stars. This is one of the leading suggested formation channels for the high incidence of blue stragglers in GCs (e.g., Davies et al., 2004; Mapelli et al., 2006; Knigge et al., 2009), and possibly also from triple-star interactions (Perets & Fabrycky, 2009; Antonini et al., 2016). (Blue stragglers are main-sequence stars above the turn-off point of the GC main-sequence. Almost every GC has several of such blue stragglers.) Because collisions are most likely to occur among the most massive main-sequence stars, such a merger result will then typically be a blue straggler. Alternatively, blue stragglers are produced via stable mass transfer in binaries, that is, post-Algol systems, as we describe in detail in Chapters 9 and 16.

The direct collision between a compact star and a red (sub)giant star will lead to a quite different outcome. This case was pioneered by Sutantyo (1975b). Because of the large radius of a red giant, the collision cross section in this case is much larger than for a main-sequence target star. A low-mass star in the (sub)giant phase has already a compact degenerate helium core. It is possible that after the collision the compact star and this compact (WD-like) core will orbit each other inside the (sub)giant envelope. This is a typical common envelope (CE) situation (Chapter 4), and tidal friction may then lead to a deep spiral-in of these two components and expulsion of the envelope, leaving the two compact stars in an ultra-compact circular binary orbit (Verbunt, 1987). In this way, the 11.4 min orbital period X-ray binary 4U 1820−30 in NGC 6624 may have been formed (but see also Section 11.4). Stellar merges may be a common occurrence also in resonance encounters between a compact star and a binary consisting of two normal stars, as will be dealt with in the next section. In the course of the complicated three-body orbits, the distance of closest approach may easily become small enough for a merger to occur. The third star may either participate in the merger, stay in orbit around the merger product, or escape from the system (Hut & Bahcall, 1983; Krolik et al., 1984; McMillan et al., 1987; Portegies Zwart & McMillan, 2000; Benacquista & Downing, 2013, and references therein).

12.3.3 Exchange Encounters

We follow here closely the review by Mapelli (2018). Binaries have a much larger cross section for collisions than single stars. So, when there is a sizeable fraction of binaries in a cluster, gravitational encounters between binaries and single stars will be an important source of interactions, in which a binary component can be exchanged for a passing single star (see Fig. 12.1). If this passing star is compact, a binary with a compact component may form. As we will see below, binaries furthermore form an energy reservoir for the star cluster, which may drive its dynamical evolution. For a bound binary system holds that the total energy is negative:

$$E_{\rm tot} = \frac{1}{2}\mu v^2 - \frac{Gm_1m_2}{r} < 0, \tag{12.9}$$

where again $\mu \equiv (m_1m_2)/(m_1+m_2)$ is the reduced mass, m_1 and m_2 are the masses of the two components, G is the gravitational constant, and v and r are the relative velocity and the distance between the two stars in the binary, respectively. When a single star closely approaches a binary, within a few times the orbital radius of the binary, part of the binary's energy can be transferred to the single star, such that its kinetic energy increases while the binary's separation shrinks: the binary *hardens*. In this process, the total energy of the binary becomes more negative, and the system gets more strongly bound. Alternatively, the single star may lose part of its kinetic energy in the encounter, causing the binary orbit to get wider: it becomes *softer*, or even may be ionized if its total energy becomes positive.

The binding energy, $E_{\rm bind}$ of a binary is the energy needed to unbind the system, that is, to make $E_{\rm tot} = 0$ (i.e., $E_{\rm bind} = -E_{\rm tot}$). A binary in a cluster is called hard or soft, depending on:

$$\begin{aligned} \text{hard}: \quad & E_{\rm bind} > \frac{1}{2}\langle m\rangle\,\sigma_{\rm vel}^2 \\ \text{soft}: \quad & E_{\rm bind} < \frac{1}{2}\langle m\rangle\,\sigma_{\rm vel}^2, \end{aligned} \tag{12.10}$$

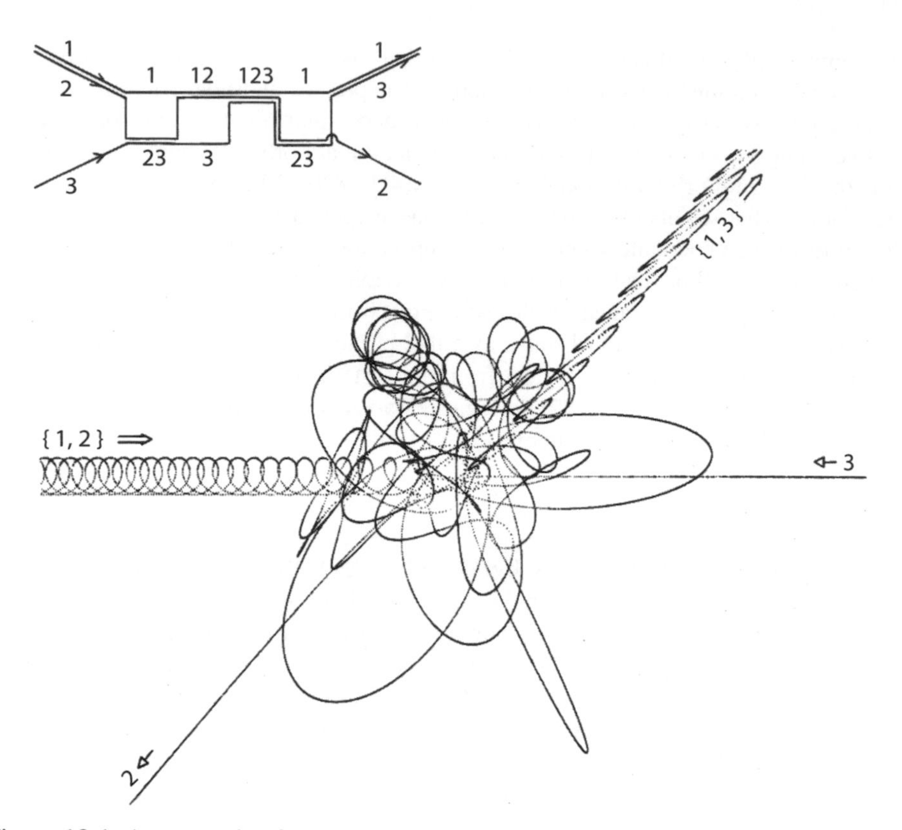

Figure 12.1. An example of a resonance scattering process involving a hard binary and a field star. The trajectories of the three equal-mass stars are plotted in the center-of-mass frame. The original binary comes in from the left and contains the stars that are labeled 1 and 2. The single star comes in from the right and is labeled 3. The orbits are represented by dots spaced at constant time intervals. The outcome of the scattering process is an exchange: star 2 escapes, leaving 1 and 3 behind as a newly formed binary. In the upper left-hand corner, the degree of mutual gravitational binding of the three stars is represented schematically in a scattering diagram. After Hut & Bahcall (1983).

where $\langle m \rangle$ is the average mass of a star in the cluster, and $\sigma_{\rm vel}$ is the velocity dispersion of the stars. According to the fundamental work of Heggie (1975), hard binaries tend to become harder due to three-body encounters, and soft binaries tend to get softer ("Heggie's law"). It was found that the hardening rate of the binary does not depend on the mass of the binary, but only on the local mass density, ρ in the cluster, and the velocity dispersion, $\sigma_{\rm vel}$ of the stars:

$$\frac{d}{dt}\left(\frac{1}{a}\right) = 2\pi\xi\,\frac{G\rho}{\sigma_{\rm vel}}, \tag{12.11}$$

where a is the semi-major axis of the binary, and $\xi \sim 0.2-3$ is a dimensionless efficiency parameter (Hills, 1983a; Quinlan, 1996; Miller & Hamilton, 2002b). According to Eq. (12.11), a hard binary hardens at a constant rate. The (slight) dependence of the hardening rate on a is absorbed in the value of the constant ξ, which is derived from numerical N-body simulations. In this way, by energy loss due to hardening, binaries can pump kinetic energy into the star cluster. The evolution of star clusters due to this "burning of binaries" resembles the evolution of a star due to nuclear burning. The binaries are the energy source of the cluster.

12.3.4 Core Collapse: Start of Dynamical Domination of NSs/BHs

As noticed by Spitzer (1987), isolated star clusters composed of a dense core and a low-density halo of stars are, like a star without nuclear energy sources, expected to undergo a continuous shrinking of the core ("core collapse" or *gravothermal catastrophe*) due to their negative heat capacity, resulting from the virial theorem (see the beginning of Chapter 8). This core collapse due to dynamical interactions in a cluster composed of single stars would—if nothing would stop it—lead to an infinite density and a zero core radius. However, at very high densities, three-body encounters will lead to the formation of binaries. And when binaries are present in the core, the onset of the "burning" of binaries will (like the onset of nuclear burning in a contracting star) halt the further collapse of the core. Here potential energy is removed from the core by hardening of the binaries, and by injecting fresh kinetic energy into the core, as was first pointed out by Hénon (1961). It appears that the net result of the hardening of binaries is that the injected kinetic energy of the stars causes the core to expand: the presence of binaries causes the core collapse to reverse. During the core collapse itself, the star density in the cluster core may become very high before the reversal due to binaries sets in. During this very high-density collapse state, all kinds of binaries may form by three-body encounters. A key example is the post-core collapse GC M15, in which a close DNS with a short orbital period is present (PSR B2127+11C: $P = 30.5$ ms, $P_{\rm orb} = 0.335$ d, $e = 0.681$), as well as many other interesting radio pulsars and one luminous X-ray binary (4U 2127+11: $L_{\rm X} = 10^{36.7}$ erg s^{-1}, $P_{\rm orb} = 8.6$ hr). The formation of the DNS system is likely to have happened during the core collapse, as it is difficult to see how otherwise such a close system consisting of two NSs could have formed. (The usual isolated binary star [standard] DNS formation scenario requires that the binary is very wide prior to the CE phase to avoid coalescing during the subsequent in-spiral; see Figs. 10.13 and 14.35.)

12.4 DYNAMICAL FORMATION OF DOUBLE BHs

12.4.1 Formation in Globular Clusters

Although many BHs are expected to have formed in the early life of a GC, observational evidence for their existence in globular clusters in our Galaxy is very scarce. Giesers et al. (2018) detected a BH candidate in a detached binary in the cluster NGC

3201, and Strader et al. (2012); Chomiuk et al. (2013) and Miller-Jones et al. (2015) reported the radio detection of likely BH-X-ray binaries in the clusters M22, M62, and 47 Tuc, respectively. However, there is indirect evidence for a large population of BHs in the GCs NGC 6397 (Vitral & Mamon, 2021) and Palomar 5 (Gieles, et al., 2021). On the other hand, solid X-ray evidence for the existence of BHs has been found in several globular clusters in external galaxies (Maccarone et al., 2007; Shih et al., 2008; Shih et al., 2010; Zepf et al., 2008; Irwin et al., 2010).

Colpi & Devecchi (2009), Downing et al. (2010), Rodriguez et al. (2016), and Ziosi et al. (2014) have shown that BHs in dense (compact) clusters are very efficient in capturing a companion by dynamical interactions because after 10^8 yr they are more massive than the other cluster stars. Exchanges of a binary member (in a binary-single star interaction) is more likely if the intruder is more massive than one of the binary members. Exchanges therefore favor the formation of more and more massive binaries, which will eventually be double BHs. The BH binaries formed in this way start out with very large eccentricities. They will be relatively hard binaries, and therefore encounters with other cluster stars will further harden them, which may lead to binaries close enough to merge within a Hubble time, and thus being detected by the LIGO-Virgo-KAGRA network of GW detectors (Chapter 15). One can make a simple analytical estimate of the evolution of the semi-major axis, a of a double BH binary due to three-body encounters in combination with GW losses, as shown by Colpi et al. (2003):

$$\left(\frac{da}{dt}\right) = -2\pi\,\xi\,\frac{G\rho}{\sigma_{\rm vel}}\,a^2 - \frac{64}{5}\frac{G^3 m_1 m_2 (m_1+m_2)}{c^5 a^3\,(1-e^2)^{7/2}}, \tag{12.12}$$

which is found by adding contributions from hardening via three-body encounters (Eq. 12.11) and an approximation for hardening due to emission of GWs in a binary with eccentricity, e (Eq. 15.7, see Chapter 15).

Figure 12.2 shows the solutions of Eq. (12.12) for six double BH binaries, numerically calculated by Michela Mapelli (private communication). In the beginning, the orbital shrinking due to the three-body interactions dominates, but during the last part of the evolution the shrinking due to GW losses dominates. For further details of the role of GW emission in binary-single stellar encounters and the formation of eccentric compact binary (double BH) in-spirals, including relativistic N-body dynamics, see, for example, Samsing et al. (2014); Samsing & D'Orazio (2018).

If the assembled double BH binaries and their merger products were to remain inside the clusters, one would expect BHs with increasing masses to be produced in clusters: for example, a merging double BH of $(40+35)\,M_\odot$ would produce a BH with mass[2] $> 70\,M_\odot$, and such a BH would easily capture another BH companion, merge, and produce a still more massive BH, and so on. As a result, in clusters more and more massive BHs would then be produced. Because of their large masses, the mergers of such very massive BHs would easily be detected by GW detectors.

[2] An energy of a few solar masses ($M_\odot c^2$) is typically lost in gravitational wave emission.

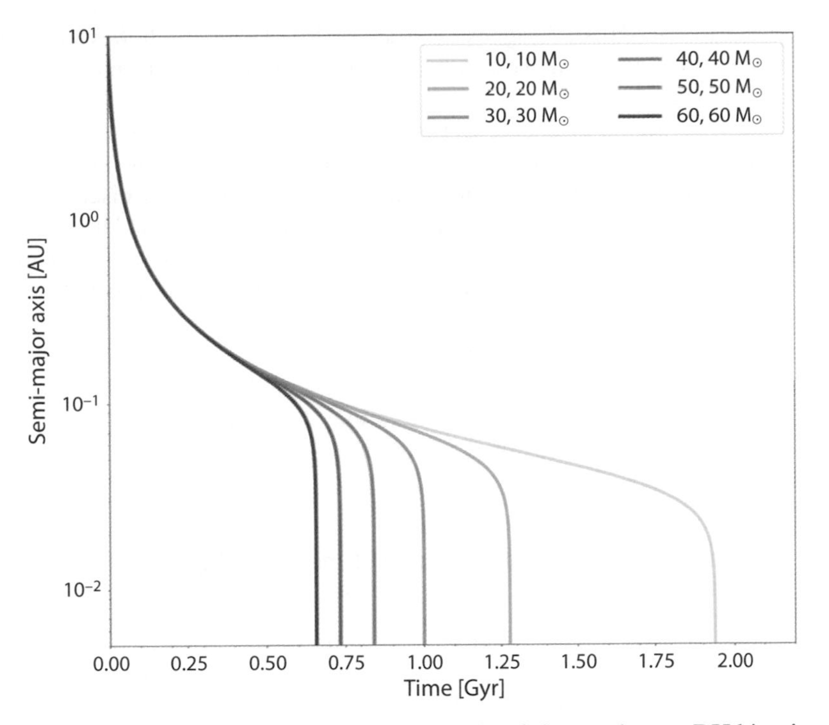

Figure 12.2. Time evolution of the semi-major axis of six equal mass BH binaries estimated from Eq. (12.12). The BH masses are indicated in the figure. For all BH binaries: $\xi = 2$, $\rho = 10^5\, M_\odot\, \mathrm{pc}^{-3}$, $\sigma_{\mathrm{vel}} = 10\,\mathrm{km\,s}^{-1}$, initial semi-major axis of $a_\mathrm{i} = 10\,\mathrm{AU}$ and eccentricity, $e = 0$. Figure courtesy: Michela Mapelli, private communication, (2021).

However, the scenario is more complicated due to momentum kicks. As shown, for example, by Kulkarni et al. (1993), Breen & Heggie (2013), and Mapelli & Bressan (2013), the three-body encounters induce kicks to the BH binaries that are likely to kick them and the interacting BH out of their clusters, because the escape velocities of clusters are generally less than a few tens of $\mathrm{km\,s}^{-1}$. One would therefore expect that the merging events of double BHs produced by N-body dynamics in star clusters take place *outside* the clusters. Furthermore, the merger process itself also induces a kick (recoil) onto the BH merger remnant (Campanelli et al., 2007; González et al., 2007), such that these remnants are often kicked out of their host clusters.

Looking at the masses of the > 85 double BH mergers detected so far by the LIGO network (Fig. 15.13), the great majority of these systems may indeed be first-generation mergers. However, an increasing population of potential second- and third-generation mergers seems to change this picture. Most noteworthy are the extremely massive double BH mergers, GW190426 and GW190521, which had BH component masses of $\sim 107 + 77\, M_\odot$ and $\sim 95 + 69\, M_\odot$, respectively (see Section 15.6.2 for discussions of these sources). These systems are therefore interesting candidates for being produced by repeated BH mergers via dynamical interactions in star clusters (Fragione et al.,

2020; Gayathri et al., 2020; Anagnostou et al., 2020). There are, however, other formation models on how to form these systems that do not involve dynamical encounters (see Section 15.6.2).

As a nice illustration of double BH formation in GCs due to dynamical interactions, Figure 12.3 depicts as an example the formation of such a system according to the numerical calculations by Rodriguez et al. (2016). For a further discussion of the formation of double BHs by dynamical interactions in star clusters, and the importance of this formation mechanism relative to formation by binary evolution, see Chapter 15.

Further interesting studies of double BH merger effects in GCs include those by Zevin et al. (2019), who studied the contribution of binary-binary interactions in GCs and found that these make a substantial (25 to 45%) contribution to all of the eccentric mergers resulting from GC binary formation processes, and the work by Ng, Vitale et al. (2021), who studied the contributions of the various double BH formation processes to double BH mergers observable out to a redshift of $z \sim 30$, which are expected to be detectable by the 3G detectors.

12.4.2 Dynamical Formation of Double BH Binaries in Active Galactic Nuclei

The supermassive BHs in Galaxy nuclei tend to be surrounded by nuclear star clusters, and in the case of active galactic nuclei (AGNs) these BHs are also surrounded by gaseous disks that can be very dense. Here dynamical processes may occur that do not occur in GCs. For example, Hamers et al. (2018) studied the evolution of BH binaries near a massive BH due to the Lidov-Kozai effect, which tends to continuously increase the eccentricity of the binaries. They found, however, that this may only lead to significant (observable) effects if the massive BH has a mass $\lesssim 10^4\ M_\odot$. This is much smaller than the masses of most galactic nuclei. On the other hand, Tagawa et al. (2020) find that in AGNs, double BHs can form and evolve due to interactions with the nuclear star cluster and the gaseous disk. Binaries may form here from two single BHs because of kinetic energy dissipation by friction in the gaseous disk. They find that even without pre-existing binaries, this environment can produce double BH mergers at a rate of $0.02 - 60\,\mathrm{Gpc^{-3}\,yr^{-1}}$. See also recent work by Samsing et al. (2022).

12.5 COMPACT OBJECTS IN GLOBULAR CLUSTERS CONSTRAIN BIRTH KICKS

We know that NSs produced by iron-core collapse SNe are imparted with large velocity kicks at birth, of the order of at least several hundreds of $\mathrm{km\,s^{-1}}$. Such NSs cannot be retained in GCs. On the other hand, NSs produced by electron-capture collapses of degenerate ONeMg cores are expected to receive hardly any kick at birth (see Chapter 13), and the same is generally expected for ultra-stripped metal cores (depending on their masses), which are typical products of close binary evolution prior to the second SN (Case BB RLO). Also, as pointed out above, the AIC of ONeMg WDs in binaries are expected to impart only small kick velocities to the newly formed

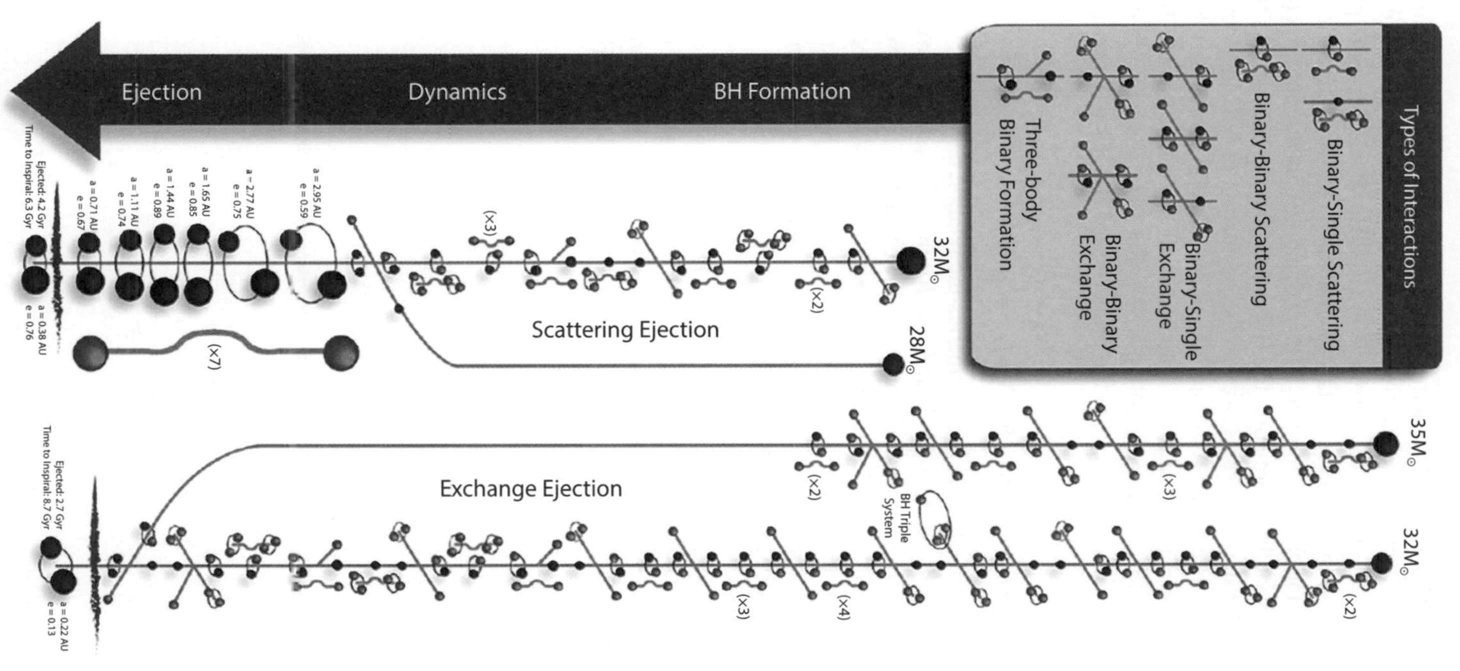

Figure 12.3. Interaction diagram showing the formation history for two GW150914 progenitors in a single GC model. From top to bottom, the history of each individual BH that will eventually comprise a GW150914-like binary is illustrated, including all binary interactions. The legend shows the various types of gravitational encounters included in the GC model (with the exception of two-body relaxation). In each interaction, the black sphere represents the GW150914 progenitor BH, while the blue and red spheres represent other BHs (and stars) in the cluster core. After Rodriguez et al. (2016).

NSs (see also Chapter 11). These facts explain that there is a population of NSs residing in GCs, and could later, by dynamical interactions, capture companions and produce the NS binaries and their products that we now observe in GCs. However, it is quite possible that, like NSs produced by iron-core collapses, BHs also receive a large kick at birth. This would make the formation of double BHs by dynamical interactions in clusters very unlikely, as already at their formation the BHs would have received kicks that made them escape from the star clusters. However, the population of NSs in globular clusters, along with the evidence from GW detections of double BHs, which may have formed dynamically, tell us that at least a fair fraction of all newborn NSs and BHs must receive small momentum kicks in core-collapse SNe. For BHs, some information about possible birth kicks can be derived from the properties of BH X-ray binaries, as follows.

Information about BH Kicks Derived from BH X-ray Binaries

Mirabel (2017) studied for Galactic BH X-ray binaries the relation between BH mass and excess space velocity of the system. He found that for the five systems for which this information is available, the two systems with the lowest BH masses (between 5 and $\sim 7\,M_{\odot}$) have the largest excess velocities ($\sim 120-180\,\mathrm{km\,s^{-1}}$), whereas the three systems with the largest BH masses (between 9 and $\sim 20\,M_{\odot}$) have the lowest space velocities (between <9 and $\sim 40\,\mathrm{km\,s^{-1}}$). Mirabel (2017) made the interesting suggestion that the lowest-mass BHs formed by fallback onto an initially-formed (proto) NS, and also suggested that in this case a high velocity might be imparted, whereas the more massive BHs formed by direct core collapse, in which case hardly any velocity is imparted. Other observational evidence about birth kick of BHs comes largely from the Galactic latitude distribution of the BH X-ray binaries but, unfortunately, the evidence is still rather inconclusive (Nelemans et al., 1999; Janka, 2013; Mandel, 2016; Repetto et al., 2017; Janka, 2017).

Chapter Thirteen

Supernovae in Binaries

13.1 INTRODUCTION

From the work of Fritz Zwicky, Walter Baade, and Rudolf Minkowski in the late 1930s, it became clear that observed supernovae (SNe) are divided into what Minkowski called Type I and Type II (Minkowski, 1941), depending on whether hydrogen is absent or present in the early-time optical spectra (see, e.g., Filippenko, 1997; Turatto, 2003). These two main types were in later years further divided in a number of subtypes. Figure 13.1 shows the spectra of some of the main SN sub-types near maximum light (approximately one week). Among the Type II SNe, the following subtypes are distinguished (see Fig. 13.2 showing the shapes of the light curves of the different SN subtypes): Type IIp (long plateau of constant luminosity in the light curve, lasting several months), Type IIL (linear light-curve decay), Type IIb (hydrogen only in the spectra of the first few weeks, then disappearing), and Type IIn (very narrow H-emission lines, doppler-broadening $< 100\,\mathrm{km\,s^{-1}}$, ascribed to interaction of the SN ejecta with circumstellar matter).

Among the Type I SNe, Type Ia constitute a special class. They show no hydrogen and helium but strong Si-absorption lines, which is the defining characteristic of this type. These SNe are different from all the other types of SNe, as it is the only type of SN that occurs in elliptical galaxies, which contain only low-mass stars ($< 1.5\,M_\odot$). All other types of SNe occur only in star-forming galaxies that contain massive stars (Type Ia SNe occur in these galaxies as well). It is generally accepted that Type Ia SNe are thermonuclear explosions resulting from the runaway carbon fusion in a carbon-oxygen (CO) WD, triggered by mass transfer from a binary companion or a merger event between two WDs. This implies that they are definitely associated with the evolution of binary systems. Their nature, physics, and proposed formation models are covered in the next section of this chapter.

The other hydrogen-poor SNe are divided into Types Ib and Ic, which are distinguished by the presence or absence of helium in their spectra, respectively. Among the helium-poor Type Ic SNe, one further distinguishes a narrow-line class with typical ejecta outflow velocities of order $10^4\,\mathrm{km\,s^{-1}}$, and a *broad-line* class Type Ic-BL, discovered by Galama et al. (1998), with outflow velocities $(3-5)\times10^4\,\mathrm{km\,s^{-1}}$. Part of the SNe in the latter class were found to be the birth events of long gamma ray bursts (GRBs; see Galama et al., 1998; Hjorth et al., 2003; Kouveliotou et al., 2012). Apart from Type Ia, all other SN types are associated with the core collapses of massive stars

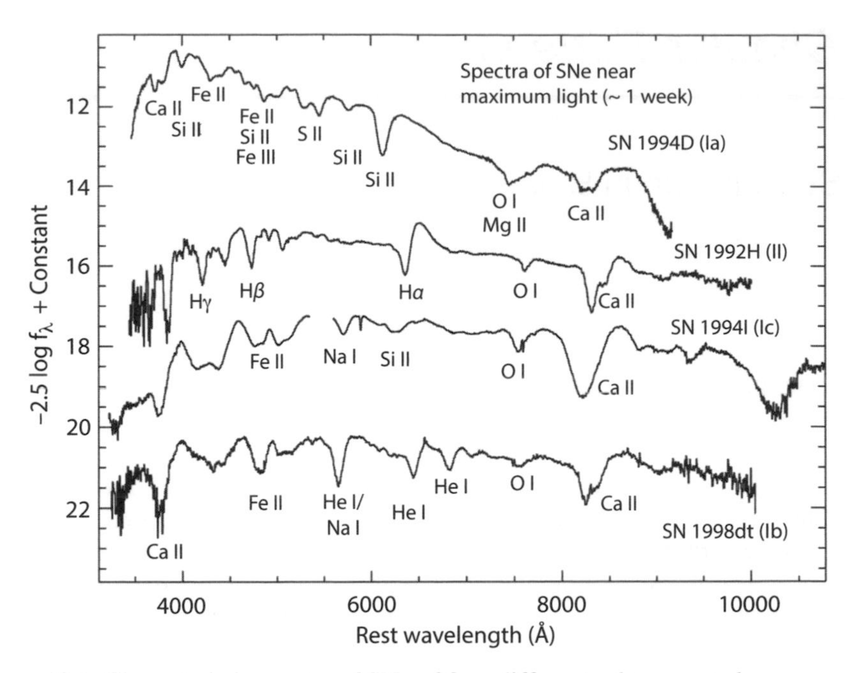

Figure 13.1. Characteristic spectra of SNe of four different subtypes, at about one week after maximum light. Figure courtesy: Thomas Matheson and Alex Filippenko.

($M \gtrsim 8\,M_\odot$). They are, therefore, the birth events of NSs and BHs (see Chapter 8). Figure 13.3 depicts the observational classification of SN types.

As to the SNe of Type Ib and Ic: the absence of hydrogen in their spectra indicates that they are core collapses of stars that have been stripped of their hydrogen-rich envelopes. For this reason, they are also designated as *stripped-envelope SNe* (SE-SNe). Together they make up some 30% of all SNe.

Envelope stripping can for massive stars ($> 30\,M_\odot$) be achieved by strong stellar wind mass loss (see Chapter 8), which may turn a massive star into a (WR) star. For this reason, initially Type Ib/c SNe were thought to be associated with WR-stars (e.g., Begelman & Sarazin, 1986; Gaskell et al., 1986). However, stars more massive than $30\,M_\odot$ are far too rare to produce 30% of all SNe. In stars less massive than $30\,M_\odot$, the winds are too weak to cause envelope stripping, and this stripping can only be achieved by mass transfer in a binary system. Indeed, observations show that some 70% of massive stars are born in interacting binary systems (Sana et al., 2012). Due to the shape of the initial mass function (IMF) and the star formation rate (SFR), the death rate of stars with masses between 8 and $30\,M_\odot$ is about six times larger than the death rate of stars $> 30\,M_\odot$. For this reason, it is now thought that the vast majority of the SE-SNe is associated with envelope stripping by mass transfer in binary systems with initial primary masses $\leq 30\,M_\odot$ (see, e.g., Dessart et al., 2012).

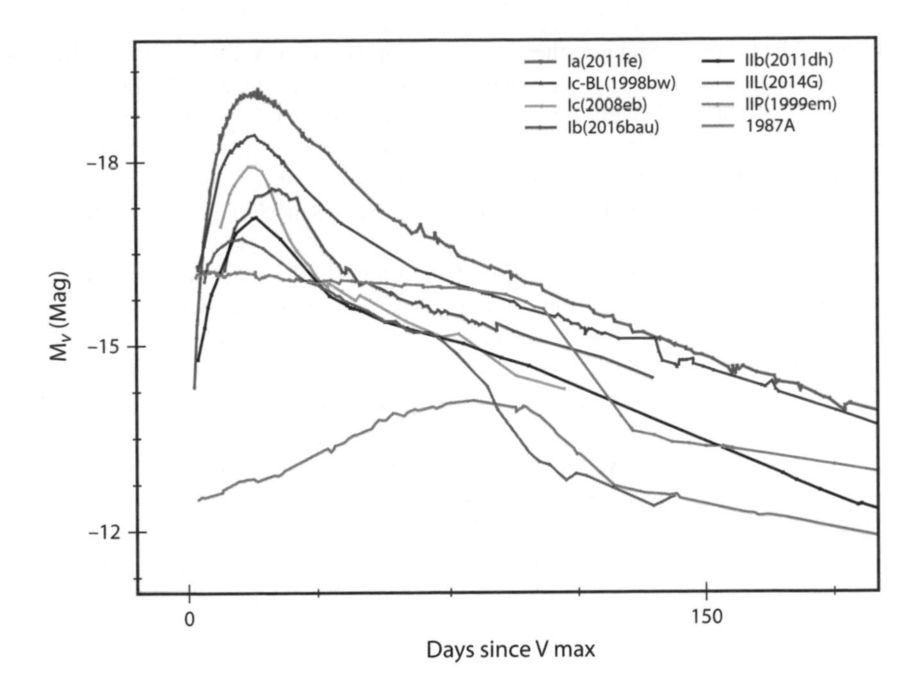

Figure 13.2. Shapes of the light curves of different SN subtypes. Figure courtesy: WeiKang Zheng and Alex Filippenko.

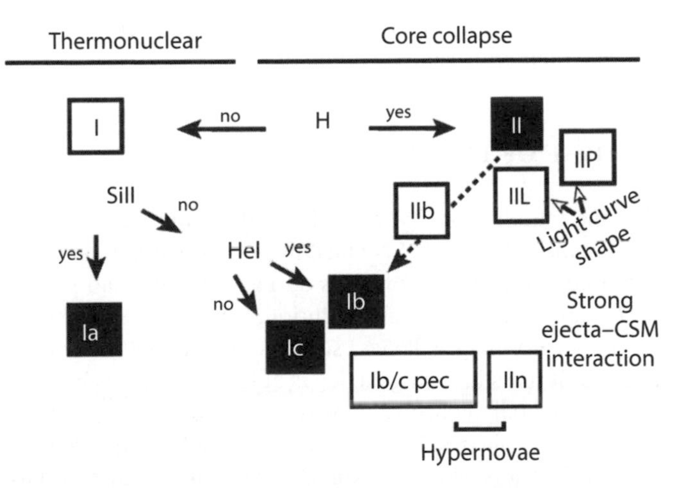

Figure 13.3. Observational classification scheme of the different subtypes of SNe. The Type Ia SNe, which have no H and He in their spectra but strong lines of Si, are different from all others: they are thermonuclear SNe, powered by the explosion of a CO WD, triggered by accretion or a merger event between two WDs. All the other subtypes are core-collapse SNe. Further explanation is in the text. After Turatto (2003).

Lucy (1991) showed that non-thermal excitation and ionization are key for the production of He-I lines in the spectrum, which lead to the classification Ib. Dessart et al. (2012) showed that moderately-mixed SN models of stripped stars (originating from ZAMS masses below 30 $M_{\odot}$), which still have helium, fail to non-thermally excite helium atoms and may therefore lead to classification Ic. For this reason, classification Ic of SE-SN spectra does not necessarily imply the absence of helium in the direct progenitor of the SN.

Since the progenitors of Type II SNe retained their hydrogen-rich envelopes, most of these SNe are thought to be due to single stars or the components of wide non-interacting binaries. There are, however, exceptions, for example the Type II SN SN1987A, the first naked-eye SN since Kepler's star in 1604. There is strong evidence that its blue supergiant progenitor star was the product of a binary merger, as will be described later in this chapter. This is most likely the reason for its quite anomalous light curve (see Fig. 13.2). We will not further discuss Type II SNe here, as most of them are expected to be products of single star evolution.

This chapter is structured as follows: The next section is devoted to observational properties and models of Type Ia SNe. The subsequent section reviews the observational properties of stripped-envelope SNe (Types Ib and Ic) and gives a general overview of theoretical models of the evolution and fate of stripped stars. After this follow two sections devoted to special types of SNe: electron-capture SNe and ultra-stripped SNe, respectively. As these are expected to often impart only small kick velocities to NSs, these SNe are of special importance for the formation of X-ray binaries and DNSs. Then follows a section in which we discuss how the different types of theoretically predicted core collapses of stripped-envelope stars may be identified with observed types of SNe Ib and Ic. The last section of this chapter is devoted to SN kicks imparted to NSs and their effects on the orbits of binaries.

13.2 SUPERNOVAE OF TYPE Ia

As mentioned in the introduction, Type Ia SNe differ from all other types of SNe by the fact that they are the only type of SNe that also occur in elliptical galaxies, in which star formation ceased many billions of years ago and in which all stars more massive than about 1.5 $M_{\odot}$ have terminated their evolution. Like all other types of SNe, Type Ia SNe also occur in star-forming galaxies, such as spirals and irregular galaxies, and their incidence per unit mass in such galaxies is at least as high as in elliptical galaxies (Maoz et al., 2014). The only conceivable way in which, in a population of low-mass stars such as in elliptical galaxies, a star can be triggered to explode as a SN is: if we are dealing with a WD that is driven to explode due to an increase of mass (Hansen & Wheeler, 1969), which is most easily achieved by mass transfer from a companion star in a binary (Whelan & Iben, 1973) or via a merger event between two WDs (Webbink, 1984; Iben & Tutukov, 1984; van Kerkwijk et al., 2010; Pakmor et al., 2012).

It was realized long ago by Hoyle & Fowler (1960) that a degenerate star that consists of C+O still contains a large amount of combustible nuclear fuel in the form of carbon that, if it were to ignite carbon burning, would lead to a catastrophic situation.

This is because the ignition of a nuclear fuel in a degenerate star leads to runaway nuclear burning, since the pressure in a degenerate gas does not respond to temperature increase with expansion and cooling (see explanation in Chapter 8). Therefore, one expects that upon carbon ignition, most or all the carbon of the degenerate star will be consumed (and also its oxygen, but that releases less energy) in a gigantic explosion. The combustion of 0.5 $M_{\odot}$ of carbon releases of order 10^{51} erg, typically the energy output of a SN Ia explosion. Hoyle & Fowler (1960) therefore proposed that Type Ia SNe are due to the thermonuclear explosion of a degenerate C+O star, which basically is a CO WD. The ratio C/O in CO WDs is expected to range from 50/50 to 30/70, depending on the main-sequence mass of the progenitor. This model for the origin of Type Ia SNe was further developed by Arnett (1969) into the C-detonation SN model, and later into the C-deflagration model of Nomoto et al. (1976), who also coined the name *deflagration*. In a deflagration explosion, the burning front moves subsonically, while in a detonation its speed is supersonic.

To trigger a CO WD to explode, its mass must be increased. One class of models requires an increase in mass up to the Chandrasekhar mass of about 1.4 $M_{\odot}$ to produce an explosion, but in recent years it has been found that models that require triggering a CO WD of lower mass to explode appear to give a good explanation of the observed light curves and spectra of SNe Ia (see, e.g., the arguments summarized earlier by Livne, 1990, and references therein).

As already mentioned, the only conceivable way to substantially increase the mass of a WD is by mass transfer from, or via collision with, a companion star (Whelan & Iben, 1973; Nomoto, 1982). For this reason, all Type Ia SNe are thought to be due to interactions in binary systems. This has become the generally accepted model of these SNe. The companion can either be a non-degenerate star or another WD. The first case is called the *single-degenerate* (SD) model, and the second case is the *double-degenerate* (DD) model, first proposed by Webbink (1984); Iben & Tutukov (1984), illustrated in Figure 2.11. For reviews of Type Ia SN models, we refer to Hillebrandt et al. (2013) and Maoz et al. (2014).

13.2.1 Spectra and Light Curves of Type Ia SNe

Type Ia SNe are observationally distinguished by their spectra and the shape of their light curves. The defining spectral classification criterion is the presence of strong Si-absorption lines. The spectra are further dominated by the typical ashes expected to be produced by the thermonuclear combustion of a CO WD, resulting mostly in ^{56}Ni plus substantial amounts of a variety of intermediate elements (IMEs) as burning products; apart from Si: Ca, Mg, Na, O, and so on. This leads to a composition structure in which the inner structure, exposed at later times, contains 0.2–1.0 $M_{\odot}$ of ^{56}Ni, while the outer layers, exposed at maximum light, are composed mainly of IMEs such as Ca-Si-O (Nomoto et al., 1984). This composition excludes pure detonation models that overproduce iron-group elements (IGEs). The radioactive ^{56}Ni decays with a half-life of 6.1 d to ^{56}Co, which with a half-life of 77.7 d decays to stable ^{56}Fe. It appears that the decay of ^{56}Co to iron explains well the shape of the late light curves of SNe Ia (e.g., see Arnett, 1979; Colgate et al., 1980; Kuchner et al., 1994, and references therein).

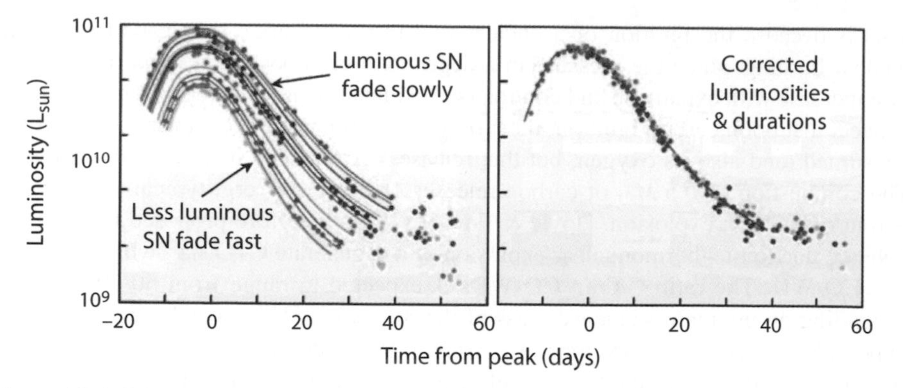

Figure 13.4. The left frame shows observed light curves of Type Ia SNe with different peak luminosities; the right frame shows that all these different light curves, with proper correction for luminosity and duration, fit to one universal shape. Figure from Durham University Department of Physics.

The required amounts of ^{56}Ni for powering the observed maximum light of SNe Ia range from 0.2 to 1 $M_{\odot}$ (Nomoto et al., 1984). These amounts are so large that Type Ia SNe are thought to be the main producers of iron in the Universe.

The light curves of Type Ia SNe are quite similar in shape, consisting of a very short rise time to maximum followed by a characteristic rather steep early decline, followed by the previously mentioned slower late-time decline. The early light-curve decline after reaching the peak luminosity presumably is powered by the decay of ^{56}Ni plus that of the intermediate elements produced in the explosion. Observations show that the exact early decline timescale is correlated with the absolute peak luminosity reached by the SN (Phillips, 1993), as is shown in Figure 13.4. This empirical *Phillips relation* between peak blue luminosity and decline timescale of the blue light following the peak, is very important for cosmology. This is because by measuring the early decline light-curve shape (the characteristic decline timescale), one immediately knows the absolute peak luminosity of the SN, which allows one to use the Type Ia SNe as "standard candles" for cosmological research. By comparing the apparent and absolute luminosity, one obtains the *light-distance* of the SN, which is based on a $1/r^2$ decrease of apparent luminosity with distance r. By comparing this light distance with the redshift of the SN one can derive the geometry of the Universe.[1] Studies of distant Type Ia SNe in this way led to the discovery in 1998 of the accelerated expansion of the Universe, indicating the presence of *dark energy* that dominates the energy balance of the Universe. The discovery of dark energy earned Saul Perlemutter, Brian Schmidt, and Adam Riess the 2011 Nobel Prize in Physics.

In recent years, a subluminous class of SNe Ia has been discovered, called SNe Iax (Fig. 13.5). They are spectroscopically similar to SN Ia at maximum light, except that the ejection velocities are lower (Foley et al., 2013). At late times, they differ

[1] Similar to the use of GW mergers (i.e., "standard sirens") when comparing the GW lumimosity distance and the redshift of the host Galaxy (see Chapter 15).

Figure 13.5. Composite image of SA(s)bc Galaxy NGC 1309, with the position of Type Iax SN 2012Z indicated. North is up, East is left. After Stritzinger et al. (2015).

spectroscopically from other SNe (McCully et al., 2014, and references therein), but their composition, dominated by iron-group elements and intermediate elements resembles the composition of a SN Ia. They are not rare and make up between 5% and 30% of all SNe Ia (Foley et al., 2013). The models for these SNe suggest that they are thermonuclear explosions, in which only part of the CO WD is consumed and the star is not completely disrupted. It thus is expected that a small remnant of this star may be left, called a "zombie star." McCully et al. (2014) discovered the progenitor of one SN of this type, SN 2012Z, in-spiral Galaxy NGC 1309 to be a luminous blue star resembling the Galactic helium nova V445 Puppis. This suggests that this SN Iax resulted from a binary in which the companion of the WD was a (helium-burning) helium star, and thus that the transferred matter was helium here, and this was an example of a SD system.

13.2.2 Binary Models for Producing Type Ia SNe

There are two main types of models for explosions of a WD that can be triggered by mass transfer from a binary companion. These models, Chandrasekhar-mass vs. sub-Chandrasekhar-mass explosions are now discussed in detail.

Chandraskehar-Mass Explosion Models

In these models, the mass transfer causes the mass of the WD to grow to the Chandrasekhar limit[2] of about 1.4 $M_\odot$. The idea among theorists working on such models is that when 99% of this limiting mass is approached, due to the increase in density and pressure, the temperature in the center of the WD begins to increase and a sub-sonic carbon burning front may slowly develop (so-called simmering C-deflagration, see Mazzali et al., 2007, and references therein). Slightly later, oxygen burning is ignited, but this is not consumed as fully as carbon. The temperature rise due to the burning under degenerate conditions would then cause the burning to accelerate, causing the burning front to develop into a supersonic detonation in which the star is disrupted. This is the *deflagration-detonation* model or *delayed detonation* model (see also Khokhlov, 1991). These processes are still not fully understood, but once ignition is started this sequence of developments of the explosion is supported by 3D supercomputer simulations. In these models, the burning is often assumed to start in a number of points close to the WD center (i.e., the distribution of burning points is artificially put in by hand). In more recent years, also pure-deflagration Chandrasekhar-mass models were developed, in which the burning did not reach detonation (e.g., see Fink et al., 2014, and references therein). Here the sub-sonic burning allows the burning region to expand (as this region no longer is fully degenerate), which prevents development into a detonation. Such models, with some randomness in the start of the ignition process, can quite well reproduce observed Type Ia SN light curves and observed spectra as well as a variety in SN Ia absolute luminosities. The weaker explosions among these are expected to leave a small remnant of the exploded star, which will be rich in ashes from the nuclear burning. These *only-deflagration* explosions are thought to be due to SD binaries, in which the mass transfer produces a steady burning layer on top of the WD, which allows the mass of the WD to gradually grow to approach the Chandrasekhar mass (e.g., Fink et al., 2014).

However, it was argued quite long ago that an alternative outcome of approaching the Chandrasekhar mass by accretion from a companion is that the accreting degenerate CO WD may not experience a thermonuclear explosion, but collapses to a NS, due to electron captures on oxygen and carbon nuclei; this is particularly expected if the mass donor is itself a CO WD (Saio & Nomoto, 1985; Ruiter et al., 2019; Liu & Wang, 2020). If this is true, Chandrasekhar-mass DD models for Type Ia SNe produced by mergers of two CO WDs (e.g., van Kerkwijk et al., 2010; Pakmor et al., 2012) will not be viable.

Finally, there is also the possibility that the merger of two CO WDs, or an ONeMg WD plus a CO WD, may leave behind a (bloated) massive WD remnant (Schwab, 2021). The remnant is expected to be a rapidly rotating, highly magnetized, meta-stable super-Chandrasekhar-mass object of order 10,000 yr, before collapsing (Schwab et al.,

[2]We remind the reader that the Chandrasekhar mass depends on the spin of the WD. For a non-rotating WD, $M_{\rm Ch} \simeq 1.37\, M_\odot$, and for a maximum rigid-body rotating WD $M_{\rm Ch} \simeq 1.48\, M_\odot$. For a differentially rotating WD, the stability limit may, in principle, reach as much as $\sim 4.0\, M_\odot$—see Yoon & Langer (2004) and references therein.

2016). The intriguing central star IRAS 00500+6713 recently discovered (Gvaramadze et al., 2019; Oskinova et al., 2020) is suggested to be such an object. Whether the final remnant is a NS or a massive WD (in case a catastrophic SN Ia is avoided) depends on the mass loss during the luminous giant phase following the merger event (Schwab, 2021). At this moment the dispute about these three possible outcomes is still not settled.

Sub-Chandrasekhar-Mass Explosions

It has long been known that pure Chandrasekhar-mass models of WDs in hydrostatic equilibrium produce mainly iron-group elements (IGEs) (Arnett et al., 1971). They cannot produce the significant amounts of intermediate elements (IMEs) that are observed to produce the dominant features of SNe Ia spectra at maximum light. To obtain these elements, pre-expansion of the WD is needed such that burning partially takes place under lower density conditions, where IMEs are produced. To achieve this, it was proposed/postulated that the flame ignites as a deflagration that produces enough energy to expand the star before the deflagration can grow into an explosion (Khokhlov, 1991; Mazzali et al., 2007). An alternative to this pre-expansion of a Chandrasekhar-mass WD is the detonation of a sub-Chandrasekhar-mass WD starting in a hydrostatic configuration. The density profile here depends on the WD mass. Close to M_{Ch}, the detonation mostly produces IGEs and a few IMEs, but at lower WD masses, more IMEs are produced and fewer IGEs.

Detonation of a sub-Chandrasekhar-mass WD cannot occur spontaneously but needs a trigger. This has led to the double-detonation model, which has a long history starting with Taam (1980), followed by, for example, Nomoto (1982); Nomoto et al. (1984); Woosley et al. (1986); and Livne (1990). In this model, the CO WD accreted from a companion and built up a degenerate helium-rich outer layer on top of the WD. This can result by direct accretion from a helium-rich companion or from accretion of hydrogen that steadily, or in flashes, is burned to helium. Helium will ignite with a flash, which is a detonation. Compression of the core by the inward propagating detonation shock may then produce a secondary C-detonation that blows up the WD (e.g., Woosley & Weaver, 1994; Livne & Arnett, 1995; Hoeflich & Khokhlov, 1996; Fink et al., 2007). Here one thinks of helium envelopes of about $0.2\,M_{\odot}$ or larger. These thick helium-envelope models synthesized significant amounts of ^{56}Ni in the outer ejecta, which conflicts with the spectra of SNe Ia (see, however, Nomoto et al., 1984, for a way to avoid this by introducing pure C-deflagration models). Bildsten et al. (2007) then made the important suggestion for the double-detonation model that already a helium layer of low mass ($\sim 0.05\,M_{\odot}$) may detonate and thus produce only a small amount of ^{56}Ni ($0.012\,M_{\odot}$) in the outer layers. Subsequently, Guillochon et al. (2010) found that rapid dynamical mass transfer from a He WD to a CO WD may also trigger a thin low-mass helium layer deposited on the WD to already detonate and cause the CO WD to detonate. They also found that intermediate elements are produced in this case. It should be noticed here that the helium-transferring WD does not need to be a pure He WD, since many CO WDs also have a helium-rich envelope and can therefore supply the thin helium layer required for triggering the explosion.

These results produced the ingredients on which Sim et al. (2010), and later Shen et al. (2018a) and Shen et al. (2021), built their sub-Chandrasekhar double-detonation models for SNe Ia. As the required helium layer can have a very low mass, Sim et al. (2010) ignored it completely and studied just the light curves and spectra produced by detonating CO WDs with a range of masses and obtained results that globally agree very well with the observed SNe Ia light curves and spectra. They even were able to qualitatively reproduce the trends of the Phillips relation between maximum luminosity and decline timescale after maximum in blue light. These results represented a major breakthrough in our understanding of Type Ia SNe. Shen et al. (2018a) basically took the same double-detonation model, but now also included the detonation of the low-mass helium layer in their simulations. Subsequently Shen et al. (2021) refined these results by using non-LTE modelling of the light curves. Their results are a refinement and improvement on the breakthrough results of Sim et al. (2010), and we present here some of their results as an illustration.

Figure 13.6 shows computations of theoretical light curves of SNe Ia, and Figure 13.7 shows the theoretically predicted Phillips relation resulting from these computations of Shen et al. (2021) as compared with the observed relation. These two figures show that the D^6 model for sub-Chandrasekhar explosions (Shen et al., 2021) appears to give results that agree at least qualitatively and, to a good degree, also quantitatively well with the observations. This is very encouraging and gives strong credence to the sub-Chandrasekhar model for Type Ia SNe, as had already emerged from the work of Sim et al. (2010).

An important point mentioned by Shen et al. (2018a) is that the double-degenerate merger rate recently derived from observations by Maoz & Hallakoun (2017) is one order of magnitude larger than the SN Ia rate and in global agreement with the double WD interaction rate derived from population synthesis calculations by Ruiter et al. (2011); Toonen et al. (2017). So, only 10% of the double-degenerate mergers are required to produce the observed Type Ia SNe. It seems therefore perfectly reasonable to assume that only a sub-population of the double-degenerate mergers, that is, the ones with He WD donors or with CO WD donors with a helium-rich envelope, produce the Type Ia SNe. The mergers of double pure CO WDs might then possibly produce other objects, for example, NSs produced by electron-capture collapse, as proposed by Saio & Nomoto (1985); Ruiter et al. (2019). An interesting point here is that the latter might produce a population of slow-moving NSs, since AIC is not expected to impart a kick to a NS (see Sections 14.5 and 13.7). The rate of this process may be substantial, and a sizeable fraction of all young pulsars may be produced in this way. We notice here that some 30% of the young pulsars have low space velocities (Verbunt et al., 2017). Possibly, among these could be products of double pure CO WD mergers.

We already mentioned that quite some time ago, using sub-Chandrasekhar models with helium accretion and assuming detonation of the accreted helium to trigger a deflagration explosion of the WD, Nomoto et al. (1984) were able to explain the domination of intermediate elements in the spectra of SNe Ia at peak luminosity. At that time, the Phillips relation was not yet known. It would be interesting to investigate if these models can also explain this relation, as an alternative to the now very

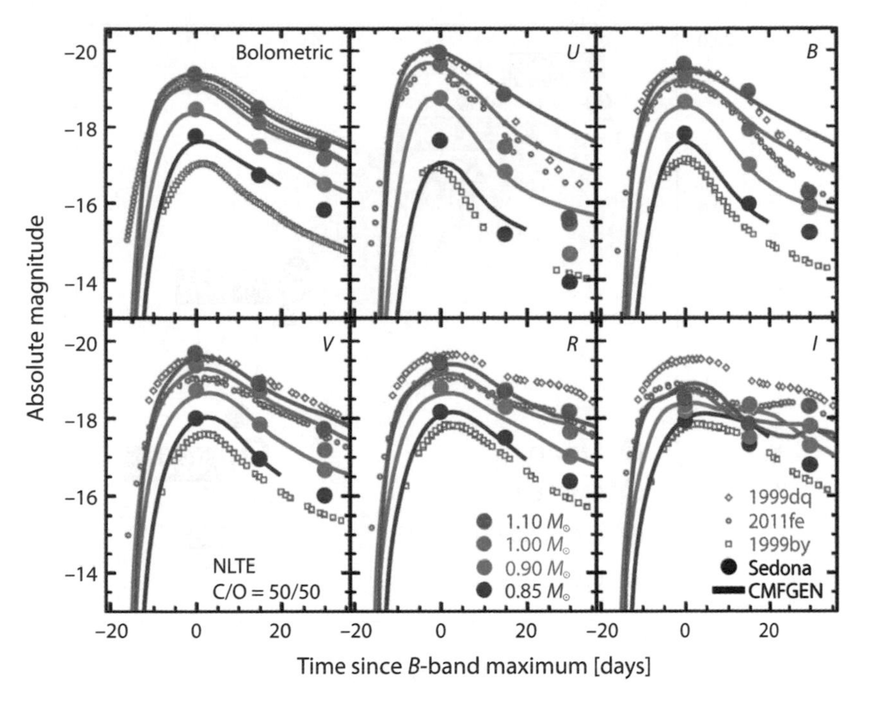

Figure 13.6. Multi-band non-LTE light curves of 50/50 C/O mass ratio WD detonations of sub-Chandrasekhar-mass WDs. Masses are as labeled with brighter B-band magnitudes as mass increases. CMFGEN and Sedona are two different non-LTE codes. Solid lines are CMFGEN results, and circles are Sedona snapshots. Open diamonds, circles, and squares show observed curves of SNe 1999dq, 2011fe, and 1999by, respectively. After Shen et al. (2021).

successful *dynamically-driven double-degenerate double-detonation* scenario (abbreviated D^6 scenario; Shen et al., 2018a, 2021).

Weaker Thermonuclear Explosions with Only Partly Consumed WDs

If the donor star is a (helium-burning) helium star, the helium deposited on top of the WD may burn steadily, producing C and O, such that the WD mass gradually grows to the Chandrasekhar mass. Several authors (Jordan et al., 2012; Kromer et al., 2013; Fink et al., 2014) have considered pure deflagration explosion models of such WDs with masses close to the Chandrasekhar mass, as was already described earlier. These subsonic explosions (deflagration) may result in a peculiar weaker SN, of Type Iax, with lower explosion energies than a SN Ia, and in which not the entire WD is consumed. 3D hydrodynamical computations by the previously mentioned authors show that such pure deflagration models may lead to survival of part of the WD as a bound remnant. This partly burned object would temporarily expand to almost solar size and then settle to a state resembling an expanded WD, containing nuclear burning ashes such as

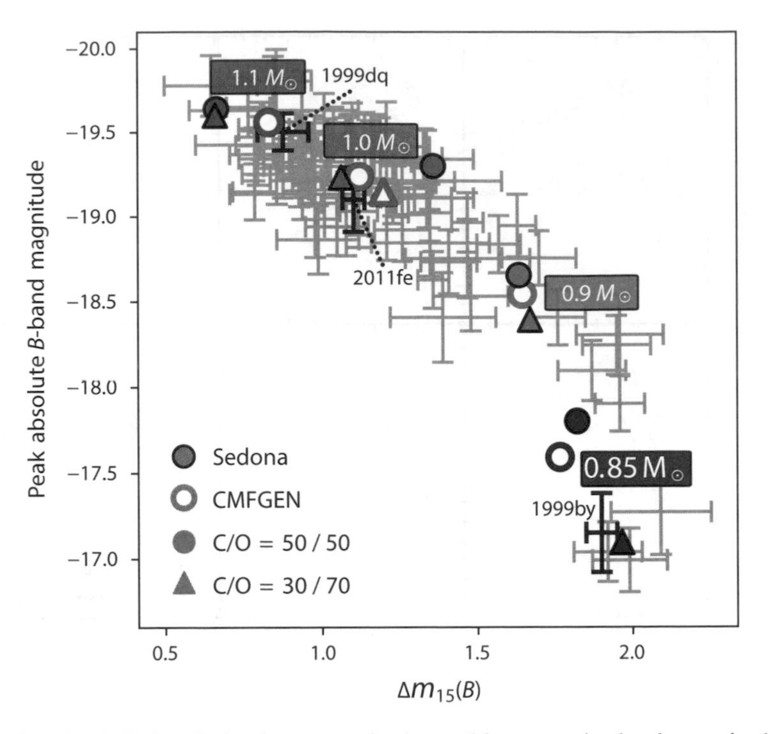

Figure 13.7. Peak B-band absolute magnitude vs. blue magnitude change in 15 days after the peak, $\Delta m_{15}(B)$: the Phillips relation (see text). Solid symbols show the results from Sedona non-LTE snapshots, and open symbols show CMFGEN results (see Fig. 13.6). Circles (triangles) represent initial C/O fractions 50/50 (30/70). Masses of WDs are as labeled. Grey error bars are values from a sample of SNe Ia, and black error bars are values for SNe 1999by, 2011fe, and 1999dq. After Shen et al. (2021).

silicon, iron, and nickel, and having a mass between 0.2 and 1.0 $M_\odot$. Such objects were later dubbed zombie stars (Vennes et al., 2017). Observations suggest that the Type Iax SNe in-spiral galaxies show a close association with sites of continuing star formation (Foley et al., 2013; Lyman et al., 2018), which might indicate that their ages do not exceed 10^8 yr. Their favored donor stars are then helium stars with masses of about 0.5 to 2 $M_\odot$ (helium cores of stars with initial ZAMS masses up to $\sim 7\, M_\odot$). The hydrogen envelope may have been lost here during a preceding CE-phase, in which the WD spiraled down into the envelope of its companion (Section 4.8). Indeed, the bright blue progenitor star of the Type Iax SN in the Galaxy NGC 1309 has the properties of a helium star of about 2 $M_\odot$ (McCully et al., 2014).

Types of Donor Stars

As mentioned previously, there are several conceivable types of mass donors in the SN Ia progenitor binary systems (e.g., see the review by Maoz et al., 2014). They can be divided into:

- non-degenerate companions; these are the SD models:
  - main-sequence stars (hydrogen-rich envelope)
  - sub-giants and giants stars (hydrogen-rich envelope)
  - helium stars
- degenerate companions; these are the DD models:
  - He WDs (with no, or just a tiny, hydrogen envelope)
  - CO WDs (which often may have a substantial helium mantle).

To proceed to a successful explosion in the SD case, the mass-transfer rate provided by the companion must enable the CO WD to grow in mass.[3] With hydrogen-rich companions, this requires a mass-transfer rate of $10^{-7} \leq |\dot{M}_2| \leq 10^{-6}\, M_\odot\,\mathrm{yr}^{-1}$ (see Figs. 5.13 and 11.5, and Section 11.1.3.). Main-sequence and sub-giant companions can provide such a rate if they are more massive than the WD (typical donor masses of $M_2 \simeq 1.4 - 2.6\, M_\odot$; Tauris et al., 2013), with the lower-end masses for low-metallicity stars, and even up to $M_2 \simeq 3.5\, M_\odot$ (Li & van den Heuvel, 1997; Han & Podsiadlowski, 2004; Wang et al., 2010), if one allows for the optically thick wind model (Kato & Iben, 1992; Kato & Hachisu, 1994) to operate for high values of $|\dot{M}_2|$. These binaries are observed as super soft X-ray sources (SSS; van den Heuvel et al., 1992), as described in Chapter 5. In the case of low-mass giant companions, donors less massive than the WD can supply a sufficiently high rate by nuclear evolution, provided the initial binary period is $> 100\,\mathrm{d}$ (e.g., Joss & Rappaport, 1983; Li & van den Heuvel, 1997; Tauris et al., 2013), but see also Liu et al. (2019)—and see general discussions on the evolution of wide-orbit LMXBs described in Chapter 11.

For helium-star donors, evolution towards a Type Ia SN (or an AIC event) is possible in binaries with donor masses $M_{\mathrm{He}} \simeq 1.0 - 3.0\, M_\odot$ with the optically thick wind assumption (Wang & Han, 2010), but only $M_{\mathrm{He}} \simeq 1.1 - 1.5\, M_\odot$ without accepting this approximation as valid (Tauris et al., 2013).

In the DD case, the systems must start out sufficiently close to drive the binary to merge due to emission of GWs within a Hubble time. This requires initial binary periods, at the time of formation of the double WD, of less than half a day. Here the lower-mass WD has the largest radius and the smallest Roche lobe of the two, such that the mass will be transferred from the lower-mass to the higher-mass WD, leading to the explosion of the latter one (Webbink, 1984; Iben & Tutukov, 1984). van den Heuvel & Bonsema (1984) found that mass transfer from a WD with mass larger than $0.66\, M_\odot$ cannot proceed in a stable way and always leads to complete disruption of the WD donor. Only WD donors with masses below $0.66\, M_\odot$ are, in principle, able to transfer mass to a more massive WD in a stable way although, if the transferred matter forms a disk around the accretor, the mass transfer may still become unstable (Bonsema & van den Heuvel, 1985).

[3] In analogy for an accreting ONeMg WD to reach M_{Ch} and undergo an AIC event.

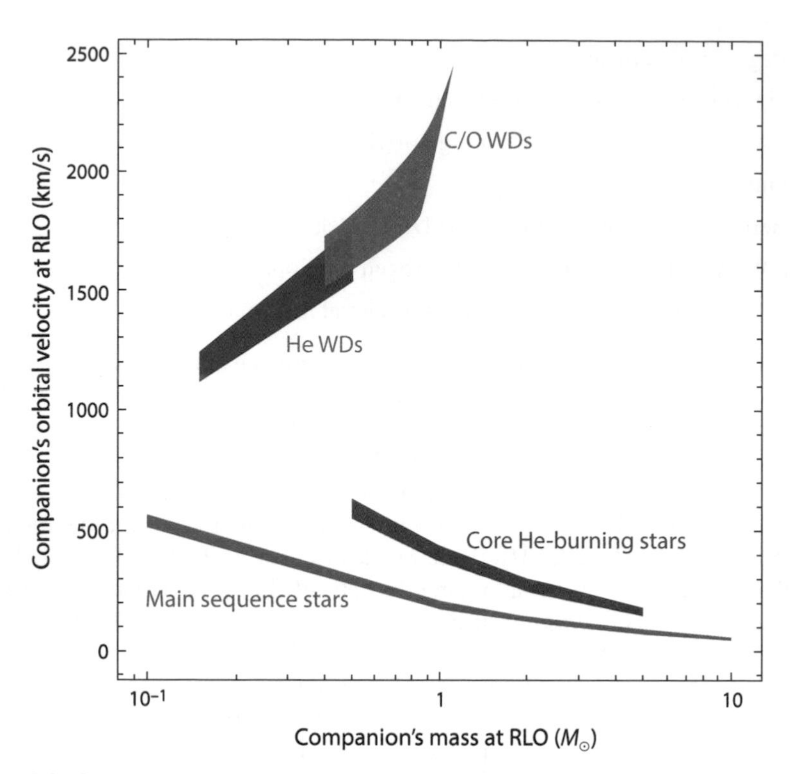

Figure 13.8. Companion star orbital velocity vs. its mass at Roche-lobe overflow. The upper boundary of each region is for a 1.1 $M_{\odot}$ WD, while the lower boundary corresponds to a 0.85 $M_{\odot}$ WD. After Shen et al. (2018b).

13.2.3 Recognition of High-velocity Remnants of Type Ia SN Explosions in Binaries

If a CO WD in a close binary is triggered by mass transfer to be completely destroyed in a Type Ia SN explosion, its mass-donor companion star will roughly[4] be launched into space with a velocity equal to its orbital velocity in the binary at the time of the explosion. If the mass donor is itself a WD or a helium star, these orbital velocities can be very large, of the order of 500 to 2,000 km s^{-1}. Figure 13.8 shows these orbital velocities for different types of companions, as calculated by Shen et al. (2018b)—see also Bauer et al. (2019); Neunteufel et al. (2021). The figure shows that He WD donors have orbital velocities of 1,200 to 1,600 km s^{-1}, CO WD donors have velocities of 1,500 to 2,400 km s^{-1}, and helium stars have velocities between 200 and 500 km s^{-1}. For general modelling of the interaction between the SN ejecta and the companion star in Typa Ia SNe, see, for example, Marietta et al. (2000); Liu et al. (2012).

[4]Disregarding dynamical effects from the SN ejecta impact on the companion star (e.g., Wheeler et al. [1975]; Tauris & Takens [1998]).

As already mentioned, it also is possible, in the SN Iax models of Fink et al. (2014) that the exploding CO WD is not completely destroyed in the explosion but leaves a small WD-like remnant of some 0.2–0.7 $M_\odot$ (a zombie star). If in the explosion 1 $M_\odot$ of material is ejected, this ejected mass may be larger than half the pre-explosion system mass, such that the system is disrupted in the explosion and the zombie becomes a runaway star, again with a velocity of the order 1,000 km s^{-1}. In recent years, thanks to ESA's *Gaia* astrometry satellite, interesting evidence on likely remnants of Type Ia SN explosions in binaries has been uncovered, in the form of stars with highly peculiar chemical compositions that are moving in our Galaxy with extremely high space velocities (i.e., *hypervelocity stars*, see also Section 13.8.6). We here summarize this evidence.

Possible High-velocity Remnants of Type Ia SNe in Binaries

Raddi et al. (2019) summarized the recent evidence for the existence of different types of runaway star remnants of SNe in our Galaxy, discovered thanks to the proper motion and parallax measurements with *Gaia*. This evidence consists of: (1) the hypervelocity hot subdwarf star US 708, which shows evidence of being the remnant of the released companion star of a completely exploded star; (2) four stars of the type of the subdwarf LP 40–365, suggested to be zombie remnants, three of which have extremely large velocities; and (3) three hypervelocity CO WDs that may be remnants from the D^6 scenario (Shen et al., 2018b; Shen et al., 2021). We now briefly describe the properties of these three types of high-velocity peculiar compact stars. Their positions in the HR diagram are shown in Figure 13.9.

US 708 is a hot subdwarf star (Hirsch et al., 2005; Geier et al., 2015; Neunteufel, 2020) that moves with a space velocity of 1,160 km s^{-1} and shows in its spectrum a photospheric composition of thermonuclear ashes, indicating that in its close neighborhood a SN explosion took place. The most likely explanation is that here the exploding star was completely disrupted when mass transfer from the progenitor of this subdwarf triggered it to explode.

LP 40–365 is a unique object whose peculiar nature was discovered by Vennes et al. (2017). Already long ago, it was suggested to be a WD (e.g., Giclas et al., 1970). Vennes et al. (2017) discovered that its spectrum does not show hydrogen or helium, although its temperature is 10^4 K, which makes it different from other WDs of the same temperature. The spectrum shows that the atmosphere is dominated by Ne (60%–65%), O(30%), Mg(3%–9%). The remaining 1%–2% consists of heavier elements: Si, Na, Ni, and Fe (0.1%). Its measured surface gravity is an order of magnitude larger than that of the Sun but three orders of magnitude lower than that of a WD. Its large radial velocity of 500 km s^{-1} and peculiar composition led Vennes et al. (2017) to suggest that it survived a low-luminosity Type Iax SN, in which only part of the WD exploded. Subsequently, Raddi et al. (2018) found that in combination with its *Gaia* parallax and proper motion, its real space velocity is 850 km s^{-1} in the Galactic rest frame. Further, its parallax allowed for determining its radius (0.16 $R_\odot$) and, by using its surface gravity, its mass, which is only 0.3 $M_\odot$. Its space velocity exceeds the local escape velocity from the Galaxy (520 km s^{-1}). It must therefore have formed relatively recently, less than a

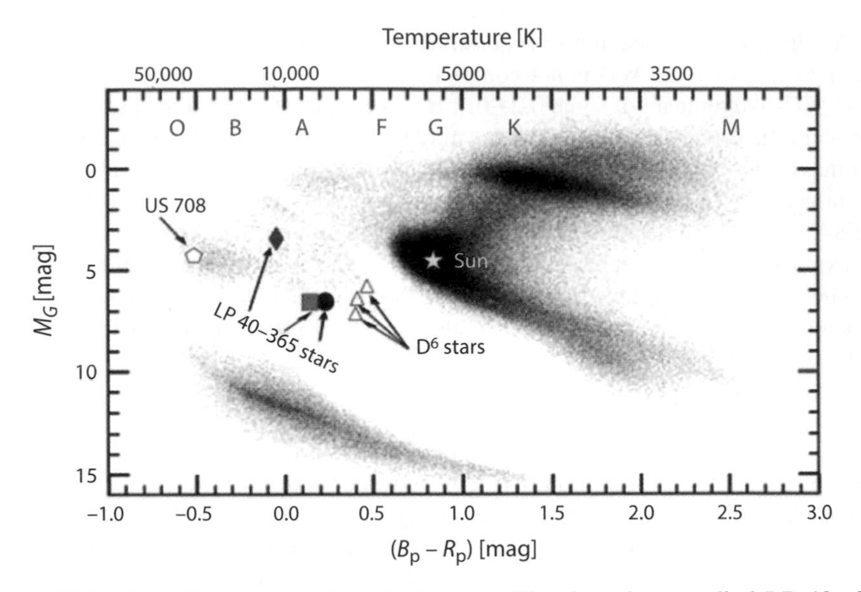

Figure 13.9. *Gaia* Hertzsprung-Russell diagram. The three best studied LP 40−365 stars are displayed along with the other runaways suspected to be produced by SNe Ia, as mentioned in the text. After Raddi et al. (2019).

few times 10^7 yr ago; otherwise it would have escaped from the Galaxy by now. If the star exploded with a mass close to the Chandrasekhar limit, it must have ejected about 1 $M_\odot$ which, since this was likely more than half the total progenitor binary mass, will have disrupted the binary. To obtain the large space velocity from binary disruption requires a very tight pre-SN orbit, and thus its companion must have been a WD.

Recently, Raddi et al. (2019) found three more LP 40−365-like stars, two of which also have very high space velocities, exceeding the local escape velocity of the Galaxy. These stars have known parallaxes, thanks to *Gaia*. The three escaping stars all have similar masses and chemical compositions, and all must have been born less than a few times 10^7 yr ago. A problem is, however, that these stars contain very little C and too much O and Ne, relative to the theoretical predictions for zombie stars (Fink et al., 2014). Also, theoretical models disfavor that the remaining remnants have the low observed masses of these stars. This means that most systems will not be disrupted by the explosive mass ejection; systems in which the remnant is bound to its former donor star may be hard to recognize (see next paragraph). Raddi et al. (2019) remarked that the SN Iax rate in the Galaxy could produce of the order of 50,000 such remaining remnants in 10^8 yr. Since numerical simulations show that the majority of the LP 40−365 stars would not have extraordinarily large space velocities (see also the next paragraph), most of them will be bound to our Galaxy, and of the order of 20 of them would be expected to be found within 2 kpc from us.

Three hypervelocity CO WDs, of the order of 1,000 km s^{-1}, were discovered by Shen et al. (2018) and were suggested to have been released from a close double WD binary when their companion WD was completely disrupted in a SN Ia explosion—similar to the case of the progenitor system of US 708. Shen et al. (2018) suggested that these explosions were of the D^6 type. They speculated that US 708 and these three objects are different evolutionary states of similar objects, US 708 being the youngest of them. These stars must have been the remnants of donor stars in double-degenerate systems. The explosion must have been triggered here while the mass transfer was going on, and the donor was not yet completely consumed. The reason why these stars—from their position in the HR diagram, as depicted in Figure 13.9—have considerably larger radii than WDs could be frictional tidal heating in the very close double WD binary prior to the start of RLO. These stars are then later in life, after cooling, expected to settle in the WD region of the HR diagram.

The existence of these different observed classes of runaway objects is the first clear observational evidence that different classes of thermonuclear SNe occur in nature. The apparent discrepancy between the observed low masses of the LP 40−365-like stars (0.2–0.3 $M_\odot$) and the typical theoretically predicted remnant masses (say 0.5–0.7 $M_\odot$, for the lower-luminosity SN Iax models by Fink et al., 2014) may well be explained by an observational selection effect. The explosions leaving relatively high-mass remnants eject less material, and these systems are therefore more likely to remain bound (Hills, 1983b) compared to the explosions that leave behind low-mass remnants, which eject more material and are thus more likely to disrupt the binaries and produce hypervelocity runaway stars. In fact, if indeed SN Iax explosions tend to leave relatively high-mass zombie remnants, the bulk of these remnants will be in relatively low-velocity post-SN binary systems with their predecessor donors. It may be difficult then, due to the presence of the companion, which may be a helium star (McCully et al., 2014), to see this remnant, so it may go undetected. The LP 40−365-like stars are therefore expected to be a very small tip of a large iceberg of zombies (Raddi et al., 2019).

13.2.4 The Relative Importance of the Single Degenerate and Double-degenerate Scenarios for the Formation of Type Ia SNe

To examine the relative contributions of the SD and DD models to the Type Ia SN rates in different galaxies, several groups have carried out population synthesis evolution calculations, assuming a realistic initial fraction of binary systems (usually between 50% and 100%). Among the SD models, one has to distinguish between SD systems in which the donor star is a hydrogen-rich star (SD, H) and a helium star (SD, He). The latter systems are the result of CE evolution, where the WD spiraled down through the envelope of a giant star with a helium core. The DD systems underwent two episodes of mass transfer, in which at least the second one involved CE evolution and spiral-in (see, e.g., Webbink, 1984; Iben & Tutukov, 1984). In such calculations, one starts from a burst of star formation and follows the evolution of the entire stellar population, including the evolution of all the various types of binary systems (for information about population synthesis techniques, see Chapter 16). Here, a number of assumptions has to be made about what happens in certain stages in the evolution of binaries for which the outcome is presently still quite uncertain. These are particularly the stages in which the binary loses much mass and orbital angular momentum, for example, in

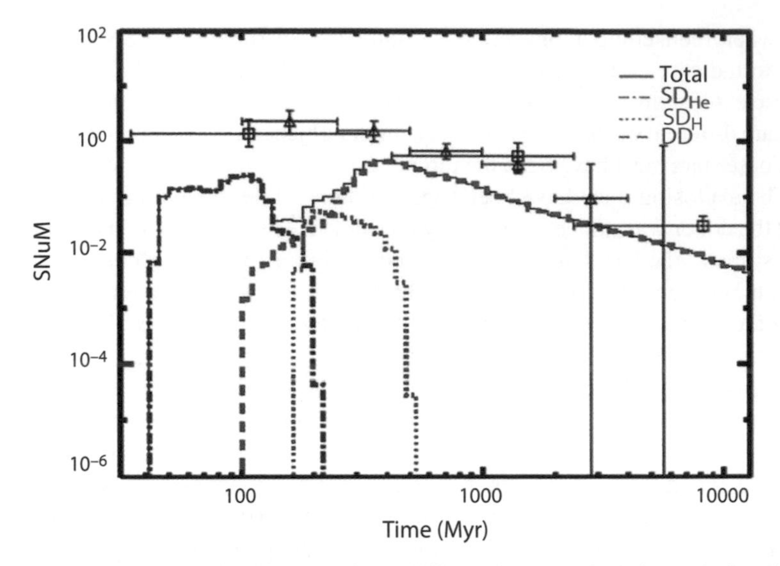

Figure 13.10. Delay-time distribution of the different Type Ia SN channels for the standard models: (SD, He), (SD, H), and DD channels are plotted in dark blue dash-dotted, red dotted, and purple dashed, respectively. The overall SN Ia rate (black thin full line) is the sum of the three channels. The rate is presented in units of SNuM, which is the SN Ia rate per 100 yr per $10^{10}\,M_\odot$ in stars. Data points represent the observed DTD from Totani et al. (2008) (triangles) and Maoz et al. (2011) (squares). Figure from Claeys et al. (2014).

the important common-envelope (CE) phases (Section 4.8). Here the outcome depends critically on the precise formalism that is used, particularly on the value of the CE *efficiency parameter* $\alpha_{\rm CE}$, which is rarely calculated for each binary from detailed stellar structure analysis in a rapid code. Different assumptions may lead to final orbital dimensions differing by up to an order of magnitude. Still, it turns out that the results obtained by different groups, for example, Yungelson (2005); Mennekens et al. (2010); Toonen et al. (2012); Claeys et al. (2014), globally make similar predictions.

In an important study, Claeys et al. (2014) examined the influence of different input assumptions of the models by different authors on the SN rates they predict as a function of time—or more specifically, the so-called delay-time distribution (DTD). This is the SN Ia rate vs. time following a short burst of star formation. For elliptical galaxies (that have not had star formation for several billion years), the SN Ia rates follow a DTD which is a $1/t$ distribution as a function of Galaxy age, since the last star formation event (see, e.g., Maoz et al., 2012, and references therein). A key aspect here is that different progenitor scenarios for Type Ia SNe predict different DTDs.

As an example, we show in Figure 13.10 one of these simulations. It is assumed that the SN Ia occurs when one of the WDs reaches the Chandrasekhar mass limit of $1.4\,M_\odot$. (For the sub-Chandrasekhar models, where the explosion occurs at a lower WD mass, the results will be qualitatively similar.)

One observes from this simulation (and the same is roughly true for the other simulations) that when the starburst is still young ($0.4-2 \times 10^8$ yr), the (SD, He) process dominates; and only from $2-5 \times 10^8$ yr, the (SD, H) process makes a significant contribution. The DD process becomes important from about 10^8 yr on, peaks at about 4×10^8 yr, and after this remains the most important SN Ia process—decaying with time as $1/t$, like the observed DTD for elliptical galaxies, which is also indicated in the figure. (Among the first ones to predict, from population-synthesis calculations, that the SN Ia rate at late times is dominated by the DD process, and will decay as $1/t$, were Jørgensen et al., 1997.) The simulations predict that the absolute Type Ia rate is highest in young star-forming galaxies, such as spirals and irregulars, and is much higher than in ellipticals, which show the $1/t$ delay rates, which in absolute terms are much lower than the rates in star-forming galaxies (in accordance with the observations, e.g., Dilday et al., 2010). We thus conclude that the simulations represent the overall character of the observed DTD quite well and that in the older galaxies the DD model dominates the production of Type Ia SNe.

A current open question relates to the outcome of a merger event of two WDs of lower mass, that is, two He WDs. It has been suggested by Webbink (1984) and Han et al. (2002) that a hot subdwarf (sdO or sdB) may form from the merger event of two He WDs. This scenario is particularly attractive for explaining highly magnetic sdO stars (Dorsch et al., 2022, and references therein). Another possibility is that the RLO between two He WDs will be stable and not result in a merger event as such—similarly to the outcome of many AM CVns, as will be discussed in Section (15.12).

13.3 STRIPPED-ENVELOPE CORE-COLLAPSE SNe

13.3.1 Overview of Observational Properties

As mentioned at the beginning of this chapter, in contrast to the thermonuclear SNe of Type Ia, SNe of Types Ib and Ic are core-collapse SNe which produce neutron stars (NSs) and black holes (BHs). Since Type Ib/c SNe show no hydrogen in their spectra, the hydrogen-rich envelopes of their progenitors were removed. For this reason, these SNe are also called stripped-envelope (SE) SNe. For stars more massive than about 30 $M_\odot$ with solar-like metallicity, the envelope stripping may have been achieved by strong stellar winds. For stars below this mass limit, the only way to remove the envelope is by binary mass-transfer. As also mentioned in Section 13.1, stars in the mass range 8–30 $M_\odot$ are some six times more abundant than stars more massive than 30 $M_\odot$, and therefore the far majority of the Type Ib/c SNe is expected to be produced by binary evolution. For excellent overviews of the observations of Type Ib and Ic SNe, we refer to Eldridge et al. (2013); Prentice et al. (2019).

In addition to the SNe Ib/c, also a fraction of the Type II SNe may be the result of binary evolution, as the case of naked-eye SN1987A has shown. Contrary to the expectation that the progenitors of Type II SNe must be red supergiants, the progenitor here was a well-studied blue supergiant with a radius of about 40 $R_\odot$. A clue to the mystery of why the progenitor was not a red supergiant came from the discovery with the Hubble Space Telescope of a triple ring structure in the SN remnant, which is clearly axi-symmetric. This suggests that the progenitor was the result of a binary merger, some 20,000 yr before the explosion. A merger origin of the blue supergiant had already been suggested (Hillebrandt & Meyer, 1989; Podsiadlowski et al., 1990) long before the nice discovery of the remnant structure. For details of the merger model, see, for example, Morris & Podsiadlowski (2007). Since a binary origin for Type II SNe is not expected to be a common phenomenon, we will in this chapter only concentrate on core-collapse explosions in binaries that give rise to Type Ib and Ic SNe.

Prentice et al. (2019) made a study of the properties of 18 SE-SNe of different types observed during 2013–2018. They found that in all 18 cases, the initial ZAMS progenitor mass was $< 20\,M_{\odot}$. They were able to determine the ejected masses from the light curves and spectral evolution. Combining these results with those of other SE-SNe in the literature, they were able to collect the ejecta masses of 80 SE-SNe and found that SN Ib have the smallest ejecta masses $M_{\rm ej}$, followed by the narrow-line SN Ic, hydrogen/helium-rich SNe, broad-line SNe, and finally GRB SNe. The average $M_{\rm ej}$ for all SE-SNe was found to be $2.8 \pm 1.5\,M_{\odot}$, which strengthens the evidence that these SNe arise from progenitors that typically have masses $\leq 5\,M_{\odot}$ at the time of the explosion, and with initial ZAMS progenitor masses of $\leq 20-25\,M_{\odot}$. The authors therefore concluded that the envelope stripping in binaries is the dominant evolutionary pathway of these SNe.

While normally in SN Ib/c, typical mass ejection velocities are of the order of 10^4 km s^{-1}, several long GRBs have been observed to coincide with a very energetic peculiar broad-line Type Ic SN, with outflow velocities of some 30,000–40,000 km s^{-1} (Galama et al., 1998; Hjorth et al., 2003). These are thought to be the sign of the collapse of a very rapidly spinning, naked, massive CO-star to a BH. This CO-star is thought to be the stripped core of an initially massive star. Such a collapse event as the cause of a GRB was predicted by the *collapsar* model of Woosley (1993); MacFadyen & Woosley (1999).

Long GRB/broad-line Type Ic SNe are very rare events, occurring perhaps at a rate of the order of 10^{-5} or 10^{-6} yr^{-1} for a Milky Way–size Galaxy. They must, therefore, be the result of a very rare type of evolution. As envelope stripping most easily can take place in a binary, and since this also can explain the required rapid rotation (by tidal spin-up), it seems not unlikely that these long GRBs are the products of a very special kind of binary evolution (see, e.g., van den Heuvel & Yoon, 2007; Cantiello et al., 2007). Because of their rareness, we will not further discuss them here.

13.3.2 Theoretically Expected pre-SN Structure of Stripped Stars of Different Masses

Some classic and more modern review papers on the theory of core-collapse SNe include Bethe (1990); Woosley & Weaver (1995); Janka (2012); Burrows (2013); Burrows & Vartanyan (2021). There are often significant differences between the nature of SNe (observational characteristics and remnant masses) resulting from similar ZAMS stars, depending on whether they are evolving in isolation or in close binaries (e.g.,

Paczyński, 1971b; van den Heuvel, 1976; Podsiadlowski et al., 1992; Wellstein & Langer, 1999; Brown et al., 2001; Podsiadlowski et al., 2004; Smartt, 2009; Yoon et al., 2010; Langer, 2012; Tauris et al., 2013, 2015; Müller et al., 2018, 2019; Laplace et al., 2020, 2021; Vartanyan et al., 2021) as discussed in further detail in Section 9.7.

As shown in Chapter 9, the stripping in binaries can follow the binary-evolution pathways with RLO Cases A, B, and C. Figure 10.14 shows an example of the evolution of a massive Case A system that produces a NS, but in general for primary masses $\leq 20\,M_{\odot}$, from where most of the SNe Ib/c originate, Case A is expected to be rare. In the following, we therefore focus on evolution via RLO Cases B and C.

The (large) differences between the expected final remnants (NS vs. BH) as a function of ZAMS mass of the primary star, and between Case B and Case C evolution, were already summarized at the end of Chapter 9. The question of whether the production of these final remnants is accompanied by a SN Ib or Ic was not discussed there. As mentioned in the introductory section of this chapter, for stripped stars with ZAMS masses $\leq 30\,M_{\odot}$, the absence of He-lines in the spectrum does not necessarily mean absence of helium in the progenitor's envelope. Dessart et al. (2012) showed that moderately mixed SN models of stripped stars originating in this mass range may fail to non-thermally excite helium atoms and may therefore lead to classification Ic. Hachinger et al. (2012) argue that $> 0.06\,M_{\odot}$ helium must be present in the SN ejecta to secure a detection of Type Ib. For this reason, progenitors in this mass range with a helium envelope around their CO core may produce SNe Ib as well as Ic. Before we go into the details of the final evolution, we first start with an overview of the late evolution and expected remnants as a function of the mass of the stripped star.

In Case B RLO, the stripped remnant of the initial primary star, immediately after the mass transfer, is practically a pure helium star at core-helium ignition (i.e., on the He-ZAMS). In Case C RLO, it is the core of the primary star near carbon ignition. This is a CO core with a helium envelope. In view of the very short stellar lifetime after carbon ignition, the helium envelope is expected to still be present at core collapse. Depending on the degree of mixing of the ejected material, the spectrum may then be either of Type Ib or Ic, as already discussed.

Yoon et al. (2010) investigated SN Ib/c progenitors in binary systems with initial ZAMS masses of $12-25\,M_{\odot}$ (at solar metallicity) and predicted them to have a wide range of final pre-SN masses up to about $7\,M_{\odot}$, with helium envelopes of $M_{\rm He} \simeq 0.2-1.5\,M_{\odot}$. Furthermore, they find that a thin hydrogen layer ($0.001\,M_{\odot} \leq M_{\rm H} \leq 0.01\,M_{\odot}$) is expected to be present in many SN Ib progenitors at the pre-SN stage. Concerning the late evolution and remnants expected in Case B evolution, there are many more studies available compared to those of Case C evolution. We note that binary evolution leading to the second SN explosion in tight binaries, in which already one NS or BH is present, will often lead to *ultra-stripped* SNe; see the discussions in Section 13.5.

To get a first general overview of the expected evolution and fates of helium stars resulting from this Case B RLO, we follow the results by Woosley (2019) for helium stars in the mass range 1.6 to $120\,M_{\odot}$, evolved including stellar wind mass loss. It appears from this study (and many other studies in this field) that with respect to the type of late and final evolution, one can distinguish six different initial helium-star mass regimes[5] where $M_{\rm He,i}$ is a function of the ZAMS mass of the primary star, given roughly by Eq. (9.8). These six different evolutionary paths and outcomes of Case B RLO are as follows:

1. $M_{\rm He,i} \lesssim 1.6\,M_\odot$: leading to a CO WD.
   Examples of such evolution are found, for example, in intermediate-mass X-ray binaries, which evolve via early-Case B RLO to produce binary pulsars with CO WD companions (see, e.g., Fig. 6.9 and Section 14.4). Note that very-low-mass helium stars with initial masses $\lesssim 0.33\,M_\odot$ will never ignite helium and will leave instead a He WD remnant.
2. $M_{\rm He,i} \simeq 1.6 - 3.2\,M_\odot$: leading to a massive WD or an unusual SN.
   This helium-star mass range corresponds roughly to ZAMS primary stars with masses of $M_{\rm ZAMS} \simeq 10{-}16\,M_\odot$. In this mass range, the physics of the cores is complex due to the occurrence of degeneracy, leading to various kinds of flashes and off-center ignition of carbon in the lower-mass part of this range, and of oxygen, neon, and silicon at the higher mass part. Also, during helium-shell burning, the envelopes of these helium stars expand to giant dimensions, which in binaries leads to further stages of mass loss by RLO (i.e., Case BB RLO; see Section 10.13). In many cases, in this mass range, the final products can be NSs, produced in quite complex ways. And the collapsing cores can be almost naked, due to further mass transfer phases, leading to ultra-stripped SNe (Section 13.5). The helium stars in this mass range develop degenerate cores of CO, ONeMg, or Si for $1.6 \leq M_{\rm He,i}/M_\odot \leq 1.8$, $1.9 \leq M_{\rm He,i}/M_\odot \leq 2.4$, or $2.5 \leq M_{\rm He,i}/M_\odot \leq 3.2$, respectively. For the lower-mass range, these helium stars failed carbon burning (possibly some traces of neon are produced at the core boundary from unsuccessful off-center carbon ignition). In the mid-range of masses, all stars ignite off-center carbon ignition that migrates to the center before the envelope of the star expands to giant dimensions (see Figs. 8.28 and 8.31). These stars develop degenerate ONeMg cores, which may grow to electron-capture collapse, producing a NS. Their evolution and further mass stripping will be discussed in the next two sections. For the high-mass range, the stars ignite central carbon ignition, but off-center oxygen-, neon-, and also off-center silicon ignition, and still have considerable envelope expansion, leading to stripping by further binary mass exchange via Case BB RLO (if these stars are in tight binaries), before iron-core collapse. We also discuss their binary evolution further in the next two sections.
3. $M_{\rm He,i} \simeq 3.2$–$45\,M_\odot$: "normal" evolution leading to a SN.
   Models in this mass range (corresponding to $16 \lesssim M_{\rm ZAMS}/M_\odot \lesssim 100$) go through all burning stages, which they ignite in the center and produce NSs and BHs. The initial helium-star mass ranges that mainly produce NSs, but see Section 9.7, are roughly $M_{\rm He,i} \lesssim 11\,M_\odot$ ($M_{\rm ZAMS} \lesssim 34\,M_\odot$) and $17 \lesssim M_{\rm He,i}/M_\odot \lesssim 25$ ($45 \lesssim M_{\rm ZAMS}/M_\odot \lesssim 62$). The other mass ranges produce BHs.

[5]Notice here that we have supplemented with the results from investigations by Marchant et al. (2019); Renzo et al. (2020) to get a wider picture. See also Chapters 8 and 15.

4. $M_{\text{He,i}} \simeq 45$–(80)120 $M_\odot$: pulsational pair-instability SN (PISN) and a BH remnant.
   The helium stars in this mass range undergo pulsational pair-instability SNe (PPISNe), leading to a BH remnant with a maximum mass of about 45 $M_\odot$. The upper limit of $M_{\text{He,i}} \simeq 120\, M_\odot$ for PPISNe is still debated, with other works finding values close to only 80 or 90 $M_\odot$, depending on various assumed physical conditions (Marchant et al., 2019; Renzo et al., 2020). The ZAMS masses of the hydrogen-rich progenitor stars leading to PPISNe are roughly between 100–200 $M_\odot$, depending on metallicity.
5. $M_{\text{He,i}} \simeq$ (80)120–(200)240 $M_\odot$: PISN and no remnant.
   These very massive helium stars undergo regular pair-instability SNe (PISNe), leading to total destruction of the star in a thermonuclear SN, and no remnant is left behind. This phenomenon results in a mass gap of stellar-mass BHs (unless they are produced by hierarchical merger events, cf. Sections 12.4 and 15.6.2). For further discussions of (P)PISNe and the BH mass gap, see Chapters 9 and 15.
6. $M_{\text{He,i}} \gtrsim$ (200)240 $M_\odot$: leading to a PISN and a massive BH remnant.
   These extremely massive helium stars (originating from ZAMS stars with masses possibly exceeding ~400 $M_\odot$, strongly depending on metallicity), undergo a full core collapse—despite violent ignition of carbon and oxygen, following the initial collapse triggered by (e^-, e^+)–pair creation from high-energy photons, which thereby remove pressure support in the core. The BH remnants left behind have masses in excess of 120 $M_\odot$.

The exact PISN mass boundaries listed here depend on the applied input physics (see more detailed discussions in Section 15.5.1) and are therefore still a subject of debate. Factors at play that determine the various PISN mass boundaries include: (1) the $^{12}\text{C}(\alpha,\gamma)^{16}\text{O}$ nuclear reaction rate, which determines the final C/O ratio in the core and thus affects the mass limit where the star enters the pair-instability region (Takahashi, 2018; Farmer et al., 2019); (2) rotation, which may shift the mass ranges upward by as much as 20% (Marchant & Moriya, 2020; Woosley & Heger, 2021) since rotation stabilizes the star against pair instability; and furthermore, (3) metallicity, an important factor that affects wind mass loss, which also affects the precise boundary masses. It can be argued (P. Marchant 2021, private communication) that the properties of the progenitor stars at helium depletion (assuming not much mass is lost between that and core collapse) can better determine the final outcome. Thus, rather than initial helium-star mass, final helium or CO core mass is a better representative to predict the outcome. As an example, Marchant estimates that the final helium core mass limit for leaving BHs above the PISN mass gap (see above under point 6) could be as low as 120 $M_\odot$.

The discovery of superluminous SNe (SLSNe), such as SN 2005ap, SN 2006gy, and SN 2007bi (Quimby et al., 2007; Smith et al., 2007; Gal-Yam et al., 2009; Quimby et al., 2011), which have a peak luminosity ~100 times brighter than typical SNe, is well documented in the literature. These SLSNe provide another twist in the classification of SNe and are believed to originate from the collapse of massive stars where the explosion mechanism is either a core-collapse SN or a (pulsational) PISN, or a luminosity source driven by magnetar spin-down, collisions of ejected shells, or

^{56}Ni decay (see, e.g., Gal-Yam, 2019; Chen K. -J. 2021, for reviews). Some of these hydrogen-poor events (SLSNe-I), for example, SN 2007bi, are indeed believed to be associated with PISNe of very massive helium stars.

13.4 ELECTRON-CAPTURE SNe IN SINGLE AND BINARY STARS

Direct observational evidence for electron-capture SNe (EC SNe) is often ambiguous, although SN 2018zd (Hiramatsu et al., 2021) may be a robust case. In the following, we therefore focus on the theoretical predictions of progenitor evolution leading to EC SNe.

13.4.1 Evolution Leading to Electron-capture SNe

This type of evolution was already very briefly introduced in Section 8.1.4. For stars that do not ignite oxygen, but for which their ONeMg cores are more massive than $1.37\,M_{\odot}$, the final fate is generally believed to be an EC SN (Miyaji et al., 1979; Nomoto, 1987; Takahashi et al., 2013; Jones et al., 2016; Chanlaridis et al., 2022). In these cores, degenerate electrons become energetic enough to be captured by magnesium (^{24}Mg) and neon (^{20}Ne), which thereby removes the pressure supported by these energetic electrons against gravity in the core. In combination with the released energy through the γ-decay of ^{20}O, the initiated gravitational collapse is followed by explosive oxygen fusion ($^{16}\mathrm{O}+^{16}\mathrm{O}$) when the central density reaches $\sim 10^{10}\,\mathrm{g\,cm^{-3}}$ (Miyaji et al., 1979; Nomoto, 1984, 1987). The reason for the explosive burning (thermonuclear runaway) is the high degree of degeneracy of the material, which makes it so that there is little to no expansion resulting from the temperature increase. Instead, the increase in temperature accelerates the rate of fusion until the temperature becomes so high ($\sim 10^{10}$ K) that the degeneracy is lifted and the composition reaches nuclear statistical equilibrium (Raduta & Gulminelli, 2019). At that point, the core can expand in response to the energy release and complete the SN explosion.

However, the picture is more complicated as such. To be more specific, for stars that evolve to develop low-mass metal cores, the exact boundary in outcome between iron-core collapse SNe (Fe CCSNe) and EC SNe depends on the location of ignition of the off-center neon and oxygen burning (Timmes et al., 1994), and also on the propagation of the neon-oxygen flame, which is sensitive to mixing in the convective layers across the flame front (Jones et al., 2014). The latter process may affect the electron fraction and the density in the central region, and thereby the electron-capture efficiency, and thus the nature of the SN.

Following the previously mentioned runaway oxygen burning process preceding the EC SN, the core will collapse rapidly to produce a NS (more rapidly than a typical iron-core progenitor), launching a shock wave and producing a dim SN (possibly of Type IIp for an isolated progenitor star). Simulations by Kitaura et al. (2006); Dessart et al. (2006); Fischer et al. (2010) predict relatively low explosion energies and low Ni ejecta mass. The promptness of these SN explosions is caused by the steep density gradient at the edge of the core in the progenitor stars of EC SNe (similar to that of

AIC on ONeMg WDs, Section 14.5). In such rapid explosions, there is therefore little time for asymmetries to develop that would otherwise produce a significant kick onto the newborn NS. It is therefore believed that EC SNe and AIC events produce small NS kicks (Section 13.8.4).

The cross section for electron capture on ^{20}Ne (i.e., the transition between the ground states of ^{20}Ne and ^{20}F) has recently been revised experimentally (Kirsebom et al., 2019). This result has an important impact for a degenerate ONeMg stellar core. By measuring this transition, they found that its strength is exceptionally large and that it enhances the capture rate by several orders of magnitude compared to previous estimates, which causes much more heat release by explosive oxygen burning. This has a decisive impact on the evolution of the core, increasing the likelihood that the star is (partially) disrupted by a thermonuclear explosion (Jones et al., 2016, 2019) rather than collapsing to form a NS. This makes any predictions for the final fate of these stars even more uncertain (Zha et al., 2019).

13.4.2 The Mass Range of EC SN Progenitors for Single Stars

The range of ZAMS masses of isolated stars producing EC SNe is believed to be rather confined and with a width of the order of $\Delta M_{\rm EC} = 0.1 - 0.2\,M_\odot$ (Doherty et al., 2017, and references therein). Within a ZAMS interval of roughly $8-11\,M_\odot$ (Jones et al., 2013, 2014; Doherty et al., 2015; Woosley & Heger, 2015), the exact location of this mass interval, however, is rather uncertain and depends on metallicity (see Fig. 13.11) and to some degree on the treatment of convective overshooting (Siess & Pumo, 2006; Siess, 2007; Poelarends et al., 2008). The reason for the narrow mass window, $\Delta M_{\rm EC}$, for producing EC SNe is that most super-asymptotic giant branch (SAGB) stars are expected to end their lives as ONeMg WDs, rather than EC SNe, due to the ability of the second dredge-up to significantly reduce the mass of the helium core and of the efficient AGB winds to remove the stellar envelope before the degenerate core reaches the critical mass ($\sim 1.37\,M_\odot$) for the activation of electron-capture reactions.

By taking the mass range of the EC SN channel from Doherty et al. (2015) and weighting it with a Salpeter IMF, Doherty et al. (2017) demonstrated that the resulting EC SN contribution to the overall core-collapse SN rate (Type II-P SNe) of single stars is about 2%–5% when assuming the maximum mass for a Type II-P SN is $18\,M_\odot$, based on the analysis of SN observations by Smartt (2015). However, this result is strongly dependent on the poorly constrained stellar wind mass-loss rate at low metallicity. Applying, for example, the results from Poelarends (2007), the analysis of Doherty et al. (2017) shows that 5%, 17%, and 38% of all Type II-P SNe for metallicities $Z = 0.02$, 0.001, and 10^{-5}, respectively, originate from EC SNe.

From an empirical point of view, one can also compute the expected range of ZAMS masses for the progenitors of EC SNe, given the known observed fraction of EC SN-like events relative to the overall rate of core-collapse SNe, and combine this result with an IMF. A main problem in this exercise, however, is to identify the EC SN events among observed potential candidates, for example, intermediate-luminosity red transients and SN2008S-like events. Depending on the classification of SN spectra and assumptions about the lowest ZAMS mass producing EC SNe, the empirical range

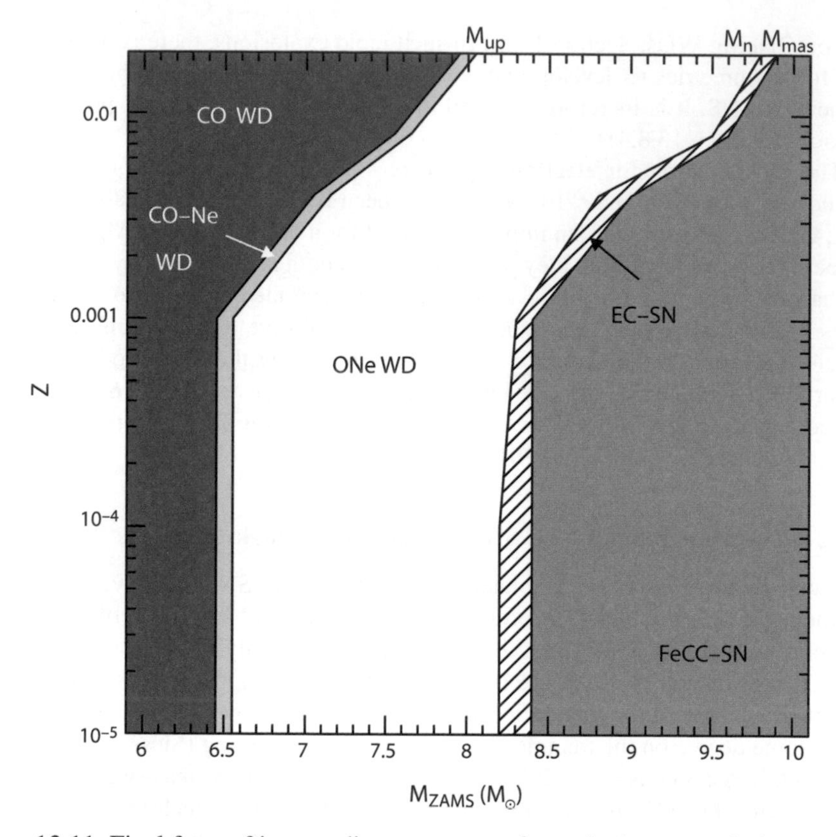

Figure 13.11. Final fates of intermediate-mass stars from single star evolution. ZAMS mass on the x-axis, metallicity (Z) on the y-axis. The hatched region represents the width of the EC SNe channel, $\Delta M_{\rm EC}$. The critical mass for producing NSs is seen to decrease with lower metallicity. After Doherty et al. (2017).

of ZAMS masses producing EC SNe seems to be somewhat larger ($\sim$0.3$-$1.0 $M_\odot$; Strittzinger et al., 2020), assuming a Salpeter IMF with isolated single stars as progenitors. This relatively wider range in ZAMS masses could be due to differences in metallicities of the SN progenitor environments, although the EC SN candidates found in the local Universe are all at approximately solar metallicity. It should be noticed, however, that the observables of SNe from stars in the ZAMS mass interval $8 - 11\,M_\odot$ are expected to be rather similar, independent of the SN being an EC SN or a low-mass Fe CCSN (Kozyreva et al., 2021).

13.4.3 The Mass Range of EC SN Progenitors in Binaries

It has been demonstrated that the width of the initial mass range for producing EC SNe is smaller in single stars compared to binary stars in close orbits (Podsiadlowski et al., 2004; Poelarends et al., 2008). The reason for this effect is that the presence in a binary can dramatically affect the structure of the core of a massive star at the time of core collapse. Stars more massive than $\sim$11 $M_{\odot}$ are generally expected to have smaller iron cores if they lose their envelopes in a close binary. Stars in the range of $8-11$ $M_{\odot}$ may explode in an EC SN if they are located in a *close* binary, while stars in *wide* binaries or single stars will experience a second dredge-up phase and are more likely to end their evolution as ONeMg WDs. Indeed, Podsiadlowski et al. (2004) argued that the minimum initial mass of a massive single star that becomes a NS may be as high as $10-12$ $M_{\odot}$, while for close binaries it may be as low as $6-8$ $M_{\odot}$. Stars less massive than 10–12 $M_{\odot}$ can end up with larger helium and metal cores if they have a close companion, since the second dredge-up phase that reduces the helium core mass dramatically in single stars does not occur once the hydrogen envelope is lost via mass transfer (Podsiadlowski et al., 2004; Siess & Lebreuilly, 2018). Therefore, these binary stars are more likely to produce ONeMg cores reaching the critical value of $\sim$1.37 $M_{\odot}$ required for EC SNe. This difference between the final evolution of single stars and primaries of interacting binaries in this mass range is illustrated in Figure 13.12. The precise width of the critical mass range depends on the applied input physics. This includes the treatment of convection, the amount of convective overshooting, and the metallicity of the star, and will generally be lower for larger amounts of convective overshooting or lower metallicity (Podsiadlowski et al., 2004; Siess & Lebreuilly, 2018).

Interestingly, since EC SNe are not expected to impart a large velocity kick to the NS, the orbits of the resulting NS binaries are expected to have small eccentricities, just as is observed in a considerable fraction of the Be/X-ray binaries (e.g., X Per), and also of the DNSs (Chapter 14). We further refer to Section 10.3.2 for the consequences of these ideas for the formation of different types of HMXBs.

Although a large fraction of main-sequence massive stars are found in close binaries (Chini et al., 2011, 2012; Sana et al., 2012), a hydrogen-rich spectrum of an EC SN points to an isolated, or wide-orbit, progenitor star. In contrast, SN Ibn progenitors have lost their hydrogen (and partly also their helium) envelopes and are embedded in dense, helium-rich circumstellar material (CSM). Their narrow spectral lines arise when the fast SN ejecta interacts with the previously ejected and slow-moving CSM. Thus, SNe Ibn may provide a connection between the "stripped-envelope"and the "interacting" SN populations (Podsiadlowski et al., 1992).

13.4.4 Numerical Modelling of EC SNe in Close Binaries

Given the difficulty in determining the final fate for single stars with a final metal core mass close to the Chandrasekhar mass, it is hardly surprising that this problem also applies to binary stars. For low-mass metal cores, the exact boundary between EC SNe and FeCC SNe depends on the location of ignition of the off-center neon and oxygen burning (Timmes et al., 1994), and also on the propagation of the neon-oxygen flame, which is sensitive to mixing in the convective layers across the flame front (Jones et al., 2014). The latter process may affect the electron fraction and the density in the central region, and thereby the electron-capture efficiency, and thus the nature of the SN. In binary star modelling, where stars often cannot be evolved onto the onset of the core collapse due to numerical issues, one can instead compare evolutionary tracks in the central mass density-central temperature diagram, that is, $(\rho_c,\ T_c)$–plane, with those available in the literature for isolated massive stars, for example, Umeda et al. (2012)

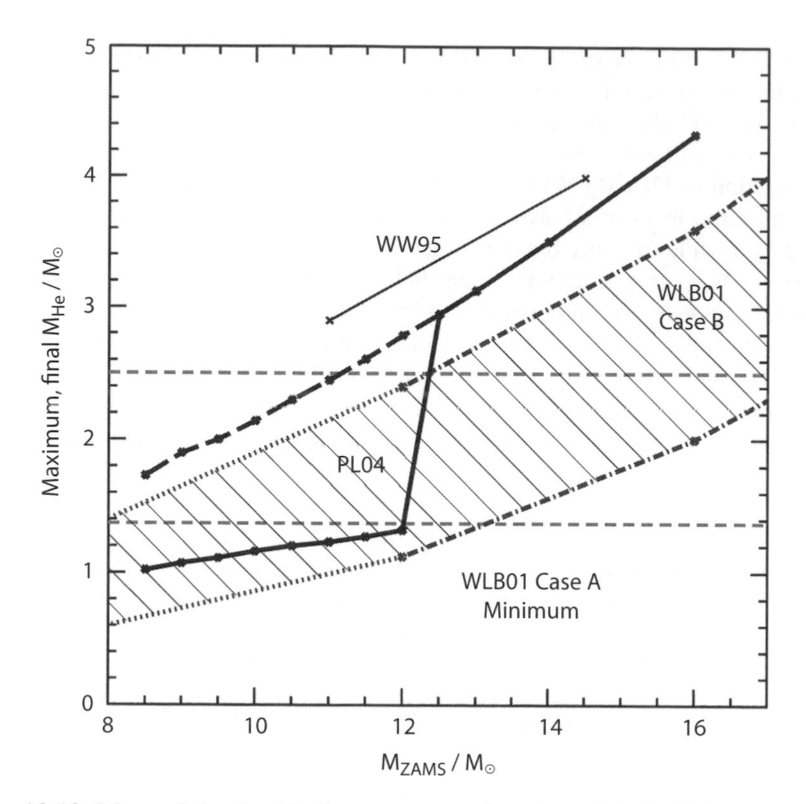

Figure 13.12. Mass of the final helium core as a function of the initial ZAMS mass, for single stars (fully drawn black curve) and binary components (upper thick black curve that is partly dashed and partly solid). In single stars in the mass range 8 to $\sim 10 - 12\, M_\odot$, on the asymptotic-giant branch, "dredge-up" erodes away the helium layers surrounding the degenerate ONeMg core, which prevents the cores of single stars in this mass range from growing to electron-capture collapse. On the other hand, in an interacting binary, the stars in this mass range lose their hydrogen-rich envelopes by mass transfer to the companion, and therefore avoid dredge-up. They become He-stars with the original mass of the helium core (thick dashed black line), which is larger than the Chandrasekhar mass, and their degenerate ONeMg cores are able to grow by helium-shell burning to electron-capture core collapse and the formation of a NS. Almost vertical line PL04 was calculated by Poelarends & Langer (2004, unpublished). WW95 give the final helium core masses calculated by Woosley & Weaver (1995); the final core masses of close binary models undergoing Case B RLO from Wellstein et al. (2001) are shown in the upper boundary of the hatched region, while those undergoing Case A RLO may be anywhere between the lower boundary of the hatched region and the Case B curve. The results from the binary calculations have been extrapolated for initial masses below $12\, M_\odot$ (dotted curves). The dashed grey horizontal lines give the range for the final helium core masses for which the star may undergo electron-capture collapse. After Podsiadlowski et al. (2004).

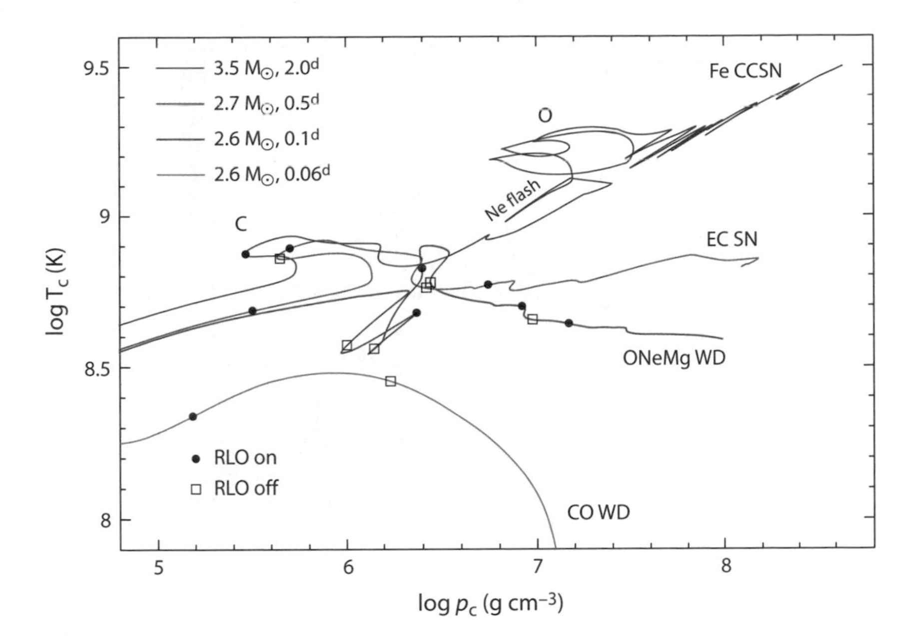

Figure 13.13. Central temperature vs. central density for four selected binary helium star–NS models that terminate their evolution as an Fe CCSN, EC SN, ONeMg WD, and CO WD, respectively, depending on the initial mass of the helium star (see legend for initial masses and orbital periods). Carbon burning (marked by C) is seen in the first three cases, and oxygen burning (marked by O) occurs only in the most massive star. Epochs of RLO onset (solid circles) and RLO termination (open squares) are shown along each track. After Tauris et al. (2015).

and Jones et al. (2013). Using this method, Tauris et al. (2015) studied the fate of naked helium donor stars in post-HMXB/post-CE systems, which undergo Case BB RLO to an accreting NS (see Section 10.13). They found, as a simple rule of thumb, that evolutionary tracks where the post-carbon-burning central temperature, T_c, rises above the value of T_c during carbon burning will eventually ignite oxygen (typically at $T_c \gtrsim 10^9$ K, depending on ρ_c) and later burn silicon to produce an iron core and finally undergo an Fe CCSN (see Fig. 13.13). This is the case for their models that have a metal core, $M_{\rm core} > 1.43\,M_\odot$, and they therefore adopted this mass as an approximate threshold limit separating Fe CCSNe from EC SNe. Hence, they argued that EC SNe will be the final fate if these stars develop metal cores (degenerate ONeMg cores) in the mass interval, $M_{\rm core} \simeq 1.37-1.43\,M_\odot$ (Tauris et al., 2015).

13.5 ULTRA-STRIPPED SUPERNOVAE

In Chapter 10, we described the final phase of mass transfer in a post-HMXB system (Case BB RLO), in which a naked helium star transfers mass to a NS in a tight orbit. As a consequence of the compact nature of the NS, it is able to strip off (almost) the entire

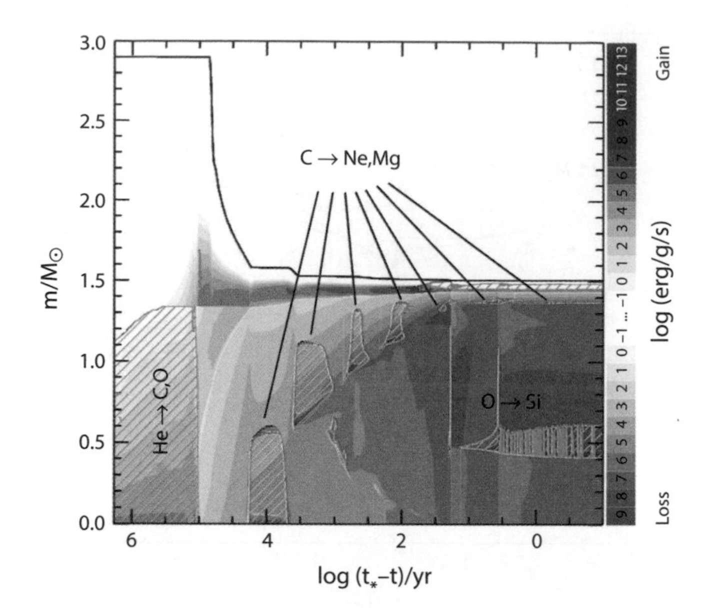

Figure 13.14. Kippenhahn diagram of a 2.9 $M_{\odot}$ helium star undergoing Case BB/BBB RLO prior to an ultra-stripped SN. The plot shows cross-sections of the helium star in mass-coordinates from the center to the surface of the star, along the y-axis, as a function of stellar age on the x-axis. The value $(t_{*} - t)/\mathrm{yr}$ is the remaining time of the calculations, spanning a total time of $t_{*} = 1.854356$ Myr. The green hatched areas denote zones with convection; red color indicates semi-convection. The intensity of the blue/purple color indicates the net energy-production rate. Shortly after off-center oxygen ignition (at $m/M_{\odot} \simeq 0.5$, when $\log(t_{*} - t) = 1.3$), the star was evolved further for another $\sim$20 yr until about 10 yr prior to core collapse. After Tauris et al. (2013).

helium envelope of its companion star while this star evolves through helium, carbon, oxygen, and silicon burning, and loses mass to the NS via Case BB RLO (Tauris et al., 2013, 2015; Müller et al., 2019; Jiang et al., 2021). Alternatively, the accretor could be a BH or a massive WD (Krishnan et al., 2020). It has been demonstrated (Tauris et al., 2015) that helium-star companions to NSs may evolve into naked metal cores with masses as low as $\sim 1.5\,M_{\odot}$, barely above the Chandrasekhar-mass limit, by the time they explode (Fig. 13.14). Depending on the initial orbital period and mass of the helium star at the onset of Case BB RLO, the final metal core left behind prior to core collapse is often embedded in an envelope with a mass of only $\lesssim 0.1 - 0.2\,M_{\odot}$, in extreme cases even $< 0.01\,M_{\odot}$ (Fig. 13.15). The explosion of such a star thus results in an *ultra-stripped SN*. This new subclass of SNe was first explored by Tauris et al. (2013), and they argued for expected optical signatures compatible with observations of rapidly decaying SN light curves, such as that of SN 2005ek (Drout et al., 2013); see Figure 13.16.

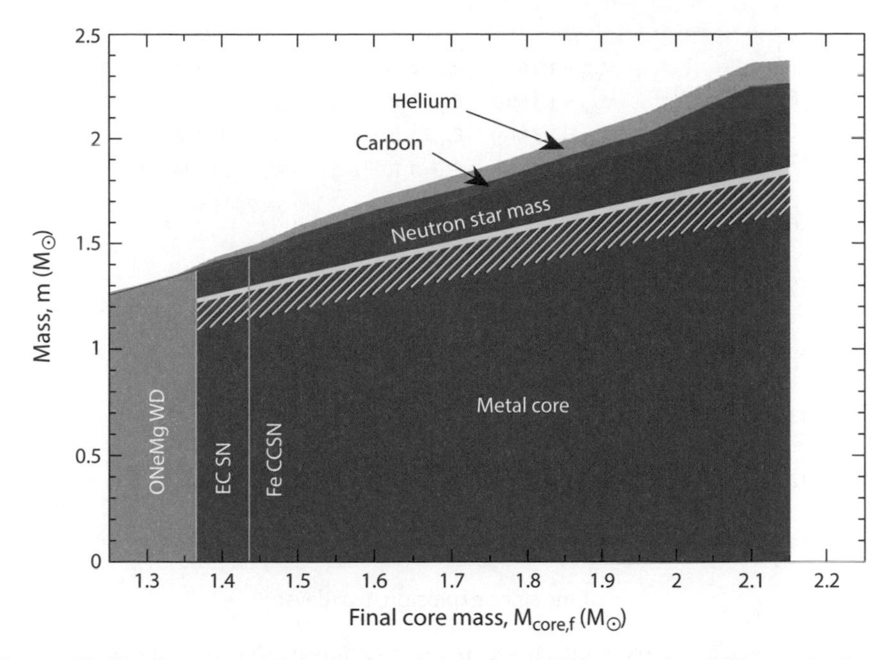

Figure 13.15. Cross sections of ultra-stripped pre-SN stars for different final core masses. The total mass is divided into three regions representing the amount of helium (orange), the remaining carbon-rich envelope (red), and the metal core (purple). The blue shaded region represents stars leaving ONeMg WDs. The yellow line indicates the gravitational masses of the NS remnants. After Tauris et al. (2015).

13.5.1 Observational Signatures of Ultra-stripped SNe

In a simple approach, one can estimate the expected light-curve properties of ultra-stripped SNe from the photon diffusion time through a homologously expanding SN envelope (Arnett, 1979, 1982; Kleiser & Kasen, 2014). The rise time and the decay time of ultra-stripped SNe is thus estimated to be roughly (Tauris et al., 2015):

$$\tau_{\rm rise} = 5.0\,{\rm days}\ M_{0.1}^{3/4}\,\kappa_{0.1}^{1/2}\,E_{50}^{-1/4} \qquad \tau_{\rm decay} = 25\,{\rm days}\ M_{0.1}\,\kappa_{0.1}^{1/2}\,E_{50}^{-1/2}, \qquad (13.1)$$

where $M_{0.1}$ is the ejecta mass in units of $0.1\,M_\odot$, E_{50} is the SN kinetic energy in units of 10^{50} erg, and $\kappa_{0.1}$ is the opacity in units of $0.1\,{\rm cm^2\,g^{-1}}$. Graphical representations of these light-curve timescales are plotted in Figure 13.17. An example of a rapidly decaying light curve expected for an ultra-stripped SN is that of SN 2005ek, which has $\Delta m_{15} \geq 3.5$ (Drout et al., 2013); see Figure 13.16. Depending on the amount of helium in the ejected envelope, and the amount of nickel mixed into this ejecta, the ultra-stripped SNe can be classified as both Type Ib or Type Ic, cf. Figure 13.18.

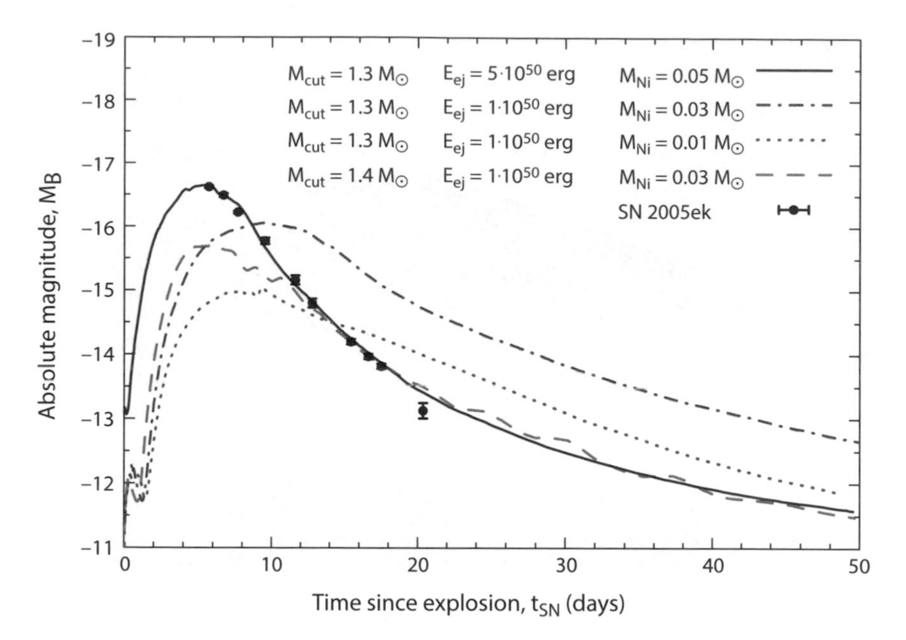

Figure 13.16. *B*-band SN light curves of ultra-stripped SNe Ic from a progenitor calculation obtained by binary evolution leading to a $1.50\,M_\odot$ exploding star with a metal core of $1.45\,M_\odot$ (Fig. 13.14). The different curves correspond to various combinations of mass cut ($M_{\rm cut}$), explosion energy ($E_{\rm ej}$), and amount of ^{56}Ni synthesized ($M_{\rm Ni}$). The total bolometric luminosities, $L_{\rm bol}$, of the four curves peak in the range $0.4 - 2.3 \times 10^{42}\,{\rm erg\,s^{-1}}$. Data for SN 2005ek is from Drout et al. (2013). The explosion date is arbitrarily chosen to match the light curve. After Tauris et al. (2013).

The expected light curves and the spectra of ultra-stripped SNe were initially discussed in detail by Moriya et al. (2017). Besides SN 2005ek (the first observed candidate of an ultra-stripped SN, [Drout et al., 2013; Tauris et al., 2013]) a number of other identified candidates are SN 2010X (Moriya et al., 2017) and a number of the puzzling Ca-rich gap transients: PTF10iuv (Moriya et al., 2017), iPTF16hgs (De et al., 2018), iPTF14gqr (De et al., 2018), SN 2019ehk (Nakaoka et al., 2021) SN 2019dge (Yao et al., 2020) and SN 2019bkc (Prentice et al., 2020), among which, in particular, iPTF14gqr and SN 2019dge are strong candidates.

Ca-rich SNe are a heterogeneous group of objects that are primarily characterized by peak magnitudes of -14 to -16.5, quickly evolving light curves, and strong calcium features in photospheric and nebular phase spectra (Taubenberger, 2017). A defining feature of the class is an integrated [Ca II]/[O I] flux ratio greater than $\sim$2, and the majority of these objects exhibit low ejecta and ^{56}Ni masses of $\lesssim 0.5\,M_\odot$ and $\lesssim 0.1\,M_\odot$, respectively (e.g., Perets et al., 2010; Lunnan et al., 2017). Ca-rich SNe have generally been suggested to arise from a progenitor system containing a WD, similar to that of thermonuclear objects. For example, it has been argued that the Ca-rich transient

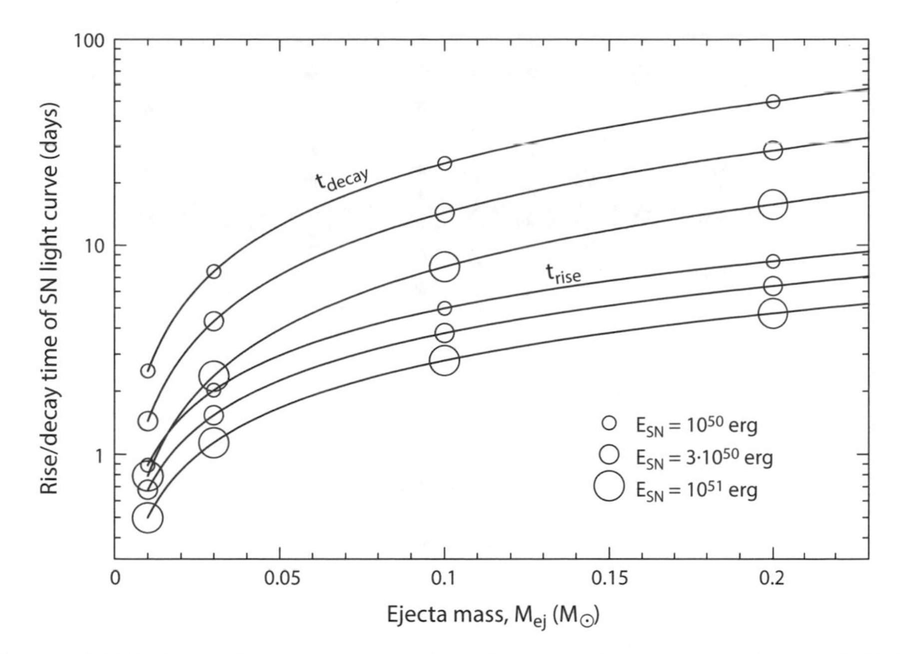

Figure 13.17. Rise times (lower, red) and decay times (upper, blue) of light curves expected from ultra-stripped SNe with different explosion energies between $10^{50}-10^{51}$ erg.

SN 2016hnk originated from the helium-shell detonation of a sub-Chandrasekhar WD (Jacobson-Galán et al., 2020). However, there are other examples of SNe (e.g., iPTF15eqv, Milisavljevic et al., 2017) that exhibit a unique combination of properties that bridge those observed in Ca-rich transients and ordinary SNe Ib/c, and which therefore challenge the notion that spectroscopically classified Ca-rich transients only originate from WD progenitor systems.

Given the old stellar environment of the sites where Ca-rich transients are often discovered (Lunnan et al., 2017), it is unlikely that ultra-stripped SNe are the origin of the bulk of Ca-rich transients. The reason for this is that ultra-stripped SNe should represent a young population (max age ~40 Myr; see Section 13.5.2). However, it is still possible that a fraction of Ca-rich transients may come from ultra-stripped SNe; see further discussions in De et al. (2020); Nakaoka et al. (2021). Hopefully, near-future and ongoing high-cadence surveys and dedicated SN searches will discover more of these exotic ultra-stripped SNe, characterized by modest luminosity with rapidly decaying light curves, and illuminate their potential link to Ca-rich transients.

NS properties and the NS equation-of-state can also help distinguish binary formation channels and SN scenarios. In principle, it might in this way be possible to obtain tight constraints on the baryonic mass and binding energy of NSs that are thought to have formed from an ultra-stripped progenitor (Holgado, 2021).

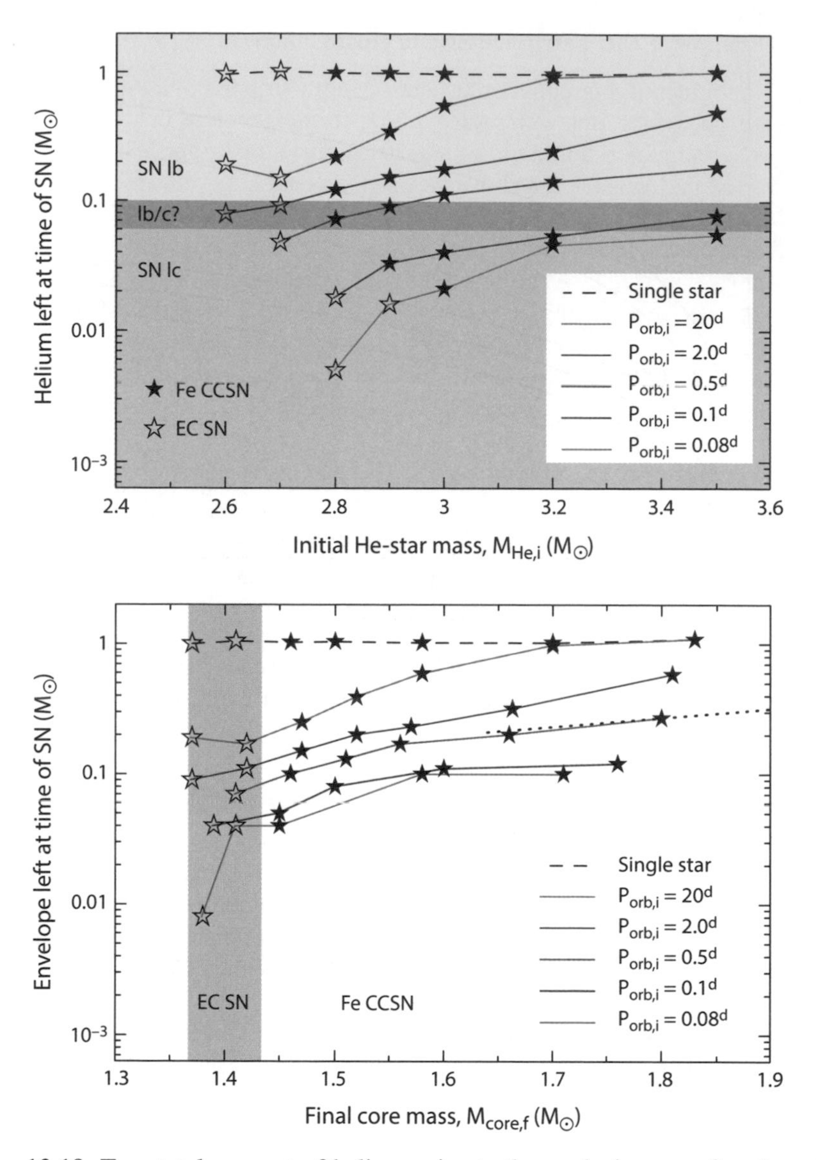

Figure 13.18. Top: total amount of helium prior to the explosion as a function of the initial helium-star mass of the progenitor before mass transfer. Different curves correspond to different initial orbital periods. The dashed curve is for isolated evolved helium stars. The expected classification of observed SNe as Type Ib or Ic are shown in different colors. Open star symbols indicate EC SNe, solid star symbols indicate Fe CCSNe. Bottom: total envelope mass as a function of final core mass prior to the SN, for the same models as in the top panel. After Tauris et al. (2015).

13.5.2 Location of Ultra-stripped SNe in Host Galaxies

To further help identifying optical transients as ultra-stripped SNe, it is important to know their expected location with respect to star-forming regions in host galaxies. Following the age argument in Tauris et al. (2015), an ultra-stripped SN is the second SN in a binary, and thus the travel distance of the system, d, with respect to its original birth location as a ZAMS binary, is the product of the systemic velocity resulting from the first SN, v_{sys}, and the lifetime of the secondary star following mass accretion from the primary star, t_2. Given that the progenitor mass of the exploding ultra-stripped star is expected within the interval $8-25\,M_{\odot}$, it is clear that $t_2 \simeq 7-40$ Myr. (This is the total lifetime of this secondary star before it loses its hydrogen envelope in a CE, which creates the naked helium star orbiting the NS, and then subsequently undergoes Case BB RLO and finally explodes.) Given their arguments for why the first SN must result in a small kick (for the system to survive in a wide orbit and successfully avoid coalescence in the subsequent CE phase and eventually produce an ultra-stripped SN, leading to tight DNS systems, see Fig. 1.1), the resulting value of only $v_{sys} \simeq 10\,\mathrm{km\,s^{-1}}$ means that the typical expected value of the travel distance is $d = v_{sys} \cdot t_2 \simeq 200-300$ pc (the entire interval probably spanning between $100-800$ pc). Therefore, ultra-stripped SNe occur relatively close to the star-forming regions in host galaxies.

13.5.3 Rate of Ultra-stripped SNe

Given that ultra-stripped SNe represent the second SN in a tight binary, they are highly relevant for LIGO sources since most (probably all) merging DNS systems have evolved through such explosions. The estimated merger rate of DNS systems in a Milky Way–equivalent Galaxy is of the order of a few $10^{-6} - 10^{-5}\,\mathrm{yr^{-1}}$ (see also Chapters 15 and 16), and the Galactic rate of core-collapse SNe is of the order of $10^{-2}\,\mathrm{yr^{-1}}$, and thus we notice that the fraction of ultra-stripped SNe to all SNe is of the order of 0.1%. Moreover, some ultra-stripped SNe result in NSs receiving large kicks (see discussions in Section 13.7.7) which may unbind a fraction of binaries that therefore never become DNS systems. In addition, also BHs and WDs are able to strip a Roche-lobe filling helium star prior to its SN, producing ultra-stripped SNe resulting in BH+NS binaries and systems containing a young pulsar orbiting a WD (Krishnan et al., 2020). Thus the total fraction of ultra-stripped SNe to all core-collapse SNe could be larger—perhaps even reaching the 1% level.

13.6 COMPARISON BETWEEN THEORY AND OBSERVATIONS OF SNe Ib AND Ic

We briefly discuss how the theoretical ideas and models for the formation of SNe Ib and Ic, presented in the last three sections, can be related to the observed properties of these SNe, and in how far theory can predict whether a SE-SN will be of Type Ib or Ic.

We saw already in the last section that ultra-stripped SNe, resulting from the second SN in a binary system, are expected to be of both Type Ib and Ic, but they are not expected to contribute more than 1% to the total core-collapse SN rate. The bulk of the much more common SE-SNe, however, are expected to be due to the first explosion in the binary, produced by the helium star resulting from the initial primary star. Here we saw that there are two main regimes of ZAMS primary star masses: (1) $M_{\rm ZAMS} \leq 30\,M_\odot$, where mass stripping can only be achieved by binary mass transfer, and (2) $M_{\rm ZAMS} > 30\,M_\odot$, where the stripping can also be achieved by strong stellar winds. We discuss these two regimes separately.

1. We saw already earlier that the bulk (more than 80%) of the SE-SNe are expected to originate from stars with $M_{\rm ZAMS} < 30\,M_\odot$. In this mass range, Case BB RLO occurs most often in systems with helium stars less massive than about $3.2\,M_\odot$ (corresponding to $M_{\rm ZAMS} \lesssim 16\,M_\odot$), but this second mass-transfer phase is not expected to entirely remove the helium envelope (unlike the situation for donor star progenitors of the second SN, which are being ultra-stripped by their NS/BH companion during Case BB RLO). This means that at the time of core collapse, there still is a helium envelope around the collapsing core. This is independent of whether it is an electron-capture collapse or an iron-core-collapse SN.

   For helium stars resulting from $M_{\rm ZAMS} \simeq 16 - 30\,M_\odot$, there is usually no second phase of mass transfer, and also wind mass loss is not substantial, so also these stars still have a helium envelope at the time of core collapse.

   Dessart et al. (2012) pointed out that moderately-mixed SN models of stripped stars with still a helium layer around the collapsing CO core may still show up as SNe Ic, in case they fail to non-thermally excite helium atoms in the ejecta. For this reason, classification Ic for SNe resulting from stripped stars with $M_{\rm ZAMS} < 30\,M_\odot$ does not give information about the presence or absence of helium in the direct progenitor of the SN. Collapsing cores with a helium envelope may thus show up as SNe Ib as well as Ic. This holds for the stripped remnants from Case B as well as Case C evolution. We are therefore—in absence of precise information of the degree of mixing in the ejecta of these SNe—unable to make a prediction about the expected ratio of SNe Ib/c for SE-SNe originating from this mass range.

2. For $M_{\rm ZAMS} > 30\,M_\odot$, which make up only about 20% of all SE-SNe, the situation may be somewhat different. Here the mass-loss remnants are helium stars more massive than about $8\,M_\odot$, which will show up as WR-stars with very strong stellar-wind mass loss. If the stellar wind removes the entire helium envelope, the star will turn into a WR-star of type WC or even WO, and the explosion will be a Type Ic SN.

   We saw in Section 9.7 that, except in the ZAMS mass range of about 35–46 $M_\odot$ (initial WR-star masses 11–17 $M_\odot$), all primaries with $M_{\rm ZAMS} \leq 62\,M_\odot$ (initial WR-star masses $\leq 25\,M_\odot$) that undergo Case B RLO leave NSs and will produce a successful SN explosion. (These quoted mass ranges are for solar metallicities.) If we assume—very roughly—half of all WR-stars to terminate as WC/WO-stars, then the Type Ic SNe from stars with $M_{\rm ZAMS} > 30\,M_\odot$ will constitute of the order of 10% of the total number of SE-SNe. As mentioned in Section 13.3, the peculiar broad-line SNe Ic connected with GRBs are extremely rare. Their contribution to the total SE-SNe rate can therefore be neglected.

As discussed in Chapter 8, helium stars with initial masses in the range[6] $35 - 80\,M_\odot$ (initial ZAMS masses about $80 - 170 M_\odot$) are expected to experience a pulsational pair-instability SN (PPISN) producing a BH. Whether or not the associated SN explosion will show helium lines will, like for ZAMS masses below $30\,M_\odot$, again depend on the non-thermal excitation of neutral helium atoms, and may possibly lead to Type Ib as well as Ic spectra. Due to the shape of the IMF, this mass range will not contribute more than 2.5% to the total SE-SNe rate. In the initial helium star mass range of about $80 - 240\,M_\odot$, the star will experience a PISN, which leaves no remnant. As here there may still be a helium mantle, this may, depending on the non-thermal excitation state of helium, again have the appearance of a Type Ib or Type Ic SN. They will contribute less than about 2% to the total SE-SNe rate.

In summary, the conclusion by the foregoing is that, if we adopt that some 10% of the SE cores terminate as WC/WO-stars, the lower limit to the expected fraction of SNe Ic among the SE-SNe is 10%. For all other cases, there is expected ejection of helium, and the non-thermal excitation of neutral helium atoms will determine whether or not helium will be observable in the spectrum of the SN. The latter depends mainly on the degree of mixing in the ejecta (Dessart et al., 2012). In the absence of knowledge about mixing of ejecta, it is therefore impossible to predict whether or not the expected fraction of SE-SNe observable as Type Ic is larger than 10%. Unfortunately, our present knowledge of the subject does not allow us to draw any stronger conclusions.

Finally, we notice that the modelling of core-collapse SNe has only recently included the combined effects of rotation and magnetic fields in 2D or 3D (e.g., Jardine et al., 2021; Bugli et al., 2021; and references therein). These effects can each have a significant impact on CC SN dynamics, and acting together, rapid rotation and strong magnetic fields can give rise to powerful magnetorotational explosions. The coming decade with enhanced computing power will reveal further progress of such advanced CC SN models and their applications (e.g., for generating γ-ray bursts and GW signals). Another important effect recently investigated (e.g., Yoshida et al., 2021; McNeill & Müller, 2022) is the effect of rapid rotation during convective oxygen-shell burning in the final evolution of a fast-rotating massive star, which may possibly affect the outcome of the subsequent implosion or SN explosion (and with implications for the NS birth spin period).

13.7 SUPERNOVA KICKS

13.7.1 Explosion Mechanism

It is well established that massive stars end their life as core-collapse SNe. However, numerical simulations have long struggled to conclusively explain the mechanism that powers these explosions. The best-explored scenario is the neutrino-driven mechanism (e.g., Mezzacappa, 2005; Janka, 2012; Burrows et al., 2012), which relies on the partial re-absorption of neutrinos emitted from the young proto-NS and the accretion layer

[6]The exact mass boundaries are strongly dependent on metallicity and assumed stellar physics applied in various theoretical models, and the quoted mass boundaries may therefore vary somewhat throughout this book.

at its surface to revive the shock and power the explosion, as briefly described in Section 8.2.2. Besides the neutrino heating (in which neutrinos revive the stalled shock by energy deposition [e.g., Colgate & White, 1966; Bethe & Wilson, 1985]), convective processes and hydrodynamic instabilities enhance the heating mechanism (e.g., Herant et al., 1994; Blondin et al., 2003; Foglizzo et al., 2007; Müller et al., 2017). The increase of computational capabilities over the last three decades, along with successively improved treatment of the microphysics and new developments for neutrino transport methods, have enabled full 3D simulations of the core-collapse SN explosion and thereby driven our understanding of the explosion mechanism to a new level (e.g., Janka, 2012; Burrows, 2013; Müller, 2016; Janka, 2017; Müller et al., 2019; Burrows et al., 2020; Stockinger et al., 2020; Bollig et al., 2021). It is possible that more than one explosion mechanism operates, depending on the physical properties of the exploding star (iron core mass, helium core size, rotation, B-field). Especially, the most energetic SNe and hypernovae seem to demand magnetorotational driving. For a discussion of alternatives to the neutrino-driven mechanism, including the magnetohydrodynamic-, acoustic-, and the QCD phase-transition mechanism, see Janka (2012).

13.7.2 Observational Evidence for SN Kicks

Following core collapse, a momentum kick is imparted onto the newborn NS remnant. The evidence for this goes back to Gunn & Ostriker (1970) who noticed that the distribution of radio pulsars around the Galactic plane is much wider than that of their progenitors, the O and B-stars with masses greater than 8 $M_{\odot}$, which are concentrated in a layer within a few hundred pc around the Galactic plane. Also the young SN remnants are concentrated in this thin layer. The much wider distribution of the pulsars in space indicates that they have much larger space velocities than their progenitor stars, which can be explained only by assuming that in their birth events NSs receive a kick velocity of order several $100\,\mathrm{km\,s^{-1}}$. These large space velocities have subsequently been confirmed by proper motion measurements of pulsars (Lyne & Lorimer, 1994; Hobbs et al., 2005), which, thanks to high-precision timing of the pulse arrival times on Earth and also by interferometric methods, can be carried out with great precision.

Work by Verbunt et al. (2017) shows a somewhat smaller velocity distribution of young pulsars, with a best fit resulting in 42% of the pulsars following a Maxwellian distribution with average velocity, $a\sqrt{8/\pi} = 120\,\mathrm{km\,s^{-1}}$ (5% have a birth velocity less than $60\,\mathrm{km\,s^{-1}}$) and 58% in a Maxwellian distribution with average velocity $540\,\mathrm{km\,s^{-1}}$ (see also Exercise 13.9).

Figure 13.19 shows the proper motion vectors of 233 pulsars measured by Hobbs et al. (2005). The average pulsar velocity of the 73 youngest pulsars in the sample is close to $400\,\mathrm{km\,s^{-1}}$, and there are NSs that have received significantly larger kicks. These include the radio pulsars B2011+38 and B2224+65, which (depending on their precise distances) both have 2D velocities exceeding $1{,}500\,\mathrm{km\,s^{-1}}$ (Hobbs et al., 2005). The latter pulsar is observed with a bow shock (the "guitar nebula"), which confirms that it is moving with a large velocity (Cordes et al., 1993). Other supersonic runaway pulsars with velocities in excess of $1{,}000\,\mathrm{km\,s^{-1}}$ are IGR J11014−6103 (Tomsick et al., 2012; Pavan et al., 2014) and PSR J0002+6216 (Schinzel et al., 2019), see Figure 13.20. Finally, B1508+55 has a fairly precisely measured velocity of

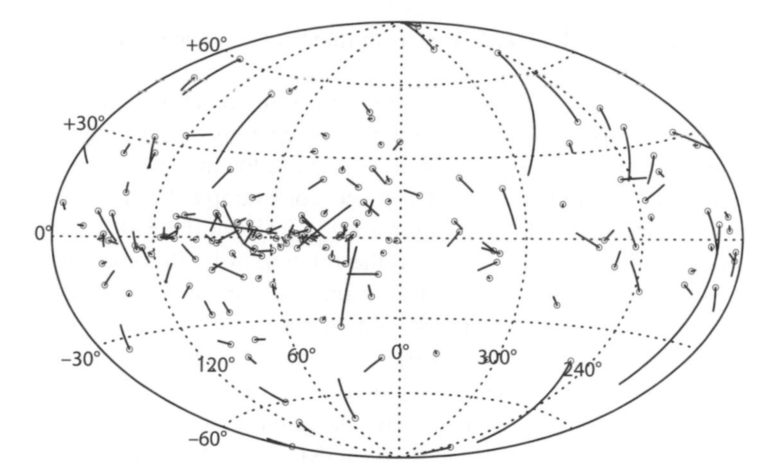

Figure 13.19. Hammer-Aitoff projection map with the proper motion vectors of 233 pulsars in Galactic coordinates. Open circles show the current positions of the pulsars, and the solid lines are the distances traveled by the pulsars on the sky in the last one million years, which are a measure of their proper motion. After Hobbs et al. (2005).

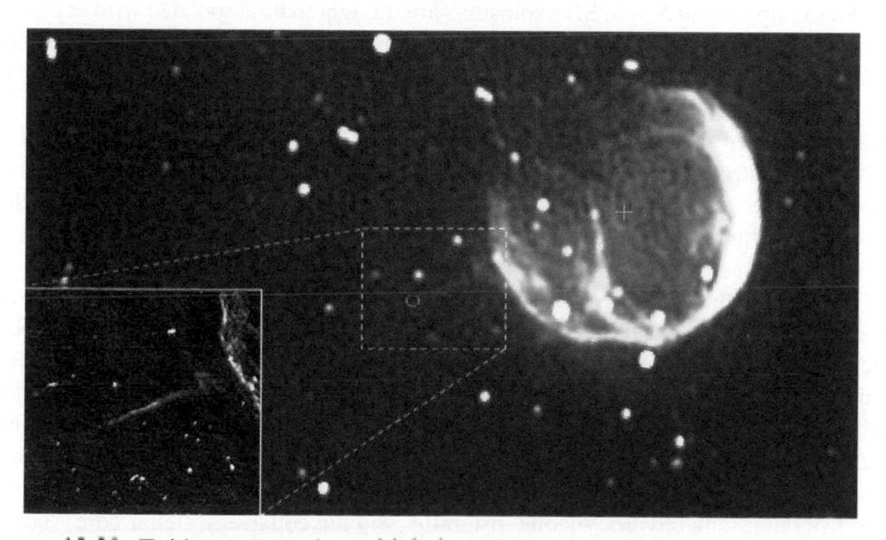

Figure 13.20. Evidence for a large kick imparted on a NS: the 115 ms γ-ray and radio pulsar J0002+6216 was shot out of the SN remnant CTB1 with a velocity of $1{,}100\,\mathrm{km\,s^{-1}}$, leaving a comet-like tail of non-thermal radio emission produced by its bow shock in the ISM. After Schinzel et al. (2019).

$\sim$1100 $\pm$ 100 km s^{-1} based on VLBA measurements of its proper motion and parallax (Chatterjee et al., 2005).

Further evidence for large kicks can be found from combining simulations of the dynamical effects of SNe (Section 13.8) with observations of NS binaries (pulsar or X-ray binaries) with large systemic velocities (e.g., following the recipes outlined by van den Heuvel, 1994a; Tauris et al., 2017); see, for example, Chapter 10. And more specifically, the fact that the binary pulsar J0045−7319 in the SMC is moving in a misaligned retrograde orbit with respect to the rotation of its Be-type companion can only be explained by a considerable birth kick imparted on the NS (Kaspi et al., 1996; Lai, 1996). In addition, it seems that the SN remnant Semeis 147 was formed in a binary that became disrupted and produced a B0V runaway star with a transverse velocity of 74 km s^{-1} and a pulsar with a transverse velocity of 357 km s^{-1} (Dinçel et al., 2015)—again, an event that can only be explained by a large birth kick, see Chapter 10.

Finally, by modelling offset distances of sGRBs from host galaxies, it has been demonstrated (Kelley et al., 2010; Zevin et al., 2020) that some binary NS systems must have had large systemic velocities, requiring kicks > 200 km s^{-1}.

13.7.3 Theoretical Evidence for SN Kicks

The kicks imparted to NSs in SN explosions are thought to be associated with explosion asymmetries and may arise from non-radial hydrodynamic instabilities in the collapsing stellar core. These instabilities lead to large-scale anisotropies of the innermost SN ejecta, which interact gravitationally with the proto-NS and accelerate the nascent NS on a timescale of several seconds (e.g., Janka, 2012; Wongwathanarat et al., 2013). In addition, a neutrino-induced kick is at work as well. For current theoretical understanding of how NS kicks come about in neutrino-driven core-collapse SN explosions, see, for example, Janka (2017); Stockinger et al. (2020). In the following, we summarize the main features.

The kick is the result of two components: a hydrodynamical kick and a neutrino kick. The hydrodynamical kick is caused by an induced asymmetry of the post-shock ejecta leading to an acceleration of the proto-NS (lasting a few seconds) in the direction of the slower ejecta, opposite of the strongest direction of the explosion, compatible with global momentum conservation. Because of this gravitational pull of the slower ejecta, the effect was simply termed the *gravitational tug-boat mechanism* (Wongwathanarat et al., 2013; Gessner & Janka, 2018). The explosion asymmetries may arise from non-radial hydrodynamic instabilities in the collapsing stellar core, such as neutrino-driven convection bubbles or the standing accretion shock instability (SASI); see, for example, Blondin & Mezzacappa (2006); Foglizzo et al. (2007). Moreover, it has been suggested that multi-dimensional effects in the convective burning shells at the onset of core collapse can have an effect on the NS kick amplitude (e.g., Burrows & Hayes, 1996; Arnett & Meakin, 2011; Müller et al., 2016), i.e., pre-collapse asymmetries in the progenitor star could influence the asymmetry parameter of Eq. (13.2) within the framework of the hydrodynamic kick mechanism.

Janka (2017) summarized the theoretical understanding of how NS kicks come about using the gravitational tug-boat mechanism in asymmetric neutrino-driven core-collapse SN explosions, and he derived a simple proportionality between the kick velocity, w, and the energy of the explosion, $E_{\rm SN}$:

$$w = 211\,{\rm km\,s^{-1}}\,C\,\left(\frac{\alpha_{\rm ej}}{0.1}\right)\left(\frac{E_{\rm SN}}{10^{51}\,{\rm erg}}\right)\left(\frac{M_{\rm NS}}{1.5\,M_{\odot}}\right)^{-1}, \tag{13.2}$$

where C is a constant typically of order unity, $M_{\rm NS}$ is the (baryonic)[7] NS mass, and $\alpha_{\rm ej}$ is the momentum-asymmetry parameter of the explosion ejecta, whose statistics needs to be determined by hydrodynamic explosion modelling. The more asymmetric and more powerful the SN explosion is, the larger the NS kick can be. For example, according to Janka (2017), and references therein, core-collapse SNe of relatively massive iron cores have $E_{\rm SN} \simeq 3\times10^{50} - 2.5\times10^{51}$ erg and $\alpha_{\rm ej} = 0 - 0.33$, yielding Fe CCSN kick values in a broad range between $w = 0 - 1{,}000\,{\rm km\,s^{-1}}$, whereas progenitors of EC SNe and low-mass Fe CCSNe—often relevant for ultra-stripped SNe, see below—typically have $E_{\rm SN} \simeq 10^{50}$ erg and $\alpha_{\rm ej} \lesssim 0.03$, resulting in very small kicks of $w \leq 10\,{\rm km\,s^{-1}}$.

The amount of liberated gravitational binding energy from the collapsing core ($\Delta E_{\rm grav} \sim G\,M^2/R \simeq 3\times10^{53}$ erg) is roughly 100 times larger than that of the typical explosion (kinetic) energy of the SN ($E_{\rm SN} \sim 10^{51}$ erg $\equiv$ 1 Bethe), and thus it constitutes a huge energy reservoir. A neutrino-induced kick arises due to anisotropic neutrino emission such that a force is exerted onto the proto-NS in the direction opposite of the most intense neutrino emission, that is, the direction of the neutrino-induced kick is affected by the neutrino-emission dipole that is associated with one-sided proto-NS accretion. While the magnitude of the hydrodynamical kicks is between a few $10\,{\rm km\,s^{-1}}$ up to more than $1{,}000\,{\rm km\,s^{-1}}$ (Scheck et al., 2006; Wongwathanarat et al., 2013), the magnitude of the neutrino kick is significantly smaller (Gessner & Janka, 2018)—typically less than $\sim 30 - 50\,{\rm km\,s^{-1}}$ (but in some cases up to $\sim 100\,{\rm km\,s^{-1}}$). Therefore, usually the neutrino kick is substantially smaller than the hydrodynamic kick from the same explosion, except for explosions of some low-mass cores where both kick components are small (Stockinger et al., 2020).

Finally, it should be mentioned that a fundamentally different mechanism for producing kicks was proposed by Harrison & Tademaru (1975), whereby a NS is accelerated after core collapse as a result of asymmetric electromagnetic (EM) dipole radiation (i.e., the so-called *EM rocket effect*). This NS kick, imparted along the spin axis of the pulsar, is attained at the expense of kinetic rotational energy on the spin down timescale. Therefore, the kick magnitude only becomes significantly larger than $100\,{\rm km\,s^{-1}}$ for initial spin periods of $\lesssim 5$ ms (Lai et al., 2001). Thus for a Crab-like pulsar, this kind of kick is expected to be less than $\sim 5\,{\rm km\,s^{-1}}$ (Gessner & Janka, 2018).

[7]The timescale of neutrino emission is about 10 sec (cf. the neutrino detections from SN 1987A). On a timescale of 1–2 sec, the newborn NS receives its kick but loses only about 1/3 of its gravitational binding energy in neutrinos, and thus it is more appropriate to apply baryonic mass rather than gravitational mass (H.-T. Janka [2020], private communication).

13.7.4 Kick Directions

The hydrodynamic and neutrino-induced kick directions are in general not parallel and should be added as vectors to yield the resultant kick. The direction of the neutrino-induced kick is affected by the neutrino-emission dipole that is associated with one-sided proto-NS accretion. The hydrodynamical kick direction has a more stochastic nature.

Some studies indicate a preference for a kick along the spin axis of isolated radio pulsars (Noutsos et al., 2013; Johnston & Lower, 2021, and references therein). However, SN explosion simulations find that the NS spins seem to be randomly oriented relative to the NS kick direction, that is, the angle between the NS spin vector and the total kick vector (hydrodynamic plus neutrino contribution) does not show any preference for spin-kick alignment, neither at early times nor days later when the fallback is complete and the simulations are terminated (Wongwathanarat et al., 2013; Müller et al., 2019; Stockinger et al., 2020). Moreover, an investigation of the kick direction in the second SN in DNS systems (Tauris et al., 2017) does not find evidence for any particularly favored direction. Stockinger et al. (2020) argue that any convincing mechanism that could provide spin-kick alignment would indeed be unexpected within current accepted physics (i.e., without invoking uncertain ingredients or extreme physical assumptions). And even if such an alignment were achieved during the first seconds of the explosion, it is hard to imagine how this initial alignment could not be overruled by the stochastic effects of the later fallback and its dominant influence on the NS spin. The only possibility to impose a preferred direction for the explosion (such as spin-kick alignment) would probably require very rapid progenitor rotation and/or in combination with strong B-fields (Stockinger et al., 2020). Recently, however, Janka et al. (2022) argue for new support to explain a spin-kick alignment as a consequence of tangential vortex flows in the SN fallback matter that is accreted in three phases on timescales of minutes to hours (blue supergiant) or days (red supergiant). In this new scenario, conclusions based on previous concepts are reversed. We conclude that the kick direction remains an open question.

When comparing to measurements of post-SN binaries, it is also uncertain whether or not the spin axis of the exploding star is tossed in a new (possibly random) direction as a result of the SN (Spruit & Phinney, 1998; Farr et al., 2011), in which case all past memory is lost. The latter seems to be the case, for example, for both of the known young pulsars in DNS systems: PSR J0737−3039 (Breton et al., 2008) and J1906+0746 (Desvignes et al., 2019). Spin tossing in BH formation is discussed in Section 13.7.9.

13.7.5 NS Kick Magnitudes in Binaries

Whereas average kicks of $400-500\,\mathrm{km\,s^{-1}}$ have been demonstrated for young isolated radio pulsars as described previously (Section 13.7.2), it has been suggested for a couple of decades that exploding stars that are stripped in close binaries (i.e., Type Ib/c SNe) may produce substantially smaller NS kicks compared to Type II SN explosions of isolated, or very wide-orbit, stars (Tauris & Bailes, 1996; Pfahl et al., 2002; Voss & Tauris, 2003; Podsiadlowski et al., 2004; van den Heuvel, 2004; Dewi et al., 2005; Bray

& Eldridge, 2016). This awareness came about from both theoretical and observational arguments—the former along the lines of stripped stars often leading to relatively fast explosions with small kicks (see Section 13.7.7), and the latter from comparison of the observed radio pulsar velocity distribution, or population synthesis studies based on this distribution, with the space velocities, orbital periods, and eccentricities of both X-ray binaries (Pfahl et al., 2002), MSPs (Tauris & Bailes, 1996), and DNS systems (Portegies Zwart & Yungelson, 1998; Podsiadlowski et al., 2004; van den Heuvel, 2004; Piran & Shaviv, 2005; Schwab et al., 2010; Ferdman et al., 2013, 2014; Beniamini & Piran, 2016; Tauris et al., 2017). It is simply not possible to reproduce the observed data if exploding stars in close binaries, in general, would receive kicks of $400-500\,\mathrm{km\,s^{-1}}$.

Moreover, for PSR J0737−3039 (the double pulsar) and PSR J1756−2251, Ferdman et al. (2013, 2014) derived small misalignment angles from observations of these systems, giving further support for small kicks.

On the other hand, we have clear evidence that even relatively large kicks can happen in close-orbit DNS progenitor systems. For example, to explain the characteristics (i.e., large proper motions) of the Hulse-Taylor pulsar (PSR B1913+16) and PSR B1534+12, a kick of at least $w \simeq 200\,\mathrm{km\,s^{-1}}$ is needed (see, e.g., Wex et al., 2000; Wong et al., 2010; Tauris et al., 2017). Furthermore, as mentioned earlier, to explain the significant misalignment of the spin axis of the B-star companion from the orbital angular momentum vector in the pulsar binary system PSR J0045−7319, a large kick is needed too (Kaspi et al., 1996; Wex, 1998). Finally, there is some evidence for binaries being disrupted in the second SN, thereby explaining the observations of isolated mildly recycled radio pulsars with similar properties to the first-born NS in DNS systems (Lorimer et al., 2004). The velocities of such ejected NSs can be large even if the kick is small in case the former binary is tight and disrupted due to a large amount of mass loss during the SN event (Tauris & Takens, 1998). However, ultra-stripping prior to the second SN event often prevents disruption due to mass loss, and, in these cases, a large kick is needed to break up the system.

13.7.6 Kicks from EC SNe

The almost spherical explosion of an EC SN progenitor yields very low hydrodynamic kick velocities of $\lesssim 1-2\,\mathrm{km\,s^{-1}}$ by the previously mentioned gravitational tug-boat mechanism (Wongwathanarat et al., 2013; Gessner & Janka, 2018; Stockinger et al., 2020). The total momentum kick imparted to a newborn NS via an EC SN is therefore expected to be small. This is hardly surprising since: (1) EC SN explosion energies are significantly smaller ($\sim 10^{50}$ erg; Kitaura et al., 2006; Dessart et al., 2006) than those inferred for standard Fe CCSNe ($\sim 10^{51}$ erg); and (2) rapid explosions means a short timescale to revive the stalled SN shock, compared to the timescales of the non-radial hydrodynamic instabilities that are required to produce strong anisotropies and significant kicks, for example, Podsiadlowski et al. (2004); Janka (2012). The resultant (hydrodynamic plus neutrino) kick velocities for EC SNe are thus expected to be (significantly) less than $50\,\mathrm{km\,s^{-1}}$, that is, substantially smaller than the average kick velocities of the order $400-500\,\mathrm{km\,s^{-1}}$ that are evidently imparted on young pulsars in general (Lyne & Lorimer, 1994; Hobbs et al., 2005). In analogue to EC SNe, the kicks imparted on NSs produced via AIC (Sections 11.1.3, 12.2 and 14.5), following similar physical conditions, are also expected to be very small.

13.7.7 Kicks from Ultra-stripped SNe

The concept of ultra-stripped SNe is addressed in detail in Section 13.5. These are SNe where a naked helium star experiences extreme stripping by mass-transfer (Case BB RLO, Section 10.13) to a close-orbit compact object prior to its explosion (Tauris et al., 2013, 2015). Ultra-stripped SNe are therefore particularly relevant for the second SN in DNS systems, for example, producing GW merger sources (Fig. 1.1).

One may ask what would be the magnitude of a kick imparted to a NS born in an ultra-stripped SN? The flavor of ultra-stripped SNe can be either an EC SN or an Fe CCSN. As described in the previous section, the momentum kick imparted on a newborn NS via an EC SN is always expected to be small. Whereas Fe CCSNe are certainly able to produce large kicks, small NS kicks have been suggested and demonstrated to originate from CCSNe with small iron cores (Podsiadlowski et al., 2004; Janka, 2017; Müller et al., 2019; Stockinger et al., 2020). For such small Fe CCSNe, the situation is somewhat similar to that of EC SNe. In both cases, SN simulations (Gessner & Janka, 2018) suggest fast explosions where non-radial hydrodynamical instabilities (convectively driven or from the standing accretion shock) are unable to grow, leading to small kick velocities. This is in agreement with expectations from Eq. (13.2).

However, ultra-stripped SNe have small and loosely bound envelopes, and their progenitors must be modelled in binary systems with severe mass transfer prior to the core collapse. Applying such modelling and using binary stellar evolution arguments, Tauris et al. (2015) identified two factors that also imply that in ultra-stripped SNe the NS kicks may be small in general (but not always, see the last paragraph of this subsection): (1) from their modelling of the progenitor stars of ultra-stripped SNe, they have demonstrated that the expected amount of ejecta is extremely small ($\sim 0.1\,M_{\odot}$) compared to standard SN explosions in which several $M_{\odot}$ of material is ejected. This may lead to a weaker gravitational tug on the proto-NS (caused by asphericity of the ejecta; e.g., Wongwathanarat et al., 2013; Janka, 2017) and thus a small kick. (2) the binding energies of the envelopes of their final progenitor star models are often only a few 10^{49} erg, such that even a weak outgoing shock can quickly lead to their ejection, potentially before large anisotropies can build up.

The work of Suwa et al. (2015) has supported this prediction using axisymmetric hydrodynamical simulations of neutrino-driven explosions of ultra-stripped Fe CCSNe using the stellar evolution outcomes of single, evolved CO-stars (to mimic the outcome, but avoiding detailed binary stellar evolution calculations of Case BB RLO). All their models exhibited successful explosions driven by neutrino heating. Their diagnostic explosion energy, ejecta mass, and nickel mass were typically 10^{50} erg, $0.1\,M_{\odot}$, and $0.01\,M_{\odot}$, respectively, that is, compatible with observations of rapidly decaying light curves such as SN 2005ek (Drout et al., 2013; Tauris et al., 2013). Moreover, their calculated kick velocities were typically less than $50\,\mathrm{km\,s^{-1}}$, and sometimes even below

$\sim$10 km s^{-1}, in agreement with the simulations discussed in Janka (2017); Müller et al. (2019), and references therein.

However, ultra-stripped SNe also occur for more massive cores (Fig. 13.15). Tauris et al. (2015) predict that ultra-stripped SNe may potentially produce young (second-born) NSs in DNS systems with a mass within the entire range $1.10 - 1.80\,M_\odot$. While the binary stellar evolution arguments presented earlier for small kicks hold best for the lowest-mass NSs formed, Tauris et al. (2017) argue that it is also expected that more massive pre-SN metal cores produce larger iron cores and thus more "normal" Fe CCSNe with larger explosion energies and therefore larger kicks. Indeed, it has been demonstrated that to explain orbital characteristics and proper motions of Galactic DNS systems, some of the kicks associated with the second-formed NS in these DNS systems (which originate from collapsing stars that could not avoid being ultra-stripped) must have had magnitudes $w \gtrsim 200$ km s^{-1} (Tauris et al., 2017, and references therein). Moreover, 3D simulations of neutrino-driven ultra-stripped SNe by Müller et al. (2019) clearly show a case where a relative massive NS obtains a substantial kick exceeding $w \gtrsim 200$ km s^{-1} (see Fig. 13.21). Although in Eq. (13.2) w appears to scale inversely proportionally to $M_{\rm NS}$, the values of $E_{\rm SN}$ and $\alpha_{\rm ej}$ are systematically smaller for explosions leading to small values of $M_{\rm NS}$, such that a correlation between w and $M_{\rm NS}$ is indeed expected in general.

13.7.8 Correlation between NS Mass and Kick Magnitude?

While it is difficult to estimate kick values, the above arguments taken together allowed Tauris et al. (2017) to speculate that ultra-stripped SNe (EC SNe and, at least, Fe CCSNe with relatively small metal cores, i.e., collapse events of stellar cores with low compactness) generally lead to the formation of $\lesssim 1.3\,M_\odot$ NSs with small kicks, whereas ultra-stripped SNe producing more massive NSs ($> 1.3\,M_\odot$) are expected to supply larger kicks, in general. Tauris et al. (2017) find empirical evidence to support this hypothesis, which would have important consequences for the formation of double NS systems. For a suggested analytical probabilistic relation between remnant mass and kick magnitude, see Mandel & Müller (2020).

13.7.9 BH Kicks and Tossing of Spin Axis

Regarding kicks on newly formed BHs, less is known from observations and theory than for NSs, and the picture is much less clear; see, for example, Nelemans et al. (1999); Janka (2013); Repetto & Nelemans (2015); Mandel (2016). As mentioned in the last paragraph of Chapter 12, observations of space velocities of Galactic BH-X-ray binaries, summarized and analyzed by Mirabel (2017), suggest that the lowest-mass BHs (possibly formed by fallback) have the highest space velocities, whereas the more massive BHs, formed by direct collapse, have hardly any excess space velocity. For population synthesis purposes, often a simple flat 3D distribution up to some value (say 200 km s^{-1}) is applied as a default distribution (e.g., Kruckow et al., 2018). Alternatively, it is popular in the literature to scale the BH kick magnitudes inversely with the BH masses (i.e., $w_{\rm BH} \propto M_{\rm NS}/M_{\rm BH}$) to equate the total momentum in the kicks

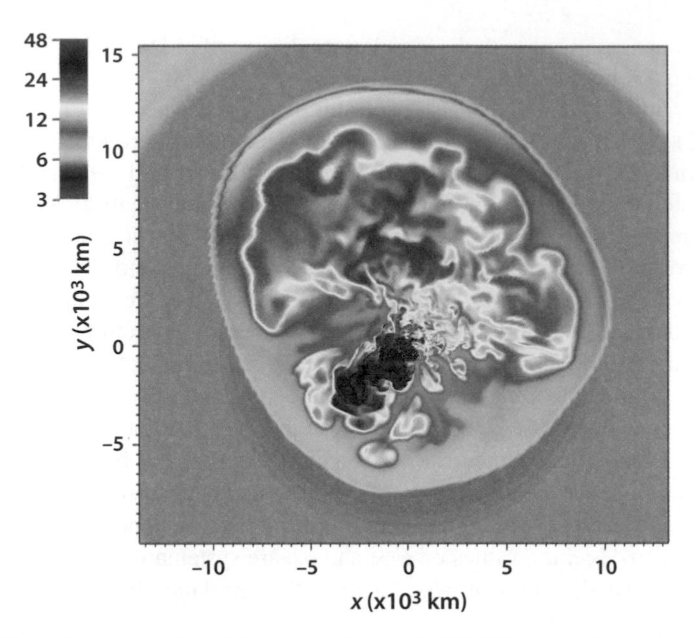

Figure 13.21. Entropy in units of k_B per nucleon of a 3D SN simulation (at $t = 1{,}023$ ms after bounce) for model he3.5. This is an ultra-stripped exploding star modelled from a 3.5 $M_\odot$ helium star in a binary with a NS and an initial orbital period of 2.0 days. The result of this simulation, with an explosion energy of $E_{\rm SN} = 2.78 \times 10^{50}$ erg, is a 1.41 $M_\odot$ NS with a kick velocity of $w = 238\,{\rm km\,s^{-1}}$, a spin period of $P = 98$ ms, and an angle between the spin and the kick vector of $\alpha = 80°$. Similar simulations of a 2.8 $M_\odot$ helium star yield a lower mass NS of 1.28 $M_\odot$ with a kick velocity of only $w = 11\,{\rm km\,s^{-1}}$. After Müller et al. (2019).

imparted on newborn NSs and BHs. In view of the results of Mirabel (2017), this may well be a reasonable approach as a description of BH kicks (however, see also Janka, 2013, who argued that the kick momentum is roughly proportional to the mass accreted from fallback, and thus to the BH mass).

There is clear evidence for spin-axis tossing in DNS systems, that is, the fact that the spin axis of the exploding star is tossed in a new (possibly random) direction as a result of the SN (Section 13.7.4). The question of whether or not such a spin-axis tossing process could also occur during BH formation is of uttermost importance for the interpretation of the formation of the double BH mergers detected by the LIGO-Virgo-KAGRA network (Tauris, 2022). It has traditionally been expected that double BH binaries that are produced via evolution of isolated binary systems would end up with more or less parallel spin axes of the two BH components (if the kick imparted onto the last-formed BH is small in magnitude), whereas double BH binaries assembled in dense stellar environments via exchange encounters would have random spin directions of the two BH components (see Chapter 15 for details of GW mergers detected by the LIGO-Virgo-KAGRA network and formation scenarios for these systems). However, if spin-axis tossing occurs during BH formation, it would complicate the ability to distinguish between these two formation channels based on GW oberservations of the effective spin parameter, $\chi_{\rm eff}$.

To produce a torque sufficiently strong to toss the spin axis during BH formation is not trivial (H.-T. Janka 2021, private communication): It requires an off-center density maximum, given that the emitted neutrinos from the core collapse (assuming they avoid being trapped within the event horizon) diffuse along the density gradient. Such a situation would require exotic conditions and ultra-fast rotation, where one could think of triaxial instabilities or break-up effects, destroying the spherical or axially symmetric density structure. Also, for a less exotic anisotropic fallback scenario, one needs to invoke non-spherical mass ejection to take place (or anisotropic neutrino or GW emission, but it has not been demonstrated that neutrinos or GWs can change the spin axis). It may be easier to obtain such hydrodynamic mechanisms that could more easily lead to spin-axis changes during formation of low-mass BHs, compared to more massive BHs (Chan et al., 2020; Janka et al., 2022). However, other work hints that spin axis tossing may also apply to formation of massive BHs (Antoni & Quataert, 2022).

13.8 KINEMATIC IMPACTS ON POST-SN BINARIES

The dynamical effects of SNe in close binaries (and hierarchical triples) have been studied both analytically and numerically in a number of papers over the last four decades, since the discovery of the Hulse-Taylor pulsar, for example, Flannery & van den Heuvel (1975); Sutantyo (1978); Hills (1983b); Brandt & Podsiadlowski (1995); Tauris & Bailes (1996); Kalogera (1996); Tauris & Takens (1998); Wex et al. (2000); Pijloo et al. (2012).

To solve for the dynamical effects of the SN explosion, the SN event can be assumed to be instantaneous given that the SN ejecta velocity is much greater than the binary orbital velocity. The mass loss reduces the absolute value of the potential energy and affects the orbital kinetic energy by decreasing the reduced mass of the system. In addition, a kick is imparted on the newborn NS mainly due to explosion asymmetrics and may arise from non-radial hydrodynamic instabilities in the collapsing stellar core (see, e.g., Janka, 2017, for the current theoretical understanding of how NS kicks come about using the gravitational tug-boat mechanism in asymmetric neutrino-driven core-collapse SN explosions). Considering the change in total energy of the system, the change in the orbital semi-major axis (ratio of final to initial value) can be expressed by (Hills, 1983b):

$$\frac{a_{\rm f}}{a_{\rm i}} = \left[\frac{1-\Delta M/M}{1-2\Delta M/M-(w/v_{\rm rel})^2-2\cos\theta\,(w/v_{\rm rel})}\right], \tag{13.3}$$

where ΔM is the amount of instantaneous mass loss from the exploding star (in our notation applied here, $\Delta M = M_{\rm He} - M_{\rm NS}$), $M = M_{\rm He} + M_2$ is the total mass of the pre-SN system, and $v_{\rm rel} = \sqrt{G(M_{\rm He}+M_2)/a_{\rm i}}$ is the relative velocity between the two stars. That is, we denote the exploding star "He" to indicate the typical nature of the

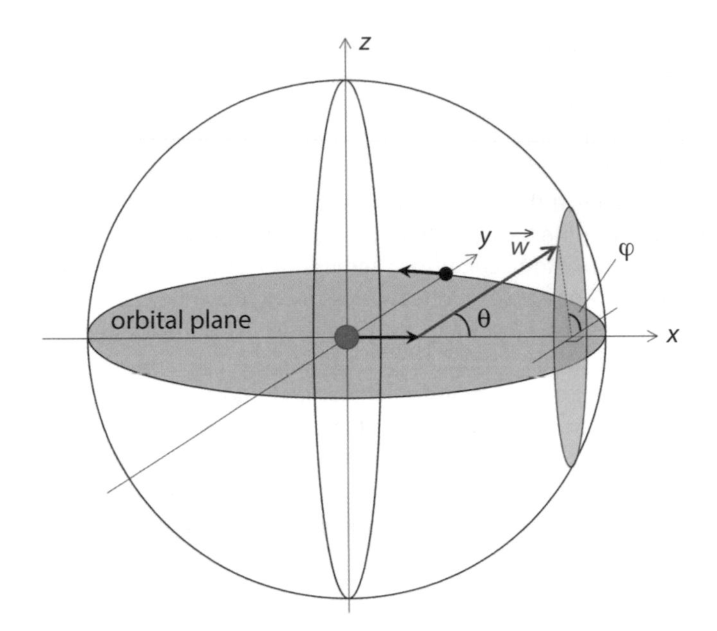

Figure 13.22. Illustration of the geometry of the two kick angles (θ, ϕ) and the kick sphere surrounding the exploding star (dark blue). The pre-SN orbital angular momentum vector is along the z-axis. The kick velocity vector ($\vec{w}$) is shown in red.

exploding star in a binary being a helium star; the resulting post-SN compact object is here denoted by "NS" (but apply just as well to a BH remnant); and the companion star is denoted as star "2". The kick angle, θ, is defined as the angle between the kick velocity vector, $\vec{w}$, and the pre-SN orbital velocity vector of the exploding star, $\vec{v}_{\rm He}$, in the pre-SN center-of-mass rest frame (see Fig. 13.22). Using Kepler's third law, the change in $P_{\rm orb}$ can be obtained. The effects of the impact on the companion star from the ejected SN shell are discussed in Section 13.8.5.

Equation (13.3) applies to a circular pre-SN binary. This is a good approximation for relatively close binaries with $P_{\rm orb}$ less than a few years, given the tidal interactions during RLO prior to the SN. For the more general case, see Hills (1983b). Solving for the denominator being equal to zero in Eq. (13.3) yields the critical angle, $\theta_{\rm crit}$, so that $\theta < \theta_{\rm crit}$ will result in the disruption of the orbit, cf. Figure 13.25. Thus the probability for a binary system to survive a SN with a kick in a random (isotropic) orientation can be found by integration and yields $P_{\rm bound} = 1 - (1 - \cos\theta_{\rm crit})/2$, or equivalently (Sutantyo, 1978; Hills, 1983b):

$$P_{\rm bound} = \frac{1}{2}\left\{1 + \left[\frac{1 - 2\Delta M/M - (w/v_{\rm rel})^2}{2\,(w/v_{\rm rel})}\right]\right\}, \tag{13.4}$$

where $P_{\rm bound}$ is restricted to [0, 1], and thus takes the value of 0 or 1, below and above this interval, respectively, when applying Eq. (13.4). For examples of probabilities

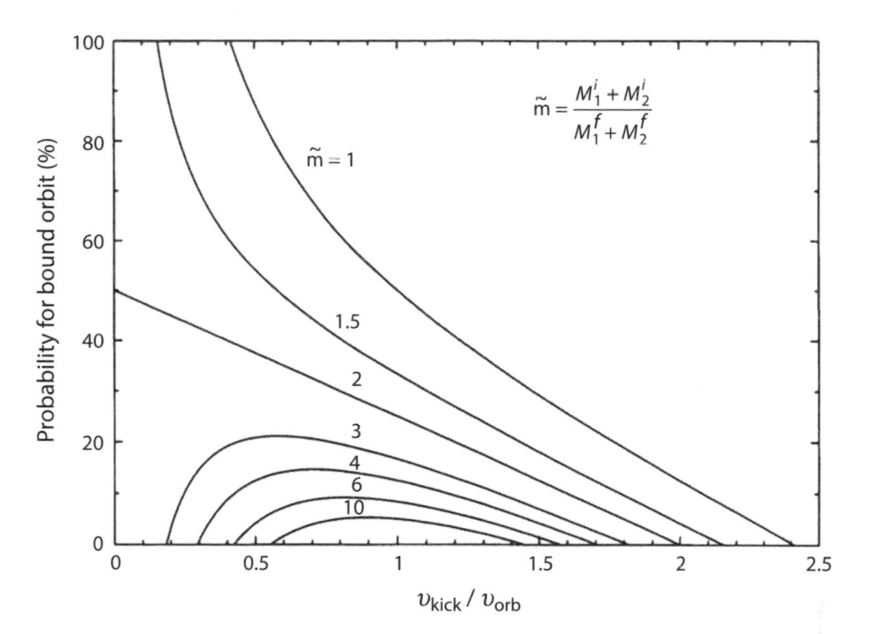

Figure 13.23. Probability of a binary to remain bound as a function of mass loss and kick. Here $\tilde{m}$ denotes the ratio between initial total mass to final total mass, and the x-axis shows the kick velocity in units of the pre-SN relative velocity, w/v_{rel}, in our notation. After Brandt & Podsiadlowski (1995).

for binaries surviving a SN, see Figure 13.23 for the general case, Figure 13.24 for explosions in pre-HMXB systems, and Figure 14.36 for the second SN in tight binaries for the formation of DNS systems.

The eccentricity of the post-SN system can be evaluated directly from the post-SN orbital angular momentum, $L_{orb,f}$, and is given by:

$$e = \sqrt{1 + \frac{2\, E_{orb,f}\, L_{orb,f}^2}{\mu_f\, G^2 M_{NS}^2 M_2^2}}\,, \tag{13.5}$$

where

$$L_{orb,f} = a_i\, \mu_f \sqrt{(v_{rel} + w\cos\theta)^2 + (w\sin\theta\,\sin\phi)^2}\,. \tag{13.6}$$

Here μ_f and $E_{orb,f} = -GM_{NS}M_2/2a_f$ are the post-SN reduced mass and orbital energy, respectively. The kick angle, ϕ, is measured in the plane perpendicular to the pre-SN velocity vector of the exploding star, $\vec{v}_{He}$ (Hills, 1983b; Tauris & Takens, 1998), such that the component of the kick velocity pointing directly toward the companion star is given by $w_y = w\ \sin\theta\ \cos\phi$ (Fig. 13.22).

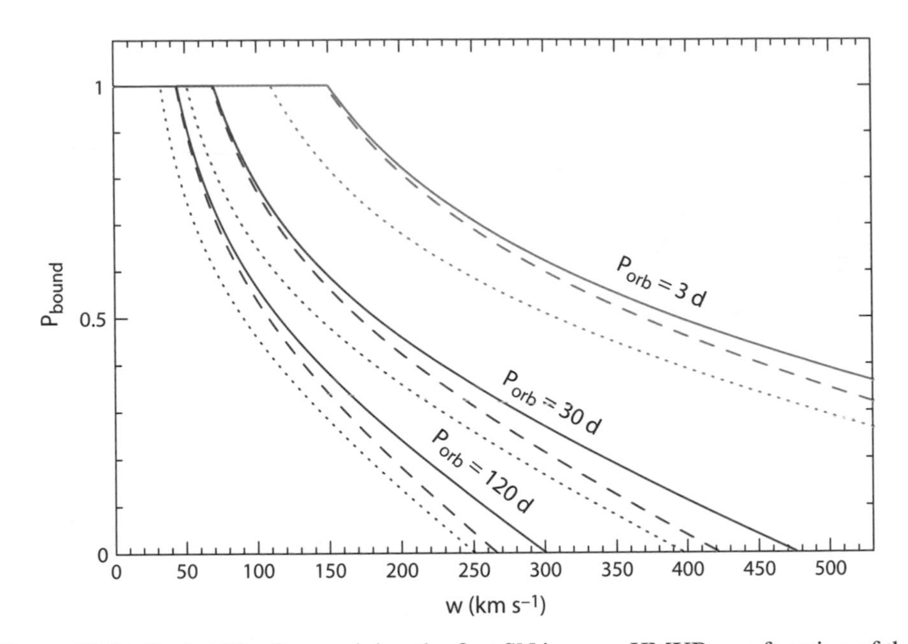

Figure 13.24. Probability for surviving the first SN in a pre-HMXB as a function of the imparted NS kick velocity, w, and for three different pre-SN orbital periods of 3, 30, and 120 days—see Eq. (13.4). The solid, dashed, and dotted lines are for Type Ib SNe with $(M_{\rm He,f}/M_\odot, M_2/M_\odot) = (4.0, 22.0)$, (2.0, 15.0), and (4.0, 12.0), respectively. A newborn NS mass of $M_{\rm NS,1} = 1.40\, M_\odot$ was assumed. After Tauris et al. (2017).

Solving for the right-hand side in Eq. (13.3) being equal to one yields another interesting angle, $\theta_{\rm a}$:

$$\theta_{\rm a} = \cos^{-1}\left(-\frac{\Delta M/M + (w/v_{\rm rel})^2}{2\,(w/v_{\rm rel})}\right), \tag{13.7}$$

which determines if the post-SN semi-major axis, $a_{\rm f}$, is larger or smaller than the pre-SN orbital radius, $a_{\rm i}$. Thus, as a consequence of the SN explosion, the orbit widens[8] if the kick angle $\theta < \theta_{\rm a}$, and it shrinks if $\theta > \theta_{\rm a}$. For random kick directions, we find the probability that the post-SN orbit shrinks ($a_{\rm f} < a_{\rm i}$) is given by:[9]

$$P_{\rm a}^{-} = \frac{1}{2} - \left[\frac{\Delta M/M + (w/v_{\rm rel})^2}{4\,(w/v_{\rm rel})}\right]. \tag{13.8}$$

A plot of this expression is demonstrated in Figure 14.37. It is convenient to use as an indicator for estimating the distribution of merger timescales of DNS systems produced

[8]It should be noted that in case the parenthesis in Eq. (13.7) takes a value less than -1, it follows that $\theta_{\rm a} = 180°$, meaning that all post-SN systems will widen.

[9]$P_{\rm a}^{-}$ is restricted to $[0, \frac{1}{2}]$ and is always 0 below this interval.

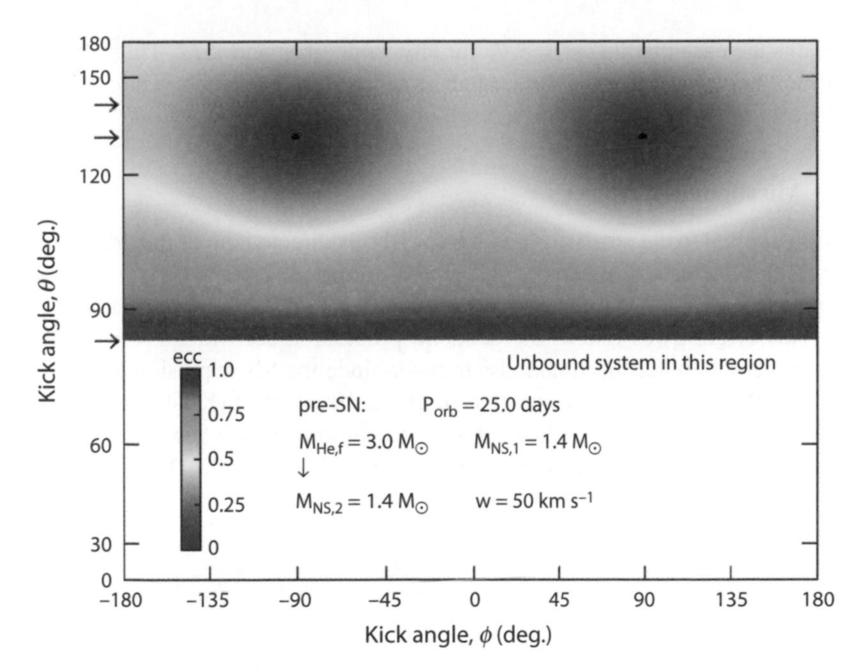

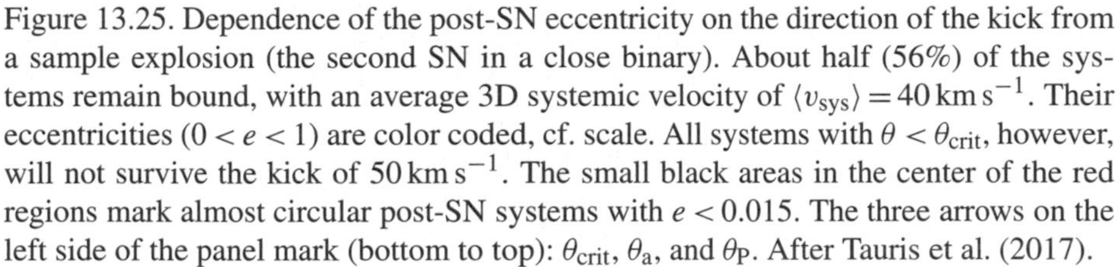

Figure 13.25. Dependence of the post-SN eccentricity on the direction of the kick from a sample explosion (the second SN in a close binary). About half (56%) of the systems remain bound, with an average 3D systemic velocity of $\langle v_{\rm sys}\rangle = 40\,{\rm km\,s^{-1}}$. Their eccentricities ($0 < e < 1$) are color coded, cf. scale. All systems with $\theta < \theta_{\rm crit}$, however, will not survive the kick of $50\,{\rm km\,s^{-1}}$. The small black areas in the center of the red regions mark almost circular post-SN systems with $e < 0.015$. The three arrows on the left side of the panel mark (bottom to top): $\theta_{\rm crit}$, $\theta_{\rm a}$, and $\theta_{\rm P}$. After Tauris et al. (2017).

by the second SN, that is, to estimate the number of GW sources that the LIGO-Virgo-KAGRA network and LISA will detect (Chapter 15).

Combining Eqs. (13.5) and (13.7) we can thus derive the two fine-tuned solutions of the kick angles for which the post-SN system is circular (i.e., $e = 0$). The result is simply: $(\theta,\ \phi) = (\theta_{\rm a},\ \pm 90°)$, see Figure 13.25 for an example.

Similarly, using Kepler's third law, one can work out a critical angle, $\theta_{\rm P}$:

$$\theta_{\rm P} = \cos^{-1}\left(\frac{1 - 2\Delta M/M - (1-\Delta M/M)^{2/3} - (w/v_{\rm rel})^2}{2\,(w/v_{\rm rel})}\right), \tag{13.9}$$

above which the post-SN orbital period is smaller than the pre-SN orbital period. The probability that $\theta > \theta_{\rm P}$ (and therefore that $P_{\rm orb}$ decreases) is thus given by:

$$P_{\rm P}^{-} = \frac{1}{2} - \left[\frac{2\Delta M/M + (1-\Delta M/M)^{2/3} + (w/v_{\rm rel})^2 - 1}{4\,(w/v_{\rm rel})}\right]. \tag{13.10}$$

It is always the case that $\theta_{\rm crit} < \theta_{\rm a} < \theta_{\rm P}$ for any value of w and ΔM. Thus, for systems with a kick angle $\theta_{\rm a} < \theta < \theta_{\rm P}$, the orbital semi-major axis shrinks while at the same time the orbital period increases as a result of the SN.

13.8.1 Misalignment Angles

If the kick applied to the newly formed NS (or BH) is directed out of the orbital plane of the pre-SN system (i.e., if the kick angle $\phi \neq 0°$ and $\phi \neq \pm 180°$), then the spin axis of the companion star will be tilted with respect to the post-SN orbital angular momentum vector. This effect gives rise to the geodetic precession seen in several radio pulsar binaries, where there has been no mass transfer since the SN explosion and where the spin axis of the main-sequence star (Kaspi et al., 1996), WD (Krishnan et al., 2020), or NS (Kramer et al., 2006; Ferdman et al., 2013) companion is measured to be tilted. This misalignment angle can be calculated as[10] (see Exercise 13.11):

$$\delta = \cos^{-1}\left(\frac{v_{\rm rel} + w\cos\theta}{\sqrt{(v_{\rm rel} + w\cos\theta)^2 + (w\sin\theta\sin\phi)^2}}\right). \tag{13.11}$$

If the misalignment angle is large ($> 90°$), the new orientation of the orbit will cause retrograde spin of one or both binary components—to the sense of orbital revolution—depending on their pre-SN spin directions and whether (and in which direction) the spin axis of the newborn NS or BH might be tilted during the SN explosion (for the case of NSs see, e.g., Desvignes et al., 2019; and for BHs, see Tauris, 2022).

Retrograde spin of a NS or BH will affect its accretion process and thus its spin period evolution. If the sense of rotation of a NS is retrograde, it may more easily accrete the material donated by its companion. The accretion torque will slow down the rate of spin and cause the spin axis to migrate toward the nearest pole of the orbit (Hills, 1983b; Biryukov & Abolmasov, 2021). Its rate of rotation will continue to slow down and will eventually stop as it interacts with the material donated by its companion. As it continues to be torqued beyond this point, its spin vector will rapidly flip around by 180° so that the spin becomes prograde. Its spin rate will then increase rapidly, and its subsequent evolution will eventually resemble that of a NS that always has had a prograde spin. For a BH accretor with a retrograde spin ($a_* < 0$) relative to its accretion disk, the last stable circular orbit, $R_{\rm ISCO}$, is greater than it would be for a non-spinning BH ($6\,GM/c^2$; see Fig. 7.19), and the fraction of the accreted mass converted to energy is less than the 5.7% value obtained by a non-spinning BH. Further discussions and kinematic constraints on retrograde vs. prograde spins of post-SN orbits are discussed further in Hills (1983b); Brandt & Podsiadlowski (1995).

To use the observed misalignment angle to constrain kick properties in, for example, DNS systems (Tauris et al., 2017), we must rely on the assumption that accretion

[10]This expression should replace Eq. (16) in Tauris et al. (2017), which is only a good approximation for small values of $(w/v_{\rm rel})$.

torques align the spin axis of the first-born NS with the orbital angular momentum vector during the recycling process (e.g., Hills, 1983b; Bhattacharya & van den Heuvel, 1991; Biryukov & Abolmasov, 2021). Observational evidence for such an alignment to actually occur in nature was demonstrated for LMXBs by Guillemot & Tauris (2014), who found agreement between the viewing angles of binary MSPs (as inferred from γ-ray light-curve modelling) and their orbital inclination angles. Although the timescale of accretion during Case BB RLO in DNS progenitor systems is substantially shorter (by two to four orders of magnitude) than in LMXBs, the torques at work will be larger due to the much higher mass-transfer rates during Case BB RLO (while the size of the magnetospheres remain roughly the same as a result of larger NS B-fields in DNS systems compared to fully recycled MSPs). Therefore, it is reasonable to assume $\delta_i = 0$ prior to the second SN explosion and thus legitimate to use the post-SN measurements of δ of the recycled NSs to constrain kicks in the second SN event.

13.8.2 Systemic Velocities

Another important diagnostic quantity, besides from $P_{\rm orb}$, $P_{\rm spin}$, eccentricity, $M_{\rm NS}$, and δ (e.g., for understanding HMXB and DNS formation), is their post-SN systemic velocity, $v_{\rm sys}$. Any bound system receives a recoil velocity relative to the center-of-mass rest frame of the pre-SN system. This is due to the combined effects of sudden mass loss (see Fig. 4.10 and Chapter 10) and a kick velocity imparted on the newborn NS. From simple conservation of momentum considerations (e.g., following Tauris & Bailes, 1996), we can write this 3D velocity as:

$$v_{\rm sys} = \sqrt{(\Delta P_x)^2 + (\Delta P_y)^2 + (\Delta P_z)^2} \,/\, (M_{\rm NS} + M_2)\,, \tag{13.12}$$

where the change in momentum is given by:

$$\begin{aligned} \Delta P_x &= M_{\rm NS}\, w \cos\theta - \Delta M\, M_2 \sqrt{G/(M a_{\rm i})}\,, \\ \Delta P_y &= M_{\rm NS}\, w \sin\theta \cos\phi\,, \\ \Delta P_z &= M_{\rm NS}\, w \sin\theta \sin\phi\,. \end{aligned} \tag{13.13}$$

Observations of the proper motion of binary radio pulsars or accreting NSs in X-ray binaries can be combined with their distance estimates to infer their systemic velocity with respect to the local frame of rest, after correcting for differences in Galactic motion relative to the Solar System barycenter. This velocity ($v_{\rm sys}$) is thought to originate from the recoil of the SN explosion producing the NS. In DNS systems, two recoils are imparted, although the first recoil is usually relatively small due to the presence of a massive companion star at that epoch (the pre-HMXB stage). Derived values of $v_{\rm sys}$ are therefore quite valuable when it comes to constraining the kicks imparted on NSs (e.g., Tauris et al., 2017, for a discussion of an applied method)—similarly to the case for BH kicks (Nelemans et al., 1999; Repetto & Nelemans, 2015; Mandel, 2016; Mirabel, 2017). Alternatively, kicks on a given population of compact object binaries can be estimated from their dispersion in location with respect to the Galactic disk (Gunn & Ostriker, 1970). The distribution of HMXBs with respect to star-forming regions in the Galactic disk (Kaper et al., 1997; Bodaghee et al., 2012; Coleiro & Chaty, 2013; Fortin et al., 2022, see also Fig. 6.4 and Chapter 10), and sGRBs to their host galaxies (Voss & Tauris, 2003; Fong & Berger, 2013), can be used in the same way to constrain kick magnitudes.[11]

13.8.3 Symmetric SNe

For purely symmetric SNe ($w=0$), the equations governing the dynamical effects of the SN explosion (Eqs. 13.3–13.5) simplify into (Flannery & van den Heuvel, 1975):

$$\frac{a_\mathrm{f}}{a_\mathrm{i}} = \left[\frac{1-\Delta M/M}{1-2\Delta M/M}\right], \tag{13.14}$$

and

$$e = \Delta M/(M_\mathrm{NS}+M_2), \tag{13.15}$$

and where the probability of remaining bound is always $P_\mathrm{bound}=1$ for $\Delta M/M < 0.5$, whereas (following the virial theorem) all systems are disrupted if more than half the total mass is lost, that is, if $\Delta M/M > 0.5$ (Exercise 4.9). There is probably always a small element of asymmetry (i.e., a kick) involved in SN explosions or AIC events. Even a tiny kick of a few $\mathrm{km\,s^{-1}}$ can change the post-SN eccentricity quite a bit in systems with little ejecta mass, which makes it difficult to map the pre-SN to post-SN orbital parameters in such cases (Freire & Tauris, 2014; Tauris & Janka, 2019).

13.8.4 Kicks Imparted on NSs in Binaries

As summarized in this chapter, there is ample empirical evidence for large kicks ($> 400\,\mathrm{km\,s^{-1}}$) received by isolated young radio pulsars, as well as in some binary pulsar systems. However, there is also good evidence for some NSs being produced with only small kicks ($< 50\,\mathrm{km\,s^{-1}}$). Based on an analysis of post-SN binaries, the evidence for a certain population of NSs receiving only relatively small kicks has already been pointed out in several studies (e.g., Tauris & Bailes, 1996; Portegies Zwart & Yungelson, 1998; van den Heuvel et al., 2000; Pfahl et al., 2002; Podsiadlowski et al., 2004; van den Heuvel, 2004; Piran & Shaviv, 2005; Schwab et al., 2010; Ferdman et al., 2013, 2014; Beniamini & Piran, 2016; Tauris et al., 2017). In addition to an origin from electron capture SNe (EC SNe), from collapsing ONeMg cores (Nomoto, 1987; Gessner & Janka, 2018), small NS kicks have also been suggested to originate from small Fe CCSNe (Podsiadlowski et al., 2004; Tauris et al., 2015, 2017; Janka, 2017). In both cases, SN simulations (Janka, 2017) suggest fast explosions where non-radial

[11]Perets & Beniamini (2021) recently argued that sGRB (and Ca-rich SN) offsets should be taken with a grain of salt as the large spatial offsets of these transients might be explained by the observations of highly extended underlying stellar populations in (mostly early type) Galaxy halos, typically missed in surveys.

hydrodynamical instabilities (convectively driven or from the standing accretion shock) are unable to grow, leading to somewhat small kick velocities. Small kicks are therefore often expected to be imparted onto the second-formed NSs in close binaries produced from ultra-stripped progenitors (Tauris et al., 2013, 2015), in particular if these are low-mass NSs (Tauris et al., 2017; Janka, 2017; Gessner & Janka, 2018), as has also been demonstrated by multi-dimensional SN explosion modelling (Suwa et al., 2015; Müller et al., 2019). This scenario is particularly relevant for the formation of DNS systems that merge and become GW sources. However, we emphasize again that in some cases even ultra-stripped SNe result in relatively large kicks (200−300 km s^{-1}), for example, in the case of the Hulse-Taylor pulsar (Tauris et al., 2017, and references therein), possibly related to more massive exploding stars producing more massive NSs.

Finally, in addition to the kinematic effects from the second SN, DNS systems also have a relic systemic velocity component from the first SN. This value, however, is expected to be small, typically $v_{\rm sys} \leq 25\,{\rm km\,s^{-1}}$ and practically always $v_{\rm sys} < 50\,{\rm km\,s^{-1}}$, if these first SN explosions of stripped stars are limited in kick values to $w < 170\,{\rm km\,s^{-1}}$ (M. Kruckow, 2019, private communication). Based on the distribution and the peculiar velocities of HMXBs, Coleiro & Chaty (2013) find that their resulting systemic velocities are often only $v_{\rm sys} \simeq 10-20\,{\rm km\,s^{-1}}$ (see, however, also Section 10.4), which supports the hypothesis that exploding stars generally produce smaller kicks compared to the distribution of, for example, Hobbs et al. (2005), if they have lost (at least) their hydrogen-rich envelope via mass transfer prior to the SN. Further arguments of a small kick imparted on the first NS to form, for those systems that later evolve to become observable DNS systems, follow from the need of a wide orbital separation before the first SN for the descendant HMXB system to survive its subsequent CE evolution without merging (Pfahl et al., 2002; Voss & Tauris, 2003; Tauris et al., 2017; Kruckow et al., 2018).

The direct observational constraints on kicks added to NSs in newly formed HMXBs is very limited as there are very few HMXBs detected inside SN remnants. Recently, a case was discovered in the LMC (Maitra et al., 2019) where a binary pulsating NS with an orbital period of 2.2 days was found at the geometrical center of the SN remnant MCSNR J0513−6724. This SN remnant is very young and has an estimated age of less than 6,000 yr. Only Circinus X-1 with an age estimate of ~4,600 yr (Heinz et al., 2013) seems to be a younger X-ray binary inside a SN remnant. It is still unclear whether Cir X-1 is a LMXB or a HMXB.

13.8.5 Disrupted Binaries and Runaway Stars

The analytical description of the kinematic effects of bound post-SN binaries was derived by Hills (1983b). The full description for the general case of disrupted binaries, including shell impact effects on the companion star, was derived by Tauris & Takens (1998). Their analytical equations describing the runaway velocities of stellar components, originating from disrupted binaries via asymmetric SNe, in cartesian coordinates in the original pre-SN center-of-mass inertial reference frame, are given by:

$$v_{\rm NS,x} = w\cos\theta\left(\frac{1}{R}+1\right)+\left(\frac{1}{R}+\frac{m_2}{1+m_{\rm shell}+m_2}\right)v_{\rm rel}, \tag{13.16}$$

$$v_{\mathrm{NS,y}} = w \sin\theta \cos\phi \left(1 - \frac{1}{S}\right) + \frac{1}{S}\, v_{\mathrm{im}} + \frac{Q\sqrt{P}}{S}\, v_{\mathrm{rel}}, \tag{13.17}$$

$$v_{\mathrm{NS,z}} = w \sin\theta \sin\phi \left(\frac{1}{R} + 1\right), \tag{13.18}$$

and

$$v_{2,\mathrm{x}} = \frac{-w\cos\theta}{m_{2\mathrm{f}}\, R} - \left(\frac{1}{m_{2\mathrm{f}}\, R} + \frac{1 + m_{\mathrm{shell}}}{1 + m_{\mathrm{shell}} + m_2}\right) v_{\mathrm{rel}}, \tag{13.19}$$

$$v_{2,\mathrm{y}} = \frac{w \sin\theta\cos\phi}{m_{2\mathrm{f}}\, S} + \left(1 - \frac{1}{m_{2\mathrm{f}}\, S}\right) v_{\mathrm{im}} - \frac{Q\sqrt{P}}{m_{2\mathrm{f}}\, S}\, v_{\mathrm{rel}}, \tag{13.20}$$

$$v_{2,\mathrm{z}} = \frac{-w\sin\theta\sin\phi}{m_{2\mathrm{f}}\, R}, \tag{13.21}$$

where the index "NS" refers to the newborn NS, and index "2" refers to its companion star, and

$$P \equiv 1 - 2\tilde{m} + \frac{w^2}{v_{\mathrm{rel}}^2} + \frac{v_{\mathrm{im}}^2}{v_{\mathrm{rel}}^2} + 2\frac{w}{v_{\mathrm{rel}}^2}\left(v_{\mathrm{rel}}\cos\theta - v_{\mathrm{im}}\sin\theta\cos\phi\right), \tag{13.22}$$

$$Q \equiv 1 + \frac{P}{\tilde{m}} - \frac{(w\sin\theta\cos\phi - v_{\mathrm{im}})^2}{\tilde{m}\, v_{\mathrm{rel}}^2}, \tag{13.23}$$

$$R \equiv \left(\frac{\sqrt{P}}{\tilde{m}\, v_{\mathrm{rel}}}\,(w\sin\theta\cos\phi - v_{\mathrm{im}}) - \frac{P}{\tilde{m}} - 1\right)\frac{1 + m_{2\mathrm{f}}}{m_{2\mathrm{f}}}, \tag{13.24}$$

$$S \equiv \left(1 + \frac{P}{\tilde{m}}\,(Q+1)\right)\frac{1 + m_{2\mathrm{f}}}{m_{2\mathrm{f}}}, \tag{13.25}$$

and where w is the kick magnitude, θ and ϕ are the two kick angles defining the kick direction (Fig. 13.22), m_{shell} is the amount of ejecta mass in units of the gravitational mass of the newborn NS (i.e., $m_{\mathrm{shell}} = \Delta M/M_{\mathrm{NS}} = (M_{\mathrm{He}} - M_{\mathrm{NS}})/M_{\mathrm{NS}}$), m_2 and $m_{2\mathrm{f}}$ denote the companion mass before and after the shell impact, and again here in units of the gravitational mass of the newborn NS.

An example of the importance of the kick magnitude and kick direction on the final fate of a binary system is plotted in Figure 13.26. Here it is demonstrated that a kick can stabilize a binary that would otherwise be disrupted in the symmetric SN case, since in this example more than half of the total system mass is lost in the SN. The binary considered could be a case of a system prior to the second SN forming a DNS system. The pre-SN mass of the exploding star was assumed to be 4.3 $M_\odot$, and the NS remnant left behind was assumed to have a gravitational mass of 1.4 $M_\odot$ (similar to the mass of the first-born NS). We thus find that the amount of mass instantly ejected ($\Delta M = 2.9\ M_\odot$) is more than half the total mass prior to the SN ($M = 4.3\ M_\odot + 1.4\ M_\odot = 5.7\ M_\odot$), and therefore the system would not survive a purely symmetric SN. This

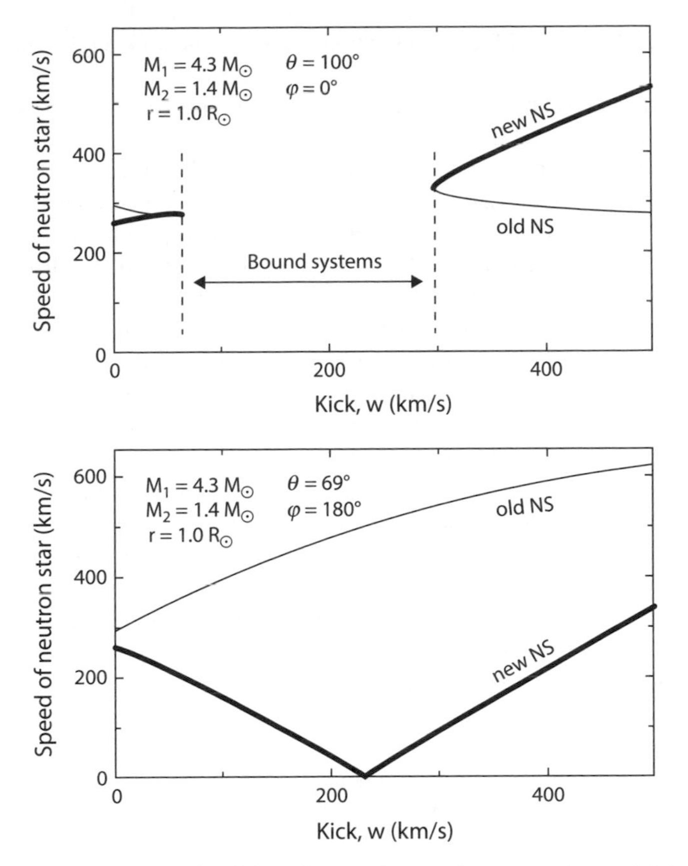

Figure 13.26. The top panel of the figure shows that a moderate kick ($65 < w < 295\,\mathrm{km\,s^{-1}}$) can stabilize a binary and prevent it from disruption though more than half of the total mass is ejected in the explosion. The bottom panel shows that it is possible to form a stationary NS (in the reference frame of the pre-SN binary) if the direction and magnitude of the kick are carefully tuned. In both cases, the newborn NS has a mass of $1.4\,M_\odot$. See Exercise 13.5. After Tauris & Takens (1998).

is indeed seen in the figure for $w = 0$, and a pre-SN orbital radius of $1.0\,R_\odot$, which leads to a disrupted system with runaway velocities of roughly 260 and $300\,\mathrm{km\,s^{-1}}$ for the newborn NS and the old (first-born) NS, respectively. However, if moderate kick of $65 < w < 295\,\mathrm{km\,s^{-1}}$ with fixed kick angles of $\theta = 100°$ and $\phi = 0°$ are applied in the SN, the post-SN system is stabilized and the system remains bound. A larger kick ($w > 295\,\mathrm{km\,s^{-1}}$), on the other hand, will always disrupt the binary. In the bottom panel, it is seen that for fixed kick angles of $\theta = 69°$ and $\phi = 180°$ there is a solution to produce a *stationary* newborn NS (in the reference frame of the pre-SN binary) if $w \simeq 230\,\mathrm{km\,s^{-1}}$. Thus we conclude that finding an isolated NS with a very small velocity is not necessarily evidence for an origin in a SN with a very small kick. Such a NS could simply be left behind from a disrupted binary (if the kick angles are sufficiently fine-tuned).

For a number of years, the Jena group (led by Ralph Neuäuser) has investigated potential runaway stars and their associations with SN disrupted binaries (e.g., Tetzlaff et al., 2010, 2011; Dinçel et al., 2015). More recently, they argued that the detection of 1.5–3.2 Myr old ^{60}Fe on Earth indicates a recent nearby core-collapse SN and managed to trace back kinematical evidence that the runaway ζ Oph and the radio pulsar PSR B1706−16 were released by a SN in a binary 1.8 Myr ago at a distance of about 100 pc (Neuhäuser et al., 2019). On the theoretical side, simulations have been performed to investigate runaway stars based on population synthesis of binary star evolution. For Galactic disk stars more massive than 15 $M_{\odot}$, Renzo et al. (2019) estimate that $\sim$10% of them are walkaways and only less than 1% are runaways, nearly all of which have accreted mass from their companion. The low runaway fraction we find is in tension with observed fractions of about 10%.

While the shell-impact effects on the companion star from exploding helium stars in tight binaries can potentially be important for the first SN explosion (e.g., Wheeler et al., 1975; Tauris, 2015; Liu et al., 2015), we disregard such effects in the second SN where the companion star is a NS (essentially a point mass).

Shell-impact effects, however, are important clues to search for evidence for post-SN Ia events in the single degenerate scenario. The former non-degenerate donor star is expected to be polluted with SN ejecta, giving rise to unusual chemical composition in its surface abundances. Additionally, the presence of the non-degenerate star should cause a shadow in the ejected SN remnant.

For Galactic SN remnants, it is interesting to notice the first evidence of a SN in a binary system, which is the case of the SN remnant Semeis 147, described in Chapter 10 (see Fig. 10.8). This SN remnant contains a runaway radio pulsar and a runaway B-star, which resulted from the disruption of a massive binary system. A second case is the source RCW 86 (Gvaramadze et al., 2017). Here, the SN explosion apparently, like in Semeis 147, occurred near the center of the hemispherical optical nebula in the southwest of RCW 86. Archival Chandra data reveals an X-ray spectrum typical of a young pulsar. At the same position, follow-up VLT observations show a G-star with radial velocity variations (with an estimated orbital period of the order of a month) and a very peculiar chemical abundance, including an overabundance of Calcium by a factor of 6.

13.8.6 Hypervelocity Stars from SN Disrupted Binaries

In the last two decades, a large number of hypervelocity star (HVS) candidates have been reported (e.g., Brown et al., 2005, 2006; Edelmann et al., 2005; Hirsch et al., 2005; Brown et al., 2009, 2012, 2014; Tillich et al., 2009; Li et al., 2012; Palladino et al., 2014; Zhong et al., 2014; Li et al., 2020, and references therein). The nature of the HVSs spans a wide range of types from OB-stars to metal-poor F-stars and G/K dwarfs, as well as a variety of dense peculiar stars, mentioned in Section 13.2, which are thought to be the remnants of SNe Ia. In this section, we will only concentrate on the normal main-sequence stars with very high velocities. While there is evidence from many late-type B HVSs in the halo to originate from near the SMBH in the center of our Galaxy (e.g., Brown et al., 2014), other HVSs seem to originate from the Galactic disk (e.g., Heber et al., 2008; Li et al., 2012; Palladino et al., 2014; Kreuzer et al., 2020; Irrgang et al., 2021). Genuine HVSs can be defined as stars that will escape the gravitational potential of our Galaxy. Depending on the location and direction of motion, this criterion typically corresponds to a stellar velocity in the Galactic rest frame $>400\,\mathrm{km\,s^{-1}}$ (Kenyon et al., 2008). Previously, more than 50 stars were classified as HVSs. The *Gaia* mission has revealed further candidates (e.g., Marchetti et al., 2019, Du et al., 2019; Li et al., 2020; Irrgang et al., 2021), while other previous candidates have been found to have sub-escape velocities (Boubert et al., 2018), in particular most of the late-type HVS candidates.

HVSs can obtain their large velocities from a number of different processes. Hills (1988) predicted the formation of HVSs via tidal disruption of tight binary stars by the central supermassive black hole (SMBH) of the Milky Way. In this process one star is captured by the SMBH while the other is ejected at high speed via the gravitational slingshot mechanism. Also, exchange encounters in dense stellar environments (e.g., Aarseth, 1974) between hard binaries and massive stars may cause stars to be ejected and escape our Galaxy (Leonard, 1991; Gvaramadze et al., 2009). A competing mechanism for producing HVSs, however, is disruption of close binaries via SN explosions (Blaauw, 1961; Boersma, 1961; Tauris & Takens, 1998; Zubovas et al., 2013; Tauris, 2015). As demonstrated in Tauris & Takens (1998), the runaway velocities of both ejected stars can reach large values when asymmetric SNe are considered, that is, when the newborn NS receives a momentum kick at birth.

Tauris (2015) investigated the maximum possible ejection velocities of HVSs with different masses originating from disrupted binaries via asymmetric core-collapse SNe and found that HVSs up to $\sim$1,280, $\sim$770, and $\sim 550\,\mathrm{km\,s^{-1}}$ are possible in the Galactic rest frame from this scenario for ejected HVSs of masses 0.9, 3.5, and 10.0 $M_{\odot}$, respectively. Although a binary origin cannot explain all the observed velocities of B-type HVSs (in agreement with their proposed central SMBH origin), it can indeed account for the far majority (if not all) of the detected G/K-dwarf HVS candidates.

The analytical formulae to calculate the velocities of stars ejected from binaries in which asymmetric SNe occur were derived in Tauris & Takens (1998), see Eqs. (13.16)–(13.21). The major component affecting the maximum possible ejection velocity of the companion star, v_2^{max}, is its pre-SN orbital velocity, $v_{2,\mathrm{orb}}$. Hence, to produce ejected HVSs it is clear that very close binaries are needed. The minimum separation between the two stars prior to the SN explosion is assumed to be limited by the Roche-lobe radius of the companion star. Another required constraint is that the post-SN trajectory of the NS will not lead to a merger event in an otherwise disrupted system, that is, if the periastron separation, q, in case of motion along an inbound leg of the hyperbolic orbit ($\gamma > 0$ and $\xi > 2$, in the notation of Tauris & Takens, 1998) is smaller than the radius of the companion star, R_2, then the system will likely merge and not produce a HVS.

In Figure 13.27, it is shown that the velocity of the ejected companion star (i.e., the value of v_2) is highly dependent on the direction of the kick imparted on the NS. The largest values of v_2 are obtained when the newborn NS is kicked into a trajectory just in front of the companion star, thereby accelerating it during its outward trajectory motion. The white area in the middle corresponds to cases where the newborn NS is

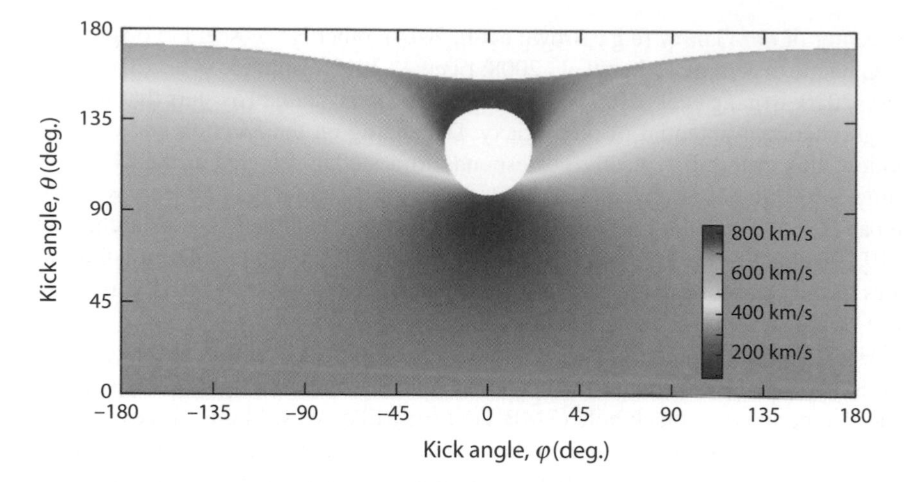

Figure 13.27. Ejection velocities of G/K-dwarf (0.9 $M_\odot$) companions plotted as a function of the kick direction angles, θ and ϕ, for a given pre-SN setup and a SN kick of $w = 1000\,\mathrm{km\,s^{-1}}$. The colors represent the resulting ejection velocities, v_2 varying between $87\,\mathrm{km\,s^{-1}}$ (red) and $839\,\mathrm{km\,s^{-1}}$ (blue). After Tauris (2015).

shot into the companion star and the system is assumed to merge. For highly retrograde kick directions ($\theta \to 180°$) the systems remain bound, hence the white area at the top of the plot (the shape of which depends on the pre-SN orbital separation and the kick velocity, w). The SN-induced HVSs with the largest values of v_2 are ejected close to the plane of the pre-SN binary ($\phi = 0$), causing their spin axis to be almost perpendicular to their velocity vector.

13.8.7 Effects of SN Shell Impact

An additional consequence of the SN explosion, besides the kinematic effects arising from instantaneous mass loss and kick, is that the companion star is affected by the impact of the shell debris ejected from the exploding star (Wheeler et al., 1975). In close-orbit systems, this impact can induce significant mass loss and heating of the companion star (Liu et al., 2015). Moreover, the companion star will be exposed to chemical enrichment. Empirical evidence for such chemical contamination is found in the Galactic SN remnant RWC 86 where a binary G-star, thought to orbit the young NS, is measured to have the highest calcium content of any known star in the Galactic disk (Gvaramadze et al., 2017). Moreover, besides Ca which is overabundant by a factor of 6 compared to solar composition, also Si, Ti, V, Cr, Mn, Fe, Co, Ni, and Ba are enhanced by a factor of about 3. These elements are typical trace elements of SN ejecta. Moreover, statistical evidence from samples of Type Ib/c SNe suggest that their progenitors are often stripped, relatively low-mass, exploding stars in binaries (Drout et al., 2011; Bianco et al., 2014; Taddia et al., 2015; Lyman et al., 2016). One example is iPTF13bvn, where pre-SN images, light-curve modelling, and late-time photometry suggest that it is in an interacting binary system (Cao et al., 2013).

After a SN explosion occurs in a binary system, the ejected debris is expected to expand freely for a few minutes to hours and eventually impact the companion star. As a result, the early SN light curve can be brightened by the collision of the SN ejecta with the companion star (Kasen, 2010; Moriya et al., 2015). Following the SN shell impact, the companion star may be significantly heated and shocked so that the envelope of the companion star is partially removed due to the stripping and ablation mechanism (Wheeler et al., 1975; Fryxell & Arnett, 1981; Taam & Fryxell, 1984). Non-degenerate companion stars are expected to survive the impact of the SN explosion, despite the fact that the kinetic energy of the SN ejecta incident upon the companion star is much greater than the total binding energy of this star. The star survives the SN impact because energy is deposited in only a small fraction of the stellar mass and excess energy is redirected into kinetic energy of the expelled material (Fryxell & Arnett, 1981; Taam & Fryxell, 1984). In addition to mass loss, the surviving companion star will show some peculiar properties, such as a high runaway velocity and probably enrichment in heavy elements, which may be identified by observations if mixing processes are not efficient on a long timescale, as in the case of RWC 86 described previously. (As mentioned in Section 6.5, the F-subgiant companion of the BH V1033 Sco [GRO J1655−40] shows a large overabundance of alpha elements [O, Mg, Si, S] due to matter that must have been ejected in the formation event of the BH.) In addition, these stars are significantly bloated and overluminous as a consequence of internal heating by the passing shock wave. The SN remnant may also show an observational feature in the form of a hole in the ejecta (García-Senz et al., 2012; Gray et al., 2016).

Many numerical simulations have been carried out of collisions of SN ejecta with the companion star in single-degenerate Type Ia SNe (e.g., Marietta et al., 2000; Pakmor et al., 2008; Pan et al., 2012; Liu et al., 2012). All these studies suggest that about 2%–30% of the companion mass, or almost the whole envelope of the companion star (depending on its structure), is removed by the SN impact. 3-D hydrodynamical simulations (Fig. 13.28) by Liu et al. (2015) of the impact of SN ejecta on 0.9 and 3.5 $M_{\odot}$ main-sequence companion stars in core-collapse SNe of Type Ib/c reveal power-law-like fitting functions for the dependence of total mass loss, resulting impact velocity, and amount of accreted ejecta on the pre-SN binary separation. All three quantities decrease significantly with increasing pre-SN binary separation. It was found that in most cases less than 5% of the MS companion mass can be removed by the SN impact and that the companion star typically receives an impact velocity of a few $10\,\mathrm{km\,s^{-1}}$. The amount of SN ejecta captured by the companion star after the explosion is most often less than $10^{-3}\,M_{\odot}$. Only in the closest pre-SN systems, the MS companion stars are affected more strongly by the SN ejecta impact, leading to mass loss of about 10%, impact velocities of the order of $100\,\mathrm{km\,s^{-1}}$, and an amount of captured SN ejecta up to $4 \times 10^{-3}\,M_{\odot}$, depending on the mass of the companion star. For SNe in binaries where at the moment of the explosion the companion is a red supergiant star, however, Hirai et al. (2020) find that up to 50%–90% of the envelope of the companion can be unbound due to ejecta impact, as long as the pre-SN orbital separation is less than 5 times the stellar radius. As a consequence, if the system is disrupted, the subsequent explosion of the secondary (now solitary) red supergiant would produce a stripped-envelope SN.

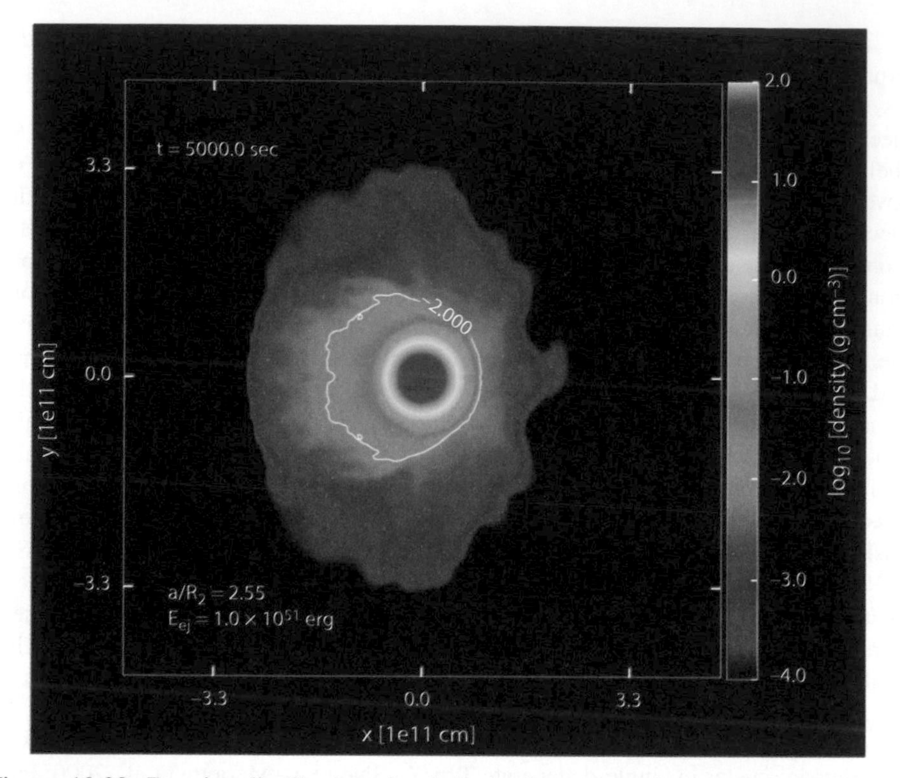

Figure 13.28. Density distribution simulations of a companion star after SN shell impact (from the right side) on a 3.5 $M_\odot$ main-sequence B-star model with a pre-SN binary separation of $a = 2.55\ R_2$, where R_2 is its radius. Only the bound material that originally belonged to the companion star and final accreted ejecta are shown. The logarithm of density is color-coded. The snapshot is taken at the moment (5,000 sec after SN shell impact) that the removed companion mass, the resulting impact velocity, and the contamination from the SN ejecta converged to (almost) constant values. The white curve shows a constant density contour ($\log \rho = -2.0$). After Liu et al. (2015).

Finally, we notice that the SN shell collision with dense circumstellar material will lead to superluminous light curves and that the Rayleigh-Taylor instability, which is expected to take place in the cool dense shell between the SN ejecta and the dense circumstellar medium, is important in understanding SNe powered by shock interaction (Moriya et al., 2013).

EXERCISES

Exercise 13.1. *Kinematic impact of the 2nd SN in producing a BH+BH system. Consider a pre-SN system composed of an exploding WR-star with a mass of $M_{\rm He} =$ 10.0 $M_\odot$ in orbit with the first-formed BH with a mass of $M_2 = 12.0\ M_\odot$ and an orbital* *period of $P_{orb} = 3.00$ d. Assume that the asymmetric SN leaves behind a second BH with a mass of $M_{BH} = 7.0\,M_\odot$ and also imparts a kick onto it with a magnitude of $w = 200$ km s^{-1} in the direction: $\theta = 114°$ and $\phi = 83°$. Using the equations from Section 13.8, derive for the post-SN binary:*

a. *the critical angle, θ_{crit} for which $\theta < \theta_{crit}$ would have disrupted the binary, and verify thereby that this system will remain bound*
b. *the ratio (a_f/a_i) of the post- to pre-SN orbital semi-major axis*
c. *the orbital period, P_{orb}*
d. *the eccentricity, e*
e. *the 3D systemic recoil velocity of the system, v_{sys}*
f. *the misalignment angle, δ*
g. *keeping the kick direction the same, what is the maximum kick velocity imparted onto the BH such that the post-SN system will remain bound?*

Exercise 13.2. *Kinematic impact of the 2nd SN in producing a NS+NS system. Consider a pre-SN system composed of an exploding helium star with a mass of $M_{He} = 3.00\,M_\odot$ in orbit with the first-formed NS with a mass of $M_2 = 1.40\,M_\odot$ and an orbital period of $P_{orb} = 25.0$ d. Assume that the asymmetric SN leaves behind a second NS also with a mass of $M_{NS} = 1.40\,M_\odot$ and in addition imparts a kick onto it with a magnitude of $w = 50$ km s^{-1} in the direction: $\theta = 100°$ and $\phi = -45°$. Using the equations from Section 13.8, derive for the post-SN binary:*

a. *the critical angle, θ_{crit} for which $\theta < \theta_{crit}$ would have disrupted the binary, and verify thereby that this system will remain bound*
b. *the ratio (a_f/a_i) of the post- to pre-SN orbital semi-major axis*
c. *the resulting orbital period, P_{orb}*
d. *the resulting eccentricity, e*
e. *the 3D recoil velocity of the system, v_{sys}*
f. *the misalignment angle δ of the new orbital plane relative to the original orbital plane.*

Compare your answers to Figure 13.25.

Exercise 13.3. *Repeat Exercise 13.2 questions (a) to (f) for a system with pre-SN parameters: $M_{He} = 1.80\,M_\odot$, $M_2 = 1.40\,M_\odot$, and $P_{orb} = 0.50$ d. Assume that the SN produces a NS with $M_{NS} = 1.50\,M_\odot$ and imparts a kick of $w = 300$ km s^{-1} in the direction: $\theta = 170°$ and $\phi = 170°$. Compare your answers to Figure 40 in Tauris et al. (2017).*

g. *Assume that the companion would instead have been a main-sequence star of $M_2 = 1.40\,M_\odot$. Would the binary still survive the SN?*
(Hint: consider the post-SN periastron separation, $q = a_f(1-e)$.)

Exercise 13.4. *Consider a pre-SN system similar to that in Exercise 13.3. Assume that all parameters are the same, except that now* $\theta = 30°$. *Demonstrate that the system will break apart after the SN, and calculate the runaway velocities at infinity of the two ejected NSs in the pre-SN c.m. reference frame.*

Exercise 13.5. *Consider a pre-SN system composed of an exploding helium star with mass* $M_{\rm He} = 4.3\,M_\odot$ *orbiting a* $1.4\,M_\odot$ *NS in a circular orbit with separation* $a_0 = 1.0\,R_\odot$. *The SN leaves behind a second NS, also with a mass of* $1.4\,M_\odot$. *Assume a fixed kick direction of* $(\theta,\ \phi) = (100°,\ 0°)$ *and four different kick values of: (a)* $w = 0\ \rm km\,s^{-1}$; *(b)* $w = 50\ \rm km\,s^{-1}$; *(c)* $w = 200\ \rm km\,s^{-1}$; *and (d)* $w = 400\ \rm km\,s^{-1}$.

Evaluate, in each case, whether the system disrupts or remains bound. If the system remains bound, calculate the post-SN orbital period, $P_{\rm orb}$; *eccentricity,* e; *and systemic velocity,* $v_{\rm sys}$. *Compare your results to Figure 13.26.*

Exercise 13.6. *Derive expressions for the runaway velocities, at infinity, for stellar components ejected from a binary system (see Eqs. 13.16—13.21) in the case where:*

a. *the kick,* $w \to \infty$
b. *the mass of the SN remnant,* $M_{\rm NS} \to 0$ *(i.e., as in a Type Ia SN)*
c. *the SN is purely symmetric* $(w = 0)$ *and neglecting shell impact.*

Exercise 13.7. *Reproduce Figure 13.24.*

Exercise 13.8. *Derive Eq. (13.8) and show that the expression for* $P_{\rm a}^{-}$ *has a peak value of:*

$$P_{\rm max} = \frac{1}{2}\left(1 - \sqrt{\Delta M/M}\right). \tag{13.26}$$

Exercise 13.9. *A Maxwellian (Maxwell-Boltzmann) probability distribution function (PDF) is often used in the literature to assign a kick magnitude to a newborn NS or BH:*

$$f(v) = \sqrt{\frac{2}{\pi}}\,\frac{v^2}{a^3}\,\exp\left(-\frac{v^2}{2a^2}\right), \tag{13.27}$$

where v *is the 3D speed and* a *is the scale parameter (1D root-mean-square value). An example of a truncated PDF is shown in Figure 13.29 where also the most probable value* $v_{\rm p} = \sqrt{2}\,a$, *the mean value* $\langle v\rangle = \sqrt{8/\pi}\,a$, *and the root-mean-square value* $v_{\rm rms} = \sqrt{3}\,a$ *are indicated. The variance is given by:* $\sigma^2 = a^2(3\pi - 8)/\pi$.

a. *Verify the values of* $v_{\rm p}$, $\langle v\rangle$ *and* $v_{\rm rms}$ *in Figure 13.29.*
b. *Derive Eq. (14.36).*

For the truncated PDF plotted in Figure 13.29, for which $a = 300\ \rm km\,s^{-1}$ *and* $0 < v < 800\ \rm km\,s^{-1}$, *calculate:*

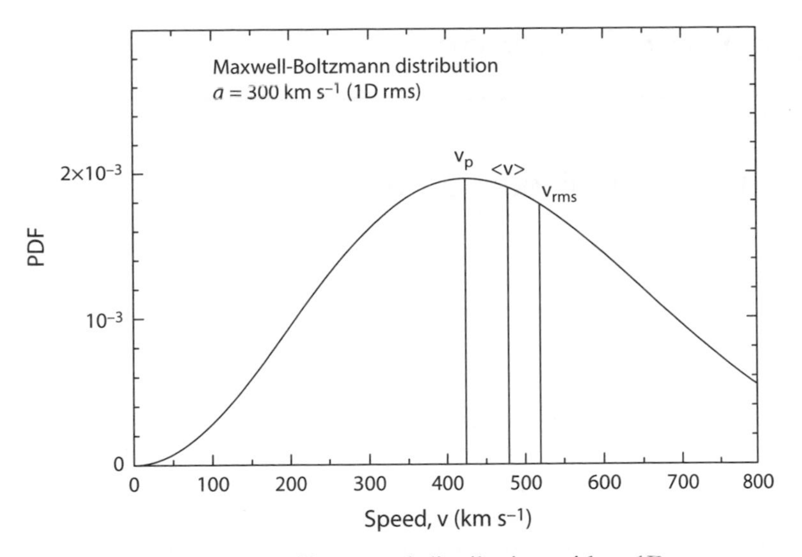

Figure 13.29. Truncated Maxwellian speed distribution with a 1D root-mean-square (rms) value of $a = 300\,\mathrm{km\,s^{-1}}$ (Eq. 13.27). The most probable value, v_{p}, the mean value, $\langle v \rangle$, and the root-mean-square value, v_{rms}, are indicated. Note that the PDF plotted here is normalized for $v \in [0;\ \infty[$.

c. *the probability that* $v > v_{\mathrm{rms}}$

d. *the probability that* $v < \frac{1}{2}\langle v \rangle$.

Exercise 13.10. *Consider a pre-double BH system composed of an exploding helium star with a mass of* $M_{\mathrm{He}} = 10\,M_{\odot}$ *in orbit with the first-formed BH with a mass of* $M_2 = 12\,M_{\odot}$ *and an orbital period of* $P_{\mathrm{orb}} = 4$ d. *Assume that the asymmetric SN leaves behind a second BH with a mass of* $M_{\mathrm{BH}} = 8\,M_{\odot}$ *and in addition imparts a kick onto it with a magnitude,* w, *and kick direction angles:* θ *and* $\phi = 60°$ *(the latter angle being fixed and also disregarding here the possibility of spin axis tossing of the second-formed BH). Plot the resulting BH+BH misalignment angle,* δ, *as a function of* w *and* θ, *treated as free parameters. This can be done in a 3D plot or by color-coding* δ *in the* (θ, w)*–plane. See also Exercise 15.4.*

Exercise 13.11. *Use the Cartesian coordinate system depicted in Figure 13.22 (or any other convenient coordinate system) to calculate the tilt of the orbital angular momentum vector* $\vec{L}_{\mathrm{orb}} = \vec{r} \times \vec{p}$ *due to an asymmetric SN, that is:*

$$\cos\delta = \frac{\vec{L}_{\mathrm{orb,i}} \cdot \vec{L}_{\mathrm{orb,f}}}{|\vec{L}_{\mathrm{orb,i}}|\,|\vec{L}_{\mathrm{orb,f}}|}\,. \tag{13.28}$$

This tilt angle, δ, *is also called the misalignment angle because it is the expected angle between the spin angular momentum vector of the companion star,* $\vec{S}_2$, *and the post-SN orbital angular momentum vector,* $\vec{L}_{\mathrm{orb,f}}$. *Compare your result with Eq. (13.11).*

Chapter Fourteen

Binary and Millisecond Pulsars

Binary and millisecond pulsars (MSPs) are known to be important objects of research for many areas of fundamental physics. This includes testing general relativity (Damour & Taylor, 1992; Kramer et al., 2006; Wex, 2014; Kramer et al., 2021), the neutron star equation-of-state (Lattimer & Prakash, 2007; Özel & Freire, 2016), and pulsar timing arrays for detecting nano-Hz gravitational waves (GWs) (Hobbs et al., 2010). Equally important, however, binary MSPs represent end points of stellar evolution, and their observed orbital and stellar properties are fossil records of their evolutionary history. Thus one can also use binary pulsar systems as key probes of stellar astrophysics. It is well established that the neutron star (NS) in binary MSP systems forms first, descending from the initially more massive of the two binary stellar components. The NS is subsequently spun up to a high spin frequency via accretion of mass and angular momentum once the secondary star evolves (Alpar et al., 1982; Radhakrishnan & Srinivasan, 1982; Fabian et al., 1983; Bhattacharya & van den Heuvel, 1991; Tauris & van den Heuvel, 2006; Tauris et al., 2012). During this *recycling* phase, the system is observable as an X-ray binary (see Chapters 6 and 7) and toward the end of this phase as an X-ray millisecond pulsar (Wijnands & van der Klis, 1998; Archibald et al., 2009; Papitto et al., 2013, 2020; Di Salvo & Sanna, 2020; Patruno & Watts, 2021), see Table 7.5. Although this formation scenario is now commonly accepted,[1] many aspects of the mass-transfer process and the accretion physics (e.g., details of non-conservative evolution, accretion efficiency, and B-field decay) are still not well understood. Some of these issues will be addressed here.

In this chapter, we first give a general introduction to radio pulsars followed by a review of the characteristics and the formation of all observed classes of the binary and MSP population. The main focus is on the stellar astrophysics of X-ray binaries leading to the production of either (1) fully recycled MSPs with white dwarf (WD) or substellar semi-degenerate companions, or (2) mildly recycled pulsars in DNS (double NS) systems. It is demonstrated that the evolutionary status of the donor star in X-ray binaries—or, equivalently, the orbital period—at the onset of the Roche-lobe overflow (RLO) is the main determining factor for the outcome of the mass-transfer phase and thus the characteristics of the recycled pulsar. We show that, depending on the nature of the companion star, MSPs are believed to form from either LMXBs or intermediate-mass X-ray binaries (IMXBs). In HMXBs, on the other hand, the timescale of mass

[1] Already proposed from the properties of the first known binary pulsar, B1913+16, see Section 15.1.

transfer from a massive star onto the NS is too short to fully recycle the pulsar to millisecond spin periods. Furthermore, the formation of binary MSPs is discussed in context of the $P\dot{P}$–diagram, as well as the Corbet diagram for radio pulsars. We summarize accretion torque calculations and the physics behind spin equilibrium and the spin-up line. Masses of recycled pulsars are discussed at the end of the chapter. Pulsars in globular clusters are, in general, not suitable as tracers of their stellar evolution history because of the frequent encounters and exchanges of companion stars in the dense environments (Ransom et al., 2005; Ridolfi et al., 2022). Hence, this chapter mainly focuses on binary pulsars formed in the Galactic disk.

14.1 INTRODUCTION TO RADIO PULSARS

Pulsars were the first population of NSs to be discovered (1967). In 1968, Jocelyn Bell Burnell and Antony Hewish announced the first discovery of a radio pulsar (Hewish et al., 1968). Later in the same year, the Crab pulsar was discovered (PSR B0531+21; Staelin & Reifenstein, 1968) in the remnant of the supernova SN 1054, which was widely observed on Earth in the year 1054. The Crab pulsar is a relatively young NS. The star is a central star in the Crab nebula and its discovery solved the puzzle of the unknown energy source of the nebula (Oort & Walraven, 1956). The related puzzle of the origin and acceleration of the relativistic electrons in the Crab Nebula later led Woltjer (1964) to predict that NSs can have magnetic field strengths up to 10^{14}–10^{16} G based on flux conservation during core collapse. Just prior to the discovery of the radio pulsars, Pacini (1967) calculated that a rotating NS with a strong magnetic field will emit much energy in the form of magnetic dipole radiation. The energy emission of the radio pulsars is indeed rotationally driven. The rate of increase of the 0.033 s pulse period of the Crab pulsar, $\dot{P} = 38\,\mathrm{ns\,day^{-1}}$, was measured within a day after its discovery in 1968. This means that the total loss rate of rotational energy is $\dot{E}_{\rm rot} = I\Omega\dot{\Omega} \sim 10^{38}\,\mathrm{erg\,s^{-1}}$. Only 10^{-8} of $\dot{E}_{\rm rot}$ is emitted as detectable radio waves. As we shall discuss later on, the braking torque is a combination of a pulsar wind and a rotating magnetic dipole emitting magnetodipole radiation (Section 14.1.3) with a frequency equal to the pulsar spin frequency. This long-wavelength radiation energy is absorbed by and powers the Crab nebula (a "megawave oven"). With the current $\dot{P}$, the Crab pulsar will double its spin period in a few 10^3 yr. For an excellent introduction to radio pulsars, we refer to Lorimer & Kramer (2012), and for the physics of rotation and accretion-driven pulsars in general, see Ghosh (2007). For an overview of the present status of ongoing and upcoming surveys for radio MSPs see, for example, Bhattacharyya & Roy (2022).

14.1.1 The Kaleidoscopic Neutron Star Population

Figure 1.2 shows a plot of all currently known radio pulsars with measured values of spin period (P) and their time derivative ($\dot{P}$). The classic *radio pulsars* (red dots) are concentrated in the region with $P \simeq 0.2 - 2$ sec and $\dot{P} \simeq 10^{-16} - 10^{-13}$. They have magnetic fields of the order of $B \simeq 10^{10} - 10^{13}$ G and lifetimes as radio sources of a few 10^7 yr. The population plotted in blue circles are binary pulsars, and they clearly indicate a connection to the rapidly spinning MSPs. The pulsars marked with stars indicate young pulsars observed inside, or near, their gaseous SN remnants. Then there are the "drama queens" of the NS population: the *magnetars* (see also Section 10.3.3), which undergo bursts and are powered by their huge magnetic energy reservoirs. Also plotted are the mysterious *rotating radio transients* (RRATs), the *isolated X-ray dim neutron stars* (XDINS), and the *central compact objects* (CCOs) of SN remnants. In addition, there are other exotic radio pulsars such as the *intermittent pulsars*, the *black widows* and *redbacks*, and pulsars in triple systems. A main challenge of the past decade was—and continues to be—to find a way to unify this variety into a coherent physical picture (e.g., GUNS, the grand unification of neutron stars; Kaspi, 2010). The NS zoo raises essential questions such as: What determines whether a NS will be born with, for example, magnetar-like properties or as a Crab-like pulsar? What are the branching ratios for the various varieties, and, given estimates of their lifetimes, how many of each are there in the Galaxy? Can individual NSs evolve from one species to another (and why)? How does a NS interact with a companion star during and near the end of mass transfer (recycling)? And what limits its final spin period? Ultimately such questions are fundamental to understanding the fate of massive stars, close binary evolution, and the nature of core-collapse SNe, while simultaneously relating to a wider variety of interesting fundamental physics and astrophysics questions, ranging from the nature of matter in extremely high magnetic fields to the equation-of-state of ultra-dense matter and details of accretion processes. The spin period distribution of NSs spans over seven orders of magnitudes—from a slow 5.3 hr of the accreting pulsar in the X-ray binary 4U 1954+319 (Enoto et al., 2014) to the rapidly spinning 1.4 ms recycled MSP J1748−2446ad in the globular cluster Terzan 5 (Hessels et al., 2006).

14.1.2 Basic Observational Properties of Radio Pulsars

The "*pulsar lighthouse model*" is illustrated in Figure 14.1 and explained in more detail in the following sub-sections. For further descriptions of the properties of radio pulsars, we refer to Manchester & Taylor (1977), Lorimer & Kramer (2012), and references therein.

Spin Periods, Energy Loss, and Lifetimes

Radio pulsars have observed spin periods between 1.4 ms and 76 s. Some pulsar periods are known with 16 significant digits, that is, down to the attosecond level (e.g., PSR J0437−4715, which has a spin period $P = 0.005\,757\,451\,924\,362\,137(2)$ s; Verbiest et al., 2008). Hence, pulsars represent excellent clocks, and it is even possible to construct a pulsar-based timescale that has a precision comparable to the best modern atomic clocks (Hobbs et al., 2012). Of course, pulsars continuously lose rotational energy (see Section 14.1.3), which steadily increases their spin period, but this can be corrected for by measuring their (stable) spin period derivative, $\dot{P}$. With an ensemble of

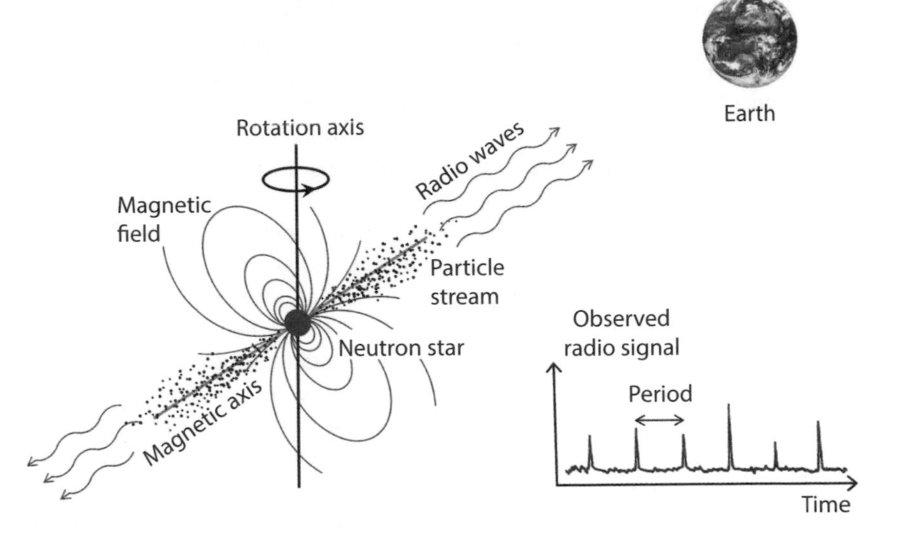

Figure 14.1. The pulsar lighthouse model. The rotating NS emits a beam of radio waves from above the polar regions of its magnetic axis, which is inclined with respect to its rotational axis. At the observatory on Earth, periodic radio signals are recorded, revealing the spin period of the pulsar.

high-precision-timing pulsars, a so-called pulsar timing array (PTA; e.g., Verbiest et al., 2021) can be constructed to detect low-frequency (nano-Hz) GWs passing through the Milky Way, originating from the in-spiral of merging super-massive BHs in distant galaxies. Globally there are three active PTAs: the Parkes PTA, the European PTA (EPTA), and the North-American NANOGrav (Hobbs et al., 2010; Manchester et al., 2013; McLaughlin, 2013; Kramer & Champion, 2013; Pol et al., 2021). These projects collaborate in the International Pulsar Timing Array (IPTA) project and may hopefully soon succeed with detections (see discussions in Antoniadis et al., 2022). Recently, also the Indian Pulsar Timing Array (InPTA) joined in (Tarafdar et al., 2022).

The interval of observed spin period derivatives spans more than ten orders of magnitude (see Fig. 1.2) with a mean value of $\sim \dot{P} \simeq 10^{-15}$ (always positive and thus a slow down of the spin rate). The loss of rotational energy is due to a combined effect of the energy loss by magnetic dipole radiation and by the pulsar wind of highly relativistic particles, which both are driven by the rotation of the highly magnetized NS. This energy loss is a function of the strength B of the surface dipole magnetic field and the rotation rate of the NS, as will be explained in Section 14.1.3. The measured spin rate and spin-down rate allow for calculation of the rotational energy loss rate and therefore enable estimation of the strength, B, of the dipole component of the pulsar's magnetic field. Using this method, radio pulsars have estimated B-fields between 10^7 and 10^{14} G (see Fig. 1.2).

Only a tiny fraction of the rotational energy loss, $|\dot{E}_{\rm rot}|$, of a spinning radio pulsar is emitted as the radio waves that cause it to be detected. For example, for the Crab pulsar, the radio flux density[2] of the detected signal at 436 MHz is $\sim$0.48 Jy corresponding to a luminosity of $L_{\rm radio} \sim 10^{31}$ erg s^{-1}, at a distance of 2.0 kpc, which is $\sim 10^8$ times smaller than $|\dot{E}_{\rm rot}| \approx 5 \times 10^{38}$ erg s^{-1}. A large fraction of $|\dot{E}_{\rm rot}|$ also goes to light up the Crab nebula via injection of the pulsar wind of highly relativistic particles. Furthermore, a small fraction is needed for the observed magnetospheric emission of optical-, X-, and γ-rays (via curvature or synchrotron radiation generated from gyration of charged particles in the strong magnetic field).

The average lifetime of a normal (i.e., *non-recycled*) pulsar is a few times 10 Myr. The radio emission process terminates once the electrostatic potential drop across the polar cap, $\Delta\phi \propto B/P^2$, decreases below a critical value for maintaining the required electron-positron pair production: $\gamma + \vec{B} \rightarrow e^- + e^+$ (i.e., when the pulsar crosses the so-called "death line" in the $P\dot{P}$–diagram [cf. Ruderman & Sutherland 1975; Beskin et al. 1988; Chen & Ruderman 1993]). As a consequence of the radio emission criterion, it is clear that the vast majority of the NSs that were formed since the birth of the Galaxy, some 13 Gyr ago, have long since disappeared into the "graveyard" area of the $P\dot{P}$–diagram and thus have become undetectable. Assuming a core-collapse SN rate of one per 30 to 50 years throughout the lifetime of the Galaxy, the total number of defunct pulsars—in the graveyard—will be $\sim (4-7) \times 10^8$. This number should, however, be corrected upward to take into account a Galactic starburst in the first 2 to 3 Gyr (Sandage, 1986). For comparison: the number of WDs in the Galaxy is probably around 10^{10}, while the total mass in the form of normal stars is $\sim (1-2) \times 10^{11}\ M_\odot$. So, NSs make up $\sim 1\%$ of the baryonic mass of the Galaxy, and (very) old NSs outnumber the active pulsars by a factor of the order of 10^4, given an estimate of $\sim 10^5$ active radio pulsars in the Milky Way (Keane et al., 2015).

Recycled pulsars and MSPs, which constitute the radio pulsars focused on in this chapter, have small B-fields and rapid spins. Hence, their ratio of $E_{\rm rot}/|\dot{E}_{\rm rot}|$ is very large, and thus they remain active radio sources on the order of a Hubble time.

Pulsar Spectra, Duty Cycles, and Beaming Fractions

Pulsars generally have rather steep radio-frequency spectra, $S_\nu \propto \nu^\alpha$, where S_ν is the flux density and the spectral index, α, is typically -1.5 and even steeper at high frequencies (> 1 GHz). Given the radio luminosity, one can calculate the surface intensity of the radio emission, I_ν, and use a Planck function to demonstrate that if the radio emission was caused by thermal black body radiation, one would obtain an extremely high brightness temperature ($T \approx 10^{28}$ K, leading to absurdly large particle energies, $E = kT \sim 10^{24}$ eV). Therefore the radiation mechanism of a radio pulsar must be *coherent*. (Most models invoke curvature radiation or a maser mechanism, e.g., Ruderman & Sutherland, 1975; Harding 2018.)

The shape of the observed radio pulse profile depends on the intersection of the line of sight across the emission region and the geometric structure of the pulsar beam (Lyne & Manchester, 1988; Manchester, 1995; Rankin, 1983, 1990; Karastergiou &

[2]The unit is Jansky, and 1 Jy $\equiv 10^{-23}$ erg s^{-1} cm^{-2} Hz^{-1} St^{-1}.

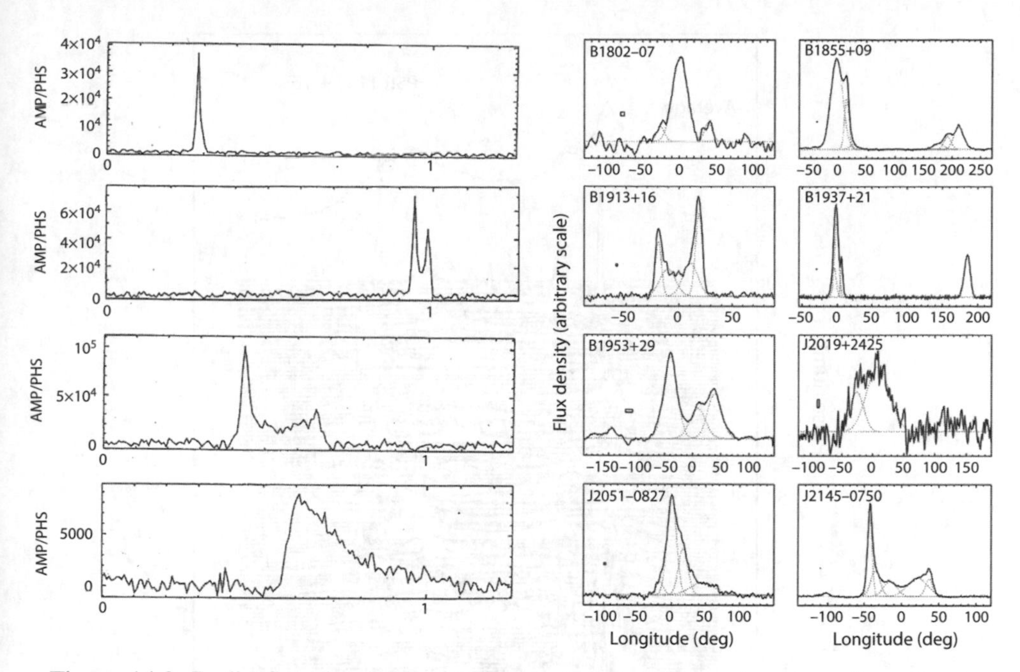

Figure 14.2. Radio flux density in arbitrary units plotted as a function of rotational phase for 12 pulsars. Left: Pulsar profiles of PSR J0206−4028 (top), J2346−0609, J2145−0750, and J1801−2304 (bottom) obtained from the Parkes Radio Telescope at a frequency of 436 MHz in the mid-1990s (Tauris, 1997). The upper two pulsars are slow ($P \sim 1$ s). The third one (PSR J2145−0750) is a MSP with $P = 16$ ms. Note the exponential tail due to interstellar scattering in the profile of J1801−2304 ($DM \sim 1{,}000\,\mathrm{cm}^{-3}$ pc, see Section 14.1.2). Right: Pulse profiles of one DNS and 7 MSPs observed at a frequency of 1.4 GHz by the Effelsberg Radio Telescope (Kramer et al., 1998). Here, one full rotation $= 360°$ in longitude of rotational phase. The pulse profiles can be quite complex with multiple components. PSR B1855+09 and B1937+21 even exhibit an interpulse, that is, a component of emission from the opposite magnetic pole (at a longitude $\sim 180°$) compared to that of the main pulse.

Johnston, 2007). Many pulsars have linear polarized profiles—up to 100%. Circular polarization is also seen but is not as frequent or as strong as the linear polarization. Polarization measurements of pulsars make it possible to determine the inclination angle between the magnetic and the rotational axes—an angle that affects the braking torque acting on the pulsar (e.g., Radhakrishnan & Cooke, 1969; Tauris & Manchester, 1998; Johnston & Karastergiou, 2017).

The duty cycle of radio pulsars (fraction of rotational phase with measurable emission) is typically 1%–5%, although substantially larger for MSPs (Kramer et al., 1998), see Figure 14.2. Similarly, the beaming fraction, the portion of the sky illuminated by a given pulsar, decreases with spin period (Tauris & Manchester, 1998). The observed pulse shapes are quite different in nature and often include two or more sub-pulses

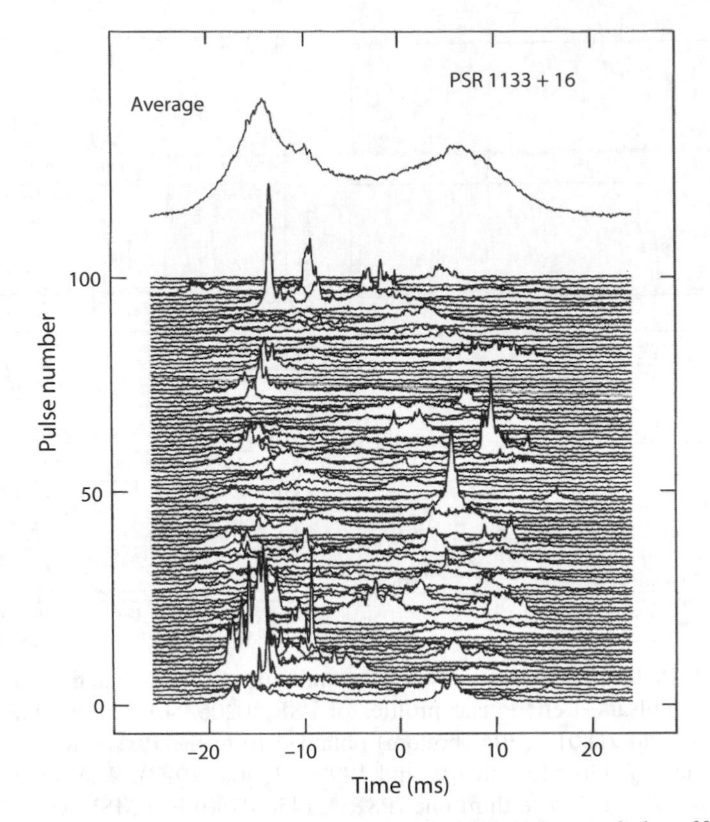

Figure 14.3. A sequence of 100 pulses from PSR 1133+16 recorded at 600 MHz by Cordes (1979). Consecutive individual pulses are plotted vertically to show their large variations. The *average* pulse behavior, however, is extremely stable. An average of 500 pulses is shown at the top.

(Figure 14.2). The micro-structure of each pulse (or sub-pulse) can be very complex, as shown in Figure 14.3. However, the *average* pulse profile is remarkably stable—a feature that is essential for precise pulsar timing.

Timing Pulsars

The concept of pulsar timing is straightforward in principle: one measures pulse times-of-arrival (TOAs) at the observatory and compares them with time kept by a stable reference. During the data analysis, the recorded TOAs must be transformed to the corresponding proper time of emission, T, in the pulsar reference frame (Taylor & Weisberg, 1989; Lorimer & Kramer, 2012). Such a transformation includes a dispersive delay from the interstellar medium, a transformation to the Solar System barycenter, and special and general relativistic time delays (Rømer, Einstein, and Shapiro delays) for both the Solar System and for the emitting pulsar system if the pulsar orbits a companion star. For some pulsars, it is the post-Keplerian parameters of the binary

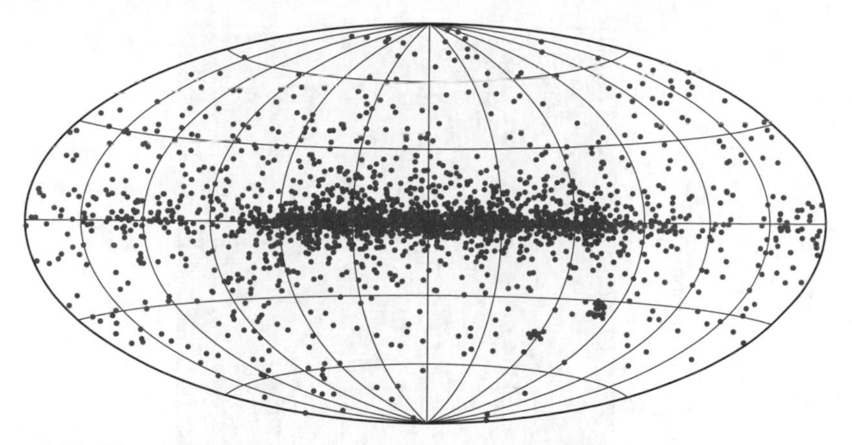

Figure 14.4. The sky distribution in Galactic coordinates of 2,781 known pulsars. Data from the ATNF Pulsar Catalogue, Manchester et al. (2005)—version 1.64, Feb. 2021. Figure provided by Norbert Wex.

system that allows for testing gravitational physics (Taylor & Weisberg, 1982, 1989; Stairs, 2003; Kramer et al., 2006; Kramer & Wex, 2009; Freire et al., 2012; Antoniadis et al., 2013; Wex, 2014; Kramer et al., 2021).

Under the assumption of a deterministic spin-down law, the rotational phase of a pulsar can be written as:

$$\phi(T) = \phi_0 + \Omega T + \frac{1}{2}\dot{\Omega}T^2 + \frac{1}{6}\ddot{\Omega}T^3 + \cdots . \tag{14.1}$$

To obtain a timing solution, it is necessary to assign a pulse number to each recorded TOA. Some of the observations can be separated by weeks, months, or even years. Hence, between two consecutive observations, the pulsar may have rotated as many as $10^7 - 10^{10}$ turns, and to extract the maximum information content from the data, these integer numbers of turns must be recovered *exactly*. The spin period derivative, $\dot{P}$, is an essential parameter for pulsars. All their physical parameters that can be extracted basically depend on this parameter: for example, B-field, energy loss, braking torque, age, and electrostatic potential gap across the polar cap region of the magnetosphere. To measure $\dot{P}$ one must assure that the time interval between two neighboring observations, multiplied by the error in the measured rotational frequency of the pulsar, is much less than one, that is, $(t_2 - t_1)\,\Delta\Omega/2\pi \ll 1$.

Distance Measurements and the Distribution of Galactic Pulsars

Figure 14.4 shows the sky distribution of pulsars in Galactic coordinates. The clear clustering of sources in the Galactic plane shows that pulsars are indeed of Galactic origin and that they are the likely remnants of massive OB stars. However, the broad scatter in their distribution is very significant. While their OB progenitor stars are found close to the Galactic plane (with a velocity dispersion of only $\sim 10 - 20\,\mathrm{km\,s^{-1}}$),

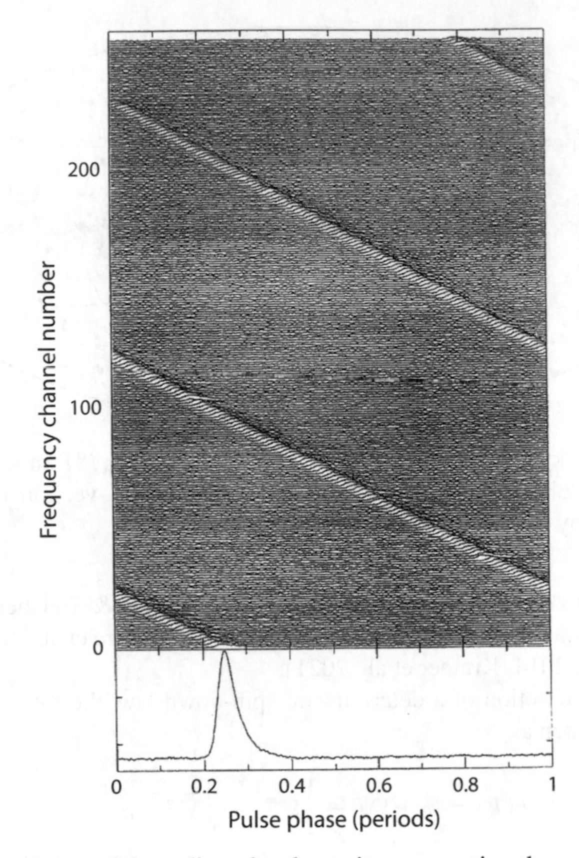

Figure 14.5. Dispersion of the radio pulse due to its propagation through the interstellar medium. Shown is the received signal at each frequency channel at the observatory (channel no.1 @ 420 MHz → channel no.256 @ 452 MHz) as a function of pulse phase for the Vela pulsar, PSR B088−45. After Lorimer (1994).

pulsars have a velocity dispersion exceeding $100\,\mathrm{km\,s^{-1}}$. The explanation for this is that (most) NSs apparently receive a significant momentum kick at birth, resulting in a mean birth velocity of $\sim 400\,\mathrm{km\,s^{-1}}$ (Lyne & Lorimer, 1994), although a fraction of these runaway velocities can also be explained by the disruption of binary systems (Tauris & Takens, 1998).

For some nearby pulsars with good timing properties, it is possible to obtain a parallax measurement. However, for the majority of pulsars, one has to rely on distance estimates obtained from the amount of dispersion of the radio waves as they propagate through the ionized interstellar medium along the line of sight, i.e., the delay in the TOA of the different frequency components of the pulse, see Figure 14.5. Since this method requires a model for the distribution of free electrons in the Milky Way (e.g., Cordes & Lazio, 2002; Yao et al., 2017), the resulting distance estimates are, in general, only accurate to within 20%–50%.

14.1.3 The Magnetized Rotating Neutron Star

Given that radio pulsars are rapidly rotating, strongly magnetized NSs that have an inclined dipole magnetic field axis with respect to their rotation axis, they radiate significant amounts of energy in the form of magnetodipole waves (electromagnetic waves with a frequency equal to the spin frequency of the pulsar). The energy-loss rate due to magnetic dipole radiation is given by (Landau & Lifshitz, 1951, 1971):

$$\dot{E}_{\rm dipole} = -\frac{2}{3c^3}|\ddot{\vec{m}}|^2 \quad \wedge \quad |\ddot{\vec{m}}| = BR^3\Omega^2 \sin\alpha, \tag{14.2}$$

where $\vec{m}$ is the magnetic moment of the NS, B is the magnetic flux density at its surface (equator), R is the radius, $\Omega = 2\pi/P$ is the angular velocity with P being the pulsar spin period, c is the speed of light in vacuum, and the magnetic inclination angle is $0 < \alpha \leq 90°$. The total loss of rotational energy of a magnetized single NS is caused by a combination of magnetic dipole radiation (first suggested by Pacini, 1967), the presence of plasma currents in the magnetosphere (the Goldreich-Julian term; Goldreich & Julian, 1969; Spitkovsky, 2006), and GW radiation (e.g., Wade et al., 2012):

$$\dot{E}_{\rm rot} = \dot{E}_{\rm dipole} + \dot{E}_{\rm GJ} + \dot{E}_{\rm GW}. \tag{14.3}$$

Theoretically, the value of $\dot{E}_{\rm GJ}$ is found by considering the outward Poynting energy flux, $S \sim c \cdot B^2/4\pi$, crossing the light cylinder, $r_{\rm lc} = c/\Omega$, and giving rise to the observed high-frequency radiation as well as emission of relativistic particles. Observations have shown that $\dot{E}_{\rm GJ}$ is of the same order as $\dot{E}_{\rm dipole}$ within a factor of a few (Kramer et al., 2006; Lorimer et al., 2012; Camilo et al., 2012), and it is usually a very good approximation that $\dot{E}_{\rm GW} \ll \dot{E}_{\rm rot}$ (Abadie et al., 2010). Hence, the dipole component of pulsar B-fields is traditionally found simply by equating the loss rate of rotational energy ($\dot{E}_{\rm rot} = I\Omega\dot{\Omega} = -4\pi^2 I\dot{P}/P^3$) to the energy loss caused by magnetic dipole radiation (Eq. 14.2) yielding:

$$B_{\rm dipole} = \sqrt{\frac{3\,c^3\,I}{8\pi^2\,R^6}\,P\dot{P}} \quad \simeq \quad 3.2 \times 10^{19}\,\sqrt{P\dot{P}}\ \text{Gauss}, \tag{14.4}$$

where the numerical constant is calculated for the equatorial B-field of an orthogonal rotator ($\alpha = 90°$), and assuming $R = 10$ km and a NS moment of inertia of $I = 10^{45}$ g cm^2. Tauris et al. (2012) discuss this equation further and derive an alternative expression (Section 14.6.2). For a discussion of Eq. (14.4) in the context of efficient γ-ray emitting MSPs, see Guillemot & Tauris (2014).

14.1.4 Pulsar Evolutionary Tracks and True Ages

The slow-down of pulsar spin is characterized by the observable braking index, $n \equiv \Omega\ddot{\Omega}/\dot{\Omega}^2$, which relates the braking torque ($N = dJ_{\rm spin}/dt = I\,\dot{\Omega}$) to the rotational angular velocity via $\dot{\Omega} \propto -\Omega^n$. By integrating this pulsar spin-deceleration equation one can obtain evolutionary tracks and isochrones in the $P\dot{P}$–diagram (e.g., Tauris & Konar, 2001; Lazarus et al., 2014; see also Section 14.6). For a pulsar evolving with a constant value of n, the kinematic solution at time t (positive in the future, negative in the past) is given by:

$$P = P_0 \left[1 + (n-1) \frac{\dot{P}_0}{P_0} t \right]^{1/(n-1)} \tag{14.5}$$

and

$$\dot{P} = \dot{P}_0 \left(\frac{P}{P_0} \right)^{2-n}, \tag{14.6}$$

where P_0 and $\dot{P}_0$ represent values at $t = 0$. Similarly, the true age can be written as:

$$t = \frac{P}{(n-1)\dot{P}} \left[1 - \left(\frac{P_0}{P} \right)^{n-1} \right]. \tag{14.7}$$

The so-called *characteristic age* (Manchester & Taylor, 1977; Shapiro & Teukolsky, 1983) is defined as: $\tau \equiv P/(2\dot{P})$. However, this expression is only a good age estimate for pulsars that have evolved with a constant $n = 3$ and for which $P_0 \ll P$. For many pulsars, and in particular for magnetars and recycled pulsars (MSPs), τ is not a good true age estimator (Tauris et al., 2012). For example, some young NSs associated with SN remnants have τ values of several Myr, although SN remnants are believed to be dissolved into the interstellar medium within 50 kyr (which is therefore an upper limit on the true age of these NSs). A specific deviant case includes PSR J0538+2817 in the supernova remnant S147, which has $\tau = 620$ kyr but a kinematic age of only 30 kyr (Kramer et al., 2003) and a similar cooling age estimate (Ng et al., 2007). Other extreme examples are the MSPs, which in some cases have $\tau > 30$ Gyr (more than a couple of Hubble times). Ironically, in many cases τ can better be thought of as an estimate of the *remaining* lifetime of a pulsar given that $\tau = E_{\rm rot}/|\dot{E}_{\rm rot}|$.

The first ideas of a so-called recycling process of old NSs date back to the mid-1970s following the discovery of the Hulse-Taylor pulsar (Bisnovatyi-Kogan & Komberg, 1976; Smarr & Blandford, 1976, and see Section 15.1). This concept of pulsar recycling was given a boost by the discovery of the first MSP (Backer et al., 1982; Alpar et al., 1982; Radhakrishnan & Srinivasan, 1982; Fabian et al., 1983). The idea is that the MSP obtains its rapid spin (and weak B-field) via a long phase of accretion of matter from a companion star in a LMXB. As a result of the high incidence of binaries found in the following years among these fast-spinning pulsars (see Fig. 1.2), this formation scenario has now become generally accepted (Bhattacharya & van den Heuvel, 1991). Furthermore, the model was beautifully confirmed with the discovery of the first millisecond X-ray pulsar in the LMXB system SAX 1808.4−3658 (Wijnands & van der Klis, 1998), and more recently by the detection of the so-called transitional MSPs which undergo changes between accretion and rotational powered states (Archibald et al., 2009; Papitto et al., 2013), see Section 14.3.1.

Roughly 12% of all known radio pulsars are MSPs (here defined as fully recycled pulsars with spin periods $P < 10\,\mathrm{ms}$) and the majority ($\sim 2/3$) of these have been found to host a companion star. Radio pulsars in general have been discovered in binary systems with a variety of companions: white dwarfs (WDs), NSs, main-sequence stars, and even planets; see Table 14.1. The vast majority of the binary pulsar systems contain a MSP with a helium WD companion. However, there is a growing number of MSPs with a non- or semi-degenerate companion star that is being ablated by the pulsar wind, the so-called *black widows* and *redbacks* (Roberts, 2013). This is evidenced by the radio signal from the pulsar being eclipsed for some fraction of the orbit (Fruchter et al., 1988; Stappers et al., 1996; Archibald et al., 2009). These companions are all low-mass stars with a mass between 0.02 and 0.3 $M_\odot$ (Roberts, 2013; Breton et al., 2013). Chen et al. (2013) argued that that black widows and redbacks are two distinct populations and that they are not linked by an evolutionary path.[3] An important long-standing question to answer is whether black widows are the progenitors of the isolated MSPs. Can ablation by the energetic pulsar flux lead to a complete evaporation of a low-mass companion (Kluzniak et al., 1988; Ruderman et al., 1989)? Or will the companion star in an ultra low-mass binary pulsar system eventually disrupt via an internal instability when its mass becomes too small (Deloye & Bildsten, 2003; Possenti, 2013)?

Based on stellar evolution theory, it is expected that pulsars can also be found with a helium star or a BH companion. These systems have not yet been discovered, but it is likely that the SKA will reveal pulsars with such companions within the next decade.

In recent years, a few binary pulsars with peculiar properties have been discovered that indicate a hierarchical triple system origin: for example, PSR J1903+0327 (Champion et al., 2008; Freire et al., 2011; Portegies Zwart et al., 2011). In 2013, the first two puzzling MSPs were discovered in eccentric binaries: PSR J2234+06 (Deneva et al., 2013) and PSR J1946+3417 (Barr et al., 2013). Since then, another few eccentric MSPs have been discovered, see Section 14.3.3. These systems might also have a triple origin. Apart from these intriguing systems, an exotic triple system MSP with two WD companions (PSR J0337+1715) was announced by Ransom et al. (2014). This amazing system must have survived three phases of mass transfer and one SN explosion and challenges current knowledge of multiple stellar system evolution (Tauris & van den Heuvel, 2014).

14.2 TO BE RECYCLED OR NOT TO BE RECYCLED

The $P\dot{P}$–diagram introduced in Figure 1.2 in Chapter 1 (and showing measured values of spin period, P, and their time derivative, $\dot{P}$), clearly demonstrates that the bulk of the pulsar population appears to fall into two broad groups that differ significantly in a number of characteristics. As mentioned previously, the classical isolated pulsars (young, strong-magnetic field pulsars that have not been recycled) are concentrated in the region with $P \simeq 0.2-2\,\mathrm{s}$ and $\dot{P} \simeq 10^{-16}-10^{-13}$. They have magnetic fields

[3] See, however, Benvenuto et al. (2014) for a different point of view.

Table 14.1. Representative Examples of the Different Classes of Binary and Millisecond Radio Pulsars

Radio pulsar	Class	P (ms)	$\dot{P}$ (10^{-20})	B $(10^8$ G)	$P_{\rm orb}$ (d)	ecc	$M_{\rm psr}$ $(M_\odot)$	$M_{\rm comp}$ $(M_\odot)$
J0453+1559[a]	DNS	45.8	18.6	9.2	4.072	0.113	1.559	1.174
J1913+1102[b]	DNS	27.3	15.7	6.3	0.206	0.090	1.62	1.27
J0737−3039A[c]	DNS	22.7	176	20	0.102	0.088	1.338	1.249
J0737−3039B[c]	DNS young	2,773.5	89,200	4,900	− ‖ −	− ‖ −	1.249	1.338
J1906+0746[d]	DNS young*	144.1	2,030,000	5,300	0.166	0.085	1.291	1.322
B1534+12[e]	DNS ecc	37.9	242	30	0.421	0.274	1.333	1.346
B1913+16[f]	DNS ecc	59.0	863	70	0.323	0.617	1.440	1.389
J2222−0137[g]	CO WD**	32.8	5.80	4.4	2.45	3.8e-4	1.76	1.29
J0621+1002[h]	CO WD	28.9	4.73	3.7	8.32	2.5e-3	~1.53	~0.76
B2303+46[i]	CO WD young	1,066	56,900	2,500	12.3	0.658	~1.34	~1.3
J1141−6545[j]	CO WD young	394	431,000	4,100	0.198	0.172	1.27	1.02
J1737+0333[k]	He WD	5.85	2.41	1.2	0.355	3.4e-7	1.47	0.181
J1713+0747[l]	He WD	4.57	0.853	0.62	67.8	7.5e-5	1.35	0.292
J2234+0611[m]	eMSP	3.58	1.20	0.66	32.0	0.129	1.35	0.298
J1946+3417[n]	eMSP	3.17	0.315	0.32	27.0	0.134	1.83	0.266
B1957+20[o]	black widow	1.61	1.69	0.52	0.382	<4e-5	>1.7	~0.03
J2051−0827[p]	black widow	4.51	1.27	0.76	0.099	5.1e-5	~1.8	~0.05
J2129−0429[q]	redback	7.63	—	—	0.635	—	~1.7	~0.44
J2339−0533[r]	redback	2.88	1.41	0.64	0.193	2.1e-4	~ 1.5	~0.32

J1023+0038[s]	tMSP	1.69	0.693	0.34	0.198	—	>1.6	>0.22
J1824−2452I[t]	tMSP	3.93	—	—	0.459	<1e-4	—	—
B1257+12[u]	planets (multi)	6.22	11.4	2.7	25-98	—	—	$\sim M_\oplus$
J1719−1438[v]	planet	5.79	0.804	0.68	0.091	8e-4	—	$\sim M_J$
B1937+21[w]	single	1.56	10.5	1.3			—	
J2053+1718[x]	single	119	28.7	18			—	
J0337+1715[y]	triple	2.73	1.77	0.70	1.6/327	≤0.035	1.438	0.20/0.41
J1740−3052[z]	non-deg.	570	2,550,000	12,000	231	0.579	—	~20
B1259−63[aa]	non-deg.	47.8	228000	1,000	1237	0.870	—	>10
J1807−2459B[ab]	GC DNS*	4.2	8.23	1.8	9.957	0.747	1.366	1.206
B2127+11C[ac]	GC DNS	30.5	499	38	0.335	0.681	1.358	1.354
J1748−2446Y[ad]	GC He WD	2.05	—	—	1.17	2e-5	—	>0.14
B1620−26[ae]	GC triple***	11.1	67.1	8.6	191	0.025	—	~0.34

$\dot{P}$ values are not corrected for the (in most cases negligible) Shklovskii effect (Shklovskii, 1970). B-field values are here estimated from $B \simeq 1.0 \times 10^{19} \sqrt{P\dot{P}}$ G—see also Eq. (14.15) assuming a magnetic inclination angle, $\alpha = 60°$. Masses are usually obtained from timing—see refs. for uncertainties.

*Not a confirmed DNS system—could also be a WD+NS binary.

**WDs with $M \gtrsim 1.05\,M_\odot$ are most likely ONeMg WDs (e.g., Lazarus et al., 2014).

***Hosts a circumbinary planet of $M \simeq 2.5\,M_J$ and $P_{orb} \simeq 90$ yr.

References for most recent parameters (and see also references therein): [a]Martinez et al. (2015). [b]Lazarus et al. (2016). [c]Ferdman et al. (2013). [d]van Leeuwen et al. (2015). [e]Fonseca et al. (2014). [f]Weisberg et al. (2010). [g]Cognard et al. (2017). [h]Kasian (2012). [i]van Kerkwijk & Kulkarni (1999). [j]Bhat et al. (2008). [k]Antoniadis et al. (2012). [l]Arzoumanian et al. (2018). [m]Stovall et al. (2019). [n]Barr et al. (2017). [o]van Kerkwijk et al. (2011). [p]Lazaridis et al. (2011). [q]Bellm et al. (2016). [r]Pletsch & Clark (2015). [s]Shahbaz et al. (2019). [t]Papitto et al. (2013). [u]Wolszczan & Frail (1992). [v]Ng et al. (2014). [w]Backer et al. (1982). [x]Brinkman et al. (2018). [y]Ransom et al. (2014). [z]Madsen et al. (2012). [aa]Johnston et al. (1992). [ab]Lynch et al. (2012). [ac]Anderson et al. (1990). [ad]Ransom et al. (2005). [ae]Sigurdsson et al. (2003).

of the order of $B \simeq 10^{10} - 10^{13}$ G and lifetimes as radio sources of a few 10^7 yr. The population of fast-spinning and weak-magnetic-field pulsars is clearly connected with the presence of a companion star in a binary system (plotted in blue circles), and there is no doubt that these pulsars are old NSs that have been rejuvenated via recycling to become MSPs (Bhattacharya & van den Heuvel, 1991, and references therein). Also the recycled pulsars can be divided into a number of sub-classes (van den Heuvel & Taam, 1984), depending on their properties, which we shall demonstrate and evolve further in more detail in this chapter. The term *recycled* for an old spun-up pulsar was coined by V. Radhakrishnan in 1978 (private communication to van den Heuvel), who on the Bandung IAU Asian-Pacific Regional Meeting in 1980, together with G. Srinivasan, also introduced the concept of a *spin-up* line, see Radhakrishnan & Srinivasan (1984).

The fact that the pulsar characteristics of the single and binary MSPs are indistinguishable strongly suggests that they are closely related and must have a similar origin. The extraordinarily large percentage of binaries among the MSPs then suggests that their formation is indeed related to the evolution of binary systems. The continuous line of observational evidence for the standard recycling scenario is now complete with several known cases of, for example, LMXBs; accreting X-ray MSPs (AXMSPs, Wijnands & van der Klis, 1998, see Chapter 7), of which a few dozen are now known (Di Salvo & Sanna, 2020; Patruno & Watts, 2021); transitional MSPs (switching between being X-ray and radio emitters, e.g., Archibald et al., 2009; Papitto et al., 2020); black widow or eclipsing MSPs (Fruchter et al., 1988); MSPs with planets (e.g., Wolszczan & Frail, 1992; Bailes et al., 2011; Spiewak et al., 2018); and, the final evolutionary endpoint, isolated MSPs (e.g., Backer et al., 1982).

In Table 14.1, we list examples of observed binary pulsars and MSPs from the various sub-classes with different characteristics. In the rest of the chapter, their formation and evolution will be further outlined in much detail. Notice that in systems where the second-formed compact object is a NS, some significant eccentricity ($e \gg 0.01$) remains from the second SN explosion. Thus double NS (DNS) systems, and systems where the NS formed *after* the WD (Tauris & Savonije, 2001), have much larger eccentricities than the usual MSP systems where the WD companion formed after a long LMXB phase of stable RLO that circularized the orbit.

14.2.1 The Observed Population of Binary MSPs

There are now more than 400 MSPs known in the Milky Way (according to the ATNF Pulsar Catalogue in September 2022, Manchester et al., 2005). In total, there are 485 fully recycled MSPs with spin periods less than 10 ms, of which 311 are located in the Galactic disk and 174 are located in a globular cluster (GC). The number of these MPSs found in binaries is 257 (173 in the disk and 84 in a GC). If we consider more broadly the number of *recycled* pulsars (i.e., here identified as those with spin periods less than 100 ms *and* a measured spin period derivative, $\dot{P} < 10^{-16}$), the total number is 381, of which 283 are located in the disk and 98 are located in a GC. The sub-population of these sources that are found in binaries is 271: 216 in the disk and 55 in a GC. It should be noticed that, for GC sources, the measured value of $\dot{P}$ may be severely affected by acceleration of the pulsar in the gravitational potential of the cluster (Phinney, 1993).

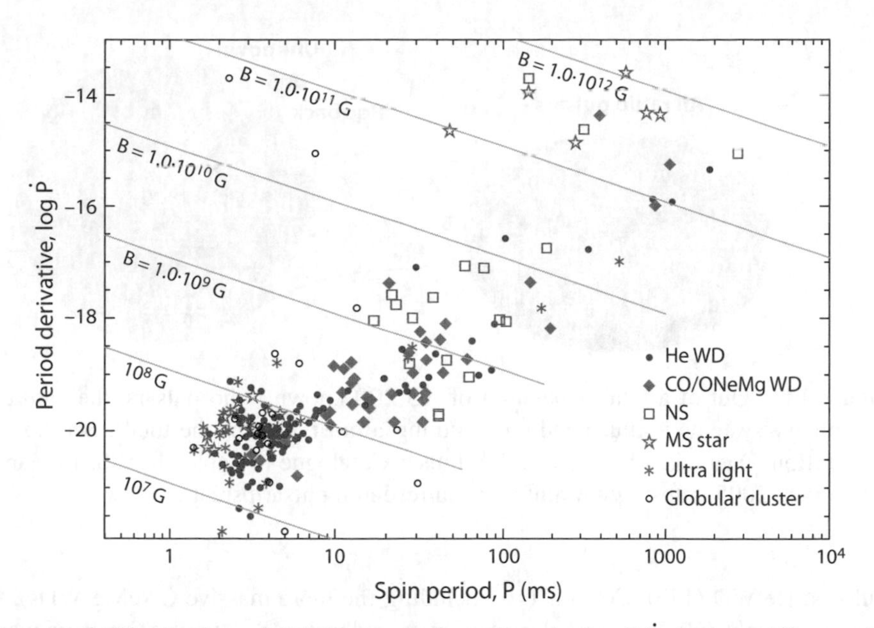

Figure 14.6. Distribution of 245 binary radio pulsars in the $P\dot{P}$-diagram. The nature of the companion stars is indicated with different symbols. Lines of constant surface dipole B-field flux density are shown. Data taken from the ATNF Pulsar Catalogue in June 2021 (Manchester et al., 2005, https://www.atnf.csiro.au/research/pulsar/psrcat).

Figure 14.6 shows the distribution of the known binary pulsars in the $P\dot{P}$–diagram, i.e., those with measured values of P as well as $\dot{P}$. Notice that this is a zoom-in of the typical $P\dot{P}$–diagram plotted in Figure 1.2. The fully recycled MSPs ($P < 10$ ms) are dominated by having mainly helium white dwarf (He WD) companions, although also ultra-light (sub-stellar) semi-degenerate companions are seen—often in eclipsing "black widow" or "redback"-like systems (see below)—as well as a few systems with the more massive carbon-oxygen (CO) WD companions. The mildly recycled MSPs ($10\,\text{ms} < P < 100\,\text{ms}$) are dominated by CO WD (or ONeMg WD) and NS companions. As we shall see, the relatively slow spin rate of the pulsars in these systems is expected from an evolutionary point of view as a consequence of the rapid mass-transfer phase from a relatively massive donor star. The few pulsars with similarly slow spin periods and He WD companions apparently also had a limited recycling phase, which may provide a hint as to their origin. The DNS systems descend from HMXBs and are discussed in Section 14.9.

Figure 14.7 illustrates the nature of 333 pulsar companions in known binary systems. These binary radio pulsars constitute almost 11% of the currently known population of $\sim 3{,}200$ radio pulsars. The sub-populations of companion types[4] are as

[4]See definitions of *BinComp* in Tauris et al. (2012) and in the ATNF Pulsar Catalogue https://www.atnf.csiro.au/research/pulsar/psrcat. The attentive reader will notice that these numbers add up to 335, as a few sources are categorized as one of two types. As of September 2022, 357 out of 3341 pulsars are in binaries.

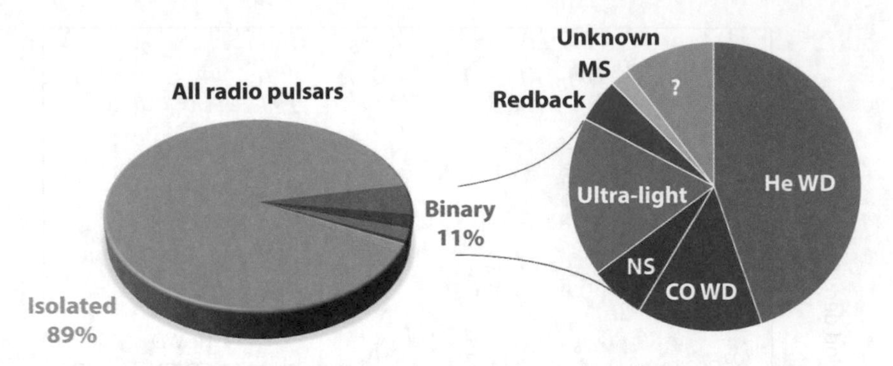

Figure 14.7. Out of a total population of ~3,200 known radio pulsars, the nature of 333 binary systems is illustrated (constituting almost 11% of the total). See text for description. Data taken from the ATNF Pulsar Catalogue in September 2021 (Manchester et al., 2005, https://www.atnf.csiro.au/research/pulsar/psrcat).

follows: He WD (149), CO WD (46, including the more massive ONeMg WDs), NS (20), ultra-light (60, typically black widows), redback (18, low-mass main-sequence stars which experienced mass loss), and MS (7, unevolved main-sequence companions), and 35 companion types are still unknown. The nature of these various companions is discussed in more detail throughout this chapter.

Figure 14.8 shows the distribution of orbital periods vs. spin periods of the known population of binary pulsars. From this Corbet diagram of radio pulsars, it is clear the MSPs can form fully recycled for $P_{\rm orb} \lesssim 200$ days. MSPs are not only characterized by a rapid spin. All of them also possess a low surface magnetic flux density, B which is typically of the order of 10^8 G, or some 3 to 5 orders of magnitude less than the B-fields of ordinary, non-recycled pulsars (see discussion in Section 14.6.4). Hence, the recycled pulsars do not suffer as much from loss of rotational energy due to emission of magnetodipole waves. For this reason, the MSPs have small period derivatives $\dot{P} < 10^{-18}$, and hence they are able to maintain a rapid spin—and thus a sufficiently large electrostatic potential across the polar caps (Section 14.1.2)—enabling the emission of radio waves, keeping them observable for billions of years.

When discussing the origin of MSPs with various companion stars, it makes sense to plot the binary orbital period as a function of the companion star mass—see Figure 14.9. This figure is essential for understanding the progenitor systems that we will now review in more detail.

14.2.2 Basics from Mass Transfer in X-ray Binaries and the Dependence on the Nature of the Donor Star

To produce a recycled radio pulsar, consider a close interacting binary system that consists of a non-degenerate donor star and a compact object, in our case a NS. If the orbital separation is small enough, the (evolved) non-degenerate star fills its inner common equipotential surface (Roche lobe, Chapter 4) and becomes a donor star for a subsequent epoch of mass transfer toward the accreting NS. In this phase, the system

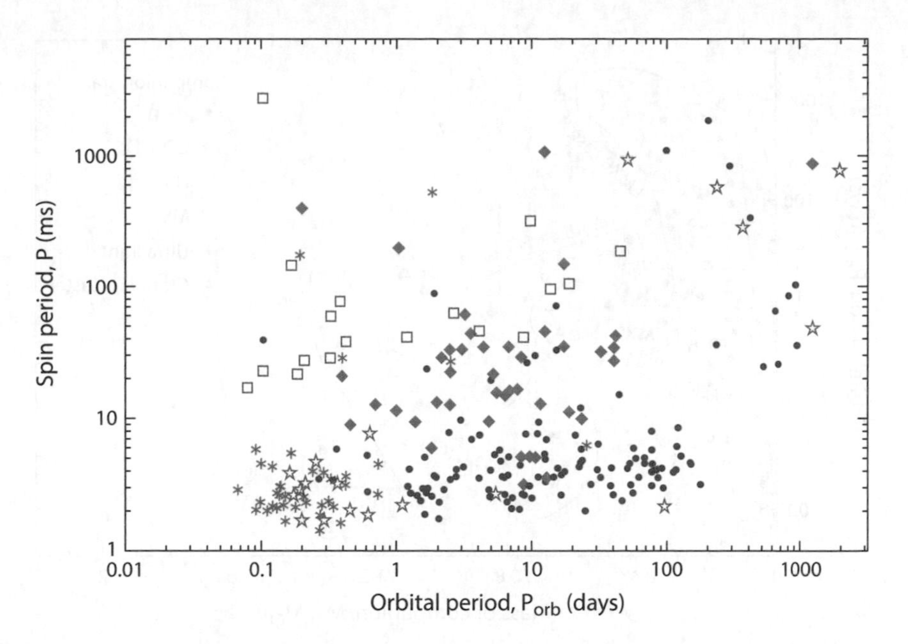

Figure 14.8. Distribution of 238 Galactic disk binary radio pulsars in the Corbet diagram (see Fig. 14.6 for explanation of the various symbols). MSPs with He WD companions can be fully recycled up to an orbital period of $P_{\rm orb} \simeq 200$ days. Data taken from the ATNF Pulsar Catalogue in June 2021 (Manchester et al., 2005, https://www.atnf.csiro.au/research/pulsar/psrcat).

is observed as an X-ray binary (Chapter 6). When the donor star fills its Roche lobe, it is perturbed by removal of mass and falls out of hydrostatic and thermal equilibrium. In the process of re-establishing equilibrium, the star will either grow or shrink—depending on the properties of its envelope layers—first on a dynamical (adiabatic) timescale and subsequently on a slower thermal timescale (Chapter 9). However, any exchange and loss of mass in such an X-ray binary system will also lead to alterations of the orbital dynamics via modifications in the orbital angular momentum (Chapter 4) and hence changes in the size of the critical Roche-lobe radius of the donor star. The stability of the mass-transfer process therefore depends on how these two radii evolve, i.e., the radius of the star and the Roche-lobe radius, see Chapters 4 and 9 for details.

As discussed in other chapters of this book, the various possible modes of mass exchange and loss include, for example, direct fast wind mass loss, Roche-lobe overflow (with or without isotropic re-emission), and CE evolution (e.g., van den Heuvel, 1994a; Soberman et al., 1997; Tauris & van den Heuvel, 2006, and references therein). The RLO mass transfer can be initiated while the donor star is still on the main sequence (Case A RLO), during hydrogen shell burning (Case B RLO), or during helium shell burning (Case C RLO); see Figure 8.5. The corresponding evolutionary timescales for these different cases will in general proceed on a nuclear, thermal, or dynamical timescale, respectively, or a combination thereof. This timescale is important for the extent to which the accreting NS can be recycled (i.e., with respect to its final spin period and B-field strength).

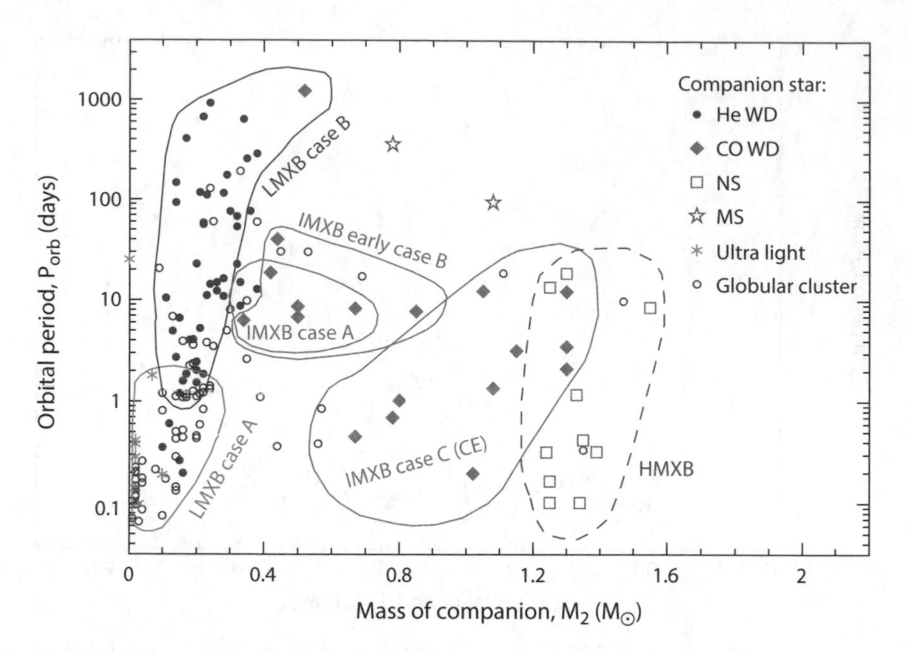

Figure 14.9. Five roads to form a MSP are shown: two roads from LMXBs (Cases A and B) and three roads from IMXBs (Cases A, B, and C). An additional possible formation path is via AIC. HMXBs leave behind the DNS systems and only partially recycle the first-born NS. The plotted boundaries are not very strict, and most of the companion masses have large error bars (not shown) due to the unknown orbital inclination angles. After Tauris (2011).

When modelling the evolution of an X-ray binary for the production of recycled pulsars, one should therefore take into account a number of issues related to the stellar evolution (Chapter 9), the stability of the mass-transfer process (Chapter 4), the ejection of matter from the system, and the accretion of material onto the NS. Ideally, all these calculations should be performed *self-consistently*. The major uncertainties here are related to the amount of, and the mode of, the specific orbital angular momentum of ejected matter—and for close systems, also the treatment of tides, spin-orbit couplings, and irradiation effects (Fruchter et al., 1988; Podsiadlowski, 1991; Tavani & Brookshaw, 1992; Benvenuto et al., 2014). In the following sections, we discuss the various sub-populations of pulsars and highlight their formation paths in an evolutionary context of X-ray binaries.

14.3 MSPs WITH He WD OR SUB-STELLAR DWARF COMPANIONS—EVOLUTION FROM LMXBs

To the left in Figure 14.9, one sees the MSPs with either ultra-light companions or He WD companions. Having low-mass companions, these systems thus descend from LMXBs. The exceptions are a few apparent carbon-rich remnants—such as the "diamond planet-pulsar" (Bailes et al., 2011)—that may be produced via mass-transfer in ultra-compact X-ray binaries (UCXBs, e.g., van Haaften et al., 2012, see also Section 11.4) or possibly by evaporation of former CO WD companions via particle and photon irradiation from energetic pulsars. MSPs with $P_{\rm orb} \leq 1$ day generally originate from LMXBs in very tight orbits where the donor star already initiated RLO while it was still on the main sequence (thus Case A RLO), whereas the MSPs with $P_{\rm orb} > 1$ day originate from wider orbit LMXBs where the donor star did not fill its Roche lobe until it had evolved and expanded to become a (sub)giant, i.e., Case B RLO. As previously discussed in Chapter 11, a critical orbital bifurcation period ($P_{\rm bif} \sim 1$ day) exists at the onset of the RLO, separating the formation of *converging* LMXBs from *diverging* LMXBs (Pylyser & Savonije, 1988, 1989; van der Sluys et al., 2005; Ma & Li, 2009; Istrate et al., 2014; Van et al., 2019) (see Chapter 11). The converging systems evolve with decreasing $P_{\rm orb}$, roughly until the mass-losing component becomes degenerate and an ultra-compact binary is formed, whereas the diverging systems expand their orbits until the mass-losing star has lost its hydrogen-rich envelope and a wide-orbit binary with a detached WD is formed. The final fate of LMXBs is thus either wide-orbit MSPs with He WD companions (or CO companions in the very widest orbits where pulsars are only partly recycled) or tight-orbit MSPs with He WDs (which may later evolve into UCXBs), ultra-light companions, planets, or solitary MSPs. The standard formation path leading to production of binary MSPs is depicted in Figure 14.10 with an example of calculated values of orbital period and age of a given system (see also Chapter 11). In the following sections, we discuss the formation of the various sub-classes of binary (an isolated) MSPs in more detail.

14.3.1 Close-Orbit LMXB Systems—Formation of Isolated, Planetary, and Black-Widow Pulsars

In observed MSP systems with $P_{\rm orb} \lesssim 1$ day, the mass transfer was driven by loss of orbital angular momentum due to magnetic braking and (for more narrow systems) emission of GWs. (The evolution of such systems is very similar to that of CVs, see, e.g., Spruit & Ritter, 1983; Knigge et al., 2011). In some LMXB systems, the donor stars never detach to form a He WD. Instead, the hydrogen-rich donors are continuously being stripped off while becoming semi-degenerate objects with a brown dwarf–like structure (Deloye & Bildsten, 2003). These are the so-called *converging* systems (Pylyser & Savonije, 1988; Istrate et al., 2014, see Chapter 11). In these binaries, the donor stars still possess a significant hydrogen content—even in their cores—and due to their very small nuclear burning rates, they still have a mixture of hydrogen and helium when they finally become degenerate near the orbital period minimum of $\sim 10-85$ min (Paczynski & Sienkiewicz, 1981; Rappaport et al., 1982; Podsiadlowski et al., 2002; van der Sluys et al., 2005). After this point, their orbits widen substantially due to the continuous mass transfer and ever decreasing mass ratio between the donor star and the accreting NS. On the widening branch, these systems are now observable as *black widow* binary radio MSPs (Fruchter et al., 1988; Stappers et al., 1996). In such systems, the radio pulsars display eclipses during their orbital motion, and in the case of PSR J2051−0827 one can even measure the effects of gravitational quadrupole

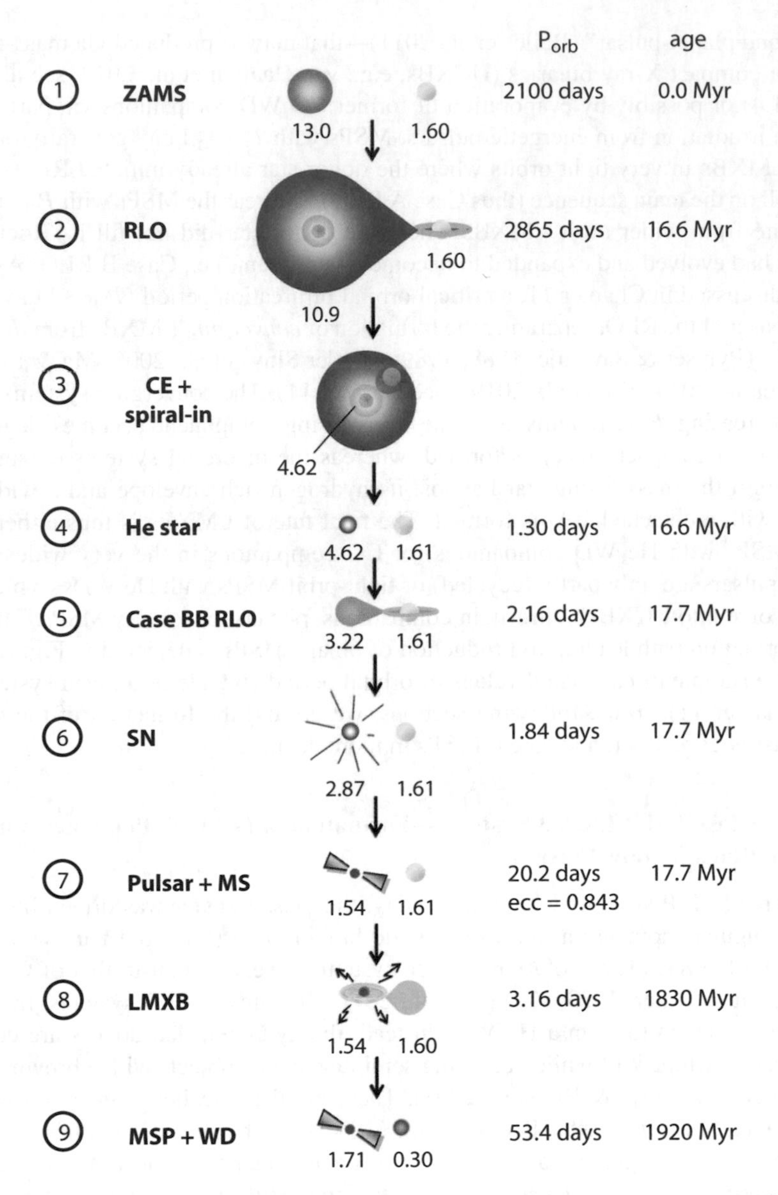

Figure 14.10. Standard formation path leading to a LMXB and a binary MSP. (See also Figs. 11.1 and 11.2.) Typical ZAMS masses of the two stars are $10-15\ M_\odot$ and $1.0-2.0\ M_\odot$.

moment changes of the ultra-light companion star (Lazaridis et al., 2011; Shaifullah et al., 2016).

In LMXB systems that detach to form a MSP with a He WD companion in a tight orbit (less than ~9 hours), an additional mass-transfer phase is initiated within a Hubble time (leading to UCXBs, see Section 11.4), driven by emission of GWs, which brings the two stellar components closer together until the WD fills its Roche lobe (for detailed calculations, see e.g., Sengar et al., 2017; Tauris, 2018; Chen et al., 2021). These MSPs in UCXB systems are often observed as X-ray transients with orbital periods less than one hour and an ultra-light companion star of $M_2 \simeq 10^{-2}\, M_\odot$ (Deloye & Bildsten, 2003; Heinke et al., 2013).

In both of the two previous cases, the final product is expected to be either a MSP with a planet of a few Jupiter masses (possibly disrupted and leaving behind more planets of much lower mass, if it undergoes a tidal instability, Ruderman & Shaham, 1985) or an isolated MSP, in case irradiation and evaporation processes are efficient in destroying the companion star (van den Heuvel & van Paradijs, 1988; Ruderman et al., 1989; Podsiadlowski, 1991; Ergma & Fedorova, 1991; Tavani & Brookshaw, 1992; Martin et al., 2016; Ginzburg & Quataert, 2020). These irradiation effects are caused first by the intense X-ray flux from the accreting NS (Podsiadlowski, 1991; Benvenuto et al., 2012, 2014) and later by the pulsar wind of relativistic particles and hard photons (Tavani & Brookshaw, 1992). Tidal dissipation of energy in the envelope (Applegate & Shaham, 1994) may cause the companion star to be thermally bloated and evaporate more easily. Observational evidence for this scenario is found in MSPs with planets (e.g., PSR B1257+12, Wolszczan & Frail, 1992), the occurrence of many solitary MSPs (the first MSP ever discovered is the isolated PSR B1937+21 spinning at 1.56 ms, Backer et al., 1982), and the presence of ultracool He WD companions (Tang & Li, 2021). Finally, it is possible that the accreting NS in some cases undergoes accretion-induced collapse (AIC) and leaves behind a BH remnant (Fig. 11.2).

Black Widows vs. Redbacks

In the last decade, a rapidly increasing number of eclipsing MSPs (the so-called *spiders*) have been found in the Galactic field and have tight orbits with $P_{\rm orb} \lesssim 24\,{\rm hr}$, see, for example, Roberts (2013); Pallanca et al. (2012); Romani et al. (2012); Kaplan et al. (2013); Breton et al. (2013); Bellm et al. (2016); Cromartie et al. (2016). A large fraction of these new MSPs are associated with Fermi gamma-ray sources (e.g., Abdo et al., 2009; Pletsch et al., 2012). As Roberts (2013) suggested, the population of eclipsing binary MSPs can be divided into two different classes: *black widows* and *redbacks*. The black widows have very low-mass companions ($M_2 \ll 0.1\, M_\odot$), while the redbacks usually have companion masses of a few tenths of a solar mass ($M_2 \simeq 0.1 - 0.4\, M_\odot$). A number of detailed optical studies of their irradiated companion stars (e.g., Breton et al., 2013; Draghis et al., 2019; Linares et al., 2018) reveal important information on the pulsar wind, gamma-ray beaming, heating models, Roche-lobe filling factors, and companion star mass densities. Detailed stellar evolution modelling has demonstrated that black widows form as the outcome of converging LMXB evolution (e.g., Ergma & Fedorova, 1992; Podsiadlowski et al., 2002; Benvenuto et al., 2012), although an origin as a population of binaries ejected from GCs has also been suggested by King et al. (2003). In view

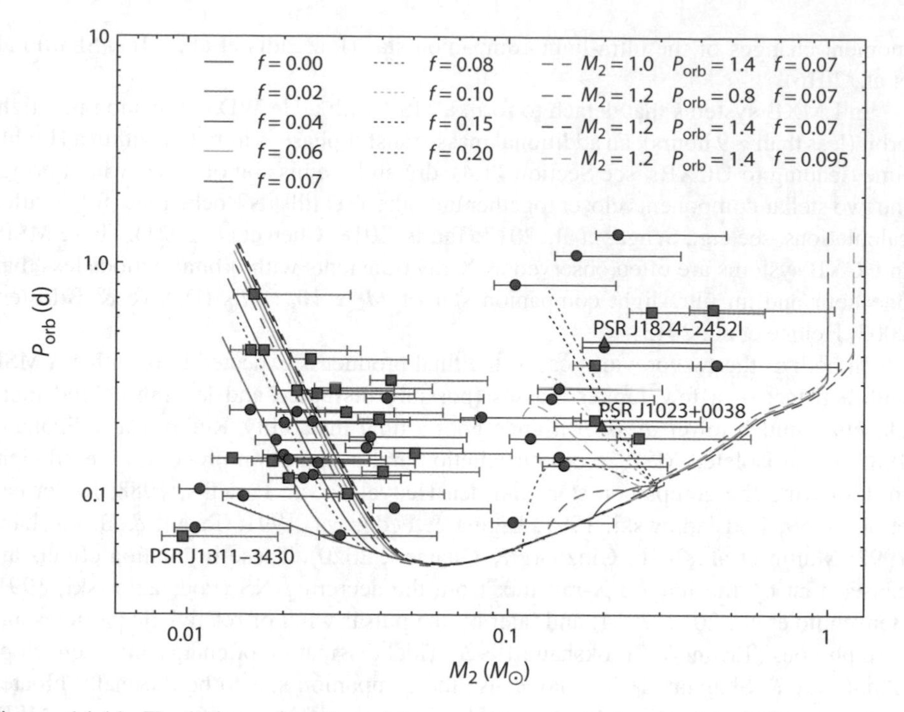

Figure 14.11. Evolution of various binary systems calculated with different values of the evaporation efficiency, f. The observed data shows the black widows ($M_2 \ll 0.1\,M_\odot$) and the redbacks ($M_2 \simeq 0.1-0.4\,M_\odot$). Galactic field sources are marked by green squares, and GC sources are red circles. For the error bars, the left and right ends correspond to an orbital inclination angle of 90° and 25.8° (the 90% probability limit), respectively, for $M_{\rm NS} = 1.4\,M_\odot$. The calculated LMXB tracks shown follow evolution from right to left; initially $M_2 = 1.0-1.2\,M_\odot$. The two branches with increasing orbital periods are due to efficient donor evaporation ($f \gtrsim 0.08$, redback tracks and $M_2 \simeq 0.1-0.4\,M_\odot$) and less efficient donor evaporation ($f \lesssim 0.07$, black-widow tracks and $M_2 \ll 0.1\,M_\odot$). After Chen et al. (2013).

of the many recent discoveries, it seems that the latter scenario may have difficulties in explaining the large number of such binary systems in the Galactic field.

Numerical calculations by Chen et al. (2013) suggest that redbacks and black widows and are not linked by an evolutionary path—they represent two distinct populations (i.e., redbacks do not evolve into black widows). They argue that these two populations represent two distinct regimes of companion-star evaporation upon (partial) detachment from the Roche lobe. The most efficient evaporation ($f \gtrsim 0.08$) leads to redback systems, with an early orbital widening as a result, whereas more inefficient evaporation ($f \lesssim 0.07$) produces binaries that are trapped by GW radiation and evolve down to the orbital period minimum before the systems evolve up the black widow branch, as shown in Figure 14.11. In the model of Chen et al. (2013), the differences in evaporation efficiency are simply thought to be due to the geometry of the system, i.e., to which extent the companion star is beamed by the pulsar outflow of relativistic particles and hard photons.

In recent works, Ginzburg & Quataert (2020, 2021) propose that the evaporative wind couples to the companion's magnetic field, removes angular momentum from the binary, and maintains stable RLO (i.e., an indirect magnetic braking mechanism). Furthermore, in their model the magnetic braking is sustained even for low-mass companions via a B-field generated by the convective luminosity (Christensen et al., 2009). They argue that strongly irradiated donor stars may retain radiative cores down to black widow masses, such that magnetic braking likely persists for the entire evolution of such systems. As a result, their evolutionary tracks link redbacks to black widows, which is in contrast to the model proposed by Chen et al. (2013) where the two spider populations evolve primarily on separate tracks. Future observations may settle this question.

Finally, complicating matters even more, Novarino et al. (2021) argue that the obtained period pattern in the redback system PSR J1723–2837 reveals the presence of a more complex phenomenology than described in the frame of the weak friction model of equilibrium tides.

Radio Ejection and Transitional Millisecond Pulsars

As the mass-transfer rate continues to decrease, the magnetospheric boundary, $r_{\rm mag}$, expands and eventually crosses the light-cylinder radius, $r_{\rm lc}$, of the NS, given by Eq. (7.7). When $r_{\rm mag} > r_{\rm lc}$ the plasma wind of the pulsar can stream out along the open field lines, providing the necessary condition for the radio emission mechanism to turn on (e.g., Michel, 1991; Spitkovsky, 2008). Once the radio MSP is activated, the presence of the plasma wind may prevent further accretion from the now weak flow of material at low $\dot{M}$ (Kluzniak et al., 1988). As discussed by Burderi et al. (2001, 2002), this will be the case when the total spin-down pressure of the pulsar (from magnetodipole radiation and the plasma wind) exceeds the inward pressure of the material from the donor star, i.e., if $\dot{E}_{\rm rot}/(4\pi r^2\, c) > P_{\rm disk}$.

However, the radio ejection phase is not necessarily the end of the accretion history. It is interesting to notice the existence of the so-called transitional MPSs, which manage to switch back and forth between X-ray emission and radio emission. The reason for this transition from an accreting X-ray source to a radio emitting and rotation-powered pulsar is somewhat unclear. The discovery of a number of *transitional MSPs* (tMSPs Archibald et al., 2009; Papitto et al., 2013; Bassa et al., 2014; Roy et al., 2015; Papitto & de Martino, 2022) indicates that the development from an X-ray source to a radio source is a gradual process where the source for a period of time is able to jump between the two emission stages as the magnetospheric boundary ($r_{\rm mag}$) is either pushed outside or inside the light-cylinder radius ($r_{\rm lc}$) of the pulsar. Interestingly, all known tMSPs are redback systems. The Roche-lobe decoupling phase is discussed in Section 14.6.8.

Low-Mass Proto-White Dwarfs

In those binaries where the companion leaves behind a low-mass He WD ($M_{\rm WD} < 0.20\,M_\odot$), pycnonuclear burning at the bottom of their thick residual hydrogen envelopes can keep the low-mass He WDs warm for $\sim 10^9$ yr (see discussions in Alberts

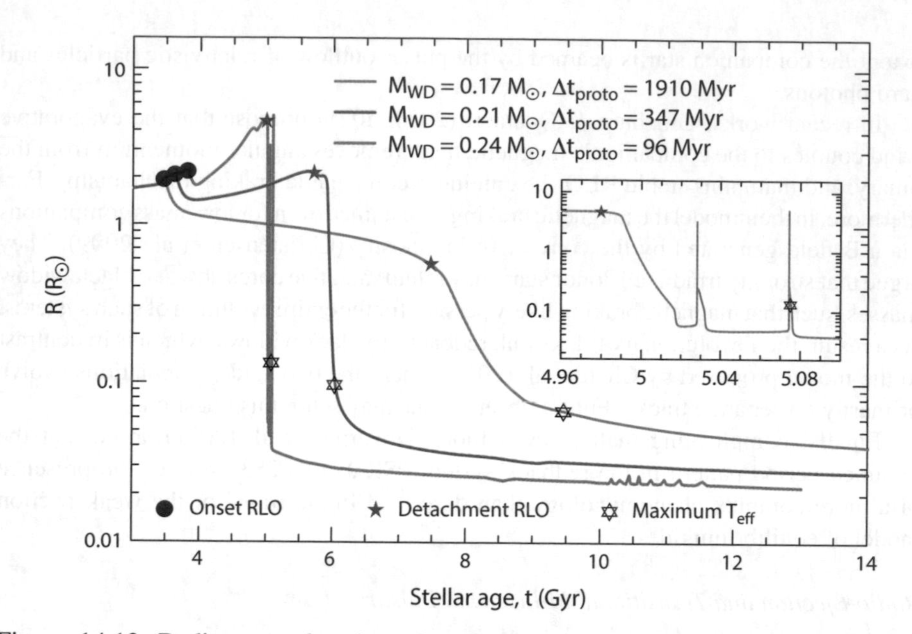

Figure 14.12. Radius as a function of stellar age for the progenitor stars of three He WDs of mass 0.17, 0.21, and 0.24 $M_\odot$, evolved in LMXB systems. The most massive proto-WD evolves with hydrogen shell flashes, see inset. The epoch between the solid red star (RLO termination) and the open black star (max. $T_{\rm eff}$) marks the contraction timescale, $\Delta t_{\rm proto} = 96-1{,}910$ Myr, strongly increasing with decreasing $M_{\rm WD}$. After Istrate et al. (2014).

et al., 1996; Driebe et al., 1998; Althaus et al., 2001; van Kerkwijk et al., 2005). Since it has been known for a long time (and clearly demonstrated in, e.g., Tauris, 2012; Tauris et al., 2012) that the characteristic ages of recycled pulsars are a bad indicator of their true ages (See Section 14.6.7), a precise cooling age determination of such WDs, and thus the true age of the MSPs since their recycling, is of utmost importance.

Numerical calculations by Istrate et al. (2014, 2016) found that the time interval between Roche-lobe detachment until the low-mass proto-He WD reaches the WD cooling track is typically $\Delta t_{\rm proto} = 0.5 - 2.5$ Gyr, depending systematically on the WD mass (Fig. 14.12) and therefore on its luminosity. Istrate et al. (2014) derived the expression:

$$\Delta t_{\rm proto} \simeq 400\,{\rm Myr}\left(\frac{0.20\,M_\odot}{M_{\rm WD}}\right)^7. \qquad (14.8)$$

Hence, the duration of this radial contraction phase decreases strongly with increasing mass of the proto-He WD. This can be understood from the well-known correlation between degenerate core mass and luminosity of an evolved low-mass star. Therefore, after Roche-lobe detachment, the rate at which the residual ($0.01 \pm 0.005\ M_\odot$) hydrogen in the envelope is consumed is directly proportional to the luminosity, and thus $\sim M_{\rm WD}^7$ as a consequence of a steep core mass-luminosity relation.

Table 14.2. Stellar parameters for giant stars with $R = 50\,R_\odot$. The core mass ($0.34 \pm 0.01\,M_\odot$) is seen to be independent of the envelope mass (here $0.2-1.2\,M_\odot$). (After Tauris & Savonije 1999.)

$M_{\rm initial}/M_\odot$	1.0*	1.6*	1.0**	1.6**
$\log L/L_\odot$	2.644	2.723	2.566	2.624
$\log T_{\rm eff}$	3.573	3.593	3.554	3.569
$M_{\rm core}/M_\odot$	0.342	0.354	0.336	0.345
$M_{\rm env}/M_\odot$	0.615	1.217	0.215	0.514

*Single star ($X = 0.70$, $Z = 0.02$, $\alpha = 2.0$ and $\delta_{\rm ov} = 0.10$).

**Binary star (at onset of RLO: $P_{\rm orb} \simeq 60$ days and $M_{\rm NS} = 1.3\,M_\odot$).

In a more detailed study, Istrate et al. (2016) investigated the combined effects of rotational mixing and element diffusion (e.g., gravitational settling, thermal and chemical diffusion) on the evolution of proto-WDs and on the cooling properties of the resulting WDs. They investigated a large number of models of different metallicities and confirmed that element diffusion plays a significant role in the evolution of proto-WDs that experience hydrogen shell flashes (e.g., Althaus et al., 2013).

The long timescale of low-mass proto-He WD evolution can explain a number of observations, including some MSP systems hosting He WD companions with very low surface gravity and high effective temperature, such as, for example, PSR J1816+4510 (Kaplan et al., 2012, 2013) which has $P_{\rm orb} = 8.7$ hr, $T_{\rm eff} = 16\,000 \pm 500$ K, a surface gravity of $\log g = 4.9 \pm 0.3$, and a companion mass of $M_{\rm WD} \simeq 0.19\,M_\odot$. A renewed interest in such low-mass proto-WDs is sparked by the observations of asteroseismic pulsations in some of these WDs (Kilic et al., 2015; Gianninas et al., 2016; Istrate et al., 2016; Calcaferro et al., 2017).

14.3.2 Wide-Orbit LMXB Systems—Formation of Classic MSPs with He WD Companions

In LMXBs with initial $P_{\rm orb} \gtrsim 2-3$ days, the mass transfer is driven by internal nuclear evolution of the donor star since it evolves into a (sub)giant before onset of the RLO. As mentioned in Section 11.5, these systems have been studied by, for example, Webbink et al. (1983); Savonije (1987); Joss et al. (1987); Rappaport et al. (1995); Tauris & Savonije (1999); Podsiadlowski et al. (2002); Lin et al. (2011). For a donor star on the red giant branch (RGB), the growth in core mass is directly related to the luminosity, as this luminosity is entirely generated by hydrogen shell burning. As such a star, with a small compact core surrounded by an extended convective envelope, is forced to move up the Hayashi track its luminosity increases strongly with only a fairly modest decrease in surface temperature. Hence, one also finds a relationship between the giant's radius and the mass of its degenerate helium core (Refsdal & Weigert, 1971)—almost entirely independent of the mass present in the hydrogen-rich envelope (see, e.g., Section 11.5 and Table 14.2). This relation is valid for all stars with initial

ZAMS masses $M_2 < 2.3\,M_\odot$ since these low-mass stars produce degenerate helium cores on the RGB (see Section 8.1). Their resultant WDs have masses of $0.15 \lesssim M_{\rm WD}/M_\odot \lesssim 0.46$.

It has also been shown that the core mass determines the rate of mass transfer (Webbink et al., 1983). This can be understood as a consequence of more evolved stars having deeper convective envelopes, thereby a negative mass-radius exponent, and thus high mass-transfer rates. In the scenario under consideration, the extended envelope of the giant is expected to fill its Roche lobe until termination of the mass transfer. Since the Roche-lobe radius, R_L, only depends on the masses and separation between the two stars, it is clear that the core mass, from the moment the star begins RLO, is uniquely correlated with $P_{\rm orb}$ of the system. Thus, also the final orbital period ($\sim$2 to 10^3 days) is expected to be a function of the mass of the resulting He WD companion (Savonije, 1987; see also Section 11.5). The expected ($P_{\rm orb}$, $M_{\rm WD}$) correlation was calculated in detail by Tauris & Savonije (1999) who derived an overall best fit:

$$M_{\rm WD} = \left(\frac{P_{\rm orb}}{b}\right)^{1/a} + c, \tag{14.9}$$

where, depending on the chemical composition of the donor,

$$(a,b,c) = \begin{cases} 4.50 & 1.2\times10^5 & 0.120 & \text{for Pop.I stars } (Z=0.02) \\ 4.75 & 1.1\times10^5 & 0.115 & \text{for Pop.I+II stars} \\ 5.00 & 1.0\times10^5 & 0.110 & \text{for Pop.II stars } (Z=0.001). \end{cases} \tag{14.10}$$

Here $M_{\rm WD}$ is in solar mass units and $P_{\rm orb}$ is measured in days. The fit is valid for binary MSPs with He WD companions and $0.18 \lesssim M_{\rm WD}/M_\odot < 0.46$. For final values of $P_{\rm orb}$ less than 2 days (leading to WD masses of $\sim 0.16 - 0.20\,M_\odot$), more recent calculations reveal a marginally steeper slope (Istrate et al., 2014, 2016). Note, if the system was initially so wide that the donor ascended the asymptotic giant branch (AGB) before initiating RLO, a CO WD companion is left behind (as observed in the binary pulsar B0820+02 which has an orbital period of 1,232 days). Actually, Eq. (14.9) is applicable to *all* binaries that produce a He WD remnant via stable RLO from a low-mass ($< 2.3\,M_\odot$) donor star, including post-Algol systems (Brogaard et al., 2018). The formula depends slightly on the adopted value of the convective mixing-length parameter. We emphasize that the correlation is *independent* of β (the fraction of the transferred material lost from the system; see Chapter 4), the mode of the mass loss, and the strength of the magnetic braking torque since the relation between giant radius and core mass of the donor star remains unaffected by the exterior stellar conditions governing the process of mass transfer. However, for the *individual* binary, $P_{\rm orb}$ and $M_{\rm WD}$ do depend on these parameters. In Figure 14.13 is shown a calculation of a theoretical ($P_{\rm orb}$, $M_{\rm WD}$) correlation and also the evolutionary tracks of four LMXBs. At the end of the calculations, the remaining mass of the hydrogen envelope ($< 0.01\,M_\odot$) is the difference between the donor's total mass ($M_{\rm WD}$) and its core mass.

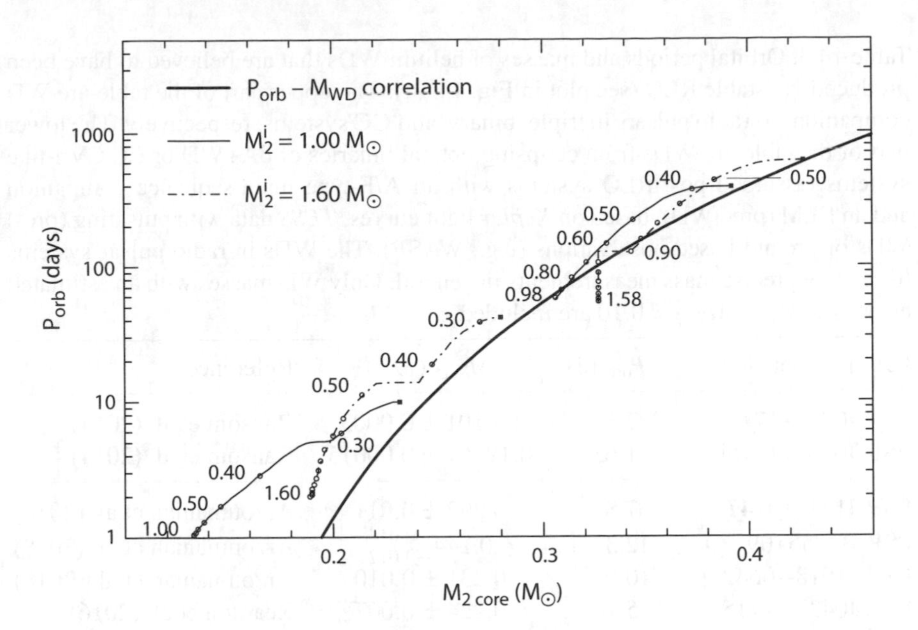

Figure 14.13. Evolutionary tracks of four LMXBs showing P_{orb} as a function of M_{core} of the donor star. The initial donor masses were 1.0 and 1.6 $M_\odot$ (each calculated at two different initial P_{orb}), and the initial NS mass was 1.3 $M_\odot$. The total mass of the donors during the evolution is written along the tracks. At the termination of the mass-transfer process, the donor only has a tiny ($\leq 0.01\ M_\odot$) hydrogen envelope, and the end-points of the evolutionary tracks are located near the (P_{orb}, M_{WD}) correlation for binary MSPs. After Tauris & Savonije (1999).

The correlation between P_{orb} and M_{WD} was previously difficult to verify observationally (van Kerkwijk et al., 2005) since very few MSPs had accurately measured masses of their companion. From the observed mass function alone, one can only make a rough estimate of the WD mass, which depends on the orbital inclination angle of the system as well as the NS mass—both of which are often unknown. However, in recent years, the number of binary WD systems with accurately measured WD masses has increased significantly due to improved optical spectroscopy and pulsar timing, see Table 14.3. It was demonstrated by Tauris & van den Heuvel (2014), that the theoretically derived relation (Eqs. 14.9 and 14.10) is indeed quite good when comparing to data (see our updated version in Fig. 14.14), if one only considers those systems for which the WD mass has been estimated fairly accurately. The empirical verification of this correlation between helium WD orbital period and mass is a major success for binary stellar evolution theory. (When the correlation of Tauris & Savonije, 1999, was published, only one WD mass [PSR B1855+09] was known.) On the other hand, there might be indications of a slight systematic deviation from the correlation for pulsars with $P_{orb} > 100$ days (Tauris, 1996; Tauris & Savonije, 1999; Stairs et al., 2005), although this claim is still based on relatively small number statistics. Unfortunately, only one WD in such a system with $P_{orb} > 100$ d has an accurate mass measurement.

Table 14.3. Orbital periods and masses of helium WDs that are believed to have been produced via stable RLO (see plot in Fig. 14.14). The upper part of the table are WD companions to radio pulsars in triple, binary, and GC systems, respectively. The lower part of the table are WDs from eclipsing optical binaries of dA+WD or EL CVn-like systems (assumed post-RLO systems with an A/F-type main-sequence companion and an ELM (pre-)WD), based on *Kepler* light curves, *TESS* data with pulsating (pre-) WDs or ground-based observations (e.g., WASP). The WDs in radio pulsar systems have more precise mass measurements in general. Only WD masses with an estimated error of $\Delta M_{\rm WD}/M_{\rm WD} < 0.10$ are included.

Pulsar companion	$P_{\rm orb}$ (d)	$M_{\rm WD}$ ($M_\odot$)	Reference
PSR J0337+1715o	327	0.4101 ± 0.0003	Ransom et al. (2014)
PSR J0337+1715i	1.63	0.19751 ± 0.00015	Ransom et al. (2014)
PSR J1713+0747	67.8	0.292 ± 0.011	Arzoumanian et al. (2018)
PSR B1855+09	12.3	$0.244^{+0.014}_{-0.012}$	Arzoumanian et al. (2018)
PSR J1918−0642	10.9	0.231 ± 0.010	Arzoumanian et al. (2018)
PSR J0437−4715	5.74	0.224 ± 0.007	Reardon et al. (2016)
PSR J0740+6620	4.77	$0.253^{+0.006}_{-0.005}$	Fonseca et al. (2021)
PSR J1909−3744	1.53	0.208 ± 0.002	Arzoumanian et al. (2018)
PSR J2043+1711	1.48	0.173 ± 0.010	Arzoumanian et al. (2018)
PSR J1012+5307	0.605	0.165 ± 0.015	Mata Sánchez et al. (2020)
PSR J1738+0333	0.355	$0.181^{+0.007}_{-0.005}$	Antoniadis et al. (2012)
PSR J0751+1807	0.263	0.16 ± 0.01	Desvignes et al. (2016)
PSR J0348+0432	0.1025	0.172 ± 0.003	Antoniadis et al. (2013)
PSR J1910−5959A	0.837	0.180 ± 0.018	Corongiu et al. (2012)

Eclipsing optical WD	$P_{\rm orb}$ (d)	$M_{\rm WD}$ ($M_\odot$)	Reference
RRLYR–02792	15.2	0.261 ± 0.015	Pietrzyński et al. (2012)
KOI 74b	5.19	0.228 ± 0.014	Bloemen et al. (2012)
KIC 2851474	2.77	0.210 ± 0.018	Faigler et al. (2015)
KOI 1224b	2.70	0.22 ± 0.02	Breton et al. (2012)
KIC 10989032	2.31	0.24 ± 0.02	Zhang et al. (2017)
KIC 7368103	2.18	0.21 ± 0.02	Wang et al. (2019)
KIC 9285587	1.81	0.191 ± 0.019	Faigler et al. (2015)
KIC 8823397	1.51	0.21 ± 0.02	Wang et al. (2019)
KIC 9164561	1.27	0.197 ± 0.005	Rappaport et al. (2015)
KIC 4169521	1.17	0.210 ± 0.015	Faigler et al. (2015)
TIC 416264037	1.16	0.18 ± 0.01	Wang et al. (2020)
TIC 149160359	1.12	0.16 ± 0.01	Wang et al. (2020)
EL CVn	0.796	0.176 ± 0.004	Wang et al. (2020)
WASP 1628+10b	0.720	0.135 ± 0.008	Luo (2020)
WASP 0247−25b	0.668	0.186 ± 0.002	Maxted et al. (2013)

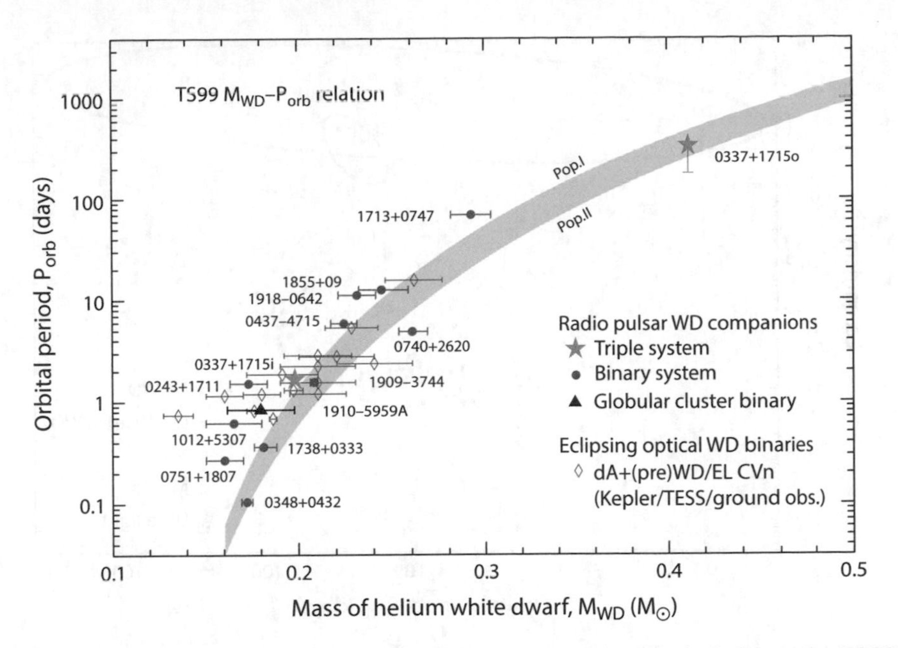

Figure 14.14. $M_{\rm WD} - P_{\rm orb}$ relation (TS99), as calculated by Tauris & Savonije (1999), along with observational data (plotted with 1σ error bars) for helium WD companions orbiting radio pulsars (blue circles), including one source in a GC (black triangle), and two WDs from a triple system (green stars [Ransom et al., 2014]). The vertical error bar plotted for the outer WD in the latter (PSR J0337+1715o) is caused by an uncertainty in the widening of the outer orbit during the inner LMXB-phase (Tauris & van den Heuvel, 2014). Red diamonds are from eclipsing optical binaries of dA+WD or EL CVn-like systems (post-RLO systems with an A/F-type main-sequence companion and an ELM (pre-)WD), based on *Kepler* light curves, *TESS* data with pulsating (pre-) WDs or ground-based observations (e.g., WASP). The width of the TS99 relation is caused by using metallicities from $Z = 0.001$ (lower boundary) to 0.02 (upper boundary). The relation becomes uncertain for $P_{\rm orb} < 1$ day (see e.g., Istrate et al., 2014). Data is taken from Table 14.3. We do not include SDSS and ELM survey sources in general in this plot as these WDs often originate from a CE phase.

Hopefully, future mass determinations of some of these wide binaries can help settle this question. As mentioned in Section 11.5 (see also Exercise 11.3), it is interesting to notice that the correlation between $M_{\rm WD}$ and $P_{\rm orb}$ is *independent* of the mass of the accreting star.

Finally, it should be mentioned that wide-orbit binary MSPs also have another fossil of the mass-transfer phase: the correlation between orbital eccentricity and orbital period. These residual eccentricities from the mass-transfer phase are, in general, very small: typically between $10^{-6} - 10^{-3}$. A ($P_{\rm orb}$, ecc) correlation was derived by Phinney (1992), see Figure 14.15, and it is related to tides and arises because density fluctuations

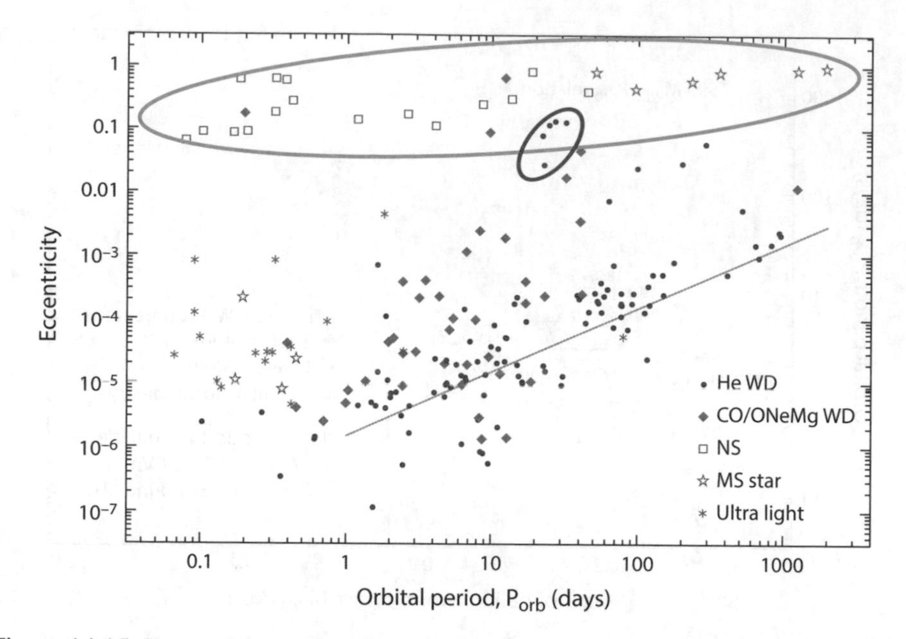

Figure 14.15. Eccentricity as function of orbital period for binary radio pulsars in the Galactic field. Except for the mysterious old eccentric MSPs within the purple ellipse (see text), systems located within the green boundary contain a young pulsar that has not experienced accretion since its formation. This explains their large eccentricity as a relic from the SN explosion. All other pulsars experienced mass accretion, which explains their more circularized orbits. The grey line shows an extrapolation of the (P_{orb}, ecc) correlation of Phinney (1992). Data taken from the ATNF Pulsar Catalogue in June 2021 (Manchester et al., 2005, https://www.atnf.csiro.au/research/pulsar/psrcat).

in the (convective) envelope increase with more evolved donor stars, which have wider orbits during the mass transfer, thus preventing perfect circularization.

14.3.3 Eccentric MSPs

Within the last decade, a new mysterious sub-class of binary MSPs has been discovered (see, e.g., Stovall et al., 2019, for characteristics of the first five sources). These are the so-called eccentric MSPs (eMSPs). Unlike the far majority of binary MSPs with He WD companions in circularized systems, found in a broad range of orbital periods, these eMSPs possess significant non-zero eccentricities between 0.027 and 0.14. Furthermore, they are only found in a limited orbital period range between 22 and 32 days (Fig. 14.15). The WD masses, however, are following the mass-period correlation discussed previously. Theoretical explanations to understand the origin of these unusual systems include a triple star origin, rotational-delayed accretion-induced collapse (RD-AIC) of a WD (Freire & Tauris, 2014), a phase transition inside the MSP resulting in the formation of a strange-star core (Jiang et al., 2015), and eccentricity pumping (see also Section 4.7) via interaction with a circumbinary disk (Antoniadis, 2014). Recently, Han & Li (2021) suggested a scenario with small kicks ($1-8\,\mathrm{km\,s^{-1}}$) imparted onto the proto-ELM WD companion due to hydrogen-shell flashes, whereas Ginzburg & Chiang (2022) proposed that the anomalously large eccentricities are driven by resonant convection. This latter idea is based on their finding that the orbital periods of the eMSPs are similar to the red-giant convective eddy turnover time of the WD progenitor stars.

A triple system origin of eMSPs (similar to what has been suggested for the peculiar binary PSR J1903+0327, Freire et al., 2011) seems unlikely given the very narrow interval of both eccentricities and orbital periods of the observed class of eMSPs. Such a fine-tuning would be difficult to explain if these binaries formed via ejection of a tertiary star. The RD-AIC of a WD, or the phase transition to a strange star, would most likely result in a low-velocity binary MSP. However, the local transverse velocity of the eMSP J1946+3417 is $200 \pm 60\,\mathrm{km\,s^{-1}}$ (Barr et al., 2017). Furthermore, the eMSP J1946+3417 is a rather massive NS ($1.83 \pm 0.02\,M_\odot$, Barr et al., 2017), which is unexpected for the RD-AIC scenario unless differential rotation was at work in the WD prior to its implosion (Freire & Tauris, 2014). The currently five known eMSPs seem to satisfy best the predictions of the Antoniadis (2014) circumbinary disk model (although also this model has been subject of some critique, see Rafikov, 2016) and the hypothesis of resonant convection by Ginzburg & Chiang (2022). However, it is still not understood why some MSPs with similar orbital periods and WD companion masses have very small eccentricities of $e \simeq 10^{-5}$.

14.4 MSPs WITH CO WD COMPANIONS—EVOLUTION FROM IMXBs

CO WDs (and ONeMg WDs) are substantially more massive than He WDs. Therefore, MSPs with these more massive WD companions must originate from binaries with more massive WD progenitor stars—that is, IMXBs with donor masses of $3-7\,M_\odot$. Observationally, these systems were identified by Camilo (1996) as an independent class. There seems to be three roads to produce such systems: RLO Cases A, B, and C, depending on the evolutionary status (or, equivalently, the orbital period) at the onset of the RLO. We now discuss each of these cases—see also Figure 14.16 and Table 14.4 for an overview of producing binary (and isolated) MSPs in LMXBs and IMXBs, and via AIC from accreting WDs in CVs and supersoft X-ray sources.

14.4.1 Wide-orbit IMXB Systems: Case C RLO and Common-envelope Evolution

It was demonstrated by van den Heuvel (1994b) that binary systems with a mildly recycled MSP and a CO WD companion could form from an IMXB with a donor star on the AGB, leading to common envelope (CE) and spiral-in evolution once the RLO sets in. This model is in particular the favored scenario for the formation of very tight binary MSPs with orbital periods $P_{\mathrm{orb}} < 3$ days. The reason for this is the ability of the

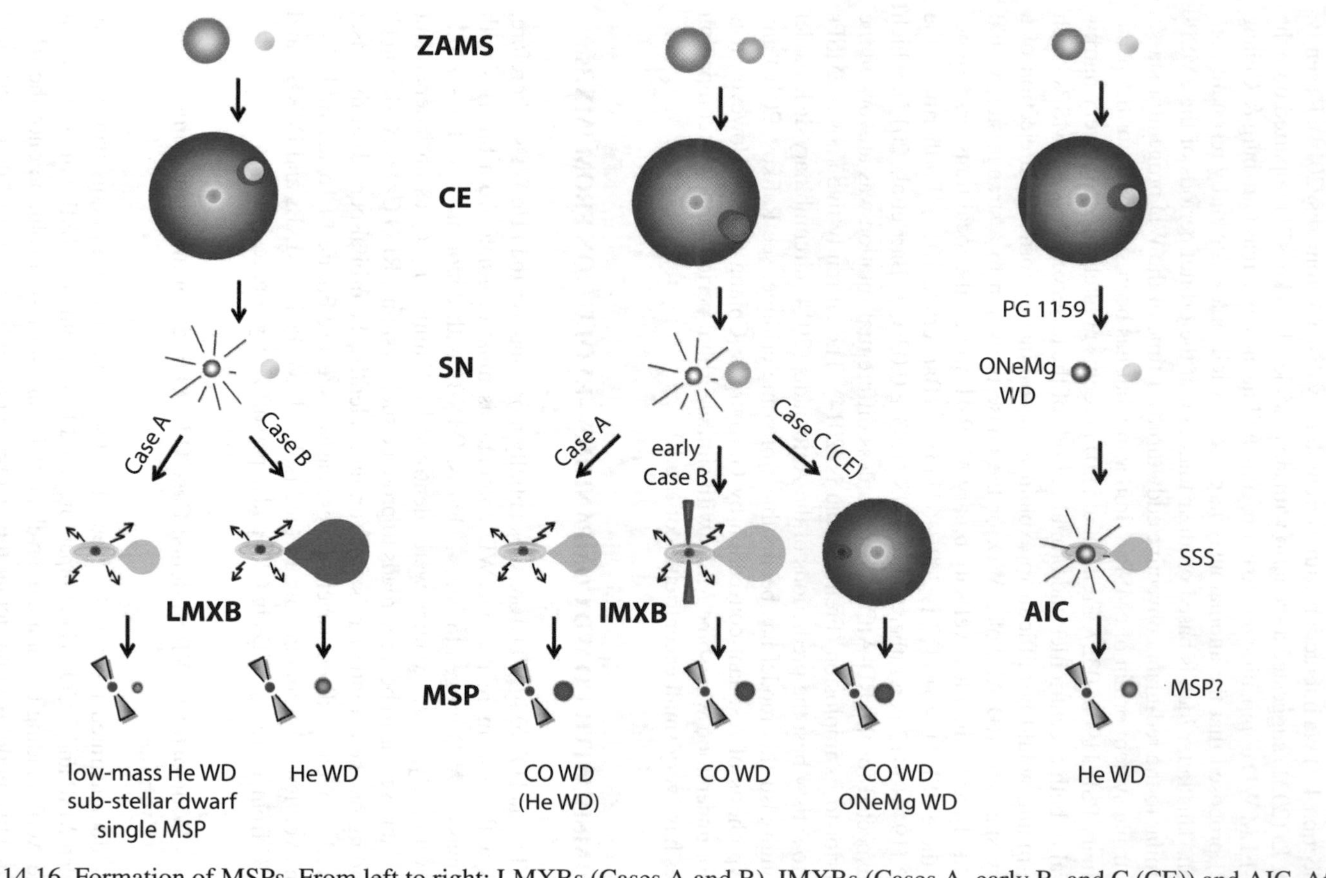

Figure 14.16. Formation of MSPs. From left to right: LMXBs (Cases A and B), IMXBs (Cases A, early B, and C (CE)) and AIC. After Tauris (2011).

Table 14.4. Progenitor systems leading to the formation of MSPs—see also Figure 14.16. All the characteristic values for both the X-ray binaries and the MSP systems are only rough indications and depend on effects that are poorly known—for example, concerning the strength of magnetic braking and other spin-orbit couplings as well as the CE and spiral-in evolution. AICs are not included in this table. After Tauris (2011).

LMXB	Case A	Case B	
Donor mass ($M_\odot$)	$1-2$	$1-2$	
$P_{\rm RLO}$ (days)	≤ 1.0*	> 1.0*	
MSP	↓	↓	
$P_{\rm orb}$ (days)	≤ 1.0	$1-1000$	
Recycling	full	full**	
Companion	single/ultra-light/He WD	He WD	
Example PSR	B1957+20	1713+0747	
IMXB	Case A	early Case B	Case C***
Donor mass ($M_\odot$)	$3-5$	$3-6$	$3-9$
$P_{\rm RLO}$ (days)	≤ 2.5	$3-10$	$100-1000$
MSP	↓	↓	↓
$P_{\rm orb}$ (days)	$3-20$	$3-50$	$0-20$****
Recycling	full	partial	partial
Companion WD	CO (He)	CO	CO/ONeMg
Example PSR	1614−2230	0621+1002	1802−2124

*Depending on the exact bifurcation period (e.g., Istrate et al., 2014).

**Often full recycling if the final $P_{\rm orb} \leq 200$ days. Otherwise, partial recycling.

***Via common-envelope evolution.

****λ and $\alpha_{\rm CE}$ are uncertain (see Section 4.8).

spiral-in phase to reduce the orbital angular momentum, and thus the orbital period, by a huge amount. One point to notice is that the CE phase is expected to be extremely short (Section 4.8). The plunge-in phase proceeds on a dynamical timescale (a few orbital periods) followed by an envelope ejection phase that may take up to 10^3 years (Ivanova, Justhan, Chen et al., 2013). However, at least ~0.01 $M_\odot$ is needed to spin up a pulsar to ~10 ms (Tauris et al., 2012), and this requires some 0.4 Myr of efficient accretion at the Eddington limit (a few times $10^{-8}\ M_\odot\ \mathrm{yr}^{-1}$). It is therefore believed that the recycling of the MSP may actually take place from either wind accretion from the naked post-CE core or, more likely, from Case BB RLO once the naked helium star expands and fills its Roche lobe (again) leading to a stable phase of mass transfer recycling the NS.

14.4.2 Hertzsprung-Gap IMXB Systems: Early Case B RLO and Isotropic Re-emission

An alternative way of producing MSPs with CO WD companions was demonstrated in detail by Tauris et al. (2000). They considered donor stars that had just left the main sequence (i.e., Hertzsprung-gap, or sub-giant stars) and applied the so-called isotropic re-emission model to IMXBs. In this model (Bhattacharya & van den Heuvel, 1991), the far majority of the matter—being transferred at a highly super-Eddington rate—is ejected (e.g., in a jet or a disk wind) with the specific orbital angular momentum of the accreting NS. This model can stabilize the RLO and prevent it from becoming dynamically unstable (Tauris et al., 2000; Podsiadlowski & Rappaport, 2000; Podsiadlowski et al., 2002; Misra et al., 2020). The typical donor star masses in this scenario are $3-5\,M_{\odot}$, and they leave behind a CO WD. Hence, the outcome of this formation scenario is mildly recycled MSPs with CO WD companions and orbital periods roughly between 3 and 50 days. Misra et al. (2020) recently expanded the investigation of IMXBs and their outcomes, see Figure 11.24 and discussions on dynamical stability in Section 11.6.

14.4.3 Close-Orbit IMXB Systems—Case A RLO and a Long-lasting, Stable Mass Transfer

The formation of MSPs from close-orbit IMXBs with main-sequence donors has been studied in detail by Podsiadlowski et al. (2002); Lin et al. (2011); Tauris et al. (2011). The outcome of these systems is mainly MSPs with CO WD companions, but He WD companions are also produced if the donor mass is at the low end, or the initial binary is very tight. The MSPs formed through this channel are expected to be fully recycled, that is, with spin periods of a few ms, since in this case the last part of the mass-transfer phase lasts $\sim 10^7$ yr—see Figure 14.17 which reproduces PSR J1614$-$2230 (Demorest et al., 2010) with a spin period of 3.15 ms. A handful of fully recycled MSPs with CO WDs are currently known in the Galactic disk that are thought to belong to this class of pulsars forming via Case A RLO. These are PSRs J1614$-$2230 (Demorest et al., 2010), J1101$-$6424 (Ng et al., 2015), J1943+2210 (Scholz et al., 2015), J1933$-$6211 (Graikou et al., 2017), J1618$-$4624 (Cameron et al., 2020), and possibly J1337$-$6423 (Keith et al., 2012).

The effective recycling in Case A RLO is opposite to the preferentially mildly recycled MSPs formed by IMXBs via early Case B or Case C RLO (described in Sections 14.4.1 and 14.4.2) where the duration of the total accretion phase is less (or significantly less) than 10^6 yr, particularly for systems evolving through a CE phase. Hence, recycled pulsars with massive WD companions evolving via these channels have longer spin periods.

In Figure 14.17, it is interesting to notice the huge drop in luminosity of the main-sequence donor star during Case A RLO. This is a direct consequence of intense mass-loss on a short timescale in which the star is driven out of thermal equilibrium. In order for the star to replace the rapid loss of the envelope mass with material from further below and to remain in hydrostatic equilibrium, an endothermic expansion of its inner region (requiring work against gravity) causes the surface luminosity to

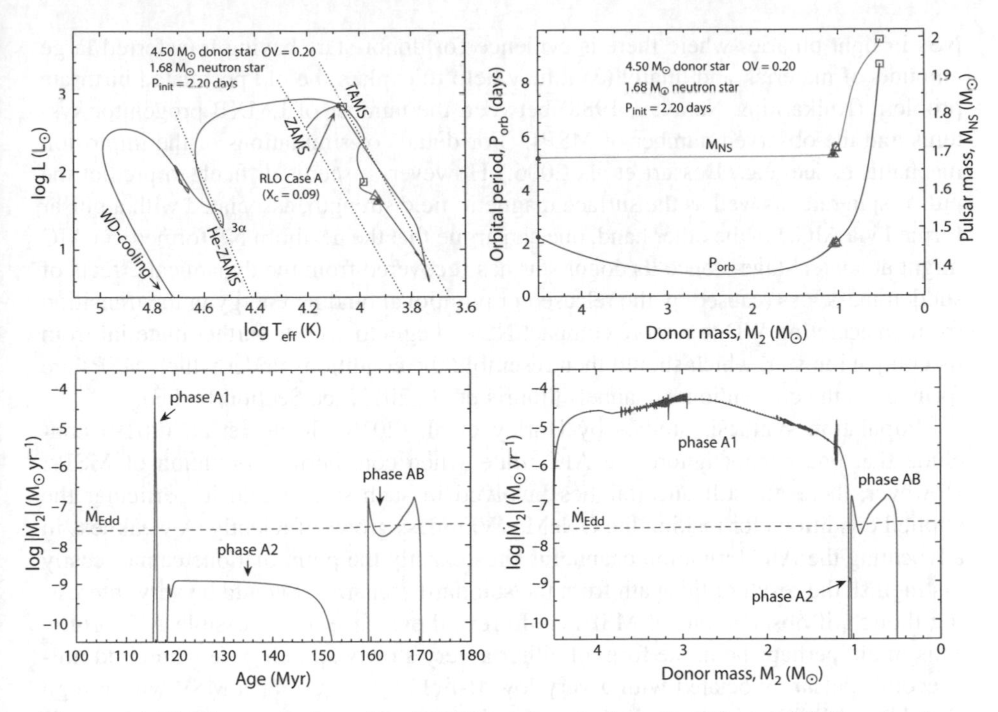

Figure 14.17. The evolution of an IMXB system undergoing Case A RLO and calculated to reproduce the properties of PSR J1614−2230. The mass transfer takes place between the blue dots (phase A1), open triangles (phase A2) and open squares (phase AB), respectively, on the donor star track in the HR diagram (upper left panel). The two bottom panels show the mass-transfer rate as a function of age and donor star mass, respectively. The NS is spun up to a MSP during phase AB (for example, upper right panel). After Tauris et al. (2011).

decrease by more than two orders of magnitude. In extreme cases of rapid mass loss, the central nuclear burning of the donor star can temporarily be completely exhausted (Tauris et al., 2015), as the nuclear-burning rate is a rather sensitive function of temperature.

14.5 FORMATION OF MSPs VIA ACCRETION-INDUCED COLLAPSE

It has been suggested that a NS can form from the implosion of a massive ONeMg WD, if such a WD accretes sufficient material from a companion star (see Section 11.1.3, and, e.g., Nomoto et al., 1979; Taam & van den Heuvel, 1986; Michel, 1987; Canal et al., 1990; Nomoto & Kondo, 1991; van den Heuvel & Bitzaraki, 1995, and references therein). This companion can either be a main-sequence star (i.e., a super-soft X-ray source), a low-mass giant (i.e., a novae-like system), or a helium star. The AIC route to form MSPs has three main advantages: (1) it can explain the existence of young NSs in an old stellar environment (like GCs), (2) it can explain the existence of high B-field

NSs in tight binaries where there is evidence for donor stars having transferred large amounts of material, and finally (3) it may help to explain the old postulated birthrate problem (Kulkarni & Narayan, 1988) between the number of LMXB progenitor systems and the observed number of MSPs. (For details of simulations of the implosion mechanism, see, e.g., Dessart et al., 2006.) However, it seems difficult to predict the pulsar spin rate, as well as the surface magnetic field strength, associated with a pulsar formed via AIC. On the other hand, one can argue that the newborn NS formed via AIC might at a later stage—once its donor star has recovered from the dynamical effects of sudden mass loss (caused by the released gravitational binding energy in the transition from an accreting WD to a more compact NS)—begin to accrete further material from its companion star, which should then resemble the conditions under which MSPs are formed via the conventional channels (Tauris et al., 2013, see Section 11.1.3).

Population synthesis studies by Hurley et al. (2010); Ruiter et al. (2019) conclude that one cannot ignore the AIC route when considering formation of MSPs. However, there are still uncertainties involved in such studies and in particular the applied conditions for making the ONeMg WD mass grow sufficiently. A weakness in advocating the AIC formation channel is thus exactly the point that one cannot easily distinguish the result of this path from the standard scenario. It would be very interesting though, if observations of MSPs could reveal evidence for a possible AIC origin. This might perhaps be in the form of either a very slowly spinning (few hundred millisecond) pulsar associated with a very low B-field ($\leq 10^8$ G), or a MSP with a high B-field ($> 10^{10}$ G). However, the former might not be operating as an active radio pulsar (as it would be located below the death line, Chen & Ruderman, 1993) and the latter kind of pulsar would be very unlikely to be detected given that its very strong magnetic dipole radiation would slow down its spin rate within a few Myr.

Although it has been suggested that in some cases AIC might lead to a thermonuclear explosion (Nomoto & Kondo, 1991; Schwab et al., 2015), there is evidence from the presence of apparently young and relatively strong-magnetic-field pulsars in GCs (see Chapter 12) that AIC may occur in nature. (However, see also discussion in Verbunt & Freire, 2014, for an alternative interpretation of the apparently young pulsars in GCs.) It is therefore quite plausible that the NSs in a fraction of the LMXBs and symbiotic X-ray binaries, and their descendants the binary MSPs, were formed also by this mechanism.

Formation of Ordinary Pulsars via AIC Revisited

In Section 11.1.3, we presented observational arguments for NS formation via AIC. The question now is if AIC is more likely to occur in wide-orbit systems. The fact that for $P_{\rm orb} > 200$ d, the pulsars with circular orbits and He WD companions tend to have relatively strong magnetic fields and normal long pulse periods, indicating that they are young, might be seen as an indication that they were recently formed during the preceding mass-transfer process that went on in these systems (van den Heuvel & Bitzaraki, 1995). The reason for this is that the progenitors of these $\lesssim 0.46\,M_\odot$ He WDs had a ZAMS mass, $M_2 < 2.3\,M_\odot$ (Section 14.3.2) and therefore must have had an age of more than 1 Gyr before they became giant stars and initiated RLO. During that time interval, the B-field of the NS, if it was formed long ago, a few times

10^7 yr after the formation of the system, may have decayed significantly below the presently observed values ($10^9 - 10^{11}$ G) in these systems, which means that the NSs must then have been produced more recently (via AIC). Alternatively, if the NS B-fields do not decay by more than a factor of $\sim$100 in a Gyr (assuming birth B-fields of the usual $10^{12} - 10^{13}$ G, then these systems could have been produced via inefficient mass transfer (mild recycling) in wide-orbit LMXBs as proposed by Tauris et al. (2012). The extent to which B-fields of NSs decay on a Gyr timescale (without any mass accretion) is still an open and important question (see, e.g., Cruces et al., 2019).

14.6 RECYCLING OF PULSARS

After outlining the formation of binary (and single) MSPs in LMXB and IMXB systems, we now proceed with a general examination of the recycling process of pulsars. We begin by analyzing the concept of spin-up lines in the $P\dot{P}$–diagram and derive an analytic expression for the amount of accreted mass needed to spin up a pulsar to a given equilibrium spin period. We use standard theory for deriving these expressions, but include it here to present a coherent and detailed derivation needed for our purposes. Furthermore, we apply the Spitkovsky (2006) solution to the pulsar spin-down torque, which combines the effect of a plasma current in the magnetosphere with the magnetic dipole model. The physics of the disk–magnetosphere interaction is still not known in detail. The interplay between the NS magnetic field and the conducting plasma in the accretion disk is a rather complex process. For further details of the accretion physics we refer to, for example, Pringle & Rees (1972); Lamb et al. (1973); Davidson & Ostriker (1973); Ghosh & Lamb (1979a,b); Shapiro & Teukolsky (1983); Ghosh & Lamb (1992); Spruit & Taam (1993); Campana et al. (1998); Frank et al. (2002); Rappaport et al. (2004); Ghosh (2007); Bozzo et al. (2009); D'Angelo & Spruit (2010); Ikhsanov & Beskrovnaya (2012); Shakura et al. (2015), and references therein. In the following sections, we follow Tauris et al. (2012).

14.6.1 The Accretion Disk

The accreted gas from a binary companion possess large specific angular momentum. For this reason the flow of gas onto the NS is not spherical, but leads to the formation of an accretion disk where excess angular momentum is transported outward by (turbulent-enhanced) viscous stresses, for example, Shapiro & Teukolsky (1983); Frank et al. (2002), and see also the discussion in Section 4.6. Depending on the mass-transfer rate, the opacity of the accreted material and the temperature of the disk, the geometric shape, and flow of the material may take a variety of forms (thick disk, thin thick, slim disk, torus-like, ADAF). Popular models of the inner disk (Ghosh & Lamb, 1992; Ghosh, 2007) include optically thin/thick disks which can be either gas (GPD) or radiation pressure dominated (RPD). The exact expression for the spin-up line in the $P\dot{P}$–diagram also depends on the assumed model for the inner disk—mainly as a result of the magnetosphere boundary, which depends on the characteristics of the inner disk. Close to the NS surface, the magnetic field is strong enough that the magnetic stresses truncate the Keplerian disk, and the plasma is channeled along field lines to accrete on to the surface of the NS. At the inner edge of the disk, the magnetic field interacts directly with the disk material over some finite region. The physics of this transition zone from Keplerian disk to magnetospheric flow is important and determines the angular momentum exchange from the differential rotation between the disk and the NS.

Interestingly enough, it seems to be the case that the resultant accretion torque, acting on the NS, calculated using detailed models of the disk–magnetosphere interaction does not deviate much from simple expressions assuming idealized, spherical accretion and Newtonian dynamics. For example, it has been pointed out by Ghosh & Lamb (1992) as a fortuitous coincidence that the equilibrium spin period calculated under simple assumptions of spherical flow resembles the more detailed models of an optically thick, gas pressure–dominated inner accretion disk. Hence, as a starting point, this allows us to expand on standard prescriptions in the literature with the aims of (1) performing a more careful analysis of the concept of a spin-up line, (2) deriving an analytic expression for the mass needed to spin up a given observed MSP, and (3) understanding the effects on the spinning NS when the donor star decouples from its Roche lobe. As we shall discuss further in Section 14.6.8, the latter issue depends on the location of the magnetospheric boundary (or the inner edge of the accretion disk) relative to the co-rotation radius and the light-cylinder radius of the pulsar. All stellar parameters listed in the next section refer to the NS unless explicitly stated otherwise.

14.6.2 The Accretion Torque—The Basics

The mass transfered from the donor star carries with it angular momentum that eventually spins up the rotating NS once its magnetic flux density, B, is low enough to allow for efficient accretion, that is, following initial phases where either the magnetodipole radiation pressure dominates or propeller effects are at work.

The accretion torque acting on the spinning NS has a contribution from both material stress (dominant term), magnetic stress, and viscous stress, and is given by $N = \dot{J}_\star \equiv (d/dt)(I\Omega_\star)$, where $J_\star$ is the pulsar spin angular momentum, $\Omega_\star$ is its angular velocity, and $I \approx 1 - 2 \times 10^{45}$ g cm^2 is its moment of inertia. The exchange of angular momentum ($\vec{J} = \vec{r} \times \vec{p}$) at the magnetospheric boundary eventually leads to a gain of NS spin angular momentum, which can approximately be expressed as:

$$\Delta J_\star = \sqrt{GMr_{\rm A}}\, \Delta M\, \xi, \tag{14.11}$$

where $\xi \simeq 1$ is a numerical factor that depends on the flow pattern (Ghosh & Lamb, 1979b, 1992), $\Delta M = \dot{M} \times \Delta t$ is the amount of mass accreted in a time interval Δt with average mass accretion rate $\dot{M}$, and

$$\begin{aligned} r_{\rm A} &\simeq \left(\frac{B^2 R^6}{4\dot{M}\sqrt{2GM}}\right)^{2/7} \\ &\simeq 15\,{\rm km} \quad B_8^{4/7} \left(\frac{\dot{M}}{0.1\,\dot{M}_{\rm Edd}}\right)^{-2/7} \left(\frac{M}{1.4\,M_\odot}\right)^{-5/7} \end{aligned} \tag{14.12}$$

is the Alfvén radius defined as the location where the magnetic energy density will begin to control the flow of matter (i.e., where the incoming material couples to the magnetic field lines and co-rotate with the NS magnetosphere), see Eq. (7.11). Here B is the surface magnetic flux density, R is the NS radius, M is the NS mass (see relation between R and M in the last paragraph of Section 14.6.2), and B_8 is B in units of 10^8 Gauss. (Note, some descriptions in the literature apply the polar B-field strength, $B_{\rm p} = 2B$, rather than the equatorial B-field strength used here. Sometimes also a factor two differs in the ram pressure when defining the location of $r_{\rm A}$. Finally, the exact NS radius is also uncertain. Thus, the various expressions in the literature for the quantity of $r_{\rm A}$ may vary a little bit.) A typical value for the Alfvén radius in accreting X-ray millisecond pulsars (AXMSPs), obtained from $B \sim 10^8$ G and $\dot{M} \sim 0.01\,\dot{M}_{\rm Edd}$, is $\sim$30 km, corresponding to $\sim 2-3\,R$. The expression in Eq. (14.12) is found (see Section 7.3) by equating the magnetic energy density $(B^2/8\pi)$ to the ram pressure of the incoming matter and using the continuity equation (e.g., Pringle & Rees, 1972). Furthermore, it assumes a scaling with distance, r, of the far-field strength of the dipole magnetic moment, μ, as: $B(r) \propto \mu/r^3$ (i.e., disregarding poorly known effects such as magnetic screening [Vasyliunas, 1979]). A more detailed estimation of the location of the inner edge of the disk, that is, the coupling radius or magnetospheric boundary, is given by: $r_{\rm mag} = \phi \cdot r_A$, where ϕ is $0.5 - 1.4$ (Ghosh & Lamb, 1992; Wang, 1997; D'Angelo & Spruit, 2010).

The Surface B-field Strength of Recycled Radio Pulsars

Before we proceed, we need an expression for B. One can estimate the B-field of recycled MSPs based on their observed spin period, P, and its time derivative $\dot{P}$. The usual assumption is to apply the vacuum magnetic dipole model in which the rate of rotational energy loss ($\dot{E}_{\rm rot} = I\Omega\dot{\Omega}$) is equated to the energy-loss rate caused by emission of magnetodipole waves (with a frequency equal to the spin frequency of the pulsar) due to an inclined axis of the magnetic dipole field with respect to the rotation axis of the pulsar:

$$\dot{E}_{\rm dipole} = (-2/3c^3)|\ddot{\mu}|^2, \tag{14.13}$$

where $\mu = BR^3$ is the magnetic moment and $\ddot{\mu} = BR^3\Omega^2 \sin\alpha$. The result is:

$$\begin{aligned} B_{\rm dipole} &= \sqrt{\frac{3c^3 I P\dot{P}}{8\pi^2 R^6}\,\frac{1}{\sin\alpha}} \\ &\simeq 1.6 \times 10^{19}\,G \quad \sqrt{P\dot{P}}\left(\frac{M}{1.4\,M_\odot}\right)^{3/2}\frac{1}{\sin\alpha}, \end{aligned} \tag{14.14}$$

where the magnetic inclination angle is $0 < \alpha \leq 90°$. This is the standard equation for evaluating the B-field of a radio pulsar. Our numerical scaling constant differs by a factor of a few from the conventional one: $B = 3.2 \times 10^{19}\,G\,\sqrt{P\dot{P}}$, which assumes $R = 10$ km and $I = 10^{45}$ g cm^2, both of which are likely to be slightly underestimated values. Note also that some descriptions in the literature apply the polar B-field strength

($B_p = 2\,B$) rather than the equatorial B-field strength used here. It is important to realize that the expression in Eq. (14.14) does not include the rotational energy loss obtained when considering the spin-down torque caused by the $\vec{j} \times \vec{B}$ force exerted by the plasma current in the magnetosphere, even in the case of an aligned rotator ($\alpha = 0°$, see e.g., Goldreich & Julian, 1969; Spitkovsky, 2008). Thus, a pure vacuum magnetic dipole model does not predict any spin-down torque for an aligned rotator which is not correct. The incompleteness of the vacuum magnetic dipole model was in particular evident after the discovery of intermittent pulsars by Kramer et al. (2006) and demonstrated the need for including the plasma term in the spin-down torque. A combined model was derived by Spitkovsky (2006) and applying his relation between B and α, one can rewrite Eq. (14.14) slightly (Tauris et al., 2012):

$$B = \sqrt{\frac{c^3 I P \dot{P}}{4\pi^2 R^6} \frac{1}{1+\sin^2\alpha}} \tag{14.15}$$

$$\simeq 1.3 \times 10^{19}\, G \quad \sqrt{P\dot{P}} \left(\frac{M}{1.4\, M_\odot}\right)^{3/2} \sqrt{\frac{1}{1+\sin^2\alpha}}.$$

This new expression leads to smaller estimated values of B by a factor of at least $\sqrt{3}$, or more precisely $\sqrt{\frac{3}{2}(2+\cot^2\alpha)}$, compared to the vacuum dipole model. As we shall shortly demonstrate, this difference in dependence on α is quite important for the location of the spin-up line in the $P\dot{P}$–diagram. Note that in deriving Eqs. (14.14) and (14.15), we have adopted expressions for $R(M)$ and $I(M)$ given in Sections 14.6.2 and 14.6.3, respectively.

The Ram Pressure

The ram pressure of the incoming matter is roughly $\frac{1}{2}\rho\, v_{\rm ff}(r_{\rm mag})^2$, where

$$\rho = \frac{\dot{M}}{4\pi\, r_{\rm mag}^2\, v_{\rm ff}(r_{\rm mag})} \tag{14.16}$$

and $v_{\rm ff} = \sqrt{2GM/r}$ is the free-fall velocity. The accretion rate is restricted by the Eddington limit , which for a NS is given by (see Section 4.9):

$$\dot{M}_{\rm Edd} \simeq 3.0 \times 10^{-8}\, M_\odot\, {\rm yr}^{-1} \quad R_{13} \left(\frac{1.3}{1+X}\right), \tag{14.17}$$

where X is the hydrogen mass fraction of the accreted material, and R_{13} is the NS radius in units of 13 km. Note that the value of $\dot{M}_{\rm Edd}$ is only a rough measure since the derivation assumes spherical symmetry, steady-state accretion, Thompson scattering opacity, and Newtonian gravity.

To estimate the NS radius, we use a mass-radius exponent following a simple non-relativistic degenerate Fermi-gas polytrope ($R \propto M^{-1/3}$) with a scaling factor such that $R = 15\,(M/M_\odot)^{-1/3}$ km, calibrated from PSR J1614−2230, cf. figure 3 in Demorest et al. (2010). Following measurements of tidal deformability in the DNS merger GW170817 (Abbott et al., 2018), and based on NICER observations of MSPs in X-rays (Bogdanov et al., 2019; Miller et al., 2021), constraints on the NS EoS were derived. However, the uncertainties of these results are still relatively large, and we proceed here with a simple polytropic NS EoS.

14.6.3 The Spin-up Line and Equilibrium Spin

The observed spin evolution of accreting NSs often shows rather stochastic variations on a short timescale (Bildsten et al., 1997). The reason for the involved dramatic torque reversals is not well known—see hypotheses listed at the beginning of Section 14.6.8. However, the long-term spin rate will eventually tend toward the *equilibrium* spin period, $P_{\rm eq}$—meaning that the pulsar spins at the same rate as the particles forced to co-rotate with the B-field at the magnetospheric boundary. The location of the associated so-called *spin-up line* (a concept introduced in 1980 by Radhakrishnan & Srinivasan, 1984, at a meeting in Bandung, Indonesia, see page 574) for the rejuvenated radio pulsar in the $P\dot{P}$–diagram can be found by considering the equilibrium configuration when the angular velocity of the NS is equal to the Keplerian angular velocity of matter at the magnetospheric boundary where the accreted matter enters the magnetosphere, that is, $\Omega_\star = \Omega_{\rm eq} = \omega_c\,\Omega_{\rm K}(r_{\rm mag})$ or:

$$
\begin{aligned}
P_{\rm eq} &= 2\pi\sqrt{\frac{r_{\rm mag}^3}{GM}}\,\frac{1}{\omega_c} \qquad (14.18)\\
&\simeq 1.40\,{\rm ms}\quad B_8^{6/7}\left(\frac{\dot{M}}{0.1\,\dot{M}_{\rm Edd}}\right)^{-3/7}\left(\frac{M}{1.4\,M_\odot}\right)^{-5/7} R_{13}^{18/7},
\end{aligned}
$$

which can be derived[5] from the equations in Section 7.3.1 (Exercise 14.6), and where R_{13} is the NS radius in units of 13 km and $0.25 < \omega_c \leq 1$ is the so-called critical fastness parameter, which is a measure of when the accretion torque vanishes (depending on the dynamical importance of the pulsar spin rate and the magnetic pitch angle, Ghosh & Lamb, 1979b). One must bear in mind that factors that may differ from unity were omitted in the numerical expression in Eq. (14.18). In all subsequent numerical expressions, we assume $\sin\alpha = \phi = \xi = \omega_c = 1$. Actually, the dependence on the NS radius, R, disappears in the full analytic formula obtained by inserting Eqs. (14.12) and (14.15) into the top expression in Eq. (14.18), and using $r_{\rm mag} = \phi\cdot r_{\rm A}$, which yields:

$$
P_{\rm eq} = \left(\frac{\pi c^9}{\sqrt{2}\,G^5}\,\frac{I^3\,\dot{P}^3}{M^5\dot{M}^3}\right)^{1/4} \quad (1+\sin^2\alpha)^{-3/4}\,\phi^{21/8}\,\omega_c^{-7/4}. \qquad (14.19)
$$

[5]Here we give the expression following Tauris et al. (2012). The numerical value may vary by a factor of approximately two, depending on assumptions on, for example, $r_{\rm A}$ and $R_{\rm NS}$.

Notice, in the above step we needed to link the B-fields of accreting NSs to the B-fields estimated for observed recycled radio pulsars (expressed by P and $\dot{P}$). This connection can be approximated in the following manner: If the radio pulsar after the recycling phase is "born" with a spin period P_0 that is somewhat close to $P_{\rm eq}$, then we can estimate the location of its magnetosphere when the source was an AXMSP just prior to the accretion turn-off during the Roche-lobe decoupling phase (RLDP), *if* this event did not significantly affect the spin period of the pulsar, cf. discussion in Section 14.6.8. (Further details of our assumptions of the B-fields of accreting NSs are given in Section 14.6.4, and in Section 14.6.7 we discuss the subsequent spin evolution of recycled radio pulsars toward larger periods, $P > P_0$.)

It is often useful to express the time derivative of the spin period as a function of the equilibrium spin period, for example, for the purpose of drawing the spin-up line in the $P\dot{P}$–diagram:

$$\dot{P} = \frac{2^{1/6}G^{5/3}}{\pi^{1/3}c^3}\frac{\dot{M}M^{5/3}P_{\rm eq}^{4/3}}{I}\,(1+\sin^2\alpha)\,\phi^{-7/2}\,\omega_c^{7/3}. \tag{14.20}$$

Given that $\dot{M}_{\rm Edd}$ is a function of the NS radius, and using the relation between M and R stated in Section 14.6.2, we need a relation between the moment of inertia and the mass of the NS. According to the equations-of-state studied by Worley et al. (2008), these quantities scale very close to linearly as $I_{45} \simeq M/M_\odot$ (see their figure 4), where I_{45} is the moment of inertia in units of 10^{45} g cm^2. Toward the end of the mass-transfer phase, the amount of hydrogen in the transfered matter is usually quite small ($X < 0.20$). The donor star left behind is basically a naked helium core (the proto WD). Hence, we can rewrite Eq. (14.20) and estimate the location of the spin-up line for a recycled pulsar in the $P\dot{P}$–diagram only as a function of its mass and the mass-accretion rate:

$$\dot{P} = 3.7\times10^{-19}\;P_{\rm ms}^{4/3}\left(\frac{M}{M_\odot}\right)^{2/3}\left(\frac{\dot{M}}{\dot{M}_{\rm Edd}}\right), \tag{14.21}$$

assuming again $\sin\alpha = \phi = \omega_{\rm c} = 1$, and where $P_{\rm ms}$ is the equilibrium spin period in units of milliseconds.

In the literature the spin-up line is almost always plotted without uncertainties. Furthermore, one should keep in mind the possible effects of the applied accretion disk model on the location of the spin-up line, cf. Section 14.6.1. Figure 14.18, taken from Tauris et al. (2012), shows a plot of Eq. (14.20) for different values of α, ϕ, and $\omega_{\rm c}$ to illustrate the uncertainties in the applied accretion physics to locate the spin-up line. The upper boundary of each band (or "line") is calculated for a NS mass $M = 2.0\,M_\odot$ and a magnetic inclination angle, $\alpha = 90°$. The lower boundary is calculated for $M = 1.0\,M_\odot$ and $\alpha = 0°$. The green hatched band corresponds to $\phi = 1$ and $\omega_{\rm c} = 1$, which are often used as default values in such calculations. The blue and red hatched bands are upper and lower limits set by reasonable choices of the two parameters (ϕ, $\omega_{\rm c}$). In all cases, a fixed accretion rate of $\dot{M} = \dot{M}_{\rm Edd}$ was assumed. The location of the spin-up line is simply shifted one order of magnitude in $\dot{P}$ down (up) for every order of magnitude $\dot{M}$ is decreased (increased).

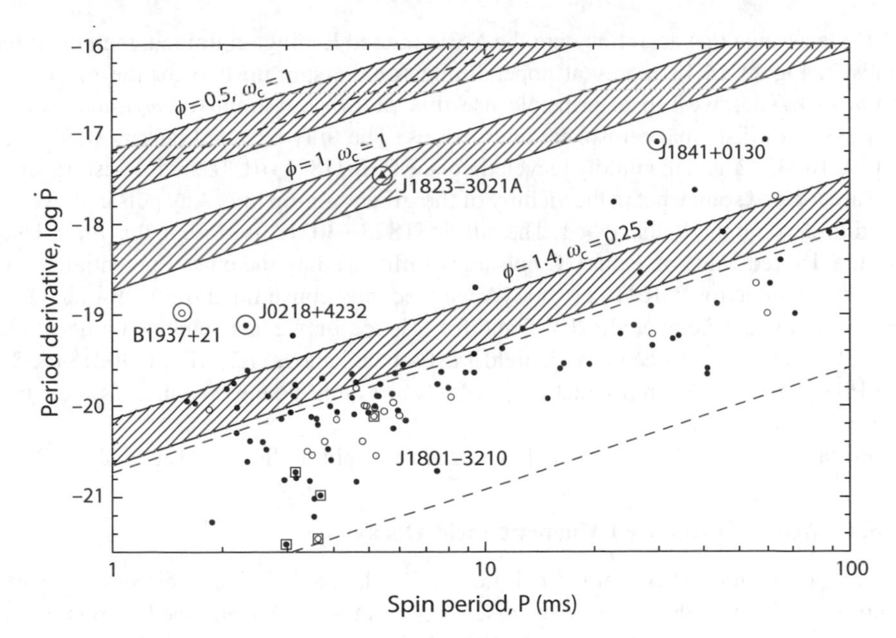

Figure 14.18. The spin-up line is shown as three colored bands depending on the parameters (ϕ, ω_c). In all three cases the spin-up line is calculated assuming accretion at the Eddington limit, $\dot{M} = \dot{M}_{\rm Edd}$, and applying the Spitkovsky torque formalism. Thus these "lines" represent upper limits for the given set of parameters. The width of each line (band) results from using a spread in NS mass and magnetic inclination angle from ($2.0\,M_\odot$, $\alpha = 90°$) to ($1.0\,M_\odot$, $\alpha = 0°$), upper and lower boundary, respectively. The dashed line within the upper blue band is calculated from ($2.0\,M_\odot$, $\alpha = 0°$). Hence, the bandwidth above this line reflects the dependence on α, whereas the bandwidth below this line shows the dependence on M. The two red dashed spin-up lines below the red band are calculated for a $1.4\,M_\odot$ NS using the vacuum magnetic dipole model for the radio pulsar torque and assuming $\phi = 1.4$, $\omega_c = 0.25$, and $\dot{M} = \dot{M}_{\rm Edd}$ for $\alpha = 90°$ (upper) and $\alpha = 10°$ (lower). The observed binary and isolated radio pulsars in the Galactic disk (i.e., outside GCs) are plotted as small, filled and open circles, respectively. Also plotted is the pulsar J1823−3021A, which is located in the GC NGC 6624. For the meaning of the other symbols, see the original work. After Tauris et al. (2012).

It is important to realize that there is no universal spin-up line in the $P\dot{P}$–diagram. Only an upper limit. Any individual pulsar has its own spin-up line/location that in particular depends on its unknown accretion history ($\dot{M}$). Also notice that the dependence on the magnetic inclination angle, α, is much less pronounced when applying the Spitkovsky formalism for estimating the B-field of the MSP compared to applying the vacuum dipole model. The difference in the location of spin-up lines using $\alpha = 90°$ and $\alpha = 0°$ is only a factor of two in the Spitkovsky formalism. For a comparison, using the vacuum dipole model with a small magnetic inclination of $\alpha = 10°$ results in a spin-up line that is translated downward by almost two orders of magnitude compared to its equivalent orthogonal rotator model, cf. the two red dashed lines in Figure 14.18.

If we assume that accretion onto the NS is indeed Eddington limited, then the three bands in Figure 14.18 represent upper limits for the spin-up line for the given sets of (ϕ, ω_c). Thus we can in principle use this plot to constrain (ϕ, ω_c) and, hence, the physics of disk–magnetosphere interactions. The fully recycled pulsars B1937+21 and J0218+4232, and the mildly recycled pulsar PSR J1841+0130, are interesting since they are located somewhat in the vicinity of the green spin-up line. Any pulsar above the green line would imply that $\phi < 1$. The pulsar J1823−3021A is close to this limit. Usually the derived value of $\dot{P}$ for GC pulsars is influenced by the cluster potential. However, the $\dot{P}$ value for PSR J1823−3021A was recently constrained from Fermi LAT γ-ray observations (Freire et al., 2011), and for this reason we have included it here. The high $\dot{P}$ values of the three Galactic field MSPs listed above, B1937+21, J0218+4232, and PSR J1841+0130, imply that these MSPs are quite young. We discuss their true ages in Section 14.6.7.

For a discussion of observational evidence of a spin-up line, see Liu et al. (2022).

14.6.4 Accretion-Induced Magnetic Field Decay

Whether or not the B-fields of normal (i.e., isolated, non-recycled) radio pulsars decay or not is a subject of debate since the discovery of pulsars. Already the first population study of radio pulsars by Gunn & Ostriker (1970) suggested magnetic field decay on a short timescale of a few Myr. However, later work showed that the timescale of B-field decay is probably very long and in excess of 100 Myr, that is, longer than the typical active lifetime of a radio pulsar (Bhattacharya et al., 1992). The question of B-field decay is further complicated by the possible alignment of the magnetic inclination angle with time (Tauris & Manchester, 1998). Such evolution gives rise to a similar decay of the braking torque (Tauris & Konar, 2001; Johnston & Karastergiou, 2017). In addition, we caution that some misleading confusion has arisen in the literature from conclusions based simply on plotting the derived B-field ($B \propto \sqrt{P\dot{P}}$) as a function of characteristic age ($\tau \propto P\dot{P}^{-1}$) of radio pulsars, and then erroneously used this result to argue in favor of B-field decay (see Exercise 14.3).

An important observation for the discussion of B-field decays in radio pulsars was the determination of a cooling age of ~2 Gyr (Kulkarni, 1986) for a WD companion star to the 0.196 s binary radio pulsar B0655+64 with an estimated B-field of $B \simeq 10^{10}$ G. This observation led Kulkarni (1986) to hypothesize that the B-field in NSs consists of two components—an exponentially decaying field in the crust and a steady (residual) field residing in the core. Further arguments for a B-field decay of MSPs on a timescale of at least several Gyr were presented in van den Heuvel et al. (1986). Observations of a large number of MSP systems have since then confirmed that the B-fields of these recycled pulsars are typically of the order of $10^8 - 10^{10}$ G (compared to the $10^{12} - 10^{13}$ G in young pulsars) and show no evidence for further decay.

The widely accepted idea of accretion-induced magnetic field decay in NSs is based on observational evidence (e.g., early studies by Taam & van den Heuvel, 1986; Shibazaki et al., 1989; van den Heuvel & Bitzaraki, 1994). During the recycling process, the surface B-field of pulsars is apparently reduced by several orders of magnitude, from values of 10^{11-12} G to 10^{7-9} G. However, it is still not understood if this is caused by spin-down induced flux expulsion of the core proton fluxoids (Srinivasan et al., 1990), or if the B-field is confined to the crustal regions and decays due to diffusion and Ohmic dissipation, as a result of a decreased electrical conductivity when heating effects set in from nuclear burning of the accreted material (Geppert & Urpin, 1994; Konar & Bhattacharya, 1997), or if the B-field decay is simply caused by a diamagnetic screening by the accreted plasma—see review by Bhattacharya (2002) and references therein.

The decay of the crustal B-field is found by solving the induction equation:

$$\frac{\partial \vec{B}}{\partial t} = -\frac{c^2}{4\pi}\vec{\nabla} \times (\frac{1}{\sigma}\vec{\nabla} \times \vec{B}) + \vec{\nabla} \times (\vec{v} \times \vec{B}), \tag{14.22}$$

where the electrical conductivity, σ, depends on the temperature of the crust (as well as the local density and lattice impurities, Geppert & Urpin, 1994; Konar & Bhattacharya, 1997).

Although there have been attempts to model or empirically fit the magnetic field evolution of accreting NSs (e.g., Shibazaki et al., 1989; Zhang & Kojima, 2006; Wang et al., 2011), the results may not be fully certain. One reason is that it is difficult to estimate how much mass a given recycled pulsar has accreted. Recently, Cruces et al. (2019) investigated whether the low B-field of radio MSPs is instead a simple consequence of ambipolar diffusion on a Gyr timescale *before* accretion of material. The pre-accretion timescale is then related to the type of companion star. The conclusion of this work is that the main distribution of B-field flux densities of recycled pulsars can be explained by this model, although a number of individual sources cannot. However, one cannot rule out the possibility that some of these odd NSs may originate from the accretion-induced collapse of a massive white dwarf, in which case they might be formed with a high B-field near the end of the mass-transfer phase.

14.6.5 Amount of Mass Needed to Spin Up a Pulsar

Here, we follow Tauris et al. (2012) and make a number of assumptions to model the B-field evolution with the aim to relate the spin period of recycled pulsars to the amount of mass accreted:

- The B-field decays rapidly in the early phases of the accretion process via some unspecified process (see beginning of Section 14.6.4).
- Accreting pulsars accumulate the majority of mass while spinning at/near equilibrium.
- The magnetospheric boundary, $r_{\rm mag}$, is approximately kept at a fixed location for most of the spin-up phase, until the mass transfer ceases during the Roche-lobe decoupling phase (RLDP).
- During the RLDP, the B-field of an AXMSP can be considered to be constant since very little envelope material ($\sim$0.01 $M_\odot$) remains to be transfered from its donor at this stage.

Accreted Mass vs. Final Spin Period Relation

The amount of spin angular momentum added to an accreting pulsar is given by:

$$\Delta J_{\star} = \int n(\omega,t)\,\dot{M}(t)\,\sqrt{GM(t)r_{\rm mag}(t)\,\xi(t)}\,dt, \qquad (14.23)$$

where $n(\omega)$ is a dimensionless torque. Assuming $n(\omega)=1$, and $M(t)$, $r_{\rm mag}(t)$ and $\xi(t)$ to be roughly constant during the major part of the spin-up phase, we can rewrite the expression and obtain a simple formula (see also Eq. 14.11) for the amount of matter needed to spin up the pulsar:

$$\Delta M \simeq \frac{2\pi I}{P\sqrt{GMr_{\rm mag}\,\xi}}. \qquad (14.24)$$

Note that the initial spin angular momentum of the pulsar prior to accretion is negligible given that $\Omega_0 \ll \Omega_{\rm eq}$. To include all numerical scaling factors properly, we follow again Tauris et al. (2012) and insert Eqs. (14.12), (14.15) and (14.20) into Eq. (14.24), recalling that $r_{\rm mag} = \phi\cdot r_{\rm A}$, and we find:

$$\Delta M_{\rm eq} = I\left(\frac{\Omega_{\rm eq}^4}{G^2M^2}\right)^{1/3} f(\alpha,\xi,\phi,\omega_c), \qquad (14.25)$$

where $f(\alpha,\xi,\phi,\omega_c)$ is some dimensionless number of order unity. Once again, one can apply the relation between moment of inertia and mass of the NS (e.g., Worley et al., 2008) and obtain a simple convenient expression to relate the amount of mass to be accreted to spin up a pulsar to a given (equilibrium) rotational period (Tauris et al., 2012):

$$\Delta M_{\rm eq} = 0.22\,M_{\odot}\,\frac{(M/M_{\odot})^{1/3}}{P_{\rm ms}^{4/3}} \qquad (14.26)$$

assuming that the numerical factor $f(\alpha,\xi,\phi,\omega_c)=1$.

In the above derivation, we have neglected minor effects related to release of gravitational binding energy of the accreted material—see, for example, Eq. (22) in Tauris & Savonije (1999), and in particular Bejger et al. (2011) and Bagchi (2011) for a more detailed general discussion including various equations of state, general relativity, and the critical mass shedding spin limit. However, since the exchange of angular momentum takes place near the magnetospheric boundary, the expression in Eq. (14.26) refers to the baryonic mass accreted from the donor star. To calculate the increase in gravitational mass of the pulsar, one must apply a reducing correction factor of $\sim$0.85 – 0.90, depending on the NS EoS (Lattimer & Yahil, 1989).

In Figure 14.19, we show the amount of mass, $\Delta M_{\rm eq}$, needed to spin up a pulsar to a given spin period. The value of $\Delta M_{\rm eq}$ is a strongly decreasing function of the pulsar spin period, $P_{\rm eq}$. For example, considering a pulsar with a final mass of 1.4 $M_{\odot}$ and a recycled spin period of either 2 ms, 5 ms, 10 ms, or 50 ms requires accretion

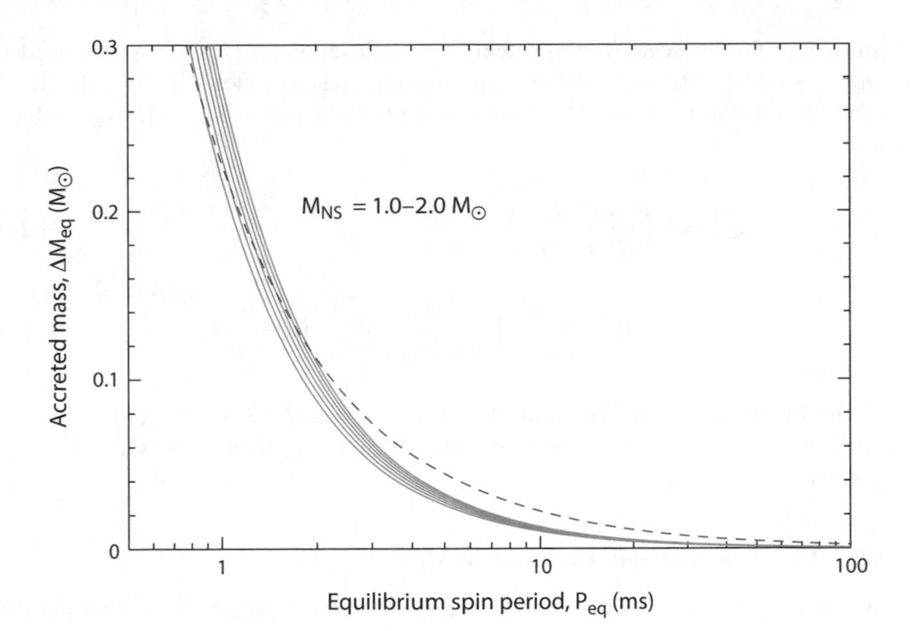

Figure 14.19. Amount of mass needed to spin up a pulsar as a function of its equilibrium spin period using Eq. (14.26). The dashed red curve is from Lipunov & Postnov (1984). The calculated green curves correspond to various NS masses in steps of 0.2 $M_\odot$, increasing upward. After Tauris et al. (2012).

of 0.10 $M_\odot$, 0.03 $M_\odot$, 0.01 $M_\odot$, or 10^{-3} $M_\odot$, respectively. Therefore, it is no surprise that observed recycled pulsars with massive companions (CO WD, ONeMg WD, or NS) in general are much slower rotators—compared to binary MSPs with He WD companions—since the progenitor of their massive companions evolved on a relatively short timescale, only allowing for very little mass to be accreted by the pulsar.

On the one hand, the value of $\Delta M_{\rm eq}$ should be regarded as a lower limit to the actual amount of material required to be transfered to the NS, even at sub-Eddington rates, since a non-negligible amount may be ejected from the pulsar magnetosphere due to magnetodipole wave pressure or the propeller effect (Illarionov & Sunyaev, 1975). Furthermore, accretion disk instabilities (Pringle, 1981; van Paradijs, 1996) are also responsible for ejecting part of the transferred material. It was demonstrated by Antoniadis et al. (2012) that the accretion efficiency in some cases is less than 40%, even in short orbital period binaries accreting at sub-Eddington levels (Fig. 6.10). On the other hand, the above derivation did not take into account the possibility of a more efficient angular momentum transfer early in the accretion phase where the value of $r_{\rm mag}$ (the lever arm of the torque) could have been larger if the B-field did not decay rapidly.

14.6.6 Spin-relaxation Timescale

In the previous section, we have reproduced a calculation of the amount of mass needed to spin up a pulsar to a given equilibrium spin period. However, one must be sure that the accretion-torque can actually transmit this acceleration on a timescale shorter than the mass-transfer timescale. To estimate the spin-relaxation timescale (the time needed to spin up a slowly-rotating NS to spin equilibrium) one can follow Tauris et al. (2012) and simply consider $t_{\rm torque} = J/N$ where $J = 2\pi I/P_{\rm eq}$ and $N = \dot{M}\sqrt{GMr_{\rm mag}}\,\xi$ which yields:

$$t_{\rm torque} = I\left(\frac{4G^2M^2}{B^8R^{24}\dot{M}^3}\right)^{1/7}\frac{\omega_c}{\phi^2\,\xi} \tag{14.27}$$

$$\simeq 50\,{\rm Myr}\quad B_8^{-8/7}\left(\frac{\dot{M}}{0.1\,\dot{M}_{\rm Edd}}\right)^{-3/7}\left(\frac{M}{1.4\,M_\odot}\right)^{17/7}$$

(the latter for $\phi = \xi = \omega_c = 1$) or equivalently, $t_{\rm torque} = \Delta M/\dot{M}$, see Eq. (14.24). If the duration of the mass-transfer phase, t_X is shorter than $t_{\rm torque}$, then the pulsar will not be fully recycled.

14.6.7 True Ages and Spin Evolution of MSPs

Knowledge of the true ages of recycled radio pulsars is important for comparing the observed population with the properties expected from the spin-up theory outlined here in Section 14.6. The following discussions are taken from Tauris et al. (2012) who provide many more details.

All radio pulsars lose rotational energy with age and the braking index, n, is given by (see Section 14.1.4):

$$\dot{\Omega} \propto -\Omega^n, \tag{14.28}$$

which yields (for n constant): $n \equiv \Omega\ddot{\Omega}/\dot{\Omega}^2$. The deceleration law can also be expressed as: $\dot{P} \propto P^{2-n}$ and hence the slope of a pulsar evolutionary track in the $P\dot{P}$–diagram is simply given by: $2-n$. Depending on the physical conditions under which the pulsar spins down, n can take different values as mentioned earlier. For example:

$$\begin{array}{ll} \text{pure gravitational wave radiation} & n=5 \\ \text{B-decay or alignment or multipoles} & n>3 \\ \text{perfect magnetic dipole} & n=3 \\ \text{B-growth/distortion or counter-alignment} & n<3. \end{array} \tag{14.29}$$

The combined magnetic dipole and plasma current spin-down torque may also result in $n \neq 3$ (Contopoulos & Spitkovsky, 2006). A simple integration of Eq. (14.28) for a *constant* braking index ($n \neq 1$) yields the well-known expression in Eq. (14.7) for the true age of the pulsar.

Isochrones in the $P\dot{P}$–Diagram

Unfortunately, one cannot directly use Eq. (14.7) to obtain evolutionary tracks in the $P\dot{P}$–diagram (even under the assumption of a constant n and a known initial spin period, P_0). For chosen values of t, n, and P_0 one can find a whole family of solutions $(P,\,\dot{P})$ to be plotted as an isochrone (Kiziltan & Thorsett, 2010; Tauris et al., 2012).

However, there is only one point that is a valid solution for a given pulsar and Eq. (14.7) does reveal which point is correct. The problem is that the variables in Eq. (14.7) are not independent, and we do not know a priori the initial spin period derivative, $\dot{P}_0$. The evolution of the spin period is a function of both t, n, the initial spin period, and its time derivative, that is, $P(t, n, P_0, \dot{P}_0)$ (Tauris et al., 2012); see also Eqs. (14.5) and (14.6).

Characteristic vs. True Ages of MSPs

Introducing the characteristic age of a pulsar, $\tau \equiv P/(2\dot{P})$, one finds the relation between τ (the observable) and t:

$$\log \tau = \log t + \log\left(\frac{n-1}{2}\right) - \log\left(1 - (\frac{P_0}{P})^{n-1}\right) \tag{14.30}$$

for evolution with a constant braking index, n. The asymptotic version of this relation (as $t \to \infty$ and $P \gg P_0$) is given by:

$$\log \tau = \begin{cases} \log t + \log 2 & \text{for } n = 5 \\ \log t & \text{for } n = 3 \\ \log t - \log 2 & \text{for } n = 2. \end{cases} \tag{14.31}$$

Figure 14.20 shows plots of τ as a function of t on log-scales for braking indices of $n = 2$, 3, and 5, respectively. Initially, when t is small and $P \simeq P_0$, τ is always greater than t. After a certain timescale, the value of τ can either remain larger or become smaller than t, depending on n. Those pulsars with smallest values of $\dot{P}_0$ for a given B_0 or P_0 are those with $\tau_0 \gg t$ (which is hardly a surprise given that $\tau \equiv P/(2\dot{P})$). The extent to which they continue evolving with $\tau \gg t$ depends on their initial conditions (P_0, B_0) and well as n. Notice, for recycled pulsars n is not measurable, and for this reason its value remains unknown.

Since it is an empirical fact that the final post-accretion B-field strength of a MSP is often as low as a few times 10^7 G—a typical value for binary MSPs with He WD companions—this clearly demonstrates why characteristic ages of MSPs are untrustworthy as true age indicators, as such MSPs are born with $\tau_0 \simeq 13$ Gyr (see also Tauris, 2012). This fact is important since the characteristic ages of recycled pulsars are often compared to the cooling ages of their white dwarf companions (e.g., see discussion in van Kerkwijk et al. [2005]).

In any discussion of the ages and the evolution of MSPs in the $P\dot{P}$–diagram, kinematic corrections must be included when considering the observed values of $\dot{P}$ ($\dot{P}_{obs}$). The kinematic corrections to the intrinsic $\dot{P}$ ($\dot{P}_{int}$) are caused by acceleration due to proper motion for nearby pulsars (Shklovskii, 1970) and to vertical (a_Z) and differential rotational acceleration in our Galaxy (a_{GDR}). The total corrections are given by:

$$\left(\frac{\dot{P}_{obs}}{P}\right) = \left(\frac{\dot{P}_{int}}{P}\right) + \left(\frac{\dot{P}_{shk}}{P}\right) + \frac{a_Z}{c} + \frac{a_{GDR}}{c}, \tag{14.32}$$

where $\dot{P}_{shk}/P = \mu^2 d/c$, d is the distance to the pulsar, and μ is its proper motion related to its transverse velocity, $v_\perp = \mu d$. The expressions for Galactic vertical and differential rotational acceleration can be found in the literature (e.g., Damour &

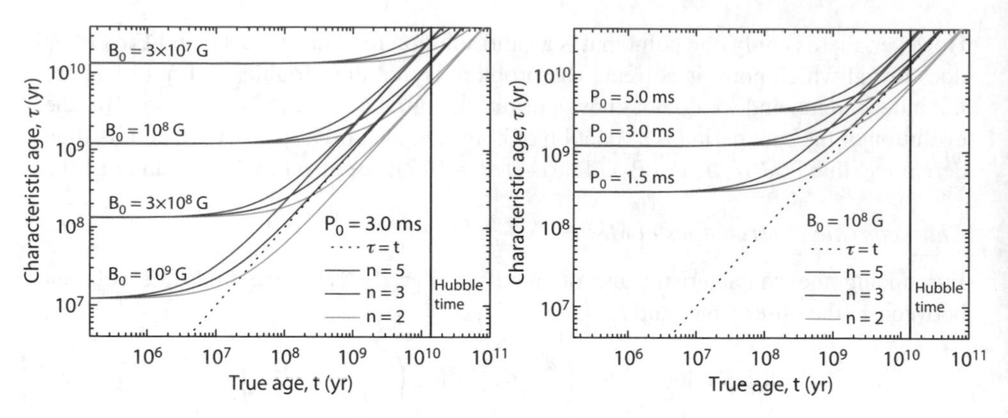

Figure 14.20. Evolutionary tracks of characteristic ages, τ, calculated as a function of true ages, t, for recycled pulsars with a constant braking index of $n = 2$, $n = 3$, and $n = 5$. In all cases for n assuming a constant $M = 1.4\,M_{\odot}$, $\sin\alpha = 1$, a constant moment of inertia, I, and for $n = 5$ a constant ellipticity, $\varepsilon \neq 0$. In the left panel, in all cases an initial spin period of $P_0 = 3.0$ ms was assumed and the value of the initial surface magnetic dipole flux density, B_0 was varied. In the right panel, it was assumed in all cases that $B_0 = 10^8$ G and P_0 was varied. The dotted line shows a graph for $\tau = t$ and thus only pulsars located on (near) this line have characteristic ages as reliable age indicators. Note, that recycled pulsars with small values of $\dot{P}_0$, resulting from either small values of B_0 (left panel) and/or large values of P_0 (right panel), tend to have $\tau \gg t$, even at times exceeding the age of the Universe. In all degeneracy splittings of the curves, the upper curve (blue) corresponds to $n = 5$, the central curve (red) corresponds to $n = 3$ and the lower curve (green) corresponds to $n = 2$. After Tauris et al. (2012).

Taylor, 1991; Wolszczan et al., 2000). The corrections to $\dot{P}$ due to these two terms are typically quite small (a few $10^{-22} \ll \dot{P}_{\rm int}$) and can be ignored, except in a few cases.

The kinematic corrections listed above also apply to the measured time derivative of the orbital period of binary systems, $\dot{P}_{\rm orb}$. Usually, this effect is irrelevant, but for some binary MSPs it may play an important role (Bell & Bailes, 1996; Lazaridis et al., 2009)—see also Exercise 14.9.

Evolutionary Tracks and True Age Isochrones

To investigate if we can understand the distribution of MSPs in the $P\dot{P}$–diagram, Tauris et al. (2012) traced the evolution of eight hypothetical recycled MSPs with different birth locations in the diagram. In each case, they traced the evolution for $2 \leq n \leq 5$ and calculated isochrones. The results are shown in Figure 14.21 together with observed data.

Three main conclusions can be drawn from this diagram:

1) The overall distribution of observed pulsars follows nicely the banana-like shape of an isochrone with multiple choices for $\dot{P}_0$ (or B_0), see the fat purple line. The

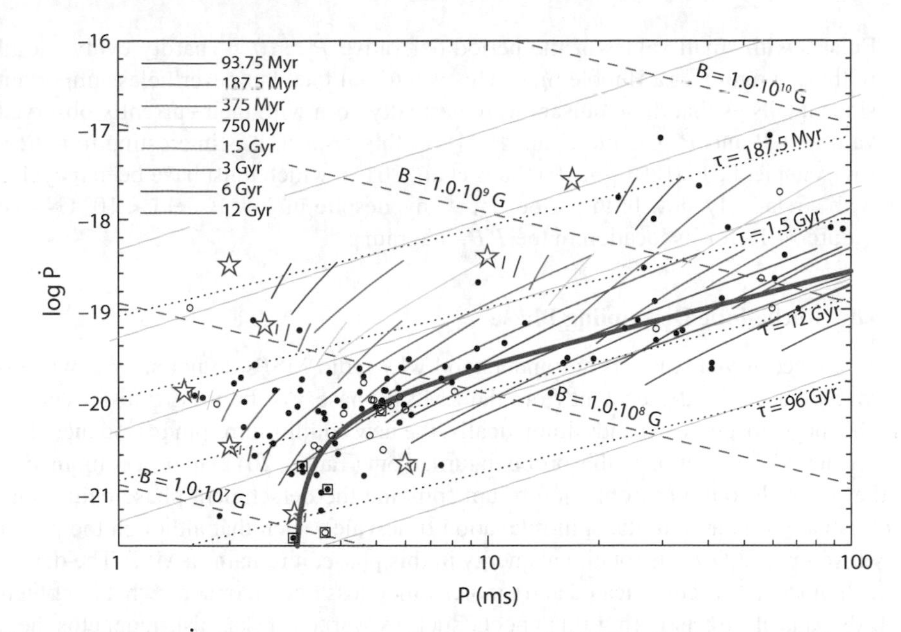

Figure 14.21. $P\dot{P}$–diagram with isochrones of eight hypothetical recycled pulsars born at the locations of the red stars. Solid and open circles mark observed binary and recycled isolated pulsars, respectively. (All values of $\dot{P}$ have been corrected for the Shklovskii effect.) The isochrones (see rainbow-colored lines in the legend for ages) were calculated for different values of the braking index, $2 \leq n \leq 5$ and using Eqs. (14.5) and (14.6). Also plotted are inferred B-field values (dashed lines) and characteristic ages, τ (dotted lines). The thin grey lines are spin-up lines with $\dot{M}/\dot{M}_{\rm Edd} = 1$, 10^{-1}, 10^{-2}, 10^{-3} and 10^{-4} (top to bottom, and assuming $\sin\alpha = \phi = \omega_{\rm c} = 1$, see Eq. 14.20). In all calculations, a pulsar mass of $1.4\,M_\odot$ was assumed. It is noted that the majority of the observed population is found near the isochrones for $t = 3 - 12$ Gyr, as expected. The fat, solid purple line indicates an example of a $t = 6$ Gyr isochrone for pulsars with any value of $\dot{P}_0$, but assuming $P_0 = 3.0$ ms and $n = 3$. It is seen how the banana shape of such a type of an isochrone fits very well with the overall distribution of observed pulsars in the Galactic disk. After Tauris et al. (2012).

chosen values of $P_0 = 3.0$ ms, $n = 3$, and $t = 6$ Gyr are for illustrative purposes only and not an attempt for a best fit to the observations. Fitting to one curve would not be a good idea given that MSPs are born with different initial spin periods, which depend on their accretion history, and given that the pulsars have different ages. The spread in the observed population is hinting that recycled pulsars are born at many different locations in the $P\dot{P}$–diagram.

2) The majority of the recycled pulsars seem to have true ages between 3 and 12 Gyr, as expected since the population accumulates and the pulsars keep emitting radio waves for a Hubble time.

3) Pulsars with small values of the period derivative $\dot{P} \simeq 10^{-21}$ hardly evolve at all in the diagram over a Hubble time. This is a trivial fact, but nevertheless important since it tells us that these pulsars were basically born with their currently observed values of P and $\dot{P}$ (Camilo et al., 1994). In this respect, it is interesting to notice, for example, PSR J1801−3210 (Bates et al., 2011), which must have been recycled with a relatively slow birth period, $P_0 \sim 7$ ms despite its low B-field $< 10^8$ G—see Figure 14.18 for its location in the $P\dot{P}$–diagram.

14.6.8 Roche-lobe Decoupling Phase

For a number of years, it was not understood why radio MSPs do not slow down significantly during the detachment from the LMXB phase. As the mass transfer ceases, and the magnetosphere expands dramatically, the new equilibrium spin period increases significantly. The solution to this problem turned out (Tauris, 2012) to be a combination of the pulsar decoupling from equilibrium spin and the detachment phase being relatively short (compared to the spin-relaxation timescale), such that although the pulsar may lose some 50% of its rotational energy in this phase, it remains a MSP. The details of such modelling is complicated and ignores other, less known or somewhat hypothetical, dynamical effects during this epoch, such as warped disks, disk-magnetosphere instabilities, variations in the mass-transfer rate caused by X-ray irradiation effects on the donor star, or transitions between, for example, a Keplerian thin disk and a sub-Keplerian, advection-dominated accretion flow (ADAF).

An example of a numerical calculation of the Roche-lobe decoupling phase (RLDP) is shown in Figure 14.22. During the equilibrium spin phase, the rapidly alternating sign changes of the torque (partly unresolved on the graph) reflect small oscillations around the semi-stable equilibrium, corresponding to successive small episodes of spin-up and spin-down. The reason is that the relative location of $r_{\rm mag}$ and $r_{\rm co}$ depends on the small fluctuations in $\dot{M}$. Despite applying an implicit coupling scheme in the code behind these calculations and ensuring that the time steps during the RLDP ($\sim 8 \times 10^4$ yr) are much smaller than the duration of the RLDP ($\sim 10^8$ yr), the calculated mass-transfer rates are subject to minor numerical oscillations. However, these oscillations could in principle be physical within the frame of the simple applied model. Examples of physical perturbations that could cause real fluctuations in $\dot{M}$, but on a much shorter timescale, are accretion disk instabilities and clumps in the transfered material from the donor star (the donor stars here have active, convective envelopes during the RLDP).

At some point ($t \sim 940$ Myr in Fig. 14.22) the equilibrium spin is broken. Initially, the spin can remain in equilibrium by adapting to the decreasing value of $\dot{M}$. However, further into the RLDP the result is that $r_{\rm mag}$ increases on a timescale faster than the spin-relaxation timescale, $t_{\rm torque}$, at which the torque can transmit the effect of deceleration to the NS, and therefore $r_{\rm mag} > r_{\rm co}$ (see Tauris, 2012, and supporting online material therein). During the propeller phase, the resultant accretion torque acting on the NS is negative, that is, a spin-down torque. When the radio pulsar is activated and the accretion has come to an end, the spin-down torque is simply caused by the magnetodipole radiation combined with the pulsar wind (see Section 14.6.2).

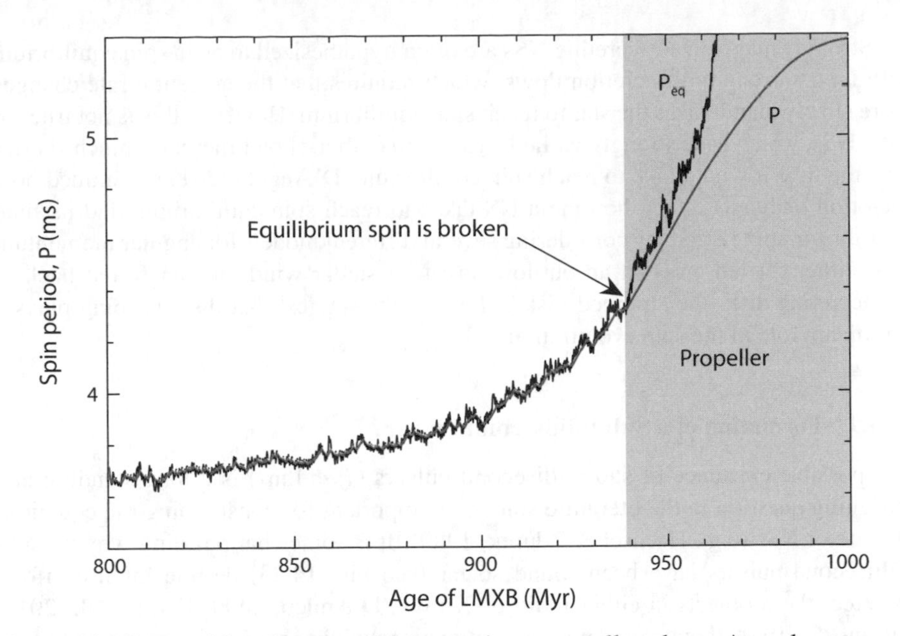

Figure 14.22. Transition from equilibrium spin to propeller phase. At early stages of the Roche-lobe decoupling phase (RLDP) the NS spin is able to remain in equilibrium despite the outward moving magnetospheric boundary caused by decreasing ram pressure. However, at a certain point (indicated by the arrow), when the mass-transfer rate decreases rapidly, the torque can no longer transmit the deceleration fast enough for the NS to remain in equilibrium. This point marks the onset of the propeller phase where $r_{\rm mag} > r_{\rm co}$ (and $P < P_{\rm eq}$) at all times. After Tauris (2012).

In this LMXB model calculation, the value of the spin period increased significantly during the RLDP. The reason for this is that the duration of the propeller phase (i.e., the RLDP) is a substantial fraction of the spin-relaxation timescale, $t_{\rm torque}$. Using Eq. (14.27), we find $t_{\rm torque} = 195$ Myr whereas the RLDP lasts for $t_{\rm RLDP} \simeq 56$ Myr, which is a significant fraction of $t_{\rm torque}$ ($t_{\rm RLDP}/t_{\rm torque} = 0.29$). Therefore, the RLDP has quite a significant impact on the spin evolution of the NS. It is seen that the spin period increases from 3.7 ms to 5.2 ms during this short time interval, that is, the pulsar loses 50% of its rotational energy during the RLDP. As shown by Tauris (2012), this RLDP effect is important for understanding the apparent difference in spin period distributions between AXMSPs and radio MSPs (Hessels, 2008).

In another paper (Tauris et al., 2012), similar RLDP calculations were applied to IMXB systems producing MSPs with massive WD companions. In these IMXB model calculations, the value of the spin period only increased very little during the RLDP. The reason for this is that here the duration of the RLDP is relatively short compared to the spin-relaxation timescale. The resulting small ratio $t_{\rm RLDP}/t_{\rm torque} \simeq 0.04$ at the onset of the RLDP causes the spin period to "freeze" at the original value of $P_{\rm eq}$ (see also Ruderman et al., 1989). Hence, the loss of rotational energy during RLDP is not significant for IMXBs and for Case BB RLO in post-HMXB systems.

Strongly magnetized accreting NSs are often hypothesized to be in spin equilibrium with their surrounding accretion flows, which requires that the accretion rate changes more slowly than it takes the star to reach spin equilibrium. However, this is not true for most NSs, which have strongly variable accretion outbursts on timescales much shorter than the time it would take to reach spin equilibrium. D'Angelo (2017) examined how accretion outbursts affect the time a NS takes to reach spin equilibrium, and its final equilibrium spin period, by considering several different models for angular momentum loss, either carried away in an outflow, lost to a stellar wind, or transferred back to the accretion disk (the "trapped disk"). Her results suggest that disk trapping plays a significant role in the spin evolution of NSs.

14.6.9 Formation of a Sub-millisecond Pulsar?

The possible existence of sub-millisecond pulsars ($P < 1$ ms) is a long-standing and intriguing question in the literature since it is important for constraining the equation-of-state of NSs (e.g., Haensel & Zdunik, 1989). It is somewhat puzzling why no sub-millisecond pulsars have been found so far (see Fig. 14.23) despite intense efforts to detect these objects in either radio or X-rays (D'Amico, 2000; Keith et al., 2010; Patruno, 2010). Although modern observational techniques are sensitive enough to pick up sub-millisecond radio pulsations, the fastest spinning radio MSP among the ~500 known sources, is J1748−2446ad (Hessels et al., 2006), has a spin frequency of only 716 Hz, corresponding to a spin period of 1.4 ms. This spin rate is far from the expected minimum equilibrium spin period and the physical mass shedding limit of ~1,500 Hz.

On the theoretical side, it has been suggested that GW radiation during the accretion phase halts the spin period above a certain level (Bildsten, 1998; Chakrabarty et al., 2003). In a detailed investigation, Patruno et al. (2017) find that different lines of evidence suggest that GWs might be playing an important role in regulating the spin of accreting NSs, although they caution that open questions remain on the exact emission mechanism. As discussed in Section 14.6.8, Tauris (2012) argued that an important issue for understanding the physics of the early spin evolution of radio MSPs is the impact of the expanding magnetosphere during the terminal stages of the mass-transfer process (i.e., the so-called Roche-lobe decoupling phase outlined in the previous section). He demonstrated using detailed binary stellar evolution calculations that the braking torque acting on a NS, when the companion star decouples from its Roche lobe, is able to dissipate more than 50% of the rotational energy of the pulsar. This effect may then explain the apparent difference in observed spin distributions between X-ray and radio MSPs (e.g., Hessels, 2008; Papitto et al., 2014).

We note from Eq. (14.18) that the equilibrium spin period, $P_{\rm eq}$, scales with the location of the inner edge of the accretion disk (roughly given by $r_{\rm mag}$), as $P_{\rm eq} \propto r_{\rm mag}^{3/2}$ and thereby $P_{\rm eq} \propto B^{6/7}\,\dot{M}^{-3/7}$. Therefore, recycling a NS to a small spin period P requires two conditions to be fulfilled: (1) a small value of $r_{\rm mag}$, and (2) a long enough timescale to allow for sufficient spin-up by the applied torque (see Section 14.6.6). However, these two conditions are practically mutually exclusive. On the one hand, in those LMXBs where the mass-accretion rate, $\dot{M}$, is large (thus creating a large ram pressure pushing the size of the magnetosphere to a small value), the evolutionary

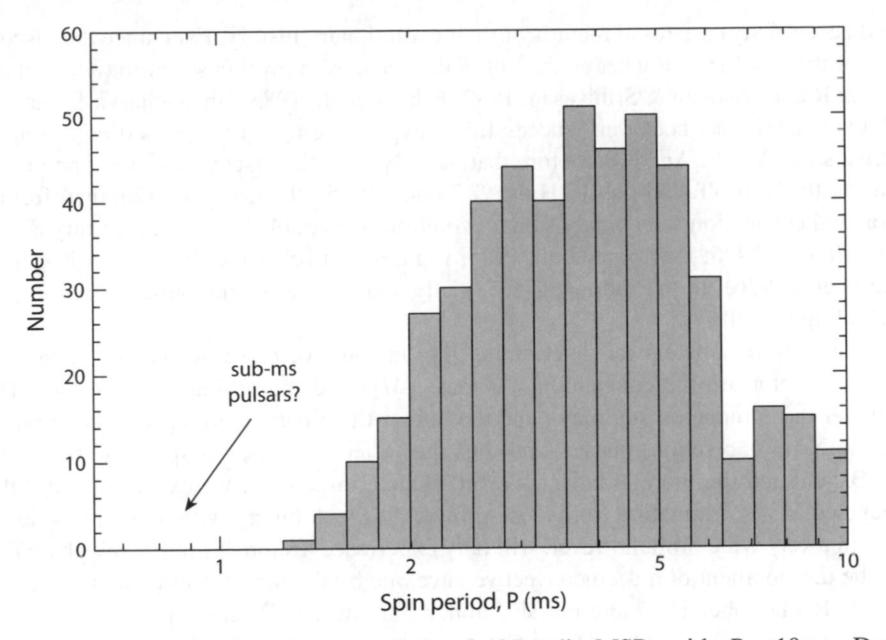

Figure 14.23. Distribution of spin periods of 485 radio MSPs with $P < 10$ ms. Data from ATNF Pulsar Catalogue in September 2022 (Manchester et al. 2005).

timescale of the donor star is small since a large mass-transfer rate can only be sustained if the donor star is either of large mass or an evolved giant. In this case, not only is the net amount of material accreted by the NS limited to a relatively small amount (especially if the Eddington-accretion limit applies), or equivalently, the mass-transfer timescale is smaller than the required spin-relaxation timescale (Section 14.6.6), the B-field of the accreting NS may not have decayed sufficiently to produce a small value of $r_{\rm mag}$, a necessary condition of a small value of P. On the other hand, in those LMXBs where the accreting NS has a small B-field strength, the mass-transfer phase is expected to have been of long duration, meaning that the mass of the donor star, and thus the mass-transfer rate, have been relatively small, whereby $r_{\rm mag}$ may not have reached a sufficiently small value. Finally, even *if* the boundary of the magnetosphere is pushed sufficiently inward and extends all the way down to the surface of the NS, the length of lever arm ($r_{\rm mag}$) for the spin-up torque would then be small, which means that the spin-relaxation timescale may exceed a Hubble time.

We therefore conclude that the reason no sub-millisecond pulsars have been discovered so far in either X-rays or radio, can be explained either by GW radiation or by simple arguments based on binary stellar evolution and magnetospheric conditions, as discussed in this chapter.

14.6.10 Strong Link between Stellar Evolution Theory and Pulsar Properties

As we have discussed in detail in this section, MSPs represent the end point of binary stellar evolution, and their observed properties (spin period, B-field, mass, orbital period, and eccentricity) are fossil records of their evolutionary history. Ever since the discovery of the first MSP (Backer et al., 1982), the standard *recycling* scenario (Alpar et al., 1982; Radhakrishnan & Srinivasan, 1982; Fabian et al., 1983; Bhattacharya & van den Heuvel, 1991) has been very successful in explaining the gross properties of binary pulsars and MSPs. And even before that, shortly after the discovery of the first binary pulsar, the Hulse-Taylor pulsar (Hulse & Taylor, 1975), the first papers on their formation and connection with binary stellar evolution were published (e.g., Flannery & van den Heuvel, 1975; De Loore et al., 1975; van den Heuvel, 1976; Bisnovatyi-Kogan & Komberg, 1976). In the following, we briefly summarize the immense success of the recycling model.

The key parameters for determining the outcome of close binary evolution with a NS accretor are the companion star mass (M_2) and the orbital period ($P_{\rm orb}$). The smaller the companion star mass, and the tighter the binary, the longer is the duration of the X-ray (accretion) phase—and thus the faster the pulsar is spinning, the lower its B-field and the more circular its orbit. Hence, in general, we expect mainly fully recycled MSPs originating from close-orbit LMXBs. A binary with a low-mass donor star in a very wide orbit, however, will only experience a short-lasting X-ray phase due to the development of a deep convective envelope by the time the low-mass giant will fill its Roche lobe. The more massive donor stars in IMXBs can only leave behind a fully recycled MSP (with a He or CO companion) if its orbital period is very small (Case A RLO). Otherwise, the outcome is a more massive WD (CO/ONeMg WD) orbiting a mildly recycled pulsar. HMXBs are even more extreme and only lead to mild (or in extreme cases, only marginal) recycling during the subsequent Case BB RLO following a CE phase, for the first-born NS in a DNS system. This is the connection between stellar evolution and pulsar properties in a nutshell (Bhattacharya & van den Heuvel, 1991; Tauris & van den Heuvel, 2006; Tauris et al., 2012, 2015).

The evidence for this strong link is best illustrated by a couple of plots—see also Figures 14.6 to 14.9. In Figure 14.24 (top panel), we show the spin period distribution of all binary Galactic disk pulsars (with $\dot{P} < 10^{-16}$ and $P_{\rm orb} < 200$ days). And in Figure 14.24 (bottom panel), we show the distribution of their B-fields. As expected, those binary MSPs with He WD companions have the fastest spins and the smallest B-fields, whereas those with more massive WD companions (here denoted CO WDs) have slower spins and stronger B-fields; and systems with two NSs have even slower spins and again stronger B-fields. So it is obvious that there is this clear connection between being either fully, mildly, or marginally recycled and the type of companion star in a given system. The average spin periods of these binary pulsars are 7.2 ms, 28.6 ms, and 57.7 ms with He WD, CO/ONeMg WD, and NS companions, respectively. Their average B-fields are: 1.8×10^8 G, 8.2×10^8 G, and 3.5×10^9 G, respectively. If we consider the spin periods of 53 isolated recycled pulsars in the Galactic disk (here identified as those with a measured spin period derivative, $\dot{P} < 10^{-16}$, and also having a spin period of less than 100 ms to avoid too many non-recycled pulsars in the sample), the average spin period is 13.7 ms and the average B-field is 4.7×10^8 G.

Further pieces of evidence for the strong link and accordance between binary stellar evolution theory and observed binary pulsar properties are shown in Figures 14.14 and 14.25. The former figure shows the correlation between He WD mass and orbital

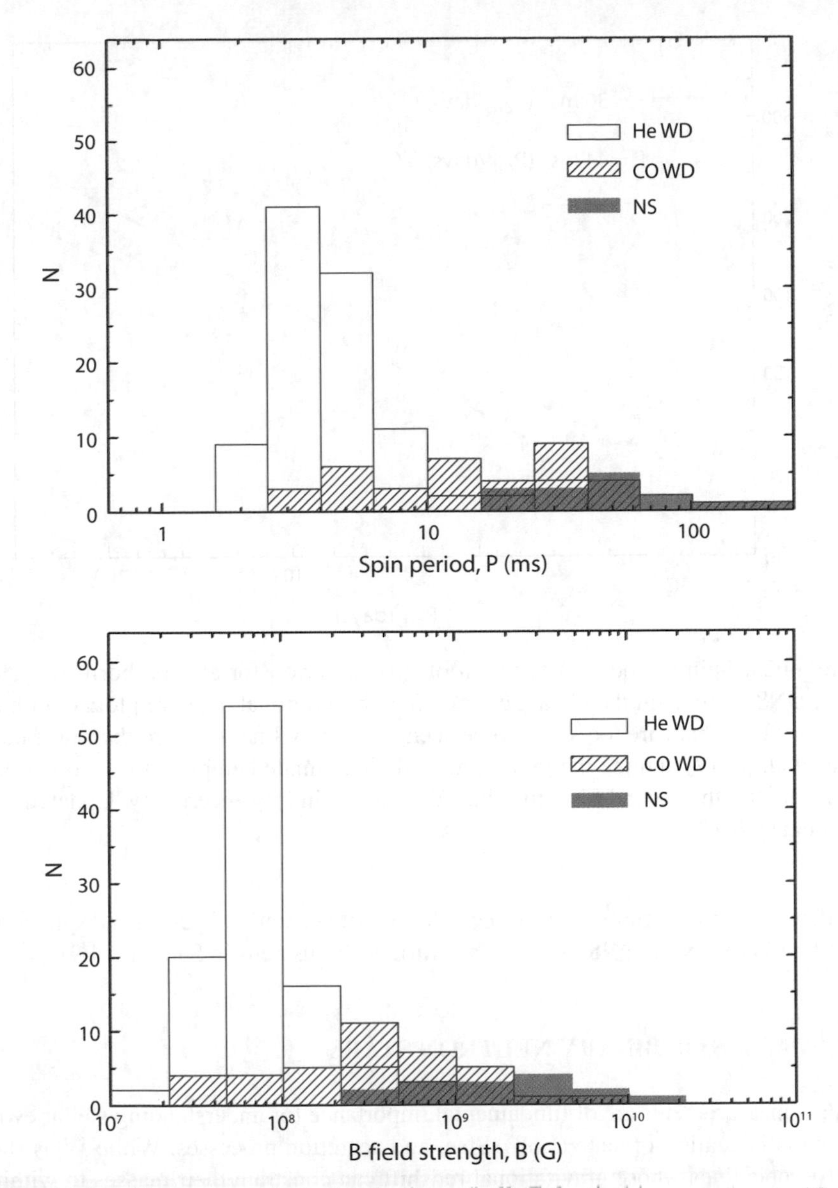

Figure 14.24. Top: spin period distribution of all Galactic binary pulsars with $\dot{P} < 10^{-16}$ and $P_{\rm orb} < 200$ days. Bottom: magnetic dipole flux density (B-field) distribution of the same pulsars. The latter values were calculated according to Eq. (14.15), assuming $\alpha = 60°$ and $M = 1.4\,M_{\odot}$ for all pulsars. The different types of pulsar companion stars are indicated. Data from ATNF Pulsar Catalogue in June 2021 (Manchester et al. 2005).

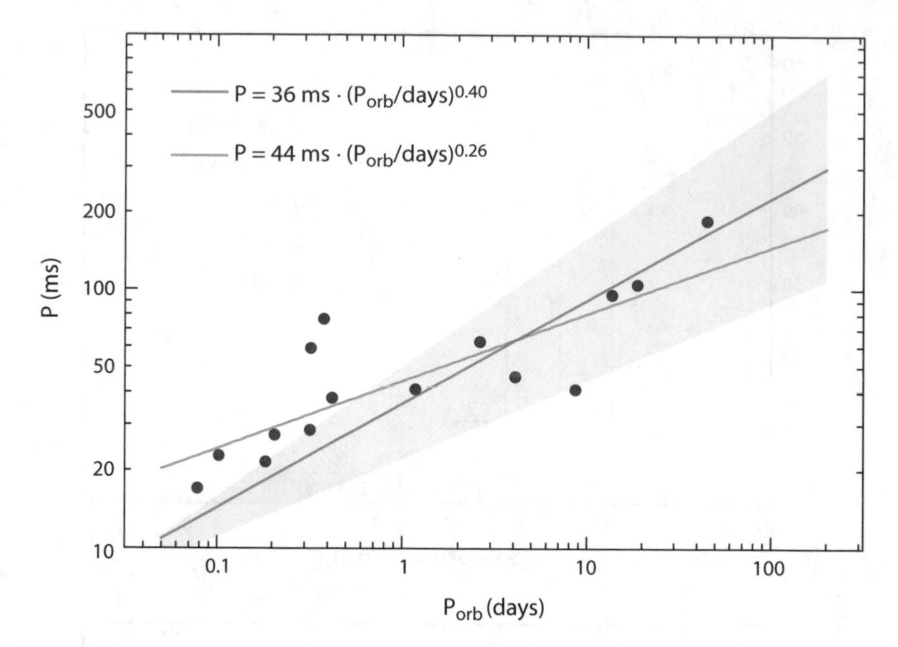

Figure 14.25. Spin period as a function of orbital period for all first-born (recycled) NSs in DNS systems in the Galactic disk. The observational data are plotted with red points (the error bars are too small to be seen). The grey line is a fit to the raw data by Tauris et al. (2017) and the the green line is their estimated empirical relation at birth (i.e., right after the second SN), the shaded region being its uncertainty. Updated after Tauris et al. (2017).

period, and the latter figure shows the correlation between orbital period and spin period of the first-born NSs in DNS systems (see further discussions in Section 14.9).

14.7 MASSES OF BINARY NEUTRON STARS

NS mass measurements are of fundamental importance for understanding stellar evolution, the NS equation-of-state (EoS), SNe, and accretion processes. While WDs show spectroscopic lines whose gravitational redshift can constrain their masses to within a few percent via M/R relations (Barstow et al., 2005; Joyce et al., 2018), NS masses can only be determined in binary systems. Figure 3.5 of Chapter 3 shows how the longitude of periastron $\omega_{\rm p}$ is determined from the shape of the observed radial velocity curve of the pulsar. For $\omega_{\rm p} = 0°$ or $180°$, the curve is symmetrical with respect to periastron passage, while for $\omega_{\rm p} = 90°$ or $270°$ it has an asymmetric sawtooth shape. The observed Doppler-delay curve yields the five orbital elements $K_1 = v_1 \sin i$ (radial velocity amplitude), $P_{\rm orb}$, e, $\omega_{\rm p}$, and T_0 (time of periastron passage), see Figure 3.3. The shape of the curve is characterized by e and $\omega_{\rm p}$. Integration of the radial velocity $v_1 \sin i$ over the orbital period yields $a_1 \sin i$, the projected semi-major axis of the

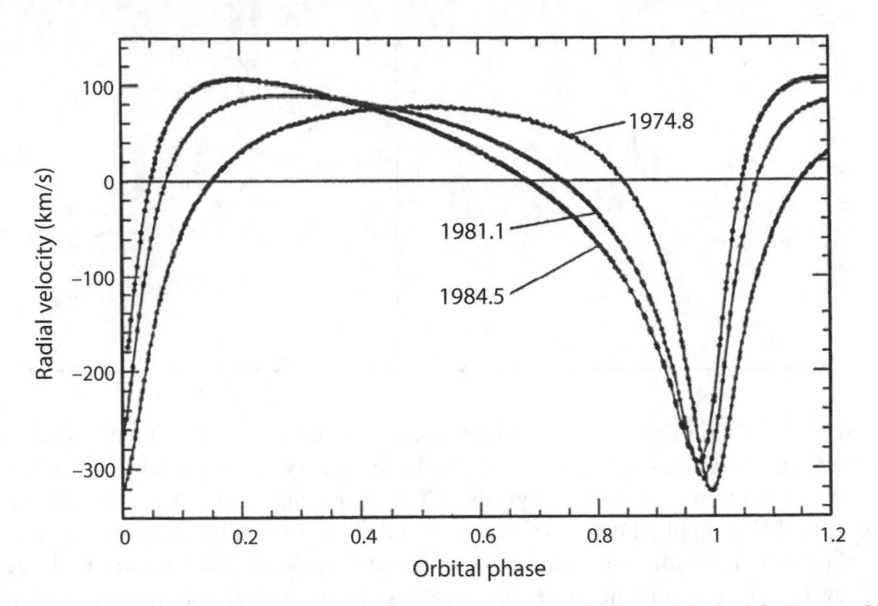

Figure 14.26. The shape of the radial velocity curve of PSR B1913+16 observed at three different epochs over 10 years (after Taylor, 1987). The changes in shape are due to the general relativistic change in the longitude of periastron, $\dot{\omega}_{\mathrm{p}}$ by $4.226598 \pm 0.000005\,\mathrm{deg\,yr^{-1}}$ (Weisberg et al., 2010), that is, some 35,000 times larger than that of Mercury in the Solar System.

observed pulsar. Furthermore, the mass function $f(M)$ can be determined as explained in Chapter 3. In some cases, even the misalignment angle between the pulsar spin axis and the orbital angular momentum vector can be measured (Ferdman et al., 2013). However, all this orbital information is still not sufficient to determine the pulsar mass.

Figure 14.26 shows how the shape of the radial velocity curve of PSR B1913+16 changed between its discovery in 1974 and the year 1984, due to the change in ω_{p}. From the comparison between these graphs, the rate of periastron advance (apsidal precession), $\dot{\omega}_{\mathrm{GR}}$ can be determined. Such relativistic effects hold the key to measuring the NS masses accurately.

14.7.1 Mass Determination of Binary Radio Pulsars Using Special and General Relativity

Pulsars are extremely accurate clocks and the DNSs with very narrow and eccentric orbits such as the Hulse-Taylor pulsar B1913+16 ($P_{\mathrm{orb}} = 7.75\,\mathrm{hr}$) and the double pulsar J0737−3039AB ($P_{\mathrm{orb}} = 2.4\,\mathrm{hr}$) are so close that up to seven post-Keplerian (PK) parameters are observable (i.e., parameters characterizing deviations from Newtonian motion and signal propagation in the timing model to accurately describe the measured pulse times of arrival; see e.g., Damour & Deruelle (1986); Lorimer & Kramer (2012); Wex (2014); Kramer et al. (2021), for more details. We can expect the theory to describe PK parameters as a function of the measured Keplerian (K) parameters and

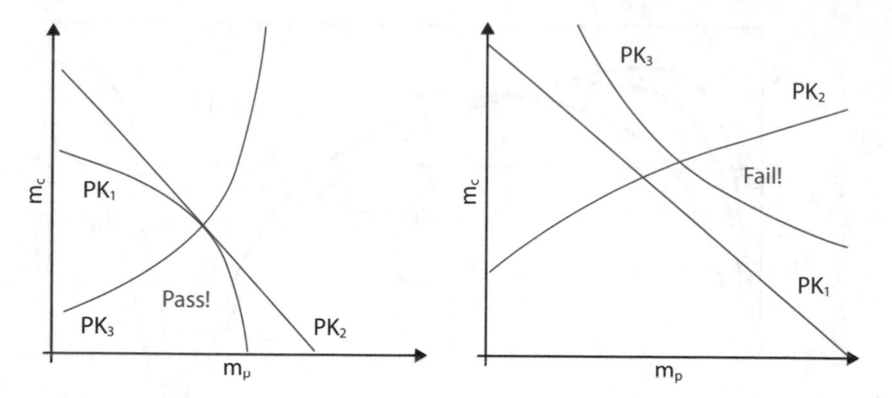

Figure 14.27. Example of three measured PK parameters in a pulsar binary. If all the PK lines intersect in a single point (left panel), the theory is capable of describing the experiment correctly. In cases where the PK lines do not intersect in a single point (right panel), the applied theory of gravity is falsified. In reality, each line has a certain thickness, indicating the measurement uncertainty of the PK parameters. In general, every pair of N measured PK parameters defines an intersection point, and thus allows $(N - 2)$ independent tests of theory. Additional parameters derived independently can also be displayed in the diagram to provide additional constraints. This usually includes a region of the mass–mass diagram excluded by the value of the measured mass-function, the mass ratio, the condition that $\sin i < 1$, or an independent measurement of the orbital inclination angle, i. After Kramer (2018).

the two (apriori unknown) masses of the binary components in the system, $M_{\rm psr}$ and $M_{\rm comp}$, that is, following the functional form $PK = f_{\rm PK}\,\{M_{\rm psr},\ M_{\rm comp},\ K\}$ (Kramer, 2018). Depending on the applied theory of gravity, these functional forms differ. However, with two PK parameters measured, one can solve for the two mass values of the pulsar and its companion. If the applied theory is a correct description of the data, these two masses should accurately predict the value for any other PK parameter in the system. Therefore, each additional PK parameter measured in a given system has the potential for falsifying the tested theory. A theory, which produces the same pair of mass values from any combination of two PK parameters, is consistent with the data. This way of testing a given theory of gravity can be displayed graphically in a so-called mass–mass diagram, where each of the measured PK parameters represents a line, showing its dependence on the two masses (Fig. 14.27). A correct theory will produce lines that all intersect in the same $(M_{\rm psr},\ M_{\rm comp})$ point. A measurement of a number (N) of PK parameters will reveal $N - 2$ independent tests, where in each case the intersection point could in principle differ.

As a result, for the double pulsar, the measurements of the seven[6] PK parameters (Table 14.6), by careful monitoring of the pulse arrival times on Earth, yield seven

[6]Including the first successful detection of the relativistic deformation of the orbit, δ_θ (Kramer et al., 2021), and geodetic precession of pulsar B (Breton et al., 2008).

different equations for the two unknowns M_{psr} and M_{comp}. Hence these unknowns are in fact overdetermined. For the binary pulsars listed in Table 14.5 and plotted in Figure 14.28, it has been found that the solutions of these PK equations are fully consistent with one another.

The first system to which this analysis of the relativistic effects was applied was the binary radio pulsar B1913+16, discovered in 1974 (Hulse & Taylor, 1975). One of the relativistic effects is the rate of decrease of the orbital period, $\dot{P}_{orb}$, due to the emission of GWs. At the time of the first measurement of this decrease by Taylor and Weisberg (Taylor & Weisberg, 1982, 1989), GWs had not yet been directly measured on Earth, so their existence had not been demonstrated experimentally.[7] Therefore, this relativistic effect could not yet (in principle) be used to determine the masses of the two compact stars. Instead, one could determine the masses from other general relativistic effects that had been well verified and established by measurements in the Solar System and on Earth (such as the Mössbauer effect).

One of these effects is the average rate of the periastron advance, $\dot{\omega}_{GR}$:

$$\langle \dot{\omega}_{GR} \rangle = \frac{3\,G^{2/3}}{c^2} \left(\frac{2\pi}{P_{orb}} \right)^{5/3} \frac{(M_{psr} + M_{comp})^{2/3}}{(1 - e^2)}, \tag{14.33}$$

which was measured to be $\sim$4.22 deg yr^{-1} for PSR B1913+16 (Taylor et al., 1976) shortly after its discovery (see Fig. 14.26). Another effect is caused by variations in gravitational redshift and second order Doppler effect, γ_{GR}:

$$\gamma_{GR} = \frac{G^{2/3}}{c^2} \left(\frac{P_{orb}}{2\pi} \right)^{1/3} e \, \frac{M_{comp}\,(M_{psr} + 2M_{comp})}{(M_{psr} + M_{comp})^{4/3}}. \tag{14.34}$$

Since all quantities in the right-hand side of the above two equations are known except for the two masses, these masses can be solved for by measuring the two PK parameters $\dot{\omega}$ and γ. In this way, accurate values of NS masses have been determined in DNS systems (Table 14.5), as well as in some NS+WD systems with close to edge-on orbital inclination or tight and/or eccentric orbits (see, e.g., Jacoby et al., 2005; Bhat et al., 2008; Demorest et al., 2010; Freire et al., 2012). Observations of PSR B1913+16 by Weisberg et al. (2010); Weisberg & Huang (2016) resulted in high-precision measurements of $\dot{\omega}$ and γ_{GR} (with error bars of 0.0001% and 0.02%, respectively) and thus revealed a total system mass of $2.828378 \pm 0.000007\,M_\odot$, and component masses of $M_{psr} = 1.4398 \pm 0.0002\,M_\odot$ and $M_{comp} = 1.3886 \pm 0.0002\,M_\odot$. These are (along with those of the double pulsar, see Section 14.7.4) the most accurately known stellar masses in all of astronomy until now, aside from the mass of the Sun.

[7]The first direct measurement of GWs was performed by LIGO in 2015, see Chapter 15.

Table 14.5. Properties of 22 DNS Systems with Published Data (including a Few Unconfirmed Candidates).

Radio Pulsar	Type	P (ms)	$\dot{P}$ (10^{-18})	B (10^9 G)	P_{orb} (days)	e	M_{psr} ($M_\odot$)	M_{comp} ($M_\odot$)	M_{total} ($M_\odot$)	δ (deg)	Dist. (kpc)	v^{LSR} ** (km s^{-1})	τ_{GW} (Myr)
J0453+1559[a]	recycled	45.8	0.186	1.1	4.072	0.113	1.559	1.174	2.734	—	1.07	82	∞
J0509+3801[b]	recycled	76.5	7.93	7.2	0.380	0.586	~1.34	~1.46	2.805	—	1.56	—	579
J0737−3039A[c]	recycled	22.7	1.76	1.8	0.102	0.088	1.338	1.249	2.587	< 3.2	1.15	32	86
J0737−3039B[c]	young	2773.5	892	410	—‖—	—‖—	1.249	1.338	—‖—	130 ± 1	—‖—	—‖—	—‖—
J1325−6253[u]	recycled	29.0	0.0480	0.37	1.815	0.064	<1.59	>0.98	2.57	—	4.4	—	∞
J1411+2551[d]	recycled	62.5	0.0956	0.66	2.616	0.170	<1.62	>0.92	2.538	—	1.13	—	∞
J1518+4904[e]	recycled	40.9	0.0272	0.33	8.634	0.249	***	***	2.718	—	0.63	30	∞
B1534+12[f]	recycled	37.9	2.42	2.8	0.421	0.274	1.333	1.346	2.678	27 ± 3	1.05	143	2730
J1753−2240[g]	recycled	95.1	0.970	2.5	13.638	0.304	—	—	—	—	3.46	—	∞
J1755−2550[h]*	young	315.2	—	270	9.696	0.089	—	>0.40	—	—	10.3	—	∞
J1756−2251[i]	recycled	28.5	1.02	1.6	0.320	0.181	1.341	1.230	2.570	< 34	0.73	39****	1660
J1757−1854[j]	recycled	21.5	2.63	2.2	0.184	0.606	1.341	1.392	2.733	—	19.6	—	76
J1759+5036[v]	recycled	176.0	0.243	2.0	2.043	0.308	<1.93	>0.701	2.63	—	0.71	—	∞
J1811−1736[k]	recycled	104.2	0.901	2.7	18.779	0.828	<1.64	>0.93	2.57	—	5.93	—	∞
J1829+2456[l]	recycled	41.0	0.0495	0.42	1.176	0.139	1.306	1.299	2.606	—	0.91	49	∞
J1906+0746[m]*	young	144.1	20300	470	0.166	0.085	1.291	1.322	2.613	—	7.40	—	309
J1913+1102[n]	recycled	27.3	0.157	0.83	0.206	0.090	1.62	1.27	2.889	—	—	—	471

B1913+16[o]	recycled	59.0	8.63	7.3	0.323	0.617	1.440	1.389	2.828	18 ± 6	9.80	241	301
J1930−1852[p]	recycled	185.5	18.0	16	45.060	0.399	<1.32	>1.30	2.59	—	1.5	—	∞
J1946+2052[q]	recycled	17.0	0.92	1.0	0.078	0.064	<1.31	>1.18	2.50	—	1.5	—	46
J0514−4002A[r]*	GC	5.0	0.00070	0.016	18.79	0.888	~1.25	~1.22	2.473	—	12.1	—	∞
J1807−2459B[s]*	GC	4.2	0.0823	0.18	9.957	0.747	1.366	1.206	2.572	—	3.0	—	∞
B2127+11C[t]	GC	30.5	4.99	3.7	0.335	0.681	1.358	1.354	2.713	—	12.9	—	217

$\dot{P}$ values are not corrected for the (in most cases negligible) Shklovskii effect (Shklovskii, 1970). Similarly for B-field values estimated from Eq. (14.15) using $\alpha = 60^\circ$. All the NS masses quoted with four significant figures have uncertainties of $\lesssim 0.010\,M_\odot$. Distances are based on the NE2001 DM distance model. All values of $\tau_{GW} > 50$ Gyr are shown with ∞.

*Not a confirmed DNS system—could also be a WD+NS binary.

**Based on the median value of v^{LSR}, (cf. Tauris et al., 2017). Note, large uncertainty in general.

***Preliminary $M_{NS,1} = 1.41\,M_\odot$ and $M_{NS,2} = 1.31\,M_\odot$ (G. Janssen 2017, private communication).

****Based on the 2D proper motions estimated in Ferdman et al. (2014).

References: [a]Martinez et al. (2015). [b]Lynch et al. (2018). [c]Kramer et al. (2006); Breton et al. (2008); Ferdman et al. (2013). [d]Martinez et al. (2017). [e]Janssen et al. (2008). [f]Fonseca et al. (2014). [g]Keith et al. (2009). [h]Ng et al. (2015, 2018). [i]Faulkner et al. (2005); Ferdman et al. (2014). [j]Cameron et al. (2018, 2022). [k]Corongiu et al. (2007). [l]Champion et al. (2004); Haniewicz et al. (2021). [m]Lorimer et al. (2006); van Leeuwen et al. (2015). [n]Lazarus et al. (2016); Ferdman et al. (2020). [o]Hulse & Taylor (1975); Kramer (1998); Weisberg et al. (2010). [p]Swiggum et al. (2015). [q]Stovall et al. (2018). [r]Ridolfi et al. (2019). [s]Lynch et al. (2012). [t]Anderson et al. (1990); Jacoby et al. (2006). [u]Sengar et al. (2022). [v]Agazie et al. (2021).

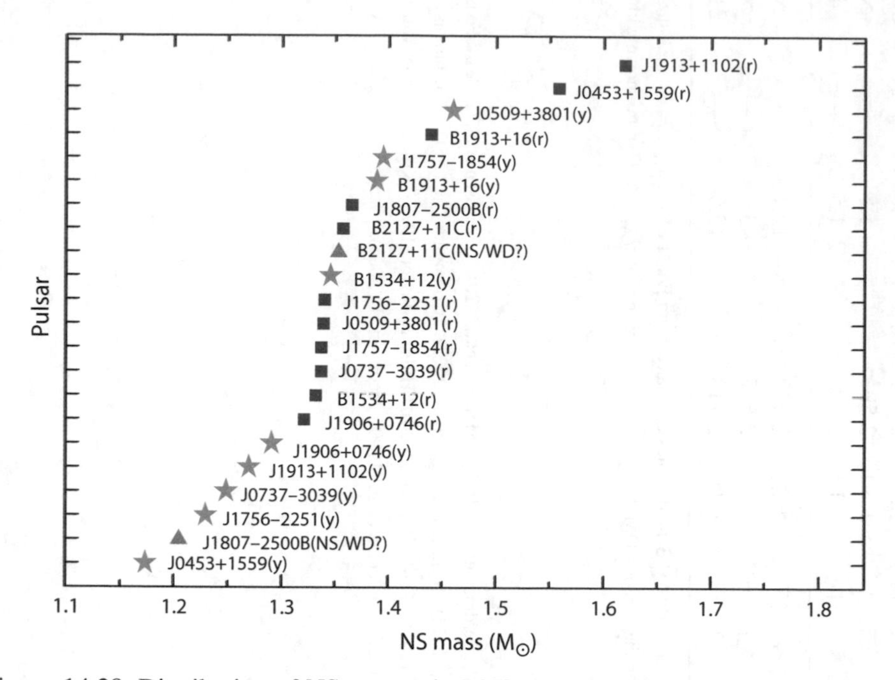

Figure 14.28. Distribution of NS masses in DNS systems (Feb. 2021). Recycled pulsars are indicated by (r) and blue squares; young pulsars are indicated by (y) and green stars. See also Figure 14.33 for further info and a comparison to MSP masses. A concentration of recycled DNS masses is noticed at $1.33 - 1.36\, M_\odot$.

14.7.2 Confirmation of the Existence of Gravitational Radiation

Inserting the thus determined values of these masses for PSR B1913+16 into the general relativistic formula for $\dot{P}_{\rm orb}$ (Peters, 1964):

$$(\dot{P}_{\rm orb})_{\rm GR} = -\frac{192\pi\, G^{5/3}}{5\, c^5} \left(\frac{P_{\rm orb}}{2\pi}\right)^{-5/3} \left(1 + \frac{73}{24}e^2 + \frac{37}{96}e^4\right) \times (1-e^2)^{-7/2}\, \frac{M_{\rm psr}\, M_{\rm comp}}{(M_{\rm psr} + M_{\rm comp})^{1/3}} \tag{14.35}$$

then yields a theoretical prediction for the rate of decrease of the orbital period of this system by the emission of GWs, according to Einstein's version of general relativity. $\dot{P}_{\rm orb}$ can be accurately measured by monitoring the times of periastron passage of the pulsar (Fig. 14.29). The value was measured first by Taylor & Weisberg (1982), and their revised value seven years later (Taylor & Weisberg, 1989) was $\dot{P}_{\rm orb} = -2.427(\pm 0.026) \times 10^{-12}$. The theoretically predicted value at the same time, obtained by inserting the masses obtained from $\dot{\omega}$ and γ into Eq. (14.35) is $(\dot{P}_{\rm orb})_{\rm GR} = -2.40216 \times 10^{-12}$.

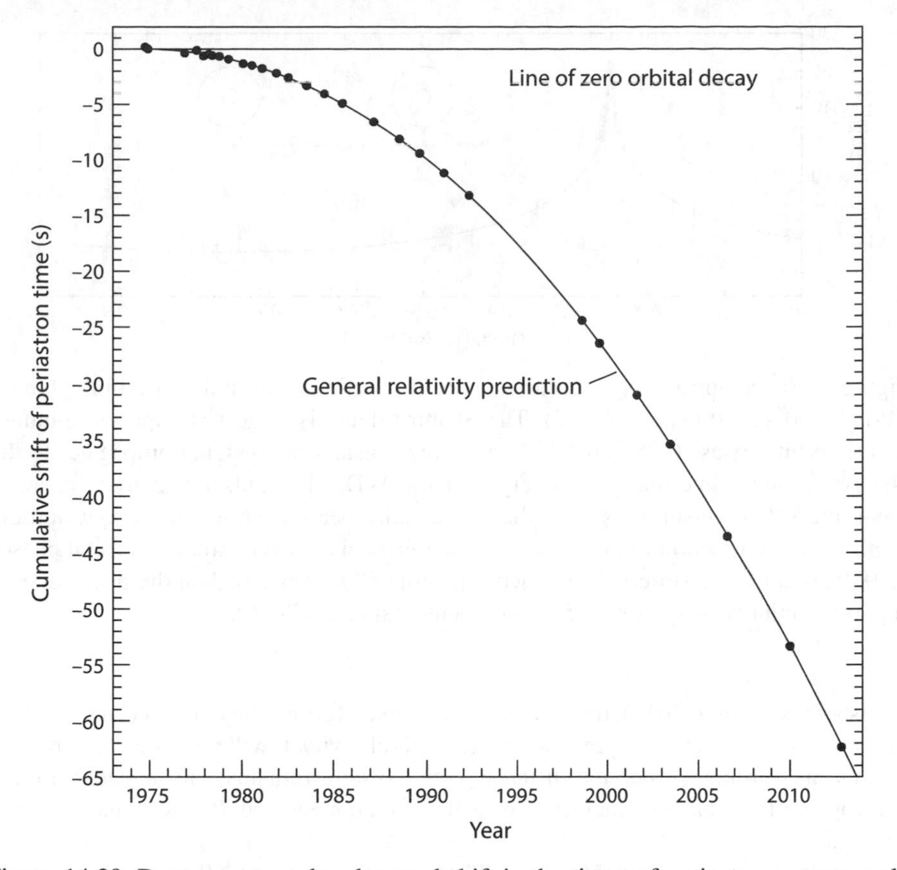

Figure 14.29. Dots represent the observed shift in the times of periastron passage relative to a constant orbital period for PSR B1913+16. The curve is the theoretically predicted shift for orbital decay due to the emission of GWs from the system, according to general relativity. The theoretically predicted and observed values of the rate of orbital decay agree now to within the observational uncertainties of 0.2% with each other. After Weisberg & Huang (2016).

The observed and theoretically predicted values of $\dot{P}_{\rm orb}$ were thus found to be in excellent agreement with each other (within the observational errors in 1989 of less than 1%), confirming in an indirect way the existence of gravitational radiation. These highly accurate confirmations of effects predicted by the general theory of relativity, by using the first-discovered binary pulsar, resulted in awarding the 1993 Nobel Prize in Physics to the discoverers of this pulsar, Taylor and Hulse. (The Nobel Prize in Physics 2017 was awarded to Weiss, Barish, and Thorne for their contribution to the LIGO detector, which discovered the first direct signals of GWs from the BH+BH merger event GW150914, see the next chapter.) Timing the Hulse-Taylor pulsar over a longer baseline now shows that the observed orbital period decrease caused by GW damping (corrected by a kinematic term) to the general relativistic prediction is 0.9983 ± 0.0016

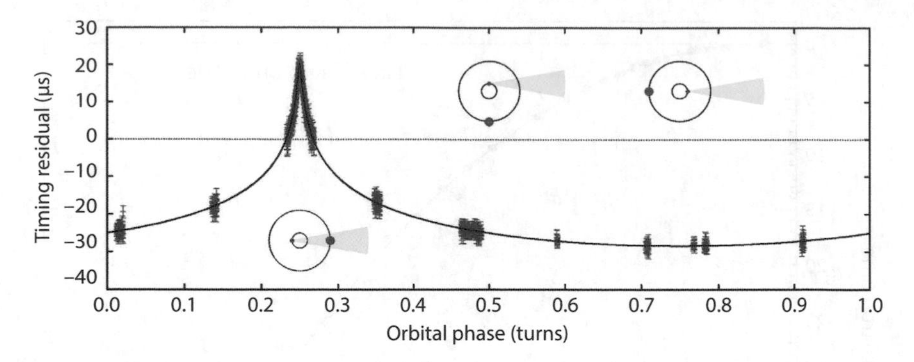

Figure 14.30. Shapiro-delay measurement as a function of orbital phase in the binary MSP J1640−2230 ($P_{\rm orb} = 8.69$ d). The Shapiro delay is largest at superior conjunction (at orbital phase 0.25, left orbit) where the pulsar's radio signal propagates to the observer along a line that passes very near the WD. (The pulsar is marked as a red point, the WD companion is blue, the pulsar radio beam is shown in yellow, and the Earth is in the direction to the right.) The Shapiro delay is very small at orbital phases 0.00, 0.50, and 0.75 (inferior conjunction), some 50 μs smaller than the peak value at superior conjunction (phase 0.25). After Demorest et al. (2010).

(Weisberg & Huang, 2016), thereby confirming even further the existence and strength of gravitational radiation as predicted by general relativity. It will be difficult to improve on this measurement since the distance to the system remains with a sizeable error bar, causing the Galactic kinematic correction to the measured $\dot{P}_{\rm orb}$ to remain slightly uncertain.

Besides from measurements of the PK parameters $\dot{\omega}_{\rm p}$, γ, and $\dot{P}_{\rm orb}$, the excess propagation delay (Shapiro, 1964) caused by the passage of pulsar signals through the curved spacetime of the companion star (largest at the pulsar's superior conjunction, see Fig. 14.30) has also been measured in a number of binary radio pulsar systems. Damour & Deruelle (1986) characterize the measurable quantities as *range*, $r = (GM_{\rm comp}/c^3)$, and *shape*, $s \equiv \sin i$, of the Shapiro delay, where i is the orbital inclination. These measurements are important for further testing relativistic gravity (e.g., Stairs et al., 1998). Another interesting, and in some cases measurable, relativistic phenomenon is geodetic precession (Damour & Ruffini, 1974; Barker & O'Connell, 1975), also known as de Sitter precession, that is, precession of the spin axis due to orbital motion. A related phenomenon is the Lense-Thirring effect (*frame dragging*) which is caused by a rotating central mass. The first detection of this effect, although in combination with the effect of the rotationally induced quadrupole, was recently made for the binary pulsar J1141−6545 (Krishnan et al., 2020), which is in a tight and eccentric orbit with a rapidly spinning WD as companion. The total precession is calculated by combining the geodetic precession with the Lense-Thirring precession.

PSR J1141−6545 is also noteworthy for the formation order of its compact object components since here the radio pulsar formed *after* the WD (Tauris & Sennels, 2000). In 99% of the ~300 known NS+WD systems, the NS formed first. But in a few cases, the primary star of the progenitor system was just not massive enough to leave behind a NS, and a WD was formed instead. However, if its companion star, the secondary star of the progenitor system, was of similar (or slightly lower) mass, it may have accreted sufficient material via RLO (while the primary star was a giant donor) to push it over the SN threshold and produce a NS as the second-formed compact object (Tutukov & Yungelson, 1993a; Portegies Zwart & Yungelson, 1999; Tauris & Sennels, 2000; Davies et al., 2002; Kalogera et al., 2005; Church et al., 2006; Zapartas et al., 2017; Ng et al., 2018; Krishnan et al., 2020). The progenitor binaries to such WD+NS systems (PSR J1141−6545, PSR B2303+46 and possibly PSR J1755−2550 [Ng et al., 2018]) would be WD-Be/X-ray binaries, that is, Be/X-ray binaries where the accreting object is a WD, rather than the usual NS (or in rare cases a BH). Out of approximately 100 known HMXBs (mostly Be/X-ray binaries) in the SMC, a few systems are indeed such WD-Be/X-ray binaries (Coe et al., 2020; Kennea et al., 2021). In our Galaxy, the naked-eye star γ Cassiopeia, which has a $\sim 1\,M_\odot$ unseen companion in a 203-day circular orbit, might also be such a system (Nemravová et al., 2012). (This was the first star ever to be designated a Be-star, see Chapter 7.)

14.7.3 Further Gravity Tests

Unlike Einstein's general relativity, in alternative theories, such as those violating the strong equivalence principle (SEP), one expects a modification in the emission of GWs. That is, in addition to quadrupolar emission and higher multipoles, dipolar GWs are also emitted due to an effective dipole term from additional gravitational "charges." One class of such alternative theories is the so-called scalar-tensor theories (Damour & Esposito-Farese, 1992; Will, 2014). In DNS systems, the masses of the two components are rather similar. Thus, the dipole term is naturally much smaller than in the case of systems where the compactness of the binary components is sufficiently different. For testing such gravity theories, one therefore seeks experiments with pulsar binaries where the compactness of the companion star is either much lower (NS+WD systems) or higher (NS+BH) than that of the NS. Since no NS+BH systems have been discovered yet, one has to rely on such experiments in NS+WD systems (Freire et al., 2012; Antoniadis et al., 2013). So far, no deviations from Einstein's general relativity have been seen.

Einstein's general relativity is based on the universality of free fall, which specifies that all objects (even bodies with strong self-gravity) accelerate identically in an external gravitational field. This strong equivalence principle (SEP), thus extending the weak equivalence principle (WEP), is in contrast to almost all alternative theories of gravity. Therefore, a detection of a SEP violation would directly falsify general relativity. The relativistic hierarchical triple MSP J0337+1715 (composed of an inner binary with a NS and a low-mass WD with an orbital period of 1.63 days, and another outer WD orbiting with a period of 327 days) provides an excellent laboratory to test whether the two compact objects of the inner binary actually fall with the same acceleration in the gravity of the outer WD. Archibald et al. (2018) carried out this direct test of the SEP and found that the pulsar and the inner WD experience accelerations that differ fractionally by $|\Delta| < 2.6 \times 10^{-6}$ (95% confidence), and thereby improved on the previous

Table 14.6. Post-Keplerian (PK) parameters measured for PSR J0737−3039AB. Figures in parentheses represent uncertainties in the last quoted digits. Six of these PK parameters are included in Figure 14.31. After Kramer et al. (2021).

$\dot{\omega}$	secular advance of periastron	16.899323(13) deg yr^{-1}
γ	amplitude of Einstein delay for pulsar A	0.384045(94) ms
$\dot{P}_{orb}$	secular change in orbital period	$-1.247920(78) \times 10^{-12}$
s	shape of Shapiro delay for pulsar A	0.999936(10)
r	range of Shapiro delay for pulsar A	6.162(21) μs
Ω_B^{geod}	rate of geodetic precession of pulsar B	4.77(66) deg yr^{-1}
δ_θ	relativistic deformation of orbit	$13(13) \times 10^{-6}$

limit by almost three orders of magnitude. Recently, this limit has been reduced even further by Voisin et al. (2020).

For a detailed treatment of the relativistic effects in binary pulsars, and for testing theories of gravity using pulsars, we refer to the books of Lorimer & Kramer (2012), Will (2014), and to the review of Wex (2014).

14.7.4 The Double Pulsar

A system in which relativistic effects have been verified even more accurately than in PSR B1913+16 is that of the double pulsar, PSR J0737−3039AB, discovered by Burgay et al. (2003), Lyne et al. (2004). This is the only binary pulsar known in which both NSs are observed as radio pulsars. Careful timing of this system (Kramer et al., 2006; Breton et al., 2008; Kramer et al., 2021) has revealed remarkable precision of measured parameters (Table 14.6) and unprecedented tests of gravity (Kramer & Wex, 2009; Kramer et al., 2021), for example, as illustrated in the mass–mass diagram in Figure 14.31. The double pulsar has an orbital period, $P_{orb} = 0.1022515592973(10)$ days, and an eccentricity, $e = 0.087777023(61)$. From the measurement of the shape, s, of the Shapiro delay, the orbital inclination of the system is constrained to be $i = 89.35 \pm 0.05°$, that is, almost completely edge-on as seen from the Solar System. An additional unique feature of the system is its nature as a dual-line source, which then yields the mass ratio between the two observable pulsars (A and B) as: $R = 1.07152(2)$ (see the red line in Fig. 14.31). The masses of the two pulsars can be derived to extreme precision based on all the measured PK parameters and the results for the recycled and the young pulsar are $m_A = 1.338185^{+0.000012}_{-0.000014}\, M_\odot$ and $m_B = 1.248868^{+0.000013}_{-0.000011}\, M_\odot$, respectively.

As a consequence of the extended pulsar magnetosphere, during the superior conjunction the signals of pulsar A pass close to pulsar B at a distance of only 20,000 km (Wex, 2014) which causes its radio signals to become eclipsed for ~30 s of the orbital period due to absorption by the extended plasma trapped in the magnetosphere of pulsar B (Lyne et al., 2004). The eclipsing pulses also provide a successful test of spin precession of pulsar B (the young and non-recycled pulsar) from the fact that the pulsar's axis wobbles around like a top as it spins. Breton et al. (2008) found that over 75 yr, the

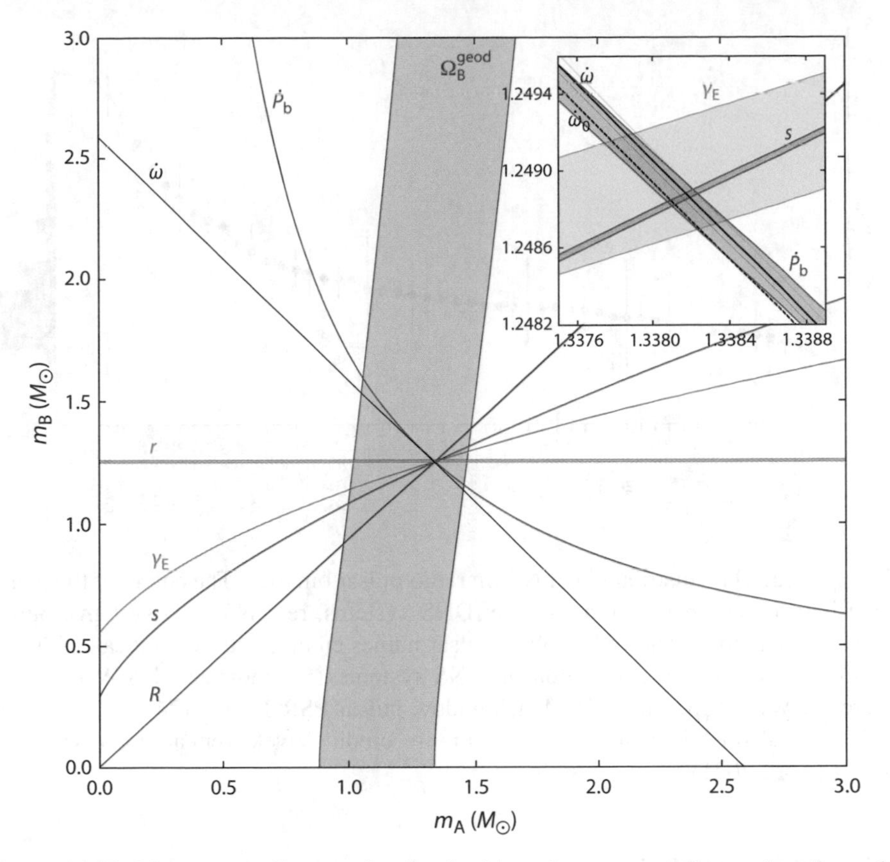

Figure 14.31. Mass–mass diagram for the double pulsar system. Constraints from the various measured relativistic effects are shown by pairs of lines representing the uncertainty of the measurement (not visible for the most precisely measured effects such as $\dot{\omega} = \dot{\omega}_{GR}$, see Table 14.6 for values). The inset shows the enlarged region around the intersection of the various constraints which yield a very precise mass determination of the two NSs: $m_A = 1.338185^{+0.000012}_{-0.000014}\ M_\odot$ and $m_B = 1.248868^{+0.000013}_{-0.000011}\ M_\odot$. See original paper for further details. After Kramer et al. (2021), adapted and provided by Norbert Wex.

radio emission beam of pulsar B will wobble in a full circle. In March 2008, pulsar B therefore disappeared since its radio beam is simply no longer pointing in the direction toward the Solar System. However, it is expected to return in the 2020s. Finally, we note that effects like the Lense-Thirring precession of the orbit will become measurable with the SKA radio telescope (Kehl et al., 2018; Hu et al., 2020). Such a measurement will enable the double pulsar to help constraining the equation-of-state at supra-nuclear densities in NSs.

In Section 15.1, we highlight the importance of the discovery of the double pulsar for a final confirmation of the suggested evolution model for producing DNS systems via recycling.

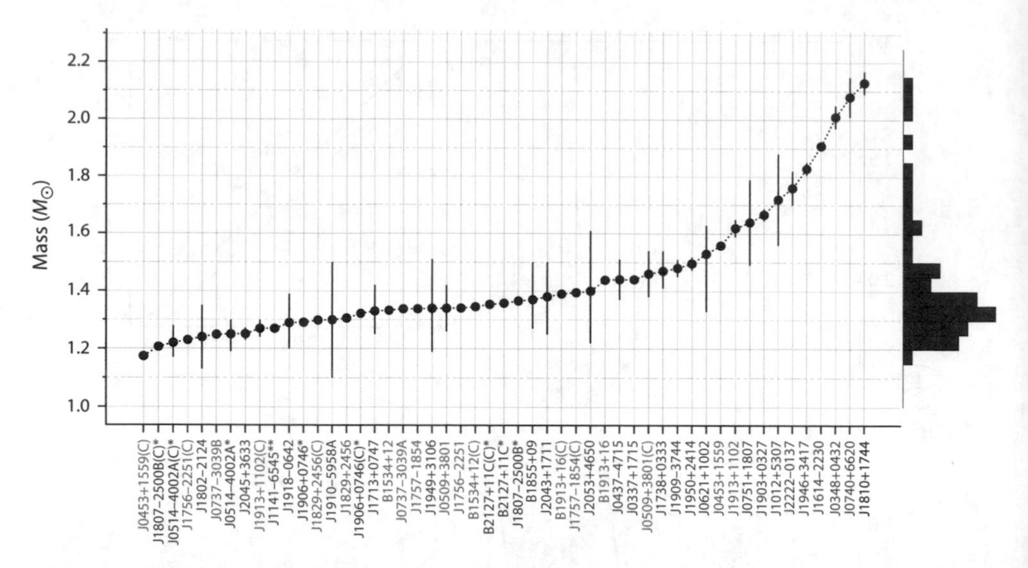

Figure 14.32. Mass distribution of NSs in radio pulsar binaries. The colors of the pulsar name indicate companion types—blue: DNS systems, red: NS+WD systems, purple and (*): unsettled if DNS or NS+WD. Pulsar names ending with (C) indicate NSs that are unseen radio pulsar companions in DNS systems, (**) indicate a NS+WD system where the WD formed first. The black widow pulsar PSR J1810+1744 is included as well. Vertical lines indicate error bars. Figure credit: Vivek Venkatraman Krishnan, MPIfR (April 2021).

14.7.5 Masses of MSPs

NS masses can also be well constrained in NS+WD binaries, using optical spectroscopy of the WD companion. Hence, these systems become dual-line systems, where the Doppler shifts in the spectral lines can be used, together with the timing observations of the pulsar, to determine the mass ratio, R. By modelling the WD, taking into account its atmosphere and cooling history, its mass can be deduced, which then also reveals the NS mass. With this method, possibly the most massive of the accurately determined NS masses has been found ($2.01 \pm 0.04\, M_\odot$ for PSR J0348+0432, Antoniadis et al., 2013). However, note also PSRs J0740+6620 and J1810+1744 which are good candidates of even more massive NSs.[8] Shapiro delay measurements of the former yields $M_{\rm NS} = 2.08 \pm 0.07\, M_\odot$ (Cromartie et al., 2020; Fonseca et al., 2021), whereas analysis of the latter "spider" yields a mass determination of $M_{\rm NS} = 2.13 \pm 0.04\, M_\odot$ (Romani et al., 2021). The distribution of derived NS masses in radio pulsar systems is shown in Figure 14.32 and discussed further in an evolutionary context in Section 14.7.6.

In MSP binary systems, the orbits become efficiently circularized during the recycling process (Section 14.6), while leaving behind a WD companion. In such systems

[8]Less precisely constrained is J2215+5135 with $M_{\rm NS} = 2.27 \pm 0.16 M_\odot$ (Linares et al. 2018).

with very small eccentricities, the PK parameters $\dot{\omega}_p$ and γ are generally impossible to measure. There are a few cases where the orbit is seen sufficiently edge-on, so that a measurement of the Shapiro delay gives access to the two PK parameters r and s with good precision (Ryba & Taylor, 1991; Demorest et al., 2010). With these two parameters, the system is then fully determined, and in principle can be used for a gravity test in combination with a measurement of, for example, $\dot{P}_{orb}$ as a third PK parameter.

14.7.6 NS Masses from an Evolutionary Point of View

A histogram comparing the masses of NSs in DNS and NS+WD systems is shown in Figure 14.33. There is a remarkable difference between the masses of the young, non-recycled pulsars and those of the old, recycled pulsars—especially for the NS+WD systems. Whereas the average NS (birth) mass for the 11 DNS systems and the 3 WD+NS systems (in which the NS formed *after* the WD) are both 1.30 $M_{\odot}$, the average mass of the recycled NSs for the 11 DNS systems and the 24 NS+WD systems (the usual MSP systems where the NS formed *before* the WD) is 1.40 $M_{\odot}$ and 1.55 $M_{\odot}$, respectively. However, this mass difference between non-recycled and recycled NS masses need not be dominated by accretion of matter after the formation of the NS. As argued by Tauris et al. (2017), the first-born NSs in DNS systems probably only accrete $\lesssim 0.02\,M_{\odot}$, based on an analysis of the various accretion phases and observational data of both mass and spin periods. There is no doubt that NSs in LMXB systems (e.g., producing MSP binaries with He WD companions) are able to accrete significantly more mass than the first-born NSs in DNS systems, since the timescale of evolution in the former systems is much longer. Yet, there are many examples of fully recycled MSPs that have small masses of $1.2-1.4\,M_{\odot}$. Since their original NS masses could not be much smaller than $1.2\,M_{\odot}$ (Tauris et al., 2015; Tauris & Janka, 2019, and references therein) these systems show clear evidence of very inefficient accretion. Therefore, we conclude that the differences in masses between non-recycled and recycled NSs are mainly caused by a difference in the birth masses of NSs born first or second in a binary system. Thus these mass differences are likely to be the result of the progenitor star evolution. As shown in Chapter 8 for Fe-core collapses, a large range of birth masses of NSs is expected. In tight binaries producing DNS systems (e.g., PSR J0737−3039) or WD+NS systems (e.g., PSR J1141−6545), ultra-stripping of the secondary star prior to the SN explosion (Tauris et al., 2015), as shown in Chapter 13, may partly explain why these NSs forming as the second compact object in a binary have lower NS masses, on average, compared to those NSs that form first and later become recycled.

An Anti-correlation between Pulsar Mass and Orbital Period?

Assuming all NSs to be born with a mass of 1.3 $M_{\odot}$, Tauris & Savonije (1999) demonstrated that an (P_{orb}, M_{NS}) anti-correlation would be expected for binary MSPs as a simple consequence of the interplay between mass-transfer rate (and thus accretion rate), orbital period, and the evolutionary status of the LMXB donor star at the onset of the RLO. However, since this model predicted rather massive ($> 2\,M_{\odot}$) NSs in binary MSP systems with $P_{orb} \lesssim 30$ days, it failed to explain, for example, the mass of

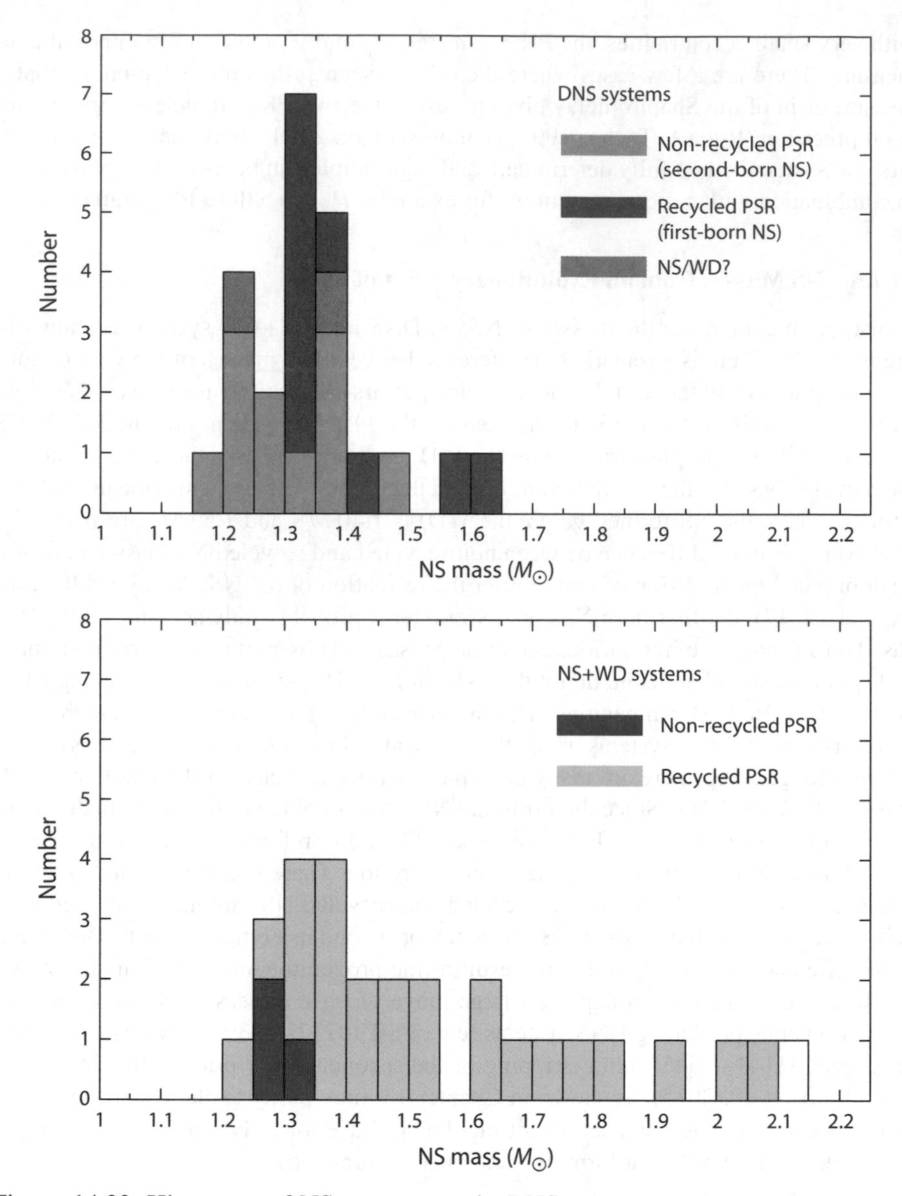

Figure 14.33. Histogram of NS star masses in DNS systems (top) and NS+WD systems (bottom). Recycled pulsars are in blue or yellow color; young pulsars are in green or red color. It is evident that NSs formed second in these binaries (i.e., the non-recycled, young pulsars) in general have a smaller mass than the recycled NSs. Data is mostly taken from Paulo Freire's webpage (April 2021): https://www3.mpifr-bonn.mpg.de/staff/pfreire/. The massive "spider" PSR J1810+1744 with a derived mass of $2.13 \pm 0.04\ M_\odot$ (Romani et al., 2021 is shown in white).

PSR B1855+09 ($P_{\rm orb} = 12.3$ days) which, at that time, was known to be $< 1.55\,M_\odot$ from constraints on its Shapiro delay (the most recent measurement yields $\sim 1.37 \pm 0.11\,M_\odot$, Arzoumanian et al., 2018). Tauris & Savonije (1999) concluded that this was a proof for the fact that a large amount of matter *must* be lost from the LMXB, even for sub-Eddington accretion rates—probably as a result of either accretion disk instabilities (Pringle, 1981; van Paradijs, 1996) or the propeller effect (Illarionov & Sunyaev, 1975). One should also keep in mind that such an anti-correlation is probably blurred by the fact that NSs are born with different masses, as shown in Figure 14.33. However, we still suggest that there should be an anti-correlation between the *amount* of mass accreted and the orbital period, due to the well-known correlation between mass-transfer rate and the radial extent of a low-mass donor star. (Evolved giants have deep convective envelopes which they lose on a very short timescale after onset of RLO, Tauris & Savonije, 1999, leading to little accretion onto the NS.)

The Accretion Efficiency

As discussed earlier, in Tauris & Savonije (1999) we noticed from a comparison between observational constraints on pulsar masses and our calculated LMXB models that the accretion efficiency is sometimes less than 50%—even for those LMXBs where the NS accretes at sub-Eddington rates. The conclusion was confirmed and constrained even further (to accretion efficiencies below 30%, and possibly even below 10%) in measurements of pulsar masses in J1738+0333 (Antoniadis et al., 2012) and J2234+0611 (Stovall et al., 2019).

Maximum NS masses

The maximum birth mass of a NS is of large interest for understanding the final stages of stellar evolution and the physics of SNe (Langer, 2012; Ugliano et al., 2012; Pejcha & Thompson, 2015). As a result of accretion, NSs may increase their mass further. Therefore, measuring the masses of recycled NSs that accreted material is also of great interest for improving our knowledge on accretion physics, as well as the equation-of-state (EoS) for dense matter (Lattimer & Prakash, 2016; Özel & Freire, 2016). Regarding the birth masses of NSs, it has been reported that NSs in HMXBs (the progenitors of DNS systems) may be as high as $\sim 2.0\,M_\odot$, for example, Vela X-1 (Barziv et al., 2001; Falanga et al., 2015), see Chapter 7. In the last decade it has become clear that all soft EoSs, including the classic kaon condensation EoSs (Brown & Bethe, 1994; Bethe & Brown, 1995), can be ruled out. This is either because they cannot explain observations of massive NSs ($\sim 2.0\,M_\odot$) and/or because they are not compatible with the inferred tidal deformability parameter from NS mergers. The coming decade will reveal many constraints on the NS EoS from detection of GW mergers, similar to GW170817 (Abbott et al., 2017c) and GW190425 (Abbott et al., 2020a), as well as from X-ray satellite missions such as eXTP, Athena, and STROBE-X (Sieniawska et al., 2018; Watts et al., 2019; Ray et al., 2019) and also from binary radio pulsars using MeerKat and SKA (Kramer et al., 2021).

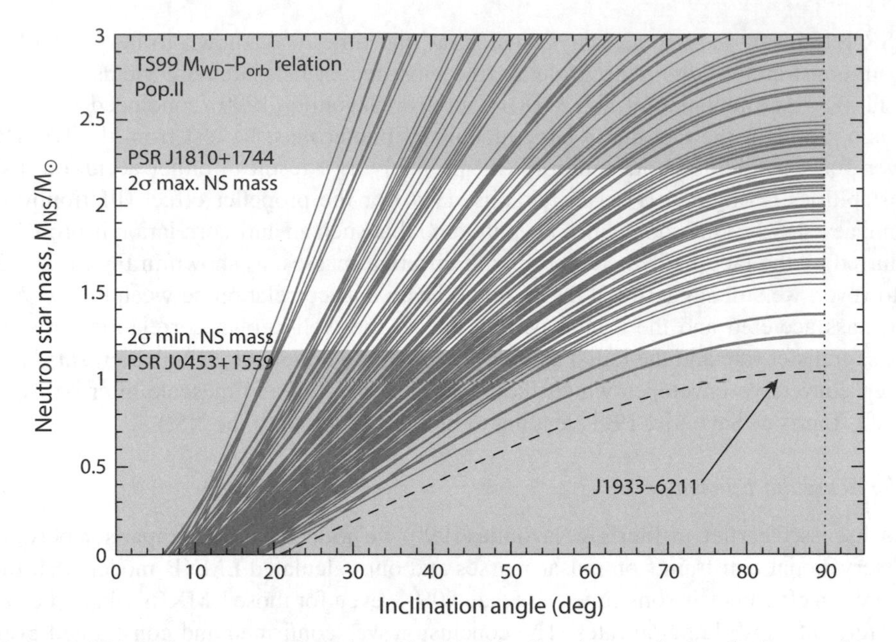

Figure 14.34. NS mass as a function of orbital inclination angle for the known binary radio pulsars with WD companions. These curves are derived from measured mass functions and assuming the WD mass in each system to be uniquely determined by the $M_{\rm WD} - P_{\rm orb}$ relation for helium WDs. The white horizontal band corresponds to the interval of known NS masses (including 2σ error bars). PSR J1933−6211 (black dashed line) is a clear outlier, and its companion is now confirmed to be a CO WD (see main text and further discussions in Antoniadis et al., 2016). Data taken from the ATNF Pulsar Catalogue in June 2021 (Manchester et al., 2005, https://www.atnf.csiro.au/research/pulsar/psrcat).

14.7.7 NS Mass and Orbital Inclination Constraints from the $M_{\rm WD}-P_{\rm orb}$ Relation of WDs

Previously in this chapter (Section 14.3.2), we discussed the $M_{\rm WD}-P_{\rm orb}$ relation of helium WDs and presented solid evidence for this correlation (Fig. 14.14). Therefore, assuming this correlation to be valid in general for all binary radio pulsars (with helium WD companions) thought to have formed from LMXB evolution, interesting constraints can be made on the pulsar masses and the orbital inclinations.

In Figure 14.34, we have plotted NS mass as a function of orbital inclination angle for binary radio pulsars with He WD companions. These curves can be drawn from measured values of the projected semi-major axis and orbital period, yielding the mass function (see Eqs. 14.40 and 14.41) for each individual binary, assuming that the WD mass is given uniquely by the $M_{\rm WD} - P_{\rm orb}$ relation. From this "spaghetti plot," two important points can be made. Firstly, the steepest curves have a very narrow interval of solutions for the orbital inclination angles. Therefore, predictions can be made on the orientation of these orbits. Such predictions have been verified (Guillemot & Tauris, 2014) from comparison to the derived viewing angles (i.e., between the pulsar spin axis and the line of sight) of binary MSPs detected in γ-rays. For example, from Figure 14.34, we can constrain the inclination angle to be roughly $20° < i < 30°$ for some of these systems with the steepest curves, assuming their NS masses to be within the full range of known NS masses (horizontal white band in this plot).

A second important point from Figure 14.34 is that the upper limit on the NS mass in a given system can be constrained from the lower curves that reach a flat plateau. As an example, some curves are seen to result in $M_{\rm NS} \lesssim 1.5\,M_{\odot}$. In one case (PSR J1933−6211, black dashed line), it has already been pointed out by Antoniadis et al. (2016) that the derived NS mass is very small ($\sim$0.90−1.05 $M_{\odot}$, depending on chemical composition of the WD progenitor star, see Eqs. 14.9 and 14.10), *if* indeed this system hosts a helium WD and was produced in a LMXB system. Given that such small NS masses are impossible to create in SNe according to current models (see, e.g., discussions in Tauris & Janka, 2019), it was concluded that this WD is most likely a CO WD and that this system therefore did not evolve from a LMXB system, and therefore does not obey the $M_{\rm WD} - P_{\rm orb}$ relation. Subsequent spectroscopy has now confirmed the CO composition of this WD (J. Antoniadis, private communication).

14.8 PULSAR KICKS

Massive stars usually end their life in a SN where the envelope is ejected while the core collapses to become a NS or a BH. In this process, a momentum kick is imparted on the NS or BH remnant. Empirical evidence for such kicks, as well as the mechanisms behind the SN explosion and the origin of the kicks according to current theoretical knowledge, was described in Chapter 13; see also Chapter 10. Thus, summarized in very few words, the evidence for kicks stems from the general distribution of radio pulsars and their height above the Galactic plane (Gunn & Ostriker, 1970), as well from the proper motions of young radio pulsars, showing average velocities of the order of 400 km s^{-1} (Lyne & Lorimer, 1994; Hobbs et al., 2005). Moreover, observations of bow shocks and supersonic runaway pulsars near SN remnants often yield kick velocities in excess of 1,000 km s^{-1} (Cordes et al., 1993; Hobbs et al., 2005; Chatterjee et al., 2005; Tomsick et al., 2012; Pavan et al., 2014; Schinzel et al., 2019). The kicks giving rise to these large pulsar velocities are mainly ascribed to ejecta asymmetries following the core collapse in which the NSs are formed (see Janka, 2012, 2017, for reviews). To be more specific, non-radial hydrodynamic instabilities in the collapsing stellar core (neutrino-driven convection and the standing accretion-shock instability) lead to large-scale anisotropies of the innermost SN ejecta, which interact gravitationally with the proto-NS and accelerate the nascent NS on a timescale of several seconds (e.g., Janka, 2012; Wongwathanarat et al., 2013). In addition, a smaller magnitude neutrino-induced kick is imparted as well (see Section 13.7 for further details on kicks).

Theoretical models suggest that, particularly if NSs are formed by iron-core collapse SNe (FeCC SNe), the resulting hydrodynamical kick velocities can be quite large (in some simulations exceeding 1,000 km s^{-1}, Scheck et al., 2006; Wongwathanarat et al., 2013). On the other hand, for electron capture SNe (EC SNe, Nomoto, 1987; Takahashi et al., 2013), which occurs only in a narrow mass interval of $\Delta M_{\rm EC} \simeq 0.1 - 0.2\,M_\odot$, for isolated progenitor stars with initial ZAMS masses between 8 and 10 $M_\odot$ (increasing with metallicity, Doherty et al., 2015, see also Chapters 8 and 13), the explosion proceeds in a more symmetric way, resulting in hardly any birth kick velocity. SN simulations show that such SNe, occurring in progenitor cores with steep density gradients, usually result in small kicks of (much) less than 50 km s^{-1} (Kitaura et al., 2006; Dessart et al., 2006; Gessner & Janka, 2018), in some cases just $1-2$ km s^{-1}.

For low-mass metal cores, the exact boundary between EC SNe and FeCC SNe depends on the location of ignition of the off-center neon and oxygen burning (Timmes et al., 1994), and also on the propagation of the neon-oxygen flame, which is sensitive to mixing in the convective layers across the flame front (Jones et al., 2014). The latter process may affect the electron fraction and the density in the central region, and thereby the electron-capture efficiency, and thus the nature of the SN. The boundary between EC SNe and FeCC SNe is roughly a metal core mass of $\sim 1.43\,M_\odot$ (Tauris et al., 2015). Based on both emerging empirical evidence and theoretical reasoning, Tauris et al. (2017) argued for a correlation between kick magnitude and NS mass. According to this picture, NSs born with masses $\lesssim 1.3\,M_\odot$ tend to receive small kicks $\lesssim 50$ km s^{-1}, whereas more massive iron cores produce NSs with masses $\gtrsim 1.3\,M_\odot$ and larger kicks $\gtrsim 200$ km s^{-1}.

For binary systems, the impact of birth kicks imparted on NSs and BHs are important, as they may lead to disruption of the binary in the SN event (Flannery & van den Heuvel, 1975; Tauris & Takens, 1998, see also Chapter 13). The fact that a large fraction of the newborn (strong magnetic field) pulsars are single is commonly ascribed to the effect of these large birth kicks.

In population synthesis simulations, the magnitude of the kick, w, is often treated as a free parameter (besides the assumption of isotropy in the kick direction) and chosen from a distribution function using Monte Carlo techniques. For simulating dynamical effects of SNe (Chapter 13), a Maxwell-Boltzmann distribution is often used for the kick magnitude, w:

$$f(w) = \sqrt{\frac{54}{\pi}\frac{w^2}{w_{\rm rms}^3}}\;\exp\left(-\frac{3}{2}\frac{w^2}{w_{\rm rms}^2}\right). \tag{14.36}$$

For FeCC SNe producing NSs, the default value for the projected 1D root-mean-square velocity is $w_{\rm rms}^{\rm 1D} = 265$ km s^{-1}, taken from Hobbs et al. (2005). The 3D $w_{\rm rms}$ of the Maxwell-Boltzmann distribution is then found by $w_{\rm rms} = \sqrt{3}\,w_{\rm rms}^{\rm 1D} \simeq 460$ km s^{-1}.

It has been suggested for a couple of decades (see discussion in Tauris et al., 2017, and references therein) that exploding stars that are stripped via mass transfer in close binaries prior to the SN (i.e., SNe Ib/c) may produce substantially smaller NS kicks compared to (SN II) explosions of isolated, or very wide-orbit, stars. As a consequence, it has been suggested for population synthesis simulations (Kruckow et al., 2018) to apply a reduced $w_{\rm rms}^{\rm 1D} = 120$ km s^{-1} for such explosions. An even more stripped progenitor star, which also loses its helium envelope prior to the core collapse (via Case BB RLO, Habets, 1986a), may result in something like $w_{\rm rms}^{\rm 1D} = 60$ km s^{-1}, unless this star has a compact object companion. If the exploding star forms the second-born compact star of the binary, it undergoes an ultra-stripped SN (Tauris et al., 2013, 2015; Suwa et al., 2015; Moriya et al., 2017) because of severe mass stripping by the nearby compact object, leaving an almost naked metal core at the time of the explosion. For such ultra-stripped SNe, one may apply, for example, $w_{\rm rms}^{\rm 1D} = 30\,{\rm km\,s^{-1}}$, based on the many observed DNS binaries with small eccentricities of $e \lesssim 0.2$ (Tauris et al., 2017) and their small derived systemic velocities, in addition to the small kicks resulting from 3D simulations of such ultra-stripped SNe (Müller et al., 2018, 2019). For population synthesis use, one may apply, for example, a flat 3D kick distribution up to $50\,{\rm km\,s^{-1}}$ for NSs produced by either EC SNe or AIC of a massive white dwarf (Section 14.5).

14.9 FORMATION OF DOUBLE NEUTRON STAR SYSTEMS

Early theoretical work on the physics of DNS formation includes (here disregarding general population synthesis studies): Flannery & van den Heuvel (1975); De Loore et al. (1975); van den Heuvel (1976); Bisnovatyi-Kogan & Komberg (1976); Srinivasan & van den Heuvel (1982); van den Heuvel (1994a); Ivanova et al. (2003); Dewi & Pols (2003); Podsiadlowski et al. (2004); van den Heuvel (2004); Dewi et al. (2005). For a more recent, detailed, and comprehensive work we refer to Tauris et al. (2017). From these papers, a *standard scenario* has emerged (e.g., van den Heuvel, 1976; Bhattacharya & van den Heuvel, 1991; Tauris & van den Heuvel, 2006), which we now summarize in more detail. An illustration of this standard formation scenario of a DNS system, and its subsequent merger event most likely producing a single BH following a short GRB, is depicted in Figure 1.1 in Chapter 1.

14.9.1 Résumé of DNS Formation

In Figure 14.35, we demonstrate a specific example of the standard scenario and provide more details, including age, stellar masses, and orbital period evolution for a given system evolving to a HMXB and a DNS system. The initial system contains a pair of OB-stars that are massive enough[9] to terminate their lives in a core-collapse SN (CCSN). To enable formation of a tight DNS system in the end, the two stars must initially be in a binary system close enough to ensure interactions via either stable or unstable mass transfer. If the binary system remains bound after the first SN explosion (which is of Type Ib/c, Yoon et al., 2010), the system eventually becomes observable as a HMXB (Chapter 10). Before this stage, the system may also be detectable as a radio pulsar orbiting an OB-star, for example, as in PSRs B1259−63 (Johnston et al., 1992), J0045−7319 (Kaspi et al., 1994) and J2108+4516 (Andersen et al., 2022). When the secondary star expands and initiates full-blown RLO during the HMXB stage, the system eventually becomes dynamical unstable. For wide systems, where the donor star

[9]The secondary (initially least massive) star could in principle be a 5−7 $M_\odot$ star that effectively accretes mass from the primary (initially most massive) star to reach the threshold limit for core collapse between $\sim 8-12\,M_\odot$ (Wellstein & Langer, 1999; Jones et al., 2013; Woosley & Heger, 2015).

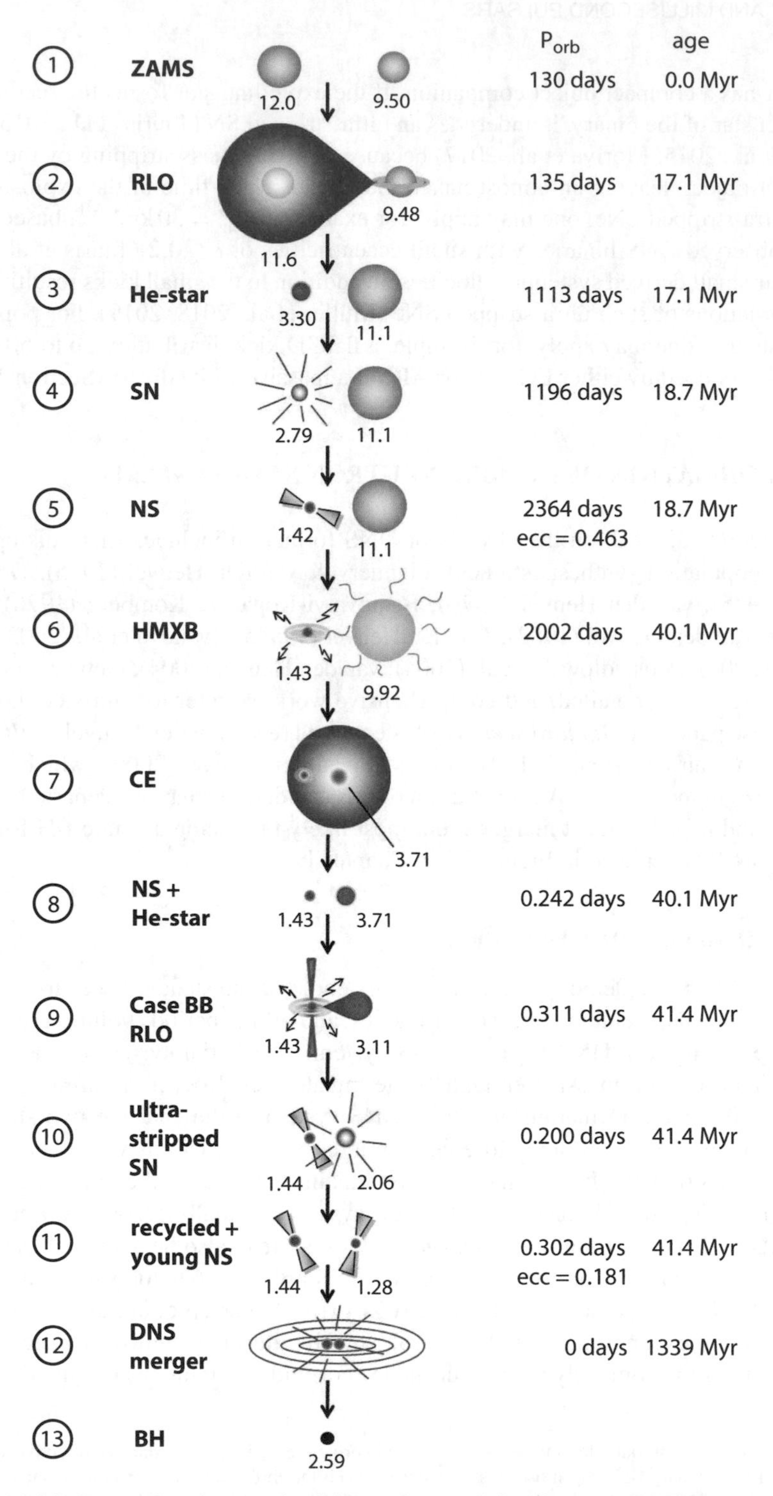

Figure 14.35. Example of a standard formation path leading to a DNS system, which later merges and leaves behind an isolated BH.

has a deep convective envelope at the onset of the mass transfer (i.e., during so-called Case B RLO, following termination of core hydrogen burning), the timescale on which the system becomes dynamically unstable might be as short as a few 100 yr (Savonije, 1978). This leads to the formation of a CE (see, e.g., Paczyński, 1976, and also discussed here in more detail in Section 4.3 and Chapter 10) where dynamical friction of the motion of the NS inside the giant star's envelope often causes extreme loss of orbital angular momentum and (in some cases) ejection of the hydrogen-rich envelope. If the binary system survives the CE phase, it consists of a NS orbiting a helium star (the naked core of the former giant star). Depending on the orbital separation and the mass of the helium star, an additional phase of mass transfer (Case BB RLO, Habets, 1986a; Tauris et al., 2015) may be initiated, see Section 10.13. This stage of mass transfer is important since it enables a relatively long phase of accretion onto the NS, whereby the NS is recycled, and it allows for extreme stripping of the helium star prior to its explosion (as a so-called *ultra-stripped* SN, Tauris et al., 2013, 2015; Suwa et al., 2015; Moriya et al., 2017; De et al., 2018; Müller et al., 2018, 2019), see Section 13.5 for further discussions. Whether or not the system survives the second SN depends on the orbital separation and the kick imparted onto the newborn NS (Flannery & van den Heuvel, 1975; Hills, 1983b; Tauris & Takens, 1998; see Section 13.8). As argued in, for example, Kruckow et al. (2018); Tauris et al. (2017), we expect most systems to survive the second SN explosion. If the post-SN orbital period is short enough (and especially if the eccentricity is large), the DNS system will eventually merge due to GW radiation and produce a strong high-frequency GW signal and possibly a short GRB (e.g., Eichler et al., 1989). The final remnant is most likely a BH, although, depending on the EoS, a NS (or, at least, a meta-stable NS) may be left behind instead (Vietri & Stella, 1998).

14.9.2 Major Uncertainties of DNS Formation

Apart from the pre-HMXB evolution, which is discussed earlier in this book, the most important and uncertain aspects of our current understanding of DNS formation are related to:

1. CE evolution and spiral-in of the NS
2. momentum kicks imparted onto newborn NSs
3. the mass distribution of NSs.

In the following sub-sections, we provide more details and briefly discuss each of these three aspects.

1. CE Evolution

In Section 10.11.1, we already discussed some of the limitations and uncertainties of CE models. A CE phase (see Section 4.8; Taam & Sandquist, 2000; Ivanova, Justhan, Chen et al., 2013 for reviews) develops when the donor star in a HMXB fills its Roche lobe. The large mass ratio between the OB-star donor and the accreting NS causes the orbit to shrink significantly upon mass transfer, whereby the NS is captured inside the envelope of the donor star. As a consequence of the resulting drag force acting on the orbiting NS, efficient loss of orbital angular momentum often leads to a huge reduction (up to a factor of $\sim$1,000) in orbital separation prior to the second SN explosion, thus explaining the tight orbits of the observed DNS systems.

In a study on CE evolution with massive stars, it was demonstrated by Kruckow et al. (2016) that an in-spiraling NS may indeed be able to eject the envelope of its massive star companion, at least from an energy budget point of view. However, they also showed that it is difficult to predict the final post-CE separation for several reasons. First of all, it is difficult to locate the bifurcation point (Tauris & Dewi, 2001) within the massive star, separating the remaining core from the ejected envelope. Secondly, while additional energy sources, such as accretion energy, may play a significant role in helping to facilitate the CE-ejection process, models of time-dependent energy transport in the convective envelope are needed to quantify this. And finally, the effect of inflated envelopes of the remaining helium core—as a consequence of the stellar luminosity reaching the Eddington limit in their interiors (e.g., Sanyal et al., 2015)—makes it non-trivial to map the core size of a massive donor star model to the size of the naked helium core (Wolf-Rayet star) left behind. Until future 3D hydrodynamical simulations might succeed in ejecting the CE of massive stars, the estimated post-CE separation will remain highly uncertain (and thus population synthesis simulations of, e.g., the LIGO-Virgo-KAGRA network detection rates of merging BH/NS binaries will remain uncertain, e.g., Voss & Tauris, 2003).

2. NS Kicks

The second major remaining issue that needs to be solved is the magnitude and direction of the momentum kick added to a NS during the SN explosion (Janka, 2012, 2017)—see also discussions in Section 14.8. In particular for the application to population synthesis modelling of DNS formation, it is important to identify any differences between the first and the second SN explosion (e.g., Bray & Eldridge, 2016, and references therein). While there is ample observational evidence for large NS kicks (typically $400-500\,\mathrm{km\,s^{-1}}$) in observations of young radio pulsars (Gott et al., 1970; Cordes et al., 1993; Lyne & Lorimer, 1994; Kaspi et al., 1996; Hobbs et al., 2005; Chatterjee et al., 2005; Verbunt et al., 2017), it also seems evident that the second SN explosion forming DNS systems involves, on average, significantly smaller kicks (Podsiadlowski et al., 2004; van den Heuvel, 2004; Schwab et al., 2010; Beniamini & Piran, 2016; Tauris et al., 2017). Furthermore, there is evidence from observations of pulsars in GCs (which have small escape velocities of less than $40\,\mathrm{km\,s^{-1}}$) that some of them are born with very small kicks. This could suggest a bimodal kick velocity distribution, which might be connected to a bimodality in the formation mechanism of NSs. This picture is supported by observations of HMXBs (Pfahl et al., 2002; Knigge et al., 2011). It was argued already in Tauris & Bailes (1996) that NSs produced from progenitors that had lost significant material prior to the SN explosion would, on average, receive smaller kicks than the typical velocities of $400-500\,\mathrm{km\,s^{-1}}$ inferred from proper motions of radio pulsars a few years earlier (e.g., Lyne & Lorimer, 1994).

Due to the continuous growth of supercomputing power and the development of efficient and highly parallelized numerical simulation tools, 2D and 3D simulations of stellar core collapse and explosions with sophisticated treatment of the complex microphysics have become feasible and demonstrate the viability of the neutrino-driven and MHD explosion mechanisms in principle (Janka, 2012). Initiating the SN blast wave with an approximate neutrino transport description in 2D and 3D simulations, it could be demonstrated (Wongwathanarat et al., 2013) that mass-ejection asymmetries imprinted by hydrodynamic instabilities in the SN core during the initiation of the explosion can lead to NS (and BH) kicks compatible with the measured velocities of young pulsars. Moreover, a first systematic attempt to establish the progenitor–explosion–remnant connection based on the neutrino-driven mechanism has revealed the interesting possibility that the NS–BH transition might occur in stars below 20 $M_\odot$ (Ugliano et al., 2012; Pejcha & Thompson, 2015; Sukhbold et al., 2016; Ebinger et al., 2019, see Section 8.2.2), which seems to be in line with conclusions drawn from observational SN-progenitor associations (Smartt, 2009) and from formation models of BH–LMXBs (see Chapter 10). This result should have consequences for the predicted ratio of the number of compact objects (NSs vs. BHs) of the GW mergers that the LIGO-Virgo-KAGRA network will detect.

As discussed earlier, it has been demonstrated (Tauris et al., 2013, 2015) that close-orbit DNS systems (i.e., those that are tight enough to merge within a Hubble time due to GWs) must form via ultra-stripped SNe when the last star explodes. The reason being that Case BB RLO, that is, mass transfer via Roche-lobe overflow from a naked helium star in the post-HMXB/post-CE system, causes the NS to significantly strip its evolved helium star companion, almost to a naked metal core prior to its explosion. This has an important effect on the number and properties of surviving DNS systems—in particular, in terms of their kinematic properties—as demonstrated in population synthesis studies (e.g., Beniamini & Piran, 2016; Chruslinska et al., 2018; Kruckow et al., 2018; Giacobbo & Mapelli, 2018; Vigna-Gómez et al., 2018).

The outcome of the second SN explosion is crucial for the survival of the binary system and the further development toward a DNS merger source. Figure 14.36 shows calculations of the probability that the binaries remain bound systems after the second SN (for various amounts of pre-SN stripping, orbital periods and kicks), whereas Figure 14.37 shows the probability that the surviving DNS systems will obtain shorter orbital periods as a consequence of the SN, assuming various kick velocities and an isotropic kick direction. It is seen that of the order of 1/3 of typical pre-SN systems will have their post-SN orbits reduced, thereby (not least, along with the resultant eccentricity) enhancing the probability of the systems evolving into a GW merger source within a Hubble time.

To learn about the progenitor binaries that formed DNS systems and the conditions of the SN explosions that produced the second-born NSs, many studies in the literature have analyzed or simulated a large number of SNe and compared the outcome with observations of DNS systems (Flannery & van den Heuvel, 1975; Hills, 1983b; Kornilov & Lipunov, 1984; Radhakrishnan & Shukre, 1985; Dewey & Cordes, 1987; Brandt & Podsiadlowski, 1995; Kalogera, 1996; Tauris & Takens, 1998; Wex et al., 2000; Andrews et al., 2015). Tauris et al. (2017) carried out a detailed analysis including

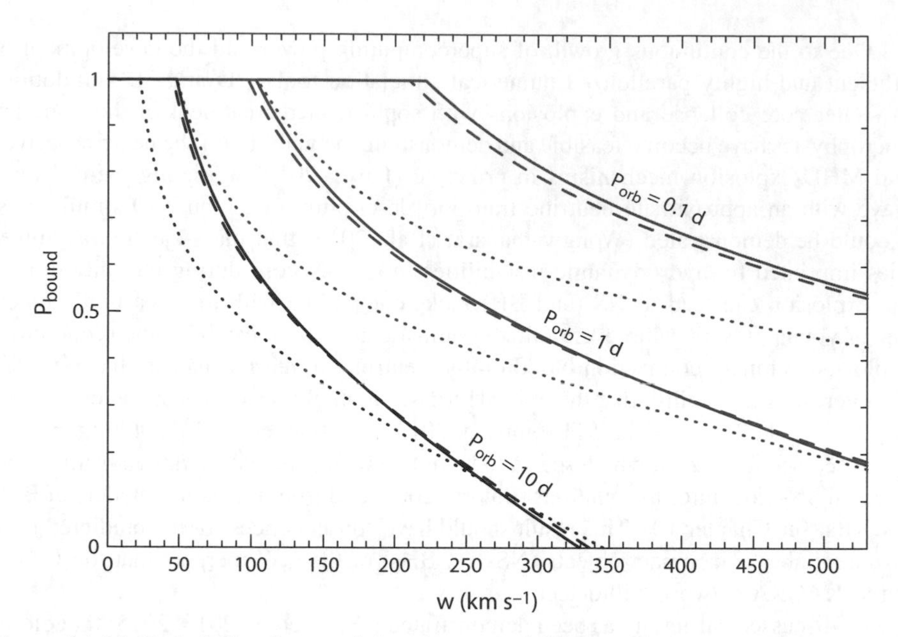

Figure 14.36. Probability for surviving the second SN in a binary as a function of the imparted NS kick velocity, w, and for three different pre-SN orbital periods of 0.1, 1, and 10 days—see Eq. (13.4). The solid and the dashed lines are for ultra-stripped SNe with $(M_{\mathrm{He,f}}/M_{\odot}, \Delta M/M_{\odot}) = (1.50, 0.30)$ and $(1.80, 0.45)$, respectively, and the dotted line is for a less stripped SN (3.00, 1.55). In all cases the first-born NS mass is $M_{\mathrm{NS,1}} = 1.40\,M_{\odot}$. After Tauris et al. (2017).

all observational constraints on known Galactic DNS systems (e.g., misalignment angles of spin axes, local rest frame proper motions, ages, etc.), leading to firm results for each individual system regarding the kick magnitude and, not least, kick direction, as well as the mass of the exploding star in the second SN event (as we summarize further in Sections 14.9.7 and 14.9.8).

3. Mass Distribution of Neutron Stars

A third main aspect of DNS systems that is far from being understood—and possibly related to the previous discussion on SNe and binary stellar evolution in general—is the mass distribution of NSs, and why it differs from the NS masses measured in systems with WD companions. Antoniadis et al. (2016) analyzed the mass distribution of 32 MSPs in orbits with (mostly) WD companions. They found evidence for a bimodal mass distribution with a low-mass component centered at $1.39 \pm 0.03\,M_{\odot}$ and dispersed by $0.06\,M_{\odot}$, and a high-mass component with a mean of $\sim 1.81 \pm 0.10\,M_{\odot}$ and dispersed by $0.18\,M_{\odot}$. The diversity in spin and orbital properties of high-mass NSs suggests that this mass difference is most likely not a result of the recycling process, but rather reflects differences in the NS birth masses. For the case of DNS systems, it was

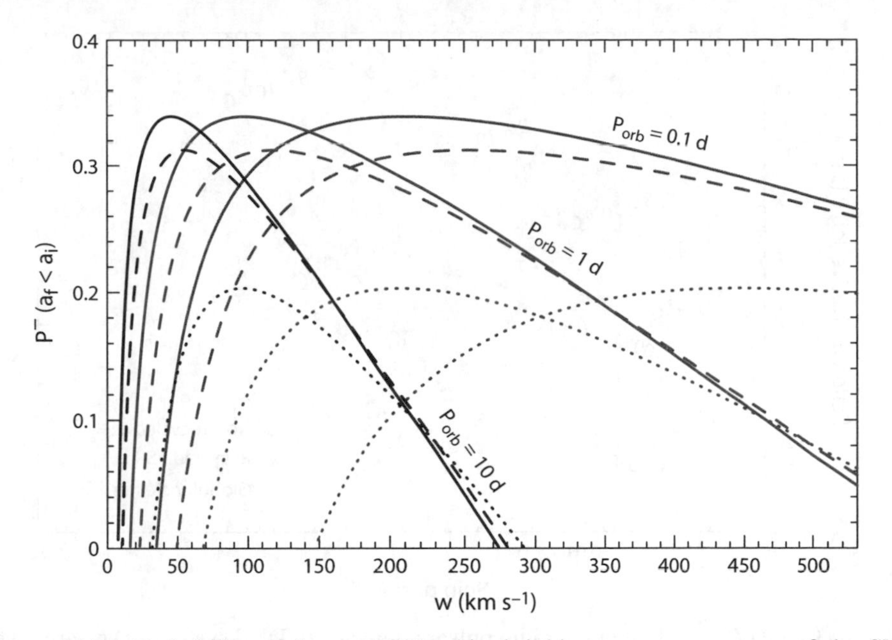

Figure 14.37. Probability of the post-SN orbit shrinking as a consequence of the SN explosion with a kick value, w (i.e., $(a_f/a_i) < 1$, cf. Eqs. 13.3 and 13.8) for the same systems plotted in Figure 14.36. These calculations are important for evaluating the number of tight DNS systems that the LIGO-Virgo-KAGRA network and LISA will detect as GW sources. After Tauris et al. (2017).

argued by Tauris et al. (2017) that the total amount of mass accreted by the first-born NS at various subsequent phases of evolution only adds up to $\lesssim 0.02\,M_{\odot}$. Thus, the measured NS masses of the first-born NSs also closely match their birth masses (similarly to the situation for the second-born NSs, which did not accrete material after their formation).

14.9.3 Observational Data of DNS Systems

In Table 14.5, we list 22 DNS systems, including a couple of sources for which the DNS nature is not confirmed. The vast majority of these DNS systems are found in the Galactic disk; only three sources are found in a GC. These three latter sources will be disregarded for further discussions related to the recycling process and impact of the second SN explosion. The reason for this is that these three systems almost certainly formed by secondary exchange encounters, that is, where already recycled pulsars change stellar companions or are being exchanged into new binaries because of close encounters. In the cases of PSRs B2127+11C, B1807−2500B, and J0514−4002A, such arguments were given by Prince et al. (1991); Lynch et al. (2012); Verbunt & Freire (2014); Ridolfi et al. (2019). In these dynamical processes, information on all

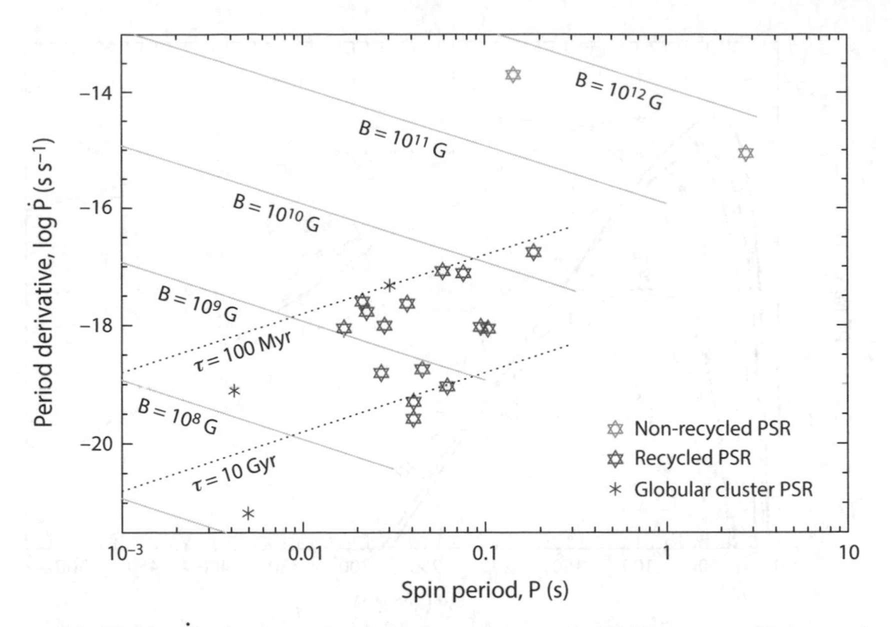

Figure 14.38. $(P, \dot{P})$–diagram of radio pulsars detected in DNS systems. Data are given in Table 14.5. The solid grey lines represent constant surface dipole B-fields (Eq. 14.15, assuming $M_{\rm NS} = 1.4\,M_{\odot}$ and $\alpha = 60°$) and the dotted black lines represent constant characteristic ages. The $\dot{P}$ values are not corrected for (mostly negligible) kinematic effects (Shklovskii, 1970). For the GC pulsars, the observed values of $\dot{P}$ must also be corrected for the (potentially significant) NS acceleration in the cluster potential. Updated after Tauris et al. (2017).

traces of the past evolutionary links to their former companions is lost and cannot be recovered from observations of the present binary systems. In Table 14.5, we indicate whether an observed NS is a recycled pulsar or a young (second-born) NS component. Only in the PSR J0737−3039 system, both NS components are detected as radio pulsars. This system is therefore known as the *double pulsar*.

14.9.4 Evidence of Mild Recycling

The $P\dot{P}$–diagram for DNS systems is plotted in Figure 14.38. The combination of recycled spin periods of typically 20 − 100 ms and B-fields of $\sim 10^9 - 10^{10}$ G provides evidence of an accretion history with mild accretion (see also Fig. 14.24), as expected from the short evolutionary timescale of their massive star progenitors. In this diagram, the two observed young (non-recycled) pulsars are clearly distinct from the recycled ones. In addition, it is seen how two of the GC DNS pulsars, J1807−2500B and J0514−4002A, have been fully recycled (probably as a result of long-term recycling in a LMXB system followed later by an encounter exchange of the MSP into the present DNS system).

14.9.5 Spin—Orbital Period Correlation

Tauris et al. (2017) demonstrated the existence of a correlation between the spin period of the recycled (i.e., first-born) NS and the orbital period of DNS systems in the Galactic disk (see Fig. 14.25), and also gave an explanation based on stellar evolution mass-transfer timescales and their dependence of the orbital period (i.e., based on the evolutionary stage of the donor at the onset of RLO): The wider the pre-SN orbit is prior to the second SN, the more evolved is the helium star donor by the time it fills its Roche lobe and initiates a recycling process via Case BB RLO (Section 10.13), and the less time it has left before it undergoes a CCSN (see Fig. 10.21). Thus, in wide-orbit systems, the recycling of the first-born NS is very inefficient and thus its spin period remains fairly large. Tauris et al. (2017) also calculated theoretical correlations based on a combination of stellar evolution and SN modelling (Fig. 14.25). They predict a wide spread in this $(P,\ P_{\rm orb})$–correlation due to, in particular, a range of possible initial helium (donor) star masses and the impact of the kick involved in the second SN explosion (even for relatively small kick magnitudes).

14.9.6 Proper Motion of DNS Systems

Measurements of the proper motion of DNS systems are important for probing their formation history. The proper motion, combined with a distance estimate (from pulsar timing and/or dispersion measure), will yield the transverse velocity with respect to the Solar System barycenter. This velocity can then be converted to the local standard of rest in the Galaxy, using a Galactic model (e.g., McMillan, 2017). Whereas pulsar binaries with an optical companion star can have their radial velocity measured, the radial component of the velocity of a DNS system remains unknown and cannot be constrained. This is usually accounted for by simulating the addition of a radial velocity component such that the 3D velocity vector has a random orientation with respect to the line of sight. From such a method, the probability distribution can be derived for the 3D velocity of a given DNS system. The results vary from less than $20\,{\rm km\,s^{-1}}$ to more than $200\,{\rm km\,s^{-1}}$ (Tauris et al., 2017). Although the proper motion of a DNS system is the sum of the impacts from both SNe acting on the binary, the kinematic effect from the first SN is usually quite limited due to the presence of a massive secondary star being able to absorb a large fraction of momentum transfer from the explosion. This explains the relatively small systemic velocities of wide (mostly B-emission) HMXBs with respect to their local standard of rest (of the order of $10–20\,{\rm km\,s^{-1}}$), compared to, for example, the motion of binary MSPs (typically of the order of $70-100\,{\rm km\,s^{-1}}$) that had a low-mass companion star at the moment of the SN explosion (Tauris & Bailes, 1996).

14.9.7 Simulating the Kinematics from the Second SN

The inferred 3D velocities of DNS systems can then be used as input parameters, together with the observed binary properties such as orbital periods, eccentricities, and misalignment angles, for simulating the second SN producing the DNS system. From such studies, one can in some cases derive strong constraints on not only the magnitude

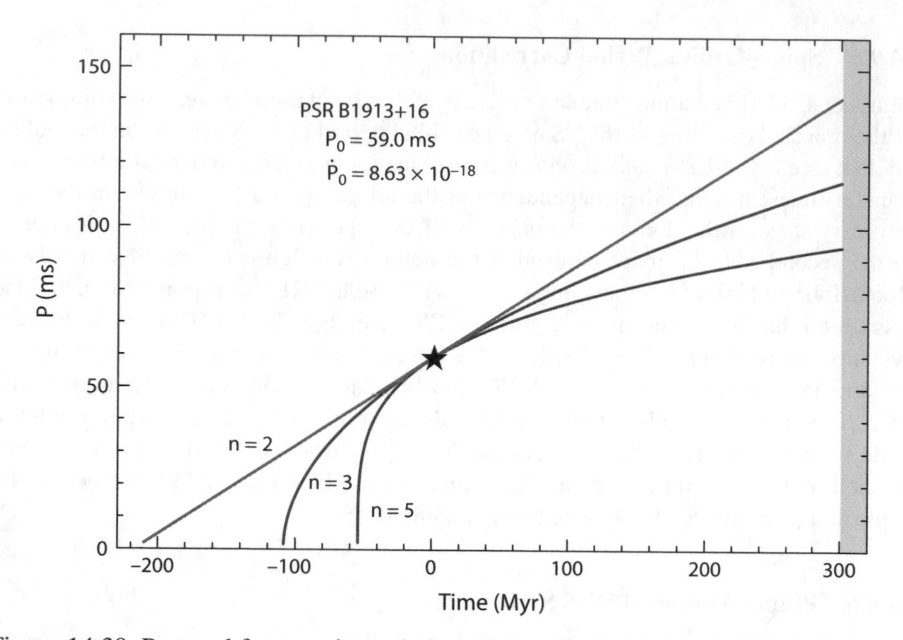

Figure 14.39. Past and future spin evolution of the recycled pulsar in PSR B1913+16, calculated for three different values of the braking index, $n = \{2, 3, 5\}$, see Section 14.1.4. The current system is shown with a solid star, located at $t = 0$. The system will merge in 301 Myr. After Tauris et al. (2017).

and direction of the SN kick but also the mass of the exploding star. This method is very useful to gain insight into the formation of DNS systems (Tauris et al., 2017).

Corrections Due to Post-SN GW Evolution

DNS systems in close orbits evolve with time, not only due to the spin-down of the observed radio pulsar but also due to GW damping of the orbit. The lowest-order secular orbital evolution is easily calculated from changes in the elements of the relative orbit of two point masses resulting from GW damping (Peters, 1964). This means that the simulated post-SN parameters applicable to DNS systems at birth (P, $P_{\rm orb}$, and eccentricity), are often somewhat different from their current observed values (Tauris et al., 2017). To illustrate the spin evolution of a DNS system since its recycling prior to the second SN, in Figure 14.39, we have plotted the past and the future spin evolution of PSR B1913+16, using three different values of the braking index, $n = \{2, 3, 5\}$, see, e.g., Tauris et al. (2012) and references therein for a discussion.

14.9.8 A Kick–Mass Correlation for NSs?

It was argued by Tauris et al. (2017) that ultra-stripped SNe often, but not always, produce relatively small kicks. The reason that some of the ultra-stripped SNe produce large kicks was ascribed to the fact that more massive pre-SN cores produce larger iron cores and thus more "normal" Fe CCSNe with larger explosion energies, and therefore larger kicks. Hence, they advocated for a correlation between kick magnitude and the resultant NS mass. Indeed, the computed models of Tauris et al. (2015) predict that ultra-stripped SNe may potentially produce young NSs in DNS systems with a mass within the entire range $1.10 - 1.80\,M_\odot$.

Based on theoretical arguments, Janka (2017) provides arguments why collapse events of stellar cores with low compactness and steep density gradients (both of which are associated with EC SNe, low-mass Fe CCSNe, and ultra-stripped SNe with small metal cores) can be expected to produce NSs with considerably smaller kicks than SNe from more ordinary iron cores. Since stellar cores with low compactness also typically produce lower-mass NSs (see Ugliano et al., 2012; Ertl et al., 2016; Sukhbold et al., 2016, see also Section 8.2.4), a relation between SN kick and NS mass seems plausible. We notice that although in Eq. (13.2) w appears to scale inversely proportionally to $M_{\rm NS}$, the values of $E_{\rm SN}$ and $\alpha_{\rm ej}$ are systematically smaller for explosions leading to small values of $M_{\rm NS}$, such that a correlation between w and $M_{\rm NS}$ is indeed expected in general.

This hypothesis was put to the test in Tauris et al. (2017) by comparing to observations of DNS systems (see their figure 16). They found that the DNS systems where the second-born NS has a relatively large mass $\gtrsim 1.33\,M_\odot$ are also those systems where it is evident from observations of proper motion that a large kick must have been at work. The hypothesis of a possible correlation between $M_{\rm NS}$ and w is still based on small number statistics. Tauris et al. (2017) noticed that there could also be a transition region between small kicks and large kicks, possibly related to some element of stochasticity involved in the development of the explosion asymmetry (Wongwathanarat et al., 2013) that may mask any correlation.

14.9.9 Alternative DNS Formation Scenarios

A very elegant alternative formation scenario (the "*double core scenario*") for DNS systems was proposed by Brown (1995); Bethe & Brown (1998), and later applied to population synthesis investigations by Dewi et al. (2006). The peculiarity (or advantage) of this scenario is that it does not involve a NS passing through a CE. (The scenario was originally developed to explain why no NSs were observed with masses greater than $1.45\,M_\odot$, the rationale being that accretion onto the NS in a CE scenario would push it above its maximum mass, at that time thought to be close to $1.45\,M_\odot$, thus producing a BH.)

In the double core scenario, the initial binary components have very similar masses (within 4% on the ZAMS), and both components have become giant stars before they evolve into contact. In the subsequent CE and in-spiral phase, the two cores lose their hydrogen-rich envelopes, creating a tight binary before both stars explode in SNe. Although most investigations on CE physics since then (Section 4.8) point to a NS in-spiral without hypercritical accretion and collapse to a BH (and thus supporting the standard formation channel of DNSs via CE evolution), this alternative double core scenario is still believed to be valid, although probably much less frequent compared to the standard scenario.

Finally, DNS systems may, of course, also be produced via exchange interactions in dense stellar environments like GCs (Chapter 12).

EXERCISES

For all exercises involving a NS: unless stated otherwise, assume a standard NS with $M = 1.4\,M_\odot$, $R = 10\,\mathrm{km}$, and $I = 10^{45}\,\mathrm{g\,cm^2}$.

Exercise 14.1. *For the Crab pulsar, the radio flux-density of the detected signal at 436 MHz is* $S_{436} \simeq 0.48\,\mathrm{Jy}$ *(Jansky), and* $1\,\mathrm{Jy} \equiv 10^{-23}\,\mathrm{erg\,s^{-1}\,cm^{-2}\,Hz^{-1}\,St^{-1}}$*. The distance to the Crab pulsar is* $d = 2.0\,\mathrm{kpc}$*.*

a. *Show that the radio luminosity of the pulsar is* $L_{radio} \simeq 10^{31}\,\mathrm{erg\,s^{-1}}$*.* *(Assume [not correct] a flat spectral index, i.e., that the radio emission is equally intense at all frequencies and assume a total frequency bandwidth of* 436 MHz*. Remember that an entire sphere has a solid angle of* 4π St *[steradian].)*
b. *The Crab pulsar has a spin period of* $P = 33.4\,\mathrm{ms}$ *and a spin period derivative of* $\dot{P} = 4.21 \times 10^{-13}$*. Given its radio luminosity,* L_{radio} *from above, calculate the fraction of spin-down energy that is emitted in the radio band.*
c. *What are the main causes of its loss of rotational energy?*
d. *Apply the Raleigh-Jeans limit of a Planck function to demonstrate that if the radio emission was caused by thermal black body radiation, one would obtain an extremely high brightness temperature (leading to absurdly large particle energies) and therefore the radiation mechanism of a radio pulsar must be coherent.*

Exercise 14.2. *Derive Eq. (14.4).*

Exercise 14.3. *Assume that the distribution of normal radio pulsars can be described in the* $P\dot{P}$*–diagram from a simple Gaussian distribution in both* $\log P$ *and* $\log \dot{P}$*.*

a. *From an eye inspection of the* $P\dot{P}$*–diagram in Figure 1.2, estimate mean and spread values of the two distributions.*
b. *Using, for example, a numerical Monte Carlo technique, choose a dataset of values for* P *and* $\dot{P}$ *and plot surface magnetic fields (here simply assumed to be given by* $B = 3.2 \times \sqrt{P\dot{P}}$*) as a function of characteristic age (*$\tau = P/2\dot{P}$*).*
c. *Is this plot providing evidence for B-field decay in pulsars? Why/why not?*

Exercise 14.4. *To solve for the magnetic field evolution of a NS, consider the induction equation stated in Eq. (14.22). Disregarding convective transport of material (the second term), and considering only the uniform conductivity case, this equation takes the simple form of a pure diffusion equation (by virtue of the divergence-free condition for the magnetic field):*

$$\frac{\partial \vec{B}}{\partial t} = -\frac{c^2}{4\pi\,\sigma}\,\nabla^2 \vec{B}, \tag{14.37}$$

where σ *is the electrical conductivity.*

a. *Show that the solution to Eq. (14.37) is an exponential decay of the surface B-field:*

$$B = B_0 \, e^{-t/\tau_D}, \tag{14.38}$$

and write an expression for the decay-time constant, τ_D.

b. *Insert this exponential expression into Eq. (14.4) and integrate to obtain the following expression for the spin period of the pulsar:*

$$P(t)^2 = P_0^2 + B_0^2 \, \tau_D \left(1 - e^{-2t/\tau_D}\right) \cdot \frac{1}{k^2}, \tag{14.39}$$

where k is a constant that depends on the NS radius and moment of inertia.

c. *Find $\lim_{t\to\infty} P(t)$ for a pulsar born with $B_0 = 10^{13}$ G and an initial spin period $P_0 =$ 20 ms, assuming a decay constant of $\tau_D = 10$ Myr and using $R = 10$ km and $I =$ 10^{45} g cm^2.*

d. *Plot the above evolutionary track in a $P\dot{P}$–diagram.*

Exercise 14.5. *The mass function of a binary pulsar system is defined by:*

$$f \equiv \frac{(M_2 \, \sin i)^3}{(M_{\rm psr} + M_2)^2}, \tag{14.40}$$

where $M_{\rm psr}$ is the pulsar mass, M_2 is the mass of its companion star, and i is the orbital inclination angle of the binary system ($i = 90°$ for an edge-on geometry, $i = 0°$ for a face-on geometry. See Chapter 3 for further details).

a. *Show that the mass function can be written in terms of observables as:*

$$f = \frac{P_{\rm orb} \, v_{\rm psr}^3}{2\pi \, G} = \frac{4\pi^2}{G} \frac{(a_{\rm psr} \sin i)^3}{P_{\rm orb}^2}, \tag{14.41}$$

where $v_{\rm psr} = 2\pi \, a_{\rm psr} \sin i \,/\, P_{\rm orb}$ is the projected orbital velocity of the pulsar (parallel to the line of sight), $a_{\rm psr} \sin i$ is the projected semi-major axis of the pulsar orbit around the common center of mass, and $P_{\rm orb}$ is the orbital period of the binary given by Kepler's third law.

b. *Consider a binary MSP with an orbital period of $P_{\rm orb} = 2.00$ d and a projected semi-major axis of $a_{\rm psr} \sin i = 1.90$ lt · s. Determine its mass function.*

c. *Assume a NS mass of $1.30 \, M_\odot$ or $1.80 \, M_\odot$ and plot the mass of its companion star as a function of the unknown orbital inclination angle, i. What is the expected nature of the companion star?*

Exercise 14.6. *a. Derive the equation for spin equilibrium, Eq. (14.18). Try also to slightly vary assumptions on, for example, $r_{\rm A}$ (see remark in Section 7.3.1) and $R_{\rm NS}$.*

b. *Estimate the required average mass-accretion rate onto a typical NS with* $B = 2.0 \times 10^8$ G *to reach an equilibrium spin period of* $P_{\rm eq} = 3.0$ ms.

Exercise 14.7. *In Sections 14.1.3, 14.1.4, and 14.6.7, we discussed pulsar evolutionary tracks within the magnetic dipole model (and often assuming with a perfect magnetic dipole, i.e., evolution with a braking index of* $n = 3$*). Here, we consider the spin evolution of a pulsar emitting gravitational waves (GWs)—see also Shapiro & Teukolsky (1983); Palomba (2005); Woan et al. (2018); Andersson (2020)—leading to a rotational energy loss rate of:*

$$\dot{E}_{\rm GW} \simeq -\frac{32\,G}{5\,c^5}\, I^2 \varepsilon^2\, \Omega^6, \tag{14.42}$$

assuming an idealized NS as a slightly deformed homogeneous ellipsoid with rotation about its principal axis of inertia (see more details in Section 15.9.1) and having a small ellipticity ($a \approx b$) *of:*

$$\varepsilon = \frac{a - b}{(a + b)/2}. \tag{14.43}$$

where a *and* b *are the semi-major and semi-minor axes in the equatorial plane of the NS.*

a. *Consider a pulsar whose loss rate of rotational energy is entirely dominated by emission of GWs. Show that its ellipticity is roughly given by:*

$$\varepsilon \simeq 6.0\sqrt{P_{\rm ms}^3\,\dot{P}}, \tag{14.44}$$

where $P_{\rm ms}$ *is the pulsar spin period in ms.*

b. *Show that the spin evolution of this pulsar follows the expression:*

$$P(t)^4 = \left(\frac{2\varepsilon}{A}\right)^2 t + P_0^4, \tag{14.45}$$

where $A = \varepsilon/\sqrt{P^3\dot{P}}$ *is a constant.*
In a combined model considering both magnetodipole radiation and GW radiation ($\dot{E}_{\rm rot} = \dot{E}_{\rm GW} + \dot{E}_{\rm dipole}$)*, the spin evolution can be expressed as:*

$$\dot{\Omega} = \gamma\Omega^5 - \beta\Omega^3. \tag{14.46}$$

c. *Derive an expression for the true age of the pulsar,* t*, in terms of* γ*,* β*,* Ω*, and its initial value,* Ω_0*.*

Exercise 14.8. *Following the equations in Exercise 14.7, consider a pulsar exclusively losing rotational energy via GW radiation. Consider a pulsar with* $\varepsilon = 10^{-7}$ *(i.e., a "1 mm mountain" with* $a - b = 1$ mm, $a \approx b = 10$ km*) and a spin period of* $P = 2$ ms.

a. *Calculate its loss rate of rotational energy due to GWs.*
b. *How much time does it take until its spin period has decreased to 10 ms?*

Recent results reveal that the value of $\varepsilon \ll 10^{-7}$.

c. *Repeat the previous two questions assuming* $\varepsilon = 10^{-9}$.

Exercise 14.9. *The contribution to the value of the measured time derivative of the binary orbital period,* $(\dot{P}_{\rm orb})_{\rm obs}$*, caused by the Shklovskii effect (Section 14.6.7) is given by:*

$$(\dot{P}_{\rm orb})_{\rm shk} = \left(\frac{v_\perp^2}{c\,d}\right) P_{\rm orb}, \tag{14.47}$$

where $v_\perp = \mu\, d$ *is the transverse velocity, determined by the measured proper motion* (μ) *and distance* (d).

Assume that $(\dot{P}_{\rm orb})_{\rm obs} = (\dot{P}_{\rm orb})_{\rm int} + (\dot{P}_{\rm orb})_{\rm shk}$ *(additional terms due to vertical and differential rotational acceleration in the Galaxy are neglected here), where the contribution to the internal evolution,* $(\dot{P}_{\rm orb})_{\rm int}$ *may be caused mainly by GW radiation, that is,* $(\dot{P}_{\rm orb})_{\rm int} \simeq (\dot{P}_{\rm orb})_{\rm GW}$.

Consider a circular $(e \simeq 0)$ *binary MSP with* $(\dot{P}_{\rm orb})_{\rm obs} = 5.0 \times 10^{-14}$ *and where using GR it is found that* $(\dot{P}_{\rm orb})_{\rm GW} = -1.1 \times 10^{-14}$. *Assume a transverse velocity of* $v_\perp = 100\,\mathrm{km\,s^{-1}}$.

a. *Use Eq. (14.35) to estimate* $P_{\rm orb}$ *for this binary MSP with a He WD companion, assuming* $M_{\rm psr} \simeq 1.5\,M_\odot$ *and* $M_{\rm comp} \simeq 0.18\,M_\odot$.
b. *Show that:*

$$v_\perp = 4.74\, d \cdot \mu, \tag{14.48}$$

if $v_\perp$ *is measured in* $\mathrm{km\,s^{-1}}$; d *in* pc; *and* μ *in* $''\,\mathrm{yr^{-1}}$.

c. *Find the distance to the binary MSP and calculate its proper motion.*

Chapter Fifteen

Gravitational Waves from Binary Compact Objects

Gravitational wave (GW) astronomy began on September 14, 2015, at 09:50:45 UTC, with LIGO's astonishing discovery of GW150914 (Abbott et al., 2016d)—the direct detection of a binary BH merger lasting ~ 0.2 s, see Figures 15.1 and 15.2. In the following half dozen years, we witnessed a scientific revolution with many further discoveries of GW mergers, including DNS and mixed BH+NS mergers (Abbott et al., 2022b). The very first GW signal ever detected (GW150914) was a merger of a pair of relatively massive ($35+30\,M_\odot$) BHs located in a distant Galaxy and for 1.3 billion years the emitted GWs propagated with the speed of light through the Universe—much like waves caused by a stone falling into a pond. The energy release was enormous, peaking at a power of some 10^{56} erg s^{-1} (a power equivalent to the complete annihilation of the mass of some 100 solar-like stars in a second; or more specifically, an energy of $3\,M_\odot c^2$ was radiated away in GWs within a few hundredths of a second).

GWs carry a new complementary source of information of space-time, energy, and matter at their origin of emission. This is of paramount importance, and it will expand (and bring new challenges to) our understanding of the Universe. The GW detections open a new major branch of research in multi-messenger astrophysics and also raise questions of more philosophical character. In popular terms, one can say that the last 400 years in astronomy were driven by observations and interpretation of electromagnetic signals (for most of this period: light) from outer space—since the 20th century, this was also complemented by some information obtained from relativistic particles such as cosmic rays and neutrinos. Now, with GW detections, a paradigm shift has taken place. We can finally start to probe deeper into the nature of gravity and, hopefully, someday succeed in uniting quantum physics with Einstein's geometric description of space-time—a holy grail in modern physics.

It is anticipated that the two American GW detectors (LIGO), combined with their European, Japanese, and, soon, Indian sisters (Virgo, KAGRA and IndIGO), within the next few years, when the detectors reach full design sensitivity, will detect up to hundreds of such GW events each year from merging binaries of BHs and/or NSs in the local Universe (Abbott et al., 2010).

The merger events of BH/NS binaries are the most powerful energy sources known in the Universe and bring new opportunities of insight to fundamental aspects of physics, including: new tests of gravity in a highly relativistic regime (Abbott et al., 2016e, 2021f), creation of energetic bursts of electromagnetic radiation (i.e., short gamma-ray

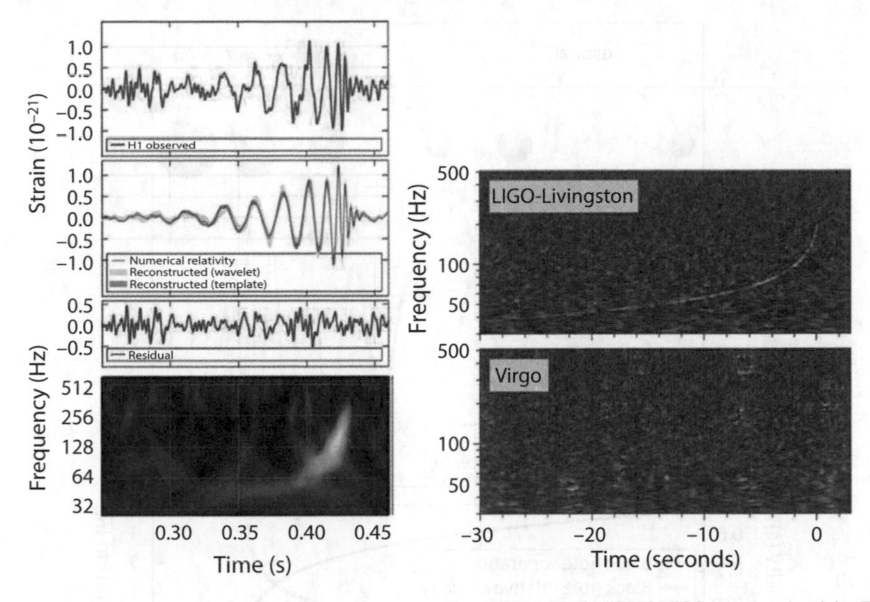

Figure 15.1. *Left:* The first ever detection of a GW event (GW150914—a double BH merger) as observed by the LIGO Hanford detector. Approximately 8 GW cycles are seen in the last ~ 0.2 s of the in-spiral (left top/central: observed signal and numerical model of strain vs. time; left bottom: frequency vs. time showing the chirp of the signal). The signal from GW150914 arrived 6.9 ms earlier at the Livingston detector, 3,002 km away. *Right:* Time-frequency representations of data containing the first DNS merger event GW170817, observed by the LIGO-Livingston detector (top). The signal was also detected by the LIGO Hanford detector. The non-detection by the Virgo detector (bottom) helped to localize GW170817 in the sky. After Abbott et al. (2016d, 2017c).

bursts, Abbott et al., 2017a), and production of heavy chemical elements (beyond iron), via so-called r-process nucleosynthesis, which decay and power an optical transient called a *kilonova* (Smartt et al., 2017; Kilpatrick et al., 2017; Rosswog et al., 2018; Giacomazzo et al., 2018; Shibata & Hotokezaka, 2019; Nakar, 2020; Metzger, 2019), see Section 15.10.

In addition to the high-frequency GW detectors LIGO-Virgo-KAGRA (and soon also the Indian IndIGO or LIGO-Aundha in ~2027), a space-borne European low-frequency GW detector (LISA; Amaro-Seoane et al., 2017; Amaro-Seoane, Andrews et al., 2022) is scheduled for launch in about a decade. This new low-frequency detector will complement and significantly increase the access to GW sources over a large frequency range, including mostly WD and NS binaries located within our own Milky Way (MW) Galaxy (see Sections 15.12 and 15.13). Similar Chinese detectors (TianQin and Taiji, see, e.g., Huang et al., 2020; Ruan et al., 2020) are planned as well. In addition, ongoing radio pulsar timing arrays (PTAs, e.g., Verbiest et al., 2021, see

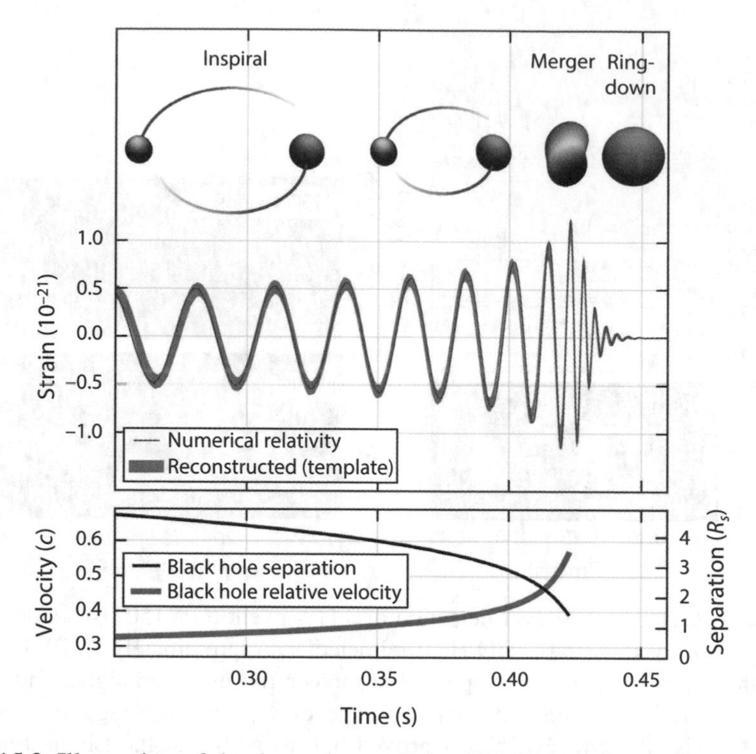

Figure 15.2. Illustration of the three phases of the GW150914 merger event: in-spiral, merger, and ring-down. The bottom panel shows the orbital velocity and the separation of the two BHs in units of the speed of light and the sum of the Schwarzschild radii, respectively. After Abbott et al. (2016d).

also Section 14.1.2) are constructed to detect low-frequency (nano-Hz) GWs from the in-spiral of merging super-massive BHs in distant galaxies. Finally, a stellar interferometry experiment (Park et al., 2021) has been proposed for detecting GWs in the lower frequency range of $10^{-6}-10^{-4}$ Hz.

To reveal in detail the origin and properties of both the merging BH/NS binaries and the low-frequency GW sources is an important science goal for the coming decades. The basics of the full evolutionary path producing these GW sources (see Fig. 1.1) has been covered in the previous chapters of this book. In this chapter, we discuss these GW sources further, starting with a historic note on the evidence of GWs based on four decades of observations of the Hulse-Taylor pulsar, PSR B1913+16. We then proceed by describing the GW luminosity, merger time, and signal strain of a given binary system before discussing merger rates in the local Universe. We highlight the classes and origin of some of the individual GW events discovered during the first three LIGO-Virgo observing runs (O1, O2, and O3), and we summarize multi-messenger astronomy as well as anticipated future discoveries in general GW astronomy. We end this chapter by discussing low-frequency GW sources and LISA sensitivity curves.

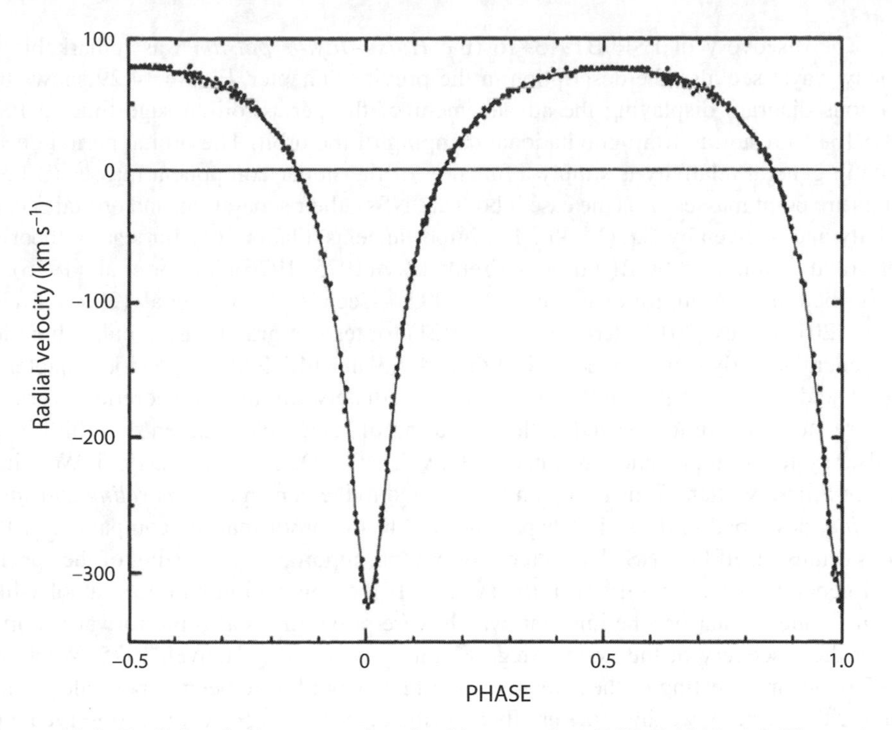

Figure 15.3. Radial velocity curve for the binary pulsar PSR B1913+16. The points represent measurements of the pulsar period distributed over parts of 10 different orbital periods. (See also Fig. 14.26.) After Hulse & Taylor (1975).

15.1 THE EVIDENCE OF GWs PRIOR TO LIGO

PSR B1913+16 was the first binary pulsar to be found. It was discovered by PhD student Russell Hulse and his supervisor Joseph Taylor, of the University of Massachusetts Amherst in the summer of 1974. They found a 59 ms radio pulsar in a highly eccentric orbit ($e = 0.61$) with an orbital period of 7.75 hours and a radial velocity semi-amplitude of 199 km s^{-1} (Fig. 15.3). At the 144th AAS Meeting in Gainesville, Florida, later in that year, Hulse and Taylor announced their discovery: "We have detected an unusual pulsar during the course of a systematic survey for new pulsars carried out at the Arecibo Observatory.... The unseen companion is apparently a neutron star or possibly a black hole. A number of interesting effects will probably be observable in forthcoming timing data on the pulsar, including the relativistic Doppler term, the gravitational redshift, and the advance of periastron." And indeed that was the case. In the following decades, this system was investigated in detail by Taylor and colleagues, and the discovery of PSR B1913+16 earned Hulse and Taylor the 1993 Nobel Prize in Physics "for the discovery of a new type of pulsar, a discovery that has opened up new possibilities for the study of gravitation."

The discovery of PSR B1913+16 (the *Hulse–Taylor pulsar*) was remarkable in many ways; see also the description in the previous chapter. Figure 14.29 shows the famous diagram displaying the advancement of the periastron passage time in PSR B1913+16 resulting from gravitational damping of the orbit. The orbital period decay rate in general relativity is simply a function of the stellar component masses (if both stars are point masses, as is the case if both are NSs), their separation, and orbital eccentricity and is given by Eq. (14.35). The importance as a laboratory for testing theories of gravity, pioneered by Blandford & Teukolsky (1975, 1976); Taylor et al. (1976), is described in detail in, for example, Will (1993)—see also Kramer et al. (2006); Freire et al. (2012); Wex (2014); Kramer et al. (2021) for tests of gravity using pulsar binaries in general (briefly summarized in Sections 14.7.3 and 14.7.3 of the previous chapter).

The discovery of PSR B1913+16 also immediately stimulated theoretical work on binary stellar evolution related to the formation of such a close eccentric-orbit binary pulsar system (e.g., Flannery & van den Heuvel, 1975; De Loore et al., 1975; Webbink, 1975; Smarr & Blandford, 1976), and it later led to the concepts of *recycling* and *spin-up line*, described in detail in Chapter 14, and to the proof that the companion of the pulsar must itself be a NS. The latter is of crucial importance for testing of the special and general relativistic effects in the system. If the companion had been a solar-like main-sequence star or a helium star, which were possibilities also put forward shortly after the discovery of the system (e.g., Flannery & van den Heuvel, 1975; Webbink, 1975), accurate testing of these relativistic effects would have been impossible. At the end of this section, we therefore briefly describe the reasoning of why the observed pulsar in this system is recycled and why the companion can only be another, younger, NS.

Furthermore, the fact that the gravitational damping of the system causes the two NSs to approach each other by 3.5 m every year, leading to a collision in 301 Myr, led to the first firm evidence of the existence of DNS mergers within the MW. The ideas of colliding NS and BH binaries had already been discussed as pure theoretical concepts before the discovery of PSR B1913+16 in a pioneering research article by Lattimer & Schramm (1974): "...although these circumstances would appear to be relatively improbable." However, the discovery of PSR B1913+16, and later several other DNS systems that will merge within a Hubble time, was essential in the late 1980s and early 1990s for the funding of the first big-science laser interferometry GW detectors, that is, LIGO. (Resonant bar detectors were built in the 1960s by Joseph Weber, but had poor sensitivity and no confirmed detections.) The expected merger rates in the MW and the local Universe were highly uncertain, however (see, e.g., Abadie et al., 2010, for a review), and at the time of writing (after the LIGO-Virgo observing runs O1–O3) the empirical rates still have large error bars. The main reason for the difficulty in determining the merger rates theoretically from population synthesis simulations is the uncertain physics related to common envelope evolution and the kicks imparted onto newborn NSs and BHs at birth (e.g., Belczynski et al., 2002; Voss & Tauris, 2003; Kruckow et al., 2018; see also Section 15.4 for discussions).

Intermezzo: Why the Hulse-Taylor Pulsar Companion Must Also Be a Neutron Star

As described in Chapter 10 (see Fig. 10.13), van den Heuvel & De Loore (1973); van den Heuvel (1974) predicted that later in life a HMXB (high-mass X-ray binary) with a NS accretor will spiral-in very deeply and—-if the system does not coalesce—form a very close binary with an orbital period of only a few hours, consisting of a helium star and a NS (see also van den Heuvel, 1976). Due to the strong friction and tidal effects during in-spiral, the post-in-spiral orbit of this system will be completely circular. When the helium star explodes, and the system is not disrupted by the explosion, a close system of two NSs will result, in an eccentric orbit (Flannery & van den Heuvel, 1975; De Loore et al., 1975). Just before the discovery of this binary pulsar, Bisnovatyi-Kogan & Komberg (1974) with great foresight had suggested that, due to the observed acceleration of the spins (spin-up) observed in X-ray binaries with RLO, such as Cen X-3 and Her X-1 (see Chapter 7), after the SN explosion of the companion, these rapidly spinning old NSs with an accretion history will be released from their binary and would be observable again as radio pulsars.

When the Hulse-Taylor binary pulsar was discovered, these authors suggested (Bisnovatyi-Kogan & Komberg, 1976) that the combination of short pulse-period (59 ms) and weak magnetic field ($B = 2 \times 10^{10}$ G) of this pulsar was due to this spin-up by accretion in an X-ray binary. (They suggested, without further arguments, the magnetic field strength having been weakened by the accretion.) Bisnovatyi-Kogan & Komberg (1976) argued that during the RLO that caused the spin-up of the old NS, the orbit must have been circularized by tidal forces, and that therefore the present eccentricity of the orbit must be due to the SN explosion of the companion star, which means that the companion must be a NS or a BH (they favored the latter possibility). Smarr & Blandford (1976) independently put forward a similar model and also argued that the observed pulsar is the spun-up first-born one.

In these papers, it was not explained why the system would have spiraled-in to such a narrow orbit, but they further had all the right arguments of why the companion must itself be a compact object and why the observed pulsar is the first-born *recycled* one in the system. Srinivasan & van den Heuvel (1982) combined these arguments with the fact that HMXBs later in life spiral-in to produce a very narrow orbit and argued that the companion of the observed pulsar must be a young non-recycled NS with a strong magnetic field: that is, a *garden variety* radio pulsar with a very short spin-down timescale, which disappeared into the pulsar graveyard within a few million years after its birth. This explains why it is not observed (although this could also be due to beaming effects). On the other hand, the recycled pulsar with its weak magnetic field has a very long spin-down timescale, which explains why, despite its old age, it is still observable (see Chapter 14 for details).

This evolution model of DNSs was confirmed by the discovery of the double radio pulsar PSR J0737−3039 (Burgay et al., 2003; Lyne et al., 2004), where it was discovered that indeed the companion of the fast-spinning weak-magnetic field (recycled) pulsar is a young, strong-magnetic field pulsar, with a pulse period of approximately one second (2.8 s in the case of PSR J0737−3039B).

We see from the previous chapters that essential elements of binary stellar evolution, including spiral-in, recycling, and SN kicks, were required for the proof that the PSR B1913+16 binary system consists of two NSs. Later on, the concepts of recycling and spin-up line—all derived from the properties of PSR B1913+16—proved essential to understand the origin of millisecond pulsars (Alpar et al., 1982; Radhakrishnan & Srinivasan, 1982; Fabian et al., 1983), see the previous chapter for details.

15.1.1 The Non-detection of GWs from Initial LIGO

The sensitivity limit of the initial LIGO instrument (LIGO I) was equivalent to detecting a NS merger within a distance of 8 Mpc. For the second-generation LIGO (Advanced LIGO), the design sensitivity corresponds to detecting a NS merger at 200 Mpc, and thus a detection volume that is larger by a factor of $(200/8)^3 \simeq 15{,}000$. Therefore, it was not surprising that LIGO I did not detect any mergers. Although it did not succeed in detecting any GW signals, the instrument was able to place an upper limit on the continuous GW emission from, for example, the Crab pulsar to be less than 3% of its total spin-down luminosity (Aasi et al., 2014); that is, emission of GWs accounted for less than 3% of its loss rate of rotational energy. The current limit is less than 0.017% (Section 15.9).

15.2 GW LUMINOSITY AND MERGER TIMESCALE

In GW astrophysics, there are multiple efforts to detect GWs over a broad range of frequencies from various astrophysical sources: orbital motion and mergers of double compact objects, mergers of super-massive BHs in galactic nuclei, tidal disruption events, supernovae, NS oscillations, and inflation of the early Universe—see Figure 15.4. Unlike electromagnetic waves, which scatter significantly in the plasma along the line of sight, GWs couple weakly to matter and arrive in almost original condition. The GW signals are weak and their interaction with spacetime can be treated as a small perturbation to the background metric tensor. The solution to Einstein's field equations is that of a plane wave—somewhat in analogue to Hooke's law: force (source) equals a constant times displacement (curvature). In the following, we state a few basic equations related to GW astrophysics of compact binary stars, disregarding their derivations and sometimes ignoring factors of order unity. For detailed textbooks on GW astrophysics, we refer to Maggiore (2018); Andersson (2020). For an excellent pedagogic review on detection of GW mergers, see also Abbott et al. (2017).

In analogue to accelerated charged particles giving rise to emission of electromagnetic waves, accelerated masses (and rotation of any objects with an asymmetric distribution of mass with respect to the spin axis) give rise to emission of GWs. The emitted GWs carry information of the change in the gravitational field of the source as a result of a change in the distribution of mass, energy, and momentum. The GWs propagate with the speed of light (the virtual particle, the graviton, mediating the force of gravity has zero rest mass), and they give rise to fluctuations in the metric where they pass through. The force field of the wave is transverse to its direction of propagation and has quadrupolar symmetry (the graviton is a spin-2 particle) with two polarizations. The GWs cause the local spacetime to oscillate when they pass through, and thus the distance, L, between two test masses in the metric will change by an amount, ΔL, which is induced by the GW strain (amplitude): $h \simeq \Delta L/L$. The GW strain is related to the second time derivative of the gravitational quadrupole moment, Q of the source:

$$h_{jk} = \frac{2G}{c^4 d}\,\ddot{Q}_{jk}, \tag{15.1}$$

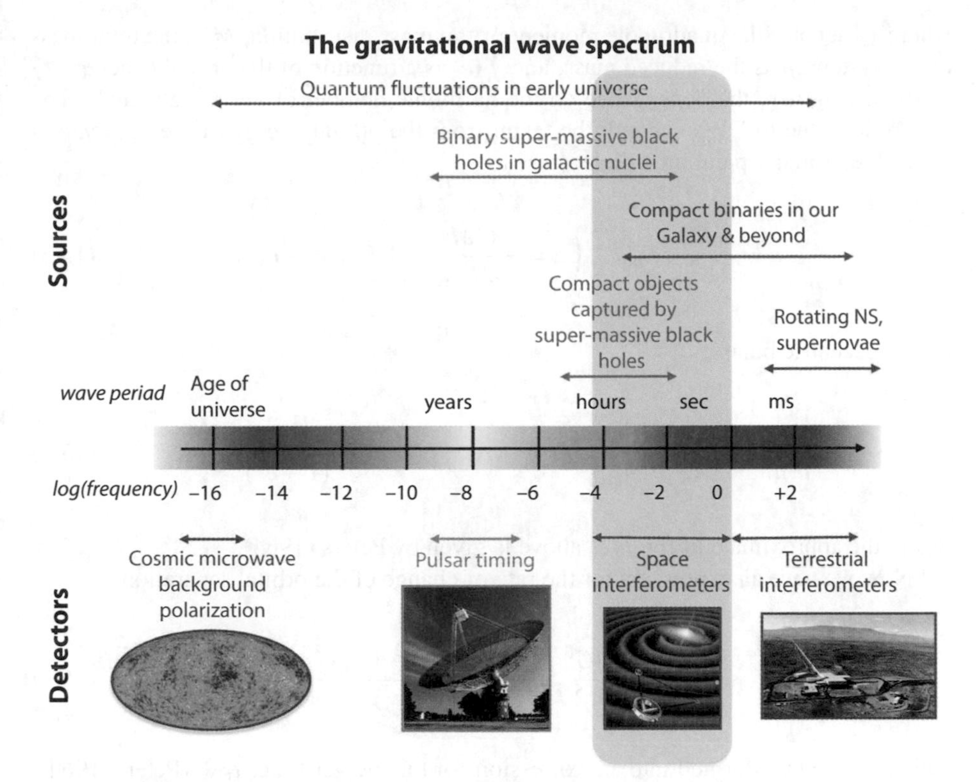

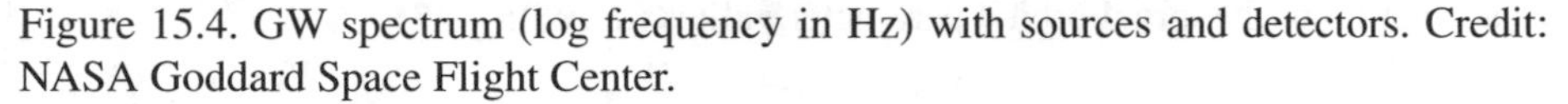

Figure 15.4. GW spectrum (log frequency in Hz) with sources and detectors. Credit: NASA Goddard Space Flight Center.

where G is the gravitational constant, c is the speed of light, and d is the distance to the source. The GW energy flux is given by:

$$F_{\rm GW} = \frac{c^3}{32\pi G} \langle \dot{h}_{jk} \dot{h}_{jk} \rangle. \tag{15.2}$$

An integration of this expression yields the GW luminosity:

$$L_{\rm GW} = d^2 \int F_{\rm GW}\, d\Omega = \frac{G}{5c^5} \langle \dddot{Q}_{jk} \dddot{Q}_{jk} \rangle \simeq \frac{4\pi d^2 c^3}{16\pi\, G} |\dot{h}|^2 . \tag{15.3}$$

For a binary compact object system, the rate of energy loss as a result of GW radiation is given by (in the quadrupole approximation, $a \ll \lambda_{\rm GW}$):

$$L_{\rm GW} = \frac{G}{5c^5} \langle \dddot{Q}_{jk} \dddot{Q}_{jk} \rangle\, g(n, e) \simeq \frac{32 G^4}{5c^5} \frac{M^3 \mu^2}{a^5}\, f(e), \tag{15.4}$$

where Q denotes the quadrupole moment of the mass distribution, M is the total mass of the system, μ is the reduced mass, and $f(e)$ is a function of the orbital eccentricity (disregarding here the dependence on the harmonic number of the wave signal). The energy loss due to GWs can only be taken from the orbital energy of the binary, and hence the orbital separation will decrease as:

$$\dot{a} = \frac{GM\mu}{2E_{\rm orb}^2}\,\dot{E}_{\rm orb} \quad \left(a = -\frac{GM\mu}{2E_{\rm orb}} \,\wedge\, \dot{E}_{\rm orb} = -L_{\rm GW}\right). \tag{15.5}$$

For an eccentric binary:

$$\frac{1}{a}\frac{da}{dt} = -\frac{1}{E}\left.\frac{dE}{dt}\right|_{e=0} f(e) \quad \wedge \quad f(e) \simeq \frac{1+\frac{73}{24}e^2+\frac{37}{96}e^4}{(1-e^2)^{7/2}}, \tag{15.6}$$

where the approximate fit for $f(e)$ above is given by Peters (1964).

Now we have an expression for the rate of change of the orbital separation:

$$\dot{a} \simeq -\frac{64G^3}{5c^5}\frac{M^2\mu}{a^3}\frac{1+\frac{73}{24}e^2+\frac{37}{96}e^4}{(1-e^2)^{7/2}}, \tag{15.7}$$

which can be transformed into an expression for the merger time, $\tau_{\rm GW}$ (Peters, 1964), as a function of the initial orbital parameter values (a_0, e_0):

$$\tau_{\rm GW}(a_0, e_0) = \frac{12}{19}\frac{C_0^4}{\beta} \times \int_0^{e_0} \frac{e^{29/19}[1+(121/304)e^2]^{1181/2299}}{(1-e^2)^{3/2}}\,de, \tag{15.8}$$

where the constants are given by:

$$C_0 = \frac{a_0(1-e_0^2)}{e_0^{12/19}}\,[1+(121/304)e_0^2]^{-870/2299} \quad \wedge \quad \beta = \frac{64G^3}{5c^5}M^2\mu. \tag{15.9}$$

This equation cannot be solved analytically and must be evaluated numerically. The timescale is very dependent on both a and e. Tight and/or eccentric orbits spiral-in much faster than wider and more circular orbits—see Figure 15.5. For example, one can show from Eq. (15.8) that the GW decay timescale decreases by a factor of 10 for a system with $e = 0.694$ compared to a circular system with the same post-SN orbital period (separation) and component masses.

For circular orbits, the merger timescale can easily be found analytically: $\tau_{\rm GW}^{\rm circ} = a_0^4/4\beta$. PSR B1913+16 ($P_{\rm orb} = 7.75$ hr, $M_{\rm NS} = 1.441$, and $1.387\ M_\odot$) with an

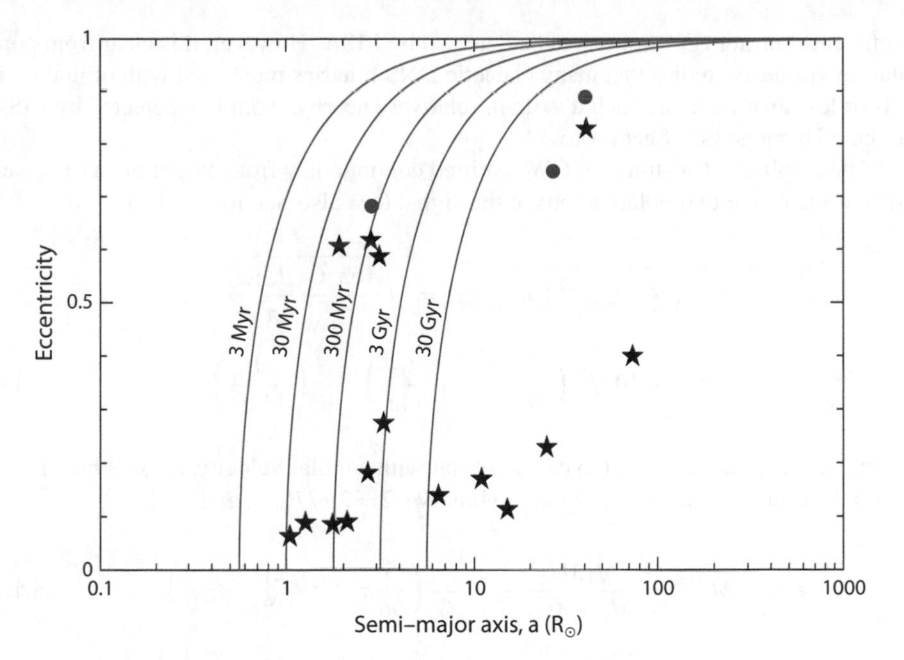

Figure 15.5. Isochrones for the merger time of DNS binaries assuming component masses of 1.4 $M_\odot$. The curves correspond to values between 3 Myr and 30 Gyr. The first 18 known Galactic DNSs are indicated with stars, including three systems in globular clusters (red circles). See Table 14.5 for data.

eccentricity of 0.617 will merge in 301 Myr; whereas if its orbit were circular, the merger time would be 1.65 Gyr.

In terms of loss of orbital angular momentum of a circular binary due to emission of GWs, one can write the resulting rate as:

$$\frac{\dot{J}_{\rm GW}}{J_{\rm orb}} = -\frac{32G^3}{5c^5}\frac{M_1M_2(M_1+M_2)}{a^4}. \tag{15.10}$$

15.3 OBSERVATIONS OF GW SIGNALS FROM BINARIES

The DNS system PSR B1913+16 has an orbital period of 7.75 hr and a distance to the Earth of ~7 kpc. Thus the continuous GW signal from this source is too weak, and the wave frequency too small, to be detected by the upcoming space-borne low-frequency GW detector LISA. Other more recently detected DNS systems, like PSRs J0737−3039 (Burgay et al., 2003), J1757−1854 (Cameron et al., 2018), and J1946+2052 (Stovall et al., 2018), have orbital periods of about a few hours or less (corresponding to GW frequencies of $f_{\rm GW} \sim 0.3$ mHz) and distances between 1 and 10 kpc, but are still not sufficiently strong GW emitters to be detected by LISA. However, it is clear from population synthesis studies that many Galactic DNS binaries must exist with orbital periods of less than an hour. Such a system, relatively nearby, would be detected by LISA in the mHz range (see Section 15.12).

The amplitude (or strain) of GWs emitted continuously from a tight binary is given by the sum of the two polarizations of the signal (see also Section 15.3.1):

$$h = \sqrt{\tfrac{1}{2}[h^2_{+,max} + h^2_{\times,max}]} = \sqrt{\frac{16\pi\, G}{c^3\, \omega^2_{\mathrm{GW}}} \frac{L_{\mathrm{GW}}}{4\pi\, d^2}}$$
$$\simeq 1 \times 10^{-21} \left(\frac{\mathcal{M}}{M_\odot}\right)^{5/3} \left(\frac{P_{\mathrm{orb}}}{1\ \mathrm{hr}}\right)^{-2/3} \left(\frac{d}{1\ \mathrm{kpc}}\right)^{-1}. \tag{15.11}$$

Here, the waves are assumed to be sinusoidal with angular velocity, ω_{GW}, which is $\simeq 2$ times the orbital angular velocity of the binary ($\Omega = 2\pi / P_{\mathrm{orb}}$), and

$$\mathcal{M} \equiv \mu^{3/5} M^{2/5} = \frac{(M_1 M_2)^{3/5}}{(M_1 + M_2)^{1/5}} = \frac{c^3}{G} \left(\frac{5}{96} \pi^{-8/3} f_{\mathrm{GW}}^{-11/3} \dot{f}_{\mathrm{GW}}\right)^{3/5} \tag{15.12}$$

is the chirp mass of the system ($\mu = M_1 M_2/(M_1 + M_2)$ being the reduced mass). The latter term in Eq. (15.12) is found by combining Eq. (15.10) with Kepler's third law ($\Omega^2 = GM/a^2$). It is important to notice that the strain amplitude (detected signal strength) is inversely proportional to d, in contrast to the case of electromagnetic signals, for which the detected signal strength is inversely proportional to d^2. The GW signal strength therefore decreases much more slowly with distance than the strength of an electromagnetic signal.

As a compact binary continues its in-spiral over millions of years after entering the LISA band ($f_{\mathrm{GW}} \sim$ mHz), the GWs will sweep upward in frequency until the system is detectable in the high-frequency LIGO[1] band from $\sim$10 to 10^3 Hz, at which point the compact objects will finally collide. It is this last phase of in-spiral (for DNSs lasting up to $\sim$ 15 minutes with $\sim$16,000 cycles of waveform oscillation) and the final merger event that the LIGO network seeks to monitor in detail.

The first DNS merger GW170817, located in the Galaxy NGC 4993 at a distance of 40 Mpc, was detected in the LIGO band for a bit more than 1 min (Abbott et al., 2017c); see Section 15.10. At design sensitivity, DNS mergers will be detected out to a distance of $\sim$200 Mpc (see Fig. 15.6). This corresponds to GW amplitudes of roughly $10^{-22} < h < 10^{-20}$, depending on inclination of the orbital plane and epoch during in-spiral. As a result of the much larger chirp mass for the double BH mergers, such binaries will be detected out to a luminosity distance, $d_L \propto \mathcal{M}^{5/6}$ (Finn, 1996) which is typically a factor of 5 to 20 larger compared to DNS binaries. Hence, mergers of double BHs are probed in a volume of the local Universe which is roughly 5^3 to 20^3

[1]From here on, we simply refer to all high-frequency interferometer detectors (LIGO, Virgo, KAGRA, IndIGO [or LIGO-Aundha]) as "LIGO" or the "LIGO network."

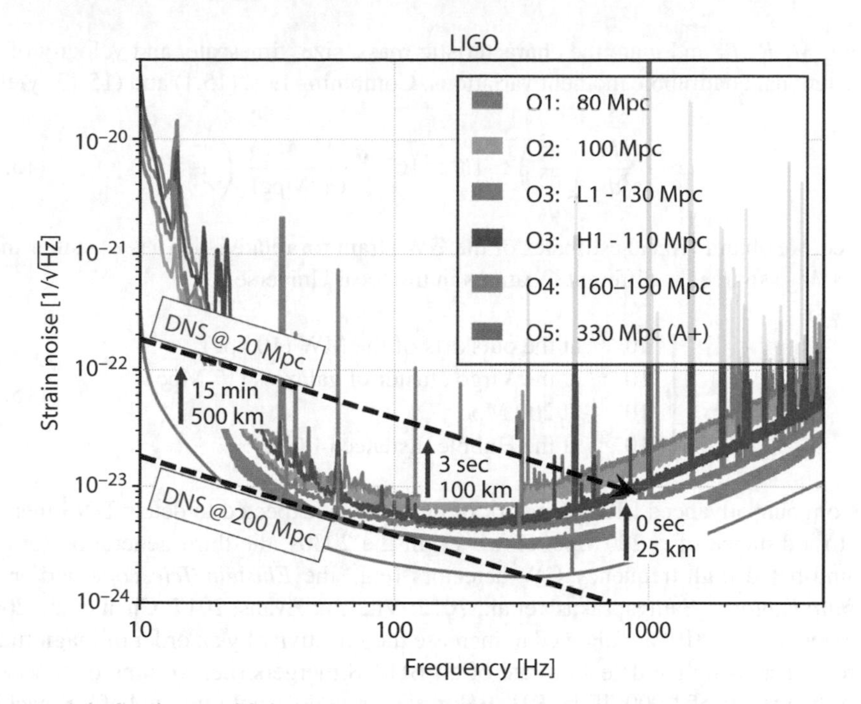

Figure 15.6. Strain sensitivity curves for LIGO from observing runs O1–O5. The dashed lines show the GW amplitude during the in-spiral of a DNS merger at a distance of 20 Mpc (upper) and 200 Mpc (lower). The red arrows indicate remaining merger time and orbital separation between the NSs. Modified after original figure by the LVK collaboration in the LIGO-P1900218-v2 document.

times larger compared to that of DNS mergers, and therefore double BH mergers are expected to completely dominate the event rate of LIGO network detections,[2] as noted in early investigations by, for example, Lipunov et al. (1997); Sipior & Sigurdsson (2002); Voss & Tauris (2003). The detected event rate of mixed BH+NS binaries is expected somewhere in between, see Section 15.4.

15.3.1 Order of Magnitude Estimates

The strain amplitude of the detected GW signal (Eq. 15.1) is proportional to the second time derivative of the gravitational quadrupole moment, Q, of the astrophysical source which can be approximated as:

$$\ddot{Q} \sim \frac{MR^2}{T^2} \simeq Mv^2, \tag{15.13}$$

[2]For monochromatic LISA sources, however, the detection volume scales with $\mathcal{M}^5$ for a constant SNR (Section 15.12).

where M, R, T, and v are the characteristic mass, size, timescale, and velocity of the gravitational quadrupole moment variations. Combining Eqs. (15.1) and (15.13) yields:

$$h \sim \frac{2GM}{c^2 d}\left(\frac{v^2}{c^2}\right) \simeq 1.0 \times 10^{-19}\,\frac{(M/M_\odot)}{(d/\mathrm{Mpc})}\left(\frac{v^2}{c^2}\right). \tag{15.14}$$

Hence, we obtain typical estimates of the GW strain for relativistic ($v \sim c$) stellar mass ($M \sim M_\odot$) sources at different locations in the local Universe:

$$h \sim \begin{cases} 10^{-17} \text{ at the outskirts of the MW (10 kpc)} \\ 10^{-20} \text{ at the Virgo cluster of galaxies (17 Mpc)} \\ 10^{-21} \text{ at 200 Mpc} \\ 10^{-22} \text{ at the Hubble distance (4 Gpc).} \end{cases} \tag{15.15}$$

The ongoing advanced LIGO network of detectors is expected to detect DNS mergers out to a distance of ~ 200 Mpc. However, in the 2030s, the third generation (3G) of ground-based high-frequency GW detectors (e.g., the *Einstein Telescope* and/or the *Cosmic Explorer*, Sathyaprakash et al., 2012; Vitale & Evans, 2017; Chan et al., 2018; Kalogera et al., 2019) are planned to increase the sensitivity by an order of magnitude,[3] thereby increasing the detection rate of all BH/NS mergers (i.e., volume of space) by roughly a factor of 1,000 (if the BH+BH merger rate does not saturate before reaching this sensitivity limit). Thus GW detections may in the best case be expected every few minutes.

Illustrations of GW strain–frequency tracks for LIGO mergers are plotted in Figure 15.6. Although the raw signal gets stronger as the frequency chirps during the final phase of in-spiral, the detectability decreases because the source progressively spends less time in each frequency bin.

15.4 GALACTIC MERGER RATES OF NEUTRON STAR/ BLACK HOLE BINARIES

It is important to simulate the local merger rate and the properties of NS/BH-binaries and compare with the empirical rates measured by the LIGO detectors (see Section 15.7). This exercise will help us to better understand the binary star interactions of their progenitor systems. For Galactic DNS systems, the theoretical merger rate can be determined either from binary population synthesis calculations (e.g., Lipunov et al., 1997; Bloom et al., 1999; Belczynski et al., 2002; Voss & Tauris, 2003; Belczynski et al., 2008; Dominik et al., 2013; Mennekens & Vanbeveren, 2014; Belczynski et al., 2016; Eldridge & Stanway, 2016; Stevenson et al., 2017; Chruslinska et al., 2018; Mapelli & Giacobbo, 2018; Kruckow et al., 2018; Vigna-Gómez et al., 2018; Neijssel et al., 2019; Tang et al., 2020; Santoliquido et al., 2021) or from observations

[3] The Advanced LIGO Plus upgrade in 2024 will already double the current sensitivity.

Table 15.1. Theoretical MW merger rates, merger-rate densities in the local Universe at redshift zero, and anticipated LIGO detection rates at design sensitivity for two different environment metallicities: $Z = Z_{MW} = 0.0088 \simeq 0.7\,Z_\odot$, and $Z = Z_{IZw18} \simeq 0.02\,Z_\odot$. The numbers are taken from Kruckow et al. (2018). The associated uncertainties are about an order of magnitude (see text).

Binary	Z ($Z_\odot$)	MW merger rate (Myr^{-1})	Merger-rate density ($\mathrm{Gpc}^{-3}\,\mathrm{yr}^{-1}$)	LIGO detection rate (yr^{-1})
NS+NS	0.7	3.0	10–35	0.3–1.0
	0.7	14*	159*	4.4*
	0.7	35**	400**	11**
	0.02		10–33	0.3–0.9
BH/NS	0.7	4.1	18–47	5.9–15
	0.02		15–53	11–39
BH+BH	0.7	0.3	0.6–3.1	2.9–15
	0.02		17–35	600–1,200

*Optimistic model for NS+NS.
**Most optimistic model for NS+NS.

of Galactic binary pulsars with NS companions (Clark et al., 1979; Kalogera et al., 2001; Kim et al., 2003, 2015; Pol et al., 2019; Grunthal et al., 2021); see, for example, Chapter 16. Both methods involve a large number of uncertainties. Nevertheless, Lipunov et al. (1997) were the first to point out, quite convincingly, that double BH mergers will be more frequently observed by Advanced LIGO than DNS mergers—see also Postnov & Prokhorov (1999). Earlier, Tutukov & Yungelson (1993) pointed out that the LIGO rates of these two types of events will be of the same order of magnitude. In the following, we discuss theoretically estimated merger rates. The empirical merger rates measured by the LIGO detectors are discussed in Section 15.7.

The current theoretical estimates for the Galactic merger rate of DNS systems are in the range $1 - 100\,\mathrm{Myr}^{-1}$ (for reviews on the merger rates, see, e.g., Abadie et al., 2010; Mandel & Broekgaarden, 2022), which surprisingly is still very similar to the very first estimate of this rate (Clark et al., 1979). As an example, see Table 15.1 for results on NS+NS, BH/NS and BH+BH rates, based on one sample investigation. The associated uncertainty in numbers for a given run with a given population synthesis code is about an order of magnitude, solely resulting from poor constraints on the input physics (mainly CE evolution and SN kicks). Thus all rates based on population synthesis codes should be taken with a large grain of salt (Chapter 16).

In order to extrapolate the Galactic merger rate out to the volume of the local Universe accessible to LIGO, one can either use a method based on star formation rates or a scaling based on the B-band luminosities of galaxies. Using the latter method, Kalogera et al. (2001) found a scaling factor of $(1.0 - 1.5) \times 10^{-2}\,\mathrm{Mpc}^{-3}$, or equivalently, $\sim$100,000 MW-equivalent galaxies (MWEGs) within the LIGO volume (out to

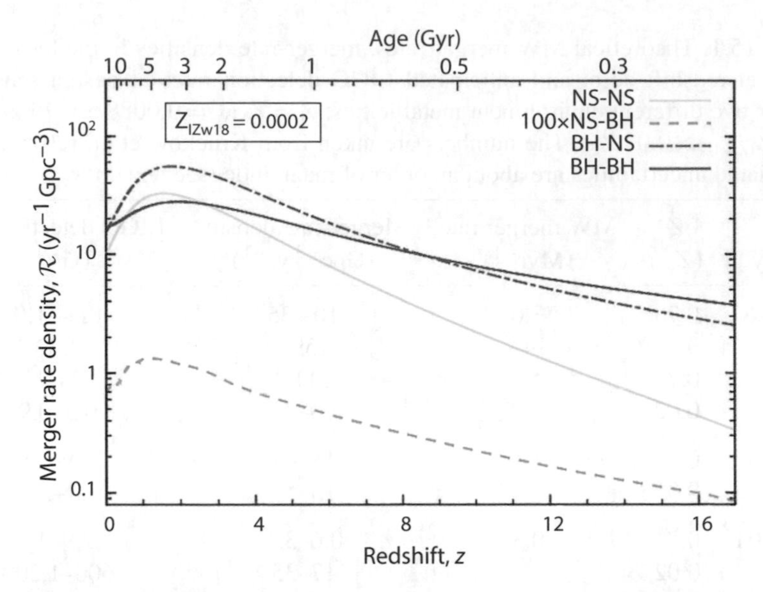

Figure 15.7. Theoretically predicted merger-rate densities of double compact object binaries at IZw18-like metallicity ($Z = 1/50\ Z_\odot$) as a function of cosmological redshift and age of the Universe. Color coding is the type of the two compact objects in their formation order. After Kruckow et al. (2018).

$d \simeq 200$ Mpc for DNS mergers at design sensitivity); see also discussions in, for example, Abadie et al. (2010); Dominik et al. (2013). For mixed BH/NS and double BH systems, the chirp masses are significantly larger and thus the LIGO detection rates, $\mathcal{R}_{\rm LIGO}$, are much higher. $\mathcal{R}_{\rm LIGO}$ scales with $\mathcal{M}^{2.5}$ and to obtain the actual detection rate at the observatories, one must apply some further geometrical factors and assume a signal-to-noise threshold (e.g., $\rho \geq 8$), see discussion in Dominik et al. (2015).

One should be aware that compact star mergers in globular clusters (Portegies Zwart & McMillan, 2000) probably also contribute significantly to the total merger rates (e.g., Rodriguez et al., 2018). It is an ongoing research question to answer which fraction of LIGO detections originate from tight binaries formed via stellar/binary encounters in dense stellar environments (see Chapter 12). The general merger rates are dependent on redshift (Fig. 15.7) through a time-varying star-forming history, as well as a metallicity dependence (e.g., Neijssel et al., 2019, Broekgaarden et al., 2021, Broekgaarden et al., 2022). The metallicity also plays an important role for the stability of RLO in the progenitor systems (Klencki et al., 2020).

Reproducing empirical merger rates using population synthesis is far from sufficient on its own to be a successful result. The physical properties (masses, spins, and orbital parameters) *must* also match those of the known populations at each stage along the complicated evolutionary path (Figs. 1.1 and 15.10) from ZAMS stars to formation of double compact objects, such as binary radio pulsars (Tauris et al., 2017). Actually, the measured properties of known binaries along the evolutionary path are very useful and important for calibrating various physical interaction parameters (Kruckow et al., 2018).

15.5 FORMATION OF DOUBLE BLACK HOLE BINARIES

Before discussing the BH masses and spins measured in GW mergers by LIGO-Virgo-KAGRA, we summarize thc outcome of the final stages of the most massive stars (see Chapter 10 for further details).

15.5.1 BHs, Pulsational Pair Instabilities, and PISNe

It has been known for many years (Fowler & Hoyle, 1964), that quite massive stars that end their main-sequence evolution with a helium (carbon-oxygen) core exceeding $\sim 60\,M_\odot$ ($57\,M_\odot$), after accounting for wind mass loss, will undergo a fatal pair-instability SN (PISN) that disrupts the entire star—see also Section 8.3.3. The reason for this is that during carbon burning the conversion of energetic photons into electron-positron pairs causes a softening of the equation-of-state, initiating the collapse of the star (Fig. 15.8). This increases the core temperature further until explosive thermonuclear oxygen burning reverts the collapse and fully disrupts the star.

Stars that are very massive, however, and leave a helium (carbon-oxygen) core exceeding $125\,M_\odot$ ($121\,M_\odot$) also experience explosive thermonuclear oxygen burning, but they are believed to survive this stage due to photodisintegration of heavy nuclei that absorbs a large fraction of the released energy. Therefore, the thermonuclear explosion is not energetic enough to reverse the collapse into a full explosion disrupting the star (Fryer et al., 2001; Heger et al., 2003), and the core collapse is speculated to leave behind a massive BH. As a result, at first sight, it is expected that there is a gap in the resulting masses of BHs roughly between[4] 60 and $125\,M_\odot$ (see Fig. 8.25).

Massive stars that produce core masses just below the threshold for PISNe ($\lesssim 60\,M_\odot$) still experience explosive burning, although somewhat weaker, and therefore they suffer from partial ejections of their envelopes. These are the so-called pulsational pair-instability (PPI) ejections (Chatzopoulos & Wheeler, 2012; Woosley, 2017, 2019; Renzo et al., 2020, see also Chapter 8), which therefore result in lowering the final core mass. The star may undergo multiple such pulses until it finally leaves behind a BH from a core that is stabilized from the combined effects of pulsational mass loss, entropy loss to neutrinos, and fuel consumption (Renzo et al., 2020). Thus the gap in the resulting masses of BHs in close binaries due to the combination of PISNe and PPI ejections is predicted roughly between $\sim 45 - 125\,M_\odot$ (Fig. 15.9). The lower edge of the mass gap is dependent on the efficiency of angular momentum transport in the progenitor star such that efficient coupling through the Spruit-Tayler dynamo shifts the lower edge of the mass-gap slightly upwards, and thus Marchant & Moriya (2020) argue that the lower edge of the upper mass-gap is dependent on BH spin. Woosley & Heger (2021) also investigated the lower and upper mass limits of the mass gap and found that they depend on a number of factors, including rotation and nuclear reaction rates. As an example, they found that rapid rotation could increase the lower mass limit

[4]The exact mass boundaries are strongly dependent on metallicity and assumed stellar physics (e.g., Woosley & Heger, 2021), and quotes may therefore vary throughout this book.

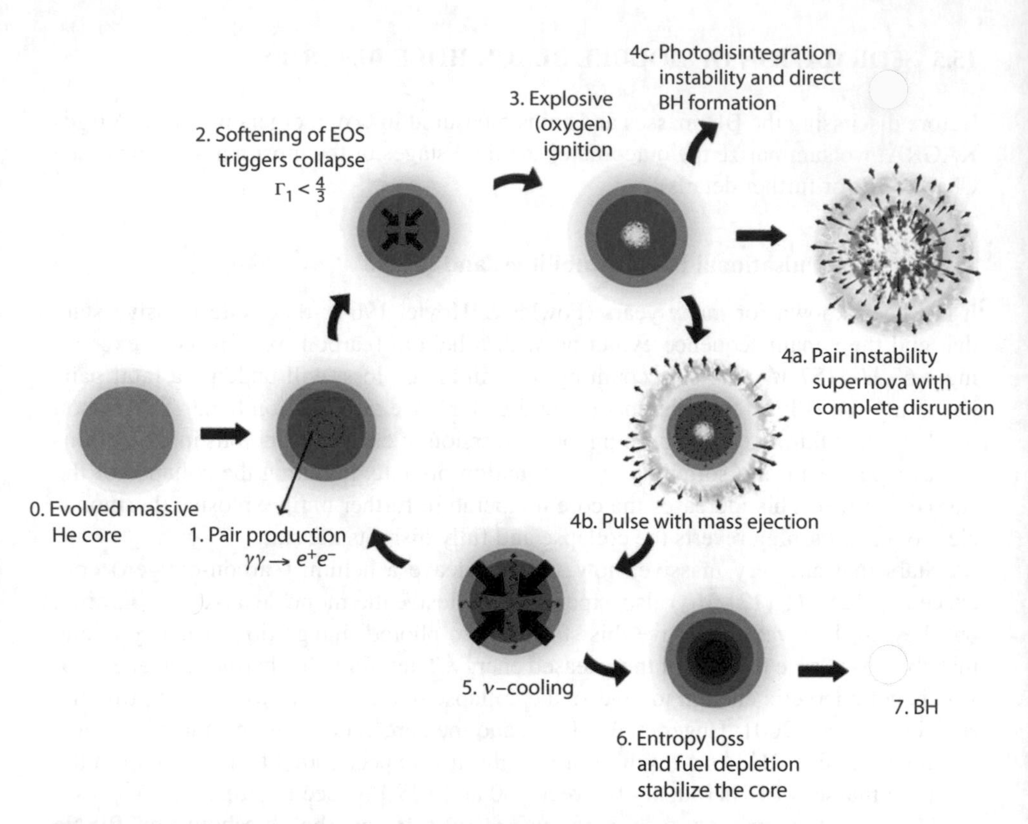

Figure 15.8. Evolution of a massive helium core in relation to pair-instability SNe (PISNe) or pulsational pair-instability (PPI) cycles. Three final outcomes are possible (above/in/below BH mass gap): a BH produced from an extremely massive core that survives, due to photodisintegration, the onset of pair instability (4c); full disruption without a compact remnant (PISN, 4a); and, for a somewhat less massive core, PPI episodic mass loss (4b) which eventually (after one or more cycles) produces a BH remnant (7). After Renzo et al. (2020).

to $\sim 70\, M_\odot$, depending on the treatment of magnetic torques. It will be interesting to see if the mass-gap prediction based on stellar evolution can be confirmed by future LIGO-Virgo-KAGRA measurements (see, e.g., discussions in Edelman et al., 2021, based on GWTC-2 data). Figure 15.8 depicts the evolution leading to PISNe and PPI cycles.

Finally, Marchant et al. (2019) demonstrated that if PPI ejections occur from a star in a close binary, the heat deposited throughout the layers of the star that remains bound causes it to expand to more than $100\, R_\odot$ for a period of 10^2 to 10^4 yr, depending on the mass of the progenitor. This may lead to subsequent RLO or a CE, thereby affecting the resulting spin of the BHs and leading to additional electromagnetic transients associated with PPI eruptions.

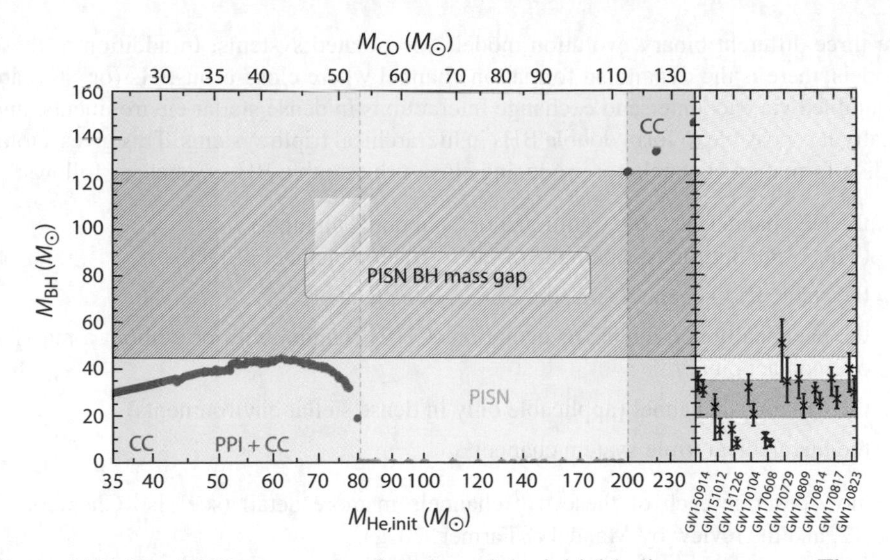

Figure 15.9. Final BH masses as a function of the initial helium-core mass. The colors in the background indicate the approximate range for each evolutionary path. (See also Figs. 8.33 and 15.8.) The right panel shows the masses inferred from the first 10 BH+BH mergers detected by LIGO. After Renzo et al. (2020). Since this plot was published, the double BH mergers GW190426 and GW190521 were announced with all four BH component masses between $\sim 69 - 107\,M_\odot$, that is, located right inside the PISN BH mass gap (see Section 15.6.2 for discussions).

15.5.2 Final Stage Prior to Formation of a Double BH System

As we will discuss in more detail in this section (15.5), in all three models for producing close-orbit double BH binaries from isolated systems, the system just prior to the formation the second BH consists of a massive helium star (i.e., a Wolf-Rayet [WR] star) and a BH, with an orbital period of less than a few days.

Presently, seven of such WR-star HMXBs are known and listed in Table 6.8. With the exception of Cyg X-3 ($P_{\rm orb} = 4.8$ hr), they are all in external galaxies, and with only one exception, their orbital periods are very short: two around 1.5 d, and four shorter than one day. For reasons derived from binary evolution (see Section 10.11), the compact stars in these systems are likely to be BHs (van den Heuvel et al., 2017). This implies that if their massive WR-star companions terminate their evolution as BHs too, these systems are expected to produce double BHs with very short orbital periods, which will merge by GW losses within a Hubble time.

15.5.3 Five Formation Channels of Double BHs

Contrary to the formation of DNSs, for which only one main binary evolution channel seems to dominate: CE evolution from a wide-orbit HMXB (see Fig. 1.1 and Sections 10.13 and 14.9), it appears that for the formation of close-orbit double BHs there are three different binary evolution models for isolated systems. In addition to these models, there is the dynamical formation channel where close-orbit BHs (or NSs) are assembled via encounter and exchange interactions in dense stellar environments, and finally it is possible to form double BHs in hierarchical triple systems. This gives a total of five formation channels for producing close-orbit double BH systems, as follows:

1. the CE channel (i.e., the traditional or "standard" channel)
   (This is the high-mass analogue of the formation channel of DNSs.)
2. the stable RLO channel (i.e., the SS433-like channel)
3. the chemically homogeneous evolution (CHE) channel with or without a massive overcontact binary
4. the dynamical channel (applicable only in dense stellar environments)
5. the hierarchical triple system channel.

We now describe each of these five channels in more detail (see also Chapters 10 and 12, and the review by Mandel & Farmer, 2022).

15.5.4 (1) Formation via CE evolution

The CE formation channel for close-orbit double BHs is depicted in Figure 15.10. This channel is similar to that which is believed to produce tight DNS systems (see Figure 1.1, and Sections 10.13 and 14.9 for discussions, and see also Tauris et al. (2017) for further details), scaled up to higher initial component masses. In this scenario, the systems always enter a CE phase following the HMXB stage, during which the O-star becomes a red supergiant (RSG) and captures its BH companion (see, e.g., Bogomazov, 2014). Belczynski et al. (2016) applied this scenario to explain the origin of the first LIGO source, the double BH merger GW150914 (however, see also Andrews et al., 2021). They found that to produce massive double BHs, with components more massive than 30 $M_{\odot}$, the progenitor binaries must have a metallicity below $\sim 0.1\, Z_{\odot}$. For higher metallicities, the stellar wind mass-loss rates from the massive progenitor stars are so high that such massive remnants can never be left behind (Vink et al., 2001; Gräfener & Hamann, 2008; Vink et al., 2011; Chen et al., 2015), see also Figure 15.14 later in this chapter. Notice, however, that recently Higgins et al. (2021) demonstrated the possibility of producing BH masses all the way up to the onset of the PISN gap ($\sim 45\, M_{\odot}$) for metallicities up to 0.5 $Z_{\odot}$ (roughly that of the LMC), based on a new wind prescription of Sander & Vink (2020).

It is generally believed that donor stars in wide-orbit HMXBs develop a deep convective envelope prior to filling their Roche lobe. They will therefore expand rapidly in response to mass loss which, in combination with the orbital shrinking due to a large mass ratio, will result in a CE (Section 4.8). The outcome of the CE depends on the orbital energy and the binding energy of donor star envelope at the onset of the RLO (see Sections 4.8.4 and 10.11.1, and Kruckow et al., 2016; Klencki et al., 2021; Marchant et al., 2021, for discussions of CE ejection in massive binaries producing double BH mergers detected by the LIGO network). The binding energy of the CE becomes significantly reduced during the red-supergiant stage of the donor star.

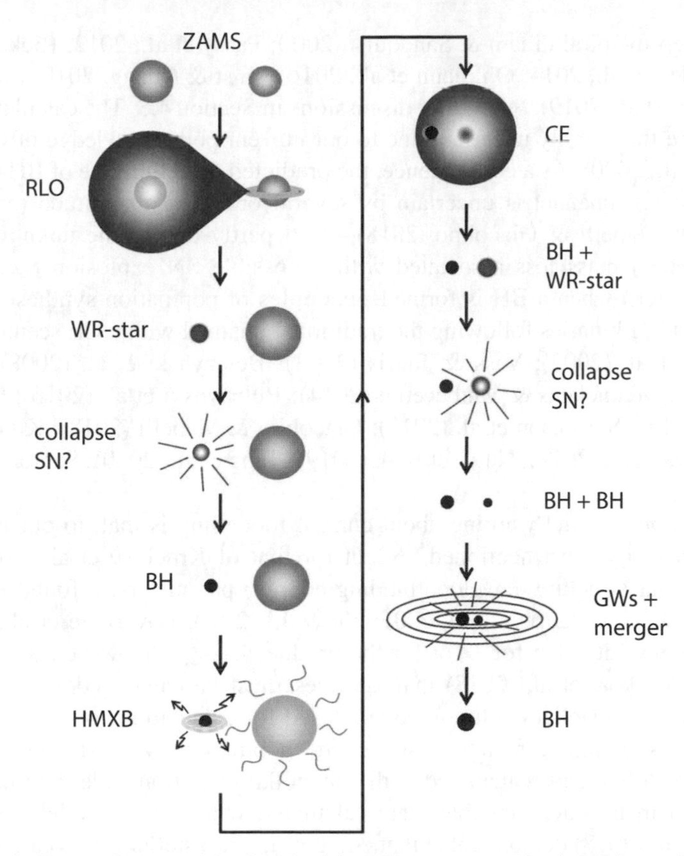

Figure 15.10. Formation model of a double BH merger as final product of close massive binary evolution with a CE. See text for explanations and discussions. Compared to formation of a DNS system (Figs. 1.1 and 14.35), ZAMS masses are larger, and resulting evolutionary timescales shorter. The BH SNe may not be observable.

Evolving the donor star to such an advanced evolutionary stage before it fills its Roche lobe and initiates RLO and CE evolution requires a wide orbit. It is therefore important to realize that the initial BH (NS) + main-sequence star binary that finally produces the close double BH (or DNS) system via CE evolution, is initially very wide with an initial period of several years. Following the in-spiral and CE ejection leaves a system resembling WR-star HMXBs with a BH accretor and an orbital period of approximately one day (Section 10.11). After the collapse of the WR-star, one is left with a double BH in a close orbit that will merge within a Hubble time.

There are many uncertainties, however, involved in calculations of the in-spiral and the subsequent CE ejection. The evolution is often tidally unstable, and the angular momentum transfer, dissipation of orbital energy, and structural changes of the donor star take place on very short timescales ($< 10^3$ yr, Podsiadlowski, 2001). A complete study of the problem requires detailed multi-dimensional hydrodynamical calculations, although early studies in this direction have difficulties ejecting the envelope and securing deep in-spiral (Taam & Sandquist, 2000; Passy et al., 2012; Ricker & Taam, 2012; Nandez et al., 2014; Ohlmann et al., 2016; Prust & Chang, 2019; Fragos et al., 2019; Ricker et al., 2019), see further discussions in Section 4.8. The calculations along this route are thus highly uncertain due to our current poor knowledge of CE physics (Ivanova et al., 2020). As a consequence, the predicted detection rate of BH+BH mergers from the CE channel is uncertain by several orders of magnitude (e.g., Abadie et al., 2010; Mapelli & Giacobbo, 2018)—also partly due to the unknown amount of (asymmetric) mass loss associated with a possible SN explosion (i.e., imparted momentum kick) when a BH is formed. Examples of population synthesis investigations of BH+BH binaries following the traditional channel with a CE scenario include Belczynski et al. (2002); Voss & Tauris (2003); Belczynski et al. (2008); Dominik et al. (2012); Mennekens & Vanbeveren (2014); Belczynski et al. (2016); Eldridge & Stanway (2016); Stevenson et al. (2017); Giacobbo & Mapelli (2018); Kruckow et al. (2018); Spera et al. (2019); Neijssel et al. (2019); Tang et al. (2020); Santoliquido et al. (2021).

A great concern and warning about current modelling is that, to our knowledge, all of the investigations mentioned, except for that of Kruckow et al. (2018), apply tabulated stellar structure (envelope binding energy) parameters, λ found in the literature (e.g., Dewi & Tauris, 2000, 2001; Xu & Li, 2010; Loveridge et al., 2011), or even use a constant value for λ (or for the product $\lambda \cdot \alpha_{\rm CE}$). However, it was demonstrated by Kruckow et al. (2018) that λ-values must be calculated *self-consistently*, such that the same stellar evolution code is used for calculations of λ and, for example, the corresponding core mass used in the population synthesis code. One *cannot* combine stellar grids calculated with one stellar evolution code with tabulated λ-values found in the literature that were calculated with another code. For example, Kruckow et al. (2018) demonstrated that combining their stellar tables calculated from the BEC code with computed λ-values based on the Eggleton code, the double BH merger rate resulting from similar population synthesis simulations increases by almost two orders of magnitude compared to the self-consistent treatment applied in their COMBINE code.

Furthermore, in CE evolution in which an extended convective envelope is involved, as is encountered when a companion enters the envelope of a red giant or red supergiant, regular pulsational mass ejection events may occur (Clayton et al., 2017), and in such circumstances it is not sure that the final outcome will be a deep in-spiral.

It is also important to notice that, as we have seen in Chapter 8, even without CE evolution, single stars more massive than $\sim 25\, M_{\odot}$ spontaneously lose their red-giant envelopes by strong wind mass loss, also at low metallicity. As the binding energy of their envelopes is apparently very low, this means that in stars of these masses very little orbital energy has to be deposited by the CE evolution to lift off the envelope. This might prevent deep in-spiral during CE evolution. Particularly for the formation of double BHs by the CE scenario, this might be a problem. See further comments on potential limitations of CE evolution in Sections 4.8.5 and 10.11.1.

In the following four formation channels of producing BH+BH binaries, the CE phase is avoided altogether. However, each of them have other issues.

15.5.5 (2) Formation via Stable RLO

The stable RLO channel (or the SS433-like channel) was discussed in detail in Section 10.11 and will briefly be summarized here. HMXBs with short orbital periods (and relatively massive BH accretors) may evolve via dynamically *stable* RLO to produce short-period WR-star X-ray binaries (e.g., van den Heuvel et al., 2017, and references therein). If these systems remain bound after the second SN, they may form double BH systems with orbital periods short enough to produce a merger event within a Hubble time. Numerically, it has been shown that RLO in intermediate-mass X-ray binaries with a radiative or slightly convective donor star can remain dynamically stable for mass ratios up to $q \sim 4$ (Tauris et al., 2000). It has even been suggested that stable RLO in X-ray binaries may proceed with larger q-values (Pavlovskii et al., 2017; Quast et al., 2019). However, further theoretical work is needed to validate this finding. Of the five known BH HMXBs, M33 X-7 is the best candidate for producing a double BH system (see review by van den Heuvel, 2019). This system has a $15.7\,M_{\odot}$ BH and a $70\,M_{\odot}$ companion in a 3.45 d orbit (Orosz et al., 2007).

The stable RLO evolution channel holds for BH HMXBs with orbital periods up to at least several months. The first clear observational indication that such stable RLO evolution can take place is provided by the peculiar X-ray binary system of SS433. In view of the crucial importance of this system for the evolution of BH-HMXBs, this system was discussed in detail in Section 10.11. Some population synthesis studies on the modelling of the formation of double BH systems (Neijssel et al., 2019; Andrews et al., 2021) indicate that the majority of LIGO-detectable double BH systems might have formed by binary evolution via the stable RLO (SS433-like) model—especially when more detailed `MESA` models are applied compared to rapid binary population synthesis (Gallegos-Garcia et al., 2021).

15.5.6 (3) Formation via Chemically Homogeneous Evolution

In the chemically homogeneous evolution (CHE) scenario for binaries (de Mink et al., 2009; Mandel & de Mink, 2016; de Mink & Mandel, 2016; Marchant et al., 2016; du Buisson et al., 2020), the stars avoid the usual strong post-main sequence expansion as a result of effective mixing enforced through the rapid rotation of the stars, which is maintained via tidal interactions (the rotational mixing prevents the build-up of the usual discontinuity in abundance composition seen at the core-envelope transition in giant stars). It is thought that this scenario works only for massive stars at low metallicity where strong angular-momentum loss due to stellar winds can be avoided—that is, a weak stellar wind will prevent significant orbital widening due to wind mass loss. (Efficient wind mass loss would widen the orbit and thus result in weaker tidal interactions and thereby loss of synchronization.)

The CHE model is based on the fact that in massive binaries with orbital periods $\lesssim 2-3$ d (such as those known in the Doradus region of the LMC), strong tidal friction will cause the rotation period of the component stars to be synchronized with the orbital period. This means that the two stars are kept in very rapid rotation. In such rapidly rotating massive stars, strong large-scale meridional currents (Eddington-Sweet circulation) will develop, which will keep the stellar material fully mixed throughout the star (Maeder, 1987). Thus, the helium produced by the hydrogen burning in the stellar core will be fully mixed through the whole star, and the star will keep a full chemically homogeneous composition throughout its entire hydrogen-burning phase and will end as a pure helium star: a WR-star. This implies that, contrary to the case of normally evolving massive stars, the stellar radius never increases during the evolution, and the two stars may avoid overflowing their Roche lobes. Very close binaries, however, may evolve into contact where both stellar components fill and even overfill their Roche-lobe volumes. The evolution during such an overcontact phase differs from a classical CE phase because co-rotation can, in principle, be maintained as long as material does not overflow the L_2 point (Section 4.3.2). This means that orbital decay and spiral-in that is due to viscous drag can be avoided, resulting in a stable system evolving on a nuclear timescale. The scenario requires, however, a mass ratio close to unity (Marchant et al., 2016).

The final helium star has a smaller radius than its original hydrogen-rich progenitor. Since the two stars will never have a mass ratio exactly equal to unity, the more massive component will be the first to become a WR-star, while its companion then still will look like an O-star with enhanced helium. By the time the WR-star collapses to a BH, the companion may itself have become a WR-star, such that for a while the system will be a short-period WR-star HMXB. After the core collapse of the second WR-star, the system terminates as a close double BH.

Marchant et al. (2016) presented the first detailed CHE models leading to the formation of BH+BH systems and demonstrated that massive overcontact binary systems are particularly suited for this channel, enabling formation of very massive stellar-mass BH+BH mergers, that is, in agreement with the detection of GW150914. Figure 15.11 depicts the evolutionary sequence via CHE producing a double BH. Lower-mass BH+BH mergers like GW151226 ($14+8\ M_{\odot}$), however, cannot be formed from this scenario (as efficient mixing requires rather massive stars), nor BH+NS binaries where the NS is recycled. (BH+NS binaries, however, where the BH is born first from a star that evolved without RLO, may possibly be formed via this CHE scenario and could produce ULX binaries as well, Marchant et al., 2017.)

Recent work by Hastings et al. (2020) shows that the effects of companion-induced circulation have strong implications for the formation rates of close binary BHs through the CHE channel: not only do the predicted detection rates increase, also double BH systems with mass ratios as low as 0.8 may be formed when companion-induced circulation is taken into account.

15.5.7 (4) Formation via Dynamical Interactions

As explained in Chapter 12, BH+BH binaries can also be formed via encounter and exchange interactions in dense star clusters, such as globular clusters or nuclear star clusters near the centers of most galaxies (for the dynamical evolution of such clusters see, e.g., Sigurdsson & Hernquist, 1993; Portegies Zwart & McMillan, 2000; Banerjee et al., 2010; Samsing et al., 2014). Such dynamically assembled pairs of BHs have also been proven to be effective BH+BH merger sources (e.g., Rodriguez et al., 2016; Antonini & Rasio, 2016; Mapelli, 2016; Askar et al., 2017; Banerjee, 2017; Fragione &

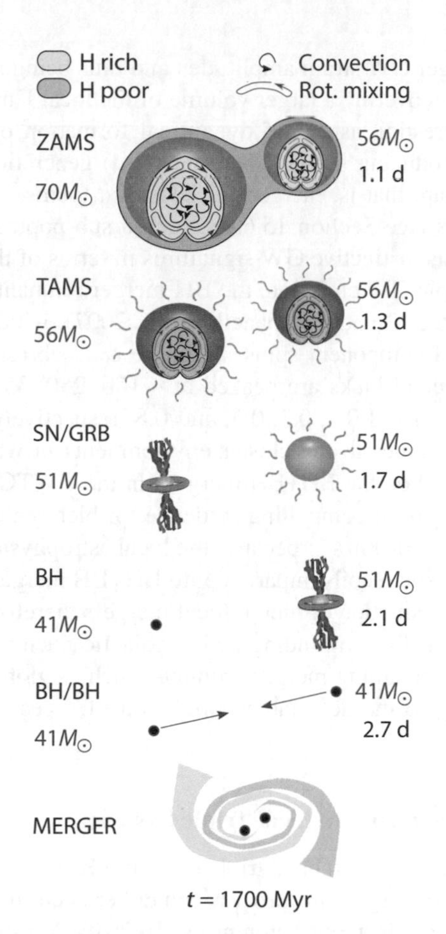

Figure 15.11. Chemically homogeneous evolution (CHE) of a very massive and close binary (initial orbital period, $P_{\rm orb} \leq 2-3$ d), into a tight double BH system which merges within the lifetime of the Universe (in this example, after 1.7 Gyr). See text for further explanation. After Marchant et al. (2017).

Banerjee, 2021; Gerosa & Fishbach, 2021). One advantage of the dynamical formation channel is that is circumvents the need for mass transfer and CE evolution, and it allows for the possibility of a chain of hierarchical mergers producing quite massive BHs (Section 15.6.2).

In analogy to the other production channels mentioned earlier in this chapter, the rate of BH+BH mergers from the dynamical formation channel is also difficult to constrain. A number of studies predict that this channel probably accounts for less than ~10% of all BH+BH mergers (e.g., Rodriguez et al., 2016), whereas other studies indicate this fraction could be larger (Rodriguez et al., 2018). It has been argued that the distribution of masses of the first 10 BH+BH mergers (Fig. 6.21) favors relatively massive BHs—more massive than the BHs typically found in X-ray binaries—which may point to an origin via a dynamical formation channel (Perna et al., 2019). However, one must carefully take into account the observational bias of more massive BH+BH mergers producing larger GW strain amplitudes and thus being more easily detectable out to longer distances, covering a larger volume of the local Universe.

A particularly interesting aspect of dynamical formation of double BH binaries is the possibility of producing second (or even third) generation BH mergers in the pair-instability mass gap, that is, mergers of BHs that themselves are the product of previous merger events (see Section 15.6.2). Such a sub-population of hierarchically assembled BHs presents distinctive GW signatures in terms of their masses and spins. As a result of anisotropic GW emission, the BH merger remnant receives a significant recoil (see, e.g., Chapter 12 and Campanelli et al., 2007), which is highly correlated with the maximum BH component spins. Fragione & Loeb (2021) find that the distributions of average recoil kicks are peaked at $\sim$150, 250, 350, and 600 km s^{-1} for maximum progenitor spins of 0.1, 0.3, 0.5, and 0.8, respectively. Therefore, we see a resulting bias when it comes to the cluster environments in which such hierarchical mergers can form. Based on the BH+BH mergers in the GWTC-2 catalogue, Kimball et al. (2021) argue in favor of compelling evidence for hierarchical mergers in clusters with escape velocities $>$ 100 km s^{-1} because the local astrophysical environments need to overcome the relativistic recoils imparted onto BH+BH merger remnants. Promising locations for efficient production of hierarchical mergers therefore include nuclear star clusters and accretion disks surrounding active galactic nuclei, though environments that are less efficient at retaining merger products, such as globular clusters, may still contribute significantly to the detectable population of repeated mergers (Gerosa & Fishbach, 2021).

15.5.8 (5) Formation in Hierarchical Triple Systems

Kozai-Lidov resonances in hierarchical triple systems (Kozai, 1962; Lidov, 1962) may bring the inner binary to merge. This process can either occur in triple systems formed in, for example, globular clusters (Antonini et al., 2016; Martinez et al., 2020) or in the Galactic field (Silsbee & Tremaine, 2017; Antonini et al., 2017; Toonen et al., 2016, 2020). This formation path to the merger of a pair of BHs most likely accounts for less than $\sim$1% of all detected BH+BH mergers. The formation of two or three compact objects in a triple system in the field is extremely complicated to model in detail given the mass-transfer processes (often involving a CE phase) and kinematic effects of SNe when NSs or BHs are formed, apart from dynamical stability criteria which limits the relative orbital sizes during the entire evolution. As an example, see our modelling (Tauris & van den Heuvel, 2014) of the formation of the triple pulsar J0337+1715 (Ransom et al., 2014) mentioned in Chapter 14. For a further description of the evolution of massive stellar triples and implications for the production of BHs and NSs, see Stegmann et al. (2022).

15.5.9 Differences in Outcomes from Formation Channels

The five formation channels outlined in Sections 15.5.4–15.5.8 differ in predictions, of for example, BH masses, mass ratios, and BH spins. The upper mass gap in the spectrum of BH masses discussed in Section 15.5.1 is one example. Another (but controversial) example is related to the origin of BH spins. Although it has very often been advocated in the current literature (e.g., Farr et al., 2017) that the CE formation channel should produce aligned spins, and the dynamical channel results in randomly oriented spins (in agreement with several LIGO network detections that seem to favor small effective spins, $\chi_{\rm eff}$, see Section 15.8), this distinction is not true if the last-formed BH had its spin axis tossed in a more or less random direction during the core collapse (Section 13.7.9). This process is known to operate during the formation of NSs and may very well also apply to newborn BHs (Tauris et al., 2017; Tauris, 2022). We discuss BH spins in further detail later in this chapter.

It is essential for our understanding of massive star evolution to identify the contribution of binary evolution to the formation of double BH binaries and how such systems differ from those formed in dense environments with dynamical exchange interactions. A promising way to progress is investigating the progenitors of double BH systems and comparing predictions with local massive star samples such as the population in 30 Doradus in the LMC (Langer et al., 2020).

Naively, a simple distinction between different formation channels would also be to look at their predictions for BH+BH merger rates. Interestingly, the literature has seen papers arguing for exclusive origins in either isolated binary systems or in dynamically assembled systems—with the claims that both channels are each being consistent with the present observed rate of BH+BH mergers of $17-45\,{\rm Gpc}^{-3}\,{\rm yr}^{-1}$ (GWTC-3, with an evolving merger rate and considering here a redshift of $z=0.2$, Abbott et al., 2022b), see, for example, discussions in Bavera et al. (2020); Roulet et al. (2021); Rodriguez et al. (2021). Given the uncertainties in current modelling, such conclusions are not surprising.

For the isolated systems, theoretical estimates of merger rates are very sensitive to assumptions regarding the CE phase and natal kicks (e.g., Voss & Tauris, 2003; Kruckow et al., 2018, and references therein) but also depend strongly on the assumed model of chemical evolution over cosmic time (Belczynski et al., 2020), while the dependence of merger rates with redshift depends mostly on the cosmic star-formation history (Madau & Fragos, 2017). For the dynamical formation channel, the major uncertainties stem from the unknown relaxation processes and star formation issues. Since all objects near, for example, a central super-massive BH (SMBH) in an AGN orbit the same thing, like in a Solar System, relaxation processes, including (vector) resonant relaxation, make it hard to predict the mass-segregation timescale for BHs to sink in the potential well of the SMBH (Hamers et al., 2018; Tagawa et al., 2020). The diversity of LIGO detections of BH+BH mergers, however, may well suggest that multiple formation channels are at play. For a recent review on evidence for multiple formation channels based on LIGO network data, see Zevin et al. (2021).

A better option to constrain the contribution from, for example, the dynamical formation channel is perhaps to consider the eccentricities of the BH+BH mergers. Romero-Shaw et al. (2019) demonstrated that all eccentricities of BH+BH events from observing runs O1+O2 of the LIGO network are smaller than 0.02 to 0.05, whereas expectations from a dynamical capture scenario point to production of merger events with higher eccentricities in 5% of the events.

15.6 PROPERTIES OF GW SOURCES DETECTED SO FAR

During the first two observing runs of LIGO-Virgo (2015–2017), a total of 11 events were reported (Abbott et al., 2019c). In addition, more than half a dozen of additional candidates were later announced using an independent analysis pipeline (Venumadhav et al., 2020; Zackay et al., 2019). Of all these events, ~ 19 were double BH mergers, and one event was a DNS merger. Selected parameters for the 11 confident detections from the first (O1) and second (O2) observing runs of the LIGO network of detectors (LIGO and Virgo) are shown in Table 15.2. Confident GW events are identified based on the statistical significance of triggers and have false-alarm rates (FAR) $\ll 1\,\mathrm{yr}^{-1}$ and network signal-to-noise ratios of typically: $10 < \mathrm{SNR} < 30$ (slightly depending on the pipeline of software used for the analysis). In addition to the confident detections, there is also a large number of marginal triggers. In observing runs O1 and O2, 14 such marginal triggers were reported. These often have $\mathrm{FAR} = 1 - 10\,\mathrm{yr}^{-1}$ and $\mathrm{SNR} \lesssim 10$. Events in this grey zone could be either real astrophysical events or simply noise.

The third observing run O3a+O3b in 2019–2020 revealed a total 74 (39+35) candidate events[5] in 11 months (Abbott et al., 2021d, 2022a). A reanalysis of the O3a data found an additional 8 BH+BH mergers (GWTC–2.1, Abbott et al., 2021) and an independent data analysis by the Princeton group revealed 10 additional BH+BH merger candidates (Olsen et al., 2022). The O3 dataset was based on a FAR threshold of typically $1.0\,\mathrm{yr}^{-1}$, corresponding to inclusion of all candidates with a likelihood of being astrophysical of $P_{\mathrm{astro}} > 50\%$. (Thus it is expected that $\sim 10\% - 15\%$ of the events are false alarms caused by instrumental noise.) Among the new sources in O3, many had spectacular properties, for example, very massive individual BH components ($\sim 100\,M_{\odot}$: GW190426 and GW190521), BH+BH mergers with significant non-zero effective spins ($\chi_{\mathrm{eff}} \simeq 0.5 - 0.7$: GW190517, GW190403, and GW200308) and binaries with small mass ratios ($q \simeq 0.1$: e.g., the mysterious lower mass-gap event GW190814). The latter gave rise to detectable GW radiation beyond the leading quadrupolar order. Besides detection of many BH+BH mergers, and a massive DNS merger (GW190425), the combined O3 sample also included a couple of mixed BH+NS merger events (GW200105 and GW200115, Abbott et al., 2021b), and an additional couple of such candidates (GW190426_152155[6] and possibly GW190814 and GW191219). The most interesting discoveries listed above are discussed further below in more detail, and their key parameters are shown in Table 15.3.

Figure 15.12 displays data of all the GW candidate events from observing runs O1–O3 that are collected in the GWTC-3 catalogue (Abbott et al., 2022b). The third observing run alone (O3, i.e., O3a+O3b) detected in average a bit more than one BH+BH merger event (candidate) every week. In total, more than 80 BH+BH mergers, besides a couple of NS+NS mergers, mixed BH+NS mergers, and mass-gap events. The double BH mergers (including the lower-mass gap event GW190917) have component

[5]See all the candidates at: https://www.gw-openscience.org/eventapi/html/GWTC/ and also https://gracedb.ligo.org/. Note, for each event the full GW name encodes the UTC date with the time of the event given after the underscore, for example, GW190924_021846. Here in this book, we abbreviate candidate names by omitting the last six digits when unambiguous.

[6]Written here with full 12 digit name to avoid confusion with the very massive GW190426.

Table 15.2. List of selected parameters for the 11 confident detections announced by LIGO-Virgo following the first two observing runs (O1 and O2). The columns show source-frame component masses m_i and chirp mass $\mathcal{M}$, dimensionless effective aligned spin χ_eff, final source-frame mass M_f, final spin a_f, radiated energy E_rad, peak luminosity ℓ_peak, luminosity distance d_L, redshift z, and sky localization $\Delta\Omega$ (the area of the 90% credible region). After Abbott et al. (2019c).

Event	$m_1/M_\odot$	$m_2/M_\odot$	$\mathcal{M}/M_\odot$	χ_eff	$M_f/M_\odot$	a_f	$E_\mathrm{rad}/(M_\odot c^2)$	$\ell_\mathrm{peak}/(\mathrm{erg\,s^{-1}})$	d_L/Mpc	z	$\Delta\Omega/\mathrm{deg}^2$
GW150914	$35.6^{+4.7}_{-3.1}$	$30.6^{+3.0}_{-4.4}$	$28.6^{+1.7}_{-1.5}$	$-0.01^{+0.12}_{-0.13}$	$63.1^{+3.4}_{-3.0}$	$0.69^{+0.05}_{-0.04}$	$3.1^{+0.4}_{-0.4}$	$3.6^{+0.4}_{-0.4}\times10^{56}$	440^{+150}_{-170}	$0.09^{+0.03}_{-0.03}$	182
GW151012	$23.2^{+14.9}_{-5.5}$	$13.6^{+4.1}_{-4.8}$	$15.2^{+2.1}_{-1.2}$	$0.05^{+0.31}_{-0.20}$	$35.6^{+10.8}_{-3.8}$	$0.67^{+0.13}_{-0.11}$	$1.6^{+0.6}_{-0.5}$	$3.2^{+0.8}_{-1.7}\times10^{56}$	1080^{+550}_{-490}	$0.21^{+0.09}_{-0.09}$	1523
GW151226	$13.7^{+8.8}_{-3.2}$	$7.7^{+2.2}_{-2.5}$	$8.9^{+0.3}_{-0.3}$	$0.18^{+0.20}_{-0.12}$	$20.5^{+6.4}_{-1.5}$	$0.74^{+0.07}_{-0.05}$	$1.0^{+0.1}_{-0.2}$	$3.4^{+0.7}_{-1.7}\times10^{56}$	450^{+180}_{-190}	$0.09^{+0.04}_{-0.04}$	1033
GW170104	$30.8^{+7.3}_{-5.6}$	$20.0^{+4.9}_{-4.6}$	$21.4^{+2.2}_{-1.8}$	$-0.04^{+0.17}_{-0.21}$	$48.9^{+5.1}_{-4.0}$	$0.66^{+0.08}_{-0.11}$	$2.2^{+0.5}_{-0.5}$	$3.3^{+0.6}_{-1.0}\times10^{56}$	990^{+440}_{-430}	$0.20^{+0.08}_{-0.08}$	921
GW170608	$11.0^{+5.5}_{-1.7}$	$7.6^{+1.4}_{-2.2}$	$7.9^{+0.2}_{-0.2}$	$0.03^{+0.19}_{-0.07}$	$17.8^{+3.4}_{-0.7}$	$0.69^{+0.04}_{-0.04}$	$0.9^{+0.0}_{-0.1}$	$3.5^{+0.4}_{-1.3}\times10^{56}$	320^{+120}_{-110}	$0.07^{+0.02}_{-0.02}$	392
GW170729	$50.2^{+16.2}_{-10.2}$	$34.0^{+9.1}_{-10.[illegible]}$	$35.4^{+6.5}_{-4.8}$	$0.37^{+0.21}_{-0.25}$	$79.5^{+14.7}_{-10.2}$	$0.81^{+0.07}_{-0.13}$	$4.8^{+1.7}_{-1.7}$	$4.2^{+0.9}_{-1.5}\times10^{56}$	2840^{+1400}_{-1360}	$0.49^{+0.19}_{-0.21}$	1041
GW170809	$35.0^{+8.3}_{-5.9}$	$23.8^{+5.1}_{-5.2}$	$24.9^{+2.1}_{-1.7}$	$0.08^{+0.17}_{-0.17}$	$56.3^{+5.2}_{-3.8}$	$0.70^{+0.08}_{-0.09}$	$2.7^{+0.6}_{-0.6}$	$3.5^{+0.6}_{-0.9}\times10^{56}$	1030^{+320}_{-390}	$0.20^{+0.05}_{-0.07}$	308
GW170814	$30.6^{+5.6}_{-3.0}$	$25.2^{+2.8}_{-4.0}$	$24.1^{+1.4}_{-1.1}$	$0.07^{+0.12}_{-0.12}$	$53.2^{+3.2}_{-2.4}$	$0.72^{+0.07}_{-0.05}$	$2.7^{+0.4}_{-0.3}$	$3.7^{+0.4}_{-0.5}\times10^{56}$	600^{+150}_{-220}	$0.12^{+0.03}_{-0.04}$	87
GW170817	$1.46^{+0.12}_{-0.10}$	$1.27^{+0.0[illegible]}_{-0.0[illegible]}$	$1.186^{+0.001}_{-0.001}$	$0.00^{+0.02}_{-0.01}$	≤ 2.8	≤ 0.89	≥ 0.04	$\geq 0.1\times10^{56}$	40^{+7}_{-15}	$0.01^{+0.00}_{-0.00}$	16
GW170818	$35.4^{+7.5}_{-4.7}$	$26.7^{+4.3}_{-5.2}$	$26.5^{+2.1}_{-1.7}$	$-0.09^{+0.18}_{-0.21}$	$59.4^{+4.9}_{-3.8}$	$0.67^{+0.07}_{-0.08}$	$2.7^{+0.5}_{-0.5}$	$3.4^{+0.5}_{-0.7}\times10^{56}$	1060^{+420}_{-380}	$0.21^{+0.07}_{-0.07}$	39
GW170823	$39.5^{+11.2}_{-6.7}$	$29.0^{+6.7}_{-7.[illegible]}$	$29.2^{+4.6}_{-3.6}$	$0.09^{+0.22}_{-0.26}$	$65.4^{+10.1}_{-7.4}$	$0.72^{+0.09}_{-0.12}$	$3.3^{+1.0}_{-0.9}$	$3.6^{+0.7}_{-1.1}\times10^{56}$	1940^{+970}_{-900}	$0.35^{+0.15}_{-0.15}$	1666

Table 15.3. A sample of confident detections announced by LIGO-Virgo-Kagra following the first three observing runs (O1, O2, O3a and O3b). The columns show source-frame component masses m_i and chirp mass $\mathcal{M}$, dimensionless effective aligned spin $\chi_{\rm eff}$, and luminosity distance d_L. Data taken from Abbott et al. (2019c, 2021d, 2022a).

Event	Type	$m_1(M_\odot)$	$m_2(M_\odot)$	$\mathcal{M}(M_\odot)$	$\chi_{\rm eff}$	d_L(Mpc)	Remark
GW150914	BHBH	$35.6^{+4.7}_{-3.1}$	$30.6^{+3.0}_{-4.4}$	$28.6^{+1.7}_{-1.5}$	$-0.01^{+0.12}_{-0.13}$	440^{+150}_{-170}	first GW detection
GW190426	BHBH	$106.9^{+41.6}_{-25.2}$	$76.6^{+26.2}_{-33.6}$	$77.1^{+19.4}_{-17.1}$	$+0.19^{+0.43}_{-0.40}$	4350^{+335}_{-215}	very massive BHBH
GW190521	BHBH	$95.3^{+28.7}_{-18.9}$	$69.0^{+22.7}_{-23.1}$	$69.2^{+17.0}_{-10.6}$	$+0.03^{+0.32}_{-0.39}$	3920^{+219}_{-195}	very massive BHBH
GW200308	BHBH	$36.4^{+11.2}_{-9.6}$	$13.8^{+7.2}_{-3.3}$	$19.0^{+4.8}_{-2.8}$	$+0.65^{+0.17}_{-0.21}$	5400^{+2700}_{-2600}	high spin
GW190403	BHBH	$88.0^{+28.2}_{-32.9}$	$22.1^{+22.7}_{-23.1}$	$36.3^{+14.4}_{-8.8}$	$+0.70^{+0.15}_{-0.27}$	8000^{+5880}_{-3990}	high spin, small mass ratio
GW190514	BHBH	$39.0^{+14.7}_{-8.2}$	$28.4^{+9.3}_{-8.8}$	$28.5^{+7.9}_{-4.8}$	$-0.19^{+0.29}_{-0.32}$	4130^{+2650}_{-2170}	likely significant negative spin
GW191109	BHBH	65^{+11}_{-11}	47^{+15}_{-13}	$47.5^{+9.6}_{-7.5}$	$-0.29^{+0.42}_{-0.31}$	1290^{+1130}_{-650}	likely significant negative spin
GW191113	BHBH	29^{+12}_{-14}	$5.9^{+4.4}_{-1.3}$	$10.7^{+1.1}_{-1.0}$	$0.00^{+0.37}_{-0.29}$	1370^{+1150}_{-620}	small mass ratio
GW190814	BHBH	$23.2^{+1.1}_{-1.0}$	$2.59^{+0.08}_{-0.09}$	$6.09^{+0.06}_{-0.06}$	$0.00^{+0.06}_{-0.06}$	240^{+40}_{-50}	extreme mass ratio
GW190924	BHBH	$8.9^{+7.0}_{-2.0}$	$5.0^{+1.4}_{-1.9}$	$5.8^{+0.2}_{-0.2}$	$+0.03^{+0.30}_{-0.09}$	570^{+220}_{-220}	low-mass BHBH
GW190917	BHBH?	$9.3^{+3.4}_{-4.4}$	$2.1^{+1.5}_{-0.5}$	$3.7^{+0.2}_{-0.2}$	$-0.11^{+0.24}_{-0.49}$	720^{+340}_{-310}	lower-mass gap secondary?
GW200210	BHBH?	$24.1^{+7.5}_{-4.6}$	$2.83^{+0.47}_{-0.42}$	$6.56^{+0.38}_{-0.40}$	$+0.02^{+0.22}_{-0.21}$	940^{+430}_{-340}	lower-mass gap secondary?
GW200105	BHNS	$8.9^{+1.1}_{-1.3}$	$1.9^{+0.2}_{-0.2}$	$3.41^{+0.08}_{-0.07}$	$-0.01^{+0.08}_{-0.12}$	280^{+110}_{-110}	mixed system
GW200115	BHNS	$5.9^{+1.4}_{-2.1}$	$1.4^{+0.6}_{-0.2}$	$2.42^{+0.05}_{-0.07}$	$-0.14^{+0.17}_{-0.34}$	310^{+150}_{-110}	mixed system
GW170817	NSNS	$1.46^{+0.12}_{-0.10}$	$1.27^{+0.09}_{-0.09}$	$1.186^{+0.001}_{-0.001}$	$0.00^{+0.02}_{-0.01}$	40^{+7}_{-15}	first DNS + sGRB + kilonova
GW190425	NSNS	$1.74^{+0.17}_{-0.09}$	$1.56^{+0.08}_{-0.014}$	$1.44^{+0.02}_{-0.02}$	$+0.012^{+0.01}_{-0.01}$	159^{+69}_{-72}	massive DNS system

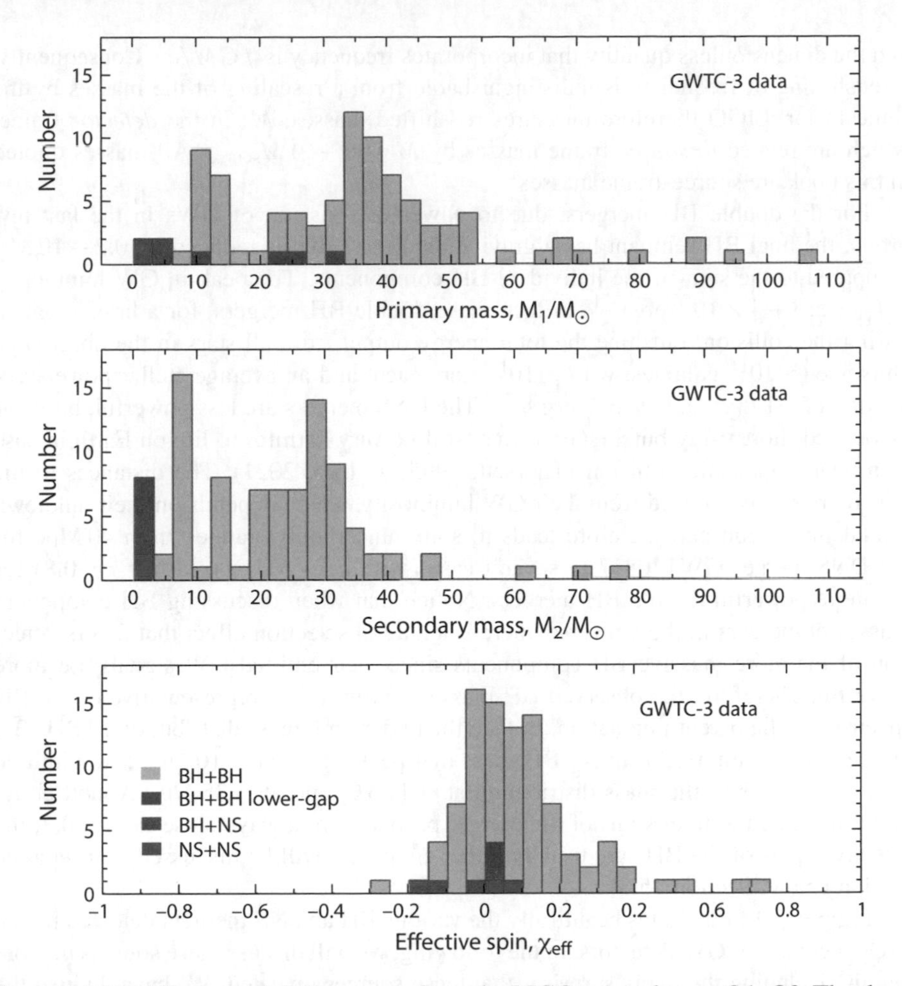

Figure 15.12. All reported GW mergers from the LIGO network runs O1–O3. The three panels show: primary mass (top), secondary mass (center), and effective spin (bottom), with color codings indicating the nature of the various sources. Plotted without error bars with data from the GWTC-3 catalogue: https://www.gw-openscience.org/eventapi/html/GWTC/.

masses from $\sim 2-107\ M_\odot$, and associated chirp masses of $4 \lesssim \mathcal{M}/M_\odot \lesssim 77$. For comparison, the DNS mergers have typical NS mass components around $1.2-2.1\ M_\odot$ and thus a chirp mass of $\mathcal{M} \simeq 1.0-1.8\ M_\odot$. Hence, as we shall discuss below, already after the third observing run (O3), interesting events were detected which have compact object masses located inside the two mass gaps: the PISN mass gap for BHs (discussed in Section 15.5.1) and the putative lower-mass gap of $\sim 2-5\ M_\odot$ between the maximum NS mass and the minimum BH mass (Bailyn et al., 1998).

We remind the reader that since we are living in an expanding Universe, the observed frequency of the GW signal is redshifted by a factor of $(1+z)$, where z is the cosmological redshift. There is no intrinsic mass or length scale in vacuum general relativity, and the dimensionless quantity that incorporates frequency is fGM/c^3. Consequently, a redshifting of frequency is indistinguishable from a rescaling of the masses by the same factor. LIGO therefore measures redshifted masses, M, in the *detector* frame, which are related to *source*-frame masses by $M=(1+z)\,M_{\rm source}$. All masses quoted in this book are source-frame masses.

For the double BH mergers, due to powerful emission of GWs in the last few orbits, the final BH remnant has a gravitational mass that is reduced by $0.2-10\,M_\odot$ compared to the sum of the individual BH components. The peak in GW luminosity is $\ell_{\rm peak}\simeq 3-4\times 10^{56}\,{\rm erg\,s^{-1}}$. In fact, these double BH mergers, for a brief moment during the collision, outshine the total energy output from all stars in the observable Universe ($\sim 10^{11}$ galaxies, with $\sim 10^{11}$ stars each, and an average stellar power production of $\sim 1\,L_\odot \simeq$ a few $10^{33}\,{\rm erg\,s^{-1}}$). The DNS mergers are less powerful, but their associated short γ-ray burst (sGRB) may still be very harmful to life on Earth in case of a nearby Galactic collision (Thorsett, 1995; Tøttrup, 2021). The distances to the GW mergers (as inferred from their GW luminosity, which depends on their unknown orbital inclination and therefore leads to some uncertainty) ranges from 40 Mpc for the DNS merger GW170817 to several cases with $d_L\sim 5$ Gpc (redshift $z\sim 0.8$) for the more powerful double BH mergers. Notice that when discussing BH component masses of mergers in the Universe, there is a natural selection effect that favors detection of the more massive BH components since their emitted GW signals are more powerful. Therefore, the observed BH mass spectrum is not representative of the BH spectrum in the parent population. In fact, the underlying mass distribution of BH+BH mergers has an inferred primary BH mass that peaks mass at $\sim 10\,M_\odot$, and with two additional peaks in the mass distribution at $\sim 17\,M_\odot$ and at $\sim 35\,M_\odot$ (Abbott et al., 2022b). While the final spins of the merger products are always close to $a_{\rm f}\simeq 0.7$, the effective spins of the BHs are usually rather small and will be discussed in a separate section below (Section 15.8).

Figure 15.13 illustrates beautifully the various BH and NS mergers detected by the LIGO network of GW detectors. In the following, we will discuss these sources in more detail—including the many surprises that these sources revealed. We have divided the different merger events into a number of classes.

In the following sections, we start by discussing the double BH mergers, subdivided into different classes depending on their chirp mass (Table 15.4). We also discuss extreme mass-ratio binaries, followed by DNS binaries, and finally GW events that are candidates for being mergers of mixed BH/NS binaries.

15.6.1 The Common Class of GW150914-like BH Binaries

Already the very first LIGO detection GW150914 (Abbott et al., 2016d,c) was a merger event of two relatively massive stellar-mass BHs ($35+30\,M_\odot$, $\mathcal{M}\simeq 29\,M_\odot$), which raises interesting questions to its origin. After subsequent data release with the GWTC-2 and GWTC-3 catalogues (Abbott et al., 2021d, 2022a), it became clear, however, that GW150914 is about of average mass compared to other BH+BH mergers detected. In fact, the bulk of the observed BH+BH merger events seem to have $20<\mathcal{M}/M_\odot<40$.

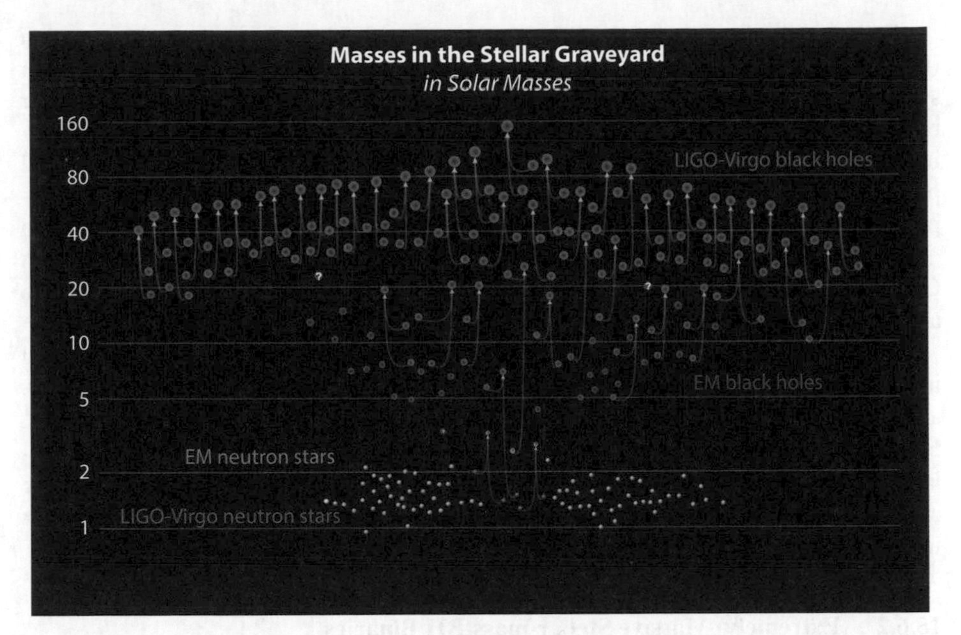

Figure 15.13. Illustration of derived compact object masses from detected LIGO-Virgo merger events in O1–O3a, together with those derived from X-ray binaries and binary radio pulsars. BHs and NSs detected by LIGO-Virgo are colored in blue and orange, respectively. The purple (BH) and yellow (NS) masses are those known from EM observations. Credit: LSC, Frank Elavsky & Aaron Geller, Northwestern University.

Table 15.4. Classes of BH+BH mergers depending on their chirp mass, $\mathcal{M}$. Notice that for equal mass binaries ($q \simeq 1$), the relation between $\mathcal{M}$ and individual component masses, m, is given by: $m \simeq 1.15\,\mathcal{M}$ (see Eq. 15.12). This scaling factor decreases only slowly for unequal mass components using $m = \sqrt{m_1 m_2}$.

BH+BH Class	Chirp mass, $\mathcal{M}/M_\odot$	Example
very-low mass	3–10	GW170608
low mass	10–20	GW151012
intermediate mass	20–40	GW150914
massive	40–60	GW190602
extremely massive	>60	GW190426

Following its discovery, GW150914 was suggested to have formed via CE evolution (Belczynski et al., 2016). In this standard scenario (see Sections 10.11 and 15.5.4), such systems always enter a CE phase following the HMXB stage, during which the O-star becomes a red supergiant and captures its BH companion. Kruckow et al. (2016) verified that, based on detailed stellar structure considerations, CE evolution may indeed produce BH+BH systems like GW150914, and even systems with individual BH component masses up to $\sim 50\,M_\odot$. (Assuming more massive ZAMS progenitor stars up to $\sim 200\,M_\odot$, Belczyński, 2020, proposed the possibility of the CE channel resulting in GW mergers with BH component masses up to $\sim 90\,M_\odot$, see below, assuming PISNe are avoided.)

It is immediately clear that even the common $25-60\,M_\odot$ stellar-mass BHs detected in, for example, GW150914, GW170729, GW170809, GW190503, and GW190929, are much more massive than the BHs known in MW X-ray binaries ($\lesssim 21\,M_\odot$, Fig. 15.13). This points to an origin in a low-metallicity environment where the stellar wind mass-loss rate is much weaker, thereby leaving more massive stars prior to the core collapse—see Figure 15.14 for the difference in resulting BH+BH masses depending on metallicity.

15.6.2 (Extremely) Massive Stellar-mass BH Binaries

GW mergers with quite massive BH components and chirp masses in the range $40 \lesssim \mathcal{M}/M_\odot \lesssim 60$ were reported for the first time in O3. These systems, e.g., GW190519, GW190602, and GW191109, often have total masses above $100\,M_\odot$ and one BH mass component $\gtrsim 65\,M_\odot$. However, in LIGO run O3, even more massive BH+BH merger candidates were reported (inside the aforementioned PISN gap in BH masses), which we now discuss.

The most extreme massive BH+BH mergers detected in LIGO run O3 were GW190426 and GW190521 (Abbott et al., 2020a, 2021), with total masses[7] of $M = 184.4^{+41.7}_{-36.6}\,M_\odot$ and $M = 163.9^{+39.2}_{-23.5}\,M_\odot$, chirp masses of $\mathcal{M} = 77.1^{+19.4}_{-17.1}\,M_\odot$ and $\mathcal{M} = 69.2^{+17.0}_{-10.6}\,M_\odot$, and estimated BH component masses of $106.9^{+41.6}_{-25.2} + 76.6^{+26.2}_{-33.6} M_\odot$ and $95.3^{+28.7}_{-18.9} + 69.0^{+22.7}_{-23.1}\,M_\odot$, respectively (Abbott et al., 2021d). Given the earlier discussion of the expected PISN gap in the mass distribution of BH binaries (Section 15.5.1), these BH component masses are clearly *inside* the mass gap. That is, these masses are above the lower threshold mass for which a PISN is expected to completely disrupt the progenitor star, and thereby not leave behind any compact remnant (Fowler & Hoyle, 1964; Heger & Woosley, 2002; Chatzopoulos & Wheeler, 2012; Woosley, 2019; Stevenson et al., 2019; Farmer et al., 2019; Renzo et al., 2020; du Buisson et al., 2020; Woosley & Heger, 2021). For this reason, the discovery of the exotic GW190521 merger (announced a year before the even more massive GW190426) has sparked a lively debate in the literature regarding its origin. The many and creative suggested formation scenarios include, for example, repeated BH mergers via dynamical interactions in star clusters (Fragione et al., 2020; Gayathri et al., 2020; Anagnostou et al., 2020),

[7] The slightly higher GW190521 masses quoted here were announced in Abbott et al. (2021d).

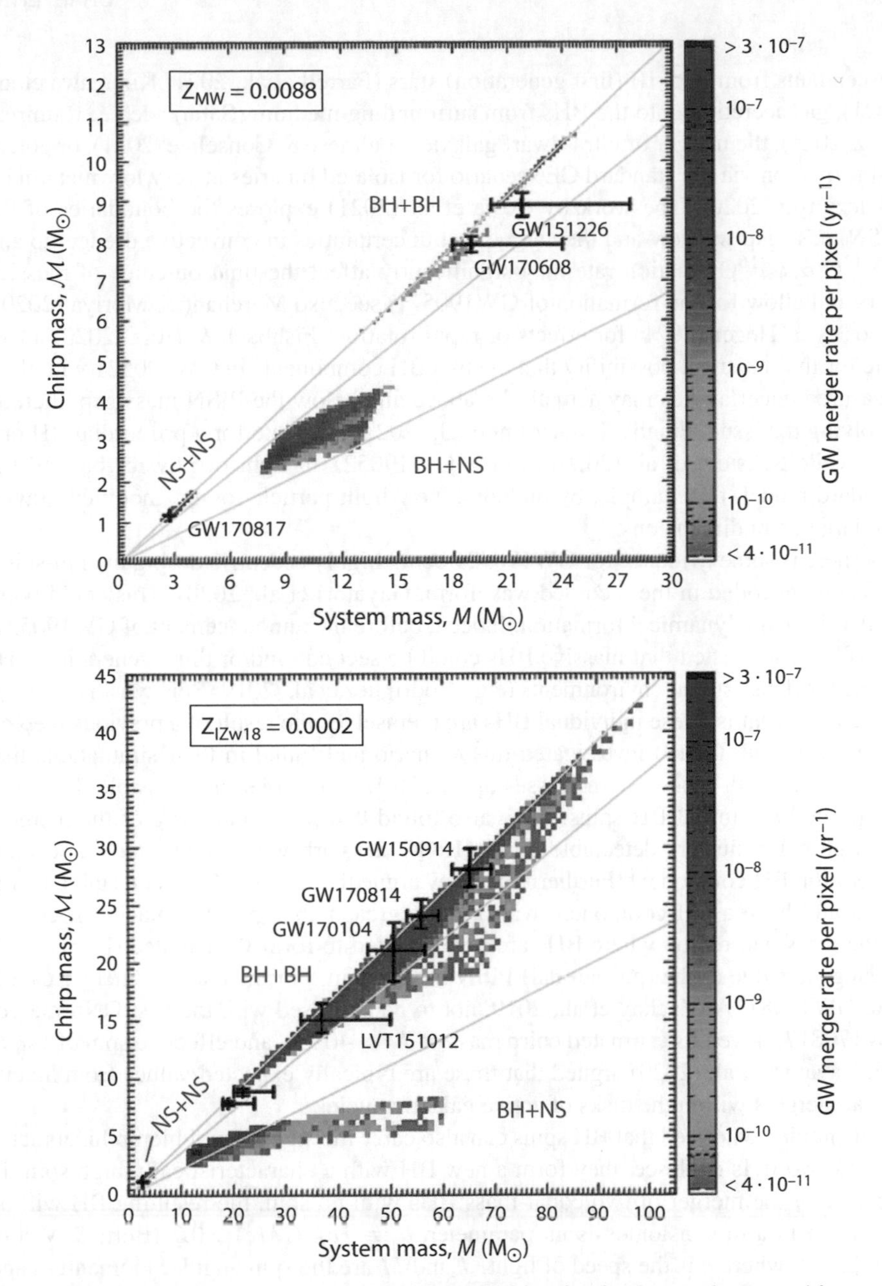

Figure 15.14. Distribution of simulated BH/NS binaries in the total mass–chirp mass plane for a metallicity of $Z = 0.0088$ (top) and $Z = 0.0002$ (bottom) formed through the CE channel. Three islands of data are visible, corresponding to BH+BH, mixed BH+NS, and NS+NS systems. The color coding indicates the merger rate per pixel in a MW-equivalent galaxy. Three solid grey lines indicate a constant mass ratio of 1, 3, and 10 (from top to bottom). Observed LIGO-Virgo sources are shown with black crosses, and event names are stated. The higher mass BH+BH mergers can only be reproduced at low metallicity (bottom). After Kruckow et al. (2018).

descendants from pop. III (first generation) stars (Farrell et al., 2021; Kinugawa et al., 2021), gas accretion onto the BHs from surrounding medium (Safarzadeh & Ramirez-Ruiz, 2021), the merger of ultra-dwarf galaxies (Palmese & Conselice, 2021), or potential formation via the standard CE scenario for isolated binaries at very low metallicity (Belczyński, 2020). The work by Costa et al. (2021) explores the boundaries of the PISN mass gap window and they argue that uncertainties in convective dredge-up and the $^{12}C(\alpha, \gamma)^{16}O$ reaction rate may significantly affect the final outcome of massive stars and allow for the formation of GW190521; see also Marchant & Moriya (2020); Woosley & Heger (2021) for effects of rapid rotation. Fishbach & Holz (2020) advocate for the statistical possibility that the two BH components in GW190521 with their measured uncertainties may actually lie above and below the PISN mass gap, thereby resolving the issue. Finally, Franciolini et al. (2022) advocate for a primordial BH origin, while Sakstein et al. (2020) discuss GW190521 in light of physics beyond the standard model, for example, by including new light particle losses, modified gravity and large extra dimensions.

There is some evidence for GW190521 being highly eccentric during final in-spiral (a feature encoded in the recorded waveform, Gayathri et al., 2020). This could point to an origin via dynamical formation. Indeed, before the announcement of GW190521, it has been suggested that massive BHs could be second- and/or third-generation BHs formed in dense stellar environments (e.g., Rodriguez et al., 2019; Belczynski & Banerjee, 2020). That is, these individual BHs are themselves the result of a previous merger. Rodriguez et al. (2019) investigated this scenario and found in their simulations that BH binaries with total system masses up to 150 $M_\odot$ (120 $M_\odot$) are possible if excluding (including) initial BH spins. They also found that more than 20% of the mergers from globular clusters detectable by the LIGO network will contain such a second-generation BH component. Furthermore, they argue that nearly 7% of detectable mergers would have a BH component with a mass greater than 55 $M_\odot$, placing it clearly in the mass-gap region where BHs are not expected to form via isolated (binary) collapsing stars due to the (pulsational) PISN mechanism. For the claimed merger candidate GW170817A (Zackay et al., 2019, not to be confused with the first DNS merger GW170817), given its estimated chirp mass of $\mathcal{M} \sim 40\,M_\odot$ and effective spin of $\chi_{\rm eff} \sim 0.5$, Gayathri et al. (2020) argued that these are typically expected values from hierarchical mergers within the disks of active galactic nuclei.

It should be noticed that BH spins can also carry the signature of hierarchical mergers. As two BHs coalesce, they form a new BH with a characteristically high spin. In the case of the merger of two equal-mass BHs with no spin, the resulting BH will be formed with a dimensionless spin parameter, $a_* \equiv Jc/(GM^2) \simeq 0.7$ (Berti & Volonteri, 2008), where c is the speed of light, J and M are the spin angular momentum and mass of the BH, respectively, and G is the gravitational constant. The spin parameter is bound to be within $-1 < a_* < 1$ for any BH (Section 4.9 and Table 7.2).

15.6.3 The Class of (Very) Low-mass BH Binaries

The class of relatively low-mass GW merger events have $\mathcal{M} \simeq 10 - 20\,M_\odot$ and typical individual component masses between $15 - 30\,M_\odot$. Members of this class are, for example, GW151012 (formerly LVT151012, Abbott et al., 2016a), GW190408, GW190828, and GW200225 from the O3 runs (Abbott et al., 2021d, 2022a).

GW151226 (Abbott et al., 2016b), GW170608 (Abbott et al., 2017b), GW190725, GW190917 (Abbott et al., 2021), GW190924, and GW200202 (Abbott et al., 2021d, 2022a) are typical representatives of *very low-mass* stellar BH mergers. These are characterized by $\mathcal{M} \simeq 3-10\,M_{\odot}$ and individual BH component masses of $\sim 6-15\,M_{\odot}$. An interesting example is GW190917, a marginally significant detection, which has a secondary component mass of only $2.1^{+1.5}_{-0.5}\,M_{\odot}$—which could be either a lower-mass gap BH or a massive NS. An additional such example of a merger with a lower mass-gap secondary component is GW200210 with $2.83^{+0.47}_{-0.42}\,M_{\odot}$. Hence, many of these above-mentioned systems are compatible with the stellar BH masses found in Galactic BH X-ray binaries (see Tables 6.5, 6.6, and 7.3 and McClintock et al. [2014]). This is particularly true since selection effects strongly favor GW detections of more massive binaries. Indeed, population synthesis favors production of such low-mass BH systems in relatively metal-rich environments (see Fig. 15.14). Kruckow et al. (2018) find that MW or LMC metallicities are best suited to reproduce such binaries (but see also García et al., 2021).

The classes of (very) low-mass double BH mergers seem to have a slight preference for a larger spread in observed mass ratios compared to the more massive classes of BH mergers which seem to have more equal mass components in general, although number statistics is still rather limited from LIGO runs O1–O3. The formation of (very) low-mass BH binaries is expected from either the standard CE channel, the stable RLO channel or the dynamical channel. CHE evolution is not expected to produce BH masses below $\sim 25\,M_{\odot}$.

15.6.4 Small Mass-ratio BH Binaries

Another interesting source from LIGO run O3 is GW190412 (Abbott et al., 2020) with a signal-to-noise ratio SNR $= 19$ across all three GW detectors and which has BH mass components of roughly $\sim 30.1 + 8.3\,M_{\odot}$, and thereby a very small mass ratio of only $q = 0.28^{+0.12}_{-0.07}$. The effective spin parameter was determined to be $\chi_{\rm eff} = 0.25^{+0.08}_{-0.11}$, which means that most likely it is one of relatively few merging BHs that had a significant spin magnitude—although the interpretation has been debated (Mandel & Fragos, 2020; Safarzadeh & Hotokezaka, 2020). Other detected mergers with small mass ratios of $q \simeq 0.20-0.25$ are GW190403, GW190917, GW191113, and GW200208.

Similar to the idea discussed above, second- or third-generation mergers in a dense stellar environment could be the explanation for the origin of these systems (Rodriguez et al., 2020; Gerosa et al., 2020), see Figure 15.15. Even more extreme mass ratio events ($q \simeq 0.1$) are discussed in Section 15.6.7.

15.6.5 The Class of Double NS Binaries

The double NS (DNS) mergers are all expected have chirp masses in the range $\mathcal{M} = 1.0-1.8\,M_{\odot}$ (with the far majority having $\mathcal{M} \simeq 1.2\,M_{\odot}$) and thus considerably smaller than those of double BH or mixed BH/NS mergers. Similar to the situation for

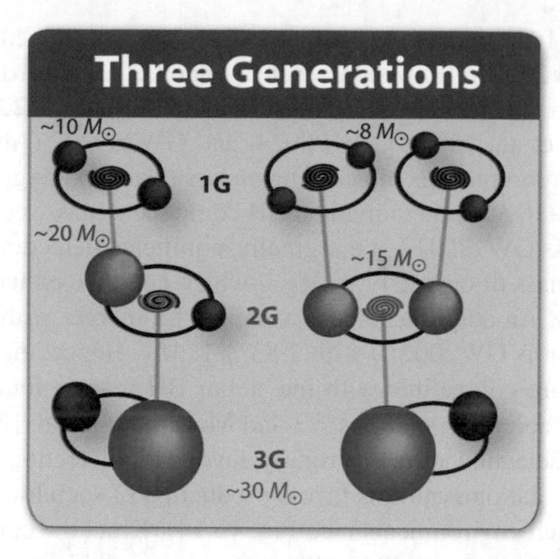

Figure 15.15. Merger tree of two possible multi-merger progenitors of GW190412. The massive primary BH (in red) is created from the merger of either one or two second-generation (2G) BHs (in blue). After Rodriguez et al. (2020).

double BH and mixed BH/NS mergers, the observed distribution of component spins and mass ratios will be of great importance for better understanding their formation process. GW170817 was detected with $\mathcal{M} = 1.188^{+0.004}_{-0.002}\,M_\odot$ and a total mass of $M = 2.74^{+0.04}_{-0.01}\,M_\odot$. The distribution of NS masses simulated to reproduce GW170817 is shown in Figure 15.16. It represents quite well the observed distribution of Galactic NS masses (Table 14.5).

GW190425 was reported (Abbott et al., 2020a) to be an unusual massive DNS with $\mathcal{M} = 1.44^{+0.02}_{-0.02}\,M_\odot$ and a total mass of $M = 3.4^{+0.3}_{-0.1}\,M_\odot$. For a more realistic low-spin prior of $\chi_{\rm eff} < 0.05$ (as expected for DNS systems, Tauris et al., 2017) the total mass is constrained to be $M = 3.3^{+0.1}_{-0.1}\,M_\odot$—still remarkably more massive than those found among known Galactic DNS systems ($M = 2.47 - 2.89\,M_\odot$). As a result, the formation of GW190425 has been speculated to be unusual. However, a Galactic NS+WD system is known with a total mass of $M = 3.15^{+0.01}_{-0.01}\,M_\odot$ (PSR J2222−0137, Guo et al., 2021). Furthermore, Kruckow (2020) demonstrated that a DNS system like GW190425 can be reproduced from the standard scenario at a reasonably frequent rate and even containing a NS mass component above $2.0\,M_\odot$ (Fig. 15.17). The simulated delay times are between less than 0.1 Gyr and up to a Hubble time.

Similar to the case for double BH mergers, DNS mergers can also be produced either via the standard CE channel (the majority of systems) or via the dynamical channel in globular clusters, as is witnessed from detections of DNS pulsars in our Galaxy (Table 14.5). The NS spins play an important role in determining the origin of the progenitor binary. While the first-formed (and thus recycled) NSs in DNS systems, with an origin in the Galactic disk via the CE channel, can only be formed with spin periods longer than ~10−12 ms (Tauris et al., 2015), DNS systems in globular clusters can be produced with fully recycled MSPs via exchange reactions. An example of

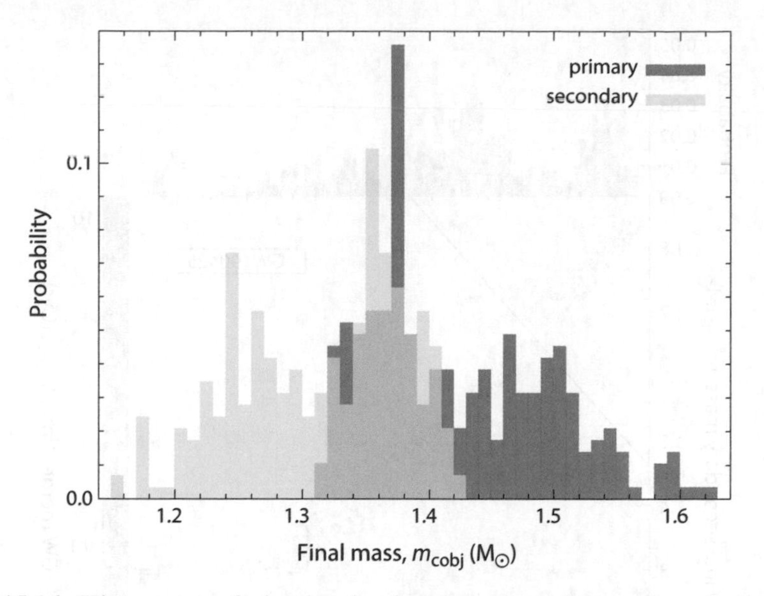

Figure 15.16. Histograms of simulated primary (red) and secondary (yellow) NS masses which are solutions to the progenitor binary of the DNS merger GW170817. After Kruckow et al. (2018).

such a DNS candidate system formed via dynamical interactions is PSR J1807−2500B (Lynch et al., 2012), which is a binary 4.2 ms pulsar found in the globular cluster NGC 6544. Unlike the situation for double BH binaries, the fraction of DNS binaries (and mixed BH/NS binaries) formed dynamically in globular clusters is probably quite insignificant. According to Ye et al. (2020), the merger-rate density in the local Universe is estimated to be $\sim 0.02\,\mathrm{Gpc}^{-3}\,\mathrm{yr}^{-1}$ for both NS+NS and BH/NS binaries in globular clusters, or a total of $\sim 0.04\,\mathrm{Gpc}^{-3}\,\mathrm{yr}^{-1}$ for both populations together. In comparison, a conservative estimate based on a simulated local merger rate of isolated (field) NS+NS and BH/NS systems combined is between 50 and 200 $\mathrm{Gpc}^{-3}\,\mathrm{yr}^{-1}$ (Kruckow et al., 2018), that is, four orders of magnitude larger.

During the final orbits of in-spiral, tides are very strong and act to deform the structure of the NS. This information is very important for extracting details of the long-sought-after NS equation-of-state (EoS). The tidal deformation is embedded in the emitted waveform and can be detected (Abbott et al., 2017c). The dimensionless tidal deformability parameter is given by (Flanagan & Hinderer, 2008; Favata, 2014):

$$\tilde{\Lambda} \equiv \frac{16}{13} \frac{(1+12q)\Lambda_1 + (12+q)q^4\Lambda_2}{(1+q)^5}, \tag{15.16}$$

where $\Lambda_{1,2}$ are the dimensionless quadrupole tidal deformability for each NS component, and $q \equiv M_2/M_1 \leq 1$ is the mass ratio. Thus we may expect an increasing number of such detections in future DNS and NS/BH mergers to provide further constraints on the NS EoS.

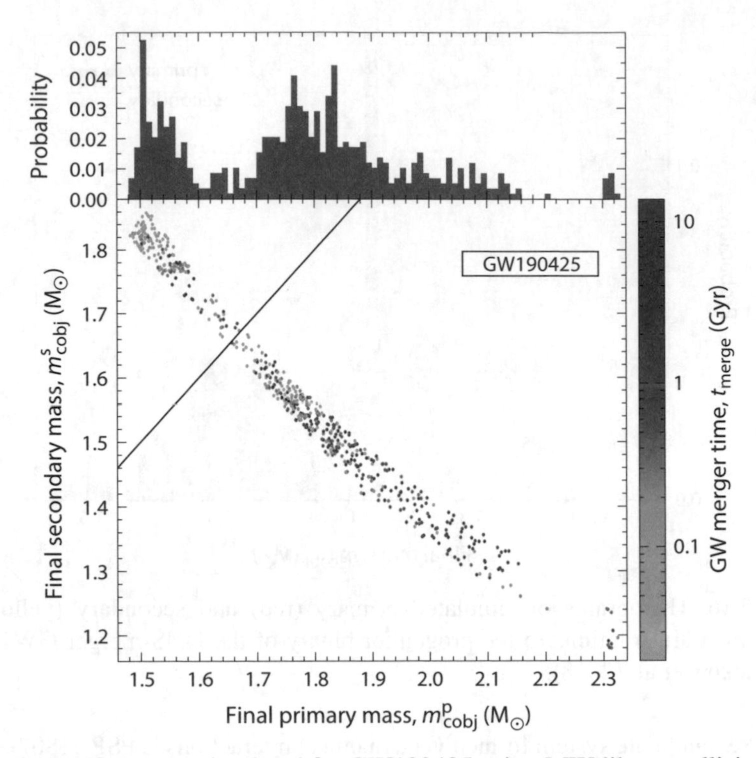

Figure 15.17. NS masses simulated for GW190425 using MW-like metallicity. Colors indicate the time from formation of the DNS system until the GW merger. The models are selected according to the reported data for GW190425 (Abbott et al., 2020a). The diagonal line indicates a mass ratio of $q = 1$. The histogram shows the distribution function for the primary NS masses. This figure shows that the standard CE channel may produce DNS systems that could, in rare cases, have very asymmetric masses with a mass ratio down to $q \sim 0.5$. After Kruckow (2020).

15.6.6 The Class of Mixed BH/NS Binaries

Many population synthesis investigations of compact object mergers predicted a relatively large number of mixed BH/NS binaries to be detected—see, for example, Chattopadhyay et al. (2021) and the example in Table 15.1 where LIGO detections of mixed BH/NS mergers were predicted to exceed that of DNS mergers by an order of magnitude. It was therefore anticipated that the LIGO network would have detected a number of such systems shortly after the beginning of operation. However, not until LIGO observing run O3b, firm cases of such mixed BH/NS mergers systems were finally detected (Abbott et al., 2021b), although a couple of good candidates had been announced earlier (see below).

The two fairly firm mixed BH/NS events found in O3b are GW200105 and GW200115 (Abbott et al., 2021b). The first source has component masses of $8.9^{+1.2}_{-1.5}\,M_\odot$ and $1.9^{+0.3}_{-0.2}\,M_\odot$, whereas the second source has component masses of $5.7^{+1.8}_{-2.1}\,M_\odot$ and

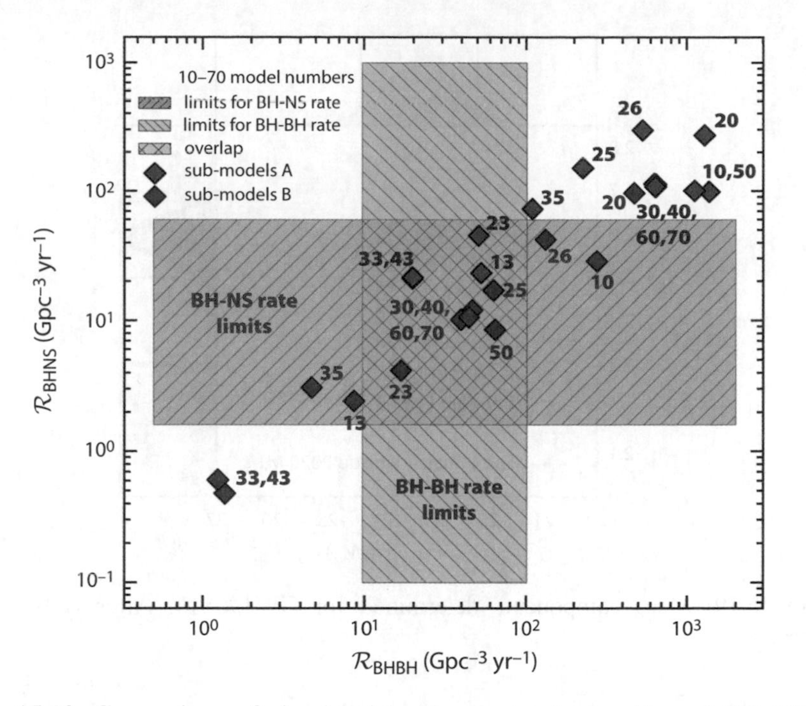

Figure 15.18. Comparison of simulated local merger-rate densities of BH+NS and BH+BH systems based on various models (diamond symbols) with the empirical limits in shaded regions. Despite theoretical predictions of relatively high merger rates of BH+NS systems, only few such candidates are detected, even when taking observational selection effects into account—see also theoretical estimates in Table 15.1 based on a different code. After Belczynski et al. (2020).

$1.5^{+0.7}_{-0.3}\,M_\odot$ (at the 90% credible level, Abbott et al., 2021b). The luminosity distances are estimated to be $d_L = 280^{+110}_{-110}$ Mpc and $d_L = 300^{+150}_{-100}$ Mpc, respectively. Whereas the BH component in GW200105 has a BH spin magnitude, $\chi_1 < 0.23$ (90% credibility), the BH component in GW200115 apparently has a negative spin (i.e., $\chi_1 < 0$ at 88% credibility).

To produce a BH+NS merger like GW200115 with a negative spin projection onto the orbital angular momentum, Fragione et al. (2021) argue for the need of large NS kicks ($\gtrsim 150\,\mathrm{km\,s^{-1}}$) and efficient CE ejection, if they were produced via isolated binary evolution (see also Broekgaarden & Berger, 2021).

It is somewhat puzzling that so few mixed BH/NS binaries are observed compared to theoretical predictions (Fig. 15.18), and this circumstance may well indicate a deficiency in our current modelling. For discussions on the large uncertainties (of one or even two orders of magnitude) in the merger-rate estimates of mixed BH/NS binaries from population synthesis due to massive binary star interactions and cosmic evolution, see, for example, Broekgaarden et al. (2021).

Additional mixed BH/NS candidate events from LIGO science runs O1–O3 are GW190426_152155 (Abbott et al., 2021d) and GW191219_163120 (Abbott et al.,

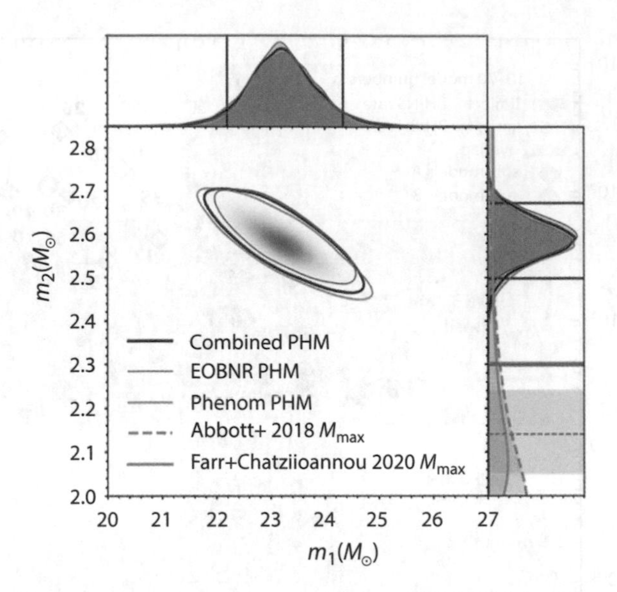

Figure 15.19. Mass components determined for GW190814. After Abbott et al. (2020b).

2022a). Unfortunately, these events had relatively large false-alarm rates. GW 190426_152155 had a FAR $= 1.4\,\mathrm{yr}^{-1}$ and is thus a marginal detection. If real, this merger had component masses of $5.7^{+4.0}_{-2.3}\,M_\odot$ and $1.5^{+0.8}_{-0.5}\,M_\odot$ (i.e., a prime example of the merger of a low-mass BH and a NS, similar to GW200115), and it is located at a distance of $d_L \simeq 380^{+190}_{-160}$ Mpc. Another potential BH/NS merger candidate is GW190814, which we will now discuss.

15.6.7 The Mysterious Mass-gap Event GW190814

GW190814 (Abbott et al., 2020b) is a unique merger consisting of a 23 $M_\odot$ BH merging with a 2.6 $M_\odot$ compact object (Fig. 15.19). This merger is the best localized event detected in LIGO–Virgo observing run O3a, with the 90% contour of 2D sky position being $\Delta\Omega = 19\,\mathrm{deg}^2$ (the 90% volume of 3D sky position is constrained to $\Delta V_{90} = 3.2 \times 10^{-5}$ Gpc). It was detected with a high level of significance (FAR $< 1.0 \times 10^{-5}\,\mathrm{yr}^{-1}$), and yet its nature remains unknown. The system has one of the most unequal mass ratios ($q = 0.11$) yet detected among all GW sources, and its secondary compact object is either one of the lightest BHs or possibly the heaviest NS ever discovered in a double compact-object system. Another similar event is GW200210 (Abbott et al., 2022a), which has component masses of 24.1 $M_\odot$ and 2.83 $M_\odot$ ($q \simeq 0.12$).

Astrophysical models predict that binaries with mass ratios similar to these events can form (see, e.g., bottom panel of Fig. 15.14 among the population of mixed BH/NS binaries), although usually without any compact object component within the so-called *lower mass gap*, roughly between 2 and 5 $M_\odot$—a topic of intense ongoing debate (see, e.g., Antoniadis et al., 2022).

Such a gap in the compact-object mass spectrum has been suggested to exist (Fryer et al., 2012) if the instability growth and launch of the SN proceed on rapid timescales ($\sim$10 and $\sim$100 ms, respectively). Alternatively, if instabilities are delayed and develop over longer timescales ($\sim$200 ms), accretion can occur on the proto-NS before the neutrino-driven SN explosion is launched and this would produce BHs in the mass gap. Interestingly enough, the mass distributions of NSs and BHs observed in our Galaxy provided initial evidence that compact object formation proceeds on rapid timescales, which therefore has led most population synthesis codes to follow the recipe leading to a mass gap. Zevin et al. (2020) has tested population modelling assuming delayed explosions (with a variety of input physics) to reproduce systems like GW190814. They find it difficult to match the empirical rate of GW190814-like systems while simultaneously being consistent with the rates of other compact binary populations inferred by GW mergers. Mandel et al. (2021) applied a probabilistic prescription for compact remnant masses to reproduce systems like GW190814, but overpredict the number of massive Galactic DNS systems.

Another possibility is that systems like GW190814 and GW200210 are of dynamical origin. However, dynamical exchanges in dense stellar environments like globular clusters tend to pair up massive compact objects with similar masses (Kulkarni et al., 1993; Sigurdsson & Hernquist, 1993), and thus most merging binary BHs have $q \simeq 1$ (Rodriguez et al., 2016), highly suppressing mixed BH/NS mergers. For the same reason, a population of second-generation BHs in the mass range between 2 and 5 $M_\odot$ (e.g., produced from mergers of DNS systems), which subsequently merges with a much more massive stellar-mass BH, are expected to be exceedingly rare (Ye et al., 2020).

The BH/NS systems formed in isolated binaries from population synthesis codes, with a prescription including the mass gap caused by rapid SN explosions, are anticipated to have chirp masses $\mathcal{M} = 2 - 4\,M_\odot$ in high-metallicity environments and $\mathcal{M} = 2 - 8\,M_\odot$ at low metallicity, see Figure 15.14. However, GW190814 is clearly demonstrating that we cannot rely exclusively on rapid explosions, even though they are consistent with the currently known population of Galactic compact object binaries.

Although there are currently no known compact object masses measured accurately between roughly 2 and 5 $M_\odot$ among Galactic X-ray binaries, it has been suggested by Zdziarski et al. (2013) that the HMXB Cyg X-3 (discussed in Chapters 6 and 10) has a compact object accretor with an estimated mass of $2.4^{+2.1}_{-1.1}\,M_\odot$. These authors argue for its companion being a WR-star with a mass of $10.3^{+3.9}_{-2.8}\,M_\odot$. If this analysis is correct, this system might be able to produce a BH (if the WR-star collapses to produce a $\sim$10 $M_\odot$ BH) in a binary system with a compact object in the lower mass gap and in a very narrow orbit (currently $P_{\rm orb} = 4.8$ hr) such that the system will later merge by GW radiation within a Hubble time. So, perhaps Cyg X-3 is in fact a progenitor for a GW190814-like system. In the latter system, the mass ratio is more extreme, but still.

In the wake of the discoveries of GW190814 and GW200210, many alternative models have been proposed to explain their unexpected component masses, including models where a newborn NS collapses to form a low-mass BH shortly after its birth, either due to instability of an extremely rapidly rotating NS (Most et al., 2020) or due to sigificant accretion of the SN ejecta (Safarzadeh & Loeb, 2020), or simply as a result of explosibility fluctuations of massive stellar cores (see Sections 8.2.3 and 9.7) such that the collapse of carbon-oxygen cores up to $\sim 30\,M_{\odot}$ may either leave behind a relatively massive BH of 20–25 $M_{\odot}$ or a lower-mass gap object (e.g., $2.5-3.0\,M_{\odot}$), and thus the possibility of a pairing of the two components in an isolated binary system (Antoniadis et al., 2022).

Combining the contradictory constraints obtained from GW170817 ($M_{\rm TOV} \lesssim 2.3\,M_{\odot}$) and GW190814 ($M_{\rm TOV} \gtrsim 2.5\,M_{\odot}$) on the maximum mass of a (non- or slowly rotating) NS, Nathanail et al. (2021) argued that the low-mass component of GW190814 is most likely to be a BH at merger. We point out that such a low-mass BH might have been produced from the accretion-induced collapse of a massive NS at an earlier epoch in the evolution of the progenitor system—perhaps in a dense cluster where the low-mass BH was later paired with the 23 $M_{\odot}$ BH component.

Finally, there are also mixed BH/NS merger candidates in which the detected secondary mass is close to the minimum mass of a NS: GW191219 (Abbott et al., 2022a) has component masses of $31.1^{+2.2}_{-2.8}\,M_{\odot}$ and $1.17^{+0.07}_{-0.06}\,M_{\odot}$ ($q \simeq 0.038$). It is of particular interest to see if future LIGO detections of BH/NS mergers will also be accompanied with a sGRB event as expected for favorable beaming geometry, and whether such systems can be found with mass ratios of $q < 0.1$.

15.7 EMPIRICAL MERGER RATES

The current (GWTC-3) empirical merger rates are stated in Table 15.5, along with those announced after the release of earlier data. The much larger sample of events in GWTC-2 and GWTC-3 is the reason for the decrease in error bars compared to GWTC-1. Despite the rapidly increasing number of detected GW mergers, these rates are still subject to small number statistics and, not least, the underlying model assumptions.

Following the data release of GWTC-3, a number of merger-rate densities were derived using different model assumptions, and with or without an evolving merger rate with redshift (Abbott et al., 2022b). As a specific example of such uncertainties, the empirical BH+BH merger rate announced in previous GWTC-2 would change from $\mathcal{R}_{\rm BHBH} = 23.9^{+14.9}_{-8.6}\,{\rm Gpc^{-3}\,yr^{-1}}$ to $\mathcal{R}_{\rm BHBH} = 58^{+54}_{-29}\,{\rm Gpc^{-3}\,yr^{-1}}$ just by lowering the minimum accepted BH mass below 3.0 $M_{\odot}$, to include the mass-gap event GW190814 ($M_2 = 2.6\,M_{\odot}$, see Section 15.6.7) in the sample. It is also due to the uncertainty of the actual nature of such lower mass-gap sources, besides the relatively sparse detection of mixed BH/NS events so far, that the merger-rate density for BH/NS events is somewhat difficult to constrain. For example, regardless of the nature of the secondary compact object in GW190814, the estimated merger-rate density for such systems alone is estimated to be $1-23\,{\rm Gpc^{-3}\,yr^{-1}}$ (Abbott et al., 2020b).

The empirical merger rate evolves with redshift as expected (see Fig. 15.7 for an example of a model). However, the merger rate evolves slower than the naive expectation of $(1+z)^{2.7}$ from the star formation rate (SFR) at 87% credibility. Abbott et al. (2021e) find that the merger rate of BH+BH events at $z=1$ differs from the merger rate at $z=0$ by a factor of $2.5^{+7.8}_{-1.9}$, compared to a factor ~ 6 in the corresponding increase of the SFR.

Table 15.5. Empirical merger-rate densities following the release of LIGO–Virgo–KAGRA observing runs O1+O2+O3 data in GWTC-1, GWTC-2, GWTC-2.1, and GWTC-3 (Abbott et al., 2019c, 2021d, 2021, 2022a). The quoted rates are dependent on the assumed threshold false-alarm rate (FAR), the underlying mass and spin models, and whether or not allowing for a model that evolves with redshift. Ranges with full uncertainties are given in parentheses.

Binaries	Merger-rate density $\mathcal{R}_{\rm LIGO}$ (Gpc^{-3} yr^{-1})	Data release
BH+BH	33^{+16}_{-10} (16 – 130) (17.3 – 45)*	GWTC-3**
BH+NS	32^{+62}_{-25} (7.4 – 320)	GWTC-3**
NS+NS	99^{+260}_{-86} (13 – 1900)	GWTC-3**
BH+BH	$23.9^{+14.9}_{-8.6}$	GWTC-2
BH+NS	130^{+112}_{-69}	GWTC-2.1
NS+NS	320^{+490}_{-240}	GWTC-2
BH+BH	$53.2^{+55.8}_{-28.2}$	GWTC-1
BH+NS	< 610	GWTC-1
NS+NS	1540^{+3200}_{-1220}	GWTC-1

*For a BH+BH merger rate evolving with redshift (here at $z = 0.2$).
**Non-parametric model BGP.

Although the LIGO network has detected a number of very high BH component masses (Section 15.6.2), it is important to realize that this is due to a selection effect that favors detection of such massive mergers as a result of their much higher GW luminosities. In fact, the merger-rate density of BH+BH systems in GWTC-3, using the non-parametric BGP model, with primary masses $> 50\,M_\odot$ is only $0.74^{+1.2}_{-0.46}$ Gpc^{-3} yr^{-1}, that is, just a few % of the total BH+BH merger-rate density (33^{+16}_{-10} Gpc^{-3} yr^{-1}). The merger-rate density with primary masses of $20 - 50\,M_\odot$ is $6.4^{+3.0}_{-2.1}$ Gpc^{-3} yr^{-1}, illustrating again that low-mass ($< 20\,M_\odot$) primary BH components strongly dominate the BH mass spectrum.

For theoreticians and modellers of the Galactic merger rate of compact object binaries, the merger rate in a MWEG is often a more useful quantity than the merger-rate density in the local Universe. Assuming,[8] for example, that there are ~0.01 MWEGs

[8] See also discussions in Dominik et al. (2013, 2015) for alternatives.

in $1\,\mathrm{Mpc}^{-3}$ of space (Kopparapu et al., 2008), the quoted empirical merger-rate density of DNS mergers of $320^{+490}_{-240}\,\mathrm{Gpc}^{-3}\,\mathrm{yr}^{-1}$ would amount to roughly $32^{+49}_{-24}\,\mathrm{Myr}^{-1}$ in our Galaxy. In Chapter 16, we discuss the Galactic DNS merger rate (theoretical and empirical from radio pulsars) in more detail.

15.8 BH SPINS—EXPECTATIONS AND OBSERVATIONS

Besides mass, the other fundamental parameter describing an astrophysical BH is spin (a third parameter, electric charge, is not relevant as astrophysical BHs are neutralized by the interstellar medium). In the following sections, we discuss expectations and observations of BH spins in GW detections of BH+BH mergers. For further discussions of BH spins in general and their measurements in X-ray binaries, see Section 7.6.

15.8.1 Expectations of BH Spins from Stellar Evolution

The spin magnitude of a newborn BH reflects primarily the spin of the core of its collapsing progenitor star, which itself depends on the angular momentum transport during its previous phases of stellar evolution (Meynet & Maeder, 2000; Aerts et al., 2019). Additional effects from the SN may play a minor role as well and cannot be excluded: spin-up from an off-centered kick (which has been suggested in the context of NSs, Spruit & Phinney, 1998), loss of rotational energy due to the Blandford-Znajek mechanism (Blandford & Znajek, 1977), and spin-up from accretion of fallback material due to slow-moving SN ejecta. Regarding the latter possibility, Schrøder et al. (2018) showed that realistic velocity profiles of the SN ejecta can only lead to mild spin-up of the BH.

Without any angular momentum transport within the progenitor star, most compact objects would be born with maximum spin (Heger et al., 2000), but efficient angular momentum transport by viscosity will couple the stellar core to its envelope, thereby slowing the spin of the core as the envelope expands when it becomes a giant star. (The enhanced spin of the contracting core during the crossing of the Hertzsprung gap and early hydrogen shell-burning is lost in the same process as internal shear activates the Tayler instability [Spruit, 2002] counteracting the core spin-up.) The outer layers of the star may then later be lost due to stellar winds or mass transfer if in a binary system. Evidence from observations of asteroseismology (although for low-mass giants, Fuller et al., 2014; Cantiello et al., 2014) as well as NS and WD spins (Heger et al., 2005; Suijs et al., 2008), verifies that efficient angular momentum transport must be at work inside stars. Applying the magnetic torques from such Tayler instability, Fuller & Ma (2019) argued that BHs born from single stars should all have very slow spins of $a_* \sim 0.01$, where the dimensionless spin parameter for a rotating (Kerr) BH is defined by (see also Section 4.9):

$$a_* \equiv \frac{Jc}{GM^2} \tag{15.17}$$

and where J is the spin angular momentum of the BH with mass, M. (Notice that it is no problem that $a_* \gg 1$ for the collapsing core. The resulting BH remnant will always have $a_* \leq 1$ since the excess angular momentum is lost with mass ejecta which is regulated to prevent a non-physical value of a_*.)

Similar conclusions on slow spin of isolated BHs or first-born BHs in binaries have been reached by Heger et al. (2005) and Qin et al. (2018), based on the classic Tayler-Spruit dynamo (Spruit, 2002), which provides somewhat weaker magnetic torques compared to the Tayler instability explored by Fuller et al. (2019). Exceptions from anticipated slow spins of the first-born BHs stem only from evolution via the CHE channel (Marchant et al., 2016; Fuller & Ma, 2019) since rotational mixing and tidal locking in these tight massive systems prevent the stars from expanding during their evolution.

It is interesting to notice that the estimated spins of BHs observed in HMXBs, however, are apparently all quite large (McClintock et al., 2014; Miller & Miller, 2015; Miller-Jones et al., 2021). The spins for Cyg X-1, LMC X-1, LMC X-3, and M33 X-7 are estimated to be $0.84 \lesssim a_* \lesssim 0.99$. These values are clearly in tension with the aforementioned predictions of slow spins for such (first-born) BHs. To explain this discrepancy, one can argue that the BH spin measurements in HMXBs are uncertain (e.g., Kawano et al., 2017) or that these BH–HMXBs form a population different from those that eventually produce BH+BH mergers (Belczynski et al., 2012; Gallegos-Garcia et al., 2022). For example, Qin et al. (2019) suggest that these known BH–HMXBs might have formed via CHE or Case A RLO with a weak coupling between the core and its envelope following the main-sequence stage. This is, nevertheless, in contrast with the idea that the BH in Cyg X-1, and probably also in other BH–HMXBs, was presumably formed through Case C evolution, although this picture has recently been challenged as argued in Chapter 10.

Finally, we notice that although the first-born BH may accrete material from its companion star, the timescale and the Eddington-limited poor efficiency of the spin-up process would prevent any significant increase in its spin value. Here we follow the heuristic arguments given by Mandel & Fragos (2020): Accretion of a mass element, δM, that is moving at Keplerian velocity, $v = \sqrt{GM/R}$, as it approaches the BH's equator with radius R, leads to a gain in spin angular momentum of $\delta J = \delta M v R$ and thereby a gain in the dimensionless spin parameter of $\delta a_* = (\delta M/M)\sqrt{Rc^2/GM}$. One would therefore expect that the BH mass must be roughly doubled to produce a rapidly spinning BH. Accurate calculations (Thorne, 1974; King & Kolb, 1999) show that 20% of the BH mass must be accreted to increase the spin of an initially non-rotating BH to a spin value of 0.5, and the BH mass must be more than doubled to produce a spin value of 0.99. The mass doubling timescale of a BH, however, is of the order of ~100 Myr—far longer than typical lifetime of HMXBs ($0.1 - 1$ Myr). Moreover, only a small fraction of mass is anticipated to be accreted during the subsequent CE phase (MacLeod & Ramirez-Ruiz, 2015a; Cruz-Osorio & Rezzolla, 2020).

Therefore, we conclude that any spin angular momentum of a first-born BH in a binary must come from the progenitor star, or possibly the SN itself via an off-centered kick or accretion of fallback material from slow-moving ejecta. Hence, Cyg X-1, LMC X-1, LMC X-3, and M33 X-7 (see Section 7.6 for their BH spin measurements) were born with rapid spins, in contradiction with the aforementioned predictions. As we shall discuss in Section 15.8.2, GW detections of BH+BH mergers also point to

BH spins with significant, although generally small, magnitudes. In our view, it may well be the case that we are still missing some key input physics in our understanding of producing BH spins, or the observational estimates of the spins of BHs in HMXBs (Section 7.6) are more uncertain than thought (Kawano et al., 2017). While the first-born BHs are thus predicted to be slowly rotating from current theoretical models, the expectation of the spin of the second-born BH is quite different.

The spin of the second-born BH is also determined by angular momentum transport, stellar evolution, and mass loss of its progenitor star. Particularly important for the second BH, however, are tidal interactions in the post-HMXB (post-CE or post-RLO) phase where the naked helium-star (WR-star) is orbiting the first-born BH in a tight orbit. Before this stage, similarly to the case of the primary star, any initial or acquired rotation during evolution of the secondary star is expected to be diminished in most cases. Therefore, using analytical arguments and semi-analytical calculations, one can derive approximate estimates of the spin-up of the secondary naked core prior to its collapse and thereby constrain the spin of the second-born BH (van den Heuvel & Yoon, 2007; Kushnir et al., 2016; Hotokezaka & Piran, 2017; Zaldarriaga et al., 2018). In a follow-up paper with numerical computations, Fuller & Lu (2022) demonstrated that indeed in close-orbit BH+He-star systems with $P_{\rm orb} \lesssim 1$ d, for which tides are strong, the spin angular momentum of the helium star is greatly increased, thereby leaving a rapidly spinning second-born BH after core collapse.

The orbital separation of the two BHs after their formation (besides eccentricity and masses) determines their coalescence time and also, if assuming only a minor change in orbital separation during core collapse, whether or not the secondary WR-star was synchronized by tides prior to its collapse (Section 4.7). Thus, there should be an anti-correlation between the spin of the second-born BH and orbital separation (and thus coalescence time). That is, one expects systems with initially short coalescence times to host secondary BHs with high spins (Hotokezaka & Piran, 2017)—see, for example, Eq. (15) in Belczynski et al. (2020) for a fit between resulting BH spin magnitude and orbital period of the progenitor BH+WR-star binary, under the simple assumption of rigid WR-star rotation and ignoring orbital evolution.

From simple arguments, the pre-SN synchronization timescale of the WR-star can be related to the coalescence timescale, $\tau_{\rm GW}$, of the descendant BH+BH binary roughly as (Kushnir et al., 2016):

$$\tau_{\rm sync} \simeq 10\,{\rm Myr}\quad q^{-1/8}\left(\frac{1+q}{2q}\right)^{31/24}\left(\frac{\tau_{\rm GW}}{1\,{\rm Gyr}}\right)^{17/8}, \tag{15.18}$$

where $q \leq 1$ is the mass ratio of the two BH components.

Piran & Hotokezaka (2018) concluded that WR-stars formed at high redshifts ($z \geq 2$) are the best candidates for being progenitors of BH+BH mergers with low values of $\chi_{\rm eff}$ (see Eq. 15.19). These binaries have a long merger time, implying large initial separations and thus weak synchronization. This prediction, however, depends strongly on the assumed strength of the WR-star wind which removes spin angular momentum and also widens the orbit, thereby weakening the synchronization ability of the binary. Another uncertainty is the applied BH+BH formation rate in such semi-analytical investigations.

More detailed studies than the previously mentioned semi-analytical investigations were recently presented, for example, by Belczynski et al. (2020); Bavera et al. (2020), and they include a more self-consistent modelling of the final post-CE stages by calculating simultaneously the orbital evolution and the combined effects of stellar winds and tides. In addition, changes in stellar structure and spin-orbit couplings during the evolution are included. Both studies find that efficient angular momentum transport is favored as it results in distributions of $\chi_{\rm eff}$ and BH component masses consistent with observations (see in Section 15.8.2 and Fig. 15.12), while inefficient angular momentum transport would lead to rapidly spinning post-merger BHs, which are not observed at present. Bavera et al. (2020) conclude that if angular momentum transport in massive stars is efficient, then any (electromagnetic or GW) observation of a rapidly spinning BH would indicate the following possibilities: (1) a very effective tidal spin-up of the progenitor star (either via CHE or stable RLO HMXB evolution with Case A mass transfer, or via spin-up of a WR-star in a tight binary by a close companion); (2) significant mass accretion by the BH; or as a third possibility (3) a BH origin as a second-generation BH produced via the merger of two or more BHs in a dense stellar environment. Similarly to these studies, Piran & Piran (2020) also showed that the observed $\chi_{\rm eff}$ distribution favors an origin from isolated field binaries over a dynamical formation scenario.

In general, the standard CE channel (i.e., classic isolated binary channel) predicts a bimodal distribution of spins of the secondary BHs with low and high spin peaks (e.g., Zaldarriaga et al., 2018). The high spin peak corresponds to synchronized binaries with strong tidal couplings. Bavera et al. (2020) find that only 20% of all BH+BH binaries produced via the CE channel will have $\chi_{\rm eff} > 0.1$. However, if a significant fraction of BH+BH mergers were formed via stable RLO (see, e.g., population synthesis by Giacobbo et al., 2018; Neijssel et al., 2019), which tend to form BH+BH binaries with wider separations and thus smaller secondary BH spins due to weaker tidal interactions in their progenitor systems, then the fraction of all BH+BH mergers formed via isolated (field) evolution that have $\chi_{\rm eff} > 0.1$ is even smaller. Perhaps of the order of $\sim 10\%$. Assuming Eddington-limited accretion efficiency and that the first-born BH is formed with a negligible spin, Bavera et al. (2021) find that all non-zero $\chi_{\rm eff}$ systems among the detected BH+BH mergers (here neglecting the possibility of dynamical assembly of the binaries) can come only from the CE channel, as the stable RLO HMXB channel cannot shrink the orbits sufficiently for efficient tidal spin-up to take place. An uncertain factor in all such calculations is the strength of the angular momentum transport inside massive stars. A similarly important issue is related to a number of uncertainties in the late stages of close binary evolution of massive stars, such as orbital angular momentum loss due to mass loss through the outer Lagrangian points. Apart from BH kicks, also the outcome of the CE phase remains crucial (the post-CE orbital separation and the spin of the two stellar components), and the strength of the stellar winds of WR-stars (e.g., Higgins et al., 2021).

Given that the event rate of BH+BH mergers inferred by the LIGO network of GW detectors is similar to the rate of long gamma-ray bursts (LGRBs), after beaming corrections with a reasonable value (Wanderman & Piran, 2010), it is tempting to suggest that LGRBs are produced exactly in such close binaries that eventually evolve to produce BH+BH mergers (Hotokezaka & Piran, 2017). The rationale being that LGRBs are produced from the core collapse of massive stars (hypernovae/collapsars), which form a spinning BH surrounded by an accretion disk (i.e., the central engine powering the burst, Woosley, 1993; Woosley et al., 2002; Piran, 2004). Indeed, as argued by Cantiello et al. (2007); van den Heuvel & Yoon (2007), helium stars in close X-ray binaries can be tidally kept in sufficiently rapid rotation to produce a collapsar.

15.8.2 Observations of BH Spins in BH+BH Mergers

The fact that the final spins of the BH merger products are all close to $a_{*,f} \simeq 0.7$ (Table 15.2) is indirect evidence that the spin components of the two original BHs were either small or not aligned. Had the two BH spin components been large and close to aligned, the final spin of the BH merger remnant would have been close to unity (Rezzolla et al., 2008), leading to a longer ring-down phase.

Unfortunately, it is impossible to directly measure the magnitudes of the individual BH spins or their tilt (misalignment) angles with respect to the orbital angular momentum vector of the binary. Only the sum of their projected spins along the orbital angular momentum is measurable, i.e., the effective in-spiral spin:[9]

$$\chi_{\rm eff} \equiv \frac{c}{GM}\left(\frac{\vec{S}_1}{M_1}+\frac{\vec{S}_2}{M_2}\right)\cdot\frac{\vec{L}}{|\vec{L}|} \tag{15.19}$$

or, alternatively, written in the form:

$$\chi_{\rm eff} \equiv \frac{(M_1\vec{\chi}_1 + M_2\vec{\chi}_2)}{M}\cdot\frac{\vec{L}}{|\vec{L}|} = \frac{\chi_1\cos\theta_1 + q\,\chi_2\cos\theta_2}{1+q}, \tag{15.20}$$

where $M = M_1 + M_2$ is the total mass of the two BHs; $q \equiv M_2/M_1 \leq 1$ is the mass ratio; $\vec{S}_{1,2}$, $\vec{\chi}_{1,2}$, and $\theta_{1,2}$ are the spin angular momentum vectors, dimensionless spins, and tilt (misalignment) angles, respectively, of the BH components; and $\vec{L}$ is the orbital angular momentum vector. The spin tilt angle for each BH is simply given by:

$$\theta_{1,2} = \cos^{-1}\left(\frac{\vec{\chi}_{1,2}\cdot\vec{L}}{|\vec{\chi}_{1,2}|\,|\vec{L}|}\right). \tag{15.21}$$

As a consequence, the spin components in the orbital plane are very poorly constrained. These spin components drive relativistic procession of the orbital plane (Apostolatos et al., 1994), and this effect can be quantified by the effective precession spin parameter (Schmidt et al., 2015):

$$\chi_{\rm p} \equiv \max\left[\chi_1\sin\theta_1,\ \left(\frac{4q+3}{4+3q}\right)q\,\chi_2\sin\theta_2\right]. \tag{15.22}$$

[9]Constraints on $\chi_{\rm eff}$ stem from observations of the GW signals before and after the merger.

Notice that, although the spin tilts of each BH may evolve during final in-spiral, the values of $\chi_{\rm eff}$ and χ_p are approximately constant (Gerosa et al., 2015). If the system precesses, by convention the spin parameters are usually defined at a reference frequency of $f_{\rm GW} = 20$ Hz (Abbott et al., 2021d).

The effective BH spins from the sources in O1 and O2 are depicted in Figure 15.20. Somewhat surprisingly, all the measured values of $\chi_{\rm eff}$ are more or less consistent with $\chi_{\rm eff} = 0$, perhaps except GW151226 and in particular GW170729 where $\chi_{\rm eff} = 0.18^{+0.20}_{-0.12}$ and $\chi_{\rm eff} = 0.37^{+0.21}_{-0.25}$, respectively.

From observing run O3a (Fig. 15.21), this picture changed slightly. Here, 10 systems showed signs of non-zero $\chi_{\rm eff}$ at the 95% credibility level. No individual sources were confidently found to have $\chi_{\rm eff} < 0$ (BH spin, $\vec{\chi}$ in a direction opposite to that of the orbital angular momentum vector, $\vec{L}$; i.e., *retrograde* spin with a tilt angle, $\theta > 90°$). An analysis of all systems in O1–O3a by Abbott et al. (2021e) yielded that a fraction of 0.67 ± 0.16 of all BH+BH mergers have $\chi_{\rm eff} > 0$, a fraction of $0.27^{+0.17}_{-0.15}$ have $\chi_{\rm eff} < 0$, and, finally, a fraction of $0.05^{+0.02}_{-0.01}$ have vanishing effective spins of $|\chi_{\rm eff}| < 0.01$. The above conclusions were more or less verified in Abbott et al. (2022b), following the addition of 35 systems in observing run O3b and released in GWTC-3 (see Fig. 15.12). They found, based on the combined data, that BH+BH mergers with $\chi_{\rm eff} < 0$ do exist at between 88% and 99% credibility, depending on the modelling assumptions.

In particular, six events with a high BH spin component are noticeable: GW190403 with $\chi_{\rm eff} = 0.70^{+0.15}_{-0.27}$, GW200308 with $\chi_{\rm eff} = 0.65^{+0.17}_{-0.21}$, GW190517 with $\chi_{\rm eff} = 0.53^{+0.20}_{-0.19}$, GW200208 with $\chi_{\rm eff} = 0.45^{+0.43}_{-0.44}$, GW190719 with $\chi_{\rm eff} = 0.35^{+0.28}_{-0.32}$, and GW190412 with $\chi_{\rm eff} = 0.28^{+0.07}_{-0.08}$. The BH+BH events with the lowest effective spin values are probably GW191109 with $\chi_{\rm eff} = -0.29^{+0.42}_{-0.31}$ and GW190514 with $\chi_{\rm eff} = -0.19^{+0.29}_{-0.32}$. There is no strong evidence for variation in the spin distribution with mass. However, it was found by Callister et al. (2021) that high spin correlates with asymmetric binaries. In some cases, the joint $\chi_{\rm eff}$ and q measurements for an event enables a tighter measurement of the spin magnitude of the more massive component; especially if $q \not\simeq 1$, although the resulting error bars are still large.

The uncertainty on the measurements of $\chi_{\rm eff}$ in GWTC-3 covers a range from $|\Delta\chi_{\rm eff}| < 0.1$ (GW191204) to $|\Delta\chi_{\rm eff}| > 0.4$ (GW200322), and typically $|\Delta\chi_{\rm eff}| = 0.1 - 0.25$. Therefore, the *individual* values of $\chi_{\rm eff}$ should be taken with a grain of salt, whereas the shape of the $\chi_{\rm eff}$ distribution for the *general* population is presumably reliable. Selection effects (Gerosa et al., 2018) from masses or spins, internal relations (Callister et al., 2021) and priors used in the LIGO-Virgo-KAGRA analysis (Mandel & Smith, 2021) imply that caution should still be taken in present analysis of the data. In addition, it is expected that $\chi_{\rm eff}$ depends on the redshift of the mergers (Bavera et al., 2022).

Whereas tilt (mislignment) angles are generally difficult to determine, the tilt angle of, for example, the more massive component in GW190412 is particularly well constrained to $\theta_1 \simeq 46°^{+30}_{-21}$, as seen in Figure 15.22. See further below for a discussion of the interpretation of this unusual GW event.

A detection of ~ 9 sources was found by Zackay et al. (2019), and references therein, in addition to the confident merger sources announced from the LIGO–Virgo

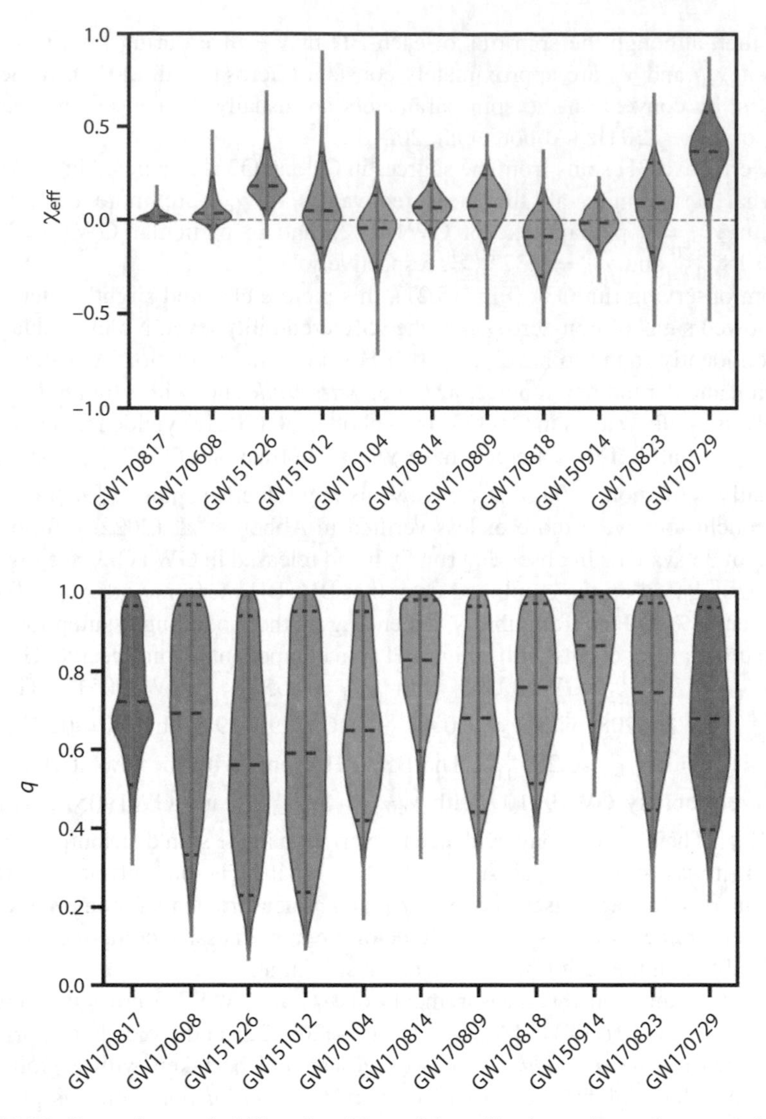

Figure 15.20. Posterior probability densities of the effective spin and mass ratio parameters of the GW events from LIGO-Virgo observing runs O1 and O2 (GWTC-1). The shaded probability distributions have equal maximum widths, and horizontal lines indicate the medians and 90% credible intervals of the distributions. Events are ordered by source-frame chirp mass. Top: effective aligned spin magnitude, $\chi_{\rm eff}$. Bottom: mass ratio, $q = M_2/M_1$. After Abbott et al. (2019c).

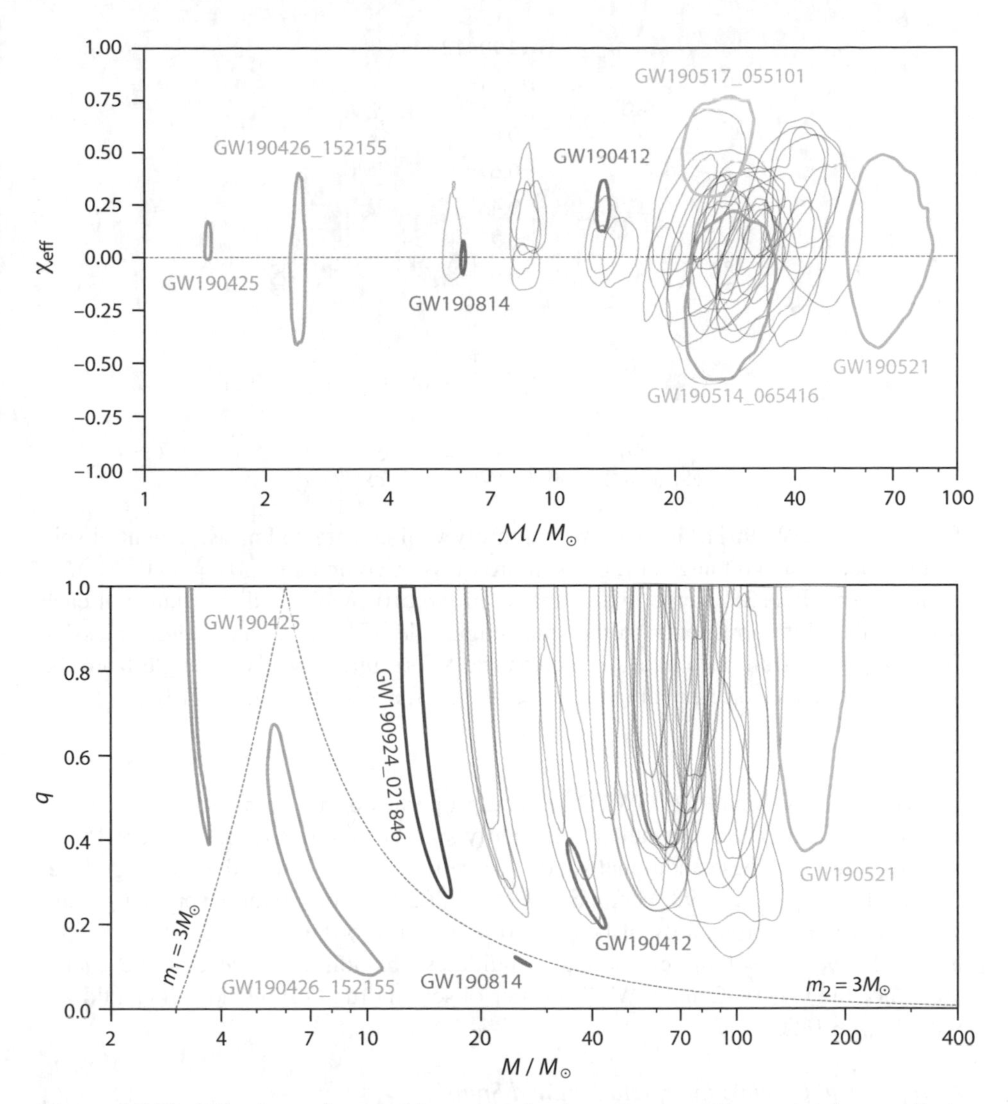

Figure 15.21. Effective spin vs. chirp mass (top), and mass ratio vs. total mass (bottom), of the candidate GW events from LIGO-Virgo observing run O3a. Each contour represents the 90% credible region for any given event. A sub-sample of the more peculiar evens are highlighted. The dashed lines in the bottom panel delineate regions where zero (DNS events), one (most likely mixed BH/NS events) or two components (BH+BH events) have masses above 3.0 $M_\odot$. See discussions in text. After Abbott et al. (2021d).

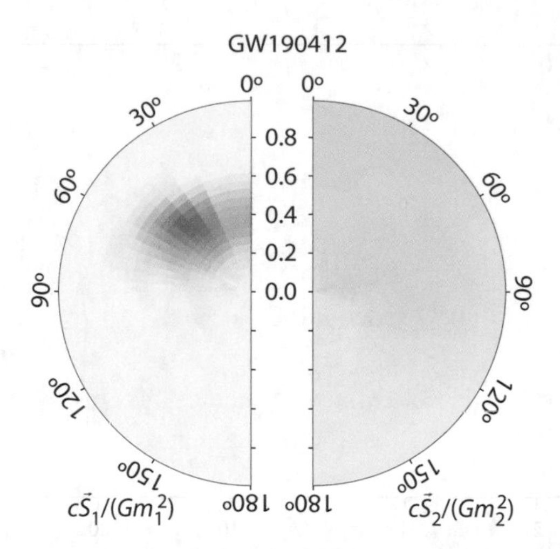

Figure 15.22. GW190412 is a case with a rarely well-constrained measurement of spin magnitude, χ_1, and tilt angle, θ_1, of the more massive component ($M_1 = 30.0^{+4.7}_{-5.1}\, M_\odot$) of an unusual BH+BH merger with a mass ratio, $q \simeq 0.28$. The radial distance of each pixel in the plot corresponds to the spin magnitude, $|\vec{\chi}|$, of the more (less) massive BH component from the circle's center on the left (right). The pixel's angle from the vertical axis is the corresponding tilt angle, $\theta_{1,2}$. After Abbott et al. (2021d).

observing runs O1 and O2 (GWTC-1). The properties of all claimed BH+BH mergers from O1 and O2 (confident and marginally significant) are shown in Figure 15.23. Although some of these additional sources were found to be detected with a high level of confidence ($P_{\rm astro} > 0.98$, Venumadhav et al., 2020), others are only very marginally significant with a probability of being astrophysical in origin of barely $P_{\rm astro} > 0.50$. Most likely, some of these marginal candidates are noise. Notice that the events announced in GWTC-2 and GWTC-3 (after observing run O3) include all candidates with $P_{\rm astro} > 0.50$.

Astrophysical Origins: Interpretations and Spins

Stevenson et al. (2017) argued that both the mass and $\chi_{\rm eff}$ of GW170729 are consistent with a hierarchical merger occurring in an active galactic nucleus (AGN). However, according to Gayathri et al. (2020), the reconstructed parameters of GW170729 are consistent with being part of the same population as the rest of the observed BH mergers if the prior probability of hierarchical mergers is low (Abbott et al., 2019b; Fishbach et al., 2020).

Given that the far majority of all BH+BH mergers reported so far have near-zero effective spins leads to only three potential explanations (e.g., Belczynski et al., 2020): If the individual BH spin magnitudes are large, then (1) either both BH spin vectors must be nearly in the orbital plane, or (2) the spin angular momenta of the BHs must be oppositely directed and similar in magnitude. Finally, there is also the possibility that

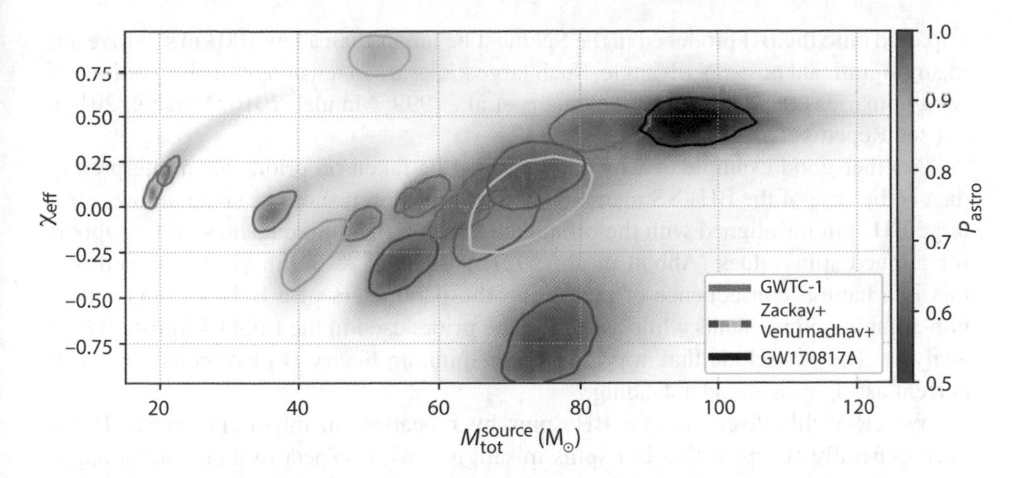

Figure 15.23. BH+BH merger events reported from LIGO-Virgo observing runs O1 and O2 (GWTC-1), in the plane of effective spin vs. total mass. In blue are shown the 10 BH+BH events reported in Abbott et al. (2019c) which are all certainly astrophysical in origin ($P_{\rm astro} \simeq 1$). Color coding by $P_{\rm astro}$ (the estimated probability of an astrophysical origin, rather than noise) are shown seven additional events with $P_{\rm astro} > 0.5$. In black is shown GW170817A ($P_{\rm astro} = 0.86$). Displayed are $1\,\sigma$ probability contours. After Zackay et al. (2019).

(3) the BH spin magnitudes are small. Belczynski et al. (2020) demonstrated that they can reproduce the observed distribution of low $\chi_{\rm eff}$ values within the classical isolated binary evolution scenario (the CE channel) of BH+BH formation assuming efficient angular momentum transport.

How to interpret the measured value of $\chi_{\rm eff}$ in terms of individual BH component spin magnitudes is indeed controversial as it is very sensitive to a priori assumptions about the plausible distributions of these values that enter the analysis as priors. An excellent example is given by Mandel & Fragos (2020) regarding GW190412 (Abbott et al., 2020) for which they demonstrate a very different conclusion on the results for the estimated spin magnitudes of the two BH components compared to that found by the LIGO–Virgo Collaboration (but see Zevin et al., 2020, for further discussions). Mandel & Fragos (2020) argue for a dimensionless spin component of the secondary BH in GW190412 between 0.64 and 0.99 along the orbital angular momentum. This scenario might provide evidence for the possibility of efficient spin-up of a tidally locked post-CE helium star in a tight orbit with a BH companion.

It should be noticed that even in cases where the spins of the first-born BH and the collapsing WR-star are both parallel to the orbital angular momentum, a kick added to the second-born BH in a direction out of the orbital plane will result in tilting the angle between the spin of the first-born BH and the post-SN orbital angular momentum vector (see Eq. 13.11). However, as noticed by Mandel & Fragos (2020), in tight orbits (and thus particularly in systems where the secondary star is tidally spun up to high spin) the pre-SN orbital velocity of the WR-star may approach $1{,}000\,{\rm km\,s^{-1}}$, so any kick imparted onto the BH produced in the SN must be larger than a few $100\,\mathrm{km\,s^{-1}}$ to result in any significant post-SN tilt angle. Such large kicks are in tension with observations of X-ray binaries containing BHs (Nelemans et al., 1999; Mandel, 2016; Mirabel, 2017), but see Repetto et al. (2012).

Another good example of why caution must be taken on priors when interpreting data is the case of the BH+NS merger GW200115. Inference on the signal allows for a large BH spin misaligned with the orbital angular momentum, but shows little support for aligned spin values (Abbott et al., 2021b). Mandel & Smith (2021) argued that this is a natural consequence of measuring the parameters of a BH+NS binary with non-spinning components while assuming the priors used in the LIGO-Virgo-KAGRA analysis. They conclude that, a priori, a non-spinning binary is more consistent with current astrophysical understanding.

We close this discussion on BH spins by repeating an important caveat. It has been generally accepted that BH spins misaligned with respect to their orbital angular momenta is a "smoking gun" signature of dynamical binary formation inside dense stellar clusters (e.g., Callister et al., 2020). However, if the last-formed BH had its spin axis tossed in a more or less random direction during the core collapse (see Section 13.7.9 and Tauris, 2022) it would completely jeopardize all current views and interpretations of the observed values of χ_{eff} for BH+BH mergers. In particular, such a scenario could easily produce $\chi_{\mathrm{eff}} < 0$ for systems originating from isolated binary systems.

In a recent investigation, Tauris (2022) demonstrated that empirical data (GWTC-3) strongly disfavor a main origin of BH+BH mergers from isolated binary star evolution *without* BH spin-axis tossing. Including (isotropic) BH tossing, however, provides a simple solution and an excellent fit to the bulk data for a large range of input distributions of relevant physical parameters. More statistical investigations (including selection effects) are needed to firmly verify and quantify the effect of BH spin-axis tossing, as well as for deriving constraints on the spin magnitudes of the first- and second-born BHs more robustly.

The physical tossing mechanism for NSs and low-mass BHs may be caused by fallback of material (Janka et al., 2022), whereas 3D convection in the envelope of a massive progenitor star (Antoni & Quataert, 2022), that is expected to collapse and directly produce a BH, may apply to more massive BHs. There could thus be a possible mass (and possibly spin) dependence on the BH tossing mechanism—a feature that may be revealed from future data of BH+BH mergers.

15.9 ANTICIPATED OTHER SOURCES TO BE DETECTED IN THE GW ERA

Following the successful detection of many double BH mergers, DNS mergers, as well as a few promising mixed BH/NS candidates found in LIGO–Virgo observing runs 01–O3, all combinations of double compact object mergers anticipated to show up in the high-frequency LIGO band (few 10 Hz to ~ 1 kHz) have now been discovered. However, there are still other exciting GW sources out there that await detection in the coming years.

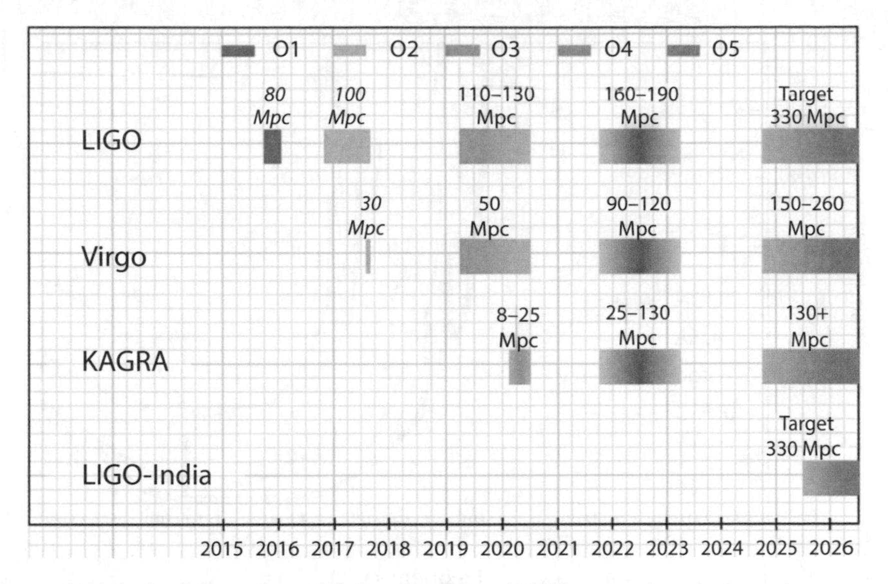

Figure 15.24. Anticipated sensitivity evolution (DNS ranges) and observing runs of advanced LIGO, advanced Virgo, KAGRA, and LIGO-India detectors over the coming years. See text for explanation. After Abbott et al. (2020b).

Figure 15.24 displays the planned sensitivity evolution and observing runs of the advanced LIGO, advanced Virgo, and KAGRA detectors over the coming years. The *ranges*, d_R, quoted are average distances for which DNS mergers can be detected with $\mathrm{SNR} = 8$. Mergers can be detected further away if they are orientated in a favorable way and are in the right part of the sky. The furthest distance the LIGO network can detect a source is the *horizon*, $d_H \simeq 2.264\, d_R$ (Finn & Chernoff, 1993). The number of detected sources scales roughly with $(4\pi/3)d_R^3$ (the redshifted volume), such that doubling the sensitivity (d_R) of the detectors will increase the expected detection number by a factor of 8.

Figure 15.25 displays a comparison of current and (potential) future GW detectors across a wide GW frequency band. As an example of the potential performance of a 3G ground-based high-frequency GW detector, Singh et al. (2021) estimate that the Einstein Telescope may record up to $10^5 - 10^6$ double BH mergers and $\sim 7 \times 10^4$ DNS mergers in just one year.

15.9.1 High-frequency GW Sources

GWs from supernovae

One class of potential transient high-frequency GW sources awaiting discovery include core-collapse SNe (CCSNe) in the MW or neighboring galaxies (Andresen et al., 2017; Morozova et al., 2018; Powell & Müller, 2019), although statistically expected to be rare events. (The CCSN rate in the MW is less than two per century.) Holgado & Ricker (2019) argue for the interesting possibility that GWs may be produced not only from

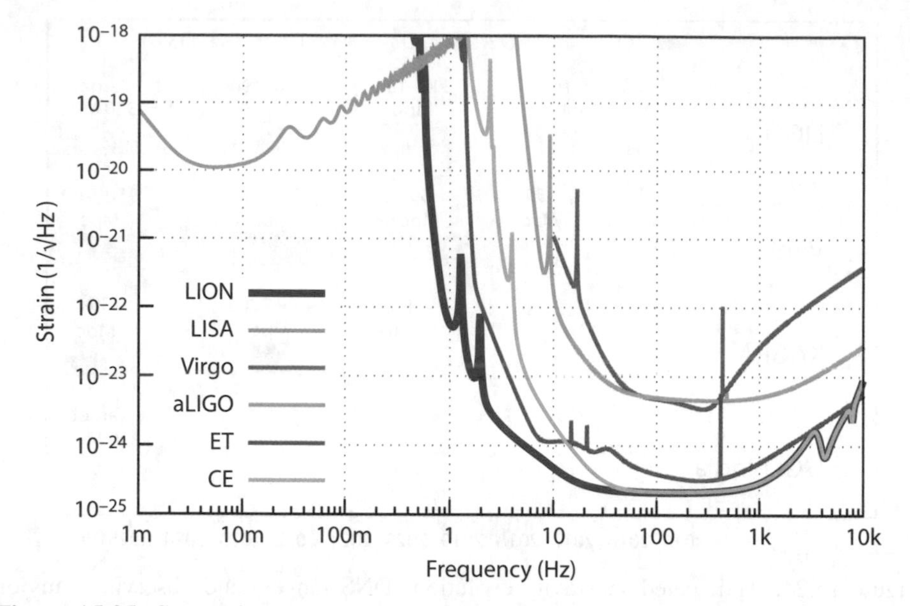

Figure 15.25. Comparison of the design sensitivity of several current and future GW detectors. Detectors shown here include at low frequencies: LISA, and at high frequencies: aLIGO and Virgo, as well as the proposed 3G detectors: ET (Einstein Telescope) and CE (Cosmic Explorer), and finally the suggested laser interferometer on the Moon (LION). After Amaro-Seoane et al. (2021).

the core-collapse SN process but also from the SN mass loss and SN natal kick during the transition from a pre- to post-SN binary. The reason why we can only expect to detect GWs from Galactic (or nearby) SNe is that the integrated GW luminosity over time from a given event is small: of the order of $E_{\rm GW} \lesssim 10^{-8}\, M_\odot\, c^2$ (e.g., Radice et al., 2019)—see Szczepanczyk et al. (2021) for a comprehensive review. Thus Advanced LIGO would only detect a typical CCSN with a signal-to-noise ratio, $\mathrm{SNR} \simeq 1-10$ for a SN distance out to 10 kpc. (For comparison, the SNR from GW150914 was 24.) For the 3G detectors (Fig. 15.25), the prospects are brighter with a corresponding SNR expected in the range from 20 to 110.

A small fraction of SNe, however, may produce metastable millisecond magnetars that survive of the order of one minute before collapsing to a BH. During this short time interval, such a magnetar may accrete sufficient material ($\lesssim 1\, M_\odot$) to produce a “mountain” and thereby emit GWs at kHz frequencies, for example, with a strain of $h_c \simeq 10^{-23}$ for an object located at 1 Mpc (Sur & Haskell, 2021).

Continuous GWs from Spinning NSs

More promising in terms of potential detection rates is continuous high-frequency GWs emitted from Galactic accreting MSPs or young rapidly spinning NSs. These sources are anticipated to be measurable if their ellipticity is large enough (Haskell et al., 2015;

Konar et al., 2016; Woan et al., 2018; Chen, 2020). The non-detection so far (Abbott et al., 2019a; Dergachev & Papa, 2020; Steltner et al., 2021; Abbott et al., 2021a) can be used to constrain the mass quadrupole moment and ellipticity. For the Crab and Vela pulsars, the non-detection now results in constraining the GW emission to account for less than 0.017% and 0.18% of their spin-down luminosity, respectively (Abbott et al., 2019d). A search for detection of continuous GWs in the LIGO O2 public data (Steltner et al., 2021) excludes NSs rotating faster than 5 ms with equatorial ellipticities (see below) larger than 10^{-7} and located closer than 100 pc. A recent search for GWs from the promising source Sco X-1 (the brightest point X-ray source in the sky) using LIGO O3 data also failed a detection, resulting in a GW amplitude constraint of $h_0 < 10^{-25}$ (Abbott et al., 2022).

The expected GW signal from a spinning NS is studied in several works (e.g., Aasi et al., 2014; Haskell et al., 2015; Woan et al., 2018; Abbott et al., 2019c). A quadrupolar GW signal from a triaxial NS (i.e., a NS with some asymmetry with respect to its rotation axis and therefore possessing a triaxial moment of inertia ellipsoid), steadily spinning about one of its principal axes of inertia, has a GW frequency twice the rotation frequency ($f_{\rm GW} = 2 f_{\rm spin}$), and a strain of:

$$h(t) = \frac{1}{2}\, F_{+}(t,\psi)\, h_0\, (1+\cos^2\iota)\cos\phi(t) + F_{\times}(t,\psi)\, h_0\, \cos\iota \sin\phi(t) \qquad (15.23)$$

in the detector, where

$$h_0 = h_{\rm spin} = \frac{16\pi^2 G}{c^4} \frac{f_{\rm spin}^2\, I_{zz}\, \varepsilon}{d_L} \qquad (15.24)$$

is the dimensionless GW strain amplitude. This strain amplitude is dependent on I_{zz}, the fiducial equatorial ellipticity defined as $\varepsilon = (I_{xx} - I_{yy})/I_{zz}$ in terms of principal moments of inertia, the rotational frequency, $f_{\rm spin}$, and the distance to the source, d_L. The signal amplitudes in the two polarizations (+ and $\times$) depend on the inclination of the NS's rotation axis with respect to the line-of-sight, ι, while the detector antenna pattern responses for the two polarization states, $F_{+}(t,\psi)$ and $F_{\times}(t,\psi)$, depend on the GW polarization angle, ψ, as well as the detector location, orientation, and source sky position. The GW phase evolution, $\phi(t)$, depends on both the intrinsic spin frequency of the NS and and its time derivative, as well as on Doppler and propagation effects (Aasi et al., 2014). These extrinsic effects include relativistic modulations caused by the orbital and rotational motion of Earth, the presence of massive bodies in the Solar System close to the line-of-sight to the NS, the proper motion of the NS, and its orbital motion if in a binary system.

The value of the ellipticity (due to asymmetries in the NS crust) remains unknown and is probably $\varepsilon < 10^{-7}$. The input physics behind the "NS mountain" calculations and, in particular, modelling the force creating non-sphericity remains challenging (Gittins et al., 2021). Non-detection by a 3G detector of GWs from spinning NSs, would even push the limit as far as $\varepsilon < 10^{-9}$. For discussion and details of NS asteroseismology, we refer to Andersson (2020). Exercises 14.7 and 14.8 deal with the spin evolution of a pulsar emitting GWs.

Transient GWs from Glitching Pulsars

Ho et al. (2020) investigated the potential for detecting GWs from NS fluid oscillations, which have been excited to energies typical of a pulsar glitch. They find that current GW detectors may observe nearby pulsars undergoing large events with Vela pulsar-like glitch energies, while next-generation detectors (3G) could observe a significant number of such events.

15.9.2 Low-frequency GW Sources

A broad palette of low-frequency GW sources is expected to be detected by LISA (Amaro-Seoane et al., 2017; Amaro-Seoane, Andrews et al., 2022). These are briefly summarized here and discussed in more detail in Section 15.12. The low-frequency GW sources include first of all continuous GWs emitted from Galactic compact object binaries (Nelemans, Yungelson, Portegies Zwart et al., 2001). The main sub-class is constituted by tight binaries of double WDs, but also mixed NS/WD, DNS and possibly a few BH binaries will be detected. As soon as the detector turns on, it is guaranteed to detect at least a dozen sources (Kupfer et al., 2018). A large number of transient sources are also expected. These are evolving sources (i.e., binaries for which the GW frequency, f_{GW}, changes significantly over the observation time) and include mainly the slow merger process of super-massive black hole (SMBH) binaries in distant galaxies. Furthermore, a number of extragalactic stellar-mass BH mergers are expected to be detected in the low-frequency LISA band a few years prior to their final merger event at high frequencies (Sesana, 2016)—see example plotted in Figure 15.26. The LISA sensitivity curve will be summarized in much more detail in Section 15.13.

LISA will measure the low-frequency GW signal ($f_{\mathrm{GW}} = 2\, f_{\mathrm{orb}}$) arising from the orbital motion, h_{orb}, of a large number of Galactic binaries with a strain amplitude given by (Evans et al., 1987):

$$h_{\mathrm{orb}} = \left(\frac{32}{5}\right)^{1/2} \frac{(2\pi)^{2/3}\, G^{5/3}\, f_{\mathrm{orb}}^{2/3}\, \mathcal{M}^{5/3}}{c^4\, d_L} \tag{15.25}$$

for binaries with an average orbital orientation and polarization.

It is possible that other Galactic sources whose evolution is not directly driven by GW radiation may be detectable as well. Renzo et al. (2021) investigated the emission of low-frequency GWs from CE events of in-spiraling binaries and predict the detection of $\sim 0.1 - 100$ such sources in the MW during the duration of the LISA mission. Detecting such a GW signal would provide valuable direct insight on the gas-driven physics.

15.9.3 A Dual-line GW Source

A very interesting possibility in upcoming GW astronomy will be the potential detection of a *dual-line* Galactic GW source (Tauris, 2018), where the LIGO network, or a 3G detector, would detect the continuous high-frequency GW emission from the

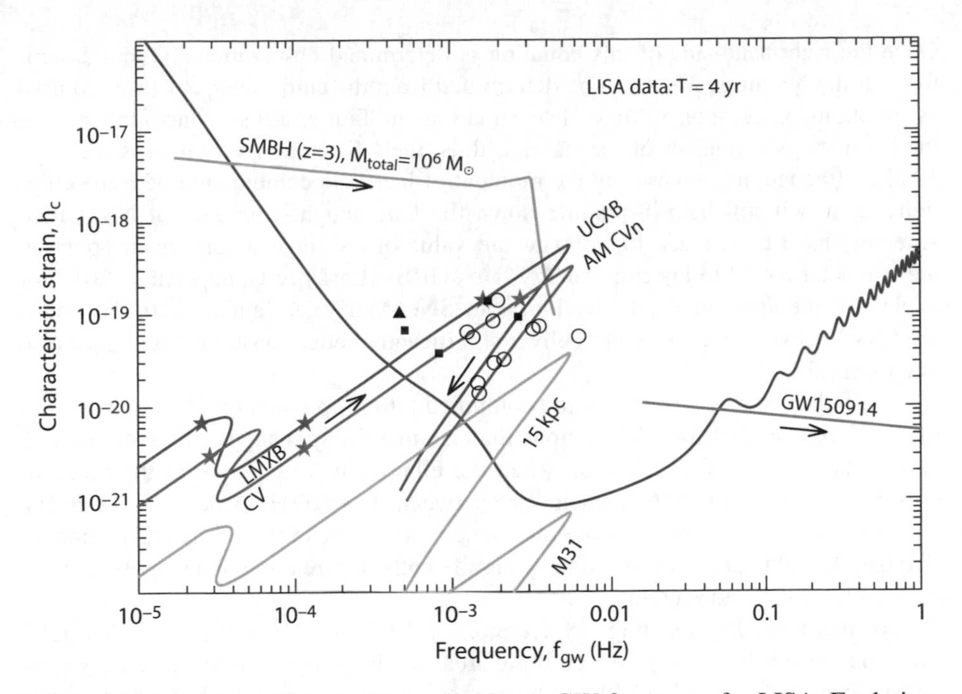

Figure 15.26. Characteristic strain amplitude vs. GW frequency for LISA. Evolutionary tracks for UCXB (blue) and AM CVn systems (magenta) are plotted assuming a distance of $d_L = 1\,\mathrm{kpc}$. The grey curves are for the same UCXB located at $d_L = 15\,\mathrm{kpc}$ and $d_L = 780\,\mathrm{kpc}$ (M31), respectively. The three star symbols represent (with increasing f_{GW}): onset of the LMXB/CV stage, termination of the LMXB/CV stage, and onset of the UCXB/AM CVn stage. These sources evolve slowly on a Myr timescale (Fig. 15.35) and are *monochromatic* during the lifetime of LISA. The LISA sensitivity curve (shown as a red line) is based on four years of observations. Comparison tracks for *evolving* sources are shown for: a super-massive BH (SMBH) merger with $M = 10^6\,M_\odot$ at $z = 3$ (green), and the last four years of in-spiral of GW150914 (orange). Data from LISA verification sources include detached double WD binaries (solid squares), AM CVn systems (open circles), and a hot subdwarf binary (solid triangle). After Tauris (2018).

spinning (recycled) NS, and LISA would simultaneously detect the gravitational damping of the system's orbital motion via continuous low-frequency GW emission. Such a system could very well be an UCXB (Sections 7.8.4 and 11.4).

The strain amplitude of a LIGO/3G measurement of the high-frequency NS spin would be given by Eq. (15.24) and LISA may then measure the low-frequency GW signal arising from the orbital motion expressed by Eq. (15.25). Combining these two expressions yields (Tauris, 2018):

$$I_{zz}\,\varepsilon = \sqrt{\frac{2}{5}}\left(\frac{\sqrt{G}}{2\pi}\right)^{4/3}\left(\frac{f_{\mathrm{orb}}^{1/3}}{f_{\mathrm{spin}}}\right)^{2}\mathcal{M}^{5/3}\left(\frac{h_{\mathrm{spin}}}{h_{\mathrm{orb}}}\right). \tag{15.26}$$

Once the right-hand-side of this equation is determined observationally, and assuming that the NS mass, M_{NS}, can be determined from the chirp mass, $\mathcal{M}$ (see required assumptions on component mass determinations in Tauris, 2018), constraints can be made on the NS moment of inertia and, thus, the NS radius (Ravenhall & Pethick, 1994). Although only measuring the moment of inertia in combination with the ellipticity, ε, it will still help in pinning down the long sought-after EoS of NS matter. Attempts have been made to estimate the value of ε—for example, from applying the magnetar model to injecting energy into sGRBs (Lasky & Glampedakis, 2016) or explaining the light curves of ultra-luminous SNe (Moriya & Tauris, 2016). However, the LIGO network is expected to deliver significantly better constraints on ε (see also Section 15.9.1).

The pulsar spin is affected by the combined torques producing GW radiation and magnetic dipole radiation. Calculations indicate that the spin angular momentum loss is dominated by the GW radiation when the ellipticities of NSs are in the range of $(1-50)\times 10^{-7}$ and GW frequencies are between $10-100\,\mathrm{Hz}$ (Chen, W.-C., 2021). However, such large values of ε are questionable (Section 15.9.1). Finally, we emphasize that the value of ε is expected to be significantly different in young newborn magnetars, old MSPs, and accreting NSs.

For thermonuclear bursting NS accretors in UCXBs, however, mass-radius relations (obtained from the apparent emitting area and the peak flux achieved during photospheric bursts, see Section 7.8), can be used to further constrain the EoS of NS matter (e.g., van Paradijs, 1979; Özel et al., 2009). Based on this argument, Suvorov (2021) identified two Galactic systems (4U 1728−34 and 4U 1820−30) as being potentially detectable as dual-line GW sources that could provide percent-level constraints on the mass, radius, and even internal magnetic field strength of the NS.

15.10 GW FOLLOW-UP MULTI-MESSENGER ASTRONOMY

The detection of GW170817 represents a grand milestone in multi-messenger astronomy (Abbott et al., 2017d; Margutti & Chornock, 2020; Hajela et al., 2021). It was a very loud GW signal from a DNS merger with a SNR of 32. Following only $1.74\pm 0.05\,\mathrm{s}$ after the GW merger event, a short ($\sim 2\,\mathrm{s}$ duration) γ-ray burst (sGRB, denoted GRB 170817A) was detected by the Fermi and INTEGRAL satellite telescopes (Abbott et al., 2017a; Savchenko et al., 2017; Ajello et al., 2018). This is quite remarkable and fortunate given that this sGRB occurred more than 10 times closer to Earth (40 Mpc $\simeq$ 130 million light years) than any previously known GRB. The short time delay between the GW and the γ-ray signal proves that GWs propagate with the speed of light to at least 15 significant figures. Some delay is expected in any case by a combination of the jet-launch timescale and the fireball propagation out to the photospheric radius.

An astronomical transient designated AT 2017gfo (originally, SSS17a) was found 11 hr after the GW signal (Abbott et al., 2017d; Coulter et al., 2017; Drout et al., 2017; Smartt et al., 2017; Soares-Santos et al., 2017). It was discovered in the Galaxy NGC 4993 during a search of the sky region indicated by the GW and γ-ray detection

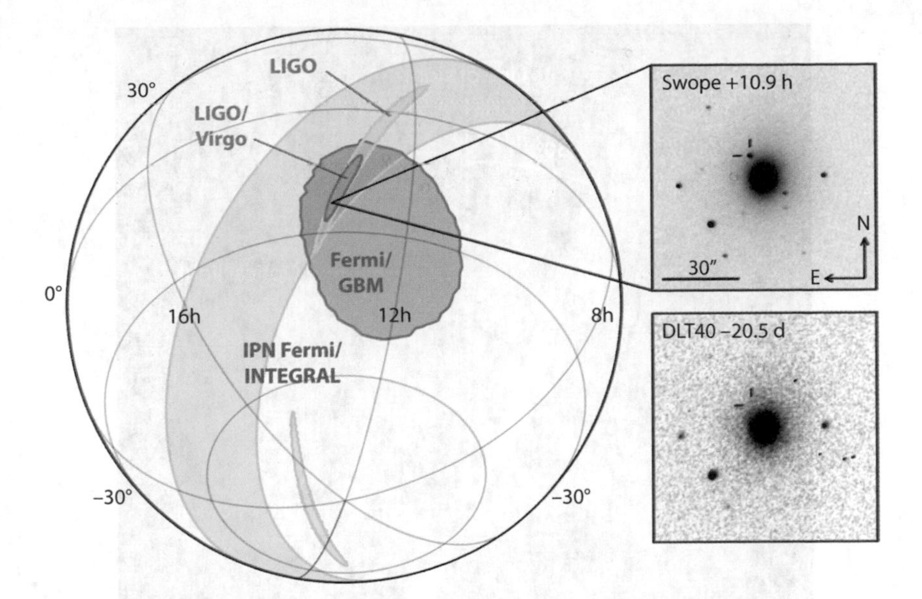

Figure 15.27. Localization of the GW, γ-ray, and optical signals related to GW170817. Shown are the projection of the 90% credible regions from LIGO (190 deg^2; light green), combined LIGO–Virgo localization (31 deg^2; dark green), IPN triangulation from the time delay between Fermi and INTEGRAL (light blue), and Fermi-GBM (dark blue). The insets on the right-hand side show the location of GW170817 in the apparent host Galaxy NGC 4993 in the Swope optical discovery image at 10.9 hr after the merger (top right) and the DLT40 pre-discovery image from 20.5 days prior to merger (bottom right). The reticle marks the position of the transient in both images. After Abbott et al. (2017d).

(Fig. 15.27). The transient was observed by numerous electromagnetic (EM) telescopes, from radio to X-ray wavelengths, over the following days and weeks, and it was shown to be a fast-moving, rapidly-cooling cloud of neutron-rich material, as expected for debris ejected from a NS merger (i.e., a *kilonova*, or alternatively called a *macronova*, see Figs. 15.28, 15.29, and 15.30). The kilonova is powered by radioactive decay of heavy r-process elements (Fig. 15.31) produced from ejected iron-group nuclei seeds that are bombarded by the intense flux of free neutrons. (For general reviews on nuclear synthesis, dynamics, and observations of kilonovae, see e.g., Giacomazzo et al., 2018; Shibata & Hotokezaka, 2019; Nakar, 2020; Metzger, 2019; Radice et al., 2020).

The relatively massive S0 Galaxy NGC 4993 was identified as the host Galaxy of GW170817 (e.g., Coulter et al., 2017; Soares-Santos et al., 2017). It is located ~40 Mpc away and has a metallicity between 0.2 and 1.0 $Z_{\odot}$, that is, similar to that of the MW (e.g., Im et al., 2017). NGC 4993 is less massive than the MW ($M_{4993} \simeq 10^{10.5}\ M_{\odot}$ and $M_{\rm MW} \simeq 10^{11.8}\ M_{\odot}$, respectively) and shows negligible recent star formation (Pan et al., 2017). Therefore, GW170817 is expected to have an old progenitor system and a delay time of at least a few Gyr. Such DNS systems can be produced with population

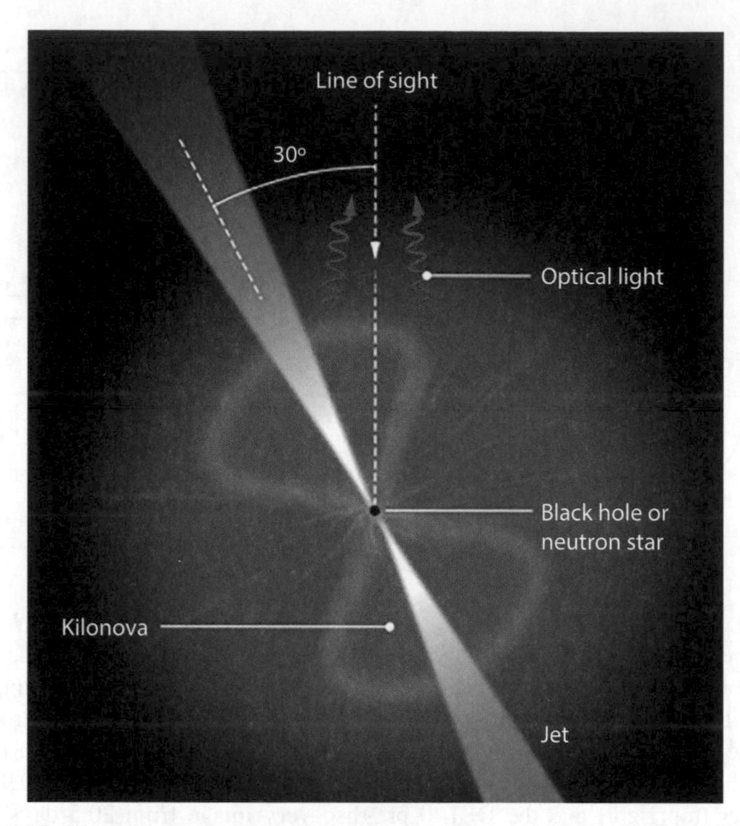

Figure 15.28. Illustration of a binary NS merger leading to a short γ-ray burst (confined to a jet) and an optical kilonova. Credit: C. Bickel/Science.

synthesis (e.g., Kruckow et al., 2018; Belczynski et al., 2018) and the measured offsets of observed sGRBs from the star-forming regions in their host galaxies (Fong & Berger, 2013; Zevin et al., 2020) can be explained by their spread in systemic recoil velocities (most often $v_{\rm sys} < 200\,{\rm km\,s^{-1}}$) resulting from the second SN explosion (Section 13.8).

Interestingly, following the GW170817/GRB 170817A kilonova event, a number of archival sGRB events have been re-analyzed (in optical, near-infrared, and radio) to search for other kilonova transients. A prime candidate is GRB 160821B ($z = 0.162$), which shows evidence of being a merger event with dynamic ejecta mass $M_{\rm dyn} = (1.0 \pm 0.6) \times 10^{-3}\,M_\odot$ and a secular (post-merger) ejecta mass with $M_{\rm pm} = (1.0 \pm 0.6) \times 10^{-2}\,M_\odot$, consistent with a DNS merger resulting in a short-lived massive NS (Lamb et al., 2019). Theoretical models of the amount of ejecta mass from both components resulting from DNS mergers are shown in Figure 15.32. The amount of ejected matter in DNS and NS/BH mergers, and thus the kilonova light curve and jet launching, depends on the mass ratio as well as the spin of the NS/BH components (Rosswog et al., 1999; Lehner et al., 2016; Cowperthwaite et al., 2017; Ruiz et al., 2020). Note

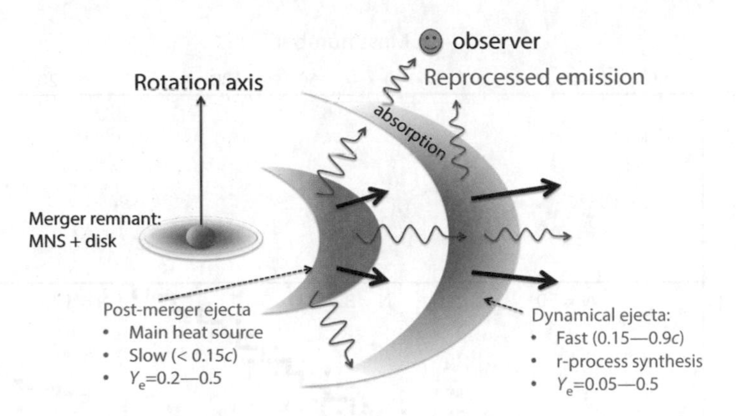

Figure 15.29. Schematic picture of the ejecta profile for the case that a long-lived massive NS is formed as a remnant of a NS+NS merger. The outer falcate component denotes the neutron-rich dynamical ejecta. The inner falcate component denotes the less neutron-rich post-merger ejecta which is slower than the dynamical ejecta. GW data indicates that GW170817 is observed within 30° along the direction of the rotation axis. After Shibata & Hotokezaka (2019).

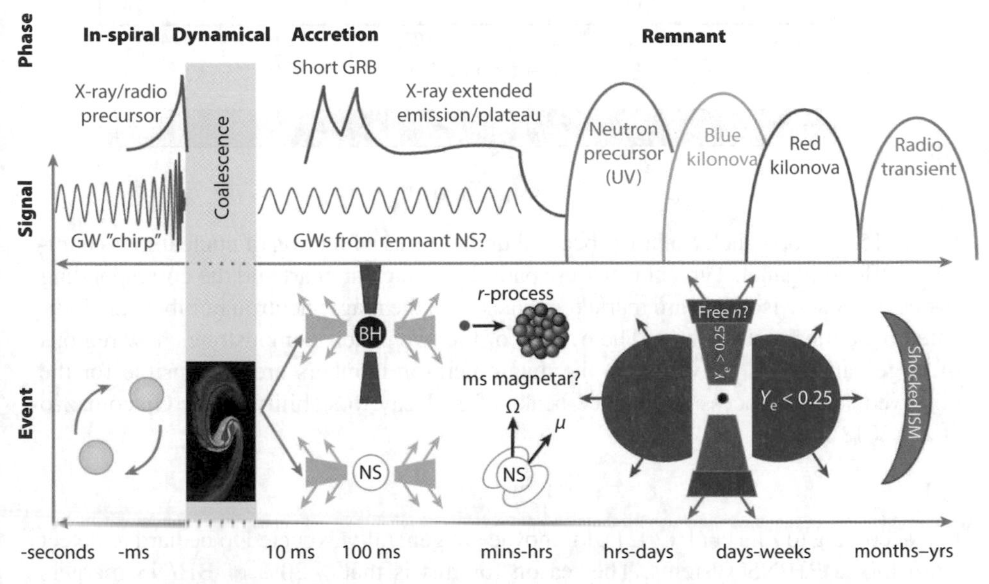

Figure 15.30. Phases of a NS+NS merger as a function of time, showing the associated rich observational signatures and underlying physical phenomena. Abbreviations are ISM: interstellar medium; n: neutron; Y_e: electron fraction. After Fernández & Metzger (2016).

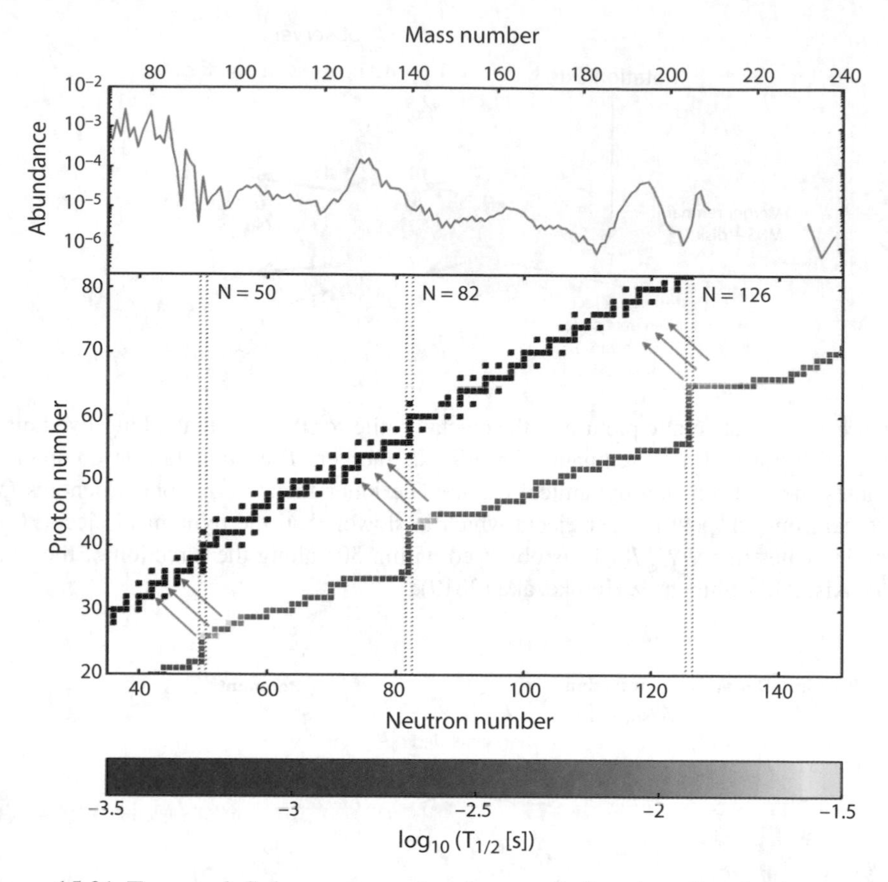

Figure 15.31. Top panel: Solar r-process abundances as a function of nuclear mass number, A. Bottom panel: Typical r-process path in the nuclear chart and the corresponding β-decays. Stable isotopes are marked in black, and the magic neutron numbers are indicated by vertical dotted lines. The overlay of the two panels demonstrates how regions of large half-life ($T_{1/2}$) values at the magic neutron numbers are responsible for the observed solar r-process abundance peaks after decay to stability. After Giacomazzo et al. (2018).

that according to Zhu et al. (2021), kilonovae are generally expected to be hard to detect from mixed BH/NS systems. The reason for this is that $\lesssim 20\%$ of BH/NS mergers are tidally disrupted since most of the primary BHs are expected to have low spins. Furthermore, the associated kilonovae for those disrupted events are still difficult to detect because of their low brightness and large distances.

Unlike the situation for GW170817, no clear EM counterpart (sGRB nor optical kilonova) was reported for the DNS merger event GW190425. A good reason for this is that the 90% credible event sky area for GW190425 was more than 8,000 deg^2 (compared to only 28 deg^2 for GW170817). Furthermore, GW190425 has an estimated luminosity distance of $d_L \simeq 160 \pm 70$ Mpc, compared to $d_L = 40 \pm 3$ Mpc for

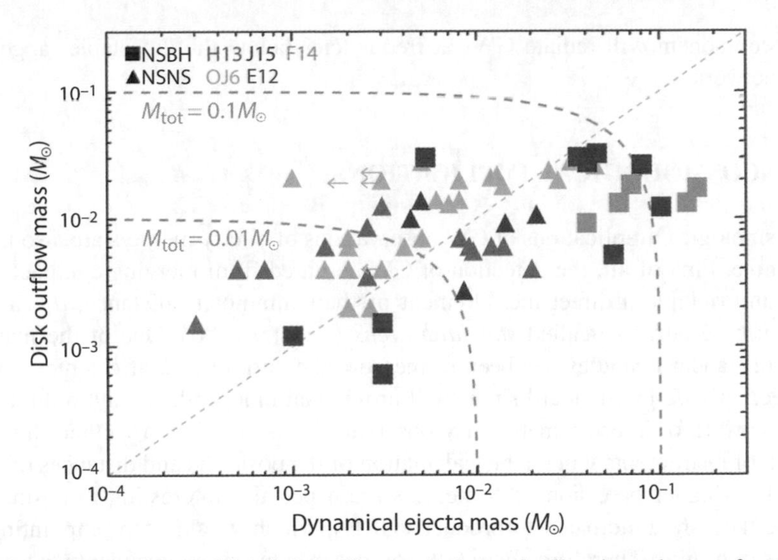

Figure 15.32. Mass ejected in disk outflows vs. dynamically ejected mass for various model simulations of NS+NS mergers. Each point corresponds to the results of a single simulation. After Fernández & Metzger (2016).

GW170817. Finally, it may be the case that any associated sGRB is simply not beaming in our direction of the sky.

We still await EM emission to be detected from mixed BH/NS mergers. In the case of the BH/NS candidate merger GW190814 (Abbott et al., 2020b), no optical nor infrared emission was detected from follow-up campaigns (Ackley et al., 2020; Vieira et al., 2020).

Binary BH mergers are not expected to be directly associated with EM counterparts as such, due to the lack of baryonic matter involved in such a merger event between two BHs. However, de Mink & King (2017) pointed out that the progenitor binary may have shed a significant amount of gas during its prior evolution which could be retained in a circumbinary disk. The recoil of the BH merger (Campanelli et al., 2007; González et al., 2007) may then shock and heat this material. Gravitationally bound gas attempting to follow the kicked binary BH merger product may thus collide with the surrounding disk gas of an AGN, thereby producing an EM counterpart, such as the one suggested to be associated with the binary BH merger event GW190521 (Graham et al., 2020). In fact, production of BH+BH binaries in an AGN disk may be a viable road to grow the masses of the BHs significantly (into or beyond the mass-gap, see Section 15.5) via accretion from the gaseous environment (Kaaz et al., 2021).

It is quite possible, however, that wide-field all-sky synoptic surveys will soon detect optical transients from massive WDs (CO/ONeMg WDs) merging with NSs or BHs (Zenati et al., 2020; Bobrick et al., 2022). SN AT2018kzr has been suggested to be such a transient (McBrien et al., 2019; Gillanders et al., 2020) based on its rapidly-evolving light curve, large oxygen, magnesium, and silicon content, and high inferred $^{54}Fe/^{56}Ni$ ratio. Notice that since WDs have radii that are two orders of magnitudes larger than NS radii and Schwarzschild radii of $10\,M_\odot$ BHs, a binary merger with a WD component will radiate GWs at frequencies below the detectable range of the LIGO network.

15.11 COSMOLOGICAL IMPLICATIONS

The cosmological implications of GW observations of binary mergers are also interesting to note. First of all, the detection of GWs emitted from merging compact objects allows in principle a direct measurement of their luminosity distance, d_L, and thus they can be used as so-called *standard sirens* (Schutz, 1986). One of the main uses of SN Ia standard candles has been to measure the current rate of cosmic expansion (see Section 13.2.1). Standard sirens will provide an independent way to do this. The global network of interferometric GW observatories is growing. Together, this LIGO-network of instruments will enable calculation of the positions and distances of merger events to a higher precision. (NS mergers are especially interesting to cosmologists because they are anticipated to produce sGRBs, which would help pinpointing their galaxies of origin.) Therefore, these GW sources provide an independent measurement of the Hubble constant, H_0, in analogue to standard candles from Type Ia SNe. Adding further EM information to GW data will be even better. Hotokezaka et al. (2019) estimated that if combining GW data, radio images, and light curves of DNS mergers similar to GW170817, it would only require 15 more events, as compared with 50 to 100 GW events without such additional data, to potentially resolve the tension between the Planck and Cepheid-SN measurements on the value of H_0.

The expression of the luminosity distance as a function of redshift, $d_L(z)$, is:

$$d_L(z) = \frac{1+z}{H_0} \int_0^z \frac{d\tilde{z}}{\sqrt{\Omega_R(1+\tilde{z})^4 + \Omega_M(1+\tilde{z})^3 + \rho_{\rm DE}(\tilde{z})/\rho_0}}, \qquad (15.27)$$

where $\rho_0 = 3H_0^2/(8\pi G)$, $\rho_{\rm DE}$ is the dark energy density, and Ω_M and Ω_R are the present matter and radiation density fractions (the latter contribution is negligible at the redshifts relevant for standard sirens). In the limit $z \ll 1$, the Hubble law $d_L(z) = H_0^{-1}\, z$ is recovered, and from measurements at such small redshifts we can only obtain information on H_0. Third-generation (3G) interferometers, however, may measure standard sirens up to large redshifts, $z \sim 8$, allowing for investigations of the effects of dark matter (Pierce et al., 2018) and dark energy (Ezquiaga & Zumalacárregui, 2017). Space-based interferometers (LISA) can also be used to track dark energy (Amaro-Seoane, Andrews et al., 2022). Finally, a measurement of the order of $10^2 - 10^3$ standard sirens will also allow for testing modified gravity theories in which the propagation of GWs differs from general relativity (Belgacem et al., 2018).

15.12 LISA SOURCES

As already briefly mentioned in Section 15.9.2, detection of GW emission from binary compact stars is one of the key drivers for the LISA mission. In this section, we first summarize some of the scientific advantages and goals for LISA. Thereafter, we discuss the binary compact objects with detached components, as well as those systems which undergo mass-transfer interactions (RLO) from a WD donor star to a WD or NS accretor. For further details and a comprehensive review of the Galactic binary compact objects that LISA is expected to detect, we refer to Amaro-Seoane, Andrews et al. (2022).

At the time of writing (March 2022), there are already two dozen of known Galactic sources that are guaranteed to be detected with LISA after a few years of its operation. These "verification sources" are tight binaries (typically with orbital periods of $P_{\rm orb} \simeq 5-30\,{\rm min}$) of double WDs (DWDs), which give rise to continuous emission of GWs. Unlike binaries containing NSs and BHs, binaries with WD components (due to their larger compact star radii and thus orbital periods prior to merger) are not readily detectable by ground-based high-frequency (Hz–kHz) GW observatories, such as LIGO-Virgo-KAGRA, nor by the 3G of such proposed detectors, for example, the Einstein Telescope and the Cosmic Explorer. These high-frequency detectors can observe the final few orbits of in-spiral (lasting from a fraction of a second to a few minutes) and the merger event itself. As we have discussed earlier in this chapter, however, such merger events are rare (of the order of a dozen events $\rm Myr^{-1}$ for a MWEG) and therefore they are anticipated to be detected only as extragalactic sources. A major advantage of LISA, with its low-frequency ($\sim$ mHz) GW window, is that it can follow the in-spiral of the numerous population of tight Galactic double WDs, NSs, and BHs (due to orbital GW damping in these compact binaries)—up to $\sim 10^6$ yr prior to their merger event (see Section 15.12.4 below). Thus a significant number of such local sources are anticipated to be detected in GWs by LISA, even though their emitted GW luminosity is relatively small compared to that of the final merger process.

Binary population synthesis studies predict of the order of 10^4 resolved Galactic DWDs that may be detected with LISA. This population includes both detached DWDs (Section 15.12.1) and systems undergoing mass transfer (AM CVn systems, Section 15.12.3). DNS systems are also expected to be detected by LISA (Section 15.12.4). Based on the known Galactic population of tight-orbit radio pulsar binaries in combination with population synthesis predictions, an estimated number of $10^1 - 10^2$ DNS systems with a significant signal-to-noise ratio (SNR) may be detected by LISA within a 4 yr mission. An even larger number of NS+WD systems is expected to be detected too, including ultra-compact X-ray binaries (UCXBs). Double BHs detectable by LISA are strong candidates to become the first discoveries of such systems in the MW. Given that LISA's volume sensitivity for a constant SNR scales with the fifth power of the chirp mass, $\mathcal{M}^5$, double BH sources can easily be be detected in distant galaxies, located several Gpc away. Interestingly enough, this fortuitous condition will therefore allow LISA to discover extragalactic double BHs several years before their final merger events that the LIGO network will detect (Sesana, 2016; Toubiana et al., 2021). Moreover, LISA is also expected to detect exotic Galactic systems like triple stellar systems, tight systems of WDs with exoplanets, and possibly helium star binaries (Amaro-Seoane, Andrews et al., 2022).

The LISA mission will provide great opportunities to explore physical interactions and answer key scientific questions related to formation and evolutionary processes of tight binary (and multiple) stellar systems containing compact objects. This includes questions related to the stability and efficiency of mass transfer, CEs, tides, and stellar angular momentum transport, irradiation effects; as well as details of their formation and destruction in CCSNe and Type Ia SNe, respectively. Furthermore, information about the environments of these sources will be available too. The number and Galactic distribution of LISA sources are excellent probes to gain new knowledge on the star formation history and the structure of the MW. Finally, the sheer numbers and rates of LISA sources will provide crucial knowledge on their formation and evolution processes and help to place constraints on key physical parameters related to binary (and triple-star) interactions.

In Figures 15.36–15.40, we illustrate formation channels for producing GW sources detectable by LISA. The concentric ellipses indicate stages with emission of significant GWs. All interaction processes are discussed (and acronyms defined) throughout various chapters in this book, in particular in Chapters 9 through 12.

Regarding LISA sources, it is often useful to state the characteristic merger timescale due to orbital decay from emission of GWs. For a circular binary orbit, it is given by (see Exercise 15.1):

$$\tau_{\rm GW} = \frac{3}{8}\left|\frac{P_{\rm orb}}{\dot{P}_{\rm orb}}\right| \quad \simeq 0.0713\,{\rm Myr}\,\left|\frac{P_{\rm min}}{\dot{P}_{-11}}\right|, \tag{15.28}$$

where $P_{\rm min}$ is the orbital period in minutes and $\dot{P}_{-11}$ is the orbital period decay rate in units of $10^{-11}\,{\rm s\,s^{-1}}$ (see Eq. 14.35). For binaries with one or two WD components, a tidal contribution may affect $\dot{P}_{\rm orb}$ by a few percent (Burdge et al., 2019)—see further discussions in Section 15.12.1. Note also, that such systems will become UCXB or AM CVn sources when the WD fills its Roche lobe after a time, $t < \tau_{\rm GW}$ (depending on the WD radius, i.e., mass and temperature).

15.12.1 Detached Double WD Systems

It has been known for several decades that nearby Galactic double WD (DWD) binaries will be the dominant contributor to signals detected by LISA (Hils et al., 1990). As a consequence of DWDs being so numerous, however, a large portion of LISA's sensitivity range will be swamped with GWs from millions of DWDs located in the Galactic disk and bulge, thereby producing a GW foreground (or background) "confusion noise" of unresolved binaries (Nelemans, Yungelson, Portegies Zwart, 2001)—see the bend in the sensitivity curve in Figure 15.26 near a frequency of $f_{\rm GW} \sim 1$ mHz. Nevertheless, any binary with a significant SNR will be resolved, and already now an increasing number of secure "verification sources" are identified (i.e., known binaries that are so bright in the LISA GW band that they will be detected with a high SNR of $\sim 10-100$ within a few years of LISA operation [Kupfer et al., 2018], see Fig. 15.33). Based on these sources along with population synthesis, it is anticipated that LISA will detect a large number ($\sim 10^4$) of resolved detached DWDs (Nelemans, Yungelson et al., 2001; Farmer & Phinney, 2003; Ruiter et al., 2010; Korol et al., 2017).

Over the past couple of years, observational campaigns using the Zwicky Transient Facility (ZTF) have discovered a rapid growth in the population of known DWDs

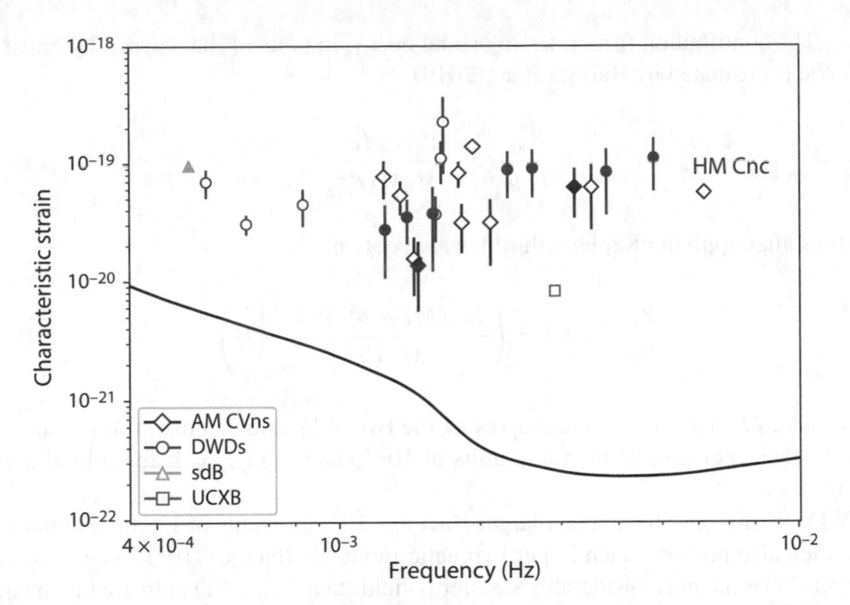

Figure 15.33. Known LISA verification sources as of June 2021. Many sources are seen to have a high signal-to-noise ratio, SNR > 10. The symbols represent various Galactic sources: blue diamonds (AM CVn); red circles (DWD); yellow triangle (sdB); and purple square (UCXB). Notice that here the sensitivity curve is based on 10 yr of observations with a duty cycle of 80%, slightly different than the one applied in Figure 15.26. Filled symbols represent eclipsing sources. Figure credit: T. Kupfer (2021, private communication).

with orbital periods less than an hour. Three of the sources discovered by ZTF so far (Burdge et al., 2020), the eclipsing DWDs: ZTF J1539+5027 ($P_{\rm orb} = 6.91$ min), ZTF J2243+5242 ($P_{\rm orb} = 8.80$ min), and ZTF J0538+1953 ($P_{\rm orb} = 14.4$ min), should all be detected by LISA with a high SNR (Fig. 15.33), enabling precise parameter estimation using GWs (Littenberg & Cornish, 2019). An observational example along the formation path via a single CE of such tight GW sources (Fig. 15.36) is the post-CV system LAMOST J0140355+392651 (El-Badry et al., 2021), which is composed of a bloated (proto)-He ELM WD in orbit with a massive WD. (ELM is an acronym for "extremely-low mass" and refers to a He WD with $M \lesssim 0.20\,M_\odot$.)

The detached DWDs may consist of any combination pair of He WDs, CO WDs and ONeMg WDs. LISA measurements of the orbital-decay rate will yield the chirp mass for a given system, which can be combined with EM observations to reveal individual WD component masses. With a large number of sources detected, the distribution of WD masses will provide important information to help understand their formation history and constrain physical parameters related to, for example, mass-transfer efficiencies and CE physics.

Beware, as mentioned earlier, tides may well contribute to orbital decay as well. Besides producing thermal energy within the WDs, tidal dissipation can transfer orbital energy into rotational energy, causing the orbit to decay slightly faster than via GWs alone. The contribution from tides to orbital decay, in units of that caused by emission of GWs, is estimated by Burdge et al. (2019):

$$\frac{\dot{P}_{\rm tide}}{\dot{P}_{\rm GW}} \simeq \frac{12\pi^2 I a}{G M_1 M_2 P_{\rm orb}^2}, \tag{15.29}$$

and thus after applying Kepler's third law, we obtain:

$$\frac{\dot{P}_{\rm tide}}{\dot{P}_{\rm GW}} \simeq 0.036 \left(\frac{I_{50}\,(M_1+M_2)^{1/3}}{M_1 M_2} \frac{1}{P_{\rm min}^{4/3}} \right), \tag{15.30}$$

where here M_1 and M_2 are the masses of the two WD components in $M_\odot$, I_{50} is the sum the their moments of inertia in units of 10^{50} g cm^2, and $P_{\rm min}$ is the orbital period in minutes.

WDs do not only have much larger (factor $\sim 10^5$) moments of inertia compared to NSs, they also possess much larger magnetic moments (factor $\sim 10^3$) since $\mu = BR^3$. (Despite WDs having considerably smaller B-fields than NSs, WD radii are much larger that those of NSs.) It is therefore possible that tight WD+WD systems may, in some cases, generate a secular electromagnetic contribution to the orbital evolution, leading to modifications of the gravitational waveforms that are potentially detectable by LISA (Bourgoin et al., 2021).

15.12.2 Detached NS+WD and BH+WD Systems

There are two main classes in the known population of Galactic NS+WD systems. There are those systems with: (1) massive WDs (ONeMg or CO WDs, typically more massive than $0.7\,M_\odot$) and those with: (2) low-mass He WDs (typically less massive than $0.3\,M_\odot$). The massive NS+WD systems can be further subdivided into two populations, depending on the formation order of the WD and NS components. The known populations of NS+WD systems are observed as binary radio pulsars, and the formation order of the two compact objects can easily be distinguished by the properties of the pulsar: if the pulsar has a strong B-field and is orbiting in an eccentric orbit (e.g., Tauris & Sennels, 2000; Church et al., 2006; Zapartas et al., 2017; Krishnan et al., 2020), it is the last-formed compact object; whereas if the pulsar is (mildly) recycled with a low-B-field and possess a fairly rapid spin, and in a near-circular orbit, it is the first-formed compact object (van den Heuvel and Taam, 1984; Tauris et al., 2012).

Figure 15.34 displays an example of the rare formation of a system where the WD forms *before* the NS. This may occur in binaries where the two ZAMS stars are fairly equal in mass, but just below the threshold mass for undergoing a SN (e.g., a $7+6\,M_\odot$ system; see also Fig. 16.1), such that the secondary star at a later stage may accrete sufficient mass via RLO to end in a SN. (Similar cases of such *mass reversal* between the stellar components may happen in more massive binaries producing NS+BH systems where the NS forms before the BH.) For LISA detections, the formation order of the NS and the WD is irrelevant, and among both types of systems there are observed

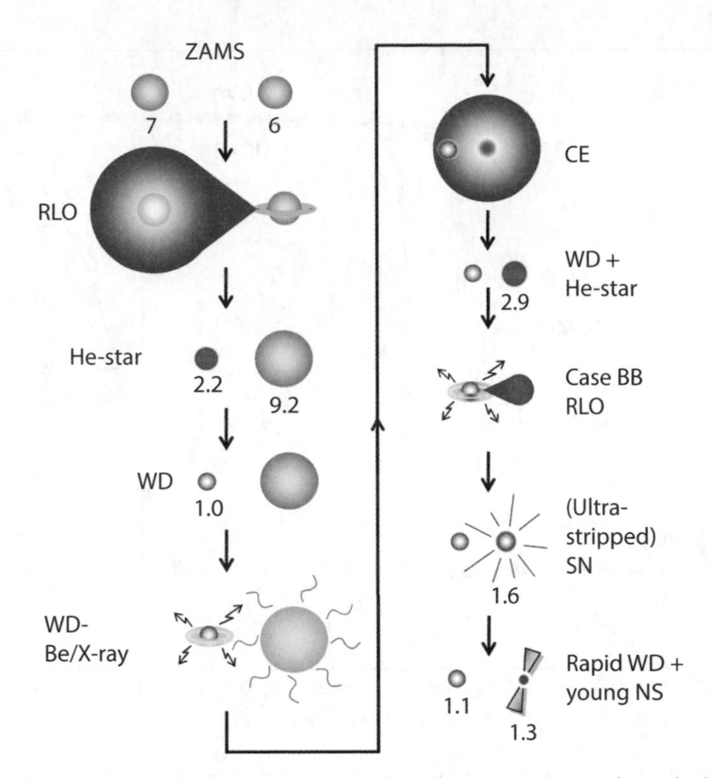

Figure 15.34. Example of binary system undergoing mass reversal, such that the primary star produces a WD whereas the secondary star receives sufficient mass to later produce a NS. Such evolution can explain the formation of the observed WD-Be/X-ray binaries (Section 7.1.1) as well as the eccentric WD+NS systems with a young and non-recycled NS (e.g., PSRs B2303+46 and J1141−6545, see Section 14.7.2). Indicative masses (in units of $M_{\odot}$) are shown at selected evolutionary stages.

examples that are known to merge within a Hubble time, thus producing a bright LISA source well before their final merger.

Binary MSPs detected in radio are completely dominating the detached low-mass He WDs with NS companions in relatively tight orbits. According to the ATNF Pulsar Catalogue (Manchester et al., 2005), there are approximately 120 such systems known in the MW disk, a handful of which will merge within a Hubble time, producing a bright detached LISA source (depending on their distance) for the last 10−20 Myr, or so, of the in-spiral, before the WD fills its Roche lobe and an UCXB is formed (Section 11.4). Based on the observed population of radio pulsars and their selection effects, we have argued (Tauris, 2018) that LISA could detect ~50 of these systems while still being detached, that is, before they become UCXBs and widen their orbits again, resulting in a negative chirp of the GW signal (see Figs. 15.26 and 15.35).

There are no BH+WD binaries detected yet. This is not surprising since these systems are difficult to observe. The only EM radiation we would expect from such

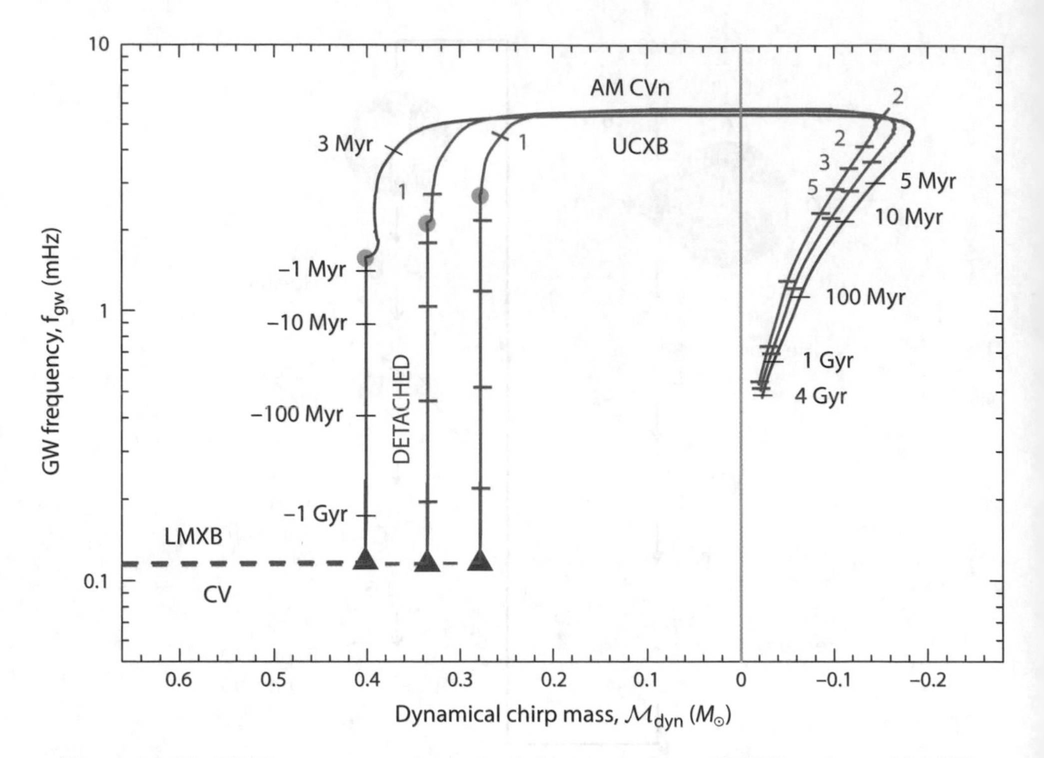

Figure 15.35. GW frequency vs. dynamical chirp mass for an UCXB and two AM CVn systems, based on detailed mass-transfer calculations (including finite-temperature effects of the WD donor star) using the MESA code. The end points of the first mass-transfer phases (LMXB and CV) are indicated by red triangles; the starting points of the second mass-transfer phases (UCXB and AM CVn) are indicated by green circles. The time stamps along the AM CVn tracks are for the same values as indicated for the UCXB system, unless stated otherwise (in Myr). Time zero is defined at the onset of the second mass-transfer phase. The maximum GW frequencies (i.e., strongest LISA signal) in these three examples are 5.45 mHz (UCXB), 5.64 mHz (AM CVn1), and 5.72 mHz (AM CVn2) corresponding to orbital periods of 6.12 min, 5.91 min, and 5.83 min, respectively. The frequency at the onset of the RLO (green circles) depends on the temperature of the low-mass He WD donor ($T_{\rm eff} = 10{,}850$ K, 9,965 K, and 8,999 K, respectively). After Tauris (2018).

detached systems, would be from the cooling of the WD companion—unlike the situation for semi-detached systems or systems containing NSs, which can be detected in X-rays and radio waves. Nevertheless, several Galactic LMXBs are known with low-mass donor stars and BH accretors (see Chapters 6 and 7, and McClintock & Remillard [2006]), and some of these systems may eventually leave detached BH+WD systems—although it is still to be proven that these systems would produce tight orbits that LISA can detect. A more viable formation channel for more massive WDs in tight orbits with BHs is formation via a CE. The BH mass in these systems can be

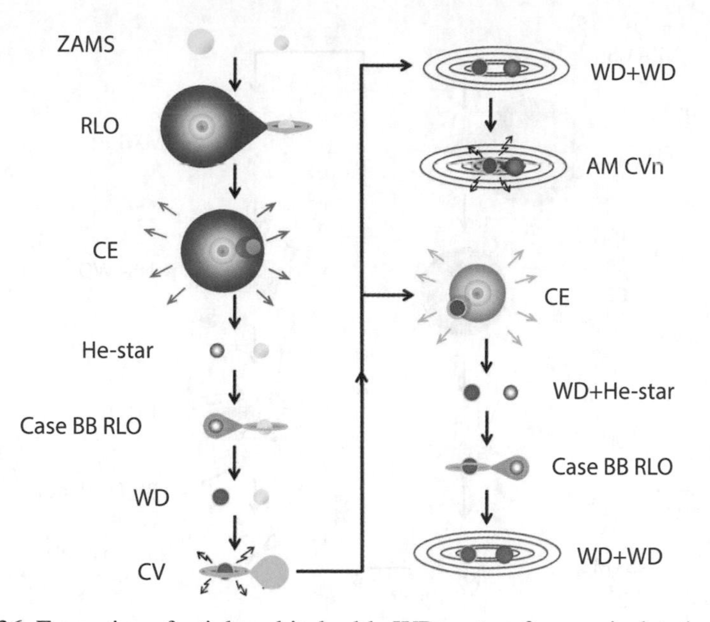

Figure 15.36. Formation of a tight-orbit double WD system from an isolated progenitor binary requires one or two epochs of CE evolution.

constrained from a combination of the measured chirp mass and optical follow-up observations of the WD companion. Simulations by Nelemans, Yungelson, Portegies Zwart (2001) predicted a Galactic merger rate of BH+WD binaries of the order of $\sim 100\,\mathrm{Myr}^{-1}$ and thus it is expected that roughly ~ 100 such systems will be detectable by LISA.

15.12.3 The Interacting Systems: AM CVns and UCXBs

It has been known for many years that tight-orbit post-CV/LMXB systems, leaving behind a DWD or NS+WD binary (often a binary MSP) that spirals-in due to GW radiation, may avoid a catastrophic event once the WD fills its Roche lobe. The outcome is instead expected to be a long-lived AM CVn or UCXB source. AM CVn (or AM Canum Venaticorum star) systems were introduced in Section 5.4 and consist of a WD accreting from a hydrogen-deficient star (or WD) companion in a close orbit. Their formation history includes one or two episodes of CE evolution (see Figs. 15.36 and 15.37). UCXBs and their formation process were discussed in Sections 7.8.4 and 11.4. These systems descend from tight-orbit MSPs and are quite analogus to AM CVns, except that their accreting compact object is a NS. Because AM CVns and UCXBs are interacting binaries, they are excellent multi-messenger sources for combined EM (optical and X-rays) and GW observations (Nelemans et al., 2004). Depending on the nature and the temperature of the companion star (and thus orbital period), the active mass transfer in current AM CVns/UCXBs is initiated due to orbital damping caused by GW radiation, at orbital periods of typically 5–20 min.

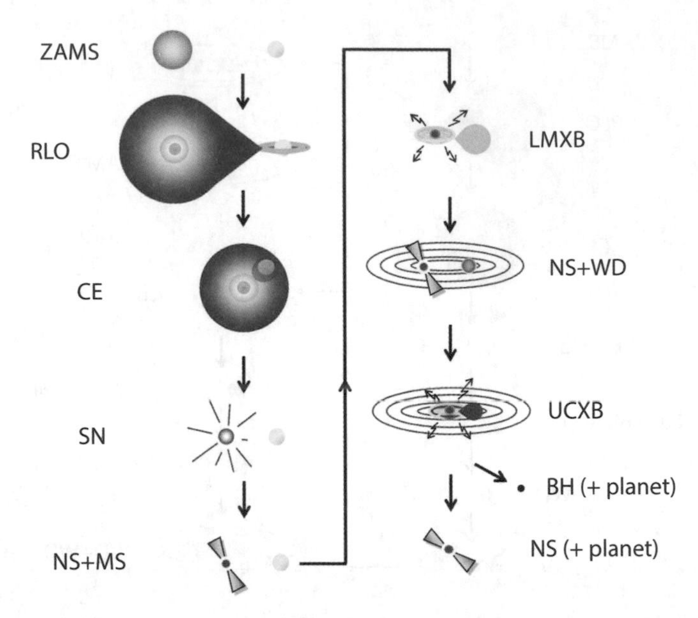

Figure 15.37. UCXBs descend from LMXBs that detach in tight orbits (producing MSPs), decaying due to GW radiation until the WD fills its Roche lobe.

For systems with a low-mass He WD donor, the cooling of such a WD determines its radius (and thus orbital period) at the onset of RLO. Systems that detach from the previous CV/LMXB stage in relatively wide orbits will allow for the He WD to cool and shrink in size, such that it will only initiate RLO at a very short orbital period ($\sim 5 - 15$ min), compared to systems that detach from the CV/LMXB stage in a relatively tight orbit, thereby initiating RLO shortly thereafter while the (proto) He WD is still bloated and thus in a relatively wide orbit ($P_{\rm orb} \simeq 30 - 50$ min), as demonstrated by Sengar et al. (2017). In other words, the wider the post-CV/LMXB orbit, the longer it takes to reach the AM CVn/UCXB stage via GW radiation, the cooler and smaller in size the WD is, and thus the tighter the orbit when it becomes an AM CVn/UCXB source. For DWDs with two CO WDs, the cooling is rapid.

The mass-transfer rate is determined by a competition between orbital angular momentum loss through emission of GWs and orbital widening due to RLO from the less massive WD donor star to the more massive WD or NS accretor. If the system survives the onset of the semi-detached phase, a stable accreting AM CVn/UCXB binary is formed in which the orbital separation widens shortly after onset of RLO, and the system evolves to longer orbital periods on a Gyr timescale (Tauris, 2018; Chen et al., 2021), see Figures 15.26 and 15.35. For this reason, in particular the numerous AM CVn sources are prime candidates for being bright sources in GWs, and thus there are high expectations for LISA to detect many of these sources. Searches for deep eclipses of Zwicky Transient Facility light curves of WDs selected using *Gaia* parallaxes have recently proven to be a succesful way to discover new eclipsing AM CVn systems (van Roestel et al., 2021). AM CVns/UCXBs will be extremely

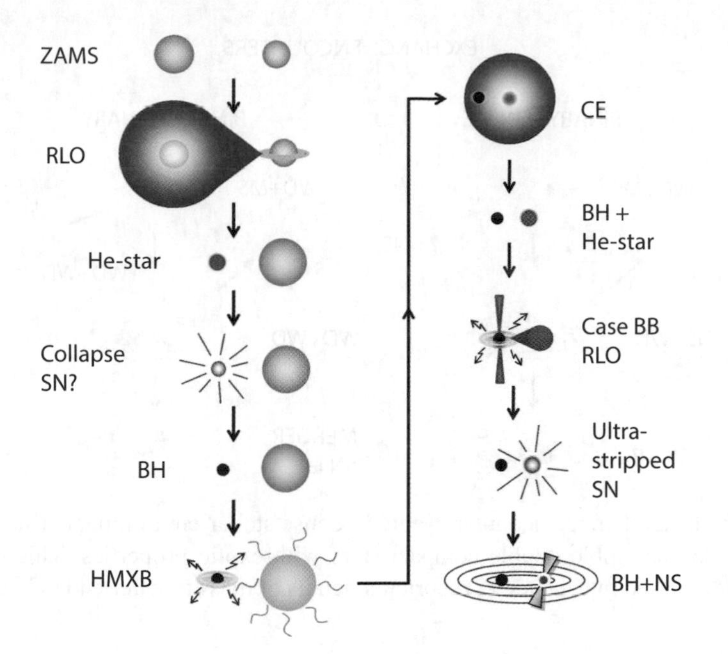

Figure 15.38. Progenitors of BH/NS binaries experience two SNe, the latter often being an ultra-stripped SN. A final phase of mass-transfer (Case BB RLO) and/or tides determine the spins of the first/second-formed BH/NS.

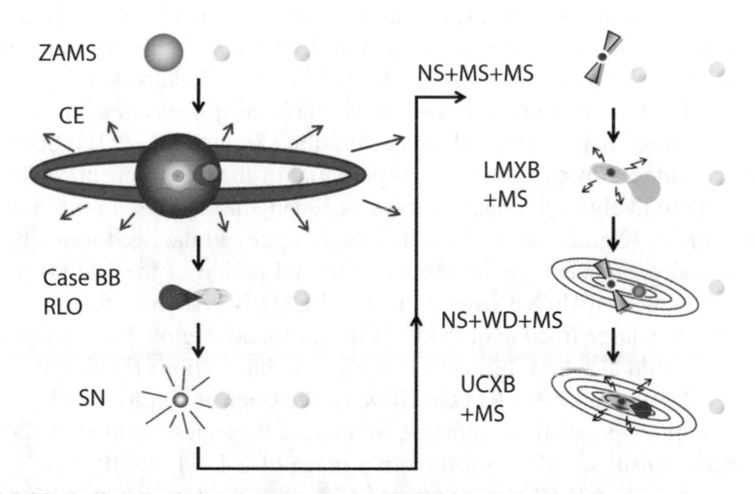

Figure 15.39. Isolated triple stars may eventually produce a tight inner binary, e.g., an UCXB, detectable in GWs. Modelling these systems is complicated due to a triple-star CE, a SN, and the requirement of three-body dynamical stability.

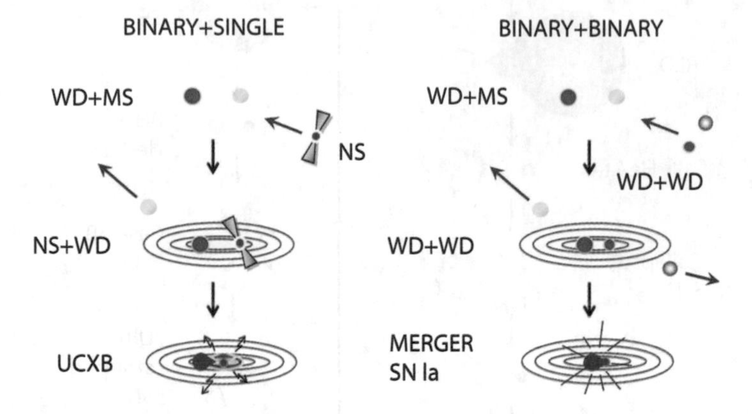

Figure 15.40. Exchange encounter events in dense stellar environments (like globular clusters) may assemble double compact stars with exotic properties. This formation channel (discussed in Chapter 12) shortcuts details of mass transfer and effects of SNe.

useful for LISA because they simultaneously provide EM information across different wavelengths, as well as being observable in LISA's GW frequency range.

It should be noticed that the measured time derivative of the GW frequency ($\dot{f}_{\rm GW}$) will be a combination of a pure general relativistic and an astrophysical contribution. The latter contribution arises from mass transfer and tides, and thus the combined signal leads to a significant deviation from the contribution from GR alone (Breivik et al., 2018). The decoupling of the astrophysical chirp from the GR chirp will in many cases be possible with combined measurements from LISA and Gaia. Note that effects from acceleration of the LISA sources may affect the GW chirp measurements and lead to degeneracy between mass and peculiar acceleration (Xuan et al., 2021). Such a variation could be induced by the gravitational potential of the environment, such as a star cluster, and more likely by a tertiary object since many binaries are in triple systems.

Earlier studies (e.g., Istrate et al., 2014) have suggested the need for extreme fine-tuning of initial parameters (stellar mass and orbital period of the LMXB progenitor systems) to produce an UCXB from an isolated LMXB system in the disk (see Section 11.3). Since a large fraction of the UCXBs are found in globular clusters, some of these systems could also have formed via dynamical interactions (Fabian et al., 1975), see Chapter 12. For stability of RLO in UCXBs, see Chen et al. (2022).

Based on binary population synthesis, Nelemans, Portegies Zwart et al. (2001) predicted a space density of AM CVn stars in a range of $0.4 - 1.7 \times 10^{-4}\,\rm pc^{-3}$, and the number of resolvable AM CVn systems for LISA was found to be equal to the number of detached DWDs (Nelemans et al., 2004). More recently, Kremer et al. (2017) predict that $\sim$2700 systems will be observable by LISA with a negative chirp of $0.1\,\rm yr^{-2}$ (i.e., $\dot{f}_{\rm GW} < 0$, resulting from orbital expansion due to mass transfer). We have argued that the number of UCXBs detected by LISA is expected to be approximately 50 to 100 sources (Tauris, 2018). Somewhat more optimistic numbers (up to 320 sources) were predicted by Chen et al. (2020).

15.12.4 DNS Systems

As discussed in detail in Chapter 14, the known DNS systems so far only manifest themselves as radio pulsars. There are currently $\sim$22 DNS systems detected in our Galaxy. Except for one case, the *double pulsar* PSR J0737$-$3039 (Lyne et al., 2004), only one of the two NSs is detected—usually the recycled pulsar (Tauris et al., 2017). The other pulsar is either not beaming in our direction, or, more likely, it is not an active radio pulsar anymore since non-recycled pulsars fade away on a timescale of $\sim 10 - 100$ Myr (Srinivasan and van den Heuvel, 1982; Chen & Ruderman, 1993).

About half of the 22 known DNS systems have orbital periods small enough (or eccentricities sufficiently large) to merge within a Hubble time. As an example, a typical DNS system with NS masses of $1.35\,M_\odot$ and, for example, an orbital period of 16 hr will merge in 11.8 Gyr, 4.4 Gyr, or 0.35 Gyr for an initial eccentricity, e_0 of 0.1, 0.5, or 0.8, respectively. The number of DNS sources that LISA will detect can be evaluated, approximately to first order, simply from a combination of the Galactic DNS merger rate and the distribution of these sources within the MW. The three above-mentioned cases of DNS systems will have a remaining lifetime of $\sim$247 kyr (or slightly less, $\sim$243 kyr for $e_0 = 0.8$ as this system circularizes the least) by the time they enter the LISA band, if this occurs at a GW frequency of 2 mHz. If they enter the LISA band already at 1 mHz, the three systems will have remaining lifetimes of between 1.48–1.57 Myr. Thus, if the Galactic merger rate is, say, $10\,\mathrm{Myr}^{-1}$, we can roughly expect to detect between a few and a dozen LISA sources. Of course, the details depend on the Galactic distribution of these DNS sources, the SNR required for a detection and the duration of the LISA mission. The merger rate can be estimated from population synthesis (e.g., Table 15.1), but its value is uncertain by, at least, one or two orders of magnitude (Abadie et al., 2010; Kruckow et al., 2018); see discussions in Section 16.3. The merger rate derived from an extrapolation of the LIGO–Virgo empirical merger rate of DNS systems still has very large error bars due to small number statistics (see Table 15.5).

Korol & Safarzadeh (2021) explored in detail how we can probe a population of Galactic GW190425-like DNSs with LISA and investigate their origin. Other works by Lau et al. (2020); Andrews et al. (2020) suggest that LISA may even detect up to $\sim 50 - 200$ DNS sources with a SNR greater than 7 within a 4 yr mission. Applying a more conservative number for the merger rate of Galactic DNS systems of $\sim 3 - 14\,\mathrm{Myr}^{-1}$ (Kruckow et al., 2018) would lead to a substantial reduction in the predicted number of LISA detections. The Galactic merger rate is expected to be significantly better constrained in the coming decade such that we will have a clear idea about the expected number of DNS sources detected by LISA prior to its operation.

15.12.5 BH+NS and Double BH Systems

Predictions for BH+BH and BH+NS GW sources in our Galactic backyard detectable with LISA were recently carried out by Wagg et al. (2022). They estimate up to a few

100 detections of each kind of sources (and up to about 50 DNS systems) in the most optimistic case. A number of HMXBs containing BH accretors have been identified in the MW and nearby galaxies (e.g., Cyg X-1, LMC X-1, LMC X-3, MCW 656, M33 X-7, see Chapters 6 and 10), and we have discussed that a fraction of these known wind-accreting HMXBs may eventually produce double BHs or BH+NS systems, while other will merge in an upcoming CE phase (Section 2.7 and Chapter 10) once the companion star evolves to giant-star dimensions and initiates dynamically unstable RLO (depending on its stellar structure and the mass ratio between the two binary components). The masses of compact objects in X-ray binaries can be estimated by a combination of spectroscopy and photometry of their donor stars (Chapter 6), and the Galactic stellar-mass BHs are found to have masses roughly in the range $5-21\ M_\odot$.

Finally, non-interacting binaries with a BH component can also, in principle, be discovered by combining radial velocity measurements with photometric variability data (e.g., Thompson et al., 2019; Rivinius et al., 2020), although their interpretations can in some cases be subject to alternative explanations, as we have demonstrated (van den Heuvel & Tauris, 2020)—see also Section 6.6.2.

As discussed earlier in this chapter, the LIGO network of high-frequency detectors has discovered double BH mergers in distant galaxies, out to ~8 Gpc (Abbott et al., 2021d), with inferred BH masses ranging from $\sim 2.6\ M_\odot$ to $\sim 107\ M_\odot$, thus significantly more massive than the known Galactic stellar-mass BHs (see Tables 6.5 and 6.6). This difference is mainly attributed to the relatively high metallicity content of the MW. As mentioned earlier, although the detected GW sources are located far away in distant galaxies, their low-frequency GW radiation during the last few years of in-spiral prior to the merger event is often so luminous that it allows for detection with LISA (Sesana, 2016), see Figure 15.26 for GW150914. The extreme events GW190426 (Abbott et al., 2021) and GW190521 (Abbott et al., 2020a), with total BH component masses of ~184 and 164 $M_\odot$, respectively, and both located at distances of ~4 Gpc, would possibly also have been marginally detected during their in-spiral in the LISA band.

15.13 LISA SENSITIVITY CURVE AND SOURCE STRAIN

The sensitivity of GW detectors is limited by various sources of noise. Ground-based detectors are limited by seismic and suspension noise, test mass thermal noise, and photon shot noise. A space-borne interferometer detector (LISA) is limited by the optical metrology noise, test mass acceleration noise, and the confusion noise from unresolved Galactic binaries (Robson et al., 2019). The internal instrument noise can be non-Gaussian or Gaussian in nature. The non-Gaussian noise is, for example, strain releases in the suspension systems that isolate the detector from environmental mechanical noise. This happens several times per day and can only be removed from comparison to other detectors. The Gaussian noise (obeying the probability distribution of Gaussian statistics) can be characterized by a *spectral density*, $S_n(f)$.

The observed signal (the strain data stream, $x(t)$) at the detector is a combination of a true GW strain, $h(t)$ and noise, $n(t)$, that is, $x(t)=h(t)+n(t)$. (For an estimate of $h(t)$ for astrophysical sources, see Sections 15.3.1 and 15.13.1.) The signal-to-noise ratio is given by Maggiore (2008):

$$\rho^2 = \left(\frac{S}{N}\right)^2 = 4\int_0^\infty \frac{|\tilde{h}(f)|^2}{S_n(f)}\,df = 4\int_{f=0}^\infty \frac{f|\tilde{h}(f)|^2}{S_n(f)}\,d(\ln f), \tag{15.31}$$

where $\tilde{h}(f)$ is the Fourier transform of the signal waveform given by:

$$\tilde{h}(f) \equiv \int_{-\infty}^{\infty} h(t)\,e^{-i\omega t}\,dt, \tag{15.32}$$

and the noise power spectral density can be expressed as:

$$S_n(f) = |\tilde{n}(f)|^2. \tag{15.33}$$

In this section, we focus on the LISA sensitivity curves. In the LISA community, the above $S_n(f)$ is called $P_n(f)$, and hence the full LISA sensitivity curve, or strain spectral density, is given by (see, e.g., Robson et al., 2019, for a review):

$$S_n(f) = \frac{P_n(f)}{\mathcal{R}(f)} + S_c(f), \tag{15.34}$$

where $\mathcal{R}(f)$ is the signal response (transfer) function that relates the power spectral density of the incident GW signal to the power spectral density of the signal recorded in the detector. Hence, $\mathcal{R}(f)$ is a function of detector geometry and antenna patterns. The last term in the above equation, $S_c(f)$, is the Galactic confusion noise from unresolved binaries (Nelemans, Yungelson et al., 2001), which will decrease over mission time.

For the design of a 3-arm LISA (Amaro-Seoane et al., 2017), there are two and three independent channels for $f < f_*$ and $f > f_*$, respectively. The transfer frequency, $f_* = c/(2\pi L)$ and for current LISA design, $L = 2.5 \times 10^9$ m, yielding $f_* = 19.09$ mHz.

The response function can be fitted by:

$$\mathcal{R}(f) = \frac{3}{10}\frac{1}{(1 + 0.6\,(f/f_*)^2)}, \tag{15.35}$$

while the numerically computed signal response functions for the combination of two Michelson-style LISA data channels are provided in the literature, for example, Cornish & Robson (2018). Notice that $\mathcal{R}^{\rm LIGO} = 1/5$ and $\mathcal{R}^{\rm LISA} = 1/5 \times 2 \times (\sqrt{3}/2)^2 = 3/10$, where the factor of 2 is from the number of channels, and $\sin 60° = \sqrt{3}/2$ is from a triangular detector formation.

The power spectral density of the LISA noise, $P_n(f)$, arises from a combination of so-called metrology noise, $P_{\rm OMS}$, and the single test-mass acceleration noise, $P_{\rm acc}$, yielding a total noise:

$$P_n(f) = \frac{P_{\rm OMS}}{L^2} + 2\,(1 + \cos^2(f/f_*))\,\frac{P_{\rm acc}}{(2\pi f)^4\,L^2}, \tag{15.36}$$

where

$$P_{\rm OMS} = (1.5 \times 10^{-11}\,{\rm m})^2 \left(1 + \left(\frac{2\,{\rm mHz}}{f}\right)^4\right) {\rm Hz}^{-1} \tag{15.37}$$

and

$$P_{\rm acc} = (3 \times 10^{-15}\,{\rm m\,s^{-2}})^2 \left(1 + \left(\frac{0.4\,{\rm mHz}}{f}\right)^2\right)\left(1 + \left(\frac{f}{8\,{\rm mHz}}\right)^4\right) {\rm Hz}^{-1}. \tag{15.38}$$

Following Berti et al. (2005), the Fourier transform of the waveform for one Michelson LISA detector, in the stationary phase approximation and averaged over the pattern functions, is given by:

$$\tilde{h}_\alpha(f) = \frac{\sqrt{3}}{2}\,\mathcal{A}\,f^{-7/6}\,e^{i\psi(f)}, \tag{15.39}$$

where

$$\mathcal{A} = \sqrt{\frac{1}{30}}\,\frac{(G\mathcal{M}/c^3)^{5/6}}{\pi^{2/3}\,(d_L/c)}, \tag{15.40}$$

and where f is the GW frequency during in-spiral, $\mathcal{M}$ is the chirp mass, $M = M_1 + M_2$ is the total mass, and d_L is the luminosity distance to the source. The phasing function, $\psi(f)$, is known for point masses up to 3.5 post-Newtonian (PN) order (including spin terms up to 2 PN order).

The values of $\mathcal{M}$ and M are in the detector reference frame. They are related to the values in the source rest frame by:

$$\mathcal{M} = (1+z)\,\mathcal{M}_{\rm source} \qquad \text{and} \qquad M = (1+z)\,M_{\rm source}, \tag{15.41}$$

where z is the cosmological redshift. For a zero-spatial-curvature Universe ($\Omega_k = 0$, $\Omega_\Lambda + \Omega_M = 1$), the luminosity distance is given by (see Eq. 15.27):

$$d_L = \frac{1+z}{H_0}\int_0^z \frac{dz'}{\sqrt{\Omega_M\,(1+z')^3 + \Omega_\Lambda}}, \tag{15.42}$$

where we can assume $H_0 = 72\,{\rm km\,s^{-1}\,Mpc^{-1}}$ for the present value of the Hubble constant.

With a given noise spectral density for the detector, $S_n(f)$, one defines the inner product between the two signals (one for each of the two independent Michelson outputs for a three-arm detector with symmetric noise):

$$\langle h_1, h_2\rangle \equiv 2\int_0^\infty \frac{\tilde{h}_1^*\tilde{h}_2 + \tilde{h}_2^*\tilde{h}_1}{S_n(f)}\,df, \tag{15.43}$$

where $\tilde{h}_1(f)$ and $\tilde{h}_2(f)$ are the Fourier transforms of the respective gravitational waveforms, $h(t)$.

The signal-to-noise ratio (SNR) for a given h is then given by:

$$\sqrt{\langle \rho^2 \rangle} \equiv \langle h, h \rangle. \tag{15.44}$$

15.13.1 Characteristic Strain

The dimensionless characteristic strain amplitude, h_c, is introduced to include the effect of integrating an in-spiraling signal. One can distinguish the cases of a stationary (monochromatic) source and an evolving source.

Stationary Source—Galactic Binary

If the change in GW frequency, $f_{\rm GW}$, is small (compared to the duration of the observation time, T), one can simply estimate the characteristic strain as:

$$h_c = \sqrt{N_{\rm cycles}}\, h, \tag{15.45}$$

where h is the dimensionless GW source amplitude and the number of detected cycles is $N_{\rm cycles} = f_{\rm GW} T$. For LISA, it is often expected that $T = 4\,{\rm yr}$. The dimensionless GW amplitude generated by a binary at luminosity distance, d_L, for an average orbital orientation and polarization is given by:

$$h = \left(\frac{32}{5}\right)^{1/2} \frac{\pi^{2/3} G^{5/3} f_{\rm GW}^{2/3} \mathcal{M}^{5/3}}{c^4\, d_L}. \tag{15.46}$$

This expression can be derived (e.g., Evans et al., 1987) by combining:

$$h = \sqrt{\frac{4 G F_{\rm GW}}{\pi\, c^3\, f_{\rm GW}^2}}, \tag{15.47}$$

where the GW energy flux is given by:

$$F_{\rm GW} = \frac{L_{\rm GW}}{4\pi\, d_L^2}, \tag{15.48}$$

and the luminosity, $L_{\rm GW}$, is given by Eq. (15.4); and applying Kepler's third law, as well as $f_{\rm GW} = 2\, f_{\rm orb}$ and $\mu\, M^{2/3} = \mathcal{M}^{5/3}$ and assuming a circular orbit, that is, $f(e) = 1$ (see Exercise 15.5).

Another often plotted quantity in the literature is the source equivalent of the noise amplitude spectral density ($\sqrt{S_n(f)}$, which is the square root of the power spectral density, see Eq. (15.33), and has the unit of ${\rm Hz}^{-1/2}$), i.e., a source amplitude spectral density, $\sqrt{S_h(f)}$ (Moore et al., 2015):

$$\sqrt{S_h(f)} = h_c\, f^{1/2}. \tag{15.49}$$

Considering a one-sided power spectrum rather than two-sided, a factor $\sqrt{2T}$ can be multiplied with Eq. (15.46) for a monochromatic GW source to yield the strain spectral density for a Galactic binary (Robson et al., 2019):

$$h_{\rm GB} = \left(\frac{64}{5}\right)^{1/2} \frac{\pi^{2/3} G^{5/3} f_{\rm GW}^{2/3} \mathcal{M}^{5/3}}{c^4\, d_L} \sqrt{T}. \tag{15.50}$$

The advantage of plotting the dimensionless $h_c = \sqrt{f\, h_{\rm GB}^2}$ in a characteristic strain vs. frequency plot (Figs. 15.26 and 15.33) is that the SNR of the GW source (i.e., how loud the signal is) is simply given by the ratio of the height of the point and the height of the sensitivity curve (Exercise 15.6). The downside of plotting characteristic strain is that the values on the strain axis do not directly relate to the amplitude of the waves from the source (Moore et al., 2015).

Notice that LISA is able to measure a chirp magnitude, $\dot{f}_{\rm GW}$, greater than $\sim$0.1 bin yr^{-1} (1 bin $= T^{-1}$), corresponding to a minimum value of $\dot{f}_{\rm GW} \simeq 8 \times 10^{-10}$ Hz yr^{-1} for a four years mission (Kremer et al., 2017).

Finally, it should be mentioned that we only considered circular binaries when estimating the GW strain. For eccentric binaries like DNS systems, GWs are emitted at multiple harmonics, which makes the GW strain estimate more complicated (Randall et al., 2022; Korol & Safarzadeh, 2021).

For (almost) monochromatic binaries with little chirping during the observation time, meaning that the peak frequency, $f_{\rm p}$ and the eccentricity, e are relatively constant, Randall & Xianyu (2021); Randall et al. (2022) argued that simple formulae apply to approximate the sum over harmonics for any value of eccentricity ($0 \leq e < 1$) such that for the characteristic strain and the SNR:

$$\frac{h^2_{c(f_{\rm p},\, e)}}{h^2_{c(f_{\rm p},\, e=0)}} \simeq (1-e)^{3/2}, \qquad \frac{\rho_{(f_{\rm p},\, e)}}{\rho_{(f_{\rm p},\, e=0)}} \simeq (1-e)^{3/4}. \tag{15.51}$$

Evolving Source

A couple of evolving sources (i.e., binaries for which $f_{\rm GW}$ changes significantly over the observation time) are plotted in Figure 15.26 for comparison. Two examples are given for the last 4 yr of in-spiral. The first example is a hypothetical super-massive black hole (SMBH) binary with a total mass of $10^6\, M_\odot$ at redshift $z = 3$. For such systems, after correcting for the redshift, the waveform model of the original phenomenological model called PhenomA (Ajith et al., 2007) can be implemented (Robson et al., 2019) for the full in-spiral / merger / ring-down evolution. The other example plotted is the last four years of in-spiral of GW150914. It is indeed interesting to notice that a number of extragalactic stellar-mass double BH mergers (such as GW150914 or GW190521) are detectable in the LISA band several years before they enter and merge in the high-frequency GW band detectable by LIGO (Sesana, 2016; Toubiana et al., 2021).

Following Robson et al. (2019), one can use Eq. (15.31) and average over sky location, inclination and polarization, to obtain:

$$\overline{\rho^2} = \frac{16}{5} \int_0^{\infty} \frac{f\,\mathcal{A}^2}{S_n(f)}\, d(\ln f) = \frac{16}{5} \int_0^{\infty} \frac{f^2\,\mathcal{A}^2}{f\,S_n(f)}\, d(\ln f), \tag{15.52}$$

and thus a characteristic strain:

$$h_c = \sqrt{\frac{16}{5}}\, f\,\mathcal{A}, \tag{15.53}$$

given that:

$$\rho^2 = \int \frac{|h_c|^2}{|h_n|^2}\, d(\ln f) \tag{15.54}$$

and

$$h_c = 2f\,|\tilde{h}| \qquad \text{and} \qquad h_n = \sqrt{f\,S_n}. \tag{15.55}$$

The convention is often to plot:

$$h_c = \sqrt{f\,h_{\rm eff}^2}, \tag{15.56}$$

where

$$h_{\rm eff}^2 = \frac{16\,(2fT)\,S_h(f)}{5} \tag{15.57}$$

and

$$S_h(f) = \frac{2\langle \tilde{h}(f)\,,\tilde{h}^*(f)\rangle}{T} \tag{15.58}$$

and where

$$\langle \tilde{h}(f)\,,\tilde{h}^*(f)\rangle = \frac{4}{5}\,\mathcal{R}(f)\,\mathcal{A}^2(f). \tag{15.59}$$

In the low-frequency approximation ($f \ll f_*$, and hence $\mathcal{R} = 3/10$), this yields:

$$h_{\rm eff}^2 = \frac{256}{25}\,\mathcal{R}\,f\,\mathcal{A}^2, \tag{15.60}$$

and thus:

$$h_c = \sqrt{\frac{382}{125}}\, f\, \mathcal{A}. \tag{15.61}$$

We notice that following Berti et al. (2005):

$$h_c = 2\sqrt{\left(\frac{4}{5}\mathcal{R}\right)}\, \mathcal{A}, \tag{15.62}$$

where $\mathcal{R} = 3/20$ since there is no summing over two data channels, and hence:

$$h_c = \sqrt{\frac{12}{25}}\, f\, \mathcal{A}. \tag{15.63}$$

The difference between these two expressions for h_c in Eqs. (15.61) and (15.63) is a factor: $\sqrt{32/5}$. This matches the difference of the h_c curves plotted in Figures 1 and 6 in Robson et al. (2019).

Recently, a convenient python package tool (LEGWORK) for computing the evolution and LISA detectablity of stellar-origin GW sources has been made public (Wagg et al. (2022).

EXERCISES

Exercise 15.1. *The relative rate of change of the orbital angular momentum in a circular binary due to emission of GWs is given by Eq. (15.10).*

a. *Show that this equation can be combined with the expression for orbital angular momentum (J) and Kepler's third law, and then integrated to reveal a merger timescale:*

$$\tau_{\rm GW} = \frac{1}{8}\left|\frac{J}{\dot{J}}\right|. \tag{15.64}$$

b. *Similarly, using Eq. (14.35) for $e = 0$, show that:*

$$\tau_{\rm GW} = \frac{3}{8}\left|\frac{P_{\rm orb}}{\dot{P}_{\rm orb}}\right| \simeq 0.0713\,{\rm Myr}\left|\frac{P_{\rm min}}{\dot{P}_{-11}}\right|, \tag{15.65}$$

where $P_{\rm min}$ is the orbital period in minutes and $\dot{P}_{-11}$ is the orbital period decay rate in units of $10^{-11}\,{\rm s\,s^{-1}}$. This expression is particularly useful for DWD ($e \simeq 0$) LISA binaries.

c. *For a circular binary, rewrite the expressions in Section 15.2 and show:*

$$\tau_{\rm GW} \simeq 47.1\,{\rm Gyr}\left(\frac{M_\odot}{\mathcal{M}}\right)^{5/3}\left(\frac{P_{\rm orb}}{\rm day}\right)^{8/3}. \qquad (15.66)$$

Exercise 15.2. *Calculate the GW merger time of the Hulse-Taylor pulsar (PSR B1913 +16 with $P_{\rm orb} = 7.75$ hr, $e = 0.617$, $M_{\rm NS,1} = 1.441\,M_\odot$, and $M_{\rm NS,2} = 1.387\,M_\odot$) and compare to a system, with similar NS masses and $P_{\rm orb}$, in a circular orbit ($e = 0$) and a system with $e = 0.90$.*

Exercise 15.3. *Consider the system in Exercise 13.1.*

a. *Calculate the GW merger time of this BH+BH system.*
b. *Calculate the GW merger time of this BH+BH system if the post-SN $P_{\rm orb} = 2.0$ d.*
c. *Calculate the GW merger time of this BH+BH system if the post-SN $P_{\rm orb} = 1.0$ d.*
d. *Discuss the implications of large misalignment angles, δ, of BH+BH systems on the measurement of effective spin, $\chi_{\rm eff}$, of BH+BH mergers.*

Exercise 15.4. *Consider the system in Exercise 13.10 and assume a kick magnitude of $w = 375\,{\rm km\,s^{-1}}$ and $\theta = 150°$. Using the equations from Section 13.8, calculate for the post-SN binary:*

a. *the post-SN orbital period, $P_{\rm orb}$*
b. *the eccentricity, e*
c. *the 3D systemic recoil velocity of the system, $v_{\rm sys}$*
d. *the misalignment angle, δ*
e. *the GW merger timescale, $\tau_{\rm GW}$*
f. *Discuss the implications for the effective spin parameter, $\chi_{\rm eff}$, of this BH+BH system in relation to the ongoing debate on the origin of GW mergers (i.e., an isolated binary origin vs. a dynamical origin). Can one rule out BH+BH binaries having an isolated binary origin if the LIGO network detects such a merger with a significant value of $\chi_{\rm eff}$? (Hint: consider the calculated misalignment angle, δ.)*

Exercise 15.5. *Derive Eq. (15.46) and show that the GW amplitude, h, is dimensionless.*

Exercise 15.6. *Estimate the expected SNR detected by LISA, during a 4 yr mission, of the 6.91 min orbital period double WD system ZTF J1539+5027 (Burdge et al., 2019), which has component masses of 0.61 and $0.21\,M_\odot$ and a distance of 2.3 kpc.*
(Hint: calculate the amplitude of the integrated GW signal and compare to the sensitivity curve in Fig. 15.26.)

Exercise 15.7. *Estimate the minimum orbital period of a Galactic circular DNS system located at a distance of 1 kpc to be detected by LISA in a* $T = 4\,\mathrm{yr}$ *mission, assuming a SNR of at least 10 and 4, respectively.*

Exercise 15.8. *Estimate the loss of SNR (signal-to-noise ratio) for an eccentric DNS system with* $e = 0.5$*, 0.7 and 0.9, compared to a circular (*$e = 0$*) DNS system with the same orbital period and NS masses.*

Exercise 15.9. *Following the detection of GW170817 in observing run O2, LIGO announced an empirical GW merger-rate density of DNS systems in the local Universe of an astonishing* $\mathcal{R}_{\mathrm{DNS}} = 1540^{+3200}_{-1220}\,\mathrm{Gpc^{-3}\,yr^{-1}}$*. The central value corresponds to a rate of the order of* $\sim 150 - 450\,\mathrm{Myr^{-1}}$ *for a MWEG. In our Galaxy, the DNS system with the shortest known merger time is J1946+2052 (Stovall et al., 2018) with* $\tau_{\mathrm{GW}} = 46\,\mathrm{Myr}$*. If the announced central value for the DNS merger rate is correct, at least* $\sim$*7,000 DNS binaries should exist in the MW with a merger time shorter than that of J1946+2052.*

a. *Verify the above number and discuss whether it is realistic that all these systems should have escaped detection as radio pulsar sources?*
   (Notice, following LIGO O3 (GWTC-3) the new empirical rate was estimated to be $\mathcal{R}_{\mathrm{DNS}} = 99^{+260}_{-86}\,\mathrm{Gpc^{-3}\,yr^{-1}}$*, see Table 15.5).*
b. *The LIGO network observing run O3a had detector sensitivity ranges at Livingston and Hanford of roughly 100 Mpc for a DNS merger. The survey lasted* $\sim$*1 yr. Show that the DNS merger-rate density based on only one firm detection of a DNS merger (GW190425) in O3a, results in a significantly lower value of* $\mathcal{R}_{\mathrm{DNS}} \simeq 250\,\mathrm{Gpc^{-3}\,yr^{-1}}$ *compared to the value of* $\mathcal{R}_{\mathrm{DNS}} = 1540^{+3200}_{-1220}\,\mathrm{Gpc^{-3}\,yr^{-1}}$ *from LIGO O2.*
   (Hint: the merger-rate density is the ratio between the number of events detected and the volume-time [VT] of the survey. Note, in fact the DNS in-spiral ranges were slightly larger than 100 Mpc, but on the other hand the duty cycles of the two LIGO detectors were only 71%–76%.)
c. *Assuming that there are 0.01 MWEGs in* $1\,\mathrm{Mpc^{-3}}$ *of space (Kopparapu et al., 2008), state the expected local merger rate in our Galaxy, based on the empirical number from question (b) above. Discuss how reliable the number is.*

Exercise 15.10. *What is the slope of the evolving source tracks for in-spiral in Figure 15.26 (e.g., GW150914)?*

Exercise 15.11. *Verify the last term in Eq. (15.12). (Remember:* $f_{\mathrm{GW}} = 2\,f_{\mathrm{orb}}$*).*

Chapter Sixteen

Binary Population Synthesis and Statistics

16.1 INTRODUCTION

As we have seen in Chapter 3, the incidence of interacting binaries among stars of all masses is high, ranging from ∼40% among solar-like stars to over 80% for O-type stars. It is therefore clear that remnants of close binary evolution will be produced in large quantities. The effects of mass transfer in a close binary may often cause the post-mass-transfer accretor to be more massive than the original primary star of the system. For this reason, in those cases, the original secondary star is sometimes referred to as the new primary star.

In view of the steepness of the mass function of stars (the IMF [Ψ] as well as the observed mass function [Φ], see Chapter 3), stars of lower mass are much more abundant than stars of higher mass. Consequently, effective mass transfer between stellar components in binaries may produce new primaries of higher mass, that is, among stars that are less abundant (Fig. 16.1). Therefore, close binary evolution may make a relatively large contribution to the formation rate of stars at this higher mass, as was pointed out by van den Heuvel (1969). It was already known in the late 1960s that a high fraction of the stars are found in interacting binaries (van Albada, 1968). For more recent work on how binary evolution may inject higher-mass stars in a population of stars of lower mass, and thus change the shape of the mass function and produce blue stragglers, see Schneider et al. (2014, 2015, 2018).

Due to this same effect, close binary mass transfer may cause a close binary consisting of components that both have too low a mass to explode as a SN and produce a NS, to be transformed into a system in which the new primary is massive enough to explode as a SN (Fig. 16.1). Since, due to the shape of the IMF, stars of lower mass have a higher birthrate than star of higher mass, close binary evolution may result in a sizeable increase of the SN rate and NS-formation rate. Examples of NSs formed in this way, originating from binary stellar components that on their own could not have formed NSs, are the binary radio pulsars J1141−6545[1] and B2303+46[2], which both are young, strong-magnetic-field pulsars ($B \simeq 10^{12}$ G) with a massive WD companion ($M_{\rm WD} = 1.02\,M_\odot$ and $M_{\rm WD} \gtrsim 1.13\,M_\odot$, respectively). In both of these systems, the

[1] $P_{\rm orb} = 0.198$ d, $e = 0.172$, $M_{\rm NS} = 1.27\,M_\odot$, $M_{\rm WD} = 1.02\,M_\odot$.
[2] $P_{\rm orb} = 12.3$ d, $e = 0.658$, $M_{\rm NS} \gtrsim 1.3\,M_\odot$, $M_{\rm WD} \gtrsim 1.13\,M_\odot$.

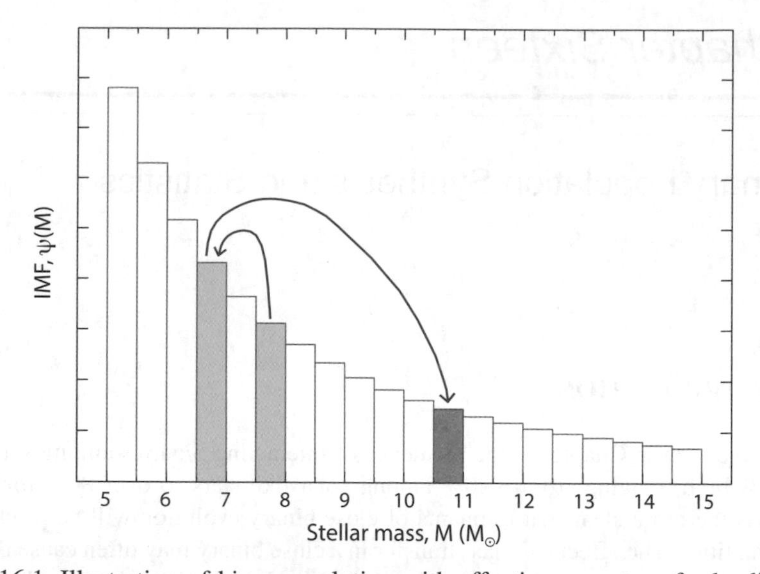

Figure 16.1. Illustration of binary evolution with effective mass transfer leading to a more massive secondary star (the new primary star), which is massive enough to undergo a SN explosion. The background distribution of stellar masses follows the initial mass function (IMF, $\Psi \propto M^{-2.35}$), which is very steep (see Chapter 3). Since the less massive stars are more numerous than the more massive ones, binary evolution can make a relatively large contribution to the formation of massive stars, as noted by van den Heuvel (1969).

WD is the first-formed stellar remnant, originating from the original primary star, and the NS is the remnant of the post-mass-transfer secondary, which was more massive than the original primary and exploded as a SN, leaving the strong-magnetic field NS in an eccentric orbit (see Tauris & Sennels, 2000, and Fig. 15.34). In these systems, the WD is *recycled* instead of the NS. That the WD in the PSR J1141−6545 system is indeed recycled was demonstrated by the estimate, thanks to the measurement of Lense-Thirring precession in this system, of its rotation period of only ~2 min, which is extremely fast for a WD (Krishnan et al., 2020).

As these examples show, the formation of NSs will be even more favorably biased toward membership in a binary, and the same will be true for SNe, and thus one expects the fraction of NSs formed in binaries to be larger than the already high fraction (> 70%) of potentially interacting binaries among massive stars in general (Chini et al., 2011, 2012; Sana et al., 2012). This effect was pointed out, for example, by Vanbeveren (1988) and Meurs & van den Heuvel (1989). See also work by Zapartas et al. (2017).

We know from the foregoing chapters that many types of astrophysical objects are products of binary evolution. If one wishes to make reliable estimates of the production rates of all these types of products of close binary evolution, one has to study the evolution of an entire stellar population that contains a realistic fraction of interacting binaries, with realistic distributions of their masses and orbital properties. This is what is called binary population synthesis (BPS). Population synthesis is a very strong tool for studying astrophysical events or populations of objects. As an example of a complicated case that may involve both single star and binary star evolution, consider the origin of measured pulsar velocities. Figure 16.2 illustrates many evolutionary paths leading to a single pulsar. It may have formed directly from a single star, or it may have originated in binary system that was later disrupted (in either the first or the second SN). It may also have been produced via an AIC event or the merger of two WDs. Or it could have evaporated its former companion star from intense irradiation effects; or it might originate from a triple/quadruple-star system.... Hence, we see that to probe the origin of single pulsar velocities, one must evaluate each of these many formation paths. BPS is a useful tool for this purpose and an example of how one must include all possible formation channels if one wishes to understand the observed single NS population.

Further examples of BPS applications are many and include: open cluster binary stars, Type Ia SN rates, dynamical effects of SNe in binaries (e.g., for production of hypervelocity stars), CE events, properties of X-ray binaries, CVs, MSPs, or, very popular recently, compact object merger rates for the LIGO network of GW detectors as well as other GW sources for LISA and other instruments. However, without the correct input physics, or by applying incorrect assumptions or (too many) free variables in a given investigation, the results can be misleading. In the following section, we describe and discuss the methodology of population synthesis based on Monte Carlo techniques, including critical assessments and pitfalls. After this, we will briefly discuss simple statistical considerations for deriving properties such as birth rates of various binary star systems, in particular double neutron stars (DNSs). We end this chapter with a short summary of historical developments regarding population synthesis, and we demonstrate one sample case of studying properties of open cluster binaries using BPS. We will not discuss the scientific results from specific population synthesis investigations here, as this has been done throughout the book. Further introductions to population synthesis of binary stars can be found in the reviews of Eldridge (2017); Izzard & Halabi (2018).

To our knowledge, the first attempt of population synthesis of binary stars was carried out by one of us (van den Heuvel) in 1969 to calculate the evolution of two synthetic star clusters with binaries, as will be briefly described in Section 16.4.

16.2 METHODOLOGY OF POPULATION SYNTHESIS

If one wishes to make reliable estimates of, for example, the production rate and properties of a certain type of outcome of close binary evolution, one first of all needs to know some basic starting ingredients (i.e., input probability density functions, PDFs) of the unevolved population, such as:

1. The fraction, F_{bin}, of binaries (and multiple stars) as a function of primary star mass in the general newborn stellar population.
2. The distribution at birth of the masses of primary stars, $\Psi(M_1)$.

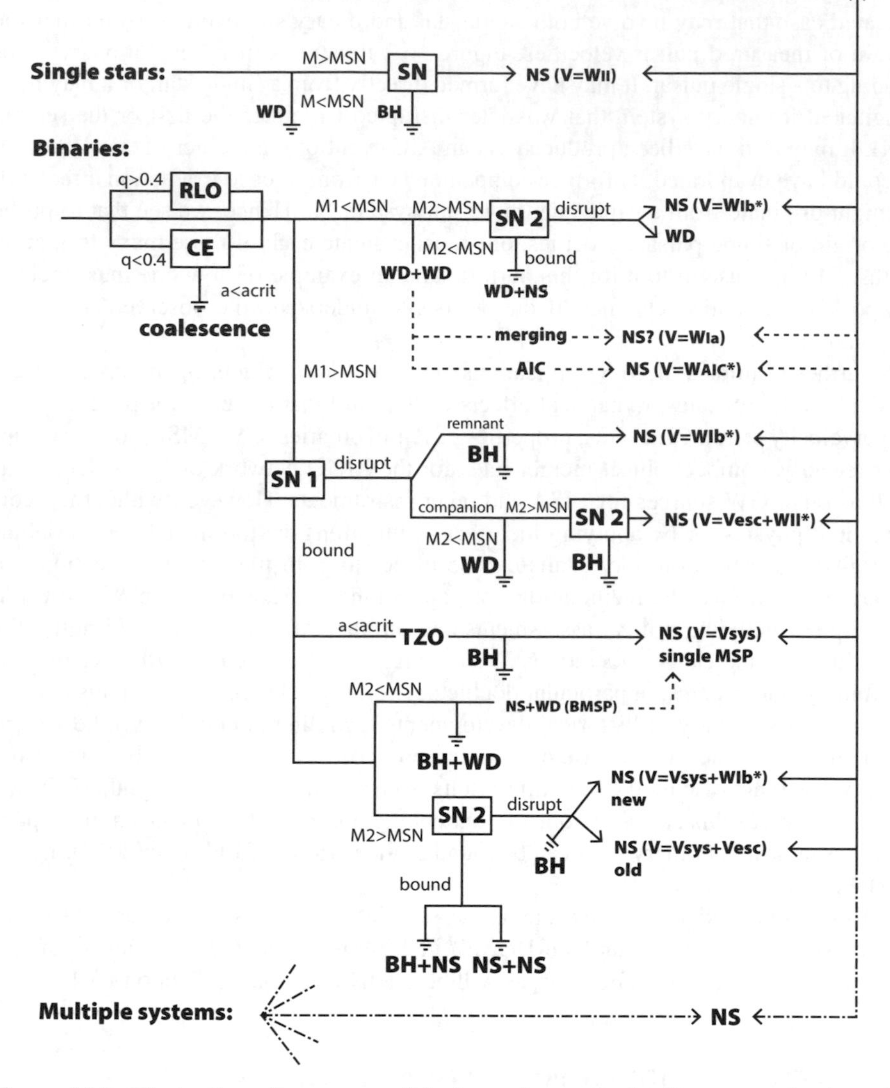

Figure 16.2. Illustration of many possible formation paths leading to production of a single radio pulsar. Progenitors include single, binary, and multiple-star systems, see text. It can be quite a complex exercise trying to simulate theoretical expectation values for comparison with observations of, e.g., pulsar velocities. Symbols: v is pulsar velocity; w indicates SN kick velocity with the index referring to SN Type: Ia, Ib, II, or AIC; $v_{\rm sys}$ and $v_{\rm esc}$ refer to the binary systemic and escape velocity, respectively; q is the mass ratio of the pre-SN stellar components; $M_{\rm SN}$ indicates the threshold mass for undergoing a SN; $a_{\rm crit}$ is the critical separation below which a CE results in a merger; and TZO refers to Thorne-Żytkow Object. After Tauris (1999).

3. The distribution $f(q)$ of mass ratios $q = M_2/M_1$ of potentially interacting binaries.
4. The distribution $f_{\log}(P_{\rm orb})$ of orbital periods $P_{\rm orb}$, or semi-major axes, $f_{\log}(a)$, of unevolved binaries.

These input distributions are derived from the observations of binary systems and were given is Section 3.7. The input values needed for each binary are most often chosen using simple Monte Carlo techniques (Section 16.2.3), which select, using a random number generator based on the CPU, a value from a given weighted distribution. Because of the very large number of possible combinations of input parameters, the synthetic binary stellar population will, to be a realistic representation of the stellar population in, for example, a Galaxy need to contain at least of the order of $N = 10^6 - 10^{10}$ binaries—depending on the purpose and the required detailed level of stellar evolution and binary interactions in a given run. A minimum number of $N = 10^6$ binary systems is needed if varying systematically only three basic parameters (e.g., M_1, q and a) over a range of 100 initial values for each.

Once one has constructed an initial (typically ZAMS) stellar population, one has to evolve each binary by combining binary interactions with a stellar evolution code that is able to simulate the evolution of stars in all kinds of binaries with different component masses, rotation rates, and metallicities. For fast calculation of the evolution of the entire population, interpolation methods have been developed for determining the evolution of single stars and binary systems, as will be described in the next section.

It is clear that if one would like to compute the complete detailed evolution of each binary in the synthetic stellar population, this will require an unrealistically large amount of computer time, although an impressive number of $\mathcal{O}(10^5)$ MESA runs using high-performance computing (HPC) with many cores are now possible (see early examples in, e.g., Lin et al., 2011; Marchant et al., 2017; Wang et al., 2020; du Buisson et al., 2020; Fragos et al., 2022). A typical binary system calculated in detail with, for example, MESA takes at least 10 hr of present-day CPU time (Paxton et al. 2019). Thus, access to, for example, a 10,000-core HPC cluster would still require more than one month to calculate just $N = 10^6$ binaries. Nevertheless, we foresee such full population synthesis studies in the coming decade.

16.2.1 Fast Binary Population Synthesis

In the past, such HPC calculations were impossible, and different ways have been invented to very much speed up the calculation of the evolution of single stars and binaries. Two commonly used methods for accelerating stellar evolution calculations involve applying:

1. interpolations in a library with grids of detailed pre-calculated models
2. analytical fitting formulae to detailed stellar evolution models

The first method is using as input a grid of a limited number of detailed pre-calculated single star and/or binary evolutionary tracks, covering the full range of primary masses, mass ratios, and orbital periods, and applying an interpolation algorithm for computing the evolution of binaries with any combinations of input parameters. This method has been applied, for example, in Pols & Marinus (1994); Han et al. (1995); Bagot (1997); Voss & Tauris (2003); Wang et al. (2010); Vanbeveren et al. (2012); Spera et al. (2015); Eldridge et al. (2017); Kruckow et al. (2018); Giacobbo & Mapelli (2018). The second method applies analytical fitting models to pre-calculated stellar models. Examples of population synthesis studies based on this method include Dewey & Cordes (1987); Yungelson et al. (1993); Kornilov & Lipunov (1983a,b); Lipunov et al. (1987, 1996); Portegies Zwart & Verbunt (1996); Nelemans Yungelson, Portegies Zwart, Verbunt (2001); Hurley et al. (2002); Belczynski et al. (2002, 2008); Izzard et al. (2004); de Mink et al. (2013); Stevenson et al. (2017); Neijssel et al. (2019); Breivik et al. (2020); Shao & Li (2021). A widely used set of analytic fitting models for the latter method are those of Hurley et al. (2000), based on the work of Pols et al. (1998). Population synthesis based on a hybrid approach, combining a rapid BPS code with detailed stellar evolution modelling for a given evolutionary stage is also a viable approach (e.g., Bavera et al., 2020; Wu et al., 2020; Bavera et al., 2021). In all cases, the backbone of the applied stellar models should be state of the art. The names of the many various BPS codes are very creative, and some of the acronyms[3] include: binary_c, BPASS, Brussels, ComBinE, COMPAS, MOBSE, POSYDON, Scenario Machine, SeBa, SEVN, StarTrack.

An interesting question arises in terms of convergence of the various codes on the market. A study (PopCORN) by Toonen et al. (2014) was launched to answer exactly this question in the particular case of investigating the progenitors of Type Ia SNe (double WDs and WD + main-sequence star binaries). The outcome of this comparison was that the outcomes of the four different BPS codes considered were in remarkably good agreement. We suspect, however, that a similar study of massive binary stars leading the a GW merger event (i.e., binaries experiencing two SNe and a CE with a massive donor star) may produce rather different outcomes. One of the reasons for this may well be that the CE evolution in the case of SNe Ia progenitors occurs in giants with a degenerate core, where the core boundary is clearly defined, contrary to the case of massive giants with a more extended non-degenerate core, where the definition of the core boundary is much less clear (Section 4.8.4).

16.2.2 Binary Interactions in Binary Population Synthesis

Apart from all the required effects of internal stellar evolution (nuclear reactions, winds, mixing, angular momentum transport, and rotation), the code should also include (depending on the masses of the stars) a treatment of the following effects:

- tides
- mass transfer (RLO, Cases A, B, C, BB, etc.)
- CE evolution (one or two phases)
- two SN explosions (including kinematic effects of kicks, spin-axis tossing?)

[3]Possibly a major concern of the developer (some names being more successful than others).

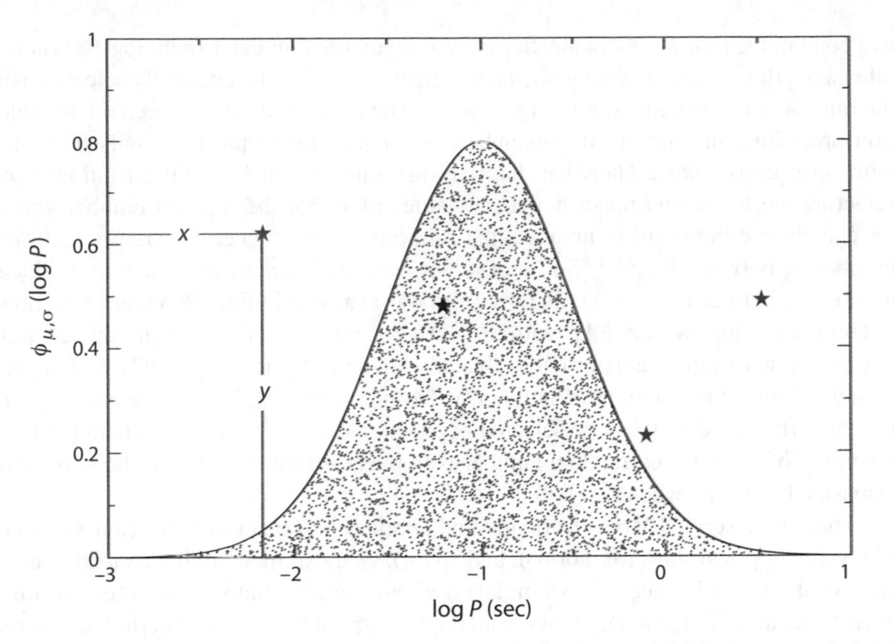

Figure 16.3. Illustration of the Monte Carlo method to select a variable from a weighted probability density function, ϕ, here a Gaussian (normal) function with $\sigma = 0.5$ and $\mu = -1.0$. The example deals with assigning a random initial spin period, P, to a newborn NS. The random number generator selects in each trial two numbers (x, y), here corresponding to a point in the $(\log P, \phi)$–plane. In case $y > \phi(x)$, the number x (here $\log P$) is disregarded (red star symbols), and a new pair of numbers is drawn. In case $y < \phi(x)$, the number is kept (black star symbol). The plot shows $N = 5{,}000$ selected points (blue dots).

- spin-orbit couplings (e.g., magnetic braking)
- irradiation effects (feedback on donor star).

16.2.3 Monte Carlo Technique in a Nutshell

The Monte Carlo method can simply be compared to a blindfolded monkey (no offense) throwing darts at a board. The procedure is shown in Figure 16.3. Here we consider an example where we want to assign a randomly-drawn spin period, P, to a newborn NS. We assume the probability density function (PDF) for $\log P$ is given by a Gaussian (normal) function with:

$$\phi(\log P) = \frac{1}{\sigma\sqrt{2\pi}}\, e^{-\frac{1}{2}\left(\frac{\log P - \mu}{\sigma}\right)^2}, \tag{16.1}$$

where the mean (or expectation) value is $\mu = -1.0$ (i.e., $\overline{P} = 0.1$ s) and $\sigma = 0.5$. The random number generator selects in each trial two numbers (x, y), here corresponding to a point in the (log P, ϕ)–plane. In case $y < \phi(x)$, the number x (here log P) is kept—otherwise, the number is disregarded and a new pair (x, y) is repeatedly selected using the random number generator until $y < \phi(x)$. There are methods to speed up this selection procedure and the given explanation is for illustrative purposes only. For PDFs with open ends (like a Gaussian distribution), one should be a little cautious when selecting the lower and upper boundary values of x that the random number generator can choose between. In practice, the random number generator simply chooses a number, r, between 0 and 1. So, in case we want the random generator to produce a number, x, between, say, -3.0 and $+1.0$ (in our example in Fig. 16.3 that corresponds to choosing values within $\pm 4\sigma$ with respect to μ, but in general one should be careful not to disregard rare events or choose values beyond physically possible boundaries), then we simply find the random number, x, from: $x = 4r - 3$. In the case of a Gaussian function, the value of y can be selected between $0 \leq y < 1/(\sigma\sqrt{2\pi})$, where the upper boundary of the selected random number is the upper limit of $\phi(x)$ for the normalized Gaussian PDF (i.e., when $x = \mu$).

The here described Monte Carlo method applies to *all* types of PDFs (flat, Gaussian, Maxwellian, power-law, log-normal, and so on). One can then apply this technique for each of the variables needed to simulate a given sample binary system (e.g., primary mass from an IMF, secondary mass from the mass-ratio PDF, orbital period, and eccentricity). As the system evolves with time, several situations will be encountered for which a variable is drawn from a stochastic process. One example deals with a SN explosion where one must select a kick magnitude and two angles to determine the kick direction (Fig. 13.22).

In population synthesis, the Markov chain Monte Carlo (MCMC) method is an algorithm for obtaining a sequence of random samples from consecutive (evolutionary) nested PDFs. By constructing a Markov chain that has the desired combined distribution as its equilibrium distribution, one can in this way obtain a sample of the desired population by recording states from the chain (see, e.g., Andrews et al., 2018, for a true MCMC method applied to binary population synthesis). Metropolis-Hastings and other MCMC algorithms are often used for sampling from multi-dimensional distributions, especially when the number of dimensions (variables) is high. MCMC methods may be used for calculating numerical approximations of multi-dimensional integrals in Bayesian statistics.

The more repetitions (number of initial binaries, N) that are included, the more closely the distribution of the sample matches the actual desired distribution. If there are many random variables describing the expected evolution of each system (and in particular if the parameter space for each variable is wide), then a very large number, N, of binaries need to be sampled. Often in binary star astrophysics, only a small fraction of all initial systems make it to the final stage under investigation. For example, for simulating a population of merging DNS systems with a reasonable number of final systems of order 10^5, such that their distributions of investigated physical properties will converge, that is, be independent of numerical noise along the MCMC path, an initial number of $N = 10^9$ may have to be generated. (In this case, the great majority of the systems will merge in a CE, be disrupted in a SN, or produce final systems that do not merge within a Hubble time.) Sampling algorithms that improve the computational efficiency of population studies of rare events have also been developed (e.g., STROOP-WAFEL, Broekgaarden et al., 2019).

16.2.4 Pitfalls in Population Synthesis

Monte Carlo simulation is a powerful tool for investigating complicated interactions that depend on many parameters. Often the difficult task is not to produce a reasonable code but to analyze the results, pinpoint the important physical parameters behind the major trends in a simulated population, and try to understand these results in terms of relatively simple physics.

As an example, a great concern and warning about current BPS modelling that includes a CE phase is that all the previously mentioned investigations (to our knowledge, except for those of Kruckow et al., 2018; Marchant et al., 2021), apply tabulated stellar structure (envelope binding energy) parameters, λ, found in the literature (e.g., Dewi & Tauris, 2000, 2001; Xu & Li, 2010; Loveridge et al., 2011), or even use a constant value for λ (or the product $\lambda\,\alpha_{\rm CE}$, see Section 4.8). However, it has been demonstrated by Kruckow et al. (2018) that λ-values must be calculated *self-consistently*, such that the same stellar evolution code is used for calculations of λ and, for example, the corresponding core mass used in the population synthesis code. One *should not* combine stellar grids calculated with one stellar evolution code with tabulated λ-values found in the literature that were calculated with another code. For example, Kruckow et al. (2018) demonstrated that combining their stellar tables calculated from the BEC code with computed λ-values based on the Eggleton code, the double BH merger rate resulting from population synthesis simulations increases by almost two orders of magnitude compared to the outcomes of the self-consistent treatment applied in their COMBINE code. Marchant et al. (2021) have refined this improvement one step further by also calculating the bifurcation point (core-envelope boundary) in a self-consistent manner.

16.3 EMPIRICAL VS. BINARY POPULATION SYNTHESIS-BASED ESTIMATES OF DOUBLE COMPACT OBJECT MERGER RATES

We consider here as an example of BPS the estimated merger rates of double compact objects in galaxies. The mergers of double neutron star (DNS) systems were early recognized as the most certain sources of detectable GWs (Clark & Eardley, 1977). It therefore became important to compute the expected merger rate of DNSs in galaxies. The first estimate of this rate, based on binary evolution considerations but neglecting, for example, the effect of birth kicks to NSs, was by Clark et al. (1979). Their result was $\sim 10^{-5}\,{\rm yr}^{-1}$ for the Milky Way, with an uncertainty of one order of magnitude in both directions, so 10^{-6} to $10^{-4}\,{\rm yr}^{-1}$. Amazingly, as we will see further along in this chapter, later determinations of this rate, determined with much more sophisticated methods, still come out in the same range.

There are basically two ways in which an estimate of the Galactic DNS merger rate can be made:

1. Empirical (observation-based), by comparing the properties of observed DNSs (detected as binary radio pulsars) with those of the populations of single pulsars found in the same surveys in which each of these DNSs were discovered (including selection effects).
2. Theory-based, by computing the outcome of BPS studies taking into account all known evolutionary effects involved, particularly CE evolution and the effects of kicks imparted onto NSs at birth.

In view of the large uncertainties particularly in the treatment of CE evolution which, so far, have not been satisfactorily resolved (Section 4.8), observation-based estimates may, at this moment, be at least as reliable as those obtained from BPS. Furthermore, the empirical merger-rate density, $\mathcal{R}_{\rm DNS}$, of DNS systems in the local Universe derived from observations by the LIGO network of GW detectors (currently at $\mathcal{R}_{\rm DNS} = 99^{+260}_{-86}\,{\rm Gpc}^{-3}\,{\rm yr}^{-1}$, see Section 15.7) will be substantially improved in the coming decade. Here, for now, we focus on the Galactic empirical rate based on observations of radio pulsars. We will start with describing the determination of observation-based merger rates, and subsequently give some examples of BPS-based estimates. The latter method for determining merger rates is particularly important for estimating double BH merger rates since, contrary to the case of DNSs, no double BHs have been observed in our Galaxy so far, and no Galactic background population of single BHs is known. Thus, currently there is no way to obtain an observation-based merger rate for Galactic double BHs. However, this may well change after the launch of LISA (Chapter 15).

16.3.1 Empirical Estimates of the DNS Merger Rate

These are based on comparing the observed properties of DNSs with those of single radio pulsars discovered in the same survey in which these DNSs were discovered. The earliest of such observation-based estimates yielded Galactic merger-rate values of $\sim 10^{-6}\,{\rm yr}^{-1}$ (Phinney, 1991; Narayan et al., 1991). Later Bailes (1996) estimated the upper limit to be $10^{-5}\,{\rm yr}^{-1}$, close to the rate determined by Curran & Lorimer (1995), after a correction by van den Heuvel & Lorimer (1996), that came out as $8 \times 10^{-6}\,{\rm yr}^{-1}$.

In making these observational estimates one has to take into account that, in general, one observes in the DNS system a weak-magnetic-field recycled pulsar, while the single pulsars detected in the same survey are strong-field non-recycled pulsars that live much shorter than the recycled pulsars in DNS systems (Chapter 14). Also the beaming fraction of recycled pulsars is different from that of newborn pulsars (Gould, 1994), and one will have to correct for these differences. Furthermore, one should correct for the different star-formation rates in galaxies of different types. Including all these effects, Phinney (1991) gave a best-guess merger rate in galaxies with a mass similar to ours, of $\sim 10\,{\rm Myr}^{-1}$, corresponding to a conservative limit on the rate for Advanced LIGO GW detections of $\sim 3\,{\rm yr}^{-1}$ within 200 Mpc. Subsequently, much important work in this field has been done, with increasing refinement of the methods, by Vicky Kalogera and her collaborators at Northwestern University, with the aim to study the GW event rate expected to be observed by the LIGO detectors. The work of Kalogera et al. (2001)

yielded a lower limit of $2\,\text{yr}^{-1}$ and an upper limit of between 300 and $1{,}000\,\text{yr}^{-1}$; the huge span in values reflecting the large uncertainties involved.

In subsequent studies, such as the one by Kim et al. (2003), for each binary pulsar, the properties of the survey in which this pulsar was discovered are used to calculate the entire Galactic population of similar binary pulsars. By then applying the theorem of Bayes to these results, the authors were able to calculate the probability density of the merger rate for each binary pulsar. By using in such studies the three binary radio pulsars B1913+16, B1534+12, and J0737−3039A, Kalogera et al. (2004), assuming a beaming correction of 6 for PSR J0737−3039A (similar to the average ones for PSR B1913+16 and PSR B1534+12, based on their polarization measurements), derived a Galactic DNS merger rate of $\sim$90 Myr^{-1}; see also Burgay et al. (2003) for an earlier estimate using PSR J0737−3039A, and Kim (2006) for a somewhat later update. O'Shaughnessy & Kim (2010) further refined these estimates by calculating for each binary pulsar the beaming fraction from polarization measurements and analysis of the pulse shape. They derived a rate of $\sim$60 Myr^{-1}.

Applying the same method, but now also including the unrecycled component PSR J0737−3039B of the double pulsar, Kim et al. (2015) derived a Galactic merger rate of $21^{+28}_{-14}\,\text{Myr}^{-1}$ at 95% confidence level, leading to a predicted ground-based GW detection rate of the Advanced LIGO-VIRGO network of $8^{+10}_{-5}\,\text{yr}^{-1}$ detections at 95% confidence. This derived merger rate was dominated by the three short-lived DNSs: B1913+16, B1534+12, and J0737−3039AB (of which B1534+12 contributed only a few percent to the merger rate), as depicted in Figure 16.4.

It is clear from the above that the empirical Galactic DNS merger rate derived from observations of radio pulsars is strongly dependent on individual systems. Hence, ongoing discoveries keep changing the derived merger rate. One example is PSR J1906+0746 (van Leeuwen et al., 2015), the first discovery of a (most likely) DNS system in which only the non-recycled strong-magnetic-field pulsar is seen.[4] Kim et al. (2015) estimated that if this system is included in the calculations, the merger rate doubles to $\sim$40 Myr^{-1} and the GW detection rate increases to $16^{+20}_{-10}\,\text{yr}^{-1}$. An independent analysis by Grunthal et al. (2021) including beam shape and viewing geometry, reveals $32\,\text{Myr}^{-1}$ and $3.5^{+2.1}_{-1.0}\,\text{yr}^{-1}$, respectively. Another example, is the discovery of the highly eccentric DNS system J0509+3801 (Lynch et al., 2018) which again, according to Pol et al. (2020), changes the Galactic merger rate to $37^{+24}_{-11}\,\text{Myr}^{-1}$, where the errors represent 90% confidence intervals. In our opinion, these derived rates are likely to be too high, see Exercise 15.9. In the following section, we discuss theoretical merger rates based on analytic calculations and using BPS codes.

16.3.2 Theoretical Estimates of the DNS Merger Rate

As mentioned earlier, several codes have been used to predict the formation rates of DNSs and double BHs, and their GW mergers. This is the case, for example, with the codes: *StarTrack* (Belczynski et al., 2008), ComBinE (Kruckow et al., 2018), BPASS

[4] Alternatively, this system could potentially also be a WD+NS binary (Ng et al., 2018).

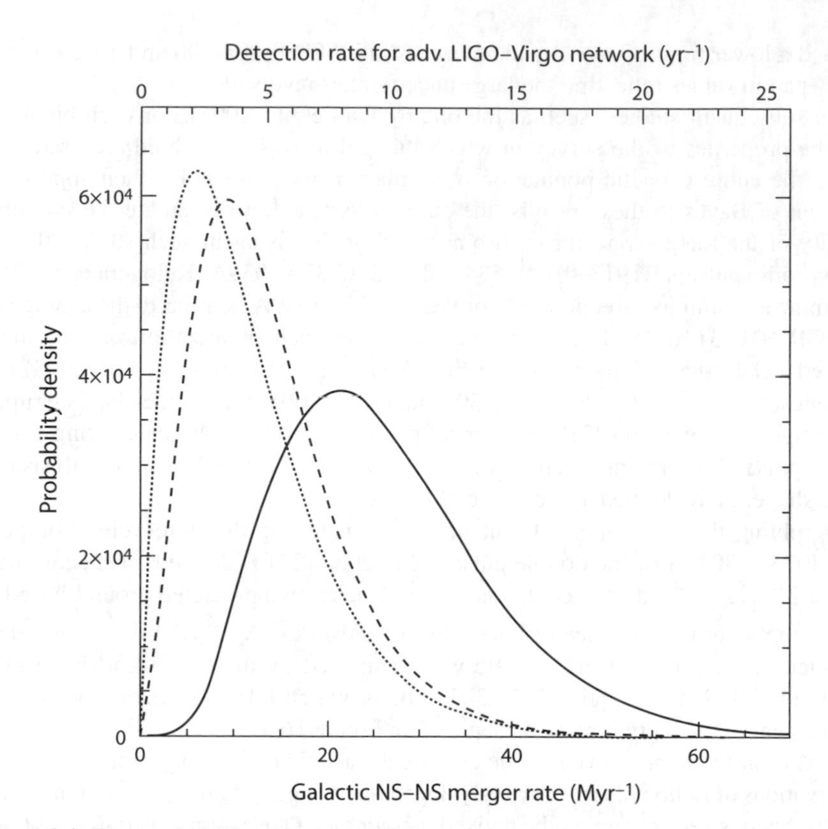

Figure 16.4. The probability density (solid curve) of the Galactic merger rate (bottom scale) and LIGO detection rate (top scale) overlaid with the contributions obtained individually from PSR B1913+16 (dotted) and PSR J0737-3039AB (dashed). After Kim et al. (2015).

(Eldridge & Stanway, 2016), COMPAS (Stevenson et al., 2017), MOBSE (Giacobbo et al., 2018), SEVN (Spera et al., 2019), and *Scenario Machine* (Lipunov et al., 2009). Another important application of the BPS codes is the prediction of the formation rate of close double WDs and/or candidate progenitors of Type Ia SNe. Applications of BPS codes for these calculations include, for example, SeBa (Nelemans, Yungelson, Portegies Zwart, Verbunt, 2001) and binary_c (Izzard et al., 2006).

A rough estimate mimicking the calculation of the DNS (or double BH) merger rate with a BPS code proceeds along the following lines. The Galactic merger rate, $\mathcal{R}_{\rm MW}$, can be simply written as a product (inspired by, e.g., Lipunov et al., 2018) as follows:

$$\mathcal{R}_{\rm MW} = f \cdot B \cdot \beta \cdot \gamma \cdot \delta \cdot \epsilon, \tag{16.2}$$

where: f is the star formation rate in the Galaxy, B is the fraction of stars in interacting binaries with a primary star massive enough to explode in a SN ($\sim 9\, M_\odot$, and a total binary mass exceeding $\sim$15 $M_\odot$), β is the fraction of high-mass binaries that survive the first SN explosion (this fraction depends strongly on the assumed kick-velocity distribution, see, e.g., Renzo et al., 2019), γ is the fraction of HMXBs that survive the CE evolution, δ is the fraction of the surviving post-CE binaries that survive the second SN explosion, and ϵ is the fraction of the latter systems that will merge within a Hubble time. The value of f for our Galaxy is roughly 1.0 $M_\odot$ yr^{-1}, and to obtain the value of B one must include a Salpeter-like IMF and the fact that $\sim$70% of massive binaries will interact via mass transfer (Sana et al., 2012). In relation to the value of δ, we discussed in Section 4.8.4 that only HMXBs with $P_{\rm orb} \gtrsim 1$ yr are able to survive the CE phase. The values of the other quantities are less trivial to evaluate analytically but can be derived following Monte Carlo simulations of, for example, the effects of kicks (methods that are included in BPS codes). For a recent example of a BPS code and its many input parameters, and parametrized assumptions, see, for example, Riley et al. (2021).

The first computations of this type were made by Kornilov & Lipunov (1983a,b) with an early version of the *Scenario Machine* that already included kicks and by Dewey & Cordes (1987). The first estimate with the fully developed *Scenario Machine* by Lipunov et al. (1987) was a Galactic DNS merger rate of 100 Myr^{-1}. Tutukov & Yungelson (1993b) who assumed that NSs do not get birth kicks, obtained even a rate of 3×100 Myr^{-1}. With a newer version of the *Scenario Machine*, Lipunov et al. (1996, 1997) again obtained a rate of $\sim$100 Myr^{-1}. With the SeBa code Portegies Zwart & Spreeuw (1996), including kicks, obtained a rate of $\sim$10 Myr^{-1}, and Portegies Zwart & Yungelson (1998) obtained a rate of 20 Myr^{-1}. Other early results on the Galactic DNS merger rate include Bloom et al. (1999) with $\sim$ 10 Myr^{-1}, Voss & Tauris (2003) with $\sim$ 1.5 Myr^{-1}, and the work of Belczynski et al. (2002) that gave $\sim$ 50 Myr^{-1}. As is already clear from these results, and summarized by Kalogera et al. (2001) and in further detail by Abadie et al. (2010), the theoretically estimated Galactic DNS merger rate is highly uncertain and the value is found to be anywhere between 1 and 100 Myr^{-1}. The large spread of a factor of 100 in the derived rates (often even when applying the same code by adjusting assumptions on input physics, e.g., Voss & Tauris, 2003), is due to uncertainties in the input physics.

In the last two decades, several new and more refined studies on the DNS merger rate yielded results within the same range ($1 - 100$ Myr^{-1}) and, perhaps surprisingly, with the same large span of uncertainties. These works include the more recent ones of Chruslinska et al. (2018); Belczynski et al. (2018); Lipunov et al. (2018); Kruckow et al. (2018); Vigna-Gómez et al. (2018); Neijssel et al. (2019); Tang et al. (2020); Santoliquido et al. (2021). A common feature here is that the models have to be stretched to the very optimistic limits to reach the local merger-rate density of DNS systems reported in LIGO GWTC-1 and GWTC-2 of $\mathcal{R}_{\rm DNS} = 1540^{+3200}_{-1220}$ Gpc^{-3} yr^{-1} (Abbott et al., 2019c) and $\mathcal{R}_{\rm DNS} = 320^{+490}_{-240}$ Gpc^{-3} yr^{-1} (Abbott et al., 2021d), respectively. These numbers, depending on the number scaling of galaxies, correspond roughly to $\sim$ 150 Myr^{-1} and $\sim$30 Myr^{-1} for a Milky Way–equivalent Galaxy. If the empirical LIGO network merger-rate density for DNS events continues to decrease further in the coming decade, it would be encouraging for many theoreticians (at the time of writing, it is $\sim$10 Myr^{-1}, see Table 15.5). Finally, we remark that the here quoted uncertainties in simulated DNS merger rates of two orders of magnitude is similar to the situation for the simulations of BH+BH binaries, and probably even larger for the mixed BH+NS binaries (Broekgaarden et al., 2021). While the kicks probably play a larger role for DNS systems, the metallicity dependence is much more important for the BH+BH systems. For the merger rates of these systems calculated with various BPS codes, see Sections 15.4, 15.5.9 and 15.6.

Problems with BPS Codes and Possible Directions for Future Work

Apart from the common pitfalls of population synthesis mentioned in Section 16.2.4, another major issue is related to the *calibration* of the codes (or the lack thereof). As discussed in Kruckow et al. (2018), binary interaction parameters must be calibrated to match, for example, the observed properties of Galactic DNS systems. Any binary population synthesis code on double compact object binaries must be able to reproduce the known and precise data on masses and orbital characteristics of binary radio pulsars and for example, mass-transfer efficiency of X-ray binaries (Kruckow et al., 2018; Bavera et al., 2021; Marchant et al., 2021). Similarly, one may also use properties of massive progenitor star binaries as important probes (Langer et al., 2020).

The importance of calibration also applies to modelling WD binaries (e.g., Camacho et al., 2014) and, in particular, double WD (DWD) binaries. As remarked by Toonen et al. (2014) from comparing the outcomes of four BPS codes on the predicted distributions of orbital separations (and also WD masses) of DWDs: the general outcome is similar, but the precise distributions are quite uncertain. This is because in most cases at least three phases of RLO and at least one and often two phases of CE evolution are involved. Since each step has its own uncertainty, the cumulative uncertainty in the final distribution of orbital separations will be large. Although Toonen et al. (2014) made no comparison between the codes for the outcomes of the production of the distributions of orbital separations of DNSs and BHs, here the uncertainties will be even larger, since apart from the many binary evolution steps required for producing these systems, one has in addition still the effects of two SN explosions involving distributions of kick velocities imparted to the compact objects in their formation.

Our conclusion can therefore only be that the total uncertainly in the outcome of the distributions of orbital separations and masses of DNSs and double BHs derived from BPS studies is very large, and that the same will hold for the predicted merger rates of DNSs and double BHs based on BPS studies. A recent discussion and comparison of remnant masses determined in BPS codes using core mass or pre-SN core structure can be found in Patton et al. (2021).

As already remarked, one of the main bottlenecks in BPS calculations with all these codes is that the detailed physics of CE evolution is still poorly understood, as discussed in Chapter 4 and in Sections 10.11.1 and 15.5.4 of this book. Before further progress can be made in that respect, it is therefore essential to first get a better insight into the physics of CE evolution. The second major issue relates to the poorly known details that distinguish between formation of NS and BH remnants of stars with initial masses roughly between ~ 15 and $30\,M_{\odot}$ (see discussions in Chapter 8). The refined physics involved is at the moment very difficult (if not impossible) to incorporate in BPS codes.

16.4 SOME HISTORY OF EARLY BINARY POPULATION SYNTHESIS: EVOLUTION OF OPEN STAR CLUSTERS WITH BINARIES

A star cluster in which all stars formed at the same time by fragmentation of a molecular cloud is a simple example of a stellar population. The evolution of a star cluster with a realistic fraction of binaries is therefore a good and simple form of the evolution of a stellar population with binaries. Studies of the evolution of such clusters with binaries were already made since the late 1960s and showed that, in a natural way, blue stragglers in star clusters are produced by binary evolution (Pols, 1993; Pols & Marinus, 1993, 1994; van den Heuvel, 1994a).

A realistic synthetic open cluster of N stars with binaries is constructed in the same way as outlined in the foregoing section for constructing synthetic stellar populations with binaries in general. The evolution code to be used must be able to handle the different types of evolution of binaries.

To our knowledge, the first one to calculate the evolution of a star cluster with binaries was one of us (van den Heuvel) in 1969. The aim of this study was to explain the formation of blue stragglers in open star clusters, such as the α Per cluster, Praesepe and M7 (see Fig. 16.5). Working at Lick Observatory, he used the grids of stripped-star models by Giannone et al. (1968), which can be used to calculate the evolution in the binary evolution Cases A and B, and the single star evolutionary tracks kindly provided by I. Iben. Evolutionary tracks for arbitrary stellar masses were (logarithmically) interpolated between these single star tracks and binary grids. The evolution with time of the HR diagrams of two clusters of stars was calculated. Both clusters were assigned 33% interacting binaries ($a < 10$ AU) and 33% visual binaries, a flat distribution of $\log a$ and a q-distribution as given by Kuiper (1935). It was found that the effects of mass transfer will often cause the post-RLO secondary (new primary) star of a close binary to be more massive than the original primary star of the system, thus producing blue stragglers and new stars above the turn-off of the main sequence. The results were not published, but they were presented at colloquia at Lick Observatory in 1969, and at the Universities of Brussels and Geneva, and at the Institut d'Astrophysique (Paris) in 1970 and 1971. Some of the original figures from 1969 were published as figure 87 in van den Heuvel (1994a), where a more extensive description of the applied evolutionary models is given.

An important similar work was independently made by Collier & Jenkins (1984) who calculated the evolution of star clusters with Case B binaries, again with the aim to explain the existence of blue stragglers in star clusters, and showed that indeed binary evolution successfully explains the existence of these objects.

A refined work, which one may now call *a classic*, on the same subject of star cluster evolution, was carried out by Pols (1993) and Pols & Marinus (1993, 1994). We present the results of these authors here as a nice example of how the HR diagrams of star clusters evolve with time if one takes binary evolution into account. (For a recent similar study, using MESA and taking also the rotation of stars into account, see Wang et al., 2020). The code constructed by Pols used the single star evolutionary tracks of Maeder & Meynet (1989) and took full account of all the various possible types of close binary evolution described in this book, including CE evolution, reverse mass transfer, and SN effects. Also, the calculated luminosities and effective temperatures

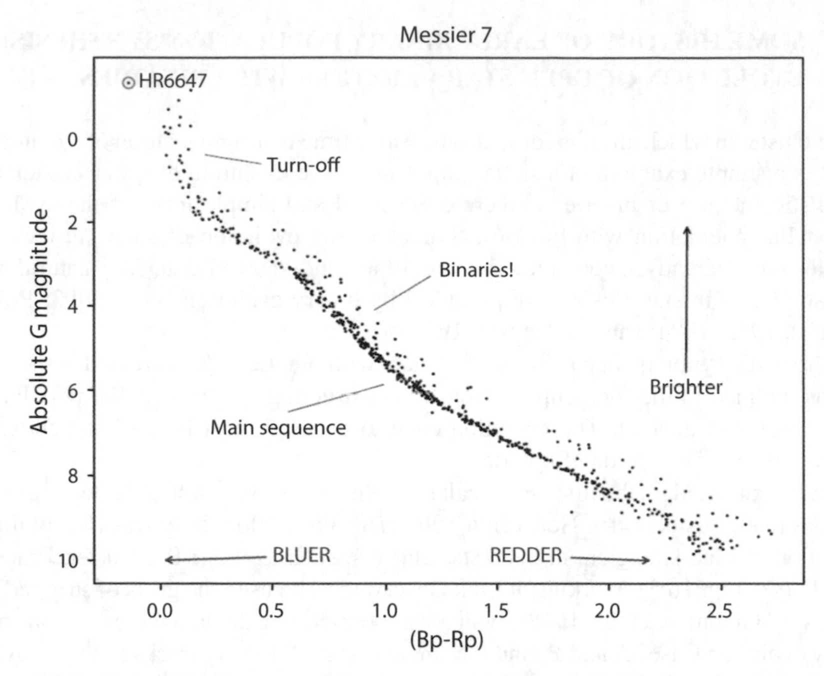

Figure 16.5. *Gaia* HR diagram of the open star cluster M7 (Ptolemy's cluster, with an age of $\sim 2 \times 10^8$ yr), clearly showing the position of the blue straggler HR6647 (with circle), a naked-eye star, as well as the second main sequence, due to unresolved binaries. The data in this diagram were derived by Henri Boffin from the early third *Gaia* data release (*Gaia* Collaboration et al., 2021). Figure courtesy of Henri Boffin.

were converted into U, B, and V magnitudes and $(U-B)$, $(B-V)$ colors, using the best transformations at the time (Coté, 1993), as is necessary for comparison of the obtained theoretical HR diagrams of the clusters at different times with observed HR diagrams of star clusters.

Figures 16.6, 16.7, and 16.8 show some of the results of the calculations by Pols & Marinus (1993, 1994) for a cluster of 2,000 stars with 50% binaries, with the q-distribution of Kuiper (1935). In a number of the CE cases, the two stars coalesced, and their evolution was then continued as that of a single star. A couple of the results of these calculations merit special attention, as follows.

a) The Occurrence of a Second Main Sequence

If an unresolved binary consists of two stars of the same mass, its luminosity will be twice that of a single star of that mass, but its color will be exactly the same. A twice as large luminosity means a magnitude difference of 0.75 mag. In the Hertzsprung-Russell diagram (i.e., HR or color–magnitude diagram), such a binary will be located 0.75 mag. above the main sequence of the single stars. In the HR diagrams of open star clusters, such as the Hyades, Praesepe, and the Pleiades, one observes indeed a second main sequence (Haffner & Heckmann, 1937) located some 0.75 mag. above the

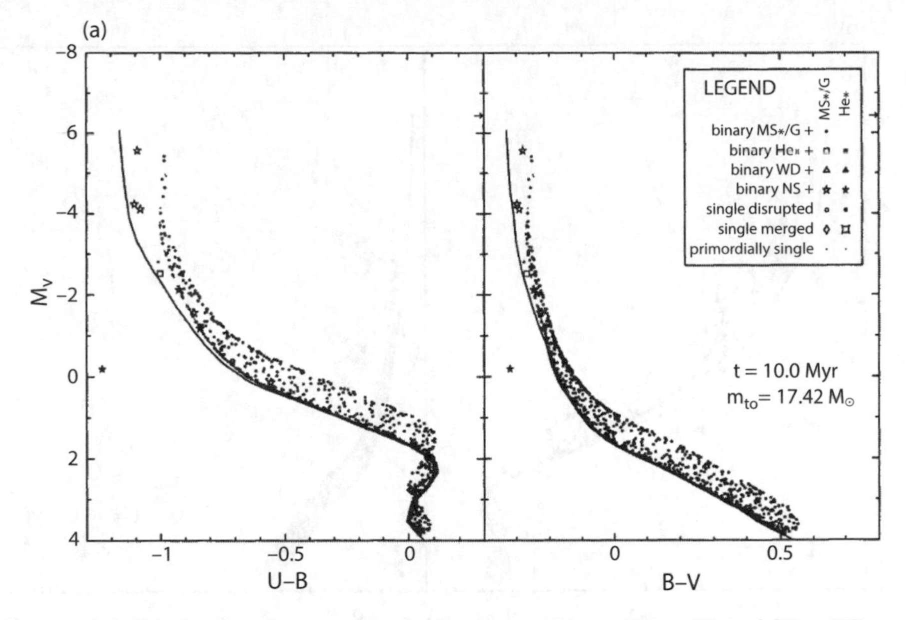

Figure 16.6. Synthetic color–magnitude diagrams (M_V vs. $[U - B]$ and $[B - V]$) at a cluster age $t = 10$ Myr corresponding to a turn-off mass of 17.42 $M_\odot$, resulting from the simulations by Pols & Marinus (1993, 1994) of a star cluster model containing 2,000 binary systems. The various symbols depict different types of stellar systems (see also legend). For binary systems: • indicate main sequence or giant stars; □ are helium stars plus a normal star; △ are WDs plus a normal star; and open stars indicate NSs—in all cases the binary companion is a normal star (or a helium star if the symbols are filled). Arrows on the right edges indicate red (super)giants that fall off the scale. After Pols & Marinus (1994).

main sequence, and this second one must indeed consist of unresolved binaries of about equal-mass stars. As an example, Figure 16.5 shows the second main sequence clearly for the cluster M7.

The three figures of the HR diagrams of the synthetic star clusters (Figs. 16.6–16.8) show that, although there is no peak at $q = 1$ in the q-distribution adopted in the simulations, the unevolved binaries spread in a band above the single star main sequence, which has a densely populated upper boundary that resembles the *second main-sequence* observed in open star clusters. It is located ∼0.75 mag. above the ordinary main sequence, which in these figures, except for the most massive stars, follows the ZAMS plotted here as a thin solid line. In synthetic clusters with an order of magnitude fewer stars than 2,000, there will be proportionally fewer binaries between the two main sequences, but the second one will, as it is relatively densely populated, still be quite pronounced.

At first glance, the presence of this second main sequence appears to suggest that in nature, for unevolved binaries, there is a preference for having a mass ratio close to unity. It was, however, realized long ago by Haffner & Heckmann (1937) and independently

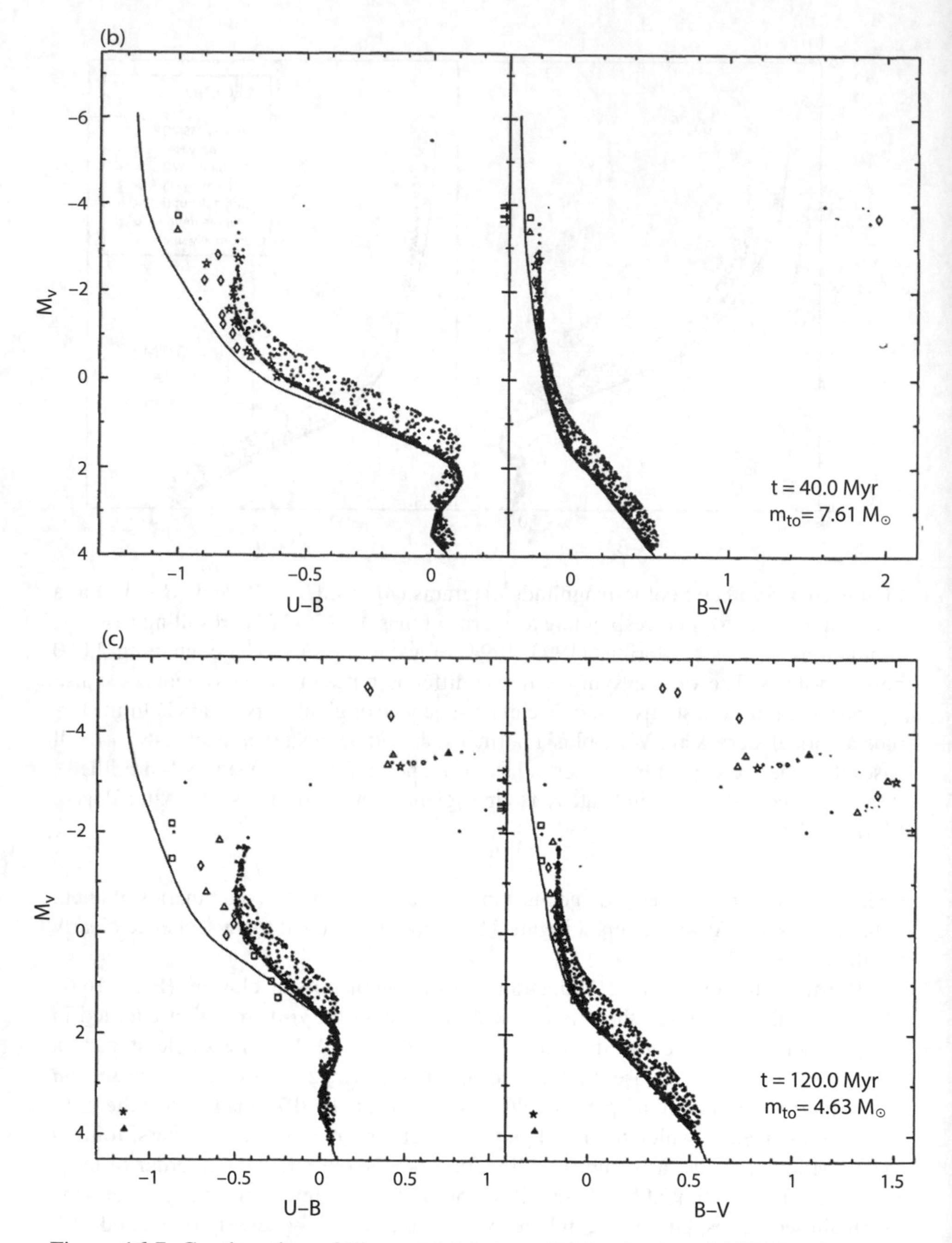

Figure 16.7. Continuation of Figure 16.6 for $t = 40$ Myr (top) and 120 Myr (bottom) and resulting turn-off masses of 7.61 $M_\odot$ and 4.63 $M_\odot$, respectively. Notice that even at $t = 120$ Myr there still is one NS+giant star binary and one NS+helium star binary left in the cluster. Because of the SN kicks to the NSs, these systems will usually have moved out of the cluster.

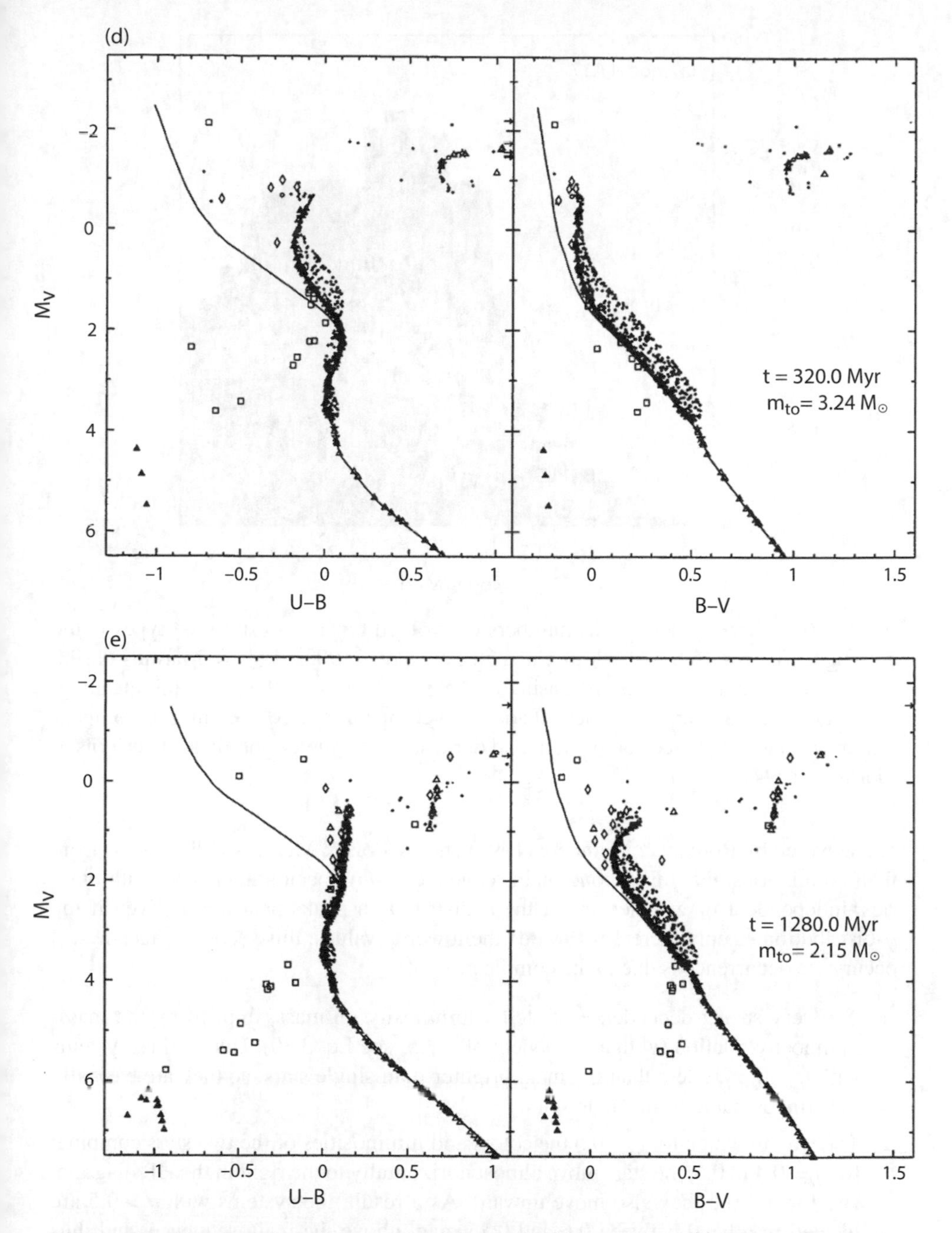

Figure 16.8. Continuation of Figure 16.6 for $t = 320$ Myr (top) and 1,280 Myr (bottom) and resulting turn-off masses of 3.24 $M_{\odot}$ and 2.15 $M_{\odot}$, respectively. The diagrams for $t = 1{,}280$ Myr resemble those of the Hyades and Praesepe clusters, although these have far fewer stars than this synthetic cluster, and therefore have fewer blue stragglers and giants.

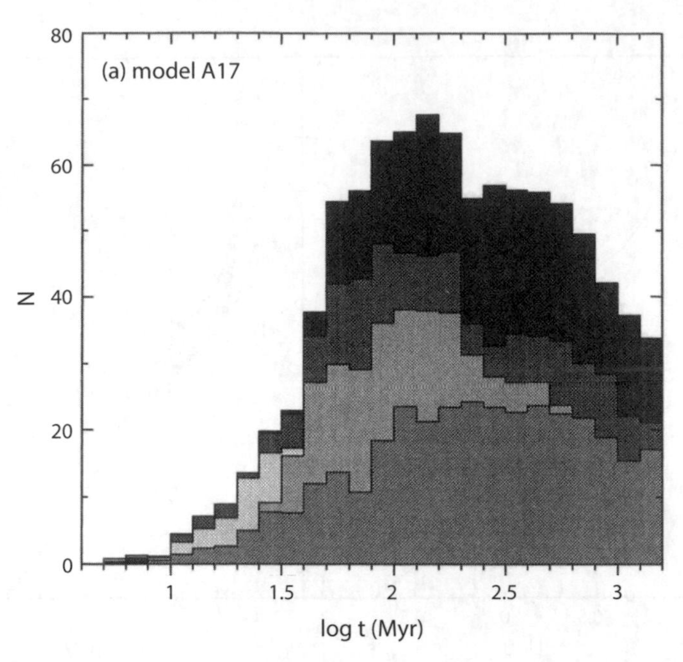

Figure 16.9. Development of the numbers of evolved binaries of different types in the cluster of Figures 16.6 to 16.8, as calculated by Pols (1993); Pols & Marinus (1993, 1994), indicated in shades of increasing darkness, which have the following meaning: lightest color: NS companion; next shade: WD companion; next: helium star companion; then: main-sequence companion; and darkest shade: merger product. After Pols & Marinus (1994).

rediscovered by Romani & Weinberg (1991) and by Pols & Marinus (1993, 1994) from their simulations, that this second main sequence *always* occurs in clusters with binaries, independent of whether or not the q-distribution peaks near $q = 1$. Even if the q-distribution steeply increases towards the lower q-values, this second sequence still occurs. Its occurrence is due to the combination of:

1. The very strong dependence of stellar luminosity on mass, implied by the mass-luminosity relation (with an exponent of $\sim$3.5, see Eq. 3.30). Due to this, systems with $q < 0.5$ are less than 0.1 mag. brighter than single stars, so they are basically indistinguishable from single stars.
2. The way in which for $q > 0.5$ the colors and luminosities of the two stars combine: for $q = 0.4$ to 0.6, the stars move almost horizontally to the right in the HR diagram, and for $q > 0.6$ they also move upward. As a result, the systems with $q > 0.5$ are located in a band between 0.6 and 0.75 mag. above the main sequence, and thus appear to form a second main sequence.

Thus, to conclude: the existence of the second main sequence observed in star clusters does not tell us anything about the shape of the q-distribution.

b) The Occurrence of Blue Stragglers, Be-Stars, Runaway OB-Stars and X-ray binaries

Figures 16.6 through 16.8 show that at almost any age there are a number of blue stragglers in the cluster, produced by close binary evolution. The new primary is a rejuvenated star: as it received a large amount of hydrogen-rich material from the envelope of the companion, it grew in mass, and, consequently, also the mass of its convective core increased. Since the convective core boundary moved outward into a region where no hydrogen burning had occurred, fresh hydrogen from this region was mixed into the core, lowering its helium content, causing the star to be rejuvenated. Thus, after the mass transfer, this rejuvenated star moved closer to the ZAMS, whereas the most massive single stars in the cluster are close to core-hydrogen exhaustion and are located to the right of the ZAMS. For this reason, the new primary star is bluer and, if q was sufficiently large (and assuming more or less conservative mass transfer), also brighter than the single stars near the main-sequence turn-off. See also Schneider et al. (2014).

One notices that at age 10^7 yr there are also blue stragglers with NS companions. Such systems will have runaway velocities of several $10\,\mathrm{km\,s^{-1}}$ as indicated in the figure. Blue stragglers in young clusters are often B-emission stars. This is quite understandable, since the preceding mass transfer took place through an accretion disk around the mass-receiving star (see Chapter 10). This star is therefore expected to become a very rapid rotator. The NS binaries, which will often be runaways from the cluster due to recoils from the SN mass loss and kick (Section 13.8), may later evolve into HMXBs.

Figure 16.9 depicts the development as a function of time of the numbers of different kinds of evolved binaries in the cluster, with NS, WD, helium-star, and main-sequence star companions, respectively, as well as resulting mergers.

Acknowledgments

We would like to express our deepest acknowledgments to all our colleagues over the years: collaborators, competitors, and anonymous referees. But most of all, we would like to express our gratitude to all the students we have had the pleasure to teach and work with in the field of binary star evolution. Thanks for your interest, wisdom, optimism, challenging questions, and huge inspiration! Without you, this book would never have been written. We hope in return that your students, as well as our many professional colleagues in the field, will benefit from this book.

We thank Princeton University Press for the invitation to write this book and, in particular, its staff members Ingrid Gnerlich, Whitney Rauenhorst, and Elizabeth Byrd for their continuous support and advice. We thank the staff of Westchester Publishing Services, in particular Christine Marra, for their great care in converting our manuscript into a real book. Without all of you, this book would not have existed.

Thomas would like to thank the following collaborators in particular. From ATNF: Dick Manchester (my splendid PhD supervisor), Matthew Bailes, Simon Johnston, Andrew Lyne. Amsterdam: Gertjan Savonije, Roelf Takens, Gijs Nelemans, Jasinta Dewi. In Bonn: Norbert Langer, Michael Kramer, Paulo Freire, Norbert Wex (thanks for commenting on Chapter 14), David Champion, John Antoniadis, Takashi Moriya, Sung-Chul Yoon, Lucas Guillemot, Joris Verbiest, Alina Istrate, Matthias Kruckow, Pablo Marchant, Rahul Sengar, Cherry Ng. In Oxford: Philipp Podsiadlowski. McGill: Vicky Kaspi. Yunnan Observatories: Hai-Liang Chen, Zheng-Wei Liu, Zhanwen Han. Monash: Bernhard Müller, Alexander Heger. From MPA: Hans-Thomas Janka. In addition, Thomas would like to thank: Gerry Brown, Simon Portegies Zwart, Sylvain Chaty, Ilya Mandel, Nanda Rea, Sushan Konar, Vasilii Gvaramadze, Ulrik Uggerhøj, and, not least, Kjeld Pedersen. Aarhus Institute of Advanced Studies (AIAS) and the physicists at the Dept. of Physics and Astronomy, Aarhus University are acknowledged for financial support. And finally, I would like to thank the long list of wonderful students that I had the pleasure to teach, work with, and learn from.

Edward would like to thank, first and foremost: my unforgettable Utrecht astrophysics teacher Marcel Minnaert; my two wonderful Utrecht PhD thesis advisors, Kees de Jager, who taught me stellar evolution, and Anne Underhill, who taught me the physics of early-type stars; and my Lick Observatory post-doctoral supervisor Peter Conti, who taught me about massive stars, as well as observing with large optical telescopes. Furthermore, my colleagues from all over the world who taught me very much, either by collaborating with me in research, or in the organization of conferences, workshops and summer schools. A non-exhaustive list of those whom I wish to thank personally: In the Netherlands: Max Kuperus, John Heise, Gertjan Savonije, Jan van Paradijs,

Michiel van der Klis, Godelieve Hammerschlag-Hensberge, Giel Habets, Vincent Icke, Frank Verbunt, Piet Hut, Lex Kaper, Ralph Wijers, Marten van Kerkwijk, Simon Portegies Zwart, Onno Pols, Gijs Nelemans, Selma de Mink, Paul Groot, Titus Galama, Paul Vreeswijk, Evert Rol, Henk Spruit, Johan Bleeker, Wim Hermsen, Erik Kuulkers, Rens Waters, Alex de Koter, Henny Lamers, Huib Henrichs and Martin Heemskerk. In Belgium: Bert De Loore, Jean-Pierre de Greve, Walter van Rensbergen, Henri Boffin, Jean-Louis Tassoul, Christoffel Waelkens, and Danny van Beveren. From the rest of the world: Dipankar Bhattacharya, Ganesan Srinivasan, Venkatraman Radhakrishnan, Pranab Ghosh, Rajagopalan Ramachandran, Mahendra Vardya, Xiang-Dong Li, Qing-Zhong Liu, Ken Nomoto, Yasuo Tanaka, Ene Ergma, Lev Yungelson, Sasha Tutukov, Konstantin Postnov, Anatol Cherepashchuk, Sergei Fabrika, Rashid Sunyaev, Sergei Grebenev, Eugene Churazov, Marat Gilfanov, Vladimir Lipunov, Jacob Shaham, Mario Livio, Ali Alpar, Hakki Ögelman, Michael Kramer, Norbert Langer, Joachim Trümper, Chris Winkler, Peter Kretschmar, Monica Colpi, Steve Shore, Pietro Ubertini, Guillaume Bélanger, Martin Rees, Jim Pringle, Philipp Podsiadlowski, Andrew Lyne, Phil Charles, Andy Fabian, Roger Blandford, Ben Stappers, Lorraine Hanlon, Riccardo Giacconi, Ethan Schreier, Jerry Ostriker, John Bahcall, Joseph Taylor, David Pines, Fred Lamb, Mal Ruderman, Brian Flannery, Ron Webbink, Ron Taam, Saul Rappaport, Josh Grindlay, Icko Iben, Shri Kulkarni, Chryssa Kouveliotou, Joel Weisberg, Joanna Rankin, Miller Goss, David Helfand, Duncan Lorimer, Lars Bildsten, Vicky Kalogera, Rosanne DiStefano, Vicky Kaspi, Walter Lewin, Natasha Ivanova, Tony Moffat, Dick Manchester, Matthew Bailes, and Ilya Mandel.

Last, but not least, we are grateful to our dear wives for their wonderful indulgence and patience, and we hereby apologize for the many cold dinners and absence from family moments due to the work on this book (lasting more than five years).

The authors (Ed van den Heuvel and Thomas Tauris) enjoying a relaxed evening in Thomas' house in Djursland, Denmark, after a long day's work on the book in February 2020. [Photo by Birgitte Tauris.]

Answers to Exercises

Here we have collected answers to selected exercises. For solutions to derivations of equations, general discussions, and so forth, please consult your course instructor.

Exercise 2.1: a) $d = 2.65$ pc. b) $M_{\rm bol,A} = +1.33$ and $M_{\rm bol,B} = +8.57$. c) $L_{\rm A} = 23.1\,L_\odot$ and $L_{\rm B} = 0.0293\,L_\odot$. d) $a = 20.2$ AU. e) $M_{\rm A}/M_{\rm B} = 2.15$. f) $M_{\rm total} = 3.31\,M_\odot$. g) $M_{\rm A} = 2.26\,M_\odot$ and $M_{\rm B} = 1.05\,M_\odot$. h) $(M/L)_{\rm A} = 0.0978\,M_\odot/L_\odot$ and $(M/L)_{\rm B} = 35.8\,M_\odot/L_\odot$.

Exercise 2.2: b) $P = 2.8$ hr. c) $P = 7.3$ s. d) $P = 0.61$ ms.

Exercise 3.4: HD11613: $f = 0.00842\,M_\odot$ and $M_2 \geq 0.18\,M_\odot$. HD14214: $f = 0.04267\,M_\odot$ and $M_2 \geq 0.46\,M_\odot$.

Exercise 3.5: a) 32%.

Exercise 4.1: b) $\langle \rho_2 \rangle \simeq 23\,{\rm g\,cm^{-1}}$.

Exercise 4.3: a) $\frac{6}{5}$. c) $\frac{3}{5}$. e) $\frac{9}{10}$. f) The least massive star has the largest specific (per unit mass) orbital angular momentum.

Exercise 4.20: b) $(a/a_0) = 0.254$.

Exercise 5.1: a) β Per. Star 1: $R_{\rm L} = 7.2\,R_\odot$ and Star 2: $R_{\rm L} = 3.6\,R_\odot$. TX UMa. Star 1: $R_{\rm L} = 8.1\,R_\odot$ and Star 2: $R_{\rm L} = 4.3\,R_\odot$. In both systems: star 2 fills its Roche lobe since $R_2 \simeq R_{\rm L}$. (i.e., the least massive star; hence the original *Algol paradox*). b) $a_1 = 2.52\,R_\odot$, $v_1 = 44.5\,{\rm km\,s^{-1}}$, $f \simeq 0.026\,M_\odot$.

Exercise 6.1: a) $\dot{M} = 1.68 \times 10^{-9}\,M_\odot\,{\rm yr^{-1}}$.

Exercise 6.2: $M_2 \geq 17.3\,M_\odot$ ($\geq 18.9\,M_\odot$) for a 1 $M_\odot$ (2 $M_\odot$) X-ray pulsar.

Exercise 6.3: e) $M_{\rm opt}/M_\odot = \{18.1,\ 19.8,\ 26.6\}$, for $i = \{90°,\ 75°,\ 60°\}$. f) $R_{\rm opt}/R_\odot = \{12.4,\ 13.3,\ 16.0\}$, for $i = \{90°,\ 75°,\ 60°\}$.

Exercise 6.4: HMXB: a) 1.88 $R_\odot$. b) $R_{\rm L} = 14.9\,R_\odot$. c) $L_1 = 18.3\,R_\odot$. LMXB: a) 1.36 $R_\odot$. b) $R_{\rm L} = 0.576\,R_\odot$. c) $L_1 = 0.773\,R_\odot$.

Exercise 7.1: b) $\dot{M} = 2.11 \times 10^{-15}\,M_\odot\,{\rm yr^{-1}}$ ($1.33 \times 10^{11}\,{\rm g\,s^{-1}}$).

Exercise 7.4: $\Delta M_{\rm NS} < 5.2 \times 10^{-4}\,M_\odot$.

Exercise 9.3: a) $\Delta E = 6.40 \times 10^{21}$ erg. b) 0.71%. c) $\Delta E = 1.55 \times 10^{23}$ erg. d) 17%. e) (Sun, WD, BH): $\Delta E = (1.91 \times 10^{18}$ erg 9.29, $\times 10^{19}$ erg, 4.42×10^{23} erg) (0.00021%, 0.010%, 49%).

Exercise 10.2: a) $a = 22.1\,R_\odot$. c) $Z \sim 0.1$ (e.g., 55 $M_\odot$ / 72.5 $M_\odot$ tracks).

Exercise 10.3: a) $P_{\rm orb} = 2210$ d. b) For example: $\beta = 0.183$, $\delta = 0.300$ (with $\gamma = 1.50$) yields $P_{\rm orb,0} = 109$ d. c) $P_{\rm orb} = 101$ d.

Exercise 11.1: a) $a = 4.46\,R_\odot$. b) 2.55×10^{48} erg.
c) $R_{\rm donor} = 1227\,R_\odot$. d) $R_{\rm donor} = 4089\,R_\odot$.
e) $R_{\rm donor} = 1044\,R_\odot$ (for $\alpha_{\rm CE} = 1$).

Exercise 13.1: a) $\theta_{\rm crit} = 59.31°$. b) $(a_{\rm f}/a_{\rm i}) = 0.9739$. c) $P_{\rm orb} = 3.102$ d.
d) $e = 0.06454$. e) $v_{\rm sys} = 93.99\,{\rm km\,s^{-1}}$. f) $\delta = 28.62°$.
g) $w_{\rm max} = 559.1\,{\rm km\,s^{-1}}$.

Exercise 13.2: a) $\theta_{\rm crit} = 83.34°$. b) $(a_{\rm f}/a_{\rm i}) = 2.622$. c) $P_{\rm orb} = 133.1$ d.
d) $e = 0.6586$. e) $v_{\rm sys} = 35.83\,{\rm km\,s^{-1}}$. f) $\delta = 17.47°$.

Exercise 13.3: a) $\theta_{\rm crit} = 81.03°$. b) $(a_{\rm f}/a_{\rm i}) = 0.5235$. c) $P_{\rm orb} = 0.1989$ d.
d) $e = 0.9297$. e) $v_{\rm sys} = 172.8\,{\rm km\,s^{-1}}$. f) $\delta = 5.174°$.
g) $q = 0.1437\,R_\odot$ (Hence, $q \ll R_2 \simeq 1.20\,R_\odot$).

Exercise 13.4: $v_{\rm NS,new} = 315.8\,{\rm km\,s^{-1}}$, $v_{\rm NS,old} = 144.3\,{\rm km\,s^{-1}}$.

Exercise 13.5: c) $P_{\rm orb} = 17.5$ d; $e = 0.976$; $v_{\rm sys} = 299\,{\rm km\,s^{-1}}$.

Exercise 13.9: c) $P(v > v_{\rm rms}) = 0.428$. d) $P(v < \frac{1}{2}\langle v \rangle) = 0.120$.

Exercise 14.1: b) 2.8×10^{-8}. d) $T = 3.9 \times 10^{28}$ K ($kT = 3.3 \times 10^{24}$ eV).

Exercise 14.4: c) $P_\infty = 5.54$ s.

Exercise 14.5: b) $f = 1.85 \times 10^{-3}\,M_\odot$.

Exercise 14.6: b) $\dot{M} = 1.4 \times 10^{-9}\,M_\odot\,{\rm yr^{-1}}$.

Exercise 14.8: a) $\dot{E}_{\rm GW} = 1.69 \times 10^{38}\,{\rm erg\,s^{-1}}$. b) $t = 289$ Myr.
c) $\dot{E}_{\rm GW} = 1.69 \times 10^{34}\,{\rm erg\,s^{-1}}$ and $t = 2.89 \times 10^{12}$ yr.

Exercise 14.9: a) $P_{\rm orb} = 14.8$ hr. c) $d = 946$ pc and $\mu = 0.0223''\,{\rm yr^{-1}}$
(22.3 mas $\rm yr^{-1}$).

Exercise 15.2: $\tau_{\rm GW} = 301$ Myr (1.65 Gyr for $e = 0$. 5.57 Myr for $e = 0.90$).

Exercise 15.3: a) $\tau_{\rm GW} = 30.2$ Gyr. b) 9.36 Gyr. c) 1.47 Gyr.

Exercise 15.4: a) $P_{\rm orb} = 1.883$ d. b) $e = 0.7843$. c) $v_{\rm sys} = 168.1\,{\rm km\,s^{-1}}$.
d) $\delta = 56.67°$. e) $\tau_{\rm GW} = 7.200$ Gyr.

Exercise 15.6: SNR$\sim$100 ($h_c = 1.40 \times 10^{-19}$, $h_{\rm GB} = 2.02 \times 10^{-18}\,{\rm Hz^{-1/2}}$)

Exercise 15.7: $P_{\rm orb} = 67$ min ($h_c = 2.33 \times 10^{-19}$, $h_n \simeq 2 \times 10^{-20}$, SNR ~ 10)
$P_{\rm orb} = 111$ min ($h_c = 1.29 \times 10^{-19}$, $h_n \simeq 3.5 \times 10^{-20}$,
SNR ~ 3.7)

Exercise 15.8: 19.4% ($e = 0.5$), 39.7% ($e = 0.7$), 71.2% ($e = 0.9$).

Exercise 15.9: c) 25 $\rm MWEG^{-1}\,Myr^{-1}$.

Exercise 15.10: $-1/6$.

List of Acronyms

AIC	accretion-induced collapse
aLIGO	advanced LIGO
AM CVn	AM Canum Venaticorum
AXMSP	accreting X-ray millisecond pulsar
Be-star	(pre-) main-sequence B-star with emission lines
B[e] star	B supergiant star with forbidden emission lines
Be/X-ray binary	B-emission X-ray binary
BH	black hole
BPS	binary population synthesis
CCO	central compact object (in a supernova remnant)
CCSN	core-collapse supernova
CE	common envelope
CE	Cosmic Explorer (future American GW telescope)
CEMP star	carbon-enhanced metal-poor star
CH star	carbon-enriched red giant
CHE	chemically homogeneous evolution
CHEMP	chemically peculiar
CO	carbon-oxygen
CV	cataclysmic variable
DD	double degenerate
DM	dispersion measure
DNS	double neutron star
DTD	delay-time distribution
DWD	double white dwarf
EC SN	electron capture supernova
ELM	extremely low mass
EM	electromagnetic
ET	Einstein Telescope (future European GW Telescope)
FAST	five-hundred-meter Aperture Spherical (radio) Telescope
Fe CCSN	iron-core collapse supernova
FRB	fast radio burst
GC	globular cluster
GUNS	grand unification of neutron stars
GW	gravitational wave
Gyr	gigayear
He	helium

HMXB	high-mass X-ray binary
HPC	high performance computing
IGE	iron-group elements
IME	intermediate-mass elements
IMF	initial mass function
IMXB	intermediate-mass X-ray binary
IndIGO	Indian initiative in gravitational-wave observations
ISM	interstellar medium
KAGRA	Kamioka Gravitational Wave Detector
LBV	luminous blue variable
LGRB	long gamma-ray burst
LIGO	Laser Interferometer Gravitational-Wave Observatory
LION	laser interferometer on the Moon
LISA	Laser Interferometter Space Antenna
LMC	Large Magellanic Cloud
LMXB	low-mass X-ray binary
MS	main sequence
MSP	millisecond pulsar
MW	Milky Way
MWEG	Milky Way equivalent galaxy
Myr	megayear
NS	neutron star
OGLE	optical gravitational lensing experiment
ONeMg	oxygen-neon-magnesium
PDF	probability density function
PISN	pair-instability supernova
PN	planetary nebula
PPISN	pulsational pair-instability supernova
PSR	pulsar
QPO	quasi-periodic oscillation
R Cor Bor (R CrB)	R Coronae Borealis
RG	red giant
RLO	Roche-lobe overflow
RRATs	rotating radio transients
RS CVn	RS Canum Venaticorum
RSG	red supergiant
SE-SN	stripped-envelope supernova
SD	single degenerate
SDSS	Sloane digital sky survey
SFXT	supergiant fast X-ray transient
sg-HMXB	supergiant high-mass X-ray binary
sGRB	short gamma-ray burst
SMC	Small Magellanic Cloud
SN	supernova
SNR	supernova remnant; alternative meaning: signal-to-noise ratio

SSS	super-soft (X-ray) sources
TAMS	terminal main sequence
TZO	Thorne-Żytkow object
UCXB	ultra-compact X-ray binary
ULX	ultra-luminous X-ray binary
VLBI	very long baseline interferometry
WC	carbon-type Wolf-Rayet star
WD	white dwarf
WN	nitrogen-type Wolf-Rayet star
WO	oxygen-type Wolf-Rayet star
WR-star	Wolf-Rayet star
YSG	yellow supergiant
XDIN	X-ray dim neutron star
ZAMS	zero-age main sequence
ZTF	Zwicky transient facility

References

Aadland E., Massey P., Neugent K. F., Drout M. R., 2018, *Astronomical Journal*, 156, 294.
Aarseth S. J., 1974, *Astronomy and Astrophysics*, 35, 237.
Aasi J., Abadie J., Abbott B. P., Abbott R., Abbott T., Abernathy M. R., et al. 2014, *Astrophysical Journal*, 785, 119.
Abadie J., Abbott B. P., Abbott R., Abernathy M., Accadia T., Acernese F., Adams C., Adhikari R., Ajith P., Allen B., et al. 2010, *Classical and Quantum Gravity*, 27, 173001.
Abate C., Pols O. R., Izzard R. G., Karakas A. I., 2015, *Astronomy and Astrophysics*, 581, A22.
Abate C., Pols O. R., Stancliffe R. J., 2018, *Astronomy and Astrophysics*, 620, A63.
Abbott B. P., Abbott R., Abbott T. D., Abernathy et al. 2017, *Annalen der Physik*, 529, 1600209.
Abbott B. P., Abbott R., Abbott T. D., Abernathy M. R., Acernese F., Ackley K., et al. 2016a, *Physical Review X*, 6, 041015.
Abbott B. P., Abbott R., Abbott T. D., Abernathy M. R., Acernese F., Ackley, K., et al. 2016b, *Physical Review Letters*, 116, 241103.
Abbott B. P., Abbott R., Abbott T. D., Abernathy M. R., Acernese F., Ackley K., et al. 2016c, *Physical Review X*, 6, 041014.
Abbott B. P., Abbott R., Abbott T. D., Abernathy M. R., Acernese F., Ackley K., Adams C., Adams T., et al. 2016d, *Physical Review Letters*, 116, 061102.
Abbott B. P., Abbott R., Abbott T. D., Abernathy M. R., Acernese F., Ackley K., et al. 2016e, *Physical Review Letters*, 116, 221101.
Abbott B. P., Abbott R., Abbott T. D., Abraham S., Acernese F., Ackley K., et al. 2020a, *Astrophysical Journal*, 892, L3.
Abbott B. P., Abbott R., Abbott T. D., Abraham S., Acernese F., Ackley K., et al. 2020b, *Living Reviews in Relativity*, 23, 3.
Abbott B. P., Abbott R., Abbott T. D., Abraham S., Acernese F., et al. 2019a, *Physical Review D*, 100, 024004.
Abbott B. P., Abbott R., Abbott T. D., Abraham S., Acernese F., et al. 2019b, *Astrophysical Journal*, 882, L24.
Abbott B. P., Abbott R., Abbott T. D., Abraham S., Acernese F., et al. 2019c, *Physical Review X*, 9, 031040.
Abbott B. P., Abbott R., Abbott T. D., Abraham S., Acernese F., et al. 2019d, *Astrophysical Journal*, 879, 10.
Abbott B. P., Abbott R., Abbott T. D., Acernese F., Ackley K., Adams C., et al. 2017a, *Astrophysical Journal*, 848, L13.
Abbott B. P., Abbott R., Abbott T. D., Acernese F., Ackley K., Adams C., et al. 2017b, *Astrophysical Journal*, 851, L35.
Abbott B. P., Abbott R., Abbott T. D., Acernese F., Ackley K., Adams C., et al. 2017c, *Physical Review Letters*, 119, 161101.
Abbott B. P., Abbott R., Abbott T. D., Acernese F., Ackley K., Adams C., et al. 2017d, *Astrophysical Journal*, 848, L12.

Abbott B. P., Abbott R., Abbott T. D., Acernese F., Ackley K., Adams C., et al. 2017e, *Astrophysical Journal*, 850, L40.
Abbott B. P., Abbott R., Abbott T. D., Acernese F., Ackley K., Adams C., et al. 2018, *Physical Review Letters*, 121, 161101.
Abbott B. P., Abbott R., Acernese F., Adhikari R., Ajith P., Allen B., et al. 2010, *Astrophysical Journal*, 713, 671.
Abbott D. C., Conti P. S., 1987, *Annual Review of Astronomy and Astrophysics*, 25, 113.
Abbott R., Abbott T. D., Abraham S., Acernese et al. 2021a, *Astrophysical Journal*, 922, 71
Abbott R., Abbott T. D., Abraham S., Acernese et al. 2021b, *Astrophysical Journal*, 915, L5.
Abbott R., Abbott T. D., Abraham S., Acernese F., et al 2020, *Physical Review D*, 102, 043015.
Abbott R., Abbott T. D., Abraham S., Acernese F., et al. 2020a, *Physical Review Letters*, 125, 101102.
Abbott R., Abbott T. D., Abraham S., Acernese F., et al. 2020b, *Astrophysical Journal*, 896, L44.
Abbott R., Abbott T. D., Abraham S., Acernese F., et al. 2021d, *Physical Review X*, 11, 021053.
Abbott R., Abbott T. D., Abraham S., Acernese F., et al. 2021e, *Astrophysical Journal*, 913, L7.
Abbott R., Abbott T. D., Abraham S., Acernese F., et al. 2021f, *Physical Review D*, 103, 122002.
Abbott R., Abbott T. D., Acernese et al. 2022a, arXiv:2111.03606.
Abbott R., Abbott T. D., Acernese et al. 2022b, arXiv:2111.03634.
Abbott R., Abe H., Acernese F., Ackley K., Adhicary S., Adhikari N., et al. 2022, arXiv:2209.02863.
Abbott R., Abbott T. D., Acernese F., Ackley K., Adams et al. 2021, arXiv:2108.01045.
Abdo A. A., Ackermann M., Ajello M., Atwood W. B., et al. 2009, *Science*, 325, 848.
Abramowicz M. A., Calvani M., Nobili L., 1980, *Astrophysical Journal*, 242, 772.
Abt H. A., Conti P. S., Deutsch A. J., Wallerstein G., 1968, *Astrophysical Journal*, 153, 177.
Abt H. A., Gomez A. E., Levy S. G., 1990, *Astrophysical Journal Supplement Series*, 74, 551.
Abt H. A., Levy S. G., 1976, *Astrophysical Journal Supplement Series*, 30, 273.
Abt H. A., Levy S. G., 1978, *Astrophysical Journal Supplement Series*, 36, 241.
Ackley K., Amati L., Barbieri C., Bauer F. E., Benetti S., Bernardini M. G., et al. 2020, *Astronomy and Astrophysics*, 643, A113.
Adams S. M., Kochanek C. S., Gerke J. R., Stanek K. Z., 2017, *Monthly Notices of the Royal Astronomical Society*, 469, 1445.
Aerts C., Mathis S., Rogers T. M., 2019, *Annual Review of Astronomy and Astrophysics*, 57, 35.
Agazie G. Y., Mingyar M. G., McLaughlin M. A., Swiggum J. K., Kaplan D. L., et al. 2021, *Astrophysical Journal*, 922, 35.
Ajello M., Allafort A., Axelsson M., Baldini L., Barbiellini G., et al. 2018, *Astrophysical Journal*, 861, 85.
Ajith P., Babak S., Chen Y., et al. 2007, *Classical and Quantum Gravity*, 24, S689.
Ake T. B., Griffin E., 2015, *Giants of Eclipse: The ζ Aurigae Stars and Other Binary Systems* Vol. 408 of Astrophysics and Space Science Library, Springer, Dordrecht.
Alberts F., Savonije G. J., van den Heuvel E. P. J., Pols O. R., 1996, *Nature*, 380, 676.
Alexander M. E., 1973, *Astrophysics and Space Science*, 23, 459.
Alpar M. A., Cheng A. F., Ruderman M. A., Shaham J., 1982, *Nature*, 300, 728.
Althaus L. G., Miller Bertolami M. M., Córsico A. H., 2013, *Astronomy and Astrophysics*, 557, A19.
Althaus L. G., Serenelli A. M., Benvenuto O. G., 2001, *Monthly Notices of the Royal Astronomical Society*, 324, 617.

Amaro-Seoane P., Andrews J., Arca Sedda M., Askar A., Balasov R., Bartos I., Bavera S. S., et al. 2022, arXiv:2203.06016
Amaro-Seoane P., Audley H., Babak S., Baker J., Barausse E., et al. 2017, arXiv:1702.00786.
Amaro-Seoane P., Bischof L., Carter J. J., Hartig M.-S., Wilken D., 2021, *Classical and Quantum Gravity*, 38, 125008.
Anagnostou O., Trenti M., Melatos A., 2020, arXiv:2010.06161.
Andersen B. C., Bandura K., Bhardwaj M., Boubel P., Boyce M. M., Boyle P. J., et al. 2019, *Astrophysical Journal*, 885, L24.
Andersen B. C., Fonseca E., McKee J. W., Meyers B. W., Luo J., Tan C. M., et al. 2022, arXiv:2209.06895.
Andersen B. C., Bandura K., Bhardwaj M., Boyle P. J., Brar C., Breitman D., et al. 2022, *Nature*, 607, 256.
Andersen J., 1991, *Astronomy and Astrophysics Review*, 3, 91.
Andersen J., Clausen J. V., Gimenez A., Nordstroem B., 1983, *Astronomy and Astrophysics*, 128, 17.
Anderson S. B., Gorham P. W., Kulkarni S. R., Prince T. A., Wolszczan A., 1990, *Nature*, 346, 42.
Andersson N., 2020, *Gravitational-Wave Astronomy: Exploring the Dark Side of the Universe*, Oxford University Press.
Andresen H., Müller B., Müller E., Janka H.-T., 2017, *Monthly Notices of the Royal Astronomical Society*, 468, 2032.
Andrews J. J., Breivik K., Pankow C., D'Orazio D. J., Safarzadeh M., 2020, *Astrophysical Journal*, 892, L9.
Andrews J. J., Cronin J., Kalogera V., Berry C. P. L., Zezas A., 2021, *Astrophysical Journal*, 914, L32.
Andrews J. J., Farr W. M., Kalogera V., Willems B., 2015, *Astrophysical Journal*, 801, 32.
Andrews J. J., Zezas A., Fragos T., 2018, *Astrophysical Journal Supplement Series*, 237, 1.
Ankay A., Kaper L., de Bruijne J. H. J., Dewi J., Hoogerwerf R., Savonije G. J., 2001, *Astronomy and Astrophysics*, 370, 170.
Antoni A., Quataert E., 2022, *Monthly Notices of the RAS*, 511, 176.
Antoniadis J., 2014, *Astrophysical Journal*, 797, L24.
Antoniadis J., Aguilera-Dena D. R., Vigna-Gomez A., Kramer M., Langer N., MŁuller B., Tauris T. M., Wang C., Xu X.-T., 2022, *Astronomy and Astrophysics*, 657, L6.
Antoniadis J., Arzoumanian Z., Babak S., Bailes M., Bak Nielsen A. S., et al. 2022, *Monthly Notices of the Royal Astronomical Society*, 510, 4873.
Antoniadis J., Freire P. C. C., Wex N., Tauris T. M., Lynch R. S., van Kerkwijk M. H., et al. 2013, *Science*, 340, 448.
Antoniadis J., Tauris T. M., Ozel F., Barr E., Champion D. J., Freire P. C. C., 2016, arXiv:1605.01665.
Antoniadis J., van Kerkwijk M. H., Koester D., Freire P. C. C., Wex N., Tauris T. M., ... Bassa C. G., 2012, *Monthly Notices of the Royal Astronomical Society*, 423, 3316.
Antonini F., Chatterjee S., Rodriguez C. L., Morscher M., Pattabiraman B., Kalogera V., Rasio F. A., 2016, *Astrophysical Journal*, 816, 65.
Antonini F., Rasio F. A., 2016, *Astrophysical Journal*, 831, 187.
Antonini F., Toonen S., Hamers A. S., 2017, *Astrophysical Journal*, 841, 77.
Apostolatos T. A., Cutler C., Sussman G. J., Thorne K. S., 1994, *Physical Review D*, 49, 6274.
Applegate J. H., 1992, *Astrophysical Journal*, 385, 621.
Applegate J. H., Shaham J., 1994, *Astrophysical Journal*, 436, 312.
Archibald A. M., Gusinskaia N. V., Hessels J. W. T., Deller A. T., Kaplan D. L., Lorimer D. R., ... Stairs I. H., 2018, *Nature*, 559, 73.

Archibald A. M., Stairs I. H., Ransom S. M., Kaspi V. M., Kondratiev V. I., Lorimer D. R., ... Remillard R. A., 2009, *Science*, 324, 1411.
Aret A., Kraus M., Kolka I., Maravelias G., 2017, in Balega Y. Y., Kudryavtsev D. O., Romanyuk I. I., Yakunin I. A., eds, *Stars: From Collapse to Collapse*, Vol. 510 of Astronomical Society of the Pacific Conference Series, The Yellow Hypergiant—B[e] Supergiant Connection, p. 162.
Armitage P. J., Livio M., 2000, *Astrophysical Journal*, 532, 540.
Armitage P. J., Natarajan P., 1999, *Astrophysical Journal*, 525, 909.
Arnett W. D., 1969, *Astrophysics and Space Science*, 5, 180.
Arnett W. D., 1972a, *Astrophysical Journal*, 176, 681.
Arnett W. D., 1972b, *Astrophysical Journal*, 176, 699.
Arnett W. D., 1974, *Astrophysical Journal*, 193, 169.
Arnett W. D., 1978, in Giacconi R., Ruffini R., eds, *Physics and Astrophysics of Neutron Stars and Black Holes North Holland Publishing Company, Amsterdam: The Final Stages of the Evolution of Single Stars*, pp. 356–436.
Arnett W. D., 1979, *Astrophysical Journal*, 230, L37.
Arnett W. D., 1982, *Astrophysical Journal*, 253, 785.
Arnett W. D., Meakin C., 2011, *Astrophysical Journal*, 733, 78.
Arnett W. D., Truran J. W., Woosley S. E., 1971, *Astrophysical Journal*, 165, 87.
Arons J., Lea S. M., 1976, *Astrophysical Journal*, 207, 914.
Artymowicz P., Lubow S. H., 1994, *Astrophysical Journal*, 421, 651.
Arzoumanian Z., Brazier A., Burke-Spolaor S., et al. 2018, *Astrophysical Journal Supplement Series*, 235, 37.
Arzoumanian Z., Fruchter A. S., Taylor J. H., 1994, *Astrophysical Journal*, 426, L85.
Ashman K. M., Zepf S. E., 1998, Globular Cluster Systems, Cambridge University Press.
Askar A., Szkudlarek M., Gondek-Rosińska D., Giersz M., Bulik T., 2017, *Monthly Notices of the Royal Astronomical Society*, 464, L36.
Baade D., Pigulski A., Rivinius T., Wang L., Martayan C., Handler G., ... Zwintz K., 2018, *Astronomy and Astrophysics*, 620, A145.
Baade W., Zwicky F., 1934, *Physical Review*, 46, 76.
Bachetti M., Harrison F. A., Walton D. J., Grefenstette B. W., Chakrabarty D., Fürst F., et al. 2014, *Nature*, 514, 202.
Backer D. C., Kulkarni S. R., Heiles C., Davis M. M., Goss W. M., 1982, *Nature*, 300, 615.
Bagchi M., 2011, *Monthly Notices of the Royal Astronomical Society*, 413, L47.
Bagot P., 1997, *Astronomy and Astrophysics*, 322, 533.
Bailes M., 1996, in van Paradijs J., van den Heuvel E. P. J., Kuulkers E., eds, *Compact Stars in Binaries*, proceedings of IAU Symposium 165, Pulsar Velocities. p. 213.
Bailes M., Bates S. D., Bhalerao V., Bhat N. D. R., Burgay M., Burke-Spolaor S., ... van Straten W., 2011, *Science*, 333, 1717.
Bailyn C. D., Jain R. K., Coppi P., Orosz J. A., 1998, *Astrophysical Journal*, 499, 367.
Banerjee S., 2017, *Monthly Notices of the Royal Astronomical Society*, 467, 524.
Banerjee S., Baumgardt H., Kroupa P., 2010, *Monthly Notices of the Royal Astronomical Society*, 402, 371.
Barai P., Gies D. R., Choi E., Das V., Deo R., Huang W., ... Peters G. J., 2004, *Astrophysical Journal*, 608, 989.
Bardeen J. M., Petterson J. A., 1975, *Astrophysical Journal*, 195, L65.
Bardeen J. M., Press W. H., Teukolsky S. A., 1972, *Astrophysical Journal*, 178, 347.
Barker B. M., O'Connell R. F., 1975, *Physical Review D*, 12, 329.
Barnard R., Garcia M. R., Primini F., Murray S. S., 2014, *Astrophysical Journal*, 791, 33.

Pedretti E., Zhao M., Schaefer G., Parks R., Che X., Thureau N., ten Brummelaar T. A., McAlister H. A., Ridgway S. T., Farrington C., Sturmann J., Sturmann L., Turner N., 2012, *Astrophysical Journal*, 752, 20.

Barr E. D., Champion D. J., Kramer M., Eatough R. P., Freire P. C. C., Karuppusamy R., ... Klein B., 2013, *Monthly Notices of the Royal Astronomical Society*, 435, 2234.

Barr E. D., Freire P. C. C., Kramer M., Champion D. J., Berezina M., Bassa C. G., ... Stappers B. W., 2017, *Monthly Notices of the Royal Astronomical Society*, 465, 1711.

Barstow M. A., Bond H. E., Holberg J. B., Burleigh M. R., Hubeny I., Koester D., 2005, *Monthly Notices of the Royal Astronomical Society*, 362, 1134.

Bartlett E. S., Clark J. S., Coe M. J., Garcia M. R., Uttley P., 2013, *Monthly Notices of the Royal Astronomical Society*, 429, 1213.

Barziv O., Kaper L., Van Kerkwijk M. H., Telting J. H., Van Paradijs J., 2001, *Astronomy and Astrophysics*, 377, 925.

Basko M. M., Sunyaev R. A., 1976, *Monthly Notices of the Royal Astronomical Society*, 175, 395.

Bassa C. G., Patruno A., Hessels J. W. T., Keane E. F., Monard B., Mahony E. K., ... Tendulkar S., 2014, *Monthly Notices of the Royal Astronomical Society*, 441, 1825.

Bates S. D., Bailes M., Bhat N. D. R., Burgay M., Burke-Spolaor S., D'Amico N., ... van Straten W., 2011, *Monthly Notices of the Royal Astronomical Society*, 416, 2455.

Batten A. H., Fletcher J. M., MacCarthy D. G., 1989, *Publications of the Dominion Astrophysical Observatory Victoria*, 17, 1.

Bauer E. B., White C. J., Bildsten L., 2019, *Astrophysical Journal*, 887, 68.

Bavera S. S., Fragos T., Qin Y., Zapartas E., Neijssel C. J., Mandel I., ... Stevenson S., 2020, *Astronomy and Astrophysics*, 635, A97.

Bavera S. S., Fragos T., Zevin M., Berry C. P. L., Marchant P., Andrews J. J., ... Zapartas E., 2021, *Astronomy and Astrophysics*, 647, A153.

Bavera S. S., Fishbach M., Zevin M., Zapartas E., Fragos T., 2022, *Astronomy and Astrophysics*, 665, A59.

Becker P. A., Wolff M. T., 2007, *Astrophysical Journal*, 654, 435.

Becklin E. E., Kristian J., Neugebauer G., Wynn-Williams C. G., 1972, *Nature Physical Science*, 239, 130.

Begelman M. C., 2002, *Astrophysical Journal*, 568, L97.

Begelman M. C., King A. R., Pringle J. E., 2006, *Monthly Notices of the Royal Astronomical Society*, 370, 399.

Begelman M. C., Sarazin C. L., 1986, *Astrophysical Journal*, 302, L59.

Bejger M., Zdunik J. L., Haensel P., Fortin M., 2011, *Astronomy and Astrophysics*, 536, A92.

Belczyński K., 2020, *Astrophysical Journal*, 905, L15.

Belczynski K., Askar A., Arca-Sedda M., Chruslinska M., Donnari M., Giersz M., ... Belloni D., 2018, *Astronomy and Astrophysics*, 615, A91.

Belczynski K., Banerjee S., 2020, *Astronomy and Astrophysics*, 640, L20.

Belczynski K., Bulik T., Fryer C. L., 2012, arXiv:1208.2422.

Belczynski K., Holz D. E., Bulik T., O'Shaughnessy R., 2016, *Nature*, 534, 512.

Belczynski K., Kalogera V., Bulik T., 2002, *Astrophysical Journal*, 572, 407.

Belczynski K., Kalogera V., Rasio F. A., Taam R. E., Zezas A., Bulik T., ... Ivanova N., 2008, *Astrophysical Journal Supplement Series*, 174, 223.

Belczynski K., Klencki J., Fields C. E., Olejak A., Berti E., Meynet G., ... Bulik T., et al. 2020, *Astronomy and Astrophysics*, 636, A104.

Belczyński K., Mikołajewska J., Munari U., Ivison R. J., Friedjung M., 2000, *Astronomy and Astrophysics Supplement Series*, 146, 407.

Belgacem E., Dirian Y., Foffa S., Maggiore M., 2018, *Physical Review D*, 97, 104066.
Bell J. F., Bailes M., 1996, *Astrophysical Journal*, 456, L33.
Bellm E. C., Kaplan D. L., Breton R. P., et al. 2016, *Astrophysical Journal*, 816, 74.
Belloni T. M., Stella L., 2014, *Space Science Reviews*, 183, 43.
Beloborodov A. M., Li X., 2016, *Astrophysical Journal*, 833, 261.
Benacquista M. J., Downing J. M. B., 2013, *Living Reviews in Relativity*, 16, 4.
Beniamini P., Piran T., 2016, *Monthly Notices of the Royal Astronomical Society*, 456, 4089.
Beniamini P., Wadiasingh Z., Metzger B. D., 2020, *Monthly Notices of the Royal Astronomical Society*, 496, 3390.
Bennett P. D., Bauer W. H., 2015, in *Giants of Eclipse: The ζ Aurigae Stars and Other Binary Systems* eds. T. B. Ake and E. Griffin, Vol. 408 of Astrophysics and Space Science Library, *The Special Case of VV Cephei*, p. 85.
Benvenuto O. G., De Vito M. A., Horvath J. E., 2012, *Astrophysical Journal*, 753, L33.
Benvenuto O. G., De Vito M. A., Horvath J. E., 2014, *Astrophysical Journal*, 786, L7.
Benz W., Hills J. G., 1987, *Astrophysical Journal*, 323, 614.
Berger E., 2014, *Annual Review of Astronomy and Astrophysics*, 52, 43.
Berti E., Buonanno A., Will C. M., 2005, *Physical Review D*, 71, 084025.
Berti E., Volonteri M., 2008, *Astrophysical Journal*, 684, 822.
Beskin V. S., Gurevich A. V., Istomin I. N., 1988, *Astrophysics and Space Science*, 146, 205.
Bethe H. A., 1990, *Reviews of Modern Physics*, 62, 801.
Bethe H. A., Brown G. E., 1995, *Astrophysical Journal*, 445, L129.
Bethe H. A., Brown G. E., 1998, *Astrophysical Journal*, 506, 780.
Bethe H. A., Wilson J. R., 1985, *Astrophysical Journal*, 295, 14.
Bhardwaj M., Gaensler B. M., Kaspi V. M., Landecker T. L., Mckinven R., Michilli D., et al. 2021, *Astrophysical Journal*, 910, L18.
Bhat N. D. R., Bailes M., Verbiest J. P. W., 2008, *Physical Review D*, 77, 124017.
Bhattacharya D., 2002, *Journal of Astrophysics and Astronomy*, 23, 67.
Bhattacharya D., van den Heuvel E. P. J., 1991, *Physics Reports*, 203, 1.
Bhattacharya D., Wijers R. A. M. J., Hartman J. W., Verbunt F., 1992, *Astronomy and Astrophysics*, 254, 198.
Bhattacharyya B., Roy J., 2022, in Bhattacharyya S., Papitto A., Bhattacharya D., eds, Astrophysics and Space Science Library, Vol. 465, *Radio Millisecond Pulsars*. p. 1
Bianco F. B., Modjaz M., Hicken M., Friedman A., Kirshner R. P., Bloom J. S., ... Rest A., 2014, *Astrophysical Journal Supplement Series*, 213, 19.
Bidelman W. P., Keenan P. C., 1951, *Astrophysical Journal*, 114, 473.
Biehle G. T., 1991, *Astrophysical Journal*, 380, 167.
Biermann P., Kippenhahn R., 1971, *Astronomy and Astrophysics*, 14, 32.
Biggs J. D., Bailes M., Lyne A. G., Goss W. M., Fruchter A. S., 1994, *Monthly Notices of the Royal Astronomical Society*, 267, 125.
Bildsten L., 1998, *Astrophysical Journal*, 501, L89.
Bildsten L., Chakrabarty D., Chiu J., Finger M. H., Koh D. T., Nelson R. W., ... Wilson R. B., 1997, *Astrophysical Journal Supplement Series*, 113, 367.
Bildsten L., Shen K. J., Weinberg N. N., Nelemans G., 2007, *Astrophysical Journal*, 662, L95.
Binder B. A., Sy J. M., Eracleous M., Christodoulou D. M., Bhattacharya S., Cappallo R., ... Williams B. F., 2021, *Astrophysical Journal*, 910, 74.
Binnendijk L., 1960, *Properties of Double Stars; A Survey of Parallaxes and Orbits.* University of Pennsylvania Press.
Binney J., Tremaine S., 2008, *Galactic Dynamics: Second Edition*, Princeton University Press.

Biryukov A., Abolmasov P., 2021, *Monthly Notices of the Royal Astronomical Society*, 505, 1775.
Bisnovatyi-Kogan G. S., Komberg B. V., 1974, AZh, 51, 373.
Bisnovatyi-Kogan G. S., Komberg B. V., 1976, *Soviet Astronomy Letters*, 2, 130.
Bitzaraki O. M., van den Heuvel E. P. J., 1993, in Regev O., Shaviv G., eds, *Cataclysmic Variables and Related Physics*, Annals of the Israel Physical Society, Jerusalem, Vol. 10, *Mass Transfer in Black-Hole Low-Mass X-Ray Binaries*, p. 277.
Blaauw A., 1956, *Scientific American*, 194, 36.
Blaauw A., 1961, *Bulletin of the Astronomical Institutes of the Netherlands*, 15, 265.
Blaauw A., 1993, in Cassinelli J. P., Churchwell E. B., eds, *Massive Stars: Their Lives in the Interstellar Medium* Vol. 35 of Astronomical Society of the Pacific Conference Series, *Massive Runaway Stars*, p. 207.
Blaauw A., Morgan W. W., 1954, *Astrophysical Journal*, 119, 625.
Blagorodnova N., Kotak R., Polshaw J., Kasliwal M. M., et al. 2017, *Astrophysical Journal*, 834, 107.
Blandford R., Teukolsky S. A., 1975, *Astrophysical Journal*, 198, L27.
Blandford R., Teukolsky S. A., 1976, *Astrophysical Journal*, 205, 580.
Blandford R. D., 1987, "Astrophysical Black Holes," in *Three Hundred Years of Gravitation*, editors S. W. Hawking and W. Israel, Cambridge University Press, pp. 277–329.
Blandford R. D., Znajek R. L., 1977, *Monthly Notices of the Royal Astronomical Society*, 179, 433.
Bleach J. N., Wood J. H., Smalley B., Catalán M. S., 2002a, *Monthly Notices of the Royal Astronomical Society*, 335, 593.
Bleach J. N., Wood J. H., Smalley B., Catalán M. S., 2002b, *Monthly Notices of the Royal Astronomical Society*, 336, 611.
Bloemen S., Marsh T. R., Degroote P., Østensen R. H., Pápics P. I., Aerts C., et al. 2012, *Monthly Notices of the Royal Astronomical Society*, 422, 2600.
Blondin J. M., Mezzacappa A., 2006, *Astrophysical Journal*, 642, 401.
Blondin J. M., Mezzacappa A., DeMarino C., 2003, *Astrophysical Journal*, 584, 971.
Bloom J. S., Sigurdsson S., Pols O. R., 1999, *Monthly Notices of the Royal Astronomical Society*, 305, 763.
Blundell K. M., Bowler M. G., 2004, *Astrophysical Journal*, 616, L159.
Bobrick A., Davies M. B., Church R. P., 2017, *Monthly Notices of the Royal Astronomical Society*, 467, 3556.
Bobrick A., Zenati Y., Perets H. B., Davies M. B., Church R., 2022, *Monthly Notices of the Royal Astronomical Society*, 510, 3758.
Bochenek C. D., Ravi V., Belov K. V., Hallinan G., Kocz J., Kulkarni S. R., McKenna D. L., 2020, *Nature*, 587, 59.
Bodaghee A., Courvoisier T. J. L., Rodriguez J., Beckmann V., Produit N., Hannikainen D., . . . Wendt G., 2007, *Astronomy and Astrophysics*, 467, 585.
Bodaghee A., Tomsick J. A., Rodriguez J., James J. B., 2012, *Astrophysical Journal*, 744, 108.
Bodenheimer P., Taam R. E., 1984, *Astrophysical Journal*, 280, 771.
Bodensteiner J., Shenar T., Mahy L., Fabry M., Marchant P., Abdul-Masih M., . . . Sana H., 2020, *Astronomy and Astrophysics*, 641, A43.
Boersma J., 1961, *Bulletin of the Astronomical Institutes of the Netherlands*, 15, 291.
Boffin H. M. J., 2015, *Mass Transfer by Stellar Wind*, p. 153.
Boffin H. M. J., Beccari G., Petr-Gotzens M. G., 2017, *The Messenger*, 169, 61.
Boffin H. M. J., Jorissen A., 1988, *Astronomy and Astrophysics*, 205, 155.

Bogdanov S., Guillot S., Ray P. S., Wolff M. T., Chakrabarty D., Ho W. et al. 2019, *Astrophysical Journal*, 887, L25.

Bogomazov A. I., 2014, *Astronomy Reports*, 58, 126.

Bogomazov A. I., Cherepashchuk A. M., Lipunov V. M., Tutukov A. V., 2018, *New Astronomy*, 58, 33.

Bohannan B., Conti P. S., 1976, *Astrophysical Journal*, 204, 797.

Bollig R., Yadav N., Kresse D., Janka H.-T., Müller B., Heger A., 2021, *Astrophysical Journal*, 915, 28.

Bolton C. T., 1971, in *Bulletin of the American Astronomical Society*, Vol. 3 of BAAS, The Spectrum of HDE 226868 = Cyg X-1 (?).. p. 458.

Bonanos A. Z., Stanek K. Z., Udalski A., Wyrzykowski L., Żebruń K., Kubiak M., . . . Soszyński I., 2004, *Astrophysical Journal*, 611, L33.

Bond H. E., 2011, *Astrophysical Journal*, 737, 17.

Bond H. E., Bedin L. R., Bonanos A. Z., Humphreys R. M., Monard L. A. G. B., Prieto J. L., Walter F. M., 2009, *Astrophysical Journal*, 695, L154.

Bond H. E., Henden A., Levay Z. G., Panagia N., Sparks W. B., Starrfield S., . . . Munari U., 2003, *Nature*, 422, 405.

Bondi H., Hoyle F., 1944, *Monthly Notices of the Royal Astronomical Society*, 104, 273.

Bonnell I. A., Bate M. R., Vine S. G., 2003, *Monthly Notices of the Royal Astronomical Society*, 343, 413.

Bonnet-Bidaud J. M., van der Klis M., 1981, *Astronomy and Astrophysics*, 101, 299.

Bonsema P. F. J., van den Heuvel E. P. J., 1985, *Astronomy and Astrophysics*, 146, L3.

Boubert D., Guillochon J., Hawkins K., Ginsburg I., Evans N. W., Strader J., 2018, *Monthly Notices of the Royal Astronomical Society*, 479, 2789.

Bourgoin A., Le Poncin-Lafitte C., Mathis S., Angonin M. C., 2021, in Siebert A., Baillié K., Lagadec E., Lagarde N., Malzac J., Marquette J. B., N'Diaye M., Richard J., Venot O., eds, SF2A-2021: Proceedings of the Annual Meeting of the French Society of Astronomy and Astrophysics; Dipolar magnetic fields in binaries and gravitational waves, pp. 101–104.

Bours M. C. P., Marsh T. R., Parsons S. G., Copperwheat C. M., Dhillon V. S., Littlefair S. P., . . . Tremblay P.-E., 2014, *Monthly Notices of the Royal Astronomical Society*, 438, 3399.

Boyer R. H., Lindquist R. W., 1967, *Journal of Mathematical Physics*, 8, 265.

Boyles J., Lorimer D. R., Turk P. J., Mnatsakanov R., Lynch R. S., Ransom S. M., . . . Belczynski K., 2011, *Astrophysical Journal*, 742, 51.

Bozzo E., Ducci L., Falanga M., 2021, *Monthly Notices of the Royal Astronomical Society*, 501, 2403.

Bozzo E., Falanga M., Stella L., 2008, *Astrophysical Journal*, 683, 1031.

Bozzo E., Oskinova L., Feldmeier A., Falanga M., 2016, *Astronomy and Astrophysics*, 589, A102.

Bozzo E., Stella L., Vietri M., Ghosh P., 2009, *Astronomy and Astrophysics*, 493, 809.

Bradt H., Levine A. M., Remillard R. A., Smith D. A., 2000. arXiv:astro-ph/0001460.

Braes L. L. E., Miley G. K., 1971, *Nature*, 232, 246.

Bragaglia A., Greggio L., Renzini A., D'Odorico S., 1990, *Astrophysical Journal*, 365, L13.

Brandt N., Podsiadlowski P., 1995, *Monthly Notices of the Royal Astronomical Society*, 274, 461.

Brandt W. N., Podsiadlowski P., Sigurdsson S., 1995, *Monthly Notices of the Royal Astronomical Society*, 277, L35.

Braun H., Langer N., 1995, *Astronomy and Astrophysics*, 297, 483.

Bray J. C., Eldridge J. J., 2016, *Monthly Notices of the Royal Astronomical Society*, 461, 3747.

Breen P. G., Heggie D. C., 2013, *Monthly Notices of the Royal Astronomical Society*, 432, 2779.

Breivik K., Coughlin S., Zevin M., Rodriguez C. L., Kremer K., Ye C. S., . . . Rasio F. A., 2020, *Astrophysical Journal*, 898, 71.

Breivik K., Kremer K., Bueno M., Larson S. L., Coughlin S., Kalogera V., 2018, *Astrophysical Journal*, 854, L1.

Breton R. P., Kaspi V. M., Kramer M., McLaughlin M. A., Lyutikov M., Ransom S. M., . . . Possenti A., 2008, *Science*, 321, 104.

Breton R. P., Rappaport S. A., van Kerkwijk M. H., Carter J. A., 2012, *Astrophysical Journal*, 748, 115.

Breton R. P., Roberts M. S. E., Ransom S. M., Kaspi V. M., Durant M., Bergeron P., Faulkner A. J., 2007, *Astrophysical Journal*, 661, 1073.

Breton R. P., van Kerkwijk M. H., Roberts M. S. E., Hessels J. W. T., Camilo F., McLaughlin M. A., . . . Ray P. S., Stairs I. H., 2013, *Astrophysical Journal*, 769, 108.

Brinkman C., Freire P. C. C., Rankin J., Stovall K., 2018, *Monthly Notices of the Royal Astronomical Society*, 474, 2012.

Broekgaarden F. S., Berger E., 2021, *Astrophysical Journal*, 920, L13.

Broekgaarden F. S., Berger E., Neijssel C. J., Vigna-Gomez A., Chattopadhyay D., Stevenson S., Chruslinska M., Justhan S., de Mink S. E., Mandel I., 2021, *Monthly Notices of the Royal Astronomical Society*, 508, 5028.

Broekgaarden F. S., Berger E., Neijssel C. J., Vigna-Gómez A., Chattopadhyay D., Stevenson S., . . . Mandel I., 2021, Monthly Notices of the Royal Astronomical Society, 508, 5028.

Broekgaarden F. S., Justhan S., de Mink S. E., Gair J., Mandel I., Stevenson S., . . . Neijssel C. J., 2019, *Monthly Notices of the Royal Astronomical Society*, 490, 5228.

Brogaard K., Christiansen S. M., Grundahl F., Miglio A., Izzard R. G., Tauris T. M., et al. 2018, *Monthly Notices of the Royal Astronomical Society*, 481, 5062.

Brooks J., Schwab J., Bildsten L., Quataert E., Paxton B., 2017, *Astrophysical Journal*, 843, 151.

Brott I., de Mink S. E., Cantiello M., Langer N., de Koter A., Evans C. J., . . . Vink J. S., 2011, *Astronomy and Astrophysics*, 530, A115.

Brown G. E., 1995, *Astrophysical Journal*, 440, 270.

Brown G. E., Bethe H. A., 1994, *Astrophysical Journal*, 423, 659.

Brown G. E., Heger A., Langer N., Lee C., Wellstein S., Bethe H. A., 2001, *New Astronomy*, 6, 457.

Brown G. E., Lee C.-H., Bethe H. A., 1999, *Nature*, 4, 313.

Brown G. E., Lee C. H., Tauris T. M., 2001, *New Astronomy*, 6, 331.

Brown W. R., Geller M. J., Kenyon S. J., 2009, *Astrophysical Journal*, 690, 1639.

Brown W. R., Geller M. J., Kenyon S. J., 2012, *Astrophysical Journal*, 751, 55.

Brown W. R., Geller M. J., Kenyon S. J., 2014, *Astrophysical Journal*, 787, 89.

Brown W. R., Geller M. J., Kenyon S. J., Kurtz M. J., 2005, *Astrophysical Journal*, 622, L33.

Brown W. R., Geller M. J., Kenyon S. J., Kurtz M. J., 2006, *Astrophysical Journal*, 640, L35.

Brown W. R., Kilic M., Allende Prieto C., Kenyon S. J., 2010, *Astrophysical Journal*, 723, 1072.

Brown W. R., Kilic M., Kosakowski A., Andrews J. J., Heinke C. O., Agüeros M. A., Camilo F., Gianninas A., Hermes J. J., Kenyon S. J., 2020, *Astrophysical Journal*, 889, 49.

Bugli M., Guilet J., Obergaulinger M., 2021, *Monthly Notices of the Royal Astronomical Society*, 507, 443.

Büning A., Ritter H., 2004, *Astronomy and Astrophysics*, 423, 281.

Burderi L., D'Antona F., Burgay M., 2002, *Astrophysical Journal*, 574, 325.

Burderi L., Possenti A., D'Antona F., Di Salvo T., Burgay M., Stella L., . . . d'Amico N., 2001, *Astrophysical Journal*, 560, L71.

Burdge K. B., Coughlin M. W., Fuller J., Kaplan D. L., Kulkarni S. R., Marsh T. R., Prince T. A., 2020, *Astrophysical Journal*, 905, L7.

Burdge K. B., Coughlin M. W., Fuller J., Kupfer T., Bellm E. C., Bildsten L., et al. 2019, *Nature*, 571, 528.

Burdge K. B., Prince T. A., Fuller J., Kaplan D. L., Marsh T. R., Tremblay P.-E., et al. 2020, *Astrophysical Journal*, 905, 32.

Burgay M., D'Amico N., Possenti A., Manchester R. N., Lyne A. G., Joshi B. C., ... Lorimer D. R., 2003, *Nature*, 426, 531.

Burrows A., 2013, *Reviews of Modern Physics*, 85, 245.

Burrows A., Dessart L., Livne E., Ott C. D., Murphy J., 2007, *Astrophysical Journal*, 664, 416.

Burrows A., Dolence J. C., Murphy J. W., 2012, *Astrophysical Journal*, 759, 5.

Burrows A., Hayes J., 1996, *Physical Review Letters*, 76, 352.

Burrows A., Radice D., Vartanyan D., 2019, *Monthly Notices of the Royal Astronomical Society*, 485, 3153.

Burrows A., Radice D., Vartanyan D., Nagakura H., Skinner M. A., Dolence J. C., 2020, *Monthly Notices of the Royal Astronomical Society*, 491, 2715.

Burrows A., Vartanyan D., 2021, *Nature*, 589, 29.

Busso M., Gallino R., Wasserburg G. J., 1999, *Annual Review of Astronomy and Astrophysics*, 37, 239.

Cadelano M., Ransom S. M., Freire P. C. C., Ferraro F. R., Hessels J. W. T., Lanzoni B., ... Stairs I. H., 2018, *Astrophysical Journal*, 855, 125.

Calcaferro L. M., Córsico A. H., Althaus L. G., 2017, *Astronomy and Astrophysics*, 607, A33.

Callister T. A., Farr W. M., Renzo M., 2021, *Astrophysical Journal*, 920, 157.

Callister T. A., Haster C.-J., Ng K. K. Y., Vitale S., Farr W. M., 2021, *Astrophysical Journal*, 922, L5.

Camacho J., Torres S., García-Berro E., Zorotovic M., Schreiber M. R., Rebassa-Mansergas A., ... Gänsicke B. T., 2014, *Astronomy and Astrophysics*, 566, A86.

Cameron A. D., Champion D. J., Bailes M., Balakrishnan V., Barr E. D., Bassa C. G., et al. 2020, *Monthly Notices of the Royal Astronomical Society*, 493, 1063.

Cameron A. D., Champion D. J., Kramer M., Bailes M., Barr E. D., et al. 2018, *Monthly Notices of the Royal Astronomical Society*, 475, L57.

Camilo F., 1995, PhD thesis, Princeton University.

Camilo F., 1996, in Johnston S., Walker M. A., & Bailes, M., eds, IAU Colloq. 160: *Pulsars: Problems and Progress* Vol. 105 of Astronomical Society of the Pacific Conference Series, *Intermediate-Mass Binary Pulsars: a New Class of Objects?* pp 539.

Camilo F., Lorimer D. R., Freire P., Lyne A. G., Manchester R. N., 2000, *Astrophysical Journal*, 535, 975.

Camilo F., Ransom S. M., Chatterjee S., Johnston S., Demorest P., 2012, *Astrophysical Journal*, 746, 63.

Camilo F., Thorsett S. E., Kulkarni S. R., 1994, *Astrophysical Journal*, 421, L15.

Campana S., Colpi M., Mereghetti S., Stella L., Tavani M., 1998, *Astronomy and Astrophysics Review*, 8, 279.

Campanelli M., Lousto C., Zlochower Y., Merritt D., 2007, *Astrophysical Journal*, 659, L5.

Canal R., Isern J., Labay J., 1990, *Annual Review of Astronomy and Astrophysics*, 28, 183.

Cannon R. C., 1993, *Monthly Notices of the Royal Astronomical Society*, 263, 817.

Cantiello M., Mankovich C., Bildsten L., Christensen-Dalsgaard J., Paxton B., 2014, *Astrophysical Journal*, 788, 93.

Cantiello M., Yoon S.-C., Langer N., Livio M., 2007, *Astronomy and Astrophysics*, 465, L29.

Cao Y., Kasliwal M. M., Arcavi I., et al. 2013, *Astrophysical Journal*, 775, L7.

Carpano S., Haberl F., Maitra C., Vasilopoulos G., 2018, *Monthly Notices of the Royal Astronomical Society*, 476, L45.

Cartwright T. F., Engel M. C., Heinke C. O., Sivakoff G. R., Berger J. J., Gladstone J. C., Ivanova N., 2013, *Astrophysical Journal*, 768, 183.

Casares J., Negueruela I., Ribó M., Ribas I., Paredes J. M., Herrero A., Simón-Díaz S., 2014, *Nature*, 505, 378.

Cassisi S., Iben Jr. I., Tornambc A., 1998, *Astrophysical Journal*, 496, 376.

Castro N., Fossati L., Langer N., Simón-Díaz S., Schneider F. R. N., Izzard R. G., 2014, *Astronomy and Astrophysics*, 570, L13.

Chakrabarty D., Morgan E. H., Muno M. P., Galloway D. K., Wijnands R., van der Klis M., Markwardt C. B., 2003, *Nature*, 424, 42.

Chakrabarty D., Morgan E. H., Wijnands R., van der Klis M., Galloway D. K., Muno M. P., Markwardt C. B., 2003, in AAS/High Energy Astrophysics Division #7, Thermonuclear X-Ray Burst Oscillations at the Spin Frequency of the Accretion-Powered Millisecond Pulsar SAX J1808.4-3658, p. 45.01.

Chamandy L., Frank A., Blackman E. G., Carroll-Nellenback J., Liu B., Tu Y., . . . Peng B., 2018, *Monthly Notices of the Royal Astronomical Society*, 480, 1898.

Champion D. J., Lorimer D. R., McLaughlin M. A., Cordes J. M., Arzoumanian Z., Weisberg J. M., Taylor J. H., 2004, *Monthly Notices of the Royal Astronomical Society*, 350, L61.

Champion D. J., Ransom S. M., Lazarus P., et al. 2008, *Science*, 320, 1309.

Chan M. L., Messenger C., Heng I. S., Hendry M., 2018, *Physical Review D*, 97, 123014.

Chandrasekhar S., 1933, *Monthly Notices of the Royal Astronomical Society*, 93, 390.

Chanlaridis S., Antoniadis J., Aguilera-Dena D. R., GrŁafener G., Langer N., Stergioulas N., 2022, arXiv:2201.00871.

Charles P. A., Coe M. J., 2010, *Optical, Ultraviolet and Infrared Observations of X-ray Binaries*, in W.H.G. Lewin and M. van der Klis, eds, *Compact Stellar X-ray Sources*, Cambridge University Press, p. 215.

Chashkina A., Abolmasov P., Poutanen J., 2017, *Monthly Notices of the Royal Astronomical Society*, 470, 2799.

Chatterjee S., Vlemmings W. H. T., Brisken W. F., Lazio T. J. W., Cordes J. M., Goss W. M., . . . Kramer M., 2005, *Astrophysical Journal*, 630, L61.

Chattopadhyay D., Stevenson S., Hurley J. R., Bailes M., Broekgaarden F., 2021, *Monthly Notices of the Royal Astronomical Society*, 504, 3682.

Chaty S., 2011, in Schmidtobreick L., Schreiber M. R., Tappert C., eds, *Evolution of Compact Binaries*, Vol. 447 of Astronomical Society of the Pacific Conference Series, *Nature, Formation, and Evolution of High Mass X-Ray Binaries*, p. 29.

Chaty S., 2013, *Advances in Space Research*, 52, 2132.

Chaty S., Rahoui F., 2012, *Astrophysical Journal*, 751, 150.

Chatzopoulos E., Wheeler J. C., 2012, *Astrophysical Journal*, 748, 42.

Chen H.-L., Chen X., Tauris T. M., Han Z., 2013, *Astrophysical Journal*, 775, 27.

Chen H.-L., Tauris T. M., Chen X., Han Z., 2022, *Astrophysical Journal*, 930, 134.

Chen H.-L., Tauris T. M., Han Z., Chen X., 2021, *Monthly Notices of the Royal Astronomical Society*, 503, 3540.

Chen K., Ruderman M., 1993, *Astrophysical Journal*, 402, 264.

Chen K.-J., 2021, *International Journal of Modern Physics D*, 30, 2130001.

Chen W.-C., 2020, *Physical Review D*, 102, 043020.

Chen W.-C., 2021, *Physical Review D*, 103, 103004.

Chen W.-C., Li X.-D., 2015, *Astronomy and Astrophysics*, 583, A108.

Chen W.-C., Liu D.-D., Wang B., 2020, *Astrophysical Journal*, 900, L8.

Chen W.-C., Podsiadlowski P., 2017, *Astrophysical Journal*, 837, L19.

Chen X., Maxted P. F. L., Li J., Han Z., 2017, *Monthly Notices of the Royal Astronomical Society*, 467, 1874.
Chen X., Wang W., Tong H., 2021, *Journal of High Energy Astrophysics*, 31, 1.
Chen Y., Bressan A., Girardi L., Marigo P., Kong X., Lanza A., 2015, *Monthly Notices of the Royal Astronomical Society*, 452, 1068.
Cherepashchuk A., Postnov K., Molkov S., Antokhina E., Belinski A., 2020, *New Astronomy Reviews*, 89, 101542.
Cherepashchuk A. M., 1981, *Monthly Notices of the Royal Astronomical Society*, 194, 761.
Cherepashchuk A. M., Belinski A. A., Dodin A. V., Postnov K. A., 2021, *Monthly Notices of the Royal Astronomical Society*, 507, L19.
Chevalier R. A., 1993, *Astrophysical Journal*, 411, L33.
Chevalier R. A., 1996, *Astrophysical Journal*, 459, 322.
Chini R., Hoffmeister V. H., Nasseri A., Stahl O., Zinnecker H., 2012, *Monthly Notices of the Royal Astronomical Society*, 424, 1925.
Chini R., Nasseri A., Hoffmeister V. H., Buda L. S., Barr A., 2011, in Schmidtobreick L., Schreiber M. R., Tappert C., eds, *Evolution of Compact Binaries*, Vol. 447 of Astronomical Society of the Pacific Conference Series, *Most High-Mass Stars are Born as Twins*, p. 67.
Chojnowski S. D., Labadie-Bartz J., Rivinius T., Gies D., Panoglou D., Borges Fernandes M., et al. 2018, *Astrophysical Journal*, 865, 76.
Chomiuk L., Metzger B. D., Shen K. J., 2021, *Annual Review of Astronomy and Astrophysics*, 59.
Chomiuk L., Soderberg A. M., Moe M., Chevalier R. A., Rupen M. P., Badenes C., . . . Dittmann J. A., 2012, *Astrophysical Journal*, 750, 164.
Chomiuk L., Strader J., Maccarone T. J., Miller-Jones J. C. A., Heinke C., Noyola E., . . . Ransom S., 2013, *Astrophysical Journal*, 777, 69.
Chou Y., Grindlay J. E., 2001, *Astrophysical Journal*, 563, 934.
Christensen U. R., Holzwarth V., Reiners A., 2009, *Nature*, 457, 167.
Christodoulou D. M., Laycock S. G. T., Kazanas D., Cappallo R., Contopoulos I., 2017, *Research in Astronomy and Astrophysics*, 17, 063.
Chruslinska M., Belczynski K., Klencki J., Benacquista M., 2018, *Monthly Notices of the Royal Astronomical Society*, 474, 2937.
Church R. P., Bush S. J., Tout C. A., Davies M. B., 2006, *Monthly Notices of the Royal Astronomical Society*, 372, 715.
Claeys J. S. W., Pols O. R., Izzard R. G., Vink J., Verbunt F. W. M., 2014, *Astronomy and Astrophysics*, 563, A83.
Claret A., Cunha N. C. S., 1997, *Astronomy and Astrophysics*, 318, 187.
Clark G. W., 1975, *Astrophysical Journal*, 199, L143.
Clark J. P. A., Eardley D. M., 1977, *Astrophysical Journal*, 215, 311.
Clark J. P. A., van den Heuvel E. P. J., Sutantyo W., 1979, *Astronomy and Astrophysics*, 72, 120.
Clark J. S., Goodwin S. P., Crowther P. A., Kaper L., Fairbairn M., Langer N., Brocksopp C., 2002, *Astronomy and Astrophysics*, 392, 909.
Clark J. S., Najarro F., Negueruela I., Ritchie B. W., Urbaneja M. A., Howarth I. D., 2012, *Astronomy and Astrophysics*, 541, A145.
Clavel M., Dubus G., Casares J., Babusiaux C., 2021, *Astronomy and Astrophysics*, 645, A72.
Clayton D. D., 1968, *Principles of Stellar Evolution and Nucleosynthesis*. McGraw-Hill, New York, pp. 612.
Clayton M., Podsiadlowski P., Ivanova N., Justhan S., 2017, *Monthly Notices of the Royal Astronomical Society*, 470, 1788.

Coe M. J., Kennea J. A., Evans P. A., Udalski A., 2020, *Monthly Notices of the Royal Astronomical Society*, 497, L50.
Coe M. J., Kirk J., 2015, *Monthly Notices of the Royal Astronomical Society*, 452, 969.
Cognard I., Freire P. C. C., Guillemot L., Theureau G., Tauris T. M., Wex N., et al. 2017, *Astrophysical Journal*, 844, 128.
Coleiro A., Chaty S., 2013, *Astrophysical Journal*, 764, 185.
Colgate S. A., Petschek A. G., Kriese J. T., 1980, *Astrophysical Journal*, 237, L81.
Colgate S. A., White R. H., 1966, *Astrophysical Journal*, 143, 626.
Collado A., Gamen R., Barbá R. H., Morrell N., 2015, *Astronomy and Astrophysics*, 581, A49.
Collier A. C., Jenkins C. R., 1984, *Monthly Notices of the Royal Astronomical Society*, 211, 391.
Colpi M., Casella P., Gorini V., Moschella U., Possenti A., 2009, *Physics of Relativistic Objects in Compact Binaries: From Birth to Coalescence*. Astrophysics and Space Science Library, Springer, Dordrecht, Vol. 359.
Colpi M., Devecchi B., 2009, *Dynamical Formation and Evolution of Neutron Star and Black Hole Binaries in Globular Clusters*, in Astrophysics and Space Science Library, Springer Dordrecht, Vol. 359, p. 199.
Colpi M., Mapelli M., Possenti A., 2003, *Astrophysical Journal*, 599, 1260.
Conti P. S., Crowther P. A., Leitherer C., 2008, *From Luminous Hot Stars to Starburst Galaxies*.
Conti P. S., Ebbets D., 1977, *Astrophysical Journal*, 213, 438.
Contopoulos I., Spitkovsky A., 2006, *Astrophysical Journal*, 643, 1139.
Copperwheat C. M., Marsh T. R., Littlefair S. P., Dhillon V. S., Ramsay G., Drake A. J., . . . Tulloch S., 2011, *Monthly Notices of the Royal Astronomical Society*, 410, 1113.
Corbet R. H. D., 1984, *Astronomy and Astrophysics*, 141, 91.
Corbet R. H. D., Sokoloski J. L., Mukai K., Markwardt C. B., Tueller J., 2008, *Astrophysical Journal*, 675, 1424.
Cordes J. M., 1979, *Space Science Reviews*, 24, 567.
Cordes J. M., Chatterjee S., 2019, *Annual Review of Astronomy and Astrophysics*, 57, 417.
Cordes J. M., Lazio T. J. W., 2002, ArXiv: astro-ph/0207156.
Cordes J. M., Romani R. W., Lundgren S. C., 1993, *Nature*, 362, 133.
Coriat M., Fender R. P., Dubus G., 2012, *Monthly Notices of the Royal Astronomical Society*, 424, 1991.
Corongiu A., Burgay M., Possenti A., Camilo F., D'Amico N., Lyne A. G., . . . van Straten W., 2012, *Astrophysical Journal*, 760, 100.
Corongiu A., Kramer M., Stappers B. W., Lyne A. G., Jessner A., Possenti A., . . . Löhmer O., 2007, *Astronomy and Astrophysics*, 462, 703.
Corradi R. L. M., Sabin L., Miszalski B., et al. 2011, *Monthly Notices of the Royal Astronomical Society*, 410, 1349.
Corral-Santana J. M., Casares J., Muñoz-Darias T., Bauer F. E., Martínez-Pais I. G., Russell D. M., 2016, *Astronomy and Astrophysics*, 587, A61.
Costa G., Bressan A., Mapelli M., Marigo P., Iorio G., Spera M., 2021, *Monthly Notices of the Royal Astronomical Society*, 501, 4514.
Coté J., 1993, PhD thesis, University of Amsterdam.
Coti Zelati F., Rea N., Pons J. A., Campana S., Esposito P., 2018, *Monthly Notices of the Royal Astronomical Society*, 474, 961.
Coughlin E. R., Quataert E., Fernández R., Kasen D., 2018, *Monthly Notices of the Royal Astronomical Society*, 477, 1225.
Coulter D. A., Foley R. J., Kilpatrick C. D., et al. 2017, *Science*, 358, 1556.
Counselman Charles C. I., 1973, *Astrophysical Journal*, 180, 307.

Cowperthwaite P. S., Berger E., Villar V. A., Metzger B. D., Nicholl M., Chornock R., et al. 2017, *Astrophysical Journal*, 848, L17.

Cox J. P., Giuli R. T., 1968, Principles of Stellar Structure, Vol. 1 and 2, Gordon and Breach, New York.

Crampton D., Cowley A. P., Hutchings J. B., 1980, *Astrophysical Journal*, 235, L131.

Crawford J. A., 1955, *Astrophysical Journal*, 121, 71.

Cromartie H. T., Camilo F., Kerr M., Deneva J. S., Ransom S. M., Ray P. S., ... Wood K. S., 2016, *Astrophysical Journal*, 819, 34.

Cromartie H. T., Fonseca E., Ransom S. M., Demorest P. B., Arzoumanian Z., Blumer H., et al. 2020, *Nature Astronomy*, 4, 72.

Cropper M., Harrop-Allin M. K., Mason K. O., Mittaz J. P. D., Potter S. B., Ramsay G., 1998, *Monthly Notices of the Royal Astronomical Society*, 293, L57.

Crowther P. A., 2007, *Annual Review of Astronomy and Astrophysics*, 45, 177.

Crowther P. A., Lennon D. J., Walborn N. R., 2006, *Astronomy and Astrophysics*, 446, 279.

Cruces M., Reisenegger A., Tauris T. M., 2019, *Monthly Notices of the Royal Astronomical Society*, 490, 2013.

Cruz-Osorio A., Rezzolla L., 2020, *Astrophysical Journal*, 894, 147.

Cumming A., Zweibel E., Bildsten L., 2001, *Astrophysical Journal*, 557, 958.

Curran S. J., Lorimer D. R., 1995, *Monthly Notices of the Royal Astronomical Society*, 276, 347.

Curtis S., Ebinger K., Fröhlich C., Hempel M., Perego A., Liebendörfer M., Thielemann F.-K., 2019, *Astrophysical Journal*, 870, 2.

Dall'Osso S., Perna R., Stella L., 2015, *Monthly Notices of the Royal Astronomical Society*, 449, 2144.

Dall'Osso S., Stella L., 2022, in Bhattacharyya S., Papitto A., Bhattacharya D., eds, *Astrophysics and Space Science Library*, Vol. 465 of Astrophysics and Space Science Library, Millisecond Magnetars, pp. 245–280.

D'Amico N., 2000, in Kramer M., Wex N., & Wielebinsk, R., eds, IAU Colloq. 177: Pulsar Astronomy—2000 and Beyond Vol. 202 of Astronomical Society of the Pacific Conference Series, *The Bologna Submillisecond Pulsar Survey*, p. 27.

Damour T., Deruelle N., 1986, *Annales de l'Institut Henri Poincaré Physique Théorique*, Vol. 44, No. 3, pp. 263–292.

Damour T., Esposito-Farese G., 1992, *Classical and Quantum Gravity*, 9, 2093.

Damour T., Ruffini R., 1974, *Academie des Sciences Paris Comptes Rendus Serie Sciences Mathematiques*, 279, 971.

Damour T., Taylor J. H., 1991, *Astrophysical Journal*, 366, 501.

Damour T., Taylor J. H., 1992, *Physical Review D*, 45, 1840.

D'Angelo C. R., 2017, *Monthly Notices of the Royal Astronomical Society*, 470, 3316.

D'Angelo C. R., Spruit H. C., 2010, *Monthly Notices of the Royal Astronomical Society*, 406, 1208.

D'Angelo C. R., Spruit H. C., 2012, *Monthly Notices of the Royal Astronomical Society*, 420, 416.

D'Antona F., Mazzitelli I., Ritter H., 1989, *Astronomy and Astrophysics*, 225, 391.

D'Antona F., Ventura P., Burderi L., Di Salvo T., Lavagetto G., Possenti A., Teodorescu A., 2006, *Astrophysical Journal*, 640, 950.

Darwin G. H., 1879, *Proceedings of the Royal Society of London*, 29, 168.

Davidson K., Ostriker J. P., 1973, *Astrophysical Journal*, 179, 585.

Davies M. B., Piotto G., de Angeli F., 2004, *Monthly Notices of the Royal Astronomical Society*, 349, 129.

Davies M. B., Ritter H., King A., 2002, *Monthly Notices of the Royal Astronomical Society*, 335, 369.

Davies R. E., Pringle J. E., 1980, *Monthly Notices of the Royal Astronomical Society*, 191, 599.

Davis P. J., Kolb U., Knigge C., 2012, *Monthly Notices of the Royal Astronomical Society*, 419, 287.

De K., Kasliwal M. M., Cantwell T., Cao Y., Cenko S. B., Gal-Yam A., et al. 2018, *Astrophysical Journal*, 866, 72.

De K., Kasliwal M. M., Ofek E. O., Moriya T. J., et al. 2018, *Science*, 362, 201.

De K., Kasliwal M. M., Tzanidakis A., Fremling U. C., et al. 2020, *Astrophysical Journal*, 905, 58.

De S., MacLeod M., Everson R. W., Antoni A., Mandel I., Ramirez-Ruiz E., 2020, *Astrophysical Journal*, 897, 130.

de Freitas Pacheco J. A., 1998, *Mass Loss Rates from B[e] Stars*, in Anne Marie Hubert and Carlos Jaschek, eds, B[e] stars: Proceedings of the Paris workshop held 9-12 June, 1997. Springer, Dordrecht. p. 221.

De Greve J. P., 1986, *Space Science Reviews*, 43, 139.

De Greve J. P., De Loore C., 1976, *Astrophysics and Space Science*, 43, 35.

de Jager C., Lobel A., Nieuwenhuijzen H., Stothers R., 2001, *Monthly Notices of the Royal Astronomical Society*, 327, 452.

de Jager C., Nieuwenhuijzen H., van der Hucht K. A., 1988, *Astronomy and Astrophysics Supplement Series*, 72, 259.

de Kool M., 1990, *Astrophysical Journal*, 358, 189.

de Kool M., van den Heuvel E. P. J., Rappaport S. A., 1986, *Astronomy and Astrophysics*, 164, 73.

de La Chevrotière A., Moffat A. F. J., Chené A.-N., 2011, *Monthly Notices of the Royal Astronomical Society*, 411, 635.

De Loore C., De Greve J. P., 1975, *Astrophysics and Space Science*, 35, 241.

De Loore C., de Grève J. P., 1976, in Eggleton P., Mitton S., Whelan J., eds, *Structure and Evolution of Close Binary Systems*, proceedings of IAU Symposium 73, *Two Types of Evolution of Massive Close Binary Systems*, p. 27.

De Loore C., De Greve J. P., de Cuyper J. P., 1975, *Astrophysics and Space Science*, 36, 219.

De Loore C., Sutantyo W., 1984, *Astrophysics and Space Science*, 99, 335.

de Marco O., Passy J., Moe M., Herwig F., Mac Low M., Paxton B., 2011, *Monthly Notices of the Royal Astronomical Society*, 411, 2277.

de Mink S. E., Cantiello M., Langer N., Pols O. R., Brott I., Yoon S.-C., 2009, *Astronomy and Astrophysics*, 497, 243.

de Mink S. E., King A., 2017, *Astrophysical Journal*, 839, L7.

de Mink S. E., Langer N., Izzard R. G., Sana H., de Koter A., 2013, *Astrophysical Journal*, 764, 166.

de Mink S. E., Mandel I., 2016, *Monthly Notices of the Royal Astronomical Society*, 460, 3545.

De Rosa R. J., Patience J., Wilson P. A., Schneider A., Wiktorowicz S. J., Vigan A., . . . Lai O., 2014, *Monthly Notices of the Royal Astronomical Society*, 437, 1216.

de Val-Borro M., Karovska M., Sasselov D. D., Stone J. M., 2017, *Monthly Notices of the Royal Astronomical Society*, 468, 3408.

Dehman C., Viganò D., Rea N., Pons J. A., Perna R., Garcia-Garcia A., 2020, *Astrophysical Journal*, 902, L32.

Delgado A. J., Thomas H.-C., 1981, *Astronomy and Astrophysics*, 96, 142.

Deloye C. J., Bildsten L., 2003, *Astrophysical Journal*, 598, 1217.

Demorest P. B., Pennucci T., Ransom S. M., Roberts M. S. E., Hessels J. W. T., 2010, *Nature*, 467, 1081.

Deneva J. S., Stovall K., McLaughlin M. A., Bates S. D., Freire P. C. C., Martinez J. G., ... Bagchi M., 2013, *Astrophysical Journal*, 775, 51.

Deng Z.-L., Li X.-D., Gao Z.-F., Shao Y., 2021, *Astrophysical Journal*, 909, 174.

Dergachev V., Papa M. A., 2020, *Physical Review Letters*, 125, 171101.

Dermine T., Izzard R. G., Jorissen A., Van Winckel H., 2013, *Astronomy and Astrophysics*, 551, A50.

Dessart L., Burrows A., Ott C. D., Livne E., Yoon S.-C., Langer N., 2006, *Astrophysical Journal*, 644, 1063.

Dessart L., Hillier D. J., Li C., Woosley S., 2012, *Monthly Notices of the Royal Astronomical Society*, 424, 2139.

Desvignes G., Caballero R. N., Lentati L., Verbiest J. P. W., Champion D. J., Stappers B. W., et al. 2016, *Monthly Notices of the Royal Astronomical Society*, 458, 3341.

Desvignes G., Kramer M., Lee K., van Leeuwen J., Stairs I., Jessner A., ... Stappers B. W., 2019, *Science*, 365, 1013.

Dewey R. J., Cordes J. M., 1987, *Astrophysical Journal*, 321, 780.

Dewi J. D. M., Podsiadlowski P., Pols O. R., 2005, *Monthly Notices of the Royal Astronomical Society*, 363, L71.

Dewi J. D. M., Podsiadlowski P., Sena A., 2006, *Monthly Notices of the Royal Astronomical Society*, 368, 1742.

Dewi J. D. M., Pols O. R., 2003, *Monthly Notices of the Royal Astronomical Society*, 344, 629.

Dewi J. D. M., Pols O. R., Savonije G. J., van den Heuvel E. P. J., 2002, *Monthly Notices of the Royal Astronomical Society*, 331, 1027.

Dewi J. D. M., Tauris T. M., 2000, *Astronomy and Astrophysics*, 360, 1043.

Dewi J. D. M., Tauris T. M., 2001, in Podsiadlowski P., Rappaport S., King A. R., D'Antona F., & Burderi L., eds, *Evolution of Binary and Multiple Star Systems*, Vol. 229 of Astronomical Society of the Pacific Conference Series, *On the λ-Parameter of the Common Envelope Evolution*. p. 255.

Di Salvo T., Sanna A., 2020, arXiv:2010.09005.

Diehl R., Halloin H., Kretschmer K., Lichti G. G., et al. 2006, *Nature*, 439, 45.

Dilday B., Bassett B., Becker A., Bender R., Castander F., Cinabro et al. 2010, *Astrophysical Journal*, 715, 1021.

Dinçel B., Neuhäuser R., Yerli S. K., Ankay A., Tetzlaff N., Torres G., Mugrauer M., 2015, *Monthly Notices of the Royal Astronomical Society*, 448, 3196.

Doherty C. L., Gil-Pons P., Siess L., Lattanzio J. C., 2017, *Publications of the Astronomical Society of Australia*, 34, e056.

Doherty C. L., Gil-Pons P., Siess L., Lattanzio J. C., Lau H. H. B., 2015, *Monthly Notices of the Royal Astronomical Society*, 446, 2599.

Dominik M., Belczynski K., Fryer C., Holz D. E., Berti E., Bulik T., Mandel I., O'Shaughnessy R., 2012, *Astrophysical Journal*, 759, 52.

Dominik M., Belczynski K., Fryer C., Holz D. E., Berti E., Bulik T., Mandel I., O'Shaughnessy R., 2013, *Astrophysical Journal*, 779, 72.

Dominik M., Berti E., O'Shaughnessy R., Mandel I., Belczynski K., Fryer C., ... Pannarale F., 2015, *Astrophysical Journal*, 806, 263.

Dominis D., Mimica P., Pavlovski K., Tamajo E., 2005, *Astrophysics and Space Science*, 296, 189.

Doroshenko O., Löhmer O., Kramer M., Jessner A., Wielebinski R., Lyne A. G., Lange C., 2001, *Astronomy and Astrophysics*, 379, 579.

Doroshenko V., Santangelo A., Suleimanov V., 2011, *Astronomy and Astrophysics*, 529, A52.
Doroshenko V., Tsygankov S., Santangelo A., 2016, *Astronomy and Astrophysics*, 589, A72.
Dorsch M., Reindl N., Pelisoli I., Heber U., Geier S., Istrate A. G., Justhan S., 2022, *Astronomy and Astrophysics*, 658, L9.
Downing J. M. B., Benacquista M. J., Giersz M., Spurzem R., 2010, *Monthly Notices of the Royal Astronomical Society*, 499, 5941.
Draghis P., Romani R. W., Filippenko A. V., Brink T. G., Zheng W., Halpern J. P., Camilo F., 2019, *Astrophysical Journal*, 883, 108.
Driebe T., Schoenberner D., Bloecker T., Herwig F., 1998, *Astronomy and Astrophysics*, 339, 123.
Drout M. R., Piro A. L., Shappee B. J., Kilpatrick C. D., Simon J. D., et al. 2017, *Science*, 358, 1570.
Drout M. R., Soderberg A. M., Gal-Yam A., Cenko S. B., Fox D. B., Leonard D. C., . . . Green Y., 2011, *Astrophysical Journal*, 741, 97.
Drout M. R., Soderberg A. M., Mazzali P. A., et al. 2013, *Astrophysical Journal*, 774, 58.
Du C., Li H., Yan Y., Newberg H. J., Shi J., Ma J., Chen Y., Wu Z., 2019, *Astrophysical Journal Supplement Series*, 244, 4.
Du S., Wang W., Wu X., Xu R., 2021, *Monthly Notices of the Royal Astronomical Society*, 500, 4678.
du Buisson L., Marchant P., Podsiadlowski P., Kobayashi C., Abdalla F. B., Taylor P., . . . Langer N., 2020, *Monthly Notices of the Royal Astronomical Society*, 499, 5941.
Dubner G. M., Holdaway M., Goss W. M., Mirabel I. F., 1998, *Astronomical Journal*, 116, 1842.
Dubus G., 2013, *Astronomy and Astrophysics Review*, 21, 64.
Dubus G., Hameury J.-M., Lasota J.-P., 2001, *Astronomy and Astrophysics*, 373, 251.
Duchêne G., Kraus A., 2013, *Annual Review of Astronomy and Astrophysics*, 51, 269.
Duquennoy A., Mayor M., 1991, *Astronomy and Astrophysics*, 500, 337.
Ebinger K., Curtis S., Fröhlich C., Hempel M., Perego A., Liebendörfer M., Thielemann F.-K., 2019, *Astrophysical Journal*, 870, 1.
Ebinger K., Curtis S., Ghosh S., Fröhlich C., Hempel M., Perego A., . . . Thielemann F.-K., 2020, *Astrophysical Journal*, 888, 91.
Echevarría J., de la Fuente E., Costero R., 2007, *Astronomical Journal*, 134, 262.
Echevarría J., Smith R. C., Costero R., Zharikov S., Michel R., 2008, *Monthly Notices of the Royal Astronomical Society*, 387, 1563.
Edelman B., Doctor Z., Farr B., 2021, *Astrophysical Journal*, 913, L23.
Edelmann H., Napiwotzki R., Heber U., Christlieb N., Reimers D., 2005, *Astrophysical Journal*, 634, L181.
Edwards D. A., Pringle J. E., 1987, *Monthly Notices of the Royal Astronomical Society*, 229, 383.
Eggleton P., 2006, *Evolutionary Processes in Binary and Multiple Stars*, Cambridge University Press.
Eggleton P. P., 1983, *Astrophysical Journal*, 268, 368.
Eggleton P. P., Pringle J. E., 1985, *Astrophysical Journal*, 288, 275.
Eggleton P. P., Verbunt F., 1986, *Monthly Notices of the Royal Astronomical Society*, 220, 13P.
Eichler D., Livio M., Piran T., Schramm D. N., 1989, *Nature*, 340, 126.
Eksi K. Y., Andac I. C., Cikintoglu S., Gencali A. A., Gungor C., Oztekin F., 2015, *Monthly Notices of the Royal Astronomical Society*, 448, L40.
Ekström S., Georgy C., Eggenberger P., Meynet G., Mowlavi N., Wyttenbach A., . . . Frischknecht U., 2012, *Astronomy and Astrophysics*, 537, A146.
Ekström S., Georgy C., Meynet G., Groh J., Granada A., 2013, in Kervella P., Le Bertre T., Perrin G., eds, EAS Publications Series Vol. 60, *Red supergiants and stellar evolution*, pp. 31–41.

El-Badry K., Quataert E., 2021, *Monthly Notices of the Royal Astronomical Society*, 502, 3436.
El-Badry K., Quataert E., Rix H.-W., Weisz D. R., Kupfer T., Shen K. J., ... Liu X., 2021, *Monthly Notices of the Royal Astronomical Society*, 505, 2051.
El Mellah I., Casse F., 2015, in SF2A-2015: Proceedings of the Annual meeting of the French Society of Astronomy and Astrophysics, *Numerical Simulations of Axisymmetric Bondi-Hoyle Accretion onto a Compact Object*, pp, 325–331.
El Mellah I., Sundqvist J. O., Keppens R., 2019, *Astronomy and Astrophysics*, 622, L3.
Eldridge J. J., 2017, *Population Synthesis of Massive Close Binary Evolution*, in Handbook of Supernovae, Springer International Publishing AG, p. 671.
Eldridge J. J., Fraser M., Smartt S. J., Maund J. R., Crockett R. M., 2013, *Monthly Notices of the Royal Astronomical Society*, 436, 774.
Eldridge J. J., Izzard R. G., Tout C. A., 2008, *Monthly Notices of the Royal Astronomical Society*, 384, 1109.
Eldridge J. J., Stanway E. R., 2016, *Monthly Notices of the Royal Astronomical Society*, 462, 3302.
Eldridge J. J., Stanway E. R., Xiao L., McClelland L. A. S., Taylor G., Ng M., ... Bray J. C., 2017, *Publications of the Astronomical Society of Australia*, 34, e058.
Eldridge J. J., Tout C. A., 2004, *Monthly Notices of the Royal Astronomical Society*, 353, 87.
Eldridge J. J., Tout C. A., 2019, *The Structure and Evolution of Stars*, World Scientic Publishing Company, Singapore.
Enoto T., Sasano M., Yamada S., Tamagawa T., Makishima K., Pottschmidt K., ... Wilms J., 2014, *Astrophysical Journal*, 786, 127.
Ergma E., Fedorova A. V., 1992, *Astronomy and Astrophysics*, 265, 65.
Ergma E., Sarna M. J., Antipova J., 1998, *Monthly Notices of the Royal Astronomical Society*, 300, 352.
Ergma E., van den Heuvel E. P. J., 1998, *Astronomy and Astrophysics*, 331, L29.
Ergma E. V., Fedorova A. V., 1991, *Astronomy and Astrophysics*, 242, 125.
Ertl T., Janka H.-T., Woosley S. E., Sukhbold T., Ugliano M., 2016, *Astrophysical Journal*, 818, 124.
Esposito P., Israel G. L., Milisavljevic D., Mapelli M., Zampieri L., Sidoli L., Fabbiano G., Rodríguez Castillo G. A., 2015, *Monthly Notices of the Royal Astronomical Society*, 452, 1112.
Evans C. R., Iben Jr. I., Smarr L., 1987, *Astrophysical Journal*, 323, 129.
Evans N. R., Berdnikov L., Lauer J., Morgan D., Nichols J., Günther H. M., ... Moskalik P., 2015, *Astronomical Journal*, 150, 13.
Exter K. M., Pollacco D. L., Maxted P. F. L., Napiwotzki R., Bell S. A., 2005, *Monthly Notices of the Royal Astronomical Society*, 359, 315.
Ezquiaga J. M., Zumalacárregui M., 2017, *Physical Review Letters*, 119, 251304.
Fabbiano G., 2006, *Annual Review of Astronomy and Astrophysics*, 44, 323.
Fabian A. C., Pringle J. E., Rees M. J., 1975, *Monthly Notices of the Royal Astronomical Society*, 172, 15p.
Fabian A. C., Pringle J. E., Verbunt F., Wade R. A., 1983, *Nature*, 301, 222.
Fabian A. C., Rees M. J., 1979, *Monthly Notices of the Royal Astronomical Society*, 187, 13P.
Fabian A. C., Rees M. J., Stella L., White N. E., 1989, *Monthly Notices of the Royal Astronomical Society*, 238, 729.
Fabrika S., 2004, *Astrophysics and Space Physics Reviews*, 12, 1.
Faigler S., Kull I., Mazeh T., Kiefer F., Latham D. W., Bloemen S., 2015, *Astrophysical Journal*, 815, 26.

Falanga M., Bozzo E., Lutovinov A., Bonnet-Bidaud J. M., Fetisova Y., Puls J., 2015, *Astronomy and Astrophysics*, 577, A130.
Falanga M., Kuiper L., Poutanen J., Galloway D. K., Bonning E. W., Bozzo E., ... Stella L., 2011, *Astronomy and Astrophysics*, 529, A68.
Farmer A. J., Phinney E. S., 2003, *Monthly Notices of the Royal Astronomical Society*, 346, 1197.
Farmer R., Renzo M., de Mink S. E., Marchant P., Justhan S., 2019, *Astrophysical Journal*, 887, 53.
Farr W. M., Kremer K., Lyutikov M., Kalogera V., 2011, *Astrophysical Journal*, 742, 81.
Farr W. M., Stevenson S., Miller M. C., Mandel I., Farr B., Vecchio A., 2017, *Nature*, 548, 426.
Farrell E., Groh J. H., Hirschi R., Murphy L., Kaiser E., Ekström S., ... Meynet G., 2021, *Monthly Notices of the Royal Astronomical Society*, 502, L40.
Faulkner A. J., Kramer M., Lyne A. G., Manchester R. N., McLaughlin M. A., Stairs I. H., ... Burgay M., 2005, *Astrophysical Journal*, 618, L119.
Favata M., 2014, *Physical Review Letters*, 112, 101101.
Fender R., 2006, *Jets from X-Ray Binaries*, in W. H. G. Lewin and M. van der Klis, ed., *Compact Stellar X-ray Sources*, Cambridge University Press, pp. 381–419.
Fender R., Belloni T., 2004, *Annual Review of Astronomy and Astrophysics*, 42, 317.
Feng Y., Zhao X., Li Y., Gou L., Jia N., Liao Z., Wang Y., 2022, Monthly Notices of the Royal Astronomical Society, 516, 2074.
Ferdman R. D., Freire P. C. C., Perera B. B. P., Pol N., Camilo F., Chatterjee S., ... van Leeuwen J., 2020, *Nature*, 583, 211.
Ferdman R. D., Stairs I. H., Kramer M., Breton R. P., McLaughlin M. A., Freire P. C. C., ... Lyne A. G., 2013, *Astrophysical Journal*, 767, 85.
Ferdman R. D., Stairs I. H., Kramer M., Janssen G. H., Bassa C. G., Stappers B. W., ... Possenti A., 2014, *Monthly Notices of the Royal Astronomical Society*, 443, 2183.
Fernández R., Metzger B. D., 2016, *Annual Review of Nuclear and Particle Science*, 66, 23.
Ferrario L., de Martino D., Gänsicke B. T., 2015, *Space Science Reviews*, 191, 111.
Ferrario L., Pringle J. E., Tout C. A., Wickramasinghe D. T., 2009, *Monthly Notices of the Royal Astronomical Society*, 400, L71.
Figer D. F., 2005, *Nature*, 434, 192.
Figer D. F., Najarro F., Morris M., McLean I. S., Geballe T. R., Ghez A. M., Langer N., 1998, *Astrophysical Journal*, 506, 384.
Filippenko A. V., 1997, *Annual Review of Astronomy and Astrophysics*, 35, 309.
Fink M., Hillebrandt W., Röpke F. K., 2007, *Astronomy and Astrophysics*, 476, 1133.
Fink M., Kromer M., Seitenzahl I. R., Ciaraldi-Schoolmann F., Röpke F. K., Sim S. A., ... A. J., Hillebrandt W., 2014, *Monthly Notices of the Royal Astronomical Society*, 438, 1762.
Finn L. S., 1996, *Physical Review D*, 53, 2878.
Finn L. S., Chernoff D. F., 1993, *Physical Review D*, 47, 2198.
Fischer T., Whitehouse S. C., Mezzacappa A., Thielemann F. K., Liebendörfer M., 2010, *Astronomy and Astrophysics*, 517, A80.
Fishbach M., Farr W. M., Holz D. E., 2020, *Astrophysical Journal*, 891, L31.
Fishbach M., Holz D. E., 2020, *Astrophysical Journal*, 904, L26.
Fitzpatrick E. L., Garmany C. D., 1990, *Astrophysical Journal*, 363, 119.
Flanagan É. É., Hinderer T., 2008, *Physical Review D*, 77, 021502.
Flannery B. P., 1975, *Monthly Notices of the Royal Astronomical Society*, 170, 325.
Flannery B. P., van den Heuvel E. P. J., 1975, *Astronomy and Astrophysics*, 39, 61.
Flowers E., Ruderman M. A., 1977, *Astrophysical Journal*, 215, 302.
Foglizzo T., Galletti P., Scheck L., Janka H.-T., 2007, *Astrophysical Journal*, 654, 1006.

Foley R. J., Challis P. J., Chornock R., Ganeshalingam M., Li W., Marion G. H., et al. 2013, *Astrophysical Journal*, 767, 57.

Fong W., Berger E., 2013, *Astrophysical Journal*, 776, 18.

Fonseca E., Cromartie H. T., Pennucci T. T., Ray P. S., Kirichenko A. Y., Ransom S. M., 2021, *Astrophysical Journal*, 915, L12.

Fonseca E., Stairs I. H., Thorsett S. E., 2014, *Astrophysical Journal*, 787, 82.

Fortin F., Chaty S., Sander A., 2020, *Astrophysical Journal*, 894, 86.

Fortin F., Garcia F., Chaty S. (2022) arXiv: 2207.02114

Fossati L., Mochnacki S., Landstreet J., Weiss W., 2010, *Astronomy and Astrophysics*, 510, A8.

Fossati L., Schneider F. R. N., Castro N., Langer N., Simón-Díaz S., Müller A., . . . Wade G. A., 2016, *Astronomy and Astrophysics*, 592, A84.

Fowler W. A., Hoyle F., 1964, *Astrophysical Journal Supplement Series*, 9, 201.

Fragione G., Banerjee S., 2021, *Astrophysical Journal*, 913, L29.

Fragione G., Loeb A., 2021, *Monthly Notices of the Royal Astronomical Society*, 502, 3879.

Fragione G., Loeb A., Rasio F. A., 2020, *Astrophysical Journal*, 902, L26.

Fragione G., Loeb A., Rasio F. A., 2021, *Astrophysical Journal*, 918, L38.

Fragos T., Andrews J. J., Bavera S. S., Berry C. P. L., Coughlin S., et al. 2022, arXiv:2202.05892.

Fragos T., Andrews J. J., Ramirez-Ruiz E., Meynet G., Kalogera V., Taam R. E., Zezas A., 2019, *Astrophysical Journal*, 883, L45.

Franciolini G., Baibhav V., De Luca V., Ng K. K. Y., Wong K. W. K., Berti E., . . . Vitale S., 2022, Physical Review D, 105, 083526.

Frank J., King A., Raine D. J., 2002, *Accretion Power in Astrophysics*, Third Edition, Cambridge University Press.

Fregeau J. M., Cheung P., Portegies Zwart S. F., Rasio F. A., 2004, *Monthly Notices of the Royal Astronomical Society*, 352, 1.

Freire P. C. C., Abdo A. A., Ajello M., et al. 2011, *Science*, 334, 1107.

Freire P. C. C., Bassa C. G., Wex N., Stairs I. H., Champion D. J., Ransom S. M., 2011, *Monthly Notices of the Royal Astronomical Society*, 412, 2763.

Freire P. C. C., Kramer M., Wex N., 2012, *Classical and Quantum Gravity*, 29, 184007.

Freire P. C. C., Ransom S. M., Bégin S., Stairs I. H., Hessels J. W. T., Frey L. H., Camilo F., 2008, *Astrophysical Journal*, 675, 670.

Freire P. C. C., Tauris T. M., 2014, *Monthly Notices of the Royal Astronomical Society*, 438, L86.

Freire P. C. C., Wex N., Esposito-Farèse G., Verbiest J. P. W., Bailes M., Jacoby B. A., . . . Janssen G. H., 2012, *Monthly Notices of the Royal Astronomical Society*, 423, 3328.

Fruchter A. S., Berman G., Bower G., Convery M., Goss W. M., Hankins T. H., . . . Weisberg J. M., 1990, *Astrophysical Journal*, 351, 642.

Fruchter A. S., Stinebring D. R., Taylor J. H., 1988, *Nature*, 333, 237.

Fryer C. L., Belczynski K., Wiktorowicz G., Dominik M., Kalogera V., Holz D. E., 2012, *Astrophysical Journal*, 749, 91.

Fryer C. L., Heger A., Langer N., Wellstein S., 2002, *Astrophysical Journal*, 578, 335.

Fryer C. L., Kalogera V., 2001, *Astrophysical Journal*, 554, 548.

Fryer C. L., Woosley S. E., Heger A., 2001, *Astrophysical Journal*, 550, 372.

Fryxell B. A., Arnett W. D., 1981, *Astrophysical Journal*, 243, 994.

Fryxell B. A., Taam R. E., 1988, *Astrophysical Journal*, 335, 862.

Fuller J., Lecoanet D., Cantiello M., Brown B., 2014, *Astrophysical Journal*, 796, 17.

Fuller J., Lu W., 2022, *Monthly Notices of the Royal Astronomical Society*, 511, 3951.

Fuller J., Ma L., 2019, *Astrophysical Journal*, 881, L1.

Fuller J., Piro A. L., Jermyn A. S., 2019, *Monthly Notices of the Royal Astronomical Society*, 485, 3661.

Fürst F., Pottschmidt K., Wilms J., Tomsick J. A., Bachetti M., Boggs S. E., ... Zhang W., 2014, *Astrophysical Journal*, 780, 133.
Fürst F., Walton D. J., Harrison F. A., Stern D., Barret D., Brightman M., ... Middleton M. J., 2016, *Astrophysical Journal*, 831, L14.
Gaia Collaboration, Smart R. L., Sarro L. M., Rybizki J., Reylé C., Robin A. C., Hambly N. C., 2021, *Astronomy and Astrophysics*, 649, A6.
Gal-Yam A., 2019, *Annual Review of Astronomy and Astrophysics*, 57, 305.
Gal-Yam A., Mazzali P., Ofek E. O., Nugent P. E., Kulkarni S. R., Kasliwal M. M., 2009, *Nature*, 462, 624.
Galama T. J., Vreeswijk P. M., van Paradijs J., Kouveliotou C., Augusteijn T., Böhnhardt H., 1998, *Nature*, 395, 670.
Gallegos-Garcia M., Berry C. P. L., Marchant P., Kalogera V., 2021, *Astronomy and Astrophysics*, 922, 110.
Gallegos-Garcia M., Fishbach M., Kalogera V., Berry C. P. L., Doctor Z., 2022, arXiv: 2207.14290.
Gallo E., 2010, in Belloni T., ed, Lecture Notes in Physics, Berlin Springer Verlag Vol. 794, Radio Emission and Jets from Microquasars, p. 85.
Gallo E., Fender R., Kaiser C., Russell D., Morganti R., Oosterloo T., Heinz S., 2005, *Nature*, 436, 819.
Gänsicke B. T., Dillon M., Southworth J., Thorstensen J. R., Rodríguez-Gil P., Aungwerojwit A., 2009, *Monthly Notices of the Royal Astronomical Society*, 397, 2170.
Gao S.-J., Li X.-D., 2021, *Research in Astronomy and Astrophysics*, 21, 196.
Gao Y., Wang Q. D., Appleton P. N., Lucas R. A., 2003, *Astrophysical Journal*, 596, L171.
García F., Simaz Bunzel A., Chaty S., Porter E., Chassande-Mottin E., 2021, *Astronomy and Astrophysics*, 649, A114.
Garcia M., Herrero A., Najarro F., Lennon D. J., Alejandro Urbaneja M., 2014, *Astrophysical Journal*, 788, 64.
García-Senz D., Badenes C., Serichol N., 2012, *Astrophysical Journal*, 745, 75.
Gaskell C. M., Cappellaro E., Dinerstein H. L., Garnett D. R., Harkness R. P., Wheeler J. C., 1986, *Astrophysical Journal*, 306, L77.
Gayathri V., Bartos I., Haiman Z., Klimenko S., Kocsis B., Márka S., Yang Y., 2020, *Astrophysical Journal*, 890, L20.
Gayathri V., Healy J., Lange J., O'Brien B., Szczepanczyk M., Bartos I., ... O'Shaughnessy R., 2020, arXiv:2009.05461.
Ge H., Webbink R. F., Chen X., Han Z., 2015, *Astrophysical Journal*, 812, 40.
Ge H., Webbink R. F., Chen X., Han Z., 2020, *Astrophysical Journal*, 899, 132.
Ge H., Webbink R. F., Han Z., 2020, *Astrophysical Journal Supplement Series*, 249, 9.
Geier S., Fürst F., Ziegerer E., Kupfer T., Heber U., ... 2015, *Science*, 347, 1126.
Georgy C., Ekström S., Eggenberger P., Meynet G., Haemmerlé L., Maeder A., ... Barblan F., 2013, *Astronomy and Astrophysics*, 558, A103.
Georgy C., Ekström S., Saio H., Meynet G., Groh J., Granada A., 2013, in Kervella P., Le Bertre T., Perrin G., eds, *EAS Publications Series*, Vol. 60, *How the Mass-Loss Rates of Red-Supergiants Determine the Fate of Massive Stars*, pp. 43–50.
Geppert U., Urpin V., 1994, *Monthly Notices of the Royal Astronomical Society*, 271, 490.
Gerosa D., Kesden M., Sperhake U., Berti E., O'Shaughnessy R., 2015, *Phys. Rev. D*, 92, 064016.
Gerosa D., Berti E., O'Shaughnessy R., Belczynski K., Kesden M., Wysocki D., Gladysz, W., 2018, *Phys. Rev. D*, 98, 084036.
Gerosa D., Fishbach M., 2021, *Nature Astronomy*, 5, 749.
Gerosa D., Vitale S., Berti E., 2020, *Physical Review Letters*, 125, 101103.

Gessner A., Janka H.-T., 2018, *Astrophysical Journal*, 865, 61.
Ghosh P., 2007, *Rotation and Accretion Powered Pulsars*, World Scientific Series in Astronomy and Astrophysics, Vol. 10, World Scientific Publishing Company.
Ghosh P., Lamb F. K., 1979a, *Astrophysical Journal*, 232, 259.
Ghosh P., Lamb F. K., 1979b, *Astrophysical Journal*, 234, 296.
Ghosh P., Lamb F. K., 1992, in van den Heuvel E. P. J., Rappaport S. A., eds, *X-Ray Binaries and Recycled Pulsars, Diagnostics of Disk-Magnetosphere Interaction in Neutron Star Binaries*, Springer, pp. 487–510.
Giacconi R., 1975, in Bergman P. G., Fenyves E. J., Motz L., eds, Seventh Texas Symposium on Relativistic Astrophysics Vol. 262, *Her X-1 and Cen X-3 Revisited*, 312–330.
Giacconi R., 2005, *Annual Review of Astronomy and Astrophysics*, 43, 1.
Giacconi R., Gursky H., Paolini F. R., Rossi B. B., 1962, *Physical Review Letters*, 9, 439.
Giacobbo N., Mapelli M., 2018, *Monthly Notices of the Royal Astronomical Society*, 480, 2011.
Giacobbo N., Mapelli M., Spera M., 2018, *Monthly Notices of the Royal Astronomical Society*, 474, 2959.
Giacomazzo B., Eichler M., Arcones A., 2018, *Electromagnetic Emission and Nucleosynthesis from Neutron Star Binary Mergers*, in *The Physics and Astrophysics of Neutron Stars*, Astrophysics and Space Science Library, Volume 457, Springer Nature Switzerland AG, 2018, p. 637.
Gianninas A., Curd B., Fontaine G., Brown W. R., Kilic M., 2016, *Astrophysical Journal*, 822, L27.
Giannone P., Kohl K., Weigert A., 1968, *Zeitschrift für Astrophysik*, 68, 107.
Giclas H. L., Burnham R., Thomas N. G., 1970, *Lowell Observatory Bulletin*, 7, 183.
Gieles M., Erkal D., Antonini F., Balbinot E., Peñarrubia J., 2021, *Nature Astronomy*, 5, 957.
Gies D. R., 1987, *Astrophysical Journal Supplement Series*, 64, 545.
Gies D. R., Bolton C. T., 1986, *Astrophysical Journal Supplement Series*, 61, 419.
Giesers B., Dreizler S., Husser T.-O., Kamann S., Anglada Escudé G., Brinchmann J., ... Wisotzki L., 2018, *Monthly Notices of the Royal Astronomical Society*, 475, L15.
Gillanders J. H., Sim S. A., Smartt S. J., 2020, *Monthly Notices of the Royal Astronomical Society*, 497, 246.
Ginzburg S., Chiang E., 2022, *Monthly Notices of the Royal Astronomical Society*, 509, L1.
Ginzburg S., Quataert E., 2020, *Monthly Notices of the Royal Astronomical Society*, 495, 3656.
Ginzburg S., Quataert E., 2021, *Monthly Notices of the Royal Astronomical Society*, 500, 1592.
Gittins F., Andersson N., Jones D. I., 2021, *Monthly Notices of the Royal Astronomical Society*, 500, 5570.
Glanz H., Perets H. B., 2021, *Monthly Notices of the Royal Astronomical Society*, 507, 2659.
Glebbeek E., Gaburov E., Portegies Zwart S., Pols O. R., 2013, *Monthly Notices of the Royal Astronomical Society*, 434, 3497.
Goldreich P., Julian W. H., 1969, *Astrophysical Journal*, 157, 869.
González J. A., Hannam M., Sperhake U., Brügmann B., Husa S., 2007, *Physical Review Letters*, 98, 231101.
González-Galán A., Kuulkers E., Kretschmar P., Larsson S., Postnov K., Kochetkova A., Finger M. H., 2012, *Astronomy and Astrophysics*, 537, A66.
Gordon M. S., Humphreys R. M., 2019, *Galaxies*, 7, 92.
Gordon M. S., Humphreys R. M., Jones T. J., 2016, *Astrophysical Journal*, 825, 50.
Götberg Y., 2018, PhD thesis, University of Amsterdam.
Götberg Y., de Mink S. E., Groh J. H., Kupfer T., Crowther P. A., Zapartas E., Renzo M., 2018, *Astronomy and Astrophysics*, 615, A78.
Gott III J. R., Gunn J. E., Ostriker J. P., 1970, *Astrophysical Journal*, 160, L91.

Gottlieb E. W., Wright E. L., Liller W., 1975, *Astrophysical Journal*, 195, L33.
Gould D. M., 1994, PhD thesis, University of Manchester, UK.
Gräfener G., Hamann W. R., 2008, *Astronomy and Astrophysics*, 482, 945.
Graham M. J., Ford K. E. S., McKernan B., Ross N. P., Stern D., Burdge K., . . . 2020, *Physical Review Letters*, 124, 251102.
Graikou E., Verbiest J. P. W., Osłowski S., Champion D. J., Tauris T. M., Jankowski F., Kramer M., 2017, *Monthly Notices of the Royal Astronomical Society*, 471, 4579.
Gray W. J., Raskin C., Owen J. M., 2016, *Astrophysical Journal*, 833, 62.
Green M. J., Marsh T. R., Steeghs D. T. H., Kupfer T., Ashley R. P., Bloemen S., . . . 2018, *Monthly Notices of the Royal Astronomical Society*, 476, 1663.
Greiner J., 2000a, *New Astronomy*, 5, 137.
Greiner J., 2000b, *New Astronomy Reviews*, 44, 149.
Greiner J., Hasinger G., Kahabka P., 1991, *Astronomy and Astrophysics*, 246, L17.
Grichener A., Sabach E., Soker N., 2018, *Monthly Notices of the Royal Astronomical Society*, 478, 1818.
Griffin R. E., Ake T. B., 2015, in *Giants of Eclipse: The ζ Aurigae Stars and Other Binary Systems*, Vol. 408 of Astrophysics and Space Science Library, *The zeta Aurigae Binaries*, 1.
Griffin R. E., Eaton J. A., Ake T. B., Schröder K.-P., 2015, in *Giants of Eclipse: The ζ Aurigae Stars and Other Binary Systems*, Vol. 408 of Astrophysics and Space Science Library, *Observing and Analyzing the zeta Aurigae Systems*, 15.
Grimm H.-J., Gilfanov M., Sunyaev R., 2002, *Astronomy and Astrophysics*, 391, 923.
Grindlay J., Gursky H., Schnopper H., Parsignault D. R., Heise J., Brinkman A. C., Schrijver J., 1976, *Astrophysical Journal*, 205, L127.
Grindlay J. E., Bailyn C. D., 1988, *Nature*, 336, 48.
Groh J. H., Oliveira A. S., Steiner J. E., 2008, *Astronomy and Astrophysics*, 485, 245.
Grudzinska M., Belczynski K., Casares J., de Mink S. E., Ziolkowski J., Negueruela I., . . . Benacquista M., 2015, *Monthly Notices of the Royal Astronomical Society*, 452, 2773.
Grunthal K., Kramer M., Desvignes G., 2021, *Monthly Notices of the Royal Astronomical Society*, 507, 5658.
Guillemot L., Tauris T. M., 2014, *Monthly Notices of the Royal Astronomical Society*, 439, 2033.
Guillochon J., Dan M., Ramirez-Ruiz E., Rosswog S., 2010, *Astrophysical Journal*, 709, L64.
Gullikson K., Kraus A., Dodson-Robinson S., 2016, *Astronomical Journal*, 152, 40.
Gunn J. E., Ostriker J. P., 1970, *Astrophysical Journal*, 160, 979.
Guo Y. J., Freire P. C. C., Guillemot L., . . . Theureau G., 2021, *Astronomy and Astrophysics*, 654, A16.
Gursky H., 1973, in Proceedings of Cambridge University Conference, "Physics and Astrophysics of Compact Objects," July 1973 (unpublished): X-ray Sources in the Galaxy.
Gursky H., van den Heuvel E. P. J., 1975, *Scientific American*, 232, 24.
Guseinov O. K., Zel'dovich Y. B., 1966, *Soviet Astronomy*, 10, 251.
Gvaramadze V. V., Gräfener G., Langer N., Maryeva O. V., Kniazev A. Y., Moskvitin A. S., Spiridonova O. I., 2019, *Nature*, 569, 684.
Gvaramadze V. V., Gualandris A., Portegies Zwart S., 2009, *Monthly Notices of the Royal Astronomical Society*, 396, 570.
Gvaramadze V. V., Kniazev A. Y., Gallagher J. S., Oskinova L. M., Chu Y. H., Gruendl R. A., Katkov I. Y., 2021, *Monthly Notices of the Royal Astronomical Society*, 503, 3856.
Gvaramadze V. V., Langer N., Fossati L., Bock D. C. J., Castro N., Georgiev I. Y., . . . Tauris T. M., 2017, *Nature Astronomy*, 1, 0116.
Haberl F., Sturm R., 2016, *Astronomy and Astrophysics*, 586, A81.
Habets G. M. H. J., 1985, PhD thesis, University of Amsterdam.

Habets G. M. H. J., 1986a, *Astronomy and Astrophysics*, 165, 95.
Habets G. M. H. J., 1986b, *Astronomy and Astrophysics*, 167, 61.
Habets G. M. H. J., 1987, *Astronomy and Astrophysics*, 184, 209.
Hachinger S., Mazzali P. A., Taubenberger S., Hillebrandt W., Nomoto K., Sauer D. N., 2012, *Monthly Notices of the Royal Astronomical Society*, 422, 70.
Hachisu I., Kato M., 2001, *Astrophysical Journal*, 558, 323.
Hachisu I., Kato M., Nomoto K., 1996, *Astrophysical Journal*, 470, L97+.
Hachisu I., Kato M., Nomoto K., 1999, *Astrophysical Journal*, 522, 487.
Hachisu I., Kato M., Saio H., Nomoto K., 2012, *Astrophysical Journal*, 744, 69.
Haensel P., Zdunik J. L., 1989, *Nature*, 340, 617.
Haffner H., Heckmann O., 1937, Veroeffentlichungen der Universitaets-Sternwarte zu Goettingen, 0004, 77.
Hainich R., Oskinova L. M., Torrejón J. M., Fuerst F., Bodaghee A., Shenar T., ... Hamann W. R., 2020, *Astronomy and Astrophysics*, 634, A49.
Hajela A., Margutti R., Bright J. S., ... Berger E., et al. 2022, *Astrophysical Journal*, 927, L17.
Hakobyan A. A., Nazaryan T. A., Adibekyan V. Z., Petrosian A. R., Aramyan L. S., Kunth D., ... Turatto M., 2014, *Monthly Notices of the Royal Astronomical Society*, 444, 2428.
Halbwachs J. L., Mayor M., Udry S., Arenou F., 2003, *Astronomy and Astrophysics*, 397, 159.
Hamers A. S., Bar-Or B., Petrovich C., Antonini F., 2018, *Astrophysical Journal*, 865, 2.
Hamilton R. T., Harrison T. E., Tappert C., Howell S. B., 2011, *Astrophysical Journal*, 728, 16.
Han Q., Li X.-D., 2021, *Astrophysical Journal*, 909, 161.
Han Z., Podsiadlowski P., 2004, *Monthly Notices of the Royal Astronomical Society*, 350, 1301.
Han Z., Podsiadlowski P., Eggleton P. P., 1994, *Monthly Notices of the Royal Astronomical Society*, 270, 121.
Han Z., Podsiadlowski P., Eggleton P. P., 1995, *Monthly Notices of the Royal Astronomical Society*, 272, 800.
Han Z., Podsiadlowski P., Maxted P. F. L., Marsh T. R., Ivanova N., 2002, *Monthly Notices of the Royal Astronomical Society*, 336, 449.
Haniewicz H. T., Ferdman R. D., Freire P. C. C., Champion D. J., Bunting K. A., Lorimer D. R., McLaughlin M. A., 2021, *Monthly Notices of the Royal Astronomical Society*, 500, 4620.
Hansen C. J., Kawaler S. D., Trimble V., 2004, *Stellar Interiors: Physical Principles, Structure, and Evolution*. Springer Verlag, New York.
Hansen C. J., van Horn H. M., 1975, *Astrophysical Journal*, 195, 735.
Hansen C. J., Wheeler J. C., 1969, *Astrophysics and Space Science*, 3, 464.
Hansen T. T., Andersen J., Nordström B., Beers T. C., Placco V. M., Yoon J., Buchhave L. A., 2016, *Astronomy and Astrophysics*, 588, A3.
Hardie R. H., 1950, *Astrophysical Journal*, 112, 542.
Harding A. K., 2018, in Weltevrede P., Perera B. B. P., Preston L. L., Sanidas S., eds, *Pulsar Astrophysics the Next Fifty Years*, Vol. 337, *Pulsar Emission Physics: The First Fifty Years*, pp. 52–57.
Harmanec P., 2002, *Astronomische Nachrichten*, 323, 87.
Harris J., Zaritsky D., 2004, *Astronomical Journal*, 127, 1531.
Harrison E. R., Tademaru E., 1975, *Astrophysical Journal*, 201, 447.
Harrison T. E., Bornak J., McArthur B. E., Benedict G. F., 2013, *Astrophysical Journal*, 767, 7.
Hartmann L., 1998, *Accretion Processes in Star Formation*, Cambridge Univeristy Press.
Haskell B., Priymak M., Patruno A., Oppenoorth M., Melatos A., Lasky P. D., 2015, *Monthly Notices of the Royal Astronomical Society*, 450, 2393.
Hastings B., Langer N., Koenigsberger G., 2020, *Astronomy and Astrophysics*, 641, A86.
Heber U., 2016, *Publications of the Astronomical Society of the Pacific*, 128, 082001.

Heber U., Edelmann H., Napiwotzki R., Altmann M., Scholz R.-D., 2008, *Astronomy and Astrophysics*, 483, L21.
Heger A., Fryer C. L., Woosley S. E., Langer N., Hartmann D. H., 2003, *Astrophysical Journal*, 591, 288.
Heger A., Langer N., 2000, *Astrophysical Journal*, 544, 1016.
Heger A., Langer N., Woosley S. E., 2000, *Astrophysical Journal*, 528, 368.
Heger A., Woosley S. E., 2002, *Astrophysical Journal*, 567, 532.
Heger A., Woosley S. E., Spruit H. C., 2005, *Astrophysical Journal*, 626, 350.
Heggie D. C., 1975, *Monthly Notices of the Royal Astronomical Society*, 173, 729.
Heida M., Lau R. M., Davies B., Brightman M., Fürst F., Grefenstette B. W., . . . Harrison F. A., 2019, *Astrophysical Journal*, 883, L34.
Heinke C. O., Ivanova N., Engel M. C., Pavlovskii K., Sivakoff G. R., Cartwright T. F., Gladstone J. C., 2013, *Astrophysical Journal*, 768, 184.
Heinz S., Sell P., Fender R. P., Jonker P. G., Brandt W. N., Calvelo-Santos D. E., . . . van der Klis M., 2013, *Astrophysical Journal*, 779, 171.
Helfand D. J., 1987, *Nature*, 329, 285.
Hénon M., 1961, *Annales d'Astrophysique*, 24, 369.
Herant M., Benz W., Hix W. R., Fryer C. L., Colgate S. A., 1994, *Astrophysical Journal*, 435, 339.
Hermes J. J., Kilic M., Brown W. R., Winget D. E., Allende Prieto C., Gianninas A., . . . Kenyon S. J., 2012, *Astrophysical Journal*, 757, L21.
Hessels J. W. T., 2008, in Wijnands R., Altamirano D., Soleri P., Degenaar N., Rea N., Casella P., Patruno A., Linares M., eds, *American Institute of Physics Conference Series*, Vol. 1068, *The Observed Spin Distributions of Millisecond Radio and X-ray Pulsars*, pp. 130–134.
Hessels J. W. T., Ransom S. M., Stairs I. H., Freire P. C. C., Kaspi V. M., Camilo F., 2006, *Science*, 311, 1901.
Hessman F. V., Gänsicke B. T., Mattei J. A., 2000, *Astronomy and Astrophysics*, 361, 952.
Hewish A., Bell S. J., Pilkington J. D. H., Scott P. F., Collins R. A., 1968, *Nature*, 217, 709.
Higgins E. R., Sander A. A. C., Vink J. S., Hirschi R., 2021, *Monthly Notices of the Royal Astronomical Society*, 505, 4874.
Hilditch, Ronald W., *An Introduction to Close Binary Stars*, 2001. Cambridge University Press.
Hilditch R. W., Harries T. J., Hill G., 1996, *Monthly Notices of the Royal Astronomical Society*, 279, 1380.
Hillebrandt W., 1987, in Pacini F., ed, NATO Advanced Science Institutes (ASI) Series C, Vol. 195, Kluwer Academic Publishers, Dordrecht *Stellar Collapse and Supernova Explosions*, pp. 73–104.
Hillebrandt W., Kromer M., Röpke F. K., Ruiter A. J., 2013, *Frontiers of Physics*, 8, 116.
Hillebrandt W., Meyer F., 1989, *Astronomy and Astrophysics*, 219, L3.
Hills J. G., 1983a, *Astronomical Journal*, 88, 1269.
Hills J. G., 1983b, *Astrophysical Journal*, 267, 322.
Hills J. G., 1988, *Nature*, 331, 687.
Hillwig T. C., Gies D. R., 2008, *Astrophysical Journal*, 676, L37.
Hillwig T. C., Gies D. R., Huang W., McSwain M. V., Stark M. A., van der Meer A., Kaper L., 2004, *Astrophysical Journal*, 615, 422.
Hils D., Bender P. L., Webbink R. F., 1990, *Astrophysical Journal*, 360, 75.
Hinkle K. H., Lebzelter T., Fekel F. C., Straniero O., Joyce R. R., Prato L., Karnath N., Habel N., 2020, *Astrophysical Journal*, 904, 143.
Hirai R., Podsiadlowski P., Owocki S. P., Schneider F. R. N., Smith N., 2021, *Monthly Notices of the Royal Astronomical Society*, 503, 4276.
Hirai R., Podsiadlowski P., Yamada S., 2018, *Astrophysical Journal*, 864, 119.

Hirai R., Sato T., Podsiadlowski P., Vigna-Gómez A., Mandel I., 2020, *Monthly Notices of the Royal Astronomical Society*, 499, 1154.
Hiramatsu D., Howell D. A., Van Dyk S. D., Goldberg J. A., Maeda K., Moriya T. J., . . . 2021, *Nature Astronomy*.
Hirsch H. A., Heber U., O'Toole S. J., Bresolin F., 2005, *Astronomy and Astrophysics*, 444, L61.
Hirschi R., Meynet G., Maeder A., 2005, *Astronomy and Astrophysics*, 443, 581.
Hirv A., Annuk K., Eenmäe T., Liimets T., Pelt J., Puss A., Tempel M., 2006, *Baltic Astronomy*, 15, 405.
Hjellming M. S., 1989, PhD thesis, Illinois, Univ. at Urbana-Champaign, Savoy.
Hjellming R. M., 1973, *Science*, 182, 1089.
Hjellming R. M., Balick B., 1972a, *Nature Physical Science*, 239, 135.
Hjellming R. M., Balick B., 1972b, *Nature*, 239, 443.
Hjellming R. M., Wade C. M., 1971, *Nature*, 234, 138.
Hjorth J., Sollerman J., Møller P., Fynbo J. P. U., Woosley S. E., Kouveliotou C., Tanvir N. R., . . . 2003, *Nature*, 423, 847.
Ho W. C. G., Jones D. I., Andersson N., Espinoza C. M., 2020, *Physical Review D*, 101, 103009.
Ho W. C. G., Wijngaarden M. J. P., Andersson N., Tauris T. M., Haberl F., 2020, *Monthly Notices of the Royal Astronomical Society*, 494, 44.
Hobbs G., Archibald A., Arzoumanian Z., et al. 2010, *Classical and Quantum Gravity*, 27, 084013.
Hobbs G., Coles W., Manchester R. N., et al. 2012, *Monthly Notices of the Royal Astronomical Society*, 427, 2780.
Hobbs G., Lorimer D. R., Lyne A. G., Kramer M., 2005, *Monthly Notices of the Royal Astronomical Society*, 360, 974.
Hoeflich P., Khokhlov A., 1996, *Astrophysical Journal*, 457, 500.
Hoffleit D., Jaschek C., 1991, *The Bright Star Catalogue*, 5th edition, Yale University Observatory, New Haven, USA.
Hogeveen S. J., 1992a, *Astrophysics and Space Science*, 195, 359.
Hogeveen S. J., 1992b, *Astrophysics and Space Science*, 194, 143.
Hogeveen S. J., 1992c, *Astrophysics and Space Science*, 196, 299.
Holgado A. M., 2021, arXiv:2103.13605.
Holgado A. M., Ricker P. M., 2019, *Monthly Notices of the Royal Astronomical Society*, 490, 5560.
Holmgren D., 1989, *Space Science Reviews*, 50, 347.
Hotokezaka K., Nakar E., Gottlieb O., Nissanke S., Masuda K., Hallinan G., Mooley K. P., Deller A. T., 2019, *Nature Astronomy*, 3, 940.
Hotokezaka K., Piran T., 2017, *Astrophysical Journal*, 842, 111.
Howitt G., Stevenson S., Vigna-Gómez A., Justhan S., Ivanova N., Woods T. E., Neijssel C. J., Mandel I., 2020, *Monthly Notices of the Royal Astronomical Society*, 492, 3229.
Hoyle F., Fowler W. A., 1960, *Astrophysical Journal*, 132, 565.
Hoyle F., Lyttleton R. A., 1939, Proceedings of the Cambridge Philosophical Society, 35, 405.
Hu H., Kramer M., Wex N., Champion D. J., Kehl M. S., 2020, *Monthly Notices of the Royal Astronomical Society*, 497, 3118.
Huang S.-J., Hu Y.-M., Korol V., Li P.-C., Liang Z.-C., Lu Y., . . . Mei J., 2020, *Physical Review D*, 102, 063021.
Huang S.-S., 1963, *Astrophysical Journal*, 138, 471.
Hulse R. A., Taylor J. H., 1975, *Astrophysical Journal*, 195, L51.

Humphreys R. M., Davidson K., 1994, *Publications of the Astronomical Society of the Pacific*, 106, 1025.
Humphreys R. M., Davidson K., Hahn D., Martin J. C., Weis K., 2017, *Astrophysical Journal*, 844, 40.
Hunter I., Brott I., Lennon D. J., Langer N., Dufton P. L., Trundle C., . . . Ryans R. S. I., 2008, *Astrophysical Journal*, 676, L29.
Hunter I., Dufton P. L., Smartt S. J., Ryans R. S. I., Evans C. J., Lennon D. J., Trundle C., Hubeny I., Lanz T., 2007, *Astronomy and Astrophysics*, 466, 277.
Hurley J. R., Pols O. R., Tout C. A., 2000, *Monthly Notices of the Royal Astronomical Society*, 315, 543.
Hurley J. R., Tout C. A., Pols O. R., 2002, *Monthly Notices of the Royal Astronomical Society*, 329, 897.
Hurley J. R., Tout C. A., Wickramasinghe D. T., Ferrario L., Kiel P. D., 2010, *Monthly Notices of the Royal Astronomical Society*, 402, 1437.
Hut P., 1980, *Astronomy and Astrophysics*, 92, 167.
Hut P., 1981, *Astronomy and Astrophysics*, 99, 126.
Hut P., 1983, *Astronomical Journal*, 88, 1549.
Hut P., Bahcall J. N., 1983, *Astrophysical Journal*, 268, 319.
Hutton K., Henden A., Terrell D., 2009, *Publications of the Astronomical Society of the Pacific*, 121, 708.
Iaconi R., De Marco O., Passy J.-C., Staff J., 2018, *Monthly Notices of the Royal Astronomical Society*, 477, 2349.
Iaconi R., Reichardt T., Staff J., De Marco O., Passy J.-C., Price D., . . . Herwig F., 2017, *Monthly Notices of the Royal Astronomical Society*, 464, 4028.
Ibeling D., Heger A., 2013, *Astrophysical Journal*, 765, L43.
Iben Jr. I., Livio M., 1993, *Publications of the Astronomical Society of the Pacific*, 105, 1373.
Iben Jr. I., Tutukov A. V., 1985, *Astrophysical Journal Supplement Series*, 58, 661.
Iben Jr. I., Tutukov A. V., 1999, *Astrophysical Journal*, 511, 324.
Iben I. J., Tutukov A. V., 1984, *Astrophysical Journal Supplement Series*, 54, 335.
Iben Icko J., 2013a, *Stellar Evolution Physics*, Volume 1: *Physical Processes in Stellar Interiors*, Cambridge University Press.
Iben Icko J., 2013b, *Stellar Evolution Physics*, Volume 2: *Advanced Evolution of Single Stars*, Cambridge University Press.
Ikhsanov N. R., Beskrovnaya N. G., 2012, *Astronomy Reports*, 56, 589.
Ikhsanov N. R., Mereghetti S., 2015, *Monthly Notices of the Royal Astronomical Society*, 454, 3760.
Illarionov A. F., Sunyaev R. A., 1975, *Astronomy and Astrophysics*, 39, 185.
Ilovaisky S. A., Chevalier C., Motch C., Pakull M., van Paradijs J., Lub J., 1984, *Astronomy and Astrophysics*, 140, 251.
Im M., Yoon Y., Lee S.-K. J., Lee H. M., Kim J., Lee C. U., . . . Shim H., 2017, *Astrophysical Journal*, 849, L16.
Ioka K., Zhang B., 2020, *Astrophysical Journal*, 893, L26.
Irrgang A., Dimpel M., Heber U., Raddi R., 2021, *Astronomy and Astrophysics*, 646, L4.
Irrgang A., Geier S., Kreuzer S., Pelisoli I., Heber U., 2020, *Astronomy and Astrophysics*, 633, L5.
Irwin J. A., Brink T. G., Bregman J. N., Roberts T. P., 2010, *Astrophysical Journal*, 712, L1.
Israel G. L., 1996, PhD thesis, Scuola Int. Superiore Stud. Avanzati, Trieste, 1996.
Israel G. L., Belfiore A., Stella L., Esposito P., Casella P., De Luca A., . . . Puccetti S., 2017, *Science*, 355, 817.

Israel G. L., Papitto A., Esposito P., Stella L., Zampieri L., Belfiore A., ... Lisini G., 2017, *Monthly Notices of the Royal Astronomical Society*, 466, L48.

Israelian G., Rebolo R., Basri G., Casares J., Martín E. L., 1999, *Nature*, 401, 142.

Istrate A. G., Fontaine G., Gianninas A., Grassitelli L., Marchant P., Tauris T. M., Langer N., 2016, *Astronomy and Astrophysics*, 595, L12.

Istrate A. G., Marchant P., Tauris T. M., Langer N., Stancliffe R. J., Grassitelli L., 2016, *Astronomy and Astrophysics*, 595, A35.

Istrate A. G., Tauris T. M., Langer N., 2014, *Astronomy and Astrophysics*, 571, A45.

Istrate A. G., Tauris T. M., Langer N., Antoniadis J., 2014, *Astronomy and Astrophysics*, 571, L3.

Ivanova N., 2011, *Astrophysical Journal*, 730, 76.

Ivanova N., Belczynski K., Kalogera V., Rasio F. A., Taam R. E., 2003, *Astrophysical Journal*, 592, 475.

Ivanova N., Chaichenets S., Fregeau J., Heinke C. O., Lombardi J. C. J., Woods T. E., 2010, *Astrophysical Journal*, 717, 948.

Ivanova N., Heinke C. O., Rasio F. A., Belczynski K., Fregeau J. M., 2008, *Monthly Notices of the Royal Astronomical Society*, 386, 553.

Ivanova N., Justhan S., Avendano Nandez J. L., Lombardi J. C., 2013, *Science*, 339, 433.

Ivanova N., Justhan S., Chen X., De Marco O., Fryer C. L., Gaburov E., ... Webbink R. F., 2013, *Astronomy and Astrophysics Review*, 21, 59.

Ivanova N., Justhan S., Ricker P., 2020, *Common Envelope Evolution*, AAS-IOP Astronomy Book Series, IOP Publishing, Online ISBN: 978-0-7503-1563-0 Print ISBN: 978-0-7503-1561-6.

Ivanova N., Podsiadlowski P., Spruit H., 2002, *Monthly Notices of the Royal Astronomical Society*, 334, 819.

Iyer N., Paul B., 2017, *Monthly Notices of the Royal Astronomical Society*, 471, 355.

Izzard R. G., Dermine T., Church R. P., 2010, *Astronomy and Astrophysics*, 523, A10.

Izzard R. G., Dray L. M., Karakas A. I., Lugaro M., Tout C. A., 2006, *Astronomy and Astrophysics*, 460, 565.

Izzard R. G., Halabi G. M., 2018, arXiv:1808.06883.

Izzard R. G., Tout C. A., Karakas A. I., Pols O. R., 2004, *Monthly Notices of the Royal Astronomical Society*, 350, 407.

Jacobson-Galán W. V., Polin A., Foley R. J., Dimitriadis G., Kilpatrick C. D., Margutti R., ... Rojas-Bravo C., 2020, *Astrophysical Journal*, 896, 165.

Jacoby B. A., Cameron P. B., Jenet F. A., Anderson S. B., Murty R. N., Kulkarni S. R., 2006, *Astrophysical Journal*, 644, L113.

Jacoby B. A., Hotan A., Bailes M., Ord S., Kulkarni S. R., 2005, *Astrophysical Journal*, 629, L113.

Jager R., Mels W. A., Brinkman A. C., Galama M. Y., Goulooze H., Heise J., ... Wiersma G., 1997, *Astronomy and Astrophysics Supplement Series*, 125, 557.

Jahanara B., Mitsumoto M., Oka K., Matsuda T., Hachisu I., Boffin H. M. J., 2005, *Astronomy and Astrophysics*, 441, 589.

Janka H.-T., 2012, *Annual Review of Nuclear and Particle Science*, 62, 407.

Janka H.-T., 2013, *Monthly Notices of the Royal Astronomical Society*, 434, 1355.

Janka H.-T., 2017, *Astrophysical Journal*, 837, 84.

Janka H.-T., Melson T., Summa A., 2016, *Annual Review of Nuclear and Particle Science*, 66, 341.

Janka H.-T., Wongwathanarat A., Kramer M., 2022, *Astrophysical Journal*, 926, 9.

Janssen G. H., Stappers B. W., Kramer M., Nice D. J., Jessner A., Cognard I., Purver M. B., 2008, *Astronomy and Astrophysics*, 490, 753.

Jardine R., Powell J., Müller B., 2022, *Monthly Notices of the Royal Astronomical Society*, 510, 5535.
Jawor J. A., Tauris T. M., 2022, *Monthly Notices of the Royal Astronomical Society*, 509, 634.
Jedrzjec E., 1969, Master's thesis, Warsaw University, Poland.
Jiang L., Chen W.-C., Li X.-D., 2017, *Astrophysical Journal*, 837, 64.
Jiang L., Li X.-D., Dey J., Dey M., 2015, *Astrophysical Journal*, 807, 41.
Johnston H. M., Fender R., Wu K., 1999, *Monthly Notices of the Royal Astronomical Society*, 308, 415.
Johnston H. M., Soria R., Gibson J., 2016, *Monthly Notices of the Royal Astronomical Society*, 456, 347.
Johnston S., Karastergiou A., 2017, *Monthly Notices of the Royal Astronomical Society*, 467, 3493.
Johnston S., Lower M. E., 2021, *Monthly Notices of the Royal Astronomical Society*, 507, L41.
Johnston S., Manchester R. N., Lyne A. G., Bailes M., Kaspi V. M., Qiao G., D'Amico N., 1992, *Astrophysical Journal*, 387, L37.
Jones D., 2019, "The Importance of Binarity in the Formation and Evolution of Planetary Nebulae," in G. Beccari and M. J. Boffin, eds, *The Impact of Binary Stars on Stellar Evolution*, Cambridge University Press, pp. 106–127.
Jones S., Hirschi R., Nomoto K., 2014, *Astrophysical Journal*, 797, 83.
Jones S., Hirschi R., Nomoto K., Fischer T., Timmes F. X., Herwig F., ... Bertolli M. G., 2013, *Astrophysical Journal*, 772, 150.
Jones S., Röpke F. K., Fryer C., Ruiter A. J., Seitenzahl I. R., Nittler L. R., ... Belczynski K., 2019, *Astronomy and Astrophysics*, 622, A74.
Jones S., Röpke F. K., Pakmor R., Seitenzahl I. R., Ohlmann S. T., Edelmann P. V. F., 2016, *Astronomy and Astrophysics*, 593, A72.
Jonker P. G., Kaur K., Stone N., Torres M. A. P., 2021, *Astrophysical Journal*, 921, 131.
Jordan George C. I., Perets H. B., Fisher R. T., van Rossum D. R., 2012, *Astrophysical Journal*, 761, L23.
Jørgensen H. E., Lipunov V. M., Panchenko I. E., Postnov K. A., Prokhorov M. E., 1997, *Astrophysical Journal*, 486, 110.
Jorissen A., Van Eck S., Mayor M., Udry S., 1998, *Astronomy and Astrophysics*, 332, 877.
Joss P. C., 1977, *Nature*, 270, 310.
Joss P. C., 1978, *Astrophysical Journal*, 225, L123.
Joss P. C., Rappaport S., Lewis W., 1987, *Astrophysical Journal*, 319, 180.
Joss P. C., Rappaport S. A., 1983, *Nature*, 304, 419.
Joyce S. R. G., Barstow M. A., Holberg J. B., Bond H. E., Casewell S. L., Burleigh M. R., 2018, *Monthly Notices of the Royal Astronomical Society*, 481, 2361.
Justesen A. B., Albrecht S., 2021, *Astrophysical Journal*, 912, 123.
Kaaret P., Feng H., Roberts T. P., 2017, *Annual Review of Astronomy and Astrophysics*, 55, 303.
Kaaz N., Schrøder S. L., Andrews J. J., Antoni A., Ramirez-Ruiz E., 2021, arXiv:2103.12088.
Kahabka P., van den Heuvel E. P. J., 1997, *Annual Review of Astronomy and Astrophysics*, 35, 69.
Kahabka P., van den Heuvel E. P. J., 2006, *Super-Soft Sources*, in W. H. G. Lewin and M. van der Klis, eds, *Compact Stellar X-ray Sources*, Cambridge University Press, 461–474.
Kalirai J. S., Hansen B. M. S., Kelson D. D., Reitzel D. B., Rich R. M., Richer H. B., 2008, *Astrophysical Journal*, 676, 594.
Kalogera V., 1996, *Astrophysical Journal*, 471, 352.
Kalogera V., 1997, PhD thesis, University of Illinois at Urbana-Champaign.

Kalogera V., 1998, *Astrophysical Journal*, 493, 368.
Kalogera V., 2001, in Kaper L., Heuvel E. P. J. V. D., Woudt P. A., eds, *Black Holes in Binaries and Galactic Nuclei, ESO Astrophysics Symposia, Springer Verlag, Heidelberg: Formation of Black-Hole X-Ray Binaries with Low-Mass Donors*, 299.
Kalogera V., Baym G., 1996, *Astrophysical Journal*, 470, L61.
Kalogera V., Berry C. P. L., Colpi M., Fairhurst S., Justhan S., Mandel I., ... Valiante R., 2019, BAAS, 51, 242.
Kalogera V., Kim C., Lorimer D. R., Burgay M., D'Amico N., Possenti A., ... Camilo F., 2004, *Astrophysical Journal*, 601, L179.
Kalogera V., Kim C., Lorimer D. R., Ihm M., Belczynski K., 2005, *The Galactic Formation Rate of Eccentric Neutron Star—White Dwarf Binaries*, ASP Conference Series, Vol. 328, 261.
Kalogera V., Narayan R., Spergel D. N., Taylor J. H., 2001, *Astrophysical Journal*, 556, 340.
Kalogera V., Webbink R. F., 1996, *Astrophysical Journal*, 458, 301.
Kalogera V., Webbink R. F., 1998, *Astrophysical Journal*, 493, 351.
Kaper L., van der Meer A., van Kerkwijk M., van den Heuvel E., 2006, ESO, *The Messenger*, 126, 27.
Kaper L., van Loon J. T., Augusteijn T., Goudfrooij P., Patat F., Waters L. B. F. M., Zijlstra A. A., 1997, *Astrophysical Journal*, 475, L37.
Kaplan D. L., Bhalerao V. B., van Kerkwijk M. H., Koester D., Kulkarni S. R., Stovall K., 2013, *Astrophysical Journal*, 765, 158.
Kaplan D. L., Bildsten L., Steinfadt J. D. R., 2012, *Astrophysical Journal*, 758, 64.
Kaplan D. L., Stovall K., Ransom S. M., Roberts M. S. E., ... 2012, *Astrophysical Journal*, 753, 174.
Karakas A. I., Tout C. A., Lattanzio J. C., 2000, *Monthly Notices of the Royal Astronomical Society*, 316, 689.
Karastergiou A., Johnston S., 2007, *Monthly Notices of the Royal Astronomical Society*, 380, 1678.
Karino S., Miller J. C., 2016, *Monthly Notices of the Royal Astronomical Society*, 462, 3476.
Karino S., Nakamura K., Taani A., 2019, *Publications of the Astronomical Society of Japan*, 71, 58.
Kasen D., 2010, *Astrophysical Journal*, 708, 1025.
Kashiyama K., Quataert E., 2015, *Monthly Notices of the Royal Astronomical Society*, 451, 2656.
Kasian L. E., 2012, PhD thesis, The University of British Columbia, Vancouver.
Kasliwal M. M., Kulkarni S. R., Arcavi I., 2011, *Astrophysical Journal*, 730, 134.
Kasliwal M. M., Kulkarni S. R., Gal-Yam A., Nugent P. E., Sullivan M., Bildsten L., ... 2012, *Astrophysical Journal*, 755, 161.
Kaspi V. M., 2010, *Proceedings of the National Academy of Science*, 107, 7147.
Kaspi V. M., Bailes M., Manchester R. N., Stappers B. W., Bell J. F., 1996, *Nature*, 381, 584.
Kaspi V. M., Beloborodov A. M., 2017, *Annual Review of Astronomy and Astrophysics*, 55, 261.
Kaspi V. M., Johnston S., Bell J. F., Manchester R. N., Bailes M., Bessell M., ... D'Amico N., 1994, *Astrophysical Journal*, 423, L43.
Kaspi V. M., Kramer M., 2016, ArXiv:1602.07738; paper published in R. Blandford, D. Gross, A. Sevrin, eds, *Astrophysics and Cosmology*, Proceedings of the 26th Solvay Conference on Physics, World Scientific Publishing Comp. 21–70.
Kato M., 2010, *Astronomische Nachrichten*, 331, 140.
Kato M., Hachisu I., 1994, *Astrophysical Journal*, 437, 802.
Kato M., Iben Jr. I., 1992, *Astrophysical Journal*, 394, 305.
Kato S., Machida M., 2020, *Publications of the Astronomical Society of Japan*, 72, 38.

Kato T., Nogami D., Baba H., 2001, *Publications of the Astronomical Society of Japan*, 53, 901.
Katoh N., Itoh Y., Toyota E., Sato B., 2013, *Astronomical Journal*, 145, 41.
Katz J. I., 1975, *Nature*, 253, 698.
Kawano T., Done C., Yamada S., Takahashi H., Axelsson M., Fukazawa Y., 2017, *Publications of the Astronomical Society of Japan*, 69, 36.
Keane E., Bhattacharyya B., Kramer M., et al. 2015, *Advancing Astrophysics with the Square Kilometre Array (AASKA14)*, 40.
Keane E. F., Kramer M., 2008, *Monthly Notices of the Royal Astronomical Society*, 391, 2009.
Kehl M. S., Wex N., Kramer M., Liu K., 2018, in Bianchi M., Jansen R. T., Ruffini R., eds, *Fourteenth Marcel Grossmann Meeting—MG14 Future Measurements of the Lense-Thirring Effect in the Double Pulsar*, pp. 1860–1865.
Keith M. J., Jameson A., van Straten W., Bailes M., Johnston S., Kramer M., et al. Stappers B. W., 2010, *Monthly Notices of the Royal Astronomical Society*, 409, 619.
Keith M. J., Johnston S., Bailes M., . . . 2012, *Monthly Notices of the Royal Astronomical Society*, 419, 1752.
Keith M. J., Kramer M., Lyne A. G., Eatough R. P., Stairs I. H., Possenti A., . . . Manchester R. N., 2009, *Monthly Notices of the Royal Astronomical Society*, 393, 623.
Kelley L. Z., Ramirez-Ruiz E., Zemp M., Diemand J., Mandel I., 2010, *Astrophysical Journal*, 725, L91.
Kennea J. A., Coe M. J., Evans P. A., Townsend L. J., Campbell Z. A., Udalski A., 2021, *Monthly Notices of the Royal Astronomical Society*, 508, 781.
Kenyon S. J., 1986, *The Symbiotic Stars*, Cambridge University Press.
Kenyon S. J., Bromley B. C., Geller M. J., Brown W. R., 2008, *Astrophysical Journal*, 680, 312.
Kerr R. P., 1963, *Physical Review Letters*, 11, 237.
Keszthelyi Z., Meynet G., Shultz M. E., David-Uraz A., ud-Doula A., Townsend R. H. D., Wade G. A., Georgy C., Petit V., Owocki S. P., 2020, *Monthly Notices of the Royal Astronomical Society*, 493, 518.
Khokhlov A. M., 1991, *Astronomy and Astrophysics*, 245, 114.
Khokhlov S. A., Miroshnichenko A. S., Zharikov S. V., Manset N., Arkharov A. A., Efimova N., . . . 2018, *Astrophysical Journal*, 856, 158.
Khorrami Z., Vakili F., Lanz T., Langlois M., Lagadec E., Meyer M. R., . . . Ramos J., 2017, *Astronomy and Astrophysics*, 602, A56.
Khouri T., Vlemmings W. H. T., Tafoya D., Perez-Sanchez A. F., Sanchez Contreras C., Gomez J. F., Imai H., Sahai R., 2021, *Nature Astronomy*, 6, 275.
Kilic M., Hermes J. J., Gianninas A., Brown W. R., 2015, *Monthly Notices of the Royal Astronomical Society*, 446, L26.
Kilpatrick C. D., Foley R. J., Kasen D., et al. 2017, *Science*, 358, 1583.
Kim C., 2006, PhD thesis, Northwestern University, Illinois.
Kim C., Kalogera V., Lorimer D. R., 2003, *Astrophysical Journal*, 584, 985.
Kim C., Perera B. B. P., McLaughlin M. A., 2015, *Monthly Notices of the Royal Astronomical Society*, 448, 928.
Kim S. S., Figer D. F., Kudritzki R. P., Najarro F., 2006, *Astrophysical Journal*, 653, L113.
Kimball C., Talbot C., Berry C. P. L., Zevin M., Thrane E., Kalogera V., Buscicchio R., Carney M., Dent T., Middleton H., Payne E., Veitch J., Williams D., 2021, *Astrophysical Journal*, 915, L35.
King A., Lasota J.-P., 2019, *Monthly Notices of the Royal Astronomical Society*, 485, 3588.
King A., Lasota J.-P., Kluźniak W., 2017, *Monthly Notices of the Royal Astronomical Society*, 468, L59.

King A. R., 2004, *Monthly Notices of the Royal Astronomical Society*, 347, L18.
King A. R., Begelman M. C., 1999, *Astrophysical Journal*, 519, L169.
King A. R., Davies M. B., Beer M. E., 2003, *Monthly Notices of the Royal Astronomical Society*, 345, 678.
King A. R., Davies M. B., Ward M. J., Fabbiano G., Elvis M., 2001, *Astrophysical Journal*, 552, L109.
King A. R., Kolb U., 1999, *Monthly Notices of the Royal Astronomical Society*, 305, 654.
King A. R., Schenker K., Kolb U., Davies M. B., 2001, *Monthly Notices of the Royal Astronomical Society*, 321, 327.
King A. R., Taam R. E., Begelman M. C., 2000, *Astrophysical Journal*, 530, L25.
Kinugawa T., Nakamura T., Nakano H., 2021, *Monthly Notices of the Royal Astronomical Society*, 501, L49.
Kippenhahn R., Kohl K., Weigert A., 1967, *Zeitschrift für Astrophysik*, 66, 58.
Kippenhahn R., Weigert A., 1967, *Zeitschrift für Astrophysik*, 65, 251.
Kippenhahn R., Weigert A., 1990, *Stellar Structure and Evolution*. Springer, Berlin
Kirsebom O. S., Jones S., Strömberg D. F., Martínez-Pinedo G., Langanke K., Röpke F. K., . . . 2019, *Physical Review Letters*, 123, 262701.
Kirsten F., Marcote B., Nimmo K., Hessels J. W. T., Bhardwaj M., Tendulkar S. P., Keimpema A., Yang J., et al. 2022, *Nature Astronomy*, 602, 585.
Kirsten F., Snelders M. P., Jenkins M., Nimmo K., van den Eijnden J., Hessels J. W. T., . . . Yang J., 2020, *Nature Astronomy*.
Kiss L. L., Kasza J., Borza S., 2000, *Information Bulletin on Variable Stars*, 4962.
Kitamoto S., Miyamoto S., Yamamoto T., 1989, *Publications of the Astronomical Society of Japan*, 41, 81.
Kitaura F. S., Janka H.-T., Hillebrandt W., 2006, *Astronomy and Astrophysics*, 450, 345.
Kiziltan B., Thorsett S. E., 2010, *Astrophysical Journal*, 715, 335.
Kleiser I. K. W., Kasen D., 2014, *Monthly Notices of the Royal Astronomical Society*, 438, 318.
Klencki J., Nelemans G., Istrate A. G., Chruslinska M., 2021, *Astronomy and Astrophysics*, 645, A54.
Klencki J., Nelemans G., Istrate A. G., Pols O., 2020, *Astronomy and Astrophysics*, 638, A55.
Kloppenborg B. K., Hopkins J. L., Stencel R. E., 2012, *Journal of the American Association of Variable Star Observers (JAAVSO)*, 40, 647.
Klus H., Ho W. C. G., Coe M. J., Corbet R. H. D., Townsend L. J., 2014, *Monthly Notices of the Royal Astronomical Society*, 437, 3863.
Kluzniak W., Lasota J. P., 2015, *Monthly Notices of the Royal Astronomical Society*, 448, L43.
Kluzniak W., Ruderman M., Shaham J., Tavani M., 1988, *Nature*, 334, 225.
Knigge C., Baraffe I., Patterson J., 2011, *Astrophysical Journal Supplement Series*, 194, 28.
Knigge C., Coe M. J., Podsiadlowski P., 2011, *Nature*, 479, 372.
Knigge C., Leigh N., Sills A., 2009, *Nature*, 457, 288.
Kobulnicky H. A., Fryer C. L., 2007, *Astrophysical Journal*, 670, 747.
Kobulnicky H. A., Kiminki D. C., Lundquist M. J., Burke J., Chapman J., Keller E., . . . Brotherton M. M., 2014, *Astrophysical Journal Supplement Series*, 213, 34.
Kolb U., Ritter H., 1990, *Astronomy and Astrophysics*, 236, 385.
Konar S., Bhattacharya D., 1997, *Monthly Notices of the Royal Astronomical Society*, 284, 311.
Konar S., Mukherjee D., Bhattacharya D., Sarkar P., 2016, *Physical Review D*, 94, 104036.
Kopparapu R. K., Hanna C., Kalogera V., O'Shaughnessy R., González G., Brady P. R., Fairhurst S., 2008, *Astrophysical Journal*, 675, 1459.
Körding E., Falcke H., Markoff S., 2002, *Astronomy and Astrophysics*, 382, L13.
Kornilov V. G., Lipunov V. M., 1983a, *Soviet Astronomy*, 27, 163.

Kornilov V. G., Lipunov V. M., 1983b, *Soviet Astronomy*, 27, 334.
Kornilov V. G., Lipunov V. M., 1984, *Soviet Astronomy*, 28, 402.
Korol V., Rossi E. M., Groot P. J., Nelemans G., Toonen S., Brown A. G. A., 2017, *Monthly Notices of the Royal Astronomical Society*, 470, 1894.
Korol V., Safarzadeh M., 2021, *Monthly Notices of the Royal Astronomical Society*, 502, 5576.
Kourniotis M., Kraus M., Arias M. L., Cidale L., Torres A. F., 2018, *Monthly Notices of the Royal Astronomical Society*, 480, 3706.
Kouveliotou C., Dieters S., Strohmayer T., van Paradijs J., Fishman G. J., Meegan C. A., ... Murakami T., 1998, *Nature*, 393, 235.
Kouveliotou C., Wijers R. A. M. J., Woosley S., 2012, *Gamma-Ray Bursts*, Cambridge University Press.
Kouwenhoven M. B. N., Brown A. G. A., Portegies Zwart S. F., Kaper L., 2007, *Astronomy and Astrophysics*, 474, 77.
Kozai Y., 1962, *Astronomical Journal*, 67, 591.
Kozyreva A., Baklanov P., Jones S., Stockinger G., Janka H.-T., 2021, *Monthly Notices of the Royal Astronomical Society*, 503, 797.
Kramer M., 1998, *Astrophysical Journal*, 509, 856.
Kramer M., 2018, in Weltevrede P., Perera B. B. P., Preston L. L., Sanidas S., eds, *Pulsar Astrophysics the Next Fifty Years*, proceedings of IAU Symposium 337, *Gravity Tests with Pulsars*, 128–133.
Kramer M., Champion D. J., 2013, *Classical and Quantum Gravity*, 30, 224009
Kramer M., Lyne A. G., Hobbs G., Löhmer O., Carr P., Jordan C., Wolszczan A., 2003, *Astrophysical Journal*, 593, L31.
Kramer M., Lyne A. G., O'Brien J. T., Jordan C. A., Lorimer D. R., 2006, *Science*, 312, 549.
Kramer M., Stairs I. H., Manchester R. N., McLaughlin M. A., Lyne A. G., Ferdman R. D., ... Freire P. C. C., Camilo F., 2006, *Science*, 314, 97.
Kramer M., Stairs I. H., Manchester R. N., Wex N., ... 2021, *Physical Review X*, 11, 041050.
Kramer M., Stairs I. H., Venkatraman Krishnan V., Freire P. C. C., Abbate F., Bailes M., ... 2021, *Monthly Notices of the Royal Astronomical Society*, 504, 2094.
Kramer M., Wex N., 2009, *Classical and Quantum Gravity*, 26, 073001.
Kramer M., Xilouris K. M., Lorimer D. R., Doroshenko O., Jessner A., Wielebinski R., ... Camilo F., 1998, *Astrophysical Journal*, 501, 270.
Kraus M., 2016, *Boletin de la Asociacion Argentina de Astronomia La Plata Argentina*, 58, 70.
Kreiner J. M., 1981, in Rudnicki K., Flin P., eds, *Rocznik Astronomiczny Obserwatorium Krakowskiego 1982, Ephemerides of Eclipsing Binaries among Cataclysmic Variables for the Year 1982*, 107–111.
Kremer K., Breivik K., Larson S. L., Kalogera V., 2017, *Astrophysical Journal*, 846, 95.
Kretschmar P., Fürst F., Sidoli L., Bozzo E., Alfonso-Garzon J., Bodaghee A., Chaty S., Chernyakova M., Ferrigno C., et al. 2019, *New Astronomy Reviews*, 86, 101546.
Kreuzer S., Irrgang A., Heber U., 2020, *Astronomy and Astrophysics*, 637, A53.
Krishnan V. V., Bailes M., van Straten W., Wex N., Freire P. C. C., Keane E. F., ... Osłowski S., 2020, Science, 367, 577.
Krolik J. H., Meiksin A., Joss P. C., 1984, *Astrophysical Journal*, 282, 466.
Kromer M., Fink M., Stanishev V., Taubenberger S., Ciaraldi-Schoolman F., Pakmor R., ... Hillebrandt W., 2013, *Monthly Notices of the Royal Astronomical Society*, 429, 2287.
Kroupa P., 2001, *Monthly Notices of the Royal Astronomical Society*, 322, 231.
Kruckow M. U., 2020, *Astronomy and Astrophysics*, 639, A123.
Kruckow M. U., Tauris T. M., Langer N., Kramer M., Izzard R. G., 2018, *Monthly Notices of the Royal Astronomical Society*, 481, 1908.

Kruckow M. U., Tauris T. M., Langer N., Szécsi D., Marchant P., Podsiadlowski P., 2016, *Astronomy and Astrophysics*, 596, A58.

Krumholz M. R., Klein R. I., McKee C. F., Offner S. S. R., Cunningham A. J., 2009, *Science*, 323, 754.

Kruszewski A., 1964, *Acta Astronomica*, 14, 241.

Kuchner M. J., Kirshner R. P., Pinto P. A., Leibundgut B., 1994, *Astrophysical Journal*, 426, L89.

Kudritzki R.-P., Puls J., 2000, *Annual Review of Astronomy and Astrophysics*, 38, 613.

Kuiper G. P., 1935, *Publications of the Astronomical Society of the Pacific*, 47, 15.

Kuiper G. P., 1941, *Astrophysical Journal*, 93, 133.

Kulkarni S. R., 1986, *Astrophysical Journal*, 306, L85.

Kulkarni S. R., Hester J. J., 1988, *Nature*, 335, 801.

Kulkarni S. R., Hut P., McMillan S., 1993, *Nature*, 364, 421.

Kulkarni S. R., Narayan R., 1988, *Astrophysical Journal*, 335, 755.

Kulkarni S. R., Ofek E. O., Rau A., Cenko S. B., Soderberg A. M., Fox D. B., . . . Sanders D. B., 2007, *Nature*, 447, 458.

Kundra E., Hric L., 2011, *Astrophysics and Space Science*, 331, 121.

Kundt W., 1976, *Physics Letters A*, 57, 195.

Kunze S., Speith R., Hessman F. V., 2001, *Monthly Notices of the Royal Astronomical Society*, 322, 499.

Kupfer T., Korol V., Shah S., Nelemans G., Marsh T. R., Ramsay G., et al. Rossi E. M., 2018, *Monthly Notices of the Royal Astronomical Society*, 480, 302.

Kushnir D., Zaldarriaga M., Kollmeier J. A., Waldman R., 2016, *Monthly Notices of the Royal Astronomical Society*, 462, 844.

Lagos F., Schreiber M. R., Parsons S. G., Gänsicke B. T., Godoy N., 2020, *Monthly Notices of the Royal Astronomical Society*, 499, L121.

Lai D., 1996, *Astrophysical Journal*, 466, L35.

Lai D., Chernoff D. F., Cordes J. M., 2001, *Astrophysical Journal*, 549, 1111.

Lamb D. Q., Lamb F. K., 1978, *Astrophysical Journal*, 220, 291.

Lamb F. K., Pethick C. J., Pines D., 1973, *Astrophysical Journal*, 184, 271.

Lamb G. P., Tanvir N. R., Levan A. J., de Ugarte Postigo A., et al. 2019, *Astrophysical Journal*, 883, 48.

Landau L. D., Lifshitz E. M., 1951, *The Classical Theory of Fields*. Addison-Wesley Press, USA.

Landau L. D., Lifshitz E. M., 1971, The Classical Theory of Fields. Pergamon Press, Oxford.

Langanke K., Martínez-Pinedo G., Zegers R. G. T., 2021, *Reports on Progress in Physics*, 84, 066301.

Langer N., 2012, *Annual Review of Astronomy and Astrophysics*, 50, 107.

Langer N., Deutschmann A., Wellstein S., Höflich P., 2000, *Astronomy and Astrophysics*, 362, 1046.

Langer N., Schürmann C., Stoll K., Marchant P., et al. 2020, *Astronomy and Astrophysics*, 638, A39.

Lanza A. F., Rodonò M., 1999, *Astronomy and Astrophysics*, 349, 887.

Lanza A. F., Rodono M., Rosner R., 1998, *Monthly Notices of the Royal Astronomical Society*, 296, 893.

Laor A., 1991, *Astrophysical Journal*, 376, 90.

Laplace E., Götberg Y., de Mink S. E., Justhan S., Farmer R., 2020, *Astronomy and Astrophysics*, 637, A6.

Laplace E., Justhan S., Renzo M., Götberg Y., Farmer R., Vartanyan D., de Mink S. E., 2021, *Astronomy and Astrophysics*, 656, A58.

Lasky P. D., Glampedakis K., 2016, *Monthly Notices of the Royal Astronomical Society*, 458, 1660.

Lasota J.-P., Dubus G., Kruk K., 2008, *Astronomy and Astrophysics*, 486, 523.

Lattimer J. M., Prakash M., 2007, *Physics Reports*, 442, 109.

Lattimer J. M., Prakash M., 2016, *Physics Reports*, 621, 127

Lattimer J. M., Schramm D. N., 1974, *Astrophysical Journal*, 192, L145.

Lattimer J. M., Schramm D. N., 1976, *Astrophysical Journal*, 210, 549.

Lattimer J. M., Yahil A., 1989, *Astrophysical Journal*, 340, 426.

Lau M. Y. M., Mandel I., Vigna-Gómez A., Neijssel C. J., Stevenson S., Sesana A., 2020, *Monthly Notices of the Royal Astronomical Society*, 492, 3061.

Law-Smith J. A. P., Everson R. W., Ramirez-Ruiz E., de Mink S. E., van Son L. A. C., Götberg Y., ... Hutchinson-Smith T., 2020, arXiv:2011.06630.

Laycock S. G. T., Maccarone T. J., Christodoulou D. M., 2015, *Monthly Notices of the Royal Astronomical Society*, 452, L31.

Lazaridis K., Verbiest J. P. W., Tauris T. M., Stappers B. W., Kramer M., Wex N., ... Smits R., 2011, *Monthly Notices of the Royal Astronomical Society*, 414, 3134.

Lazaridis K., Wex N., Jessner A., Kramer M., Stappers B. W., Janssen G. H., ... Zensus J. A., 2009, *Monthly Notices of the Royal Astronomical Society*, 400, 805.

Lazarus P., Freire P. C. C., Allen B., Aulbert C., Bock O., et al. 2016, *Astrophysical Journal*, 831, 150.

Lazarus P., Tauris T. M., Knispel B., Freire P. C. C., Deneva J. S., Kaspi V. M., ... Zhu W. W., 2014, *Monthly Notices of the Royal Astronomical Society*, 437, 1485.

Leahy D. A., Leahy J. C., 2015, *Computational Astrophysics and Cosmology*, 2, 4.

Leavitt H. S., Pickering E. C., 1912, *Harvard College Observatory Circular*, 173, 1.

Lee H. M., Ostriker J. P., 1986, *Astrophysical Journal*, 310, 176.

Lehner L., Liebling S. L., Palenzuela C., Caballero O. L., O'Connor E., Anderson M., Neilsen D., 2016, *Classical and Quantum Gravity*, 33, 184002.

Lennon D. J., Maíz Apellániz J., Irrgang A., Bohlin R., Deustua S., Dufton P. L., ... de Burgos A., 2021, *Astronomy and Astrophysics*, 649, A167.

Leonard P. J. T., 1991, *Astronomical Journal*, 101, 562.

Lépine S., Bongiorno B., 2007, *Astronomical Journal*, 133, 889.

Leung S.-C., Zha S., Chu M.-C., Lin L.-M., Nomoto K., 2019, *Astrophysical Journal*, 884, 9.

Levesque E. M., Massey P., Olsen K. A. G., Plez B., Josselin E., Maeder A., Meynet G., 2005, *Astrophysical Journal*, 628, 973.

Levesque E. M., Massey P., Olsen K. A. G., Plez B., Meynet G., Maeder A., 2006, *Astrophysical Journal*, 645, 1102.

Levesque E. M., Massey P., Żytkow A. N., Morrell N., 2014, *Monthly Notices of the Royal Astronomical Society*, 443, L94.

Levine A. M., Bradt H., Cui W., Jernigan J. G., Morgan E. H., Remillard R., ... Smith D. A., 1996, *Astrophysical Journal*, 469, L33.

Levitan D., Groot P. J., Prince T. A., Kulkarni S. R., Laher R., Ofek E. O., ... Surace J., 2015, *Monthly Notices of the Royal Astronomical Society*, 446, 391.

Lewin W. H. G., van der Klis M., 2006, *Compact Stellar X-Ray Sources*. Cambridge University Press.

Lewin W. H. G., van Paradijs J., Taam R. E., 1993, *Space Science Reviews*, 62, 223.

Lewin W. H. G., van Paradijs J., Taam R. E., 1995, in *X-ray Binaries*, W. H. G. Lewin, J. A. van Paradijs and E. P. J. van den Heuvel, eds, Cambridge University Press, 175–232.

Li X.-D., 2015, *New Astronomy Reviews*, 64, 1.

Li X.-D., van den Heuvel E. P. J., 1997, *Astronomy and Astrophysics*, 322, L9.

Li Y., Luo A., Zhao G., Lu Y., Ren J., Zuo F., 2012, *Astrophysical Journal*, 744, L24.

Liao Z., Liu J., Zheng X., Gou L., 2020, *Monthly Notices of the Royal Astronomical Society*, 492, 5922.

Lidov M. L., 1962, *Planetary and Space Science*, 9, 719.

Lin J., Rappaport S., Podsiadlowski P., Nelson L., Paxton B., Todorov P., 2011, *Astrophysical Journal*, 732, 70.

Linares M., Shahbaz T., Casares J., 2018, *Astrophysical Journal*, 859, 54.

Lippincott S. L., 1961, Leaflet of the Astronomical Society of the Pacific, 8, 311.

Lipunov V., Kornilov V., Gorbovskoy E., Lipunova G., Vlasenko D., Panchenko I., . . . Grinshpun V., 2018, *New Astronomy*, 63, 48.

Lipunov V. M., Blinnikov S., Gorbovskoy E., Tutukov A., Baklanov P., et al. 2017, *Monthly Notices of the Royal Astronomical Society*, 470, 2339.

Lipunov V. M., Postnov K. A., 1984, *Astrophysics and Space Science*, 106, 103.

Lipunov V. M., Postnov K. A., Prokhorov M. E., 1987, *Astronomy and Astrophysics*, 176, L1.

Lipunov V. M., Postnov K. A., Prokhorov M. E., 1996, *The Scenario Machine: Binary Star Population Synthesis*, *Astronomy and Astrophysics*, 310, 489.

Lipunov V. M., Postnov K. A., Prokhorov M. E., 1997, *New Astronomy*, 2, 43.

Lipunov V. M., Postnov K. A., Prokhorov M. E., Bogomazov A. I., 2009, *Astronomy Reports*, 53, 915.

Lissauer J. J., Backman D. E., 1984, *Astrophysical Journal*, 286, L39.

Littenberg T. B., Cornish N. J., 2019, *Astrophysical Journal*, 881, L43.

Liu D., Wang B., 2020, *Monthly Notices of the Royal Astronomical Society*, 494, 3422.

Liu D., Wang B., Ge H., Chen X., Han Z., 2019, *Astronomy and Astrophysics*, 622, A35.

Liu J., Zhang H., Howard A. W., Bai Z., Lu Y., Soria R., . . . Casares J., et al. 2019, *Nature*, 575, 618.

Liu Q. Z., van Paradijs J., van den Heuvel E. P. J., 2000, *Astronomy and Astrophysics Supplement Series*, 147, 25.

Liu Q. Z., van Paradijs J., van den Heuvel E. P. J., 2001, *Astronomy and Astrophysics*, 368, 1021.

Liu Q. Z., van Paradijs J., van den Heuvel E. P. J., 2006, *Astronomy and Astrophysics*, 455, 1165.

Liu Q. Z., van Paradijs J., van den Heuvel E. P. J., 2007, *Astronomy and Astrophysics*, 469, 807.

Liu X.-J., You Z.-Q., Zhu X.-J., 2022, *Astrophysical Journal, Letters*, 934, L2.

Liu Z. W., Pakmor R., Röpke F. K., Edelmann P., Wang B., Kromer M., . . . Han Z. W., 2012, *Astronomy and Astrophysics*, 548, A2.

Liu Z.-W., Tauris T. M., Röpke F. K., Moriya T. J., Kruckow M., Stancliffe R. J., Izzard R. G., 2015, *Astronomy and Astrophysics*, 584, A11.

Livio M., 1994, in Shore S. N., Livio M., van den Heuvel E. P. J., Nussbaumer H., Orr A., eds, *Saas-Fee Advanced Course 22: Interacting Binaries: Topics in the Theory of Cataclysmic Variables and X-ray Binaries*, 135–262.

Livne E., 1990, *Astrophysical Journal*, 354, L53.

Livne E., Arnett D., 1995, *Astrophysical Journal*, 452, 62.

Lombardi Jr. J. C., Proulx Z. F., Dooley K. L., Theriault E. M., Ivanova N., Rasio F. A., 2006, *Astrophysical Journal*, 640, 441.

Long K. S., Helfand D. J., Grabelsky D. A., 1981, *Astrophysical Journal*, 248, 925.

Long K. S., van Speybroeck L. P., 1983, in Lewin W. H. G., van den Heuvel E. P. J., eds, *Accretion-Driven Stellar X-ray Sources X-ray Emission from Normal Galaxies*, 117–146.

Lorimer D. R., 1994, PhD thesis, The University of Manchester, UK.

Lorimer D. R., Bailes M., McLaughlin M. A., Narkevic D. J., Crawford F., 2007, *Science*, 318, 777.

Lorimer D. R., Kramer M., 2004, *Handbook of Pulsar Astronomy*, Cambridge University Press.

Lorimer D. R., Kramer M., 2012, *Handbook of Pulsar Astronomy*, Cambridge University Press.
Lorimer D. R., Lyne A. G., McLaughlin M. A., Kramer M., Pavlov G. G., Chang C., 2012, *Astrophysical Journal*, 758, 141.
Lorimer D. R., McLaughlin M. A., Arzoumanian Z., Xilouris K. M., Cordes J. M., Lommen A. N., ... Backer D. C., 2004, *Monthly Notices of the Royal Astronomical Society*, 347, L21.
Lorimer D. R., Stairs I. H., Freire P. C., et al. 2006, *Astrophysical Journal*, 640, 428.
Lovegrove E., Woosley S. E., 2013, *Astrophysical Journal*, 769, 109.
Loveridge A. J., van der Sluys M. V., Kalogera V., 2011, *Astrophysical Journal*, 743, 49.
Lubow S. H., Artymowicz P., 1996, in Wijers R. A. M. J., Davies M. B., Tout C. A., eds, NATO Advanced Science Institutes (ASI) Series C Vol. 477, Kluwer Academic Publishers, Dordrecht, *Young Binary Star/Disk Interactions*, 53.
Lubow S. H., Shu F. H., 1975, *Astrophysical Journal*, 198, 383.
Lucy L. B., 1991, *Astrophysical Journal*, 383, 308.
Lucy L. B., Ricco E., 1979, *Astronomical Journal*, 84, 401.
Lunnan R., Kasliwal M. M., Cao Y., Hangard L., Yaron O., Parrent J. T., et al. 2017, *Astrophysical Journal*, 836, 60.
Luo Y., 2020, *New Astronomy*, 78, 101363.
Lyman J. D., Bersier D., James P. A., Mazzali P. A., Eldridge J. J., Fraser M., Pian E., 2016, *Monthly Notices of the Royal Astronomical Society*, 457, 328.
Lyman J. D., Taddia F., Stritzinger M. D., Galbany L., Leloudas G., Anderson J. P., ... Stanway E. R., 2018, *Monthly Notices of the Royal Astronomical Society*, 473, 1359.
Lynch R. S., Freire P. C. C., Ransom S. M., Jacoby B. A., 2012, *Astrophysical Journal*, 745, 109.
Lynch R. S., Swiggum J. K., Kondratiev V. I., Kaplan D. L., Stovall K., et al. 2018, *Astrophysical Journal*, 859, 93.
Lynden-Bell D., Pringle J. E., 1974, *Monthly Notices of the Royal Astronomical Society*, 168, 603
Lyne A. G., Biggs J. D., Harrison P. A., Bailes M., 1993, *Nature*, 361, 47.
Lyne A. G., Brinklow A., Middleditch J., Kulkarni S. R., Backer D. C., 1987, *Nature*, 328, 399.
Lyne A. G., Burgay M., Kramer M., Possenti A., Manchester R. N., Camilo F., ... Freire P. C. C., 2004, Science, 303, 1153.
Lyne A. G., Lorimer D. R., 1994, *Nature*, 369, 127.
Lyne A. G., Manchester R. N., 1988, *Monthly Notices of the Royal Astronomical Society*, 234, 477.
Lyne A. G., Manchester R. N., D'Amico N., 1996, *Astrophysical Journal*, 460, L41.
Lyne A. G., Manchester R. N., Taylor J. H., 1985, *Monthly Notices of the Royal Astronomical Society*, 213, 613.
Lyutikov M., Barkov M. V., Giannios D., 2020, *Astrophysical Journal*, 893, L39.
Lyutikov M., Popov S., 2020, arXiv:2005.05093.
Lyutikov M., Toonen S., 2017, arXiv:1709.02221.
Ma B., Li X.-D., 2009, *Astrophysical Journal*, 691, 1611.
Maccarone T., Knigge C., 2007, *Astronomy and Geophysics*, 48, 5.12
Maccarone T., Steiner J., McClintock J., Orosz J., Walton D., 2014, *Determining the Mass of a Putative Heavy Stellar Black Hole*, NOAO Proposal.
Maccarone T. J., de Mink S. E., 2016, *Monthly Notices of the Royal Astronomical Society*, 458, L1.
Maccarone T. J., Fender R. P., Knigge C., Tzioumis A. K., 2009, *Monthly Notices of the Royal Astronomical Society*, 393, 1070.
Maccarone T. J., Kundu A., Zepf S. E., Rhode K. L., 2007, *Nature*, 445, 183.
MacFadyen A. I., Woosley S. E., 1999, *Astrophysical Journal*, 524, 262.

MacLeod M., Antoni A., Murguia-Berthier A., Macias P., Ramirez-Ruiz E., 2017, *Astrophysical Journal*, 838, 56.
MacLeod M., Macias P., Ramirez-Ruiz E., Grindlay J., Batta A., Montes G., 2017, *Astrophysical Journal*, 835, 282.
MacLeod M., Ostriker E. C., Stone J. M., 2018, *Astrophysical Journal*, 863, 5.
MacLeod M., Ramirez-Ruiz E., 2015a, *Astrophysical Journal*, 803, 41.
MacLeod M., Ramirez-Ruiz E., 2015b, *Astrophysical Journal*, 798, L19.
Madau P., Fragos T., 2017, *Astrophysical Journal*, 840, 39.
Madau P., Rees M. J., 2001, *Astrophysical Journal*, 551, L27.
Madsen E. C., Stairs I. H., Kramer M., Camilo F., Hobbs G. B., Janssen G. H., . . . Stappers B. W., 2012, *Monthly Notices of the Royal Astronomical Society*, 425, 2378.
Maeder A., 1987, *Astronomy and Astrophysics*, 178, 159.
Maeder A., Lequeux J., Azzopardi M., 1980, *Astronomy and Astrophysics*, 90, L17.
Maeder A., Meynet G., 1989, *Astronomy and Astrophysics*, 210, 155.
Maeder A., Meynet G., 2000a, *Astronomy and Astrophysics*, 361, 159.
Maeder A., Meynet G., 2000b, *Annual Review of Astronomy and Astrophysics*, 38, 143.
Maggiore M., 2008, *Gravitational Waves: Volume 1: Theory and Experiments*, Oxford University Press.
Maggiore M., 2018, *Gravitational Waves: Volume 2: Astrophysics and Cosmology*, Oxford University Press.
Maitra C., Haberl F., Filipović M. D., Udalski A., Kavanagh P. J., Carpano S., et al. 2019, *Monthly Notices of the Royal Astronomical Society*, 490, 5494.
Maitra C., Haberl F., Maggi P., Kavanagh P. J., Vasilopoulos G., Sasaki M., . . . Udalski A., 2021, *Monthly Notices of the Royal Astronomical Society*, 504, 326.
Maíz Apellániz J., Sota A., Arias J. I., Barbá R. H., Walborn N. R., Simón-Díaz S., . . . Alfaro E. J., 2016, *Astrophysical Journal Supplement Series*, 224, 4.
Manchester R. N., 1995, *Journal of Astrophysics and Astronomy*, 16, 107.
Manchester R. N., Hobbs G., Bailes M., et al. 2013, *Publications of the Astronomical Society of Australia*, 30, e017.
Manchester R. N., Hobbs G. B., Teoh A., Hobbs M., 2005, *Astronomical Journal*, 129, 1993.
Manchester R. N., Lyne A. G., Robinson C., D'Amico N., Bailes M., Lim J., 1991, *Nature*, 352, 219.
Manchester R. N., Taylor J. H., 1977, *Pulsars*. W. H. Freeman, San Francisco.
Mandel I., 2016, *Monthly Notices of the Royal Astronomical Society*, 456, 578.
Mandel I., Broekgaarden F. S., 2022, *Living Reviews in Relativity*, 25, 1, 1
Mandel I., de Mink S. E., 2016, *Monthly Notices of the Royal Astronomical Society*, 458, 2634.
Mandel I., Farmer A., 2022, *Physics Reports*, 955, 1.
Mandel I., Fragos T., 2020, *Astrophysical Journal*, 895, L28.
Mandel I., Müller B., 2020, *Monthly Notices of the Royal Astronomical Society*, 499, 3214.
Mandel I., Smith J. E., 2021, *Astrophysical Journal, Letters*, 922, L14.
Mandel I., Müller B., Riley J., de Mink S. E., Vigna-Gómez A., Chattopadhyay D., 2021, *Monthly Notices of the Royal Astronomical Society*, 500, 1380.
Manousakis A., Walter R., Blondin J. M., 2012, *Astronomy and Astrophysics*, 547, A20.
Maoz D., Hallakoun N., 2017, *Monthly Notices of the Royal Astronomical Society*, 467, 1414.
Maoz D., Mannucci F., Brandt T. D., 2012, *Monthly Notices of the Royal Astronomical Society*, 426, 3282.
Maoz D., Mannucci F., Li W., Filippenko A. V., Della Valle M., Panagia N., 2011, *Monthly Notices of the Royal Astronomical Society*, 412, 1508.

Maoz D., Mannucci F., Nelemans G., 2014, *Annual Review of Astronomy and Astrophysics*, 52, 107.
Mapelli M., 2016, *Monthly Notices of the Royal Astronomical Society*, 459, 3432.
Mapelli M., 2018, arXiv:1807.07944.
Mapelli M., Bressan A., 2013, *Monthly Notices of the Royal Astronomical Society*, 430, 3120.
Mapelli M., Giacobbo N., 2018, *Monthly Notices of the Royal Astronomical Society*, 479, 4391.
Mapelli M., Santoliquido F., Bouanais Y., Arca Sedda M. A., Artale M. C., Ballone A., 2021, *Symmetry*, 13, 1678.
Mapelli M., Sigurdsson S., Ferraro F. R., Colpi M., Possenti A., Lanzoni B., 2006, *Monthly Notices of the Royal Astronomical Society*, 373, 361.
Maraschi L., Cavaliere A., 1977, in *X-ray Binaries and Compact Objects*, Joint Discussion Nr. 2 at the 16th General Assembly of the IAU (ed. E. P. J. van den Heuvel), *Highlights in Astronomy*, Vol 4, part 1: *X-ray Bursts of Nuclear Origin*, pp. 127–128.
Maraschi L., Treves A., van den Heuvel E. P. J., 1976, *Nature*, 259, 292.
Marchant P., Langer N., Podsiadlowski P., Tauris T. M., de Mink S., Mandel I., Moriya T. J., 2017, *Astronomy and Astrophysics*, 604, A55.
Marchant P., Langer N., Podsiadlowski P., Tauris T. M., Moriya T. J., 2016, *Astronomy and Astrophysics*, 588, A50.
Marchant P., Moriya T. J., 2020, *Astronomy and Astrophysics*, 640, L18.
Marchant P., Pappas K. M. W., Gallegos-Garcia M., Berry C. P. L., Taam R. E., Kalogera V., Podsiadlowski P., 2021, *Astronomy and Astrophysics*, 650, A107.
Marchant P., Renzo M., Farmer R., Pappas K. M. W., Taam R. E., de Mink S. E., Kalogera V., 2019, *Astrophysical Journal*, 882, 36.
Marchetti T., Rossi E. M., Brown A. G. A., 2019, *Monthly Notices of the Royal Astronomical Society*, 490, 157.
Margalit B., Berger E., Metzger B. D., 2019, *Astrophysical Journal*, 886, 110.
Margon B., 1983, in Lewin W. H. G., van den Heuvel E. P. J., eds, *Accretion-Driven Stellar X-ray Sources: SS433*, pp. 287–301.
Margon B., 1984, *Annual Review of Astronomy and Astrophysics*, 22, 507.
Margutti R., Chornock R., 2021, *Annual Review of Astronomy and Astrophysics*, 59, 155.
Marietta E., Burrows A., Fryxell B., 2000, *Astrophysical Journal Supplement Series*, 128, 615.
Markova N., Puls J., 2008, *Astronomy and Astrophysics*, 478, 823.
Marschall L. A., 1988, *The Supernova Story*, Plenum Press, New York.
Marsh T. R., 2000, *New Astronomy Reviews*, 44, 119.
Marsh T. R., Dhillon V. S., Duck S. R., 1995, *Monthly Notices of the Royal Astronomical Society*, 275, 828.
Marsh T. R., Nelemans G., Steeghs D., 2004, *Monthly Notices of the Royal Astronomical Society*, 350, 113.
Marsh T. R., Gänsicke B. T., Hümmerich S., Hambsch F. J., Bernhard K., Lloyd C., ... 2016, *Nature*, 537, 374.
Marston A. P., McCollum B., 2008, *Astronomy and Astrophysics*, 477, 193.
Martin R. G., Livio M., Palaniswamy D., 2016, *Astrophysical Journal*, 832, 122.
Martinez J. G., Stovall K., Freire P. C. C., Deneva J. S., Jenet F. A., McLaughlin M. A., ... Ridolfi A., 2015, *Astrophysical Journal*, 812, 143.
Martinez J. G., Stovall K., Freire P. C. C., Deneva J. S., Tauris T. M., Ridolfi A., ... Bagchi M., 2017, *Astrophysical Journal*, 851, L29.
Martinez M. A. S., Fragione G., Kremer K., Chatterjee S., Rodriguez C. L., Samsing J., ... Rasio F. A., 2020, *Astrophysical Journal*, 903, 67.
Martinez-Nunez S., Kretschmar P., Bozzo E., et al. 2017, *Space Science Reviews*, 212, 59.

Martins F., 2015, in Vink J. S., ed, *Very Massive Stars in the Local Universe*, Vol. 412 of Astrophysics and Space Science Library, Springer: *Empirical Properties of Very Massive Stars*, 9.

Mason B. D., Hartkopf W. I., Gies D. R., Henry T. J., Helsel J. W., 2009, *Astronomical Journal*, 137, 3358.

Massey P., 2002, *Astrophysical Journal Supplement Series*, 141, 81.

Massey P., Levesque E., Neugent K., Evans K., Drout M., Beck M., 2017, in Eldridge J. J., Bray J. C., McClelland L. A. S., Xiao L., eds, *The Lives and Death-Throes of Massive Stars*, proceedings IAU Symposium 329, *The Red Supergiant Content of the Local Group*, pp. 161–165.

Massey P., Neugent K. F., Dorn-Wallenstein T. Z., Eldridge J. J., Stanway E. R., Levesque E. M., 2021, *Astrophysical Journal*, 922, 177.

Massey P., Neugent K. F., Morrell N., 2015, *Astrophysical Journal*, 807, 81.

Massey P., Neugent K. F., Morrell N., 2017, *Astrophysical Journal*, 837, 122.

Massey P., Neugent K. F., Morrell N., Hillier D. J., 2014, *Astrophysical Journal*, 788, 83.

Massey P., Plez B., Levesque E. M., Olsen K. A. G., Clayton G. C., Josselin E., 2005, *Astrophysical Journal*, 634, 1286.

Mata Sánchez D., Istrate A. G., van Kerkwijk M. H., Breton R. P., Kaplan D. L., 2020, *Monthly Notices of the Royal Astronomical Society*, 494, 4031.

Mathieu R. D., Meibom S., Dolan C. J., 2004, *Astrophysical Journal*, 602, L121.

Mathys G., Khalack V., Landstreet J. D., 2020, *Astronomy and Astrophysics*, 636, A6.

Matsuda T., Inoue M., Sawada K., 1987, *Monthly Notices of the Royal Astronomical Society*, 226, 785.

Matsuda T., Ishii T., Sekino N., Sawada K., Shima E., Livio M., Anzer U., 1992, *Monthly Notices of the Royal Astronomical Society*, 255, 183.

Mauerhan J. C., Van Dyk S. D., Johansson J., Fox O. D., Filippenko A. V., Graham M. L., 2018, *Monthly Notices of the Royal Astronomical Society*, 473, 3765.

Mauron N., Josselin E., 2011, *Astronomy and Astrophysics*, 526, A156.

Maxted P. F. L., Marsh T. R., 1998, *Monthly Notices of the Royal Astronomical Society*, 296, L34.

Maxted P. F. L., Serenelli A. M., Miglio A., Marsh T. R., Heber U., Dhillon V. S., . . . Schaffenroth V., 2013, *Nature*, 498, 463.

Mazeh T., 2008, in Goupil M. J., Zahn J. P., eds, Vol. 29 of EAS Publications Series, *Observational Evidence for Tidal Interaction in Close Binary Systems*, pp. 1–65.

Mazurek T. J., Wheeler J. C., 1980, *Fundamental Cosmic Physics*, 5, 193.

Mazzali P. A., Röpke F. K., Benetti S., Hillebrandt W., 2007, *Science*, 315, 825.

McBrien O. R., Smartt S. J., Chen T.-W., Inserra C., Gillanders J. H., Sim S. A., . . . 2019, *Astrophysical Journal*, 885, L23.

McClintock J. E., Narayan R., Steiner J. F., 2014, *Space Science Reviews*, 183, 295.

McClintock J. E., Remillard R. A., 1986, *Astrophysical Journal*, 308, 110.

McClintock J. E., Remillard R. A., 2006, in W. H. G. Lewin and M. van der Klis, eds, *Compact Stellar X-ray Sources*, Cambridge University Press: *Black Hole Binaries*, 157–213.

McCully C., Jha S. W., Foley R. J., Bildsten L., Fong W.-F., Kirshner R. P., . . . Stritzinger M. D., 2014, *Nature*, 512, 54.

McCully C., Jha S. W., Foley R. J., Chornock R., Holtzman J. A., Balam D. D., . . . 2014, *Astrophysical Journal*, 786, 134.

McHardy I. M., Lawrence A., Pye J. P., Pounds K. A., 1981, *Monthly Notices of the Royal Astronomical Society*, 197, 893.

McLaughlin D. B., 1924, *Astrophysical Journal*, 60, 22.

McLaughlin M. A., 2013, *Classical and Quantum Gravity*, 30, 224008.

McMillan P. J., 2017, *Monthly Notices of the Royal Astronomical Society*, 465, 76.

McMillan S. L. W., McDermott P. N., Taam R. E., 1987, *Astrophysical Journal*, 318, 261.

McNeill L. O., Müller B., 2022, *Monthly Notices of the Royal Astronomical Society*, 509, 818.

Meibom S., Barnes S. A., Platais I., Gilliland R. L., Latham D. W., Mathieu R. D., 2015, *Nature*, 517, 589.

Méndez M., van der Klis M., Wijnands R., Ford E. C., van Paradijs J., Vaughan B. A., 1998, *Astrophysical Journal*, 505, L23.

Mennekens N., Vanbeveren D., 2014, *Astronomy and Astrophysics*, 564, A134.

Mennekens N., Vanbeveren D., De Greve J. P., De Donder E., 2010, *Astronomy and Astrophysics*, 515, A89.

Mestel L., 1968, *Monthly Notices of the Royal Astronomical Society*, 138, 359.

Metzger B. D., 2019, *Living Reviews in Relativity*, 23, 1.

Metzger B. D., Martínez-Pinedo G., Darbha S., Quataert E., Arcones A., Kasen D., ... Zinner N. T., 2010, *Monthly Notices of the Royal Astronomical Society*, 406, 2650.

Metzger B. D., Pejcha O., 2017, *Monthly Notices of the Royal Astronomical Society*, 471, 3200.

Metzger B. D., Zenati Y., Chomiuk L., Shen K. J., Strader J., 2021, *Astrophysical Journal*, 923, 100.

Meurs E. J. A., van den Heuvel E. P. J., 1989, *Astronomy and Astrophysics*, 226, 88.

Meyer F., Meyer-Hofmeister E., 1979, *Astronomy and Astrophysics*, 78, 167.

Meyer F., Meyer-Hofmeister E., 1983, *Astronomy and Astrophysics*, 121, 29.

Meynet G., Georgy C., Hirschi R., Maeder A., Massey P., Przybilla N., Nieva M. F., 2011, *Bulletin de la Societe Royale des Sciences de Liege*, 80, 266.

Meynet G., Maeder A., 2000, *Astronomy and Astrophysics*, 361, 101.

Meynet G., Maeder A., 2002, *Astronomy and Astrophysics*, 390, 561.

Meynet G., Maeder A., 2003, *Astronomy and Astrophysics*, 404, 975.

Meynet G., Maeder A., 2005, *Astronomy and Astrophysics*, 429, 581.

Mezzacappa A., 2005, *Annual Review of Nuclear and Particle Science*, 55, 467.

Michel F. C., 1987, *Nature*, 329, 310.

Michel F. C., 1991, *Theory of Neutron Star Magnetospheres*, The University of Chicago Press.

Middleton M., 2016, in Bambi C., ed, *Astrophysics of Black Holes: From Fundamental Aspects to Latest Developments* Vol. 440 of Astrophysics and Space Science Library, Springer, *Black Hole Spin: Theory and Observation*, 99.

Mikkola S., 2008, in Hubrig S., Petr-Gotzens M., Tokovinin A., eds, *Multiple Stars Across the H-R Diagram Dynamics and Stability of Triple Stars*, 11.

Milgrom M., 1979, *Astronomy and Astrophysics*, 76, L3.

Milgrom M., Anderson S. F., Margon B., 1982, *Astrophysical Journal*, 256, 222.

Milisavljevic D., Patnaude D. J., Raymond J. C., Drout M. R., Margutti R., Kamble A., et al. 2017, *Astrophysical Journal*, 846, 50.

Miller M. C., Hamilton D. P., 2002a, *Monthly Notices of the Royal Astronomical Society*, 330, 232.

Miller M. C., Hamilton D. P., 2002b, *Monthly Notices of the Royal Astronomical Society*, 330, 232.

Miller M. C., Lamb F. K., Dittmann A. J., Bogdanov S., Arzoumanian Z., Gendreau K. C., et al. 2021, *Astrophysical Journal*, 918, L28.

Miller M. C., Miller J. M., 2015, *Physics Reports*, 548, 1.

Miller-Jones J. C. A., Bahramian A., Orosz J. A., Mandel I., Gou L., Maccarone T. J., et al. 2021, *Science*, 371, 1046.

Minkowski R., 1941, *Publications of the Astronomical Society of the Pacific*, 53, 224.
Mirabel F., 2017, *New Astronomy Reviews*, 78, 1.
Mirabel I. F., Rodríguez L. F., 1994, *Nature*, 371, 46.
Mirabel I. F., Rodriguez L. F., Cordier B., Paul J., Lebrun F., 1992, *Nature*, 358, 215.
Misner C. W., Thorne K. S., Wheeler J. A., 1973, *Gravitation* W. H. Freeman and Company, San Francisco.
Misra D., Fragos T., Tauris T. M., Zapartas E., Aguilera-Dena D. R., 2020, *Astronomy and Astrophysics*, 642, A174.
Miyaji S., Nomoto K., Yokoi K., Sugimoto D., 1979, International Cosmic Ray Conference, Published by the Institute for Cosmic Ray Research, University of Tokyo, Vol. 2, 13.
Moe M., 2019, *Memorie della Societa Astronomica Italiana*, 90, 347.
Moe M., Di Stefano R., 2013, *Astrophysical Journal*, 778, 95.
Moe M., Di Stefano R., 2015, *Astrophysical Journal*, 810, 61.
Moe M., Di Stefano R., 2017, *Astrophysical Journal Supplement Series*, 230, 15.
Mokiem M. R., de Koter A., Vink J. S., Puls J., Evans C. J., Smartt S. J., . . . Villamariz M. R., 2007, *Astronomy and Astrophysics*, 473, 603.
Montez Jr. R., De Marco O., Kastner J. H., Chu Y.-H., 2010, *Astrophysical Journal*, 721, 1820.
Moore C. J., Cole R. H., Berry C. P. L., 2015, *Classical and Quantum Gravity*, 32, 015014.
Moreno Méndez E., López-Cámara D., De Colle F., 2017, *Monthly Notices of the Royal Astronomical Society*, 470, 2929.
Moreno M. M., Schneider F. R. N., Roepke F. K., Ohlmann S. T., Pakmor R., Podsiadlowski P., Sand C., 2021, arXiv:2111.12112.
Mori K., Gotthelf E. V., Hailey C. J., Hord B. J., de Oña Wilhelmi E., Rahoui F., . . . Zhang W. W., 2017, *Astrophysical Journal*, 848, 80.
Moriya T. J., 2019, *Monthly Notices of the Royal Astronomical Society*, 490, 1166.
Moriya T. J., Blinnikov S. I., 2021, *Monthly Notices of the Royal Astronomical Society*, 508, 74.
Moriya T. J., Blinnikov S. I., Tominaga N., Yoshida N., Tanaka M., Maeda K., Nomoto K., 2013, *Monthly Notices of the Royal Astronomical Society*, 428, 1020.
Moriya T. J., Liu Z.-W., Izzard R. G., 2015, *Monthly Notices of the Royal Astronomical Society*, 450, 3264.
Moriya T. J., Mazzali P. A., Tominaga N., Hachinger S., Blinnikov S. I., Tauris T. M., . . . Podsiadlowski P., 2017, *Monthly Notices of the Royal Astronomical Society*, 466, 2085.
Moriya T. J., Tauris T. M., 2016, *Monthly Notices of the Royal Astronomical Society*, 460, L55.
Morozova V., Radice D., Burrows A., Vartanyan D., 2018, *Astrophysical Journal*, 861, 10.
Morris P. W., Voors R. H. M., Lamers H. J. G. L. M., Eenens P. R. J., 1997, in Nota A., Lamers H., eds, *Luminous Blue Variables: Massive Stars in Transition* Vol. 120 of Astronomical Society of the Pacific Conference Series, *Near-Infrared Spectra of LBV; Be and B[e] Stars: Does Axisymmetry Provide a Morphological Link?* 20.
Morris T., Podsiadlowski P., 2007, *Science*, 315, 1103.
Morton D. C., 1960, *Astrophysical Journal*, 132, 146.
Most E. R., Papenfort L. J., Weih L. R., Rezzolla L., 2020, *Monthly Notices of the Royal Astronomical Society*, 499, L82.
Motch C., Pakull M. W., Soria R., Grisé F., Pietrzyński G., 2014, *Nature*, 514, 198.
Motta S. E., Belloni T. M., Stella L., Muñoz-Darias T., Fender R., 2014, *Monthly Notices of the Royal Astronomical Society*, 437, 2554.
Motta S. E., Rodriguez J., Jourdain E., Del Santo M., Belanger G., Cangemi F., . . . Wilms J., 2021, *New Astronomy Reviews*, 93, 101618.
Motz L., 1952, *Astrophysical Journal*, 115, 562.

Müller B., 2016, *Publications of the Astronomical Society of Australia*, 33, e048.
Müller B., 2020, *Living Reviews in Computational Astrophysics*, 6, 3.
Müller B., Gay D. W., Heger A., Tauris T. M., Sim S. A., 2018, *Monthly Notices of the Royal Astronomical Society*, 479, 3675.
Müller B., Heger A., Liptai D., Cameron J. B., 2016, *Monthly Notices of the Royal Astronomical Society*, 460, 742.
Müller B., Melson T., Heger A., Janka H.-T., 2017, *Monthly Notices of the Royal Astronomical Society*, 472, 491.
Müller B., Tauris T. M., Heger A., Banerjee P., Qian Y.-Z., Powell J., ... Langer N., 2019, *Monthly Notices of the Royal Astronomical Society*, 484, 3307.
Müller B., Viallet M., Heger A., Janka H.-T., 2016, *Astrophysical Journal*, 833, 124.
Murphy S. J., Moe M., Kurtz D. W., Bedding T. R., Shibahashi H., Boffin H. M. J., 2018, *Monthly Notices of the Royal Astronomical Society*, 474, 4322.
Mushtukov A. A., Suleimanov V. F., Tsygankov S. S., Poutanen J., 2015, *Monthly Notices of the Royal Astronomical Society*, 454, 2539.
Nadezhin D. K., 1980, *Astrophysics and Space Science*, 69, 115.
Nagase F., 1989, *Publications of the Astronomical Society of Japan*, 41, 1.
Nakaoka T., Maeda K., Yamanaka M., Tanaka M., Kawabata M., Moriya T. J., et al. 2021, *Astrophysical Journal*, 912, 30.
Nakar E., 2020, *Physics Reports*, 886, 1.
Nan R., Li D., Jin C., Wang Q., Zhu L., Zhu W., Zhang H., Yue Y., Qian L., 2011, *International Journal of Modern Physics D*, 20, 989.
Nandez J. L. A., Ivanova N., 2016, *Monthly Notices of the Royal Astronomical Society*, 460, 3992.
Nandez J. L. A., Ivanova N., Lombardi J. C., 2015, *Monthly Notices of the Royal Astronomical Society*, 450, L39.
Nandez J. L. A., Ivanova N., Lombardi Jr. J. C., 2014, *Astrophysical Journal*, 786, 39.
Napiwotzki R., Karl C. A., Lisker T., Catalán S., Drechsel H., Heber U., ... Yungelson L., 2020, *Astronomy and Astrophysics*, 638, A131.
Narayan R., Piran T., Shemi A., 1991, *Astrophysical Journal*, 379, L17.
Nathanail A., Most E. R., Rezzolla L., 2021, *Astrophysical Journal*, 908, L28.
Nauenberg M., Chapline Jr. G., 1973, *Astrophysical Journal*, 179, 277.
Neijssel C. J., Vigna-Gómez A., Stevenson S., Barrett J. W., Gaebel S. M., Broekgaarden F. S., ... Mandel I., 2019, *Monthly Notices of the Royal Astronomical Society*, 490, 3740.
Neijssel C. J., Vinciguerra S., Vigna-Gómez A., Hirai R., Miller-Jones J. C. A., Bahramian A., ... Mandel I., 2021, *Astrophysical Journal*, 908, 118.
Nelemans G., Portegies Zwart S. F., Verbunt F., Yungelson L. R., 2001, *Astronomy and Astrophysics*, 368, 939.
Nelemans G., Tauris T. M., 1998, *Astronomy and Astrophysics*, 335, L85.
Nelemans G., Tauris T. M., van den Heuvel E. P. J., 1999, *Astronomy and Astrophysics*, 352, L87.
Nelemans G., Tout C. A., 2005, *Monthly Notices of the Royal Astronomical Society*, 356, 753.
Nelemans G., Verbunt F., Yungelson L. R., Portegies Zwart S. F., 2000, *Astronomy and Astrophysics*, 360, 1011.
Nelemans G., Yungelson L. R., Portegies Zwart S. F., 2001, *Astronomy and Astrophysics*, 375, 890.
Nelemans G., Yungelson L. R., Portegies Zwart S. F., 2004, *Monthly Notices of the Royal Astronomical Society*, 349, 181.
Nelemans G., Yungelson L. R., Portegies Zwart S. F., Verbunt F., 2001, *Astronomy and Astrophysics*, 365, 491.

Nelemans G., Yungelson L. R., van der Sluys M. V., Tout C. A., 2010, *Monthly Notices of the Royal Astronomical Society*, 401, 1347.
Nelson L., Schwab J., Ristic M., Rappaport S., 2018, *Astrophysical Journal*, 866, 88.
Nelson L. A., Rappaport S. A., Joss P. C., 1986, *Astrophysical Journal*, 304, 231.
Nemravová J., Harmanec P., Koubský P., Miroshnichenko A., Yang S., Šlechta M., ... Votruba V., 2012, *Astronomy and Astrophysics*, 537, A59.
Neugent K. F., Massey P., Hillier D. J., Morrell N., 2017, *Astrophysical Journal*, 841, 20.
Neuhäuser R., Gießler F., Hambaryan V. V., 2019, *Monthly Notices of the Royal Astronomical Society*, 2261.
Neunteufel P., 2020, *Astronomy and Astrophysics*, 641, A52.
Neunteufel P., Kruckow M., Geier S., Hamers A. S., 2021, *Astronomy and Astrophysics*, 646, L8.
Ng C., Bailes M., Bates S. D., Bhat N. D. R., Burgay M., et al. 2014, *Monthly Notices of the Royal Astronomical Society*, 439, 1865.
Ng C., Champion D. J., Bailes M., Barr E. D., Bates S. D., et al. 2015, *Monthly Notices of the Royal Astronomical Society*, 450, 2922.
Ng C., Kruckow M. U., Tauris T. M., Lyne A. G., Freire P. C. C., Ridolfi A., ... Stappers B., 2018, *Monthly Notices of the Royal Astronomical Society*, 476, 4315.
Ng C. Y., Romani R. W., Brisken W. F., Chatterjee S., Kramer M., 2007, *Astrophysical Journal*, 654, 487.
Ng K. K. Y., Vitale S., Farr W. M., Rodriguez C. L., 2021, *Astrophysical Journal*, 913, L5.
Nimmo K., Hessels J. W. T., Kirsten F., Keimpema A., Cordes J. M., Snelders M. P., Hewitt D. M., Karuppusamy R., et al. 2022, *Nature Astronomy*, 6, 393.
Nobili L., Turolla R., Zampieri L., 1991, *Astrophysical Journal*, 383, 250.
Nomoto K., 1982, *Astrophysical Journal*, 253, 798.
Nomoto K., 1984, *Astrophysical Journal*, 277, 791.
Nomoto K., 1987, *Astrophysical Journal*, 322, 206.
Nomoto K., Kondo Y., 1991, *Astrophysical Journal*, 367, L19.
Nomoto K., Miyaji S., Sugimoto D., Yokoi K., 1979, in van Horn H. M., Weidemann V., eds, IAU Colloq. 53: *White Dwarfs and Variable Degenerate Stars: Collapse of Accreting White Dwarf to Form a Neutron Star*, pp. 56–60.
Nomoto K., Shigeyama T., Kumagai S., Yamaoka H., Tsujimoto T., 1991, in Ventura J., Pines D., eds, NATO Advanced Science Institutes (ASI) Series C Vol. 344, Kluwer Academic Publishers, Dordrecht, *Neutron Star Formation in Close Binary Systems*, 143.
Nomoto K., Sugimoto D., Neo S., 1976, *Astrophysics and Space Science*, 39, L37.
Nomoto K., Thielemann F.-K., Yokoi K., 1984, *Astrophysical Journal*, 286, 644.
Nomoto K., Yamaoka H., 1992, in, eds.: E. P. J. van den Heuvel, S. A. Rappaport, eds, *X-Ray Binaries and Recyled Pulsars*, Kluwer Academic Publishers, Dordrecht: *Accretion-Induced Collapse of White Dwarfs*, 189–205.
Noutsos A., Schnitzeler D. H. F. M., Keane E. F., Kramer M., Johnston S., 2013, *Monthly Notices of the Royal Astronomical Society*, 430, 2281.
Novarino M. L., Echeveste M., Benvenuto O. G., De Vito M. A., Ferrero G. A., 2021, MNRAS, 508, 3812.
Nugis T., Lamers H. J. G. L. M., 2000, *Astronomy and Astrophysics*, 360, 227.
O'Connor E., Bollig R., Burrows A., Couch S., Fischer T., Janka H.-T., ... Vartanyan D., 2018, *Journal of Physics G Nuclear Physics*, 45, 104001.
O'Connor E., Ott C. D., 2011, *Astrophysical Journal*, 730, 70.
Ofek E. O., Kulkarni S. R., Rau A., Cenko S. B., Peng E. W., Blakeslee J. P., ... Bouwens R., 2008, *Astrophysical Journal*, 674, 447.

Ogilvie G. I., 2014, *Annual Review of Astronomy and Astrophysics*, 52, 171.
Ohlmann S. T., Röpke F. K., Pakmor R., Springel V., 2016, *Astrophysical Journal*, 816, L9.
Ohlmann S. T., Röpke F. K., Pakmor R., Springel V., 2017, *Astronomy and Astrophysics*, 599, A5.
Olausen S. A., Kaspi V. M., 2014, *Astrophysical Journal Supplement Series*, 212, 6.
Olejak A., Belczynski K., Bulik T., Sobolewska M., 2020, *Astronomy and Astrophysics*, 638, A94.
Olsen S., Venumadhav T., Mushkin J., Roulet J., Zackay B., Zaldarriaga M., 2022, *Physical Review D*, 10, 043009.
Olsen S., Venumadhav T., Mushkin J., Roulet J., Zackay B., Zaldarriaga M. 2022, *Phys. Rev. D*, 92, 064016.
Oort J. H., Walraven T., 1956, *Bulletin of the Astronomical Institutes of the Netherlands*, 12, 285.
Öpik E., 1924, Publications of the Tartu Astrofizica Observatory, 25, 1.
Orosz J. A., McClintock J. E., Aufdenberg J. P., Remillard R. A., Reid M. J., Narayan R., Gou L., 2011, *Astrophysical Journal*, 742, 84.
Orosz J. A., McClintock J. E., Narayan R., Bailyn C. D., Hartman J. D., Macri L., . . . Mazeh T., 2007, *Nature*, 449, 872.
Orosz J. A., Steeghs D., McClintock J. E., Torres M. A. P., et al. 2009, *Astrophysical Journal*, 697, 573.
Orosz J. A., Steiner J. F., McClintock J. E., Buxton M. M., Bailyn C. D., Steeghs D., . . . Torres M. A. P., 2014, *Astrophysical Journal*, 794, 154.
Osaki Y., 1974, *Publications of the Astronomical Society of Japan*, 26, 429.
O'Shaughnessy R., Kim C., 2010, *Astrophysical Journal*, 715, 230.
Oskinova L. M., Gvaramadze V. V., Gräfener G., Langer N., Todt H., 2020, *Astronomy and Astrophysics*, 644, L8.
Ostriker J. P., Gunn J. E., 1971, *Astrophysical Journal*, 164, L95.
Ostriker J. P., McCray R., Weaver R., Yahil A., 1976, *Astrophysical Journal*, 208, L61.
Özel F., Freire P., 2016, *Annual Review of Astronomy and Astrophysics*, 54, 401.
Özel F., Güver T., Psaltis D., 2009, *Astrophysical Journal*, 693, 1775.
Özel F., Psaltis D., Narayan R., McClintock J. E., 2010, *Astrophysical Journal*, 725, 1918.
Pacini F., 1967, *Nature*, 216, 567.
Packet W., 1981, *Astronomy and Astrophysics*, 102, 17.
Packet W., De Greve J. P., 1979, *Astronomy and Astrophysics*, 75, 255.
Paczyński B., 1966, *Acta Astronomica*, 16, 231.
Paczyński B., 1967a, *Acta Astronomica*, 17, 1.
Paczyński B., 1967b, *Acta Astronomica*, 17, 193
Paczyński B., 1967c, *Acta Astronomica*, 17, 355.
Paczyński B., 1971a, *Acta Astronomica*, 21, 1.
Paczyński B., 1971b, *Annual Review of Astronomy and Astrophysics*, 9, 183.
Paczyński B., 1976, in P. Eggleton, S. Mitton, & J. Whelan, eds, *Structure and Evolution of Close Binary Systems* proceedings of IAU Symposium 73, Common Envelope Binaries, Dordrecht, Holland, 75.
Paczynski B., 1977, *Astrophysical Journal*, 216, 822.
Paczynski B., 1986, *Astrophysical Journal*, 308, L43.
Paczynski B., 1992, *Acta Astronomica*, 42, 145.
Paczyński B., Sienkiewicz R., 1972, *Acta Astronomica*, 22, 73.
Paczynski B., Sienkiewicz R., 1981, *Astrophysical Journal*, 248, L27.
Pakmor R., Kromer M., Taubenberger S., Sim S. A., Röpke F. K., Hillebrandt W., 2012, *Astrophysical Journal*, 747, L10.

Pakmor R., Röpke F. K., Weiss A., Hillebrandt W., 2008, *Astronomy and Astrophysics*, 489, 943.
Pakull M. W., Mirioni L., 2003, in Arthur J., Henney W. J., eds, Revista Mexicana de Astronomia y Astrofisica Conference Series Vol. 15: Bubble Nebulae around Ultraluminous X-Ray Sources, pp. 197–199.
Pala A. F., Gänsicke B. T., Breedt E., Knigge C., Hermes J. J., Gentile Fusillo N. P., et al. 2020, *Monthly Notices of the Royal Astronomical Society*, 494, 3799.
Palladino L. E., Schlesinger K. J., Holley-Bockelmann K., Allende Prieto C., Beers T. C., Lee Y. S., Schneider D. P., 2014, *Astrophysical Journal*, 780, 7.
Pallanca C., Mignani R. P., Dalessandro E., Ferraro F. R., Lanzoni B., Possenti A., ... Sabbi E., 2012, *Astrophysical Journal*, 755, 180.
Palmese A., Conselice C. J., 2021, *Physical Review Letters*, 126, 181103.
Palomba C., 2005, *Monthly Notices of the Royal Astronomical Society*, 359, 1150.
Pan K.-C., Liebendörfer M., Couch S. M., Thielemann F.-K., 2021, *Astrophysical Journal*, 914, 140.
Pan K.-C., Ricker P. M., Taam R. E., 2012, *Astrophysical Journal*, 750, 151.
Pan Y.-C., Kilpatrick C. D., Simon J. D., et al. 2017, *Astrophysical Journal*, 848, L30.
Pan Z., Qian L., Ma X., Liu K., Wang L., Luo J., Yan Z., Ransom S., Lorimer D., Li D., Jiang P., 2021, *Astrophysical Journal*, 915, L28.
Papitto A., de Martino D., 2022, in Bhattacharyya S., Papitto A., Bhattacharya D., eds, of Astrophysics and Space Science Library, Vol. 465, *Transitional Millisecond Pulsars*, p. 157.
Papitto A., Falanga M., Hermsen W., Mereghetti S., Kuiper L., Poutanen J., Bozzo E., et al. 2020, *New Astronomy Reviews*, 91, 101544.
Papitto A., Ferrigno C., Bozzo E., et al. 2013, *Nature*, 501, 517.
Papitto A., Torres D. F., Rea N., Tauris T. M., 2014, *Astronomy and Astrophysics*, 566, A64.
Park I. H., Choi K. Y., Hwang J., Jung S., Kim D. H., Kim M. H., ... Won E., 2021, *Journal of Cosmology and Astroparticle Physics*, 11, 008, 1.
Park M.-G., 1990a, *Astrophysical Journal*, 354, 64.
Park M.-G., 1990b, *Astrophysical Journal*, 354, 83.
Parker R. J., Goodwin S. P., Wright N. J., Meyer M. R., Quanz S. P., 2016, *Monthly Notices of the Royal Astronomical Society*, 459, L119.
Parsons S. G., Gänsicke B. T., Schreiber M. R., Marsh T. R., Ashley R. P., Breedt E., ... Meusinger H., 2021, *Monthly Notices of the Royal Astronomical Society*, 502, 4305.
Parsons S. G., Hermes J. J., Marsh T. R., ... 2017, *Monthly Notices of the Royal Astronomical Society*, 471, 976.
Parsons S. G., Marsh T. R., Copperwheat C. M., Dhillon V. S., Littlefair S. P., Gänsicke B. T., Hickman R., 2010, *Monthly Notices of the Royal Astronomical Society*, 402, 2591.
Passy J.-C., De Marco O., Fryer C. L., Herwig F., Diehl S., Oishi J. S., ... Rockefeller G., 2012, *Astrophysical Journal*, 744, 52.
Pastetter L., Ritter H., 1989, *Astronomy and Astrophysics*, 214, 186.
Pastorello A., Mason E., Taubenberger S., Fraser M., Cortini G., et al. 2019, *Astronomy and Astrophysics*, 630, A75.
Patruno A., 2010, ArXiv astro.ph/1007.1108.
Patruno A., Haskell B., Andersson N., 2017, *Astrophysical Journal*, 850, 106.
Patruno A., Watts A. L., 2021, *Accreting Millisecond X-ray Pulsars*, in Belloni, Tomaso M.; Méndez, Mariano; Zhang, Chengmin, eds, *Timing Neutron Stars: Pulsations, Oscillations and Explosions*, for Astrophysics and Space Science Library, Vol. 461, Springer, 143–208.
Patton R. A., Sukhbold T., 2020, *Monthly Notices of the Royal Astronomical Society*, 499, 2803.
Patton R. A., Sukhbold T., Eldridge J. J., 2022, *Monthly Notices of the Royal Astronomical Society*, 511, 903.

Pavan L., Bordas P., Pühlhofer G., Filipović M. D., De Horta A., O'Brien A., . . . Stella L., 2014, *Astronomy and Astrophysics*, 562, A122.
Pavlovskii K., Ivanova N., 2015, *Monthly Notices of the Royal Astronomical Society*, 449, 4415.
Pavlovskii K., Ivanova N., Belczynski K., Van K. X., 2017, *Monthly Notices of the Royal Astronomical Society*, 465, 2092.
Paxton B., Cantiello M., Arras P., Bildsten L., Brown E. F., Dotter A., Mankovich C., Montgomery M. H., Stello D., Timmes F. X., Townsend R., 2013, *Astrophysical Journal Supplement Series*, 208, 4.
Paxton B., Marchant P., Schwab J., Bauer E. B., Bildsten L., Cantiello M., Dessart L., Farmer R., Hu H., Langer N., Townsend R. H. D., Townsley D. M., Timmes F. X., 2015, *Astrophysical Journal Supplement Series*, 220, 15.
Paxton B., Smolec R., Schwab J., Gautschy A., Bildsten L., Cantiello M., et al. 2019, *Astrophysical Journal Supplement Series*, 24312, 10.
Payne D. J. B., Melatos A., 2007, *Monthly Notices of the Royal Astronomical Society*, 376, 609.
Payne-Gaposchkin C., Haramundanis H., 1970, *Introduction to Astronomy*, Prentice-Hall, Engelwood Cliffs NJ, USA.
Pejcha O., Thompson T. A., 2015, *Astrophysical Journal*, 801, 90.
Penrose R., 1969, *Nuovo Cimento Rivista Serie*, 1, 252.
Perets H. B., Beniamini P., 2021, *Monthly Notices of the Royal Astronomical Society*, 503, 5997.
Perets H. B., Fabrycky D. C., 2009, *Astrophysical Journal*, 697, 1048.
Perets H. B., Gal-Yam A., Mazzali P. A., . . . 2010, *Nature*, 465, 322.
Perlmutter S., Turner M. S., White M., 1999, *Physical Review Letters*, 83, 670.
Perna R., Wang Y.-H., Farr W. M., Leigh N., Cantiello M., 2019, *Astrophysical Journal*, 878, L1.
Peter D., Feldt M., Henning T., Hormuth F., 2012, *Astronomy and Astrophysics*, 538, A74.
Peters P. C., 1964, *Physical Review*, 136, 1224.
Petroff E., Hessels J. W. T., Lorimer D. R., 2019, *Astronomy and Astrophysics Review*, 27, 4.
Petro E., Hessels J. W. T., Lorimer D. R., 2022, *Astronomy and Astrophysics Review*, 30, 2.
Petrovic J., Langer N., van der Hucht K. A., 2005, *Astronomy and Astrophysics*, 435, 1013.
Petrovic J., Langer N., Yoon S. C., Heger A., 2005, *Astronomy and Astrophysics*, 435, 247.
Peuten M., Brockamp M., Küpper A. H. W., Kroupa P., 2014, *Astrophysical Journal*, 795, 116.
Pfahl E., Rappaport S., Podsiadlowski P., Spruit H., 2002, *Astrophysical Journal*, 574, 364.
Phillips J. A., Thorsett S. E., Kulkarni S. R., eds, 1993, *Planets around Pulsars*; Proceedings of the Conference, California Inst. of Technology, Pasadena, Apr. 30–May 1, 1992, Vol. 36 of Astronomical Society of the Pacific Conference Series.
Phillips M. M., 1993, *Astrophysical Journal*, 413, L105.
Phinney E. S., 1991, *Astrophysical Journal*, 380, L17.
Phinney E. S., 1992, *Royal Society of London Philosophical Transactions Series A*, 341, 39.
Phinney E. S., 1993, in Djorgovski S. G., Meylan G., eds, *Structure and Dynamics of Globular Clusters* Vol. 50 of Astronomical Society of the Pacific Conference Series, *Pulsars as Probes of Globular Cluster Dynamics*, 141.
Phinney E. S., Kulkarni S. R., 1994, *Annual Review of Astronomy and Astrophysics*, 32, 591.
Pierce A., Riles K., Zhao Y., 2018, *Physical Review Letters*, 121, 061102.
Pietrzyński G., Thompson I. B., Gieren W., Graczyk D., Stępień K., et al. 2012, *Nature*, 484, 75.
Pijloo J. T., Caputo D. P., Portegies Zwart S. F., 2012, *Monthly Notices of the Royal Astronomical Society*, 424, 2914.
Pinsonneault M. H., Stanek K. Z., 2006, *Astrophysical Journal*, 639, L67.
Piran T., 2004, *Reviews of Modern Physics*, 76, 1143.
Piran T., Hotokezaka K., 2018, arXiv:1807.01336.
Piran T., Shaviv N. J., 2005, *Physical Review Letters*, 94, 051102.

Piran Z., Piran T., 2020, *Astrophysical Journal*, 892, 64.
Piro A. L., Thompson T. A., 2014, *Astrophysical Journal*, 794, 28.
Plavec M., 1967, *Bulletin of the Astronomical Institutes of Czechoslovakia*, 18, 253.
Plavec M., Kratochvil P., 1964, *Bulletin of the Astronomical Institutes of Czechoslovakia*, 15, 165.
Pletsch H. J., Clark C. J., 2015, *Astrophysical Journal*, 807, 18.
Pletsch H. J., Guillemot L., Fehrmann H., et al. 2012, *Science*, 338, 1314.
Pleunis Z., Michilli D., Bassa C. G., Hessels J. W. T., Naidu A., ... 2021, *Astrophysical Journal*, 911, L3.
Podsiadlowski P., 1991, *Nature*, 350, 136.
Podsiadlowski P., 2001, in P. Podsiadlowski, S. Rappaport, A. R. King, F. D'Antona, & L. Burderi, eds, *Evolution of Binary and Multiple Star Systems* Vol. 229 of Astronomical Society of the Pacific Conference Series, *Common-Envelope Evolution and Stellar Mergers*, 239.
Podsiadlowski P., Cannon R. C., Rees M. J., 1995, *Monthly Notices of the Royal Astronomical Society*, 274, 485.
Podsiadlowski P., Joss P. C., Hsu J. J. L., 1992, *Astrophysical Journal*, 391, 246.
Podsiadlowski P., Joss P. C., Rappaport S., 1990, *Astronomy and Astrophysics*, 227, L9.
Podsiadlowski P., Langer N., Poelarends A. J. T., Rappaport S., Heger A., Pfahl E., 2004, *Astrophysical Journal*, 612, 1044.
Podsiadlowski P., Rappaport S., 2000, *Astrophysical Journal*, 529, 946.
Podsiadlowski P., Rappaport S., Han Z., 2003, *Monthly Notices of the Royal Astronomical Society*, 341, 385.
Podsiadlowski P., Rappaport S., Pfahl E. D., 2002, *Astrophysical Journal*, 565, 1107.
Poelarends A. J. T., 2007, PhD thesis, Utrecht University.
Poelarends A. J. T., Herwig F., Langer N., Heger A., 2008, *Astrophysical Journal*, 675, 614.
Pojmanski G., 2002, *Acta Astronomica*, 52, 397.
Pol N., McLaughlin M., Lorimer D. R., 2019, *Astrophysical Journal*, 870, 71.
Pol N., McLaughlin M., Lorimer D. R., 2020, *Research Notes of the American Astronomical Society*, 4, 22.
Pol N., Taylor S., Vigeland S., Kelley L., Simon J., Chen S., Nanograv Collaboration 2021, in *American Astronomical Society Meeting Abstracts*, Vol. 53, *Astrophysics Milestones For Pulsar Timing Array Gravitational Wave Detection*, 433.01.
Politano M., 2021, *Astronomy and Astrophysics*, 648, L6.
Pols O. R., 1993, PhD thesis, University of Amsterdam.
Pols O. R., Marinus M., 1993, in Saffer R. A., ed, *Blue Stragglers* Vol. 53 of Astronomical Society of the Pacific Conference Series, *Monte Carlo Simulations of Close Binary Evolution in Young Open Clusters*, 126.
Pols O. R., Marinus M., 1994, *Astronomy and Astrophysics*, 288, 475.
Pols O. R., Schröder K.-P., Hurley J. R., Tout C. A., Eggleton P. P., 1998, *Monthly Notices of the Royal Astronomical Society*, 298, 525.
Popper D. M., 1980, *Annual Review of Astronomy and Astrophysics*, 18, 115,
Portegies Zwart S., van den Heuvel E. P. J., van Leeuwen J., Nelemans G., 2011, *Astrophysical Journal*, 734, 55.
Portegies Zwart S. F., McMillan S. L. W., 2000, *Astrophysical Journal*, 528, L17.
Portegies Zwart S. F., McMillan S. L. W., Gieles M., 2010, *Annual Review of Astronomy and Astrophysics*, 48, 431.
Portegies Zwart S. F., Spreeuw H. N., 1996, *Astronomy and Astrophysics*, 312, 670.
Portegies Zwart S. F., van den Heuvel E. P. J., 2016, *Monthly Notices of the Royal Astronomical Society*, 456, 3401.

Portegies Zwart S. F., Verbunt F., 1996, *Astronomy and Astrophysics*, 309, 179.

Portegies Zwart S. F., Verbunt F., Ergma E., 1997, *Astronomy and Astrophysics*, 321, 207.

Portegies Zwart S. F., Yungelson L. R., 1998, *Astronomy and Astrophysics*, 332, 173.

Portegies Zwart S. F., Yungelson L. R., 1999, *Monthly Notices of the Royal Astronomical Society*, 309, 26.

Possenti A., 2013, in van Leeuwen J., ed, IAU Symposium, Vol. 291: *Binary Pulsar Evolution: Unveiled Links and New Species*, 121–126.

Postnov K., Shakura N. I., Kochetkova A. Y., Hjalmarsdotter L., 2012, in Proceedings of "An INTEGRAL view of the high-energy sky (the first 10 years)"—9th INTEGRAL Workshop and Celebration Of the 10th Anniversary of the Launch (INTEGRAL 2012), 15-19 October 2012. Bibliotheque Nationale de France, *Quasi-Spherical Accretion in Low-Luminosity X-ray Pulsars: Theory vs Observations*, 22.

Postnov K. A., Prokhorov M. E., 1999, arXiv astroph/9903193.

Postnov K. A., Yungelson L. R., 2014, *Living Reviews in Relativity*, 17, 3.

Pounds K., 2020, *Astronomy and Geophysics*, 61, 1.32.

Pourbaix D., Tokovinin A. A., Batten A. H., Fekel F. C., Hartkopf W. I., Levato H., . . . Udry S., 2004, *Astronomy and Astrophysics*, 424, 727.

Poutanen J., Lipunova G., Fabrika S., Butkevich A. G., Abolmasov P., 2007, *Monthly Notices of the Royal Astronomical Society*, 377, 1187.

Poveda A., Ruiz J., Allen C., 1967, *Boletin de los Observatorios Tonantzintla y Tacubaya*, 4, 86.

Powell J., Müller B., 2019, *Monthly Notices of the Royal Astronomical Society*, 487, 1178.

Prentice S. J., Ashall C., James P. A., Short L., Mazzali P. A., Bersier D., Crowther P. A., Barbarino C., Chen T. W., et al. 2019, *Monthly Notices of the Royal Astronomical Society*, 485, 1559.

Prentice S. J., Maguire K., Flörs A., Taubenberger S., Inserra C., Frohmaier C., Chen T. W., et al. 2020, *Astronomy and Astrophysics*, 635, A186.

Priedhorsky W. C., Holt S. S., 1987, *Space Science Reviews*, 45, 291.

Priedhorsky W. C., Verbunt F., 1988, *Astrophysical Journal*, 333, 895.

Prince T. A., Anderson S. B., Kulkarni S. R., Wolszczan A., 1991, *Astrophysical Journal*, 374, L41.

Pringle J. E., 1981, *Annual Review of Astronomy and Astrophysics*, 19, 137.

Pringle J. E., Rees M. J., 1972, *Astronomy and Astrophysics*, 21, 1.

Prodan S., Murray N., 2012, *Astrophysical Journal*, 747, 4.

Prust L. J., Chang P., 2019, *Monthly Notices of the Royal Astronomical Society*, 486, 5809.

Pylyser E., Savonije G. J., 1988, *Astronomy and Astrophysics*, 191, 57.

Pylyser E. H. P., Savonije G. J., 1989, *Astronomy and Astrophysics*, 208, 52.

Qian S.-B., Liao W.-P., Fernández Lajús E., 2008, *Astrophysical Journal*, 687, 466.

Qin Y., Fragos T., Meynet G., Andrews J., Sørensen M., Song H. F., 2018, *Astronomy and Astrophysics*, 616, A28.

Qin Y., Marchant P., Fragos T., Meynet G., Kalogera V., 2019, *Astrophysical Journal*, 870, L18.

Quast M., Langer N., Tauris T. M., 2019, *Astronomy and Astrophysics*, 628, A19.

Quataert E., Lecoanet D., Coughlin E. R., 2019, *Monthly Notices of the Royal Astronomical Society*, 485, L83.

Quimby R. M., Aldering G., Wheeler J. C., Höflich P., Akerlof C. W., Rykoff E. S., 2007, *Astrophysical Journal*, 668, L99.

Quimby R. M., Kulkarni S. R., Kasliwal M. M., Gal-Yam A., Arcavi I., Sullivan M., et al. 2011, *Nature*, 474, 487.

Quinlan G. D., 1996, *New Astronomy*, 1, 255.

Raddi R., Heber U., Hollands M., 2019, *Astronomy and Geophysics*, 60, 5.34.
Raddi R., Hollands M. A., Gänsicke B. T., Townsley D. M., Hermes J. J., Gentile Fusillo N. P., Koester D., 2018, *Monthly Notices of the Royal Astronomical Society*, 479, L96.
Raddi R., Hollands M. A., Koester D., Hermes J. J., Gänsicke B. T., Heber U., . . . Strader J., 2019, *Monthly Notices of the Royal Astronomical Society*, 489, 1489.
Radhakrishnan V., Cooke D. J., 1969, *Astrophysical Letter*, 3, 225.
Radhakrishnan V., Shukre C. S., 1985, in Srinivasan G., Radhakrishnan V., eds, *Supernovae, Their Progenitors and Remnants On the Meaning of Pulsar Velocities*, 155.
Radhakrishnan V., Srinivasan G., 1982, *Current Science*, 51, 1096.
Radhakrishnan V., Srinivasan G., 1984, in proceedings of the Second Asian-Pacific Regional Meeting on Astronomy, B. Hidayat and M. W. Feast, eds, Tira Pustaka Publishing House, Jakarta: *Are Many Pulsars Processed in Binaries*, 423.
Radice D., Bernuzzi S., Perego A., 2020, *Annual Review of Nuclear and Particle Science*, 70, 95.
Radice D., Morozova V., Burrows A., Vartanyan D., Nagakura H., 2019, *Astrophysical Journal*, 876, L9.
Raduta A. R., Gulminelli F., 2019, *Nuclear Physics A*, 983, 252.
Rafikov R. R., 2016, *Astrophysical Journal*, 830, 8.
Raghavan D., McAlister H. A., Henry T. J., Latham D. W., Marcy G. W., Mason B. D., . . . Brummelaar T. A., 2010, *Astrophysical Journal Supplement Series*, 190, 1.
Raithel C. A., Sukhbold T., Özel F., 2018, *Astrophysical Journal*, 856, 35.
Ramírez-Tannus M. C., Backs F., de Koter A., Sana H., Beuther H., Bik A., . . . Poorta J., 2021, *Astronomy and Astrophysics*, 645, L10.
Ramsay G., Green M. J., Marsh T. R., Kupfer T., Breedt E., Korol V., . . . Aungwerojwit A., 2018, *Astronomy and Astrophysics*, 620, A141.
Randall L., Shelest A., Xianyu Z.-Z., 2022, *Astrophysical Journal*, 924, 102.
Randall L., Xianyu Z.-Z., 2021, *Eccentricity without Measuring Eccentricity: Discriminating among Stellar Mass Black Hole Binary Formation Channels*, 914, 75.
Rankin J. M., 1983, *Astrophysical Journal*, 274, 333.
Rankin J. M., 1990, *Astrophysical Journal*, 352, 247.
Ransom S. M., Hessels J. W. T., Stairs I. H., Freire P. C. C., Camilo F., Kaspi V. M., Kaplan D. L., 2005, *Science*, 307, 892.
Ransom S. M., Stairs I. H., Archibald A. M., Hessels J., Kaplan D. L., van Kerkwijk M. H., et al. 2014, *Nature*, 505, 520.
Rappaport S., Deck K., Levine A., Borkovits T., Carter J., El Mellah I., Sanchis-Ojeda R., Kalomeni B., 2013, *Astrophysical Journal*, 768, 33.
Rappaport S., Di Stefano R., Smith J. D., 1994, *Astrophysical Journal*, 426, 692.
Rappaport S., Joss P. C., Webbink R. F., 1982, *Astrophysical Journal*, 254, 616.
Rappaport S., Nelson L., Levine A., Sanchis-Ojeda R., Gandolfi D., Nowak G., . . . Prsa A., 2015, *Astrophysical Journal*, 803, 82.
Rappaport S., Nelson L. A., Ma C. P., Joss P. C., 1987, *Astrophysical Journal*, 322, 842.
Rappaport S., Podsiadlowski P., Joss P. C., Di Stefano R., Han Z., 1995, *Monthly Notices of the Royal Astronomical Society*, 273, 731.
Rappaport S., van den Heuvel E. P. J., 1982, in Jaschek M., Groth H.-G., eds, *Be Stars*, proceedings of IAU Symposium 98, *X-ray Observations of Be Stars*, 327–344.
Rappaport S., Vanderburg A., Schwab J., Nelson L., 2021, *Astrophysical Journal*, 913, 118.
Rappaport S., Verbunt F., Joss P. C., 1983, *Astrophysical Journal*, 275, 713.
Rappaport S. A., Fregeau J. M., Spruit H., 2004, *Astrophysical Journal*, 606, 436.
Rasio F. A., Livio M., 1996, *Astrophysical Journal*, 471, 366.

Rauw G., De Becker M., Nazé Y., Crowther P. A., Gosset E., Sana H., . . . Williams P. M., 2004, *Astronomy and Astrophysics*, 420, L9.

Rauw G., Vreux J.-M., Bohannan B., 1999, *Astrophysical Journal*, 517, 416.

Rauw G., Vreux J.-M., Gosset E., Hutsemekers D., Magain P., Rochowicz K., 1996, *Astronomy and Astrophysics*, 306, 771.

Ravenhall D. G., Pethick C. J., 1994, *Astrophysical Journal*, 424, 846.

Rawls M. L., Orosz J. A., McClintock J. E., Torres M. A. P., Bailyn C. D., Buxton M. M., 2011, *Astrophysical Journal*, 730, 25.

Ray P., Arzoumanian Z., Ballantyne D., Bozzo E., Brandt S., et al. 2019, in *Bulletin of the American Astronomical Society*, Vol. 51, *STROBE-X: X-ray Timing and Spectroscopy on Dynamical Timescales from Microseconds to Years*, p. 231.

Rea N., Esposito P., Turolla R., Israel G. L., Zane S., Stella L., . . . Kouveliotou C., 2010, *Science*, 330, 944.

Reardon D. J., Hobbs G., Coles W., Levin Y., Keith M. J., Bailes M., et al. 2016, *Monthly Notices of the Royal Astronomical Society*, 455, 1751.

Reeves H., 1968, *Stellar Evolution and Nucleosynthesis*, Gordon and Breach, New York.

Refsdal S., Weigert A., 1971, *Astronomy and Astrophysics*, 13, 367.

Reichardt T. A., De Marco O., Iaconi R., Tout C. A., Price D. J., 2019, *Monthly Notices of the Royal Astronomical Society*, 484, 631.

Reimers D., 1975, *Circumstellar Envelopes and Mass Loss of Red Giant Stars*, in *Problems in Stellar Atmospheres and Envelopes*. Springer-Verlag, New York, 229–256.

Reinsch K., Beuermann K., Gänsicke B. T., 2002, *Optical Spectroscopy of the Supersoft X-ray Source RX J0439.8–6809, Astrophysics and Space Science*, 261, 653.

Remillard R. A., McClintock J. E., 2006, *Annual Review of Astronomy and Astrophysics*, 44, 49.

Renzo M., Callister T., Chatziioannou K., van Son L. A. C., Mingarelli C. M. F., Cantiello M., Ford K. E. S., McKernan B., Ashton G., 2021, *Astrophysical Journal*, 919, 128.

Renzo M., Farmer R., Justhan S., Götberg Y., de Mink S. E., Zapartas E., . . . Smith N., 2020, *Astronomy and Astrophysics*, 640, A56.

Renzo M., Zapartas E., de Mink S. E., Götberg Y., Justhan S., Farmer R. J., . . . Sana H., 2019, *Astronomy and Astrophysics*, 624, A66.

Repetto S., Davies M. B., Sigurdsson S., 2012, *Monthly Notices of the Royal Astronomical Society*, 425, 2799.

Repetto S., Igoshev A. P., Nelemans G., 2017, *Monthly Notices of the Royal Astronomical Society*, 467, 298.

Repetto S., Nelemans G., 2015, *Monthly Notices of the Royal Astronomical Society*, 453, 3341.

Repolust T., Puls J., Herrero A., 2004, *Astronomy and Astrophysics*, 415, 349.

Reynolds C., 2021, in 43rd COSPAR Scientific Assembly. Held 28 January–4 February Vol. 43, *Observational Constraints on Black Hole Spin*, 1412.

Rezzolla L., Barausse E., Dorband E. N., Pollney D., Reisswig C., Seiler J., Husa S., 2008, *Physical Review D*, 78, 044002.

Rice J. B., 1988, *Astronomy and Astrophysics*, 199, 299.

Richards M. T., Sharova O. I., Agafonov M. I., 2010, *Astrophysical Journal*, 720, 996.

Ricker P. M., Taam R. E., 2008, *Astrophysical Journal*, 672, L41.

Ricker P. M., Taam R. E., 2012, *Astrophysical Journal*, 746, 74.

Ricker P. M., Timmes F. X., Taam R. E., Webbink R. F., 2019, in Oskinova L. M., Bozzo E., Bulik T., Gies D. R., eds, IAU Symposium Vol. 346, *Common Envelope Evolution of Massive Stars*, 449–454.

Ridolfi A., Freire P. C. C., Gupta Y., Ransom S. M., 2019, *Monthly Notices of the Royal Astronomical Society*, 490, 3860.

Ridolfi A., Gautam T., Freire P. C. C., Ransom S. M., Buchner S. J., Possenti A., Venkatraman Krishnan V., Bailes M., Kramer M., Stappers B. W., et al. 2021, *Monthly Notices of the Royal Astronomical Society*, 504, 1407.

Ridolfi A., Freire P. C. C., Gautam T., Ransom S. M., Barr E. D., et al. 2022, *Astronomy and Astrophysics*, 664, A27.

Riess A. G., Filippenko A. V., Challis P., et al. 1998, *Astronomical Journal*, 116, 1009.

Riley J., Agrawal P., Barrett J. W., Boyett K. N. K., Broekgaarden F. S., et al. 2022, *Astrophysical Journal Supplement Series*, 258, 34.

Ritter H., 1988, *Astronomy and Astrophysics*, 202, 93.

Ritter H., 2008, *New Astronomy Reviews*, 51, 869.

Ritter H., Kolb U., 2003, VizieR Online Data Catalog, 5113.

Rivinius T., Baade D., Hadrava P., Heida M., Klement R., 2020, *Astronomy and Astrophysics*, 637, L3.

Rizzuto A. C., Ireland M. J., Robertson J. G., Kok Y., Tuthill P. G., Warrington B. A., ... Laliberte-Houdeville C., 2013, *Monthly Notices of the Royal Astronomical Society*, 436, 1694.

Roberts M. S. E., 2013, in van Leeuwen J., ed, proceedings of IAU Symposium 291, *Surrounded by Spiders! New Black Widows and Redbacks in the Galactic Field*, 127–132.

Robson T., Cornish N. J., Liu C., 2019, *Classical and Quantum Gravity*, 36, 105011.

Rodriguez C. L., Amaro-Seoane P., Chatterjee S., Kremer K., Rasio F. A., Samsing J., ... Zevin M., 2018, *Physical Review D*, 98, 123005.

Rodriguez C. L., Chatterjee S., Rasio F. A., 2016, *Physical Review D*, 93, 084029.

Rodriguez C. L., Haster C.-J., Chatterjee S., Kalogera V., Rasio F. A., 2016, *Astrophysical Journal*, 824, L8.

Rodriguez C. L., Kremer K., Chatterjee S., Fragione G., Loeb A., Rasio F. A., ... Ye C. S., 2021, *Research Notes of the American Astronomical Society*, 5, 19.

Rodriguez C. L., Kremer K., Grudić M. Y., Hafen Z., Chatterjee S., Fragione G., ... Ye C. S., 2020, *Astrophysical Journal*, 896, L10.

Rodriguez C. L., Zevin M., Amaro-Seoane P., Chatterjee S., Kremer K., Rasio F. A., Ye C. S., 2019, *Physical Review D*, 100, 043027.

Rodríguez Castillo G. A., Israel G. L., Belfiore A., Bernardini F., Esposito P., Pintore F., De Luca A., Papitto A., Stella L., Tiengo A., Zampieri L., et al. 2020, *Astrophysical Journal*, 895, 60.

Roelofs G. H. A., Rau A., Marsh T. R., Steeghs D., Groot P. J., Nelemans G., 2010, *Astrophysical Journal*, 711, L138.

Romani R. W., 1990, *Nature*, 347, 741.

Romani R. W., Filippenko A. V., Cenko S. B., 2015, *Astrophysical Journal*, 804, 115.

Romani R. W., Filippenko A. V., Silverman J. M., Cenko S. B., Greiner J., Rau A., ... Pletsch H. J., 2012, *Astrophysical Journal*, 760, L36.

Romani R. W., Kandel D., Filippenko A. V., Brink T. G., Zheng W., 2021, *Astrophysical Journal*, 908, L46.

Romani R. W., Weinberg M. D., 1991, *Astrophysical Journal*, 372, 487.

Romanova M. M., Blinova A. A., Ustyugova G. V., Koldoba A. V., Lovelace R. V. E., 2018, *New Astronomy*, 62, 94.

Romero-Shaw I. M., Lasky P. D., Thrane E., 2019, *Monthly Notices of the Royal Astronomical Society*, 490, 5210.

Rossiter R. A., 1924, *Astrophysical Journal*, 60, 15.

Rosswog S., Liebendörfer M., Thielemann F. K., Davies M. B., Benz W., Piran T., 1999, *Astronomy and Astrophysics*, 341, 499.

Rosswog S., Sollerman J., Feindt U., Goobar A., Korobkin O., Wollaeger R., ... Kasliwal M. M., 2018, *Astronomy and Astrophysics*, 615, A132.

Roulet J., Chia H. S., Olsen S., Dai L., Venumadhav T., Zackay B., Zaldarriaga M., 2021, *Physical Review D*, 104, 083010.
Roy J., Ray P. S., Bhattacharyya B., Stappers B., Chengalur J. N., Deneva J., Camilo F., et al. 2015, *Astrophysical Journal*, 800, L12.
Ruan W.-H., Guo Z.-K., Cai R.-G., Zhang Y. Z., 2020, *International Journal of Modern Physics A*, 35, 2050075.
Ruderman M., Shaham J., Tavani M., 1989, *Astrophysical Journal*, 336, 507.
Ruderman M., Shaham J., Tavani M., Eichler D., 1989, *Astrophysical Journal*, 343, 292.
Ruderman M. A., Shaham J., 1985, *Astrophysical Journal*, 289, 244.
Ruderman M. A., Sutherland P. G., 1975, *Astrophysical Journal*, 196, 51.
Ruiter A. J., Belczynski K., Benacquista M., Larson S. L., Williams G., 2010, *Astrophysical Journal*, 717, 1006.
Ruiter A. J., Belczynski K., Sim S. A., Hillebrandt W., Fryer C. L., Fink M., Kromer M., 2011, *Monthly Notices of the Royal Astronomical Society*, 417, 408.
Ruiter A. J., Ferrario L., Belczynski K., Seitenzahl I. R., Crocker R. M., Karakas A. I., 2019, *Monthly Notices of the Royal Astronomical Society*, 484, 698.
Ruiz M., Paschalidis V., Tsokaros A., Shapiro S. L., 2020, *Physical Review D*, 102, 124077.
Ruszkowski M., Begelman M. C., 2003, *Astrophysical Journal*, 586, 384.
Ryba M. F., Taylor J. H., 1991, *Astrophysical Journal*, 380, 557.
Sabach E., Hillel S., Schreier R., Soker N., 2017, *Monthly Notices of the Royal Astronomical Society*, 472, 4361.
Safarzadeh M., Hotokezaka K., 2020, *Astrophysical Journal*, 897, L7.
Safarzadeh M., Loeb A., 2020, *Astrophysical Journal*, 899, L15.
Safarzadeh M., Ramirez-Ruiz E., 2021, arXiv:2105.08746.
Saffer R. A., Liebert J., Olszewski E. W., 1988, *Astrophysical Journal*, 334, 947.
Saio H., 2011, *Monthly Notices of the Royal Astronomical Society*, 412, 1814.
Saio H., Nomoto K., 1985, *Astronomy and Astrophysics*, 150, L21.
Sakstein J., Croon D., McDermott S. D., Straight M. C., Baxter E. J., 2020, *Physical Review Letters*, 125, 261105.
Saladino M. I., Pols O. R., 2019, *Astronomy and Astrophysics*, 629, A103.
Saladino M. I., Pols O. R., Abate C., 2019, *Astronomy and Astrophysics*, 626, A68.
Saladino M. I., Pols O. R., van der Helm E., Pelupessy I., Portegies Zwart S., 2018, *Astronomy and Astrophysics*, 618, A50.
Salpeter E. E., 1955, *Astrophysical Journal*, 121, 161.
Salpeter E. E., 1964, *Astrophysical Journal*, 140, 796.
Salvesen G., Miller J. M., 2021, *Monthly Notices of the Royal Astronomical Society*, 500, 3640.
Samsing J., D'Orazio D. J., 2018, *Monthly Notices of the Royal Astronomical Society*, 481, 5445.
Samsing J., MacLeod M., Ramirez-Ruiz E., 2014, *Astrophysical Journal*, 784, 71.
Sana H., de Mink S. E., de Koter A., Langer N., Evans C. J., Gieles M., ... Schneider F. R. N., 2012, *Science*, 337, 444.
Sana H., Le Bouquin J. B., Lacour S., Berger J. P., Duvert G., Gauchet L., ... Zinnecker H., 2014, *Astrophysical Journal Supplement Series*, 215, 15.
Sandage A., 1986, *Annual Review of Astronomy and Astrophysics*, 24, 421.
Sandage A., Osmer P., Giacconi R., Gorenstein P., Gursky H., Waters J., ... Jugaku J., 1966, *Astrophysical Journal*, 146, 316.
Sander A. A. C., Vink J. S., 2020, *Monthly Notices of the Royal Astronomical Society*, 499, 873.
Sandquist E. L., Taam R. E., Chen X., Bodenheimer P., Burkert A., 1998, *Astrophysical Journal*, 500, 909.

Santander-García M., Rodríguez-Gil P., Corradi R. L. M., Jones D., Miszalski B., Boffin H. M. J., ... Kotze M. M., 2015, *Nature*, 519, 63.
Santoliquido F., Mapelli M., Giacobbo N., Bouffanais Y., Artale M. C., 2021, *Monthly Notices of the Royal Astronomical Society*, 502, 4877.
Sanyal D., Grassitelli L., Langer N., Bestenlehner J. M., 2015, *Astronomy and Astrophysics*, 580, A20.
Sathyaprakash B., Abernathy M., Acernese F., Ajith P., Allen B., Amaro-Seoane P., Andersson N., et al. 2012, *Classical and Quantum Gravity*, 29, 124013.
Sathyaprakash R., Roberts T. P., Walton D. J., Fuerst F., Bachetti M., Pinto C., ... Soria R., 2019, *Monthly Notices of the Royal Astronomical Society*, 488, L35.
Savchenko V., Ferrigno C., Kuulkers E., et al. 2017, *Astrophysical Journal*, 848, L15.
Savonije G. J., 1978, *Astronomy and Astrophysics*, 62, 317.
Savonije G. J., 1979, *Astronomy and Astrophysics*, 71, 352.
Savonije G. J., 1987, *Nature*, 325, 416.
Savonije G. J., 2008, in Goupil M. J., Zahn J. P., eds, Vol. 29 of EAS Publications Series, *The Dynamical Tide and Resonance Locking*, 91–125.
Savonije G. J., de Kool M., van den Heuvel E. P. J., 1986, *Astronomy and Astrophysics*, 155, 51.
Savonije G. J., Witte M. G., 2002, *Astronomy and Astrophysics*, 386, 211.
Savonije J., 1983, in Lewin W. H. G., van den Heuvel E. P. J., eds, *Accretion-Driven Stellar X-ray Sources* Cambridge University Press: *Evolution and Mass Transfer in X-ray Binaries*, 343–366.
Sazonov S., Paizis A., Bazzano A., Chelovekov I., Khabibullin I., Postnov K., ..., Lund, N., 2020, *New Astronomy Reviews*, 88, 101536.
Scarfe C. D., 1986, *Journal of the Royal Astronomical Society of Canada*, 80, 257.
Schaller G., Schaerer D., Meynet G., Maeder A., 1992, *Astronomy and Astrophysics Supplement Series*, 96, 269.
Schatzman E., 1962, *Annales d'Astrophysique*, 25, 18.
Scheck L., Kifonidis K., Janka H.-T., Müller E., 2006, *Astronomy and Astrophysics*, 457, 963.
Schinzel F. K., Kerr M., Rau U., Bhatnagar S., Frail D. A., 2019, *Astrophysical Journal*, 876, L17.
Schmidt P., Ohme F., Hannam M., 2015, *Physical Review D*, 91, 024043.
Schneider F. R. N., Izzard R. G., de Mink S. E., Langer N., Stolte A., de Koter A., ... Sana H., 2014, *Astrophysical Journal*, 780, 117.
Schneider F. R. N., Izzard R. G., Langer N., de Mink S. E., 2015, *Astrophysical Journal*, 805, 20.
Schneider F. R. N., Ohlmann S. T., Podsiadlowski P., Röpke F. K., Balbus S. A., Pakmor R., Springel V., 2019, *Nature*, 574, 211.
Schneider F. R. N., Podsiadlowski P., Langer N., Castro N., Fossati L., 2016, *Monthly Notices of the Royal Astronomical Society*, 457, 2355.
Schneider F. R. N., Podsiadlowski P., Müller B., 2021, *Astronomy and Astrophysics*, 645, A5.
Schneider F. R. N., Sana H., Evans C. J., et al. 2018, *Science*, 359, 69.
Schnurr O., Casoli J., Chené A.-N., Moffat A. F. J., St-Louis N., 2008, *Monthly Notices of the Royal Astronomical Society*, 389, L38.
Schnurr O., Moffat A. F. J., Villar-Sbaffi A., St-Louis N., Morrell N. I., 2009, *Monthly Notices of the Royal Astronomical Society*, 395, 823.
Scholz P., Kaspi V. M., Lyne A. G., et al. 2015, *Astrophysical Journal*, 800, 123.
Schootemeijer A., Langer N., Lennon D., Evans C. J., Crowther P. A., Geen S., ... Vink J. S., 2021, *Astronomy and Astrophysics*, 646, A106.

Schreiber M. R., Belloni D., Gänsicke B. T., Parsons S. G., Zorotovic M., 2021, *Nature Astronomy*, 5, 648.
Schreiber M. R., Lasota J.-P., 2007, *Astronomy and Astrophysics*, 473, 897.
Schreiber M. R., Zorotovic M., Wijnen T. P. G., 2016, *Monthly Notices of the Royal Astronomical Society*, 455, L16.
Schreier E., Levinson R., Gursky H., Kellogg E., Tananbaum H., Giacconi R., 1972, *Astrophysical Journal*, 172, L79.
Schreier E. J., 1977, in Papagiannis M. D., ed, Eighth Texas Symposium on Relativistic Astrophysics Vol. 302 of Annals of the New York Academy of Sciences, *Timing Effects in Rotating Neutron Stars*, 445.
Schrøder S. L., Batta A., Ramirez-Ruiz E., 2018, *Astrophysical Journal*, 862, L3.
Schrøder S. L., MacLeod M., Loeb A., Vigna-Gómez A., Mandel I., 2020, *Astrophysical Journal*, 892, 13.
Schrøder S. L., MacLeod M., Ramirez-Ruiz E., Mandel I., Fragos T., Loeb A., Everson R. W., 2021, arXiv:2107.09675.
Schutz B. F., 1986, *Nature*, 323, 310.
Schwab J., 2019, *Astrophysical Journal*, 885, 27.
Schwab J., 2021, *Astrophysical Journal*, 906, 53.
Schwab J., Bildsten L., Quataert E., 2017, *Monthly Notices of the Royal Astronomical Society*, 472, 3390.
Schwab J., Podsiadlowski P., Rappaport S., 2010, *Astrophysical Journal*, 719, 722.
Schwab J., Quataert E., Bildsten L., 2015, *Monthly Notices of the Royal Astronomical Society*, 453, 1910.
Schwab J., Quataert E., Kasen D., 2016, *Monthly Notices of the Royal Astronomical Society*, 463, 3461.
Schwarzschild M., 1958, *Structure and Evolution of the Stars*, Princeton University Press.
Schwarzschild M., Härm R., 1965, *Astrophysical Journal*, 142, 855.
Selam S. O., Demircan O., 1999, *Turkish Journal of Physics*, 23, 301.
Sen K., Langer N., Marchant P., Menon A., de Mink S. E., Schootemeijer A., Schürmann C., Mahy L., Hastings B., Nathaniel K., Sana H., Wang C., Xu X. T., 2022, *Astronomy and Astrophysics*, 659, A98.
Sengar R., Tauris T. M., Langer N., Istrate A. G., 2017, *Monthly Notices of the Royal Astronomical Society*, 470, L6.
Sengar R., Balakrishnan V., Stevenson S., Bailes M., Barr E. D., et al. 2022, *Monthly Notices of the RAS*, 512, 4.
Sepinsky J. F., Willems B., Kalogera V., Rasio F. A., 2010, *Astrophysical Journal*, 724, 546.
Sesana A., 2016, *Physical Review Letters*, 116, 231102.
Seward F. D., Mitchell M., 1981, *Astrophysical Journal*, 243, 736.
Shahbaz T., Linares M., Rodríguez-Gil P., Casares J., 2019, *Monthly Notices of the Royal Astronomical Society*, 488, 198.
Shaifullah G., Verbiest J. P. W., Freire P. C. C., et al. 2016, *Monthly Notices of the Royal Astronomical Society*, 462, 1029.
Shakura N., Postnov K., 2017, arXiv:1702.03393.
Shakura N., Postnov K., Kochetkova A., Hjalmarsdotter L., 2012, *Monthly Notices of the Royal Astronomical Society*, 420, 216.
Shakura N., Postnov K., Kochetkova A., Hjalmarsdotter L., 2018, in Shakura N., ed, Vol. 454 of Astrophysics and Space Science Library, *Quasi-Spherical Subsonic Accretion onto Magnetized Neutron Stars*, 331.

Shakura N. I., Postnov K. A., Kochetkova A. Y., Hjalmarsdotter L., 2014, Vol. 64 of European Physical Journal Web of Conferences, *Theory of Wind Accretion*, 02001.

Shakura N. I., Postnov K. A., Kochetkova A. Y., Hjalmarsdotter L., Sidoli L., Paizis A., 2015, Astronomy Reports, 59, 645.

Shakura N. I., Sunyaev R. A., 1973, *Astronomy and Astrophysics*, 24, 337.

Shao Y., Li X.-D., 2012, *Astrophysical Journal*, 756, 85.

Shao Y., Li X.-D., 2015, *Astrophysical Journal*, 802, 131.

Shao Y., Li X.-D., 2020, *Astrophysical Journal*, 898, 143.

Shao Y., Li X.-D., 2021, *Astrophysical Journal*, 908, 67.

Shapiro I. I., 1964, *Physical Review Letters*, 13, 789.

Shapiro S. L., Lightman A. P., 1976, *Astrophysical Journal*, 204, 555.

Shapiro S. L., Teukolsky S. A., 1983, *Black Holes, White Dwarfs, and Neutron Stars: The Physics of Compact Objects*, Wiley, New York.

Shara M. M., Regev O., 1986, *Astrophysical Journal*, 306, 543.

Shatsky N., Tokovinin A., 2002, *Astronomy and Astrophysics*, 382, 92.

Shen K. J., Bildsten L., 2007, *Astrophysical Journal*, 660, 1444.

Shen K. J., Blondin S., Kasen D., Dessart L., Townsley D. M., Boos S., Hillier D. J., 2021, *Astrophysical Journal*, 909, L18.

Shen K. J., Boubert D., Gänsicke B. T., Jha S. W., Andrews J. E., Chomiuk L., et al. 2018, *Astrophysical Journal*, 865, 15.

Shen K. J., Kasen D., Miles B. J., Townsley D. M., 2018, *Astrophysical Journal*, 854, 52.

Shen K. J., Quataert E., Pakmor R., 2019, *Astrophysical Journal*, 887, 180.

Shenar T., Bodensteiner J., Abdul-Masih M., Fabry M., Mahy L., Marchant P., . . . Sana H., 2020, *Astronomy and Astrophysics*, 639, L6.

Shenar T., Gilkis A., Vink J. S., Sana H., Sander A. A. C., 2020, *Astronomy and Astrophysics*, 634, A79.

Shi C.-S., Zhang S.-N., Li X.-D., 2015, *Astrophysical Journal*, 813, 91.

Shih I. C., Kundu A., Maccarone T. J., Zepf S. E., Joseph T. D., 2010, *Astrophysical Journal*, 721, 323.

Shih I. C., Maccarone T. J., Kundu A., Zepf S. E., 2008, *Monthly Notices of the Royal Astronomical Society*, 386, 2075.

Shibata M., Hotokezaka K., 2019, *Annual Review of Nuclear and Particle Science*, 69, 41.

Shibazaki N., Murakami T., Shaham J., Nomoto K., 1989, *Nature*, 342, 656.

Shih, I. C. and Maccarone, T. J. and Kundu, A. and Zepf, S. E., 2008, *Monthly Notices of the Royal Astronomical Society*, 386.

Shih, I. Chun and Kundu, Arunav and Maccarone, Thomas J. and Zepf, Stephen E. & Joseph, Tana D., 2010, *Astrophysical Journal*, 721, 323–328.

Shirke P., Bala S., Roy J., Bhattacharya D., 2021, *Journal of Astrophysics and Astronomy*, 42, 58.

Shklovskii I. S., 1967, *Astrophysical Journal*, 148, L1.

Shklovskii I. S., 1970, *Soviet Astronomy*, 13, 562.

Shore S. N., Livio M., van den Heuvel E. P. J., 1994, *Interacting Binaries*, H. Nussbaumer and A. Orr, eds, Springer Verlag, Heidelberg.

Sidoli L., 2012, in Proceedings of "An INTEGRAL View of the High-Energy Sky (the First 10 Years)"—9th INTEGRAL Workshop and Celebration of the 10th Anniversary of the Launch (INTEGRAL 2012), 15-19 October 2012. Bibliotheque Nationale de France, *Supergiant Fast X-ray Transients: A Review*, 11.

Sieniawska M., Bejger M., Haskell B., 2018, *Astronomy and Astrophysics*, 616, A105.

Siess L., 2006, *Astronomy and Astrophysics*, 448, 717.

Siess L., 2007, *Astronomy and Astrophysics*, 476, 893.
Siess L., Lebreuilly U., 2018, *Astronomy and Astrophysics*, 614, A99.
Siess L., Pumo M. L., 2006, *Memorie della Societa Astronomica Italiana*, 77, 822.
Sigurdsson S., Hernquist L., 1993, *Nature*, 364, 423.
Sigurdsson S., Richer H. B., Hansen B. M., Stairs I. H., Thorsett S. E., 2003, *Science*, 301, 193.
Silber A. D., 1992, PhD thesis, Massachusetts Institute of Technology.
Sills A., Pinsonneault M. H., Terndrup D. M., 2000, *Astrophysical Journal*, 534, 335.
Silsbee K., Tremaine S., 2017, *Astrophysical Journal*, 836, 39.
Sim S. A., Röpke F. K., Hillebrandt W., Kromer M., Pakmor R., Fink M., Ruiter A. J., Seitenzahl I. R., 2010, *Astrophysical Journal*, 714, L52.
Simón-Díaz S., Maíz Apellániz J., Lennon D. J., González Hernández J. I., Allende Prieto C., Castro N., ... Smartt S. J., 2020, *Astronomy and Astrophysics*, 634, L7.
Singh N., Bulik T., Belczynski K., Askar A., 2021, arXiv:2112.04058.
Singh N. S., Naik S., Paul B., Agrawal P. C., Rao A. R., Singh K. Y., 2002, *Astronomy and Astrophysics*, 392, 161.
Sipior M. S., Sigurdsson S., 2002, *Astrophysical Journal*, 572, 962.
Skumanich A., 1972, *Astrophysical Journal*, 171, 565.
Slettebak A., 1988, *Publications of the Astronomical Society of the Pacific*, 100, 770.
Smarr L. L., Blandford R., 1976, *Astrophysical Journal*, 207, 574.
Smartt S. J., 2009, *Annual Review of Astronomy and Astrophysics*, 47, 63.
Smartt S. J., 2015, *Publications of the Astronomical Society of Australia*, 32, e016.
Smartt S. J., Chen T.-W., Jerkstrand A., ... 2017, *Nature*, 551, 75.
Smith M. A., 1979, *Publications of the Astronomical Society of the Pacific*, 91, 737.
Smith N., 2016, *Monthly Notices of the Royal Astronomical Society*, 461, 3353.
Smith N., 2019, *Monthly Notices of the Royal Astronomical Society*, 489, 4378.
Smith N., Andrews J. E., Van Dyk S. D., Mauerhan J. C., Kasliwal M. M., Bond H. E., ... Sabbi E., 2016, *Monthly Notices of the Royal Astronomical Society*, 458, 950.
Smith N., Götberg Y., de Mink S. E., 2018, *Monthly Notices of the Royal Astronomical Society*, 475, 772.
Smith N., Li W., Foley R. J., Wheeler J. C., Pooley D., Chornock R., ... Hansen C., 2007, *Astrophysical Journal*, 666, 1116.
Smith N., Rest A., Andrews J. E., Matheson T., Bianco F. B., Prieto J. L., ... Zenteno A., 2018, *Monthly Notices of the Royal Astronomical Society*, 480, 1457.
Smith N., Tombleson R., 2015, *Monthly Notices of the Royal Astronomical Society*, 447, 598.
Smith N., Vink J. S., de Koter A., 2004, *Astrophysical Journal*, 615, 475.
Smits R., Lorimer D. R., Kramer M., Manchester R., Stappers B., Jin C. J., ... Li D., 2009, *Astronomy and Astrophysics*, 505, 919.
Soares-Santos M., Holz D. E., Annis J., et al. 2017, *Astrophysical Journal*, 848, L16.
Sobel D., 2016, *The Glass Universe*, New York: Viking Press
Soberman G. E., Phinney E. S., van den Heuvel E. P. J., 1997, *Astronomy and Astrophysics*, 327, 620.
Socrates A., Davis S. W., 2006, *Astrophysical Journal*, 651, 1049.
Soker N., 1998, *Astronomical Journal*, 116, 1308.
Soker N., 2004, *New Astronomy*, 9, 399.
Soker N., 2015, *Astrophysical Journal*, 800, 114.
Soker N., 2016, *New Astronomy Reviews*, 75, 1.
Soker N., Livio M., 1984, *Monthly Notices of the Royal Astronomical Society*, 211, 927.
Soker N., Regev O., Livio M., Shara M. M., 1987, *Astrophysical Journal*, 318, 760.

Sokoloski J., Lawrence S., Crotts A. P. S., Mukai K., 2016, in *Accretion Processes in Cosmic Sources Flows and Shocks*, Proceedings of Science (Saint Petersburg) *Some Recent Developments in Symbiotic Star and Nova Research*, 21.
Solheim J.-E., 2010, *Publications of the Astronomical Society of the Pacific*, 122, 1133.
Southworth J., Gänsicke B. T., Breedt E., 2012, in Richards M. T., Hubeny I., eds, *From Interacting Binaries to Exoplanets: Essential Modeling Tools*, proceedings of IAU Symposium 282, *The Orbital Period Distribution of Cataclysmic Variables Found by the SDSS*, 123–124.
Spera M., Mapelli M., Bressan A., 2015, *Monthly Notices of the Royal Astronomical Society*, 451, 4086.
Spera M., Mapelli M., Giacobbo N., Trani A. A., Bressan A., Costa G., 2019, *Monthly Notices of the Royal Astronomical Society*, 485, 889.
Spiewak R., Bailes M., Barr E. D., et al. 2018, *Monthly Notices of the Royal Astronomical Society*, 475, 469.
Spitkovsky A., 2006, *Astrophysical Journal*, 648, L51.
Spitkovsky A., 2008, in C. Bassa, Z. Wang, A. Cumming, & V. M. Kaspi eds, *40 Years of Pulsars: Millisecond Pulsars, Magnetars and More*, Vol. 983 of American Institute of Physics Conference Series, *Pulsar Magnetosphere: The Incredible Machine*, 20–28.
Spitkovsky A., Levin Y., Ushomirsky G., 2002, *Astrophysical Journal*, 566, 1018.
Spitler L. G., Scholz P., Hessels J. W. T., Bogdanov S., . . . 2016, *Nature*, 531, 202.
Spitzer L., 1987, *Dynamical Evolution of Globular Clusters*, Princeton University Press.
Spruit H., Phinney E. S., 1998, *Nature*, 393, 139.
Spruit H. C., 2002, *Astronomy and Astrophysics*, 381, 923.
Spruit H. C., 2018, arXiv:1810.06106.
Spruit H. C., Ritter H., 1983, *Astronomy and Astrophysics*, 124, 267.
Spruit H. C., Taam R. E., 1993, *Astrophysical Journal*, 402, 593.
Spruit H. C., Taam R. E., 2001, *Astrophysical Journal*, 548, 900.
Srinivasan G., Bhattacharya D., Muslimov A. G., Tsygan A. J., 1990, *Current Science*, 59, 31.
Srinivasan G., van den Heuvel E. P. J., 1982, *Astronomy and Astrophysics*, 108, 143.
Staelin D. H., Reifenstein Edward C. I., 1968, *Science*, 162, 1481.
Staff J. E., De Marco O., Macdonald D., Galaviz P., Passy J.-C., Iaconi R., Low M.-M. M., 2016, *Monthly Notices of the Royal Astronomical Society*, 455, 3511.
Stairs I. H., 2003, *Living Reviews in Relativity*, 6, 5.
Stairs I. H., Arzoumanian Z., Camilo F., Lyne A. G., Nice D. J., Taylor J. H., . . . Wolszczan A., 1998, *Astrophysical Journal*, 505, 352.
Stairs I. H., Faulkner A. J., Lyne A. G., Kramer M., Lorimer D. R., McLaughlin M. A., . . . Freire P. C., Gregory P. C., 2005, *Astrophysical Journal*, 632, 1060.
Stanway E. R., Eldridge J. J., Chrimes A. A., 2020, *Monthly Notices of the Royal Astronomical Society*, 497, 2201.
Stappers B. W., Bailes M., Lyne A. G., Manchester R. N., D'Amico N., Tauris T. M., . . . Sandhu J. S., 1996, *Astrophysical Journal*, 465, L119.
Stappers B. W., van Kerkwijk M. H., Bell J. F., Kulkarni S. R., 2001, *Astrophysical Journal*, 548, L183.
Starrfield S., Iliadis C., Hix W. R., 2016, *Publications of the Astronomical Society of the Pacific*, 128, 051001.
Staubert R., Trümper J., Kendziorra E., Klochkov D., Postnov K., Kretschmar P., . . . Fürst F., 2019, *Astronomy and Astrophysics*, 622, A61.
Steeghs D., Howell S. B., Knigge C., Gänsicke B. T., Sion E. M., Welsh W. F., 2007, *Astrophysical Journal*, 667, 442.

Stegmann J., Antonini F., Moe, M., et al. 2022, *Monthly Notices of the RAS*, 516, 1.
Steiner J. E., Oliveira A. S., 2005, *Astronomy and Astrophysics*, 444, 895.
Stella L., Priedhorsky W., White N. E., 1987, *Astrophysical Journal*, 312, L17.
Steltner B., Papa M. A., Eggenstein H. B., Allen B., Dergachev V., Prix R., ... Kwang S., 2021, *Astrophysical Journal*, 909, 79.
Stencel R. E., 2015, in *Giants of Eclipse: The ζ Aurigae Stars and Other Binary Systems*, Vol. 408 of Astrophysics and Space Science Library, T. B. Ake and E. Griffin, eds, Springer: *Epsilon Aurigae: A Two Century Long Dilemma Persists*, 107.
Stevenson S., Berry C. P. L., Mandel I., 2017, *Monthly Notices of the Royal Astronomical Society*, 471, 2801.
Stevenson S., Sampson M., Powell J., Vigna-Gómez A., Neijssel C. J., Szécsi D., Mandel I., 2019, *Astrophysical Journal*, 882, 121.
Stevenson S., Vigna-Gómez A., Mandel I., Barrett J. W., Neijssel C. J., Perkins D., de Mink S. E., 2017, *Nature Communications*, 8, 14906.
Stirling A. M., Spencer R. E., de la Force C. J., Garrett M. A., Fender R. P., Ogley R. N., 2001, *Monthly Notices of the Royal Astronomical Society*, 327, 1273.
Stockinger G., Janka H. T., Kresse D., Melson T., Ertl T., Gabler M., ... Heger A., 2020, *Monthly Notices of the Royal Astronomical Society*, 496, 2039.
Stone R. C., 1991, *Astronomical Journal*, 102, 333.
Stothers R. B., Chin C.-W., 1996, *Astrophysical Journal*, 468, 842.
Stovall K., Freire P. C. C., Antoniadis J., et al. 2019, *Astrophysical Journal*, 870, 74.
Stovall K., Freire P. C. C., Chatterjee S., Demorest P. B., Lorimer D. R., ... 2018, *Astrophysical Journal*, 854, L22.
Strader J., Chomiuk L., Maccarone T. J., Miller-Jones J. C. A., Seth A. C., 2012, *Nature*, 490, 71.
Stritzinger M. D., Taddia F., Fraser M., Tauris T. M., Contreras C., Drybye S., Galbany L., Holmbo S., Morrell N., Pastorello A., et al. 2020, *Astronomy and Astrophysics*, 639, A104.
Stritzinger M. D., Valenti S., Hoeflich P., Baron E., Phillips M. M., Taddia F., Foley R. J., et al. 2015, *Astronomy and Astrophysics*, 573, A2.
Strohmayer T., Bildsten L., 2006, *New Views of Thermonuclear Bursts*, in: W. H. G. Lewin and M. van der Klis, eds, *Compact Stellar X-ray Sources*, Cambridge University Press, 113–156.
Strohmayer T. E., Markwardt C. B., 1999, *Astrophysical Journal*, 516, L81.
Struve O., Huang S. S., 1958, *Handbuch der Physik: Astrophysics I: Stellar Surfaces-Binaries*, 243–274. Springer Verlag OHG, Berlin, Goettingen, Heidelberg.
Suijs M. P. L., Langer N., Poelarends A. J., Yoon S. C., Heger A., Herwig F., 2008, *Astronomy and Astrophysics*, 481, L87.
Sukhbold T., Ertl T., Woosley S. E., Brown J. M., Janka H.-T., 2016, *Astrophysical Journal*, 821, 38.
Sukhbold T., Woosley S. E., Heger A., 2018, *Astrophysical Journal*, 860, 93.
Summa A., Hanke F., Janka H.-T., Melson T., Marek A., Müller B., 2016, *Astrophysical Journal*, 825, 6.
Summa A., Janka H.-T., Melson T., Marek A., 2018, *Astrophysical Journal*, 852, 28.
Sundqvist J. O., Simón-Díaz S., Puls J., Markova N., 2013, *Astronomy and Astrophysics*, 559, L10.
Sur A., Haskell B., 2021, *Monthly Notices of the Royal Astronomical Society*, 502, 4680.
Šurlan B., Hamann W.-R., Aret A., Kubát J., Oskinova L. M., Torres A. F., 2013, *Astronomy and Astrophysics*, 559, A130.
Sutantyo W., 1974, *Astronomy and Astrophysics*, 35, 251.
Sutantyo W., 1975a, *Astronomy and Astrophysics*, 41, 47.

Sutantyo W., 1975b, *Astronomy and Astrophysics*, 44, 227.
Sutantyo W., 1978, *Astrophysics and Space Science*, 54, 479.
Sutantyo W., 1992, in *X-Ray Binaries and Recycled Pulsars*, E. P. J. van den Heuvel, S. A. Rappaport, eds, Kluwer Academic Publishers, Dordrecht: *The Evolution of Her X-1 and Why Her X-1-Like Systems Are Very Rare*, 293–309.
Sutantyo W., Li X.-D., 2000, *Astronomy and Astrophysics*, 360, 633.
Suvorov A. G., 2021, *Monthly Notices of the Royal Astronomical Society*, 503, 5495.
Suwa Y., Yoshida T., Shibata M., Umeda H., Takahashi K., 2015, *Monthly Notices of the Royal Astronomical Society*, 454, 3073.
Suzuki T., Zha S., Leung S.-C., Nomoto K., 2019, *Astrophysical Journal*, 881, 64.
Swank J. H., Becker R. H., Boldt E. A., Holt S. S., Pravdo S. H., Serlemitsos P. J., 1977, *Astrophysical Journal*, 212, L73.
Swiggum J. K., Rosen R., McLaughlin M. A., Lorimer D. R., Heatherly S., Lynch R., . . . et al. 2015, *Astrophysical Journal*, 805, 156.
Szczepanczyk M., Antelis J., Benjamin M., Cavaglia M., Gondek-Rosinska D., Hansen T., Klimenko S., et al. 2021, *Physical Review D*, 104, 102002.
Taam R. E., 1980, *Astrophysical Journal*, 242, 749.
Taam R. E., 1983, *Astrophysical Journal*, 270, 694.
Taam R. E., 1996, in van Paradijs J., van den Heuvel E. P. J., Kuulkers E., eds, *Compact Stars in Binaries*, proceedings of IAU Symposium 165, *Common-Envelope Evolution, the Formation of CVs, LMXBs, and the Fate of HMXBs*, 3.
Taam R. E., Bodenheimer P., Ostriker J. P., 1978, *Astrophysical Journal*, 222, 269.
Taam R. E., Fryxell B. A., 1984, *Astrophysical Journal*, 279, 166.
Taam R. E., Fryxell B. A., 1988, *Astrophysical Journal*, 327, L73.
Taam R. E., Ricker P. M., 2010, *New Astronomy Reviews*, 54, 65.
Taam R. E., Sandquist E. L., 2000, *Annual Review of Astronomy and Astrophysics*, 38, 113.
Taam R. E., van den Heuvel E. P. J., 1986, *Astrophysical Journal*, 305, 235.
Taani A., Karino S., Song L., Zhang C., Chaty S., 2018, arXiv:1808.05345.
Taddia F., Sollerman J., Leloudas G., Stritzinger M. D., Valenti S., Galbany L., . . . Wheeler J. C., 2015, *Astronomy and Astrophysics*, 574, A60.
Tagawa H., Haiman Z., Kocsis B., 2020, *Astrophysical Journal*, 898, 25.
Takahashi K., 2018, *Astrophysical Journal*, 863, 153.
Takahashi K., Yoshida T., Umeda H., 2013, *Astrophysical Journal*, 771, 28.
Tanaka Y., 1983, IAU Circ., 3891.
Tanaka Y., Lewin W. H. G., 1995, in X-ray Binaries, W. H. G. Lewin, J. A. van Paradijs, E. P. J. van den Heuvel, eds, Cambridge University Press: Black Hole Binaries, 126–174.
Tananbaum H., Gursky H., Kellogg E., Giacconi R., Jones C., 1972, *Astrophysical Journal*, 177, L5.
Tananbaum H. D., 1973, in Bradt H., Giacconi R., eds, *X- and Gamma-Ray Astronomy*, proceedings of IAU Symposium 55, *UHURU Results on Galactic X-Ray Sources*, 9.
Tang P. N., Eldridge J. J., Stanway E. R., Bray J. C., 2020, *Monthly Notices of the Royal Astronomical Society*, 493, L6.
Tang W., Li X.-D., 2021, *Monthly Notices of the Royal Astronomical Society*, 506, 3323.
Tarafdar P., Nobleson K., Rana P., Singha J., Krishnakumar M. A., Joshi B. C., et al. 2022, arXiv:2206.09289.
Taubenberger S., 2017, *The Extremes of Thermonuclear Supernovae*, in *Handbook of Supernovae*, Springer International Publishing AG, 317.
Tauris T. M., 1996, *Astronomy and Astrophysics*, 315, 453.
Tauris T. M., 1997, PhD thesis, Aarhus University.

Tauris T. M., 1999, in Arzoumanian Z., Van der Hooft F., van den Heuvel E. P. J., eds, *Pulsar Timing, General Relativity and the Internal Structure of Neutron Stars: Pulsar Velocities*, 315.

Tauris T. M., 2001, in Podsiadlowski P., Rappaport S., King A. R., D'Antona F., Burderi L., eds, *Evolution of Binary and Multiple Star Systems*, Vol. 229 of Astronomical Society of the Pacific Conference Series, *On the Pre-RLO Spin-Orbit Couplings in LMXBs*, 145.

Tauris T. M., 2011, in Schmidtobreick L., Schreiber M. R., Tappert C., eds, *Evolution of Compact Binaries*, Vol. 447 of Astronomical Society of the Pacific Conference Series, *Five and a Half Roads to Form a Millisecond Pulsar*, 285.

Tauris T. M., 2012, *Science*, 335, 561.

Tauris T. M., 2015, *Monthly Notices of the Royal Astronomical Society*, 448, L6.

Tauris T. M., 2018, *Physical Review Letters*, 121, 131105.

Tauris T. M., 2022, *Astrophysical Journal*, 938, 66.

Tauris T. M., Bailes M., 1996, *Astronomy and Astrophysics*, 315, 432.

Tauris T. M., Dewi J. D. M., 2001, *Astronomy and Astrophysics*, 369, 170.

Tauris T. M., Fender R. P., van den Heuvel E. P. J., Johnston H. M., Wu K., 1999, *Monthly Notices of the Royal Astronomical Society*, 310, 1165.

Tauris T. M., Janka H.-T., 2019, *Astrophysical Journal*, 886, L20.

Tauris T. M., Konar S., 2001, *Astronomy and Astrophysics*, 376, 543.

Tauris T. M., Kramer M., Freire P. C. C., Wex N., Janka H.-T., Langer N., ... Champion D. J., 2017, *Astrophysical Journal*, 846, 170.

Tauris T. M., Langer N., Kramer M., 2011, *Monthly Notices of the Royal Astronomical Society*, 416, 2130.

Tauris T. M., Langer N., Kramer M., 2012, *Monthly Notices of the Royal Astronomical Society*, 425, 1601.

Tauris T. M., Langer N., Moriya T. J., Podsiadlowski P., Yoon S.-C., Blinnikov S. I., 2013, *Astrophysical Journal*, 778, L23.

Tauris T. M., Langer N., Podsiadlowski P., 2015, *Monthly Notices of the Royal Astronomical Society*, 451, 2123.

Tauris T. M., Manchester R. N., 1998, *Monthly Notices of the Royal Astronomical Society*, 298, 625.

Tauris T. M., Sanyal D., Yoon S.-C., Langer N., 2013, *Astronomy and Astrophysics*, 558, A39.

Tauris T. M., Savonije G. J., 1999, *Astronomy and Astrophysics*, 350, 928.

Tauris T. M., Savonije G. J., 2001, in Kouveliotou C., Ventura J. & van den Heuvel E. P. J. eds, *The Neutron Star—Black Hole Connection*, Kluwer Academic Publishers, Dordrecht: *Spin-Orbit Coupling in X-ray Binaries*, 337.

Tauris T. M., Sennels T., 2000, *Astronomy and Astrophysics*, 355, 236.

Tauris T. M., Takens R. J., 1998, *Astronomy and Astrophysics*, 330, 1047.

Tauris T. M., van den Heuvel E. P. J., 2006, *Formation and Evolution of Compact Stellar X-ray Sources*, in W. H. G. Lewin, M. van der Klis, eds, *Compact Stellar X-ray Sources*, Cambridge University Press, 623–665.

Tauris T. M., van den Heuvel E. P. J., 2014, *Astrophysical Journal*, 781, L13.

Tauris T. M., van den Heuvel E. P. J., Savonije G. J., 2000, *Astrophysical Journal*, 530, L93.

Tavani M., Brookshaw L., 1992, *Nature*, 356, 320.

Taylor J. H., 1987, in Ulmer M. P., ed, *13th Texas Symposium on Relativistic Astrophysics, World Scientific Publishing Co: Pulsars—An Overview of Recent Developments*, 467–477.

Taylor J. H., 1992, *Royal Society of London Philosophical Transactions Series A*, 341, 117.

Taylor J. H., Hulse R. A., Fowler L. A., Gullahorn G. E., Rankin J. M., 1976, *Astrophysical Journal*, 206, L53.

Taylor J. H., Weisberg J. M., 1982, *Astrophysical Journal*, 253, 908.

Taylor J. H., Weisberg J. M., 1989, *Astrophysical Journal*, 345, 434.
Tchekhovskoy A., Narayan R., McKinney J. C., 2010, *Astrophysical Journal*, 711, 50.
Tendulkar S. P., Gil de Paz A., Kirichenko A. Y., Hessels J. W. T., et al. 2021, *Astrophysical Journal*, 908, L12.
Terman J. L., Taam R. E., Hernquist L., 1995, *Astrophysical Journal*, 445, 367.
Terquem C., Papaloizou J. C. B., Nelson R. P., Lin D. N. C., 1998, *Astrophysical Journal*, 502, 788.
Terrell D., Munari U., Zwitter T., Wolf G., 2005, *Monthly Notices of the Royal Astronomical Society*, 360, 583.
Terrell J., Priedhorsky W. C., 1984, *Astrophysical Journal*, 285, L15.
Tetzlaff N., Neuhäuser R., Hohle M. M., 2011, *Monthly Notices of the Royal Astronomical Society*, 410, 190.
Tetzlaff N., Neuhäuser R., Hohle M. M., Maciejewski G., 2010, *Monthly Notices of the Royal Astronomical Society*, 402, 2369.
Thomas H. C., 1967, *Zeitschrift für Astrophysik*, 67, 420.
Thompson C., Duncan R. C., 1995, *Monthly Notices of the Royal Astronomical Society*, 275, 255.
Thompson C., Duncan R. C., 1996, *Astrophysical Journal*, 473, 322.
Thompson T. A., Kochanek C. S., Stanek K. Z., Badenes C., Post R. S., Jayasinghe T., . . . Covey K., 2019, *Science*, 366, 637.
Thompson T. A., Prieto J. L., Stanek K. Z., Kistler M. D., Beacom J. F., Kochanek C. S., 2009, *Astrophysical Journal*, 705, 1364.
Thorne K. S., 1974, *Astrophysical Journal*, 191, 507.
Thorne K. S., Żytkow A. N., 1975, *Astrophysical Journal*, 199, L19.
Thorne K. S., Żytkow A. N., 1977, *Astrophysical Journal*, 212, 832.
Thorsett S. E., 1995, *Astrophysical Journal*, 444, L53.
Tillich A., Przybilla N., Scholz R.-D., Heber U., 2009, *Astronomy and Astrophysics*, 507, L37.
Timmes F. X., Woosley S. E., 1992, *Astrophysical Journal*, 396, 649.
Timmes F. X., Woosley S. E., Taam R. E., 1994, *Astrophysical Journal*, 420, 348.
Timmes F. X., Woosley S. E., Weaver T. A., 1996, *Astrophysical Journal*, 457, 834.
Tjemkes S. A., Zuiderwijk E. J., van Paradijs J., 1986, *Astronomy and Astrophysics*, 154, 77.
Tjin A Djie H. R. E., The P. S., Hack M., Selvelli P. L., 1982, *Astronomy and Astrophysics*, 106, 98.
Tokovinin A., 2014, *Astronomical Journal*, 147, 87.
Tokovinin A., Thomas S., Sterzik M., Udry S., 2006, *Astronomy and Astrophysics*, 450, 681.
Tokovinin A. A., 2000, *Astronomy and Astrophysics*, 360, 997.
Tomkin J., Lambert D. L., 1987, Sky and Telescope, 74, 354.
Tomsick J. A., Bodaghee A., Rodriguez J., Chaty S., Camilo F., Fornasini F., Rahoui F., 2012, *Astrophysical Journal*, 750, L39.
Toonen S., Boekholt T. C. N., Portegies Zwart S., 2022, *Astronomy and Astrophysics*, 661, A61.
Toonen S., Claeys J. S. W., Mennekens N., Ruiter A. J., 2014, *Astronomy and Astrophysics*, 562, A14.
Toonen S., Hamers A., Portegies Zwart S., 2016, *Computational Astrophysics and Cosmology*, 3, 6.
Toonen S., Hollands M., Gänsicke B. T., Boekholt T., 2017, *Astronomy and Astrophysics*, 602, A16.
Toonen S., Nelemans G., Portegies Zwart S., 2012, *Astronomy and Astrophysics*, 546, A70.
Toonen S., Portegies Zwart S., Hamers A. S., Bandopadhyay D., 2020, *Astronomy and Astrophysics*, 640, A16.

Torres G., Andersen J., Giménez A., 2010, *Astronomy and Astrophysics Review*, 18, 67.
Totani T., Morokuma T., Oda T., Doi M., Yasuda N., 2008, *Publications of the Astronomical Society of Japan*, 60, 1327.
Tøttrup M., 2021, Master's thesis, Department of Physics and Astronomy, Aarhus University, Denmark.
Toubiana A., Sberna L., Caputo A., Cusin G., Marsat S., Jani K., . . . Tamanini N., 2021, *Physical Review Letters*, 126, 101105.
Tout C. A., 1991, *Monthly Notices of the Royal Astronomical Society*, 250, 701.
Tout C. A., Aarseth S. J., Pols O. R., Eggleton P. P., 1997, *Monthly Notices of the Royal Astronomical Society*, 291, 732.
Tramper F., Sana H., Fitzsimons N. E., de Koter A., Kaper L., Mahy L., Moffat A., 2016, *Monthly Notices of the Royal Astronomical Society*, 455, 1275.
Trimble V., 1974, *Astronomical Journal*, 79, 967.
Trimble V., 1982, *Reviews of Modern Physics*, 54, 1183.
Trimble V., 1990, *Monthly Notices of the Royal Astronomical Society*, 242, 79.
Trimble V. L., Thorne K. S., 1969, *Astrophysical Journal*, 156, 1013.
Trümper J., Hasinger G., Aschenbach B., Bräuninger H., Briel U. G., Burkert W., . . . Beuermann K., 1991, *Nature*, 349, 579.
Tsesevich V. P., 1973, *Eclipsing Variable Stars* Wiley, New York.
Tsuna D., Ishii A., Kuriyama N., Kashiyama K., Shigeyama T., 2020, *Astrophysical Journal*, 897, L44.
Tudor V., Miller-Jones J. C. A., Knigge C., Maccarone T. J., Tauris T. M., Bahramian A., Chomiuk L., Heinke C. O., et al. 2018, *Monthly Notices of the Royal Astronomical Society*, 476, 1889.
Turatto M., 2003, *Classification of Supernovae*, in K. Weiler, ed, *Supernovae and Gamma-Ray Bursters, Lecture Notes in Physics*, Vol. 598, 21–36.
Turolla R., Zane S., Watts A. L., 2015, *Reports on Progress in Physics*, 78, 116901.
Tutukov A., Yungelson L., 1973a, *Nauchnye Informatsii*, 27, 86.
Tutukov A., Yungelson L., 1973b, *Nauchnye Informatsii*, 27, 70.
Tutukov A. V., Yungelson L. R., 1993a, *Astronomy Reports*, 37, 411.
Tutukov A. V., Yungelson L. R., 1993b, *Monthly Notices of the Royal Astronomical Society*, 260, 675.
Tylenda R., Hajduk M., Kamiński T., Udalski A., Soszyński I., Szymański M. K., . . . Ulaczyk K., 2011, *Astronomy and Astrophysics*, 528, A114.
Ubertini P., Bazzano A., Cocchi M., Natalucci L., Heise J., Muller J. M., in 't Zand J. J. M., 1999, *Astrophysical Journal*, 514, L27.
Udalski A., Szymanski M. K., Soszynski I., Poleski R., 2008, *Acta Astronomica*, 58, 69.
Ugliano M., Janka H.-T., Marek A., Arcones A., 2012, *Astrophysical Journal*, 757, 69.
Umeda H., Yoshida T., Takahashi K., 2012, Prog. Theor. Exp. Phys., DOI: 10.1093/ptep/pts017.
Underhill A. B., ed, 1966, The Early Type Stars, Vol. 6 of Astrophysics and Space Science Library, Reidel, Dordrecht.
Urry C. M., Padovani P., 1995, *Publications of the Astronomical Society of the Pacific*, 107, 803.
Vacca W. D., Conti P. S., 1992, *Astrophysical Journal*, 401, 543.
Valsecchi F., Glebbeek E., Farr W. M., Fragos T., Willems B., Orosz J. A., . . . Kalogera V., 2010, *Nature*, 468, 77.
Van K. X., Ivanova N., Heinke C. O., 2019, *Monthly Notices of the Royal Astronomical Society*, 483, 5595.
van Albada T. S., 1968, *Bulletin of the Astronomical Institutes of the Netherlands*, 20, 47.

van de Kamp P., ed. 1981, Stellar Paths: Photographic Astrometry with Long-Focus Instruments, Vol. 85 of Astrophysics and Space Science Library, Springer, Dordrecht.
van den Heuvel E. P., van Paradijs J. A., Taam R. E., 1986, *Nature*, 322, 153.
van den Heuvel E. P. J., 1966, *Proceedings of the Royal Netherlands Academy of Arts and Sciences*, 69, 357.
van den Heuvel E. P. J., 1968a, *Bulletin of the Astronomical Institutes of the Netherlands*, 19, 309.
van den Heuvel E. P. J., 1968b, *Bulletin of the Astronomical Institutes of the Netherlands*, 19, 326.
van den Heuvel E. P. J., 1969, *Astronomical Journal*, 74, 1095
van den Heuvel E. P. J., 1973, *Nature Physical Science*, 242, 71.
van den Heuvel E. P. J., 1974, in *Astrophysics and Gravitation, Proceedings of the 16th Solvay Conference on Physics, Editions de l'Universite de Bruxelles: The Evolutionary Origin of Massive X-ray Binaries and the Total Number of Massive Stars with a Collapsed Close Companion in the Galaxy*, 119–130.
van den Heuvel E. P. J., 1975, *Astrophysical Journal*, 198, L109.
van den Heuvel E. P. J., 1976, in Eggleton P., Mitton S., Whelan J., eds, *Structure and Evolution of Close Binary Systems*, proceedings of IAU Symposium 73, *Late Stages of Close Binary Systems*, 35.
van den Heuvel E. P. J., 1977, in Papagiannis M. D., ed, *Eighth Texas Symposium on Relativistic Astrophysics*, Vol. 302 of Annals of the New York Academy of Sciences, *Evolutionary Processes in X-Ray Binaries and Their Progenitor Systems*, 14.
van den Heuvel E. P. J., 1981, in Sugimoto D., Lamb D. Q., Schramm D. N., eds, *Fundamental Problems in the Theory of Stellar Evolution*, proceedings of IAU Symposium 93, *The Formation of Compact Objects in Binary Systems*, 155–173.
van den Heuvel E. P. J., 1985, in Boland W., van Woerden H., eds, *Birth and Evolution of Massive Stars and Stellar Groups*, Vol. 120 of Astrophysics and Space Science Library, Springer: *Evolution of Binary Stars and the Problem of the Runaway Stars*, 107–122.
van den Heuvel E. P. J., 1992, *Endpoints of Stellar Evolution: The Incidence of Stellar Mass Black Holes in the Galaxy*, in *ESA, Environment Observation and Climate Modelling Through International Space Projects. Space Sciences with Particular Emphasis on High-Energy Astrophysics*, 29–36.
van den Heuvel E. P. J., 1994a, in Shore S. N., Livio M., van den Heuvel E. P. J., eds,: Nussbaumer H., Orr A., Saas-Fee Advanced Course 22: *Interacting Binaries: Topics in Close Binary Evolution*, 263–474.
van den Heuvel E. P. J., 1994b, *Astronomy and Astrophysics*, 291, L39.
van den Heuvel E. P. J., 2004, in Schoenfelder V., Lichti G., & Winkler C., eds, 5th INTEGRAL Workshop on the INTEGRAL Universe, Vol. 552 of ESA Special Publication, *X-Ray Binaries and Their Descendants: Binary Radio Pulsars; Evidence for Three Classes of Neutron Stars?* 185.
van den Heuvel E. P. J., 2019, Proceedings of the International Astronomical Union Symposium, 346, pp. 1–13.
van den Heuvel E. P. J., Bhattacharya D., Nomoto K., Rappaport S. A., 1992, *Astronomy and Astrophysics*, 262, 97.
van den Heuvel E. P. J., Bitzaraki O., 1994, *Memorie della Societa Astronomica Italiana*, 65, 237.
van den Heuvel E. P. J., Bitzaraki O., 1995, *Astronomy and Astrophysics*, 297, L41.
van den Heuvel E. P. J., Bonsema P. T. J., 1984, *Astronomy and Astrophysics*, 139, L16.
van den Heuvel E. P. J., De Loore C., 1973, *Astronomy and Astrophysics*, 25, 387.

van den Heuvel E. P. J., Heise J., 1972, *Nature Physical Science*, 239, 67.
van den Heuvel E. P. J., Lorimer D. R., 1996, *Monthly Notices of the Royal Astronomical Society*, 283, L37.
van den Heuvel E. P. J., Ostriker J. P., Petterson J. A., 1980, *Astronomy and Astrophysics*, 81, L7.
van den Heuvel E. P. J., Portegies Zwart S. F., Bhattacharya D., Kaper L., 2000, *Astronomy and Astrophysics*, 364, 563.
van den Heuvel E. P. J., Portegies Zwart S. F., de Mink S. E., 2017, *Monthly Notices of the Royal Astronomical Society*, 471, 4256.
van den Heuvel E. P. J., Rappaport S., 1987, in Slettebak A., Snow T. P., eds, IAU Colloq. 92: *Physics of Be Stars, Cambridge University Press: X-ray Observations of B-emission Stars*, 291–307.
van den Heuvel E. P. J., Taam R. E., 1984, *Nature*, 309, 235.
van den Heuvel E. P. J., Tauris T. M., 2020, *Science*, 368, eaba3282.
van den Heuvel E. P. J., van Paradijs J., 1988, *Nature*, 334, 227.
van den Heuvel E. P. J., Yoon S. C., 2007, *Astrophysics and Space Science*, 311, 177.
van der Hucht K. A., 2001, *New Astronomy Reviews*, 45, 135.
van der Klis M., 2000, *Annual Review of Astronomy and Astrophysics*, 38, 717.
van der Klis M., 2006, *Rapid X-ray Variability*, in W. H. G. Lewin and M. van der Klis, eds, *Compact Stellar X-ray Sources*, Cambridge University Press, 39–112.
van der Klis M., Hasinger G., Verbunt F., van Paradijs J., Belloni T., Lewin W. H. G., 1993, *Astronomy and Astrophysics*, 279, L21.
van der Klis M., Swank J. H., Zhang W., Jahoda K., Morgan E. H., Lewin W. H. G., Vaughan B., van Paradijs J., 1996, *Astrophysical Journal*, 469, L1.
van der Klis M., Wijnands R. A. D., Horne K., Chen W., 1997, *Astrophysical Journal*, 481, L97.
van der Linden T. J., 1982, PhD thesis, University of Amsterdam.
van der Linden T. J., 1987, *Astronomy and Astrophysics*, 178, 170.
van der Meer A., Kaper L., van Kerkwijk M. H., Heemskerk M. H. M., van den Heuvel E. P. J., 2007, *Astronomy and Astrophysics*, 473, 523.
van der Meij V., Guo D., Kaper L., Renzo M., 2021, *Astronomy and Astrophysics*, 655, A31.
van der Sluys M. V., Verbunt F., Pols O. R., 2005, *Astronomy and Astrophysics*, 431, 647.
van Doesburgh M., van der Klis M., 2017, *Monthly Notices of the Royal Astronomical Society*, 465, 3581.
van Haaften L. M., Nelemans G., Voss R., Jonker P. G., 2012, *Astronomy and Astrophysics*, 541, A22.
van Haaften L. M., Nelemans G., Voss R., Wood M. A., Kuijpers J., 2012, *Astronomy and Astrophysics*, 537, A104.
van Kerkwijk M. H., Bassa C. G., Jacoby B. A., Jonker P. G., 2005, in Rasio F. A. & Stairs I. H., eds, *Binary Radio Pulsars*, Vol. 328 of *Astronomical Society of the Pacific Conference Series, Optical Studies of Companions to Millisecond Pulsars*, 357.
van Kerkwijk M. H., Breton R. P., Kulkarni S. R., 2011, *Astrophysical Journal*, 728, 95.
van Kerkwijk M. H., Chang P., Justhan S., 2010, *Astrophysical Journal*, 722, L157.
van Kerkwijk M. H., Charles P. A., Geballe T. R., King D. L., Miley G. K., Molnar L. A., . . . van Paradijs J., 1992, *Nature*, 355, 703.
van Kerkwijk M. H., Kulkarni S. R., 1999, *Astrophysical Journal*, 516, L25.
van Leeuwen J., Kasian L., Stairs I. H., Lorimer D. R., Camilo F., Chatterjee S., . . . Weisberg J. M., 2015, *Astrophysical Journal*, 798, 118.
van Loon J. T., Kaper L., Hammerschlag-Hensberge G., 2001, *Astronomy and Astrophysics*, 375, 498.
van Oijen J. G. J., 1989, *Astronomy and Astrophysics*, 217, 115.

van Paradijs J., 1978, *Nature*, 274, 650.
van Paradijs J., 1979, *Astrophysical Journal*, 234, 609.
van Paradijs J., 1995, in *X-ray Binaries*, W. H. G. Lewin, J. van Paradijs and E. P. J. van den Heuvel, eds, Cambridge University Press: A Catalogue of X-ray Binaries, 536–577.
van Paradijs J., 1996, *Astrophysical Journal*, 464, L139.
van Paradijs J., 1998, in Buccheri R., van Paradijs J., Alpar A., eds, NATO Advanced Science Institutes (ASI) Series C Vol. 515, Kluwer Academic Publishers, Dordrecht, *Neutron Stars and Black Holes in X-Ray Binaries*, 279.
van Paradijs J., Allington-Smith J., Callanan P., Charles P. A., Hassall B. J. M., Machin G., Mason K. O., Naylor T., Smale A. P., 1988, *Nature*, 334, 684.
van Paradijs J., Lub J., Pel J. W., Pakull M., van Amerongen S., 1983, *Astronomy and Astrophysics*, 124, 294.
van Paradijs J., van den Heuvel E. P. J., Kouveliotou C., Fishman G. J., Finger M. H., Lewin W. H. G., 1997, *Astronomy and Astrophysics*, 317, L9.
van Paradijs J. A., Hammerschlag-Hensberge G., van den Heuvel E. P. J., Takens R. J., Zuiderwijk E. J., de Loore C., 1976, *Nature*, 259, 547.
van Rensbergen W., De Greve J. P., 2016, *Astronomy and Astrophysics*, 592, A151.
van Rensbergen W., De Greve J. P., 2020, *Astronomy and Astrophysics*, 642, A183.
van Roestel J., Kupfer T., Green M. J., Wong S., Bildsten L., Burdge K., Prince T., Marsh T. R., Szkody P., et al. 2021, *Monthly Notices of the Royal Astronomical Society*.
Vanbeveren D., 1988, *Astrophysics and Space Science*, 149, 1.
Vanbeveren D., Mennekens N., De Greve J. P., 2012, *Astronomy and Astrophysics*, 543, A4.
Vanbeveren D., Mennekens N., van den Heuvel E. P. J., Van Bever J., 2020, *Astronomy and Astrophysics*, 636, A99.
Vartanyan D., Laplace E., Renzo M., GŁotberg Y., Burrows A., de Mink S. E., 2021, *Astrophysical Journal*, 916, L5.
Vasyliunas V. M., 1979, *Space Science Reviews*, 24, 609.
Vennes S., Nemeth P., Kawka A., Thorstensen J. R., Khalack V., Ferrario L., Alper E. H., 2017, *Science*, 357, 680.
Venumadhav T., Zackay B., Roulet J., Dai L., Zaldarriaga M., 2020, *Physical Review D*, 101, 083030.
Verbiest J. P. W., Bailes M., van Straten W., Hobbs G. B., Edwards R. T., Manchester R. N., ... Kulkarni S. R., 2008, *Astrophysical Journal*, 679, 675.
Verbiest J. P. W., Oslowski S., Burke-Spolaor S., 2021, in Bambi C., Katsanevas S., Kokkotas K. D., eds, *Handbook of Gravitational Wave Astronomy*, 4, 1.
Verbunt F., 1984, *Monthly Notices of the Royal Astronomical Society*, 209, 227.
Verbunt F., 1987, *Astrophysical Journal*, 312, L23.
Verbunt F., 1989, in Ögelman H., van den Heuvel E. P. J., eds, *Timing Neutron Stars*, Vol. 262 of NATO Advanced Science Institutes (ASI) Series C, *X-ray Binaries and Radiopulsars in Globular Clusters*, 593.
Verbunt F., 1990, in Kundt W., ed, *Neutron Stars and Their Birth Events*, Vol. 300 of NATO Advanced Science Institutes (ASI) Series C, *The Origin and Evolution of X-Ray Binaries and Low-Magnetic Radio Pulsars*, 179.
Verbunt F., 1993, *Annual Review of Astronomy and Astrophysics*, 31, 93.
Verbunt F., Freire P. C. C., 2014, *Astronomy and Astrophysics*, 561, A11.
Verbunt F., Hut P., 1987, in Helfand D. J., Huang J. H., eds, *The Origin and Evolution of Neutron Stars*, proceedings of IAU Symposium 125, *The Globular Cluster Population of X-Ray Binaries*, 187.
Verbunt F., Igoshev A., Cator E., 2017, *Astronomy and Astrophysics*, 608, A57.

Verbunt F., Lewin W. H. G., 2006, in W. H. G. Lewin, M. van der Klis, eds, *Compact Stellar X-ray Sources*, Cambridge University Press: *Globular Cluster X-ray Sources*, 341–379.
Verbunt F., Phinney E. S., 1995, *Astronomy and Astrophysics*, 296, 709.
Verbunt F., Wijers R. A. M. J., Burm H. M. G., 1990, *Astronomy and Astrophysics*, 234, 195.
Verbunt F., Zwaan C., 1981, *Astronomy and Astrophysics*, 100, L7.
Vieira N., Ruan J. J., Haggard D., Drout M. R., Nynka M. C., ... 2020, *Astrophysical Journal*, 895, 96.
Vietri M., Stella L., 1998, *Astrophysical Journal*, 507, L45.
Viganò D., Rea N., Pons J. A., Perna R., Aguilera D. N., Miralles J. A., 2013, *Monthly Notices of the Royal Astronomical Society*, 434, 123.
Vigna-Gómez A., Neijssel C. J., Stevenson S., Barrett J. W., Belczynski K., Justhan S., ... Mandel I., 2018, *Monthly Notices of the Royal Astronomical Society*, 481, 4009.
Vigna-Gomez A., Wassink M., Klencki J., Istrate A., Nelemans G., Mandel I., 2022, *Monthly Notices of the Royal Astronomical Society*, 511, 2326.
Vink J. S., de Koter A., Lamers H. J. G. L. M., 2001, *Astronomy and Astrophysics*, 369, 574.
Vink J. S., Muijres L. E., Anthonisse B., de Koter A., Gräfener G., Langer N., 2011, *Astronomy and Astrophysics*, 531, A132.
Vitale S., Evans M., 2017, *Physical Review D*, 95, 064052.
Voisin G., Cognard I., Freire P. C. C., Wex N., Guillemot L., Desvignes G., Kramer M., Theureau G., 2020, *Astronomy and Astrophysics*, 638, A24.
Voss R., Tauris T. M., 2003, *Monthly Notices of the Royal Astronomical Society*, 342, 1169.
Wade L., Siemens X., Kaplan D. L., Knispel B., Allen B., 2012, *Physical Review D*, 86, 124011.
Wagg T., Breivik K., de Mink S. E., 2022, *The Journal of Open Source Software*, 7, 3998.
Wagg T., Broekgaarden F. S., de Mink S. E., Frankel N., van Son L. A. C., Justham, S., 2022, *Astrophysical Journal*, 937, 118.
Walker M. F., 1954, *Publications of the Astronomical Society of the Pacific*, 66, 230.
Walter R., Lutovinov A. A., Bozzo E., Tsygankov S. S., 2015, *Astronomy and Astrophysics Review*, 23, 2.
Walter R., Rodriguez J., Foschini L., de Plaa J., Corbel S., Courvoisier T. J. L., ... Ubertini P., 2003, *Astronomy and Astrophysics*, 411, L427.
Wandel A., Yahil A., Milgrom M., 1984, *Astrophysical Journal*, 282, 53.
Wanderman D., Piran T., 2010, *Monthly Notices of the Royal Astronomical Society*, 406, 1944.
Wang B., 2018, *Monthly Notices of the Royal Astronomical Society*, 481, 439.
Wang B., Han Z., 2010, *Astronomy and Astrophysics*, 515, A88.
Wang B., Li X.-D., Han Z.-W., 2010, *Monthly Notices of the Royal Astronomical Society*, 401, 2729.
Wang B., Liu D., 2020, *Research in Astronomy and Astrophysics*, 20, 135.
Wang C., Langer N., Schootemeijer A., Castro N., Adscheid S., Marchant P., Hastings B., 2020, *Astrophysical Journal*, 888, L12.
Wang J., 2016, *International Journal of Astronomy and Astrophysics*, 6, 82.
Wang J., Zhang C. M., Zhao Y. H., Kojima Y., Yin H. X., Song L. M., 2011, *Astronomy and Astrophysics*, 526, A88.
Wang K., Zhang X., Dai M., 2020, *Astrophysical Journal*, 888, 49.
Wang K., Zhang X., Luo Y., Luo C., 2019, *Monthly Notices of the Royal Astronomical Society*, 486, 2462.
Wang L., Gies D. R., Lester K. V., Guo Z., Matson R. A., et al. 2020, *Astronomical Journal*, 159, 4.
Wang Y. M., 1981, *Astronomy and Astrophysics*, 102, 36.

Wang Y.-M., 1997, *Astrophysical Journal*, 475, L135+.

Warner B., 1995, *Astrophysics and Space Science*, 225, 249.

Warwick R. S., Marshall N., Fraser G. W., Watson M. G., Lawrence A., Page C. G., ... Smith A., 1981, *Monthly Notices of the Royal Astronomical Society*, 197, 865.

Watson C. A., Steeghs D., Shahbaz T., Dhillon V. S., 2007, *Monthly Notices of the Royal Astronomical Society*, 382, 1105.

Watts A. L., Yu W., Poutanen J., Zhang S., Bhattacharyya S., et al. 2019, *Science China Physics, Mechanics, and Astronomy*, 62, 29503.

Weaver T. A., Woosley S. E., 1980, in Meyerott R., Gillespie H. G., eds, *Supernovae Spectra*, Vol. 63 of American Institute of Physics Conference Series, *Supernova Models and Light Curves*, 15–32.

Weaver T. A., Zimmerman G. B., Woosley S. E., 1978, *Astrophysical Journal*, 225, 1021.

Webbink R. F., 1975, *Astronomy and Astrophysics*, 41, 1.

Webbink R. F., 1976, *Astrophysical Journal*, 209, 829.

Webbink R. F., 1979, *Astrophysical Journal*, 227, 178.

Webbink R. F., 1984, *Astrophysical Journal*, 277, 355.

Webbink R. F., Rappaport S., Savonije G. J., 1983, *Astrophysical Journal*, 270, 678.

Webster B. L., Murdin P., 1972, *Nature*, 235, 37.

Weidemann V., 1990, *Annual Review of Astronomy and Astrophysics*, 28, 103.

Weinberg D. H., Mortonson M. J., Eisenstein D. J., Hirata C., Riess A. G., Rozo E., 2013, *Physics Reports*, 530, 87.

Weisberg J. M., Huang Y., 2016, *Astrophysical Journal*, 829, 55.

Weisberg J. M., Nice D. J., Taylor J. H., 2010, *Astrophysical Journal*, 722, 1030.

Weiss A., Hillebrandt W., Thomas H. C., Ritter H., 2004, *Cox and Giuli's Principles of Stellar Structure*, Cambridge Scientific Publishers Ltd, Cambridge UK.

Wellstein S., Langer N., 1999, *Astronomy and Astrophysics*, 350, 148.

Wellstein S., Langer N., 2001, in Kaper L., van den Heuvel E. P. J., Woudt P. A., eds, *Black Holes in Binaries and Galactic Nuclei*, ESO Astrophysics Symposia, Springer, Heidelberg: *Constraints on the Initial Mass Limit for Black-Hole Formation from the Massive X-Ray Binary Wray*, 977, 295.

Wellstein S., Langer N., Braun H., 2001, *Astronomy and Astrophysics*, 369, 939.

Wex N., 1998, *Monthly Notices of the Royal Astronomical Society*, 298, 67.

Wex N., 2014, in *Brumberg Festschrift*, S. M. Kopeikein, ed: to be published by de Gruyter, Berlin, arXiv:1402.5594.

Wex N., Kalogera V., Kramer M., 2000, *Astrophysical Journal*, 528, 401.

Wheeler J. C., Lecar M., McKee C. F., 1975, *Astrophysical Journal*, 200, 145.

Whelan J., Iben Jr. I., 1973, *Astrophysical Journal*, 186, 1007.

Wickramasinghe D. T., Tout C. A., Ferrario L., 2014, *Monthly Notices of the Royal Astronomical Society*, 437, 675.

Wijers R. A. M. J., 1997, *Monthly Notices of the Royal Astronomical Society*, 287, 607.

Wijnands R., van der Klis M., 1998, *Nature*, 394, 344.

Wiktorowicz G., Lu Y., Wyrzykowski Ł., Zhang H., Liu J., Justhan S., Belczynski K., 2020, *Astrophysical Journal*, 905, 134.

Will C. M., 1993, *Theory and Experiment in Gravitational Physics*, Cambridge University Press.

Will C. M., 2014, *Living Reviews in Relativity*, 17, 4.

Winn J. N., 2010, arXiv:1001.2010.

Witte M. G., 2001, PhD thesis, University of Amsterdam.

Witte M. G., Savonije G. J., 1999a, *Astronomy and Astrophysics*, 341, 842.

Witte M. G., Savonije G. J., 1999b, *Astronomy and Astrophysics*, 350, 129.

Witte M. G., Savonije G. J., 2002, *Astronomy and Astrophysics*, 386, 222.
Woan G., Pitkin M. D., Haskell B., Jones D. I., Lasky P. D., 2018, *Astrophysical Journal*, 863, L40.
Wolszczan A., Doroshenko O., Konacki M., Kramer M., Jessner A., Wielebinski R., . . . Taylor J. H., 2000, *Astrophysical Journal*, 528, 907.
Wolszczan A., Frail D. A., 1992, *Nature*, 355, 145.
Woltjer L., 1964, *Astrophysical Journal*, 140, 1309.
Wong T.-W., Willems B., Kalogera V., 2010, *Astrophysical Journal*, 721, 1689.
Wongwathanarat A., Janka H.-T., Müller E., 2013, *Astronomy and Astrophysics*, 552, A126.
Woosley S. E., 1993, *Astrophysical Journal*, 405, 273.
Woosley S. E., 2017, *Astrophysical Journal*, 836, 244.
Woosley S. E., 2019, *Astrophysical Journal*, 878, 49.
Woosley S. E., Heger A., 2007, *Physics Reports*, 442, 269.
Woosley S. E., Heger A., 2015, *Astrophysical Journal*, 810, 34.
Woosley S. E., Heger A., 2021, *Astrophysical Journal*, 912, L31.
Woosley S. E., Heger A., Weaver T. A., 2002, *Reviews of Modern Physics*, 74, 1015.
Woosley S. E., Taam R. E., 1976, *Nature*, 263, 101.
Woosley S. E., Taam R. E., Weaver T. A., 1986, *Astrophysical Journal*, 301, 601.
Woosley S. E., Timmes F. X., Baron E., 1992, in *X-Ray Binaries and Recycled Pulsars*, E. P. J. van den Heuvel and S. A. Rappaport, eds, Kluwer Academic Publishers, Dordrecht: *Accretion Induced Collapse*, 167–187.
Woosley S. E., Weaver T. A., 1994, *Astrophysical Journal*, 423, 371.
Woosley S. E., Weaver T. A., 1995, *Astrophysical Journal Supplement Series*, 101, 181.
Worley A., Krastev P. G., Li B., 2008, *Astrophysical Journal*, 685, 390.
Wu Y., Chen X., Chen H., Li Z., Han Z., 2020, *Astronomy and Astrophysics*, 634, A126.
Xu K., Li X.-D., Cui Z., Li Q.-C., Shao Y., Liang X., Liu J., 2022, *Research in Astronomy and Astrophysics*, 22, 015005.
Xu Q., Li T., Li X.-D., 2012, *Research in Astronomy and Astrophysics*, 12, 1417.
Xu X., Li X., 2010, *Astrophysical Journal*, 716, 114.
Xuan Z., Peng P., Chen X., 2021, *Monthly Notices of the Royal Astronomical Society*, 502, 4199.
Yao J. M., Manchester R. N., Wang N., 2017, *Astrophysical Journal*, 835, 29.
Yao Y., De K., Kasliwal M. M., Ho A. Y. Q., Schulze S., Li Z., Kulkarni S. R., Fruchter A., et al. 2020, *Astrophysical Journal*, 900, 46.
Ye C. S., Fong W.-f., Kremer K., Rodriguez C. L., Chatterjee S., Fragione G., Rasio F. A., 2020, *Astrophysical Journal*, 888, L10.
Yoon S., Woosley S. E., Langer N., 2010, *Astrophysical Journal*, 725, 940.
Yoon S.-C., 2017, *Monthly Notices of the Royal Astronomical Society*, 470, 3970.
Yoon S.-C., Dessart L., Clocchiatti A., 2017, *Astrophysical Journal*, 840, 10.
Yoon S.-C., Langer N., 2004, *Astronomy and Astrophysics*, 419, 623.
Yoon S.-C., Langer N., 2005, *Astronomy and Astrophysics*, 443, 643.
Yoshida T., Takiwaki T., Aguilera-Dena D. R., Kotake K., Takahashi K., Nakamura K., Umeda H., Langer N., 2021, *Monthly Notices of the Royal Astronomical Society*, 506, L20.
Yu Y.-W., Chen A., Li X.-D., 2019, *Astrophysical Journal*, 877, L21.
Yungelson L., Livio M., 1998, *Astrophysical Journal*, 497, 168.
Yungelson L. R., 2005, *Population Synthesis for Progenitors of Type Ia Supernovae*, in E. M. Sion, S. Vennes, H. L. Shipman, eds, *White Dwarfs: Cosmological and Galactic Probes*. Astrophysics and Space Science Library, Springer, Dordrecht, 163–173.
Yungelson L. R., Kuranov A. G., Postnov K. A., 2019, *Monthly Notices of the Royal Astronomical Society*, 485, 851.

Yungelson L. R., Kuranov A. G., Postnov K. A., Kolesnikov D. A., 2020, *Monthly Notices of the Royal Astronomical Society*, 496, L6.
Yungelson L. R., Nelemans G., van den Heuvel E. P. J., 2002, *Astronomy and Astrophysics*, 388, 546.
Yungelson L. R., Tutukov A. V., Livio M., 1993, *Astrophysical Journal*, 418, 794.
Zackay B., Dai L., Venumadhav T., Roulet J., Zaldarriaga M., 2019, arXiv:1910.09528.
Zahn J. P., 1966a, *Annales d'Astrophysique*, 29, 313.
Zahn J. P., 1966b, *Annales d'Astrophysique*, 29, 489.
Zahn J. P., 1975, *Astronomy and Astrophysics*, 41, 329.
Zahn J. P., 1977, *Astronomy and Astrophysics*, 500, 121.
Zahn J. P., 1989, *Astronomy and Astrophysics*, 220, 112.
Zahn J. P., 2008, in Goupil M. J., Zahn J. P., eds, *Tidal Effects in Stars, Planets and Disks*, Vol. 29 of EAS Publications Series, *Tidal Dissipation in Binary Systems*, 67–90.
Zaldarriaga M., Kushnir D., Kollmeier J. A., 2018, *Monthly Notices of the Royal Astronomical Society*, 473, 4174.
Zanazzi J. J., Lai D., 2020, *Astrophysical Journal*, 892, L15.
Zapartas E., de Mink S. E., Izzard R. G., Yoon S. C., Badenes C., Götberg Y., . . . Shrotriya T. S., 2017, *Astronomy and Astrophysics*, 601, A29.
Zapolsky H. S., Salpeter E. E., 1969, *Astrophysical Journal*, 158, 809.
Zasche P., Wolf M., Kučáková H., Kára J., Merc J., Zejda M., Skarka M., Janík J., Kurfürst P., 2020, *Astronomy and Astrophysics*, 640, A33.
Zdziarski A. A., Mikołajewska J., Belczyński K., 2013, *Monthly Notices of the Royal Astronomical Society*, 429, L104.
Zel'dovich Y. B., 1964, *Soviet Physics Doklady*, 9, 195.
Zeldovich Y. B., Guseynov O. H., 1966, *Astrophysical Journal*, 144, 840.
Zenati Y., Bobrick A., Perets H. B., 2020, *Monthly Notices of the Royal Astronomical Society*, 493, 3956.
Zepf S. E., Stern D., Maccarone T. J., Kundu A., Kamionkowski M., Rhode K. L., Salzer J. J., Ciardullo R., Gronwall C., 2008, *Astrophysical Journal*, 683, L139.
Zevin M., Bavera S. S., Berry C. P. L., Kalogera V., Fragos T., Marchant P., Rodriguez C. L., Antonini F., Holz D. E., Pankow C., 2021, *Astrophysical Journal*, 910, 152.
Zevin M., Berry C. P. L., Coughlin S., Chatziioannou K., Vitale S., 2020, *Astrophysical Journal*, 899, L17.
Zevin M., Kelley L. Z., Nugent A., Fong W.-f., Berry C. P. L., Kalogera V., 2020, *Astrophysical Journal*, 904, 190.
Zevin M., Samsing J., Rodriguez C., Haster C.-J., Ramirez-Ruiz E., 2019, *Astrophysical Journal*, 871, 91.
Zevin M., Spera M., Berry C. P. L., Kalogera V., 2020, *Astrophysical Journal*, 899, L1.
Zha S., Leung S.-C., Suzuki T., Nomoto K., 2019, *Astrophysical Journal*, 886, 22.
Zhang C. M., 1998, *Astronomy and Astrophysics*, 330, 195.
Zhang C. M., Kojima Y., 2006, *Monthly Notices of the Royal Astronomical Society*, 366, 137.
Zhang S. N., Cui W., Chen W., 1997, *Astrophysical Journal*, 482, L155.
Zhang W., Smale A. P., Strohmayer T. E., Swank J. H., 1998, *Astrophysical Journal*, 500, L171.
Zhang X. B., Fu J. N., Liu N., Luo C. Q., Ren A. B., 2017, *Astrophysical Journal*, 850, 125.
Zhao X., Gou L., Dong Y., Tuo Y., Liao Z., Li Y., . . . Steiner J. F., 2021, *Astrophysical Journal*, 916, 108.
Zh1o X., Gou L., Dong Y., Tuo Y., Liao Z., Li Y., Jia N., Feng Y., Steiner J. F., 2021, *Astrophysical Journal*, 916, 108.

Zhong J., Chen L., Liu C., de Grijs R., Hou J., Shen S., . . . Zhang Z., 2014, *Astrophysical Journal*, 789, L2.

Zhong Y., Kashiyama K., Shigeyama T., Takasao S., 2021, *Astrophysical Journal*, 917, 71.

Zhu J.-P., Wu S., Yang Y.-P., Zhang B., Yu Y.-W., Gao H., Cao Z., Liu L.-D., 2021, *Astrophysical Journal*, 921, 156.

Zickgraf F. J., 1998, *Current Definition of B[e] Stars*, in A. M. Hubert, C. Jaschek, eds, B[e] stars : Proceedings of the Paris workshop held 9-12 June, 1997. Kluwer Academic Publishers, Dordrecht, 1.

Ziolkowski J., 2012, in Proceedings of "An INTEGRAL view of the high-energy sky (the first 10 years)"—9th INTEGRAL Workshop and Celebration of the 10th Anniversary of the Launch (INTEGRAL 2012), 15-19 October 2012. Bibliotheque Nationale de France, *Determination of the Masses of the Components of the HDE 226868/Cyg X-1 Binary System*, 54.

Ziosi B. M., Mapelli M., Branchesi M., Tormen G., 2014, *Monthly Notices of the Royal Astronomical Society*, 441, 3703.

Zorotovic M., Schreiber M. R., 2020, *Advances in Space Research*, 66, 1080.

Zorotovic M., Schreiber M. R., Gänsicke B. T., Nebot Gómez-Morán A., 2010, *Astronomy and Astrophysics*, 520, A86.

Zubovas K., Wynn G. A., Gualandris A., 2013, *Astrophysical Journal*, 771, 118.

Index